T0225282

Marine Biomedicine

FROM BEACH TO BEDSIDE

Marine Biomedicine

FROM BEACH TO BEDSIDE

EDITED BY

Bill J. Baker

University of South Florida
Tampa, USA
and
School of Chemistry
National University of Ireland Galway
Galway, Republic of Ireland

CRC Press
Taylor & Francis Group
Boca Raton London New York

CRC Press is an imprint of the
Taylor & Francis Group, an **informa** business

CRC Press
Taylor & Francis Group
6000 Broken Sound Parkway NW, Suite 300
Boca Raton, FL 33487-2742

First issued in paperback 2020

© 2016 by Taylor & Francis Group, LLC
CRC Press is an imprint of Taylor & Francis Group, an Informa business

No claim to original U.S. Government works

ISBN 13: 978-0-367-57530-4 (pbk)
ISBN 13: 978-1-4665-8212-5 (hbk)

This book contains information obtained from authentic and highly regarded sources. Reasonable efforts have been made to publish reliable data and information, but the author and publisher cannot assume responsibility for the valid-ity of all materials or the consequences of their use. The authors and publishers have attempted to trace the copyright holders of all material reproduced in this publication and apologize to copyright holders if permission to publish in this form has not been obtained. If any copyright material has not been acknowledged please write and let us know so we may rectify in any future reprint.

Except as permitted under U.S. Copyright Law, no part of this book may be reprinted, reproduced, transmitted, or uti-lized in any form by any electronic, mechanical, or other means, now known or hereafter invented, including photocopy-ing, microfilming, and recording, or in any information storage or retrieval system, without written permission from the publishers.

For permission to photocopy or use material electronically from this work, please access www.copyright.com (http://www.copyright.com/) or contact the Copyright Clearance Center, Inc. (CCC), 222 Rosewood Drive, Danvers, MA 01923, 978-750-8400. CCC is a not-for-profit organization that provides licenses and registration for a variety of users. For organizations that have been granted a photocopy license by the CCC, a separate system of payment has been arranged.

Trademark Notice: Product or corporate names may be trademarks or registered trademarks, and are used only for identification and explanation without intent to infringe.

Library of Congress Cataloging-in-Publication Data

Marine biomedicine : from beach to bedside / editor, Bill J. Baker.
  p. ; cm.
  Includes bibliographical references and index.
  ISBN 978-1-4665-8212-5 (alk. paper)
  I. Baker, B. J. (Bill J.), editor.
  [DNLM: 1. Drug Discovery. 2. Aquatic Organisms--chemistry. 3. Biological Products--therapeutic use. QV 241]

  RS161
  615.3'209162--dc23   2015017252

Visit the Taylor & Francis Web site at
http://www.taylorandfrancis.com

and the CRC Press Web site at
http://www.crcpress.com

Contents

Preface

Marine natural product research has been an active area of study for more than 50 years, yet notable examples of clinically useful drugs are only now emerging. The recent arrival of clinically used marine drugs is largely a reflection of the multidecade timeline of drug development. Chemical studies of marine organisms only began in earnest with ready access to marine environments, corresponding to the advent of safe and reliable scuba equipment in the 1960s. Much of what we know of marine natural products therefore trended early on with advances in field equipment. Recognizing such limitations, with concomitant issues in supply and regulatory access, the drug development industry was slow to embrace marine products. It is safe to say that those days are past. Progress in the technology of small-molecule drug development has brought innovations to biological as well as chemical techniques that make marine biomedicine one of the most promising arenas for drug discovery. We highlight those advances in this volume and consider the future potential of marine-derived drugs. While the U.S. National Institutes of Health has promoted translational research as "bench to bedside," marine biomedicine extends those efforts to the seashore, producing clinical successes that are fully "beach to bedside" translational efforts.

The framework of this book recognizes four topical areas. Part 1, "Marine Biotechnology: Informing the Search," addresses much of the technology that has recently been brought to bear on the study of marine natural products, including *omic* and bioinformatic techniques. The bulk of the chapters appear in Part 2, "Marine Biodiscovery: Lead Discovery," where the products and their activities are reviewed. Examples of clinically successful marine products and discussions of approaches to the clinic are presented in several chapters in Part 3, "Marine Products in Biomedicine." Finally, for Part 4, "Prospects for the Future of Marine Biodiscovery," we turn to a longtime and successful practitioner of the science, William Fenical, for advice and direction.

In Part 1, several chapters develop metabolomics and genomics as techniques that are significantly impacting pharmacological studies of marine natural products. The omic revolution that has swept many fields of science promises more fully informed studies of marine organisms, including metagenomic data that not only clarify the origin of certain metabolites, but also promise the potential for natural product gene identification and, with it, heterologous expression through genetic engineering and synthetic biology. The topic and promise are discussed by Hill and his group. Similarly, Schmidt demonstrates how the knowledge of biosynthetic pathways can be useful for the advancement of pharmacologically relevant marine natural products. Metabolomics, on the other hand, provide data to identify biomarkers in both producing organisms and disease states. The Verpoorte group brings this technique to the forefront in Chapter 2. Interestingly, the marine chemistry community has yet to embrace other omic technologies in the pursuit of new chemodiversity, but this is likely to change as proteomes and transcriptomes of marine producers yield biomarkers key to understanding the regulation of those pathways. Besides the bioinformatic analyses behind big data sets from omic techniques, marine natural product chemists use similar techniques in their day-to-day spectroscopic analyses and in dereplication, as demonstrated in Chapter 4 from Hartmut Laatsch.

Reviews of marine natural products and their biomedical potential abound. What we have tried to achieve in Part 2 of the book is to highlight areas with intense effort (anti-infectives, neglected diseases, niche environments)—and approaches to their investigation—that were due for evaluation. As such, Quinn discusses macroinvertebrate libraries as a screening technology, while the Baker group highlights the strengths and weaknesses of microbial libraries for screening. Much of that screening activity, historically, has been phenotypic. Luesch summarizes the use of targeted screens for discovery of bioactive marine natural products. Marine compounds active in anti-infective and neglected diseases are discussed by Pearce and Young, respectively, and Tidgewell presents the wealth of research taking place in the neurodegenerative disease area. The Pezzuto group, on the other hand, reflects on marine products that have the potential to prevent disease.

Sourcing samples for biodiscovery is critically important. The loss of time and resources that results from rediscovery of previously reported compounds significantly detracts from screening program progress. While effective dereplication (see Laatsch) can minimize these distractions, there remain many niche environments that have not been fully exploited. Vrijmoed reviews the marine–terrestrial margin as a rich source of robust and unique biodiversity to feed screening programs, while Balunas finds similar rewards in other niche environments.

Bringing marine products forward as drugs is the primary objective for many researchers in the field, highlighted in Part 3. While natural products are privileged compounds, designed as they are by nature for physiological relevance, receptor binding, and generally, bioactivity, there are many ways medicinal chemistry can benefit developmental studies. For example, structure–activity relationship (SAR) and structure–property relationship (SPR) studies can result in improvement of physiochemical properties such as solubility or pharmacological properties, including potency and resistance to clearance mechanisms. Leahy discusses the role of medicinal chemistry in advancing marine products.

The marine natural product community has been pleasantly receptive to conotoxin's and halichondrin's advancing through regulatory approval. Those achievements, including their paths and the promise they represent for future advancements, are highlighted in Chapters 16 and 17 by Olivera's and Orr's groups, respectively. Valeriote offers a promising approach to screening for neurotoxicity among cytotoxic agents. The difficult path taken by other promising marine candidates is developed by Newman and Cragg, while Carter offers an insider's view of the industry's difficulties (*vide supra*) with marine drug discovery.

Words of wisdom from Bill Fenical provide a contemporary vision of the future in Part 4. Bill has been such an ardent advocate of marine biodiscovery, leading us into the marine microbiome, into marine biotechnology, and generally arguing persuasively that marine natural products are the answer to new chemodiversity to drive drug discovery programs, that his advice is sought by many. Bill reflects here on where he sees the field advancing.

Marine natural products as a science has matured significantly in the last 10 years, with reports of new compounds trailing off as the low-hanging fruit of our oceans have been harvested. Nonetheless, the promise of marine-derived drugs has never been greater. New techniques to inform drug discovery, new efforts in niche environments, and recent successes keep the community invigorated. The next 50 years of marine natural products will be much more product oriented as the efforts of the pioneers are brought forward. We hope the material in this volume will inspire the next generation in the translation of their research from beach to bedside.

Bill J. Baker
Galway, Republic of Ireland

Acknowledgments

This book started with an agreement between my CRC editor, John Sulzycki, and me that there were aspects of marine natural product biomedicine that had not been critically evaluated, that new technology being applied in the field was ready for critical analysis, and that the field had reached a certain level of maturity that it was time to look forward and assess where we should be going. I am indebted to John for his perseverance and commitment to the project in light of the journey that the book represents.

I am overwhelmingly indebted to the many authors who made the journey with John and me. I think we all recognize that such review chapters are a labor of love and that the time, situation, and expectations all play a role in committing to such an endeavor. There are topic areas that are missing from this book, primarily because the key contributors could not make the commitment. However, 20 senior authors and their teams stepped up and not only agreed to review a topic they were expert in, but also agreed to spin it with a view to the future, to look out over the coming years and assess the needs and aspirations of the field. I believe you'll find the authors capture many of the highlights as well as the promise of marine natural product drug discovery.

The role of editor is an enviable one, with the honor of selecting topics and authors, encouraging adventurous avenues, and directing the focus of topics. The more mundane duties of reviewing, formatting, and tracking wordsmithery are often overlooked. However, I had the benefit of the services of Dr. Ryan Young, my associate at National University of Ireland Galway and research assistant professor of chemistry at University of South Florida, who organized many of those activities. Without his capable attention to detail, this project would have suffered.

I'm greatly appreciative of my staff, students, and family, all of whom have patiently forgiven my lapses of attention when I was distracted by this project. I hope, now that they have the book in their hands, it makes up in droves for the delayed attention.

During the preparation of the manuscript, my research and efforts were generously supported by several U.S. funding agencies, including the National Science Foundation (PLR-1341339), the National Institutes of Health (AI103715 and AI103673), and the Gates Foundation (1017870). In addition, the Beaufort Marine Research Award, a component of the Sea Change Strategy and the Strategy for Science Technology and Innovation (2006–2013), with the support of the Irish Marine Institute and funded under the Marine Research Sub-Programme of the Irish National Development Plan 2007–2013, currently supports both Dr. Ryan Young and myself in the development of Irish resources as marine natural product drugs.

Editor

Bill J. Baker obtained his BS in chemistry from Cal Poly in San Luis Obispo, California, then studied under Paul Scheuer at the University of Hawaii, where he earned his PhD. He conducted postdoctoral research with Ron Parry at Rice University and Carl Djerassi at Stanford University. He moved to the University of South Florida (USF) in 2001, where he is now professor of chemistry and director of USF's Center for Drug Discovery and Innovation. He is currently visiting Beaufort professor of marine biodiscovery at the National University of Ireland Galway. His research interests in marine natural product chemistry have taken him all over the globe, including 12 field seasons of research in Antarctica.

Contributors

Kh. Tanvir Ahmed
Division of Medicinal Chemistry
Graduate School of Pharmaceutical Sciences
Duquesne University
Pittsburgh, Pennsylvania

Fatma H. Al-Awadhi
Department of Medicinal Chemistry
University of Florida
Gainesville, Florida

Kashif Ali
Natural Products Laboratory
Institute of Biology
Leiden University
Leiden, The Netherlands

Bill J. Baker
Department of Chemistry and Center for Drug
    Discovery and Innovation
University of South Florida
Tampa, Florida
and
School of Chemistry
National University of Ireland Galway
Galway, Republic of Ireland

Marcy J. Balunas
Division of Medicinal Chemistry
Department of Pharmaceutical Sciences
University of Connecticut
Storrs, Connecticut

Valerie S. Bernan
Carter-Bernan Consulting, LLC
New City, New York

Leah C. Blasiak
Institute of Marine and Environmental
    Technology
University of Maryland Center for
    Environmental Science
Baltimore, Maryland

Laurent Calcul
Department of Chemistry and Center for Drug
    Discovery and Innovation
University of South Florida
Tampa, Florida

David Camp
Eskitis Institute for Drug Discovery
Griffith University
Brisbane, Queensland, Australia

Guy T. Carter
Carter-Bernan Consulting, LLC
New City, New York

Charles E. Chase
Integrated Chemistry
Eisai, Inc.
Andover, Massachusetts

Hyeong-wook Choi
Integrated Chemistry
Eisai, Inc.
Andover, Massachusetts

Young Hae Choi
Natural Products Laboratory
Institute of Biology
Leiden University
Leiden, The Netherlands

Gordon M. Cragg
Natural Products Branch
Developmental Therapeutics Program
Division of Cancer Treatment and Diagnosis
Frederick National Laboratory for Cancer
    Research
Frederick, Maryland

Danielle H. Demers
Department of Chemistry and Center for Drug
    Discovery and Innovation
University of South Florida
Tampa, Florida

Atsushi Endo
Integrated Chemistry
Eisai, Inc.
Andover, Massachusetts

Francis G. Fang
Integrated Chemistry
Eisai, Inc.
Andover, Massachusetts

William Fenical
Center for Marine Biotechnology and
    Biomedicine
Scripps Institution of Oceanography
University of California, San Diego
La Jolla, California

Samantha M. Gromek
Division of Medicinal Chemistry
Department of Pharmaceutical Sciences
University of Connecticut
Storrs, Connecticut

Russell T. Hill
Institute of Marine and Environmental
    Technology
University of Maryland Center for
    Environmental Science
Baltimore, Maryland

Martin P. Horvath
Department of Biology
University of Utah
Salt Lake City, Utah

E. B. Gareth Jones
Department of Botany and Microbiology
College of Science
King Saud University
Riyadh, Kingdom of Saudi Arabia

Hartmut Laatsch
Institute for Organic and Biomolecular
    Chemistry
University of Göttingen, Tammannstrasse
Göttingen, Germany

Neil Lax
Department of Biological Sciences
Bayer School of Natural and Environmental
    Sciences
Duquesne University
Pittsburgh, Pennsylvania

James W. Leahy
Department of Chemistry and Center for Drug
    Discovery and Innovation
University of South Florida
Tampa, Florida

Hendrik Luesch
Department of Medicinal Chemistry and
    Center for Natural Products
Drug Discovery and Development (CNPD3)
University of Florida
Gainesville, Florida

Joseph E. Media
Henry Ford Health System
Department of Internal Medicine
Division of Hematology and Oncology and the
    Josephine Ford Cancer Center
Detroit, Michigan

Brian T. Murphy
Department of Medicinal Chemistry and
    Pharmacognosy
University of Illinois at Chicago
Chicago, Illinois

David J. Newman
Natural Products Branch
Developmental Therapeutics Program
Division of Cancer Treatment and Diagnosis
Frederick National Laboratory for Cancer
    Research
Frederick, Maryland

Baldomero M. Olivera
Department of Biology
University of Utah
Salt Lake City, Utah

John D. Orr
Analytical Chemistry
Eisai, Inc.
Andover, Massachusetts

Ka-Lai Pang
Institute of Marine Biology and Centre of
    Excellence for the Oceans
National Taiwan Ocean University
Keelung, Taiwan (ROC)

Eun-Jung Park
Daniel K. Inouye College of Pharmacy
University of Hawaii at Hilo
Hilo, Hawaii

Cedric Pearce
Mycosynthetix, Inc.
Hillsborough, North Carolina

Eva-Rachele Pesce
College of Health and Biomedicine
Victoria University
Melbourne, Victoria, Australia

John M. Pezzuto
Daniel K. Inouye College of Pharmacy
University of Hawaii at Hilo
Hilo, Hawaii

Ngoc B. Pham
Eskitis Institute for Drug Discovery
Griffith University
Brisbane, Queensland, Australia

Halina Pietraszkiewicz
Henry Ford Health System
Department of Internal Medicine
Division of Hematology and Oncology and the
    Josephine Ford Cancer Center
Detroit, Michigan

Ronald J. Quinn
Eskitis Institute for Drug Discovery
Griffith University
Brisbane, Queensland, Australia

Helena Safavi-Hemami
Department of Biology
University of Utah
Salt Lake City, Utah

Jacqueline L. von Salm
Department of Chemistry and Center for Drug
    Discovery and Innovation
University of South Florida
Tampa, Florida

Lilibeth A. Salvador
Department of Medicinal Chemistry
University of Florida
Gainesville, Florida
and
Marine Science Institute
University of the Philippines, Diliman
Quezon City, Philippines

Eric W. Schmidt
Department of Medicinal Chemistry
L. S. Skaggs Pharmacy Institute
University of Utah
Salt Lake City, Utah

Anam Shaikh
Department of Medicinal Chemistry and
    Pharmacognosy
University of Illinois at Chicago
Chicago, Illinois

Balanehru Subramanian
Central Inter-Disciplinary Research Facility
Pillayarkuppam, Puducherry, India

Russell W. Teichert
Department of Biology
University of Utah
Salt Lake City, Utah

Kevin Tidgewell
Division of Medicinal Chemistry
Graduate School of Pharmaceutical Sciences
Duquesne University
Pittsburgh, Pennsylvania

Stephen Toms
Eskitis Institute for Drug Discovery
Griffith University
Brisbane, Queensland, Australia

Clement K. M. Tsui
Department of Pathology and Laboratory
    Medicine
University of British Columbia
Vancouver, British Columbia, Canada

Frederick A. Valeriote
Henry Ford Health System
Department of Internal Medicine
Division of Hematology and Oncology and the
    Josephine Ford Cancer Center
Detroit, Michigan

Robert Verpoorte
Natural Products Laboratory
Institute of Biology
Leiden University
Leiden, The Netherlands

Lilian L. P. Vrijmoed
Department of Biology and Chemistry
City University of Hong Kong
Kowloon, Hong Kong SAR
and
Beijing Normal University–Hong Kong Baptist
    University
United International College
Tangjiawan, Zhuhai, Guangdong
People's Republic of China

Ashley M. West
Division of Medicinal Chemistry
Department of Pharmaceutical Sciences
University of Connecticut
Storrs, Connecticut

Erica G. Wilson
Natural Products Laboratory
Institute of Biology
Leiden University
Leiden, The Netherlands

Christopher G. Witowski
Department of Chemistry and Center for Drug
    Discovery and Innovation
University of South Florida
Tampa, Florida

Ryan M. Young
Department of Chemistry and Center for Drug
    Discovery and Innovation
University of South Florida
Tampa, Florida
and
School of Chemistry
National University of Ireland Galway
Galway, Republic of Ireland

PART 1

Marine Biotechnology: Informing the Search

CHAPTER 1

Accessing Chemical Diversity through Metagenomics

Leah C. Blasiak and Russell T. Hill

Institute of Marine and Environmental Technology, University of Maryland
Center for Environmental Science, Baltimore, Maryland

CONTENTS

1.1 INTRODUCTION

The field of metagenomics, which is based on the analysis of DNA extracted from a community of microbes, has grown rapidly to address the challenge that the majority of microorganisms are not yet culturable in the laboratory and that those that are amenable to culture express only a fraction of their encoded natural products. Screening of metagenomic DNA libraries has emerged as a promising method to access unculturable or unexpressed chemical diversity. The methods for natural product discovery from metagenomic libraries can be divided into two main conceptual frameworks: activity-based screening and sequence-based screening. The first approach, activity-based screening, is similar to traditional bioassay-guided fractionation in that fragments of DNA are expressed in a heterologous host and assayed for production of a desired bioactivity. The alternative strategy, sequence-based screening, involves identifying fragments of DNA encoding biosynthetic genes based on sequence homology, followed by targeted heterologous expression and detection of the predicted small-molecule product. Both approaches rely on the tendency of microbial biosynthetic pathways to be clustered in a single region of the genome, and therefore a reasonable likelihood that all of the genes necessary for the production of a given natural product will be present in a contiguous fragment of DNA. Here we illustrate the two main strategies for discovery of natural products from metagenomic DNA libraries with case studies from the literature, including activity-based screening for the discovery of marine siderophores and sequence-based screening for the identification of the biosynthetic genes encoding the onnamides and polytheonamides. The gene clusters for the patellamides and ecteinascidins were recently found by direct sequencing of metagenomes. We highlight several of the practical challenges for efficient drug discovery from metagenomic libraries, including fractionation of complex communities, successful heterologous expression, and detection of the expressed products. We discuss current limitations to this approach that have resulted in the paucity of successes to date and make the case for a dramatic increase in the success rate of metagenomic approaches over the next 5 years. In addition, we explore some of the recently developed tools for metagenomic sequence assembly and detection of biosynthetic genes and make predictions about the future directions for *in silico* natural product discovery.

1.2 UNDERSTANDING METAGENOMICS

1.2.1 Definition and Origins

A metagenome is simply the total DNA present in a community of organisms or in an environmental sample. Thus, metagenomics refers to the sequencing or analysis of the total DNA extracted from a given sample. Some authors use the term *metagenomics* to include narrower, more circumscribed amplicon-based studies that examine the diversity of a particular gene from a community DNA sample (i.e., using the 16S ribosomal RNA [rRNA] gene as a phylogenetic marker sequence to compare community composition and structure). Here, we focus on the application of metagenomics to marine natural product discovery and emphasize the future goal of metagenomic methods to allow analysis of the total genetic potential of a community of organisms.

Jo Handelsman and colleagues first used the term *metagenome* in 1998 to describe the soil metagenome, or DNA from the mixed community of organisms in soil.[1] The soil metagenome contains many bacterial groups that are recalcitrant to culture and were thus an untapped source of natural products. Their initial paper outlined one of the central goals of metagenomics, accessing chemical and enzymatic diversity from uncultured bacteria, although it was published long before the advent of next-generation sequencing technologies that have brought large, complex metagenomes within reach. They also predicted some of the pitfalls for function-based screening of metagenomic libraries, including difficulty in obtaining robust heterologous expression of DNA from diverse sources; these issues remain key challenges today. In a proof-of-principle study, Handelsman and her coworkers described preparation of two soil metagenome bacterial artificial chromosome (BAC) libraries and confirmed heterologous expression of active enzymes from the clones.[2]

1.2.2 Motivation: Why Metagenomics?

1.2.2.1 Challenges of Bacterial Isolation and Culture

Marine invertebrates have been rich sources of natural products since the 1950s. A remarkable diversity of marine invertebrates and associated microbes remain unexplored for marine natural product discovery. The Census of Marine Life[3] estimated the number of known marine macro-organism species at 250,000, with ca. 750,000 remaining to be described, and the vast majority of these organisms are invertebrates. The estimate of the number of marine microbes is much higher, at tens of millions. Scientists have only fairly recently shown that many compounds discovered from marine invertebrate extracts are actually produced by host-associated microbes.[4–6] The microbial production of natural products has many important implications. A major challenge in moving forward promising compounds from marine invertebrates to become successful pharmaceuticals is the supply problem. In many cases, these compounds are present at very low concentrations and large numbers of invertebrates must be collected just for initial screening and structural elucidation of compounds of interest. For a marine drug to progress to market, a sustainable supply is necessary. In some cases, it may be possible to obtain in culture the microbes that are the true producers of the compounds, but this has proven to be difficult in practice. Metagenomic approaches hold great promise in solving the supply problem, and this can be an important factor in moving forward many of the bioactive natural products that have been identified from marine sources.[7] The cost of drug discovery is controversial and estimates vary widely, but figures in the range of $800 million to $1.8 billion have been proposed based on different underlying assumptions.[8] It is therefore not surprising that pharmaceutical companies are reluctant to invest in classes of compounds from marine sources when supply may be a problem if the compounds move forward to market. As more success stories emerge from these metagenomic approaches, pharmaceutical companies may revisit their low investments in marine natural product discovery.

An example of a promising compound isolated from a marine invertebrate but produced by symbiotic bacteria is the anticancer compound bryostatin. Bryostatin was initially isolated from the bryozoan *Bugula neritina*,[9] but was later shown to be produced by the extracellular bacterial symbiont "*Candidatus* Endobugula sertula."[10] One obstacle to determining the true producers of marine natural products is that the suspected bacterial producers are often recalcitrant to culture in the laboratory. The designation *Candidatus* for the bacterium that produces bryostatin indicates that this organism has never been cultured, and thus its taxonomic description is incomplete. In 1985, after Carl Woese first began to use rRNA genes as phylogenetic marker sequences to construct the evolutionary relatedness of all living things, a team led by Norm Pace pioneered the use of rRNA sequences to identify and quantify uncultured microorganisms from the environment.[11,12] Their work led to the remarkable discovery that the majority of environmental rRNA gene sequences came from organisms that had never been cultured in the laboratory.[12] Since then, methods to culture some of this uncultured majority have been developed, but it is still generally accepted that the majority of bacterial diversity has not yet been isolated in a laboratory setting (reviewed in Rappe and Giovannoni[13]). To put this expansion of our understanding of bacterial diversity in perspective, in 1987 only 11 phyla of bacteria were known, and in the past 25 years, scientists have cultured representatives of an additional 14 phyla and used 16S rRNA gene sequences from environmental samples to estimate that there are at least 60 major lineages of bacteria, with the majority remaining uncultured.[13–15] This uncultured bacterial diversity is likely to be a great source of novel chemical diversity. In summary, many marine bacteria are not yet amenable to culture in the laboratory, and metagenomics has emerged as a method to access the chemical diversity and biosynthetic genes of these elusive marine microbes.

1.2.2.2 Marine Invertebrates Host Complex Communities

In addition to the problem of laboratory culture of marine bacterial symbionts, the astonishing complexity and diversity of host-associated bacterial communities complicate linking bacterial producers to their natural products. Three marine invertebrates of interest in natural product research, all of which harbor symbiotic microbes, are shown in Figure 1.1. Most marine invertebrates live in association with microbes that are different from those in the surrounding seawater. For example, a high-microbial-abundance marine sponge such as *Xestospongia muta* hosts around 10^8 to 10^{10} bacterial cells per gram of wet weight,[16,17] and these bacteria are estimated to comprise around 200 operational taxonomic units (OTUs), or species-level groups.[18] Some marine sponges can have 30% of their weight comprised of symbiotic microbes, and the high density and morphological diversity of bacteria within a sponge are illustrated in Figure 1.2. It is perhaps not a coincidence that marine sponges have the highest proportion of their weight comprised of microbes and are also among the most prolific sources of novel bioactive compounds. In many cases, the bioactive compounds found on marine sponges are likely synthesized by the symbiotic microbes present in those sponges.[19,20] The amazing richness of host-associated bacterial communities makes it difficult to sort out which member of the consortium is responsible for producing an associated natural product. Further, some bacterial community members may produce low levels of compounds that are not easily detectable from within a complex metabolome. Therefore, marine invertebrate metagenomes likely encode considerably more natural product diversity than has been detected through screening bulk extracts. Metagenomics can be applied to these complex marine bacterial communities in order to identify the biosynthetic genes responsible for production of a given natural product or to identify new compounds not detected from whole-organism extracts.

(a)   (b)   (c)

Figure 1.1   Marine invertebrates as sources of natural products. (a) The marine sponge *Axinella corrugata* is chemically defended by the compound stevensine.[21] (Photo courtesy of Russell Hill.) (b) *Aplysia dactylomela*, also known as the spotted sea hare, is the source of the marine natural product dactylyne.[22] (Photo courtesy of Jan Vicente.) (c) The opistobranch *Elysia rufescens* and its algal diet *Bryopsis* sp. are both sources of the anticancer compound kahalalide F.[23] (Photo courtesy of Jeanette Davis.)

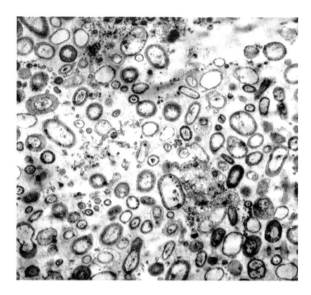

Figure 1.2 Marine invertebrates host complex microbial communities. A transmission electron micrograph through the mesohyl tissue of the marine sponge *Rhopaloeides odorabile* reveals a high density of microbial cells and a diverse array of bacterial morphotypes. (Photo courtesy of Nicole Webster.)

1.2.3 Uncultured Symbionts as Producers of Bioactive Compounds: Bryostatin Case Study

Bryostatins are cyclic peptides that have exciting biomedical potential because of their ability to modulate protein kinase C, an anticancer target.[24,25] Bryostatin 1, on which most of the pharmacological research has been focused, has been in more than 30 phase 1 and phase 2 clinical trials for cancer treatment, but has not yet progressed to phase 3 trials because of insufficient activity. Despite these setbacks, bryostatin 1 still shows promise for other biomedical applications, including treatment of neurological diseases.[26] Progress of this group of compounds, as for many other marine-derived compounds, would be greatly facilitated by an abundant and economic supply that would enable production of structural derivatives and also provide comfort to pharmaceutical companies contemplating development of this compound class.

Bryostatins were first discovered in extracts of the marine bryozoan *Bugula neritina*.[24] The research group of Margo Haygood has elucidated the role of bacterial symbionts present within this small marine invertebrate in production of the bryostatins. The structure of bryostatins indicates biosynthesis by a polyketide synthase (PKS), and ketosynthase (KS) gene fragments were successfully amplified from the holobiont comprising *B. neritina* larvae and an abundant bacterial symbiont located in the pallial sinus within the larvae. A probe based on these PKS gene fragments hybridized to the same region of the larvae as a 16S rRNA gene probe designed to target the "*Candidatus* Endobugula sertula" symbiont. Furthermore, when the numbers of bacterial symbionts were reduced by antibiotic treatment, there was a concomitant reduction in the signal from the PKS probe and in bryostatin levels.[27] Taken together, these data provide compelling circumstantial evidence that the true producer of bryostatins is the gamma-proteobacterium "*Candidatus* Endobugula sertula" that remains uncultured.

Following a procedure to enrich the bacterial fraction present in *B. neritina*, a cosmid library of ca. 14,000 clones was constructed and screened by using a probe targeting a KS gene fragment. This approach was successful in cloning the *bryA* gene proposed to encode part of the bryostatin molecule.[28] Subsequently, the entire putative biosynthetic gene cluster for byrostatin was cloned and sequenced,[10] and heterologous expression of the entire bry cluster was proposed as a way forward for producing bryostatins in sufficiently large quantities for pharmaceutical development. The substantial technical hurdles in heterologous expression of this very large gene cluster of ca. 80 kb and the possibility that some of the genes for bryostatin biosynthesis remained unidentified were acknowledged.[10] Although considerable progress was made in this interesting project, large-scale heterologous expression of bryostatins has to our knowledge not yet been achieved. This case study illustrates that even when biosynthetic gene clusters are successfully obtained, whether by the cloning approach described here or by the metagenomic approaches that are the focus of this chapter, heterologous expression can remain a formidable challenge.

1.2.4 Methods: Overview

Metagenomic approaches begin with extracting total DNA from an environmental sample or a marine organism and its associated microbes (Figure 1.3). Many companies, including Qiagen and MoBio, offer excellent DNA extraction kits that provide high-quality DNA and remove polymerase chain reaction (PCR) and sequencing inhibitors. DNA preparation can be a significant challenge for marine invertebrate samples that often contain abundant mucous and inhibiting chemicals. For approaches that require large DNA fragments (>30 kb), gentle lysis and traditional phenol-chloroform DNA extraction can be followed by pulsed-field gel electrophoresis (PFGE) to yield high-quality DNA in defined sizes.[29] In some cases, researchers use prefractionation of host tissues or cell populations to provide a less complex sample prior to DNA extraction (see Section 1.6.1).

Once DNA is available, it can be either introduced into a metagenomic DNA library for screening or directly sequenced. A metagenomic DNA library is a random set of DNA fragments cloned into a vector and maintained in a heterologous host, usually a bacterial strain. Each bacterial cell and resulting colony contains a single DNA sequence from the metagenome. The enzymes and resistance genes for natural products are usually clustered in bacterial genomes, although there are exceptions.[30] The natural grouping of biosynthetic pathways onto contiguous DNA sequence stretches facilitates identification of new pathways and means that randomly cloned fragments of DNA are likely to contain all the required biosynthetic and resistance genes for their respective product. Metagenomic libraries can then be screened by either activity-based (Section 1.3) or sequence-based (Section 1.4) methods to identify clones encoding biosynthetic gene clusters.

Alternatively, the total community DNA can be directly sequenced (Section 1.5) using next-generation sequencing methods, including 454 pyrosequencing (Roche), dye-terminator sequencing (Illumina), and single-molecule real-time (SMRT) sequencing (Pacific Biosciences). The details of these sequencing methodologies are outside the scope of this chapter, and we direct those with interest to Metzker.[31] A key advantage of all of the next-generation sequencing methods is that they do not require prepurification of many copies of each section of DNA prior to sequencing; rather, they can directly sequence random single-copy fragments of DNA. Thus, a complex mixture of DNA from a community of organisms can be directly sequenced, and then the metagenome can be assembled and searched *in silico*.

1.3 ACTIVITY-BASED SCREENING OF METAGENOMIC LIBRARIES

1.3.1 Methods and Challenges for Activity-Based Screening

Activity-based screening approaches detect a small-molecule product or enzyme activity from a heterologous host expressing a cloned metagenome fragment. For example, researchers might look for clones that display antibacterial activity. Once active clones are detected, they can identify

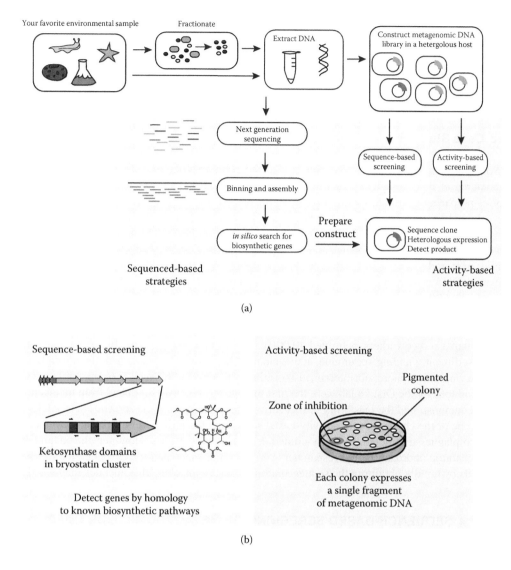

Figure 1.3 Methods for metagenomics.

the responsible small-molecule antibiotic or antibacterial protein. The activity-based screening approach requires functional expression of an entire biosynthetic gene cluster within the heterologous host, and so it will miss many clones containing partial gene clusters or clones that are inactive in the foreign expression environment. This greatly lowers the hit rate for activity-based screens but guarantees that the detected natural products can be robustly expressed to provide a supply of the compound. The activity-based screening method also links discovered natural products to biosynthetic pathways, since the genes encoding biosynthesis can be easily sequenced from the clone that encodes the bioactivity.

Although any assay type could in theory be used for activity-based screening of metagenomic libraries, in practice it is most efficient to use assays that produce visual changes to bacterial colonies on plates. Colony-based assays avoid the need to array clones in plates or prepare extracts from each clone, which would be prohibitive given the typical hit rates for activity-based metagenomic screens of around 0.01%, or 1 in 10,000 clones.[32] In colony-based assays, a metagenomic library is plated out at reasonable density and then visually inspected for active clones. The simplest clone-based, phenotypic assay is clone-dependent pigment production, where the heterologous expression of a metagenomic DNA fragment results in biosynthesis of a small-molecule pigment that changes the color of the host colony.[33] Researchers have also developed functional, colony-based screens for metagenomic DNA libraries that assay antibiotic and antifungal activity, antinematode activity, and siderophore production, as well as many assays for enzymatic activities.[2,34–37]

1.3.2 Case Study: Marine Siderophores

Some bacteria secrete small molecules called siderophores to bind and capture Fe(III), providing a source of this biologically required metal. An elegant, colony-based assay for siderophores involves preparing agar plates with Fe(III) and the dye chrome azurol S (CAS), which has a bright blue color when complexed to iron.[38] When a siderophore chelates the iron, the iron-free CAS changes color to orange. Thus, colonies that produce siderophores will have a distinctive orange halo on the blue assay plates. Otsuka and colleagues used the Fe-CAS assay to screen metagenomic libraries prepared from marine sediments for siderophore biosynthesis.[36,39] In one experiment, environmental DNA was prepared from deep-sea sediments collected in the East China Sea and a library of 60,000 clones was screened in an *Escherichia coli* heterologous host.[39] Otsuka's team found several positive clones in this assay, including one encoding production of the hydroxamate siderophore bisucaberin, a known siderophore from *Alteromonas* and *Vibrio* spp.[40,41] In a second screen, Otsuka and coworkers examined 40,000 clones from a metagenomic library prepared from tidal-flat sediment, yielding six unique, active clones. One active clone encoded the biosynthesis of the carboxylate siderophore vibrioferrin, also originally discovered from a *Vibrio* sp.[36,42] The siderophores produced by the Otsuka lab were the first to be heterologously produced from metagenomic DNA and demonstrated the feasibility of this activity-based metagenomic screen. Their results also illustrate one of the challenges for activity-based screening, the rediscovery of known compounds, which also plagues traditional, bioassay-guided drug discovery efforts. However, the advantage of the metagenomic approach is that researchers can limit the time spent on dereplication of known compounds by rapidly obtaining the sequence of the insert in the clone that encodes the bioactivity.

1.4 SEQUENCE-BASED SCREENING OF METAGENOMIC LIBRARIES

1.4.1 Methods and Challenges for Sequence-Based Screening

Another method to find natural product gene clusters within metagenomic DNA libraries involves screening for sequences with homology to known biosynthetic genes. Using degenerate primers, a mixture of primers that bind conserved sequences, researchers can amplify sequences related to the target gene. Clones that give a PCR product contain the biosynthetic gene of the desired type. For example, Schirmer et al. used degenerate PCR primers to amplify KS domains of type 1 PKSs from the sponge *Discodermia dissoluta*, the most abundant of which appeared to be encoded by a bacterial symbiont that remains uncultured.[43] Piel and coworkers used primers targeting KS domains to screen a metagenomic library from the marine sponge *Theonella swinhoei* to identify the gene clusters encoding the pederin-type polyketides, discussed in more detail below.[44] To screen a large library, Hrvatin and Piel developed a pooled screening method where many clones are screened in a single PCR and then positive pools are serially diluted and rescreened to identify the target clone.[45] Recently, a combined single-cell- and metagenomics-based approach revealed that two phylotypes of the candidate genus *Entotheonella* are present in the sponge *T. swinhoei* and are widely distributed in sponges. "*Entotheonella*" spp. are likely to include producers of diverse and unique bioactive compounds.[46]

Sequence-based screens often have higher hit rates than activity-based screens, and the PCR amplification step allows detection of even very rare clones. However, a significant problem with the sequence-based approach has been the difficulty in obtaining heterologous expression of detected clusters in order to directly link the discovered biosynthetic genes to their encoded small-molecule product.

1.4.2 Case Study: Onnamide Biosynthesis

Onnamide A was originally discovered from a Japanese marine sponge species in the genus *Theonella*.[47] The sponge *T. swinhoei* is a prodigious source of bioactive natural products, including peptides like the polytheonamides (discussed in Section 1.4.3); the thrombin inhibitors, the cyclotheonamides[48] and the cytotoxic orbiculamides[49]; and diverse other structural groups, including the aurantosides.[50] While onnamide A was isolated on the basis of its antiviral activity, it was later shown to be a potent antitumor agent that acts by inhibiting protein synthesis.[51] The onnamide A structure showed striking similarities to other known natural products, including mycalamide A[52] from a marine sponge *Mycale* sp. and the toxin pederin from rove beetles in the genus *Paederus*.[53,54] All of these natural products contain a core structure characterized by two tetrahydropyran moieties and an exomethylene group. The class of pederin-like natural products has now grown to include more than 30 structurally related compounds from 7 genera of marine sponges.[55] The core structures, along with stable isotope feeding experiments on pederin production, implied a polyketide synthase (PKS)-based biosynthetic mechanism.[56,57] Isolation of such closely related compounds from a variety of unrelated organisms suggested that they might be biosynthesized by an associated bacterial symbiont.

The original member of the structure class was the inflammatory toxin pederin, found in the hemolymph of a subset of female *Paederus* beetles.[58] Kellner first demonstrated that bacteria are the likely producers of pederin by showing that aposymbiotic beetles could acquire pederin by feeding with endosymbiont-containing eggs, and that antibiotic treatment of the eggs prevented transfer of pederin production.[59] Kellner then identified the bacterial symbiont as a *Pseudomonas* sp. by sequencing 16S ribosomal RNA genes from pederin positive and negative beetles, but he was unable to culture this putative pederin producer in the laboratory.[60] Because the symbiont could not be cultured, metagenomic methods were required to confirm bacterial production of pederin. Piel used sequence-based screening to identify PKS ketosynthase (KS) domains in a cosmid library of metagenomic DNA from beetles and their associated bacteria. He identified and sequenced a 54 kb region containing a mixed PKS-NRPS (nonribosomal polypeptide synthetase) biosynthetic gene cluster that was a strong candidate for pederin biosynthesis.[61] Sequence features in and around the gene cluster indicated that the DNA was of bacterial origin, and the surrounding genes shared high sequence identity to genes from sequenced *Pseudomonas* sp., consistent with the symbiont identified by Kellner.

The relatively simple bacterial community associated with *Paederus* beetles, which have only one dominant bacterial symbiont, facilitated Piel's successful identification of the pederin biosynthetic gene cluster. Direct PCR screening for KS genes from beetle metagenomic DNA yielded only sequences involved in pederin biosynthesis.[61] In contrast, initial screening and sequencing of 60 KS genes from the onnamide producer *T. swinhoei* produced 21 unique KS sequences, most of which were not closely related to the pederin KS sequences. By first tackling the relatively simple beetle metagenome, Piel and his team were able to sort through the complex mixture of potential biosynthetic genes in the sponge metagenome and identify the KS genes most similar to the pederin genes and therefore most likely to produce the related onnamides.[62] A distinctive feature of both the pederin and onnamide PKS clusters is that the acyltransferase (AT) domains, which select the acyl-CoA substrates, are not located within the main PKS gene but instead are encoded as separate genes. This type of "trans-AT" PKS, so-called because the AT domains act in *trans*, has since been shown to make up a large subclass of bacterial PKS pathways that are important for marine natural product biosynthesis.[63] The KS domains from trans-AT PKS pathways form a clade based on sequence similarity, which allows homology-based identification of trans-AT PKS clusters from only the KS sequence.[57]

Sequence-based screening of metagenomic libraries led to the identification of the bacterial biosynthetic gene clusters responsible for pederin and onnamide biosynthesis, without requiring culture of the bacterial symbiont. However, heterologous production of both compounds to directly link the biosynthetic genes with their small-molecule products has remained elusive. This challenge is especially frustrating given that one major goal of metagenomic studies is to produce a reliable supply of the natural product. Despite continued challenges for heterologous expression of full pathways, expression of individual biosynthetic genes can explicitly demonstrate their expected role in the biosynthetic mechanism. For example, Piel and coworkers showed that pyran synthase domains from the pederin biosynthetic gene cluster catalyze oxa-conjugate addition in to the formation of the product cyclic ether.[64]

1.4.3 Case Study: Polytheonamide Biosynthesis

The polytheonamides are large polypeptide natural products with strong cytotoxic activity. They were first isolated from the marine sponge *T. swinhoei* by bioassay-guided fraction by Hamada et al. in 1994.[65–67] Polytheonamides contain 48 amino acid residues and include many unusual amino acids, which raised the question of whether they originated from ribosomally synthesized and posttranslationally modified peptides (RiPPs) or from a nonribosomal polypeptide synthetase (NRPS).

In order to find the biosynthetic gene cluster responsible for production of the polytheonamides, Freeman et al. employed a sequenced-based strategy to screen metagenomic DNA.[68] Working under the hypothesis that the polytheonamides were produced ribosomally, they designed degenerate primers to target the presumed precursor peptide sequence. This strategy was feasible given the relatively long peptide product. Primers targeted for the ends of the 48-amino acid polytheonamide sequence yielded a ~150 bp product, and a second set of nested PCR primers was designed to provide a 100 bp product. Initially, Freeman and colleagues screened community metagenomic DNA isolated from *T. swinhoei* using their nested PCR design, and they succeeded in obtaining a DNA sequence that corresponded to the hypothesized precursor peptide. With this exact DNA sequence in hand, the Piel laboratory used PCR to screen a metagenomic library of sponge DNA containing ~60,000 cosmid clones.[44] From this enormous library, they detected a single positive clone that contained only a portion of the polytheonamide biosynthetic gene cluster. In order to obtain the surrounding DNA sequence, Freeman et al. first used PCR primers targeting a portion of the known precursor peptide gene and the vector sequence to rescreen the library. They then screened an even larger ~860,000 clone fosmid library and were finally able to identify a pool of ~1,000 clones that contained a single positive clone. However, for unknown reasons they were unable to isolate the positive clone from the pool. Instead, an additional ~7 kb sequence fragment was produced from the pool by PCR using a high-fidelity DNA polymerase and primers for the cosmid vector and targeted polytheonamide sequence. This determined effort illustrates how challenging it can be to obtain a large, intact gene cluster in a single clone.

To demonstrate that the biosynthetic genes isolated from the sponge metagenome were involved in polytheonamide biosynthesis, individual genes from the cluster were co-expressed along with the gene for the precursor peptide in *E. coli*. The authors found that co-expression with some of the tailoring enzymes was critical for production of a soluble peptide product. Surprisingly, they identified only six candidate tailoring enzymes encoded upstream of the precursor peptide gene that could explain all 48 posttranslational modifications observed in polytheonamides. The tailoring enzyme promiscuity observed for polytheonamide biosynthesis, along with the potential for mutation of the core precursor peptide, could be exploited in the future to manufacture diverse peptide libraries.

1.5 DIRECT SEQUENCING AND *DE NOVO* METAGENOME ASSEMBLY

1.5.1 Methods and Challenges for Direct Metagenome Sequencing

Next-generation sequencing technology has now matured to the point where it is possible to sequence and assemble partial genomes from complex mixtures of DNA. A single Illumina HiSeq 2500 run routinely produces 600 Gb of DNA sequence, albeit in short fragments, and the costs for sequencing are rapidly decreasing. The main obstacle for direct sequencing is *de novo* metagenome assembly, or extracting and assembling the unknown genomes of individual organisms from a mixture of short DNA sequences. Researchers are actively developing new bioinformatic software to bin sequences from different organisms and assemble complex metagenomes. Others are developing sequencing technologies that deliver very long reads, such as the Pac-Bio system, that will also help researchers assemble metagenomes. Scientists can search even partially assembled metagenomes for biosynthetic genes and then use PCR and genome walking techniques to assemble complete clusters. Alternatively, they can screen a metagenomic library with specific primers designed from the metagenome to pull out a targeted clone. As with the sequence-based library screening approach, the new challenge for direct metagenome sequencing will be heterologous expression of detected biosynthetic gene clusters in order to link the genes with natural products.

1.5.2 Case Study: Patellamides, Members of the Cyanobactin Class of Peptides

Schmidt and colleagues discovered the first biosynthetic gene cluster for a member of the cyanobactin class of natural products, patellamides A and C.[69] The patellamides were originally identified in extracts from the colonial tunicate *Lissoclinum patella*.[70] A single strain of cyanobacterium, *Prochloron didemni*, dominates the *L. patella* microbial community, and this symbiont was hypothesized to produce the patellamides.[71] Researchers were unable to culture *P. didemni* and so turned to metagenomic methods to access the patellamide biosynthetic gene cluster. Schmidt and his team used a prefractionation strategy to selectively enrich *Prochloron* DNA from the tunicate host. The relatively large size of *Prochloron* cells (10–20 μm diameter) and simple microbial community facilitated this approach; simply squeezing the tunicates released *Prochloron* cells in >95% purity.[71] Initially, a nonribosomal polypeptide synthetase (NRPS) mechanism was proposed for patellamide biosynthesis, but sequence-based searching for NRPS adenylation domains in *L. patella* metagenomic DNA yielded only one NRPS sequence that did not correlate with patellamide biosynthesis.[71]

Using the prefractionated *Prochloron* cells to generate DNA, around 10%–20% of the resulting metagenome sequence came from other tunicate-associated bacteria or the host.[72] Schmidt et al. assembled a draft genome sequence for the dominant cyanobacterial symbiont and used a simple BLAST search for the predicted core patellamide peptide to identify the biosynthetic gene cluster.[69] With the gene cluster sequence in hand, the team amplified the ~12 kb region of interest directly from a metagenomic DNA sample and cloned the sequence into an expression vector. They expressed this construct in *E. coli* and detected patellamide A production by liquid chromatography–mass spectrometry (LC-MS).

The patellamide study proved that directly sequencing a metagenome could provide access to natural product biosynthetic pathways. One advantage of the direct sequencing approach is that all the possible biosynthetic routes can be examined in one experiment, whereas PCR-based methods rely on sequence homology and a correct biosynthetic hypothesis to pull out the relevant genes. In the case of the patellamides, metagenome sequencing identified a ribosomal biosynthetic mechanism for a class of natural products that researchers previously thought were made by NRPS machinery. The cyanobactin class of ribosomally synthesized and posttranslationally modified peptides (RiPPs) continues to grow, with more than 200 members currently known.[73]

1.5.3 Case Study: ET-743 (Yondelis)

The anticancer drug ET-743 was originally isolated from the colonial tunicate *Ecteinascidia turbinata*, and structural similarity to microbial natural products saframycin and safracin suggested that a bacterial symbiont could be the actual source.[74] PharmaMar developed ET-743 as Yondelis®, which has been approved to treat relapsed soft tissue sarcomas in Europe and many countries worldwide.[75] Rath et al. sequenced the metagenome of the *E. turbinata* holobiont and identified the ET-743 gene cluster by comparison to known pathways for saframycin and safracin biosynthesis.[76] A conserved three-module NRPS encodes the production of the tetrahydroisoquinoline core for these closely related structures.[77] Although researchers have not yet produced ET-743 in a heterologous host, they have purified the ET-743 reductase domain and confirmed *in vitro* termination activity with the related saframycin substrate analogue.[76,77] In addition, Rath and colleagues showed expression of gene cluster proteins in ET-743-containing *E. turbinata* samples by mass spectrometry.[76]

The tunicate *E. turbinata* hosts a more complex bacterial community than the patellamide producer *L. patella*, with an estimated 30 bacterial community members. However, only three of these bacterial groups dominated the sequence dataset. The remaining bacterial groups contributed less than 1% each of the total sequence reads, and the tunicate host accounted for around 40% of the sequencing data. Obtaining good sequence coverage for abundant community members within a complex metagenome requires much less total sequence than accessing rare community members. Assuming an average bacterial genome size of 5 Mb and a tunicate genome around 200 Mb, the total metagenome of *E. turbinata* would be around 350 Mb. Rath et al. were able to assemble enough sequence to identify fragments of the ET-743 gene cluster from only around 280 Mb of raw sequence obtained from two 454 pyrosequencing shotgun runs. They then used specific PCR and genome walking to assemble a 35 kb contig that seems to contain the majority of the required biosynthetic genes. The team tentatively assigned the ET-743 cluster to the second most abundant bacterial symbiont, the gamma-proteobacterium "*Candidatus* Endoecteinascidia frumentensis," based on codon usage.

1.6 PRACTICAL CHALLENGES FOR METAGENOMICS

1.6.1 Fractionation

DNA from abundant community members can swamp metagenomic libraries and metagenome sequences, so methods to prefractionate communities can provide deeper coverage. For example, in the patellamide case study (Section 1.5.2), Schmidt and colleagues separated the cyanobacterial symbiont cells from the tunicate host prior to DNA extraction in order to simplify the resulting metagenome.[69] Other potential cell separation methods include physical dissection of different host tissues, density gradient centrifugation, filtration, or fluorescence-activated cell sorting (FACS). Whatever method is used, care must be taken to preserve the quality of the resulting DNA. An extreme method to fractionate a community is single-cell genomics. This technique uses cell sorting to obtain single bacterial cells, followed by whole-genome amplification (WGA) to randomly amplify the entire bacterial genome. Single-cell genomics allows sequencing and assembly of the genome from a single cell extracted from a complex community.[78]

1.6.2 Metagenome Assembly and Pathway Discovery

Rapid decreases in sequencing cost have made it possible to directly sequence large metagenomes; however, researchers are actively developing new software to improve metagenome assembly and identify natural product biosynthetic pathways. Current next-generation sequencing technologies produce relatively short reads (100–400 bp), making assembly very difficult (reviewed in Scholz et al.[79] and Schatz et al.[80]). Sequencing errors, short reads, and repetitive regions can all complicate correct genome assembly. NRPS and PKS gene clusters often contain repeating, closely related domains that can be challenging to assemble. Additionally, the sequences in a metagenome come from an unknown number of community members, and there may be large variations in the abundance of sequences from dominant and rare organisms. This biased coverage can also result in fragmented assemblies. If a typical genome project can be compared to assembling the text of a book from several shredded copies, then a metagenome assembly is like piecing back together a shredded library, with many copies of some books and only a few pages of others. Fortunately, assembly of large contigs—or long fragments of DNA—can be sufficient to identify biosynthetic genes, and a complete metagenome assembly is not required for pathway discovery. Improvements in sequencing technology, including the addition of paired-end reads and availability of much longer reads, are also expected to greatly simplify metagenome assembly in the future.

Once a metagenome is assembled into contigs, the next hurdle is to detect genes involved in natural product biosynthesis. Several tools have been developed to identify genes from common types of biosynthetic gene clusters, including KS domains from PKS clusters and adenylation and condensation domains from NRPS clusters.[81–83] The program antiSMASH 2.0, developed by the Fischbach laboratory, searches sequences for biosynthetic genes, compares those genes to characterized pathways and sequenced genomes, and even attempts to identify the boundaries of the gene cluster.[84] These tools will only improve as the number of characterized biosynthetic enzymes increases. The development of software that can mine sequences for biosynthetic pathways and then predict their encoded products will be critical to increase the success rate of metagenomic marine natural product discovery.

1.6.3 Heterologous Expression

One of the main challenges for metagenomic drug discovery is heterologous expression of biosynthetic gene clusters. Metagenomic methods rely on heterologous expression to provide access to new natural products and solve the supply problem for known products. Screening metagenomic libraries with activity-based assays depends on successful heterologous expression of the gene cluster and the small-molecule product. Heterologous expression also links the cloned genes to their small-molecule product for sequence-based screens. Of the gene clusters described in Sections 1.3 and 1.5, only the siderophores and patellamides were successfully expressed in heterologous hosts, highlighting the need for better expression systems to improve to the utility of metagenomic approaches.

The degree to which failed heterologous expression has limited metagenomic discovery efforts is debated, but researchers agree that it is a key step that requires further optimization and new approaches. Heterologous expression could fail because of incompatibilities between the cloned DNA and the host, poor transcription or regulation, or lack of required enzyme substrates. Problems with heterologous expression might explain the low hit rate observed in many function-based assays (~0.01%)[32] compared to the number of biosynthetic genes detected with sequence-based screens using degenerate PCR, with around 0.5%–0.7% of clones from a sponge metagenomic library positive for PKS genes in one study.[43] To test the ability of functional assays to detect bioactive gene clusters from metagenomic libraries, researchers screened DNA from six bacterial isolates that were known antibiotic producers. The authors obtained a hit rate of ~0.4% from this enriched and defined metagenomic sample, higher than most function-based assays, indicating that the low abundance of bioactive gene clusters contributes to low hit rates.[32] However, only DNA from proteobacterial hosts yielded active clones when expressed in their *E. coli* host, although they also assayed DNA from *Actinobacteria* and *Bacteroides*. A close relationship between the heterologous host and the original source of the biosynthetic genes seems to greatly increase the likelihood of successful expression. Others have observed that by screening the same metagenomic DNA library in multiple hosts, they detect many more natural products, with different clusters expressed by different hosts.[35]

The large size of biosynthetic operons means that many traditional cloning vectors can only capture a portion of a gene cluster on a single clone. The use of bacterial artificial chromosomes (BACs), or vectors that can handle very large inserts, now permits cloning of DNA fragments in the range of 100–300 kb. As seen in the large-scale study of the heterologous expression of proteins from *Plasmodium falciparum*, codon optimization is not the only barrier to heterologous expression, and many factors may need to be optimized to successfully express a pathway in an alternate host.[85] Some work has gone into the development of broad-host-range vectors that can be easily shuttled between a variety of hosts.[35] More commercially available vectors and expression systems are needed, in particular for facile cloning of large-insert libraries in *Streptomyces* and other hosts that will facilitate expression of genes from bacterial groups that are rich in biosynthetic genes.

The decreasing cost of synthetic DNA has made it possible to order vectors with gene clusters that are codon optimized for a given host. Until recently, cost has been a serious limitation since many biosynthetic gene clusters can be >50 kb, with some more than 100 kb. The steadily decreasing cost of DNA synthesis, with some manufacturers now offering DNA synthesis at prices under $0.30 per bp up to 10 kb genes, makes this approach increasingly attractive. A synthetic biology approach was successfully used to activate a cryptic polycyclic tetramate macrolactam biosynthetic gene cluster from *Streptomyces griseus*.[86]

An important related technical development to DNA synthesis is the ability to efficiently join linear DNA molecules by homologous recombination. Overlapping DNA fragments captured on smaller clones can also be assembled using recombination in yeast to yield a large gene cluster in a single vector.[87] Cobb and Zhao summarize the sophisticated DNA manipulation methods that are currently available and point out that the new approach devised by Fu and coworkers has several significant advantages.[88,89] Fu and coworkers used restriction digestions to obtain clusters of interest on single DNA fragments and then used their linear plus linear homologous recombination (LLHR) method to successfully clone large (10–52 kb) clusters. Two of these clusters from *Photorhabdus luminescens* were expressed in *E. coli* to identify the metabolites encoded by these pathways.

1.7 CONCLUSIONS AND FUTURE PROSPECTS

The case studies presented above highlight the successful application of metagenomics to biosynthetic pathway discovery. However, researchers have not yet translated this success to the discovery of new natural products from marine metagenomes. Put another way, while marine metagenomics has effectively connected products to gene clusters, it has proven much more challenging to obtain entirely new compounds from community DNA. A few new products, including the borregomycins,[90] arimetamycin A,[91] and the tetrarimycins,[92] have now been discovered from expression of metagenomic DNA libraries from soil, but so far, marine metagenomics has mainly been applied to the discovery of pathways encoding known compounds. In one marine success story, Donia and coworkers engineered a new cyclic peptide product, eptidemnamide, by mutating the sequence of the patellamide precursor peptide.[93]

In contrast to metagenomics, genome mining of cultured bacteria has already led to great advances in marine natural product discovery. Researchers search cultured bacterial genomes for cryptic, or uncharacterized, pathways and then detect the pathway product by induction in the native host or expression in a heterologous host. Recently, Yamanaka et al. cloned and expressed an NRPS gene cluster from the marine actinobacterium *Saccharomonospora* sp. CNQ-490 in *Streptomyces coelicolor* and discovered the new antibiotic taromycin A.[94] Genome-guided screening of extracts from *Salinispora* spp. has led to the discovery of salinilactam A,[95] salinisporamide K,[96] lomaiviticin,[97] and a new desferrioxamine siderophore.[98] Discovery from genomes of culturable bacteria circumvents some of the main challenges for metagenomic drug discovery: obtaining sufficient high-quality DNA and expression in a closely related host. Although this is an important approach in the short term, the next challenge for marine natural product discovery is the ability to obtain a sustainable supply of the many promising compounds already known from marine invertebrates and their largely uncultured associated microbes. Rapid advances in metagenomics make this an increasingly tractable problem. Even more promising is the potential to directly access the chemical diversity encoded in the huge untapped resource that exists in the genomes of marine invertebrates and their associated microbes. Overcoming the challenges of obtaining high-quality DNA and successfully expressing biosynthetic gene clusters from this DNA should make marine metagenomics a prolific source of new compounds for drug discovery.

REFERENCES

1. Handelsman, J., Rondon, M. R., Brady, S. F., Clardy, J., Goodman, R. M. Molecular biological access to the chemistry of unknown soil microbes: A new frontier for natural products. *Chem Biol* 1998, 5 (10), R245–49.
2. Rondon, M. R., August, P. R., Bettermann, A. D., Brady, S. F., Grossman, T. H., Liles, M. R., Loiacono, K. A., Lynch, B. A., MacNeil, I. A., Minor, C., Tiong, C. L., Gilman, M., Osburne, M. S., Clardy, J., Handelsman, J., Goodman, R. M. Cloning the soil metagenome: A strategy for accessing the genetic and functional diversity of uncultured microorganisms. *Appl Environ Microbiol* 2000, 66 (6), 2541–47.
3. Ausubel, J. H., Crist, D. T., Waggoner, P. E., eds. *First Census of Marine Life 2010: Highlights of a Decade of Discovery.* Census of Marine Life, Washington DC, 2010. www.coml.org.
4. Newman, D. J., Cragg, G. M. Marine natural products and related compounds in clinical and advanced preclinical trials. *J Nat Prod* 2004, 67 (8), 1216–38.
5. Salomon, C. E., Magarvey, N. A., Sherman, D. H. Merging the potential of microbial genetics with biological and chemical diversity: An even brighter future for marine natural product drug discovery. *Nat Prod Rep* 2004, 21 (1), 105–21.
6. Haygood, M. G., Schmidt, E. W., Davidson, S. K., Faulkner, D. J. Microbial symbionts of marine invertebrates: Opportunities for microbial biotechnology. *J Mol Microbiol Biotechnol* 1999, 1 (1), 33–43.
7. Newman, D. J., Hill, R. T. New drugs from marine microbes: The tide is turning. *J Ind Microbiol Biot* 2006, 33 (7), 539–44.
8. Paul, S. M., Mytelka, D. S., Dunwiddie, C. T., Persinger, C. C., Munos, B. H., Lindborg, S. R., Schacht, A. L. How to improve R&D productivity: The pharmaceutical industry's grand challenge. *Nat Rev Drug Discov* 2010, 9 (3), 203–14.
9. Pettit, G. R., Herald, C. L., Doubek, D. L., Herald, D. L., Arnold, E., Clardy, J. Anti-neoplastic agents. 86. Isolation and structure of bryostatin 1. *J Am Chem Soc* 1982, 104 (24), 6846–48.
10. Sudek, S., Lopanik, N. B., Waggoner, L. E., Hildebrand, M., Anderson, C., Liu, H., Patel, A., Sherman, D. H., Haygood, M. G. Identification of the putative bryostatin polyketide synthase gene cluster from "*Candidatus* Endobugula sertula", the uncultivated microbial symbiont of the marine bryozoan *Bugula neritina*. *J Nat Prod* 2007, 70 (1), 67–74.
11. Stahl, D. A., Lane, D. J., Olsen, G. J., Pace, N. R. Characterization of a Yellowstone hot spring microbial community by 5S rRNA sequences. *Appl Environ Microbiol* 1985, 49 (6), 1379–84.
12. Olsen, G. J., Lane, D. J., Giovannoni, S. J., Pace, N. R., Stahl, D. A. Microbial ecology and evolution: A ribosomal RNA approach. *Annu Rev Microbiol* 1986, 40, 337–65.
13. Rappe, M. S., Giovannoni, S. J. The uncultured microbial majority. *Annu Rev Microbiol* 2003, 57, 369–94.
14. Rinke, C., Schwientek, P., Sczyrba, A., Ivanova, N. N., Anderson, I. J., Cheng, J. F., Darling, A., Malfatti, S., Swan, B. K., Gies, E. A., Dodsworth, J. A., Hedlund, B. P., Tsiamis, G., Sievert, S. M., Liu, W. T., Eisen, J. A., Hallam, S. J., Kyrpides, N. C., Stepanauskas, R., Rubin, E. M., Hugenholtz, P., Woyke, T. Insights into the phylogeny and coding potential of microbial dark matter. *Nature* 2013, 499 (7459), 431–37.
15. Quast, C., Pruesse, E., Yilmaz, P., Gerken, J., Schweer, T., Yarza, P., Peplies, J., Glockner, F. O. The SILVA ribosomal RNA gene database project: Improved data processing and web-based tools. *Nucleic Acids Res* 2013, 41 (D1), D590–96.

16. Hentschel, U., Usher, K. M., Taylor, M. W. Marine sponges as microbial fermenters. *FEMS Microbiol Ecol* 2006, 55 (2), 167–77.

17. Hill, R. T. Microbes from marine sponges: A treasure trove of biodiversity for natural products discovery. In *Microbial Diversity and Bioprospecting*, ed. A. T. Bull. ASM Press, Washington, DC, 2004, 177–190.

18. Montalvo, N. F., Hill, R. T. Sponge-associated bacteria are strictly maintained in two closely related but geographically distant sponge hosts. *Appl Environ Microb* 2011, 77 (20), 7207–16.

19. Piel, J. Metabolites from symbiotic bacteria. *Nat Prod Rep* 2004, 21 (4), 519–38.

20. Waters, A. L., Hill, R. T., Place, A. R., Hamann, M. T. The expanding role of marine microbes in pharmaceutical development. *Curr Opin Biotechnol* 2010, 21 (6), 780–86.

21. Wilson, D. M., Puyana, M., Fenical, W., Pawlik, J. R. Chemical defense of the Caribbean reef sponge *Axinella corrugata* against predatory fishes. *J Chem Ecol* 1999, 25 (12), 2811–23.

22. Mcdonald, F. J., Campbell, D. C., Vanderah, D. J., Schmitz, F. J., Washecheck, D. M., Burks, J. E., Vanderhelm, D. Marine natural products: Dactylyne, an acetylenic dibromochloro ether from sea hare *Aplysia dactylomela*. *J Org Chem* 1975, 40 (5), 665–66.

23. Hamann, M. T., Scheuer, P. J. Kahalalide F: A bioactive depsipeptide from the sacoglossan mollusk *Elysia rufescens* and the green alga *Bryopsis* sp. *J Am Chem Soc* 1993, 115 (13), 5825–26.

24. Pettit, G. R., Herald, C. L., Doubek, D. L., Herald, D. L., Arnold, E., Clardy, J. Anti-neoplastic agents. 86. Isolation and structure of bryostatin-1. *J Am Chem Soc* 1982, 104 (24), 6846–48.

25. Kraft, A. S., Smith, J. B., Berkow, R. L. Bryostatin, an activator of the calcium phospholipid-dependent protein-kinase, blocks phorbol ester-induced differentiation of human promyelocytic leukemia-cells Hl-60. *Proc Natl Acad Sci USA* 1986, 83 (5), 1334–38.

26. Trindade-Silva, A. E., Lim-Fong, G. E., Sharp, K. H., Haygood, M. G. Bryostatins: Biological context and biotechnological prospects. *Curr Opin Biotechnol* 2010, 21 (6), 834–42.

27. Davidson, S. K., Allen, S. W., Lim, G. E., Anderson, C. M., Haygood, M. G. Evidence for the biosynthesis of bryostatins by the bacterial symbiont "*Candidatus* Endobugula sertula" of the bryozoan *Bugula neritina*. *Appl Environ Microb* 2001, 67 (10), 4531–37.

28. Hildebrand, M., Waggoner, L. E., Liu, H. B., Sudek, S., Allen, S., Anderson, C., Sherman, D. H., Haygood, M. bryA: An unusual modular polyketide synthase gene from the uncultivated bacterial symbiont of the marine bryozoan *Bugula neritina*. *Chem Biol* 2004, 11 (11), 1543–1552.

29. Gillespie, D. E., Rondon, M. R., Williamson, L. L., Handelsman, J. Metagenomic libraries from uncultured microorganisms. In *Molecular Microbial Ecology*. Taylor & Francis Group, New York, 2005, pp. 261–79.

30. Bosello, M., Robbel, L., Linne, U., Xie, X., Marahiel, M. A. Biosynthesis of the siderophore rhodochelin requires the coordinated expression of three independent gene clusters in *Rhodococcus jostii* RHA1. *J Am Chem Soc* 2011, 133 (12), 4587–95.

31. Metzker, M. L. Applications of next-generation sequencing technologies: The next generation. *Nat Rev Genet* 2010, 11 (1), 31–46.

32. Penesyan, A., Ballestriero, F., Daim, M., Kjelleberg, S., Thomas, T., Egan, S. Assessing the effectiveness of functional genetic screens for the identification of bioactive metabolites. *Mar Drugs* 2013, 11 (1), 40–49.

33. Brady, S. F., Chao, C. J., Handelsman, J., Clardy, J. Cloning and heterologous expression of a natural product biosynthetic gene cluster from eDNA. *Org Lett* 2001, 3 (13), 1981–84.

34. Craig, J. W., Chang, F. Y., Brady, S. F. Natural products from environmental DNA hosted in *Ralstonia metallidurans*. *ACS Chem Biol* 2009, 4 (1), 23–28.

35. Craig, J. W., Chang, F. Y., Kim, J. H., Obiajulu, S. C., Brady, S. F. Expanding small-molecule functional metagenomics through parallel screening of broad-host-range cosmid environmental DNA libraries in diverse proteobacteria. *Appl Environ Microbiol* 2010, 76 (5), 1633–41.

36. Fujita, M. J., Kimura, N., Sakai, A., Ichikawa, Y., Hanyu, T., Otsuka, M. Cloning and heterologous expression of the vibrioferrin biosynthetic gene cluster from a marine metagenomic library. *Biosci Biotech Biochem* 2011, 75 (12), 2283–87.

37. Iqbal, H. A., Feng, Z. Y., Brady, S. F. Biocatalysts and small molecule products from metagenomic studies. *Curr Opin Chem Biol* 2012, 16 (1–2), 109–16.

38. Schwyn, B., Neilands, J. B. Universal chemical-assay for the detection and determination of siderophores. *Anal Biochem* 1987, 160 (1), 47–56.

39. Fujita, M. J., Kimura, N., Yokose, H., Otsuka, M. Heterologous production of bisucaberin using a biosynthetic gene cluster cloned from a deep sea metagenome. *Mol Biosyst* 2012, 8 (2), 482–85.

40. Kadi, N., Song, L. J., Challis, G. L. Bisucaberin biosynthesis: An adenylating domain of the BibC multienzyme catalyzes cyclodimerization of N-hydroxy-N-succinylcadaverine. *Chem Commun* 2008 (41), 5119–21.

41. Takahashi, A., Nakamura, H., Kameyama, T., Kurasawa, S., Naganawa, H., Okami, Y., Takeuchi, T., Umezawa, H. Bisucaberin, a new siderophore, sensitizing tumor-cells to macrophage-mediated cytolysis. 2. Physicochemical properties and structure determination. *J Antibiot* 1987, 40 (12), 1671–76.

42. Yamamoto, S., Okujo, N., Yoshida, T., Matsuura, S., Shinoda, S. Structure and iron transport activity of vibrioferrin, a new siderophore of *Vibrio parahaemolyticus*. *J Biochem* 1994, 115 (5), 868–74.

43. Schirmer, A., Gadkari, R., Reeves, C. D., Ibrahim, F., DeLong, E. F., Hutchinson, C. R. Metagenomic analysis reveals diverse polyketide synthase gene clusters in microorganisms associated with the marine sponge *Discodermia dissoluta*. *Appl Environ Microb* 2005, 71 (8), 4840–49.

44. Piel, J., Hui, D. Q., Wen, G. P., Butzke, D., Platzer, M., Fusetani, N., Matsunaga, S. Antitumor polyketide biosynthesis by an uncultivated bacterial symbiont of the marine sponge *Theonella swinhoei*. *Proc Natl Acad Sci USA* 2004, 101 (46), 16222–27.

45. Hrvatin, S., Piel, J. Rapid isolation of rare clones from highly complex DNA libraries by PCR analysis of liquid gel pools. *J Microbiol Methods* 2007, 68 (2), 434–36.

46. Wilson, M. C., Mori, T., Ruckert, C., Uria, A. R., Helf, M. J., Takada, K., Gernert, C., Steffens, U. A., Heycke, N., Schmitt, S., Rinke, C., Helfrich, E. J., Brachmann, A. O., Gurgui, C., Wakimoto, T., Kracht, M., Crusemann, M., Hentschel, U., Abe, I., Matsunaga, S., Kalinowski, J., Takeyama, H., Piel, J. An environmental bacterial taxon with a large and distinct metabolic repertoire. *Nature* 2014, 506 (7486), 58–62.

47. Sakemi, S., Ichiba, T., Kohmoto, S., Saucy, G., Higa, T. Isolation and structure elucidation of onnamide A, a new bioactive metabolite of a marine sponge, *Theonella* sp. *J Am Chem Soc* 1988, 110 (14), 4851–53.

48. Fusetani, N., Matsunaga, S., Matsumoto, H., Takebayashi, Y. Bioactive marine metabolites. 33. Cyclotheonamides, potent thrombin inhibitors, from a marine sponge *Theonella* sp. *J Am Chem Soc* 1990, 112 (19), 7053–54.

49. Fusetani, N., Sugawara, T., Matsunaga, S., Hirota, H. Bioactive marine metabolites series. 36. Orbiculamide A: A novel cytotoxic cyclic peptide from a marine sponge *Theonella* sp. *J Am Chem Soc* 1991, 113 (20), 7811–12.

50. Matsunaga, S., Fusetani, N., Kato, Y., Hirota, H. Aurantosides A and B: Cytotoxic tetramic acid glycosides from the marine sponge *Theonella* sp. *J Am Chem Soc* 1991, 113 (25), 9690–92.

51. Burres, N. S., Clement, J. J. Antitumor-activity and mechanism of action of the novel marine natural-products mycalamide-A and mycalamide-B and onnamide. *Cancer Res* 1989, 49 (11), 2935–40.

52. Perry, N. B., Blunt, J. W., Munro, M. H. G., Pannell, L. K. Mycalamide A, an antiviral compound from a New Zealand sponge of the genus *Mycale*. *J Am Chem Soc* 1988, 110 (14), 4850–51.

53. Cardani, C., Ghiringh, D., Mondelli, R., Quilico, A. Structure of pederin. *Tetrahedron Lett* 1965 (29), 2537–45.

54. Furusaki, A., Watanabe, T., Matsumot, T., Yanagiya, M. Crystal and molecular structure of pederin di-p-bromobenzoate. *Tetrahedron Lett* 1968 (60), 6301–4.

55. Mosey, R. A., Floreancig, P. E. Isolation, biological activity, synthesis, and medicinal chemistry of the pederin/mycalamide family of natural products. *Nat Prod Rep* 2012, 29 (9), 980–95.

56. Cardani, C., Fuganti, C., Ghiringh, D., Grassell, P., Pavan, M., Valcuron, Md. Biosynthesis of pederin. *Tetrahedron Lett* 1973 (30), 2815–18.

57. Piel, J., Butzke, D., Fusetani, N., Hui, D. Q., Platzer, M., Wen, G. P., Matsunaga, S. Exploring the chemistry of uncultivated bacterial symbionts: Antitumor polyketides of the pederin family. *J Nat Prod* 2005, 68 (3), 472–79.

58. Kellner, R. L. L., Dettner, K. Allocation of pederin during lifetime of *Paederus* rove beetles (Coleoptera: Staphylinidae): Evidence for polymorphism of hemolymph toxin. *J Chem Ecol* 1995, 21 (11), 1719–33.

59. Kellner, R. L. L. Suppression of pederin biosynthesis through antibiotic elimination of endosymbionts in *Paederus sabaeus*. *J Insect Physiol* 2001, 47 (4–5), 475–83.

60. Kellner, R. L. L. Molecular identification of an endosymbiotic bacterium associated with pederin biosynthesis in *Paederus* sabaeus (Coleoptera: Staphylinidae). *Insect Biochem Mol* 2002, 32 (4), 389–95.

61. Piel, J. A polyketide synthase-peptide synthetase gene cluster from an uncultured bacterial symbiont of *Paederus* beetles. *Proc Natl Acad Sci USA* 2002, 99 (22), 14002–7.

62. Piel, J., Hui, D. Q., Fusetani, N., Matsunaga, S. Targeting modular polyketide synthases with iteratively acting acyltransferases from metagenomes of uncultured bacterial consortia. *Environ Microbiol* 2004, 6 (9), 921–27.

63. Piel, J. Biosynthesis of polyketides by trans-AT polyketide synthases. *Nat Prod Rep* 2010, 27 (7), 996–1047.

64. Poplau, P., Frank, S., Morinaka, B. I., Piel, J. An enzymatic domain for the formation of cyclic ethers in complex polyketides. *Angew Chem Int Ed Engl* 2013, 52 (50), 13215–18.

65. Hamada, T., Matsunaga, S., Yano, G., Fusetani, N. Polytheonamides A and B, highly cytotoxic, linear polypeptides with unprecedented structural features, from the marine sponge, *Theonella swinhoei*. *J Am Chem Soc* 2005, 127 (1), 110–18.

66. Hamada, T., Sugawara, T., Matsunaga, S., Fusetani, N. Polytheonamides, unprecedented highly cytotoxic polypeptides, from the marine sponge *Theonella swinhoei*. 1. Isolation and component amino-acids. *Tetrahedron Lett* 1994, 35 (5), 719–20.

67. Hamada, T., Sugawara, T., Matsunaga, S., Fusetani, N. Bioactive marine metabolism. 56. Polytheonamides, unprecedented highly cytotoxic polypeptides from the marine sponge *Theonella swinhoei*. 2. Structure elucidation. *Tetrahedron Lett* 1994, 35 (4), 609–12.

68. Freeman, M. F., Gurgui, C., Helf, M. J., Morinaka, B. I., Uria, A. R., Oldham, N. J., Sahl, H. G., Matsunaga, S., Piel, J. Metagenome mining reveals polytheonamides as posttranslationally modified ribosomal peptides. *Science* 2012, 338 (6105), 387–90.

69. Schmidt, E. W., Nelson, J. T., Rasko, D. A., Sudek, S., Eisen, J. A., Haygood, M. G., Ravel, J. Patellamide A and C biosynthesis by a microcin-like pathway in *Prochloron didemni*, the cyanobacterial symbiont of *Lissoclinum patella*. *Proc Natl Acad Sci USA* 2005, 102 (20), 7315–20.

70. Degnan, B. M., Hawkins, C. J., Lavin, M. F., McCaffrey, E. J., Parry, D. L., van den Brenk, A. L., Watters, D. J. New cyclic peptides with cytotoxic activity from the ascidian *Lissoclinum patella*. *J Med Chem* 1989, 32 (6), 1349–54.

71. Schmidt, E. W., Sudek, S., Haygood, M. G. Genetic evidence supports secondary metabolic diversity in *Prochloron* spp., the cyanobacterial symbiont of a tropical ascidian. *J Nat Prod* 2004, 67 (8), 1341–45.

72. Donia, M. S., Fricke, W. F., Partensky, F., Cox, J., Elshahawi, S. I., White, J. R., Phillippy, A. M., Schatz, M. C., Piel, J., Haygood, M. G., Ravel, J., Schmidt, E. W. Complex microbiome underlying secondary and primary metabolism in the tunicate-*Prochloron* symbiosis. *Proc Natl Acad Sci USA* 2011, 108 (51), E1423–32.

73. Arnison, P. G., Bibb, M. J., Bierbaum, G., Bowers, A. A., Bugni, T. S., Bulaj, G., Camarero, J. A., Campopiano, D. J., Challis, G. L., Clardy, J., Cotter, P. D., Craik, D. J., Dawson, M., Dittmann, E., Donadio, S., Dorrestein, P. C., Entian, K. D., Fischbach, M. A., Garavelli, J. S., Goransson, U., Gruber, C. W., Haft, D. H., Hemscheidt, T. K., Hertweck, C., Hill, C., Horswill, A. R., Jaspars, M., Kelly, W. L., Klinman, J. P., Kuipers, O. P., Link, A. J., Liu, W., Marahiel, M. A., Mitchell, D. A., Moll, G. N., Moore, B. S., Muller, R., Nair, S. K., Nes, I. F., Norris, G. E., Olivera, B. M., Onaka, H., Patchett, M. L., Piel, J., Reaney, M. J., Rebuffat, S., Ross, R. P., Sahl, H. G., Schmidt, E. W., Selsted, M. E., Severinov, K., Shen, B., Sivonen, K., Smith, L., Stein, T., Sussmuth, R. D., Tagg, J. R., Tang, G. L., Truman, A. W., Vederas, J. C., Walsh, C. T., Walton, J. D., Wenzel, S. C., Willey, J. M., van der Donk, W. A. Ribosomally synthesized and post-translationally modified peptide natural products: Overview and recommendations for a universal nomenclature. *Nat Prod Rep* 2013, 30 (1), 108–60.

74. Piel, J. Bacterial symbionts: Prospects for the sustainable production of invertebrate-derived pharmaceuticals. *Curr Med Chem* 2006, 13 (1), 39–50.

75. Schuler, M. K., Richter, S., Platzek, I., Beuthien-Baumann, B., Wieczorek, K., Hamann, C., Mohm, J., Ehninger, G. Trabectedin in the neoadjuvant treatment of high-grade pleomorphic sarcoma: Report of a rare case and literature review. *Case Rep Oncol Med* 2013, 2013, 320797.

76. Rath, C. M., Janto, B., Earl, J., Ahmed, A., Hu, F. Z., Hiller, L., Dahlgren, M., Kreft, R., Yu, F., Wolff, J. J., Kweon, H. K., Christiansen, M. A., Hakansson, K., Williams, R. M., Ehrlich, G. D., Sherman, D. H. Meta-omic characterization of the marine invertebrate microbial consortium that produces the chemotherapeutic natural product ET-743. *ACS Chem Biol* 2011, 6 (11), 1244–56.

77. Koketsu, K., Watanabe, K., Suda, H., Oguri, H., Oikawa, H. Reconstruction of the saframycin core scaffold defines dual Pictet-Spengler mechanisms. *Nat Chem Biol* 2010, 6 (6), 408–10.

78. Walker, A., Parkhill, J. Genome watch: Single-cell genomics. *Nat Rev Microbiol* 2008, 6 (3), 176–77.

79. Scholz, M. B., Lo, C. C., Chain, P. S. G. Next generation sequencing and bioinformatic bottlenecks: The current state of metagenomic data analysis. *Curr Opin Biotechnol* 2012, 23 (1), 9–15.

80. Schatz, M. C., Delcher, A. L., Salzberg, S. L. Assembly of large genomes using second-generation sequencing. *Genome Res* 2010, 20 (9), 1165–73.

81. Ziemert, N., Podell, S., Penn, K., Badger, J. H., Allen, E., Jensen, P. R. The natural product domain seeker NaPDoS: A phylogeny based bioinformatic tool to classify secondary metabolite gene diversity. *PLoS One* 2012, 7 (3).

82. Bachmann, B. O., Ravel, J. Chapter 8. Methods for *in silico* prediction of microbial polyketide and nonribosomal peptide biosynthetic pathways from DNA sequence data. *Methods Enzymol* 2009, 458, 181–217.

83. Ichikawa, N., Sasagawa, M., Yamamoto, M., Komaki, H., Yoshida, Y., Yamazaki, S., Fujita, N. DoBISCUIT: A database of secondary metabolite biosynthetic gene clusters. *Nucleic Acids Res* 2013, 41 (D1), D408–14.

84. Blin, K., Medema, M. H., Kazempour, D., Fischbach, M. A., Breitling, R., Takano, E., Weber, T. antiSMASH 2.0: A versatile platform for genome mining of secondary metabolite producers. *Nucleic Acids Res* 2013, 41 (W1), W204–12.

85. Mehlin, C., Boni, E., Buckner, F. S., Engel, L., Feist, T., Gelb, M. H., Haji, L., Kim, D., Liu, C., Mueller, N., Myler, P. J., Reddy, J. T., Sampson, J. N., Subramanian, E., Van Voorhis, W. C., Worthey, E., Zucker, F., Hol, W. G. J. Heterologous expression of proteins from *Plasmodium falciparum*: Results from 1000 genes. *Mol Biochem Parasit* 2006, 148 (2), 144–60.

86. Luo, Y. Z., Huang, H., Liang, J., Wang, M., Lu, L., Shao, Z. Y., Cobb, R. E., Zhao, H. M. Activation and characterization of a cryptic polycyclic tetramate macrolactam biosynthetic gene cluster. *Nat Commun* 2013, 4, 2894.

87. Kim, J. H., Feng, Z., Bauer, J. D., Kallifidas, D., Calle, P. Y., Brady, S. F. Cloning large natural product gene clusters from the environment: Piecing environmental DNA gene clusters back together with TAR. *Biopolymers* 2010, 93 (9), 833–44.

88. Cobb, R. E., Zhao, H. M. Direct cloning of large genomic sequences. *Nat Biotechnol* 2012, 30 (5), 405–6.

89. Fu, J., Bian, X. Y., Hu, S. B., Wang, H. L., Huang, F., Seibert, P. M., Plaza, A., Xia, L. Q., Muller, R., Stewart, A. F., Zhang, Y. M. Full-length RecE enhances linear-linear homologous recombination and facilitates direct cloning for bioprospecting. *Nat Biotechnol* 2012, 30 (5), 440–46.

90. Chang, F. Y., Brady, S. F. Discovery of indolotryptoline antiproliferative agents by homology-guided metagenomic screening. *Proc Natl Acad Sci USA* 2013, 110 (7), 2478–83.

91. Kang, H. S., Brady, S. F. Arimetamycin A: Improving clinically relevant families of natural products through sequence-guided screening of soil metagenomes. *Angew Chem Int Ed Engl* 2013, 52 (42), 11063–67.

92. Kallifidas, D., Kang, H. S., Brady, S. F. Tetarimycin A, an MRSA-active antibiotic identified through induced expression of environmental DNA gene clusters. *J Am Chem Soc* 2012, 134 (48), 19552–55.

93. Donia, M. S., Hathaway, B. J., Sudek, S., Haygood, M. G., Rosovitz, M. J., Ravel, J., Schmidt, E. W. Natural combinatorial peptide libraries in cyanobacterial symbionts of marine ascidians. *Nat Chem Biol* 2006, 2 (12), 729–35.

94. Yamanaka, K., Reynolds, K. A., Kersten, R. D., Ryan, K. S., Gonzalez, D. J., Nizet, V., Dorrestein, P. C., Moore, B. S. Direct cloning and refactoring of a silent lipopeptide biosynthetic gene cluster yields the antibiotic taromycin A. *Proc Natl Acad Sci USA* 2014, 111 (5), 1957–62.

95. Udwary, D. W., Zeigler, L., Asolkar, R. N., Singan, V., Lapidus, A., Fenical, W., Jensen, P. R., Moore, B. S. Genome sequencing reveals complex secondary metabolome in the marine actinomycete *Salinispora tropica*. *Proc Natl Acad Sci USA* 2007, 104 (25), 10376–81.

96. Eustaquio, A. S., Nam, S. J., Penn, K., Lechner, A., Wilson, M. C., Fenical, W., Jensen, P. R., Moore, B. S. The discovery of salinosporamide K from the marine bacterium "*Salinispora pacifica*" by genome mining gives insight into pathway evolution. *Chembiochem* 2011, 12 (1), 61–64.

97. Kersten, R. D., Lane, A. L., Nett, M., Richter, T. K., Duggan, B. M., Dorrestein, P. C., Moore, B. S. Bioactivity-guided genome mining reveals the lomaiviticin biosynthetic gene cluster in *Salinispora tropica*. *Chembiochem* 2013, 14 (8), 955–62.

98. Ejje, N., Soe, C. Z., Gu, J., Codd, R. The variable hydroxamic acid siderophore metabolome of the marine actinomycete *Salinispora tropica* CNB-440. *Metallomics* 2013, 5 (11), 1519–28.

CHAPTER 2

Chemical Characterization of Marine Organisms Using a Metabolomic Approach

Kashif Ali,[1,2] **Robert Verpoorte,**[1] **Erica G. Wilson,**[1] **and Young Hae Choi**[1]

[1]Natural Products Laboratory, Institute of Biology, Leiden University, Leiden, The Netherlands
[2]Biosciences Department, Shaheed Zulfikar Ali Bhutto Institute of Science and Technology (SZABIST), Karachi, Pakistan

CONTENTS

2.1 INTRODUCTION

Water covers more than 70% of the surface of the earth and oceans hold about 96.5% of this water. This leads to a great environmental diversity resulting from the very different physical and chemical conditions throughout the body of water. There is a vast number of eukaryotic organisms, and though more than 212,000 species have been accepted according to the World Register of Marine Species (WoRMS), it is estimated that this number only accounts for a quarter of the estimated 0.7 million to 1 million marine species,[1] or even less if considering other predictions of a total 2.2 million species.[2] According to WoRMS, the amount of described and accepted species increases by about 2000 per year, revealing the great interest in exploring this environment.

Marine organisms contain a wide range of structurally diverse chemical compounds related to diverse biological processes such as metabolic pathways, reproduction, development patterns, and defense mechanisms.[3] In recent years, marine biologists have been applying genomics to organisms and ecosystems, and as a result, there is a substantial amount of genomic information that can be found in several genomic resource databanks, such as Megx, CAMERA, JGI, and GOLD.[4] With this large amount of genomic data, it is possible to predict the potential of marine organisms to synthesize a great number of chemicals. Thus, while terrestrial plants seem to have monopolized the attention of novel compound researchers during the last century, the focus of investigations has now turned to the marine environment as a relatively unexplored source of new potentially bioactive molecules. Taking advantage of advances in technologies for collection, cultivation, and analysis, investigators of different disciplines, such as chemistry, pharmacology, biology, and ecology,[5–7] have directed their efforts to marine products with the hope of finding novel bioactive compounds or lead molecule candidates for drug development. According to the database Marinlit (database of the marine natural product literature, http://pubs.rsc.org/marinlit/), the number of compounds isolated over a 10-year period, starting in 1995, has increased from 6,500 to above 19,000.[8] This increase in marine natural product bioprospecting has yielded a considerable number of drug candidates. Two marine natural products have recently been approved as new medicinal drugs: Prialt (also known as ziconotide), a potent analgesic for severe chronic pain by the Food and Drug Administration (FDA), and Yondelis (known also as trabectedin or E-743), an oncologic drug for the treatment of advanced soft tissue sarcoma by the European Commission (EC).

The utility of marine organisms is not limited, however, to providing potentially new chemical entities. There is a new trend in the study of marine ecosystems that consists of investigating how metabolites reflect and affect cell functions and then integrating these results with data produced at different levels of cell organization. This shift in research trends in the marine organism field resembles what is happening in the study of terrestrial organisms and is globally known as the systems biology approach, in which the focus of investigations is on the interactions between genes, proteins, and metabolites, rather than on the identification or study of a single gene, protein, or metabolite. This holistic approach requires the input of data that reflect situations integrally, and technologies known as *omics* can provide this. Among these omic technologies, metabolomics can provide very valuable information as it rapidly shows changes in the status of an organism, reflecting its responses to any change in its environment, for example.

The complete set of metabolites that exist in an organism in a given moment are referred to as the metabolome, and the study of the metabolome is known as metabolomics. The term *metabolomics* is analogous to genomics, transcriptomics, and proteomics, and aims to provide the unbiased identification and quantification of the whole set of metabolites or, more accurately, a snapshot of the metabolome of a cell, tissue, or an organism.[9,10] In comparison with other omics, metabolomics is believed to be closest to the phenotype of an organism (Figure 2.1). However, because the current analytical platforms are neither sensitive nor comprehensive enough to detect and measure all metabolites, it is not technically possible to perform a metabolomic study in a strict sense.[11] What is actually done is metabolite fingerprinting or metabolite profiling. Metabolite fingerprinting of a sample is the high-throughput qualitative screening of its metabolic composition. This information can then be processed using a discriminant analysis that allows the comparison of different samples. Metabolite profiling, on the other hand, allows the identification and quantification of a selected and limited number of metabolites. Both these approaches are more realistic than metabolomics in their aim, and the studies referred to in this chapter correspond to either one of these.

This chapter will provide an overview of the current analytical platforms and statistical methods available for metabolomics, followed by a detailed account of recent applications.

2.2 PREANALYTICAL METHODS AND ANALYTICAL PLATFORMS FOR METABOLOMICS OF MARINE ORGANISMS

The methods used for the extraction and analysis in marine organisms metabolomics are basically the same as those used for terrestrial organisms. The workflow of a metabolomic experiment generally has four major steps: sample harvesting and extraction, analysis of the resulting extracts, data reduction and statistical processing, and identification of the relevant metabolites. The collection of samples, the first step, is harder for marine samples than for terrestrial organisms. It is

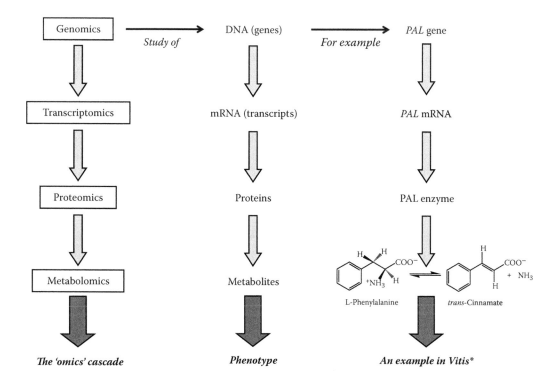

Figure 2.1   The omic cascade is comprised of genomics, transcriptomics, proteomics, and metabolomics, which are the building blocks of the systems biology approach. Transcriptomics, proteomics, and metabolomics can provide comprehensive information and better understanding about the response of a biological system toward different stresses (phenotyping). *The transcription, translation, and function of the phenylalanine ammonia-lyase (PAL) gene is a nice example, and the GenBank® IDs for the *PAL* gene, mRNA, and enzyme in *Vitis* are 100233012, DQ887093.1, and ABL74865.1, respectively.

often difficult for researchers to control the practical conditions because in most cases, samples are collected in the wild and the quality of the material is highly dependent on the collectors and their instruments. Thus, taxonomical issues and the scarcity of biological material add to the inherent difficulties associated with collecting in a marine environment. A further complication is finding the adequate conditions for growth and cultivation of the marine organisms, some of which, invertebrates, for example, are particularly difficult.[8]

Once the samples have been collected, it is usually necessary to grind them to facilitate their subsequent extraction. Though apparently simple, this step is often more complex than expected. This is particularly true in the case of macroalgal samples, such as seaweeds that contain a large amount of polysaccharides, for example, carragenaans that confer a gelling nature to the tissue, making reproducible manual grinding very difficult. It is often necessary to freeze-dry samples before grinding or directly grind them in liquid nitrogen to avoid metabolite degradation, as was suggested by Goulitquer et al.[4] In the field of animal science, microwave heating has been used to inactivate brain enzymes,[12] and this could be useful to avoid possible metabolic degradation of marine organisms.

Once samples have been collected and processed appropriately as described, it is necessary to extract the metabolites. Though vital for the success of the study, this step has received relatively less attention than other ones. In many cases, researchers claim to be using optimum or at least good extraction methods when these are actually just based on their experience and there is little cogent evidence of their efficacy, much less even minimum validation. However, as metabolomics is intended to deal with all the intact metabolites in organisms, the failure to extract any of them from the matrix or protect their integrity conspires against the overall quality of the metabolomic study. Clearly, the enormous number of chemically diverse metabolites with a wide range of polarity present in one single organism is just as hard a challenge for the design of preanalytical steps as detection and data mining.

The mechanism of extraction involves not only solute–solvent but also solute–matrix interactions. It is greatly influenced in general by chemical characteristics of the matrix, solvent, and metabolites, and in particular by specific features of the matrix, such as its swelling degree response to medium pH, contact time, and localization of the metabolites in tissues. To achieve an efficient extraction of intracellular metabolites, the cell wall (if present) and membrane must be permeabilized to allow the solvent to come in contact with the metabolites and solubilize them.[13] Though these problems are not exclusive to marine organisms, they are worsened by the relatively scarce available information on their chemical composition. In the case of metabolomics of terrestrial organisms and their biofluids, such as urine, plasma, serum, and saps, there are several validated sample pretreatment protocols, some of which are being used officially. In the case of marine organisms, as they are collected in the wild from a media that is rich in salts, protocols must include steps to remove these salts before their analysis.

Another issue to consider is that the sample pretreatment method must not only extract the metabolites efficiently, but also remove any matrix or sample components that will interfere with the analytical method. For example, when extracting algae, the presence of high amounts of carragenaans and other polysaccharides will interfere with the ^{1}H nuclear magnetic resonance (NMR) spectrum, broadening the signals.

According to the few reports on the sample preparation required for fresh marine organism metabolomics, methanol and ethanol appear to be the solvents of choice for extraction directly after their collection from natural habitats or after freeze-drying. Pure ethanol or hydroalcoholic solutions are preferred to pure water as extraction solvents because this avoids the extraction of unwanted salts as well as polysaccharides. In some cases, however, extraction of fresh, seawater-containing material with organic solvents may lead to chemical alterations of compounds due to catalytic conversions of the metabolites aided by enzymes that are released from their storage compartments during the extraction process.[14] Freeze-drying the biological material immediately after collection before extraction avoids this type of problem. The contact of freshly collected organisms with methanol or ethanol should not be unnecessarily prolonged to avoid alkylation or esterification of the secondary metabolites, yielding alkylated artifacts or esters.[8]

Primary metabolites such as amino acids, fatty acids, steroids and terpenoids (lupanes) were extracted efficiently from the spiny sea star (*Marthasterias glacialis*) with ethanol at 40 °C with magnetic stirring for their subsequent analysis with multitarget gas chromatography–mass spectrometry (GC-MS).[15]

A protocol for the collection and extraction of hard coral and its algal symbionts, as well as the subsequent NMR and MS methods, was developed by Gordon et al.[16] to study the coral–*Sybiodinium* relationship. The coral and symbionts were treated with 70% MeOH for a nontargeted extraction and analyzed by liquid chromatography (LC)-MS and ^{1}H NMR spectroscopy. In this case, the researchers stressed the importance of the choice of an appropriate quenching method.

The next step in the metabolomic workflow is the analysis of the samples. There are several potential analytical platforms that can be used, but the *sine qua non* condition for any method is that it should be capable of detecting a great number of metabolites with sufficient sensitivity. To gain a first impression of a metabolite profile, a carefully chosen analytical method could be an excellent starting point, followed by the identification of key biochemical leads for further or more focused studies.

There are a great number of reports that compare the advantages and limitations of the available spectroscopic methods, such as mass (MS), nuclear magnetic resonance (NMR), infrared (IR), and ultraviolet (UV), hyphenated or not to some type of chromatography (reviewed in Sumner et al.[10] and Kim and coworkers[17–19]). A number of protocols in which the aptitude of the analytical method has been maximized to achieve the best sensitivity, coverage of metabolites, and suitability of target tissues have been published by several authors for NMR,[17,20] GC-MS,[21] and LC-MS.[22]

Among the various available analytical platforms, the most frequently used are NMR spectroscopy and MS because they possess all or most of the above-mentioned properties. Gas or liquid chromatography, hyphenated with MS,[23,24] is very widely used, while capillary electrophoresis–mass spectrometry (CE-MS)[25] and Fourier transform–ion cyclotron–mass spectrometry (FT-ICR-MS)[26] are other less frequently used but efficient methods for sample comparison and metabolic characterization.

Each analytical method provides a different level of robustness, sensitivity, selectivity, and quantitative information. Broad groups of metabolites can be detected in one single run by NMR, and it is thus useful for macroscopic metabolomics. In terms of structure elucidation, NMR has unbeatable advantages to confirm chemical structures. However, its inherent low sensitivity still poses a problem for the detection of metabolites that are present in low concentrations, although new technologies such as cryo (cold)- and microprobes can increase sensitivity. Current NMR methods can detect up to 100 metabolites simultaneously, while MS can detect approximately 10 times more.[27] On the other hand, NMR can detect a larger diversity of metabolites. This advantage, together with signal robustness and its powerful structure elucidation capabilities, still makes NMR a preferable metabolomic tool for a macroscopic approach.

There are a large number of NMR-based metabolomic applications for marine organisms, mostly related to toxicological studies. Flounder fish (*Platichthys flesus*) were studied for molecular responses to the Tyne estuarine sediment, which contained significantly high concentrations of key environmental pollutants, particularly heavy metals and the biocide tributyltin compounds, and to reference sediment from the Ythan estuary. The ^{1}H NMR metabolomic analyses of the fish liver tissues revealed small but statistically significant alterations in fish exposed to different sediments.[28] In this experiment, samples of fish liver were extracted with the biphasic system composed of methanol–chloroform–water (2:2:1.8, v/v). The aqueous fraction was then dissolved in 0.1 M phosphate buffer (pH 7.0, 10% D_2O) containing 0.5 mM trimethylsilyl-propionate (TMSP) and measured by one-dimensional nuclear Overhauser effect spectroscopy (NOESY) ^{1}H NMR for water suppression.[29] There is a wide range of toxicological studies, such as the evaluation of environmental stress on shrimps[30]; environmental pollution on *Mytilus galloprovincialis*, sedentary filter feeders[31]; benzo(α)pyrene-induced metabolic responses in *Ruditapes philippinarum*[32]; and the effect of increased concentrations of carbon dioxide on green shore crab *Carcinus maenas*.[33] The consequences of heavy metal exposure were studied on marine mussels,[34] specifically of arsenic on the clam *Ruditapes philippinarum*[35] and cadmium on tissues of green mussels.[36] The ^{1}H NMR methods used in all cases were basically the same, differing mainly in the choice of the buffer, though phosphate buffers in D_2O are the most popular. Buffers are required to reduce signal fluctuations. However, it is possible to calibrate the signals to remove the pH effect on chemical shifts using pH modifiers such as formic acid or imidazole (www.chenomx.com). Protic solvents such as D_2O or CH_3OH-d_4 are preferred to nonprotic solvents like dimethyl sulfoxide (DMSO) or $CDCl_3$ because nonprotic solvents are largely affected by pH and temperature, producing peak broadening. Trimethylsilyl propionic acid sodium salt (TMSP)-d_4 or 4,4-dimethyl-4-silapentane-1-sulfonic acid (DSS) are used as internal standards for quantitation and signal calibration in spectra recorded in D_2O, while tetramethylsilane (TMS), hexamethyldisiloxane (HMDS), or residual solvents are used with organic solvents such as $CDCl_3$ or CH_3OH-d_4. However, it is important to note that great care should be taken when using TMSP-d_4 and DSS because they may interact with residual proteins.[37]

Other suggested internal standards as well as NMR parameters for quantitative features are well listed in several review papers.[38,39]

Apart from the previously mentioned internal standards, it is possible to use an external standard such as the Electronic Reference To Access *In Vivo* Concentrations (ERETIC).[40,41] The ERETIC experiment consists in adding an electronic signal to previously acquired spectra. The electronic signal is calibrated and integrated to a known concentration of an analyte of choice and inserted into the previously acquired spectra, where it can be used for the quantification of all other signals.[16]

Another NMR technique, two-dimensional high-resolution magic angle spinning (MAS) ^{1}H NMR spectroscopy-based metabolomics has been applied to investigate the response of MCF7 breast cancer cells to ascididemin, a marine alkaloid and lead molecule for anticancer treatment.[42] The magic angle spinning method allows semisolid samples to be handled without any extraction. High-resolution MAS ^{1}H NMR spectroscopy has also been successfully employed for whole-cell metabolomics of a microalga, *Thlassiosira pseudonana*.[43]

Mass spectrometry is more sensitive and has more resolution than NMR, and the results of metabolomic studies with MS and NMR analytical platforms reflect this. To date, there are more MS-based metabolomic studies of marine organisms, probably due to the higher sensitivity, but also because of the option of using automatized metabolite identification. Goulitquer et al.[4] published an extensive review on MS-based metabolomic studies of marine organisms.

Most MS-based metabolomic studies use hyphenated techniques in which a mass spectrometer is connected to a gas chromatography (GC-MS) or to some type of liquid chromatography (LC-MS), which might be a high-performance liquid chromatography (HPLC) or, increasingly, an ultra-high-performance liquid chromatography (UHPLC, UPLC, etc., according to the brand). Gas chromatography is most often applied for primary metabolites profiling after derivatization with trimethysilylated reagents (TMS) such as n-ethyl-N-trimethylsilytrifluoroacetamide (MSTFA) or n-butyl-N-trimethylsilytrifluoroacetamide (BSTFA). The identification of metabolites is performed using an in-house metabolic database such as BinBase,[44] among others.

Electron impact (EI) is the most commonly used ion source for GC applications. It has been applied to the metabolic profiling of fatty acids, amino acids, and oxylipins in the diatom *Cocconels scutellum*, for example, where it allowed the identification of more than 100 metabolites.[45] Another ion source, negative ion chemical isonization (NICI), has sometimes given better results, as when profiling the pentafluorbenzyle oxime–derivatized polyunsaturated aldehydes (PUAs) from *Skeletonma marinoi*.[46]

Two-dimensional gas chromatography (GC×GC or 2D-GC) has been used to increase resolution for the fatty acid profiling of *Cylindrotheca closterium* and *Seminavis robusta*.[47] However, although resolution was significantly increased, the generated data size was very large, posing a problem for further statistical or multivariate data analysis.

Headspace solid-phase microextraction (HS-SPME) has been used for volatile compound analysis. Volatile organic compounds (VOCs) of *Nitzschia* cf. *pellucida* collected by HS-SPME were analyzed by GC-MS and showed 18 different brominated and iodinated volatiles in the culture of biofilm-forming organisms, particularly cyanogen bromide that exhibits a high allelopathic activity.[48] Other interesting techniques, time-of-flight aerosol mass spectrometry (TOF-AMS) and thermo-desorption (TD), have been applied to the analysis of halogenated compounds in red and brown macroalgae after extraction by microwave-assisted tetramethylammonium hydroxide extraction.[49]

Though widely used, electron impact ionization has some limitations, the foremost being that the molecular ions are fragmented and thus provide a complex pattern. Although the fragmentation pattern can provide important information for structure identification, the lack of information on the molecular formula makes it difficult to search the compounds in conventional databases. To overcome these problems, GC can now be coupled to an atmospheric pressure chemical ionization (APCI) source, and this is the most commonly used setup nowadays for metabolic profiling of terrestrial plants.[50–53] So far there have been no reports of the use of this technique in marine organism metabolomics, but this could be expected to change in the near future. However, GC-APCI is still far from being a routine technique, one of the reasons being the lack of the spectral databases needed for compound identification. Hurtado-Fernández and his coworkers built a library of 100 compounds identified from avocado that is now publicly available.[51] The applications of gas chromatography, however, are restricted to volatile or thermo-stable compounds, and those that are not must be derivatized to acquire these properties. But even then, molecules above 500 m/z cannot be well analyzed.

Liquid chromatography hyphenated with some type of MS technique (LC-MS) is increasingly used. It is generally coupled to an electron spray ion (ESI) source or APCI, both of which are softer ionization methods that are more adequate for larger molecules. LC-MS has been extensively used for profiling secondary metabolites of marine organisms, including bacteria, micro- and macroalgae, and animals (vertebrates and invertebrates), and has been extensively reviewed by Goulitquer et al.[4] Phylogeny and bioactive secondary metabolites of marine Vibrionaceae bacteria have been extensively studied by LC-MS, and diverse groups of metabolites such as catechols, dipeptides, and phenolics were discovered in the strains.[54,55] Yan et al.[56] reported a lipidomic application consisting of profiling photosynthetic lipids in three strains of *Skeletonema* by UHPLC-Q-TOF.

A comparison of NMR- and LC-MS-based metabolomics was done by Karakach et al.[57] when evaluating salmon long-term stress handling through analysis of their plasma. They reported that while ^{1}H NMR data indicated a change in the metabolic profile after 1 week of stress detecting major metabolites, LC-MS showed a clearer metabolic change 2 weeks after stress treatment. Clearly this is due to differences in levels of detected metabolites and shows the importance of the analytical platform considering their inherent limitations and advantages.

The ionization methods used in LC-MS are mostly ESI and APCI, and both provide less fragmented ions but more molecular ion-adducted ions. This feature is quite useful to identify the molecules based on their molecular formula. The lack of fragmentation provided by these methods is partially compensated by LC-(MS)[n] technology for the profiling of certain compounds. For example, LC-MS-MS allowed the identification of prodigiosins and cycloprodigiosins, two major groups of metabolites of the marine bacteria *Zooshikella*.[58] It has also been applied to the detection of albendazole and its metabolites in fish muscle tissue,[59] benzotriazole ultraviolet stabilizers in fish,[60] and lipidomics of *Engraulis* (Peruvian anchovy).[61]

It is also possible to use MS without any previous chromatographic separation by direct infusion into the mass spectrometer. Coupled to a high-field ion cyclotron Fourier transform mass spectrometer (FT-ICR-MS), Brito-Echeverría et al.[62] used this high-resolution MS to study the fatty acid metabolism and particularly the glycerolipids and glycerophospholipids of the halophilic bacterium *Salinibacter ruber*.

Current metabolomic techniques basically use extracts, which means all the intact metabolites are pooled and spatial information, that is, the exact localization of the metabolites within the sample, is lost. This has been solved using matrix-assisted laser desorption ionization (MALDI)-TOF imaging, desorption electrospray ionization (DESI)-MS or laser ablation electrospray ionization (LAESI)-MS. For example, the spatial distribution of secondary metabolites in several cyanobacteria[63] and in *Dysidea herbaceae* sponge[64] has been revealed using MALDI-TOF imaging. Other similar techniques, such as LAESI-MS, were applied directly for the localization of small metabolites and lipids[65] and DESI-MS imaging to red alga (*Callophycus serratus*), revealing that brominated diterpene-benzoic acids were located exclusively in distinct surface patches.[66]

As in all studies, the quality of their results is heavily dependent on the reliability and accuracy of the measurements provided by the analytical instruments. Thus, all the before-mentioned platforms are affected by inherent limitations, and none of them is clearly better than all others. The choice is thus often guided by sample and metabolite characteristics. There are a number of reviews, which include the discussion of the basics and applications of different analytical platforms.[18,67]

2.2.1 Data Processing, Multivariate Data Analysis, and Database of Metabolites

Whichever the analytical platform, a metabolomic study will generate a large amount of data, and the efficiency of the subsequent data processing and its analysis determines the degree of success of the study. It requires the use of specialized informatic tools for the *in silico* multivariate data analysis (MVDA) of the dataset.

The NMR- or MS-generated raw data must be processed before they are submitted to MVDA. Extensive reviews of NMR data processing for MVDA have been published by Kim et al.[17] and Izquierdo-García et al.[68]

Once NMR data have been acquired, they have to be digitalized to numeric values for further statistical analysis. In this procedure, the NMR spectrum is divided into a series of small bins (buckets). The sum of the intensities of the signals in each bin is calculated relative to either reference areas or the sum of total intensities after the removal of unwanted signals from residual solvents or water. Scaling to the sum of the total intensities helps to minimize the effect caused by intersample variation resulting from the amount of extracted tissue. In most cases, the size of bins is 0.02 or 0.04 ppm. Binning may cause a loss of spectral resolution with the resulting difficulty in the interpretation of data (e.g., loading plot), but this can be avoided by using full-resolution NMR data for the statistical analysis, as shown by Rasmussen et al.[69] Full-resolution NMR datasets generate about 30,000 variables (0.15 Hz per data point), against the 200 variables that can result from the integrated binned NMR dataset (0.04 ppm bucketing). Consequently, they can provide more information, contributing to finding significant signals in principal component analysis (PCA) that could otherwise be lost.

It is possible to use peak alignment of ^{1}H NMR spectra, and there are a great number of available methods for this.[70–74] However, due to their inaccuracy, binning is still preferable for ^{1}H NMR data processing.

There are various commercially available methods for NMR data bucketing, including AMIX-TOOLS (Bruker Biospin GmbH, Karlsruhe, Germany), ACD NMR Manager (Advanced Chemistry Development, Toronto, Ontario, Canada), Mnova NMR Suite (MestReLab Research, Santiago de Composteia, Spain), and Chenomx NMR Suite (Chenomx, Edmonton, Alberta, Canada).

The processing of MS data has been extensively reviewed by Katajamaa and Oresic.[75] The basic aim of data processing of MS data is to transform raw data files into a type of representation that facilitates easy access to the characteristics of each observed ion. These characteristics include the mass-to-charge ratio (m/z), retention time, and intensity of each ion in each raw data file. Apart from these basic features, data processing can extract additional information, such as the isotope distribution of the ion. Since each instrument manufacturer uses its proprietary data formats, a preliminary step for data processing is required to convert such raw proprietary data to a common raw data format, such as netCDF (ASTM E2078-00, "Standard Guide for Analytical Data Interchange Protocol for Mass Spectrometric Data") or mzXML.[76] Vendor software packages usually contain scripts that can perform data conversion to netCDF or ASCII formats. Converters to more recent mzXML formats have been developed by both research groups and companies. However, many MS manufacturers also provide metabolomic software applications that process the data acquired by their instruments, avoiding the need of data file conversion.

A typical data processing workflow consists basically of four stages: filtering, feature detection, alignment, and normalization. Filtering methods process the raw measurement signal with the aim of removing effects such as baseline noise drift. Feature detection is used to detect representations of measured ions from the raw signal. Alignment methods cluster measurements across different samples, and normalization removes unwanted systematic variation between samples. Some software packages for MS data processing, including XCMS, MEtAlign, and Mzmine 2, are described in review papers.[4,75]

Having prepared the generated raw data, it is necessary to process the enormous amount of information (variables or features) acquired for the large number of analyzed samples (objects). This leads to the creation of multivariate data matrices that require the use of mathematical and statistical procedures to efficiently extract the maximum useful information from the data.[77] There are two types of MVDA methods: unsupervised and supervised. The most common unsupervised MVDA is principal component analysis (PCA).[78]

Principal component analysis is a technique that, by the reduction of data dimensionality, allows its visualization while retaining as much as possible the information present in the original data. Thus, PCA transforms the original measured variables into new uncorrelated variables called principal components. Each principal component is a linear combination of the original measured variables. This technique produces a group of orthogonal axes that represent the directions of greatest variance in the data. The first principal component (PC1) accounts for the maximum of the total variance, the second (PC2) is uncorrelated with the first and accounts for the maximum of the residual variance, and so on, until the total variance is accounted for (Figure 2.2). In most cases, the first two or three components are sufficient to achieve an adequate separation.

There are two other unsupervised methods, cluster analysis (CA) and factor analysis (FA), but neither is used as frequently as PCA because of their limitations for handling large datasets.[77,78]

Although unsupervised MVDA reveals hidden information by data reduction, useful information provided by minor changes is often lost because it is buried among data considered to be produced by unwanted sample variation. In this context, a second dataset, such as classification or biological activity, is required to build a model. Various supervised MVDA algorithms are available and routinely applied according to the type of data and the objective of the study. Some examples of these MVDAs are partial least squares (PLS), PLS-discriminant analysis (PLS-DA), orthogonal-PLS (OPLS), and bidirectional-OPLS (O2PLS). These analyses are very helpful for pattern recognition and grouping or clustering among samples.

Thus, it is clear that a considerable amount of sample processing, followed by a suitable statistical analysis of the resulting data, is required to extract useful information that can result in knowledge. There are a number of publications that explain the underlying principles and applications of various multivariate data analysis methods used in metabolomics.[77–80]

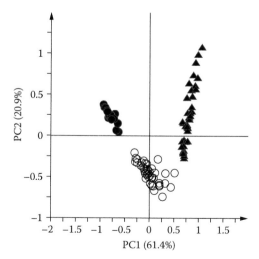

Figure 2.2   Principal component analysis (PCA) score plot of solid-phase extraction (SPE) fractions of all the wine samples. All three fractions are clearly separated from each other. Samples denoted by filled circles, open circles, and filled triangles represent water, methanol–water (1:1), and methanol fractions, respectively. (Adapted from Ali, K., et al., *Food Chemistry* 141 (3), 3124–3130, 2013.)

The last step of metabolomics is the elucidation of signals selected by the applied MVDA as relevant to the investigation. Whichever the analytical platform, researchers rely on databases for the identification of metabolites. Goulitquer et al.,[4] Sumner et al.,[10] and Tohge and Fernie[81] published reviews that list the available databases. As metabolites from marine organisms are generally included as natural products, it is possible to use diverse natural product databases such as PubChem (http://pubchem.ncbi.nlm.nih.gov/), Chemspider (www.chemspider.com/), and the combined chemical dictionary (http://ccd.chemnetbase.com) for chemical information of metabolites and KEGG (www.genome.jp/kegg/) and BioCyc (http://biocyc.org) for metabolic pathways. There are also specific databases of marine organisms, such as SWMD (www.swmd.co.in/) and Marinlit (http://pubs.rsc.org/marinlit/).

For MS data there are a few general MS spectral databases that are used to identify metabolites, such as NIST (www.nist.gov/index.html), Metlin (http://metlin.scripps.edu), and MassBank (www.massbank.jp/). The Metlin database also offers XCMS to process MS data for further MVDA and identification from MS spectra. There are also a few metabolomic databases, such as GMD (Golm Metabolome Database, http://gmd.mpimp-golm.mpg.de/) and FiehnLib (http://fiehnlab.ucdavis.edu/project/FiehnLib/index.html), that are often used as well. The Madison–Qingdao Metabolomics Consortium Database (MMCD) database (http://mmcd.nmrfam.wisc.edu/) provides quite extensive information on metabolites, including their MS spectra and chemical and NMR data information.

To use the MS database for identification of metabolites from marine organisms, it is important to bear in mind that the presence of large amounts of NaCl in marine organisms can interfere with the deconvolution of molecular ions,[4] especially because, in general, the $[M+Na]^+$ adduct shows higher intensity than the molecular ion $[M+H]^+$. Additionally, the sodium adducts $[M+Na]^+$ are less reproducible and more resistant to fragmentation.[82]

There are fewer available NMR databases. MMCD provides spectra of a relatively large number of metabolites, but it is mostly limited to primary metabolites. There are commercially available NMR spectrum databases of natural products, such as the Advanced Chemistry Development (ACD) NMR library (Toronto, Ontario, Canada) and Chenomx NMR library (Edmonton, Alberta, Canada). The ACD library is extensive, having more than 40,000 NMR spectra, including all the Marinlit NMR data, thanks to which it is possible to search for ^{13}C NMR spectra of marine products. The Chenomx NMR library contains the ^1H NMR spectra of more than 250 metabolites, mostly from primary metabolites in biofluids, but this database gives very detailed information, including spectra measured in different magnetic fields (from 400 to 800 MHz) and the pH sensitivity of each signal. The Chenomx NMR library also allows ^1H NMR spectra of extracts to be matched with spectra stored in the library.

2.3 APPLICATIONS

Despite the obvious advantages of applying the metabolomic approach to an investigation, there are fewer metabolomic papers on marine organisms than could be expected, and in most cases, they are limited to chemical profiling. Most of them are toxicological studies, bioactivity screening, and metabolic profiling or fingerprinting. In the following sections we provide some recent cases of these applications to marine organisms.

2.3.1 Lead Finding or Bioactivity Screening

Most metabolomic applications to marine natural products are related to this field. Lead finding or bioactivity screening is often considered a daunting task because not only is it labor-intensive, but also it is quite usual to lose the monitored activity during the isolation process because this activity may be the result of the synergistic effect of several compounds rather than just one or because of compound degradation. The implementation of a metabolomic approach has been very successful as an alternative for bioactivity screening since it is very suitable for handling the structural diversity of metabolites.[83,84] Applying this approach, Yuliana et al.[85] developed a comprehensive extraction method as an alternative to the classical bioguided isolation process. This extraction procedure, coupled to ^{1}H NMR analysis of the resulting extracts, allowed adenosine A1 receptor binding activity to be detected in *Orthosiphon stamineus* and proved to be very appropriate for a metabolomic approach. This method was also used to determine tumor necrosis factor alpha (TNFα) inhibition activity of different grape cultivars and their wines.[86,87]

Although the terrestrial resources for the isolation of bioactive compounds are not exhausted, marine organisms are emerging as an alternative as they provide a large pool of chemically diverse and active natural products. Marine natural products have already proved to be effective against inflammatory disorders and for the treatment of tumors, viral infections, and malaria. These uses and others are reviewed in detail by Sipkema et al.[88] and Keyzers and Davies-Coleman.[89]

We recently reported the screening of marine sponges for their adenosine A1 receptor binding activity. Based on the results of the activity profiling, a number of samples were studied to identify the bioactive ingredients in the crude extract (Figure 2.3). The NMR-based studies allowed the identification of sesterterpenes (halisulfates and suvanine). The paper revealed the potential of metabolomics for lead finding since, unlike traditional bioactivity-guided fractionation, it allowed not only the chemical characterization of the whole samples, but also the identification of the active ingredients in the crude extract.[90]

2.3.2 Toxicology

Marine organisms, especially mussels and oysters, are known to accumulate heavy metals from the environment and hence can be used to monitor heavy metal pollution.[34] Several biochemical markers related to the damage resulting from this contamination have been used to measure the level of acetylcholinesterase and antioxidants as an estimation of neurotoxicity and oxidative stress, respectively.[91,92] Unlike traditional approaches, metabolomics has proved to be effective, providing information on a large number of low-molecular-weight metabolites and affording a more comprehensive view of physiological stress.

A report on Manila clam exposure to copper contamination revealed that osmotic regulation and energy metabolism were severely disturbed.[93] Another study focusing on tissue-specific toxicology in Eastern oysters showed the time-dependent effect of heavy metal pollutants.[94] In this case, the use of NMR spectroscopy and chemometrics allowed the identification of 32 different metabolites that were affected by the pollutants. Similarly, the effect of contamination with cadmium (Cd) in *Perna viridis* was investigated in its adductor muscle, revealing changes that were reflected in characteristic low levels of branched-chain amino acids and high levels of acidic amino acids.[36] Another report described the accumulation of the osmolytes betaine and taurine with free amino acids in marine mussels (Figure 2.4) following heavy metal contamination.[34] These examples indicate the potential of metabolomics to identify metabolic biomarkers, resulting in a better understanding of heavy metal contaminant toxicology.

2.3.3 Metabolic Profiling and Chemical Fingerprinting

As discussed earlier in this chapter, most of the metabolomics-based studies actually consists in metabolic profiling or fingerprinting. The information provided by either type of study allows comparison of the metabolite composition or the content of a targeted set of metabolites in samples, explaining differences among the studied samples. Diverse marine organisms have also been investigated using metabolic profiling and fingerprinting.

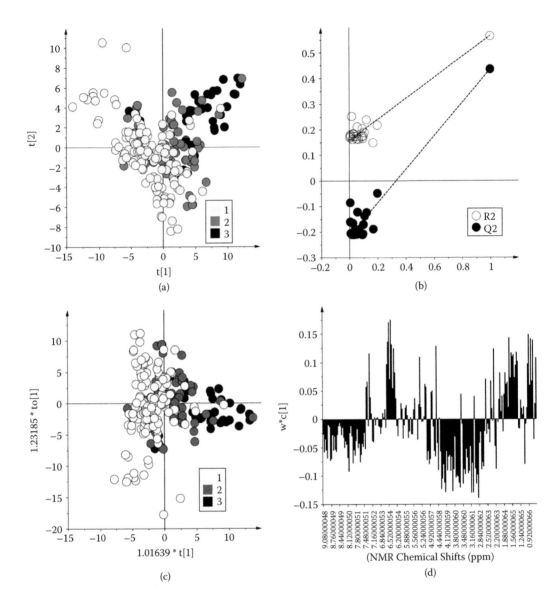

Figure 2.3 Multivariate data analyses for sponge metabolic profiling and activity profiling data correlation. (a) The PLS score plot shows grouping the samples based on their activity. (b) The permutation test plot for the PLS modeling. (c) The OPLS score plot shows the distinction among the samples based on the predictive component (x-axis). (d) The loading coefficient plot shows the positively (on positive y-axis) and negatively (negative y-axis) correlated signals to the adenosine receptor binding activity. (Adapted from Ali, K., et al., *Metabolomics* 9 (4), 778–785, 2013.)

Diatoms are one of the most studied marine organisms. A number of metabolic pathways and regulatory mechanisms can be predicted based on the genomic sequences and metabolomic information. In particular, information obtained with GC-MS-based metabolomics was able to explain the operation of metabolic pathways and define their function and regulation in those organisms (reviewed by Fernie et al.[95]). In this paper, Fernie and coworkers stated their expectations of the potential of metabolomics as a tool to reveal biochemical and physiological alterations that are reflected by the metabolome, such as nitrogen acquisition, light harvesting, and iron uptake.[95]

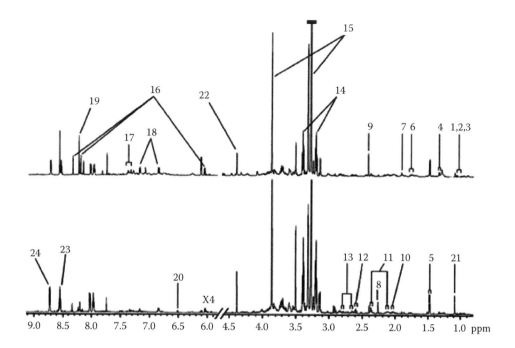

Figure 2.4   Representative 600 MHz ^{1}H NMR spectra for mantle tissues from marine mussels (*Mytilus*) from Onsan Bay and the Dokdo area. Key: (1) leucine, (2) isoleucine, (3) valine, (4) threonine, (5) alanine, (6) arginine, (7) acetate, (8) acetoacetate, (9) succinate, (10) glutamate, (11) glutamine, (12) b-alanine, (13) aspartate, (14) taurine, (15) betaine, (16) inosine, (17) phenylalanine, (18) tyrosine, (19) adenine, (20) fumarate, and (21–24) unknown. (Adapted from Kwon, Y.-K., et al., *Marine Pollution Bulletin* 64 (9), 1874–1879, 2012.)

^{1}H NMR spectroscopy has also been used for metabolic discrimination of hagfish dental and somatic skeletal muscles[96] and for gender determination of marine mussels.[97]

These studies included the one-carbon metabolism in bacteria,[98] pigment profiling and the study of the defense response in micro- and macroalgae,[99] stress handling in salmon,[57] intra- and interspecific metabolic variability to identify the taxonomic markers in sponges[100] (Figure 2.5), and tidal cycle effects on mollusks.[101] The versatility of these tools is clearly revealed by the diversity of organisms, processes, and objectives pursued in the studies, as further illustrated by a recent review.[4]

2.4 FUTURE PROSPECTS AND LIMITATIONS

With the advances in analytical instrumentation capabilities, metabolomic applications to living systems are increasing and marine organisms are not an exception. As shown by the many examples above, metabolomics can be a very useful tool to provide long-overdue answers to several questions related to the ocean or marine world, including the study of viable but nonculturable (VBNC) marine organisms,[102] biosynthetic pathways of known bioactive metabolites from marine organisms,[103] the isolation of novel bioactive compounds,[90] and quality control in marine-derived products.

However, so far there are fewer reports of marine metabolomic studies than expected. The reasons for this are diverse and range from the high concentration of interfering chemicals in marine media such as salts, the difficulties encountered in metabolite extraction, the failure to achieve correct underwater taxonomic identification of samples, and possible metabolic interference from the symbionts. Other reasons include the choice of inappropriate data analysis methods or even the total

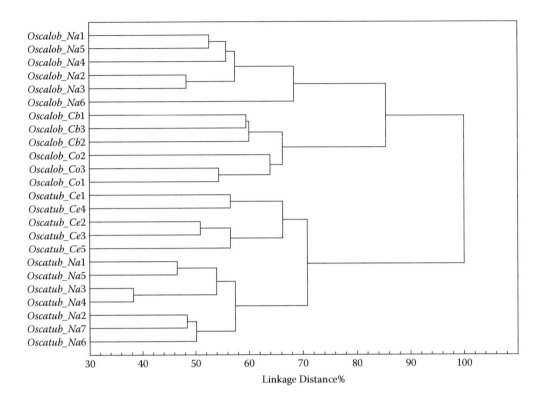

Figure 2.5 Hierarchical cluster analysis of specimens belonging to the *Oscarella* sister species. (Adapted from Ivanišević, J., et al., *Metabolomics* 7 (2), 289–304, 2011.)

lack of any method at all, and inadequate analytical capabilities that are severe limitations often mistakenly attributed to the metabolomics itself. It is to be expected that new applications in the field of marine biology will be achieved thanks to the improvements in analytical methods, the development of new data analysis algorithms, and the optimization of extraction and sampling procedures.

In recent years, marine organism research groups have adopted a new line of investigations in which the metabolomic approach is applied. These topics respond to two key terms: *environmental metabolomics* and *footprinting*. An important characteristic that in many situations distinguishes the metabolome from the genome, transcriptome, and proteome is its susceptibility to changes in the environment, for example, as a response to stressful conditions.[27] While the metabolome may be the last to reflect biological processes, it will be the first to reflect exogenous alterations in the shape of responses to anthropogenic stressors and natural daily events, such as pollutant exposure and feeding, respectively. Naturally, the most appropriate way to evaluate the changes in the whole metabolome is to apply a metabolomic approach due to its holistic nature. This application has now come to be considered a subdiscipline known as environmental metabololomics. Its goals are to monitor the prevalence of diseases in certain species of wildlife identified as sentinel species as indicators of the health of the environment; assess chemical risks of pharmaceuticals, pesticides, and other household and industrial chemicals; and provide information for the maintenance of healthy stocks of animals, including fish, for farming and industry. Within the field of environmental metabolomics, another of the most promising applications is the footprinting of the excretion of a cell or system under certain conditions.[104] All these applications arise from the evidence that metabolomics can provide valuable information to reveal the interaction between organisms and environmental conditions.[27]

2.5 CONCLUSION

The chemical characterization of an organism can provide insight into several key processes, as its metabolome is known to be the closest to its phenotype, compared to its genome, transcriptome, or even proteome. To date, the metabolism of marine organisms has been less explored than that of terrestrial organisms due mainly to the technical difficulties for sampling and cultivating and maintaining marine organisms. However, this is changing as evidence of the wealth of novel compounds available in these relatively unexplored resources increases and more funding is available to develop solutions for these difficulties. Metabolomics provides a unique opportunity to acquire knowledge about previously unknown facts, and together with other *omics* technologies, it plays an integral role in systems biology. Several key applications of the metabolic characterization of marine organisms have already been published, as mentioned above, and with more analytical capabilities and understanding, an exponential growth in further applications of this tool in marine biology is to be expected.

REFERENCES

1. Appeltans, W., Ahyong, S. T., Anderson, G., Angel, M. V., Artois, T., Bailly, N., Bamber, R., Barber, A., Bartsch, I., Berta, A., Błażewicz-Paszkowycz, M., Bock, P., Boxshall, G., Boyko, C. B., Brandão, S. N., Bray, R. A., Bruce, N. L., Cairns, S. D., Chan, T.-Y., Cheng, L., Collins, A. G., Cribb, T., Curini-Galletti, M., Dahdouh-Guebas, F., Davie, P. J. F., Dawson, M. N., De Clerck, O., Decock, W., De Grave, S., de Voogd, N. J., Domning, D. P., Emig, C. C., Erséus, C., Eschmeyer, W., Fauchald, K., Fautin, D. G., Feist, S. W., Fransen, C. H. J. M., Furuya, H., Garcia-Alvarez, O., Gerken, S., Gibson, D., Gittenberger, A., Gofas, S., Gómez-Daglio, L., Gordon, D. P., Guiry, M. D., Hernandez, F., Hoeksema, B. W., Hopcroft, R. R., Jaume, D., Kirk, P., Koedam, N., Koenemann, S., Kolb, J. B., Kristensen, R. M., Kroh, A., Lambert, G., Lazarus, D. B., Lemaitre, R., Longshaw, M., Lowry, J., Macpherson, E., Madin, L. P., Mah, C., Mapstone, G., McLaughlin, P. A., Mees, J., Meland, K., Messing, C. G., Mills, C. E., Molodtsova, T. N., Mooi, R., Neuhaus, B., Ng, P. K. L., Nielsen, C., Norenburg, J., Opresko, D. M., Osawa, M., Paulay, G., Perrin, W., Pilger, J. F., Poore, G. C. B., Pugh, P., Read, G. B., Reimer, J. D., Rius, M., Rocha, R. M., Saiz-Salinas, J. I., Scarabino, V., Schierwater, B., Schmidt-Rhaesa, A., Schnabel, K. E., Schotte, M., Schuchert, P., Schwabe, E., Segers, H., Self-Sullivan, C., Shenkar, N., Siegel, V., Sterrer, W., Stöhr, S., Swalla, B., Tasker, M. L., Thuesen, E. V., Timm, T., Todaro, M. A., Turon, X., Tyler, S., Uetz, P., van der Land, J., Vanhoorne, B., van Ofwegen, L. P., van Soest, R. W. M., Vanaverbeke, J., Walker-Smith, G., Walter, T. C., Warren, A., Williams, G. C., Wilson, S. P., Costello, M. J. The magnitude of global marine species diversity. *Current Biology* 2012, 22 (23), 2189–2202.
2. Mora, C., Tittensor, D. P., Adl, S., Simpson, A. G. B., Worm, B. How many species are there on earth and in the ocean? *PLoS Biol* 2011, 9 (8), e1001127.
3. Blunt, J. W., Copp, B. R., Keyzers, R. A., Munro, M. H. G., Prinsep, M. R. Marine natural products. *Natural Product Reports* 2012, 29 (2), 144–222.
4. Goulitquer, S., Potin, P., Tonon, T. Mass spectrometry-based metabolomics to elucidate functions in marine organisms and ecosystems. *Marine Drugs* 2012, 10 (4), 849–880.
5. Newman, D. J., Cragg, G. M., Snader, K. M. The influence of natural products upon drug discovery. *Natural Product Reports* 2000, 17 (3), 215–234.
6. König, G. M., Wright, A. D. Marine natural products research: Current directions and future potential. *Planta Medica* 1996, 62 (03), 193–211.
7. Claeson, P., Bohlin, L. Some aspects of bioassay methods in natural-product research aimed at drug lead discovery. *Trends in Biotechnology* 1997, 15 (7), 245–248.
8. Ebada, S. S., Edrada, R. A., Lin, W., Proksch, P. Methods for isolation, purification and structural elucidation of bioactive secondary metabolites from marine invertebrates. *Nature Protocols* 2008, 3 (12), 1820–1831.
9. Oliver, S. G., Wilson, M. K., Kell, D. B., Baganz, F. Systematic functional analysis in yeast genome. *Trends in Biotechnology* 1998, 16, 373–378.

10. Sumner, L. W., Mendes, P., Dixon, R. A. Plant metabolomics: Large-scale phytochemistry in the functional genomics era. *Phytochemistry* 2003, 62 (6), 817–836.

11. Weckwerth, W. Metabolomics in systems biology. *Annual Review of Plant Biology* 2003, 54 (1), 669–689.

12. Ikarashi, Y., Maruyama, Y., Stavinoha, W. B. Study of the use of the microwave magnetic field for the rapid inactivation of brain enzymes. *Japanese Journal of Pharmacology* 1984, 35 (4), 371–387.

13. Villas-Bôas, S. G., Villas-Bôas, S. G., Roessner, U., Hansen, M. A. E., Smedsgaard, J., Nielsen, J. Sampling and sample preparation. In *Metabolome Analysis*, ed. S. G. Villas-Bôas, U. Roessner, M. E. Hansen, J. Smedsgaard, J. Nielsen. John Wiley & Sons, Hoboken, NJ, 2007, pp. 39–82.

14. Thoms, C., Ebel, R., Proksch, P. Activated chemical defense in aplysina sponges revisited. *Journal of Chemical Ecology* 2006, 32 (1), 97–123.

15. Pereira, D. M., Vinholes, J., de Pinho, P. G., Valentão, P., Mouga, T., Teixeira, N., Andrade, P. B. A gas chromatography–mass spectrometry multi-target method for the simultaneous analysis of three classes of metabolites in marine organisms. *Talanta* 2012, 100 (0), 391–400.

16. Gordon, B., Leggat, W., Motti, C. Extraction protocol for nontargeted NMR and LC-MS metabolomics-based analysis of hard coral and their algal symbionts. In *Metabolomics Tools for Natural Product Discovery*, ed. U. Roessner, D. A. Dias. Humana Press, Dordrecht, The Netherlands, 2013, vol. 1055, pp. 129–147.

17. Kim, H. K., Saifullah, Khan, S., Wilson, E. G., Kricun, S. D. P., Meissner, A., Goraler, S., Deelder, A. M., Choi, Y. H., Verpoorte, R. Metabolic classification of South American *Ilex* species by NMR-based metabolomics. *Phytochemistry* 2010, 71 (7), 773–784.

18. Verpoorte, R., Choi, Y. H., Mustafa, N. R., Kim, H. K. Metabolomics: Back to basics. *Phytochemistry Reviews* 2008, 7 (3), 525–537.

19. Wolfender, J. L., Rudaz, S., Choi, Y. H., Kim, H. K. Plant metabolomics: From holistic data to relevant biomarkers. *Current Medicinal Chemistry* 2013, 20, 1506–1590.

20. Kruger, N. J., Troncoso-Ponce, M. A., Ratcliffe, R. G. ^{1}H NMR metabolite fingerprinting and metabolomic analysis of perchloric acid extracts from plant tissues. *Nature Protocols* 2008, 3 (6), 1001–1012.

21. Lisec, J., Schauer, N., Kopka, J., Willmitzer, L., Fernie, A. R. Gas chromatography mass spectrometry-based metabolite profiling in plants. *Nature Protocols* 2006, 1 (1), 387–396.

22. De Vos, R. C. H., Moco, S., Lommen, A., Keurentjes, J. J. B., Bino, R. J., Hall, R. D. Untargeted large-scale plant metabolomics using liquid chromatography coupled to mass spectrometry. *Nature Protocols* 2007, 2 (4), 778–791.

23. Koek, M. M., Jellema, R. H., van der Greef, J., Tas, A. C., Hankemeier, T. Quantitative metabolomics based on gas chromatography mass spectrometry: Status and perspectives. *Metabolomics* 2011, 7 (3), 307–328.

24. Lu, W., Bennett, B. D., Rabinowitz, J. D. Analytical strategies for LC–MS-based targeted metabolomics. *Journal of Chromatography B* 2008, 871 (2), 236–242.

25. Ramautar, R., Mayboroda, O. A., Somsen, G. W., de Jong, G. J. CE-MS for metabolomics: Developments and applications in the period 2008–2010. *Electrophoresis* 2011, 32 (1), 52–65.

26. Gidman, E., Goodacre, R., Emmett, B., Sheppard, L., Leith, I., Gwynn-Jones, D. Applying metabolic fingerprinting to ecology: The use of Fourier-transform infrared spectroscopy for the rapid screening of plant responses to N deposition. *Water, Air, and Soil Pollution: Focus* 2004, 4 (6), 251–258.

27. Viant, M. R. Metabolomics of aquatic organisms: The new omics on the block. *Marine Ecology Progress Series* 2007, 332, 301–306.

28. Williams, T. D., Davies, I. M., Wu, H., Diab, A. M., Webster, L., Viant, M. R., Chipman, J. K., Leaver, M. J., George, S. G., Moffat, C. F., Robinson, C. D. Molecular responses of European flounder (*Platichthys flesus*) chronically exposed to contaminated estuarine sediments. *Chemosphere* 2014, 108 (0), 152–158.

29. Katsiadaki, I., Williams, T. D., Ball, J. S., Bean, T. P., Sanders, M. B., Wu, H., Santos, E. M., Brown, M. M., Baker, P., Ortega, F., Falciani, F., Craft, J. A., Tyler, C. R., Viant, M. R., Chipman, J. K. Hepatic transcriptomic and metabolomic responses in the stickleback (*Gasterosteus aculeatus*) exposed to ethinyl-estradiol. *Aquatic Toxicology* 2010, 97 (3), 174–187.

30. Schock, T. B., Duke, J., Goodson, A., Weldon, D., Brunson, J., Leffler, J. W., Bearden, D. W. Evaluation of pacific white shrimp (*Litopenaeus vannamei*) health during a superintensive aquaculture growout using NMR-based metabolomics. *PLoS ONE* 2013, 8 (3), e59521.

31. Fasulo, S., Iacono, F., Cappello, T., Corsaro, C., Maisano, M., D'Agata, A., Giannetto, A., De Domenico, E., Parrino, V., Lo Paro, G., Mauceri, A. Metabolomic investigation of *Mytilus galloprovincialis* (Lamarck 1819) caged in aquatic environments. *Ecotoxicology and Environmental Safety* 2012, 84 (0), 139–146.

32. Zhang, L., Liu, X., You, L., Zhou, D., Wang, Q., Li, F., Cong, M., Li, L., Zhao, J., Liu, D., Yu, J., Wu, H. Benzo(a)pyrene-induced metabolic responses in Manila clam *Ruditapes philippinarum* by proton nuclear magnetic resonance (^{1}H NMR) based metabolomics. *Environmental Toxicology and Pharmacology* 2011, 32 (2), 218–225.

33. Hammer, K. M., Pedersen, S. A., Størseth, T. R. Elevated seawater levels of CO_2 change the metabolic fingerprint of tissues and hemolymph from the green shore crab *Carcinus maenas*. *Comparative Biochemistry and Physiology Part D: Genomics and Proteomics* 2012, 7 (3), 292–302.

34. Kwon, Y.-K., Jung, Y.-S., Park, J.-C., Seo, J., Choi, M.-S., Hwang, G.-S. Characterizing the effect of heavy metal contamination on marine mussels using metabolomics. *Marine Pollution Bulletin* 2012, 64 (9), 1874–1879.

35. Wu, H., Zhang, X., Wang, Q., Li, L., Ji, C., Liu, X., Zhao, J., Yin, X. A metabolomic investigation on arsenic-induced toxicological effects in the clam *Ruditapes philippinarum* under different salinities. *Ecotoxicology and Environmental Safety* 2013, 90 (0), 1–6.

36. Wu, H., Wang, W.-X. Tissue-specific toxicological effects of cadmium in green mussels (*Perna viridis*): Nuclear magnetic resonance-based metabolomics study. *Environmental Toxicology and Chemistry* 2011, 30 (4), 806–812.

37. Shimizu, A., Ikeguchi, M., Sugai, S. Appropriateness of DSS and TSP as internal references for ^{1}H NMR studies of molten globule proteins in aqueous media. *Journal of Biomolecular NMR* 1994, 4 (6), 859–862.

38. Pauli, G. F., Jaki, B. U., Lankin, D. C. Quantitative ^{1}H NMR: Development and potential of a method for natural products analysis. *Journal of Natural Products* 2005, 68 (1), 133–149.

39. Pauli, G. F., Gödecke, T., Jaki, B. U., Lankin, D. C. Quantitative ^{1}H NMR. Development and potential of an analytical method: An update. *Journal of Natural Products* 2012, 75 (4), 834–851.

40. Akoka, S., Barantin, L., Trierweiler, M. Concentration measurement by proton NMR using the ERETIC method. *Analytical Chemistry* 1999, 71 (13), 2554–2557.

41. Tapiolas, D. M., Raina, J.-B., Lutz, A., Willis, B. L., Motti, C. A. Direct measurement of dimethylsulfoniopropionate (DMSP) in reef-building corals using quantitative nuclear magnetic resonance (qNMR) spectroscopy. *Journal of Experimental Marine Biology and Ecology* 2013, 443 (0), 85–89.

42. Morvan, D. Functional metabolomics uncovers metabolic alterations associated to severe oxidative stress in MCF7 breast cancer cells exposed to ascididemin. *Marine Drugs* 2013, 11, 3846–3860.

43. Chauton, M. S., Størseth, T. R., Johnsen, G. High-resolution magic angle spinning ^{1}H NMR analysis of whole cells of *Thalassiosira pseudonana* (Bacillariophyceae): Broad range analysis of metabolic composition and nutritional value. *Journal of Applied Phycology* 2003, 15 (6), 533–542.

44. Lee, D., Fiehn, O. High quality metabolomic data for *Chlamydomonas reinhardtii*. *Plant Methods* 2008, 4 (1), 7.

45. Nappo, M., Berkov, S., Codina, C., Avila, C., Messina, P., Zupo, V., Bastida, J. Metabolite profiling of the benthic diatom *Cocconeis scutellum* by GC-MS. *Journal of Applied Phycology* 2009, 21 (3), 295–306.

46. Vidoudez, C., Pohnert, G. Growth phase-specific release of polyunsaturated aldehydes by the diatom *Skeletonema marinoi*. *Journal of Plankton Research* 2008, 30 (11), 1305–1313.

47. Gu, Q., David, F., Lynen, F., Vanormelingen, P., Vyverman, W., Rumpel, K., Xu, G., Sandra, P. Evaluation of ionic liquid stationary phases for one dimensional gas chromatography–mass spectrometry and comprehensive two dimensional gas chromatographic analyses of fatty acids in marine biota. *Journal of Chromatography A* 2011, 1218 (20), 3056–3063.

48. Vanelslander, B., Paul, C., Grueneberg, J., Prince, E. K., Gillard, J., Sabbe, K., Pohnert, G., Vyverman, W. Daily bursts of biogenic cyanogen bromide (BrCN) control biofilm formation around a marine benthic diatom. *Proceedings of the National Academy of Sciences* 2012, 109 (7), 2412–2417.

49. Kundel, M., Thorenz, U., Petersen, J., Huang, R.-J., Bings, N., Hoffmann, T. Application of mass spectrometric techniques for the trace analysis of short-lived iodine-containing volatiles emitted by seaweed. *Analytical and Bioanalytical Chemistry* 2012, 402 (10), 3345–3357.

50. Pacchiarotta, T., Nevedomskaya, E., Carrasco-Pancorbo, A., Deelder, A. M., Mayboroda, O. A. Evaluation of GC-APCI/MS and GC-FID as a complementary platform. *Journal of Biomolecular Techniques: JBT* 2010, 21 (4), 205–213.

51. Hurtado-Fernández, E., Pacchiarotta, T., Longueira-Suárez, E., Mayboroda, O. A., Fernández-Gutiérrez, A., Carrasco-Pancorbo, A. Evaluation of gas chromatography-atmospheric pressure chemical ionization-mass spectrometry as an alternative to gas chromatography-electron ionization-mass spectrometry: Avocado fruit as example. *Journal of Chromatography A* 2013, 1313 (0), 228–244.

52. Garratt, L. C., Linforth, R., Taylor, A. J., Lowe, K. C., Power, J. B., Davey, M. R. Metabolite fingerprinting in transgenic lettuce. *Plant Biotechnology Journal* 2005, 3 (2), 165–174.

53. Nácher-Mestre, J., Serrano, R., Portolés, T., Berntssen, M. H. G., Pérez-Sánchez, J., Hernández, F. Screening of pesticides and polycyclic aromatic hydrocarbons in feeds and fish tissues by gas chromatography coupled to high-resolution mass spectrometry using atmospheric pressure chemical ionization. *Journal of Agricultural and Food Chemistry* 2014, 62 (10), 2165–2174.

54. Mansson, M., Gram, L., Larsen, T. O. Production of bioactive secondary metabolites by marine Vibrionaceae. *Marine Drugs* 2011, 9, 1440–1468.

55. Wietz, M., Gotfredsen, C. H., Larsen, T. O., Gram, L. Antibacterial compounds from marine Vibionaceae isolated on a global expedition. *Marine Drugs* 2010, 8, 2946–2960.

56. Yan, X., Chen, D., Xu, J., Zhou, C. Profiles of photosynthetic glycerolipids in three strains of *Skeletonema* determined by UPLC-Q-TOF-MS. *Journal of Applied Phycology* 2011, 23 (2), 271–282.

57. Karakach, T., Huenupi, E., Soo, E., Walter, J., Afonso, L. B. ^{1}H-NMR and mass spectrometric characterization of the metabolic response of juvenile Atlantic salmon (*Salmo salar*) to long-term handling stress. *Metabolomics* 2009, 5 (1), 123–137.

58. Lee, J. S., Kim, Y.-S., Park, S., Kim, J., Kang, S.-J., Lee, M.-H., Ryu, S., Choi, J. M., Oh, T.-K., Yoon, J.-H. Exceptional production of both prodigiosin and cycloprodigiosin as major metabolic constituents by a novel marine bacterium, *Zooshikella rubidus* S1-1. *Applied and Environmental Microbiology* 2011, 77 (14), 4967–4973.

59. Zhang, X., Xu, H., Zhang, H., Guo, Y., Dai, Z., Chen, X. Simultaneous determination of albendazole and its metabolites in fish muscle tissue by stable isotope dilution ultra-performance liquid chromatography tandem mass spectrometry. *Analytical and Bioanalytical Chemistry* 2011, 401 (2), 727–734.

60. Kim, J.-W., Isobe, T., Ramaswamy, B. R., Chang, K.-H., Amano, A., Miller, T. M., Siringan, F. P., Tanabe, S. Contamination and bioaccumulation of benzotriazole ultraviolet stabilizers in fish from Manila Bay, the Philippines using an ultra-fast liquid chromatography–tandem mass spectrometry. *Chemosphere* 2011, 85 (5), 751–758.

61. Oh, S. F., Vickery, T. W., Serhan, C. N. Chiral lipidomics of E-series resolvins: Aspirin and the biosynthesis of novel mediators. *Biochimica et Biophysica Acta (BBA)—Molecular and Cell Biology of Lipids* 2011, 1811 (11), 737–747.

62. Brito-Echeverría, J., Lucio, M., López-López, A., Antón, J., Schmitt-Kopplin, P., Rosselló-Móra, R. Response to adverse conditions in two strains of the extremely halophilic species *Salinibacter ruber*. *Extremophiles* 2011, 15 (3), 379–389.

63. Esquenazi, E., Coates, C., Simmons, L., Gonzalez, D., Gerwick, W. H., Dorrestein, P. C. Visualizing the spatial distribution of secondary metabolites produced by marine cyanobacteria and sponges via MALDI-TOF imaging. *Molecular Biosystems* 2008, 4 (6), 562–570.

64. Simmons, T. L., Coates, R. C., Clark, B. R., Engene, N., Gonzalez, D., Esquenazi, E., Dorrestein, P. C., Gerwick, W. H. Biosynthetic origin of natural products isolated from marine microorganism–invertebrate assemblages. *Proceedings of the National Academy of Sciences* 2008, 105 (12), 4587–4594.

65. Shrestha, B., Vertes, A. *In situ* metabolic profiling of single cells by laser ablation electrospray ionization mass spectrometry. *Analytical Chemistry* 2009, 81 (20), 8265–8271.

66. Lane, A. L., Nyadong, L., Galhena, A. S., Shearer, T. L., Stout, E. P., Parry, R. M., Kwasnik, M., Wang, M. D., Hay, M. E., Fernandez, F. M., Kubanek, J. Desorption electrospray ionization mass spectrometry reveals surface-mediated antifungal chemical defense of a tropical seaweed. *Proceedings of the National Academy of Sciences* 2009, 106, 7314–7319.

67. Hall, R. D. Plant metabolomics: From holistic hope, to hype, to hot topic. *New Phytologist* 2006, 169 (3), 453–468.

68. Izquierdo-García, J. L., Villa, P., Kyriazis, A., del Puerto-Nevado, L., Pérez-Rial, S., Rodriguez, I., Hernandez, N., Ruiz-Cabello, J. Descriptive review of current NMR-based metabolomic data analysis packages. *Progress in Nuclear Magnetic Resonance Spectroscopy* 2011, 59 (3), 263–270.

69. Rasmussen, B., Cloarec, O., Tang, H., Stærk, D., Jaroszewski, J. W. Multivariate analysis of integrated and full-resolution ^{1}H-NMR spectral data from complex pharmaceutical preparations: St. John's wort. *Planta Medica* 2006, 72 (06), 556–563.

70. Yang, J., Xu, G., Zheng, Y., Kong, H., Wang, C., Zhao, X., Pang, T. Strategy for metabonomics research based on high-performance liquid chromatography and liquid chromatography coupled with tandem mass spectrometry. *Journal of Chromatography A* 2005, 1084 (1–2), 214–221.

71. Wu, W., Daszykowski, M., Walczak, B., Sweatman, B. C., Connor, S. C., Haselden, J. N., Crowther, D. J., Gill, R. W., Lutz, M. W. Peak alignment of urine NMR spectra using fuzzy warping. *Journal of Chemical Information and Modeling* 2006, 46 (2), 863–875.

72. Torgrip, R. J. O., Åberg, K. M., Alm, E., Schuppe-Koistinen, I., Lindberg, J. A note on normalization of biofluid 1D ^{1}H-NMR data. *Metabolomics* 2008, 4 (2), 114–121.

73. Forshed, J., Schuppe-Koistinen, I., Jacobsson, S. P. Peak alignment of NMR signals by means of a genetic algorithm. *Analytica Chimica Acta* 2003, 487 (2), 189–199.

74. Veselkov, K. A., Lindon, J. C., Ebbels, T. M. D., Crockford, D., Volynkin, V. V., Holmes, E., Davies, D. B., Nicholson, J. K. Recursive segment-wise peak alignment of biological ^{1}H NMR spectra for improved metabolic biomarker recovery. *Analytical Chemistry* 2008, 81 (1), 56–66.

75. Katajamaa, M., Oresic, M. Data processing for mass spectrometry-based metabolomics. *Journal of Chromatography A* 2007, 1158 (1–2), 318–328.

76. Pedrioli, P. G. A., Eng, J. K., Hubley, R., Vogelzang, M., Deutsch, E. W., Raught, B., Pratt, B., Nilsson, E., Angeletti, R. H., Apweiler, R., Cheung, K., Costello, C. E., Hermjakob, H., Huang, S., Julian Jr., R. K., Kapp, E., McComb, M. E., Oliver, S. G., Omenn, G., Paton, N. W., Simpson, R., Smith, R., Taylor, C. F., Zhu, W., Aebersold, R. A common open representation of mass spectrometry data and its application to proteomics research. *Nature Biotechnology* 2004, 22 (11), 1459–1466.

77. Berrueta, L. A., Alonso-Salces, R. M., Héberger, K. Supervised pattern recognition in food analysis. *Journal of Chromatography A* 2007, 1158 (1–2), 196–214.

78. Møller, S. F., von Frese, J., Bro, R. Robust methods for multivariate data analysis. *Journal of Chemometrics* 2005, 19 (10), 549–563.

79. Worley, B., Powers, R. Multivariate analysis in metabolomics. *Current Metabolomics* 2013, 1 (1), 92–107.

80. Stevens, J. R. 3—Statistical methods in metabolomics. In *Metabolomics in Food and Nutrition*, ed. B. C. Weimer, C. Slupsky. Woodhead Publishing, Cambridge, UK, 2013, pp 44–67.

81. Tohge, T., Fernie, A. R. Web-based resources for mass-spectrometry-based metabolomics: A user's guide. *Phytochemistry* 2009, 70 (4), 450–456.

82. Nguyen, K. T. N., Scapolla, C., Di Carro, M., Magi, E. Rapid and selective determination of UV filters in seawater by liquid chromatography–tandem mass spectrometry combined with stir bar sorptive extraction. *Talanta* 2011, 85 (5), 2375–2384.

83. Roos, G., Röseler, C., Büter, K. B., Simmen, U. Classification and correlation of St. John's wort extracts by nuclear magnetic resonance spectroscopy, multivariate data analysis and pharmacological activity. *Planta Medica* 2004, 70 (08), 771–777.

84. Cardoso-Taketa, A. T., Pereda-Miranda, R., Choi, Y. H., Verpoorte, R., Villarreal, M. L. Metabolic profiling of the Mexican anxiolytic and sedative plant *Galphimia glauca* using nuclear magnetic resonance spectroscopy and multivariate data analysis. *Planta Medica* 2008, 74 (10), 1295–1301.

85. Yuliana, N. D., Khatib, A., Verpoorte, R., Choi, Y. H. Comprehensive extraction method integrated with NMR metabolomics: A new bioactivity screening method for plants, adenosine A1 receptor binding compounds in *Orthosiphon stamineus* Benth. *Analytical Chemistry* 2011, 83 (17), 6902–6906.

86. Ali, K., Iqbal, M., Fortes, A. M., Pais, M. S., Korthout, H. A. A. J., Verpoorte, R., Choi, Y. H. Red wines attenuate TNFα production in human histiocytic lymphoma cell line: An NMR spectroscopy and chemometrics based study. *Food Chemistry* 2013, 141 (3), 3124–3130.

87. Ali, K., Iqbal, M., Korthout, H. A. A. J., Maltese, F., Fortes, A. M., Pais, M. S., Verpoorte, R., Choi, Y. H. NMR spectroscopy and chemometrics as a tool for anti-TNFα activity screening in crude extracts of grapes and other berries. *Metabolomics* 2012, 8 (6), 1148–1161.

88. Sipkema, D., Franssen, M. R., Osinga, R., Tramper, J., Wijffels, R. Marine sponges as pharmacy. *Marine Biotechnology* 2005, 7 (3), 142–162.

89. Keyzers, R. A., Davies-Coleman, M. T. Anti-inflammatory metabolites from marine sponges. *Chemical Society Reviews* 2005, 34 (4), 355–365.

90. Ali, K., Iqbal, M., Yuliana, N., Lee, Y.-J., Park, S., Han, S., Lee, J.-W., Lee, H.-S., Verpoorte, R., Choi, Y. Identification of bioactive metabolites against adenosine A1 receptor using NMR-based metabolomics. *Metabolomics* 2013, 9 (4), 778–785.

91. Matozzo, V., Tomei, A., Marin, M. G. Acetylcholinesterase as a biomarker of exposure to neurotoxic compounds in the clam *Tapes philippinarum* from the Lagoon of Venice. *Marine Pollution Bulletin* 2005, 50 (12), 1686–1693.

92. Elbaz, A., Wei, Y., Meng, Q., Zheng, Q., Yang, Z. Mercury-induced oxidative stress and impact on anti-oxidant enzymes in *Chlamydomonas reinhardtii*. *Ecotoxicology* 2010, 19 (7), 1285–1293.

93. Zhang, L., Liu, X., You, L., Zhou, D., Wu, H., Li, L., Zhao, J., Feng, J., Yu, J. Metabolic responses in gills of Manila clam *Ruditapes philippinarum* exposed to copper using NMR-based metabolomics. *Marine Environmental Research* 2011, 72 (1–2), 33–39.

94. Tikunov, A. P., Johnson, C. B., Lee, H., Stoskopf, M. K., Macdonald, J. M. Metabolomic investigations of American oysters using ^{1}H-NMR spectroscopy. *Marine Drugs* 2010, 8 (10), 2578–2596.

95. Fernie, A. R., Obata, T., Allen, A. E., Araújo, W. L., Bowler, C. Leveraging metabolomics for functional investigations in sequenced marine diatoms. *Trends in Plant Science* 2012, 17 (7), 395–403.

96. Chiu, K.-H., Ding, S., Chen, Y.-W., Lee, C.-H., Mok, H.-K. A NMR-based metabolomic approach for differentiation of hagfish dental and somatic skeletal muscles. *Fish Physiology and Biochemistry* 2011, 37 (3), 701–707.

97. Hines, A., Yeung, W. H., Craft, J., Brown, M., Kennedy, J., Bignell, J., Stentiford, G. D., Viant, M. R. Comparison of histological, genetic, metabolomics, and lipid-based methods for sex determination in marine mussels. *Analytical Biochemistry* 2007, 369 (2), 175–186.

98. Shin, M. H., Lee, D. Y., Skogerson, K., Wohlgemuth, G., Choi, I.-G., Fiehn, O., Kim, K. H. Global metabolic profiling of plant cell wall polysaccharide degradation by *Saccharophagus degradans*. *Biotechnology and Bioengineering* 2010, 105 (3), 477–488.

99. Nylund, G. M., Weinberger, F., Rempt, M., Pohnert, G. Metabolomic assessment of induced and activated chemical defence in the invasive red alga *Gracilaria vermiculophylla*. *PLoS ONE* 2011, 6 (12), e29359.

100. Ivanišević, J., Thomas, O., Lejeusne, C., Chevaldonné, P., Pérez, T. Metabolic fingerprinting as an indicator of biodiversity: Towards understanding inter-specific relationships among Homoscleromorpha sponges. *Metabolomics* 2011, 7 (2), 289–304.

101. Connor, K. M., Gracey, A. Y. High-resolution analysis of metabolic cycles in the intertidal mussel *Mytilus californianus*. *American Journal of Physiology: Regulatory, Integrative and Comparative Physiology* 2012, 302 (1), R103–R111.

102. Colwell, R. R. Global climate and infectious disease: The cholera paradigm. *Science* 1996, 274 (5295), 2025–2031.

103. Davidson, S. K., Allen, S. W., Lim, G. E., Anderson, C. M., Haygood, M. G. Evidence for the biosynthesis of bryostatins by the bacterial symbiont "*Candidatus* Endobugula sertula" of the Bryozoan *Bugula neritina*. *Applied and Environmental Microbiology* 2001, 67 (10), 4531–4537.

104. Kell, D. B., Brown, M., Davey, H. M., Dunn, W. B., Spasic, I., Oliver, S. G. Metabolic footprinting and systems biology: The medium is the message. *Nature Reviews Microbiology* 2005, 3 (7), 557–565.

CHAPTER 3

Biosynthetic Approaches to Marine Drug Discovery and Development

Eric W. Schmidt
Department of Medicinal Chemistry, L.S. Skaggs Pharmacy Institute,
University of Utah, Salt Lake City, Utah

CONTENTS

Biosynthetic approaches have been used to discover new marine compounds with medicinal potential, to supply clinically used marine drugs, and to synthesize natural product analogs for drug development. This review aims to summarize essential biosynthetic approaches behind modern marine drug research using examples of their successful application.

3.1 INTRODUCTION

The ocean is home to the greatest diversity of life on earth. This genetic diversity is reflected in the wealth of biosynthetic processes first described in marine life. For example, widely important pathways and proteins such as quorum sensing and green fluorescent protein were obtained through the study of marine biodiversity and biosynthesis.[1,2] Processes of global ecological importance, such as production of halocarbons, are known to result largely from marine biosynthetic processes.[3] The main routes to potential cyanobacterial biofuels were first found in marine organisms.[4–6] Some marine biosynthetic pathways have led to water-compatible glues and surgical products.[7] A major route to alkyl phosphonates is ubiquitous in the ocean.[8] Similarly, novel biosynthetic processes have led to a wealth of unique compounds that are of demonstrated importance to medicine: the marine natural products.

Virtually all forms of marine life contain natural products with biomedical potential. These natural products enable organisms to survive in diverse ecological niches, and thus the actual compounds are extremely diverse. Chemicals that have been so far explored, and that include Food and Drug Administration (FDA)–approved drugs and drug leads, fall roughly into three categories: small molecules, venom peptides, and other large polymers. The small-molecule natural products (used here synonymously with *secondary metabolites*) are found everywhere, but in marine research they are particularly associated with soft-bodied marine animals, algae, and microorganisms. Many compounds are synthesized to defend these organisms against predation and other competitive interactions. Venom peptides are also extremely widespread; research has focused on predatory interactions such as venoms used by anemones or cone snails to ensnare prey, although defensive venom peptides are also common. The final category includes polymers such as carbohydrates and will not be further discussed here.

The study of biosynthesis addresses several fundamental technical and scientific questions in marine natural product research. On the technical side, panning the seas for new drugs has been very fruitful, and yet key problems have been identified over the years. While bioactive compounds are readily discovered, their supply for drug development is usually not easy. Making analogs of compounds from coral reef organisms can be a challenge. Dereplication of previously defined compounds, and especially of previously defined scaffolds, also poses a problem. The discipline of biosynthesis has solved these problems in several cases. For example, compounds have been discovered and supplied and analogs synthesized using the biosynthetic approach.

Basic questions arise from the diverse chemical scaffolds discovered in the sea: What compounds are out there? How are they made? What roles do they play in nature? Biosynthetic studies address these questions and, more broadly, provide fundamental information about how organisms interact, which is critical in understanding ecology and biodiversity.

Marine biosynthesis is a vast topic. The biodiversity of the oceans is immense, and the resulting types of biosynthetic pathways and the organisms that produce them are correspondingly highly diverse. Some compounds or chemical scaffolds are identical or similar to those found on land, while others are uniquely marine. With the widespread use of molecular tools, biosynthetic genes have been described for hundreds or possibly even thousands of marine compounds. Here, the goal is not to describe exhaustively the breadth of marine natural product biosynthesis. Instead, basic considerations that are specific to marine biosynthetic studies will be described, followed by a few representative examples of direct importance to marine drug discovery and development. Finally, although marine biosynthesis has a long and storied tradition prior to the genomic era,[9–11] here the focus will be on postgenomic approaches to biosynthesis.[12]

3.2 BASIC CONSIDERATIONS IN MODERN MARINE NATURAL PRODUCT BIOSYNTHESIS

Biosynthetic studies of samples from any source, marine, terrestrial, or otherwise, are fundamentally similar. However, there are a few practical differences in terms of how to best access and study samples, which will be highlighted below. An understanding of the major sources of marine natural products is also essential. The field is fundamentally about translating chemical knowledge about natural product structures and enzymatic mechanisms into biological data (bioinformatics), and vice versa. Although the breadth of marine chemistry is beyond the scope of this chapter, it forms the essential core knowledge for practitioners. The most comprehensive source of this information is in the form of the "Marine Natural Products" reviews, which have been published annually for more than 30 years.[13]

3.2.1 Which Methods Should Be Used for the Diverse Marine Sample Types?

The ocean contains many different types of organisms, which are treated in many different ways. For example, purified cultures of marine bacteria or fungi have provided a wealth of novel and interesting metabolites, yet they are similar or identical to their terrestrial relatives and are studied by identical methods. Similarly, for marine animals such as cone snails that produce peptide natural products or venoms, methods resemble those for venom-containing animals from any environment. Some kinds of algae make important products such as biofuels or polyketide-derived polyunsaturated fatty acids; molecular tools have been developed for many of these organisms.

For some types of organisms, progress has been slow. The ocean contains a diverse array of eukaryotic microbes, which have surprising and fascinatingly diverse genetics. Therefore, genetic methods developed for animals, bacteria, and other well-studied organisms are not readily applied to some of these microbes. For example, dinoflagellates contain many important and interesting polyketides, especially polyethers.[14] Feeding studies suggest that their biosynthesis may be fundamentally different than polyketide polyether biosynthesis in bacteria. However, dinoflagellate genetics is challenging, greatly delaying progress despite substantial interest in the problem.

Many types of organisms either are uncultivable or cannot be studied as pure organisms and instead consist of mixtures of organisms. (Note: *Uncultivable* is not to be taken literally, but simply indicates organisms that have been hard to cultivate or purify so far.) Metagenomic methods are universally applicable to these mixed samples.

Examples of such organisms include the vast reservoir of uncultivated microbes in seawater, sediments, and symbioses; cyanobacterial mats; many cultivated cyanobacteria, dinoflagellates, and other organisms that are not easily purified; and many others. In the context of bioactive marine natural products, interest has focused on sponges and tunicates, which among the animals have been most prolific as sources of drugs or drug leads. Sponges and tunicates are often rich in bacteria, making it challenging to define the true producer of a marine natural product. This challenge is present for all of the mixed samples described above.

3.2.2 In a Mixture, Who Makes What?

In mixed samples, compounds could originate from any source organism living within the mixture. This necessitates a comprehensive and careful approach to identify biosynthetic genes, without undue assumptions. As described in Chapter 1, many marine animal natural products originate in symbiotic bacteria and not in the animals themselves. Often, these compounds are similar or identical to those found in cultivated bacteria. Such compounds are likely to be bacterial in origin, but since most have yet to be studied, caution is warranted. By contrast, many structural classes are so far unique to marine animals, including many alkaloids and terpenoids, and their biosynthetic sources cannot be anticipated. Some classes of compounds, such as certain demethylated sterols and sphingosine derivatives, are only known from eukaryotes, so they are more likely to originate in the animals. Most marine venoms are made by animals and not by symbiotic bacteria. Some compounds may even result from collaborations between multiple organisms, in analogy to precursor-directed biosynthesis in culture. Finally, many known marine natural products originate from the diet, and their genomic source may not exist within the sample containing the chemistry. In short, since most marine natural products are of unknown origin, it pays to expect the unexpected.

3.2.3 Localizing the Biosynthetic Genes

When researchers were first trying to nail down the origin of marine natural products in animals, a popular method was to physically separate different cell types, and then to determine where the compound was localized. The assumption was that the compounds would be found (or most abundant) in their producing organisms. Unfortunately, in the metagenomic era, this assumption has been proven to be false. For example, mycosporine amino acids are produced by the cyanobacteria *Prochloron didemni*,[5] and yet they are localized within specific types of host animal (ascidian) cells, where they protect the animal from ultraviolet irradiation.[15] As another example, bryostatins are made by symbiotic bacteria, *Candidatus* Endobugula sertula, within their host bryozoans.[16] However, bryostatins are not localized to the same spot as their producing symbiotic bacteria, but instead form a protective halo around the larvae.[17] Although in some cases compounds are co-localized with their producers, there are many contrary examples.

The location of production in space is thus a critical consideration in biosynthetic studies. In sponges, *Candidatus* Entotheonella is known to be a producer of many important sponge metabolites.[18,19] In some cases, it is found in specific regions of the sponge.[20] *Ca.* Entotheonella compounds are seemingly not always co-localized with the producer, since swinholide may be produced by the symbiont but was localized to another region of the sponge.[19,20] *Candidatus* Endolissoclinum faulkneri produces patellazole polyketides in ascidians.[21] It is found in specific blood cells inside the tiny zooid region of the animals, which are physically too small to account for the large patellazole concentration found in the animals. Sequencing approaches to obtaining the patellazole biosynthetic gene cluster completely missed the target, until a microdissection approach was employed to examine the zooid habitat specifically. Symbiotic actinomycetes in cone snails likely synthesize polyketides in the mucus-producing foot region of the animal.[22]

Symbiotic bacteria, *Teredinibacter turneri*, living in the gills of shipworms produce antibiotic metabolites. At least some of these may be transported to the gut, where they may help to keep the cellulose digestion chamber axenic.[23] This is a rare example of a cellulose-digesting animal in which digestive enzymes are produced by symbionts, but the symbionts do not reside in the gut.[24,25] Instead, digestive proteins are transferred to the gut, enabling the host to eat the resulting sugar prior to microbial sugar consumption.[26] By contrast, if bacteria are present, glucose liberated from cellulose is first consumed by bacteria. By producing antibiotics that are transferred to the gut, *T. turneri* may help the host to kill such bacteria.

Compounds that are produced by animals (and not their symbionts) can be highly localized. As one well-defined example, in production of cone snail venom peptides, the pattern of biosynthesis is quite complicated and is distributed in different parts of the cone snail venom ducts.[27]

It is quite easy to entirely miss a biosynthetic pathway by looking in the wrong place. Because compound localization is not sufficient to define the source, the best approach involves surveying several different regions of the organism using genomic methods.

3.2.4 Variation in Time and Space

There are many anecdotal stories about finding a promising drug in a source, such as a cyanobacterial mat or a sponge, and then returning the following year to find the same organism, but lacking the active compound. In short, there is a variation in chemistry in both time and space that can make biosynthetic studies challenging. A good example of this effect is in the ascidian *Didemnum molle*, which often contains potently active metabolites, but in low yields. A survey of *D. molle* samples showed that neighboring colonies can have different natural products, so that the compounds are highly concentrated within the producing ascidian but absent in nonproducing colonies.[28,29] In another study focusing on the asicidan *Lissoclinum patella*, it was shown that the variation in chemistry was highly correlated with a previously unobserved, cryptic speciation in the animals.[30]

Similarly, cyanobacterial mats containing the supergroup formerly known as *Lyngbya* vary extensively in the observed metabolites; this has been traced to speciation and strain variation in the producing cyanobacteria.[31] In the widely distributed biosynthetic powerhouses, *Salinispora* spp., variation in chemistry can similarly be traced to speciation and strain differences.[32] Such variation does not always indicate speciation, as seen in the case of sporadically varying metabolites that exist within otherwise identical *Prochloron* cyanobacteria.[28] The main lesson of this is that it is essential to look at a single colony using both chemical and genomic tools, whether that colony is a cultivated bacterial strain, a wild animal, or a microbial mat. Adjacent, otherwise identical colonies may lack the compounds entirely.

3.2.5 There Can Be Multiple Producers

In single organisms, there are often multiple families of natural products. In some sponges, *Ca.* Entotheonella is a superproducing bacterial symbiont that seems to make many of the observed classes of compounds (although it also coexists with bacteria that synthesize unusual lipids).[18,19,33] By contrast, in the ascidian *L. patella*, *Prochloron* and *Ca.* E. faulkneri are two totally different symbionts that live in different parts of the animal; one makes cyclic peptides, while the other makes polyketides.[5,21] There are certain to be found even more complex situations.

3.2.6 Genes Are Not Always Clustered

Bacterial biosynthetic pathways are usually thought to be clustered. This is convenient for biosynthetic studies because the identification of one gene in a pathway leads to identification of all genes. In order to find clustered genes, a good genome assembly is necessary, since in fragmented assemblies contiguous stretches of DNA are too small to contain large portions of biosynthetic gene clusters. Some types of genomes, such as many cyanobacterial genomes, pose assembly problems even for cultivated isolates. This problem is more difficult in genomes with many repetitive elements, such as found in cyanobacteria, especially when found in a mixed and complicated metagenomic background. Care needs to be taken to obtain reasonable assemblies in these types of genomes. Despite careful studies, there are now many examples where only partial gene clusters have been identified from marine sources. Improved sequencing methods and assembly algorithms continually reduce this problem. In addition, single-cell sequencing has been extremely useful in marine metagenomics, especially in discovering gene clusters from cyanobacteria[31] and *Ca.* Entotheonella and other bacteria in sponges.[18,19,33–35]

Sometimes, assembly is not the problem. Instead, biosynthetic genes are not actually clustered in the target organism. For example, in *Ca.* E. faulkneri the genome is undergoing degradation, and contrary to expectation, the genes are not clustered at all, but instead occupy seven different regions of the chromosome.[21] Therefore, identifying all of the needed genes for a biosynthetic pathway requires accurate assembly of the complete genome. This lack of clustering also indicates that clone library-based methods will fail for many types of samples, and that good genome assembly in bioinformatic protocols is required. This phenomenon is not unique to *Ca.* E. faulkneri, and seems to occur in other endosymbionts as well.

3.2.7 Assigning Function of Identified Genes

To make it worth investing in biochemistry or synthetic biology for a particular pathway, it is helpful to rapidly assign the function of the pathway in question. In the case of *Prochloron*, the biosynthetic pathway function was assigned by heterologous expression of the whole gene cluster.[36] This gene cluster was small and readily cloned, and its simplicity was helpful in providing proof of concept. This standard of function has traditionally been harder to accomplish with larger biosynthetic pathways. For example, for more complex symbiotic pathways, such as those to ecteinascidin (Et)-743, bryostatin, and pederin (similar to onnamide), evidence supporting the role of the identified genes was provided in part by expressing small portions of the biosynthetic pathways, among other experiments.[37–39] Although these experiments were well done, those individual proteins often have predictable functions, so that the argument runs the danger of being somewhat circular. To avoid this problem, examination of multiple proteins can be useful. When multiple proteins are expressed and analyzed individually, and all of the resulting data match the chemistry of the natural product, this brings exceptionally good confidence that the right cluster has been identified. A good example of this approach involved the curacin biosynthetic cluster, from cyanobacteria.[40] Although the curacin pathway is very large and was difficult to express using older methods, an exhaustive interrogation of the function of many individual proteins *in vitro* provided exceptionally clear and convincing evidence of function.[41,42] As synthetic biology improves, it will become routine to express whole clusters.

Another potential way around this problem is based on bioinformatics, and taking advantage of the colony variation described above. Two neighboring colonies of *L. patella* were collected, one of which contained patellazoles and the other of which did not.[21] The main metagenomic difference between these two samples was that one contained the putative patellazole-producing symbiont and the patellazole gene cluster, whereas the other did not. This type of "virtual knockout" comparative analysis is fairly easy to do with marine organisms, since chemical variation is a commonly observed phenomenon.

3.3 CASE STUDIES: BIOSYNTHESIS IN THE DISCOVERY AND DEVELOPMENT OF MARINE DRUGS

Marine drug discovery involves identifying hit compounds, supplying them for further study, and optimizing their pharmacological properties. Marine biosynthetic studies have succeeded in all three of these steps.

3.3.1 Before the Application: Discovery of Biosynthetic Pathways

The modern application of biosynthesis to marine drugs has met with substantial success, but it rests upon a long series of basic research studies. Initially, feeding studies involving putative biosynthetic precursors were most common. Subsequently, attempts were made to identify producing organisms by looking for which organism contained the natural products. Many of these studies have not stood the test of time because of the fact that, in many cases, compounds do not co-localize with their producers. However, many of the studies have correctly identified the true producing organism, as later validated by molecular approaches. In addition, these groundbreaking studies brought exceptional evidence that symbiotic bacteria might actually produce marine natural products.[43]

At the start of the molecular era, approaches mainly focused on identifying these difficult-to-cultivate symbiotic bacteria and sequencing putative biosynthetic genes.

The most impactful studies are those that identified the bryostatin polyketide-producing symbiont in the bryozoan *Bugula neritina* and the identification of onnamide-type biosynthetic genes in sponges.[44,45] Pathway identification continues to be a major focus today, as highlighted in Chapter 1. As mentioned above, pathway identification methods have evolved from initial homology cloning and DNA library approaches to the pure sequencing approaches that are employed today. Below, applications subsequent to pathway identification will be described.

Within cultivated marine bacteria, the approaches and history have tracked well with the general trends of natural product biosynthesis, although marine samples have led to some stunning new compounds and some innovative applications. These will be highlighted below.

3.3.2 Supplying Marine Natural Products

Many marine natural products originate in animals, bacterial mats, or other environmental samples. It has been difficult to translate these rare samples into drugs in part because of the large barrier between discovering a potentially useful compound and obtaining a sufficient compound even for preclinical investigation. Biosynthesis now provides methods that are increasingly powerful in overcoming this supply problem. Although supply can also be an issue for natural products from cultivated organisms, it will not be discussed here, as methods are identical to those used in terrestrial strains.

Many environmental marine biosynthetic pathways have been discovered through metagenomic and biosynthetic studies, as described in Chapter 1. Of particular note, substantial, foundational progress has been made with the production of natural products by the sponge symbiont *Ca.* Entotheonella (Chapter 1).[19] In order to go from pathway discovery to compound supply, it is essential to express the biosynthetic pathway in sufficient quantity for further study. One method is to clone or synthesize the pathway in question, and then to express it in a heterologous host. An alternative method is to identify closely related pathways or compounds in cultivated bacteria, either by genome mining or by chemical examination, and then to produce the compounds or their precursors in culture. Both methods have yielded successes.

Probably the best-known example of the latter approach is in the supply of the clinically used anticancer compound Et-743 (described in more detail in Chapter 14) (Figure 3.1).[46] Although ecteinascidins are isolated from tunicate animals, their core structures are very similar to those of the bacterial metabolites, safracins and saframycins. Et-743 is currently supplied by using a bacterial metabolite as a precursor, with semisynthesis from the fermentation product affording the

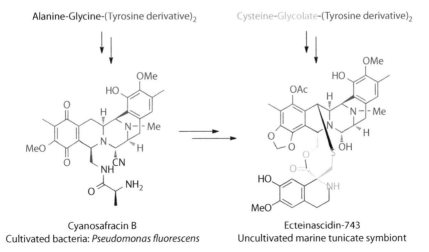

Alanine-Glycine-(Tyrosine derivative)$_2$ Cysteine-Glycolate-(Tyrosine derivative)$_2$

Cyanosafracin B
Cultivated bacteria: *Pseudomonas fluorescens*

Ecteinascidin-743
Uncultivated marine tunicate symbiont

Figure 3.1 Biosynthetic and semisynthetic origin of Et-743. Ecteinascidin-743 was isolated from a marine tunicate. It is obtained semisynthetically from cyanosafracin B. The biosynthesis of both compounds has been defined by sequencing and analysis.

desired metabolite.[46,47] The synthesis begins with cyanosafracin B, a derivative of the natural bacterial fermentation product safracin B obtained by adding cyanide during fermentation or isolation. This product is modified to yield Et-743. The discovery of a partial ecteinascidin biosynthetic gene cluster from the marine animal may serve to further improve this method.[37] This gene cluster is very similar to others in the series, such as safracin,[48] although the nonribosomal peptide synthetase genes use cysteine and glycolate in place of alanine glycine, meaning that potentially very simple molecular engineering could be used to provide an improved precursor. The biosynthesis of these compounds has many interesting and noncanonical features,[49] but in a general sense, the isolated natural products are derived from four amino acids, including a tyrosine derivative added twice by a single module of the nonribosomal peptide synthetase.

A second example relevant to clinically used drugs is aplidine, an anticancer agent supplied by synthesis. Like ecteinascidins, aplidine and relatives are found in tunicates. Free-living marine α-proteobacteria *Tistrella mobilis* and *Tistrella bauzanensis* were found that produce the close relative didemnin B (Figure 3.2). Both aplidine and didemnin B are tunicate products, which differ solely by the oxidation state at one position. The biosynthetic gene cluster was sequenced and analyzed, providing a potential source of the natural products.[50]

Heterologous expression has succeeded in providing two groups of marine natural products that were otherwise difficult to obtain. Cyanobactins are N-C circular peptides found in some tunicates and cyanobacteria.[51] In tunicates, they are biosynthesized by symbiotic cyanobacteria, *Prochloron* spp.[36,52] Although cyanobactins are not now drugs, some of the ascidian cyanobactins are fairly cytotoxic, and there was for a time some interest in the potential of trunkamide as an anticancer agent. The first cyanobactin locus to be described was a short, ~11 kbp biosynthetic gene cluster for patellamides A and C.[36] These compounds were shown to be produced on a ribosomally synthesized precursor peptide, via a ribosomally synthesized and posttranslationally modified peptide (RiPP) pathway. This precursor is enzymatically modified to produce the mature natural products. Initial attempts to express the biosynthetic pathway in *Escherichia coli* led to relatively low yields (now estimated at low nanograms per liter). Subsequently, a second cyanobactin pathway type to trunkamide and its relatives was described (Figure 3.3).[53] This pathway was expressed, again at a very low level. Over a 6-year period, the trunkamide yield from *E. coli* expression was gradually improved in a stepwise manner to now be >1 mg L^{-1}.[29,54,55] This provides the first example of a metagenome sequence-to-synthetic biology method to produce a natural product from any source and provides proof of concept for the overall method.

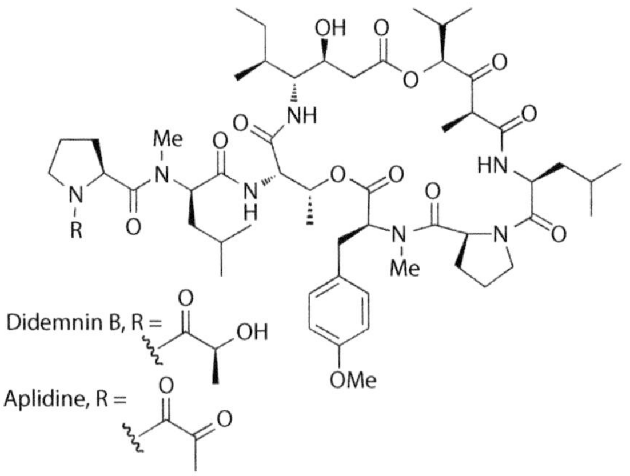

Figure 3.2 The didemnin family is found in free-living bacteria and tunicates. The discovery of a biosynthetic gene cluster in free-living marine bacteria may provide an improved supply of the compound.

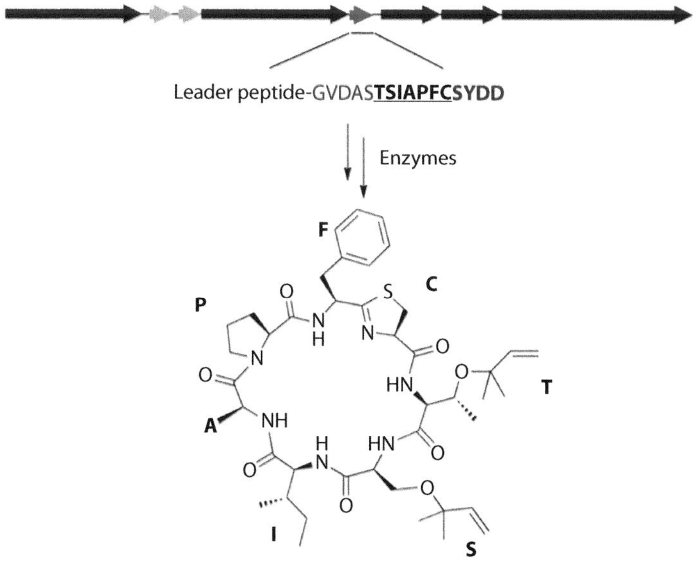

Leader peptide-GVDAS**TSIAPFCS**YDD

Enzymes

Trunkamide, from a tunicate metagenome
Expressed in *E. coli*

Figure 3.3 Heterologous expression to produce marine tunicate products. Trunkamide is an anticancer compound from tunicates. The *tru* gene cluster (top) was identified by metagenome sequencing expressed in *E. coli* to provide trunkamide in the milligrams per liter range.

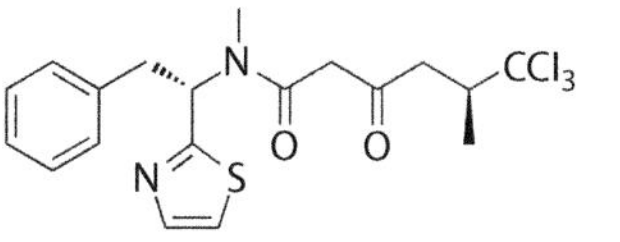

O-demethylbarbamide
Expressed in *Streptomyces*

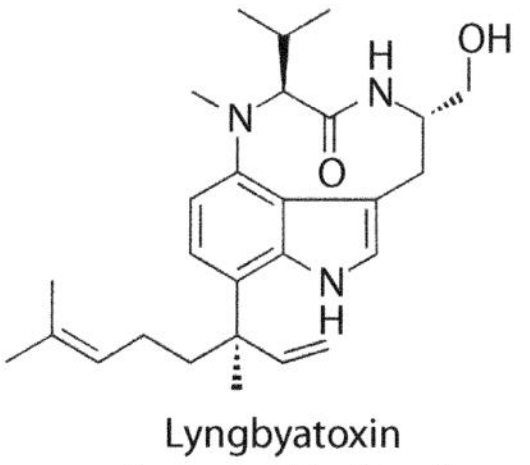

Lyngbyatoxin
Expressed in *E. coli*

Figure 3.4 Marine cyanobacterial products obtained by heterologous expression.

An early step toward the widespread application of the metagenomes-to-synthetic biology method for larger pathways is found in the barbamide cluster from the marine cyanobacterium *Moorea producens*.[56] Although this organism is cultivated, it is not a pure culture, and the genome is complicated. Therefore, the strategy used is conceptually somewhat similar to that used for marine animals. The barbamide cluster is a hybrid polyketide–nonribosomal peptide biosynthetic pathway, and it also features some unique tailoring events, such as leucine trichlorination. Interestingly, a close structural homolog and barbamide-like biosynthetic genes are also found in sponges.[57,58] To produce the compound, the gene cluster was cloned into a *Streptomyces* vector. Two putative operons are present in the 26 kbp cluster; each was placed under control of a *Streptomyces* promoter, while the remainder of the cluster was left intact. Although the natural product was not produced, a very close homolog simply lacking the *O*-methyl group was found in low (<μg L⁻¹) yields.[59]

More recently, the lyngbyatoxin cluster (involving a nonribosomal peptide synthetase) was successfully produced in high yield in *E. coli* (Figure 3.4).[60] One of the intermediates to lyngbyatoxin

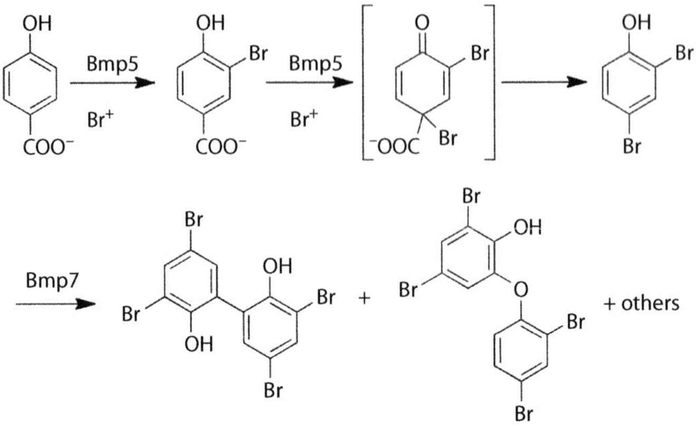

Figure 3.5 Expression of pathways to polybrominated aromatic compounds. Examples of key steps include aromatic bromination catalyzed by flavoprotein Bmp5, followed by oxidative coupling catalyzed by the cytochrome P450, Bmp7.

was reported to be present in ~150 mg L^{-1} yields, and the final product was present at about 50 mg L^{-1}. Like barbamide, lyngbyatoxin originates in a marine *M. producens*; unlike barbamide, it is found in an uncultivated sample. Like the cyanobactins, the gene cluster is fairly small (<12 kbp), and like both cyanobactins and barbamide, it was expressed as an operon under control of a single heterologous promoter.

A potential route to supply was achieved using biosynthetic genes to polybrominated aromatic compounds (Figure 3.5). This compound class has been found in many environments. Intriguingly, derivatives were identified in marine sponges,[43] where in a now classic study they were localized to symbiotic cyanobacteria. Using a genome mining approach, biosynthetic genes to polybrominated phenyl ethers and pyrroles were identified in a series of phylogenetically unrelated marine bacteria.[61] A whole biosynthetic pathway operon was transferred for inducible expression in *E. coli*, leading to functional synthesis of these compounds *in vivo*. In addition, *in vitro* studies enabled stepwise synthesis of the compounds.

The above provides examples of successful heterologous expression of targeted gene clusters from marine environmental samples, including animals and cyanobacterial mats. They provide proof of concept for the sequencing-to-expression paradigm that will become prevalent in marine natural products. A caveat is that there are some substantial unknowns, especially for biosynthetic genes and pathways from genome-reduced bacteria and from novel and understudied eukaryotic taxa.

In addition to solving the uncultivated supply problem, biosynthetic approaches have also been used to improve the supply of marine natural products from cultivated bacteria. For example, salinisporamide A is a product of a hybrid polyketide–nonribosomal peptide pathway.[62] The compound is produced by *Salinispora tropica*, an actinomycete that has so far only been found in the oceans, and the compound is currently in phase II clinical trials.[63] Manipulation of regulatory elements doubled the production of this compound without doubling the production of competing side products.[64]

3.3.3 Discovering Marine Natural Products

Genome mining provides one of the major methods to discover new natural products, and it has been applied extensively to the discovery of metabolites from diverse marine organisms, including animals, cultivated bacteria, and metagenomes. Here, specific applications to the discovery of new marine natural products will be described.

Probably the earliest example of successful application of biosynthesis to drug discovery is in the field of conotoxins/conopeptides, which are described in detail in Chapter 16. These compounds are components of cone snail venoms, and they are specifically found in and synthesized by cells in the animal venom duct.[65,66] "Cabals" of up to thousands of individual venom peptides are synthesized in each snail and injected via a harpoon into target prey organisms. Because the peptides often exhibit exquisite selectivity for receptors and ion channels, they have been developed into clinically applied agents, including the FDA-approved analgesic ziconotide.

Like other RiPPs, conotoxins were shown to originate as precursor peptides.[67] The precursor peptides contain different elements, including the N-terminal leader peptide, which is cleaved prior to toxin maturation, and the core peptide, which directly encodes the sequence of the natural product. While the core peptide has a hypervariable sequence, the leader peptide is highly conserved and falls into only a few distinct sequence categories.

Conotoxins have several distinct structural families, each with potentially tens of thousands of representatives.[66] Families are defined based on their pattern of disulfide bond formation. In addition, there are many other known posttranslational modifications, such as carboxylation of glutamate, hydroxylation of proline, halogenation of tryptophan, and C-terminal amidation, among others.[68] All of these modifications are synthesized by discrete posttranslational modifying enzymes, many of which are already known and described (with the exception of halogenation). The protein disulfide isomerases (PDIs) should also be considered in this category, since they appear to catalyze the formation of specific disulfide combinations, out of several possible structures.[69]

The biosynthetic understanding of conotoxins has greatly impacted their discovery. As for other natural products, discovery initially followed a process of purification and painstaking chemical analysis. In 1992, for the first time, a biosynthetic approach was applied to marine natural product discovery (Figure 3.6).[70] The conotoxin ω-MVIIC was discovered by probing a cDNA library and sequencing the resulting hits. Probes were designed based on the sequences of known conotoxins, ω-MVIIA and ω-MVIIB. The resulting sequence, in isolation, would not provide sufficient information to synthesize the native ω-MVIIC. Instead, based on biosynthetic insights from its

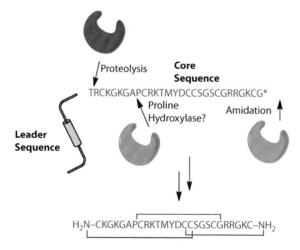

Figure 3.6 Biosynthetic approaches to conotoxin discovery. The first example was ω-MVIIC, an important Ca^{2+}-channel blocker. The peptide sequence was obtained from cDNA, and based on a biosynthetic hypothesis, the mature peptide was synthesized chemically (disulfide bonds shown schematically). Proline hydroxylation was not required for activity, so the commercially available compound lacks this modification. Similar approaches have since been used with other conotoxins. Note that this figure also shows the general route to all RiPPs: a ribosomally encoded peptide that is modified by enzymes and loses its leader sequence upon maturation.

relatives, the N-terminal cleavage site was predicted. It was also predicted that the C-terminal glycine would be a substrate for oxidative terminal amide formation. In addition, possibly the proline residue would be hydroxylated. Finally, the pattern of disulfide bond formation was inferred from knowledge of the specific conotoxin family. These predictions enabled the chemical synthesis of an active Ca^{2+}-channel blocker that remains an important and widely applied research tool. A later, but very important example is the discovery of α-RgIA by a similar method.[71] This peptide is a potent inhibitor of the α9α10 nicotinic acetylcholine receptor, an important target in drug discovery (see Chapter 10).

Discovery today is shifting further toward genome mining, often paired with proteomic approaches. For example, the *Conus bullatus* venom duct transcriptome was characterized, leading to simultaneous identification of putative genes for precursor peptides and all of the modifying enzymes that would be necessary for conotoxin maturation.[72] It may be possible to simultaneously apply these proteins to the *in vitro* or *in vivo* synthesis of conotoxin variants. In addition, such approaches are providing exceptional insights into the origin, extent of diversity, and potential for discovery of novel conopeptides.[73,74] Currently, a limiting factor is that conotoxins can be difficult to synthesize, despite stunning advances in recombinant and chemical synthetic technologies.[75]

RiPPs have also been extensively mined from marine bacteria, including cultivated bacteria and metagenomic samples (Figure 3.7).[76] An early example was the discovery of trichamide, a cyanobactin cyclic peptide in the marine cyanobacterium *Trichodesmium erythraeum*.[77] Trichamide was characterized by genome mining, in which the biosynthetic pathway was found and analyzed *in silico*. Based on knowledge of cyanobactin biosynthetic modifications, a putative structure was predicted. The predicted compound was found by mass spectrometry (MS) and confirmed by tandem MS, but it was present in too low an amount for nuclear magnetic resonance (NMR) confirmation. Since *T. erythraeum* forms enormous blooms in the ocean (the largest such cyanobacterial blooms), potential metric tons of this new metabolite are present in such blooms. The biological activity of trichamide remains undetermined. Many other cyanobactins have since been discovered by genome mining and biosynthetic insight, and they may be present in ~30% of all cyanobacteria

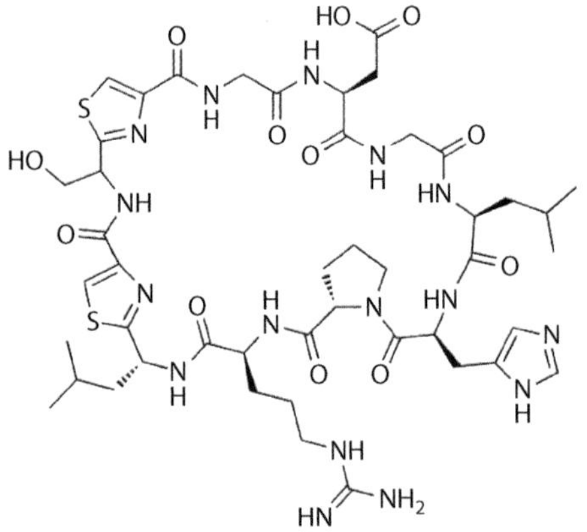

Trichamide, from the cyanobacterium
Trichodesmium erythraeum

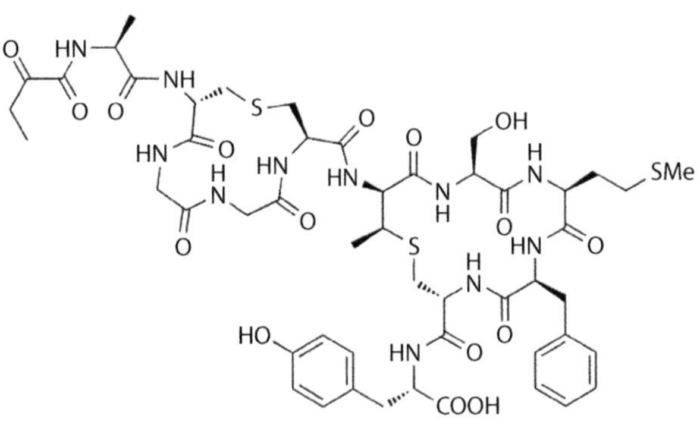

Prochlorosin 4.3, from the cyanobacterium
Prochlorococcus MIT913

Figure 3.7 Discovery of bacterial RiPPs using genome and metagenome mining and biosynthetic information. These compounds were identified by combining bioinformatics of genome sequences with chemical and biochemical methods, based on biosynthetic knowledge or experiments. More than 60 cyanobactins have been discovered by genome and metagenome mining, including the first example, trichamide. These compounds are found in free-living cyanobacteria and in cyanobacteria that live symbiotically with tunicates. Lanthipeptide prochlorosins were found in free-living cyanobacteria by genome mining. A single pathway naturally processes 29 precursor peptides to create diverse derivatives.

on earth. Cyanobactins also provide an early example of metagenome mining from marine animals. By combining mass spectrometry and gene sequencing, new cyanobactins were discovered in the ascidian *D. molle*.[5,29] Later chemical analysis and pharmacological study further confirmed the structure and demonstrated that the compounds exhibit modest inhibition of HIV replication.[78] In total, pathways for about 60 ascidian cyanobactins and hundreds of other cyanobacterial cyanobactins have been described, largely through a genome mining approach.[51,53,79–87]

RiPPs known as prochlorosins were discovered in the global planktonic cyanobacterium *Prochlorococcus* MIT913 using a genome mining and chemical approach.[88,89] Amazingly, a single pathway processes 29 different precursors within a single organism. These RiPPs have been found throughout the oceans, wherever *Prochlorococcus* and relatives exist. A large series of interesting RiPPs has been characterized from other marine bacteria by using a novel combination of genomics and proteomics.[90] It should be emphasized that there are also many other very interesting and bioactive RiPPs that have been found in the marine environment, but here the emphasis is on compounds discovered by the biosynthetic approach.

RiPPs were one of the early targets for marine genome mining because of their accessibility. By examining precursor peptide sequences in tandem with understanding enzymatic function, it is easy to guess possible structures of the mature natural products. These structures can then be rigorously examined by methods that are basically technical extensions of mass spectrometry-based proteomics. Other pathway types remain more difficult, primarily because it is hard to go from the sequence of genes to a precise structure of a product.

One of the premier examples of genome mining for the discovery of natural products has developed with the marine actinomycetes, *Salinispora* spp. These cultivated bacteria are amenable to methods commonly used for terrestrial actinomycetes (and thus are well developed), but they contain a wealth of novel and abundant biosynthetic pathways.[91] This combination of traits has made it possible to make rapid progress in discovering exceptionally novel compounds and gene clusters. For example, the macrolactam salinilactam A was identified using a combination chemical and genome mining approach (Figure 3.8).[92] The *S. tropica* genome was sequenced, leading to identification of a potential polyene macrolactam biosynthetic gene cluster. The compound was unstable in the isolation conditions, and both discovery and structure elucidation would have been difficult absent genome sequencing information. In the event, spectroscopy and genome sequencing were coupled in the structural analysis. In particular, configurational analysis was not possible, so the configuration of three OH centers was predicted using genome analysis. This method is proving particularly valuable in challenging cases. Inspired by this and other work in the field, stereocenters in the tunicate products, patellazoles, were recently anticipated by genome analysis.[21] Genome mining of this and other *Salinispora* genomes is leading to discovery of novel compounds and gene clusters. For example, novel salinosporamide derivatives have been uncovered in this way.[93]

Calyculin and its relatives are well-known monophosphorylated natural products isolated from sponges (Figure 3.9). The discovery of the calyculin biosynthetic pathway in symbiotic *Ca.* Entotheonella sp. paved the way to new compound discovery.[34] Three phosphotransferases were discovered in the calyculin pathway. One of these, CalQ, converted monophosphorylated calyculin

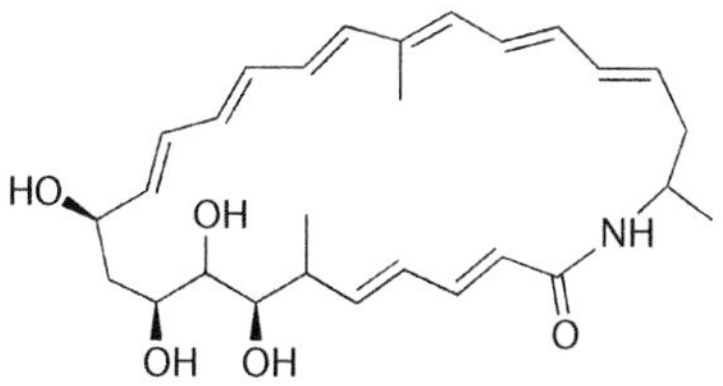

Figure 3.8 Salinilactam. The compound was identified by genome mining and chemical analysis in *Salinispora tropica*.

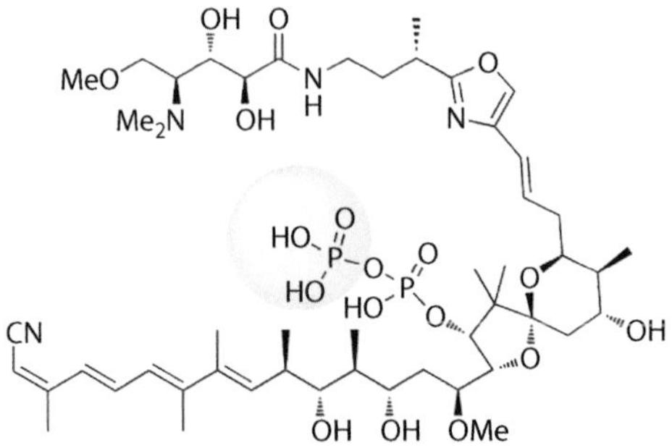

Figure 3.9 Discovery of phosphocalyculin A via metagenome mining. The highlighted phosphate was not present in the initially identified natural product, but through a metagenomic approach, it was discovered that the natural compound within a sponge is a natural protoxin, where dephosphorylation affords the active form of the compound.

derivatives into diphosphates. This was a very surprising finding since such a derivative was previously neither observed nor suspected. This led to the hypothesis that the diphosphate may be enzymatically degraded to produce isolated calyculin. Indeed, a careful study in which sponges were freshly extracted led to the identification of the diphosphate derivative as the major product existing within sponges. Calyculin is a cytotoxic phosphatase inhibitor, and its diphosphate derivative was about 1,000-fold less active than calyculin itself. Thus, calyculin may exist as a prodrug that is enzymatically liberated in response to challenge. The discovery of this new compound would likely not have been made absent genomic data, and indeed, chemical research over decades overlooked the existence of the diphosphate.

3.3.4 Making Analogs of Marine Natural Products

A key advantage of obtaining a functional biosynthetic pathway is that it becomes possible to make analogs. Thus, it enables a medicinal chemistry approach to determine structure–activity relationships and optimize pharmacological properties, the early preclinical steps in drug development. Beyond analogs, a major goal of synthetic biology is to make completely artificial compounds (similar to the combinatorial library idea) that can be screened in extremely high throughput. There are good marine examples of both of these processes. *Salinispora* spp., which are cultivated, have been used in the extensive modification of natural products to find potent new analogs. From the animal world, cyanobactins have been extensively modified in synthetic biology approaches, mainly aimed at making artificial compounds with new activities.

Salinosporamide A, from *S. tropica*, is in clinical trials as an anticancer agent and is a subject of Chapter 6 (Figure 3.10). Salinosporamide A is made from acetate, phenylalanine, and a novel chlorinated polyketide precursor derived from S-adenosyl methione (SAM).[62,94,95] The phenylalanine subunit is modified to a unique 3-R-cyclohex-2-enyl-L-alanine moiety. Salinosporamide A analogs have been made using mutasynthesis by two methods. In one example, the *salX* gene was knocked out of the *S. tropica* genome.[96] SalX is a key enzyme in the phenylalanine modification pathway, and abolishing the protein led to a loss of production of salinisporamide A. Instead, a novel derivative was produced incorporating a leucine derivative in place of the phenylalanine derivative, to produce the known compound antiprotealide. This discovery was further exploited by feeding a series of synthetic amino acid analogs to the *salX* knockout strain, including those with cyclobutyl and cyclohexyl side chains.[97] One compound, with a cyclopentene side chain, had a lower IC_{50} than the parent salinosporamide A.

Mutasynthesis has also been employed successfully to modify the chlorinated polyketide subunit of salinosporamide A.[98] SalL is a chlorinase that modifies SAM en route to salinosporamide A.

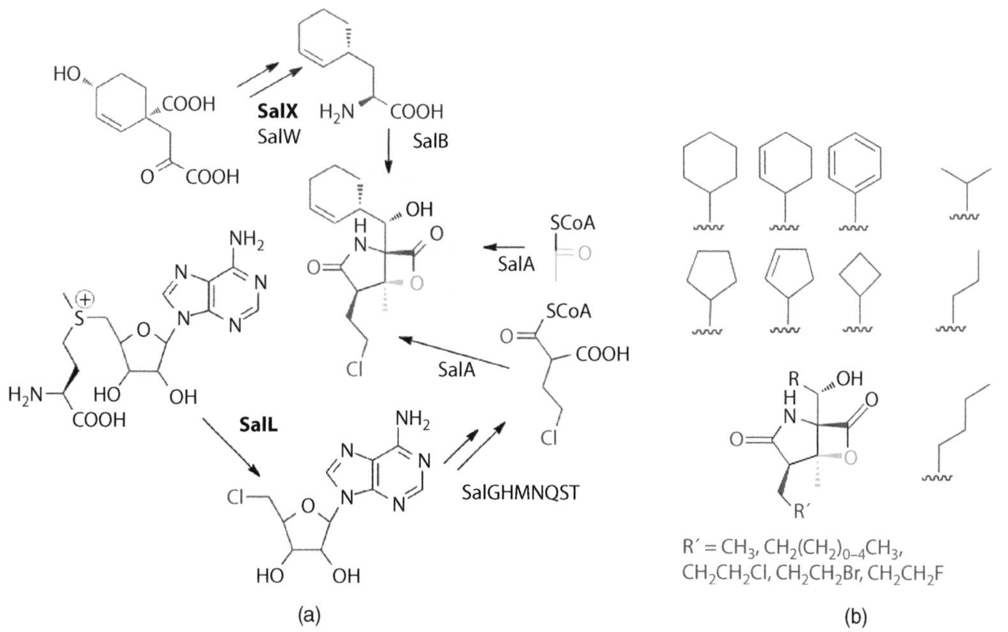

Figure 3.10 Biosynthetic approaches to salinosporamide. (a) Salinisporamide A (center) is comprised of novel polyketide and amino acid subunits, which are linked via the action of polyketide synthase–nonribosomal peptide synthetases SalA and SalB. Enzymes for precursor biosynthesis are shown and are in bold when they have been deleted in analog synthesis. (b) Natural and unnatural derivatives created by manipulating SalX or SalW and feeding advanced precursors.

A strain in which *salL* had been knocked out was supplemented with synthetic 5′-fluoro-5′-deoxyadenosine and 5-fluoro-5-deoxyribose. The strain accepted both subunits to produce a fluorinated analog of salinisporamide A. Since fluorine is a common substituent of synthetic drugs, but is rarely found in natural products, this study represented a novel mechanism to combine the strengths of both approaches in drug development.

It should be clear that there are many reports of biosynthetic studies using marine bacteria, as variously defined, so that the above represents the most illustrative example that is directly drug relevant and involves the deliberate synthesis of analogs for medicinal chemistry. For example, the first complete marine biosynthetic pathway is that to enterocin, from *Streptomyces maritimus* bacteria,[99] and there have been many subsequent reports of biosynthesis in marine bacteria.

It is desirable to use biosynthesis to create analogs of marine natural products from animals, which are otherwise inaccessible except via complex synthetic chemistry approaches. A relatively simple, purely synthetic biology approach has been employed with cyanobactin pathways. As RiPP-derived products, one would expect that the cyanobactin biosynthetic machinery would tolerate mutations. A surprising thing was the degree of mutations that are allowed. For example, *pat* and *tru* pathways have been cloned and identified in more than 30 different ascidian samples.[53,79] Wherever a *pat*-like pathway has been found in ascidians, it is nearly 100% DNA sequence identical to the first *pat* pathway that was found. The same is true with *tru* pathways. The only exception to this rule is that the short core peptide—the exact region that winds up in the mature natural product—is hypervariable. Another way of saying this is that in nature, identical enzymes synthesize a wealth of products by accepting many different substrates. Usually, it is thought that enzymes evolve to accept new substrates; here, the reverse is true, and substrates evolve while enzymes are constant. This mechanism has been named substrate evolution. Although this may not be true with all cyanobactin pathways, at least the *pat* and *tru* pathways from marine ascidians are among the broadest-substrate pathways known.

This feature has been exploited to create many new derivatives using a purely synthetic biology-based approach, in which all enzymes and substrates are derived from wholly synthetic DNA sequences and are expressed in *E. coli*. In the first proof-of-principle paper, a *pat* pathway core peptide was manipulated to contain only amino acids that are not found in any known *pat* product.[79] The resulting compound, eptidemnamide, was designed to imitate the clinically approved anticoagulant agent, eptifibatide. Since the initial experiment, many different derivatives have been made, exploiting the relaxed substrate specificity of the parent pathway (Figure 3.11).[29,54] Each position in the peptide chain can be substituted, and tandem mutations of all positions are accepted. One of the key advances was the ability to include nonproteinogenic amino acids in the process, enabling access to highly unique compounds.[54] Such an approach is also amenable to *in vitro* synthesis of derivatives.[100–103]

The technology behind cyanobactin manipulation is being continuously improved. For example, the heterocyclase enzyme has been paired with *in vitro* transcription and translation systems to produce highly diverse derivatives.[104] Several different strategies exist to manipulate peptides *in vitro*,[29,100,102,103,105,106] including producing compounds in a scalable fashion. Finally, large *in vivo* libraries have been created; currently, the ability to make large libraries of new compounds encoding millions of products is being exploited to hit targets that are not hit by the native sequences.[54,107]

3.4 CONCLUSIONS

As with terrestrial natural products, biosynthesis is beginning to have a large impact on the discovery and development of marine natural products. Proof of concept has been achieved for discovering, supplying, and manipulating compounds, both from cultivated bacteria and from environmental samples such as ascidians. The goal of going directly from metagenome sequencing to compound heterologous expression via synthetic biology has also been achieved. As costs come down and technologies improve, biosynthetic methods are anticipated to be at the center of marine natural product research. In particular, they are expected to afford access to natural products from tiny or rare organisms that are present in too small a quantity for the traditional chemical methods for natural products. Such studies are already beginning to show that perhaps the hidden majority of marine animal natural products remain to be discovered.

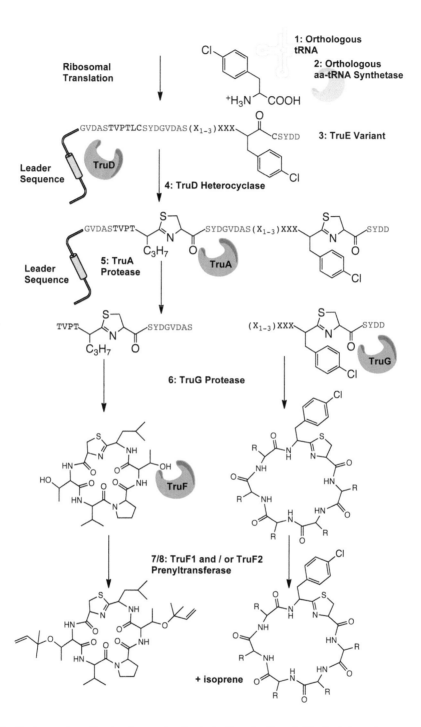

Figure 3.11 Engineered synthesis of cyanobactin derivatives in *E. coli*. Manipulation of precursor peptide sequences has led to diverse products, including those containing unnatural amino acids. (Reproduced from Tianero MD et al., *Journal of the American Chemical Society* 134:41825, 2012. With permission.)

REFERENCES

1. Fuqua WC, Winans SC, Greenberg EP. Quorum sensing in bacteria: The LuxR-LuxI family of cell density-responsive transcriptional regulators. *Journal of Bacteriology* 1994;176:269–75.
2. Morise H, Shimomura O, Johnson FH, Winant J. Intermolecular energy transfer in the bioluminescent system of *Aequorea*. *Biochemistry* 1974;13:2656–62.
3. Theiler R, Cook JC, Hager LP, Siuda JF. Halohydrocarbon synthesis by bromoperoxidase. *Science* 1978;202:1094–96.
4. Schirmer A, Rude MA, Li X, Popova E, del Cardayre SB. Microbial biosynthesis of alkanes. *Science* 2010;329:559–62.
5. Donia MS, Fricke WF, Partensky F, et al. Complex microbiome underlying secondary and primary metabolism in the tunicate-*Prochloron* symbiosis. *Proceedings of the National Academy of Sciences of the United States of America* 2011;108:E1423–32.
6. Mendez-Perez D, Begemann MB, Pfleger BF. Modular synthase-encoding gene involved in alpha-olefin biosynthesis in *Synechococcus* sp. strain PCC 7002. *Applied and Environmental Microbiology* 2011;77:4264–67.
7. Stewart RJ, Ransom TC, Hlady V. Natural underwater adhesives. *Journal of Polymer Science Part B: Polymer Physics* 2011;49:757–71.
8. Metcalf WW, Griffin BM, Cicchillo RM, et al. Synthesis of methylphosphonic acid by marine microbes: A source for methane in the aerobic ocean. *Science* 2012;337:1104–7.
9. Garson MJ. The biosynthesis of marine natural products. *Chemical Reviews* 1993;93:1699–733.
10. Moore BS. Biosynthesis of marine natural products: Microorganisms (Part A). *Natural Product Reports* 2005;22:580–93.
11. Moore BS. Biosynthesis of marine natural products: Macroorganisms (Part B). *Natural Product Reports* 2006;23:615–29.
12. Lane AL, Moore BS. A sea of biosynthesis: Marine natural products meet the molecular age. *Natural Product Reports* 2011;28:411–28.
13. Blunt JW, Copp BR, Keyzers RA, Munro MH, Prinsep MR. Marine natural products. *Natural Product Reports* 2013;30:237–323.
14. Kellmann R, Stuken A, Orr RJ, Svendsen HM, Jakobsen KS. Biosynthesis and molecular genetics of polyketides in marine dinoflagellates. *Marine Drugs* 2010;8:1011–48.
15. Maruyama T, Hirose E, Ishikura M. Ultraviolet-light-absorbing tunic cells in didemnid ascidians hosting a symbiotic photo-oxygenic prokaryote, *Prochloron*. *Biology Bulletin* 2003;204:109–13.
16. Davidson SK, Allen SW, Lim GE, Anderson CM, Haygood MG. Evidence for the biosynthesis of bryostatins by the bacterial symbiont "*Candidatus* Endobugula sertula" of the bryozoan *Bugula neritina*. *Applied and Environmental Microbiology* 2001;67:4531–37.
17. Sharp KH, Davidson SK, Haygood MG. Localization of '*Candidatus* Endobugula sertula' and the bryostatins throughout the life cycle of the bryozoan *Bugula neritina*. *ISME Journal* 2007;1:693–702.
18. Freeman MF, Gurgui C, Helf MJ, et al. Metagenome mining reveals polytheonamides as posttranslationally modified ribosomal peptides. *Science* 2012;338:387–90.
19. Wilson MC, Mori T, Ruckert C, et al. An environmental bacterial taxon with a large and distinct metabolic repertoire. *Nature* 2014;506:58–62.
20. Bewley CA, Holland ND, Faulkner DJ. Two classes of metabolites from *Theonella swinhoei* are localized in distinct populations of bacterial symbionts. *Experientia* 1996;52:716–22.
21. Kwan JC, Donia MS, Han AW, Hirose E, Haygood MG, Schmidt EW. Genome streamlining and chemical defense in a coral reef symbiosis. *Proceedings of the National Academy of Sciences of the United States of America* 2012;109:20655–60.
22. Lin Z, Torres JP, Ammon MA, et al. A bacterial source for mollusk pyrone polyketides. *Chemistry and Biology* 2013;20:73–81.
23. Elshahawi SI, Trindade-Silva AE, Hanora A, et al. Boronated tartrolon antibiotic produced by symbiotic cellulose-degrading bacteria in shipworm gills. *Proceedings of the National Academy of Sciences of the United States of America* 2013;110:E295–304.
24. Yang JC, Madupu R, Durkin AS, et al. The complete genome of *Teredinibacter turnerae* T7901: An intracellular endosymbiont of marine wood-boring bivalves (shipworms). *PloS One* 2009;4:e6085.

25. Lechene CP, Luyten Y, McMahon G, Distel DL. Quantitative imaging of nitrogen fixation by individual bacteria within animal cells. *Science* 2007;317:1563–66.
26. O'Connor RM, Fung JM, Sharp KH, et al. Gill bacteria enable a novel digestive strategy in a wood-feeding mollusk. *Proceedings of the National Academy of Sciences of the United States of America* 2014;111:E5096–104.
27. Hu H, Bandyopadhyay PK, Olivera BM, Yandell M. Elucidation of the molecular envenomation strategy of the cone snail *Conus geographus* through transcriptome sequencing of its venom duct. *BMC Genomics* 2012;13:284.
28. Donia MS, Fricke WF, Ravel J, Schmidt EW. Variation in tropical reef symbiont metagenomes defined by secondary metabolism. *PloS One* 2011;6:e17897.
29. Donia MS, Ruffner DE, Cao S, Schmidt EW. Accessing the hidden majority of marine natural products through metagenomics. *Chembiochem: A European Journal of Chemical Biology* 2011;12:1230–36.
30. Kwan JC, Tianero MD, Donia MS, Wyche TP, Bugni TS, Schmidt EW. Host control of symbiont natural product chemistry in cryptic populations of the tunicate *Lissoclinum patella*. *PLoS One* 2014;9:e95850.
31. Jones AC, Monroe EA, Podell S, et al. Genomic insights into the physiology and ecology of the marine filamentous cyanobacterium *Lyngbya majuscula*. *Proceedings of the National Academy of Sciences of the United States of America* 2011;108:8815–20.
32. Edlund A, Loesgen S, Fenical W, Jensen PR. Geographic distribution of secondary metabolite genes in the marine actinomycete *Salinispora arenicola*. *Applied and Environmental Microbiology* 2011;77:5916–25.
33. Hochmuth T, Niederkruger H, Gernert C, et al. Linking chemical and microbial diversity in marine sponges: Possible role for poribacteria as producers of methyl-branched fatty acids. *Chembiochem: A European Journal of Chemical Biology* 2010;11:2572–78.
34. Wakimoto T, Egami Y, Nakashima Y, et al. Calyculin biogenesis from a pyrophosphate protoxin produced by a sponge symbiont. *Nature Chemical Biology* 2014;10:648–55.
35. Hentschel U, Piel J, Degnan SM, Taylor MW. Genomic insights into the marine sponge microbiome. *Nature Reviews Microbiology* 2012;10:641–54.
36. Schmidt EW, Nelson JT, Rasko DA, et al. Patellamide A and C biosynthesis by a microcin-like pathway in *Prochloron didemni*, the cyanobacterial symbiont of *Lissoclinum patella*. *Proceedings of the National Academy of Sciences of the United States of America* 2005;102:7315–20.
37. Rath CM, Janto B, Earl J, et al. Meta-omic characterization of the marine invertebrate microbial consortium that produces the chemotherapeutic natural product ET-743. *ACS Chemical Biology* 2011;6:1244–56.
38. Lopanik NB, Shields JA, Buchholz TJ, et al. *In vivo* and *in vitro* trans-acylation by BryP, the putative bryostatin pathway acyltransferase derived from an uncultured marine symbiont. *Chemistry and Biology* 2008;15:1175–86.
39. Zimmermann K, Engeser M, Blunt JW, Munro MH, Piel J. Pederin-type pathways of uncultivated bacterial symbionts: Analysis of O-methyltransferases and generation of a biosynthetic hybrid. *Journal of the American Chemical Society* 2009;131:2780–81.
40. Chang Z, Sitachitta N, Rossi JV, et al. Biosynthetic pathway and gene cluster analysis of curacin A, an antitubulin natural product from the tropical marine cyanobacterium *Lyngbya majuscula*. *Journal of Natural Products* 2004;67:1356–67.
41. Gu L, Geders TW, Wang B, et al. GNAT-like strategy for polyketide chain initiation. *Science* 2007;318:970–74.
42. Gu L, Wang B, Kulkarni A, et al. Metamorphic enzyme assembly in polyketide diversification. *Nature* 2009;459:731–35.
43. Unson MD, Holland ND, Faulkner DJ. A brominated secondary metabolite synthesized by the cyanobacterial symbiont of a marine sponge and accumulation of the crystalline metabolite in the sponge tissue. *Marine Biology* 1994;119:1–12.
44. Haygood MG, Davidson SK. Small-subunit rRNA genes and *in situ* hybridization with oligonucleotides specific for the bacterial symbionts in the larvae of the bryozoan *Bugula neritina* and proposal of "*Candidatus* endobugula sertula". *Applied and Environmental Microbiology* 1997;63:4612–16.
45. Piel J, Hui D, Wen G, et al. Antitumor polyketide biosynthesis by an uncultivated bacterial symbiont of the marine sponge *Theonella swinhoei*. *Proceedings of the National Academy of Sciences of the United States of America* 2004;101:16222–27.

46. Cuevas C, Francesch A. Development of Yondelis (trabectedin, ET-743). A semisynthetic process solves the supply problem. *Natural Product Reports* 2009;26:322–37.

47. Cuevas C, Perez M, Martin MJ, et al. Synthesis of ecteinascidin ET-743 and phthalascidin Pt-650 from cyanosafracin B. *Organic Letters* 2000;2:2545–48.

48. Velasco A, Acebo P, Gomez A, et al. Molecular characterization of the safracin biosynthetic pathway from *Pseudomonas fluorescens* A2-2: Designing new cytotoxic compounds. *Molecular Microbiology* 2005;56:144–54.

49. Koketsu K, Watanabe K, Suda H, Oguri H, Oikawa H. Reconstruction of the saframycin core scaffold defines dual Pictet-Spengler mechanisms. *Nature Chemical Biology* 2010;6:408–10.

50. Xu Y, Kersten RD, Nam SJ, et al. Bacterial biosynthesis and maturation of the didemnin anti-cancer agents. *Journal of the American Chemical Society* 2012;134:8625–32.

51. Donia MS, Schmidt EW. Cyanobactins–ubiquitous cyanobacterial ribosomal peptide metabolites. In *Comprehensive Natural Products II: Chemistry and Biology*. Elsevier, Oxford, 2010, pp. 539–58.

52. Long PF, Dunlap WC, Battershill CN, Jaspars M. Shotgun cloning and heterologous expression of the patellamide gene cluster as a strategy to achieving sustained metabolite production. *Chembiochem: A European Journal of Chemical Biology* 2005;6:1760–65.

53. Donia MS, Ravel J, Schmidt EW. A global assembly line for cyanobactins. *Nature Chemical Biology* 2008;4:341–43.

54. Tianero MD, Donia MS, Young TS, Schultz PG, Schmidt EW. Ribosomal route to small-molecule diversity. *Journal of the American Chemical Society* 2012;134:418–25.

55. Pierce E, Tianero MD, Schmidt EW. Unpublished data.

56. Chang Z, Flatt P, Gerwick WH, Nguyen VA, Willis CL, Sherman DH. The barbamide biosynthetic gene cluster: A novel marine cyanobacterial system of mixed polyketide synthase (PKS)-non-ribosomal peptide synthetase (NRPS) origin involving an unusual trichloroleucyl starter unit. *Gene* 2002;296:235–47.

57. Flatt PM, Gautschi JT, Thacker RW, Musafija-Girt M, Crews P, Gerwick WH. Identification of the cellular site of polychlorinated peptide biosynthesis in the marine sponge *Dysidea* (*Lamellodysidea*) herbacea and symbiotic cyanobacterium *Oscillatoria spongeliae* by CARD-FISH analysis. *Marine Biology* 2005;147:761–74.

58. Ridley CP, Bergquist PR, Harper MK, Faulkner DJ, Hooper JN, Haygood MG. Speciation and biosynthetic variation in four dictyoceratid sponges and their cyanobacterial symbiont, *Oscillatoria spongeliae*. *Chemistry and Biology* 2005;12:397–406.

59. Kim EJ, Lee JH, Choi H, et al. Heterologous production of 4-O-demethylbarbamide, a marine cyanobacterial natural product. *Organic Letters* 2012;14:5824–27.

60. Ongley SE, Bian X, Zhang Y, et al. High-titer heterologous production in *E. coli* of lyngbyatoxin, a protein kinase C activator from an uncultured marine cyanobacterium. *ACS Chemical Biology* 2013;8:1888–93.

61. Agarwal V, El Gamal AA, Yamanaka K, et al. Biosynthesis of polybrominated aromatic organic compounds by marine bacteria. *Nature Chemical Biology* 2014;10:640–47.

62. Beer LL, Moore BS. Biosynthetic convergence of salinosporamides A and B in the marine actinomycete *Salinispora tropica*. *Organic Letters* 2007;9:845–48.

63. Feling RH, Buchanan GO, Mincer TJ, Kauffman CA, Jensen PR, Fenical W. Salinosporamide A: A highly cytotoxic proteasome inhibitor from a novel microbial source, a marine bacterium of the new genus *Salinospora*. *Angewandte Chemie (International Edition in English)* 2003;42:355–57.

64. Lechner A, Eustaquio AS, Gulder TA, Hafner M, Moore BS. Selective overproduction of the proteasome inhibitor salinosporamide A via precursor pathway regulation. *Chemistry and Biology* 2011;18:1527–36.

65. Olivera BM. E.E. Just Lecture, 1996. *Conus* venom peptides, receptor and ion channel targets, and drug design: 50 million years of neuropharmacology. *Molecular Biology of the Cell* 1997;8:2101–9.

66. Olivera BM. *Conus* peptides: Biodiversity-based discovery and exogenomics. *Journal of Biological Chemistry* 2006;281:31173–77.

67. Woodward SR, Cruz LJ, Olivera BM, Hillyard DR. Constant and hypervariable regions in conotoxin propeptides. *EMBO Journal* 1990;9:1015–20.

68. Myers RA, Cruz LJ, Rivier JE, Olivera BM. *Conus* peptides as chemical probes for receptors and ion channels. *Chemical Reviews* 1993;93:1923–36.

69. Bulaj G, Buczek O, Goodsell I, et al. Efficient oxidative folding of conotoxins and the radiation of venomous cone snails. *Proceedings of the National Academy of Sciences of the United States of America* 2003;100(Suppl 2):14562–68.

70. Hillyard DR, Monje VD, Mintz IM, et al. A new *Conus* peptide ligand for mammalian presynaptic Ca2+ channels. *Neuron* 1992;9:69–77.

71. Ellison M, Haberlandt C, Gomez-Casati ME, et al. Alpha-RgIA: A novel conotoxin that specifically and potently blocks the alpha9alpha10 nAChR. *Biochemistry* 2006;45:1511–17.

72. Hu H, Bandyopadhyay PK, Olivera BM, Yandell M. Characterization of the *Conus bullatus* genome and its venom-duct transcriptome. *BMC Genomics* 2011;12:60.

73. Dutertre S, Jin AH, Kaas Q, Jones A, Alewood PF, Lewis RJ. Deep venomics reveals the mechanism for expanded peptide diversity in cone snail venom. *Molecular and Cellular Proteomics: MCP* 2013;12:312–29.

74. Safavi-Hemami H, Siero WA, Gorasia DG, et al. Specialisation of the venom gland proteome in predatory cone snails reveals functional diversification of the conotoxin biosynthetic pathway. *Journal of Proteome Research* 2011;10:3904–19.

75. Walewska A, Zhang MM, Skalicky JJ, Yoshikami D, Olivera BM, Bulaj G. Integrated oxidative folding of cysteine/selenocysteine containing peptides: Improving chemical synthesis of conotoxins. *Angewandte Chemie (International Edition in English)* 2009;48:2221–24.

76. Arnison PG, Bibb MJ, Bierbaum G, et al. Ribosomally synthesized and post-translationally modified peptide natural products: Overview and recommendations for a universal nomenclature. *Natural Product Reports* 2013;30:108–60.

77. Sudek S, Haygood MG, Youssef DT, Schmidt EW. Structure of trichamide, a cyclic peptide from the bloom-forming cyanobacterium *Trichodesmium erythraeum*, predicted from the genome sequence. *Applied and Environmental Microbiology* 2006;72:4382–87.

78. Lu Z, Harper MK, Pond CD, Barrows LR, Ireland CM, Van Wagoner RM. Thiazoline peptides and a tris-phenethyl urea from *Didemnum molle* with anti-HIV activity. *Journal of Natural Products* 2012;75:1436–40.

79. Donia MS, Hathaway BJ, Sudek S, et al. Natural combinatorial peptide libraries in cyanobacterial symbionts of marine ascidians. *Nature Chemical Biology* 2006;2:729–35.

80. Ziemert N, Ishida K, Quillardet P, et al. Microcyclamide biosynthesis in two strains of *Microcystis aeruginosa*: From structure to genes and vice versa. *Applied and Environmental Microbiology* 2008;74:1791–97.

81. Sivonen K, Leikoski N, Fewer DP, Jokela J. Cyanobactins-ribosomal cyclic peptides produced by cyanobacteria. *Applied Microbiology and Biotechnology* 2010;86:1213–25.

82. Leikoski N, Fewer DP, Sivonen K. Widespread occurrence and lateral transfer of the cyanobactin biosynthesis gene cluster in cyanobacteria. *Applied and Environmental Microbiology* 2009;75:853–57.

83. Donia MS, Schmidt EW. Linking chemistry and genetics in the growing cyanobactin natural products family. *Chemistry and Biology* 2011;18:508–19.

84. Leikoski N, Fewer DP, Jokela J, Alakoski P, Wahlsten M, Sivonen K. Analysis of an inactive cyanobactin biosynthetic gene cluster leads to discovery of new natural products from strains of the genus *Microcystis*. *PloS One* 2012;7:e43002.

85. Houssen WE, Koehnke J, Zollman D, et al. The discovery of new cyanobactins from Cyanothece PCC 7425 defines a new signature for processing of patellamides. *Chembiochem: A European Journal of Chemical Biology* 2012;13:2683–89.

86. McIntosh JA, Lin Z, Tianero MD, Schmidt EW. Aestuaramides, a natural library of cyanobactin cyclic peptides resulting from isoprene-derived Claisen rearrangements. *ACS Chemical Biology* 2013;8:877–83.

87. Leikoski N, Liu L, Jokela J, et al. Genome mining expands the chemical diversity of the cyanobactin family to include highly modified linear peptides. *Chemistry and Biology* 2013;20:1033–43.

88. Li B, Sher D, Kelly L, et al. Catalytic promiscuity in the biosynthesis of cyclic peptide secondary metabolites in planktonic marine cyanobacteria. *Proceedings of the National Academy of Sciences of the United States of America* 2010;107:10430–35.

89. Tang W, van der Donk WA. Structural characterization of four prochlorosins: A novel class of lantipeptides produced by planktonic marine cyanobacteria. *Biochemistry* 2012;51:4271–79.

90. Kersten RD, Yang YL, Xu Y, et al. A mass spectrometry-guided genome mining approach for natural product peptidogenomics. *Nature Chemical Biology* 2011;7:794–802.

91. Nett M, Moore BS. Exploration and engineering of biosynthetic pathways in the marine actinomycete *Salinispora tropica*. *Pure and Applied Chemistry* 2009;81:1075–84.

92. Udwary DW, Zeigler L, Asolkar RN, et al. Genome sequencing reveals complex secondary metabolome in the marine actinomycete *Salinispora tropica*. *Proceedings of the National Academy of Sciences of the United States of America* 2007;104:10376–81.

93. Eustaquio AS, Nam SJ, Penn K, et al. The discovery of salinosporamide K from the marine bacterium "Salinispora pacifica" by genome mining gives insight into pathway evolution. *Chembiochem: A European Journal of Chemical Biology* 2011;12:61–64.

94. Eustaquio AS, Pojer F, Noel JP, Moore BS. Discovery and characterization of a marine bacterial SAM-dependent chlorinase. *Nature Chemical Biology* 2008;4:69–74.

95. Eustaquio AS, McGlinchey RP, Liu Y, et al. Biosynthesis of the salinosporamide A polyketide synthase substrate chloroethylmalonyl-coenzyme A from S-adenosyl-L-methionine. *Proceedings of the National Academy of Sciences of the United States of America* 2009;106:12295–300.

96. McGlinchey RP, Nett M, Eustaquio AS, Asolkar RN, Fenical W, Moore BS. Engineered biosynthesis of antiprotealide and other unnatural salinosporamide proteasome inhibitors. *Journal of the American Chemical Society* 2008;130:7822–23.

97. Nett M, Gulder TA, Kale AJ, Hughes CC, Moore BS. Function-oriented biosynthesis of beta-lactone proteasome inhibitors in *Salinispora tropica*. *Journal of Medicinal Chemistry* 2009;52:6163–67.

98. Eustaquio AS, Moore BS. Mutasynthesis of fluorosalinosporamide, a potent and reversible inhibitor of the proteasome. *Angewandte Chemie (International Edition in English)* 2008;47:3936–38.

99. Piel J, Hertweck C, Shipley PR, Hunt DM, Newman MS, Moore BS. Cloning, sequencing and analysis of the enterocin biosynthesis gene cluster from the marine isolate 'Streptomyces maritimus': Evidence for the derailment of an aromatic polyketide synthase. *Chemistry and Biology* 2000;7:943–55.

100. McIntosh JA, Robertson CR, Agarwal V, Nair SK, Bulaj GW, Schmidt EW. Circular logic: Nonribosomal peptide-like macrocyclization with a ribosomal peptide catalyst. *Journal of the American Chemical Society* 2010;132:15499–501.

101. McIntosh JA, Donia MS, Schmidt EW. Insights into heterocyclization from two highly similar enzymes. *Journal of the American Chemical Society* 2010;132:4089–91.

102. Lee J, McIntosh J, Hathaway BJ, Schmidt EW. Using marine natural products to discover a protease that catalyzes peptide macrocyclization of diverse substrates. *Journal of the American Chemical Society* 2009;131:2122–24.

103. McIntosh JA, Donia MS, Nair SK, Schmidt EW. Enzymatic basis of ribosomal peptide prenylation in cyanobacteria. *Journal of the American Chemical Society* 2011;133:13698–705.

104. Goto Y, Ito Y, Kato Y, Tsunoda S, Suga H. One-pot synthesis of azoline-containing peptides in a cell-free translation system integrated with a posttranslational cyclodehydratase. *Chemistry and Biology* 2014;21:766–74.

105. Sardar D, Pierce E, McIntosh JA, Schmidt EW. Recognition sequences and substrate evolution in cyanobactin biosynthesis. *ACS Synthetic Biology* 2015;4:482–92.

106. Houssen WE, Bent AF, McEwan AR, et al. An efficient method for the *in vitro* production of azol(in)e-based cyclic peptides. *Angewandte Chemie (International Edition in English)*. 2014;126:14395–98.

107. Ruffner DE, Schmidt EW, Heemstra JR. Assessing the combinatorial potential of the RiPP cyanobactin *tru* pathway. *ACS Synthetic Biology* 2015;53:14171–74.

CHAPTER 4

Dereplication of Natural Products Using Databases

Hartmut Laatsch

Institute for Organic and Biomolecular Chemistry, University of Göttingen,
Tammannstrasse, Göttingen, Germany

CONTENTS

4.1 INTRODUCTION

An important reason for the worldwide interest in natural products lies in their often surprisingly strong and selective bioactivity. Many of them are therefore valuable as drugs in medicine, but also as diagnostic tools or for agricultural and even technical purposes: depending on the field of application, 30%–60% of all medical drugs on the market are natural products or were inspired by structures derived from nature.[1–3] It therefore makes sense to proceed with the isolation and identification of new natural products, even if they are usually not applied in their original form, but may require optimization over many years before they are suitable for human utilization.

At present, at least 260,000 low-molecular-weight natural products have been described. Although their number could be nearly unlimited, with respect to the capacity of the enzymes involved in their synthesis, in reality their occurrence is certainly restricted. The isolation of each new product will therefore reduce the chances of finding a further new compound in nature. Depending on the source organisms and the test systems in use, usually between 80% and nearly 100% of all metabolites isolated thereof are already known. Even if it may not be difficult to find a new compound, tens of thousands of extracts have to be screened until a *new clinically versatile drug* or even a *new lead*—a new pharmaceutical principle—is detected.

The reisolation of known compounds is not necessarily the end of a promising project, as previously undiscovered properties or activities of known compounds are often brought to light, although new structures showing these new properties are certainly better, at least from the view of being patented. This situation has been coined the "traffic light in natural product chemistry." New structures *and* interesting properties are like the green light: proceed—the road is free. The yellow light may stand for a *new structure* targeting well-covered applications (e.g., a new polyene-macrolide with antifungal activity) or for a known structure showing *new indications* (such as a pentabromo-pseudiline targeting the nonmevalonate pathway).[4] A careful continuation may make sense here. If these conditions are not fulfilled, further investment will just waste resources. This is the red traffic light. It is obvious that the clear identification of this situation as early as possible will prevent a loss of money, time, and motivation.

It is therefore desirable to have an affordable and, if possible, automatic *ab initio* technique for rapid structure elucidations or, as this is not yet available, other methods to establish that a given compound has been published previously. In chemistry such procedures are known as dereplication, although this term was initially used in biochemistry to sort metabolites on the basis of their activities.

Ab initio structure elucidation treats known and new compounds in the same way with the same technical methods. In suitable cases, it is possible indeed to elucidate a structure by x-ray diffraction in a fully automated process,[5] and the same is possible in other cases by a computer-aided nuclear magnetic resonance (NMR) interpretation by means of, for example, the ACD structure elucidator.[6]

The drawback of these *ab initio* methods is that the traffic light comes at the very end of the isolation and separation steps, which actually take more time than the structure elucidation itself. In principle, this problem can be ameliorated by direct investigation of the source organism by matrix-assisted laser desorption ionization (MALDI) mass spectrometry (MS), followed by analysis of the resulting signal pattern by means of statistical methods. This technique has been successfully applied to the taxonomic identification of cyanobacteria using the Spectral Archiving and Microbial Identification System (SARAMIS), developed by AnagnosTec, Potsdam, Germany[7] (Figure 4.1), but is useful also for the identification of metabolites directly from microorganisms, at least in suitable cases such as when peptides are present.[8]

A more conventionally oriented method uses the automatic separation of the crude extracts, for example, by the sep-box technique.[9] It seems, however, that this combination of automatic separation and structure elucidation is not yet fully optimized, particularly with respect to the costs. Ito and Masubuchi have published a summary of this recent development.[10]

The alternative technique of dereplication applies a dragnet search, where data of the compounds in question are compared in a computer-aided search with *published* compound properties compiled in a database. This database should therefore be as large and as complete as possible. The limitation of this method is obvious: the dereplication may fail if data sets are missing for any reason and if published descriptions are incomplete or even wrong. And per se, this method is not suitable for the elucidation of new compounds, although valuable hints from databases will usually facilitate their identification, as we will see later.

Data suitable for such searches are all experimentally accessible molecular descriptors in the widest sense, such as molecular masses, NMR data, and substructures, as well as the taxonomic origin, biological activities, color, melting points, optical rotations, and so forth. It is obvious that some of these descriptors are predictable (e.g., from the chemical structure). Others, however, are not. We

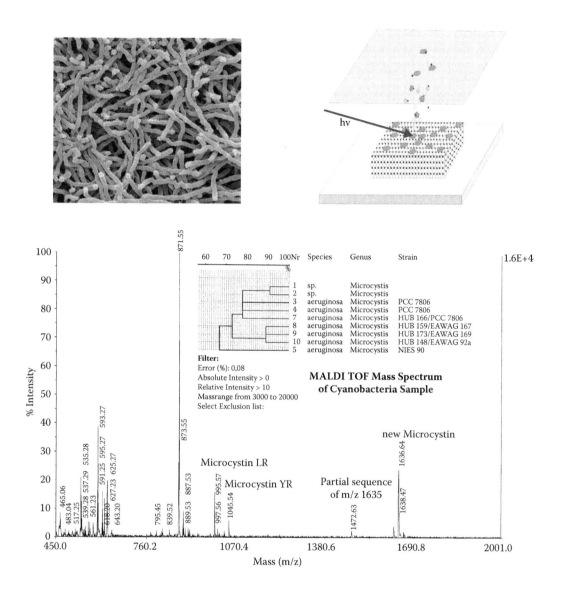

Figure 4.1 MALDI MS of cyanobacteria for their taxonomic identification using SARAMIS.

can therefore divide the information into soft and hard data. The soft data are empirical values, such as the various biological activities, thin layer chromatography (TLC) behavior (including R_f values and color reactions), melting points, and solubilities. They will never be completely accessible for *all* natural products, as their determination depends strongly on the interest of the authors and the availability and use of the respective test systems.

In contrast, the hard data are predictable and can be derived from a chemical structure for those cases where the respective values have not been published, or even if the assumed structure does not exist (e.g., in the discussion of structure alternatives). Well-known examples are NMR data, which can be predicted with surprising accuracy from the structure. In principle, other properties, such as ultraviolet (UV), circular dichroism (CD), and infrared (IR) spectra, or optical rotation data, can also be calculated if the molecules are not too big and if high-performance computers are accessible.

The technique of dereplication therefore requires a comprehensive collection of data, consisting of molecular descriptors (mainly the chemical structure) and further properties that have to be extracted continuously from the primary literature. In addition, a special program is required to handle these data.

The most complete structure-based data collections are certainly the *Chemical Abstracts*, followed by PubChem (pubchem.ncbi.nlm.nih.gov) and ChemSpider (www.chemspider.com). The *Chemical Abstracts* cover nearly 100% of all published inorganic and organic compounds, of both synthetic and natural origin, so that this database could be considered the highest authority when deciding on the novelty of a chemical entity. However, the advantage, and also the limitation, of this database is its size. The search for a compound with, for example, exactly three methyl groups requires at present a search for three (or more) and for four (or more) methyl groups, followed by subtraction of the latter answer set from the first. Due to the huge size of the database, such a search of small and frequently occurring fragments usually ends with big answer sets that cause a system overflow. Besides the high usage price, a further limitation is the absence of searchable spectroscopic data. Good substitutes for the *Chemical Abstracts* are PubChem and ChemSpider: they are metadatabases covering the content of many other databases. They are available for free and are very valuable for reference searches and the identification of properties of fully defined compounds. As they were not designed for dereplications, the capacity for substructure searches is limited, but available.

For these reasons, a number of specialized databases for structure dereplication are on the market. After the early period, when border-punched cards were used (~1963), from 1975 to 1993 Hungarian chemist Janos Bérdy developed and merchandised the first computer-aided in-house database: Bioactive Natural Products Database (BNPD).[11] This data collection was also the basis of his *CRC Handbook of Antibiotic Compounds*.[12] Unfortunately, this DOS-based system could not handle structures. As it was not further developed, it disappeared from the market ca. 1995.

The *Dictionary of Natural Products on CD-ROM* (DNP)[13] resulted from the book series "Dictionary of …" (Antibiotics, Alkaloids, Steroids, etc.; seven volumes), published by Chapman & Hall, and has been distributed electronically by Chapman & Hall since 1992, and now by CRC Press. The DNP has found a wide audience, especially for phytochemical research, as it covers nearly all natural products and contains at present about 260,000 entries. AntiBase[14] (on the market since 1993 and containing >43,000 terrestrial and marine compounds) covers microbial products and metabolites from algae and higher fungi, while MarinLit[15] (>26,000 compounds) is focused especially on marine natural products. AntiMarin is a combination of AntiBase and MarinLit, with more than 50,000 entries. The great advantage of this combined database is the inclusion of all types of spectroscopic data, especially of ^{13}C NMR spectra. It should be mentioned that MarinLit is available online only, since the Royal Chemical Society started to market the program in 2014. Also, AntiBase is available via the Internet, and AntiMarin is expected to be accessible in the same way in the near future.

The Antibiotics Database (>5400 compounds) is a public collection of references from the *Journal of Antibiotics* until 2008. The database does not contain graphical elements, but is available free of charge from the Internet.[16] Further databases have been summarized by Füllbeck et al.[17] Most of them have been designed for special applications or are limited in other ways (see Figure 4.2).

All of these databases consist of data collections and a separate program to handle the data. The data sets differ in focus, coverage, and size, but are based mostly on published and therefore generally accessible data. The tools for catenation and logical combination of these data, however, are highly specialized programs with the capability to search chemical structures and substructures and handle data ranges (e.g., of spectroscopic data), other spectroscopic information, and text fragments, as well as combinations of these data via hit lists by Boolean algebra. In addition, the program must be fast enough to handle large data sets, and usually it protects the data against being copied as well. In the *Chemical Abstracts* and the DNP, data and database programs are merged.

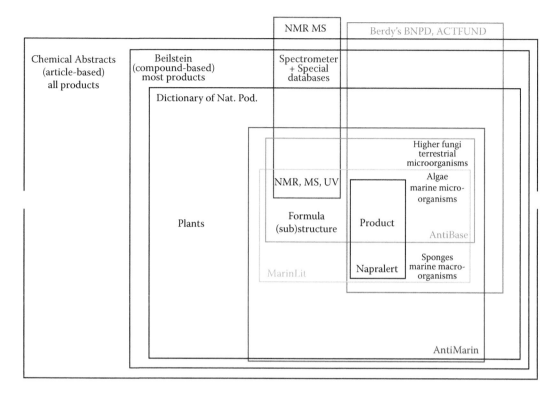

Figure 4.2 Most databases cover only an excerpt of the entirety of all natural products and their properties.

MarinLit previously used Chemfinder, while AntiBase is based at present on Chemfinder or Scidex, respectively, but in principle works also with ISIS or ORACLE.

Herein, examples for dereplication of natural products using databases will be given mostly for AntiBase and AntiMarin, as these databases are issued or co-edited in the lab of the author, thus giving the author extensive experience in their use. It should be stated, however, that similar results could also be obtained with other programs mentioned in this chapter.

4.2 SEARCH TECHNIQUES FOR DEREPLICATION

4.2.1 Dereplication Using Sensory Discrimination

Nowadays, for the structure elucidation of natural products, x-ray diffraction analysis and spectroscopic methods are applied nearly exclusively. Derivatives or degradation reactions are used only for special questions, for example, for the determination of the absolute configuration. Nevertheless, it should be emphasized that already a few simple tests will allow crucial conclusions and may speed up the dereplication substantially. The following tests are, of course, not an alternative for spectroscopic analyses, but they may help to find the best direction for the further search.

In most cases, the reduction of an initial list of candidates requires expensive equipment. We should not forget, however, that our nose and eyes are highly sensitive instruments as well, if we know how to apply them, at least in special cases and on a qualitative basis. To distinguish between the isomeric aldehydes **1** and **2** requires a detailed analysis of two-dimensional (2D) NMR data, but the nose needs only 1 sec: aldehyde **1** is known as vanillin and has a well-known typical and pleasant odor, while isovanillin (**2**) does not have any smell.

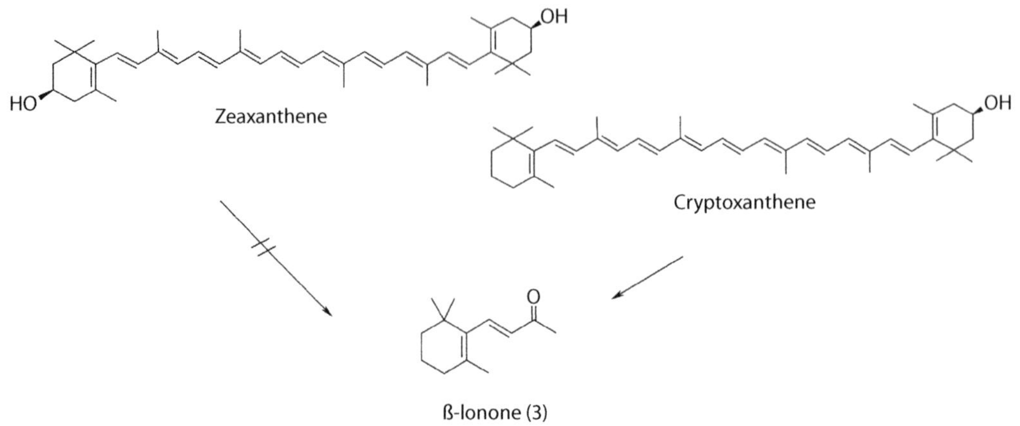

Figure 4.3 Oxidative cleavage only of carotenoids with at least one oxygen-free ß-end group delivering ß-ionone (**3**) with the typical smell of violets.

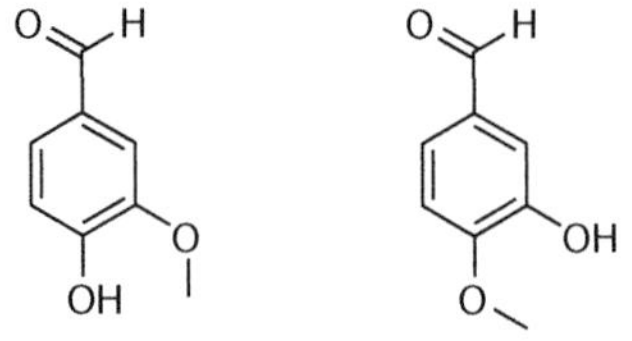

Another example is the bacterial metabolite phenylacetic acid. The smell of the pure compound is unpleasant, acidic, and sweat-like; in high dilution, however, it smells strongly and very characteristically like honey. A further application of sensory analysis is found in the following test: if traces of a carotenoid are moistened in a test tube with a drop of water and warmed, the characteristic smell of violets due to ß-ionone (**3**) is developed if the carotenoid has at least one unsubstituted ß-end group. The principle of this reaction is the slow oxidative cleavage of carotenoids in the presence of oxygen. Ionone formed only from respective carotenoids is perceptible, while the oxidation products from other carotenoids have a higher odor threshold and are not observable (Figure 4.3).

In practice, a droplet of a diluted carotenoid solution in chloroform and a drop of water are warmed in a test tube with the hands. The sensoric test is performed when the chloroform has disappeared. This solvent combination makes the jonon better perceptible, perhaps by azeotropic distillation.

4.2.2 Chromatographic Identification

Of the natural products presently listed in AntiBase, ~37% contain double bonds, 43% contain aromatic systems, and 22% include both. Silica gel TLC cards are usually dealt with a pigment that gives a strong green fluorescence on irradiation with UV light at 254 nm. Separated extracts may show on these TLC plates under UV_{254} light dark spots on a green background, giving by fluorescence quenching a hint for the presence of aromatic systems or double bonds. Additionally, fluorescence at 365 nm may indicate a rigid chromophoric system, as this is mostly found in aromatic compounds. In this way, the inspection of a developed TLC card under the two UV wavelengths allows the easy discrimination of compounds with or without π-systems. It should be mentioned, however, that a few compounds are giving at 254 nm the same green fluorescence as the background and are therefore easily overlooked.

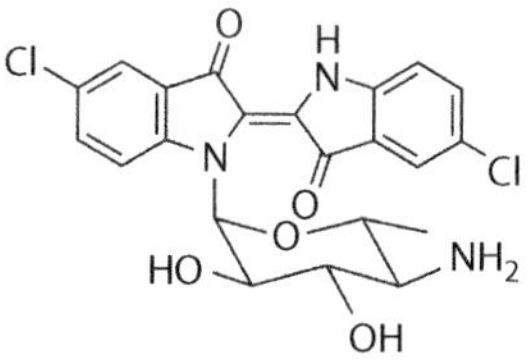

Figure 4.4 Some pH-independent blue pigments: akashin A (**4**), indigoidin (**5**), violacein (**6**), and candidin (**7**).

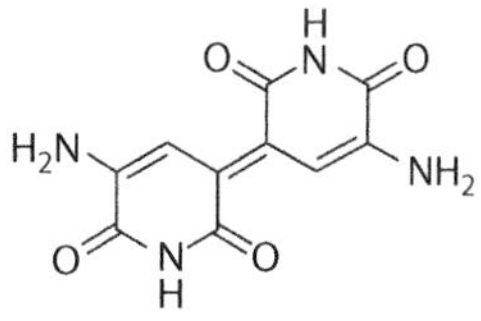

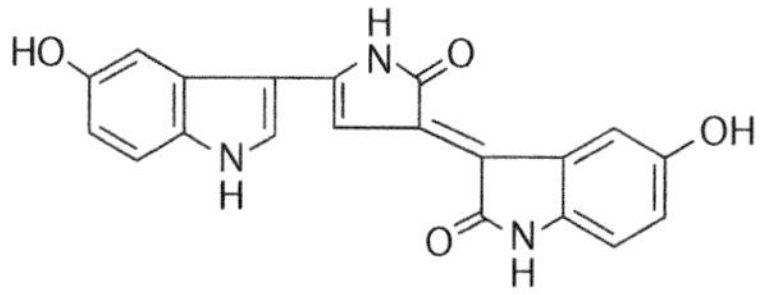

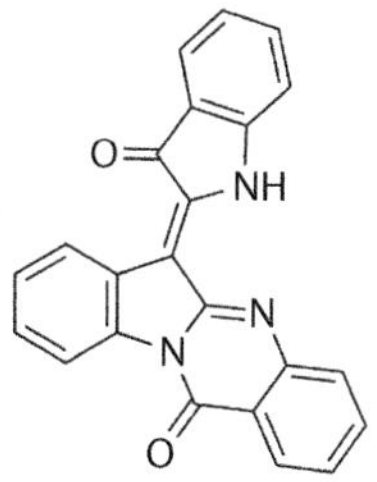

The color of a compound in solution or in a chromatogram is of even higher diagnostic value: green and blue natural products are extremely rare in nature, especially if the color is pH independent (test with diluted HCl and NaOH on TLC). In addition to the chlorophylls, further green compounds are the phenacine-derived esmeraldines or the iron-containing viridomycins and some nickel complexes. Blue (and blue–violet) compounds include the few natural indigo derivatives (such as **4**), as well as some related chromophores (e.g., **5**) and, additionally, the bacterial violaceins (like **6**), candidin (**7**) (Figure 4.4), and about only 100 further pigments.

The group of yellow to red compounds is very large and chemically heterogeneous. Compounds in this group that do *not* change color with diluted sodium hydroxide solution may be nitro compounds, polyenes or carotenoids, phenazines, phenoxazinones (including actinomycins), or certain sulfur compounds. With concentrated sulfuric acid, polyenes (like the important antifungal drug amphothericin B) turn brown to black, while the related carotenoids will give a dark blue color or, if carbonyl groups are in conjugation with double bonds, a green coloration. Under these conditions, the orange color of certain aromatic nitro compounds or the highly cytotoxic actinomycins is not changed but often intensified.

The slightly yellow flavones and isoflavones sometimes turn brown with NaOH on the chromatogram. Most yellow to red compounds are, however, *peri*-hydroxyquinones, which develop a red (rare) to violet or deep blue color with NaOH. More than 1600 of these compounds are known from microorganisms, and important examples are the antitumor antibiotics doxorubicin (adriamycin) and daunorubicin.

The behavior of red quinones is clearly distinguished from the raspberry-red prodigiosins (prodiginin salts), which develop the yellow color of the free base with NaOH (Figure 4.5). The red prodigiosins are of low polarity ($R_f \sim 0.8$ on silica gel with $CHCl_3/MeOH$ 95:5) and are widespread in yeasts and streptomycetes. They show, among others, strong immunosuppressive and antimalarial activities, but may cause problems due to their high lipophilicity and the high affinity to adsorber resins and reversed phase (RP) materials.

Among the many other spray reagents, some have very useful selectivity. An example is Ehrlich's reagent (*p*-dimethylamino-benzaldehyde), which gives intensive red, blue, or violet colors with most indole derivatives. In contrast, the widely applied anisaldehyde/sulfuric acid reagent and related compositions (vanillin, orcin, naphthoresorcin, etc., instead of anisaldehyde) are not group selective, but are useful for the detection of a wide range of metabolites, including the many

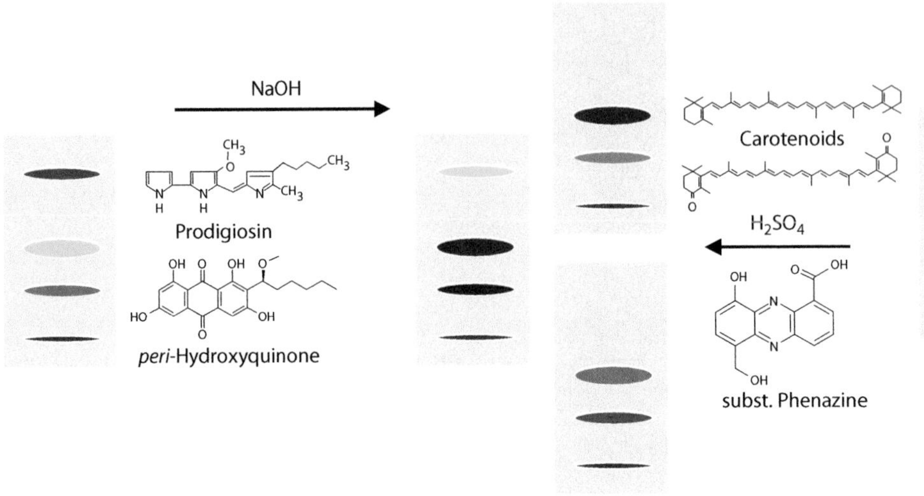

Figure 4.5 Some color reactions with sodium hydroxide and concentrated sulfuric acid.

trivial compounds (like anthranilic acid, uracil, genistein) or the xenobiotics introduced easily, for example, as solvent impurities (phthalic esters etc.).

4.2.3 Dereplication by HPLC-MS

The purity of a compound is the most important precondition for a successful and rapid dereplication by data mining. For milligram amounts, the required purity is achieved in a time- and cost-intensive separation process, resulting in a chemically defined material that is ready for all types of measurements. It is obvious, however, that dereplication at the beginning of any isolation procedure must be much more effective if data mining methods are applied on enriched fractions or even crude extracts, together with powerful separation methods in a combination termed hyphenated techniques.

Such a technique is usually the combination of gas chromatography (GC) or high-pressure liquid chromatography (HPLC) with UV spectroscopy or mass spectrometry. In particular, mass spectrometry using electrospray ionization (ESI) is indispensable now in natural product chemistry. ESI is a soft ionization method that works in the positive and negative modes and provides pseudomolecular ions, usually without noteworthy fragmentation. The assignment of these ions as $[M+H]^+$, $[M+Na]^+$, $[M-H]^-$ ions, and so on, is mostly not difficult, so that the respective masses can be searched directly in databases. A corresponding technique has even been patented.[18,19] As each mass between 100 and 2000 does not occur more than 200 times in microbial natural products, even a low-resolved mass value is a very powerful discriminator. Between 100 and 1200, all masses are allocated (see Figure 4.6). Below and above this range (up to 2000), about 200 mass values have not been found in natural products so far, so that molecules with masses of, for example, $m/z = 85$, 95, 1235, and 1412, may indicate at present a new natural product.

A further improvement is obtained by measuring high-resolution masses. This allows the determination of the empirical formula with high reliability. Even a search with only three valid digits is usually sufficient for this purpose, as nature uses only selected elemental compositions.

Nevertheless, in most cases, ESI mass spectra will only determine that the compound under investigation is either (1) a new compound (mass or elemental composition not found in databases), (2) one

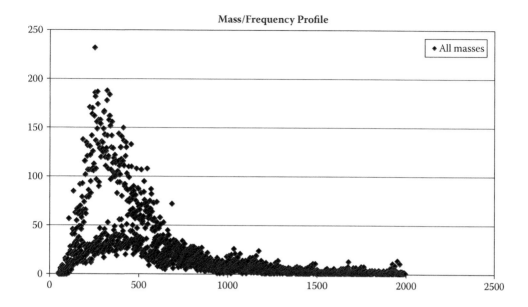

Figure 4.6 Mass distribution of natural products in AntiBase with MW = 70–2000 Da.

of xy known compounds with the same mass, or (3) perhaps a new compound with the same mass as the xy known isomers. For a further reduction of the hit number, additional properties have to be used.

An obvious extension of mass spectrometry as a tool for dereplication is the inclusion of fragmentation patterns by MS^n. In some cases, at least some steps of this fragmentation are predictable, such as for peptides or glycosides. In most other cases, however, the data for authentic compounds need to be available for comparison. As the fragmentation pattern depends strongly on the fragmentation technique and the hardware used (ion trap or quadrupole MS), and of course on the number of accessible compounds, usually rather limited data sets are available. In spite of the resulting restriction, this method is very powerful, especially in combination with retention times and UV data.[20]

4.2.4 Dereplication by HPLC-UV or HPLC-MS/UV

HPLC instruments and HPLC-MS combinations are usually coupled with a diode array detector (DAD), allowing the UV/vis spectra of a whole separation run to be measured in flight. Specialized databases compare not only the shape of the spectrum, but also the retention time, so that in many cases a clear identification of the chromophore and even of the individual compound is possible (see also Figure 4.18).

Peripheral substituents usually do not strongly influence the chromophore of a metabolite. An important advantage of these HPLC-UV and HPLC-MS/UV techniques is therefore the capability to also assign *new* compounds to a known chromophore, if present in the database. A major drawback of this technique is the fact that as many reliable data sets as possible must be available for comparison. A serious problem is the fact that retention times in particular, but also MS/MS spectra, are strongly dependent on the experimental conditions, including the hardware, so that the reproduction of results in another lab is difficult. Therefore, a number of standards should be available in order to measure the reference values of retention time and UV data for direct comparison. Nevertheless, it has been shown that carefully established in-house HPLC-MS/UV databases are a very powerful tool in natural product dereplication and structure elucidation (see Figure 4.18).[10,20,21]

The same technique can be applied to measure CD spectra. In addition, IR spectra can be registered during HPLC separations, although this method is less frequently used in natural product dereplication. Instead of HPLC, the faster and cheaper TLC techniques can be applied, using direct electrospray ionization (DESI) MS techniques[22] and reflection UV/vis spectra directly from the chromatogram,[23] although this is typically less successful and has lower sensitivity and resolution.

4.2.5 Dereplication by NMR

The most important technique for structure elucidation and also dereplication is certainly nuclear magnetic resonance (NMR) spectroscopy. In contrast to MS data, an NMR spectrum not only provides properties of the molecule as a complete unit, but also gives details of the structure and substructure down to the relative and absolute stereochemistry when using special experimental conditions. A further advantage is the amazing accuracy of predicted NMR data, especially concerning carbon shifts. A number of very fast programs for empirical and even *ab initio* predictions are available for this purpose.[6,24,25] In AntiBase and MarinLit/AntiMarin, searchable ^{13}C NMR data of high reliability are included.

As NMR spectra can be predicted directly from the formula, this method is not restricted to the availability of published data or even reference compounds. Together with its widespread accessibility, this is the reason that NMR data are among the most important dereplication tools.

In comparison with MS and UV techniques, conventional NMR spectroscopy is less sensitive by several orders of magnitude. For routine spectra, about 1–5 mg of pure compound is needed. Using cryoprobes and capillary techniques, however, even 2D spectra can be measured with a few micrograms, thus making dereplication of the constituents in crude extracts by HPLC/NMR possible.[26] However, due to the high costs, ^{13}C NMR spectra are not often applied in the early phases of dereplication.

4.3 APPLICATION OF DATABASES

Ideally, a single HPLC separation will deliver a full set of UV, CD, MS, and NMR spectra for the individual components of a crude extract. There may not be very many research laboratories that have all these techniques available simultaneously at present, however. Additionally, the different research interests in the laboratories may result in data that highlight only certain facets of the material; thus, the experimental data sets available for dereplication will remain fragmentary in most cases (Figure 4.2).

This limited number of complete data sets results in an important, but often underestimated, fact: if a database with m entries is screened for a certain property (e.g., the wavelength of a UV maximum), but does not include data for all database entries in the respective property field ($n < m$), then the $m - n$ incomplete data sets cannot be excluded automatically from further searches; this makes the dereplication process confusing.

It is strongly recommended, therefore, that investigators start with search properties where an answer can be expected in most, if not all, cases. In AntiBase, (sub)structures and structure-derived data are available in >93% of cases. These data consist of masses (including high-resolution data), empirical formulas, and calculated (and in an increasing number of cases, experimental) ^{13}C NMR data. For the remaining 7% of cases, structure proposals have not been published, and the structure fields are therefore empty in the database. These poorly defined compounds were excluded from MarinLit and AntiMarin, so that the structure availability is 100% here. In contrast, UV data (15%, AntiBase data), optical rotations (14%), or melting points (12%) are given for a substantially smaller number of cases. The benefits of using MS and NMR data for dereplication as early as possible are evident from these statistics.

4.4 SEARCH EXAMPLES

For the actual structure elucidation, nowadays numerous spectroscopic procedures are available; however, prior to their application, the sample should have a purity of ≥95%. This can be achieved by means of classical techniques (distillation, crystallization, chromatography) or, better, by HPLC, the latter often in combination with an analytical method such as a hyphenated technique (HPLC-MS/UV, etc.).

It is advisable to initially determine the molecular mass, preferably by high resolution, in order to make the empirical formula accessible. Small and volatile molecules can often be measured best by electron impact mass spectrometry (EI MS), but the method of choice for natural products of higher complexity, polarity, or sensitivity is certainly electrospray ionization (ESI), or another soft ionization technique that can be combined with HPLC.

^{1}H NMR spectra can be measured with short acquisition times and high sensitivity. They allow an easy determination of the number and type of protons, including information on their chemical environments, by means of signal shifts, intensities, and coupling patterns. In difficult cases, the dereplication is complemented by ^{13}C NMR spectra. The more expensive 2D spectra are typically not needed for dereplication and are instead reserved for the structure elucidation of new metabolites. Some examples will be outlined in the following sections.

4.4.1 Additive Search: Example 1

In an additive search, certain properties must be fulfilled simultaneously in a positive sense, as in the following example: a cytotoxic and antifungal compound isolated from a *Streptomyces* sp. formed colorless crystals *and* showed by low-resolution ESI mass spectrometry pseudomolecular ions at *m/z* 813.5 (positive mode, [M+Na]$^+$) *and* 789.6 (negative mode, [M-H]$^-$), so that an isotopic mass of ca. 790.5 could be assumed.

A search in AntiBase for [M-H]$^-$ ions with a mass in the range of 789–790 (a mass of 789.49 counts as 789, but 789.51 as 790) delivered 35 hits, among them macrolides, quinones, peptides, and glycosides. A high-resolution measurement was not available.

Additionally, the ^{1}H NMR spectrum (Figure 4.7) showed six or seven olefinic methine multiplets, several (partially overlapped) oxymethine/methylene signals, and in the range of δ 1.3–0.8 ppm, at least eight aliphatic methyl doublets with typical coupling constants of ~7 Hz. The signal pattern at δ 0.83 was resolved by 2D measurements as overlapping doublet and triplet signals. However, the one-dimensional (1D) proton spectrum did not allow a clear assignment, resulting in that signal being ignored at this step. A search using only the clearly identified 8 CH–CH$_3$ groups yielded 471 hits. Only three of these hits had [M-H]$^-$ masses of 789 Da. Of these three, only one had double bonds. This was oligomycin A (**8**), whose published NMR data were close enough to the values found here to be classified as identical.

It is certainly not unexpected that increased accuracy of the experimental data usually shortens the dereplication process dramatically. For example, the oligomycin A (**8**) sample gave a high-resolution value for the [M-H]$^-$ ion at *m/z* 789.5154. A search using this mass in a range of 789.515–789.517 resulted in a single hit, again oligomycin A (**8**).

In a similar way, the spectroscopic data could be complemented (or substituted) by ^{13}C NMR data (Figure 4.8). The carbon spectrum showed 45 C signals, among them 2 keto carbonyls, 1 high-field shifted ester carbonyl, and 1 acetal signal. Methoxy signals were visible neither in the ^{1}H nor in the ^{13}C NMR spectrum. Out of the 471 compounds with 8 or more CHCH$_3$ groups, 237 had at least 2 keto carbonyls and 1 ester group, but 206 had 4 or more carbonyl groups, so that only 31 fitting candidates were left. Only 10 of these had substructures with 44–46 C atoms (4 with 45 C's), and

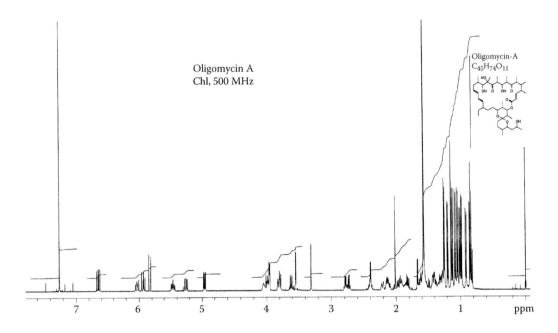

Figure 4.7 ¹H NMR spectrum (500 MHz) of oligomycin A (**8**) in CDCl₃.

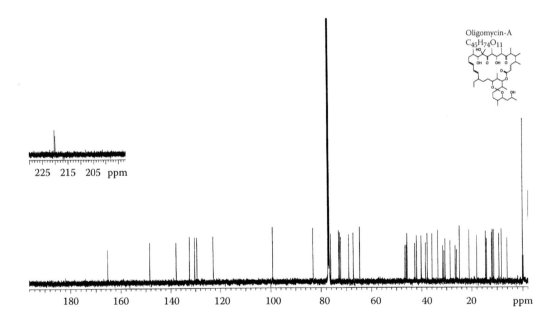

Figure 4.8 ¹³C NMR spectrum (125 MHz) of oligomycin A (**8**) in CDCl₃.

all of them were oligomycin derivatives, including rutamycin B. It should be stated that in this way, identification was possible also without MS data.

It should be confessed, however, that the dereplication as described above has a serious flaw. It includes an interpretation of the data that may be wrong. A signal at δ 165 may be an unsaturated ester, but could also be a conjugated amide, a phenolic carbon, or even an imide or guanidine residue.

A further potential source of error is highlighted in the following example: The diketopiperazine *cyclo*-(Pro,Val) (**9**) shows two clear methyl doublets in the ^{1}H NMR spectrum at δ 0.8 and 1.0. A substructure search using two CH–CH$_3$ groups and the expected neutral mass (*m/z* 196) afforded seven hits; however, *cyclo*-(Pro,Val) (**9**) was not among them. The reason is obvious: the compound in question contains only one isopropyl group, which cannot be found in a search for two CH–CH$_3$ groups with two (instead of one) methine fragments.

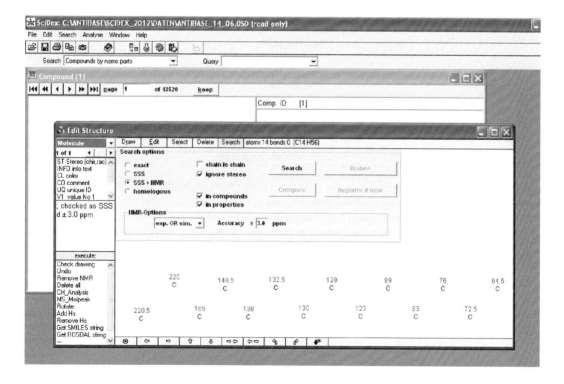

cyclo-(Pro,Val) (**9**)

It is clear, therefore, that any search strategy should omit one's own interpretations (experience) as far as possible. Such a technique could be based, for example, on the carbon shifts. In principle, all shifts should be searched simultaneously. In the case of oligomycin A (**8**), this involves 45 values. In practice, the computer capacity available usually defines the practical limits of the search, as an increasing number of parallel search parameters slow down the process dramatically. This technical problem can be easily overcome, however. An initial collection of values is searched in a first run. In the resulting answer set, a second batch is searched, and so on. In the present case, a search with 14 selected carbon shifts (220.5, 220, 165, 148.5, 138, 132.5, 130, 129, 123, 99, 83, 76, 72.5, 64.5) in the experimental spectra within a confidence range of ±2 ppm resulted in only four hits. All of these were oligomycins, and three of them had 45 C atoms. Only oligomycin A (**8**) matched the observed molecular weight (Figure 4.9).

Figure 4.9 Example of the simultaneous search for 14 ^{13}C NMR shift values in AntiBase within the collection of simulated and experimental spectra, using a confidence range of ≤±3 ppm.

The limitations of this method are obvious. It can only work perfectly if the experimental spectrum of the respective compound is present in the database, and if the shifts are not influenced strongly by a potentially different solvent. Both can never be assured for all entries. At present, the database contains only about 3000 experimental spectra, although their number is growing continuously. However, AntiBase contains high-quality simulated ^{13}C NMR spectra for all compounds. Experience to date shows that the simulation program SpecInfo[24] used in AntiBase is able to predict the spectra with a sufficient accuracy to allow successful dereplication. The influence of the solvent on ^{13}C NMR shifts is usually low and can be compensated by a slight expansion of the confidence range in the search.

A search using only the experimental values above in the simulated spectra set did not give any hits within ±2 ppm. However, in the range of ±3 ppm, again four oligomycins were found, although these were not the same compounds obtained using the experimental spectra. A further search in a confidence range of ±5 ppm gave seven hits, all of which were again oligomycins. Oligomycin A (**8**), however, was included in all results and was the only hit with the correct mass.

4.4.2 Additive Search: Example 2

A highly cytotoxic orange compound gave a UV absorbing yellow spot on TLC that turned orange with diluted sodium hydroxide, tentatively indicative for a *peri*-hydroxyquinone. This was confirmed by the signal of a chelated OH group in the ^{1}H NMR spectrum (Figure 4.10). By (+)-ESI MS, an [M+H]$^+$ ion at *m/z* 691.2 was obtained. From the proton spectrum, the presence of an additional two benzene rings with two *ortho*-coupled protons each, five methoxy groups, and 2 CHCH$_3$ fragments (or one isopropyl group) was established.

A search using a mass of *m/z* 691 for [M+H] resulted in 20 compounds, of which only 3 also had the substructure of a chelated OH group. IB-00208 (**10**) was easily identified as the correct structure (Figure 4.11).

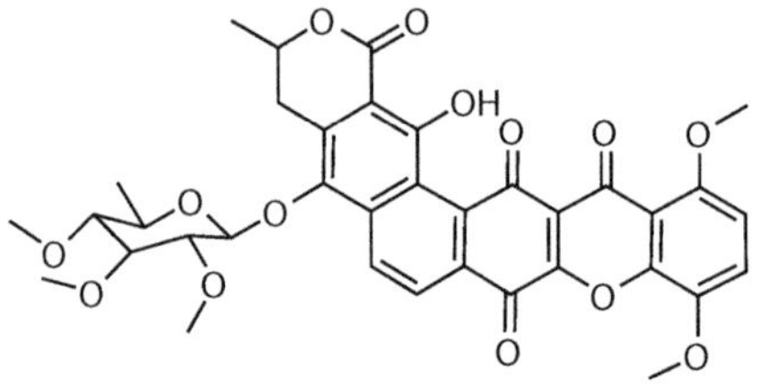

IB-00208 (**10**)

The same result (with IB-00208 as the only possibility) was obtained when five methoxy groups, two benzene rings with *ortho*-coupled protons, one CHCH$_3$ fragment, and an additional methyl were searched (the latter two to avoid the isopropyl trap) (Figure 4.12).

As in the previous example, the carbon shifts can also be used. Using the nine shifts above 140 ppm, only IB-00208 (**10**) is found in the experimental list within a search range of ±2 ppm. The list of calculated data needs a range of ±5 ppm in the search parameter in order to pinpoint this compound among 32 others as the only one with the correct mass (Figure 4.13).

When the other carbon values are also included (stepwise), only two compounds are found in the set of simulated spectra. These are IB-00208 (**10**) and viriplanin D, the latter of which has more than twice the required number of carbon atoms.

4.4.3 Subtractive and Combined Searches: Examples 3 and 4

In the previous examples, dereplication was performed with a number of characteristics that should *all* be present in the expected compound. These were additive properties. In a similar way,

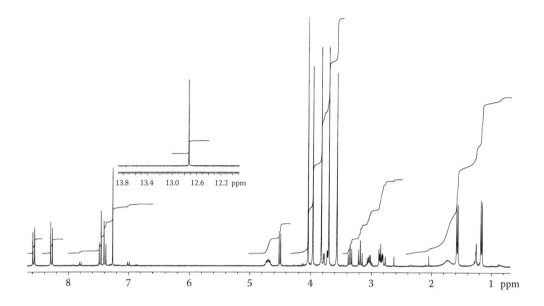

Figure 4.10 ¹H NMR spectrum (CDCl₃, 300 MHz) of IB-00208 (**10**).

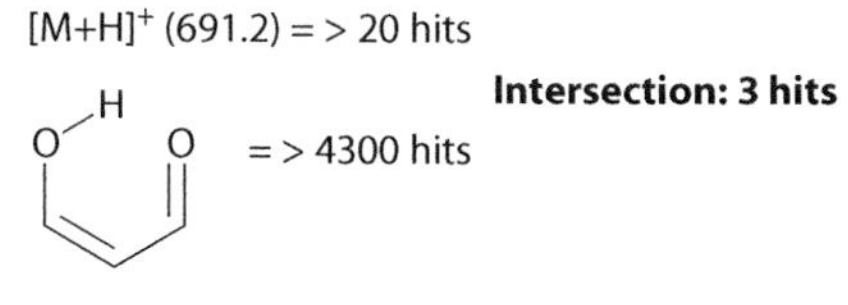

[M+H]⁺ (691.2) = > 20 hits

Intersection: 3 hits

= > 4300 hits

Figure 4.11 Search for the mass of IB-00208 (**10**) and for the presence of a chelated OH group.

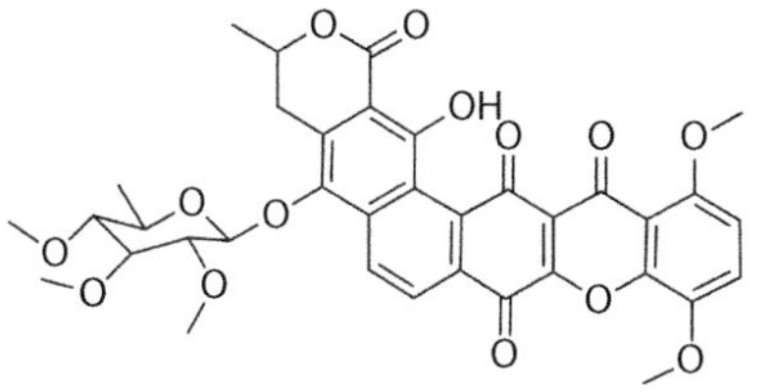

IB–00208 (**10**)

Figure 4.12 An alternative search for IB-00208 (**10**).

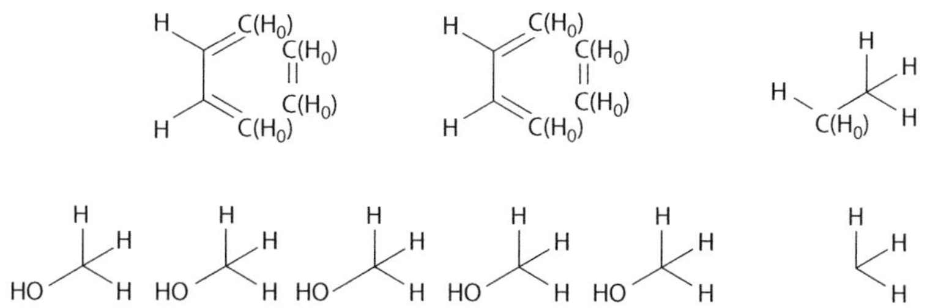

Figure 4.13 A search for IB-00208 (**10**) with nine ¹³C shift values.

we can also *exclude* certain groups of compounds, if typical characteristics are absent. If, for example, proton signals above 6.5 ppm (or carbon signals between 110 and 160) are missing, then aromatic compounds can most likely be excluded.

We isolated some time ago a UV absorbing solid that gave just one signal in the ^{1}H NMR spectrum, a singlet at δ 8.54, and only one signal, at δ 170.5, in the ^{13}C NMR spectrum. We did not obtain mass spectra under EI or ESI conditions. We later realized that this was probably due to the molecular weight below 100 Da, which needs special measurement conditions.

Obviously, C_q, CH_2, and CH_3 groups can be excluded, so that the signals must be due to one or several methine groups in a symmetrical surrounding. On searching for one CH, subtracting C_q, CH_2, and CH_3, 24 compounds were left, among them a number of liquids (halomethanes, HCN, formic acid), some UV_{254} nonabsorbing metabolites (e.g., inositols), and a few compounds with different CH types (which should show coupling patterns and more than one carbon signal, e.g., 1,1,3,3-tetrachloropropanol-2, imidazole). If these compounds were subtracted, only pyrazine remained, which gave, however, a completely different chemical shift (δ 145.3). The correct compound (1,3,5-triazene, **12**) was not known as a natural product before our investigation, but its structure was confirmed by direct comparison with the data from a synthetic sample.

Pyrazine (**11**)   1,3,5-Triazene (**12**)

In the new AntiBase version, the search for a carbon atom with δ 170.5 ± 5 ppm, with subtraction of all compounds with any shift between 1–160 ppm and 180–250 ppm, now delivers only triazene (**12**).

While an additive or subtractive search in AntiBase requires a sequence of combined substructure searches using the structure editor, these search modes can be performed in a very elegant way using MarinLit or AntiMarin: these databases are providing an extensive list of potential substituents or substitution patterns, where the respective fragment count only needs to be entered. As the substituent count can also be set to zero, compounds with absent substitution patterns or other properties can easily be excluded.

The ^{1}H NMR spectrum of an orange *peri*-hydroxyquinone (blue color reaction with sodium hydroxide) showed, among other signals, one methoxy singlet, and in the aliphatic region, one methyl triplet (= CH_2CH_3) and two methyl doublets (2 $CHCH_3$) (Figure 4.14).

A search with these three numbers only delivered ~150 hits, as all open fields remained undefined and may have any value in the results. As no C_q-Me singlet was visible, the field "Singlet CH3" was now set to 0, resulting in only 70 answers. To further reduce the answer set, the number of "all_sp2H" was set to 3, according to the integral of the ^{1}H NMR spectrum in the aromatic region: all the 10 resulting hits were anthracyclinone–glycosides, but a further reduction was possible with the number of N-methyl groups (2) or the substituent pattern (aromatic protons in 1,2-position ≡ B1234), so that finally only four musettamycin derivatives were left; two of them had the correct mass and were diastereomers. This search was performed in a few minutes (Figure 4.15).

4.4.4 Complex Data

A straightforward identification as in the previous examples is not always possible. In the case of triterpenes, many closely related compounds may be known, so that the assignment of signal patterns or even the differentiation between isomers or homologues is difficult. The dereplication will need some training and phantasm in these cases. As an example, the following spectra of a commercial nystatin (**13**) sample are given, which also for the expert may not be very meaningful (Figures 4.16 and 4.17).

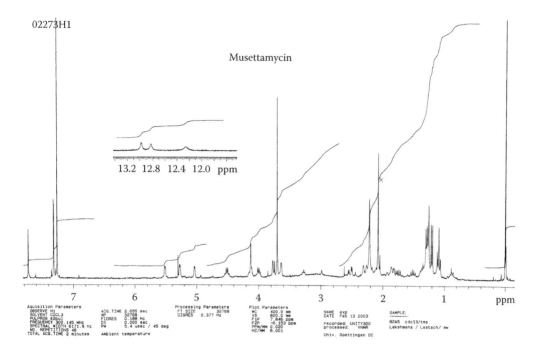

02273H1

Musettamycin

13.2 12.8 12.4 12.0 ppm

7   6   5   4   3   2   1   ppm

Aquisition Parameters
OBSERVE H1
SOLVENT CDCl3
PULPROG s2pul
FREQUENCY 300.145 MHz
SPECTRAL WIDTH 6171.9 Hz
NO. REPETITIONS 48
TOTAL ACQ.TIME 2 minutes

ACQ.TIME 2.655 sec
NP 32768
FIDRES 0.188 Hz
D1 0.000 sec
PW 5.4 usec / 45 deg
Ambient temperature

Processing Parameters
FT SIZE 32768
GIGRES 0.377 Hz

Plot Parameters
WC 400.0 mm
VS 800.0 mm
F1P 7.846 ppm
F2P -0.153 ppm
PPM/MM 0.020
HZ/MM 6.001

NAME exp
DATE Feb 13 2003
recorded: UNITY300
processed: VNMR
Univ. Goettingen OC

SAMPLE:
B2A5 cdcl3/tms
Lakshmana / Laatsch/ mw

Figure 4.14 ¹H NMR spectrum of musettamycin in CDCl₃.

The proton signals between 5.4 and 6.3 ppm are probably due to double bonds (32,388 compounds). Aromatic systems can be excluded. In highly oxygenated benzene derivatives, proton shifts may be found near δ 6, but in these cases, carbon signals between 140 and 160 are expected (14,057 compounds are left). Methyl groups at hetero atoms, double bonds, or carbonyl groups are also missing (no 3H singlets beyond 1.8 ppm; 3311 compounds). The carbon spectrum shows that at least one ester group is present, but no ketone or aldehyde (1775 compounds left). The proton spectrum shows four methyl doublets, due to four CHCH₃ fragments (or better, 2 CHCH₃ + 2 CH₃ groups; see the isopropyl trap; 205 compounds). The signal at δ 97 indicates an acetal group, perhaps of a sugar (39 compounds).

The yellow color of the sample, the carbon signals between δ 129 and 136, and the missing color reaction with sodium hydroxide, but a brown to black color change with concentrated sulfuric acid, indicate double bonds, which should be in conjugation according to the UV maxima at 290, 300, and 320 nm. This reduces the selection to 11 closely related polyene macrolides. A further refinement of the search using the molecular mass reduces the selection to nystatin (**13**) and amphotericin A. A final identification requires authentic spectra or 2D NMR measurements.

4.5 IDENTIFICATION OF NEW COMPOUNDS BY MEANS OF DATABASES

The dereplication of new compounds is a contradiction in terms, as new compounds can never be listed in a database of *known* metabolites. However, in most cases we will be able to find closely related compounds using dereplication techniques that will facilitate the structure elucidation of new relatives.

By scanning and extracting the international literature, AntiBase is growing every year by 700–900 compounds. Only about 1% of these (<10 natural products from microorganisms per year) are really novel, with a new skeleton or other unprecedented features. The remaining structures are

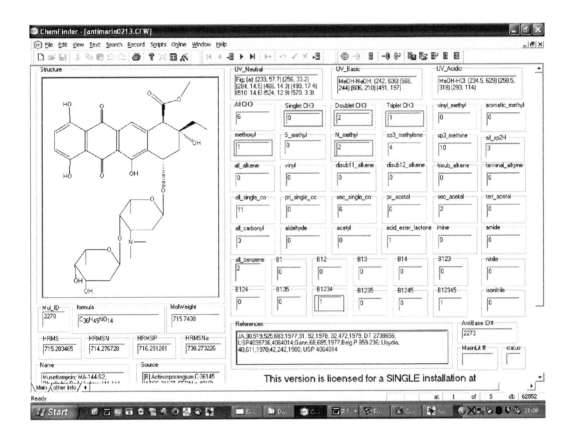

Figure 4.15 Result of a search in AntiMarin. Query values were enframed; the other fragment values are system generated.

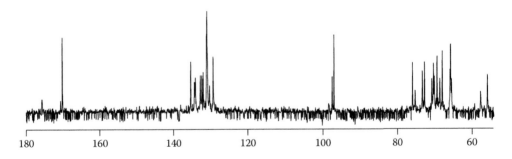

Figure 4.16 ¹³C NMR spectrum (75 MHz) of a commercial nystatin (**13**) sample in DMSO-d_6 in the range of δ > 60 ppm.

new in the sense that they were not listed in *Chemical Abstracts* before, but are simple derivatives of already existing compounds. Examples may be results of minor biosynthetic variations (homologues due to various chain lengths or the insertion of different acids), functional derivatives (esters, amides, acids, and their decarboxylation products), hydrogenation and dehydrogenation products, glycosides with variations in the number and attachment position of the sugars, and so on.

On the other hand, this means that in most new compounds, a substructure is identical with that of a known compound. If we can identify this structural core, a respective substructure search will open the way for the identification of related structures, if they exist.

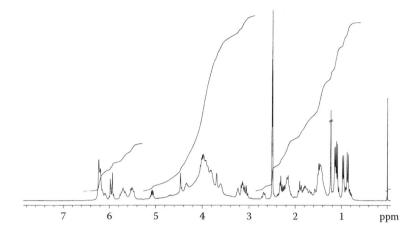

Figure 4.17 ¹H NMR spectrum (300 MHz) of a commercial nystatin (**13**) sample in DMSO-d_6.

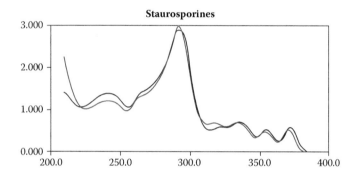

Figure 4.18 UV spectra of staurosporine (**14**, black line) and *N*-carboxamido-staurosporine (**15**, red line) in methanol.

It is obvious that the molecular weight (or pseudomolecular ions) will not be helpful in most cases. UV data, however, may identify compounds with related chromophores easily or exclude others.

A colorless crystalline solid showed a very typical UV spectrum. A search using the dominant maxima in the ranges of 240–249, 290–299, 330–339, 350–359, and 370–379 nm afforded eight candidates. Staurosporine (**14**) was identified as the best hit among them (Figure 4.18). However, according to the molecular mass, the investigated substance was not identical with the latter.

A literature search showed that other staurosporine derivatives also had closely related absorption spectra, indicating that the compound in question was most probably again a staurosporine derivative. The compound was later identified as the (at that time) new *N*-carboxamido-staurosporine (**15**).[27]

Staurosporine (**14**)   Carboxamido-staurosporine (**15**)

It is obvious, however, that this strategy has a serious weak point. At least one (or even better, several) compound with the same chromophore needs to be listed in the database. In other words, the search for UV data (or related properties) should only be used in the case of positive hits.

4.6 SUMMARY

Databases are powerful tools for the dereplication of natural products, for the classification of metabolites as new (and even novel) or already known, through the comparison of experimental data with literature values. The best way to do this is to use original data without trying to interpret these data or making any assumptions.

When ranking the various commonly encountered structure elucidation techniques used for dereplication, NMR comes out on top. NMR data are (in most cases) easily accessible and also predictable; this allows a countercheck of any conclusion. ^{1}H NMR spectra are strongly influenced by solvent effects and are also less tolerant to impurities, but can be measured quickly with high sensitivity. The crude interpretation is easy and delivers information on substructures, including the chemical neighborhood. It is advisable to search in the NMR data first for the presence or absence of common structural elements, such as benzene rings; methoxy, methyl, or carbonyl groups; double bonds; sugars; or chelated OH groups. In the resulting hit list, the observed properties (color, aggregate state, etc.) should be compared with the expected values to allow a further reduction of the hit list.

^{13}C NMR data are reasonably independent of solvent influences; however, the measurement efforts (time, costs) are higher and the sensitivity is lower, so that special techniques are requested (cryoprobes, capillary techniques). It is therefore difficult to use ^{13}C NMR data in the early separation phases. On the other side, the ^{13}C NMR spectrum allows conclusions on the number of carbon atoms and their type (APT spectrum) and on the presence or absence of carbonyl groups, benzene rings or double bonds, oxymethine/methylene groups, sugars (acetal carbons), and so on.

Correlation spectroscopy (COSY) or other 2D data are normally not needed. A search with the observed ^{13}C NMR shifts in the experimental data (in a range of <±5 ppm) or the calculated spectra (in a range of ~±10 ppm) will usually give hit lists that contain the target molecules.

In parallel, the mass data (ESI MS or another soft ionization technique is usually better than EI MS) are extremely helpful. If HRMS is available, then the experimental mass values should be used straightaway for the search, instead of the empirical formulas suggested by the spectrometer program.

A hit found in this way must be confirmed generally by comparison with the original literature. This is best done using spectra provided in the supplementary material of the respective publication or from suitable databases (see below). If the experimental spectra are not accessible, ^{1}H or ^{13}C NMR prediction programs[6] are a good alternative for the direct comparison.

If dereplication results in a structure where observed data are in good agreement with predicted values or with authentic data from the literature, then there is little doubt that the structure has been identified correctly. On the other hand, if no hit is found, great care is advisable. Instead of being novel, the proposed structure may be just wrong. It is strongly advised, therefore, to repeat the search, ideally with different search parameters or in a different sequence (see Figure 4.19).

Some reasons for a misinterpretation have already been discussed. Other reasons for a failed dereplication of known compounds may be among the following:

- The compound in question had been published previously; however, it was incompletely defined (e.g., without a structure), as was the case for sohbumycin/piperazimycin[28] and for more than 2800 other microbial products without published structures.

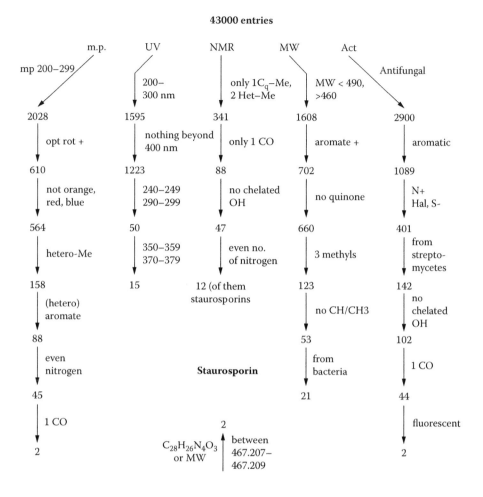

Figure 4.19 Different ways for dereplication with AntiBase using data of staurosporin (**14**). It should be emphasized that a search with high-resolution MS data **is** always the first choice for the initial step.

- The expected compound was published with a wrong structure; 1%–5% of all published structures of natural products may be wrong.[29]
- The data set in the database may be incomplete, as it is often the case for UV and IR data, melting points, biological activities, or many other soft data.
- Due to a wrong search strategy, the expected hit has been lost (see examples above).
- Databases are never complete, although all authors and editors do try their best. In the worst case, just that respective entry is missing.

Finally, the structure of a new natural product should always be checked in *Chemical Abstracts* to exclude gaps in the database in use. This would also reveal if the new natural product had been obtained previously by synthesis or from other sources not covered by the respective database. For example, many plant products have also been isolated from microorganisms, sometimes due to the activity of endophytic microorganisms.

Complete databases are the cornerstone for a successful dereplication of natural products. The best success in natural product chemistry is, however, not to find a compound isolated in one's own lab in any of these databases.

4.7 OUTLOOK

The process of dereplication as described in this chapter consists of the measurement of certain compound parameters (properties) and their comparison with values from the literature, to confirm or deny the similarity or identity by visual inspection. Also, computer programs for the comparison of experimental spectra with databases have been known for a long time. However, what we urgently need is a link of the spectrometer and measurement software with the dereplication programs and further with data (e.g., spectra) collections, so that finally, very few or, better, just one result is obtained in a fully automated way in the case of known compounds. This is obviously the only way to combine dereplication and high-throughput screening techniques. It will hopefully not take too much time until such a hardware handshake has been programmed, and local approaches have already been successful.

Another problem is even more obvious: a reliable dereplication needs reliable reference data like experimental UV, NMR, IR, or mass spectra. As mentioned before, *calculated* NMR data are provided for most structures stored in AntiBase, MarinLit, and some other databases mentioned here. However, an authentic unambiguous identification needs spectra for a direct comparison. The Spectral Database for Organic Compounds (SDBS) (http://sdbs.db.aist.go.jp/sdbs/cgi-bin/cre_index. cgi?lang = eng) is such a spectra collection and allows not only the visual comparison, but also the digital search for spectral data. The Madison–Qingdao Metabolomics Consortium Database (MMCD) (mmcd.nmrfam.wisc.edu) operates in a similar way, but contains additionally the calculated 1D and 2D NMR spectra of >20,000 compounds and a growing number of experimental spectra and *ab initio* calculations. The advantage of this database is the availability of many further compound properties and the presence of links to other databases, like nmrshiftdb2 (http:// nmrshiftdb.nmr.uni-koeln.de/).

Worldwide, every day the spectra of probably hundreds of natural products are measured and perhaps *locally* archived, but they are not published and may be discarded if the structures are known. The need is obvious to establish a central collection of spectra on the Internet that can be extended and improved by the community, similar to Wikipedia or the Cambridge Crystallographic Data Centre. In this way, the spectra from the supplementary material of thousands of publications could also be made accessible in a computer-readable form, and not only as pictures. It is desirable that authors of papers on natural products, but also scientists in general, support these efforts and make their spectra available to the public.

REFERENCES

1. Henkel, T., Brunne, R. M., Muller, H., Reichel, F. Statistical investigation into the structural complementarity of natural products and synthetic compounds. *Angew Chem Int Ed* 1999, 38 (5), 643–647.
2. Butler, M. S. The role of natural product chemistry in drug discovery. *J Nat Prod* 2004, 67 (12), 2141–2153.
3. Newman, D. J., Cragg, G. M. Natural products as sources of new drugs over the last 25 years. *J Nat Prod* 2007, 70 (3), 461–477.
4. Kunfermann, A., Witschel, M., Illarionov, B., Martin, R., Rottmann, M., Hoffken, H. W., Seet, M., Eisenreich, W., Knolker, H. J., Fischer, M., Bacher, A., Groll, M., Diederich, F. Pseudilins: Halogenated, allosteric inhibitors of the non-mevalonate pathway enzyme IspD. *Angew Chem Int Ed* 2014, 53 (8), 2235–2239.
5. Panjikar, S., Parthasarathy, V., Lamzin, V. S., Weiss, M. S., Tucker, P. A. Auto-Rickshaw: An automated crystal structure determination platform as an efficient tool for the validation of an x-ray diffraction experiment. *Acta Crystallogr D* 2005, 61, 449–457.
6. ACDLabs ACD NMR simulation programs. Advanced Chemistry Development, Toronto, 2014.

7. Anagnostec GmbH. AnagnosTec—MALDI-TOF MS microbial identification and antibiotic susceptibility testing. www.pharmaceutical-technology.com/contractors/imaging-analysis/anagnostec/.

8. Hindré, T., Didelot, S., Le Pennec, J. P., Haras, D., Dufour, A., Vallée-Réhel, K. Bacteriocin detection from whole bacteria by matrix-assisted laser desorption ionization-time of flight mass spectrometry. *Appl Environ Microbiol* 2003, 69 (2), 1051–1058.

9. Bhandari, M., Bhandari, A., Bhandari, A. Sepbox technique in natural products. *J Young Pharm* 2011, 3 (3), 226–231.

10. Ito, T., Masubuchi, M. Dereplication of microbial extracts and related analytical technologies. *J Antibiot* 2014, 67 (5), 353–360.

11. Bioactive Natural Products Database (BNPD). Not further available after 2000.

12. Bérdy, J. *CRC Handbook of Antibiotic Compounds*. Vols. 1–16. CRC Press, Boca Raton, FL, 1980–1988.

13. *Dictionary of Natural Products on CD-ROM*. Chapman & Hall/CRC Press, Boca Raton, FL, 2014. A detailed description is available on the Internet at http://dnp.chemnetbase.com/intro/DNPIntroduction.pdf (accessed August 2014).

14. Laatsch, H. AntiBase 2014: The Natural Compound Identifier. Wiley-VCH, Weinheim, Germany, 2014.

15. Munro, M., Blunt, J. MarinLit. Royal Chemical Society, University of Canterbury, Christchurch, Canterbury, New Zealand, September 2013.

16. http://www0.nih.go.jp/~jun/NADB/search.html and http://www.antibiotics.or.jp/journal/database/database-top.htm.

17. Füllbeck, M., Michalsky, E., Dunkel, M., Preissner, R. Natural products: Sources and databases. *Nat Prod Rep* 2006, 23 (3), 347–356.

18. Jakupovic, J., Binkele, H., Wolf, D., Siems, K. Identification of substance mixtures using substance libraries. 2003.

19. Jakupovic, J., Binkele, H., Wolf, D., Siems, K. Identification of substance mixtures using substance libraries. 2007.

20. Fiedler, H.-P. Screening for secondary metabolites by HPLC and UV-visible absorbance spectral libraries. *Nat Prod Lett* 1993, 2, 119–128.

21. Nielsen, K. F., Mansson, M., Rank, C., Frisvad, J. C., Larsen, T. O. Dereplication of microbial natural products by LC-DAD-TOFMS. *J Nat Prod* 2011, 74 (11), 2338–2348.

22. Takats, Z., Wiseman, J. M., Cooks, R. G. Ambient mass spectrometry using desorption electrospray ionization (DESI): Instrumentation, mechanisms and applications in forensics, chemistry, and biology. *J Mass Spectrom* 2005, 40 (10), 1261–1275. For a movie, see http://www.prosolia.com/resources/videos/desi-mass-spectrometry (accessed August 2014).

23. J&M Analytische Mess- und Regeltechnik GmbH. TLC 2010 diodenarray scanner, D-3431. http://www.j-m.de/1/hauptnavigation/j-m/j-m-home.htm.

24. SpecInfo on the Internet. Weinheim, Germany, 2014.

25. SPARTAN '14. Wavefunction, Irvine, CA, 2014.

26. Lang, G., Mayhudin, N. A., Mitova, M. I., Sun, L., van der Sar, S., Blunt, J. W., Cole, A. L. J., Ellis, G., Laatsch, H., Munro, M. H. G. Evolving trends in the dereplication of natural product extracts: New methodology for rapid, small-scale investigation of natural product extracts. *J Nat Prod* 2008, 71 (9), 1595–1599.

27. Wu, S. J., Fotso, S., Li, F., Qin, S., Kelter, G., Fiebig, H. H., Laatsch, H. N-Carboxamido-staurosporine and selina-4(14),7(11)-diene-8,9-diol, new metabolites from a marine *Streptomyces* sp. *J Antibiot* 2006, 59 (6), 331–337.

28. Umezawa, I., Tronquet, C., Funayama, S., Okada, K., Komiyama, K. A novel antibiotic, sohbumycin taxonomy, fermentation, isolation and physicochemical and biological characteristics. *J Antibiot* 1985, 38 (8), 967–971.

29. Nicolaou, K. C., Snyder, S. A. Chasing molecules that were never there: Misassigned natural products and the role of chemical synthesis in modern structure elucidation. *Angew Chem Int Ed* 2005, 44 (7), 1012–1044.

PART 2

Marine Biodiscovery: Lead Discovery

CHAPTER 5

Logistic Considerations to Deliver Natural Product Libraries for High-Throughput Screening

Ngoc B. Pham, Stephen Toms, David Camp, and Ronald J. Quinn
Eskitis Institute for Drug Discovery, Griffith University, Brisbane, Queensland, Australia

CONTENTS

5.1 INTRODUCTION

Natural product research has made great contributions to drug discovery, with nearly 50% of drugs on the market being natural products or derivatives of natural products.[1] Screening crude extracts was common in pre-1980s natural product drug discovery when screening technology was still low throughput. With the advent of high-throughput screening (HTS) in the late 1980s,[2] the automated screening of larger sample sets or libraries became possible. The discovery of bioactive natural products involving initial HTS requires the generation of libraries of crude extracts, pre-purified fractions, or pure compounds. A screening campaign can be commenced on any of these three libraries as long as the chosen library can deliver the requirements in a fast and cost-effective manner. Building a library further down the pipeline involves much more up-front investment in infrastructure and a greater expenditure on labor and consumables. This investment can shorten subsequent efforts to isolate the pure bioactive compound and needs to be considered in the context of cost-effectiveness. Along the line of discovery, continually recurring limitations of screening crude extracts slowly hinder the enthusiasm of the pharmaceutical industry. For crude extracts, nonspecific interference from fatty acids and polyphenols can cause false-positive or false-negative results. Minor components can be missed in a complex mixture because their concentrations are below the detection threshold of biological screening or masked by other compounds. Chemically unattractive compounds with poor physicochemical properties for oral drugs are often isolated. Finally, it takes a great deal of labor and time to identify and isolate bioactive components from crude extracts using bioassay-guided isolation due to the complexity of crude extracts and the need for multiple rounds of fractionation and reassay. The lower initial cost of a crude extract library needs to be evaluated against failure in screening; see below for retrospective analysis by two companies. Several prefractionation methods have been described using single- or multistep solid-phase extraction (SPE),[3–5] flash column chromatography or preparative high-performance liquid chromatography (prep-HPLC),[6,7] countercurrent chromatography (CCC),[8–10] centrifugal partition chromatography (CPC),[11,12] and ultra-performance liquid chromatography (UPLC).[13] Several pharmaceutical companies and research groups have reported their methods for building prefractionated libraries in natural product drug discovery.[4,7,13–19] These approaches have been reviewed previously.[20] These libraries were generated using reversed-phase HPLC coupled with photodiode array (PDA), mass spectroscopy (MS), or evaporative light scattering (ELS) to identify, dereplicate, and isolate biologically active constituents. The number of fractions per biota varied from 200 to 4.[6,19,21–23] Extreme hydrophilic and hydrophobic components (sugars, salts, nucleotides, fatty acids, etc.) are mostly eliminated when an extract is subjected to these methods. MerLion Pharmaceuticals reported a 12-fold increase in the number of hits in fractions compared to extracts.[22] They also found that 80% of the hit fractions from primary screening were active, while their associated crude extracts showed no activity even at four times the screening dose. A similar improvement from screening a prefractionated library was also reported by Wyeth Pharmaceuticals; around 80% of the activity was found only in the fractions, while activity only observed in the crude extract accounted for approximately 10% of the total activity.[21] Further fine-tuning this fractionation process to select lead-like compounds would make screening natural products an innovative avenue for drug discovery.[24,25] In a recent report,[26] we presented the principles of front-loading a screening library with molecules that possess desirable physicochemical properties with the flexibility to incorporate a downstream molecular weight filter. The physiochemical profiles of the natural products active against two neglected disease targets, malaria and African sleeping sickness, confirmed the feasibility of lead-like natural product fraction libraries.[26] Data from liquid chromatography–ultraviolet (LC-UV) absorption, collected in the fractionation process, have facilitated the identification of bioactive components.

In this chapter, we aim to provide practical guidelines and details based on the construction of a fraction library at the Eskitis Institute, Griffith University, as other approaches have been reviewed previously.[26] We focus on the issues of speed, reproducibility, and data tracking to achieve high quality, reproducibility, format flexibility, and cost efficiency for the whole process.

5.2 LEAD-LIKE ENHANCED EXTRACT AND FRACTION LIBRARIES

5.2.1 Quality: Basic Pillar of the Process

Taxonomic diversity is necessary to ensure the collection has sufficient chemical diversity.[27,28] The Eskitis Nature Bank houses 45,000 biota, consisting of 3,950 distinct genera. Nature Bank contains plants from Australia (~20,000), China (~6,300), and Papua New Guinea (PNG) (~5,700); marine invertebrates from Australia (~9,200) and Malaysia (~100); and Australian terrestrial invertebrates (~1,500). It is noted that Australia, China, Malaysia, and PNG are 4 of the 17 megadiverse countries of the world, according to Conservation International 1998.[29] All samples were collected in accordance with the UN Convention on Biological Diversity[30] and the principles of the Nagoya Protocol.[31]

5.2.2 Workflow Management

The fraction library was generated via a multiple-step process that can be broken into three broad stages: prefractionation of the crude extracts, HPLC fractionation, and sample storage. To optimize the entire process, careful consideration was given to each of these stages and involved instrument selection, operator (manual) workflow, and data (tracking) management. We concentrated on the total throughput of the process and considered each aspect of the process, with respect to its effect on the total throughput of the process. We found that operational management was essential for coordination of the workflow involving chemistry, drug discovery, instrument engineering, and IT. The model we used brought together a wide group of people with very different skills and abilities to provide solutions to various challenges of building a high-throughput process. Once the challenging scientific innovation was in place and the running process was optimized, the repetitive, infrequently changing, and quantifiable processes were straightforwardly managed by technical staff.

The library-building metrics were divided into four main performance dimensions, namely quality, reproducibility, flexibility, and cost (Figure 5.1). The quality of the library was judged by its enriched chemical diversity, having compounds with $\log P$ less than 5 and being devoid of nuisance compounds. Reproducibility was assessed by the consistency of the HPLC chromatogram obtained from the different samples of the prefractionated extract, and this also served to confirm

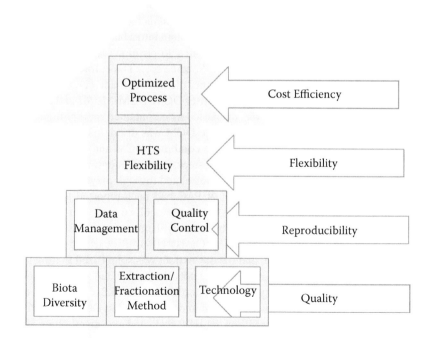

Figure 5.1 Natural product library building logistics.

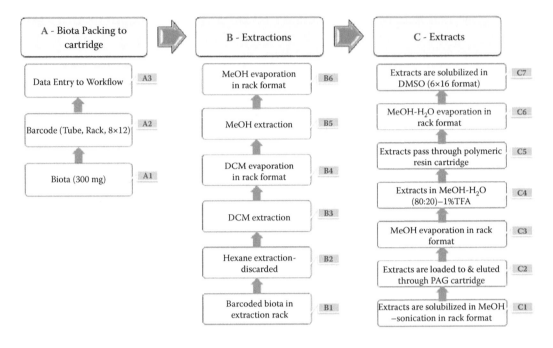

Scheme 5.1   Flow diagram of the extraction process.

the reproducibility of the prefractionation step. Ultimately, reproducibility was assessed by comparing the HPLC data obtained from a new sample of dried biota. The library storage format was designed to be easily adapted to any screen requirement. All steps were planned with a sustainability focus on labor, instrumentation load, and materials. We emphasized technology improvement, as it was the basis for all improvements of the above performance metrics.

The biota preparation process consisted of scooping biota into a barcoded solid-phase extraction (SPE) cartridge held by a matching barcoded test tube in a 96-position rack. Biota information entry was maintained by a Microsoft worksheet (step A, Scheme 5.1). The extraction process had three extraction steps with three different solvents (hexane, dichloromethane [DCM or DCM-MeOH], and methanol [MeOH]), and two evaporation steps after DCM/DCM-MeOH and MeOH extraction (step B, Scheme 5.1). The hexane extract was discarded as it contained mainly highly lipophilic or non-lead-like compounds (step B2, Scheme 5.1). The DCM was dried, and the same test tube was used to collect the MeOH extract. Transforming the combined DCM and MeOH crude extracts to lead-like enhanced extract required three sonication steps, two SPE steps, and two solvent evaporations (step C, Scheme 5.1). The crude plant extract was eluted through polyamide gel (PAG) to yield a nonpolyphenol extract. An amount of 700 mg of PAG had been reported to be able to remove all polyphenols from a 100 mg extract sample.[23] In our protocol we used 800–900 mg of PAG for extracts ranging from 10 to 50 mg in weight. The nonpolyphenol extract was dried, reconstituted in MeOH-H$_2$O (80:20), and was passed through an Oasis® HLB SPE cartridge. The eluted extract was collected, its solvent was evaporated, and the dry extract was solubilized in dimethyl sulfoxide (DMSO). The lead-like enhanced extract in DMSO was then subjected to a reversed-phase C18 high-performance liquid chromatography (HPLC) (step D, Scheme 5.2). The number of fractions collected was either 5 or 11. The fractions were dried and solubilized in DMSO. The fractions were produced by two chromatography steps: a SPE using HLB resin and a reversed-phase HPLC with an optimized protocol in which only compounds of log$P$ < 5 were in the collected region.[26] Fractions were transferred to two-dimensional (2D) barcoded microtubes on completion (step E, Scheme 5.2). These 2D microtubes allow a preparation of a variety of plate formats (step F, Scheme 5.2).

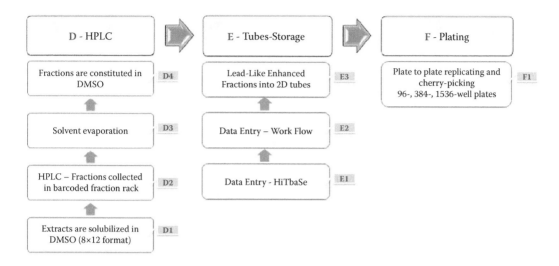

Scheme 5.2 Flow diagram of the fractionation process.

5.2.3 Technology Improvements

5.2.3.1 Extraction Rack Format

The movement of crude extracts through the prefractionation process starts with the dry extracts being collected from the Eskitis Nature Bank and placed in 10 ml glass test tubes. To allow for effective data tracking, these sample tubes are individually barcoded and racked for processing. An analysis of the sample test tube movement through the prefractionation process found that laboratory staff were required to rerack the samples multiple times depending on which instrument was used. This constant need for reracking introduced not only time delays into the workflow, but also, more importantly, the possibility of errors being introduced if samples were inadvertently returned to the wrong position in preparation for the next step in the process. To prevent these positional-related errors, it was important for laboratory staff to regularly confirm the location of each sample, by comparing the barcodes against the original pick lists—which introduced another delay. A solution was to manufacture a single "master" rack that could hold the same set of twenty-four 10 ml sample test tubes for the entire process and could be used across all instrument types. Rather than hand-barcoding each test tube at the time of sample collection, an individual sample would be identified by a unique rack number and the position of the tube within the rack via a modified Cartesian coordinate system (e.g., rack 001_A_4). Care was taken when designing a master rack to ensure that it met the design requirements for all the instruments employed within the prefractionation process, including liquid handlers, multiple evaporators, sonicators, and HPLC. Each of these devices has unique requirements (e.g., hardware features employed as alignment keys, allowances for temperature sensors, positioning of samples for liquid handing, maximum loaded mass for rotatory evaporation systems, etc.) that need to be accommodated in the final master rack design. The resultant master rack has proven itself to be effective with the samples remaining in the rack for the full prefractionation process.

5.2.3.2 Extraction Liquid Handler

The automated SPE platform employed within the institute was the Gilson ASPEC XL 4, a four-probe platform. The platform is able to undertake all the steps of the SPE process, including three-step solvent extraction, conditioning the SPE column, loading extracts onto the SPE column, washing out nuisance compounds, and eluting interesting compounds with a moderate degree of automation via a user-defined control protocol. Regardless of the level of automation, the limitation to any improvement in the extraction workflow was the instrument's four-channel head. The need to push the extraction process to higher efficiencies meant that there was a need to examine how it would be possible to increase the number of samples that were processed at any one time.

A crude way to improve the workflow of any given process is to simply run similar processes in parallel; one would expect that two parallel processes would give a twofold improvement, three parallel processes a threefold improvement, and so on. The main disadvantage of such parallelism is primarily cost, but also site requirements, and ongoing technical support needs to be considered. A goal of a 12-fold to 16-fold improvement in the overall extraction process was set, and in such a case, basic parallelism of purchasing more Gilson ASPEC XL 4's was not a financial or practical solution. As a result, our decision was to examine if it was possible to improve the number of samples that could be processed at one time by manufacturing a new pumping system with the ability to process 48 samples simultaneously. This would give the required minimum of a 12-fold improvement in throughput, while still keeping compatibility in other elements in the extraction system (sample drying and HPLC where batch processes could also be undertaken in groups of 48 samples).

We approached a manufacturer in the SPE field (J-Kem Scientific, St. Louis, Missouri) to construct a bespoke pumping system that would meet our unique requirements. J-Kem met the challenge by expanding one of its existing pumping platforms to contain sixteen 10 ml syringes, and through the use of a modified central controller, we were able to connect three of the 16-way pumping platforms in parallel to give the necessary capacity of 48. The pumps were then connected to a custom in-house manifold that interfaced directly to the biota cartridges (Figure 5.2).

Using J-Kem's intuitive programming language, it was possible to quickly rewrite our existing Gilson protocols to suit the J-Kem pump platform, thus allowing us to undertake the basic extraction processes. The loading and unloading of the 48-way manifold are currently manual processes, but given the overall improvement in workflow, this was deemed an acceptable cost to the running of the process.

After the successful testing of the 48-way manifold, a second system was built. With both systems running in parallel, we are seeing an effective 24-fold improvement over our original extraction process.

Table 5.1 details the improvements that have been achieved by moving to the J-Kem pump platform. Both platforms require a certain amount of time on preprotocol processing—preparing solvent reservoirs, priming delivery lines, and so forth. With both J-Kem systems running in parallel, an effective 24-fold improvement over our original process was achieved with a 10 min extraction time for 96 biota, compared to a 240 min extraction time for Gilson APSEC™ XL4.

Table 5.1   Time Saved for Technology Improvement of In-House Extraction Platform

Instrument	Dispense Protocol (10 ml/min for 10 min) to Process 48 Samples	Dispense Protocol (10 ml/min for 10 min) to Process 96 Samples
Gilson ASPEC XL4	120 min	240 min
J-KEM 48-way	10 min	—
J-KEM 48-way (by 2)	—	10 min
Improvement	12×	24×

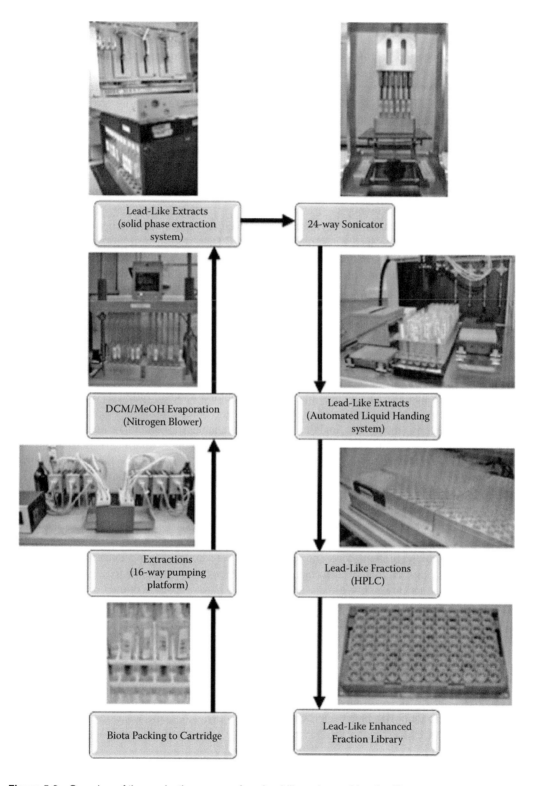

Figure 5.2 Overview of the production process for a lead-like enhanced fraction library.

5.2.3.3 Sonication

Sample resuspension originally involved having the samples manually loaded two at a time into a sonic bath for a period of 1–2 min. Thus, resuspending a rack of 24 samples would take 24–48 min. As the process was scaled up, manually handling each test tube out of the rack, into the sonicating bath, and back to the rack was an error-prone and time-consuming process. A move to a 24-way sonicating probe (a Sonics Vibra-Cell™) has allowed for the simultaneous resuspension of the rack, described above, containing 24 samples in 1–2 min (Figure 5.2). The probe controller allows for the intensity, sonication profile (including a pulsing sequence), and run time to be configured and stored, making the process a one-button action.

5.2.3.4 Solvent Evaporation

In-house designed nitrogen blowers are also used to quickly dry more volatile solvents and help reduce the workload on the larger evaporation systems. These blowers are configured to dry 48 samples simultaneously. A simple keypad allows the user to set a drying time in 30 min increments, and a custom controller then regulates the gas flow to minimize the loss of nitrogen.

Solvent evaporation is primarily undertaken by a GeneVac HT-4X evaporation system, with a Martin Christ Beta RVC unit being used when extra drying capacity is needed. These systems are able to process 96 samples (4 × 24 racks) at a time.

5.2.4 Reproducibility

The process required a capability to reproduce and resupply active fractions. Quality control of the HPLC fractionation is routinely monitored by injection of a standard once every eight injections. Data from biota, extracts, and fractions are checked and stored at each step of the process. Biota are barcoded when entering the process, and the barcode accompanies the biota to the final step of extract and fraction generation. Along the process, each biota passes through 9 (marine) or 12 (plant) different steps (3 extractions, 3 or 4 evaporations, 2 or 3 sonication steps, and 1 or 2 SPEs) on 4 instruments (automated extraction platform, 2 solvent evaporation platforms, and a sonicator). We managed to preserve biota rack coordinates for each 96-biota batch. Errors caused by biota mix-up have been minimized, if not eliminated. Extraction and fractionation processes are monitored by process workflows. The workflow application provides the robotic liquid dispenser (PerkinElmer Janus® Automated Workstation, Waltham, Massachusetts) a work list that directs extract and fraction formatting into 2D barcoded microtubes (0.5 ml) for storage. These tubes are stored at 18 °C under nitrogen and can be plated in 96-, 384-, or 1536–well formats with volumes ranging from nanoliters to microliters. Chromatographic traces are archived and used as a checkpoint when extracts or fractions are requested.

A concern of researchers when working with natural products is the biota/extract/fraction resupply. For contingency, we maintain a backup of 1 g of each biota at a separate location from the main storage. This amount of biota can last for three rounds of extraction and fractionation and can deliver at least 1000 high-throughput screens. Collaborators at the Queensland Museum and Queensland Herbarium with field collection and taxonomic expertise have eased the concern on availability of biota for large-scale fractionation follow-up. Data management was maintained by the Eskitis in-house software (HiTBaSe). This program allows storage and searching of biota/extract/fraction annotations (the unique Nature Bank ID), biota weights, biota part, family, genus, species, external ID (from our collection partners, QLD Museum, QLD Herbarium, PNG, China, and Malaysia), and biota GPS.

5.2.5 Flexibility for High-Throughput Screening: Library Size, Format, and Volume

Our process can accommodate libraries of 11 fractions or 5 fractions per biota. For a HTS campaign size of 200,000 wells, a collection of 5 fractions for each biota allows 40,000 biota to be screened, and subsequently an increase in chemical diversity for a screening campaign, compared to 11 fractions per each biota sample. Extracts and fractions are stored in tubes, which are free from any restriction of plate format. The final fractions are stored at the Compounds Australia located at the Eskitis Institute for Drug Discovery, Griffith University. Compounds Australia is Australia's only dedicated compound management facility with a capacity to store up to 400,000 × 1.0 ml microtubes, 200,000 × 0.5 ml microtubes, and 4,320 shallow-well plates in dedicated, automated stores. To preserve the sample integrity for extended periods (5+ years), the samples are kept in a dry environment at 18 °C (microtubes) or 10 °C (microplates). The samples are individually tracked using a combination of one-dimensional (1D) and 2D barcodes and an in-house database system.

5.2.6 Cost Efficiency

The focus on technology improvement in generating extracts led to a reduction in extraction time and thus improvement in cost efficiency. The effective 24-fold improvement in the extraction process has been the most cost-effective achievement. The replacement of different rack formats and multiple barcode scanning by one uniform rack format and one barcode per rack has saved time, further reduced costs, and increased reliability. Merging workflows involving reformatting of instrument platforms (Gilson XL-4, nitrogen blower, GeneVac-HT4X, 24-way sonicator) to one compatible format has simplified the complexity of a process involving thousands of biota in a pipeline of a three-step extraction and a two-step SPE prepurification.

5.3 CONCLUSION

Building a library in the hundreds of thousands of fractions has been challenging in informatics, process management, and natural product chemistry-related activities. In order to establish a large lead-like natural product library, we strategized our activity into four focused areas: library quality, reproducibility, flexibility, and cost efficiency. We generated a front-loaded library with lead-like fractions using log$P$ as a primary filter. We formulated a workflow using two steps of solid-phase extraction (PAG and HLB resin) and a reversed-phase C18 chromatography. We improved our technology to speed up the process and attain reproducibility. Challenges have led to a setup of two 48-channel extraction systems boosting a 24-fold improvement in time, a 24-way ultrasonic probe, and a uniform instrument bed platform, reducing human error compared to manual handling and also reducing process time by avoiding handling individual test tubes. The process is flexible in choosing a format of 11 fractions or 5 fractions per biota; screening plate formats of 96, 384, or 1536 wells; variable screening volumes ranging from nanoliters to microliters; and cherry-picking options for screen confirmation. Technology was the main focus for process improvement. Cost efficiency was a consequence of the technical improvements. We believe we have developed a practical approach to the preparation of a large natural product fraction library.

5.4 EXPERIMENTAL SECTION

5.4.1 Instrumentation

Extractions were performed on an in-house modified J-Kem system. Solid-phase extractions were performed on the Gilson Aspec XL4 Solid Phase Extraction system. Reconstitution of extracts into organic solvents was facilitated using a Vibra-Cell Ultrasonic Processor, Sonics & Materials, Inc. Extracts and fractions were transferred from test tubes to FluidX microtubes using a Janus Automated Liquid Handling System, PerkinElmer Life & Analytical Sciences.

5.4.2 Extracts

A freeze-dried biota sample (300 mg) was packed into an empty cartridge (3 ml volume) and was extracted with n-hexane (7 ml). The n-hexane extract was discarded. The biota was then extracted with DCM:MeOH (80:20, 7 ml) and the extract was collected in a test tube (borosilicate glass, 16 × 100 mm). The solvent was evaporated from the test tube using nitrogen gas. The biota was then extracted with methanol (13 ml) and collected in the same test tube. The solvent was then evaporated yielding a crude extract. For plant biota, the biota was extracted with DCM (100%) and then MeOH (100%). The crude extract was reconstituted in MeOH (4 ml) before being loaded onto a cartridge of polyamide gel (800–900 mg). The cartridge was washed with MeOH (2 × 4 ml) and the combined MeOH was dried. The crude extract was sonicated and solubilized in 80:20 MeOH:H_2O containing 0.1% trifluoroacetic acid (TFA) (4 ml) and loaded onto a SPE cartridge containing a functionalized polymeric reversed-phase absorbent (Oasis HLB, 400 mg). The cartridge was eluted further with the same solvent (8.5 ml). The eluent (~12 ml) was collected in a new test tube and the solvent was evaporated. The dry crude extract was solubilized in DMSO (0.6 ml) to make a stock concentration of 500 μge/μl (stock concentration = starting material [μg] per volume of DMSO [μl]). Two-thirds of the extract was stored in a 500 μl Matrix 2D tube (0.35 ml), and the rest was dispensed to a microtube for the generation of lead-like enhanced (LLE) fractions.

5.4.3 LLE Fractions

The LLE extracts were processed in batches of 96 with 100 μl injected for each sample to a C18 HPLC column (Phenomenex Onyx Monolithic C18, 4.6 × 100 mm) (Figure 5.3). HPLC fractionation conditions (Table 5.2) consisted of a linear gradient (curve 6) from 90% H_2O (0.1% TFA)/10% MeOH (0.1% TFA) to 50% H_2O (0.1% TFA)/50% MeOH (0.1% TFA) in 3 min at a flow rate of 4 ml/min; a convex gradient (curve 5) to MeOH (0.1% TFA) in 3.50 min at a flow rate of 3 ml/min,

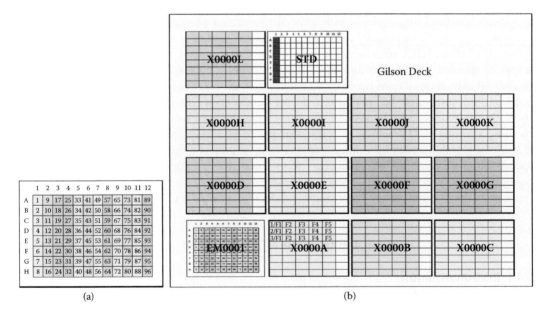

(a)   (b)

Figure 5.3 (a) Layout for the LLE extract rack; numbers signify injection order. (b) Gilson 215 liquid handler bed layout for the LLE fraction library generation. EM0001 = LLE extract track (96 samples), STD = standard rack (a standard mixture), X0000A-L = LLP fraction plates (480 fractions).

Table 5.2 Gradient Conditions for the HPLC Method Set LLE Fraction Library

Time (min)	Flow Rate (ml/min)	% Solvent B (0.1% TFA in MeOH)	% Solvent C (0.1% TFA in H₂O)	Curve
0.00	4	10	90	
3.00	4	50	50	6
3.01	3	50	50	6
6.50	3	100	0	5
7.00	3	100	0	6
7.01	4	100	0	6
8.00	4	100	0	6
9.00	4	10	90	6
11.00	4	10	90	6

held at MeOH (0.1% TFA) for 0.50 min at a flow rate of 3 ml/min, held at MeOH (0.1% TFA) for a further 1.0 min at a flow rate of 4 ml/min; and then a linear gradient (curve 6) back to 90% H_2O (0.1% TFA)/10% MeOH (0.1% TFA) in 1 min at a flow rate of 4 ml/min, held at 90% H_2O (0.1% TFA)/10% MeOH (0.1% TFA) for 2 min at a flow rate of 4 ml/min. The total run time for each LLE injection was 11 min, and five fractions were collected between 2.0 and 7.0 min: fraction 1 (time = 2.00–3.00 min), fraction 2 (3.01–4.00 min), fraction 3 (4.01–5.00 min), fraction 4 (5.01–6.00 min), and fraction 5 (6.01–7.00 min). If 11 fractions were collected, the collection times are as follows: fraction 1 (time = 2.00–2.33 min), fraction 2 (2.34–2.66 min), fraction 3 (2.67–3.00 min), fraction 4 (3.01–3.50 min), fraction 5 (3.51–4.00 min), fraction 6 (4.01–4.50 min), fraction 7 (4.51–5.00 min), fraction 8 (5.01–5.50 min), fraction 9 (5.51–6.00 min), fraction 10 (6.01–6.50 min), and fraction 11 (6.51–7.00 min). After injection of every eight LLE extracts, a standard mixture consisting of methyl 4-hydroxy benzoate, ethyl 4-hydroxy benzoate, benzophenone, and uracil (all 0.0312 mg/ml in DMSO) was injected as a positive control for the LLE fractionation process. After evaporation of solvents, each glass test tube containing an LLE fraction was resuspended in 200 µl of DMSO and transferred to a Fluid-X 500 µl microtube using a PerkinElmer Janus instrument liquid handler. The concentration of each fraction was 250 µge/µl.

ACKNOWLEDGMENTS

We thank K. Watts and R. Treers for help with preparation of the LLE extracts and the LLE fractions and P. Walve for assistance with data storage. We gratefully acknowledge Queensland Smart State Innovation Project Fund–National and International Research Alliances and Partnerships Program.

REFERENCES

1. Newman, D. J., Cragg, G. M. Natural products as sources of new drugs over the 30 years from 1981 to 2010. *J. Nat. Prod.* 2012, 75 (3), 311–335.
2. Boisclair, M. D., Egan, D. A., Huberman, K., Infantino, R. *High-Throughput Screening in Industry.* 2nd ed. Humana Press, Totowa, NJ, 2004, pp. 23–39.
3. Cardellina, J. H., Munro, M. H. G., Fuller, R. W., Manfredi, K. P., McKee, T. C., Tischler, M., Bokesch, H. R., Gustafson, K. R., Beutler, J. A., Boyd, M. R. A chemical screening strategy for the dereplication and prioritization of HIV-inhibitory aqueus natural products extracts. *J. Nat. Prod.* 1993, 56, 1123–1129.
4. Schmid, I., Sattler, I., Grabley, S., Thiericke, R. Natural products in high throughput screening: Automated high-quality sample preparation. *J. Biolmol. Screening* 1999, 4, 15–25.

5. Thiericke, R. Drug discovery from nature: automated high-quality sample preparation. *Autom. Methods Manag. Chem.* 2000, 22, 149–157.
6. Eldridge, G. R., Vervoort, H. C., Lee, C. M., Cremin, P. A., Williams, C. T., Hart, S. M., Goering, M. G., O'Neil-Johnson, M., Zeng, L. High-throughput method for the production and analysis of large natural product libraries for drug discovery. *Anal. Chem.* 2002, 74 (16), 3963–3971.
7. Abel, U., Koch, C., Speitling, M., Hansske, F. G. Modern methods to produce natural-product libraries. *Curr. Opin. Chem. Biol.* 2002, 6, 453–458.
8. Wu, S., Yang, L., Gao, Y., Liu, X., Liu, F. Multi-channel counter-current chromatography for high-throughput fractionation of natural products for drug discovery. *J. Chromatogr. A* 2008, 1180, 99–107.
9. Alvi, K. A., Peterson, J., Hofmann, B. Rapid identification of elalophylin and geldanamycin in Streptomyces fermentation broths using CPC coupled with a photodiode array detector and LC-MS methodologies. *J. Ind. Microbiol.* 1995, 15, 80–84.
10. Lu, Y., Berthod, A., Hu, R., Maa, W., Pan, Y. Screening of complex natural extracts by countercurrent chromatography using a parallel protocol. *Anal. Chem.* 2009, 81, 4048–4059.
11. Armbruster, J. A., Borris, R. P., Jiminez, Q., Zamora, N., Castillo, G. T., Harris, G. H. Separation of crude plant extracts with high speed CCC for primary screening in drug discovery. *J. Liq. Chromatogr. Relat. Technol.* 2001, 24, 1827–1840.
12. Ingkaninan, K., Hazekamp, A., Hoek, A. C., Balconi, S., Verpoorte, R. Application of centrifugal partition chromatography in a general separation and dereplication procedure for plant extracts. *J. Liq. Chromatogr. Relat. Technol.* 2000, 23, 2195–2208.
13. Tu, Y., Jeffries, C., Ruan, H., Nelson, C., Smithson, D., Shelat, A. A., Brown, K. M., Li, X. C., Hester, J. P., Smillie, T., Khan, I. A., Walker, L., Guy, K., Yan, B. Automated high-throughput system to fractionate plant natural products for drug discovery. *J. Nat. Prod.* 2010, 73, 751–754.
14. Eldridge, G. R., Vervoort, H. C., Lee, C. M., Cremin, P. A., Williams, C. T., Hart, S. M., Jonhson, M. O., Zeng, L. High-throughput method for the production and analysis of large natural product libraries for drug discovery. *Anal. Chem.* 2002, 74, 3963–3971.
15. Appleton, D. R., Buss, A. D., Butler, M. S. A simple method for high-throughput extract prefractionation for biological screening. *Chimia* 2007, 61, 327–331.
16. Jia, Q. Generating and screening a natural product library for cyclooxygenase and lipoxygenase dual inhibitors. *Stud. Nat. Prod. Chem.* 2003, 29, 643–718.
17. Wagenaar, M. M. Pre-fractionated microbial samples: The second generation natural products library at Wyeth. *Molecules* 2008, 13, 1406–1426.
18. Bitzer, J., Kopcke, B., Stadler, M., Hellwig, V., Ju, Y. M., Seip, S., Henkel, T. Accelerated dereplication of natural products, supported by reference libraries. *Chimia* 2007, 61, 332–338.
19. Bugni, T. S., Harper, M. K., McCulloch, M. W. B., Reppart, J., Ireland, C. M. Fractionated marine invertebrate extract libraries for drug discovery. *Molecules* 2008, 13, 1372–1383.
20. Camp, D., Davis, R. A., Evans-Illidge, E. A., Quinn, R. J. Guiding principles for natural product drug discovery. *Future Med. Chem.* 2012, 4 (9), 1067–1084.
21. Wagenaar, M. M. Pre-fractionated microbial samples: The second generation natural products library at Wyeth. *Molecules* 2008, 13 (6), 1406–1426.
22. Appleton, D. R., Buss, A. D., Butler, M. S. A simple method for high-throughput extract prefractionation for biological screening. *Chimia* 2007, 61 (6), 327–331.
23. Tu, Y., Jeffries, C., Ruan, H., Nelson, C., Smithson, D., Shelat, A. A., Brown, K. M., Li, X.-C., Hester, J. P., Smillie, T., Khan, I. A., Walker, L., Guy, K., Yan, B. Automated high-throughput system to fractionate plant natural products for drug discovery. *J. Nat. Prod.* 2010, 73 (4), 751–754.
24. Camp, D., Campitelli, M., Carroll, A. R., Davis, R. A., Quinn, R. J. Front-loading natural-product-screening libraries for log P: Background, development, and implementation. *Chem. Biodiv.* 2013, 10 (4), 524–537.
25. Camp, D. N., S., Newman, S., Pham, N. B., Quinn, R. J. Nature Bank and the Queensland Compound Library: Unique international resources at the Eskitis Institute for Drug Discovery. *Comb. Chem. High T. Scr.* 2013, 17, 201–209.
26. Camp, D., Davis, R. A., Campitelli, M., Ebdon, J., Quinn, R. J. Drug-like properties: Guiding principles for the design of natural product libraries. *J. Nat. Prod.* 2012, 75 (1), 72–85.

27. Ronsted, N., Symonds Matthew, R. E., Birkholm, T., Christensen Soren, B., Meerow Alan, W., Molander, M., Molgaard, P., Petersen, G., Rasmussen, N., van Staden, J., Stafford Gary, I., Jager Anna, K. Can phylogeny predict chemical diversity and potential medicinal activity of plants? A case study of Amaryllidaceae. *BMC Evol. Biol.* 2012, 12, 182.
28. Gallagher, K. A., Rauscher, K., Ioca, L. P., Jensen, P. R. Phylogenetic and chemical diversity of a hybrid-isoprenoid-producing streptomycete lineage. *Appl. Environ. Microbiol.* 2013, 79 (22), 6894–6902.
29. http://www.environment.gov.au/biodiversity/conservation/hotspots.
30. http://www.cbd.int/convention/text/.
31. http://www.cbd.int/abs/.

CHAPTER 6

Screening Marine Microbial Libraries

Jacqueline L. von Salm,[1] **Christopher G. Witowski,**[1] **Danielle H. Demers,**[1]
Ryan M. Young,[1,2] **Laurent Calcul,**[1] **and Bill J. Baker**[1,2]
[1]Department of Chemistry and Center for Drug Discovery and Innovation,
University of South Florida, Tampa, Florida
[2]School of Chemistry, National University of Ireland Galway, Galway, Republic of Ireland

CONTENTS

6.1 INTRODUCTION AND HISTORY

Synthetic, combinatorial, and computational methods for making novel scaffolds in drug discovery continue to dominate pharmaceutical research despite recent reviews describing that approximately two-thirds of all newly approved drugs are original natural products, natural product derivatives, synthetic compounds utilizing a natural product pharmacophore, or other biologicals.[1] Many of these pharmaceuticals were inspired by compounds produced by terrestrial plants, microorganisms, and the occasional marine micro- or macroorganisms. Terrestrial microorganisms continue to display pharmacological potential; however, many plants have been well investigated, rendering novel compounds and drugs harder to find. A promising resource for natural product libraries is unstudied or understudied organisms, such as the marine microbial environment. The world's oceans cover more than 70% of the earth's surface and host many diverse habitats, from deep abyssal plains to shallow coastal tide pools. Each of these environments hosts marine organisms cohabitating and interacting both physically and chemically, creating a unique ecosystem. Many of these organisms have been shown to produce biologically interesting chemistry with many other organisms still largely undiscovered.[2] Marine natural product research has largely focused on sessile or slow-moving macroorganisms such as sponges, tunicates, corals, algae, and mollusks, which often lack a physical defense against predation.[3] Studies have shown that many of these organisms host large quantities of diverse microorganisms beyond the free-living microbes within the water column and soil. This has increased efforts toward understanding host-associated marine microorganisms and their potential role in drug discovery.[4]

The history of marine microbiology is a comprehensive one, and some of the initial collection methods have proven sufficiently useful to the field that they are still being used today. Even before Jacques-Yves Cousteau provided the means for people to more closely study marine life, early marine microbiologists were studying surface microbial life in the water column. Among the first and most eminent marine microbiologists, Zobell and Allen studied the role of marine bacteria in environments by dropping glass slides off the end of a dock into the ocean and studying the bacteria and other microbe biofoulants.[5] Although their focus was primarily on bacteria and diatoms, the basic nutrient composition and isolation techniques used in the early 1930s can still be used today, and a common media used for isolation is appropriately named Zobell's Marine Agar 2216.[6]

Even before the discovery of marine microorganisms as biofoulants, there were signs of their potential as resources for drug discovery when seawater was shown to have surprising antibiotic activity against most terrestrial bacteria.[7] An explanation of this activity was hypothesized nearly 20 years later by Rosenfeld and Zobell in 1947.[8] They reasoned that the activity was due to antibacterial compounds produced by marine microbes within the seawater rather than the seawater itself. Confirmation of this hypothesis was established by Burkholder et al., who found the highly brominated antibiotic pyrrole (1) in 1966. Consisting of nearly 70% bromine by mass, this compound was produced by the common Gram-negative marine bacterium *Alteromonas luteoviolacea*.[9,10] Directly following the discovery of antibiotic-producing marine bacteria, the discoveries of the first marine-derived fungi were described, including *Cephalosporium* sp., later properly named *Acremonium chrysogenum*.[11] This fungus subsequently was found to produce the well-known β-lactam antibiotics, the cephalosporins (e.g., cephalosporin C [2]).[12,13] These compounds are representative of the marine realm's penicillins and provided a jumping-off point for the future of marine microbial drug discovery.

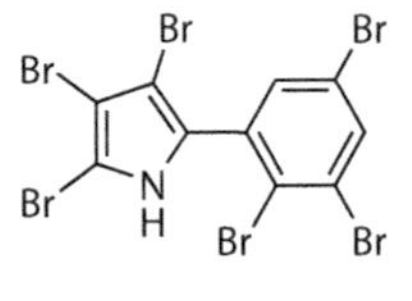

1 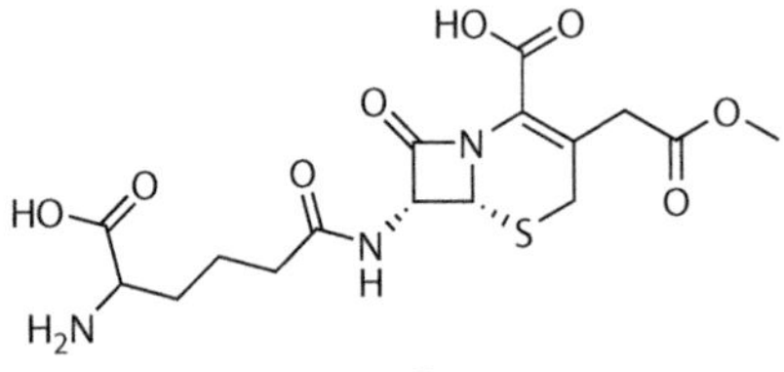   2

Currently, there are two approved drugs of marine origin on the American and European markets: ziconotide (Prialt®), from the marine snail *Conus magus*,[14] and trabectedin (Yondelis®) (3), isolated from the tunicate *Ecteinascidia turbinata*[15,16] (Figure 6.1).

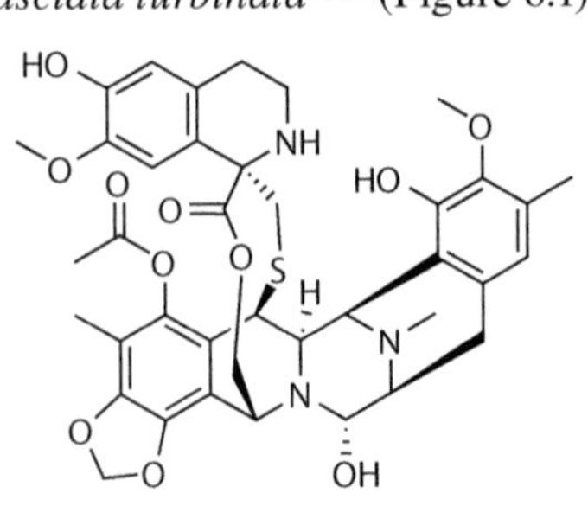

3

Figure 6.1 A photograph of *Ecteinascidia turbinata* (left) and the chemical structure of trabectedin (right). (Photo courtesy of B. J. Baker.)

Although ziconotide and trabectedin are compounds of pharmaceutical interest for their applications in medicine, supply of their respective macroorganisms is still a major concern. Isolation of bioactive compounds from a natural resource often requires tens to hundreds of kilograms of material for successful completion of preclinical studies, clinical trials, and subsequent use in treatment. The nature of macroorganisms makes it very difficult to cultivate and replicate the desired chemistry outside of their natural marine environments,[17] leaving limited options other than destroying their native ecosystems. The use of SCUBA for collection of macroorganisms is often inefficient and costly, while the use of dredges is nonspecific, resulting in a large amount of bycatch and scarring of the native habitat. In order to maintain biodiversity in oceans and related environments, other discoveries and methods like semisynthesis,[18] aquaculture,[17] and improved cultivation methods for marine microorganisms have benefited the field of marine natural product drug discovery. Advancements in the cultivation of diverse marine microorganisms have irreversibly changed the realm of marine natural products and the entire approach of drug discovery itself. Representing a renewable resource and the ability to perform simple fermentation variations, semisyntheses, and implementation of high-throughput screening (HTS), marine microbes have caught the attention of many scientists and industries. One the most advantageous properties of marine microbes is that only small amounts of biological material are required to be transported back to the laboratory, thus increasing the amount of biodiversity that can be collected and proposed for screening.

It has become increasingly clear that the world is teeming with these bacteria and fungi, many of which have yet to be identified. The most recent estimates find that 50%–78% of the world's carbon comes from prokaryotes, the majority of which originate in the open oceans.[19,20] There are an estimated $>10^5$ microbial cells per milliliter of surface seawater,[21,22] with up to 40% of a filter feeder's biomass, such as a sponge, resulting from microbes.[23] In some cases, the compounds originally isolated from marine macroorganisms have been found to originate from microbial symbionts living in association with the organism. The significant protein kinase C inhibitors, the bryostatins (e.g., bryostatin 1 [**4**]), were originally isolated from the bryozoan *Bugula neritina*, but have since been shown to come from its microbial symbiont, the bacterium *Endobugula sertula*.[24–27] In the case of trabectedin (**3**), it was first extracted from its ascidian producer *Ecteinascidia turbinata* and has demonstrated activity toward many different forms of cancer.[15,16,28] The process of extraction and isolation of trabectedin has improved significantly since the fermentation of the bacterium *Pseudomonas fluorescens* produces a similar class of compounds, the safracins, which are cultivated as semisynthetic precursors of **3**.[29] Cultivation of this microorganism is faster, easier, less expensive, and produces the majority of the pharmacophore, leaving only a small portion of the molecule left to be synthesized to yield the drug Yondelis. This produces more material for clinical studies and potentially the cure for specific cancers in the future.[30] Discoveries such as these have only come recently; however, scientists have known since the early 1900s that marine microorganisms are important as symbionts, biofoulants, and a source of nutrition for other marine organisms,[31–33] providing insight into the potential diversity of compounds being produced.

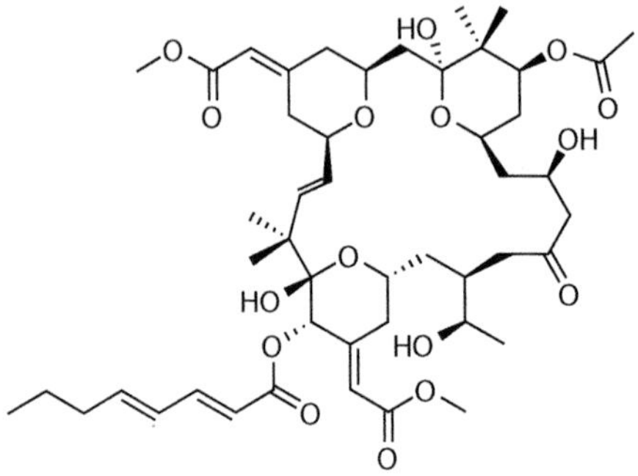

4

The vastness of their numbers and their tendency to multiply rapidly add to the benefit of targeting marine microbes for screening libraries and drug discovery.[19] A single microbial community (macroorganism, plant, or water sample) can contain hundreds of species of microorganisms. Further, a single species can produce many different compounds and often multiple scaffolds from each strain with little impact on the environment. These qualities produce an inherently diverse chemical potential.[34] For the purposes of this chapter, marine microbes will be defined as bacteria and fungi isolated from a marine source, and not just those cultivated with high-salinity media. As discussed later, however, by modifying culture conditions, one can effectively increase the amount of microbial candidates and potentially bioactive compounds produced, therefore creating a larger library for screening. With years of study now under way, marine microbial natural product drug discovery is well entrenched with a rich history providing the means to a promising future.

6.2 METHODS IN MARINE MICROBE LIBRARY DEVELOPMENT

6.2.1 Sources of Microbes

Microbes are everywhere, and developing a fermentable library requires initial investigations of ecosystems from which to isolate and cultivate the organisms. Historically, research in marine environments has focused on microbial communities in the water column or soil sediments. However, scientists have found that only a small fraction of the total amount of marine microorganisms present is culturable, and therefore, only a minute fraction of the potential drug candidates is being studied. Technological advances in genome sequencing, metagenomics, and metatranscriptomics especially have provided the evidence for this sizable diversity.[35,36] Initial estimates documented for free-living bacteria and fungi catalyzed interest in finding sources of other microflora, such as symbionts, holobionts, or other host-associated organisms.[37–39] Advanced techniques such as metabolomics and proteomics can then assist in the identification of unique natural products being produced.[40]

The potential of new species and novel compounds being overlooked has fueled research to develop better cultivation and sampling techniques. The technology for collection methods has made it possible to venture into environments such as deep coral reefs and even the Mariana Trench via SCUBA, remotely operated underwater vehicles (ROVs), or one-man submarines.[41–43] Some of the newer devices engineered for more efficient collecting and sampling include William Fenical's metal probes the "Mud Snapper" and "Mud Missile," designed to collect sediment samples of the ocean floor.[44] These developments and investigations into new ecosystems have proven the microbial biodiversity present in even the most seemingly similar marine environments[45] and have provided diverse microbes from piezophiles in oceanic trenches[46] to psychrophiles in the Antarctic.[47] Some example compounds that came from both a psychrophilic and a piezophilic Gram-positive marine bacterium are the macrolactins, macrolactinic acid, and isomacrolactinic acid, which showed antiviral activity as well as cancer cell cytotoxic properties.[48] The aforementioned methods of collection do not necessarily provide less expensive means of sampling; however, even a small sample collection can provide a large number of diverse microorganisms. Some research groups have also studied sediment and plants in coastal environments as a less expensive, although equally fruitful, endeavor (Figure 6.2). Intertidal plants such as mangroves have been shown to have a rich presence of microbial life, especially their dynamic endophytic bacterial and fungal communities, which hold a promising outlook for screening libraries.[49–51] A variety of other sources and locations have been investigated, and a summary of sources and their respective percentage of biologically active compound-producing microbes can be seen in Figure 6.2 as well as in the review by Leal et al.[52] In addition to finding new locations and strategies for collecting and cultivating microbes like extremophiles, there have been developments in culture media and isolation techniques. These techniques aim to coax out microbes and chemistry that have been missed by classical broad-spectrum isolation strategies.

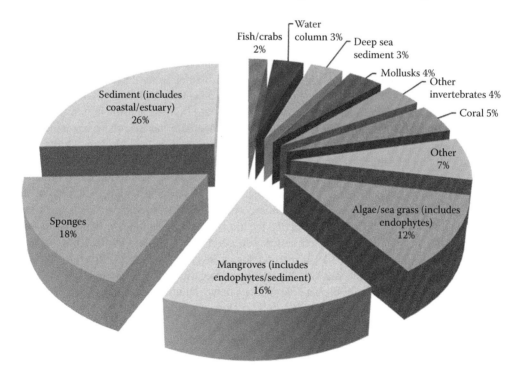

Figure 6.2 Illustration of marine microorganism sources (bacteria and fungi) from which new natural products were reported (2009–2013).[53–57] Deep-sea sediment defined as >100m. "Other invertebrates" includes tunicates, sea cucumbers, zoanthids, sea urchins, jellyfish, bryozoans, and one unknown. "Other" includes microbes isolated from submarine basalts, hydrothermal vents, sea salt fields, floating debris, and other unknown marine sources.

6.2.2 Isolation and Cultivation

The marine environment is teeming with what seems to be an infinite number of microorganism species, though the same microbes are often isolated time and time again.[58,59] Zobell's and other generic marine agars have served as a basis for much of the growth media used for common marine bacteria and fungi, but specialized culture methods have been developed to enhance the diversity of isolates. In kingdom fungi, for example, carbon makes up about 50% of dried fungal mass and is supplied in media as carbohydrates or sugars. The nitrogen needed for peptide biosynthesis is supplemented with amino acids and inorganic nitrogen. In addition, trace elements such as phosphorus, potassium, sulfur, and magnesium are required, and micronutrients such as iron, zinc, copper, manganese, and molybdenum are commonly added.[60] These specifications are generic for fungal cultivation; however, the specific nutrients necessary for diverse bacterial and fungal growth and isolation become increasingly complex as techniques are developed to diversify microbial libraries.

6.2.2.1 Targeting Free-Living Microorganisms

Marine microorganisms are chemically rich organisms with approximately 500 previously undescribed secondary metabolites reported from 1985 to 2008, many of which have pharmaceutical potential.[61] Fungi specifically are responsible for just under 300 of these natural products.[61,62] Fungi are commonly recognized for being consumers of decaying organic matter, but their roles in ecosystems are far broader than this generalization. Marine fungi are genetically diverse,[63] and current estimates suggest that only approximately 0.6% of fungi studied are marine.[64] Specific requirements are necessary for marine fungal isolation. They can be targeted with the addition of antibiotics or antifungals to isolation media, as well as variation of media content to target fast- or slow-growing organisms. Research indicates that the metabolites produced by marine fungi tend to be optimal at 20%–60% artificial seawater, with the remaining percentage consisting of distilled water. Interestingly, unlike some bacteria, terrestrial fungi tend to share these optimal growth conditions with their marine counterparts and, in some cases, produce completely different chemistry at 80%–100% artificial seawater.[62,65] By simply altering isolation media components, one can exponentially increase their screening library output by increasing the biological and, more importantly, chemical diversity of isolates. Optimal isolation conditions for marine microorganisms are under constant revision, developing innovative methods and protocols. The methodologies described here work well for isolating a diverse library of microbes; however, the isolation and cultivation of host-associated microbes have also gained a lot of attention. The techniques to isolate and characterize those microorganisms are described in the following section.

Sediments and the water column are prime locations for the isolation of marine bacteria and fungi. The isolation processes have been outlined to good effect by the Fenical laboratory, with many of the results outlined here.[66] Most methodologies involve drying the sediment first, before diluting it onto nutrient-deficient, seawater-based agar with various antibiotic treatments. Nutrient-rich agar is often detrimental to bacterial colony formation.[41] Dilutions can be done by the addition of sterile seawater to create a suspension, which is then applied systematically to media plates. In addition, dried sediment can be "stamped" using a foam plug multiple times, creating a dilution effect on the petri dish. Similarly, seawater collected from the water column can be diluted to lower individual cell densities before being inoculated onto media plates[67]; this methodology can be applied to isolate both bacteria and fungi.

Recently, research has been put into the cultivation methods utilized to target some of the terrestrial Gram-positive and Gram-negative bacteria that have gained a large amount of attention: actinobacteria and myxobacteria. These prokaryotes have promising chemical potential in marine environments as well; however, they tend to require meticulous handling and specified isolation and growth conditions.[68] Since these microorganisms are found mostly in soils, they can be found from sediments on shorelines all the way to the deep ocean sediments in the Mariana Trench.[69] Specifically for these types of bacteria, the coastal sediments have been found to produce a greater abundance of organisms and molecules similar to those of their terrestrial ancestors.[59] This probability makes alternative exploration of marine habitats such as deep oceans and extreme environments of potentially higher priority. As of 2005, more than 45% of all bioactive microbial metabolites have been discovered from the order Actinomycetales, 74% of which were just from the genera *Streptomyces* and found primarily in sediment samples.[70] Present in marine and terrestrial environments alike, the task of finding the right culture media for these chemically rich microorganisms in marine environments has become exceedingly necessary.

Marine-evolved actinomycetes, more specifically, *Streptomyces*, *Salinispora*, and *Micromonospora*, are nutrient specific and, as anticipated, always require higher ionic sodium content than terrestrial species. Taxonomically novel marine species continue to be discovered with chemistry containing unique carbon scaffolds as a result of research efforts with an increased focus on creating new isolation methods, and there are plenty more still left undiscovered.[71,72] Research groups continue to develop their own isolation and culture methods, and some of the first method alterations implemented were the inclusion of ground-up portions of marine macroorganisms in the media to select for microbial species that may have been symbionts or foulants on such organisms.[73]

More recently, the *Salinispora* species were discovered to produce salinosporamides, such as salinosporamide A (**5**), and their cultivation was found to be highly dependent on the salinity of the culture media. Currently in phase I clinical trials, salinosporamide A is a proteasome inhibitor that was originally isolated from a halophilic actinobacterium, *Salinispora tropica*, which thrives in high salt conditions. This provides an excellent example of an actinobacterium that, so far, exists exclusively in the marine environment and required specialized isolation media for its discovery.[68,74,75]

Rare marine myxobacteria also show affinity toward specific salt content in their media components and are reported to thrive at around 2%–3% salt content,[76] with their growth being completely restricted when salt is not present.[77] They have also been shown to flourish in media enriched with antibiotics and vitamins, such as cyclohexamide and cyanocobalamin, producing compounds such as the antibiotic haliangicin (**6**).[76,78] The isolation of myxobacteria is specifically intriguing since they have a different process for producing secondary metabolites than most other microorganisms. Myxobacteria have been shown to biosynthesize these compounds in the beginning of their growth cycle,[79] making the initial media composition highly important for compound variability and later stage modification less likely to be the determining factor. Despite their demonstrated promise for drug discovery in the terrestrial ecosystems, isolation from marine sources has yet to be as highly developed.

Although it has yet to be proven whether microscopic marine life will be more chemically prolific than terrestrial counterparts, the diversity of niche microbial organisms such as piezophiles, thermophiles, and halophiles points to new chemical diversity. Targeting niche environments is a good way to isolate biologically diverse microorganisms, but the environment being sampled must always be carefully considered when preparing isolation protocols. Like *S. tropica* and salinosporamide A, the macrolactins, a class of compounds discovered in Fenical's laboratory, were produced variably, depending on the culture methods used during isolation. Temperature and salinity, however, were held constant for the psychrophile to produce the desired compounds, with the liquid media consisting of 75% filtered seawater and 25% distilled water held at 20 °C.[48] The identification of the Gram-positive bacterium was not possible at the time; however, the most abundantly produced compound, macrolactin A (7), has shown activity as an antibiotic and antiviral and, more prominently, is cytotoxic against melanoma cancer cells.

6.2.2.2 Targeting Host-Associated Microorganisms

In the search for microbial diversity, host-associated fungi and bacteria represent a largely unstudied population. Hawksworth and Rossman[80] estimated that out of 1.5 million fungal species, only 70,000 have been described, most of which are hypothesized to be the "unculturables" or undiscovered endophytes. An endophyte can be bacterial or fungal and lives at least part of its life cycle within its host organisms, such as plants or algae, without causing harm to the said host.[81] A potent antileukemic and antitumor microtubule-stabilizing compound, paclitaxel (Taxol™), was originally isolated from the bark of the Pacific yew tree, *Taxus brevifolia*.[82] Low yields of the natural product required large quantities of bark to be collected, often killing the trees, thus causing destruction of the native ecosystems and outcries from activists. Exploration of the needles and leaves of *T. brevifolia* provided greater yields, but a more renewable source was needed. The fungal endophyte *Taxomyces andreanae* was isolated from *T. brevifolia* bark, and after incubation, paclitaxel was confirmed in the purified extract.[83] As the research of Taxol proves, successful isolation of endophytes can in some cases alleviate ecological stresses caused by natural product research by providing a renewable source of compounds. As previously mentioned, the pharmacological potential of diverse host-associated microorganisms is not isolated from the terrestrial environment.[4,84] Therefore, further investigations into marine symbiotic and endophytic microbes are favorable for future endeavors in drug discovery.

In order to avoid the isolation of water- or airborne microbial contaminants, surface sterilization is one of the major procedural differences between isolation of free-living and host-associated microorganisms. Kjer et al.[85] developed a sterilization protocol in which a small piece of host tissue (1 × 1 cm) is cut and rinsed with sterile seawater in triplicate and then submerged in ethanol for 30 s. One last rinse with sterile seawater is done before transferring the piece aseptically to a media plate so that the cut edge is in contact with the agar of the collection plate. This procedure is compatible with both fungi and bacteria, but the type of media chosen can selectively grow either microbe. When isolating microbes from a biologically diverse sample, sugars should be limited, which favors the growth of rapidly germinating fungi[86] that can quickly overtake a collection plate. Laboratories have had success adding sub-IC$_{50}$ (concentration that inhibits binding or activity by 50%) toxicity-level doses of antifungals and antibacterials to the media postautoclaving.[87] Antifungal examples include cyclohexamide, nystatin, and cyclosporine to suppress expeditious growing fungi, and the antibacterials (commonly chloramphenicol) can be added for higher selectivity toward fungal growth.[85]

Once colonies are observed, the isolation of fungal strains occurs through the removal and transfer of the hyphal tips, and for bacterial isolates, a streaking procedure using a sterile loop onto individual agar plates. Once a pure colony is isolated, live cells can be archived from the agar plates into tubes of glycerol (15%) at −70 °C[88] for a period of months or years, while working cultures can be maintained on agar slants for 6 months to a year.

One of the major challenges facing this approach is that it is nontargeted beyond a bacterial versus fungal distinction. This attribute tends to lead to continuous reisolation of common genera of microbes, which may lead to less chemical diversity within your library. Classical identification of microbes often involved growing the unknown microbe on various growth media and observing the microbe's morphology. Other methods involve spore, fruiting body, or mycelium identification, carbon utilization, sugar or lipid composition, and colony morphology. All these techniques are life stage dependent and may require taxonomical expertise to distinguish between structurally similar organisms. With the development of genetic techniques like polymerase chain reaction (PCR) amplification of bacterial 16S rDNA[89] or fungal 18S rDNA, 28S rDNA, or internal transcribed spacer (ITS),[90] followed by sequencing, dereplication by genetics is much more accessible. These genetic data may be compared to databases like the Basic Local Alignment Search Tool (BLAST), GenBank, European Molecular Biology Laboratory (EMBL), or DNA Data Bank of Japan (DDBJ) for a genetic match to an already described species or to an internal genetic library to avoid duplication in a collection library.

In the quest for secondary metabolite diversity to fill a screening library, genetic identification may not always be the proper dereplication method; therefore, microbial isolates may also be dereplicated via metabolite comparison.[91] A single species, isolated from different locations, may produce completely different metabolite profiles, as has been proven for marine and terrestrial macroorganisms.[92] Therefore, mass spectrometry methods that show a snapshot of the metabolite profile may be a more useful method of avoiding reisolation of known chemistry.

6.2.3 Exploiting Microbial Secondary Metabolism

Increased understanding of microbial genomics has opened the door to new possibilities for microbial fermentation and drug discovery. Genome sequencing of even a small fraction of microorganisms is enough to realize that they have the potential to produce a far greater number of secondary metabolites than have been discovered, and we now know that the genes responsible for fungal secondary metabolism are most often clustered together.[93] Further, it is clear that there are many silent or underexpressed gene clusters present within the fungal genome. These are genes whose biotic or abiotic regulation triggers are missing under normal laboratory growth conditions, causing their products to go undetected. As the field of microbial drug discovery advances, attention has been directed at discovering these gene clusters and triggering cryptic biosynthetic pathways to produce previously undescribed secondary metabolites.[94–96]

6.2.3.1 Culture-Independent Techniques to Promote Compound Discovery

According to terminology presented by Gross,[97] "silent" gene clusters are those that are not expressed or are underexpressed, and "orphan" gene clusters are those for which the metabolite product is unknown. Expressing silent gene clusters is the goal of culture-dependent genome mining strategies like epigenetics, co-culture, the OSMAC (one strain many compounds) strategy, and ribosome engineering (in prokaryotes). Elucidating the products of orphan gene clusters is the goal of culture-independent strategies like gene-directed or structure-based gene investigation, the genomisotopic approach, and proteomic approaches. Many direct genomic manipulation strategies (insertion of inducible promoters, heterologous expression, etc.) can achieve either goal, depending on which piece of the puzzle is known.[96] For all of the genomic advancements that have been made, there are an equal number of challenges that still need to be met. Nevertheless, advancements in this field continue to be at the forefront of microbial drug discovery, and when information from both genomics and culture techniques is combined, informed decisions can be made to optimize the discovery and production of novel secondary metabolites.[98]

6.2.3.1.1 Gene-Directed or Structure-Based Gene Targeting

Whole-genome sequencing of a microorganism allows a unique view of natural product biosynthesis. By knowing the gene clusters and the compounds they code for, researchers can predict the metabolites produced once that gene is transcribed and translated. This genome-up approach allows investigators to perform a directed search for a certain molecule or class of molecules in the organism's metabolome. This can be a successful tactic for uncovering silent gene clusters, as culture conditions can then be optimized with a specific molecular target (or group of targets) in mind.[94]

This genomic screening method afforded Scherlach and Hertweck[99] a suite of new prenylated quinolone alkaloids, the aspoquiniolones from *Aspergillus nidulans*. While scanning the microbe's genome sequence for noteworthy pathways, a gene cluster that looked like it, coded for a protein similar to anthranylate synthases, was uncovered. Knowing that this enzyme's primary function is the production of anthranilic acid (a precursor of the amino acid tryptophan), it was peculiar that this microbial genome featured three of these genes. Investigators hypothesized that these genes were involved in secondary metabolite production, specifically that of alkaloids such as quinazolines and quinolines. A metabolite investigation of a variety of optimized growth conditions, focusing on the presence of nitrogen-containing compounds, produced four novel prenylated quinolone-2-one alkaloids (e.g., aspoquinolones A and B [**8** and **9**]).[99]

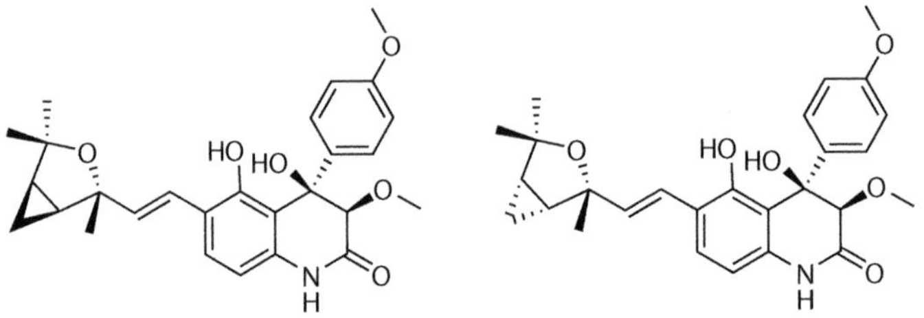

8   9

The vast majority of marine microbial genomes, however, have yet to be sequenced. This is especially true for marine symbionts, which in many cases may be responsible for the production of the elucidated secondary metabolites from marine macroorganisms, but are unknown or unculturable under laboratory conditions. One way to approach microbial drug discovery of unsequenced genomes is by searching the entire metagenome of an organism for already discovered natural product pharmacophores.[96]

Isolated from a marine sponge, psymberin (**10**) is a potent antitumor agent. Present in extremely small quantities (<1 × 10^{-4}%, wet sponge) in the macroorganism,[100] this polyketide was believed to be derived from a bacterial symbiont. Fisch and coworkers[101] demonstrated that by studying the structure of a polyketide (such as psymberin), its corresponding gene cluster can be predicted. Presenting a novel PCR-based method to target structure-correlated ketosynthase (KS) domains in a metagenome, the group successfully elucidated the entire biosynthetic pathway of psymberin. With the ability to both predict the polyketide synthase (PKS) products and identify the responsible PKS genes of an already known metabolite in a sponge symbiont metagenome, it is now possible to discover and study marine symbiont natural products even without the ability to culture them in the lab.[101] An increased understanding of the biosynthetic pathways of natural product structures can allow directed metagenome screening for genes that are similar or identical to those responsible for the structures of interest.[96] Isolation of these genes or gene clusters and expression in culturable hosts can then allow for sufficient production for drug development efforts, even in the absence of a culturable producer or fully sequenced producer genome.[102]

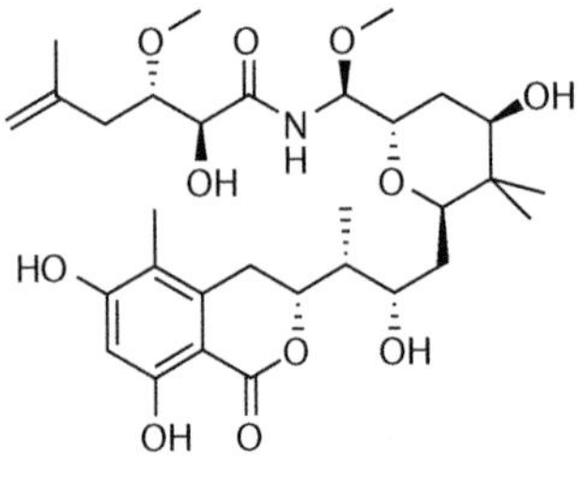

10

6.2.3.1.2 Genomisotopic Approach

As gene sequencing has become easier and cheaper and the resulting sequences have become readily available through public databases, many orphan gene clusters have been identified that have enormous potential for novel natural product discovery. To capitalize on this, a genomisotopic approach was developed by Gross et al.[103] in which bioinformation about the orphan gene cluster can be utilized to isotopically label and identify the products of the cluster.[96] In their 2007 debut of this approach, the team studied a *Pseudomonas* sp. orphan gene sequence and predicted the non-ribosomal peptide synthetase (NRPS) production of a cyclic lipopeptide (CLP). Through studying the nucleotide sequence of the NRPS, the amino acids comprising the product could be predicted. Isotope-labeled predicted amino acid precursors were then fed to the organism, and their incorporation into the biosynthetic pathway allowed the identification of the novel antimicrobial natural product orfamide A (**11**).[103] This technique, however, requires knowledge of the orphan gene cluster, as well as high enough expression of the resulting natural product to be isolated and elucidated.

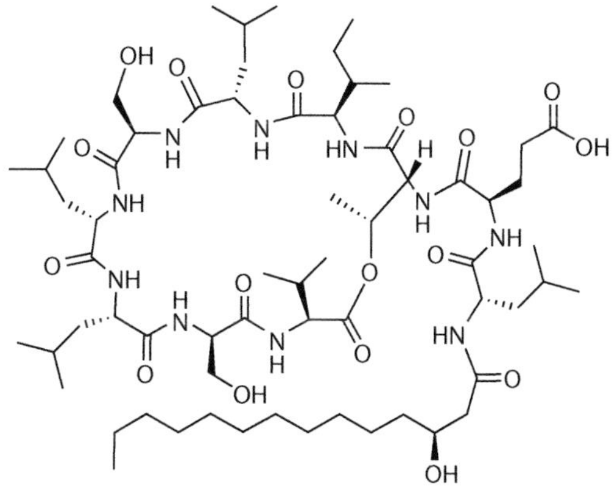

11

6.2.3.1.3 Proteomic Approach

It is well known that PKSs and NRPSs are responsible for a majority of secondary metabolite production in microorganisms. These high-molecular-weight proteins can be easily targeted by proteomic approaches. That is, in a protein-first approach, NRPSs and PKSs can be identified from an organism in an effort to help elucidate the biosynthetic pathways and products of orphan gene clusters.[96] The Proteomic Investigation of Secondary Metabolism (PrISM) method, developed in 2009, is one such approach. Complementary to bioassay-driven and genomic approaches, this method uses mass spectrometry to detect the high-molecular-weight products of expressed NRPS and PKS gene clusters. These NRPSs and PKSs can then be examined for known biosynthetic pathways and products. Screening of undescribed microbial proteomes (i.e., unsequenced environmental isolates) using this method can lead to the discovery of otherwise undetected secondary metabolites.[104] Together with the increasing number of PKS and NRPS probes, proteomic approaches are proving to be a valuable tool in microbial drug discovery.[96]

6.2.3.1.4 Direct Genomic Manipulation

A more invasive genetic approach can also be taken in fungi for which the genome has been fully sequenced. Bioinformatics can reveal if the biosynthetic gene clusters of an organism are controlled by a single transcription factor. If the organism is able to withstand genetic manipulation, an inducible promoter can be introduced to control the production of the transcription factor. The promoter can then be overinduced, leading to an abundance of transcription factors that will transcribe and upregulate the gene cluster of interest.[96,105] This technique may only be applicable to some fungi; however, if enough is known about the genome to understand the presence of the specific promoter and transcription factor, the other techniques that have been discussed may be more useful.

Gene cluster expression in a heterologous host organism is another option for elucidating the products of silent or orphan gene clusters. This approach requires well-understood genomes of both target and host organisms, but has the benefit of definitively associating the secondary metabolite with its producing gene cluster.[96]

6.2.3.2 Culture-Dependent Techniques to Promote Chemical Production

The microbial genome can be exploited with or without the full gene sequence. There are many cultivation techniques that have been proven to uncover previously silent or downregulated secondary metabolites. Often quite simple in nature, these culture methods can serve to increase the size of an extract library exponentially, often producing compounds that would have never been discovered under normal laboratory conditions.

6.2.3.2.1 Epigenetic Modification

For fungal strains (eukaryotes) for which a completed genome sequence may not be available, epigenetic modification has been shown to be an effective method of triggering previously silent biosynthetic pathways.[94,106–108] It has been well accepted that secondary metabolites found in filamentous fungi are coded for by genes that are clustered within the genome,[93] but the exact evolutionary advantage to this was initially up for debate. The discovery of the nucleic *LaeA* protein was the first example of location-based transcription regulation machinery. Through the characterization of this protein, it was realized that *Aspergillus* spp. feature chromatin-based regulation of gene expression.[109] If gene regulation is location based, it logically follows that similar genes would be located together to benefit from a single regulation mechanism. *LaeA*, a broad-based transcriptional factor, is believed to be a protein methyltransferase similar to histone methyltransferases. After the discovery of *LaeA*, it was hypothesized in 2007 by Keller and coworkers[110] that clustering in all fungal species is due to conserved regulatory machinery for these genes. With the hypothesis of chromatin remodeling being the primary method of secondary metabolite regulation, histone modification enzymes, mainly the histone deacetylases (HDACs), were studied in fungi. It was discovered that two out of three well-studied secondary metabolite gene clusters (sterigmatocystin, penicillin, and terraquinone A) exhibited increased secondary metabolite production following the genetic knockout of the gene cluster *hdaA*, which codes for the organism's HDAC. Further studies revealed that, in fact, small-molecule HDAC inhibition (rather than gene knockout) was a viable modification technique to increase secondary metabolites. Two well-studied fungi were treated with a known HDAC inhibitor, and the resulting metabolome of each included a number of upregulated and unidentified secondary metabolites. This revolutionary work showed that chromatin remodeling, specifically histone acetylation and deacetylation, plays a pivotal role in the regulation of secondary metabolite biosynthetic pathways and paved the way for epigenetic modification of other filamentous fungi.[110]

Following this discovery, Williams et al.[107] sought to prove the effectiveness of small-molecule epigenetic modifiers at revealing previously silent gene clusters and enhancing secondary metabolite production. In their first study, 12 diverse fungal strains were cultured in the presence of various epigenetic modifiers where 11 of the 12 strains studied showed responses to at least one of the small-molecule modifiers. Further proof that epigenetic modifications were an effective and exclusive producer of certain natural products was shown when fungal strains were grown under varying culture conditions. No variation of conditions altered the metabolome as notably as the epigenetic modifications. Additionally, strains were grown in the presence of amphotericin B, cycloheximide, and 5-fluorouricil. These conditions did not produce new or increased secondary metabolites, illustrating that the small-molecule modifiers were restructuring the metabolome through a mode of action that was different from a general cytotoxic response.

A marine-derived fungus, *Cladosporium cladosporioides*, was chosen for larger-scale investigation of such epigenetic metabolic manipulation. One aspect of the study included treatment with the DNA methyltransferase (DNMT) inhibitor 5-azacytidine and the other treatment with the HDAC inhibitor suberoylanilide hydroxamic acid (SAHA). This fungus demonstrated dramatic metabolite restructuring under each condition. With 5-azacytidine, several cell signaling molecules were produced, oxylipins **12–14**. The HDAC inhibitor SAHA produced a set of perylenequinones, including two new metabolites, cladochromes F (**15**) and G (**16**). This study proved that small-molecule tailoring of fungal strains is a viable strategy to express silent biosynthetic pathways and represents an easy, cheap, and quick tool for creating screening libraries by assessing the true biosynthetic potential of fungal isolates.[107]

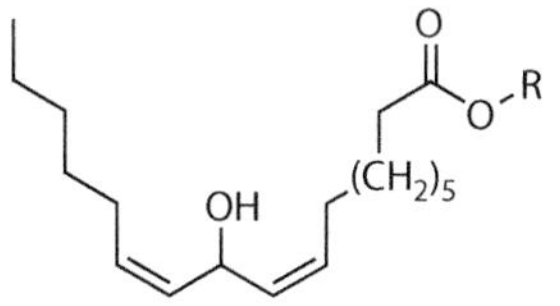

12 R = H
13 R = CH$_3$
14 R = CH$_2$-CHOH-CH$_2$OH

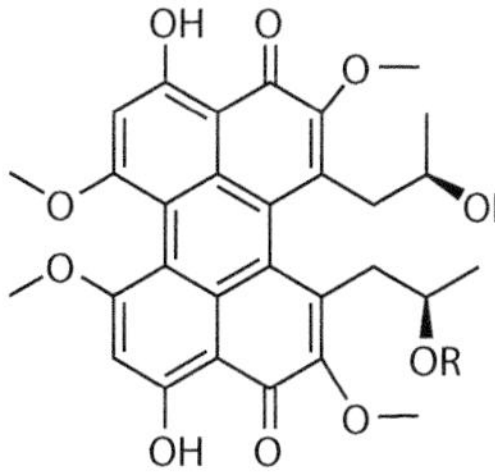

15 R = *p*-hydroxybenzoate
16 R = *p*-hydroxycarbonate

Utilizing this same strategy to uncover previously undescribed compounds, Beau et al.[106] demonstrated the importance of epigenetic tailoring of fungal metabolites for drug discovery. The mangrove-derived fungus *Leucostoma persoonii* was found to produce secondary metabolites, including cytosporones B (**17**), C (**18**), and E (**19**), which were moderately active against the malaria parasite, *Plasmodium falciparum*, and had activity against methicillin-resistant *S. aureus* (MRSA). Interested in titer improvement, and the possibility of discovering additional secondary metabolites, Beau et al. undertook an epigenetic modification study of *L. persoonii*. Sodium butyrate and 5-azacytidine were utilized as HDAC and DNMT inhibitors, respectively, which were implemented in dose–response experiments. Following optimized epigenetic treatment with the HDAC inhibitor, production of **17–19** was increased, and a new derivative, cytosporone R (**20**), was identified.[106]

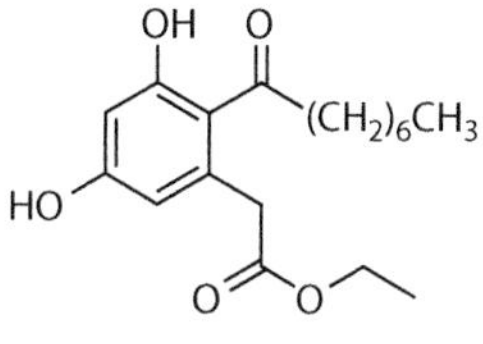

17 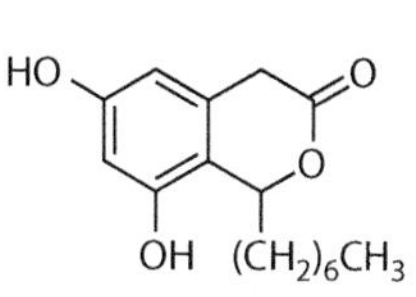   **18** 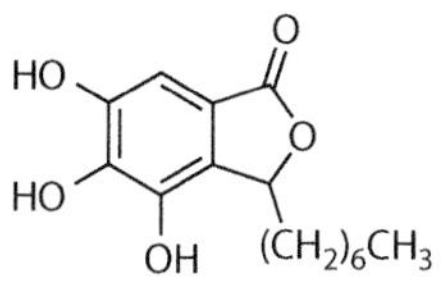   **19** 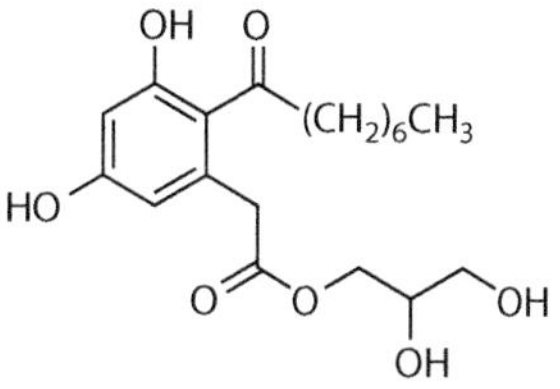   **20**

It has been shown that in lab culture conditions, more than 70% of *Aspergillus niger* PKS, NRPS, and hybrid PKS-NRPS (HPN) encoding genes are silenced under classical culture techniques. When exposed to epigenetic modifiers 5-azacytidine and SAHA, a majority of these silenced gene clusters were upregulated.[111] With this many silent or downregulated gene clusters identified

Table 6.1   Common Epigenetic Modifiers Used in Fungal Culture and Some of the Proposed Targets

Name	Proposed Targets
Sodium butyrate	HDAC2, HDAC3, HDAC4[112]
Suberoyl bis-hydroxamic acid (SBHA)	HDAC
Suberoylanilide hydroxamic acid (SAHA) (Vorinostat)	HDAC1, HDAC3[113]
Trapoxin B	HDAC1, HDAC4[114]
Trichostatin A	HDAC1, HDAC4, HDAC6[114]
5-Azacytidine	DNMT[115]
5-Aza-2′-deoxycytidine	DNMT[115]
RG-108	DNMT[116]

and successfully upregulated by epigenetic modification in one of the most studied fungi on the planet, it is easy to imagine the diversity of new and novel natural products that await epigenetic modification in the realm of marine microbes. As demonstrated, epigenetic modification is useful for the full metabolomic investigation of a single strain, but it can also be useful for large-scale screening efforts. Using just a single small-molecule modifier, fungal collections grown under both control and modified conditions effectively double the extract sample set available for bioassay. Metabolite profiling using nuclear magnetic resonance (NMR) spectroscopy and high-resolution mass spectrometry can indicate samples in which the epigenetic modification was successful in restructuring the metabolome. These profiles, combined with bioassay screening results, can then be scrutinized for scale-up candidates with the highest potential for novel compounds. Commonly used modifiers and their respective targets are outlined in Table 6.1. Although researchers continue to debate the specific mode of action followed by many of these molecules, some of the proposed targets in literature have been listed.

6.2.3.2.2 Other Culture Additives

Small-molecule manipulation of transcriptional machinery (DNMT and HDAC) is just one way to coax out new or upregulated natural products from microorganisms. Oligosaccharides, polysaccharides, enzymes, and even solvents and heavy metals have been used to manipulate microbial metabolomes.[117] Particularly with an understanding of the biosynthetic pathways of certain natural product scaffolds or precursors, any molecule that can affect these pathways can be used to elicit secondary metabolite production. Unlike the use of DNMT or HDAC inhibitors, however, sometimes the mechanism of action of such additives is not precisely known. Further, these results may be similar or identical to normal stress responses, which increases the likelihood of rediscovery.

6.2.3.2.3 Abiotic Stressers

Zeeck and coworkers[118] coined the term *OSMAC* as the practice of altering culture techniques of a single microbe to diversify its biosynthetic products. This method can be applied to small libraries to maximize output with respect to resources and financial backing. Simple physical changes such as media components, pH, temperature, aeration, and alterations to the culture vessel can lead to enhancement of the secondary metabolome.[119–121]

Fuchser and Zeeck[119] varied the aeration of culture flasks of *Aspergillus ochraceus* and isolated a suite of pentaketides, including six new compounds. The fungus was grown in broth under varying degrees of aeration, from static to shaking at 300 rpm. Static 1.5 L cultures produced aspinonene (**21**), aspyrone (**22**), triendiol (**23**), *iso*-aspinonene (**24**), dihydroaspyrone (**25**), and dientriol (**26**), whereas aerated 300 rpm exclusively produced aspinolide A (**27**) and aspinolide B (**28**). To provide insight toward the biosynthetic pathways, experiments were conducted over 8 days in an airlift-loop fermenter with ^{13}C-enriched acetate media in an $^{18}O_2$-rich environment,[122] revealing that two different PKSs (PKS_α and PKS_β) were expressed under the culture conditions. While *A. ochraceus* is not of marine origin, this method of culture variation is still a viable method for altering metabolite profiles of marine microorganisms and potentially producing novel, biologically active compounds for screening.

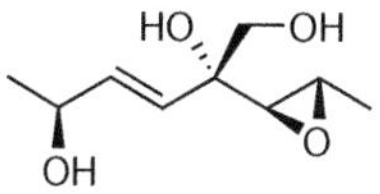

21

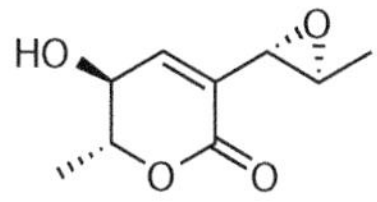

22

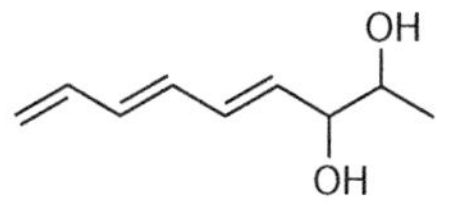

23

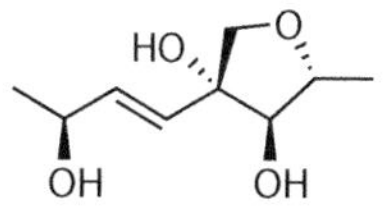

24

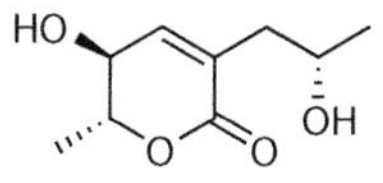

25

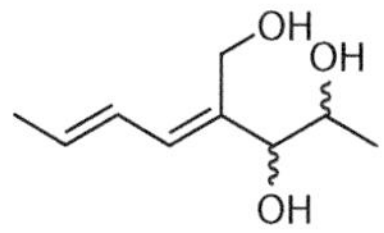

26

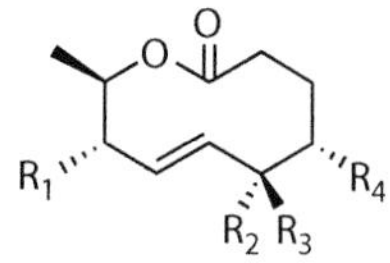

27 R_1 = H, R_2 = OH, R_3 = H, R_4 = H
28 R_1 = O-Acryloyl, R_2 = H, R_3 = OH, R_4 = OH

In an attempt to produce active halogenated metabolites, Nenkep et al.[120] added halide salts into cultures of marine-derived microbes, such as the algae-derived fungus *Fusarium tricinctum*. The fungus was incubated for 10 days on a seawater-based medium after which $CaBr_2$ (50 mM) was added and incubated for an additional 10 days. Thin-layer chromatography (TLC) was used to monitor metabolites in bromine-rich cultures and bromine-free control cultures. The $CaBr_2$ cultures showed considerable differences from the controls and displayed activity against *Staphylococcus aureus*, MRSA, and multi-drug-resistant *S. aureus* (MDRSA), thus prompting further investigation. Purification revealed two new diterpenes, bromomethylchlamydosporols A (**29**) and B (**30**), along with known compounds chlamydosporol (**31**) and fusarielin A (**32**). More recently, Crews and coworkers,[123] successfully isolated two novel chlorinated metabolites, (–)-spiromalbramide (**33**) and (+)-isomalbrancheamide B (**34**), from the marine endophyte *Malbranchea graminicola* on chlorine-rich artificial seawater. Likewise, bromine-enriched media elicited production of two new brominated analogues, (+)-malbrancheamide C (**35**) and (+)-isomalbrancheamide C (**36**).

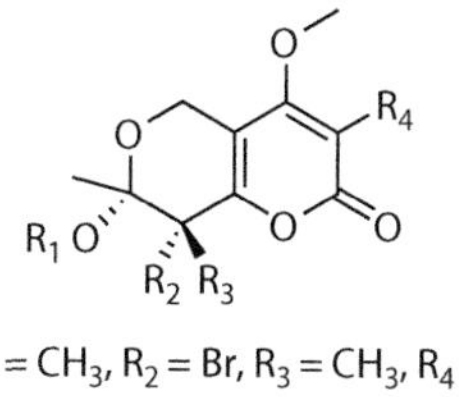

29 R_1 = CH_3, R_2 = Br, R_3 = CH_3, R_4 = H
30 R_1 = CH_3, R_2 = Br, R_3 = CH_3, R_4 = Br
31 R_1 = OH, R_2 = CH_3, R_3 = H, R_4 = H

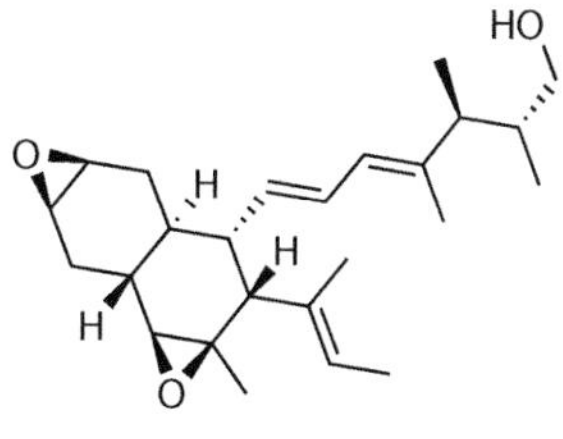

32

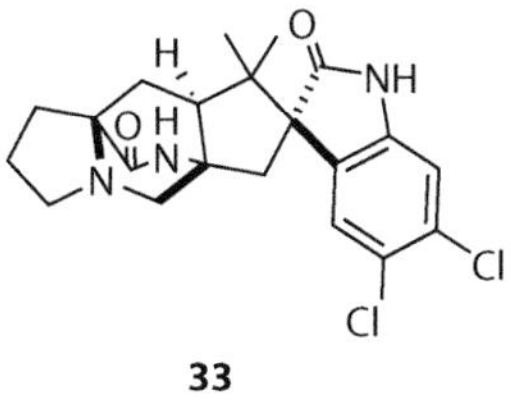

33

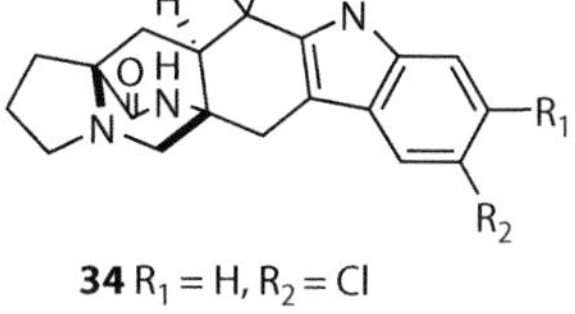

34 R_1 = H, R_2 = Cl
35 R_1 = Br, R_2 = H
36 R_1 = H, R_2 = Br

Upon varying conditions of salinity, temperature, pH, and culture medium composition, Miao et al.[121] examined the growth and bioactivity of a marine-derived fungus *Arthrinium* c.f. *saccharicola*. Growth of the cultures was measured as the weight of fungal biomass, and the extracts were bioassayed against *Pseudoalteromonas spongiae* and *Vibrio vulnificus*. Their findings showed that fungal biomass was enhanced using nonsaline conditions; however, the greatest antibacterial properties were noted at 17 and 34 parts per trillion seawater. Temperature, pH, and nutrient concentration were optimized to reveal the greatest biomass and bioactivity output. Several bacterial cultures of *Laribacter hongkongensis*, *Pseudoalteromonas piscida*, and *P. spongiae* were also sterilized by filtration and co-cultured with *A. saccharicola*, leading to increased antibiotic activity of the culture extracts.

6.2.3.2.4 Physical Culture Variations

New techniques for cultivation of fungi include the use of a sterile cellulose-based "cellophane raft," which sits atop solid media, facilitating growth without the need to extract and separate out media components. This technique was proposed by Fremlin et al.,[124] and it was noted that these rafts upregulated metabolites in two-thirds of the 100 marine fungi tested. A selected strain, *Aspergillus versicolor*, was grown on cellophane rafts on nutrient-rich agar, and the cellophane-containing mycelia were extracted with methanol and purified. The extracts afforded a new alkaloid, cottoquinazoline A (**37**), and two new cyclopentapeptides, cotteslosins A (**38**) and B (**39**), where compounds **38** and **39** were weakly cytotoxic against melanoma (MM418c5), prostate (DU145), and breast cancer (T47D) cell lines.

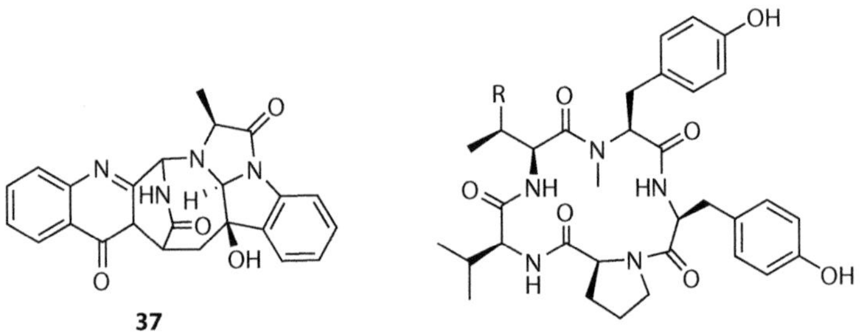

37

38 R= CH₃
39 R= CH₂CH₃

6.2.3.2.5 Co-Culturing

Microbial co-cultures have promising applications for natural product drug discovery. In an environment where competition exists, co-culturing can stimulate the production of secondary metabolites as a defense mechanism. Candidates for co-culturing are often initially discovered on isolation plates by the observation of a distinct zone of inhibition between two neighboring microorganisms. This methodology can increase yields of previously reported metabolites,[125–127] produce analogues of known metabolites,[128,129] and trigger previously unexpressed pathways to yield novel bioactive compounds.[130] Results from this and other culture methods previously mentioned are summarized in Table 6.2.

An algae-derived fungus, *Pestalotia* sp., and an unidentified Gram-negative marine bacterium isolated by Cueto et al.[131] were co-cultured in seawater-based media and monitored for their secondary metabolite production. After 24 h, 10 ml of bacterial broth was inoculated into 1 L of fungal broth and allowed to ferment for another 6 days as monitored by LC-MS. Pestalone (**40**), a potent,

Table 6.2 Summary of Microbial Culturing Methods and Their Corresponding Chemical Compounds (all percentages are quoted in m/v%)

Compound	Source	Conditions	Method
8, 9		Optimized from 40 conditions; rice media produced the most interesting extract	Genomic screening
11	Uncultivated *Psammocinia* aff. *bulbosa* symbiont	N/A	Structure-based gene targeting
12–14	*Cladosporium cladosporioides*	Addition of 5-azacytidine	Epigenetic modification
15, 16	*Cladosporium cladosporioides*	Addition of suberoylanilide hydroxamic acid	Epigenetic modification
17–20	*Leucostoma persoonii*	Addition of sodium butyrate	Epigenetic modification
21–26	*Aspergillus ochraceus*	Malt extract (2%), yeast extract (0.2%), glucose (1%), $(NH_4)_2HPO_4$ (0.05%); pH = 6.0; 21 days	Static aeration
27, 28	*Aspergillus ochraceus*	7 days	Aeration at 180 rpm
29–32	*Fusarium tricinctum*	Soluble starch (1.0%), soytone (0.1%) in seawater; 20 days	Addition of $CaBr_2$
33–36	*Malbranchea graminicola*	NaBr (2.6%), KBr (0.06%), $CaBr_2$ (0.03%), Tris base (0.2%), $MgSO_4 \cdot 7H_2O$ (0.7%), Czapek-Dox (3.5%) in artificial seawater; pH = 7; 21 days	Addition of brominated salts
37–39	*Aspergillus versicolor*	Malt extract (16%), glucose (16%), peptone (0.8%) and agar (2%); 21 days	Cellophane raft
40	*Pestalotia* sp. and unidentified bacterium	Glucose (1%), peptone (0.5%), yeast extract (0.5%), penicillin G/streptomycin sulfate (0.01%) in seawater; 6 days	Co-culture
41–44	*Libertella* sp. and *Pestalotia* sp.	Mannitol (0.4%), yeast extract (0.2%), peptone (0.2%) in seawater; 5 days	Co-culture
45, 46	*Salinispora arenicola* and *Emericella* sp.	Tryptone (0.3%), casitone (0.5%), glucose (0.4%) and mannitol (0.4%), yeast extract (0.2%), peptone (0.2%), respectively, in seawater; 2 days	Co-culture

novel benzophenone antibiotic against MRSA and vancomycin-resistant *Enterococcus faecium* (VRE), was obtained. Production of **40** does not occur in individual control cultures; however, production could be induced in low yields by the addition of ethanol (1% v/v) in a pure fungal culture that had been aged for 24 h. These observations lead the authors to believe that **40** is a product of fungal biosynthesis rather than a modification of a bacterial metabolite. Pestalone (**40**) demonstrated modest activity against National Cancer Institute's (NCI) human tumor cell line screen and displayed nanomolar potency against MRSA and VRE.

In a related study, the same marine bacterium used for the production of **40** was cultured with an ascidian-derived fungus, *Libertella* sp.[132] A 3-day-old bacterium culture was inoculated with 1 ml/1 L of marine broth into similarly aged fungal cultures and allowed to proceed for 2 additional days and was found to produce four new pimarane diterpenes: libertellenones A–D (**41-44**). Efforts to trigger libertellenone production with dead bacterial cells and bacterial extracts in fungal cultures were unsuccessful, indicating that the diterpene biosynthetic pathways of the marine fungus *Libertella* sp. are expressed through cell–cell interactions. None of the libertellenones showed antimicrobial activity; however, **44** displayed cytotoxic activity against the HCT-116 human colon carcinoma cancer cell line.

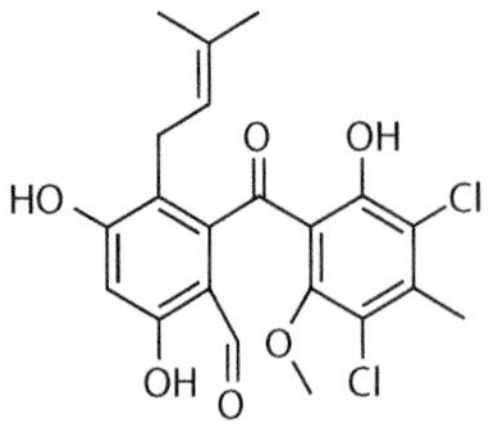

40

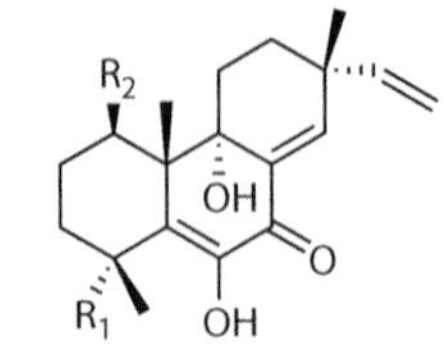

41 R$_1$ = CH$_3$, R$_2$ = OH
42 R$_1$ = CH$_2$OH, R$_2$ = H
43 R$_1$ = CH$_2$OH, R$_2$ = OH

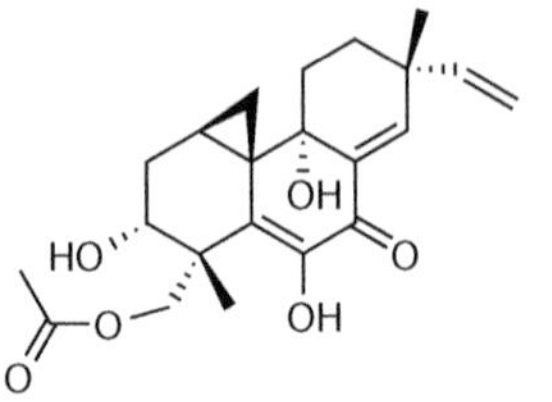

44

A mixed fermentation in marine broth was described by Oh et al.[133] with *Salinispora arenicola*, an actinomycete from a marine sediment sample, and a green algae-derived *Emericella* fungus. On day 3 of single-organism incubation, *S. arenicola* broth was added to the *Emericella* broth, and the cultures were allowed 2 additional days of growth with metabolite production screened by LC-MS. The co-culture elucidated two novel cyclic depsipeptides: emericellamindes A (**45**) and B (**46**). These depsipeptides are present in pure *Emericella* sp. cultures; however, a 100-fold yield increase aided in their characterization from the co-cultures, demonstrating the effective application of this technique in marine microbes. Moderate antimicrobial activity was reported for **45** and **46**, as well as weak cytotoxicity against the human colon carcinoma cell line (HCT-116).

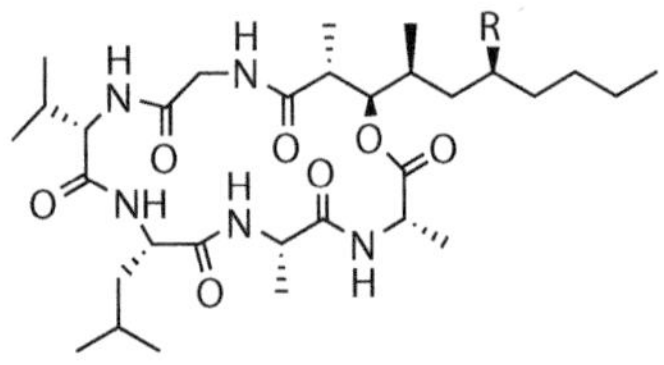

45 R = H
46 R = CH$_3$

6.2.4 Scale-Up Techniques and Screening

Once a microbial entity library has been established, a screening protocol is necessary to culture, extract, and analyze the results. Recently, much work has been done to miniaturize culture volumes, thus efficiently reducing the size and supplies needed for screening large libraries. Since large quantities of material are not needed to screen for bioactivity, these procedures call for small-scale liquid fermentations, solid culture media, or the use of microtiter plates with either liquid or solid media.[134] Academic, small biotech, and large pharmaceutical labs have successfully integrated HTS protocols of microbial libraries to establish bioactive natural products.[134,135]

The use of 96- and 384-well microtiter plates to culture microorganisms has been implemented by several laboratories.[136–138] Gao et al. prepared solid-state microcultures in 300 µl 96-deep-well plates to quantitatively screen *Streptomyces* spp. for avermectin production. Conversely, a binary cultivation system utilizing solid rice substrates and potato starch and glucose liquid media was used to culture fungi and actinomyces.[137] Miniature bioreactor (MBR) plates can be used to normalize and control culture conditions in each individual well. Isett et al.[140] showed that biomass and cellular metabolism for several fungi and bacteria were scalable from a 24-well MBR plate to a 20 L fermenter.

After sufficient incubation, cultures are submerged in an organic solvent to extract metabolites of interest. In liquid cultures, polarity-selective resins such as HP20 or XAD can be added to selectively adsorb metabolites without extracting the media components for a metabolite-specific technique that has been utilized by many researchers.[75,123,132,133,141] Natural product extracts represent a challenge for bioactivity screenings due to the complex mixtures involved, which can lead to interferences and unreliable results.[142,143] Fractionation of extracts can be performed relatively easily by liquid–liquid partition or chromatography,[137,142,144] leading to small mixtures of compounds with similar polarity. These fractions can have fewer interferences, so active compounds in minute quantities can be identified that might have otherwise been missed in a crude extract.[142] These extracts or fractions can then be screened for bioactivity against a selected target.

Eldridge and coworkers[144] developed an ultra-high-throughput method for fractionation of crude natural product extracts by subjecting the initial extract to automated flash chromatography before implementing parallel 4× channel reversed-phase (RP) preparative high-performance liquid chromatography (HPLC). The fractions are then subjected to multichannel liquid chromatography-evaporative light scattering detection-mass spectrometry (LC-ELSD-MS) for simultaneous detection of up to eight samples. These purified fractions are analyzed on a capillary (5 μl) microcoil flow probe NMR spectrometer for structure elucidation. Chromatographic instrumentation can also be coupled directly to an NMR spectrometer (LC-NMR) for even faster dereplication, providing another means for early structural data of active fractions.[145]

Biological screening can be managed in a high-throughput manner, a fact that has been exploited by many major pharmaceutical companies to determine bioactivity against a specific target.[146,147] Bristol-Myers Squibb embarked on finding compounds with distinct tubulin-binding modes, since tumors were becoming resistant to taxane (e.g., Taxol®) treatments. A myxobacterium, *Sorangium cellulosum*, displayed cytotoxic activity, which consequently inspired the semisynthetic ixabepilone, which was selected for preclinical trials.[148] The genomes of microorganisms are constantly being sequenced and made publicly available; GlaxoSmithKlein (GSK) used the complete genome sequencing of the bacterium *Haemophilus influenzae* to look at 160 essential genes as potential biological targets.[149] This prompted GSK to launch 70 HTS campaigns targeting antibiotics with novel activities, but this project was met with limited success. Payne et al.[149] noted the lack of chemical diversity in the GSK synthetic library and that a natural product screening could have generated better results. With the development of new biological assays, a "look-back" program can be implemented, as done by Wyeth, where new antibiotics were found that had been previously overlooked in their microbial library.[150]

Once a hit is established in a small-scale experiment, it requires a scale-up to undergo bioassay-guided fractionation and yield sufficient quantity of the active components to elucidate structures and for bioactivity dosing. This can be problematic for several reasons: the purification and elucidation process is time-consuming, secondary metabolites are produced in low concentrations, and reproducibility can be troublesome. According to the OSMAC approach,[118] changing any parameter, including vessel size, can alter metabolite outputs. Reproducing small-vessel experiments with many replicates is an option, but the convenience of bioreactors or fermenters will allow specific parameters to be tailored to suit experimental needs. Fermentation vessels range in size from 5 L up to 1.2 million L, where temperature, pressure, mixing, air inlet and outlet, feed inlet, and product outlet can all be programmed.[151] Once a biological hit has been identified, optimization of culture conditions and quantification of activity will reveal the proper scale-up protocol.

The production of salinosporamide A (**5**) represents an excellent example that both shake flasks and large-scale fermentations have the potential to be successful. Initially the organic extract of *Salinispora* sp. displayed anticancer activity, thus promping scale-ups in larger shake flasks, yielding **4** at 7 mg/L.[75] Nereus Pharmaceuticals implemented small shake flask experiments of *S. tropica*,[152] which induced production of **5** along with several other analogues.[141] Careful optimization of culture parameters, including adding a solid resin, which stabilizes the β-lactam ring, led to yields of 450, 350, and 260 mg/L in shake flasks, small fermenters, and industrial-sized fermenters, respectively.[153] Papagianni[154] proposed a scale-down method in which regime analysis identifies rate-limiting mechanisms at the laboratory scale that can solve issues when scaling up. The scale-up of an active culture must be exhaustively optimized before undertaking the time and resources to culture large quantities of a microbe.

6.3 CURRENTLY APPROVED DRUGS OF MICROBIAL ORIGIN

For the purposes of this section, only those approved drugs that originate from microbial sources will be considered; however, there are numerous examples of currently available drugs that are inspired by natural products, natural product mimics, or semisynthetic compounds.[1] In Table 6.3, we present the currently approved drugs of microbial origin. These organisms are largely terrestrial and primarily isolated from the genus of the group Actinobacteria, *Streptomyces*. It is clear from Table 6.3 that there are many compounds, including macrolide, aminoglycosides, and tetracyclines, produced by a range of microbes for the treatment of a variety of human afflictions. This vast range of natural products isolated from terrestrial microbes over the past 70 years suggests the microbes housed by the marine environment represent a largely untapped resource with unlimited potential for the discovery of novel structures and new drugs.

In the process of drug discovery from the natural realm, there exist three paths. The first, as shown in Table 6.3, directly uses the secondary metabolite produced by an organism in the treatment of a disease. The second entails the use of a marine natural product as a lead compound in the development of synthetic analogues, such as in the case of plinabulin (**47**). While there are currently no marine microbial natural products on the market, there is plinabulin, a fully synthetic analogue of the marine *Aspergillus* sp. metabolite halimide (**48**).[155] Compound **47** has successfully completed FDA phase II clinical trials for the treatment of cancer. Plinabulin inhibits tublin polymerization, leading to vascular disruption within the tumor, resulting in the death of the cancer cells.[156] Plinabulin provides an example of how natural products may be used as a lead compound, where bioactivity is increased via a comprehensive structural–activity relationship (SAR) study.

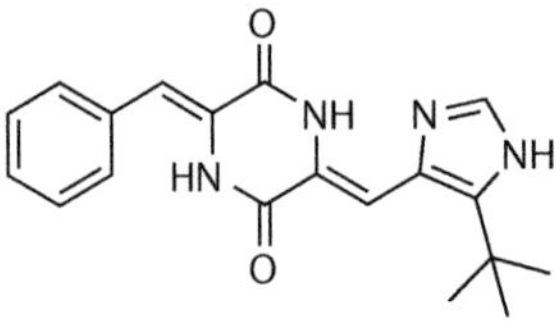

47 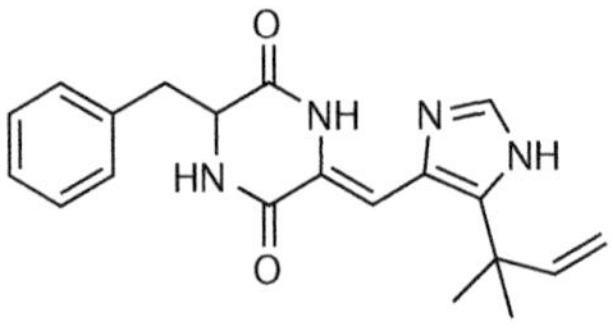   **48**

Table 6.3 List of Approved Drugs from Microbial Origin

Compound	Common Trade Names	Source	Year Approved	Treatment
Streptomycin/ tetracycline	Plantomycin®	*Streptomyces griseus*	1943	Antibacterial
Folinic acid	Leucovorin	*Leuconostoc citrovorum*	1950	Anticancer
Chlortetracycline	Aureomycin®	*Streptomyces aureofaciens*	1950	Antibiotic
Hygromycin B	Hyanthelmix	*Streptomyces hygroscopicus*	1953	Antibiotic/ antihelminthic
Carzinophilin	Carzinophilin A	*Streptomyces sahachiroi*	1954	Anticancer
Sarkomycin	Sarcomycin	*Streptomyces erythrochromogenes*	1954	Anticancer

(Continued)

Table 6.3 (Continued)   List of Approved Drugs from Microbial Origin

Compound	Common Trade Names	Source	Year Approved	Treatment
Mitomycin C	Mitomycin®	*Streptomyces lavendulae* and *Streptomyces caespitosus*	1956	Anticancer
Tetracycline	Sumycin/Hostacycline®	*Streptomyces aureofaciens*	1957	Antibiotic
Vancomycin	Vancocin®	*Amycolatopsis orientalis*	1958	Antibiotic
Paromomycin	Humatin®	*Streptomyces krestomuceticus*	1960	Antibiotic
Demeclocycline	Declomycin®	*Streptomyces aureofaciens*	1960	Antibiotic
Mithramycin	Mithracin®	*Streptomyces plicatus*	1961	Anticancer
Chromomycin A3	Toyomycin	*Streptomyces griseus*	1961	Anticancer
Gentamicin	Garamycin®/Cidomycin	*Micromonospora purpurea*	1963	Antibiotic
Actinomycin D	Dactinomycin	*Streptomyces parvullus*	1964	Anticancer
Bleomycin	Blenoxane®	*Streptomyces verticillus*	1966	Anticancer
Doxorubicin	Adriamycin®	*Streptomyces peucetius*	1966	Anticancer
Daunomycin	Cerubidine®	*Streptomyces peucetius*	1967	Anticancer
Josamycin	Josacine®	*Streptomyces narbonensis*	1970	Antibiotic
Spectinomycin	Trobicin™	*Streptomyces spectabilis*	1971	Antibiotic
Lincomycin	Lincocin®	*Streptomyces lincolnensis*	1971	Antibiotic
Neocarzinostatin	Unigli®	*Streptomyces macromomyceticus*	1976	Anticancer
Netilmicin sulfate	Netromicine	*Micromonospora* sp.	1981	Antibiotic
Peplomycin	Pepleo	*Streptomyces verticillus*	1981	Antibacterial
Aclacinomycin A	Aclarubicin	*Streptomyces galilaeus*	1981	Anticancer
Erythromycin	E-mycin™	*Saccharopolyspora erythraea*	1981	Antibiotic
Micronomicin sulfate	Sagamicin	*Micromonospora sagamiensis*	1982	Antibiotic
Cylcoserine	Seromycin®	*Streptomyces garyphalus*	1982	Antibacterial
Kanamycin	Bekanamycin/Klebicil	*Streptomyces kanamyceticus*	1983	Antibiotic
Tobramycin	Tobrex®/TOBI	*Streptomyces tenebrarius*	1984/2010	Antibiotic/cystic fibrosis
Mupirocin	Bactroban®	*Pseudomonas fluorescens*	1985	Antibiotic
Rokitamycin	Ricamycin	*Streptomyces kitasatoensis*	1986	Antibiotic
Fosfomycin trometamol	Monurol®/Fosfomycin	*Streptomyces fradiae*	1988	Antibiotic
Teicoplanin	Targocid®	*Actinoplanes teichomyceticus*	1988	Antibiotic
Pentostatin	Nipent®	*Streptomyces antibioticus*	1992	Anticancer
Neomycin sulfate	Neo-rx/Neo-Fradin	*Streptomyces fradiae*	1996/2002	Anticancer/antibiotic
Pristinamycin IA and IIA	Synercid®/Pyrostacine®	*Streptomyces pristinaespiralis*	1999	Antibiotic
Rapamycin	Rapamune®	*Streptomyces hygroscopicus*	1999	Immunosuppressant
Spiramycin	Rovamycine®	*Streptomyces ambofaciens*	2000	Antibiotic
Daptomycin	Cubicin®	*Streptomyces roseosporus*	2003	Antibiotic
Romidepsin	Istodax®	*Chromobacterium violaceum*	2010	Anticancer
Polymyxin B	Aerosporin	*Bacillus polymyxa*	2011	Antibacterial

Note: All drugs listed are approved by a regulatory body such as the Food and Drug Administration (United States), European Medicines Agency (EU), and Pharmaceuticals and Medical Devices Agency (Japan).

The final method for exploiting microbes is the use of semisynthesis. Microorganisms may not produce the complete natural product of interest; however, they may produce a structurally similar compound, which could be used as a precursor for the desired bioactive compound. For example, in the case of trabectedin (**3**), two-thirds of the complex chemical structure is biosynthesized by the bacteria producer, *P. fluorescens*, in the form of cyanosafricin B (**49**).[29] Cuevas and Francesch[18] have described a semisynthetic pathway to **3** using cyanosafricin B as a precursor, which has enabled PharmaMar to meet global demands for trabectedin (ca. 5 kg p.a.).

With these examples of natural product-inspired drugs on the market today, despite traditional hurdles in microbial drug discovery efforts, it is easy to imagine that an ever-increasing focus on marine microbe screening libraries will result in many more lead compounds and natural product drugs on the market.

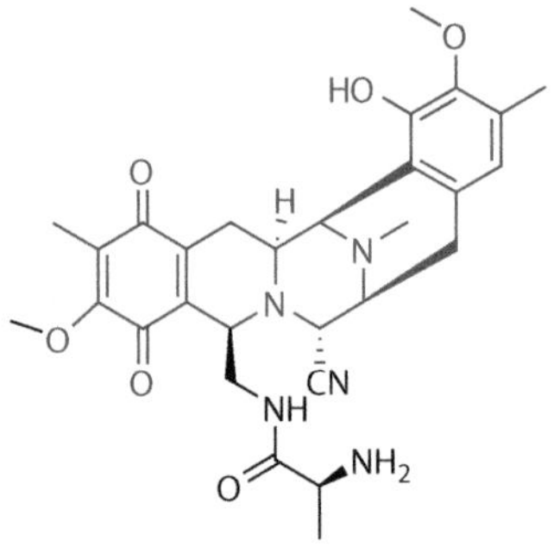

49 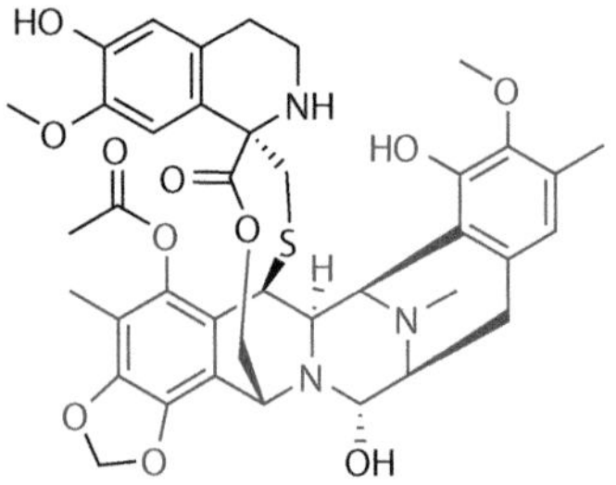   3

6.4 CONCLUSION AND FUTURE PERSPECTIVES

The daunting task of developing a screening library can be greatly diminished by choosing ideal candidates such as fermentable microbes; however, culturing diversity within the taxa isolated, and therefore compound diversity, is still difficult if improperly executed. The diversity of marine microorganisms is impressive, with investigations continuously centralized around this theme.[35,63,64,157] Many industrial and academic screening programs have been developed to combat the astronomical numbers of diseases, viruses, and other health anomalies plaguing people and other organisms around the globe.[134,136–138] Some techniques developed more recently for expanding the chemical scope of microbial libraries include epigenetic modification,[94,106–108] abiotic stressors,[118–121] and co-culturing[125–130]; however, as described in Sections 6.2.3 and 6.2.4, relatively few screening programs have focused on these techniques. As resistance rapidly develops, there is a growing need for successful screening techniques that result in new drug therapies.

Bottlenecks once keeping natural products out of pharmaceutical screening projects have been reduced by using fermentable libraries in addition to developments in chemical dereplication databases (*Dictionary of Natural Products*, Antibase, MarinLit, etc.).[151,158] With only a small number of microbial biosynthetic pathways elucidated, an even smaller fraction of the genomes fully sequenced, and relatively few secondary metabolites isolated, the opportunities are endless. The characteristics of marine microbes and the techniques described in this chapter are a few of the many reasons these organisms are desirable for screening.[151,159,160] Their abundance and record of producing biologically active novel metabolites alone prove that prosperous futures lie ahead for marine fermentable drug discovery.

REFERENCES

1. Newman, D. J., and G. M. Cragg. 2012. Natural products as sources of new drugs over the 30 years from 1981 to 2010. *J. Nat. Prod.* 75 (3):311–35.
2. Glaser, K. B., and A. M. Mayer. 2009. A renaissance in marine pharmacology: From preclinical curiosity to clinical reality. *Biochem. Pharmacol.* 78 (5):440–48.
3. Scheuer, P. J. 1990. Some marine ecological phenomena: Chemical basis and biomedical potential. *Science* 248 (4952):173–77.
4. Piel, J. 2009. Metabolites from symbiotic bacteria. *Nat. Prod. Rep.* 26 (3):338–62.
5. Zobell, C. E., and E. C. Allen. 1933. Attachment of marine bacteria to submerged slides. *Exp. Biol. Med.* 30:1409–11.
6. Zobell, C. E. 1941. Studies on marine bacteria I. The cultural requirements of heterotrophic aerobes. *J. Mar. Res.* 4:42–75.
7. Korinek, J. 1927. *Zentr. Bakt. Parasitenk. II* 71:73–79.
8. Rosenfeld, W. D., and C. E. Zobell. 1947. Antibiotic production by marine microorganisms. *J. Bacteriol.* 54 (3):393–98.
9. Lovell, F. M. 1966. The structure of a bromine-rich marine antibiotic. *J. Am. Chem. Soc.* 88 (19).
10. Burkholder, P. R., R. M. Pfister, and F. H. Leitz. 1966. Production of a pyrrole antibiotic by a marine bacterium. *Appl. Microbiol.* 14 (4):649–53.
11. Brotzu, G. 1948. Research on a new antibiotic. *Lav. Inst. Igiene Cagliari* 1–11.
12. Abraham, E. P., and G. G. Newton. 1954. Purification and some properties of cephalosporin N, a new penicillin. *Biochem. J.* 58 (1):94–102.
13. Abraham, E. P., and G. G. Newton. 1961. The structure of cephalosporin C. *Biochem. J.* 79:377–93.
14. McIntosh, J. M., G. O. Corpuz, R. T. Layer, et al. 2000. Isolation and characterization of a novel conus peptide with apparent antinociceptive activity. *J. Biol. Chem.* 275 (42):32391–97.
15. Rinehart, K. L., T. G. Holt, N. L. Fregeau, et al. 1990. Ecteinascidins 729, 743, 745, 759A, 759B, and 770: Potent antitumor agents from the Caribbean tunicate *Ecteinascidia turbinata*. *J. Org. Chem.* 55 (15):4512–15.
16. Wright, A. E., D. A. Forleo, G. P. Gunawardana, S. P. Gunasekera, F. E. Koehn, and O. J. Mcconnell. 1990. Antitumor tetrahydroisoquinoline alkaloids from the colonial ascidian *Ecteinascidia turbinata*. *J. Org. Chem.* 55 (15):4508–12.
17. Munro, M. H., J. W. Blunt, E. J. Dumdei, et al. 1999. The discovery and development of marine compounds with pharmaceutical potential. *J. Biotechnol.* 70 (1–3):15–25.
18. Cuevas, C., and A. Francesch. 2009. Development of Yondelis (trabectedin, ET-743). A semisynthetic process solves the supply problem. *Nat. Prod. Rep.* 26 (3):322–37.
19. Fuhrman, J. A. 2009. Microbial community structure and its functional implications. *Nature* 459 (7244):193–99.
20. Kallmeyer, J., R. Pockalny, R. R. Adhikari, D. C. Smith, and S. D'Hondt. 2012. Global distribution of microbial abundance and biomass in subseafloor sediment. *Proc. Natl. Acad. Sci. U.S.A.* 109 (40):16213–16.
21. Hobbie, J. E., R. J. Daley, and S. Jasper. 1977. Use of nucleopore filters for counting bacteria by fluorescence microscopy. *Appl. Environ. Microbiol.* 33 (5):1225–28.
22. Porter, K. G., and Y. S. Feig. 1980. The use of DAPI for identifying and counting aquatic microflora. *Limnol. Oceanogr.* 25 (5):943–48.
23. Proksch, P., R. A. Edrada-Ebel, and R. Ebel. 2003. Drugs from the sea: Opportunities and obstacles. *Mar. Drugs* 1:5–17.
24. Pettit, G. R., C. L. Herald, D. L. Doubek, D. L. Herald, E. Arnold, and J. Clardy. 1982. Isolation and structure of bryostatin I. *J. Am. Chem. Soc.* 104:6846–48.
25. Kortmansky, J., and G. K. Schwartz. 2003. Bryostatin-1: A novel PKC inhibitor in clinical development. *Cancer Invest.* 21 (6):924–36.
26. Woollacott, R. M. 1981. Association of bacteria with bryozoan larvae. *Mar. Biol.* 65:155–58.

27. Haygood, M. G., and S. K. Davidson. 1997. Small-subunit rRNA genes and *in situ* hybridization with oligonucleotides specific for the bacterial symbionts in the larvae of the bryozoan *Bugula neritina* and proposal of "*Candidatus* endobugula sertula". *Appl. Environ. Microbiol.* 63 (11):4612–16.

28. Molinski, T. F., D. S. Dalisay, S. L. Lievens, and J. P. Saludes. 2009. Drug development from marine natural products. *Nat. Rev. Drug Discov.* 8 (1):69–85.

29. Ikeda, Y., H. Idemoto, F. Hirayama, et al. 1983. Safracins, new antitumor antibiotics. I. Producing organism, fermentation and isolation. *J. Antibiot.* (Tokyo) 36 (10):1279–83.

30. Cuevas, C., M. Perez, M. J. Martin, et al. 2000. Synthesis of ecteinascidin ET-743 and phthalascidin Pt-650 from cyanosafracin B. *Org. Lett.* 2 (16):2545–48.

31. Schmidt, E. W. 2008. Trading molecules and tracking targets in symbiotic interactions. *Nat. Chem. Biol.* 4 (8):466–73.

32. Haygood, M. G., E. W. Schmidt, S. K. Davidson, and D. J. Faulkner. 1999. Microbial symbionts of marine invertebrates: Opportunities for microbial biotechnology. *J. Mol. Microbiol. Biotechnol.* 1 (1):33–43.

33. Zobell, C. E., and E. C. Allen. 1935. The significance of marine bacteria in the fouling of submerged surfaces. *J. Bacteriol.* 29 (3):239–51.

34. Jensen, P. R., and W. Fenical. 1996. Marine bacterial diversity as a resource for novel microbial products. *J. Ind. Microbiol.* 17:346–51.

35. Kennedy, J., B. Flemer, S. A. Jackson, et al. 2010. Marine metagenomics: New tools for the study and exploitation of marine microbial metabolism. *Mar. Drugs* 8 (3):608–28.

36. Iverson, V., R. M. Morris, C. D. Frazar, C. T. Berthiaume, R. L. Morales, and E. V. Armbrust. 2012. Untangling genomes from metagenomes: Revealing an uncultured class of marine Euryarchaeota. *Science* 335 (6068):587–90.

37. Taylor, M. W., P. J. Schupp, I. Dahllöf, S. Kjelleberg, and P. D. Steinberg. 2003. Host specificity in marine sponge-associated bacteria, and potential implications for marine microbial diversity. *Environ. Microbiol.* 6:121–130.

38. Tianero, M. D., J. C. Kwan, T. P. Wyche, et al. 2015. Species specificity of symbiosis and secondary metabolism in ascidians. *ISME J.* 9 (3):615–28.

39. Egan, S., T. Harder, C. Burke, P. Steinberg, S. Kjelleberg, and T. Thomas. 2013. The seaweed holobiont: Understanding seaweed-bacteria interactions. *FEMS Microbiol. Rev.* 37 (3):462–76.

40. Macintyre, L., T. Zhang, C. Viegelmann, et al. 2014. Metabolomic tools for secondary metabolite discovery from marine microbial symbionts. *Mar. Drugs* 12 (6):3416–48.

41. Jensen, P. R., and W. Fenical. 1994. Strategies for the discovery of secondary metabolites from marine bacteria: Ecological perspectives. *Annu. Rev. Microbiol.* 48:559–84.

42. Takami, H., A. Inoue, F. Fuji, and K. Horikoshi. 1997. Microbial flora in the deepest sea mud of the Mariana Trench. *FEMS Microbiol. Lett.* 152 (2):279–85.

43. Roberts, J. M., A. J. Wheeler, and A. Freiwald. 2006. Reefs of the deep: The biology and geology of cold-water coral ecosystems. *Science* 312 (5773):543–47.

44. Fenical, W., J. J. La Clair, C. C. Hughes, P. R. Jensen, S. P. Gaudencio, and J. B. MacMillan. 2013. The deep oceans as a source for new treatments of cancer. In *Chembiomolecular Science*, ed. M. Shibasaki, I. Masamitsu, and H. Osada. Berlin: Springer. pp. 83–91.

45. Huber, J. A., D. B. Mark Welch, H. G. Morrison, et al. 2007. Microbial population structures in the deep marine biosphere. *Science* 318 (5847):97–100.

46. Yayanos, A. A., A. S. Dietz, and R. Van Boxtel. 1981. Obligately barophilic bacterium from the Mariana Trench. *Proc. Natl. Acad. Sci. U.S.A.* 78 (8):5212–15.

47. Nichols, D., J. Bowman, K. Sanderson, et al. 1999. Developments with antarctic microorganisms: Culture collections, bioactivity screening, taxonomy, PUFA production and cold-adapted enzymes. *Curr. Opin. Biotechnol.* 10 (3):240–46.

48. Gustafson, K., M. Roman, and W. Fenical. 1989. The macrolactins, a novel class of antiviral and cytotoxic macrolides from a deep-sea marine bacterium. *J. Am. Chem. Soc.* 111 (19):7519–24.

49. Calcul, L., C. Waterman, W. S. Ma, et al. 2013. Screening mangrove endophytic fungi for antimalarial natural products. *Mar. Drugs* 11 (12):5036–50.

50. Ananda, K., and K. R. Sridhar. 2002. Diversity of endophytic fungi in the roots of mangrove species on the west coast of India. *Can. J. Microbiol.* 48 (10):871–78.

51. Schulz, B., S. Draeger, T. E. dela Cruz, et al. 2008. Screening strategies for obtaining novel, biologically active, fungal secondary metabolites from marine habitats. *Bot. Mar.* 51 (3):219–34.

52. Leal, M. C., J. Puga, J. Serodio, N. C. Gomes, and R. Calado. 2012. Trends in the discovery of new marine natural products from invertebrates over the last two decades: Where and what are we bioprospecting? *PLoS One* 7 (1):e30580.

53. Blunt, J. W., B. R. Copp, W. P. Hu, M. H. Munro, P. T. Northcote, and M. R. Prinsep. 2009. Marine natural products. *Nat. Prod. Rep.* 26 (2):170–244.

54. Blunt, J. W., B. R. Copp, M. H. Munro, P. T. Northcote, and M. R. Prinsep. 2010. Marine natural products. *Nat. Prod. Rep.* 27 (2):165–237.

55. Blunt, J. W., B. R. Copp, M. H. Munro, P. T. Northcote, and M. R. Prinsep. 2011. Marine natural products. *Nat. Prod. Rep.* 28 (2):196–268.

56. Blunt, J. W., B. R. Copp, R. A. Keyzers, M. H. Munro, and M. R. Prinsep. 2012. Marine natural products. *Nat. Prod. Rep.* 29 (2):144–222.

57. Blunt, J. W., B. R. Copp, R. A. Keyzers, M. H. Munro, and M. R. Prinsep. 2013. Marine natural products. *Nat. Prod. Rep.* 30 (2):237–323.

58. Amann, R. I., W. Ludwig, and K. H. Schleifer. 1995. Phylogenetic identification and *in situ* detection of individual microbial cells without cultivation. *Microbiol. Rev.* 59 (1):143–69.

59. Fenical, W., and P. R. Jensen. 2006. Developing a new resource for drug discovery: Marine actinomycete bacteria. *Nat. Chem. Biol.* 2 (12):666–73.

60. Cochrane, V. W. 1958. *Physiology of the Fungi.* 1st ed. New York: Wiley & Sons.

61. Hu, G. P., J. Yuan, L. Sun, et al. 2011. Statistical research on marine natural products based on data obtained between 1985 and 2008. *Mar. Drugs* 9 (4):514–25.

62. Bugni, T. S., and C. M. Ireland. 2004. Marine-derived fungi: A chemically and biologically diverse group of microorganisms. *Nat. Prod. Rep.* 21 (1):143–63.

63. Keeling, P. J., F. Burki, H. M. Wilcox, et al. 2014. The Marine Microbial Eukaryote Transcriptome Sequencing Project (MMETSP): Illuminating the functional diversity of eukaryotic life in the oceans through transcriptome sequencing. *PLoS Biol.* 12 (6):e1001889.

64. Richards, T. A., M. D. Jones, G. Leonard, and D. Bass. 2012. Marine fungi: Their ecology and molecular diversity. *Annu. Rev. Mar. Sci.* 4:495–522.

65. Masuma, R., Y. Yamaguchi, M. Noumi, S. Omura, and M. Namikoshi. 2001. Effect of sea water concentration on hyphal growth and antimicrobial metabolite production in marine fungi. *Mycoscience* 42 (5):455–59.

66. Jensen, P. R., E. Gontang, C. Mafnas, T. J. Mincer, and W. Fenical. 2005. Culturable marine actinomycete diversity from tropical Pacific Ocean sediments. *Environ. Microbiol.* 7 (7):1039–48.

67. Schut, F., E. J. de Vries, J. C. Gottschal, et al. 1993. Isolation of typical marine bacteria by dilution culture: Growth, maintenance, and characteristics of isolates under laboratory conditions. *Appl. Environ. Microbiol.* 59 (7):2150–60.

68. Williams, P. G. 2009. Panning for chemical gold: Marine bacteria as a source of new therapeutics. *Trends Biotechnol.* 27 (1):45–52.

69. Pathom-Aree, W., J. E. Stach, A. C. Ward, K. Horikoshi, A. T. Bull, and M. Goodfellow. 2006. Diversity of actinomycetes isolated from Challenger Deep sediment (10,898 m) from the Mariana Trench. *Extremophiles* 10 (3):181–89.

70. Berdy, J. 2005. Bioactive microbial metabolites. *J. Antibiot.* (Tokyo) 58 (1):1–26.

71. Jensen, P. R., T. J. Mincer, P. G. Williams, and W. Fenical. 2005. Marine actinomycete diversity and natural product discovery. *Antonie Van Leeuwenhoek* 87 (1):43–48.

72. Stach, J. E., L. A. Maldonado, D. G. Masson, A. C. Ward, M. Goodfellow, and A. T. Bull. 2003. Statistical approaches for estimating actinobacterial diversity in marine sediments. *Appl. Environ. Microbiol.* 69 (10):6189–200.

73. Okazaki, T., T. Kitahara, and Y. Okami. 1975. Studies on marine microorganisms. IV. A new antibiotic SS-228 Y produced by Chainia isolated from shallow sea mud. *J. Antibiot.* (Tokyo) 28 (3):176–84.

74. Jensen, P. R., R. Dwight, and W. Fenical. 1991. Distribution of actinomycetes in near-shore tropical marine sediments. *Appl. Environ. Microbiol.* 57 (4):1102–8.

75. Feling, R. H., G. O. Buchanan, T. J. Mincer, C. A. Kauffman, P. R. Jensen, and W. Fenical. 2003. Salinosporamide A: A highly cytotoxic proteasome inhibitor from a novel microbial source, a marine bacterium of the new genus *Salinospora. Angew. Chem. Int. Ed. Engl.* 42 (3):355–57.

76. Iizuka, T., Y. Jojima, R. Fudou, and S. Yamanaka. 1998. Isolation of myxobacteria from the marine environment. *FEMS Microbiol. Lett.* 169 (2):317–22.

77. Zhang, Y. Q., Y. Z. Li, B. Wang, et al. 2005. Characteristics and living patterns of marine myxobacterial isolates. *Appl. Environ. Microbiol.* 71 (6):3331–36.

78. Fudou, R., T. Iizuka, and S. Yamanaka. 2001. Haliangicin, a novel antifungal metabolite produced by a marine myxobacterium. 1. Fermentation and biological characteristics. *J. Antibiot.* (Tokyo) 54 (2):149–52.

79. Reichenbach, H. 2001. Myxobacteria, producers of novel bioactive substances. *J. Ind. Microbiol. Biotechnol.* 27 (3):149–56.

80. Hawksworth, D. L., and A. Y. Rossman. 1997. Where are all the undescribed fungi? *Phytopathology* 87 (9):888–91.

81. Petrini, O. 1991. Fungal endophytes of tree leaves. In *Microbial Ecology of Leaves*. Berlin: Springer.

82. Wani, M. C., H. L. Taylor, M. E. Wall, P. Coggon, and A. T. McPhail. 1971. Plant antitumor agents. VI. The isolation and structure of taxol, a novel antileukemic and antitumor agent from *Taxus brevifolia. J. Am. Chem. Soc.* 93 (9):2325–27.

83. Stierle, A., G. Strobel, and D. Stierle. 1993. Taxol and taxane production by *Taxomyces andreanae*, an endophytic fungus of Pacific yew. *Science* 260 (5105):214–6.

84. Valliappan, K., W. Sun, and Z. Li. 2014. Marine actinobacteria associated with marine organisms and their potentials in producing pharmaceutical natural products. *Appl. Microbiol. Biotechnol.* 98 (17):7365–77.

85. Kjer, J., A. Debbab, A. H. Aly, and P. Proksch. 2010. Methods for isolation of marine-derived endophytic fungi and their bioactive secondary products. *Nat. Protoc.* 5 (3):479–90.

86. Foster, M. S., G. F. Bills, and G. M. Mueller. 2004. *Biodiversity of Fungi: Inventory and Monitoring Methods*. Oxford: Elsevier.

87. Janso, J. E., and G. T. Carter. 2010. Biosynthetic potential of phylogenetically unique endophytic actinomycetes from tropical plants. *Appl. Environ. Microbiol.* 76 (13):4377–86.

88. Strobel, G., B. Daisy, U. Castillo, and J. Harper. 2004. Natural products from endophytic microorganisms. *J. Nat. Prod.* 67 (2):257–68.

89. Choi, E. J., D. S. Beatty, L. A. Paul, W. Fenical, and P. R. Jensen. 2013. *Mooreia alkaloidigena* gen. nov., sp. nov. and *Catalinimonas alkaloidigena* gen. nov., sp. nov., alkaloid-producing marine bacteria in the proposed families Mooreiaceae fam. nov. and Catalimonadaceae fam. nov. in the phylum Bacteroidetes. *Int. J. Syst. Evol. Microbiol.* 63 (Pt. 4):1219–28.

90. McDonald, L. A., L. R. Barbieri, V. S. Bernan, J. Janso, P. Lassota, and G. T. Carter. 2004. 07H239-A, a new cytotoxic eremophilane sesquiterpene from the marine-derived *Xylariaceous* fungus LL-07H239. *J. Nat. Prod.* 67 (9):1565–67.

91. El-Elimat, T., M. Figueroa, B. M. Ehrmann, N. B. Cech, C. J. Pearce, and N. H. Oberlies. 2013. High-resolution MS, MS/MS, and UV database of fungal secondary metabolites as a dereplication protocol for bioactive natural products. *J. Nat. Prod.* 76 (9):1709–16.

92. Young, R. M., J. L. von Salm, M. O. Amsler, et al. 2013. Site-specific variability in the chemical diversity of the Antarctic red alga *Plocamium cartilagineum. Mar. Drugs* 11 (6):2126–39.

93. Keller, N. P., and T. M. Hohn. 1997. Metabolic pathway gene clusters in filamentous fungi. *Fungal Genet. Biol.* 21 (1):17–29.

94. Scherlach, K., and C. Hertweck. 2009. Triggering cryptic natural product biosynthesis in microorganisms. *Org. Biomol. Chem.* 7 (9):1753–60.

95. Brakhage, A. A., and V. Schroeckh. 2011. Fungal secondary metabolites: Strategies to activate silent gene clusters. *Fungal Genet. Biol.* 48 (1):15–22.

96. Chiang, Y. M., S. L. Chang, B. R. Oakley, and C. C. Wang. 2011. Recent advances in awakening silent biosynthetic gene clusters and linking orphan clusters to natural products in microorganisms. *Curr. Opin. Chem. Biol.* 15 (1):137–43.

97. Gross, H. 2007. Strategies to unravel the function of orphan biosynthesis pathways: Recent examples and future prospects. *Appl. Microbiol. Biotechnol.* 75 (2):267–77.

98. Jensen, P. R., K. L. Chavarria, W. Fenical, B. S. Moore, and N. Ziemert. 2014. Challenges and triumphs to genomics-based natural product discovery. *J. Ind. Microbiol. Biotechnol.* 41 (2):203–9.

99. Scherlach, K., and C. Hertweck. 2006. Discovery of aspoquinolones A–D, prenylated quinoline-2-one alkaloids from *Aspergillus nidulans*, motivated by genome mining. *Org. Biomol. Chem.* 4 (18):3517–20.

100. Cichewicz, R. H., F. A. Valeriote, and P. Crews. 2004. Psymberin, a potent sponge-derived cytotoxin from *Psammocinia* distantly related to the pederin family. *Org. Lett.* 6 (12):1951–54.

101. Fisch, K. M., C. Gurgui, N. Heycke, et al. 2009. Polyketide assembly lines of uncultivated sponge symbionts from structure-based gene targeting. *Nat. Chem. Biol.* 5 (7):494–501.

102. Uria, A., and J. Piel. 2009. Cultivation-independent approaches to investigate the chemistry of marine symbiotic bacteria. *Phytochem. Rev.* 8 (2):401–414.

103. Gross, H., V. O. Stockwell, M. D. Henkels, B. Nowak-Thompson, J. E. Loper, and W. H. Gerwick. 2007. The genomisotopic approach: A systematic method to isolate products of orphan biosynthetic gene clusters. *Chem. Biol.* 14 (1):53–63.

104. Bumpus, S. B., B. S. Evans, P. M. Thomas, I. Ntai, and N. L. Kelleher. 2009. A proteomics approach to discovering natural products and their biosynthetic pathways. *Nat. Biotechnol.* 27 (10):951–956.

105. Bergmann, S., J. Schumann, K. Scherlach, C. Lange, A. A. Brakhage, and C. Hertweck. 2007. Genomics-driven discovery of PKS-NRPS hybrid metabolites from *Aspergillus nidulans*. *Nat. Chem. Biol.* 3 (4):213–17.

106. Beau, J., N. Mahid, W. N. Burda, et al. 2012. Epigenetic tailoring for the production of anti-infective cytosporones from the marine fungus *Leucostoma persoonii*. *Mar. Drugs* 10 (4):762–74.

107. Williams, R. B., J. C. Henrikson, A. R. Hoover, A. E. Lee, and R. H. Cichewicz. 2008. Epigenetic remodeling of the fungal secondary metabolome. *Org. Biomol. Chem.* 6 (11):1895–97.

108. Demers, D., K. Knestrick, R. Fleeman, et al. 2014. Chemical analysis of products from an epigenetics based fungal metabolite screening program for antibacterial and antileishmanial lead compounds. *Planta Medica* 10:PH7.

109. Bok, J. W., and N. P. Keller. 2004. LaeA, a regulator of secondary metabolism in *Aspergillus* spp. *Eukaryot. Cell* 3 (2):527–35.

110. Shwab, E. K., J. W. Bok, M. Tribus, J. Galehr, S. Graessle, and N. P. Keller. 2007. Histone deacetylase activity regulates chemical diversity in *Aspergillus*. *Eukaryot. Cell* 6 (9):1656–64.

111. Fisch, K. M., A. F. Gillaspy, M. Gipson, et al. 2009. Chemical induction of silent biosynthetic pathway transcription in *Aspergillus niger*. *J. Ind. Microbiol. Biotechnol.* 36 (9):1199–213.

112. Cousens, L. S., D. Gallwitz, and B. M. Alberts. 1979. Different accessibilities in chromatin to histone acetylase. *J. Biol. Chem.* 254 (5):1716–23.

113. Richon, V. M., S. Emiliani, E. Verdin, et al. 1998. A class of hybrid polar inducers of transformed cell differentiation inhibits histone deacetylases. *Proc. Natl. Acad. Sci. U.S.A.* 95 (6):3003–7.

114. Furumai, R., Y. Komatsu, N. Nishino, S. Khochbin, M. Yoshida, and S. Horinouchi. 2001. Potent histone deacetylase inhibitors built from trichostatin A and cyclic tetrapeptide antibiotics including trapoxin. *Proc. Natl. Acad. Sci. U.S.A.* 98 (1):87–92.

115. Creusot, F., G. Acs, and J. K. Christman. 1982. Inhibition of DNA methyltransferase and induction of friend-erythroleukemia cell-differentiation by 5-azacytidine and 5-aza-2'-deoxycytidine. *J. Biol. Chem.* 257 (4):2041–48.

116. Brueckner, B., R. G. Boy, P. Siedlecki, et al. 2005. Epigenetic reactivation of tumor suppressor genes by a novel small-molecule inhibitor of human DNA methyltransferases. *Cancer Res.* 65 (14):6305–11.

117. Pettit, R. K. 2011. Small-molecule elicitation of microbial secondary metabolites. *Microb. Biotechnol.* 4 (4):471–78.

118. Bode, H. B., B. Bethe, R. Hofs, and A. Zeeck. 2002. Big effects from small changes: Possible ways to explore nature's chemical diversity. *Chembiochem* 3 (7):619–27.

119. Fuchser, J., and A. Zeeck. 1997. Secondary metabolites by chemical screening, 34. - Aspinolides and aspinonene/aspyrone co-metabolites, new pentaketides produced by *Aspergillus ochraceus*. *Liebigs Annalen-Recueil* 1:87–95.

120. Nenkep, V., K. Yun, D. Zhang, H. D. Choi, J. S. Kang, and B. W. Son. 2010. Induced production of bromomethylchlamydosporols A and B from the marine-derived fungus *Fusarium tricinctum*. *J. Nat. Prod.* 73 (12):2061–63.

121. Miao, L., T. F. Kwong, and P. Y. Qian. 2006. Effect of culture conditions on mycelial growth, antibacterial activity, and metabolite profiles of the marine-derived fungus *Arthrinium* c.f. *saccharicola*. *Appl. Microbiol. Biotechnol.* 72 (5):1063–73.

122. Fuchser, J., R. Thiericke, and A. Zeeck. 1995. Biosynthesis of aspinonene, a branched pentaketide produced by *Aspergillus ochraceus*, related to aspyrone. *J. Chem. Soc. Perkins Trans. 1* 13:1663–66.

123. Watts, K. R., S. T. Loveridge, K. Tenney, J. Media, F. A. Valeriote, and P. Crews. 2011. Utilizing DART mass spectrometry to pinpoint halogenated metabolites from a marine invertebrate-derived fungus. *J. Org. Chem.* 76 (15):6201–8.

124. Fremlin, L. J., A. M. Piggott, E. Lacey, and R. J. Capon. 2009. Cottoquinazoline A and cotteslosins A and B, metabolites from an Australian marine-derived strain of *Aspergillus versicolor*. *J. Nat. Prod.* 72 (4):666–70.

125. Slattery, M., I. Rajbhandari, and K. Wesson. 2001. Competition-mediated antibiotic induction in the marine bacterium *Streptomyces tenjimariensis*. *Microbial Ecol.* 41 (2):90–96.

126. Trischman, J. A., R. E. Oeffner, M. G. D. Luna, and M. Kazaoka. 2004. Competitive induction and enhancement of indole and a diketopiperazine in marine bacteria. *Mar. Biotechnol.* 6 (3):215–20.

127. Dusane, D. H., P. Matkar, V. P. Venugopalan, A. R. Kumar, and S. S. Zinjarde. 2011. Cross-species induction of antimicrobial compounds, biosurfactants and quorum-sensing inhibitors in tropical marine epibiotic bacteria by pathogens and biofouling microorganisms. *Curr. Microbiol.* 62 (3):974–80.

128. Zhu, F., and Y. C. Lin. 2006. Marinamide, a novel alkaloid and its methyl ester produced by the application of mixed fermentation technique to two mangrove endophytic fungi from the South China Sea. *Chin. Sci. Bull.* 51 (12):1426–30.

129. Zhu, F., G. Y. Chen, X. Chen, M. Z. Huang, and X. Q. Wan. 2011. Aspergicin, a new antibacterial alkaloid produced by mixed fermentation of two marine-derived mangrove epiphytic fungi. *Chem. Nat. Compd.* 47 (5):767–69.

130. Pettit, R. K. 2009. Mixed fermentation for natural product drug discovery. *Appl. Microbiol. Biotechnol.* 83 (1):19–25.

131. Cueto, M., P. R. Jensen, C. Kauffman, W. Fenical, E. Lobkovsky, and J. Clardy. 2001. Pestalone, a new antibiotic produced by a marine fungus in response to bacterial challenge. *J. Nat. Prod.* 64 (11):1444–46.

132. Oh, D. C., P. R. Jensen, C. A. Kauffman, and W. Fenical. 2005. Libertellenones A–D: Induction of cytotoxic diterpenoid biosynthesis by marine microbial competition. *Bioorg. Med. Chem.* 13 (17):5267–73.

133. Oh, D. C., C. A. Kauffman, P. R. Jensen, and W. Fenical. 2007. Induced production of emericellamides A and B from the marine-derived fungus *Emericella* sp. in competing co-culture. *J. Nat. Prod.* 70 (4):515–20.

134. Waterman, C., L. Calcul, J. Beau, W. Ma, J. Fries. 2014. Miniaturized cultivation of microbiota for antimalarial drug discovery. *Med. Res. Rev.* 2014, doi: 10.1002/med.21335.

135. Waterman, C., L. Calcul, T. Mutka, D. E. Kyle, C. J. Pearce, and B. J. Baker. 2014. A potent antimalarial trichothecene from hyphomycete species. *Tetrahedron Lett.* 55 (29):3989–91.

136. Duetz, W. A., L. Ruedi, R. Hermann, K. O'Connor, J. Buchs, and B. Witholt. 2000. Methods for intense aeration, growth, storage, and replication of bacterial strains in microtiter plates. *Appl. Environ. Microbiol.* 66 (6):2641–46.

137. Gao, H., M. Liu, X. Zhou, et al. 2010. Identification of avermectin-high-producing strains by high-throughput screening methods. *Appl. Microbiol. Biotechnol.* 85 (4):1219–25.

138. Rohe, P., D. Venkanna, B. Kleine, R. Freudl, and M. Oldiges. 2012. An automated workflow for enhancing microbial bioprocess optimization on a novel microbioreactor platform. *Microb. Cell. Fact.* 11:144.

139. Ito, T., T. Odake, H. Katoh, Y. Yamaguchi, and M. Aoki. 2011. High-throughput profiling of microbial extracts. *J. Nat. Prod.* 74 (5):983–88.

140. Isett, K., H. George, W. Herber, and A. Amanullah. 2007. Twenty-four-well plate miniature bioreactor high-throughput system: Assessment for microbial cultivations. *Biotechnol. Bioeng.* 98 (5):1017–28.

141. Reed, K. A., R. R. Manam, S. S. Mitchell, et al. 2007. Salinosporamides D–J from the marine actinomycete *Salinispora tropica*, bromosalinosporamide, and thioester derivatives are potent inhibitors of the 20S proteasome. *J. Nat. Prod.* 70 (2):269–76.

142. Butler, M. S. 2004. The role of natural product chemistry in drug discovery. *J. Nat. Prod.* 67 (12):2141–53.

143. *Natural Product Chemistry for Drug Discovery*. 2010. Cambridge: Royal Society of Chemistry.

144. Eldridge, G. R., H. C. Vervoort, C. M. Lee, et al. 2002. High-throughput method for the production and analysis of large natural product libraries for drug discovery. *Anal. Chem.* 74 (16):3963–71.

145. *On-Line LC-NMR and Related Techniques*. 2002. West Sussex: John Wiley & Sons.

146. Singh, S. B., H. Jayasuriya, J. G. Ondeyka, et al. 2006. Isolation, structure, and absolute stereochemistry of platensimycin, a broad spectrum antibiotic discovered using an antisense differential sensitivity strategy. *J. Am. Chem. Soc.* 128 (48).

147. Inglese, J., D. S. Auld, A. Jadhav, et al. 2006. Quantitative high-throughput screening: A titration-based approach that efficiently identifies biological activities in large chemical libraries. *Proc. Natl. Acad. Sci. U.S.A.* 103 (31):11473–78.

148. Lee, F. Y., R. Borzilleri, C. R. Fairchild, et al. 2008. Preclinical discovery of ixabepilone, a highly active antineoplastic agent. *Cancer Chemother. Pharmacol.* 63 (1):157–66.

149. Payne, D. J., M. N. Gwynn, D. J. Holmes, and D. L. Pompliano. 2007. Drugs for bad bugs: Confronting the challenges of antibacterial discovery. *Nat. Rev. Drug Discov.* 6 (1):29–40.

150. McDonald, L. A., L. R. Barbieri, G. T. Carter, et al. 2002. Structures of the muraymycins, novel peptidoglycan biosynthesis inhibitors. *J. Am. Chem. Soc.* 124 (35):10260–61.

151. Baker, D. D., M. Chu, U. Oza, and V. Rajgarhia. 2007. The value of natural products to future pharmaceutical discovery. *Nat. Prod. Rep.* 24 (6):1225–44.

152. Maldonado, L. A., W. Fenical, P. R. Jensen, et al. 2005. *Salinispora arenicola* gen. nov., sp. nov. and *Salinispora tropica* sp. nov., obligate marine actinomycetes belonging to the family Micromonosporaceae. *Int. J. Syst. Evol. Microbiol.* 55 (Pt. 5):1759–66.

153. Fenical, W., P. R. Jensen, M. A. Palladino, K. S. Lam, G. K. Lloyd, and B. C. Potts. 2009. Discovery and development of the anticancer agent salinosporamide A (NPI-0052). *Bioorg. Med. Chem.* 17 (6):2175–80.

154. Papagianni, M. 2011. *J. Microbial Biochem. Technol.* 5 (1): doi: 10.4172/1948-5948.S5-001.

155. Fenical, W., P. R. Jensen, and X. C. Cheng. Halimide, a cytotoxic marine natural product. WO/1999/048889. Patent. USA.

156. Nicholson, B., G. K. Lloyd, B. R. Miller, et al. 2006. NPI-2358 is a tubulin-depolymerizing agent: *In-vitro* evidence for activity as a tumor vascular-disrupting agent. *Anticancer Drugs* 17 (1):25–31.

157. Gilbert, J. A., D. Field, Y. Huang, et al. 2008. Detection of large numbers of novel sequences in the metatranscriptomes of complex marine microbial communities. *PLoS One* 3 (8):e3042.

158. Harvey, A. L. 2008. Natural products in drug discovery. *Drug Discov. Today* 13 (19–20):894–901.

159. Liu, X., E. Ashforth, B. Ren, et al. 2010. Bioprospecting microbial natural product libraries from the marine environment for drug discovery. *J. Antibiot.* (Tokyo) 63 (8):415–22.

160. Xiong, Z. Q., J. F. Wang, Y. Y. Hao, and Y. Wang. 2013. Recent advances in the discovery and development of marine microbial natural products. *Mar. Drugs* 11 (3):700–17.

CHAPTER 7

Screening Strategies for Drug Discovery and Target Identification

Fatma H. Al-Awadhi,[1] **Lilibeth A. Salvador,**[1,3] **and Hendrik Luesch**[1,2]

[1]Department of Medicinal Chemistry, University of Florida, Gainesville, Florida
[2]Center for Natural Products, Drug Discovery and Development (CNPD3), University of Florida, Gainesville, Florida
[3]Marine Science Institute, College of Science, University of the Philippines, Diliman, Quezon City, Philippines

CONTENTS

7.1 INTRODUCTION

Natural product research was once limited to defining the chemistry of small molecules and looking for novel structural scaffolds. Today, natural products have successfully integrated the biology of small molecules—defining the biological activity, unraveling the cellular target, and deciphering the mechanism of action in cellular systems. Both the chemistry and biology of small molecules from natural sources helped shape the role of natural products in drug discovery. Nature has been regarded as one of the most successful producers of new small-molecule therapeutics, primarily due to the chemodiversity and potent biological activity of natural products. Aside from drug discovery, these small molecules also serve as excellent molecular probes due to the high affinity for their cellular targets arising from optimization through evolution. Natural products are recognized to be an integral part of chemical biology in defining the structure and function of proteins and signaling networks.

However, interrogation of the biological activity of natural products is limited by the complexity of the crude extract, with the bioactive component particularly present in low amounts. The presence of nuisance compounds in crude extracts, which may interfere with assay readouts, limits the applicability of natural products for high-throughput screening, particularly in assays using purified proteins. As natural products are regarded as privileged structures, these small molecules may possess multiple protein targets that define intended biological activity and unintended off-target effects, providing challenges in defining the pharmacology of these small molecules from Nature.

Significant technological improvements in fractionation and bioassay strategies have enabled rigorous evaluation of natural products in recent years, leading to a significant acceleration in their discovery and development as lead compounds. Both target- and phenotype-based screening methods have been employed, using purified proteins, cellular systems, and model organisms to define the biological activity. Bioassay methods are also being designed to provide initial clues on the mechanism of action of small molecules. Comprehensive and unbiased methods to interrogate the effects of small molecules on the genome, transcriptome, and proteome now allow for a better understanding of the effects of natural products on cellular systems. In this chapter, we highlight some recent advancements in screening strategies for drug discovery and target identification, particularly as applied to marine natural products.

7.2 BIOLOGICAL ACTIVITY PROFILING

Biological activity profiling is an integral part of the drug discovery process and crucial for defining the therapeutic potential of small molecules. Several assays and screening approaches have been developed utilizing purified enzymes, mammalian cells, or model organisms (Table 7.1). Such screening assays have been used to demonstrate the biological effects of small molecules on proteins, signaling pathways, or phenotypes (Table 7.1). The choice of assay depends on the specific question asked, and all strategies have their inherent advantages and disadvantages (Table 7.1).

7.2.1 Target-Based Assays

Target-based assays involve the direct measurement of the effect of small molecules against a purified target via *in vitro* assays (Table 7.1). The target could be a gene, gene product (RNA or protein), biomolecular complex, or protein–protein interaction. Target-based assays have several advantages, including applicability to high-throughput screening (HTS), where libraries of small molecules can be screened against a defined target. Hit compounds represent a starting point for structure–activity relationship studies that can be guided by rational drug design.[1] Knowledge about the importance of druggable targets such as proteases and kinases and their implications in several diseases necessitates sensitive assays targeting these enzymes. As target-based assays rely on

Table 7.1 Summary of Screening Methods Employed for Natural Products

Parameters	Target-based	Reporter-based	Cell-based	Simple model organism	Higher model organism
Applicability Scale: High Possible Low					
Idea of direct target					
Idea of MOA					
Phenotype monitoring					
High-throughput format					
Pharmacokinetic consideration					

disease biomarker information, they may provide a limited opportunity to discover novel proteins or genes that may present as alternative targets for therapeutic intervention.

7.2.1.1 Enzyme Assays with Purified Proteins

A substantial portion of small molecules with known molecular targets was discovered using target-based screens employing purified proteins. These assays are based on the biochemical inhibition of the protein target, for example, proteases, upon binding of the small-molecule inhibitor. These assays are usually carried out by incubating the purified protein, a tagged substrate, and the inhibitor (Figure 7.1). This method relies on the physical change accompanying substrate turnover. The change is usually monitored using fluorescence, absorbance, or luminescence measurements, following the release of the accompanying chromophore with the substrate turnover. A number of fluorescence approaches have been developed for the discovery of small-molecule inhibitors of proteases. One example involves the attachment of a chemically quenched fluorescent dye, bearing an amino group to the carboxylic end of the substrate through the formation of bound amide. The

Enzyme Assays with Purified Proteins

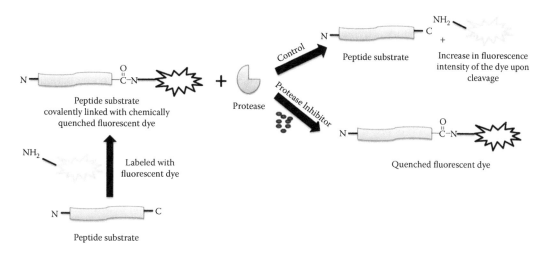

Figure 7.1 Target-based assays using purified proteins measure the biochemical inhibition of proteolytic activity of proteins, monitored through the turnover of a fluorescent substrate.

binding of the protease to the substrate results in cleavage of the quenched fluorescent dye, resulting in an increase of the fluorescence signal intensity (Figure 7.1). This approach has been widely utilized in HTS protease assays aimed at identifying inhibitors of cathepsins and caspases, which are commonly implicated in human diseases.[2] Target-based screening, aside from its utility for finding lead compounds, can also be employed to assess the selectivity of small-molecule inhibitors. This is usually achieved through evaluating the inhibitory activity against a comprehensive protease panel. This should provide information on potential off-target effects arising from a promiscuous inhibitor.

There has been increasing interest in monitoring protease activity and screening for inhibitors that target them due to their involvement in many cellular processes and abnormalities. Dysregulated proteolytic activity has been associated with several pathological conditions, such as coagulation disorders and stroke linked to thrombin, hypertension linked to the angiotensin-converting enzyme (ACE), and chronic obstructive pulmonary disease (COPD) linked to elastase,[3] as well as many others, including, but not limited to, breast cancer and Alzheimer's disease involving cathepsin D, diabetes linked to DPP4, and diseases related to the immune system involving DPP8.[4]

There are many examples from the field of marine natural products where this target-based strategy was successfully applied. One study demonstrated the discovery of potent elastase inhibitors from marine cyanobacteria.[5] These cyclic depsipeptides incorporating a characteristic 2-amino-6-hydroxypiperidone (Ahp) were tested against both bovine pancreatic and human neutrophil elastase. To determine the selectivity of this class of compounds for elastase, the representative member lyngbyastatin 7 was screened against a panel of 68 proteases at 10 µM. The results revealed preferential inhibition for serine proteases with complete inhibition of bovine elastase, bovine chymotrypsin, and proteinase K at that concentration. On the basis of these results, the related and more abundant symplostatin 5 was profiled in a dose–response manner against a panel of 26 serine proteases. The screening indicated preferential inhibition of elastase by symplostatin 5 compared with chymotrypsin, and identified an even greater selectivity index for the human variants. The selectivity and potency of this elastase inhibitor prompted the assessment of the cellular consequences of elastase inhibition by symplostatin 5. Symplostatin 5 demonstrated cytoprotective effects in bronchial epithelial cells against elastase-induced antiproliferation, apoptosis, cell detachment, and morphological changes. Furthermore, it attenuated the global transcript changes induced by elastase. Importantly, symplostatin 5 did not exhibit any cytotoxic effects on lung epithelial cells and showed superior activity in long-term assays to the clinically approved elastase inhibitor sivelestat.[5]

Target-based assay also guided mechanistic follow-up studies in physiologically relevant cellular contexts for the three marine cyanobacteria-derived linear modified peptides termed grassystatins A–C with protease inhibitory activities. The screening of grassystatin A against a panel of 59 diverse proteases revealed selective inhibitory activity against the two aspartic proteases, cathepsins D and E, with differential selectivity for cathepsin E over D. By identifying the target (cathepsin E) and suspecting its role in antigen presentation, the cellular effects of cathepsin E inhibition on peripheral blood mononuclear cells (PBMCs) were examined. The results demonstrated the ability of grassystatin A to reduce antigen stimulated T-cell proliferation, antigen presentation by dendritic cells (DCs) to T_H cells, and the levels of pro-inflammatory cytokines interleukin (IL) 17 and interferon (IFN) γ.[5a]

Grassypeptolides A–C are antiproliferative cylic depsipeptides isolated from the marine cyanobacterium *Lyngbya confervoides*. Assessment of the inhibitory potential of grassypeptolide A against a panel of proteases demonstrated a selective inhibitory activity toward dipeptidyl peptidase (DPP) 8 over DPP4. In accordance with the putative role of DPP8 in the immune system, inhibition of DPP8 by grassypeptolides attenuated IL-2 production and cell proliferation in activated T-cells.[6]

Similar comprehensive strategies to determine activity and selectivity for kinases, G protein-coupled receptors (GPCRs), or ion channels perhaps have been less extensively applied in marine natural products to date but can easily be adapted.

In general, *in vitro* assays using purified enzymes have some limitations. Since these assays are done in cell-free systems, the results of screening using purified proteins may not present cellular relevance because of the low pharmacokinetic consideration in these assay systems (Table 7.1). Also, these assays may yield false positive hits resulting from interference in assay readouts from compound chromophores, not directly related to enzyme inhibition. The use of target-based screening is also problematic for crude extracts of natural products, and hence is limited for use on fairly pure natural product samples. Since this method relies on purified proteins, it is limited by targets that can be recombinantly expressed or purified.

7.2.1.2 Enzyme Assays with Complex Cell Lysates

Chemical proteomics entails strategies involving cellular (real-life) systems as improvement of the aforementioned protease assays, one of which is named activity-based protein profiling (ABPP).[7] This strategy enables quantitative profiling of active subfractions of enzyme families within complex mixtures (lysates, cells) via site-directed probes. It involves the design of reactive small marker molecules (activity-based probes) that are supposed to selectively bind and therefore label the active site of enzymes.[8] Several different kinds of probes were designed.[4,8] In general, the probe consists of three parts: a reactive group (directed toward binding and modulating the active site of enzymes sharing similar substrate specificity or catalytic mechanism), a chemical linker, and a chemical tag (enables the detection and isolation of labeled enzymes). The basis of the strategy involves incubating the proteome with inhibitors, followed by fluorescence analysis, as the detected reduction in enzyme labeling reflects the binding of inhibitors to specific enzymes (Figure 7.2a).[7]

Enzyme Assays with Complex Cell Lysates

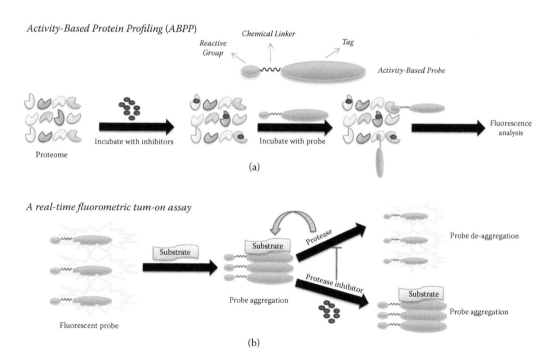

Figure 7.2 Alternative enzyme assays using cellular lysates. (a) Activity-based protein profiling (ABPP) uses site-directed probes to interrogate effects of small molecules on enzyme subfamilies. (b) Real-time fluorometric turn-on assay relies on changes in probe aggregation and, consequently, fluorescence following protamine and enzyme addition as criteria to assess the inhibitory activity of small molecules.

In one example, a real-time fluorometric turn-on assay was developed (Figure 7.2b) that is based on a designed/synthesized fluorescent benzoperylene probe and the presence of protease and substrate (e.g., trypsin and protamine, respectively). The addition of protamine to the probe results in probe aggregation, subsequently quenching its monomer fluorescence due to electrostatic interactions between the negatively charged probe and the cationic substrate. However, in the presence of a protease, the substrate will be degraded, resulting in probe de-aggregation and subsequent detection of its fluorescence signal. In the presence of an inhibitor, on the other hand, the fluorescence recovery will be diminished (Figure 7.2b).[4] Assessing protease activity by such screens enables the identification of potentially new targets by comparing the enzymatic activity profiles under both disease and healthy states. Another application of this strategy is in assessing the selectivity of a particular small-molecule inhibitor against a panel of proteases.[8] In addition to being a screening strategy, competitive ABPP experiments can be also applied to probe the target of small molecules, as demonstrated for symplostatin 4. The incubation of *Plasmodium falciparum* lysates with different concentrations of symplostatin 4, followed by labeling with the activity-based probe Cy5-DCG-04, subsequent separation of proteins by electrophoresis, and visualization via in-gel fluorescence detection, revealed falcipains as the target.[9] Falcipains FP2, FP2′, and FP3 are plasmodial papain-like cysteine proteases (PLCPs) involved in the hemoglobin degradation pathway of the parasite.[9]

ABPP approaches have been applied beyond proteases, including molecular targets involved in metabolic syndrome,[10] and are likely to be extensively utilized by the marine natural product community in years to come.

7.2.1.3 Mass Spectrometry-Based Binding Assays

Instead of relying on substrate turnover for target-based screening, mass spectrometry-based monitoring of inhibitor binding was developed, allowing for rapid and sensitive assessment, as well as compatibility for use with complex mixtures, providing a highly relevant method to natural product-based drug discovery where the starting point is a crude extract.[11] This affinity screening approach can identify ligands from natural product crude extracts based on the detection of noncovalent complexes (between active constituents and protein target) formed by incubating the extracts with a target protein.[11] The screening is performed by using bioaffinity selection coupled to electrospray ionization–Fourier transform ion cyclotron resonance–mass spectrometry (ESI-FTICR-MS). The bioaffinity selection involves separating protein complexes from small, unbound proteins to reduce mass interference. The ESI-MS involves a gentle ionization process that detects noncovalent complexes and, under certain optimized parameters, transfers the noncovalent complex from solution to the gas phase. The workflow of such a screening strategy involves analyzing the crude extract using ESI-FTICR-MS, processing the resulting data, and identifying the target protein–ligand complex. Once identified, the complex will be reanalyzed to determine the mass or molecular weight of the ligand, which can be deduced from the *m/z* of the noncovalent complex. Follow-up isolation of the hit compound from the crude extract can be subsequently carried out using mass-directed purification.

A mass spectrometric binding assay was utilized for screening marine and plant natural product (NP) extracts for potential inhibitors of bovine carbonic anhydrase II.[11] Modulators of this target have several therapeutic applications in cancer, osteoporosis, and glaucoma. A proof-of-concept experiment was performed by using NP extracts spiked with known inhibitors of the bCAII (ethoxzolamide and sulfanilamide). Spiked extracts were incubated with the target and subsequently analyzed by ESI-FTICR-MS. Noncovalent complexes of ethoxzolamide and sulfanilamide with bCAII were detected and identified in the spiked extract, thus validating this screening strategy. Next, 85 plant NP extracts were screened for active constituents against bCAII by two methods: direct infusion and online size exclusion chromatography (SEC) ESI-FTICR-MS. The screening resulted in the identification of one active extract (resulting in a noncovalent complex) in both methods. Mass-directed purification yielded 6-(1S-hydroxy-3-methylbutyl)-7-methoxy-2H-chromen-2-one, a coumarin-type ligand from the crude extract, as the active principle. Binding was confirmed, and subsequent competition experiments classified the active ligand as a specific inhibitor of bCAII.

The application of this strategy demonstrated the successful interrogation of protein–ligand complexes in NP crude extracts using FTICR-MS. Extracts must, however, be subjected to a prefractionation step to remove salts from marine-sourced samples to minimize interference with MS detection. Nonetheless, this method provides several advantages, such as having a short run time (as little as 2 min), high-throughput format capacity, sensitivity of detection, and requiring small amounts of material compared to traditional chromophore-based methods, largely owing to the utilization of a mass spectrometry-based detection method. However, since this mass spectrometry-based enzyme assay is still performed in cell-free systems, small-molecule binding may not necessarily cause a functional response. The question that remains to be addressed is whether ligand binding can indeed modulate target activity or function.

7.2.1.4 Kinobead Cellular Kinase Profiling

Protein kinases have an important role in regulating signal transductions and therefore are involved in many cellular processes. Activating mutations or overexpression of certain protein kinases is associated with several diseases, such as cancer and inflammation.[12] Therefore, kinases represent important therapeutic targets; however, exploring these targets is challenging due to the structural conservation of the adenosine triphosphate (ATP) binding pocket across the kinase family, usually resulting in multitargeted inhibitors.

Traditional *in vitro* methods used for selectivity profiling of small molecules involve the use of recombinant enzymes or protein fragments (isolated target) similarly as described above for protease inhibitors. However, the response to kinase inhibitors is dependent on the physiological context, and therefore the data generated are difficult to correlate with *in vivo* experiments measuring the compound's efficacy. This could be due to the absence of regulatory factors, protein folding, and posttranslational modification.[12–14]

From the marine environment, several small-molecule kinase inhibitors have been identified, particularly from sponges.[15] Examples include the protein kinase C (PKC) inhibitors: xestocyclamine A from the marine sponge *Xestospongia* sp. and the sesquiterpene derivatives, frondosins A–E, from the marine sponge *Dysidea frondosa*.[16,17] Penta-, hexa-, and hepta-prenylhydroquinone 4-sulfates from the marine sponge *Ircinia* sp. are tyrosine protein kinase (TPK) inhibitors.[18] However, a comprehensive assessment of the inhibition of the members of the kinome by these marine-derived compounds was not achieved.

To comprehensively assess the inhibitory profile of kinase inhibitors and also serve as a new assay system to find selective kinase inhibitors, a new chemical proteomic strategy based on kinobeads was developed. This strategy is based on immobilizing pan-selective kinase inhibitors on a resin and quantitatively monitoring the differential binding of kinases to those beads in the presence or absence of the small molecule of interest. The kinobeads are capable of assessing the inhibitory activity of small molecules against more than 500 kinases, representative of the majority of the families in the kinome. In the presence of the test compound, competition for the binding pockets of kinases may occur. As a result of the competition between the immobilized inhibitors and the compound of interest, kinobeads will have a reduced amount of captured kinases and related target proteins. Target binding profiles are then obtained through measuring the differential binding using iTRAQ, and protein identification is carried out using MS/MS (Figure 7.3).[12,19]

Bantscheff et al. demonstrated the value of kinobeads in drug discovery by applying this approach to three drugs (the tyrosine kinase abelson murine leukemia (ABL) inhibitors): imatinib, dasatinib, and bosutinib. Drugs in a range of concentrations (100 pM to 10 μM) were incubated with

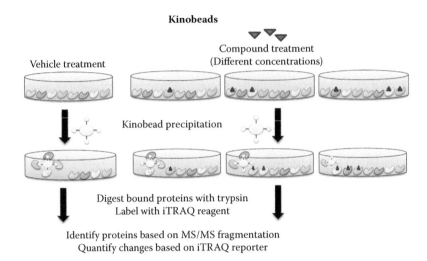

Figure 7.3   The kinobead assay is a comprehensive method to interrogate the kinome based on affinity capture of kinases using immobilized pan-selective kinase inhibitors. The identity of kinases is determined by MS/MS fragmentation, and inhibition via small-molecule treatment is related to the intensity difference in the reporter tags of iTRAQ reagents.

K562 cell lysates. Lysates were subsequently subjected to kinobead precipitation. The binding of drug to its target blocks the binding pocket of the targeted kinase, resulting in a diminished amount of targets available for capture by the kinobeads. The proteins bound to kinobeads were subjected to tryptic digestion, and the resulting peptides were labeled using iTRAQ reagent and subjected to MS for quantitation and identification (Figure 7.3). Dose–response binding profiles were obtained, which confirmed the known targets of those drugs and further identified two novel targets of imatinib: the kinase DDR1 and nonkinase (but ATP-requiring) protein NQO2.[19]

The use of the kinobead strategy enables high-throughput, quantitative, and comprehensive profiling of a small molecule of interest against hundreds of kinases or other ATP-dependent enzymes in any cell type (lysates or whole) without the need for compound modification. When used in whole cells, the kinobead approach also allows for probing the changes induced by kinase inhibitors on the phosphorylation state of the kinobead-captured proteins, thereby giving insight into the downstream signaling pathways.[19]

7.2.2 Cell-Based Assays

Cell-based assays for bioactivity screening have become a choice technique due to the versatility of the methods, adaptability to the high-throughput format, and representation of a more realistic disease model *in vitro* (Table 7.1). Cell-based assays can utilize a variety of phenotypic responses, such as proliferation, morphology, and cellular responses at the transcriptional or translational levels or signal transduction pathways, as criteria for assessing therapeutic intervention by small molecules. Changes in phenotypic or cellular responses of cells following small-molecule treatment may be assessed through microscopy or biomarker labeling that is detected by absorbance, luminescence, or fluorescence measurements.[20] Being a relevant biological surrogate, cell-based assays also provide a wealth of information about cellular permeability, stability, and toxicity of compounds.[21] In contrast to target-based screening, cell-based assays introduce complexity in target identification and mechanism of action analyses (Table 7.1). However, this setback of cell-based assays is compensated by the possibility of finding small molecules that exhibit novel mechanisms of action.

7.2.2.1 Differential Cell Proliferation and Cytotoxicity Assays

Antiproliferative activity against cancer cells is a major theme for bioactive marine natural products. As such, phenotype-based assays that measure changes in the proliferation of cancer cells are routinely done. In addition to being a screening tool, antiproliferative activity assessment is also one of the earliest methods employed to elucidate the mechanism of action of natural products through the U.S. National Cancer Institute (NCI) 60-cell line screening.[22] The NCI60 screen is based on the differential sensitivity and resistance of antiproliferative agents toward leukemia, small-cell lung, non-small-cell lung, colon, central nervous system, melanoma, ovarian, and renal cell lines. Cell growth was traditionally monitored using sulforhodamine B, which relates cell growth to protein synthesis. Other, mostly faster methods to assess cell growth can employ resazurin dye, trypan blue, methylthiazole tetrazolium (MTT) dye, or ATP lite, all of which are monitoring events that are more or less predictive of viability, although they are not identical and do not necessarily produce the same result. The characteristic similarity in the differential cytotoxicity profile of antiproliferative agents, analyzed using the COMPARE algorithm, is indicative of similarity in the biochemical mechanism of action.[22] Early work with marine natural products demonstrated that halichondrin B showed a comparable cytotoxicity profile to the known antimitotic agent maytansine.[23] The NCI60 screen has since served as a dependable tool to rapidly identify small molecules with a novel mechanism of action and those that share biological effects similar to those of other compounds with established mechanisms of action. This method is, however, limited to small molecules that cause changes in cancer cell proliferation with known mechanisms of action.

The similarity in the differential biological activity of small molecules has also been extended to identify novel antibiotics.[24] The original antibiotic mode of action profile (BioMAP) utilized 15 bacterial strains for antibiotic screening. The differential antibiotic property of small molecules from the same structural class clusters together, and hence, dereplication of novel antibiotic agents is facilitated. BioMAP rapidly identified the known antibiotics erythromycin, rifampicin, and actinomycin D from prefractionated natural product extracts. Furthermore, the new antibiotic, arromycin, was purified based on the novel BioMAP clustering.[24]

7.2.2.2 Mutant Cell Lines and Resistant Strains

Mutations in tumor suppressor proteins and oncogenes (p53 and K-Ras) have been linked to cancer development and resistance to chemotherapy. Ras mutations in particular account for 33% of all human cancers, with K-Ras being predominant in lung, colon, and pancreatic cancers, which are the top three leading cancer deaths in the United States.[25] The discovery of small-molecule modulators of K-Ras has been plagued with difficulty and was not met with success until late 2013.[25,26] Back-to-back disclosure of new classes of covalent K-Ras demonstrated the utility of targeting a guanine–nucleotide binding site as a promising strategy to inhibit Ras signaling.[26,27] Current approaches have shifted toward genome-wide unbiased genetic screens for the identification of new targets for anticancer drug discovery. This approach is based on the concept of synthetic lethality genetic interaction in which mutation in one gene (e.g., KRAS) is compatible with cell viability, whereas mutation in both genes (e.g., KRAS and the one identified by RNAi) results in cell death.[25,28] Two types of synthetic lethality screens involve screening a library of either siRNAs or small molecules to identify synthetic lethal or chemical–genetic interactions, each with its own advantages and disadvantages.[28] In general, such screens require isogenic cell lines and a functional readout to assess the viability of cells. Briefly, the methodology may involve the fluorescence tagging of isogenic cells that differ by one essential cancer gene. Those cells are mixed and sorted in 96-well plates and subsequently treated with a library consisting of either RNAi or small-molecule compounds. The viability of cells is determined based on the fluorescence readout.[28] Several RNAi screens were conducted and have identified the STK33, TBK1,[25] and PLK1[29] serine/threonine kinases, and survivin[30] as lethal partners to K-Ras mutants. A synthetic chemical lethality screen involving more than 50,000 compounds has identified lanperisone, a piperidine derivative of tolperisone (centrally acting muscle relaxant), as a selective killer of mouse embryonic fibroblasts expressing oncogenic K-Ras.[31] From marine natural products, the Philippine marine sponge *Smenospongia* sp. yielded several cytotoxic bromoindole derivatives and terpenes (5-bromo-L-tryptophan, 5-bromoabrine, 5,6-dibromoabrine, 5-bromoindole-3-acetic acid, makaluvamine O, 5,6-dibromotryptamine, aureol, and furospinulosin 1).[32] The cytotoxicity of these natural products was evaluated using wild-type (p53[+/+]) and isogenic (p53[−/−]) HCT-116 human colon cancer cell lines.[32] Aureol, furospinulosin 1, and 5,6-dibromotryptamine showed differential cytotoxicity in p53[+/+] and p53[−/−] HCT-116 cell lines, indicating the involvement of p53 in the antiproliferative activity of these compounds.[32]

Taken all together, isogenic cell lines serve as a model of cancers harboring a particular mutation and are therefore useful for better understanding the biology of tumor cells and subsequently aiding in the design of selective drugs. Such screens result in the identification of novel targets (as with siRNA screens) or potential therapeutic drugs (with small-molecule screening) that selectively kill cancer cells expressing a particular gene mutation. This approach therefore enables the discovery and development of selective anticancer drugs with a wide therapeutic index.[28]

7.2.2.3 Reversal of Transformed Phenotype

Malignant transformation that occurs in cancer is usually associated with changes in cell shape and architecture.[33] The inducers are oncogenes, which consequently represent potential therapeutic targets against cancers expressing the particular oncogene.[34] Therefore, screening compounds for their ability to revert the morphology of oncogene-transformed cells may identify potential anticancer drugs. Cell transformation assays involve the transfection of cells with oncogenes such as *Ras*, *v-sis*, and *src*. The transformed cell will undergo alteration in gene expression, culminating in changes in growth and morphology. For example, NIH3T3 cells are commonly used in transformation assays, and upon transformation, the morphology changes from flat to swirling clusters (Figure 7.4). The resulting morphological changes are usually monitored by microscopy.[35,36] Such transformed cells are often used to screen compounds for their ability to revert the altered phenotype to normal by "reprogramming" cells on the transcriptional level in order to identify compounds with potential antitumor activity. Screening based on reversion of transformed phenotypes using oncogene-transformed cells yielded several natural histone deacetylase (HDAC) inhibitors, including FK228, trichostatin A, trapoxins, and depudecin.[33,35] FK228 is a cyclic depsipeptide identified from the fermentation broth of *Chromobacterium violaceum* and is now marketed for the treatment of cutaneous T-cell lymphoma (CTCL).[35] Ha-ras-transformed NIH3T3 cells (Ras-1) treated with FK228 showed reversal of the transformed phenotype, similar to the phenotypic response of another HDAC inhibitor, trichostatin A.[37] Trichostatin A is a bacterial-derived linear HDAC inhibitor from

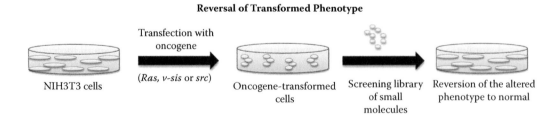

Reversal of Transformed Phenotype

Transfection with oncogene

(Ras, v-sis or src)

NIH3T3 cells   Oncogene-transformed cells   Screening library of small molecules   Reversion of the altered phenotype to normal

Figure 7.4 Anticancer activity of small molecules can be assessed by monitoring effects of oncogene-mediated phenotype alteration.

Streptomyces spp. A very low concentration of trichostatin A (1 ng/ml) was sufficient to reverse the transformed morphology of *sis*-transformed NIH3T3 cells[34] to the flat and normal morphology. The fungal metabolite depudecin was also discovered through transformation assays by its ability to reverse the rounded transformed morphology of NIH3T3 fibroblasts transfected with *v-ras* and *v-src* to the normal flattened shape.[33] These examples demonstrate the potential of such screens to identify novel HDAC inhibitors. There has been an increasing interest recently in HDAC-targeting agents due to the broad-spectrum applicability of HDAC inhibitors for a wide variety of disease indications, and the field of marine natural products is continuously providing novel structurally diverse compounds. Examples include largazole, a potent cyclic depsipeptide from the marine cyanobacterium *Symploca* sp.[38–40]; the cyclic tetrapeptides azumamides A–E from the marine sponge *Mycale izuensis*[41]; and bromotyrosine derivative psammaplin A from several marine sponge genera.[42] Santacruzamate A is another HDAC inhibitor isolated from a cyanobacterium that resembles the genus *Symploca* in terms of morphology. It shares several structural features with the clinically approved first-generation HDAC inhibitor SAHA approved for the treatment of refractory CTCL.[43]

7.2.2.4 Reporter Gene Assays

Numerous cell lines can be engineered to express a particular reporter gene product in response to stimuli, and therefore act as biodetectors of various signaling pathways[44] to be used in pathway-directed screens. Compared with phenotypic assays, the targets are narrowed down to components of a particular signaling pathway that culminates in the activity of a specific transcription factor. Normally, when cells are challenged with a stimulus (steroid hormone, cytokines, neurotransmitters), among several other ways, this agent will bind to a receptor resulting in the activation of a signal transduction pathway (second messenger cascade). This may result in the phosphorylation of certain transcriptional factors by kinases and, ultimately, the interactions or the binding of those transcriptional factors with specific response elements located in the promoter regions of responsive genes, thereby regulating transcription. Based on this, the concept of reporter gene technology arises, which involves monitoring the influence of activation of second messenger signaling cascades on gene expression reported by alterations in the activity of a particular reporter gene downstream of the promoter, with binding sites recognized by the transcription factor of interest.[45]

Reporter gene units consist of a cis-regulatory sequence (response element) that is recognized by a transcription factor, and a reporter gene (which has a measurable phenotype). Common reporter genes include chloramphenicol acetyltransferase (CAT), β-galactosidase, luciferase, alkaline phosphatase, and green fluorescent protein (GFP). Each reporter gene has its advantages and limitations, and suitability depends on the specific application (reviewed in Naylor[45]). Reporter gene assays have a variety of applications for drug discovery, including biological screens, characterization of receptors and their ligands, and signaling pathways (reviewed in Naylor[45]). Reporter gene assays are extensively utilized in primary screens since they are highly suitable for HTS. Secondary screens and follow-up studies require the examination of effects on the expression of endogenous target genes to measure a true biological activity. Although reporter gene assays are common and powerful, they have some shortcomings. The application of this assay requires the presence of gene transfer techniques; the system is rather artificial. False positive hits, as observed for HDAC inhibitors, can also arise in reporter-based assays from small molecules that act as general activators of transcription rather than targeting a specific component of the signaling cascade.[46,47]

Pathway-directed assays can be utilized to screen for both activators and inhibitors. An example for the former is a screen for activators of the antioxidant response element (ARE). ARE is a cis-acting regulatory element located in the 5′ flanking region of many phase II detoxification enzymes, such as NQO1, which plays a protective role against oxidative stress-induced damage. Oxidative stress has been implicated in aging as well as cancer, diabetes, inflammation, and neurodegenerative disorders (Alzheimer's, Parkinson's, and Huntington's diseases). Identifying activators of

ARE in the absence of oxidative stress provide a therapeutic and preventive approach for targeting those diseases.[48]

A genomic screening approach to identify novel pathway members utilized 15,000 expression cDNAs arrayed in 384-well plates that were transfected into IMR-32 human neuroblastoma cells along with the ARE–luciferase reporter construct. After 48 h incubation, the luminescence output of each well was assessed as a measure of luciferase activity. SQSTM1 and DPP3 were identified as activators of ARE and were validated by secondary assays as new players and potential drug targets.[48,49] In addition, the ARE assay has been successfully utilized to screen marine natural products for small-molecule activators from the eukaryotic alga *Ulva* sp.[50,51] Various electrophilic compounds such as isololiolide, loliolide, novel unsaturated fatty acids, and an unsaturated chlorinated aldehyde were identified as ARE inducers.[50,51] *Ulva* fractions enriched in ARE activators were able to abrogate lipopolysaccharide (LPS)-induced nitric oxide synthase (iNOS) and cyclo-oxygenase 2 (COX2) expression, thereby acting as antioxidants and anti-inflammatory natural products.[50,51]

Screening for inhibitors of pathways is somewhat more difficult since the desired decrease in signal intensity can also be affected by a reduction in cell viability. A secondary screen may involve a cell viability assay if viability cannot be assessed in parallel in the primary assay. One application of a reporter gene assay aimed at identifying pathway inhibitors was successfully demonstrated for targeting the transcription factor nuclear factor kappa B (NF-κB). NF-κB signaling is involved in cell survival and immunity, and irregularities in its activation have been implicated in the pathogenesis of cancer and inflammation.[52,53] NF-κB is regarded as an interesting target for cancer drug discovery due to its close association with carcinogenesis and inflammation. NF-κB has the ability to promote pro-survival pathways, and hence its inhibition results in the induction of apoptosis. For example, tumor necrosis factor (TNF) α can activate the NF-κB pathway, resulting in phosphorylation, degradation of IκB protein, formation and translocation of the dimer p50/p65 into the nucleus, and consequently, induction of the transcription of NF-κB target genes. New pathway players are still being identified through genomic screening approaches,[54] for example, by co-transfecting reporter gene constructs with cDNA libraries in the absence or presence of a stimulus. For inhibitor screens, cells transfected with the reporter are usually pretreated with compounds before a stimulus such as TNF-α is applied. A few marine natural products were identified as NF-κB inhibitors and are categorized based on their target. The sesterterpene lactones cacospongionolide B and petrosaspongiolide M isolated from the sponges *Fasciospongia cavernosa* and *Petrosaspongia nigra*, respectively, target IKK-dependent degradation of IκB.[55] Other compounds were found to target the proteolytic activity of the 26S proteasome, such as salinosporamide A isolated from the bacterium *Salinispora tropica*, and mycalolide A and agosterol C isolated from the marine sponges *Mycale* sp. and *Acanthodendrilla* sp., respectively.[56,57] Hymenialdisine, an alkaloid isolated from the marine sponge *Acanthella aurantiaca*, has been classified as an NF-κB-DNA binding inhibitor.[58] Moreover, there are several marine-derived NF-κB inhibitors with unknown exact mechanisms of action or molecular targets. Among these are verracurin A and ilimaquinone, isolated from the marine fungus *Myrothecium roridum* and the sponge *Hippospongia metachromia*, respectively.[52] Heteronemin, a sesterterpene isolated from the sponge *Hyrtios* sp., was also identified to target NF-κB signaling, based on DNA microarray profiling.[59] This was further confirmed by a luciferase reporter gene assay using chronic myeloid leukemia cancer cell line K562. Transfected cells were pretreated with different concentrations of heteronemin and challenged with TNF-α. A dose-dependent inhibition of the TNF-α-induced NF-κB activation was observed. In addition, heteronemin also differentially regulated genes implicated in several processes, such as apoptosis, cell cycle, and mitogen-activated protein kinase (MAPK) pathway. These results indicated heteronemin as a novel anti-inflammatory and apoptogenic marine natural product.[59] Cyclolinteinone is a sesterterpene from the Caribbean sponge *Cacospongia linteiformis* with effects on apoptosis, cell cycle, and NF-κB signaling. Anti-inflammatory properties were evident based on cyclolinteinone-mediated downregulation of iNOS and COX2 expression via inhibiting NF-κB activation. This study suggests the possible application of cyclolinteinone for inflammatory diseases that are characterized by elevated levels of nitric oxides and prostaglandins.[60]

Reporter gene assays have advantages with their sensitivity, reliability, and adaptability to HTS approaches. Moreover, such assays have the ability to monitor the cellular events before and after gene expression (i.e., signaling events and protein synthesis, respectively).[45] The continuous development of improved luminescence reporters, along with sensitive detection methods, will accelerate the discovery of novel small-molecule- and natural product-based therapies by identifying novel targets and providing high-throughput screening platforms.

7.2.2.5 High-Content Imaging

Advancements in technology, improvements in techniques to visualize the cell and subcellular compartments, the increase in computing power to quantitatively compare phenotypes across treatments, and the development of specific probes resulted in the advent of high-content imaging (HCI). HCI or high-content screening (HCS) refers to the high information content inherently present in cell images that can be compared for different conditions, such as drug treatment.[61] The basis of the technique involves the capture of image information and the automated analysis of those cellular images. Several imaging instruments are commercially available.[62] The differences between these imaging instruments is based on their throughputs, the quality versus quantity of data acquired, data analysis, and the size of the data.

This screening strategy can be applied for screening or profiling a library of natural products based on their elicited phenotypic and morphological fingerprints, therefore providing an insight into the mechanism of action (MOA). Identification of a unique fingerprint in HCI, as in the NCI60 screen, could possibly equate to a novel MOA elicited by the test compound. From the field of marine natural products there are several examples demonstrating the successful application of this screening strategy. An HCI strategy termed cytological profiling (CP) was applied to test a library of NPs by examining their cellular phenotypic features using automated fluorescence microscopy, and thus providing an insight into their MOA based on the phenotypic cellular fingerprints.[63] A library consisting of 624 marine-derived bacterial prefractions was screened, and the cytological profiles were compared to a set of commercially available compounds with known MOA. Extracts and pure compounds were successfully clustered based on MOA. Using this approach, microferrioxamines A–D were purified, and MOA prediction indicated these compounds as a previously undescribed family of iron siderophores. Moreover, based on the biological fingerprints of the prefractions, CP can also be successfully applied for dereplicating known compounds.[63]

Another study used automated HCI analysis to perform morphological profiling of a library consisting of around 400 purified natural products.[64] Nine of the compounds, within 1 h of treatment, demonstrated morphological alteration in HeLa cells characterized by marked nuclear protrusion. Although seven of those natural products were known to target actin, this strategy led to the identification of two marine natural products with previously uncharacterized targets, bisebromoamide and miuraenamide. The automated HCI analysis classified these compounds as actin filament stabilizers based on their ability to induce morphological changes similar to those induced by known actin filament stabilizers. This was further confirmed by *in vitro* polymerization and de-polymerization experiments. Furthermore, using nuclear protrusion as a morphological feature to aid the identification of actin-targeting compounds, they were able to identify and isolate pectenotoxin-2 and lyngbyabellin C from 11 marine sponge extracts. Therefore, HCI demonstrated its success by providing a sensitive method for the identification of actin-targeting molecules.[64] Observation of the changes in the morphological profile of cancer cells after small-molecule treatment was further extended to serve as an encyclopedic cell morphology database. This chemical–genetic profiling database, called Morphobase, utilizes HeLa and src^{ts}-NRK for cytological profiling of approximately 200 reference compounds with known mechanisms of action.[65] HCI analysis allowed for evaluation of 12 morphological parameters, analyzed using a multivariate statistical tool.[65] Drugs that share similar mechanisms of action cluster in the multivariate analysis.[65] Morphobase was successfully employed in identifying the mechanisms of action of three tubulin poisons.[65]

Another group has developed an RNA fluorescence-based high-content live-cell imaging (HCLCI) strategy to screen a marine NP library consisting of 2685 extracts for antimalarial activity against *P. falciparum*.[66] The extracts were added to cultures in 96-well plates and subjected to a primary screen using the standard SYBR Green I-based fluorescence growth inhibitory assay. The identified hits were subjected to a secondary screen using the HCLCI strategy, followed by purification of interesting extracts. The methodology of HCLCI involves the growth of parasites *in vitro* in the presence of extracts, followed by staining with RNA-sensitive dye, and subsequent imaging at particular time intervals using a Becton Dickinson (BD) Pathway high throughput (HT) automated confocal microscope. Using this strategy, nine extracts with potent parasite growth inhibition activity and minimal cytotoxicity to the host red blood cells (RBCs) were successfully identified by monitoring membrane integrity. The results were validated by testing the purified natural product bromophycolide A from the red macroalga *Callophycus serratus*, which induced morphological changes similar to those of the crude extracts.[66] This strategy also enables phenotypic analysis of parasite morphology incubated with extracts. Examination of the morphological changes of parasites has the potential to give an insight into the MOA of the active extract. This was demonstrated when extracts containing bromophycolide A were incubated with parasite cultures. The extract elicited a distinct phenotype of cell cycle arrest, and a lack of hemozoin (by-product of hemoglobin catabolism) formation was observed. The inhibition of hemozoin formation is therefore considered a potential MOA for bromophycolide A.[66]

The Linington group developed a 384-well high-throughput imaged-based screening approach aimed at identifying biofilm inhibitors and dispersion agents as new therapeutic options for the treatment of *P. aeruginosa* biofilm-associated infections.[67] This was accomplished by screening a library of 312 marine-derived NP (microbial) prefractions. In this approach, to measure biofilm coverage, the screen was carried out using non-z-stack epifluorescence microscopy to image a GFP-tagged strain of *P. aeruginosa*. The results were quantified using an automated image analysis script that has the ability to segment and measure the bright regions of biofilm coverage. To simulate a model for the inhibition of biofilm formation, the compounds were added to the test plates immediately after the addition of the bacterial culture. However, in order to design a model for the induction of biofilm dispersion, the cultures were incubated for 2 h first to allow biofilm formation. Then, the compounds from the screening library were added. In order to differentiate between classical antibiotics and biofilm inhibitors/dispersers, a colorimetric assay measuring the bacterial cellular metabolic activity was carried out simultaneously using the redox-sensitive dye XTT (2,3-bis(2-methoxy-4-nitro-5-sulfophenyl)-5-[(phenylamino)carbonyl]-2H-tetrazolium hydroxide). The screen resulted in the identification of five hits in the biofilm inhibition model and five hits in the biofilm dispersion model. Additionally, two hits were identified as active in both models. Further evaluation of one of the active fractions identified the cyclic depsipeptide NPs skyllamycins B and C as biofilm inhibitors with EC_{50} (half-maximal effective concentration) values of 30 and 60 µM, respectively. Skyllamycin B showed a biofilm dispersion activity with an EC_{50} value of 60 µM. Skyllamycin B was evaluated in a biofilm dispersion model assay in combination with azithromycin. Confocal images showed that the combination treatment results in the detachment of biofilms as well as the reduction in cellular metabolic activity.

HCI is a useful strategy for screening a library of NPs based on morphological changes elicited by active compounds. It can be used to identify the putative MOA of extracts early on during the drug discovery process. This will be useful in terms of hit prioritization and selecting extracts with MOA-specific fingerprints for further purification. However, one should keep in mind, as with CP, that the concentration of active compounds within crude extracts is not known, and therefore a high concentration or very low concentration could affect the profiles, resulting in death or inactivity. Another issue that needs to be considered is the presence of multiple active compounds with different MOAs. This may result in a fingerprint diverging from the MOA of each individual compound or corresponding to a compound with polypharmacology. Superimposed MOAs may be cooperative, additive, or antagonistic and not necessarily act in parallel but be in cross talk, thus complicating the interpretation. In this light, crude extracts may not be advisable for HCI screening, but this method works well in conjunction with prefractionation, where there is a simplified mixture of compounds, often containing a single major component. HCI is also limited to inhibitors that elicit a phenotype and is not applicable in finding inhibitors that cause negligible changes in phenotype.

7.2.3 Selected Model Organisms

The completion of full-genome sequencing projects highlighted the fact that many human genes are conserved through evolution from yeast to humans, including those associated with diseases.

In comparison with human genes, 43% of *Caenorhabditis elegans*, 46% of *Saccharomyces cerevisiae*, 61% of *Drosophila melanogaster*, 80% of *Danio rerio*, and up to 97% of *Mus musculus* genes show similarity to human genes,[68] which provides an opportunity to use model organisms, particularly *C. elegans* and *D. rerio*, for drug discovery and target identification. Screening using model organisms provides advantages compared with target- and cell-based assays in that it provides the physiological context for screening small molecules, and therefore yields more predictive results regarding cell permeability, bioavailability, efficacy, and toxicity (Table 7.1). The small size of simple model organisms, high fecundity, short development time, and amenability to HTS allow them to be successfully employed in a variety of applications for drug discovery, including the screening of small molecules, identification and validation of potentially novel therapeutic compounds, assessment of toxicity, and structural–activity relationship (SAR) and MOA studies. Simple model organisms have largely replaced rodent models as a primary screen for bioactivity *in vivo*.

7.2.3.1 *Caenorhabditis elegans*

C. elegans is a model organism increasingly used as a platform for drug discovery and development. In 1960, this model organism was used by Sydney Brenner in developmental biology to study animal development and their nervous system.[69–71] Since then, the use of *C. elegans* has been expanded to study a variety of biological processes, including, but not limited to, cell cycle, cell signaling, apoptosis, and gene regulation, as well as modeling of human diseases and for drug screens. Several studies and breakthrough discoveries revealed a strong relationship between this nematode and mammals in many aspects. The cellular and molecular pathways were shown to be highly conserved between these two types of organisms. Also, genome comparisons demonstrated that the majority of human disease genes were present in this invertebrate, and therefore several human diseases were modeled in *C. elegans*.[72]

C. elegans is small in size (length 1 mm, diameter 80 μm), has a short life cycle (~3 days), and can be grown on both solid agar and liquid media. It is transparent, thus allowing straightforward visualization and monitoring of cellular processes.[71–73] Owing to those features, *C. elegans* is now an attractive model for drug screening, identifying compound targets, and studying the mechanisms of their actions.

To identify novel bioactive compounds, small molecules can be screened using wild-type or mutant *C. elegans*, searching for compounds that induce or suppress a phenotype or using a transgenic reporter line.[73] The basic methodology involves the cultivation of wild-type or mutant strains of *C. elegans*, followed by usually sorting into 96- or 384-well plates (1–100 worms/well) using a worm sorter. The worms are exposed to compounds, usually one compound per well, for a certain period of time, depending on the particular assay used. The resulting phenotype is measured using a multiwell reader, which detects the fluorescence resulting from the expression of a fluorescent protein by the worms.[71] A method has been developed using fluorescent strains of worms for HTS of chemical libraries.[74] Since many inducible transcriptional pathways in *C. elegans* are well defined and have GFP transgenic reporters available, this method can be used to screen a library of small molecules for pathway modulators. In one example, hits from a pathway-directed HTS campaign were validated by testing them using a reporter strain of *C. elegans* to identify bioavailable ARE activators. This model organism possesses an inducible thiol-sensitive cytoprotective pathway similar to that seen in vertebrates. The transcription factor SKN-1 and the repressor WDR-23 are homologous to Nrf2 and Keap1, respectively.[75] When exposed to an oxidative insult, WDR-23 activates the expression of cytoprotective genes such as glutathione S-transferase 4 (*gst*-4).[75] This system enabled the identification of ARE inducer 3 (AI-3) as a bioavailable activator of cytoprotective genes in worms, but also in mice.[76] An interesting related example was the development of an ultra-HTS platform for screening small-molecule modulators of this pathway in *C. elegans*. It has been shown that genes regulated by SKN-1 are involved in multidrug resistance in parasitic nematodes, and therefore screening for small-molecule inhibitors of SKN-1 has the potential to also identify compounds that may reverse drug resistance. Leung et al. screened the National Institutes of Health (NIH) Molecular Libraries Small Molecule Repository (MLSMR) of ~364,000 compounds using a total of 272 × 1536-well plates, and the work was accomplished in 5 weeks.[77] The natural products phloretin and phloroacetophenone were among several hit compounds that were identified. This screening platform was designed to be applicable for screening small molecules targeting gene regulatory pathways.

C. elegans has, however, some limitations, including the thick cuticle surrounding the worm, which acts as barrier to the penetration and uptake of the drug. Positive results are qualitative only, and the negative ones are difficult to interpret.[78] Moreover, some of the molecular pathways relevant to human diseases might be absent in *C. elegans*.[71] Regardless of the limitations, *C. elegans* continues to be a powerful model for drug discovery.

7.2.3.2 *Danio rerio (Zebrafish)*

Over the past decade, zebrafish (*Danio rerio*) have been widely used as a cost-effective, yet powerful research tool for drug discovery and development. Unlike other model organisms, zebrafish are vertebrates containing discrete organs and tissues that are similar to those found in humans with respect to the anatomical, physiological, and molecular levels, and therefore better reflecting biological processes in humans. The small size of embryos (1–5 mm) allows for their utility in HTS campaigns. Furthermore, the transparency of the embryos enables the observation of internal organs and monitoring of the effects of small molecules on biological processes without the need for sacrificing or dissecting the animal.[79] Its closer morphological and physiological relationship to humans, in addition to the high degree of sequence similarity between human and zebrafish genes, is another advantage of this model organism[80] for small-molecule screening. Those compounds can ameliorate disease phenotypes and therefore could be potential drug candidates.[79] Several diseases have been successfully modeled in zebrafish, such as cardiomyopathy, muscular dystrophy, Alzheimer's disease, congenital dyserythropoietic anemia, and many others.[81] Their short development time enables researchers to obtain readouts on the order of hours to only several days. Developmental defects can provide clues about the well-characterized pathways affected in the zebrafish, linking phenotypes immediately to mechanism of action and target.

Several screening methods have been employed using zebrafish, which highlights the potential application of zebrafish in each of the four areas of the drug discovery process: screening and lead compound identification, target identification, validation, and assessment and optimization.

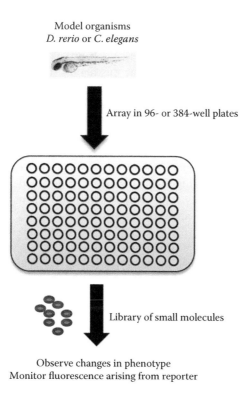

Model organisms
D. rerio or *C. elegans*

Array in 96- or 384-well plates

Library of small molecules

Observe changes in phenotype
Monitor fluorescence arising from reporter

Figure 7.5 The model organisms *C. elegans* and *D. rerio* can be utilized for high-throughput monitoring of the bioactivity of small molecules by monitoring changes in phenotype or fluorescence arising from reporter strains.

The screening approach involves arraying the embryos or larvae into 96- or 384-multiwell plates, and they are kept for hours or days postfertilization (hpf or dpf), depending on the desired time point of the assay before treatment with a library of small molecules. The uptake of those small molecules to zebrafish embryos is simply through diffusion from the surrounding media. Phenotypic changes that result from the effects of small molecules will then be analyzed (Figure 7.5).[82] The advantage of this system is that it will provide information about the effects of small molecules within the context of whole organisms since the screened molecules will act on different types of cells, whereas the results obtained from cell-based screens are specific to the type of cells being used.[82,83] The zebrafish system represents a valuable model for identifying and validating drug targets since it is amenable to large-scale genetic screenings that are conducted to identify mutations in genes that affect an organ or tissue of interest.[81] In forward genetic screens (chemical or insertional mutagenesis), novel drug targets can be discovered through the identification of novel genes underlying zebrafish disease phenotypes. Moreover, disease phenotypes can be further screened to identify drugs that can reverse them.[84] Zebrafish have been a valuable tool in studying vertebrate development through linking genes to their corresponding mutant phenotypes. This can be accomplished by conducting a saturation mutagenesis screen to identify mutants with observable phenotypes. Such mutations occur in genes affecting a particular developmental pathway. The first kind of such a screen was accomplished by the groups of Nusslein-Volhard and Driever.[85,86] The majority of mutations were identified based on morphological observation of the resultant phenotypes owing to the transparency of zebrafish. Mutations were grouped according to the elicited phenotype and were found to cover most aspects of zebrafish embryogenesis.[87] Such genetic screens are useful to maximize our understanding of vertebrate development and provide a way for dissecting the molecular basis underlying the developmental processes. This is useful for drug discovery where one can identify the target pathway based on the elicited phenotype by small molecules (phenocopy).

In one study, a zebrafish-based phenotypic screen was conducted using 954 different compounds, and the resultant phenotypes induced by this library were juxtaposed with known genetic mutations. The screen involved assaying wild-type embryos in 96-well plates and treating them with a library of small molecules at 7 hpf, followed by scoring the phenotypes at 32 hpf. The screen identified kalihinol F, a marine natural product from the sponge *Acanthella* sp. Kalihinol F elicited a characteristic phenotype that matched with the zebrafish mutant *calamity* (*cal*). The *cal* mutant displays copper deficiency as a result of mutation in copper transporter gene *atp7a*. The elicited phenotype was reversed with exogenous copper, suggesting copper chelation as the mechanism of action of kalihinol F. This study highlights the utility of this model organism in evaluating the biological effects of small molecules, with some insight into their mode of action by performing chemical screens coupled to genetic analysis.[88]

A zebrafish phenotypic assay was also utilized in screening 160 prefractionated extracts obtained from marine invertebrates.[89] Among 160 extracts, 25% showed toxicity to embryos. One of the active extracts from the sponge *Xestospongia* cf. *carbonaria* resulted in abnormalities in notochord development and death. Further purification of this extract resulted in the isolation of six pyridoacridine alkaloids, with three being novel. Amphimedine was the only one that demonstrated a phenotype following the evaluation of pure compounds using the zebrafish assay, and therefore it was identified as the agent responsible for eliciting the notochord abnormalities. Despite the low abundance in the parent extract, the compound was able to induce a phenotype in zebrafish.[89] The marine natural product apratoxin A derived from a cyanobacterium induced fibroblast growth factor (FGF)-dependent developmental defects, such as pectoral fin truncation and posterior mesoderm issues, and downregulated target genes of FGF signaling in zebrafish.[90] Those observations were similar yet not identical to those of phenotypes induced by SU5402, an FGF receptor inhibitor. The reduced expression of FGF signaling target genes was consistent with later findings that the compound reduces various RTK signaling pathways, including FGFR, by inhibiting cotranslational translocation.[91]

In summary, zebrafish, like other model organisms amenable to HTS, can be used as an effective tool to prioritize samples while obtaining useful information on biomarkers for subsequent low-throughput and expensive rodent studies.

7.3 TARGET IDENTIFICATION AND MECHANISM OF ACTION STUDIES

Target identification for small molecules is an integral and often challenging aspect of the drug discovery process. It refers to the process of identifying biomolecules (usually proteins or genes encoding them) responsible for eliciting the cellular effect of the small molecule of interest upon binding (Table 7.2). Unfortunately, the process of identifying the target is underdeveloped in the field of marine natural product research, despite the large number of compounds being discovered each year. Therefore, developing and applying new methods for identifying the target is crucial to maximize the benefit of the newly discovered compounds by revealing their pharmacological effects and possible unwanted off-target effects. Generally, two strategies are available for target identification that involve the application of either affinity-based methods or phenotype-based methods. The affinity-based methods involve the detection of direct binding between the molecule of interest and its target (usually protein), whereas phenotype-based methods rely on the detection of physiological processes elicited by the compounds of interest for the identification of the target.[92] Several strategies aimed at identifying the molecular target and mechanism of action of marine natural products interrogate the genome, transcriptome, or proteome that may be used in mammalian cells or model organisms (Table 7.2). Each method is not stand-alone and requires validation

Table 7.2   Summary of Methods for Mechanism of Action Studies and Target Identification

Parameters	Chemical Genomics	Transcriptomics	Proteomics
		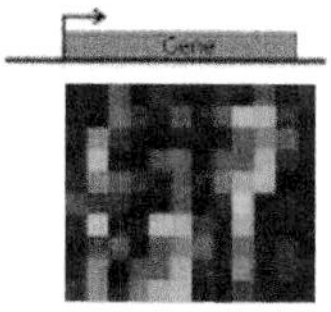	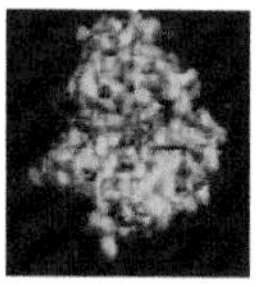
System interrogated	Genome	Transcriptome	Proteome
Methods	Haploinsufficiency-Homozygous deletion (HIP-HOP) cDNA overexpression siRNA deletion	RNASeq	Chemoproteomics DARTS
Information on mechanism of action?	Yes	Yes	No
Information of drug target?	Possible	Possible	Yes
High-throughput?	Yes	Yes	Yes
Comprehensive?	Yes	Yes	Yes
Chemical modification required?	No	No	Yes
Usable in mammalan cell system?	Limited	Yes	Yes, cell lysates not whole cells
Quantitation possible?	Yes	Yes	Yes
Interrogates mechanism of resistance?	Yes (overexpression)	Yes	No

using a complementary technique. As there are a multitude of possible molecular targets affected by compound treatment, experimental methods in target identification are geared toward comprehensive, unbiased, and high-throughput assessment (Table 7.2).

7.3.1 Affinity-Based Methods

Affinity-based approaches for target identification rely on the direct interaction between the drug and its target. These include matrix-based affinity detection (small molecules fused with solid support/moiety), matrix-free affinity labeling (small molecules labeled with radioisotope or fluorescent labels), and use of immunoaffinity fluorescent (IAF) tags and the newly developed drug affinity responsive target stability (DARTS). The general methodology for the first two involves tagging the small molecule of interest and incubating it with a complex mixture of proteins, followed by gel electrophoresis and mass spectrometry to identify the target protein (Table 7.2). IAF involves the use of an IAF probe and a corresponding anti-IAF antibody to allow both the cellular and the molecular characterization of the MOA of NPs.[93] In DARTS, on the other hand, small molecules are not labeled and incubated with protein mixtures in their native unmodified form.[92]

7.3.1.1 Chemoproteomic Approach

Among all target identification methods, affinity chromatography is considered the most widely used. This approach relies on structure–activity relationship (SAR) studies for the small molecule of interest to identify functional groups or positions that are not essential for bioactivity in order to be considered attachment sites for an affinity tag or solid matrix (Table 7.2). The success of affinity purification is dependent on both the affinity matrix and the spacer (linker), which links the solid matrix and ligand.[94] Ideally, the matrix should be macroporous, have chemical and physical stability, and be selective toward the target of interest. The spacer is preferred to be hydrophilic to reduce the nonspecific interactions. The length of the spacer should be considered during the design process. It is preferred not to be too short or too long to avoid a steric hindrance effect of the matrix or nonspecific interactions, respectively.[94]

The basic methodology involves incubating the derivatized small molecule with protein extracts, followed by subsequent washing steps to remove nonspecific binding proteins and the elution of target proteins (Figure 7.6a). Affinity chromatography can be carried out several ways, such as by using control beads or via competition with an unmodified compound. Generally, two ways of elutions can be carried out.[94] In specific elution, a competitive agent is used to challenge the ligand–protein complex, resulting in the release of a target protein. Nonspecific elution involves manipulating solvent conditions to decrease the association between the ligand and its target protein, resulting in the dissociation of the complex. This is followed by the analysis of eluted proteins by sodium dodecyl sulfate–polyacrylamide gel electrophoresis (SDS-PAGE), and protein identification is carried out using mass spectrometry (Figure 7.6a).[95] This approach, despite being widely used, has to meet certain conditions, and therefore has several limitations. It is often limited by chemistry and may not be applicable to every small molecule since the molecule of interest should have some functionality amenable to derivatization. Furthermore, the derivatization should not affect the binding/activity of the small molecule. The success of affinity chromatography is also dependent on the abundance of the protein target and the affinity of the small molecule, favoring highly abundant proteins with strong molecular interactions. Enrichment of the target protein using cellular fractionation can circumvent this limitation. Additionally, affinity chromatography of membrane-associated proteins is challenging, as these proteins are not separated very well with SDS-PAGE. Furthermore, natural products are often limited by the amount, and therefore might not be enough to undergo SAR studies. Another limitation includes some degree of nonspecific binding.[92,95] To overcome limitations related to specificity and sensitivity, affinity chromatography can be coupled with stable isotope labeling of amino acids in cell culture (SILAC), bypassing the gel electrophoresis step.[96] This approach relies as well on tandem mass spectrometry for target identification. In this quantitative proteomic approach, cell populations are cultured and labeled with light (natural isotope abundant forms) and heavy (^{13}C- and ^{15}N-bearing versions of arginine and lysine) amino acids. The lysates are then incubated with immobilized ligand (with or without unmodified ligand as competitor). The specific interacting proteins will be enriched in the heavy state over the light and will have differential ratios when analyzed by mass spectrometry. In contrast, nonspecific protein interactions will have ratios close to 1, suggesting equal enrichment in both the light and the heavy state. This approach, when coupled to affinity chromatography, has the advantages of both the identification and quantification of direct/specific interacting proteins, as well as prioritizing the target proteins based on their SILAC ratios.[96] Affinity chromatography has been successfully implicated in the discovery of the eukaryotic translation initiation factor (eIF)-4A as a target of the marine natural product pateamine A. Pateamine is a marine natural product isolated from sponge with antiproliferative and immunosuppressive activity. Its activity is therefore related to the inhibition of the eukaryotic translation initiation, which plays a role in cell survival and proliferation.[97]

The immunoaffinity fluorescent (IAF) tag methodology developed by James La Clair demonstrated its efficacy in both live cell imaging and target elucidation of natural products.[93,98] This technology involves the design of an IAF probe, usually via labeling the natural product of interest with 7-dimethylaminocoumarin-4-acetamide, to be assessed by fluorescence microscopy. Also, the design of anti-IAF antibodies with potent affinity toward the NPs labeled with 7-dimethylamino-coumarin-4-acetamide enables target elucidation via immunoaffinity assays.[93,98] This approach has been successful in elucidating the targets of several marine natural products. The extracts from the sponge *Agelas conifera* were evaluated for NPs that bind to *Escherichia coli* proteins using a bidirectional affinity approach.[99] The approach, through using affinity resins, allowed protein target identification in parallel with NP isolation. This was achieved by using resin-bound *E. coli* protein lysate to isolate NPs that have an affinity to *E. coli* proteins. This is followed by labeling the NP

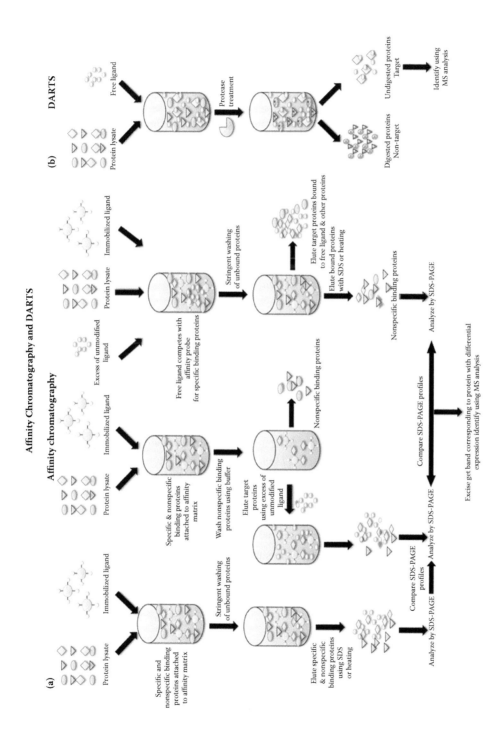

Figure 7.6 Methods for identification of direct target of small molecules. (a) The chemoproteomic approach relies on affinity purification of target proteins using small molecules immobilized on a solid support. (b) Drug affinity responsive target stability (DARTS) uses the native small molecule and is based on the proteolytic resistance of target molecules following ligand binding.

crude mixtures with IAF tag and subsequent co-immunoprecipitation of the NP binding proteins using anti-IAF mAb. LC-MS/MS protein ID analysis of the identified 40 kDa band revealed MreB as the target, which is known to play a role in bacterial cell wall assembly. NMR analysis of the material bound to the target suggested that it was sceptrin. IAF was also successful in identifying the target of napyradiomycins CNQ525.510B and A80915C, which were isolated from the marine actinomycetes.[100] Confocal microscopy images of the probes indicated the localization within the endoplasmic reticulum (ER) of HCT-116. The immunoaffinity precipitation identified a 95 kDa band that was identified by LC-MS/MS analysis to be Grp940, a heat shock protein found to localize with ER. The target of seriniquinone, a new NP isolated from a marine bacterium of the genus *Serinicoccus*, was identified by IAF tag.[101] Seriniquinone demonstrated a selective anticancer activity particularly toward melanoma cell lines. Therefore, identifying its target and the MOA is of particular interest. Confocal microscopy showed the localization within ER upon uptake. However, after 3 h, the cells underwent cell death due to the induction of autophagocytosis. Subsequent immunoaffinity assays identified the target to be dermcidin, a protein that plays a role in cancer proliferation. This technology has several advantages.[98] The use of a single probe enables studies to be conducted at both the cellular and molecular levels. Moreover, the use of a 7-dimethylaminocoumarin-4-acetic acid label is advantageous because it is nontoxic, produces minimal background, and therefore does not require extensive precleaning of the lysates.

7.3.1.2 Drug Affinity Responsive Target Stability

The interest in developing alternative strategies to overcome the limitation of affinity chromatography with respect to SAR and derivatization of small molecules led to the discovery of DARTS. The basic principle of DARTS relies on the fact that the binding of a small molecule to its target protein results in stabilization of the protein's structure. The equilibrium will shift toward the ligand-bound state. This thermodynamic stability will result in a decrease in the protein's conformational changes and unfolding and, consequently, an increase in the resistance to proteolysis. The general methodology involves incubating the native (unmodified small molecule) with protein lysates, followed by treatment with a protease (pronase) (Figure 7.6b). The nontarget proteins will be digested, whereas the target proteins will remain intact for longer as a result of being protease resistant due to their binding to their target ligands (Figure 7.6b). The key advantage of this approach is that SAR, and subsequent modification of small molecules, is not required. Since this approach is not limited by chemistry, it enables the identification of the target for natural product extracts prior to fractionation. The success of DARTS for target identification was demonstrated for didemnin B, a marine natural product whose target, Elongation factor (EF) 1α, was already identified. Lomenick et al. have demonstrated that EF-1α was protected from proteolysis in didemnin B-treated Jurkat cells. The lysates of didemnin B-treated cells were subjected to protease digestion using thermolysin and stained with Coomassie staining. The immunoblot showed a strong protected band around ~50 kDa.[92] In addition, DARTS has successfully identified a previously unknown target for the plant natural product resveratrol.[92,95]

7.3.2 Proteomic-Based Target Identification

Monitoring the effects of small molecules on the proteome has also been possible from recent development in mass spectrometry-based identification of proteins and protein purification. Such approaches have the ability to investigate the effects of compounds at the posttranslational level of proteins, unlike the chemogenomic or RNAseq methods discussed below. A successful example of this approach was demonstrated with the investigation of the molecular target of bengamides, marine natural products isolated from sponges, with potent tumor growth inhibitory activity both *in vitro* and *in vivo*. Identifying the target of bengamides has not been successful by NCI 60 cell line profiling, as it demonstrated a unique pattern of activity, suggesting a novel MOA. LAF389, the synthetic analogue of bengamide B, was tested by a several approaches and assays, but neither of them was successful in revealing the MOA. Assays for DNA binding and damage, topoisomerase binding, proteasome function, and polymerization of microtubules and actin showed no activity of LAF389. Also, studies on its transcriptional effect revealed no changes in mRNA transcription. Two-dimensional gel electrophoresis was utilized to interrogate changes in the proteome, relying on the altered mobility of proteins that interact with a synthetic variant of the natural product. A change in charge, rather than mass, was noted for such a set of proteins, which was suggestive of posttranslational modifications. Analysis led to the identification of 14-3-3 proteins that are known to modulate several processes, such as cell cycle and apoptosis via binding to specific phosphoserine-containing motifs. Further analysis revealed the retention of the N-terminal methionine upon bengamide treatment, suggesting an inhibitory effect against methionine aminopeptidases (MetAps). The target was validated by cocrystallization with the active species of bengamide. The structure shows that the binding of the inhibitor resembles that of peptide substrates. Furthermore, in the enzyme active site, the di-cobalt center is coordinated by the three hydroxy groups of bengamide.[102]

7.3.3 Chemogenomic Approach

Since gene deletion is functionally equivalent to small-molecule inhibition, the similarity between the chemical–genetic and genetic interaction profile would indicate phenocopying of the effects of gene deletion via small-molecule silencing, thus providing insights on the possible target pathway of the small molecule. For example, the MATα haploid deletion mutant of the model organism $S.$ $cerevisae$ treated with a small molecule will give a distinct chemical–genetic interaction profile.[103] This is then compared with genetic interaction profiles of double mutants of $S.$ $cerevisae.$[103]

$S.$ $cerevisae$ serves as a model organism in elucidating the mechanism of action of small molecules because of its amenability to genetic manipulations, similarity of the genome with humans, and presence of orthologues of disease-relevant human proteins. Overexpression and homozygous or heterozygous deletions of mutants of $S.$ $cerevisae$ also provide an opportunity to vary the gene dosage, leading to variation in the amount of protein target of small molecules, and consequently, either resistance or sensitivity to the small-molecule treatment is observed (Table 7.2). Increasing the gene dosage through cDNA overexpression or high-copy vector translates to an increased expression of proteins. The ordinarily lethal concentration of the small molecule will be insufficient to completely inhibit the overexpression mutants of the target protein, and hence resistance to the small molecule is observed. In contrast, haploinsufficiency mutants have one copy of the gene deleted and represent lowering of the gene dosage to 50%, while homozygous deletion mutants have both copies of the gene deleted. Deletion is facilitated by replacing the target gene with a deletion cassette containing unique barcodes for each deletion mutant, together with universal primers for amplification and KanMX4 for strain selection with successful deletion.[104] Both haploinsufficiency and homozygous deletion mutants lead to lowering and no expression of the target protein, respectively.[104] Heterozygous deletion mutants bearing the small-molecule target will then require sublethal amounts of the compound for inhibition, and hence hypersensitivity is observed. Susceptibility of homozygous deletion mutants of $S.$ $cerevisae$ to small-molecule treatment is also monitored, but rather than providing insights on the possible direct target, hits represent components of the signaling pathway perturbed by the small molecule. Homozygous deletion mutants are also limited to nonessential genes. Monitoring of the growth of multicopy suppression, haploinsufficiency, and homozygous deletion mutants of $S.$ $cerevisae$ with and without drug treatment is accomplished using either microarray analysis or barcode sequencing. Genomic or plasmid DNA is purified, amplified, and subsequently analyzed. These three methods are complementary to one another, and employing all three for profiling allows for streamlining of possible cellular targets, where the real target should confer resistance when overexpressed and sensitivity when deleted.[104]

The mechanism of action of the antiproliferative agent manzamine A from the sponge *Haliclona* sp. was probed using both haploinsufficiency and homozygous chemogenomic profiling.[105] The homozygous profile of manzamine A correlated with that of v-ATPase uncouplers.[105] This was further validated by fluorescence microscopy and comparison of the phenotype of treated cells with the known v-ATPase uncoupler bafilomycin A1.[105] While these methods are unbiased and allow for comprehensive evaluation of the mechanism of action and drug target, cellular effects of small molecules evaluated in mammalian cells may not translate into yeast, and compounds for evaluation should affect the proliferation of yeast. Mammalian counterparts of deletion and overexpression mutants have been facilitated using transfections of siRNA and cDNA, respectively. This is, however, limited by the efficiency of transfection. The mode of action of the marine natural product apratoxin A was examined through a number of approaches. Since no growth inhibitory activity of apratoxin A to yeast cells was observed, a genome-wide arrayed cDNA overexpression screen for cellular drug resistance was carried out.[90] This screen aimed to identify proteins that inhibit apratoxin A-induced apoptosis. The U2OS osteosarcoma cell line was chosen due to its high transfection efficiency and its sensitivity to apratoxin A-induced apoptosis. In this screen, 27,000 mammalian expression cDNAs along with reporter plasmid were transfected into U2OS cells. The screen was carried out in the presence and absence of apratoxin A. Following incubation and treatment with apratoxin A, cell lysis was carried out and the luciferase activity was measured. cDNAs conferring resistance to apratoxin A cytotoxic activity were selected. The apoptotic activity of apratoxin A was attenuated via the induction of fibroblast growth factor receptor (FGFR) signaling.[90] Another screen using the transfection of RNA interference-based short-hairpin RNS (shRNA) revealed the mode of action of nutlin-3. This antitumor small molecule is an inhibitor of MDM2, which can activate the p53 pathway. The application of the shRNA barcode screen identified 53BP1 as the cancer-specific mediator of nutlin-3 cytotoxicity.[106] The siRNA screen is also considered an approach to identify genes that modulate cellular responses to drugs. This was demonstrated in a study where a functional genomic approach was used to identify genes that confer resistance and sensitivity to tamoxifen.[107] Despite being widely used for the treatment of estrogen receptor positive breast cancer, resistance to tamoxifen is still considered an issue. A HTS of a library of 56,670 shRNAs targeting 16,487 human genes using an estrogen receptor positive breast cancer cell line (MCF7) was conducted. The screen, along with subsequent validation experiments, identified several genes that confer resistance (*BAP1, CLPP, GPRC5D, NAE1, NF1, NIPBL, NSD1, RAD21, RARG, SMC3*, and *UBA3*) as well as sensitivity (*C10orf72, C15orf55/NUT, EDF1, ING5, KRAS, NOC3L, PPP1R15B, RRAS2, TMPRSS2*, and *TPM4*) toward tamoxifen. Such a screen therefore allows for understanding gene expression alterations associated with diseases. Another study applied siRNA screening for the identification of the MOA of natural products through the development of Functional Signature Ontology (FUSION) maps.[108] Those maps linked the bioactive compounds within a library of natural products with their targets (proteins or biological process). In this strategy, gene expression signatures produced from a library of natural products were matched with signatures produced from libraries of siRNAs and synthetic microRNA (miRNA), through a combination of cell-based screening and a computational analysis. The natural product library consisted of 92 marine-derived bacterial strains and 20 marine invertebrate, whereas the siRNA and miRNA libraries were composed of 780 siRNAs targeting human kinases and 344 miRNA mimics. The NP library was screened for the ability to modulate reporter gene expression, and through computational analysis, the fusion map was created by computing the pairwise distances between NPs, siRNAs, and miRNAs. Several natural products were characterized in this study whose MOAs were previously underdescribed. The fraction designated SNB-019 isolated from *Streptomyces bacillaris* demonstrated a gene expression signature that correlated positively with ULK1 knockdown. Among the pure compounds isolated from this fraction, SNB-091-cmp1 was shown to inhibit autography, and structure determination revealed that this compound is identical to bafilomycin D. FUSION mapping also identified three natural products with signatures that resemble those produced by siRNA knockdown of discoidin domain receptor tyrosine kinase 2 (DDR2). One of the compounds isolated from *Bacillus hunanensis* was discoipyrrole A, which also inhibited platelet-derived growth factor (PDGF)-induced migration of fibroblasts. Those examples suggest that FUSION mapping is a successful strategy in identifying bioactive, clinically relevant natural products from natural product libraries.[108] Chemogenomics therefore continues to be a powerful approach for providing insights into the MOA of small molecules of interest.

7.3.4 RNAseq

With the advent of next-generation sequencing strategies and high computing capabilities, it is now readily possible to perform *de novo* sequencing of the genome and transcriptome of cell lines and model organisms. Transcriptome sequencing was utilized to analyze changes in the RNA of cell lines that developed resistance to small molecules upon prolonged drug treatment.[109] The resistance selected for colonies that conferred mutations to the biological target of small molecules (Table 7.2).[109] The proof of concept for this method of target identification was demonstrated for the pololike kinase 1 inhibitor BI2536 and the proteasome inhibitor bortezomib.[109] RNAseq also captured mechanisms of resistance that are independent of the drug target, such as upregulation of the expression of efflux transporters.[109]

7.3.5 X-Ray Cocrystallography and Molecular Docking

While the previously mentioned experimental methods have been successful in interrogating the mechanism of action and direct target of small molecules, no information regarding the molecular basis of target inhibition can be derived. This information may be obtained by X-ray cocrystallography and molecular docking experiments. Visualization of the molecular basis of drug action has a pivotal role in the development of second-generation inhibitors with improved potency and enables the establishment of a generalized pharmacophore of inhibitors. The utility of X-ray cocrystallization for deriving information on the chemistry of inhibitors has been successfully demonstrated for HDACs. The X-ray cocrystallization of hydroxamate inhibitors trichostatin A and SAHA with HDAC8 and histone deacetylase-like protein (HDLP) served as the basis for the generalized HDAC inhibitor pharmacophore consisting of a Zn^{2+} binding group, a four- to six-carbon spacer, and a cap moiety that interacts with the divergent regions of HDACs.[110,111] Insights on the selectivity of depsipeptide-based HDAC inhibitors were also rationalized with the cocrystallization of the marine-derived largazole thiol and HDAC8.[112] The successful visualization of the molecular interactions between largazole thiol and HDAC8 has validated the basis of the potent and selective activity of the depsipeptide-based HDAC inhibitors and provides a framework for the design of next-generation analogues. X-ray crystallography-guided improvement to inhibitor design was also successfully employed for small molecules targeting β-site APP cleaving enzyme type 1 (BACE1), using the marine natural product tasiamide B as the template.[113] Two analogues, consisting of the central phenylstatine unit critical for potent BACE1 activity, a tasiamide B-like C-terminus, and an isophthalic acid-based N-terminus, allowed for the design of inhibitors with potent activity against BACE1, both *in vitro* and *in vivo*.[113] The high affinity of natural products for their biological targets usually favors their cocrystallization with their target proteins; hence, X-ray analysis may be successful in deciphering critical molecular interactions and developing a second generation of inhibitors.

7.4 CONCLUSION

The marine environment is a rich source of structurally diverse, interesting, bioactive natural products with wide therapeutic potential. Therefore, developing new screening approaches that enable rapid and effective isolation of such valuable natural products is crucial. Several strategies have been applied to the screening of crude extracts, thus providing effective means of sample prioritization. They are also valuable for guiding and accelerating the purification process. With the large number of interesting marine natural products being discovered each year, the development and implementation of new strategies for further characterization of those discovered bioactive NPs are necessary through target identification and mechanism of action studies. Several examples from the field of marine NPs were highlighted, demonstrating the success of various methods. Several other strategies have not yet been utilized in our field that could benefit from more rigorous mechanistic and target identification studies, and which would add more value to the compounds that are being discovered. Each strategy for bioactivity screening and interrogation of the mechanism of action of small molecules has its inherent advantages and disadvantages, which must be carefully considered. Furthermore, bioactivity screening methods are also being designed to provide initial information on the mechanism of action or direct target, which will aid in deconvoluting the cellular effects of marine natural products and hopefully will accelerate the cumbersome target identification step. We believe that the successful implementation of such strategies will maximize our discoveries and ultimately can result in the development of new generations of drug candidates that are effective against a wide array of therapeutic indications.

REFERENCES

1. Hurko, O. Target-based drug discovery, genetic diseases, and biologics. *Neurochem Int* 2012, 61 (6), 892–98.
2. Zhang, G. Protease assays. In *Assay Guidance Manual*, ed. T. Holler, A. Napper. Eli Lilly & Company and the National Center for Advancing Translational Sciences, Bethesda, MD, 2012, pp. 1–14.
3. Puente, X. S., Sanchez, L. M., Overall, C. M., Lopez-Otin, C. Human and mouse proteases: A comparative genomic approach. *Nat Rev Genet* 2003, 4 (7), 544–58.
4. Zhou, C., Li, W., Chen, J., Yang, M., Li, Y., Zhu, J., Yu, C. Real-time fluorometric turn-on assay for protease activity and inhibitor screening with a benzoperylene probe. *Analyst* 2014, 139 (5), 1057–62.
5. Salvador, L. A., Taori, K., Biggs, J. S., Jakoncic, J., Ostrov, D. A., Paul, V. J., Luesch, H. Potent elastase inhibitors from cyanobacteria: Structural basis and mechanisms mediating cytoprotective and anti-inflammatory effects in bronchial epithelial cells. *J Med Chem* 2013, 56 (3), 1276–90.
5a. Kwan, J., Eksioglu, E., Liu, C., Paul, V., Luesch, H. Grassystatins A-C from marine cyanobacteria: Potent cathespin E inhibitors that reduce antigen presentation. *J. Med. Chem.* 2009, 52 (18), 5732–5747.
6. Kwan, J. C., Liu, Y., Ratnayake, R., Hatano, R., Kuribara, A., Morimoto, C., Ohnuma, K., Paul, V. J., Ye, T., Luesch, H. Grassypeptolides as natural inhibitors of dipeptidyl peptidase 8 and T-cell activation. *ChemBioChem* 2014, 15 (6), 799–804.
7. Jessani, N., Cravatt, B. F. The development and application of methods for activity-based protein profiling. *Curr Opin Chem Biol* 2004, 8 (1), 54–59.
8. Gillet, L. C., Namoto, K., Ruchti, A., Hoving, S., Boesch, D., Inverardi, B., Mueller, D., Coulot, M., Schindler, P., Schweigler, P., Bernardi, A., Gil-Parrado, S. In-cell selectivity profiling of serine protease inhibitors by activity-based proteomics. *Mol Cell Proteomics* 2008, 7 (7), 1241–53.
9. Stolze, S. C., Deu, E., Kaschani, F., Li, N., Florea, B. I., Richau, K. H., Colby, T., van der Hoorn, R. A., Overkleeft, H. S., Bogyo, M., Kaiser, M. The antimalarial natural product symplostatin 4 is a nanomolar inhibitor of the food vacuole falcipains. *Chem Biol* 2012, 19 (12), 1546–55.
10. Dominguez, E., Galmozzi, A., Chang, J. W., Hsu, K. L., Pawlak, J., Li, W., Godio, C., Thomas, J., Partida, D., Niessen, S., O'Brien, P. E., Russell, A. P., Watt, M. J., Nomura, D. K., Cravatt, B. F., Saez, E. Integrated phenotypic and activity-based profiling links Ces3 to obesity and diabetes. *Nat Chem Biol* 2014, 10 (2), 113–21.
11. Vu, H., Pham, N. B., Quinn, R. J. Direct screening of natural product extracts using mass spectrometry. *J Biomol Screen* 2008, 13 (4), 265–75.

12. Kim, D. H., Sim, T. Chemical kinomics: A powerful strategy for target deconvolution. *BMB Rep* 2010, 43 (11), 711–19.

13. Bantscheff, M., Hopf, C., Kruse, U., Drewes, G. Proteomics-based strategies in kinase drug discovery. *Ernst Schering Found Symp Proc* 2007, (3), 1–28.

14. Lemeer, S., Zorgiebel, C., Ruprecht, B., Kohl, K., Kuster, B. Comparing immobilized kinase inhibitors and covalent ATP probes for proteomic profiling of kinase expression and drug selectivity. *J Proteome Res* 2013, 12 (4), 1723–31.

15. Skropeta, D., Pastro, N., Zivanovic, A. Kinase inhibitors from marine sponges. *Mar Drugs* 2011, 9 (10), 2131–54.

16. Rodriguez, J., Peters, B. M., Kurz, L., Schatzman, R. C., McCarley, D., Lou, L., Crews, P. An alkaloid protein kinase C inhibitor, xestocyclamine A, from the marine sponge *Xestospongia* sp. *J Am Chem Soc* 1993, 115 (22), 10436–37.

17. Patil, A. D., Freyer, A. J., Killmer, L., Offen, P., Carte, B., Jurewicz, A. J., Johnson, R. K. Frondosins, five new sesquiterpene hydroquinone derivatives with novel skeletons from the sponge *Dysidea frondosa*: Inhibitors of interleukin-8 receptors. *Tetrahedron* 1997, 53 (14), 5047–5060.

18. Bifulco, G., Bruno, I., Minale, L., Riccio, R., Debitus, C., Bourdy, G., Vassas, A., Lavayre, J. Bioactive prenylhydroquinone sulfates and a novel C31 furanoterpene alcohol sulfate from the marine sponge, *Ircinia* sp. *J. Nat. Prod.* 1995, 58 (9), 1444–49.

19. Bantscheff, M., Eberhard, D., Abraham, Y., Bastuck, S., Boesche, M., Hobson, S., Mathieson, T., Perrin, J., Raida, M., Rau, C., Reader, V., Sweetman, G., Bauer, A., Bouwmeester, T., Hopf, C., Kruse, U., Neubauer, G., Ramsden, N., Rick, J., Kuster, B., Drewes, G. Quantitative chemical proteomics reveals mechanisms of action of clinical ABL kinase inhibitors. *Nat Biotechnol* 2007, 25 (9), 1035–44.

20. Zang, R., Li, D., Tang, I., Wang, J., Yang, S. Cell-based assays in high-throughput screening for drug discovery. *Int J Biotech Well Indus* 2012, 1, 31–51.

21. Riss, T. Selecting cell-based assays for drug discovery screening. *Cell Notes* 2005, 13, 16–21.

22. Shoemaker, R. H. The NCI60 human tumour cell line anticancer drug screen. *Nat Rev Cancer* 2006, 6 (10), 813–23.

23. Bai, R. L., Paull, K. D., Herald, C. L., Malspeis, L., Pettit, G. R., Hamel, E. Halichondrin B and homohalichondrin B, marine natural products binding in the vinca domain of tubulin. Discovery of tubulin-based mechanism of action by analysis of differential cytotoxicity data. *J Biol Chem* 1991, 266 (24), 15882–89.

24. Wong, W. R., Oliver, A. G., Linington, R. G. Development of antibiotic activity profile screening for the classification and discovery of natural product antibiotics. *Chem Biol* 2012, 19 (11), 1483–95.

25. Baines, A. T., Xu, D., Der, C. J. Inhibition of Ras for cancer treatment: The search continues. *Future Med Chem* 2011, 3 (14), 1787–808.

26. Lim, S. M., Westover, K. D., Ficarro, S. B., Harrison, R. A., Choi, H. G., Pacold, M. E., Carrasco, M., Hunter, J., Kim, N. D., Xie, T., Sim, T., Janne, P. A., Meyerson, M., Marto, J. A., Engen, J. R., Gray, N. S. Therapeutic targeting of oncogenic K-Ras by a covalent catalytic site inhibitor. *Angew Chem Int Ed Engl* 2014, 53 (1), 199–204.

27. Ostrem, J. M., Peters, U., Sos, M. L., Wells, J. A., Shokat, K. M. K-Ras(G12C) inhibitors allosterically control GTP affinity and effector interactions. *Nature* 2013, 503 (7477), 548–51.

28. Chan, D. A., Giaccia, A. J. Harnessing synthetic lethal interactions in anticancer drug discovery. *Nat Rev Drug Discov* 2011, 10 (5), 351–64.

29. Luo, J., Emanuele, M. J., Li, D., Creighton, C. J., Schlabach, M. R., Westbrook, T. F., Wong, K. K., Elledge, S. J. A genome-wide RNAi screen identifies multiple synthetic lethal interactions with the Ras oncogene. *Cell* 2009, 137 (5), 835–48.

30. Sarthy, A. V., Morgan-Lappe, S. E., Zakula, D., Vernetti, L., Schurdak, M., Packer, J. C., Anderson, M. G., Shirasawa, S., Sasazuki, T., Fesik, S. W. Survivin depletion preferentially reduces the survival of activated K-Ras-transformed cells. *Mol Cancer Ther* 2007, 6 (1), 269–76.

31. Shaw, A. T., Winslow, M. M., Magendantz, M., Ouyang, C., Dowdle, J., Subramanian, A., Lewis, T. A., Maglathin, R. L., Tolliday, N., Jacks, T. Selective killing of K-ras mutant cancer cells by small molecule inducers of oxidative stress. *Proc Natl Acad Sci USA* 2011, 108 (21), 8773–78.

32. Tasdemir, D., Bugni, T. S., Mangalindan, G. C., Concepcion, G. P., Harper, M. K., Ireland, C. M. Cytotoxic bromoindole derivatives and terpenes from the Philippine marine sponge *Smenospongia* sp. *Z Naturforsch* C 2002, 57 (9–10), 914–22.

33. Kwon, H. J., Owa, T., Hassig, C. A., Shimada, J., Schreiber, S. L. Depudecin induces morphological reversion of transformed fibroblasts via the inhibition of histone deacetylase. *Proc Natl Acad Sci USA* 1998, 95 (7), 3356–61.

34. Sugita, K., Koizumi, K., Yoshida, H. Morphological reversion of sis-transformed NIH3T3 cells by trichostatin A. *Cancer Res* 1992, 52 (1), 168–72.

35. Salvador, L., Luesch, H. HDAC inhibitors and other histone modifying natural products as emerging anticancer agents. In *Natural Products and Cancer Drug Discovery*, ed. F. E. Koehn. Springer, New York, 2013, pp. 59–95.

36. Salvador, L. A., Luesch, H. Discovery and mechanism of natural products as modulators of histone acetylation. *Curr Drug Targets* 2012, 13 (8), 1029–47.

37. Ueda, H., Nakajima, H., Hori, Y., Goto, T., Okuhara, M. Action of FR901228, a novel antitumor bicyclic depsipeptide produced by *Chromobacterium violaceum* no. 968, on Ha-ras transformed NIH3T3 cells. *Biosci Biotechnol Biochem* 1994, 58 (9), 1579–83.

38. Taori, K., Paul, V. J., Luesch, H. Structure and activity of largazole, a potent antiproliferative agent from the Floridian marine cyanobacterium *Symploca* sp. *J Am Chem Soc* 2008, 130 (6), 1806–7.

39. Ying, Y., Taori, K., Kim, H., Hong, J., Luesch, H. Total synthesis and molecular target of largazole, a histone deacetylase inhibitor. *J Am Chem Soc* 2008, 130 (26), 8455–59.

40. Hong, J., Luesch, H. Largazole: From discovery to broad-spectrum therapy. *Nat Prod Rep* 2012, 29 (4), 449–56.

41. Nakao, Y., Yoshida, S., Matsunaga, S., Shindoh, N., Terada, Y., Nagai, K., Yamashita, J. K., Ganesan, A., van Soest, R. W., Fusetani, N. Azumamides A–E: Histone deacetylase inhibitory cyclic tetrapeptides from the marine sponge *Mycale izuensis*. *Angew Chem Int Ed Engl* 2006, 45 (45), 7553–57.

42. Pina, I. C., Gautschi, J. T., Wang, G. Y., Sanders, M. L., Schmitz, F. J., France, D., Cornell-Kennon, S., Sambucetti, L. C., Remiszewski, S. W., Perez, L. B., Bair, K. W., Crews, P. Psammaplins from the sponge *Pseudoceratina purpurea*: Inhibition of both histone deacetylase and DNA methyltransferase. *J Org Chem* 2003, 68 (10), 3866–73.

43. Pavlik, C. M., Wong, C. Y., Ononye, S., Lopez, D. D., Engene, N., McPhail, K. L., Gerwick, W. H., Balunas, M. J. Santacruzamate A, a potent and selective histone deacetylase inhibitor from the Panamanian marine cyanobacterium cf. *Symploca* sp. *J Nat Prod* 2013, 76 (11), 2026–33.

44. Liu, A. M., New, D. C., Lo, R. K., Wong, Y. H. Reporter gene assays. *Methods Mol Biol* 2009, 486, 109–23.

45. Naylor, L. H. Reporter gene technology: The future looks bright. *Biochem Pharmacol* 1999, 58 (5), 749–57.

46. McCulloch, M. W., Coombs, G. S., Banerjee, N., Bugni, T. S., Cannon, K. M., Harper, M. K., Veltri, C. A., Virshup, D. M., Ireland, C. M. Psammaplin A as a general activator of cell-based signaling assays via HDAC inhibition and studies on some bromotyrosine derivatives. *Bioorg Med Chem* 2009, 17 (6), 2189–98.

47. Ungermannova, D., Parker, S. J., Nasveschuk, C. G., Wang, W., Quade, B., Zhang, G., Kuchta, R. D., Phillips, A. J., Liu, X. Largazole and its derivatives selectively inhibit ubiquitin activating enzyme (e1). *PLoS One* 2012, 7 (1), e29208.

48. Liu, Y., Kern, J. T., Walker, J. R., Johnson, J. A., Schultz, P. G., Luesch, H. A genomic screen for activators of the antioxidant response element. *Proc Natl Acad Sci USA* 2007, 104 (12), 5205–10.

49. Luesch, H., Liu, Y. Genome-wide overexpression screen for activators of antioxidant gene transcription. *Methods Mol Biol* 2008, 477, 343–54.

50. Ratnayake, R., Liu, Y., Paul, V. J., Luesch, H. Cultivated sea lettuce is a multiorgan protector from oxidative and inflammatory stress by enhancing the endogenous antioxidant defense system. *Cancer Prev Res* (Phila) 2013, 6 (9), 989–99.

51. Wang, R., Paul, V. J., Luesch, H. Seaweed extracts and unsaturated fatty acid constituents from the green alga *Ulva lactuca* as activators of the cytoprotective Nrf2-ARE pathway. *Free Radic Biol Med* 2013, 57, 141–53.

52. Folmer, F., Jaspars, M., Dicato, M., Diederich, M. Marine natural products as targeted modulators of the transcription factor NF-κB. *Biochem Pharmacol* 2008, 75 (3), 603–17.

53. Li, Q., Verma, I. M. NF-kappaB regulation in the immune system. *Nat Rev Immunol* 2002, 2 (10), 725–34.

54. Teo, H., Ghosh, S., Luesch, H., Ghosh, A., Wong, E. T., Malik, N., Orth, A., de Jesus, P., Perry, A. S., Oliver, J. D., Tran, N. L., Speiser, L. J., Wong, M., Saez, E., Schultz, P., Chanda, S. K., Verma, I. M., Tergaonkar, V. Telomere-independent Rap1 is an IKK adaptor and regulates NF-kappaB-dependent gene expression. *Nat Cell Biol* 2010, 12 (8), 758–67.

55. Posadas, I., De Rosa, S., Terencio, M. C., Paya, M., Alcaraz, M. J. Cacospongionolide B suppresses the expression of inflammatory enzymes and tumour necrosis factor-alpha by inhibiting nuclear factor-kappa B activation. *Br J Pharmacol* 2003, 138 (8), 1571–79.

56. Tsukamoto, S., Tatsuno, M., van Soest, R. W., Yokosawa, H., Ohta, T. New polyhydroxy sterols: Proteasome inhibitors from a marine sponge *Acanthodendrilla* sp. *J Nat Prod* 2003, 66 (9), 1181–85.

57. Tsukamoto, S., Koimaru, K., Ohta, T. Secomycalolide A: A new proteasome inhibitor isolated from a marine sponge of the genus *Mycale*. *Marine Drugs* 2005, 3 (2), 29–35.

58. Breton, J. J., Chabot-Fletcher, M. C. The natural product hymenialdisine inhibits interleukin-8 production in U937 cells by inhibition of nuclear factor-kappaB. *J Pharmacol Exp Ther* 1997, 282 (1), 459–66.

59. Schumacher, M., Cerella, C., Eifes, S., Chateauvieux, S., Morceau, F., Jaspars, M., Dicato, M., Diederich, M. Heteronemin, a spongean sesterterpene, inhibits TNF alpha-induced NF-kappa B activation through proteasome inhibition and induces apoptotic cell death. *Biochem Pharmacol* 2010, 79 (4), 610–22.

60. D'Acquisto, F., Lanzotti, V., Carnuccio, R. Cyclolinteinone, a sesterterpene from sponge *Cacospongia linteiformis*, prevents inducible nitric oxide synthase and inducible cyclo-oxygenase protein expression by blocking nuclear factor-κB activation in J774 macrophages. *Biochem J* 2000, 346 (3), 793–98.

61. Giuliano, K. A., Debiasio, R. L., Dunlay, R. T., Gough, A., Volosky, J. M., Zock, J., Pavlakis, G. N., Taylor, D. L. High-content screening: A new approach to easing key bottlenecks in the drug discovery process. *J Biomol Screening* 1997, 2 (4), 249–59.

62. McCoy, J. P., Jr. High-content screening: Getting more from less. *Nat Methods* 2011, 8 (5), 390–91.

63. Schulze, C. J., Bray, W. M., Woerhmann, M. H., Stuart, J., Lokey, R. S., Linington, R. G. "Function-first" lead discovery: Mode of action profiling of natural product libraries using image-based screening. *Chem Biol* 2013, 20 (2), 285–95.

64. Sumiya, E., Shimogawa, H., Sasaki, H., Tsutsumi, M., Yoshita, K., Ojika, M., Suenaga, K., Uesugi, M. Cell-morphology profiling of a natural product library identifies bisebromoamide and miuraenamide A as actin filament stabilizers. *ACS Chem Biol* 2011, 6 (5), 425–31.

65. Futamura, Y., Kawatani, M., Kazami, S., Tanaka, K., Muroi, M., Shimizu, T., Tomita, K., Watanabe, N., Osada, H. Morphobase, an encyclopedic cell morphology database, and its use for drug target identification. *Chem Biol* 2012, 19 (12), 1620–30.

66. Cervantes, S., Stout, P. E., Prudhomme, J., Engel, S., Bruton, M., Cervantes, M., Carter, D., Tae-Chang, Y., Hay, M. E., Aalbersberg, W., Kubanek, J., Le Roch, K. G. High content live cell imaging for the discovery of new antimalarial marine natural products. *BMC Infect Dis* 2012, 12, (1).

67. Navarro, G., Cheng, A. T., Peach, K. C., Bray, W. M., Bernan, V. S., Yildiz, F. H., Linington, R. G. Image-based 384-well high-throughput screening method for the discovery of skyllamycins A to C as biofilm inhibitors and inducers of biofilm detachment in *Pseudomonas aeruginosa*. *Antimicrob Agents Chemother* 2014, 58 (2), 1092–99.

68. Muda, M., McKenna, S. Model organisms and target discovery. *Drug Discov Today Technol* 2004, 1 (1), 55–59.

69. Brenner, S. The genetics of *Caenorhabditis elegans*. *Genetics* 1974, 77 (1), 71–94.

70. Waterston, R. H., Thomson, J. N., Brenner, S. Mutants with altered muscle structure of *Caenorhabditis elegans*. *Dev Biol* 1980, 77 (2), 271–302.

71. Artal-Sanz, M., de Jong, L., Tavernarakis, N. *Caenorhabditis elegans*: A versatile platform for drug discovery. *Biotechnol J* 2006, 1 (12), 1405–18.

72. Kaletta, T., Hengartner, M. O. Finding function in novel targets: *C. elegans* as a model organism. *Nat Rev Drug Discov* 2006, 5 (5), 387–399.

73. Segalat, L. Drug discovery: Here comes the worm. *ACS Chem Biol* 2006, 1 (5), 277–78.

74. Leung, C. K., Deonarine, A., Strange, K., Choe, K. P. High-throughput screening and biosensing with fluorescent *C. elegans* strains. *J Vis Exp* 2011, 1–5.

75. Choe, K. P., Przybysz, A. J., Strange, K. The WD40 repeat protein WDR-23 functions with the CUL4/DDB1 ubiquitin ligase to regulate nuclear abundance and activity of SKN-1 in *Caenorhabditis elegans*. *Mol Cell Biol* 2009, 29 (10), 2704–15.

76. Wang, R., Mason, D. E., Choe, K. P., Lewin, A. S., Peters, E. C., Luesch, H. *In vitro* and *in vivo* characterization of a tunable dual-reactivity probe of the Nrf2-ARE pathway. *ACS Chem Biol* 2013, 8 (8), 1764–74.

77. Leung, C. K., Wang, Y., Malany, S., Deonarine, A., Nguyen, K., Vasile, S., Choe, K. P. An ultra high-throughput, whole-animal screen for small molecule modulators of a specific genetic pathway in *Caenorhabditis elegans*. *PLoS One* 2013, 8 (4), e62166.

78. Segalat, L. Invertebrate animal models of diseases as screening tools in drug discovery. *ACS Chem Biol* 2007, 2 (4), 231–36.

79. MacRae, C. A., Peterson, R. T. Zebrafish-based small molecule discovery. *Chem Biol* 2003, 10 (10), 901–8.

80. Delvecchio, C., Tiefenbach, J., Krause, H. M. The zebrafish: A powerful platform for *in vivo*, HTS drug discovery. *Assay Drug Dev Technol* 2011, 9 (4), 354–61.

81. Sumanas, S., Lin, S. Zebrafish as a model system for drug target screening and validation. *Drug Discov Today Targets* 2004, 3 (3), 89–96.

82. Lessman, C. A. The developing zebrafish (*Danio rerio*): A vertebrate model for high-throughput screening of chemical libraries. *Birth Defects Res C Embryo Today* 2011, 93 (3), 268–80.

83. Langheinrich, U. Zebrafish: A new model on the pharmaceutical catwalk. *Bioessays* 2003, 25 (9), 904–12.

84. Zon, L. I., Peterson, R. T. *In vivo* drug discovery in the zebrafish. *Nat Rev Drug Discov* 2005, 4 (1), 35–44.

85. Haffter, P., Granato, M., Brand, M., Mullins, M. C., Hammerschmidt, M., Kane, D. A., Odenthal, J., van Eeden, F. J., Jiang, Y. J., Heisenberg, C. P., Kelsh, R. N., Furutani-Seiki, M., Vogelsang, E., Beuchle, D., Schach, U., Fabian, C., Nusslein-Volhard, C. The identification of genes with unique and essential functions in the development of the zebrafish, *Danio rerio*. *Development* 1996, 123, 1–36.

86. Driever, W., Solnica-Krezel, L., Schier, A. F., Neuhauss, S. C., Malicki, J., Stemple, D. L., Stainier, D. Y., Zwartkruis, F., Abdelilah, S., Rangini, Z., Belak, J., Boggs, C. A genetic screen for mutations affecting embryogenesis in zebrafish. *Development* 1996, 123, 37–46.

87. Ingham, P. W. Zebrafish genetics and its implications for understanding vertebrate development. *Hum Mol Genet* 1997, 6 (10), 1755–60.

88. Sandoval, I. T., Manos, E. J., Van Wagoner, R. M., Delacruz, R. G., Edes, K., Winge, D. R., Ireland, C. M., Jones, D. A. Juxtaposition of chemical and mutation-induced developmental defects in zebrafish reveal a copper-chelating activity for kalihinol F. *Chem Biol* 2013, 20 (6), 753–63.

89. Wei, X., Bugni, T. S., Harper, M. K., Sandoval, I. T., Manos, E. J., Swift, J., Van Wagoner, R. M., Jones, D. A., Ireland, C. M. Evaluation of pyridoacridine alkaloids in a zebrafish phenotypic assay. *Mar Drugs* 2010, 8 (6), 1769–78.

90. Luesch, H., Chanda, S. K., Raya, R. M., DeJesus, P. D., Orth, A. P., Walker, J. R., Izpisua Belmonte, J. C., Schultz, P. G. A functional genomics approach to the mode of action of apratoxin A. *Nat Chem Biol* 2006, 2 (3), 158–67.

91. Liu, Y., Law, B. K., Luesch, H. Apratoxin A reversibly inhibits the secretory pathway by preventing cotranslational translocation. *Mol Pharmacol* 2009, 76 (1), 91–104.

92. Lomenick, B., Hao, R., Jonai, N., Chin, R. M., Aghajan, M., Warburton, S., Wang, J., Wu, R. P., Gomez, F., Loo, J. A., Wohlschlegel, J. A., Vondriska, T. M., Pelletier, J., Herschman, H. R., Clardy, J., Clarke, C. F., Huang, J. Target identification using drug affinity responsive target stability (DARTS). *Proc Natl Acad Sci USA* 2009, 106 (51), 21984–89.

93. Yu, W. L., Jones, B. D., Kang, M., Hammons, J. C., La Clair, J. J., Burkart, M. D. Spirohexenolide A targets human macrophage migration inhibitory factor (hMIF). *J Nat Prod* 2013, 76 (5), 817–23.

94. Urh, M., Simpson, D., Zhao, K. Affinity chromatography: General methods. *Methods Enzymol* 2009, 463, 417–38.

95. Lomenick, B., Olsen, R. W., Huang, J. Identification of direct protein targets of small molecules. *ACS Chem Biol* 2011, 6 (1), 34–46.

96. Ong, S.-E., Schenone, M., Margolin, A. A., Li, X., Do, K., Doud, M. K., Mani, D. R., Kuai, L., Wang, X., Wood, J. L., Tolliday, N. J., Koehler, A. N., Marcaurelle, L. A., Golub, T. R., Gould, R. J., Schreiber, S. L., Carr, S. A. Identifying the proteins to which small-molecule probes and drugs bind in cells. *Proc Natl Acad Sci USA* 2009, 106 (12), 4617–4622.

97. Low, W. K., Dang, Y., Schneider-Poetsch, T., Shi, Z., Choi, N. S., Rzasa, R. M., Shea, H. A., Li, S., Park, K., Ma, G., Romo, D., Liu, J. O. Isolation and identification of eukaryotic initiation factor 4A as a molecular target for the marine natural product pateamine A. *Methods Enzymol* 2007, 431, 303–24.

98. Yu, W. L., Guizzunti, G., Foley, T. L., Burkart, M. D., La Clair, J. J. An optimized immunoaffinity fluorescent method for natural product target elucidation. *J Nat Prod* 2010, 73 (10), 1659–66.

99. Rodriguez, A. D., Lear, M. J., La Clair, J. J. Identification of the binding of sceptrin to MreB via a bidirectional affinity protocol. *J Am Chem Soc* 2008, 130 (23), 7256–58.

100. Farnaes, L., La Clair, J. J., Fenical, W. Napyradiomycins CNQ525.510B and A80915C target the Hsp90 paralogue Grp94. *Org Biomol Chem* 2014, 12 (3), 418–23.

101. Trzoss, L., Fukuda, T., Costa-Lotufo, L. V., Jimenez, P., La Clair, J. J., Fenical, W. Seriniquinone, a selective anticancer agent, induces cell death by autophagocytosis, targeting the cancer-protective protein dermicidin. *Proc Natl Acad Sci USA* 2014, 111 (41), 14687–92.

102. Towbin, H., Bair, K. W., DeCaprio, J. A., Eck, M. J., Kim, S., Kinder, F. R., Morollo, A., Mueller, D. R., Schindler, P., Song, H. K., van Oostrum, J., Versace, R. W., Voshol, H., Wood, J., Zabludoff, S., Phillips, P. E. Proteomics-based target identification: Bengamides as a new class of methionine aminopeptidase inhibitors. *J Biol Chem* 2003, 278 (52), 52964–71.

103. Luesch, H. Towards high-throughput characterization of small molecule mechanisms of action. *Mol Biosyst* 2006, 2 (12), 609–20.

104. Hoon, S., Smith, A. M., Wallace, I. M., Suresh, S., Miranda, M., Fung, E., Proctor, M., Shokat, K. M., Zhang, C., Davis, R. W., Giaever, G., St. Onge, R. P., Nislow, C. An integrated platform of genomic assays reveals small-molecule bioactivities. *Nat Chem Biol* 2008, 4 (8), 498–506.

105. Kallifatidis, G., Hoepfner, D., Jaeg, T., Guzman, E. A., Wright, A. E. The marine natural product manzamine A targets vacuolar ATPases and inhibits autophagy in pancreatic cancer cells. *Mar Drugs* 2013, 11 (9), 3500–16.

106. Brummelkamp, T. R., Fabius, A. W. M., Mullenders, J., Madiredjo, M., Velds, A., Kerkhoven, R. M., Bernards, R., Beijersbergen, R. L. An shRNA barcode screen provides insight into cancer cell vulnerability to MDM2 inhibitors. *Nat. Chem. Biol.* 2006, 2 (4), 202–6.

107. Mendes-Pereira, A. M., Sims, D., Dexter, T., Fenwick, K., Assiotis, I., Kozarewa, I., Mitsopoulos, C., Hakas, J., Zvelebil, M., Lord, C. J., Ashworth, A. Genome-wide functional screen identifies a compendium of genes affecting sensitivity to tamoxifen. *Proc Natl Acad Sci USA* 2012, 109 (8), 2730–35.

108. Potts, M. B., Kim, H. S., Fisher, K. W., Hu, Y., Carrasco, Y. P., Bulut, G. B., Ou, Y. H., Herrera-Herrera, M. L., Cubillos, F., Mendiratta, S., Xiao, G., Hofree, M., Ideker, T., Xie, Y., Huang, L. J., Lewis, R. E., MacMillan, J. B., White, M. A. Using Functional Signature Ontology (FUSION) to identify mechanisms of action for natural products. *Sci Signal* 2013, 6 (297), ra90.

109. Wacker, S. A., Houghtaling, B. R., Elemento, O., Kapoor, T. M. Using transcriptome sequencing to identify mechanisms of drug action and resistance. *Nat Chem Biol* 2012, 8 (3), 235–37.

110. Finnin, M. S., Donigian, J. R., Cohen, A., Richon, V. M., Rifkind, R. A., Marks, P. A., Breslow, R., Pavletich, N. P. Structures of a histone deacetylase homologue bound to the TSA and SAHA inhibitors. *Nature* 1999, 401 (6749), 188–93.

111. Vannini, A., Volpari, C., Filocamo, G., Casavola, E. C., Brunetti, M., Renzoni, D., Chakravarty, P., Paolini, C., De Francesco, R., Gallinari, P., Steinkuhler, C., Di Marco, S. Crystal structure of a eukaryotic zinc-dependent histone deacetylase, human HDAC8, complexed with a hydroxamic acid inhibitor. *Proc Natl Acad Sci USA* 2004, 101 (42), 15064–69.

112. Cole, K. E., Dowling, D. P., Boone, M. A., Phillips, A. J., Christianson, D. W. Structural basis of the antiproliferative activity of largazole, a depsipeptide inhibitor of the histone deacetylases. *J Am Chem Soc* 2011, 133 (32), 12474–77.

113. Liu, Y., Zhang, W., Li, L., Salvador, L. A., Chen, T., Chen, W., Felsenstein, K. M., Ladd, T. B., Price, A. R., Golde, T. E., He, J., Xu, Y., Li, Y., Luesch, H. Cyanobacterial peptides as a prototype for the design of potent beta-secretase inhibitors and the development of selective chemical probes for other aspartic proteases. *J Med Chem* 2012, 55 (23), 10749–65.

CHAPTER 8

Anti-Infective Agents from Marine Sources

Cedric Pearce

Mycosynthetix, Inc., Hillsborough, North Carolina

CONTENTS

8.1 INTRODUCTION

In the late 1920s, Alexander Fleming started the antibiotic movement through the discovery—at that time just an observation—of penicillin.[1] Over the course of the next 30 years, many scientists—microbiologists, chemists, and pharmacologists—worked together to survey terrestrial microorganisms for their ability to produce antibacterial substances. Their approach was to isolate microorganisms from the environment and then evaluate them for their ability to produce antibacterial or antifungal compounds. Results from this research form the backbone of the current clinically used antimicrobial agents. Antibiotics including the penicillins and cephalosporins; the aminoglycosides, such as streptomycin, gentamicin, kanamycin, and tobramycin; the polyketide-derived tetracyclines; and the macrolides, as well as various peptide derivatives, were all discovered empirically through isolation of microorganisms, evaluation for the ability to kill other microorganisms, and subsequent development. From the 1980s onward, the biggest problem with this approach was the continued rediscovery of known compounds. With the current technology to rapidly dereplicate samples, that is, identify known compounds, this is no longer such a big issue. One solution to enrich chemical novelty came to include as many unusual sources of microorganisms as possible, including those from marine samples. This latter approach has resulted in the discovery of many active anti-infective agents.

Historically, marine sources have delivered significant anti-infective agents; for example, cephalosporins, squalamine, and the antiviral vidarabine are all based directly on compounds found in marine organisms, or in the case of cephalosporin, from a fungus isolated in a marine environment.

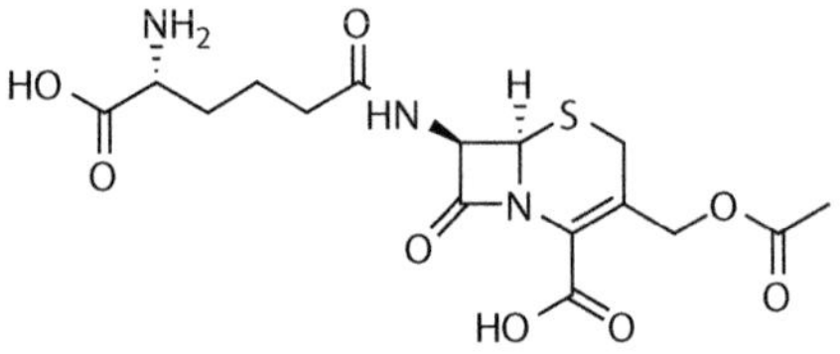

Cephalosporin C

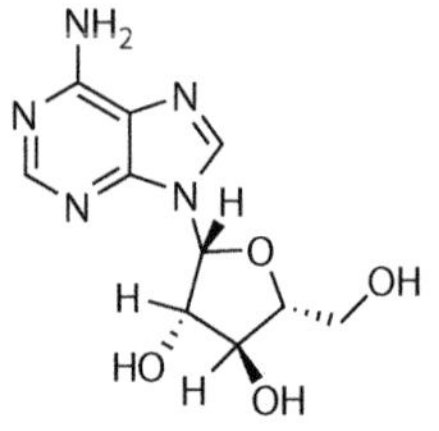

Vidarabine

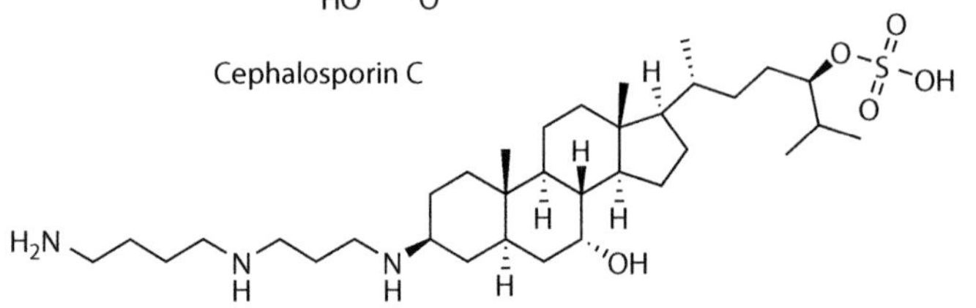

Squalamine

In this chapter are highlights of some of the more interesting and relevant marine-derived anti-infective natural products as judged both from an anti-infective and a chemistry perspective. Historical aspects are discussed, but in general, literature published between 2000 and 2014 is included in this review, with occasional deviations for significant discoveries. Compounds with antibacterial and antiviral activity are included, together with a single example of an antifungal agent illustrating the point of chemical novelty. Products from bacteria, cyanobacteria, fungi, and animals are included. General reviews on marine microbes, including those for antibacterial discovery, have been published.[2,3]

8.2 ANTIBACTERIAL COMPOUNDS

For the past 70 years, the initially marine-derived cephalosporins have been used as antibacterial agents, with the most significant compounds being cephalosporin derivatives that have been synthesized from the natural cephalosporins. More recently, other compounds have been discovered, and a selection of these are discussed below.

8.2.1 Cephalosporin

The β-lactam antibiotic cephalosporin was first isolated from a fungus *Cephalosporium acremonium* that was found in a sample of seawater collected near a sewer outlet in the Mediterranean Sea in the mid-1940s.[4] There are many detailed accounts of the discovery of cephalosporin, the structure determination, and the subsequent development of the antibiotic.[5] A brief review of these events is included here. In the study that led to this discovery, which occurred shortly after penicillin had been characterized and developed, Brotzu and his colleagues had sampled seawater in a deliberate attempt to find marine microorganisms that had the ability to produce novel antibiotics. An extract containing cephalosporin, although not highly potent compared to the antibiotics used today, was initially used to treat typhoid and *Staphylococcus aureus* infections. The structure of cephalosporin N was reported in 1956,[6] and the classic structure of cephalosporin C containing the six-membered ring was published in 1957,[7] both from Abraham's group at Oxford University in the UK. The antibacterial activity of the natural cephalosporins was not very useful, and modern cephalosporins are far superior in many ways. Through medicinal chemistry the activity and characteristics of the initial cephalosporin were improved considerably; different generations of cephalosporins vary in their abilities to treat Gram-negative and Gram-positive bacterial infections, with later generations having broad-spectrum antimicrobial properties.[8]

In the initial experiments at Oxford University, Abraham and Newton isolated from cultures of the fungus cephalosporins P, N, and C. The antibacterial activity of these compounds was fairly weak, and as such, the compounds themselves are not promising as antibiotics for human infections. Through acid hydrolysis to remove the side chain, it was possible to produce 7-aminocephalosporanic acid (7-ACA), and this was then derivatized using other side chains in a manner similar to that leading to the diversity of penicillins. Modifications at position 7 resulted in changes in antibacterial activity, and changes at position 3 resulted in changes in both human metabolism and pharmacokinetics. The new cephalosporins, again, as is the case with penicillins, showed different and improved properties. Abraham reported that the N-phenylacetyl derivative of cephalosporin C was 100 times more potent against *S. aureus* strains than the parent compound, and this observation led the way to a plethora of new and better cephalosporin antibiotics, the highlights of which are discussed below.[9]

The first cephalosporin to become available for clinical use in the United States was cephalothin, trade name Keflin and first marketed by Eli-Lilly in 1964. This and other first-generation cephalosporins, including cephapirin (Cephadyl), cephazolin (Ancef, Kefzol), cephalexin (Keflex), cephradine (Anspor, Velocef), and cefadroxil (Duricef, Ultracef), which were either orally active or given by injection, depending on the antibiotic, were active against Gram-positive bacteria with relatively modest activity against Gram-negative organisms. These compounds were used against most Gram-positive cocci but are not active against enterococci, methicillin-resistant *S. aureus* or *Staphylococcus epidermidis*. They are also active against some Gram-negative organisms, including *Klebsiella pneumonia*, *Proteus mirabilis*, and *Escherichia coli*. Second-generation cephalosporins, including cefamandole (Mandol), cefoxitin (Mefoxin), cefaclor (Ceclor), cefuroxime (Zinacef), cefonicid (Monocid), and ceforaniide (Precef), are generally more active against Gram-negative bacteria than the first-generation compounds. Third-generation cephalosprorins, which include cefotaxime (Claforam), moxalactam (Moxam), ceftizoxime (Cefizox), ceftriaxone (Rocephon), and cefoperazone (Cefobid), are all more active against Gram-negative bacteria, but in general less active against Gram-positive bacteria. In addition, changes in the compounds have led to better binding to penicillin-binding protein (PBP) and improved resistance to β-lactamases. There is a fourth generation of cephalosporins that includes ceftobiprole and ceftaroline, both with activity against methicillin-resistant *S. aureus*. These antibiotics are also more active against Gram-negative bacteria than any of the earlier generations.

Cephalosporins are bacteriocidal and kill target microorganisms through inhibition of cell wall biosynthesis, as do penicillins. Both Gram-positive and Gram-negative bacteria have a cell wall that contains as a main component peptidoglycan, a heteropolymeric network that provides a rigid structure resulting from multiple cross-links that helps the bacterium maintain its shape. The peptidoglycan comprises polymers (glycans) of N-acetylglucosamine and N-acetylmuramic acid, and these polymers are cross-linked with peptides, which vary depending upon the bacteria. In excess of 30 enzymes are involved in the process of cell wall biosynthesis. The enzyme transpeptidase is inhibited by cephalosporins, which results in terminated cell wall biosynthesis. Transpeptidase is responsible for joining the glycan to the peptide. It does this by a process in which one of two D-alanyl units at the end of the peptide chain is cleaved, followed by transferring the remainder of the peptide to the adjacent glycan. Target bacteria should be in a growth phase for the antibiotic to be active, and those in the stationary phase are much less sensitive to this group of antibiotics. The exact mechanism for the cephalosporins to inhibit cell wall biosynthesis is thought to be because these compounds are similar in shape and charge to the transition state of D-alanyl-D-alanyl that is bound to the transpeptidase, and instead of the natural substrate, the β-lactam antibiotics are bound. Following binding—the transpeptidases are also known as penicillin-binding proteins (PBPs)—the β-lactam ring is cleaved in a reaction that results in covalently linking the antibiotic to the enzyme, and thereby irreversibly inactivating it. With no further cell wall biosynthetic activity, the bacteria ultimately lyses.

8.3 ANTIBACTERIAL AGENTS FROM MARINE FISH

Fish have been the source of a number of antibacterial agents; two classes will be discussed here, the aminosterol squalamine and polypeptides found in marine fish slime.

8.3.1 Squalamine

Squalamine is an antibacterial aminosterol discovered by Zasloff et al. from the tissues of the spiny dogfish, *Squalus acanthias*, a once common marine fish found throughout the world, but now in declining numbers due to overfishing (a situation that is aggravated by the slow maturation of these animals coupled with the small number, usually less than 10, of offspring produced by mature females).[10] The initial goal for the work that Zasloff and his team conducted was to understand how animals might protect themselves in environments that contain potentially harmful microorganisms. Typically these mechanisms include peptides (see below), lipids, and alkaloids. After it had been shown that the guts of various animals, including frogs, pigs, mice, and humans,[11–16] contain antibacterial substances, the team examined other animals, including *S. acanthias*, for similar compounds. Squalamine, an aminosterol consisting of a sterol and a spermidine moiety, was initially found in the stomach, but was later shown to be distributed throughout the fish and was particularly concentrated in liver and gallbladder samples. Purified squalamine was shown to be active against a broad range of organisms, including Gram-positive and Gram-negative bacteria and fungi. The potency *in vitro* of the compound was equal to or better than that exhibited by ampicillin, the positive control used for antibacterial minimum inhibitory concentration (MIC) determinations—the MIC is defined as the minimum concentration needed to cause maximum growth inhibition. Squalamine was active against *E. coli, Pseudomonas aeruginosa, Proteus vulgaris, S. aureus, Streptococcus faecalis*, and *Candida albicans*. A number of additional metabolites were reported later.[17]

Squalamine has been found in a variety of sharks and other fish, and the synthesis has been reported from a number of groups around the world.[18–22]

As has been described above, squalamine was isolated as an antibacterial agent, possibly explaining why spiny dogfish and perhaps additional fish species do not suffer from excessive microbial infections. However, when it was evaluated *in vivo*, it showed a number of unexpected properties, including exhibiting antiangiogenic effects, and has subsequently been evaluated as a potential anticancer drug. It has also recently been shown to be broadly active against a variety of human pathogenic RNA- and DNA-enveloped viruses both *in vitro* and *in vivo*.[23] There are limited reports of the *in vivo* antibacterial activity of squalamine, but it has been shown to reduce the bacterial load in rat *P. aeruginosa* pneumonia models when administered by inhalation.[24] However, there are very few accounts of it being further pursued as a systemic antibacterial agent. A recent review suggests the possibility that squalamine could be developed as a topical anti-infective agent as a disinfectant or decontaminant.[25] The mode of action for the antibacterial effect is through membrane disruption. Recently, it has been proposed that the mechanism of action against Gram-negative organisms is through a sequence of events similar to those caused by colistin involving binding to, and subsequent disruption of, cell membranes. In contrast, it is possible that with Gram-positive bacteria, the membrane is depolarized, and this may lead to cell death. Continued work on squalamine and related compounds seems to hold promise as a route to new medicines for a variety of indications, and highlights the value of screening new marine-derived compounds against a number of targets.[25]

8.3.2 Antibacterial Peptides from Fish

It has been known for some time that the mucus and skin of fish contain antibacterial properties. Recent work on flounder[26–28] has uncovered antimicrobial peptides and enzymes. Thus, Cole et al.[26] examined winter flounder (*Pleuronectes americanus*) and discovered and characterized a novel 25-residue linear peptide with the sequence GWGSFFKKAAHVGKHVGKAALTHYL, which was named pleurocidin. This peptide shows sequence homology with tree frog (dermaseptin) and medfly (ceratoxin) antimicrobial peptides. Pleurocidin is a basic peptide, and the authors propose that it forms amphipathic α-helices and is bacteriocidal by rupturing the bacterial membrane. Pleurocidin is active against a variety of Gram-positive and Gram-negative bacteria, as well as being antifungal. The MIC values against sensitive *P. aeruginosa* range between 28 and 62 μg/ml, *Klebsiella pneumoniae* between 12.6 and 43.8, *S. aureus* between 62.5 and 82.9, and *C. albicans* between 15.4 and 24.8. Pleurocidin was also tested together with cycloserine against *Mycobacterium smegmatis*, and these were shown to be synergistic with MIC values reduced fourfold.

An interesting extension of this work was carried out by Prof. Susan Douglas at the Institute for Marine Biosciences, in Halifax, Canada,[27] in which flatfish—winter flounder, yellowtail flounder (*Pleuronectes ferruginea* Storer), American plaice (*Hippoglossoides platessoides* Fabricius), witch flounder (*Glyptocephalus cynoglossus* L.), and Atlantic halibut (*Hippoglossus hippoglossus* L.)—genes and mRNA were screened for sequences encoding potentially antimicrobial peptides. The peptide sequences were subsequently produced through synthesis and then tested for antimicrobial activity. Twenty peptides containing between 18 and 26 amino acids were evaluated against a range of Gram-positive and Gram-negative bacteria and *C. albicans*. Most of the peptides showed antimicrobial activity; the most potent had MIC values against a variety of bacteria, including methicillin-resistant *S. aureus* (MRSA), *P. aerugnosa*, *E. coli*, *Salmonella enterica*, *Aeromonas salmonicida*, and *C. albicans* in the low microgram per milliliter range, values comparable to clinically used agents. At this point, there have been no reports of the crucial *in vivo* evaluation of these compounds.

8.4 ANTIBACTERIAL AGENTS FROM MARINE MICROORGANISMS

Many advances have been made over the past decade in the drive to find new antibacterial agents from marine bacteria. These are considered valuable because of the higher probability of finding novel compounds that have not been discovered in previous industrial screening efforts using terrestrial bacteria. In addition, as marine bacteria and fungi can be cultivated in traditional fermenters (ignoring for the moment the issue of salt corrosion), it is relatively straightforward to provide the large quantities of materials needed for treating patients.

In a recent review of antibacterial agents from marine bacteria,[2] the authors claim the first reported antibacterial agent from a marine organism (bacterium), pentabromopseudilin, was reported in 1966. Since that time, many marine-derived antibacterial agents have been discovered. A further example from work using marine fungi is that of Christophersen and colleagues[29] from Denmark and Venezuela, who summarized their studies on 227 fungi isolated from Venezuelan marine samples of animal and plant origins, including those from lagoons and mangroves; this group demonstrated that 27% of these cultures were active against *Staphylococcus* or *Vibrio* test strains. Although very few of the active compounds were identified, the results illustrate the potential value of marine fungi as a source of antibacterial compounds.

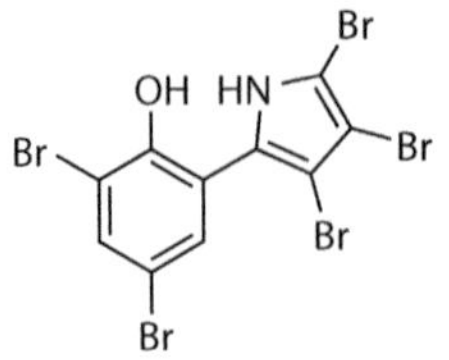

Pentabromopseudilin

Evaluating antibacterial agents *in vitro* is commonly performed using methods that generate MIC values. The inclusion of clinically used antibiotics as positive controls enables the comparison of various compounds isolated by providing a measure of activity that might vary under different laboratory conditions. Thus, in general, for Gram-positive bacteria, penicillin G might be used, and for Gram-negative bacteria, tetracycline or gentamicin could be employed. All of these commonly used antibiotics will give MIC in the low- to submicrogram per milliliter range. Antibiotic-resistant bacteria and some of the more dangerous pathogens, for example, *Mycobacterium tuberculosis*, require specialized handing, as well as careful selection of positive controls. Measuring the zones of growth inhibition on agar plates caused by some micrograms of a given compound, while demonstrating inhibitory activity, is not a desirable approach to evaluating potency.

One of the most exciting compounds discovered during the period of the review is pestalone. A chlorinated benzophenone was reported in 2001 to be produced from a marine fungus, a *Pestalotia* sp., cultured in the presence of a marine bacterium; in the absence of the bacterium, the fungus failed to produce the compound.[30] Pestalone was reported to be very potent against both MRSA and vancomycin-resistant *Enterococcus faecium* (VRE), with MIC values in the 37–78 ng/ml range. However, in 2013 pestalone was prepared by a synthetic route and the nonnatural material shown to be only modestly active, with MIC values approximately three orders of magnitude greater than the natural product.

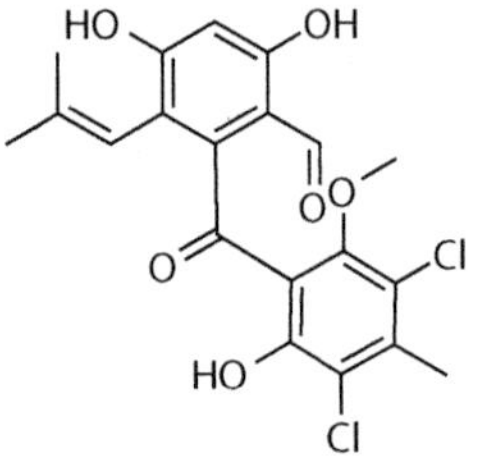

Pestalone

Marine actinomycetes are proving to be an excellent source for novel chemistry. Some examples include the marinopyrroles[31] produced by a marine actinomycete isolated from sediment collected off the Californian coast. In this case, a cultured *Streptomyces* sp. yielded two new compounds, the marinopyroles A and B, which were shown to be active again methicillin-resistant *S. aureus* with MIC_{90} values in the 2 μM range (~1 μg/ml).

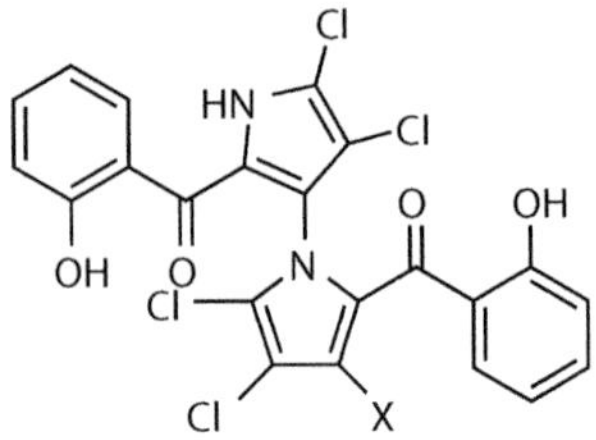

marinopyrrole A (X = H) and B (X = Br)

As is common with marine-derived products, other halogenated antibacterial agents have been isolated and characterized. Thus, from another actinomycete, CNQ-525 (possibly a new genus,

MAR4), isolated from sediment collected off the coast of California, Prof. Fenical's group also isolated and characterized three new chlorinated dihydroquinones of the general structure shown below.[32] These compounds are related to metabolites of the napyradiomycin class. Of the three new metabolites isolated, together with two previously reported compounds, MIC values against MRSA and VRE values of ~2 µg/ml were recorded for the most potent.

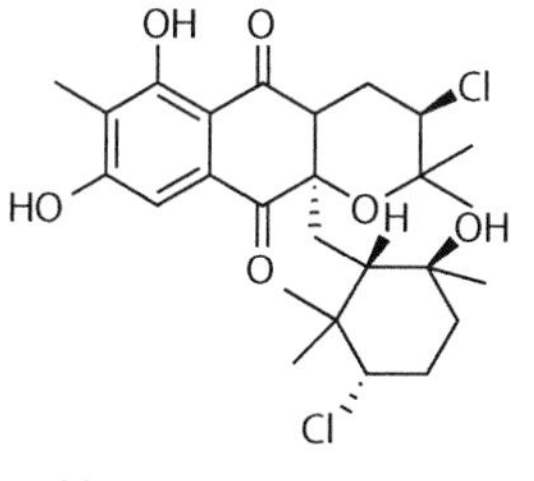

Chlorinated dihydroquinone

The discovery of the genus *Marinospora* has resulted in a number of exciting new metabolites being discovered. A recent example yielding compounds with similarities to staurosporin and rebeccamycins was reported from the Nereus Pharmaceuticals group in California.[33] In this case, lynamicins A–E, chlorinated bisindole pyrroles, were reported from a *Marinospora* isolated from a sediment sample collected off the coast of California. This new class of compounds was shown to be active against a range of Gram-positive and Gram-negative bacteria, including those that were antibiotic resistant. These compounds that are halogenated tryptophan derivatives were also shown to be active against MRSA.

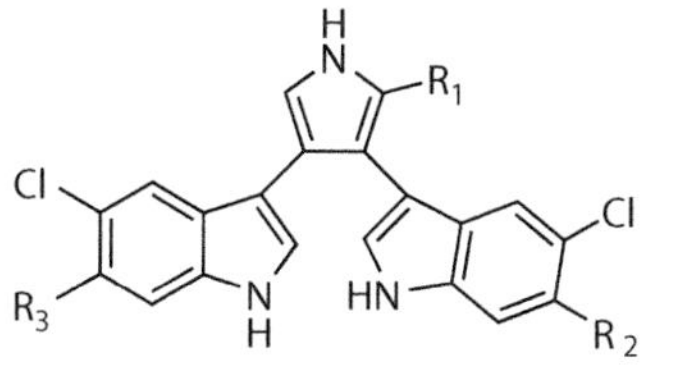

Lynamycins
R_1 = H or COOCH$_3$
R_2 = H or Cl
R_3 = H or Cl

A nucleoside antibiotic was reported from a marine actinomycete *Marinactinospora thermotolerans* SCIO 00652 that had been isolated from a deep-sea marine sediment collected from the South China Sea.[34] This previously reported compound A201A[35] was produced by the marine isolate on a large scale through a strategy involving whole-genome scanning and subsequent gene annotation, which enabled Zhu and colleagues to remove a biosynthesis suppressor. The subsequent culture of the *mtdA* gene mutant produced antibiotic A201A in 25-fold higher yield.

Marine *Nocardiopsis* species have also been the source of new antimicrobial compounds. Thus, an investigation by a European consortium[36] that surveyed isolates from sediment samples collected from Trondheim Fjord, Norway, resulted in a collection of actinomycetes that were evaluated for antimicrobial activity. Twenty-seven isolates were then screened for the presence of polyketide synthase (PKS) I, PKS II, and nonribosomal peptide synthetase (NRPS) genes, as well as the ability to produce antibacterial or antifungal activity against *Micrococcus luteus* and *C. albicans*. In order to reduce the possibility of rediscovery of known antibiotics, further antimicrobial evaluation was undertaken using multiply resistant test organisms, and based on the results from these experiments, *Nocardiopsis* sp. TFS65-07 was selected for further study; this strain was subsequently shown to produce a thiopeptide antibiotic, TP-1161. TP-1161 was active against *S. aureus*, *Staphylococcus haemolyticus*, *S. epidermidis*, *Enterococcus faecalis*, *E. faecium*, *Streptococcus pneumoniae*, and *Streptococcus* sp. and was inactive against Gram-negative bacteria, as in the case for other thiopeptide antimicrobial agents.

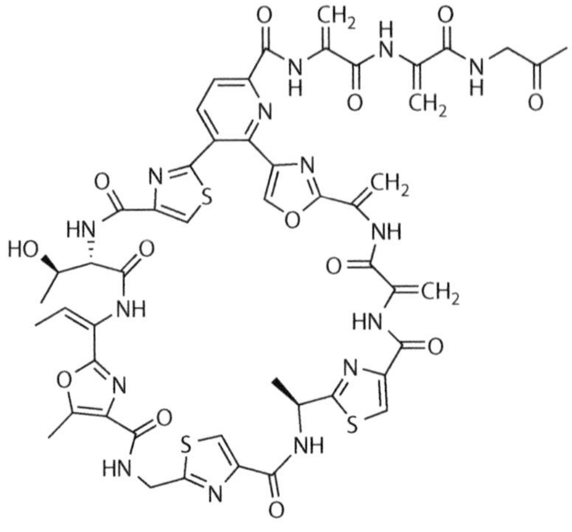

TP-1161

A further example of the use of molecular biology in the advancement of the discovery of new antibacterial agents from marine sources was reported from the Australian group of Prof. Fuerst,[37] and in this case, the discovery of a marine bacterium that produces a compound previously known from terrestrial bacteria. The application of phylogenetic analysis of the marine sponge-derived *Salinospora* strain collection targeting the polyketide biosynthesis led to rifamycin B biosynthesis motifs as found in *Amycolatopsis mediterranei* and resulted in the identification of a strain that contained similar sequences. From this observation, the group predicted that the organism might synthesize rifamycin-like compounds. By deliberately targeting this class of compounds, it was shown that the *Salinospora* produced a number of rifamycins, including rifamycins B and SV. Subsequently, using a rifamycin-specific set of primers to screen their collection of *Salinospora*, they discovered a number of other isolates containing rifamycin biosynthetic genes. This exciting discovery illustrates that marine bacteria can also be used as the source of compounds previously only known from terrestrial organisms, in this case the rifamycins. This is particularly relevant as rifamycin is a first-line antibiotic for the treatment of tuberculosis, and with the emergence of antibiotic-resistant strains of *Mycobacterium tuberculosis*, this approach could be used to generate new analogues of this class of compound, possibly with activity against the resistant strains.

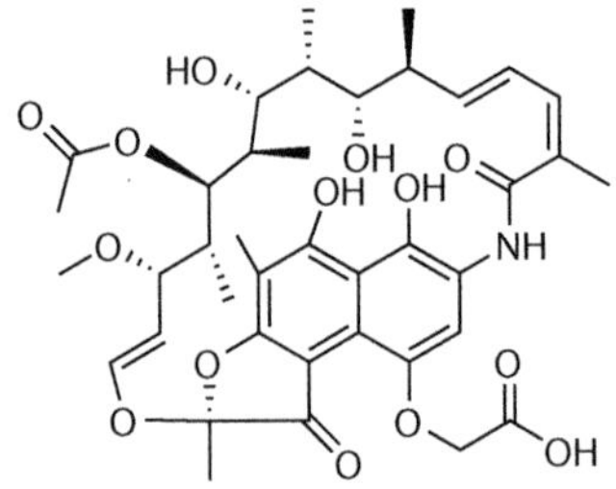

Rifamycin B

Bacteria other than actinomycetes have also been investigated for the production of antibiotics, and as an example, *Bacteroidetes*, which are Gram-negative gliding bacteria found extensively, but not exclusively, in marine environments, were targeted by a group from Japan[38] who discovered a new polyketide–peptide hybrid antibiotic from a *Rapidithrix* sp. The structure shown is of ariakemicin A, which was co-produced and inseparable from the positional isomer ariakemicin B. These compounds were active against Gram-positive bacteria and showed a MIC value of 0.46 μg/ml against a test strain of *S. aureus*; they were inactive against Gram-negative test bacteria and *C. albicans*.

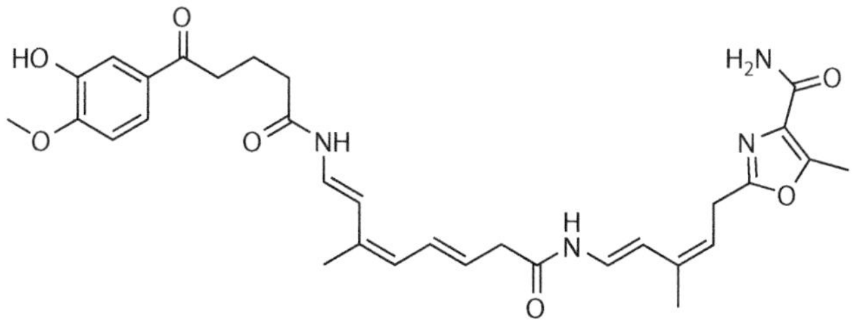

Ariakemicin A

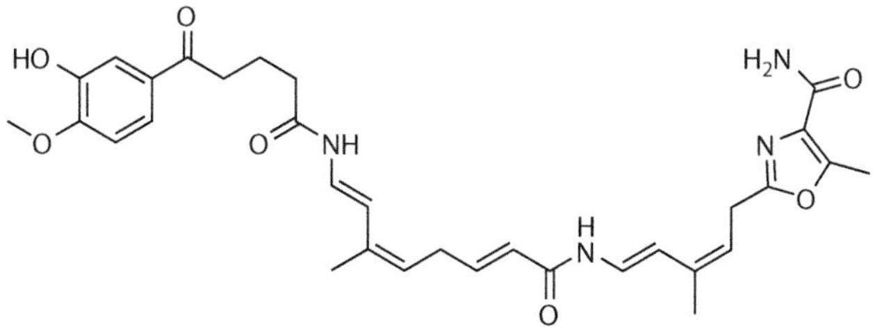

Ariakemicin B

8.5 ANTIVIRAL COMPOUNDS FROM MARINE SOURCES

There have been a number of recent reviews on antiviral marine compounds, including the extensive assessment of the field by Yasuhara-Bell and Lu,[39] Sagar et al.,[40] Uzair et al.,[41] and Dhivya et al.[42] Some very promising antiviral lead compounds and drugs have been developed over the past 50 years or so for a variety of infections and infectious agents, including human immunodeficiency virus-1 (HIV-1), herpes simplex virus-2 (HSV-2), junin virus (JV), polio virus (PV), dengue virus, severe acute respiratory syndrome (SARS) virus, measles virus, influenza virus, poxviruses, rhabdoviruses, hepadnaviruses, RNA tumor viruses, and vaccinia virus.[43–50] Recent literature illustrates the potential of this approach.[51]

8.5.1 Vidarabine

Marine sponges have been the source of many significant antiviral agents. Possibly the most important are the nucleosides spongothymidine and spongouridine that ultimately led to the synthesis of vidarabine (Ara-A) and cytarabine (Ara-C), both used clinically for viral infections and cancer, respectively. These latter two compounds differ by being the arabinoside of either adenine (Ara-A) or cytosine (Ara-C). Spongothymidine and spongouridine were discovered from a Caribbean sponge *Tethya crypta*, which belongs to the Tethyidae family found in shallow water throughout the Caribbean Sea, by Bergmann and Feeney and reported in 1951. These nucleosides contained arabinose rather than the more common ribose usually found in nucleosides.[52] Vidarabine was first synthesized by Baker et al. in 1960, primarily because it was thought such compounds would be potential anticancer medicines.[53]

Vidarabine has been isolated from a species of *Streptomyces* designated as *S. herbaceus*; the group that reported this microbial source also showed vidarabine was an herbicide.[54] Vidarabine has also been prepared through an unusual biotransformation reaction of adenosine using *Streptomyces antibioticus*. This latter reaction, which is quite selective, could also be carried out using an enzyme preparation from the bacterium, although it is not clear if either approach would lead to an industrial procedure.[55]

The antiviral activity of vidarabine was first described by Privat de Garilhe and de Rudder in 1964.[56] The mechanism of action for vidarabine leads to inhibition of DNA polymerase and therefore DNA synthesis; this occurs in both mammalian and viral systems, but to a far more significant extent with viral DNA polymerase.[57] Vidarabine is a nucleoside analogue and is phosphorylated in the presence of kinases and ATP to produce the active form of the drug, vidarabine triphosphate. This is the active form of vidarabine and is both an inhibitor and a substrate of viral DNA polymerase.[58–61] The phosphorylated form of vidarabine is incorporated into DNA in place of adenosine bases, resulting in inhibition of DNA synthesis.

The first report of the clinical use of vidarabine was from Whitley et al., who described its application for the treatment of various herpes infections.[62] Vidarabine is active against, and was used clinically against, herpes viruses, poxviruses, and certain rhabdoviruses, hepadnarviruses, and RNA tumor viruses, as well as showing *in vitro* and *in vivo* activity active against vaccinia virus.[63–65] However, for a variety of reasons, including dosing and some toxicity issues, vidarabine has been replaced by acyclovir in many cases, although the former is active against some acyclovir-resistant strains.

8.5.2 Other Antiviral Lead Compounds

There have been many interesting antiviral marine-sourced compounds reported, some of which are discussed briefly below.

Manzamine A was isolated from a *Halicona* sp. and reported by Sakai and Higa as an antitumor agent, and subsequently shown to display a variety of bioactivities, including as an anti-infective.[66–70] It has been rediscovered a number of times, and a recent survey of sponges reported from Indonesia by the group of Hamann found manzamine A and a number of related analogues. This group also reported the anti-HIV activity associated with the group of compounds.[71]

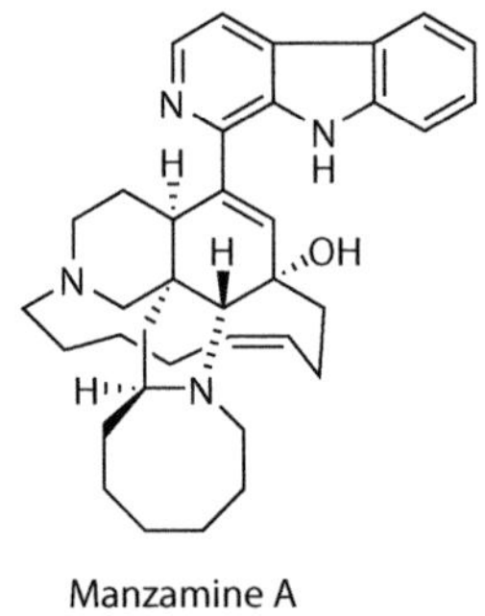

Manzamine A

The mycalamides were first reported from a marine sponge collected in New Zealand.[72] These compounds showed activity against corona virus, herpes simplex type 1, and polio type 1 viruses. These compounds also were effective in protecting infected mice. Mechanism of action studies indicate mycalamides are ultimately protein synthesis inhibitors. These compounds are also known cytotoxic agents.[73–75] The synthesis of the mycalamides has been reported by a number of groups.[76–78]

A marine-derived *Scytalidium* fungus that had been isolated from Caribbean seagrass *Halodule wrightii* by the group led by Fenical was shown to produce a series of lipophilic linear peptides, the halovirs, that inhibited herpes simplex viruses 1 and 2.[79] This discovery was made after extracts from more than 7000 marine microorganisms had been surveyed for activity. The halovirs were quite potent, with halovir A showing an ED_{50} (half-maximal effective concentration) of 280 M in a plaque reduction assay. The synthesis and some structural–activity relationships (SARs) have been reported by the same group.[80]

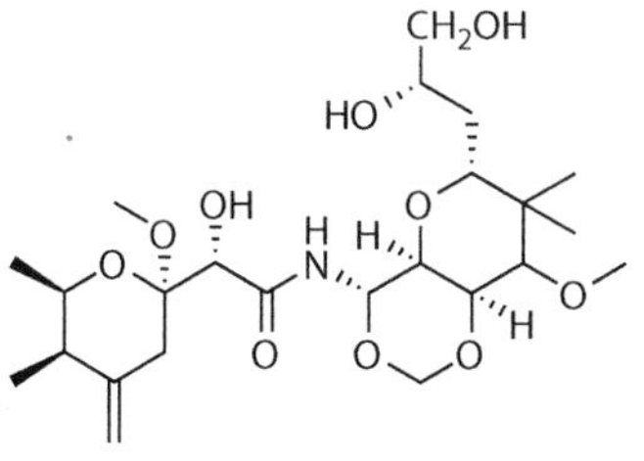

Mycalamide A

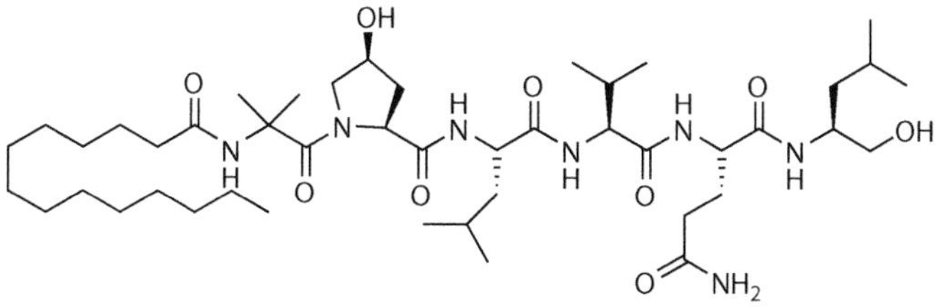

Halovir A

It has been known for some time that marine green algae are able to produce a variety of polysaccharides, comprising various sugars (e.g., rhamnose and galactose) and carbohydrate acids (e.g., glucuronic and iduronic acid), frequently being sulfated. These compounds have been shown to display antiviral activity, among a spectrum of bioactivities. One recent example was reported by Cassolato et al.,[81] who isolated a water-soluble polysaccharide from *Gayralia oxysperma*. This compound was potently active against herpes simplex type 1 and type 2 with IC_{50} (concentration that inhibits binding or activity by 50%) values in the submicrogram per milliliter range; together with the compounds' lack of cytotoxicity, this suggests these compounds may be useful candidates for clinical development.

8.6 ANTIFUNGAL COMPOUNDS

The discovery of the antifungal agents by the American Cyanamid group discussed in Section 8.6.1 is an example of how a pharmaceutical company might approach drug discovery, with an integrated team of specialists from a number of disciplines. The value of working as an integrated team has been highlighted a number of times in the literature.

8.6.1 Example of the Discovery of a Potential Drug Lead from a Marine Fungus

During the approximate period 1990–1995, the Infectious Diseases Research Division's Natural Products Section of the then pharmaceutical company American Cyanamid (subsequently acquired by Wyeth) conducted a screening program aimed at the discovery of new natural compounds with activity against the cell wall biosynthetic machinery of pathogenic fungi. In this program, approximately 250 marine fungi that had been isolated by Prof. Gareth Jones of Portsmouth Polytechnic in the United Kingdom were cultured in a variety of media, and extracts of those cultures were evaluated against a bioassay designed to detect compounds that inhibited the production of chitin. Growth media for these experiments were typically complex and included artificial seawater. Incubation was carried out at 22 °C, either stationary or with agitation. Growth periods extended to 3 months, as the marine fungi in the collection grew slowly compared to terrestrial species such as penicillium and aspergillus. The bioassay consisted of a mutated fungus with impaired cell wall biosynthetic enzymes such that the test organism produces cell wall–less cells, as well as an accumulation of orange debris on the test plate, when chitin biosynthesis is interrupted. In this case, the producing marine fungus *Hypoxylon ocenaicum* made a series of antifungal peptides, LL-15G256.[82–85] These have been patented but were not developed further. In examining the mechanism of action, it was unclear if these compounds acted through the inhibition of chitin biosynthesis.

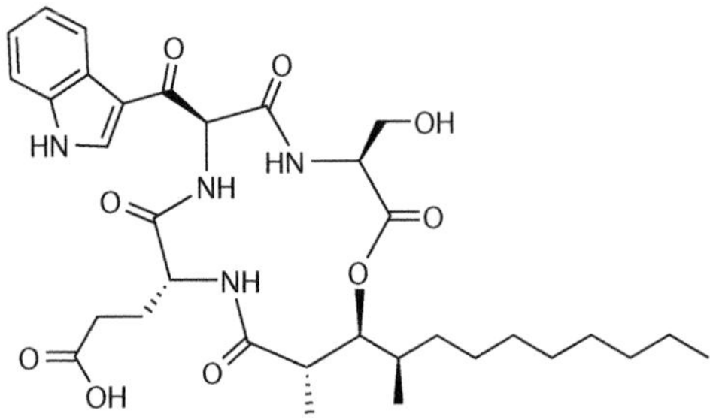

15G256γ

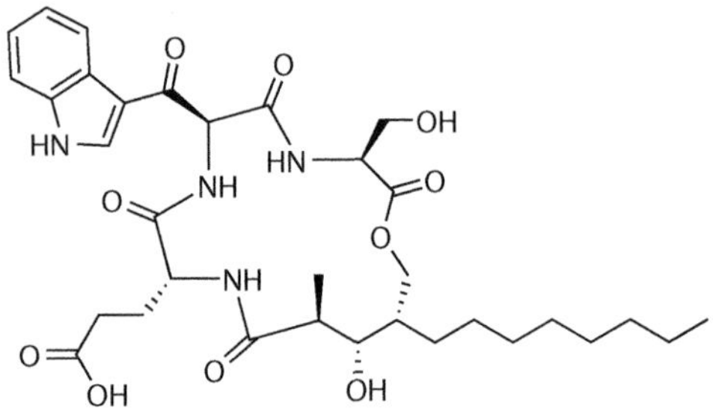

15G256δ

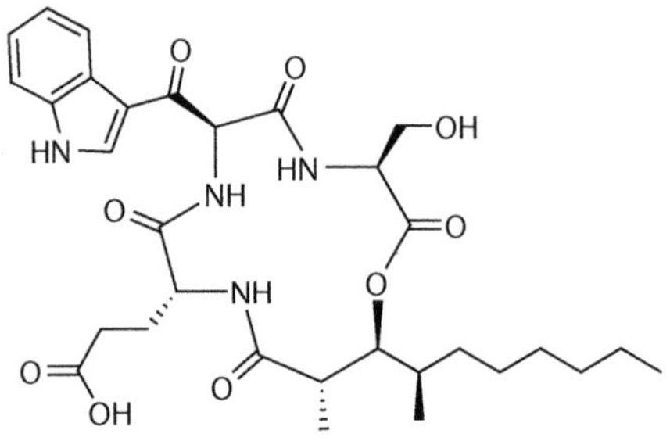

15G256ε

8.7 SUMMARY

It is clear that the marine environment is a source of molecular diversity—probably every chapter in this collection supports this conclusion. Such molecular diversity should lead to interesting and significant anti-infective activity that can relatively easily be explored from the discovery perspective in a laboratory setting. The volume and size of the global marine environment alone, compared to the terrestrial environment, could be an argument in favor of the former being a richer resource, although this could be an overly simplistic one. For example, the concentration of soil bacteria is far higher—on the order of 100 million per gram—than that in the ocean. Competition for nutrients between these organisms must exist, and may be greater in terrestrial situations. Quorum sensing may play a part in the interactions of bacteria, leading to pathways being turned on and off. Air, on the other hand, does not contain the same concentration of potential infective agents as seawater, in which marine animals live their lives. In this regard, the existence of powerful anti-infective peptides found in fish slime highlights how antibacterial properties might evolve and give an advantage to an animal.

In the golden era of antibiotic discovery, when major pharmaceutical companies were competing to discover new antibiotics, large research groups consisting of microbiologists, chemists, and pharmacologists, with virtually unlimited resources, worked around the clock to discover and develop new antibiotics. This effort resulted in the introduction of the majority of the classes of antibacterials and some significant antifungal agents used clinically today. Given the advances that have been made in so many of the scientific fields that support such work, and given the need for new lead compounds, with proper support, the field of marine-derived anti-infective agents is promising.

ACKNOWLEDGMENTS

I would like to thank Arlene Sy, Brenna Hansen, and Meagan Doss for their help in assembling this chapter, and Nancy Baker of Parlezchem, who conducted literature searches.

REFERENCES

1. A. Fleming. *Penicillin: Its Practical Application*. Butterworth & Co., London, 1946, p. 380.
2. H. Rahman et al. Novel anti-infective compounds from marine bacteria. *Mar Drugs* 8, 498–518 (2010).
3. M. Gopi et al. Antibacterial potential of sponge endosymbiont marine *Enterobacter* sp. at Kavaratti Island, Lakshadweep archipelago. *Asian Pac J Trop Med* 5, 142–146 (2012).
4. G. Brotzu. *Lavori Dell'Istituto D'Igiene di Cagliari* 1948, 6.
5. E. P. Abraham. A glimpse of the early history of the cephalosporins. *Rev Infect Dis* 1, 99–105 (1979).
6. G. G. Newton, E. P. Abraham. Degradation, structure and some derivatives of cephalosporin N. *Biochem J* 58, 103–111 (1954).
7. E. P. Abraham, G. G. Newton. The structure of cephalosporin C. *Biochem J* 79, 377–393 (1961).
8. N. C. Klein, B. A. Cunha. The selection and use of cephalosporins: A review. *Adv Ther* 12, 83–101 (1995).
9. B. Loder, G. G. Newton, E. P. Abraham. The cephalosporin C nucleus (7-aminocephalosporanic acid) and some of its derivatives. *Biochem J* 79, 408–416 (1961).
10. K. S. Moore et al. Squalamine: An aminosterol antibiotic from the shark. *Proc Natl Acad Sci USA* 90, 1354–1358 (1993).
11. K. S. Moore et al. Antimicrobial peptides in the stomach of *Xenopus laevis*. *J Biol Chem* 266, 19851–19857 (1991).
12. J. Y. Lee et al. Antibacterial peptides from pig intestine: Isolation of a mammalian cecropin. *Proc Natl Acad Sci USA* 86, 9159–9162 (1989).
13. B. Agerberth et al. Amino acid sequence of PR-39. Isolation from pig intestine of a new member of the family of proline-arginine-rich antibacterial peptides. *Eur J Biochem* 202, 849–854 (1991).
14. A. J. Ouellette et al. Developmental regulation of cryptdin, a corticostatin/defensin precursor mRNA in mouse small intestinal crypt epithelium. *J Cell Biol* 108, 1687–1695 (1989).
15. D. S. Roos, P. W. Choppin. Biochemical studies on cell fusion. I. Lipid composition of fusion-resistant cells. *J Cell Biol* 101, 1578–1590 (1985).
16. D. E. Jones, C. L. Bevins. Paneth cells of the human small intestine express an antimicrobial peptide gene. *J Biol Chem* 267, 23216–23225 (1992).
17. M. N. Rao et al. Aminosterols from the dogfish shark *Squalus acanthias*. *J Nat Prod* 63, 631–635 (2000).
18. X. Zhang et al. Synthesis of squalamine utilizing a readily accessible spermidine equivalent. *J Org Chem* 63, 8599–8603 (1998).
19. X.-D. Zhou, F. Cai, W.-S. Zhou. A stereoselective synthesis of squalamine. *Tetrahedron* 58, 10293–10299 (2002).
20. K. Okumura et al. Formal synthesis of squalamine from desmosterol. *Chem Pharm Bull* (Tokyo) 51, 1177–1182 (2003).
21. D. H. Zhang, F. Cai, X. D. Zhou, W. S. Zhou. A concise and stereoselective synthesis of squalamine. *Org Lett* 5, 3257–3259 (2003).
22. Z. Dong-Hui, C. Feng, Z. Xiang-Dong, Z. Wei-Shan. A short and highly stereoselective synthesis of squalamine from methyl chenodeoxycholanate. *Chin J Chem* 23, 176–181 (2005).
23. M. Zasloff et al. Squalamine as a broad-spectrum systemic antiviral agent with therapeutic potential (vol 108, pg 15978, 2011). *Proc Natl Acad Sci USA* 108, 18186 (2011).

24. S. Hraiech et al. Antibacterial efficacy of inhaled squalamine in a rat model of chronic *Pseudomonas aeruginosa* pneumonia. *J Antimicrob Chemother* 67, 2452–2458 (2012).

25. K. Alhanout et al. New insights into the antibacterial mechanism of action of squalamine. *J Antimicrob Chemother* 65, 1688–1693 (2010).

26. A. M. Cole, R. O. Darouiche, D. Legarda, N. Connell, G. Diamond. Characterization of a fish antimicrobial peptide: Gene expression, subcellular localization, and spectrum of activity. *Antimicrob Agents Chemother* 44, 2039–2045 (2000).

27. A. Patrzykat, J. W. Gallant, J. K. Seo, J. Pytyck, S. E. Douglas. Novel antimicrobial peptides derived from flatfish genes. *Antimicrob Agents Chemother* 47, 2464–2470 (2003).

28. K. Kasai et al. Novel L-amino acid oxidase with antibacterial activity against methicillin-resistant *Staphylococcus aureus* isolated from epidermal mucus of the flounder *Platichthys stellatus*. *FEBS J* 277, 453–465 (2010).

29. C. Christophersen et al. Antibacterial activity of marine-derived fungi. *Mycopathologia* 143, 135–138 (1998).

30. M. Cueto et al. Pestalone, a new antibiotic produced by a marine fungus in response to bacterial challenge. *J Nat Prod* 64, 1444–1446 (2001).

31. C. C. Hughes, A. Prieto-Davo, P. R. Jensen, W. Fenical. The marinopyrroles, antibiotics of an unprecedented structure class from a marine *Streptomyces* sp. *Org Lett* 10, 629–631 (2008).

32. I. E. Soria-Mercado, A. Prieto-Davo, P. R. Jensen, W. Fenical. Antibiotic terpenoid chlorodihydroquinones from a new marine actinomycete. *J Nat Prod* 68, 904–910 (2005).

33. K. A. McArthur et al. Lynamicins A–E, chlorinated bisindole pyrrole antibiotics from a novel marine actinomycete. *J Nat Prod* 71, 1732–1737 (2008).

34. Q. Zhu et al. Discovery and engineered overproduction of antimicrobial nucleoside antibiotic A201A from the deep-sea marine actinomycete *Marinactinospora thermotolerans* SCSIO 00652. *Antimicrob Agents Chemother* 56, 110–114 (2012).

35. H. A. Kirst et al. The structure of A201A, a novel nucleoside antibiotic. *J Antibiot* (Tokyo) 38, 575–586 (1985).

36. K. Engelhardt et al. Production of a new thiopeptide antibiotic, TP-1161, by a marine *Nocardiopsis* species. *Appl Environ Microbiol* 76, 4969–4976 (2010).

37. T. K. Kim, A. K. Hewavitharana, P. N. Shaw, J. A. Fuerst. Discovery of a new source of rifamycin antibiotics in marine sponge actinobacteria by phylogenetic prediction. *Appl Environ Microbiol* 72, 2118–2125 (2006).

38. N. Oku et al. Ariakemicins A and B, novel polyketide-peptide antibiotics from a marine gliding bacterium of the genus *Rapidithrix*. *Org Lett* 10, 2481–2484 (2008).

39. J. Yasuhava-Bell, Y. Lu. Marine compounds and their antiviral activities. *Antiviral Res* 86, 231–240 (2010).

40. S. Sagar, M. Kaur, K. P. Minneman. Antiviral lead compounds from marine sponges. *Mar Drugs* 8, 2619–2638 (2010).

41. B. Uzair, Z. Mahmood, S. Tabassum. Antiviral activity of natural products extracted from marine organisms. *Bioimpacts* 1, 203–211 (2011).

42. D. Dhivya, K. Alekhya, C. Nagajyothi, P. Ashok. Marine organism: Lead compounds and as a source of new antiviral agents. *Int J Novel Trends Pharm Sci* 2, 0–15 (2012).

43. M. J. Comin, M. S. Maier, A. J. Roccatagliata, C. A. Pujol, E. B. Damonte. Evaluation of the antiviral activity of natural sulfated polyhydroxysteroids and their synthetic derivatives and analogs. *Steroids* 64, 335–340 (1999).

44. Y. Hwang et al. Mechanism of inhibition of a poxvirus topoisomerase by the marine natural product sansalvamide A. *Mol Pharmacol* 55, 1049–1053 (1999).

45. A. M. S. Mayer, V. K. B. Lehmann. Marine pharmacology in 1998: Marine compounds with antibacterial, anticoagulant, antifungal, antiinflammatory, anthelmintic, antiplatelet, antiprotozoal, and antiviral activities; with actions on the cardiovascular, endocrine, immune, and nervous systems; and other miscellaneous mechanisms of action. *Pharmacologist* 42, 62–69 (2000).

46. D. C. Rowley et al. Thalassiolins A–C: New marine-derived inhibitors of HIV cDNA integrase. *Bioorg Med Chem* 10, 3619–3625 (2002).

47. M. C. Rodríguez et al. Galactans from cystocarpic plants of the red seaweed *Callophyllis variegata* (Kallymeniaceae, Gigartinales). *Carbohydr Res* 340, 2742–2751 (2005).

48. A. M. Mayer, M. T. Hamann. Marine pharmacology in 2000: Marine compounds with antibacterial, anticoagulant, antifungal, anti-inflammatory, antimalarial, antiplatelet, antituberculosis, and antiviral activities; affecting the cardiovascular, immune, and nervous systems and other miscellaneous mechanisms of action. *Mar Biotechnol* (NY) 6, 37–52 (2004).

49. N. Oku et al. Neamphamide A, a new HIV-inhibitory depsipeptide from the Papua New Guinea marine sponge *Neamphius huxleyi*. *J Nat Prod* 67, 1407–1411 (2004).

50. M. A. Rashid et al. Microspinosamide, a new HIV-inhibitory cyclic depsipeptide from the marine sponge *Sidonops microspinosa*. *J Nat Prod* 64, 117–121 (2001).

51. A. Raveh et al. Discovery of potent broad spectrum antivirals derived from marine actinobacteria. *PLoS ONE* 8, e82318 (2013).

52. W. Bergmann, R. J. Feeney. Contributions to the study of marine products. XXXII. The nucleosides of sponges. *J Org Chem* 16, 981–987 (1951).

53. E. J. Reist, A. Benitez, L. Goodman, B. R. Baker, W. W. Lee. Potential anticancer agents. 1 LXXVI. Synthesis of purine nucleosides of β-D-arabinofuranose. *J Org Chem* 27, 3274–3279 (1962).

54. J. Awaya et al. Production of 9-beta-D-arabinofuranosyladenine by a new species of *Streptomyces* and its herbicidal activity. *J Antibiot* (Tokyo) 32, 1050–1054 (1979).

55. R. J. Suhadolnik, S. Pornbanlualap, J. M. Wu, D. C. Baker, A. K. Hebbler. Biosynthesis of 9-beta-D-arabinofuranosyladenine: Hydrogen exchange at C-2′ and oxygen exchange at C-3′ of adenosine. *Arch Biochem Biophys* 270, 363–373 (1989).

56. M. Privat de Garilhe, J. de Rudder. Effect of 2 arbinose nucleosides on the multiplication of herpes virus and vaccine in cell culture. *C R Hebd Seances Acad Sci* 259, 2725–2728 (1964).

57. W. E. Muller, R. K. Zahn, K. Bittlingmaier, D. Falke. Inhibition of herpesvirus DNA synthesis by 9-beta-D-arabinofuranosyladenine in cellular and cell-free systems. *Ann NY Acad Sci* 284, 34–48 (1977).

58. A. Doering, J. Keller, S. S. Cohen. Some effects of D-arabinosyl nucleosides on polymer syntheses in mouse fibroblasts. *Cancer Res* 26, 2444–2450 (1966).

59. W. Plunkett, S. S. Cohen. Two approaches that increase the activity of analogs of adenine nucleosides in animal cells. *Cancer Res* 35, 1547–1554 (1975).

60. A. M. Mayer et al. The odyssey of marine pharmaceuticals: A current pipeline perspective. *Trends Pharmacol Sci* 31, 255–265 (2010).

61. R. A. Dicioccio, B. I. S. Srivastava. Kinetics of inhibition of deoxynucleotide-polymerizing enzyme activities from normal and leukemic human cells by 9-β-D-arabinofuranosyladenine 5′-triphosphate and 1-β-D-arabinofuranosylcytosine 5′-triphosphate. *Eur J Biochem* 79, 411–418 (1977).

62. R. J. Whitley et al. Pharmacology tolerance and anti viral activity of vidarabine mono phosphate in humans. *Antimicrob Agents Chemother* 18, 709–715 (1980).

63. D. Pavan-Langston, R. A. Buchanan, D. A. Alford. *Adenine Arabinoside: An Antiviral Agent*. Raven Press, New York, 1975.

64. O. D. White, F. J. Fenner. *Medical Virology*. Academic Press, San Diego, 1994.

65. H. J. Field, E. de Clercq. Antiviral drugs: A short history of their discovery and development. *Microbiol Today* 31, 58–61 (2004).

66. K. A. El Sayed et al. New manzamine alkaloids with potent activity against infectious diseases. *J Am Chem Soc* 123, 1804–1808 (2001).

67. K. V. Rao et al. New manzamine alkaloids with activity against infectious and tropical parasitic diseases from an Indonesian sponge. *J Nat Prod* 66, 823–828 (2003).

68. K. K. Ang, M. J. Holmes, T. Higa, M. T. Hamann, U. A. Kara. *In vivo* antimalarial activity of the beta-carboline alkaloid manzamine A. *Antimicrob Agents Chemother* 44, 1645–1649 (2000).

69. J. Peng et al. Marine natural products as prototype agrochemical agents. *J Agric Food Chem* 51, 2246–2252 (2003).

70. J. Peng, K. Rao, Y. Choo, M. Hamann. *Modern Alkaloids: Structure, Isolation, Synthesis, and Biology*. 1st ed. Wiley, Weinheim, Germany, 2007, p. 689.

71. K. V. Rao et al. Three new manzamine alkaloids from a common Indonesian sponge and their activity against infectious and tropical parasitic diseases. *J Nat Prod* 67, 1314–1318 (2004).

72. N. B. Perry, J. W. Blunt, M. H. G. Munro, L. K. Pannell. Mycalamide A, an antiviral compound from a New Zealand sponge of the genus *Mycale*. *J Am Chem Soc* 110, 4850–4851 (1988).

73. A. M. Thompson, J. W. Blunt, M. H. G. Munro, N. B. Perry. Chemistry of the mycalamides, antiviral and antitumour compounds from a marine sponge. Part 5. Acid-catalysed hydrolysis and acetal exchange, double bond additions and oxidation reactions. *J Chem Soc Perkin Trans 1* 1233–1242 (1995).

74. K. A. Hood, L. M. West, P. T. Northcote, M. V. Berridge, J. H. Miller. Induction of apoptosis by the marine sponge (*Mycale*) metabolites, mycalamide A and pateamine. *Apoptosis* 6, 207–219 (2001).

75. S. A. Dyshlovoy et al. Mycalamide A shows cytotoxic properties and prevents EGF-induced neoplastic transformation through inhibition of nuclear factors. *Mar Drugs* 10, 1212–1224 (2012).

76. N. Kagawa, M. Ihara, M. Toyota. Total synthesis of (+)-mycalamide A. *Org Lett* 8, 875–878 (2006).

77. J. H. Sohn, N. Waizumi, H. M. Zhong, V. H. Rawal. Total synthesis of mycalamide A. *J Am Chem Soc* 127, 7290–7291 (2005).

78. B. M. Trost, H. Yang, G. D. Probst. A formal synthesis of (–)-mycalamide A. *J Am Chem Soc* 126, 48–49 (2003).

79. D. C. Rowley, S. Kelly, C. A. Kauffman, P. R. Jensen, W. Fenical. Halovirs A–E, new antiviral agents from a marine-derived fungus of the genus *Scytalidium*. *Bioorg Med Chem* 11, 4263–4274 (2003).

80. D. C. Rowley, S. Kelly, P. Jensen, W. Fenical. Synthesis and structure-activity relationships of the halovirs, antiviral natural products from a marine-derived fungus. *Bioorg Med Chem* 12, 4929–4936 (2004).

81. J. E. Cassolato et al. Chemical structure and antiviral activity of the sulfated heterorhamnan isolated from the green seaweed *Gayralia oxysperma*. *Carbohydr Res* 343, 3085–3095 (2008).

82. D. Abbanat, M. Leighton, W. Maiese, E. B. Jones, C. Pearce, M. Greenstein. Cell wall active antifungal compounds produced by the marine fungus *Hypoxylon oceanicum* LL-15G256. I. Taxonomy and fermentation. *J Antibiot* (Tokyo) 51, 296–302 (1998).

83. G. Schlingmann, L. Milne, D. R. Williams, G. T. Carter. Cell wall active antifungal compounds produced by the marine fungus *Hypoxylon oceanicum* LL-15G256. II. Isolation and structure determination. *J Antibiot* (Tokyo) 51, 303–316 (1998).

84. D. Albaugh et al. Cell wall active antifungal compounds produced by the marine fungus *Hypoxylon oceanicum* LL-15G256. III. Biological properties of 15G256 gamma. *J Antibiot* (Tokyo) 51, 317–322 (1998).

85. G. Schlingmann, L. Milne, G. T. Carter. Isolation and identification of antifungal polyesters from the marine fungus *Hypoxylon oceanicum* LL-15G256. *Tetrahedron* 58, 6825–6835 (2002).

CHAPTER 9

Screening for Antiparasitic Marine Natural Products

Ryan M. Young[1,2] and Eva-Rachele Pesce[3]
[1]Department of Chemistry, University of South Florida, Tampa, Florida
[2]School of Chemistry, National University of Ireland Galway, Galway, Republic of Ireland
[3]College of Health and Biomedicine, Victoria University, Melbourne, Victoria, Australia

CONTENTS

9.1 INTRODUCTION

Infectious diseases pose a significant threat to human health. Currently, approximately 1 billion people are at risk of infection.[1] This, however, is mainly contained to the tropics, and often confined to the poorest parts of the planet; these diseases constitute the so-called neglected tropical diseases. Among these are onchocerciasis, lymphatic filariasis, leishmaniasis, malaria, toxoplasmosis, sleeping sickness, and Chagas disease. While there are several bacterial, viral, and fungal infections that may cause pathogenity in humans, for the purposes of this review, only those of a parasitic origin will be discussed.

In the past, natural products have been used in the treatment of tropical parasitic diseases. These compounds have been used either in their endogenous form or as lead compounds acting as the inspiration for many pharmaceuticals. Quinine (**1**) is the earliest of all antimalarials, and has been used in one form or another for the past 350 years. It was first discovered in South America, where it is found in the bark of the *Cinchona* tree that the local inhabitants used as an antipyretic agent.[2] In the seventeenth century, the bark of the Cinchonawas imported to Europe from Peru was used as an antimalarial.[3] Quinine is still employed today to treat uncomplicated malaria[4] and is often the only choice for life-threatening cerebral malaria.[5] Quinine has multiple side effects, most of which are not permanent, but some are severe, such as its arrhythmogenic potential and the release of insulin, which may result in hypoglycemia.[6] In the 1940s, while testing for less toxic analogues of quinine, chloroquine (**2**) was produced.[7] Until the development of resistance by the malaria parasites in the 1960s, chloroquine was the most widely used antimalarial compound.[6] Its popularity was due to chloroquine's low cost of production and lack of side effects. Mefloquine (**3**), another derivative, has been used for more than two decades and is regarded as a structurally simplified version of quinine. Mefloquine, when used as a prophylactic, can have neuropsychiatric side effects, such as insomnia, depression, and panic attacks.[8] Halofantrine is another quinine analogue, but it is linked to an increased risk of cardiac arrhythmias, as prolonged use inhibits the inward K^+ ion flow.[9] For this reason, halofantrine has been withdrawn from several countries.[6] Lumefantrine (**4**) is less effective against malaria than halofantrine, but lacks the cardiac side effects.[10]

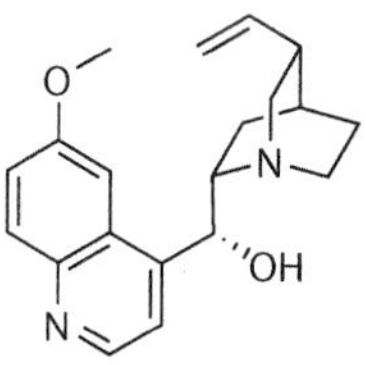

Artemisinin (**5**) is a natural product, which is active against malaria. Artemisinin is extracted from the Chinese herb qinghao (*Artemisia annua*), which was traditionally used in the treatment of fevers.[3] Many synthetic analogues of **5** have been developed, which have greatly decreased the cost of this class of antimalarials.[11] Artemether (**6**) is a more lipophilic molecule than **5** and is better absorbed from the gastrointestinal tract, allowing it to be taken orally.[12] On the contrary, due to the carboxylate fuctionality in artesunate (**7**), this drug is a hydrophilic molecule, requiring intravenous administration. This is important in the case of severe malaria, as the patient is often in an unresponsive state where no other mode of administration is possible.[13] Once in the body, all the artemisinines are metabolized to the biologically active dihydroartemisinine (**8**).

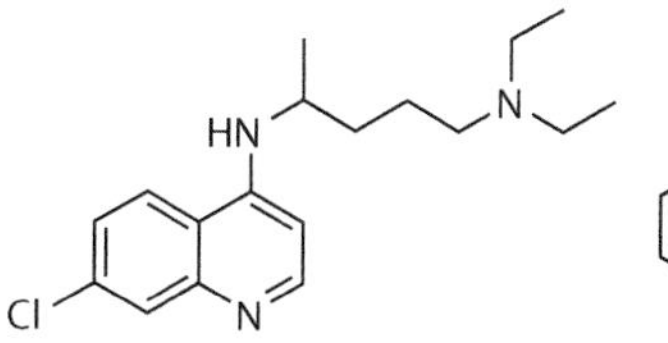

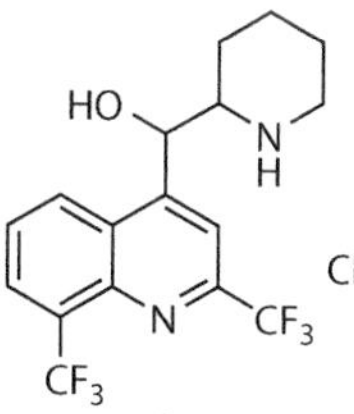

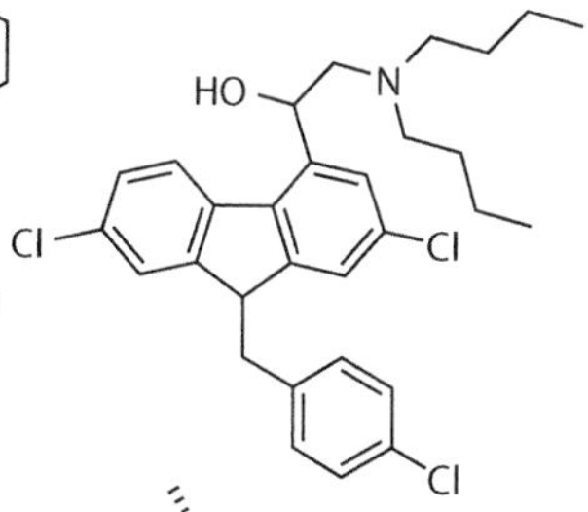

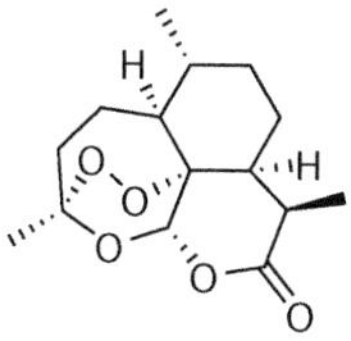

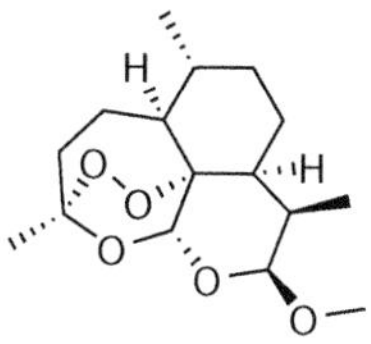

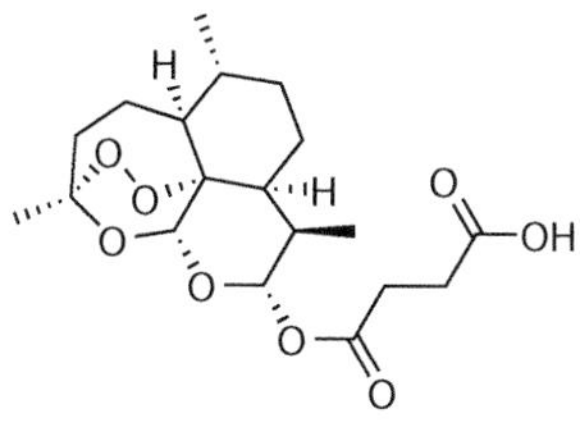

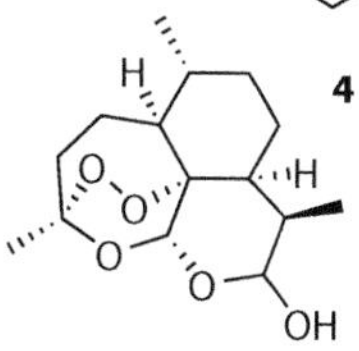

Many of the antibiotics show no visible effect on the first intraerythrocytic cycle of malaria parasites, but rather on the second cycle. The parasites are killed shortly after the second invasion of the erythrocyte. This is known as the delayed-death phenotype or the delayed-kill effect.[14,15] Antibiotics therefore show far longer fever and parasite clearance times than classic antimalarial drugs. Thus, antibiotics are used in combination with faster-acting drugs in the treatment of acute malaria. The classical antimalarials rapidly reduce the number of parasites, while the antibiotic continues to kill any remaining or resistant parasites. Doxycycline (**9**) is predominantly used as a prophylactic but may, in combination with quinine or artesunate, be used in the treatment of uncomplicated or even severe malaria.[16] Doxycycline is a semisynthetic antibiotic based on the *Streptomyces rimosus* secondary metabolite oxytetracycline (**10**).[17]

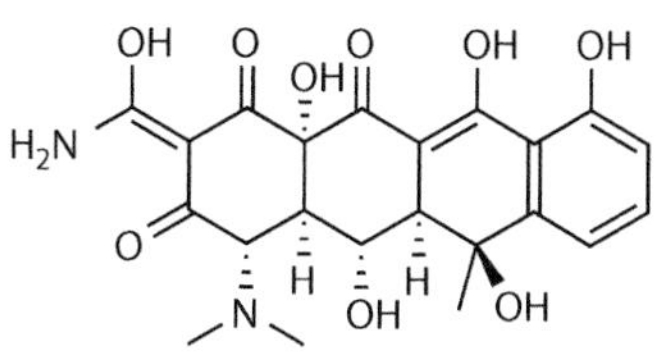

Doxycycline is currently also used in the control of *Wuchereria bancrofti*,[18] the causative parasite in lymphatic filariasis, as well as the treatment of dientamoebiasis caused by *Dientamoeba fragilis*. Dientamoebiasis may also be treated with aminoglycosides such as paromomycin[19] (**11**) isolated in the 1950s from the actinobacterium *Streptomyces krestomuceticus*.[20–23] Historically, *Streptomyces* spp. have been a great source of antiparasitic compounds; for example, *S. nodosus*,

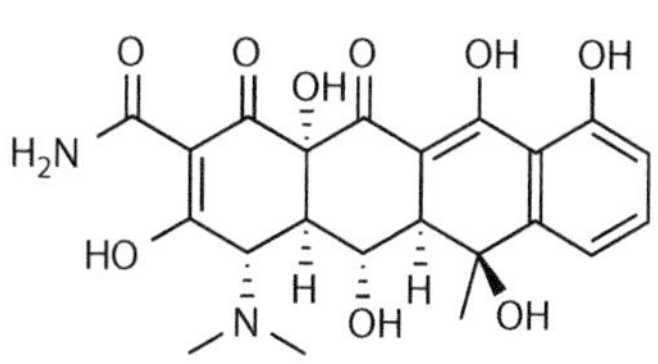

S. avermitilis, *S. lincolensis*, and *S. fradiae* have yielded amphotericin B (**12**),[24] avermectins (**13** and **14**),[25] lincomycin (**15**),[26–28] and neomycin (**16**),[29] respectively. Both **11** and **12** have been used in the treatment of leishmaniasis, while **11** has also been used in the treatment of cryptosporidiosis, amoebiasis, and dientamoebiasis. The frontline antihelminthic ivermectin (**17**) has been instrumental in the treatment and control of lymphatic filariasis, onchocerciasis, and serous cavity filariasis over the past several decades. Ivermectin was designed by the Merck Institute of Therapeutic Research based on the natural products **13** and **14** in 1981.[30] Ivermectin (**17**) acts by binding to the invertebrate glutamate-gated chloride channels, thus affecting neurological and motor function within the nematodes.[31] The semisynthetic lincosamide, clindamycin (**18**),[32] is an analogue of the biological compound **15**, which is often used in combination with **1** and **2** for the treatment of malaria, particularly in children[33] and pregnant women.[34] Neomycin (**16**), in combination with miconazole and propamidine, may be used for the treatment of Acanthamoeba keratitis.[35]

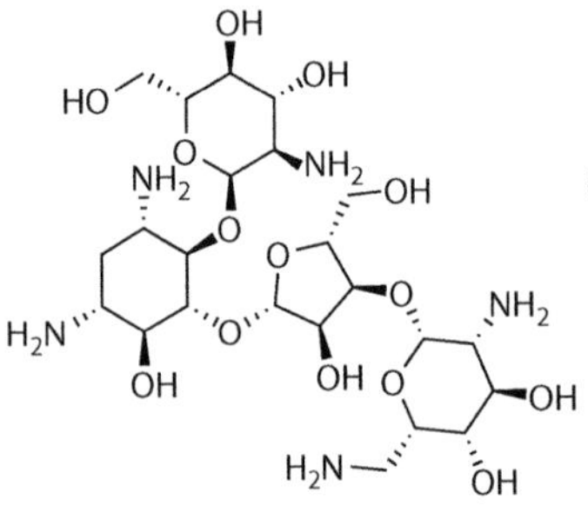

11

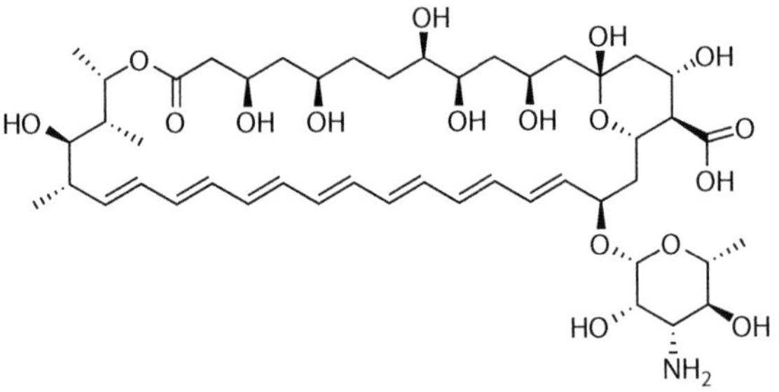

12

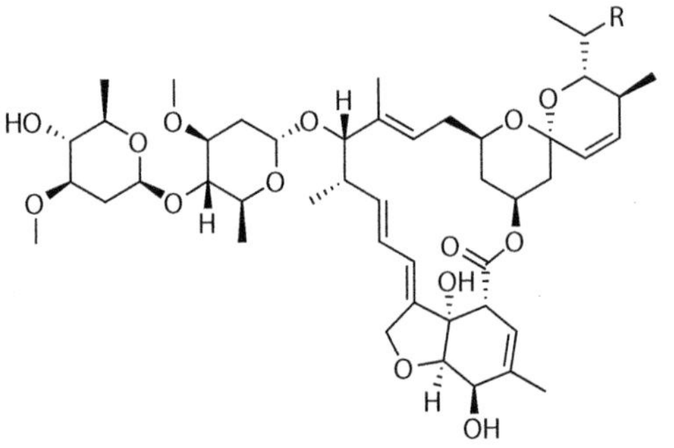

R = CH$_2$CH$_3$ Avermectin B$_{1a}$ (**13**)
R = CH$_3$ Avermectin B$_{1b}$ (**14**)

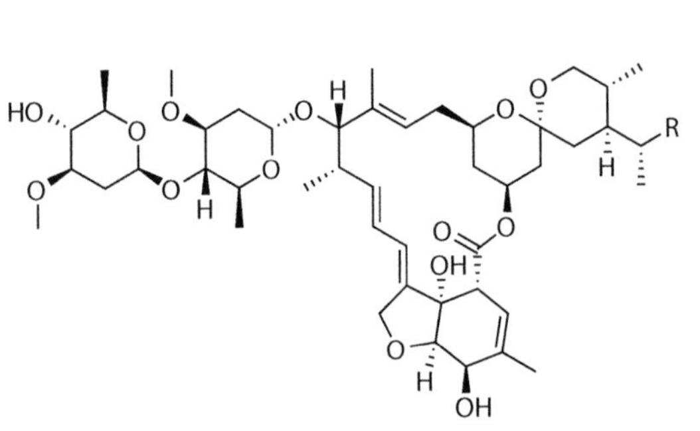

R = CH$_2$CH$_3$ Ivermectin B$_{1a}$ (**17a**)
R = CH$_3$ Ivermectin B$_{1b}$ (**17b**)

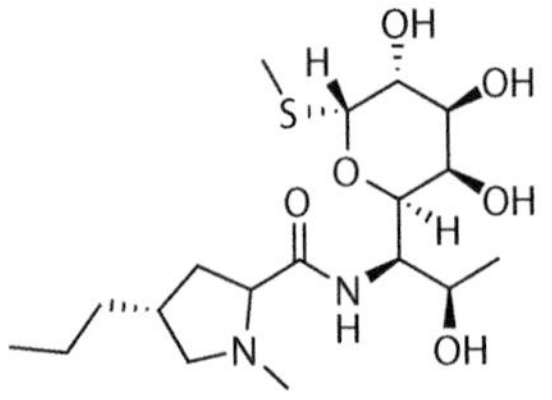

15 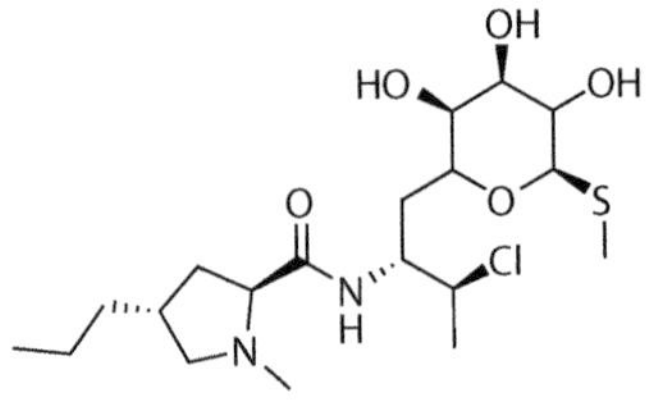   18 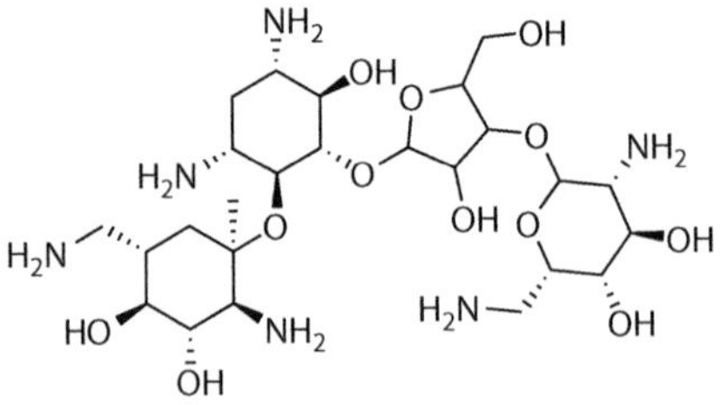   16

Rifampicin (**19**),[36] a semisynthetic compound based on the isolate of *Amycolatopsis rifmycinica*,[37] in combination with **12** and fluconazole, was administered intravenously for the successful treatment of primary amoebic meningoencephalitis (PAM) caused by *Naegleria fowleri*.[38] Another powerful antiparasitic azithromycin (**20**) was inspired by the *Saccharopolyspora erythraea* secondary metabolite, erythromycin (**21**). The combination of **20** and **12** has also shown some promise in treating *N. fowleri* in an *in vitro* and a murine *in vivo* assay.[39] Azithromycin has also been successfully used for the treatment of immune-compromised patients afflicted with cryptosporidiosis.[40,41] It should be noted that the search for medicinal natural products is not limited to microbial and botanical metabolites, but even includes the defensive egg terpenoid cantharidin (**22**) of the blister beetle, *Lytta vesicatoria*,[42] which has shown some promising results in the treatment of cutaneous leishmaniasis in animal models.[43]

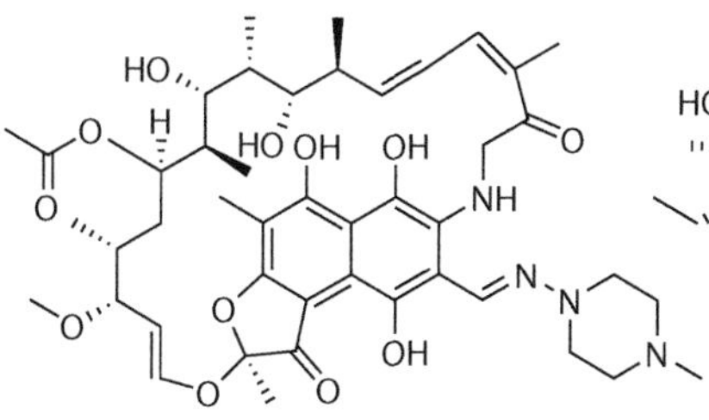

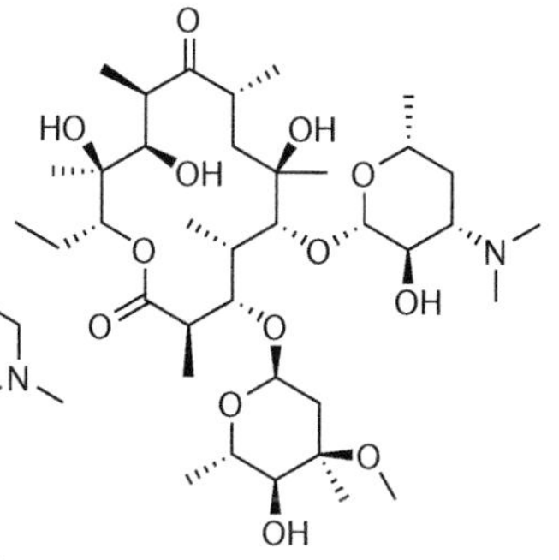

It is clear from the preceding text that the natural world holds a potentially infinite reservoir of medically relevant compounds. While traditional indigenous medicine has focused historically on terrestrial botany, the world's oceans cover more than two-thirds of the earth's surface and host approximately 100,000 species of marine invertebrates, a number that is estimated to be far higher due to numerous species still undiscovered or undescribed.[44] Many of these invertebrates belong to the Coelenterata, Porifera, Bryozoa, Mollusca, and Echinodermata phyla, which include organisms that may be sessile, slow moving, or slow growing and lack physical defenses such as spines or shells. These soft-bodied organisms may therefore contain chemical defenses to combat predation by fishes, crustacea, and so forth.[44] These diverse organisms offer a relatively untapped resource for new pharmaceuticals for the treatment of the ever-growing burden of infectious diseases.

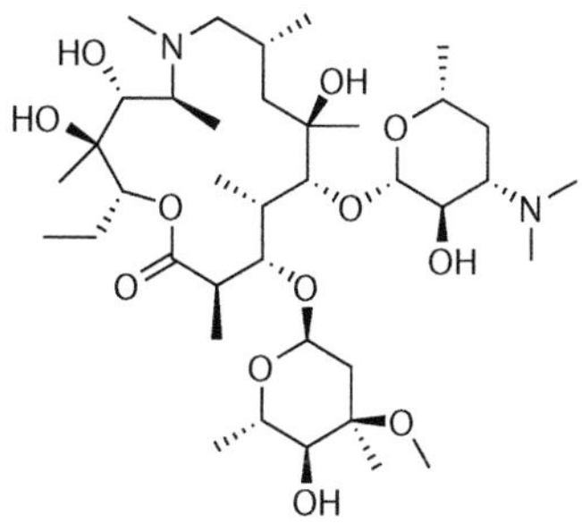

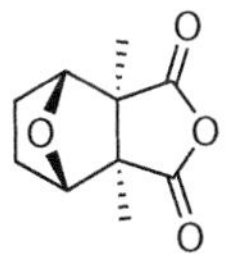

9.2 PARASITIC ORGANISMS

There exists a plethora of pathogenic parasitic organisms from a range of phyla. For the purposes of this chapter, several are discussed below; however, this list focuses on parasites, which cause pathological symptoms in humans. The full list of parasitic organisms is not limited to the ones mentioned *vide infra*.

9.2.1 Helminths

Nearly one-third of the world's population is infected with helminth parasites, which are responsible for many debilitating diseases and syndromes. There also exist a large amount of helminths responsible for the infection of livestock, resulting in large devastation to the agriculture of regions that host some of the globe's poorest people. With the current lack of pharmaceuticals, ever-growing emergence of drug resistance in the parasite, and rapid reinfection in areas where transmission has failed to be interrupted, there is a need for new pharmaceuticals and intervention strategies.

Figure 9.1   Left: The intermediate host of *S. mansoni*, *Biomphalarie* sp. Center: Adults of *Schistosoma mansoni*. Right: The intermediate for *S. haematobium* and *S. intercalatum*, *Bullinus* sp. (From DPDx Parasite Image Library, Centers for Disease Control, Atlanta, GA, 2014.)

Helminths are multicellular parasitic worms of the taxonomic groups trematodes (flukes), nematodes (roundworms), monogeneans (flatworms), and cestodes (tapeworms), each with their own life cycles and transmission. Some of these helminths are discussed in detail below.

9.2.1.1 Schistosomiasis (Bilharzia)

Schistosomiasis is a parasitic disease caused by several species of trematodes, including *Schistosoma mansoni* (Figure 9.1), *Schistosoma japaonicum*, *Schistosoma mekongi*, *Schistosoma guineensis*, *Schistosoma intercalatum*, and *Schistosoma haematobium*. The life cycle of the *Schistosoma* worm involves two stages, one in freshwater and the second within a human host. The eggs of the parasite are released into freshwater, where they hatch and become free-swimming miracidia. The miracidia penetrate the foot of a snail, thus beginning the snail stage of the parasite's life cycle. Sporocysts develop from the miracidium in the snail's hepatopancreas and finally develop into cercariae, which emerge daily from the snail, contaminating water. Once humans come into contact with cercariae-infected water, the parasite penetrates the skin and travels to the lungs and liver for further development. Mature *Schistosoma* pairs reproduce sexually, producing hundreds of eggs daily. Depending on the species, the eggs migrate to the intestines, bladder, or urethra to be expelled in the feces or urine. In areas where proper sanitation is not in place, the infected excrement often finds its way back into local water, continuing the cycle of infection when people wash, swim, or bath in contaminated waters.

While schistosomiasis is not fatal, eggs not passed by the body may develop into adult worms, becoming lodged in the bladder or intestines, resulting in inflammation or scarring, which may become more severe if not medically treated. Children who are infected by schisostomiasis may develop anemia and malnutrition, which may result in diminished cognitive functioning. Current treatment for schistosomiasis involves a single annual dose of praziquantel.[46]

9.2.1.2 Dracunculiasis (Guinea Worm Disease)

Dracunculiasis is caused by the nematode roundworm *Dracunculus medinensis*. Due to successful transmission interruption of the parasite, the incidence of Guinea worm disease has been reduced by more than 99% since 1986. Now, less than a thousand cases per year have been reported in Mali, South Sudan, Chad, and Ethiopia. Humans become infected by ingesting copepods (water fleas), which live in stagnant water. The copepod hosts are infected with Guinea worm larvae and are so small that they are invisible to the naked eye. Once ingested, the stomach acid digests the copepod but not the *Dracunculus* larvae, which is stimulated to leave the copepod host. The larvae then penetrate the host's intestine or stomach wall, where the larvae begin to mature. In the

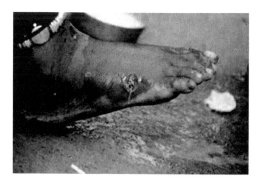

Figure 9.2   A female *Dracunculus medinensis* emerging from a human host. (From DPDx Parasite Image Library, Centers for Disease Control, Atlanta, GA, 2014.)

abdominal cavity and retroperitoneal space, the larvae mature and mate, and the fertilized female burrows deep into the subcutaneous and connective tissue. This creates large painful blisters on the host's skin, which is often perceived as a burning sensation. The host, in an attempt to relieve this burning sensation, submerges the blisters in water, allowing the female to emerge from the blister and release thousands of larvae into the water, thereby contaminating the water and renewing the cycle. While the death rate is low, this disease may induce complications due to bacterial infections of the wound generated by the parasite nematode, resulting in cellulitis, abscesses, septic arthritis, contractures, and tetanus. These symptoms can be extremely painful, often debilitating for the patient for prolonged periods. Currently there is no chemotherapy for the treatment of Guinea worm disease, and the worm is extracted from the body by physically pulling it out from the skin (Figure 9.2). Due to the female worms being up to 1 m in length, this may take several days to weeks and is extremely painful for the host.[46]

9.2.1.3 Toxocariasis (Roundworm Infection, Visceral Larva Migrans)

Toxocariasis is caused by the larvae of two species of *Toxocara*, *T. canis* and *T. cati*, which use dogs and cats as their host, respectively. While humans are not the intended host, this zoonotic helminth may infect humans who come into contact with contaminated soil. Dogs and cats transmit the parasite through their feces containing Taxocara eggs, which are therefore deposited into the soil (Figure 9.3). Although many infected people are asymptomatic, if left untreated, toxocariasis may result in eye inflammation, which could cause damage in the retina and even loss of sight. Visceral toxocariasis may spread to the organs, resulting in fever, wheezing, and abdominal pain. Toxocariasis can be treated with albendazole or mebendazole.[46]

Figure 9.3   Left: *Toxocara canis* beginning to emerge from an egg. Center: The larva of *T. canis*. Right: Cross section of liver tissue infected with larvae of *Toxocara* sp. (Tissue stained with hematoxylin and eosin.) (From DPDx Parasite Image Library, Centers for Disease Control, Atlanta, GA, 2014.)

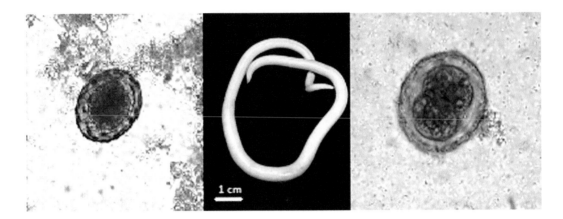

Figure 9.4 Left: The unstained stool mount showing a fertilized egg of *Ascaris lumbricoides*. Center: The adult female of *A. lumbricoides* (scale inset). Right: A wet unstained mount of a stool showing the early stage of embryo development of *A. lumbricoides*. (From DPDx Parasite Image Library, Centers for Disease Control, Atlanta, GA, 2014.)

9.2.1.4 Ascariasis (Intestinal Roundworms)

An estimated 800–1200 people are infected with the parasitic roundworm *Ascaris lumbricoides*, resulting in ascariasis. *A. lumbricoides* lives in human intestines, where the female roundworm may produce up to 200,000 eggs a day, which are deposited in the stool (Figure 9.4). Soil becomes contaminated with the parasite eggs in areas where hygiene is poor, municipal recycling of wastewater is not cleaned properly, or human feces are used as fertilizer. Humans become infected by consuming contaminated dirt or fruits and uncooked vegetables. The ingested eggs hatch into larvae in the intestines. Ascariasis often shows no symptoms, although heavy infection may result in abdominal discomfort, due to intestinal blockage, and impaired growth in children. Antihelminthic treatment of ascariasis includes albendazole and mebendazole.[46]

9.2.1.5 Hookworm Disease

Hookworm disease is primarily caused by two parasitic nematodes: *Ancylostoma duodenale* and *Necator americanus*. *A. duodenale* is predominantly confined to the Middle East, North Africa, and India, while *N. americanus* inhabits the Americas, Southeast Asia, and China (Figure 9.5). It has been estimated that 575 million–740 million people are currently infected with hookworm

Figure 9.5 Left: Hookworm egg in a wet mount. Center: Filariform (L3) hookworm larvae. Right: The adult worm of *Necator americanus*. (From DPDx Parasite Image Library, Centers for Disease Control, Atlanta, GA, 2014.)

disease.[46] Mature hookworms can penetrate the skin, and therefore the primary mode of infection of humans is due to people walking barefoot through infected soil. The soil becomes infected when a hookworm-infected person defecates outside or if feces is used as fertilizer. The hookworm eggs, deposited into the soil, hatch after 1–2 days (rhabditiform larva). The larvae continue to develop into the infective filariform larvae, which penetrate the skin of the host and travel through the blood-stream to the heart and then to the lungs. While in the lungs, the parasites burrow in the bronchial tree in order to make their way to the pharynx. Upon being swallowed, the parasites enter the gas-trointestinal tract and finally reach the small intestine. Here, the worms mature and produce eggs, continuing the cycle of infection. Due to the parasitic nature of these worms, infections may result in anemia and protein deficiency caused by blood loss at the site of attachment of the nematode (especially the small intestine). Children infected by hookworms can develop mental and growth retardation due to the reduction in available protein and iron. The general anthelminthic compounds albendazole and mebendazole are used to clear the patient of these nematodes.[46]

9.2.1.6 Trichuriasis (Whipworm Infection)

Trichuriasis is caused by an infection with the roundworm *Trichuris trichiura* (Figure 9.6). An estimated 604 million–795 million people globally are infected with whipworms. Whipworms are transmitted by the ingestion of soil containing unembryonated eggs deposited through feces. The eggs mature for 15–30 days and become infective. Once ingested by the human host, the worms hatch and establish themselves in the colon, where females may produce up to 20,000 eggs a day, which are passed through the feces. Infection may be classified as light or heavy. In light cases, many patients are asymptomatic and unaware of the parasitic infection. However, if left untreated, heavy symptoms may result, including abdominal pain. Heavy infection in children may impede physical and mental development. The current therapy for whipworm infection is a 3-day treatment with albendazole or mebendazole.[46]

9.2.1.7 Enterobiasis (Pinworm Infection)

Enterobiasis is caused by the small, thin roundworm *Enterobius vermicularis* (Figure 9.7). Although pinworms can infect anybody, enterobiasis is commonly found in children. Pinworm infection occurs when the parasite eggs are ingested from hands that have recently scratched the perianal area, by handling contaminated clothes or linen, or even inhalation of a few eggs that might have become airborne. Once ingested, the eggs hatch and the worms settle in the colon. After about 1 month, the females are fully developed and able to lay eggs. During the night, gravid females migrate out from the anus cavity, depositing eggs while crawling on the skin of the perianal area.

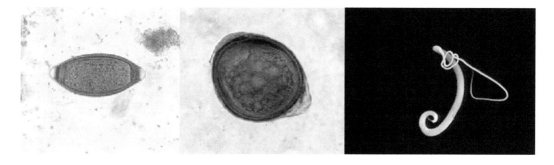

Figure 9.6 Left: Iodine-stained mount of the egg of *Trichuris trichiura*. Center: Atypical egg of *T. trichiura*. Right: Adult male of *T. trichiura*. (From DPDx Parasite Image Library, Centers for Disease Control, Atlanta, GA, 2014.)

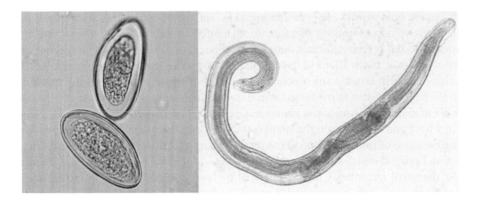

Figure 9.7 Left: Wet mount of the eggs of *Enterobius vermicularis*. Right: Adult male *E. vermicularis* from a formalin–ethyl acetate concentrated stool smear. (From DPDx Parasite Image Library, Centers for Disease Control, Atlanta, GA, 2014.)

The most common symptom of pinworm infection is an itchy anal region, which can be vulnerable to bacterial infection, particularly the female genital tract. Many patients complain of insomnia due to disturbed sleep, appendicitis, and abdominal pain. The treatment for enterobiasis is a single dose of mebendazole, pyrantel pamoate, or albendazole, followed by a second dose 2 weeks later.[46]

9.2.1.8 Gnathostomiasis

Human gnathostomiasis is caused by infection with several species of nematode from the genus *Gnathostoma* (Figure 9.8). This disease is most common in Southeast Asia; however, it has been reported in other parts of Asia as well as South and Central America. The usual mode of infection is the consumption of undercooked or raw freshwater fish, eels, frogs, birds, and reptiles that are infected with *Gnathostoma* larvae. Humans are not the natural host of this parasite, which rather proliferates in pigs, dogs, and cats that defecate unembryonated eggs. The eggs become embryonated in water, ultimately hatching into larvae. The larvae infect copepods, which are the intermediate host of the parasite. Frogs, birds, fish, and reptiles eat the infected copepods, therefore becoming carriers for the transmission of the parasite to humans. The most common symptoms of this parasitic infection are swelling under the skin and an increased level of eosinophils in the blood. Occasionally, these parasites can enter the liver or eyes, resulting in this case in loss of vision or blindness. In severe cases, the parasite enters the nervous system, spinal cord, or brain, resulting in nerve pain, paralysis, coma, and even death. Albendazole and ivermectin have been used in the treatment of cutaneous

Figure 9.8 Left and Right: The third stage larva of *Gnathostoma spinigerum*. Center: The female *G. spinigerum* head bulb. (From DPDx Parasite Image Library, Centers for Disease Control, Atlanta, GA, 2014.)

Gnathostoma infections; however, there are no published studies on the efficacy of this drug regiment in the treatment of ocular and central nervous system *Gnathostoma* infections.[46]

9.2.1.9 Lymphatic Filariasis (Elephantiasis)

Lymphatic filariasis occurs when a person is bitten by a mosquito infected with the nematodes *Wuchereria bancrofti*, *Brugia malayi*, or *Brugia timori*, with the former species resulting in 90% of all cases (Figure 9.9). Nearly 1.4 billion people within 73 countries are at risk of contracting elephantiasis, with more than 120 million currently infected.[47] The three major vectors of transmission of the Filariodidea nematodes are *Culex* spp., which are widespread in urban and semiurban areas; *Anopheles* spp., predominantly in rural areas; and *Aedes* spp., primarily in the islands of the Pacific. The parasite larvae are transmitted to humans when the insect vectors take a blood meal. The larvae then migrate to the lymphatic vessels and ultimately to the lymph nodes. Once in the lymph nodes, the larvae begin to develop into microfilariae and finally mature to the adult stage of their life cycle. The adults reside in the lymph nodes, where they continue to produce microfilariae, which circulate throughout the body via the bloodstream. It is these microfilariae that are ingested by the female mosquitoes, initiating the insect stage of the parasite's life cycle. The infection of the lymphatic system results in a diminished function, resulting in lymphedema. Lymphedema causes the swelling of the infected area due to buildup of fluid, which primarily occurs in the legs, but may also occur in the arms, breasts, and genitalia. These symptoms may only present several years after inoculation with the parasite. Due to this reduced functioning of the lymphatic system, the infected person is subject to infections by microorganisms, such as bacterial infections of the skin. This causes hardening and thickening of the skin, changing its texture and appearance, that is, elephantiasis. This deformation can be extremely painful or awkward, resulting in loss of mobility, which may result in financial hardships due to loss of income and increased medical expenses. The treatment for lymphatic filariasis is a single dose of either albendazole or ivermectin in areas where onchocerciasis is also endemic or diethylcarbamazine where river blindness is not a concern.[46]

9.2.1.10 Loiasis (African Eye Worm)

Loiasis is caused by the nematode *Loa loa* and is transmitted by the repeated bite of the vector, the deerflies *Chrysops silacea* and *Chrysops dimidiata* (Figure 9.10). These flies breed in the forest canopies of the rain forests of West and Central Africa. When the deerfly feeds on a human, the filarial larvae are deposited into the wound, through which they penetrate the skin and settle in the subcutaneous tissue layer, where they develop into adults. Adult worms reproduce, releasing

Figure 9.9 Left: Blood smear stained with Giemsa showing the microfilaria of *Wuchereria bancofti*. Center: Female *Armigeres subalbatus* after a fresh blood meal. Right: Blood smear stained with Giemsa showing the microfilaria of *Brugia malayi*. (From DPDx Parasite Image Library, Centers for Disease Control, Atlanta, GA, 2014.)

Figure 9.10   Left: Blood smears stained with Giemsa showing the microfilaria of *Loa loa*. Center: *Chrysops silacea* feeding on host. Right: Adult *L. loa* removed from a patient's eye. (From DPDx Parasite Image Library, Centers for Disease Control, Atlanta, GA, 2014.)

microfilariae into the surrounding tissue. The fly ingests microfilariae during a blood meal from an infected host, which may then be transferred to another person during a subsequent blood meal. Symptoms of this disease manifest themselves as localized Calabar swellings and eye worms. Eye worm is the migration of the adult worm across the eye, which may result in congestion, itching, pain, and light sensitivity. While these symptoms cause major discomfort, they usually cause minimal damage to the eye. A patient with a long-term infection of loiasis may manifest kidney damage, possible scarring of the heart muscle, and fluid in the lungs. Loiasis has emerged as a global health concern due to its negative impact on the control of oncocerciasis and lymphatic filariasis in areas of co-endemicity. It was shown that patients infected with *L. loa* treated with ivermectin, for onchocerciasis, developed severe neurological disorders.[48]

9.2.1.11 Ochocerciasis (River Blindness)

Ochocerciasis is caused by the parasitic filarial worm *Onchocera volvulus*, which is transmitted by blackflies of the genus *Simulium* (Figure 9.11). These flies breed in streams and rivers, often where villages are established. Onchocerciasis is a leading cause of blindness in Sub-Saharan Africa, with about half a million people blinded or visually impaired as a result of this infection. In order to induce ovulation, the female *Simulium* requires a blood meal. Therefore, blackflies will feed on humans, transferring filarial larvae to the human host through the bite wound. Once in the subcutaneous tissue, the larvae develop into the adult filariae, forming nodules. The small male worms migrate to these nodules to seek out female worms to mate with. After mating, the female may produce eggs, resulting in thousands of microfilariae, which migrate throughout the body, mature, and mate, continuing the infection. When the blackflies ingest a human blood meal, the

Figure 9.11   Left: Skin nodule showing the presence of microfilariae of *Onchocera volvulus*. Center: *Simulium ochraceum* taking a blood meal. (Photo courtesy of Dr. Nathan Burkett-Cadena.) Right: Cross section of the female *O. volvulus* showing many microfilariae within the uterus (stained with hematoxylin and eosin). (From DPDx Parasite Image Library, Centers for Disease Control, Atlanta, GA, 2014.)

microfilarae are transferred from the host into the midgut of the fly, allowing for further spread of the disease. Many people do not experience any symptoms while infected by *O. volvulus*, as the larvae can migrate throughout the body without being detected by the host's immune system. However, most of the symptoms that patients report are primarily due to an immune response to dead or dying larvae. These symptoms include skin rashes, lesions, swelling, inflammation, and intense itching, which over time may result in loss of elasticity and discoloration of the skin. As the name *river blindness* describes, onchocerciasis is the world's fourth leading cause of preventable blindness. Loss or reduction in vision occurs when microfilariae migrate to the eye and die, causing an inflammatory response. This results in the cornea becoming opaque over time, diminished eyesight, and ultimately blindness if left untreated. The current treatment for onchocerciasis is ivermectin, which kills the microfilariae but not the adult worms. Therefore, in order to break transmission of onchocerciasis, the community would require continuous treatment with ivermectin for the life span of the adult worm, which may live for up to 15 years.[46]

9.2.1.12 Taeniasis (Tapeworm Infection) and Cysticercosis

Taeniasis is a parasitic infection caused by the tapeworm species *Taenia saginata* (beef tapeworm), *Taenia solium* (pork tapeworm), and *Taenia asiatica* (Asian tapeworm) (Figure 9.12). Humans become infected by eating the raw or undercooked meat of an infected animal. Many people infected with tapeworms may be asymptomatic and therefore unaware of the infection. However, *T. solium* may cause larval cysts in the brain and muscle tissue, that is, cysticercosis. Ingested eggs hatch in the intestines, and the larvae invade the intestinal wall, from where they can migrate to the brain, liver, and other tissues, where they develop into cysticerci. The most severe form of this infection is the development of neurocysticercosis, which manifests as headaches and seizures. If untreated, the infection may result in brain swelling and hydrocephalus, which can be fatal. Praziquantel or niclosamide is the recommended treatment for taeniasis; however, should the infection progress to cysticercosis or neurocysticercosis, surgical intervention is the only course of treatment.

With the ever-growing concern of resistance, as well as serious side effects and the severe limitations of the frontline drugs ivermectin and albendazole in killing the adult worms, it is clear a new suite of anthelmintic compounds needs to be developed.[46]

9.2.2 Leishmania

Leishmaniasis is a parasitic disease that occurs when the protozoan parasite of the genus *Leishmania* (more than 20 species) is transmitted through the bite of phlebotomine sandflies (Figure 9.13). The disease is found in 88 countries located in parts of the tropics, subtropics, and southern Europe, with an estimated 1.3 million new cases and an estimated 20,000–30,000 deaths

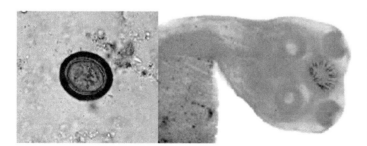
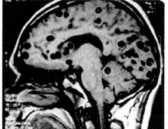

Figure 9.12 Left: The unstained wet mounts of *Taenia* spp. eggs. Center: Scolex of *T. solium*. Right: X-ray image of a patient infected with neurocysticercosis. (From DPDx Parasite Image Library, Centers for Disease Control, Atlanta, GA, 2014.)

Figure 9.13   Left: The Giemsa-stained amastigotes of *Leishmania* spp. present in a skin scraping. Center: The leishmaniasis vector *Phlebotomus papatasi* taking a blood meal. Right: The promastigotes of *Leishmania* sp. from tissue culture. (From DPDx Parasite Image Library, Centers for Disease Control, Atlanta, GA, 2014.)

occurring annually. Leishmaniasis occurs in three different forms: visceral (90% of all cases reported in Bangladesh, Brazil, India, Nepal, and Sudan), cutaneous (90% of all cases reported in Afghanistan, Brazil, Iran, Peru, Saudi Arabia, and Syria) and mucocutaneous (90% of all cases reported in Bolivia, Brazil, and Peru). The infection is initiated when a sandfly takes a blood meal, depositing promastigotes into the bite wound. The promastigotes are phagocytized by macrophages or other phagocytic cells. Once in these cells, the promastigote develops into the tissue stage of the parasite (amastigote), which undergoes multiple divisions, which produce more parasites that can infect additional macrophages and cells of various tissues. The species of parasite, host, and additional factors will determine whether visceral, cutaneous, or mucocutaneous leishmaniasis will result. Cutaneous leishmaniasis is the most common form of the disease and presents initially as papules or nodules and may develop into skin ulcers. Visceral leishmaniasis occurs when the parasitic infection spreads to the internal organs, typically the spleen, liver, and bone marrow, and can be fatal. The symptoms of this type of leishmaniasis are fever, weight loss, enlargement of the spleen and liver, and anemia. The least common form of leishmaniasis is mucocutaneous leishmaniasis, where the parasite spreads from the skin to the mucous membranes, most commonly those of the nose, mouth, and throat. Current treatment for visceral leishmaniasis is the intravenous administration of liposomal amphotericin B and meglumine antimoniate. Presently, there are no Food and Drug Administration (FDA)-approved drugs for the treatment of cutaneous and mucocutaneous leishmaniasis.[46]

9.2.3 *Plasmodium*

Although in the past century there have been many successful developments to eradicate the malarial parasite, malaria continues to be widespread. The 2014 World Health Organization (WHO) fact sheet reports that there were more than 89 million reported cases of malaria in 2012, with 660,000 cases resulting in death in 2010.[49,50] The increase in the spread of malaria has been attributed to several factors, including population movements to malaria-infected areas; changes in agricultural practices, such as the building of dams and irrigation systems or deforestation; the weakening of public health systems; global climate changes; and resistance to antimalarial drugs and insecticides.[51] With the current population growth rates in malaria-riddled regions, it has been estimated that without effective intervention, the cases of malaria will double within the next 20 years.[52]

The global distribution of malaria is centered in the tropics, although not confined to this region of the world. Many efforts have attempted to eradicate the disease; however, these only seem to have been successful in the temperate regions of the globe. This is due to, among other factors, the decreased rate of reproduction of the plasmodial parasite within the mosquito at lower temperatures. Ninety percent of all the fatal cases of malaria occurring in Africa result from infection

Figure 9.14 Left: A gametocyte of *Plasmodium falciparum*. Center: The malaria vector *Anopheles gambiae* taking a blood meal. Right: The trophozoites of *P. falciparum* from a blood smear. (From DPDx Parasite Image Library, Centers for Disease Control, Atlanta, GA, 2014.)

by *Plasmodium falciparum* (Figure 9.14).[51] Infection by other plasmodia species may also cause malaria. *Plasmodium malariae* is transmitted throughout the tropics and subtropics; although widespread, the locations are patchy. *Plasmodium vivax* is located throughout the tropics; however, this species is uncommon in most of Africa, while *Plasmodium ovale* is predominantly found in tropical Africa.[53] In more recent years, a fifth species of *Plasmodium*, *P. knowlesi*, has been reported in patients located in Southeast Asia. *P. knowlesi* usually infects macaque monkeys; however, hundreds of cases have been identified in Malaysia.[54] The widespread transmission of the malaria parasite has primarily been attributed to the vector *Anopheles gambiae* and its high preference toward anthropophily (biting of humans). The female *Anopheles* mosquito feeds by cannulating blood either from pools generated by damaged capillaries or directly from larger subcutaneous blood vessels.[55] While feeding, the mosquito vector releases sporozoites into the skin and bloodstream.

Infection in the human host occurs in two stages: the hepatic stage and the erythrocytic stage. Within minutes of entering the bloodstream, the sporozoites reach the liver cells, beginning the hepatic stage.[56] The sporozoites invade the hepatocytes and undergo multiple rounds of nuclear division and differentiate into thousands of merozoites,[57] which eventually rupture the cell with the release of the merozoites into the bloodstream, initiating the erythrocytic stage. Upon erythrocyte invasion, the parasites develop into various stages that culminate with the production of several new merozoites, which are released into the bloodstream following the burst of the infected erythrocytes. With the rupture of each erythrocyte, the parasite's waste is also released into the bloodstream, causing some of the clinical symptoms experienced by patients, including fever, chills, headaches, adnominal and back pains, nausea, diarrhea, and occasionally vomiting. *P. falciparum* can rapidly progress to severe illness, ultimately resulting in death without medical intervention. The best current treatment for malaria is an artemisinin-based combination (ACT) therapy.[58] However, with the emerging onset of artemisinin resistance in *P. falciparum*, there exists a growing need for the development of new antiplasmodials.[59]

9.2.4 Toxoplasmosis

Toxoplasmosis is caused by the parasitic protozoan *Toxoplasma gondii* (Figure 9.15). In some parts of the world the prevalence of *T. gondii* infection has been reported as high as 95% of the population, with approximately a third of the world's population infected.[46] Infection in humans occurs by ingestion of infected meat (primarily pork), water, fruit and vegetables, or soil contaminated with oocysts derived from cat feces. Felidae become infected by consuming intermediate hosts, for example, birds and rodents, which have ingested contaminated plants or soil. Once *T. gondii* oocysts enter humans by the aforementioned routes, the parasites can form cysts in the skeletal

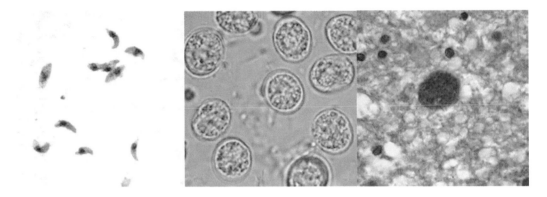

Figure 9.15 Left: *Toxoplasma gondii* tachyzoites obtained from peritoneal fluid of an inoculated mouse. Center: *T gondii* oocysts in a fecal floatation. Right: *T. gondii* cyst in brain tissue stained with hematoxylin and eosin. (From DPDx Parasite Image Library, Centers for Disease Control, Atlanta, GA, 2014.)

muscle, myocardium, brain, and eyes. In healthy humans, this usually manifests itself by flulike symptoms and may be kept at bay by the host's immune system. However, if the host is immune compromised (e.g., HIV positive), the infection may be more severe, causing fever, confusion, headache, seizures, nausea, and poor coordination. In pregnant women, the parasite may be transmitted to the fetus (congenital transmission), leading to a miscarriage, stillborn child, or deformations such as abnormal enlargement or smallness of the head. Current treatment for toxoplasmosis is a combination of drugs, such as pyrimethamine and sulfadiazine.

9.2.5 *Trypanosoma*

9.2.5.1 *Human African Trypanosomiasis (African Sleeping Sickness)*

Human African trypanosomiasis is spread by members of the genus *Glossina* (tsetse flies), which transmit the parasite *Trypanosoma brucei* (Figure 9.16). Two subspecies of *T. brucei* are responsible for the infection of humans: *Trypanosoma brucei* (*T. b.*) *gambiense* (24 countries in West and Central Africa) and *T. b. rhodesiense* (13 countries in East and southern Africa), with the former accounting for 98% of all reported cases of African sleeping sickness. At present, 36 Sub-Saharan countries are at risk of human African trypanosomiasis, with those with extended rural areas at greatest risk. While this disease continues to be a concern, due to successful interventions,

Figure 9.16 Left: *Trypanosoma brucei* ssp. in blood smear, stained with Giemsa. Center: Image of a tsetse fly, the vector of African trypanosomiasis. Right: The trypomastigote of *T. brucei* ssp. beginning to divide, stained with Giemsa. (From DPDx Parasite Image Library, Centers for Disease Control, Atlanta, GA, 2014.)

there has been a significant decrease in those infected, with 7197 reported cases in 2012 (WHO)[49]. During a blood meal, tsetse flies inject metacyclic trypomastigotes into the host's skin tissue; the parasites enter the lymphatic system and then the bloodstream. Here, *T. brucei* parasites develop into trypomastigotes that are carried to other locations throughout the host's body, including the spinal fluid and the brain, causing the neurological symptoms of the disease. *T. brucei* multiplies by binary fission, which increases the likelihood of being ingested by a host-seeking tsetse fly, continuing the cycle of infection. After the initial infected tsetse fly bite, a person will experience bouts of headaches, fever, joint pain, and itching. Usually after 1–2 years, the parasite will successfully cross the blood–brain barrier, often indicated by the infected person becoming confused, changes in personality, chronic fatigue, partial paralysis, and loss of coordination and balance. If the patient is left untreated, African trypanosomosis is often fatal. The treatment of African sleeping sickness is dependent upon the subspecies of *T. brucei*, as well as the stage to which the infection has progressed (i.e., whether the central nervous system has been invaded by the parasite). Pentamidine is the recommended drug for the initial stage of the infection; however, this drug cannot cross the blood–brain barrier and is therefore not used in the treatment of severe cases. Suramin, melarsoprol, eflornithine, and nifurtimox are used in cases where the central nervous system has been compromised by the parasite.

9.2.5.2 *Chagas Disease (American Trypanosomiasis)*

The causative agent of Chagas disease is *Trypanosoma cruzi* (Figure 9.17), which is transmitted through the infected feces of the blood-feeding triatomine bugs, the "kissing bugs." Chagas disease has been reported in North America, Europe, and western Pacific countries; however, it is most prevalent in Latin America, where the disease is endemic. It has been estimated that 7 million–8 million people are infected by this parasite worldwide. Kissing bugs usually reside in cracks of poorly constructed buildings in rural or suburban areas. Normally these insects remain hidden during the daytime and come out to feed on human blood during the night. These bugs will bite exposed skin, often defecating near the feeding site. People become infected when the feces is smeared into the skin break, eyes, or mouth when the person instinctively scratches the bite site. People may also become infected by consuming food contaminated with triatomine bug feces, contact with infected blood, and organ transplant from infected donors. Once the metacyclic trypomastigotes penetrate the cells in close proximity to the wound site, they begin to differentiate into intracellular amastigotes. The amastigotes continue to multiply via binary fission and eventually differentiate into trypomastigotes. At this point, the infected cell ruptures, releasing the trypomastigotes into the bloodstream for further cell invasion. The cycle of infection continues when the triatomine bugs consume an infected blood meal. Chagas disease presents itself in two stages: acute and chronic. The acute stage may last up to 2 months, displaying mild swelling and discomfort around the site of inoculation, enlarged lymph nodes, headaches, and muscle pain, with many people being

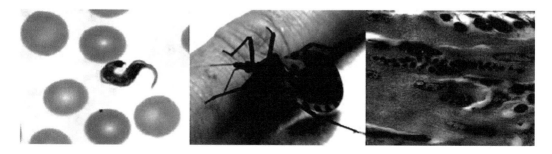

Figure 9.17 Left: *Trypanosoma cruzi* trypomastigote in blood smear stained with Giemsa. Center: Image of a triatomine, vector of *T. cruzi*. Right: The amastigotes of *T. cruzi* in heart tissue.

asymptomatic. During the chronic phase, the parasite may invade the cells of the heart and diges-tive tract, leading to cardiac disorders and dilated esophagus or colon, which results in discomfort when eating or passing stools. If the infection is left untreated, the disease may cause sudden death or heart failure due to the degradation of the heart tissue. Current treatment for Chagas disease is a course of benznidazole or nifurtimox; however, the efficacy of these treatments significantly dimin-ishes as the infection progresses.

9.2.6 Amoebozoa

9.2.6.1 Amoebiasis and Dientamoebiasis

Amoebiasis and dientamoebiasis are caused by the parasitic amoebas *Entamoeba histolytica* and *Dientamoeba fragilis*, respectively (Figure 9.18). The cysts and trophozoites of the amoebas are passed in feces, and therefore humans become infected by oral contact with contaminated hands, water, or food. Most infected people are asymptomatic; however, the infection may cause mild diarrhea. Severe amoebiasis may manifest itself in two ways: amoebic colitis and amoebic liver abscesses should the parasitic amoeba invade the intestinal lining and liver tissue, respectively. In severe cases of *D. fragilis* infections, the disease may manifest in weight loss, fatigue, stomach pain, fever, skin rashes, and pruritus. The general treatment for amoebiasis is an initial course of the amoebicide metronidazole, followed by treatment with paromomycin or diloxanide.

9.2.6.2 Primary Amoebic Meningoencephalitis

Primary amoebic meningoencephalitis (PAM) arises when the free-living amoeba *Naegleria fowleri* infects the brain (Figure 9.19). *N. fowleri* is commonly found in warm bodies of water, such as hot springs, lakes, or rivers, and is present globally. Infection primarily occurs when contami-nated water enters through the nose and the amoeba penentrates the nasal mucosa and migrates to the brain via the olfactory nerves. Currently, there are only four reported cases of patients surviving PAM, with 98% (129/132) of all U.S. cases being fatal.[60] Two of the surviving patients received a combination of amphotericin B, rifampicin, fluconazole, miconazole, dexamethasone, sulfisoxa-zole, ceftriaxone, and phenytoin. The remaining two patients' treatment involved administration of miltefosine.[61,62]

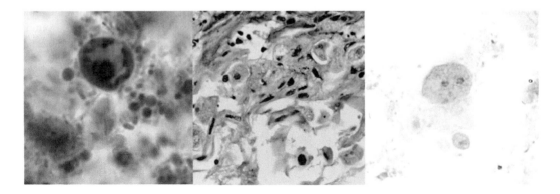

Figure 9.18 Left: The immature cyst of *Entamoeba histolytica* stained with trichrome. Center: Image of *Entamoeba histolytica* trophozoites in a perianal biopsy, stained with hematoxylin and eosin. Right: The binucleated form of trophozoites of *Dientamoeba fragilis*, stained with trichrome. (From DPDx Parasite Image Library, Centers for Disease Control, Atlanta, GA, 2014.)

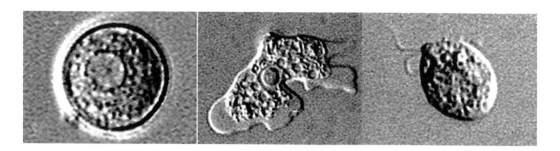

Figure 9.19 Left: The cyst of *Naegleria fowleri*. Center: *N. fowleri* in the amoeboid trophozoite stage. Right: The flagellated stage of *N. fowleri*. (From DPDx Parasite Image Library, Centers for Disease Control, Atlanta, GA, 2014.)

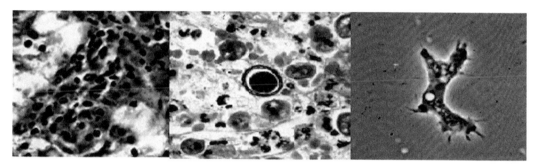

Figure 9.20 Left: *Acanthamoeba* spp. in human adrenal gland tissue, stained with hematoxylin and eosin. Center: Cysts of *Acanthamoeba* spp. in brain tissue, hematoxylin and eosin stained. Right: Trophozoite of *Acanthamoeba* spp. (From DPDx Parasite Image Library, Centers for Disease Control, Atlanta, GA, 2014.)

9.2.6.3 Acanthamoeba Keratitis and Granulomatous Amoebic Encephalitis

Acanthamoeba keratitis and granulomatous amoebic encephalitis (GAE) are serious infections of the eye and spinal cord or brain by the free-living *Acanthamoeba* spp., respectively (Figure 9.20). Infection with several species of acanthamoeba, such as *A. culbertsoni, A. polyphaga, A. castellanii, A. astronyxis, A. hatchetti, A. rhysodes, A. divionensis, A. lugdunensis,* and *A. lenticulata,* results in human disease. Acanthamoebas commonly occur in swimming pools, tap water, contact lens solution, and lakes. Acanthamoeba keratitis develops when *Acanthamoeba* spp. come into contact with the eye, while interaction of the amoeba in the nasal passages, respiratory tract, or skin lesions of immune-compromised people might result in GAE. Symptoms of acanthamoeba keratitis include eye pain, blurred vision, and photosensitivity, which, if left untreated, can lead to blindness. GAE may present in mental status changes, loss of balance and coordination, fever, double vision, muscular weakness, or partial paralysis, as well as other neurological problems. Topical treatment with chlorhexidine and polyhexamethylene is the primary therapy for acanthamoeba keratitis, although propamidine isethionate has also been successful. Several cases of GAE have been treated with pentamidine, sulfadiazine, flucytosine, and either fluconazole or itraconazole. However, most cases of brain or spinal cord infections by *Acanthamoeba* spp. are fatal.

9.2.7 Cryptosporidiosis

Cryptosporidiosis is caused by a number of parasites of the genus *Cryptosporidium*; however, the two species *Cryptosporidium parvum* and *Cryptosporidium hominis* are the major causative agents

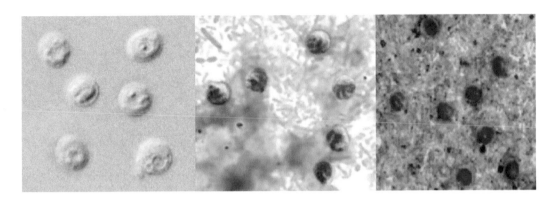

Figure 9.21   Left: The oocysts of *Cryptosporidium parvum* in a wet mount. Center: The oocysts of *Cryptosporidium* spp. stained with safranin. Right: The oocysts of *Cryptosporidium* spp. stained with Ziehl–Neelsen modified acid-fast. (From DPDx Parasite Image Library, Centers for Disease Control, Atlanta, GA, 2014.)

of the disease in humans (Figure 9.21). *Cryptosporidium* parasites are found worldwide, although developing nations are at greater risk due to poorer water treatment and sanitation. A person becomes infected when the parasite is ingested by either drinking contaminated water, eating contaminated fruits or vegetables, or touching the mouth with contaminated hands. The symptoms usually present initially as diarrhea but may develop into stomach cramps, dehydration, nausea and vomiting, fever, and weight loss. In patients with a weakened immune system, this infection might be fatal.

With many of the current drugs being discovered in the 1980s, and with some parasites[63–65] becoming resistant to the current arsenal of chemotherapies, there is a moral imperative for the discovery of new drugs for some of the poorest places in the world.

9.3 SCREENING COMPOUNDS FOR ANTIPARASITIC ACTIVITY

9.3.1 Screening Metabolites for Biological Activity

Compounds may be screened at the pure compound or crude extract level. The screening of crude extracts allows narrowing down which fraction contains the bioactive compound, that is, bioguided fractionation. Most frequently, compounds or extracts are dissolved in dimethyl sulfoxide (DMSO), ethanol, or methanol. The disadvantage of using the former two alcohols is the rapid evaporation due to their low boiling point, thus changing the concentration of the test solution. Sample solutions of 100% DMSO are the standard at a concentration of 20 mM or µg/ml for pure compounds and extracts, respectively. The advantage of using DMSO is that it is highly compatible with automation, ensuring that test compounds remain in solution during the serial dilution process. The disadvantages of using DMSO are that is highly hydroscopic, and therefore the master plate should be sealed when not in use, and the high boiling point of DMSO makes removal of the solvent to reclaim the natural product time-consuming. Storage of the master plate should be solvent-free for long-term storage or in 100% DMSO with minimal freeze–thaw cycles and humidity to reduce the dilution of the stock solution with condensation. It is crucial that the potentially active compounds are not lost, destroyed, or altered during the purification and characterization process. When screening for activity, all testing should be done with replicates, as well as contain a negative (i.e., growth media only resulting in 0% parasite growth), a positive (no inhibitor in infected growth media resulting in 100% parasite growth), and a reference control (commercially available antiparasitic compound resulting in reduction of parasite growth). Ideally, multiple reference compounds should be tested, and they should be of high purity and obtained from reputable suppliers. When preparing a microplate, the four corners of the plate should be left empty, as these wells are subject to rapid evaporation. Generally, in *in vitro* biological assays, the activity of a compound is expressed as a concentration as either the IC_{50} or IC_{90}, that is, the concentration required to kill 50% or 90% of the parasite, respectively. As a general rule, crude extracts will be deemed active at IC_{50} concentrations below 100 µg/ml, while pure compounds are considered active below 25 µM. Ideally, the IC_{50} should be in the nanomolar range.

9.3.2 Virtual Screening

The term *virtual screening* was coined in the 1990s to describe the computer algorithms used to identify bioactive molecules.[66] While there are many different algorithms to calculate the ligand conformation within a protein, there still exists much debate as to the scoring functions for virtual screening. Many of the major weaknesses of docking programs exist not in the docking algorithms, but rather in the inaccuracy of the functions used to determine the affinity between the ligand and target, that is, the scoring value. The strengths and weaknesses of these scoring functions are discussed by Stahl and Rarey in their comprehensive 2001 review.[67] Many other faults exist in the virtual screening of natural products, the major one being that the absolute structure of the natural product must be known. Natural products tend to include more chiral centers than synthetic compounds, with their stereochemical architecture being far more complex; in order to obtain accurate docking studies, the absolute confirmation should be determined. Knowledge of the three-dimensional structure of the target is equally important, which is often determined by nuclear magnetic resonance (NMR) spectroscopy or X-ray crystallography or constructed on the basis of the structure of homologous proteins. The compound may empirically show bioactivity toward the parasite but not act on a known druggable target, and therefore be omitted from testing due to being mislabeled a nonhit in a virtual screening study. The power of virtual screening lies in its ability to screen large libraries of known and proposed structures, narrowing down the list to a selected few based on the compound–target interaction. Additional filters may be added to account for physiochemical properties; however, natural products often contain more carbon, hydrogen, and oxygen atoms, while containing less nitrogen and other atoms than synthetic compounds, and natural products frequently have a molecular weight higher than 500 Da,[68] which is a clear contradiction of Lipinski's rule of five.[69]

There have been a number of virtual screening studies published in the past decade; however, for the purpose of this chapter, three recent virtual screening projects for the identification of natural products as antiparasitic agents will be used to illustrate three different approaches to virtual screening of natural products. Firstly, Venkatesan et al.[70] screened the NatProd library (Discovery Systems, Inc.) against the *Leishmania infantum* enzyme trypanothione reductase, a validated drug target in the treatment of leishmaniasis. These natural products were narrowed down to a few alkaloids and sterols as potential inhibitors of trypanothione reductase. This aforementioned study is a good example of a large library screened for a potential inhibitor of a specific protein target. The second study, by Ogungbe and Setzer,[71] virtually screened 73 known antitrypanosomal compounds against a range of validated drug targets of the parasite *T. brucei*. This screening study allowed the researchers to develop a speculative mode of action for this suite of antitrypanosomal compounds. In the final case study, Srivastava et al.[72] used artemisinin (**5**) as a lead compound and docked 144 virtual analogues against **5**'s proposed target, free heme. This approached allowed the investigators to quickly assess which structural changes to **5** would yield the greatest increase in biological activity based on the Glide score and binding free energy of the ligand and the iron center of free heme.

There have been very few reports of virtual screening driving a drug discovery project forward; however, in the world of natural products where supply is limited and the purification of additional material is costly and time-consuming, virtual screening offers insight into which enzymatic assays valuable material might be tested against. There are many other benefits to initially using virtual screening, such as a higher capacity than holistic screening and a subsequent requirement for less experimental effort for testing. Nevertheless, the information obtained in virtual screening is putative, and the efficacy of the proposed bioactive compound needs to be assessed *in vivo,* as well as absorption, distribution, metabolism, and excretion (ADME) and toxicological properties need to be considered. Virtual screening of natural products affords insight into synthetically feasible scaffolds and possible modes of action, as well as gives insight as a chromatographical scheme that targets specific predetermined compounds of a specific property, for example, size, charge, or polarity.[66,73]

9.3.3 Assay Design

There are a large variety of approaches available to identify pharmacological activity in natural products. These approaches are largely dependent upon the specific nature of the parasitic organism, as well as the availability of a practical laboratory model that closely mimics the final target, that is, the parasite within the patient. Once a compound has been deemed active in a bioassay, it should be confirmed using the model the next level up. For example, compounds that were active in the subcellular assay should be evaluated in the whole-organism assay. Those compounds that showed potential in the whole organism should be confirmed in the animal model.

The uses of the whole-organism screen have considerable advantages compared to those of the target-orientated approach. First, all available targets are included (i.e., known and unknown) and take bioavailability phenomena into account. The use of the subcellular target all too often fails to take the cell membrane permeability to the compound into account, resulting in the compound failing to achieve bioactivity when tested in the whole parasite. The whole-organism approach should be used whenever possible. The use of the whole organism is of particular interest in the screening of crude extracts, the composition of which is often unknown, and therefore the choice of screening against a subcellular target is misguided. The disadvantage of the use of whole organisms is that the mode of action is often unknown, and some crude extracts may not be considered active due to sublethal limits of the active compound or perhaps resistance from membrane barriers, thus never reaching a lethal concentration.

While the whole organism is the desired choice, the availability or ability to raise large amounts of parasites in the laboratory might be problematic or impossible, hindering screening efforts. This is frequently the case for gastrointestinal or filarial nematodes.[74] The use of surrogate or model species is often the option, for example, the use of the free-living *Caenorhabditis elegans* in the development of antihelmintic compounds. However, in recent studies these model organisms have failed to accurately represent the pharmacokinetics experienced in the adult stages of the parasitic worms.[75,76] This may lead to a large number of false positives, which are ultimately unsuccessful when tested on the desired organism.

The available resources in developing nations, where large numbers of untested marine organisms for antiparasitic activity may be located, also pose a major obstacle to the discovery of potential drugs. These regions rarely have access to high-throughput screening (HTS) platforms or highly specialized instrumentation for gauging antiparasitic activity. This problem has led to the development of model systems, such as recombinant microbes engineered to express parasite proteins,[77] or field-deployable pharmacological assays, such as screening against the free-living nematode *Panagrellus redivivus.*[78]

Animal models in laboratory are indispensable in assessing the drug efficacy in *in vivo* conditions. However, due to animal welfare considerations, this model of testing is often time-consuming and costly, and therefore the testing of compounds *in vivo* is often kept to a minimum. These models do, however, offer great insight into pharmacokinetic, metabolic, and toxicological properties of the drugs of interest.

The use of assays for bioguided fractionation is not without its challenges. As stated *supra vide*, crude extracts may contain large amounts of sugars or salts, which will skew the crude mass to become artificially high, thus diluting the potential active compounds in the crude fractions to be tested, leading to false negative results. This can be avoided by performing a liquid–liquid partition or a crude fraction scheme. As the active fraction is chased down, commonly this activity is lost. This has often been attributed to a phenomenon of synergistic effects in mixtures, and hence makes identification of the active compounds troublesome but still possible by testing each fraction in combination with the others.[79]

When conducting an antiparasitic assay, a well-characterized drug-sensitive and drug-resistant parasite strain should be used. Clinical isolates should be used at a later stage to confirm primary hits. The assay should also contain at least one reference drug to assess the test validity and accuracy. A list of reference drugs can be seen in Table 9.1. These reference drugs should be obtained from reputable suppliers to ensure purity, as well as being tested as independent replicates on the same microplate as the test solutions.

Table 9.1   List of Reference Drugs to Be Used in Biological Assays

Disease	Parasite Used in Assay	Reference Drug	IC_{50} Value (μM)
Antiprotazoal			
Malaria	*Plasmodium falciparum* (D6)	Chloroquine, atovaquone	0.05, 0.0005
	Plasmodium falciparum (W2)	Chloroquine, atovaquone	5.3, 0.002
	Plasmodium falciparum (K1)	Artemisinin	0.01
Leishmaniasis	*Leishmania donovani*	Amphotericin D, sodium stibogluconate, miltefosine	0.07, 0.5, 44.7
Toxoplasmosis	*Toxoplasma gondii*	Azithromycin, pyrimethamine, sulfamerazine	140.2, 0.64, 144
Antitrypanosomal			
Sleeping sickness	*Trypanosoma brucei brucei*	Pentamidine	0.03
	Trypanosoma brucei rhodesiense	Melarsoprol	0.01
Chagas	*Trypanosoma cruzi*	Nifurtimox, benznidazole	10, 1.2
Antihelminths			
Schistosomiasis		Niclosamide	4.6
General antihel	*Caenorhabditis elegans*		
	Heligmosomoides bakeri	Mebendazole	4.23
	Pheretima posthuma	Piperazine citrate	
Roundworm model	*Ascardia galli*	Piperazine citrate, albendazole	232.2, 75
Tapeworm model	*Raillietina spiralis*	Albendazole	75
Lymphatic filariasis	*Brugia malayi* (female adult)	Diethylcarbamazine, ivermectin	1450, 1.84
	Brugia malayi (microfilariae)	Diethylcarbamazine, ivermectin	1776, 4.14
Amoebicides			
Amoebiasis	*Entamoeba histolytica*	Metronidazole, trinidazole, nitazoxanide, iodoquinol, chloroquine, emetine, pyrvinium pamoate	1.3, <0.02, 0.65, 1.1, <7, 0.12, 0.304
Acanthamoeba keratitis	*Acanthamoeba castellanii*	Chlorhexidine, amphotericin B	8, 43
Cryptosporidiosis	*Cryptosporidium parvum*	Monensin	0.144

9.3.4 Helminths

As mentioned above, the use of model organisms such as *C. elegans* or various species of earthworms (e.g., *Pheretima posthuma*) has often been employed in the identification of antihelminthic compounds.[80–82] *C. elegans* is often an attractive model organism due to its high survival rate in the laboratory setting. However, the lifestyle of *C. elegans* is very different from that of the parasitic nematodes, resulting in much debate as to its relevance.[83] It could be argued, however, that due to the comparative physiology and pharmacology with parasitic Nematoda, *C. elegans* still remains a good model for the discovery of antihelmintics. The majority of antihelmintics target the neuromusculature, and therefore similarities between the model and target organism become very important.[84,85]

Geary et al.[86] have developed a recombinant yeast strain of *Saccharomyces cerevisae* able to express the nematode neuropeptide G protein-coupled receptors (GPCRs), which may be used to screen for nematocidal hits. Activation of the GPCR induces a cascade of signals that repairs a histidine auxotrophy, permitting growth of the yeast in histidine-rich media. The desired targets of these GPCRs are the neurotransmitters FMRFamide-like peptides, which are essential in mediating the function of the helminths. The discovery of a nonpeptide antagonist for the GPCRs would have huge application as a broad-spectrum antihelmintic.

One of the most desirable targets for the development of antifilarial drugs is inhibitors of the enzyme aminoacyl tRNA synthetase (AARS). The primary reason why this enzyme is such a desirable druggable target is the important primary and secondary functions within eukaryotes, including filarial. Of particular interest is the AARS asparaginyl-tRNA synthetase (AsnRS), due to this enzyme being expressed in both sexes, adults, and larvae of *Brugia malayi*,[87,88] as well as being well characterized in several species of nematodes. This target has also been developed in an HTS platform for the identification of AsnRS inhibitors.[89]

The visual health assessment of *B. malayi* is very easy, as healthy worms are highly motile and coiled, whereas dead worms adopt an elongated morphology devoid of motion. Differentiating between paralysis and death can be achieved through the use of 3-(4,5-dimethylthiazol-2-yl)-2,5-diphenyl-2*H*-tetrazolium bromide (MTT), which is readily reduced to formazan by live worms. This transformation may be monitored in real time by observing the absorption at 510 nm, thus assessing the filarial viability and therefore allowing for the development of an HTS assay.[90]

9.3.5 *Leishmania*

As mentioned previously, there are several species of *Leishmania* available for assays. However, the visceral species (i.e., *L. donovani* and *L. infantum*) are preferred because of their higher sensitivity to available reference drugs, and they represent a greater medical need.

The classical model of screening for antileishmanial compounds involves incubation of the test compound with the axenic promastigotes or amastigotes. The promastigotes are the extracellular form of the parasite that lives within the gut of the sandfly vector. This assay is cheap, and the easy maintenance of parasite cultures allows for screening of hundreds of compounds with unsophisticated screen equipment. The major limitations to this assay are that it only tests for growth inhibition and it is time-consuming as a result of manually counting parasites.

The use of an MTT colorimetric assay is still relatively cheap, as well as allowing for more accurate and reproducible results.[91] This method is based on the oxidative activity of the mitochondria. One limitation of this method is the possible oxidation of the MTT substrate by the test compound, leading to a false positive result.[92,93] In the case of crude extracts, the oxidative ability of the compound is rarely known. An alternative to MTT is the dye Alamar Blue, which can assess promastigote viability by monitoring wavelengths of 570 nm.[94] The use of promastigotes is discouraged due to a lower level of validation; however, it still remains a good initial screening approach. The major advantage of using axenic amastigotes in a screening assay is evaluating the potential drug candidates at a clinically relevant stage of the parasite. Preparation of the inocula for infection usually involves the isolation of amastigotes, grown in the spleen of infected hamsters. Murine peritoneal macrophages are generally used to host the amastigotes, which are added to the wells of microtiter plates containing peritoneal cells and incubated for 5 days. The parasite burden is usually microscopically assessed after Giemsa staining.

There are many subcellular targets reported in the literature, far too many to be listed here[91,95–98]; however, many have failed to be rigorously validated. The enzymatic targets or pathways that have shown potential include trypanothione metabolism,[99] cysteine peptidases,[100,101] sterol biosynthesis,[102] dihydrofolate reductases,[103] polyamine metabolism,[104,105] and tubulin.[106] All of these targets possess the ability to be screened in an HTS format.

9.3.6 *Plasmodium*

Of the five known species of malaria parasites that can infect humans, only two, *Plasmodium falciparum* and *Plasmodium knowlesi*, can be long-term cultured *in vitro*.[107] There are several strains of *P. falciparum*, and it is advised to use both drug-sensitive and drug-resistant strains in your bioassay. Drug-sensitive strains include 3D7, GHA, FCR3, NF54, and D6, while drug-resistant strains include W-2 and K1, both resistant to chloroquine, quinine, and pyrimethamine, but susceptible to mefloquine.

Evaluation of bioactive compounds against *P. falciparum* has historically involved the use of tritiated hypoxanthine,[108] which is taken up by the parasite for purine salvage and DNA synthesis.[109] Parasites are cultured in media containing ^{3}H-hypoxanthine, after which the parasites are harvested and the growth inhibition is calculated by measuring the radioactive counts. Although this is the most common method used for the screening of potential antiplasmodial compounds, this assay is costly, radioactive, and somewhat complex. An alternative assay involves the use of a transgenic *P. falciparum* clone 3D7 expressing luciferase where relative luminescence units (RLUs) are used to determine a decrease in parasite growth.[110] While this latter method is nonradioactive, it is limited by only being able to evaluate druglike molecules on the 3D7 strain and, more recently, the NF54 strain.[111] Various methodologies used to assess antiplasmodial activity have been reviewed by Fidock et al.[112]

Chloroquine has been the primary treatment for malaria for the past seven decades. Its mode of action has been studied in great detail, and the consensus is that chloroquine inhibits the polymerization of heme to hemazoin, thus poisoning the parasite with its own metabolic waste. There are several assays developed to test for new heme detoxification inhibitors, which include ^{14}C-labeled heme,[113,114] the use of trophozoites,[113] and crystallization of heme from hemin.[115–119] While numerous methods have been reported, for example, fluorometry[120] and high-performance liquid chromatography (HPLC),[121] for the quantification of hemezoin formation, spectroscopy such as UV-vis, Fourier transform infrared (FTIR), and X-ray diffraction (XRD) still remain the most convenient and widely used analytical techniques.

With ongoing research, the heat shock proteins (HSPs) of *P. falciparum* are a potentially very attractive target for new drugs. The HSPs are a family of highly conserved proteins that are generally responsible for maintaining homeostasis.[122,123] In a recent study, a number of small molecules of marine and terrestrial natural product origin, primarily prenylated alkaloids and naphthoquinones, were screened in a malate dehydrogenase-based aggregation suppression assay, adapted for the identification of inhibition of the chaperone activity of *Pf*Hsp70-1.[124–126]

The nonmevalonate pathway in *P. falciparum* consists of seven enzymatically controlled steps, resulting in the precursor isopentenyl diphosphate and its isomer dimethylallyl diphosphate.[127]

1-Deoxy-D-xylulose-5-phosphate (DOXP) reductase (DXR) is the second enzyme in the nonmevalonate cascade, and it exists as a homodimer of molecular weight 42–45 kDa. The optimum pH range is 7–8 for the function of this enzyme, with a maximum temperature range of 50 °C–60 °C.[128] DXR requires the assistance of a divalent metal ion for catalysis, usually Mg^{2+}, Mn^{2+}, or Co^{2+}. The addition of other divalent metals (Ca^{2+}, Ni^{2+}, or Zn^{2+}) *in vitro* resulted in a reduction or loss of activity.[129,130] The natural antibiotic fosmidomycin was isolated from *Streptomyces lavendulae*[129] and identified as a DXR inhibitor.[131] The enzymatic activity of DXR is monitored via the reduction of NADPH, which is essential for the function of the reaction.

Type I antifolates (e.g., dapsone) mimic *p*-aminobenzoic acid (PABA) and compete for the active site of dihydropteroate synthase (DHPS). DHPS is responsible for facilitating the conversion of hydroxymethyldihydropterin to dihydropteroate in the folate pathway. Type II antifolates (e.g., pyrimethamine) inhibit dihydrofolate reductase (DHFR), which is responsible for the reduction of H2 folate to H4 folate. Inhibition of enzymes in the folate pathway results in a decrease in the synthesis of pyrimidine, and therefore a reduction in DNA, serine, and methionine production. Due to the high level of asexual reproduction during the erythrocytic stage of the plasmodium life cycle, antifolate predominantly targets this stage.[132]

One putative target for atovaquone is the enzyme dihydroorotate dehydrogenase (DHODase),[133,134] which is crucial for electron transport as well as the *de novo* biosynthesis of pyrimidine.[6] Atovaquone has also been shown to bind to the ubiquinone binding site of the cytochrome bc_1 complex, which inhibits the movement of an iron–sulfur cluster. This iron–sulfur cluster contains a protein domain essential for electron transport within the parasite.[6] This disruption of the electron transport chain leads to the rapid disruption of the mitochondrial membrane potential.[135]

9.3.7 *Toxoplasma*

There are several methods available for the assessment of a compound or fractions of anti-*Toxoplasma* activity. At present, there are two dominant assays used for the screening of active compounds against *Toxoplasma gondii*. The first involves a strain of *T. gondii* that expresses the bacterial enzyme β-galactosidase. The parasite and potential active compound are incubated in the presence of the chlorophenol red-β-D-galactopyranoside. If the parasite is alive and healthy, there is an expression of β-galactosidase resulting in a colorimetric change from yellow–orange to a deep red, stemming from the hydrolysis of the galactoside releasing chlorophenol red. The amount of substrate converted can therefore be calculated by measuring absorbance between 570 and 595 nm.[136] The second method involves a strain of *T. gondii* transfected with a tandem yellow fluorescent protein. The amount of fluorescent protein produced was linearly correlated with parasite numbers. This method was shown to be comparable to the aforementioned method with regard to sensitivity; however, it was shown to be more accurate.[137]

9.3.8 *Trypanosoma*

Due to being nonpathogenic to humans as well as to its drug sensitivity, *Trypanosoma brucei* is the preferred strain for the identification of new drugs in the treatment of African sleeping sickness. However, if appropriate laboratory containment can be guaranteed, *Trypanosoma gambiense* or *Trypanosoma rhodesiense* can be used. The trypomastigotes are grown in a Hirumi-9[138] medium under a 5% carbon dioxide environment. After a 4-day incubation period, the growth of the parasites is assessed through the addition of a dye, followed by colorimetric or fluorimetric analysis.[139]

The current biosafety rating for *Trypanosoma cruzi* is a biohazard class 3, and therefore all laboratory work needs to be conducted in BSL-3 containment. The nifurtimox-sensitive strain

(Tulahuen) of *T. cruzi* can be cultured on MRC-5 cells. A strain that has been transfected with the β-galactosidase gene has been developed. The advantage to using this strain is the ability to conduct a colorimetric reading after the addition of chlorophenol red β-galacto-pyranoside as a substrate.[140] More recently, Doyle et al. developed an assay whereby parasites and the host cell can be automatically imaged in a 96-well plate, a format ideal for HTS platforms.[141] The assay involves the addition of a single dye, 4′,6′-diamidino-2-phenylindole (DAPI), which stains both host nuclei and the parasite kDNA, allowing for simultaneous assessment of the parasite growth inhibition, as well as a test compound's cytotoxicity based on a decrease in the number of host cells. This may easily be achieved based on the significant differences between host nuclei and parasite kinetoplasts. An additional advantage is that this assay can be used on any parasite strain, including recent clinical isolates. Currently, there are multiple subcellular targets, far too many to mention here.[142]

9.3.9 Acanthamoeba

Not until recently was there an efficient way to screen drugs for their ability to kill acanthamoebas. These classical methods were often tedious and time-consuming, involving manually assessing the amount of viable cells using a hemocytometer[143] or staining with fluorescent viability dyes and flow cytometry, using a standard plague assay,[144] and finally estimating the most probable number following plate agar plate spreading.[145] Roberts and coworkers[146] have managed to develop an HTS method exploiting the ability of live acanthamoeba to metabolize Alamar Blue, resulting in a change in color.[147]

Currently there are several subcellular targets for the development of amoebicides, which include targeting the cytoskeleton via inhibition of myosin,[148] and α- and β-tubulins.[146] The membrane may also be targeted by disrupting cellulose biosynthesis within the parasite. Another target involves the disruption of polyamine biosynthesis, which is important in the transcription and translation of genes. This is achieved primarily by inhibition of ornithine decarboxylase.[149] Finally, inhibition of the alternative oxidase and mitochondrial respiration might lead to the development of novel antimicrobial therapies.[150]

9.3.10 Cytotoxicity

Determining the cytotoxicity of the host cells is of utmost importance, as it will give a true reflection of the drug mode of action by eliminating the possibility that death of the parasite is not due to the compound of interest being generally cytotoxic, and therefore this assay should always be run in parallel with the bioactivity assay. Theoretically, many cells could be used for this. However, MRC-5 (human lung fibroblast) cells are a good example. In the case of parasitic organisms, the host cells of the parasite should be tested void of the parasite. Many natural products (and their synthetic analogues that have shown promising antiplasmodial activity against *P. falciparum*) have also altered the erythrocyte membrane shape, for example, the series of triterpenoids,[136–138] abietane-type diterpene,[139] norditerpenoids,[140] labdanes, and isopimaranes.[141] Thus, the biological activity was not attributed to the compounds killing the parasite, but rather the compounds destroying the red blood cells, resulting in the death of the parasite.

9.4 MARINE NATURAL PRODUCTS WITH BIOLOGICAL ACTIVITY

As stated above, only marine natural products with activity below 25 μM will be considered biologically active against their intended parasitic target. All compounds that fail to meet this criterion will not be included in this review.

9.4.1 Antihelminths

In the mid-1980s, Crews et al. isolated the 19-membered macrocyclic depsipeptide from the Fijian sponge *Jaspis* sp. jasplakinolide (also called jaspamide) (**23**). This cyclodepsipeptide showed an *in vitro* EC_{50} (half-maximal effective concentration) value of less than 1.4 μM against the nematode *Nippostrongylus brasiliensis*.[151] This nematode primarily infects rats; however, it has a life cycle similar to that of human hookworms. The South Australian marine brown algae *Notheia anomala* yielded dihydroxytetrahydrofuran (**24**), which showed potent nematocidal activity against the agriculturally important nematodes *Haemonchus contortus* and *Trichostrongylus colubriformis* at LD_{50} (lethal dose that kills 50%) values of 5.8 and 31.7 μM, respectively.[152] The nematocidal mode of action of **24** has been attributed to the inhibition of the development of the eggs in the L3 life stage of these two species. The sponge *Amphimedon* sp., collected off the Great Australian Bight, afforded the unusual macrolides amphilactams A–D. Amphilactam D (**25**) showed the best activity against the parasite *H. contortus* at an LD_{99} value of 0.64 μM. The macrocycle **25** inhibits the L1 life stage of the nematode, thus limiting the development of the larval stage of the parasite.[153] The Capon group isolated another nematocidal agent, geodin A magnesium salt (**26**), from the Australian sponge *Geodia* spp. This macrocyclic polyketide lactam tetrameric acid had an LD_{99} concentration of 1.1 μM against the larval stage of *H. contortus*.[154] While *H. contortus* is not a human pathogenic parasite, but a common parasite of ruminants that often results in the death of sheep and goats, these compounds may be effective against human nematodes.

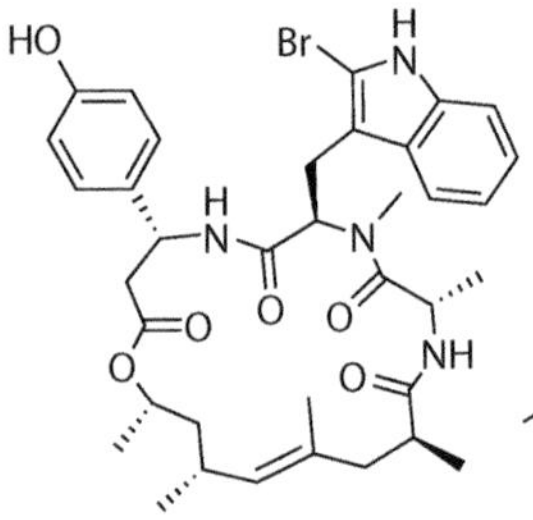

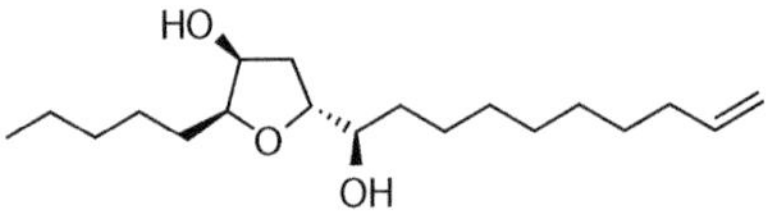

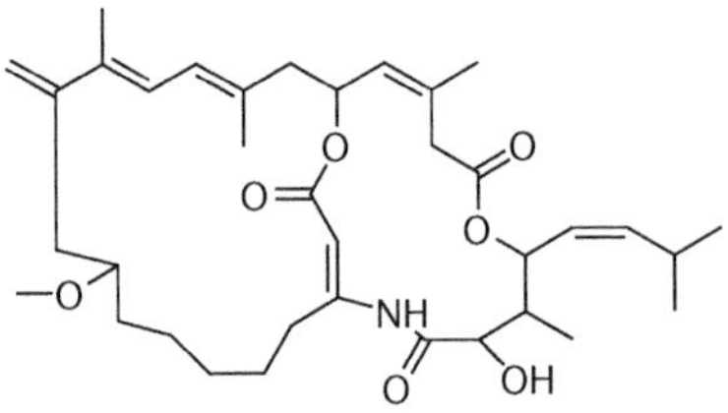

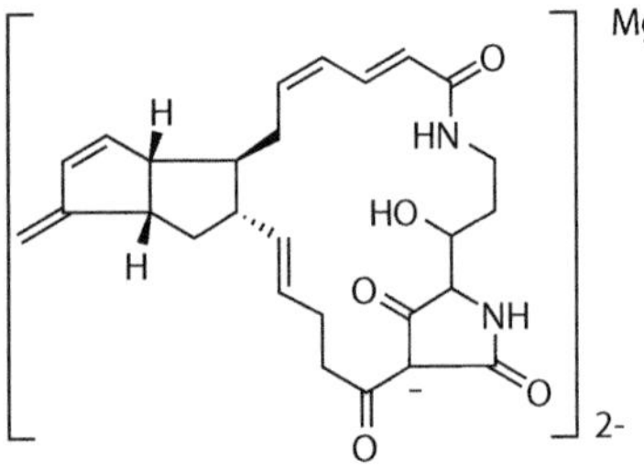

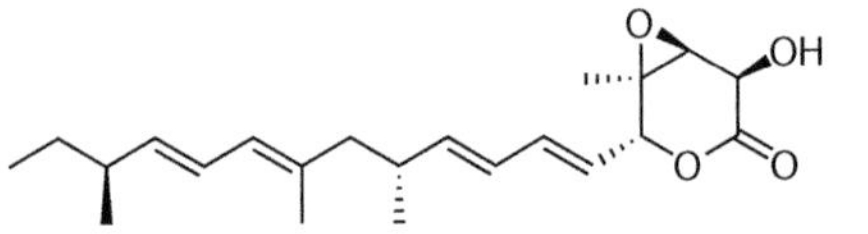

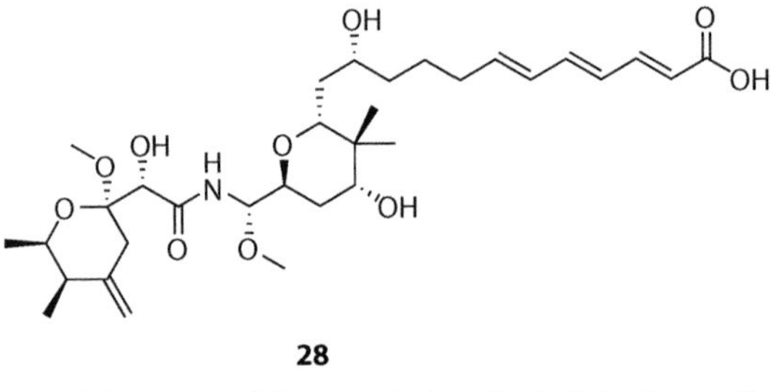

In 2001, Omura et al.[155] isolated the epoxy-δ-lactone with a methylated olefinic (prenyl) side chain, nafuredin (**27**), from the fugus *Aspergillus niger*, isolated from a marine sponge collected in Palau. This compound showed potent activity against the dwarf tapeworm *Hymenolepsis nana* in mice, as well as the pig roundworm *Ascaris suum*. The activity was determined to be due to the inhibition of the NADH-fumarate reductase (complexes I and II), which is essential to the respiratory system in helminths, at an IC_{50} value of 12 nM in adult worms. Later that year, the same group published a total synthesis of **27**.[156] A similar alkaloid, onnamide F (**28**), was isolated from the Australian marine sponge *Trachycladus laevis*, which has shown inhibition of the larval development of the parasitic nematode *H. contortus* at an *in vitro* LD_{99} of 85.3 μM.[157] Six thiocyanate-containing acyclic compounds (**29–34**) isolated from the Australian sponge *Oceanapia* spp. showed moderate nematocidal activity against the parasite *H. contortus*.[158] Recently, the simplest congeners, thiocyanatins A and B, have been prepared using microwave-assisted synthesis.[159,160] The sequiterpene isorigidol (**35**) was isolated from the Brazilian red algae *Laurencia scoparia*, which showed moderate antihelmintic activity against the parasite stage of the *N. brasilensis*.[161] In addition, this red algae yielded a halogenated β-bisabolene sesquiterepene **36**, which displayed weak antihelmintic activity against *N. brasiliensis*.[162] Another sesquiterpene active against the aforementioned parasite is ma'ilione (**37**), isolated from *L. cartilaginea* collected off the coast of Hawaii.[163]

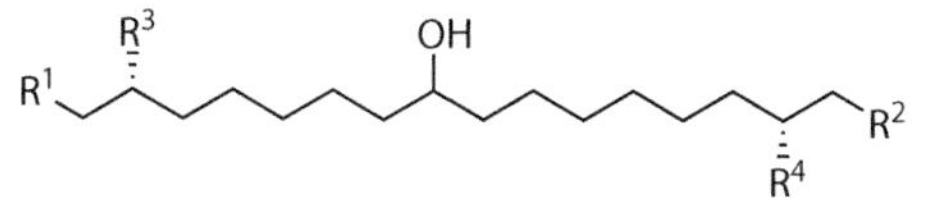

	R^1	R^2	R^3	R^4	
29	SCN	SCN	CH_3	H	thiocyanatin D_1
30	SCN	SCN	H	CH_3	thiocyanatin D_2
31	$SCON_2H$	SCN	H	H	thiocyanatin E_1
32	SCN	$SCON_2H$	H	H	thiocyanatin D

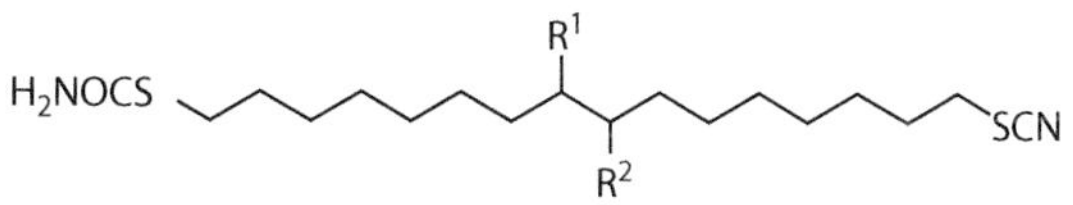

	R^1	R^2	
33	OH	H	thiocyanatin E_1
34	H	OH	thiocyanatin E_2

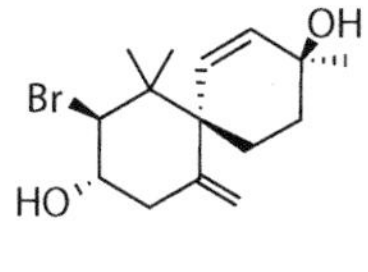

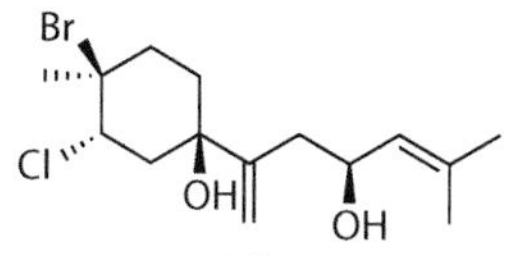

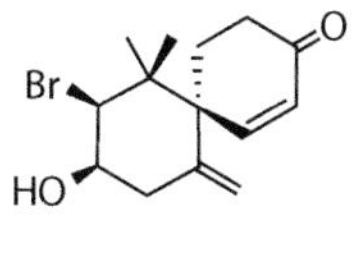

The southeastern Australia encrusting sponge, *Aplysilla sulfurea*, afforded the nematocidal compounds chromodorolides A–C (**38–40**).[164] The new tambjamine, **41**, was isolated from the marine bacterium *Pseudoalteromonas tunicate*[165] and screened and identified as having nematocidal activity against *Caenorhabditis elegans*.[166] The tetrapeptide *cyclo*-(isoleucyl-prolyl-leucyl-alanyl), **42**, isolated from a *Pseudomonas* sp. that is often associated with the sponge *Halisarca ectofibrosa* and the seaweed *Diginea* sp.,[167] showed potent antihelmintic activity against earthworms (*Megascoplex konkanensis*, *Pontoscotex corethruses*, and *Eudrilus eugeniea*).[168]

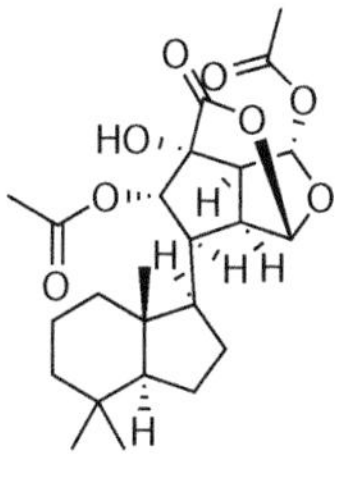

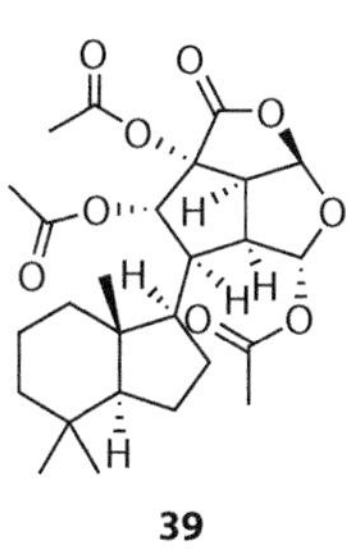

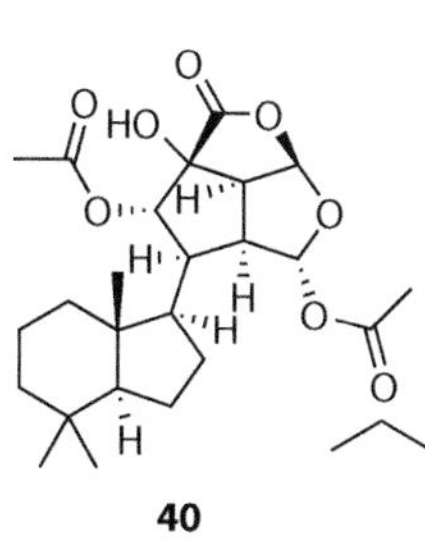

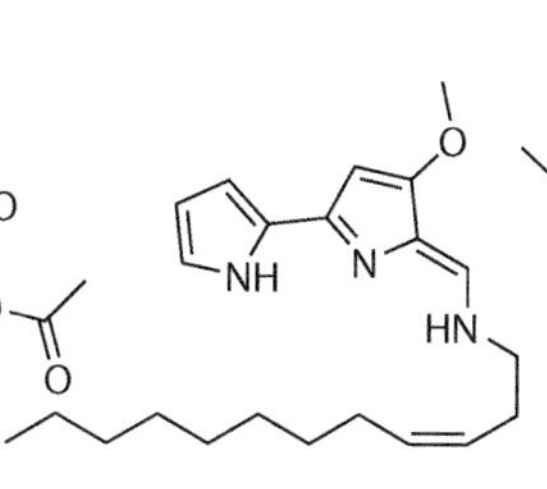

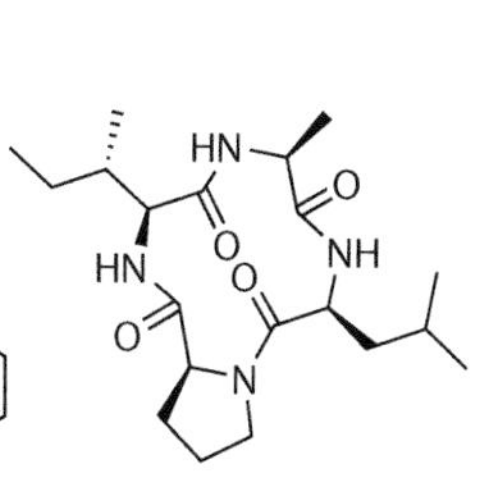

The extract of a *Zoanthus* sp. collected off the coast of India was tested for activity against the human lymphatic filarial *Brugia malayi*. The activity was determined to be in the lipophillic fraction; however, upon further purification, the pure compounds zoanthamine (**43**)[169] and gorgosterol (**44**)[170] appeared to have no activity against the parasite. This was attributed to a synergistic effect contained within the crude fraction.[171] The same group showed that the chloroform and n-hexane fractions of a red algae *Botryocladia leptopoda* had a lethal concentration of 31.25 μg/ml against *B. malayi*. Yet the active compounds were never determined.[172] The fruit of the mangrove tree, *Xylocarpus granatum*, yielded two antifilarial compounds, photogeduin (**45**) and gedunin (**46**), which were active against the adult worms/microfilaria of *B. malayi* at IC$_{50}$ values of 0.41/3.9 and 0.50/4.2 μM for **45** and **46**, respectively.[173] Two demosponges from the genus *Haliclona* yielded a series of alkaloids H27–30. Araguspongin C (**47**), was isolated from both *H. exigua* and *H. oculata*, while xestospongins C (**48**) and D (**49**), as well as mimosamycin (**50**), were only isolated from the latter. Unfortunately, only **47** was tested for antifilarial activity, with an LC$_{100}$ (lethal concentration that kills 100%) value of 32.9 μM; the other compounds were assumed to be the active components in the antifilarial crude fractions.[174,175]

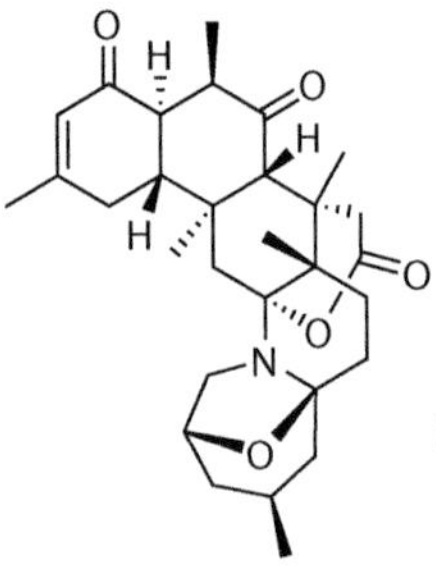

43 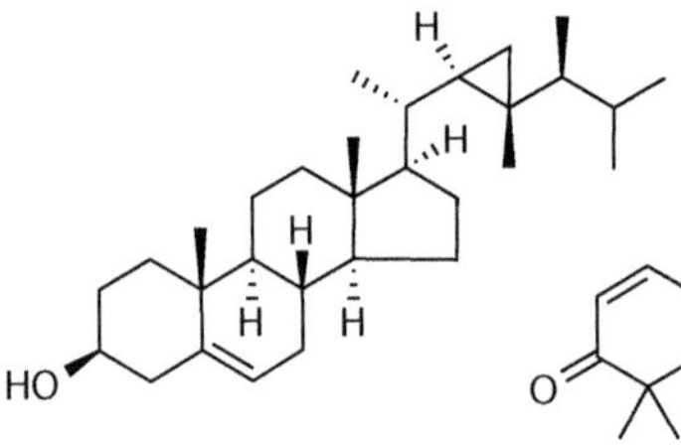   44 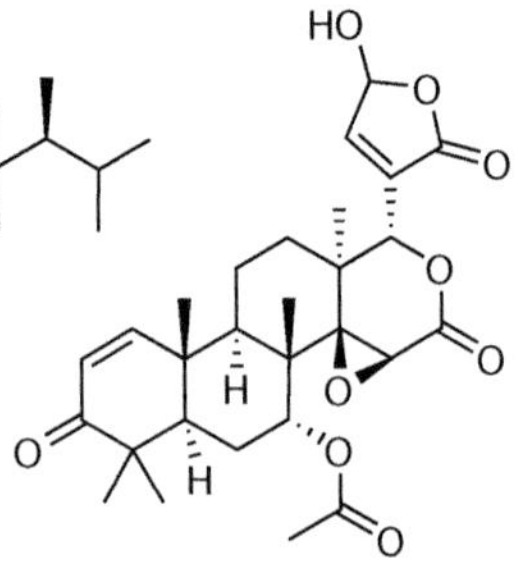   45 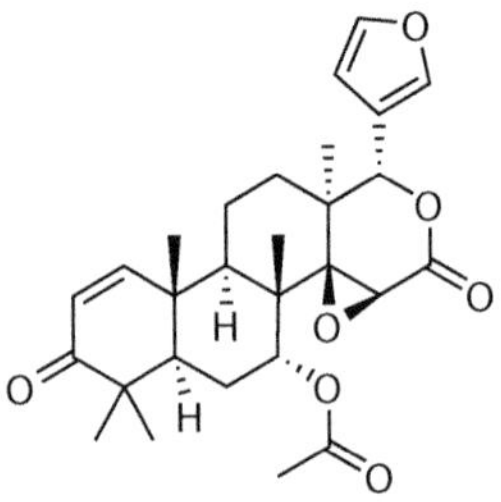   46

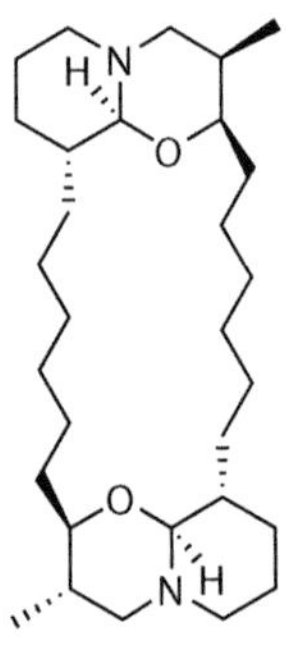

47 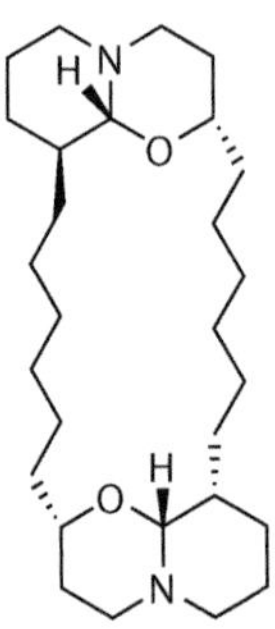   48 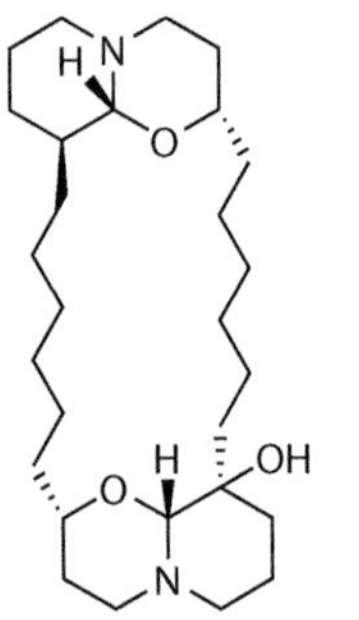   49 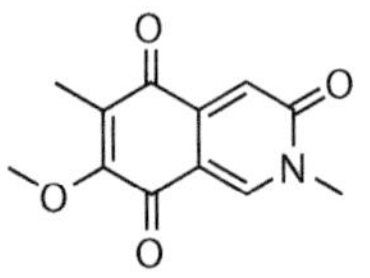   50

9.4.2 *Leishmania*

In the infancy of the marine drug discovery, Christophersen et al.[176] and Kyogoku et al.[177] isolated two triterpene glycoside metabolites, neothyosides A (**51**) and pervicoide B (**52**), from the sea cucumbers *Neothyone gibbosa* and *Holothuria pervicax*, respectively. Several years later, Christophersen et al.[178] isolated an analogue of **51**, neothyoside B (**53**), from the same Pacific sea cucumber. Compounds **51–53** would later be tested for their leishmanicidal activity.[179] A species of marine sponge of the genus *Smenospongia* yielded terpenoid ilimaquinone (**54**), which showed an EC_{50} concentration of 5.6 µM against the *Leishmania mexicana* promastigote.[180,181] A series of polyketides were isolated from the sponge *Plakortis angulospiculatus*, which displayed antiparasitic activity. The most potent were cyclic peroxides (3*S*,6*R*,8*S*)-4,6-diethyl-3,6-epidioxy-8-methyldodeca-4-enoic acid (**55**) and plakortide P (**56**), with EC_{50} values of 0.93 and 5.32 µM against *L. mexicana* promastigote cells.[182,183] The common Indo-Pacific sponge *Acanthostrongylophora* spp. yielded a suite of manzamine-type alkaloids, which showed potent activity against the promastigotes of *Leishmania donovani*, with the most active compound being manzamine A (**57**) at an EC_{50} value of 1.7 µM. Unfortunately, these compounds showed high toxicity to mammalian cells.[184,185]

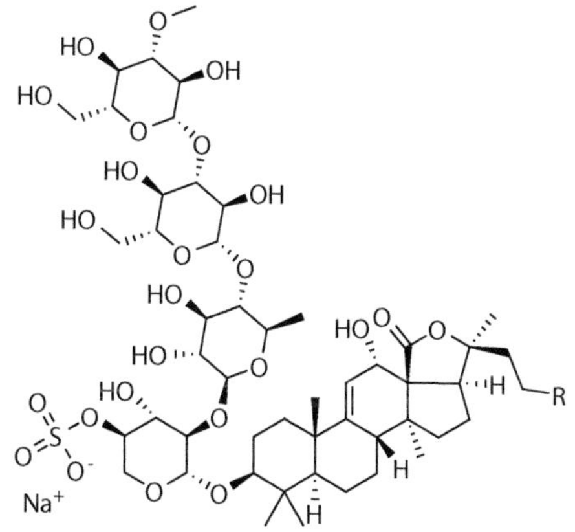

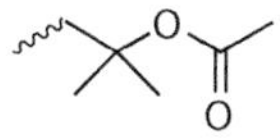

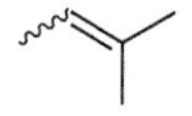

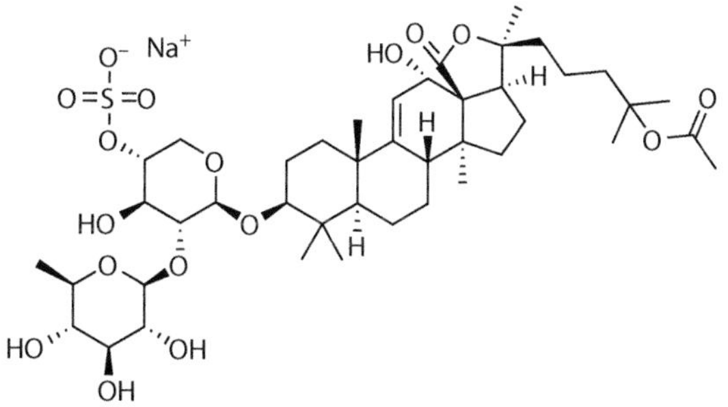

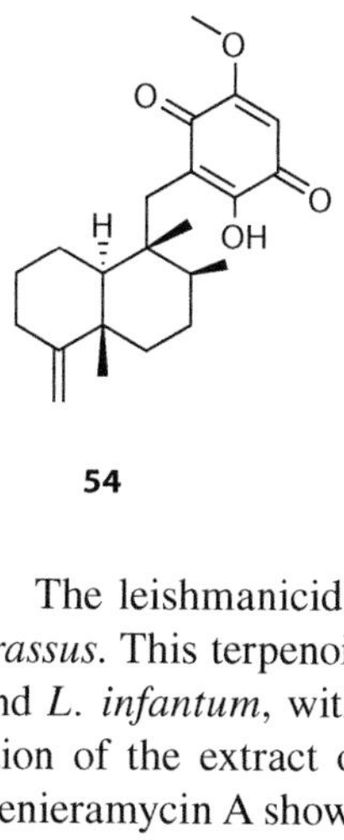

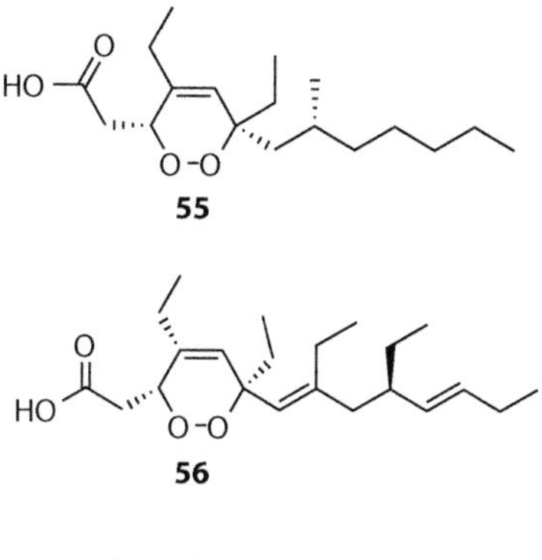

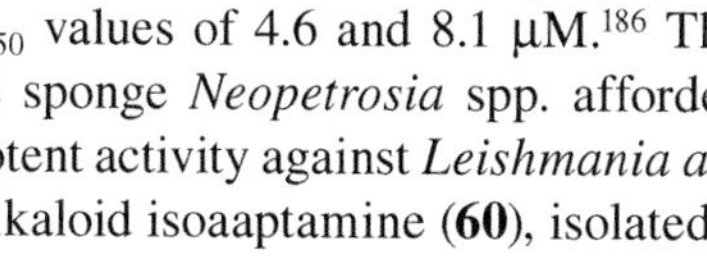

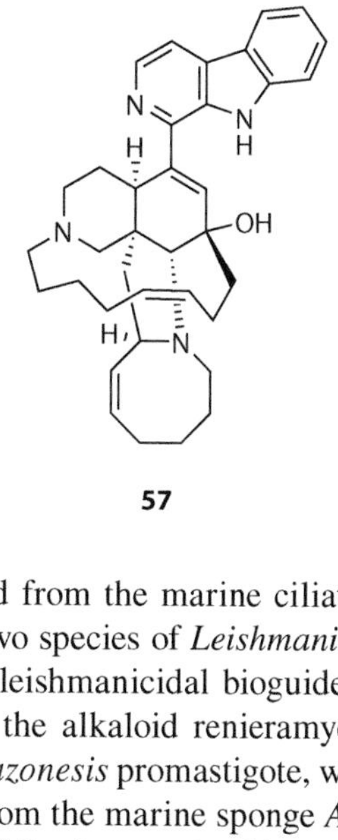

The leishmanicidal compound euplotin C (**58**) was isolated from the marine ciliate *Euplotes crassus*. This terpenoid displayed antiparasitic activity against two species of *Leishmania*, *L. major* and *L. infantum*, with EC_{50} values of 4.6 and 8.1 µM.[186] The leishmanicidal bioguided fractionation of the extract of the sponge *Neopetrosia* spp. afforded the alkaloid renieramycin A (**59**). Renieramycin A showed potent activity against *Leishmania amazonesis* promastigote, with an EC_{50} value of 0.23 µM.[187] The alkaloid isoaaptamine (**60**), isolated from the marine sponge *Aaptos* spp., displayed antileishmanial activity against the promastigotes of *L. donovani*, with an EC_{50} value of 3.1 µM.[188] That same year, Andersen and coworkers[189] isolated the meroterpenoid isoakaterpin (**61**) from the Brazilian sponge *Callyspongia* sp. This compound was identified as an inhibitor of *Leishmania tarentolae* adenine phosphoribosyl transferase (L-APRT), with an EC_{50} value of 1.05 μM. This enzyme is involved in the purine salvage pathway in many *Leishmania* species. The sterol halistanol A (**62**), isolated from the marine sponge *Petromica ciocalyptoides*, was screen against L-APRT, resulting in an EC_{50} value of 3.8 μM. Compound **62** was also tested against *L. chagasi*, but failed to display any antiparasitic activity.[190] A species of cyanobacterium from the genus *Oscillatoria* yielded two cyclic peptides, veturamides A (**63**) and B (**64**), that displayed mild activity against *L. donovani* amastigotes, both with an EC_{50} concentration of 19 μM.[80] An additional cyanobaterium from the same genus, *O. nigro-viridis*, yielded viridamide A (**65**), which showed potent antiparasitic activity against *L. mexicana* amastigotes (EC_{50} = 1.5 μM).[191] The sesquiterpene (*S*)-(+)-curcuphenol (**66**), isolated from the Jamaican sponge *Myrmekioderma styx*, displayed leishmanicidal activity, with an EC_{50} value of 11 μM against *L. donovani* promastigotes.[192]

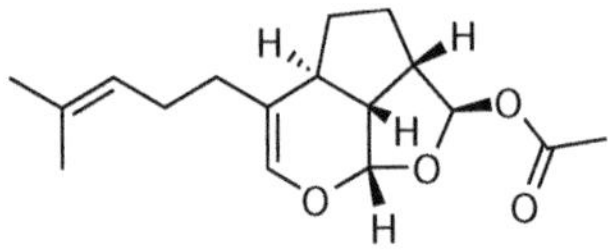

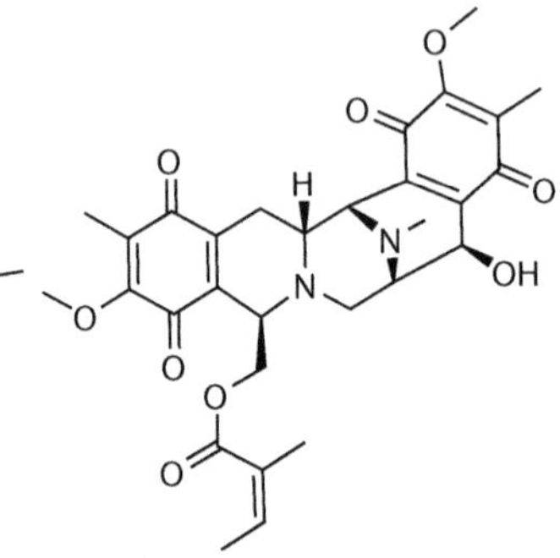

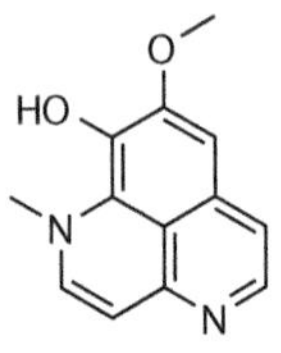

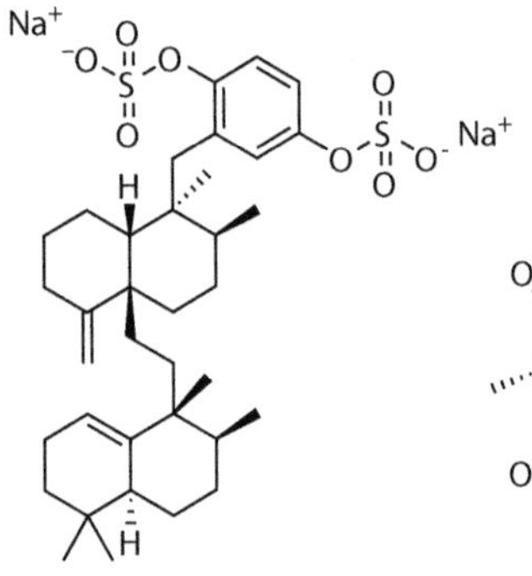

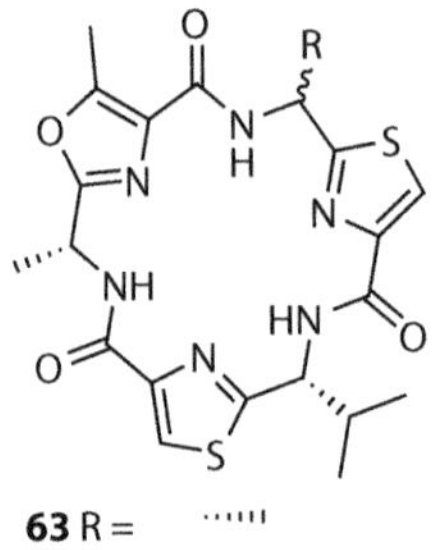

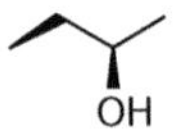

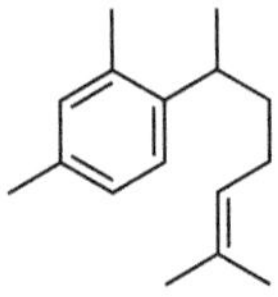

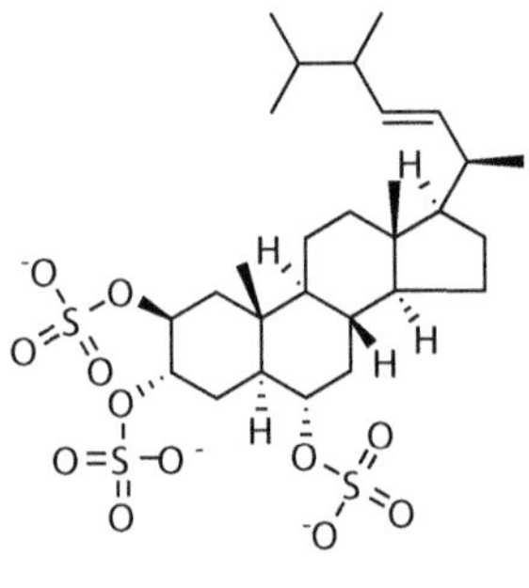

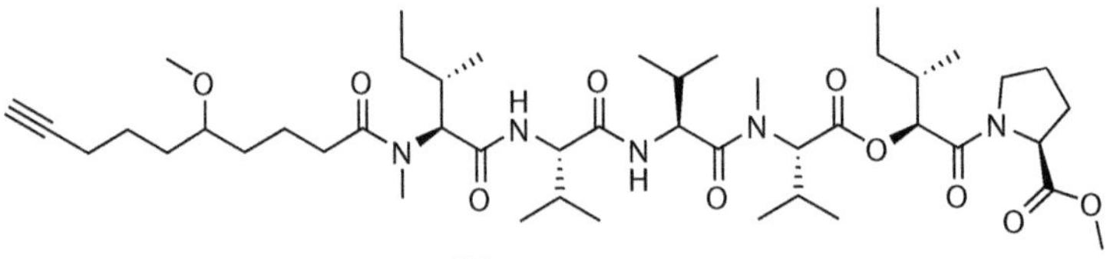

A study of the metabolites of the deep-sea sponge *Aapto ciliate* yielded series of peptides cili-atamides A–C, with only ciliatamide A (**67**) showing antiparasitic activity against *L. major* pro-mastigotes, with an IC_{50} value of 20 μM.[193] The marine bacteria *Pseudomonas fluorescens* yielded the pyrones pseudopyronines A (**68**) and B (**69**), which showed leishmanicidal activity against the *L. donovani* axenic amastigotes, with EC_{50} values of 9.9 and 4.7 μM.[194] The marine-derived fungus *Chaetomium* spp. afforded chaetoxanthones A–C, with chaetoxanthone B (**70**) showing the high-est potential as an antileishmanial, with an EC_{50} value of 9.6 μM against the intracellular amasti-gotes of *L. donovani*.[195] Two antiparasitic peptides, gallinamide A (**71**) and kahalalide F (**72**), were isolated from cyanobacteria *Schizothrix* sp. and the Hawaiian mollusk *Elysia rufescens*, respec-tively. Peptide **71** displayed antiparasitic activity at an EC_{50} value of 9.3 μM against *L. donovani* promastigotes, while **72** inhibited *L. donovani* and *Leishmania pifanoi* at concentrations of 6.1 and 8.3 μM, respectively.[196,197] The purine algelasine D (**73**), isolated from *Agelas* sp., showed leish-manicidal activity at 3.6 μM against *L. infantum*.[198]

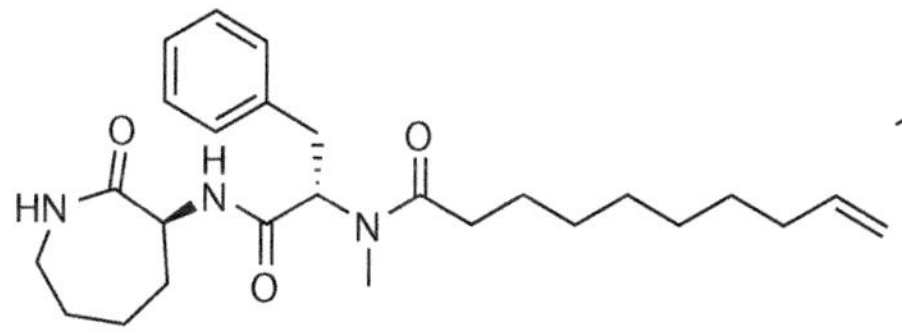

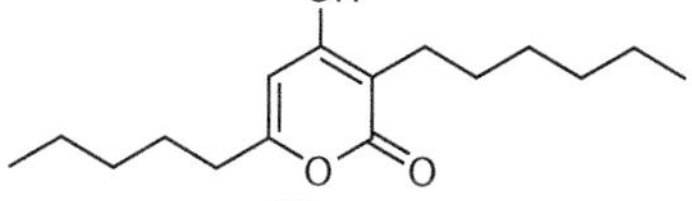

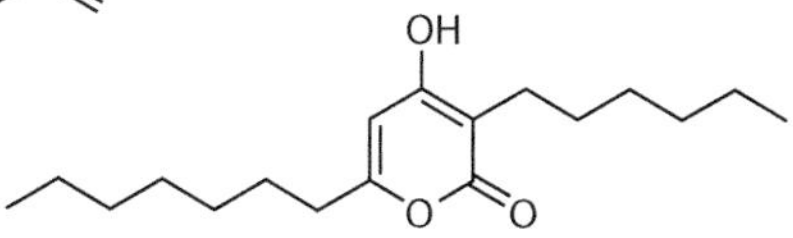

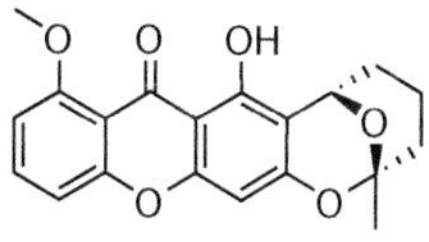

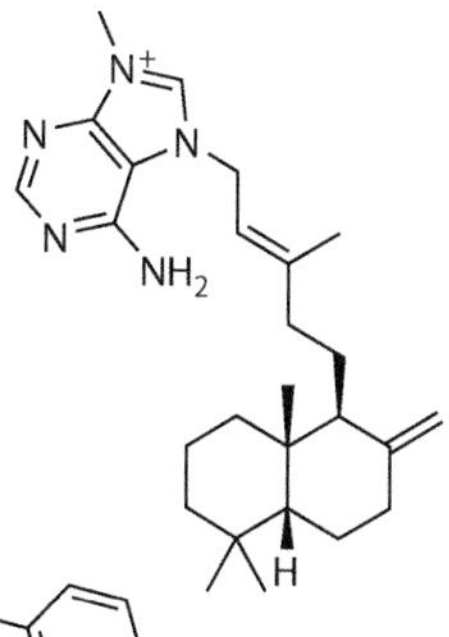

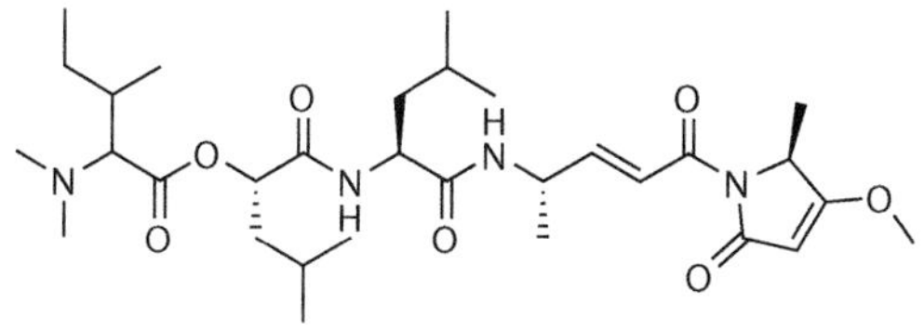

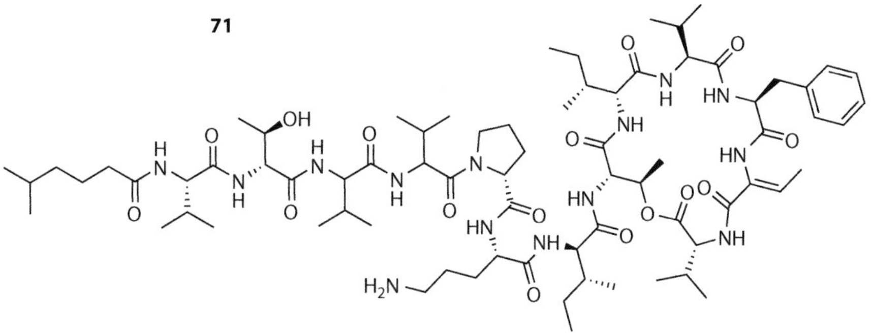

The Antarctic sponge *Crella* sp. yields a suite of sterols, norselic acids A–E (**74–78**), which showed potent activity against the leishmaniasis parasite promastigotes, displaying EC_{50} values of 2.5, 2.4, 2.6, 2.0, and 3.6 μM, respectively.[199] Three N-methylated linear lipopeptides, almiramides A–C, were isolated from the cyanobacterium *Lyngbya majuscula*. Almiramides B (**79**) and C (**80**) showed potent antiparasitic activity against *L. donovani* amasitgotes, with EC_{50} values of 2.4 and 1.9 μM.[200] In a recent study, a series of synthetic analogues of **79** and **80** were created and tested against *Trypanosoma brucei* (*T. b.*) *brucei*, with the most potent of these compounds displaying antitrypanosomal activity at 0.4 μM. Through the aid of epifluorescence microscopy, the authors proposed that the activity of the almiramides against *T. b brucei* lay in the disruption of the glycosome.[201] This same cyanobacterium yielded dragonamides A (**81**) and E (**82**) and herbamide B (**83**), which displayed *in vitro* activity against the *L. donovani* axenic amastigotes, with EC_{50} values of 6.5, 51, and 5.9 μM, respectively.[202]

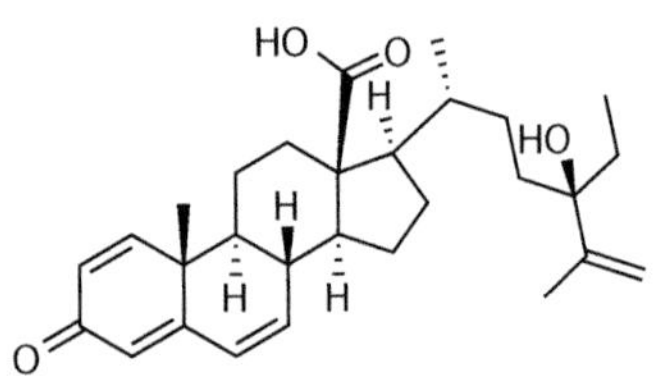

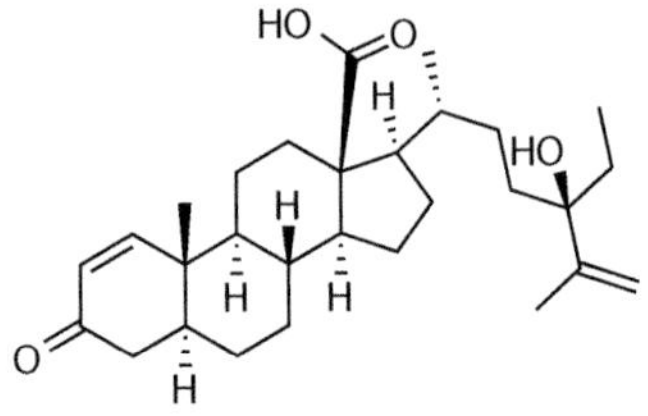

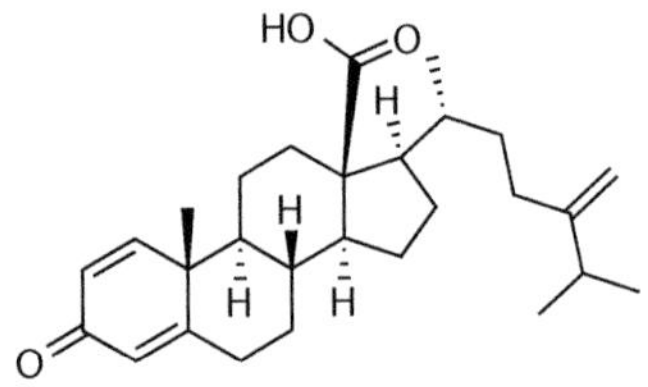

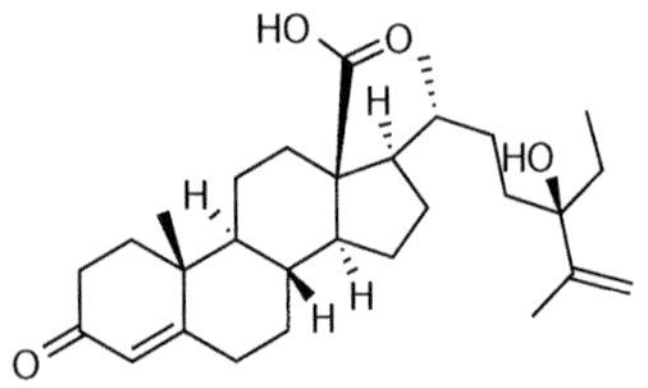

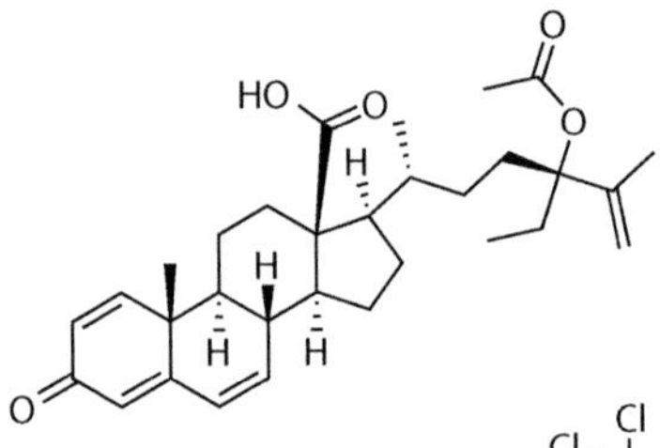

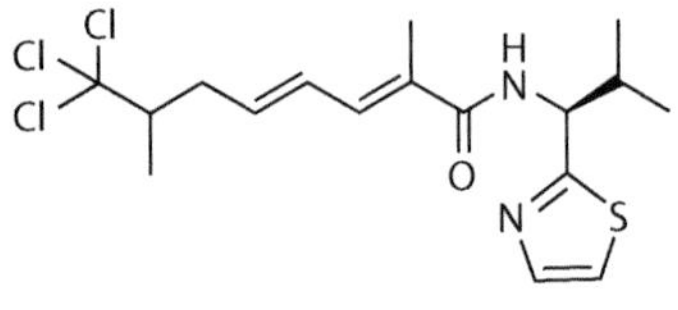

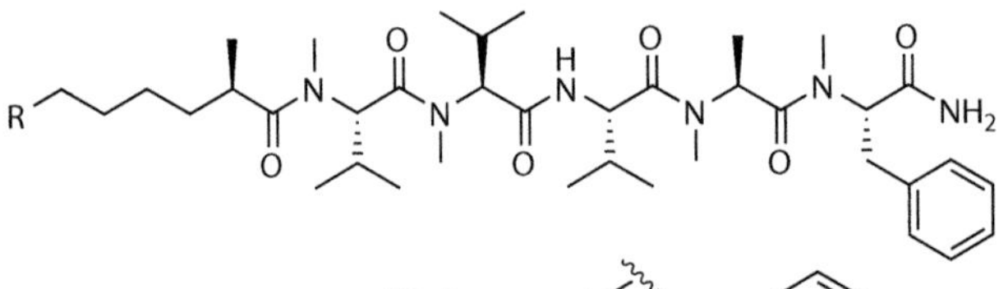

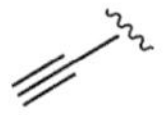

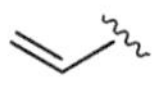

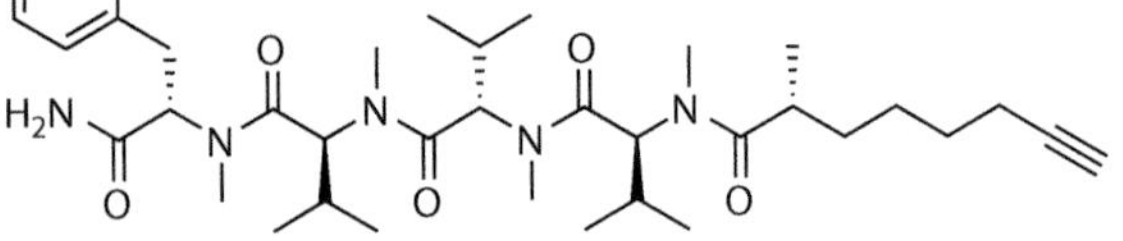

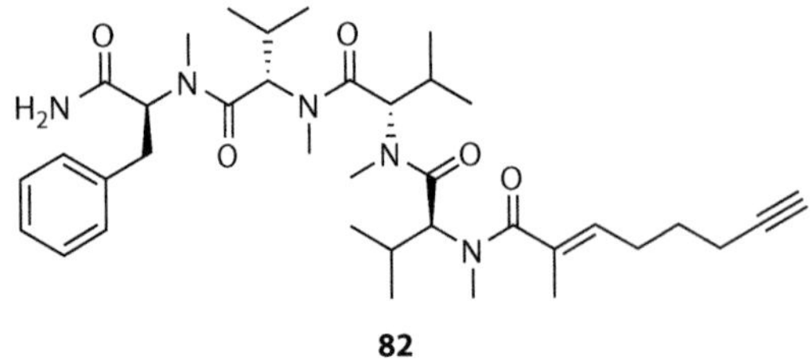

The endobiotic microbe *Streptomyces* sp., of the Mediterranean sponges *Axinella polypoides* and *Aplysina aerophoba*, afforded valinomycin (**84**). Biological evaluation of this cyclic depsipeptide showed an EC_{50} value of less than 0.11 µM against the promastigotes of *L. major*, but also exhibited cytotoxicity against 293T kidney epithelial cells and J744.1 macrophages. An additional *Streptomyces*, inhabiting an unidentified Mediterranean sponge, yielded the alkaloid staurosporine (**85**), which was active against *L. major* promastigotes, with an EC_{50} value of 5.3 µM. However, this alkaloid also showed cytoxicity toward 293T kidney epithelial cells and J774.1 macrophages.[203] The sponges, *Spongia* sp. and *Ircinia* sp., collected off the Turkish coastline, yielded a series of terpenoids. The most potent of these, **86**, displayed antileishmanial activity against *L. donovani*, with an EC_{50} value of 2.1 µM.[204] Pandaroside G (**87**) and its methyl ester (**88**) were isolated from the Caribbean sponge *Pandaros acanthifolium*. These glyocosides, **87** and **88**, inhibited the growth of axenic amastigotes of *L. donovani*, with EC_{50} values of 1.3 and 0.05 µM, respectively.[205] A series of bromopyrrole alkaloids were isolated from the marine sponges belonging to the genera *Agelas* and *Axinella*. The most potent of these compounds were longamide B (**89**) and dibromopalau'amine (**90**), with EC_{50} values of 11 and 1.9 µM, respectively, in an *in vitro* assay against the axenic amastigotes of *L. donovani*.[206]

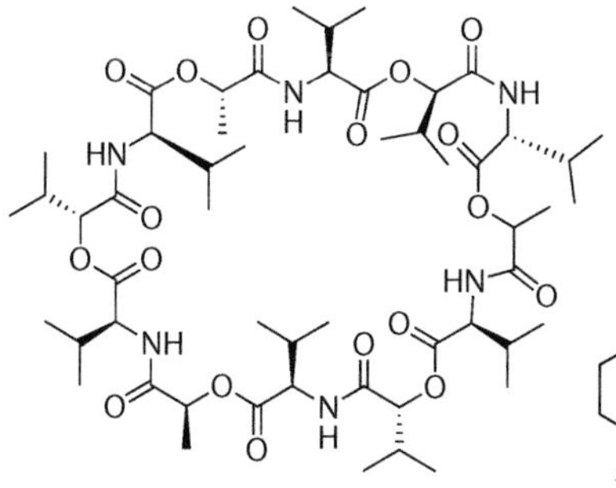

84 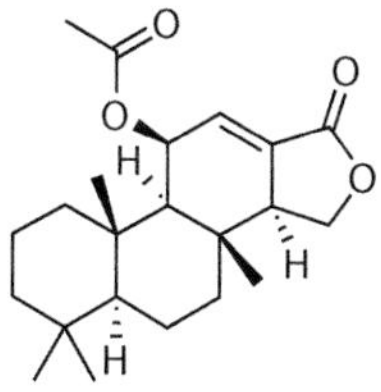   86 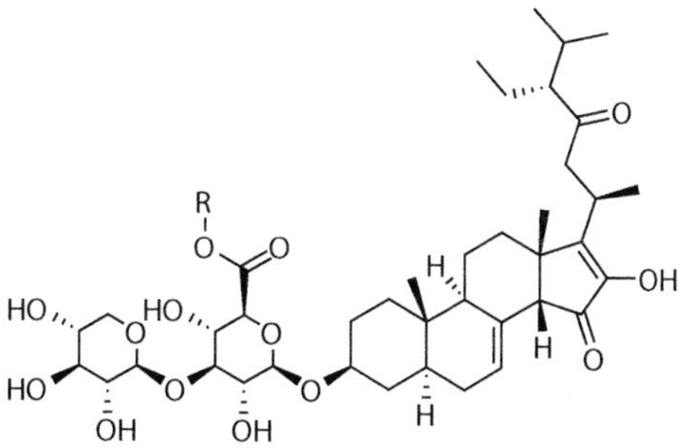   87 R = H
              88 R = CH₃

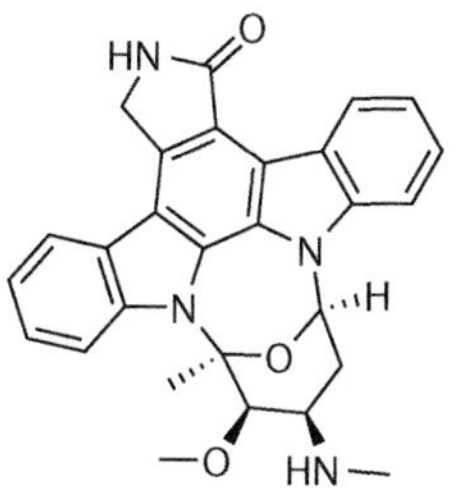

85 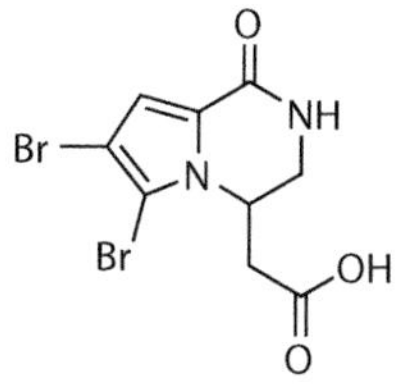   89 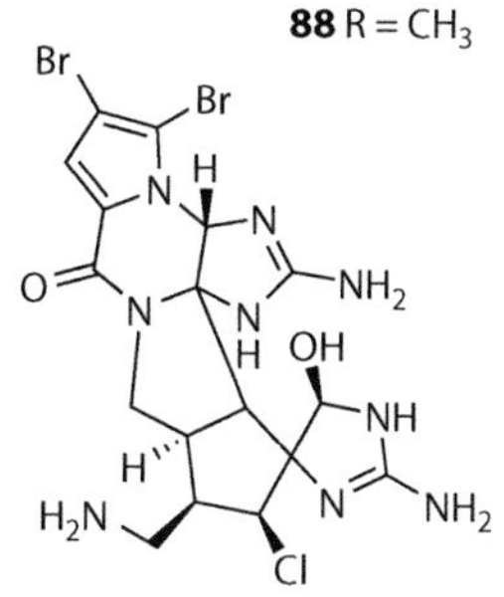   90

91 was isolated from the brown alga *Canistrocarpus cervicornis*, which proved active against the promastigotes, axenic amastigotes, and intracellular amastigote forms of *L. amazonesis*, with IC_{50} values of 5.5, 33, and 11 μM, respectively.[207] Coibacins A–D (**92–95**) were extracted from the cyanobacterium *Oscillatoria* sp. and assessed for activity against the axenic amastigotes of *L. donovani*. These polyketide lactones displayed IC_{50} values of 2.4, 7.2, 18.7, and 7.8 μM against *L. donovani* for **92–95**, respectively.[208] The potent xenican diterpenoid cristaxenicin A (**96**) was isolated from the gorgonian *Acanthoprimnoa cristata* collected in the deep seas of southern Japan. This compound inhibited 50% of the growth of the parasite *L. amazonesis* at a concentration of 0.09 μM.[209] In a recent study, an undescribed soft coral collected off the coast of Shag Rocks in the Scotia Arc yielded two new tricyclic sesquiterpenoids, shagenes A (**97**) and B (**98**). Of these two terpenes, **97** showed bioactivity against *L. donovani* (IC_{50} = 5 μM) while showing no cytotoxicity against the mammalian host (J774.A1 macrophages).[210]

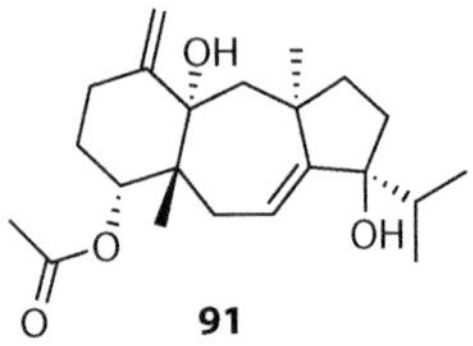

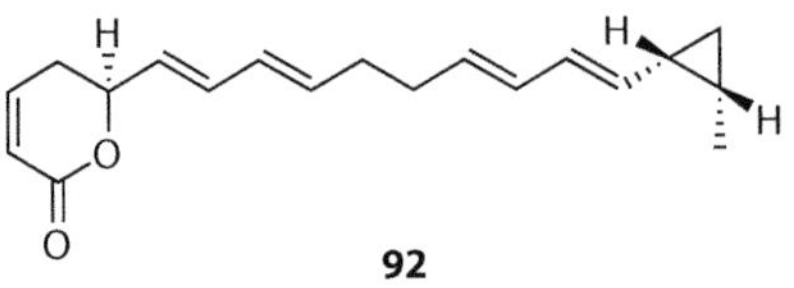

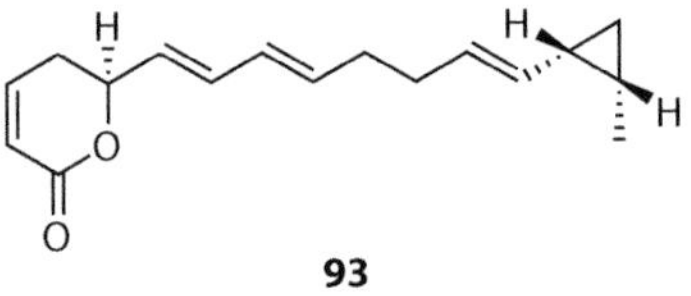

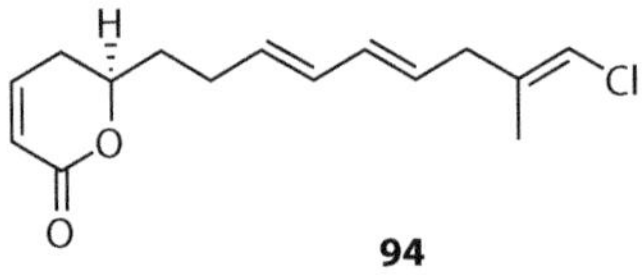

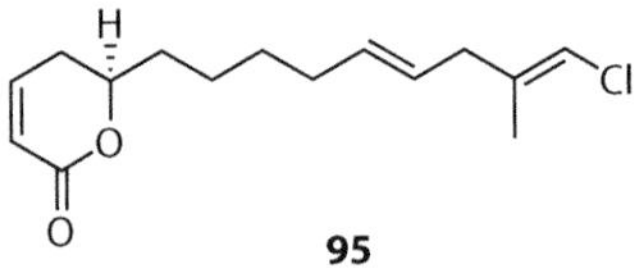

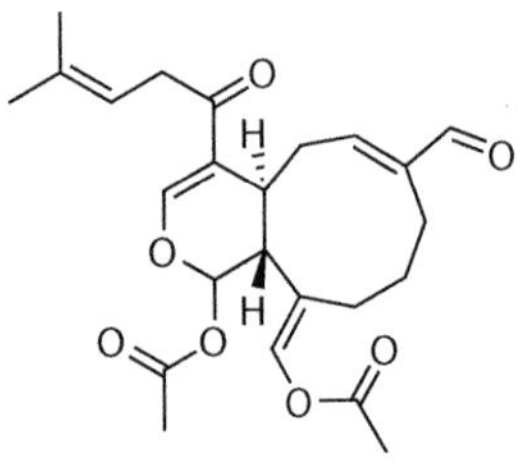

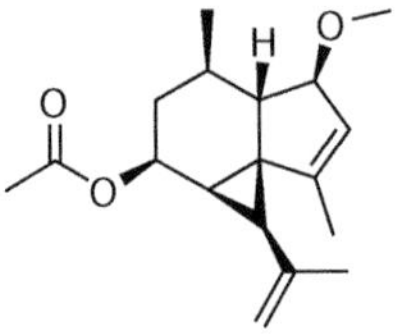

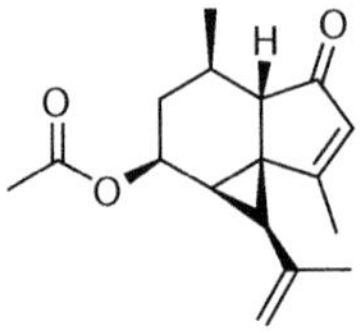

9.4.3 *Plasmodium*

In 1973 the first marine isonitrile, axisonitrile-1 (**99**), was isolated from the marine sponge *Axinella cannabina*.[211] This same sponge yielded the potent antiplasmodial axisonitrile-3 (**100**),[212] which showed activity against both chloroquine-sensitive (IC_{50}, D6 = 0.6 μM) and chloroquine-resistant (IC_{50}, W2 = 73 nM) strains.[213] A series of isonitrile, isothiocyanate, and isocyanate diterpenes consisting of isocycloamphilectane (**101–105**, with IC_{50} values of D6 = 15, 130, 220, 9, and 210 nM and W2 = 13, 8, 160, 7, and 66 nM, respectively, for each strain of *Plasmodium falciparum*), cycloamphilectane (**106** and **107**, with IC_{50} values of D6 = 290 and 250 nM and W2 = 80 and 81, respectively), amphilectane (**108–113**, with IC_{50} values of D6 = 1.02, 1.75, 1.32, 0.047, 0.2, and 2.31 μM and W2 = 0.45, 0.81, 0.31, 0.031, 0.086, and 1.23 μM, respectively), and isoamphilectane (**114**, with IC_{50} values of D6 = 0.3 μM and W2 = 0.1 μM) backbones were isolated from the halichondrid sponge *Cymbastela hooperi*.[214] Generally, all these molecules showed good antimalarial activity, with the isocycloamphilectane compound **104** showing the most potent antiplasmodial activity at an IC_{50} value of 7 nM.[215]

99

100

101

102

103

104

105

106

107

108

109

110

111

112

113

114

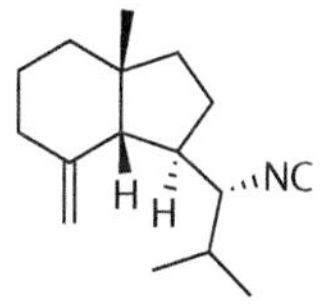

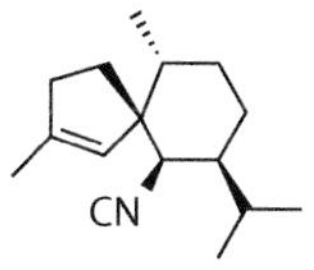

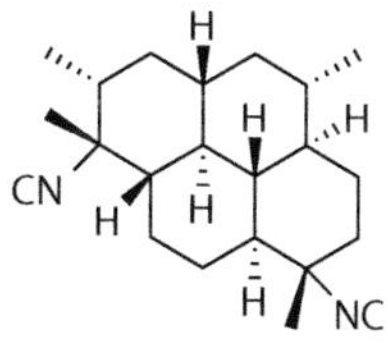

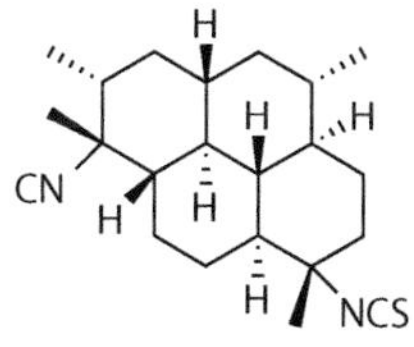

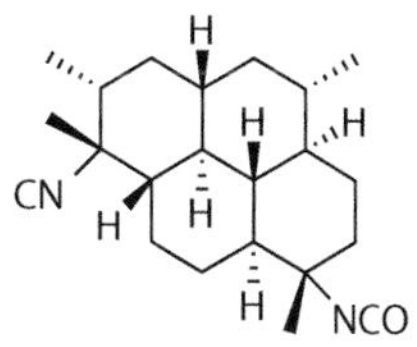

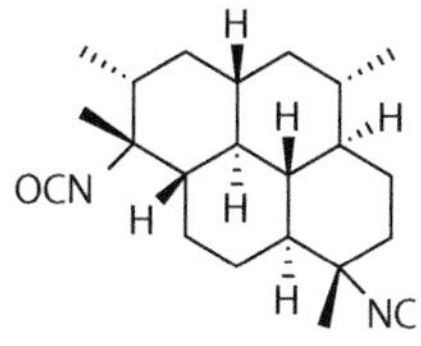

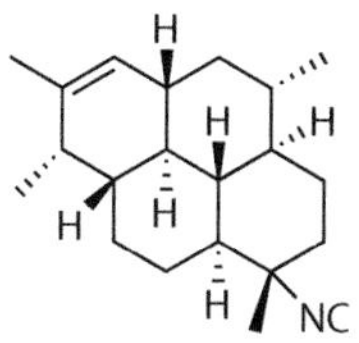

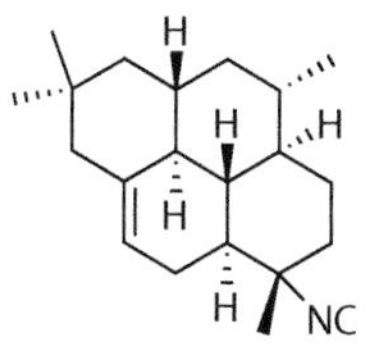

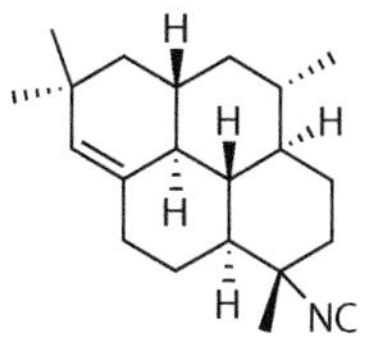

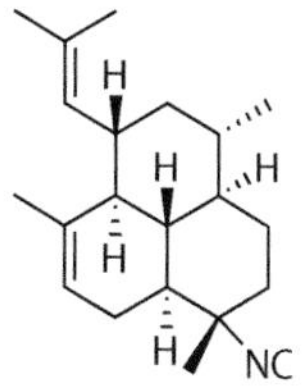

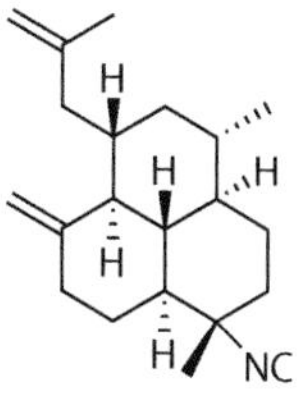

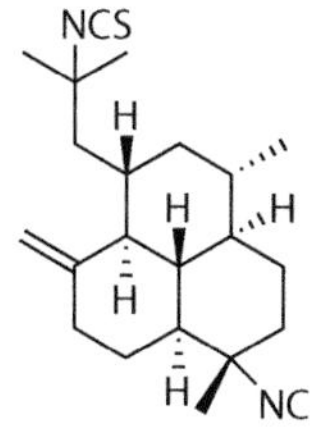

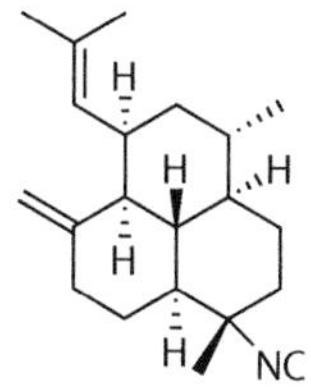

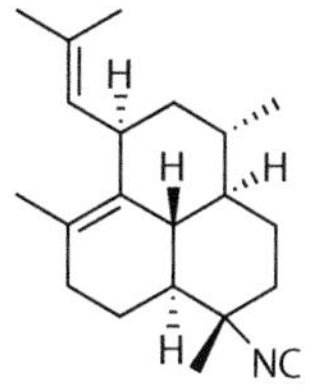

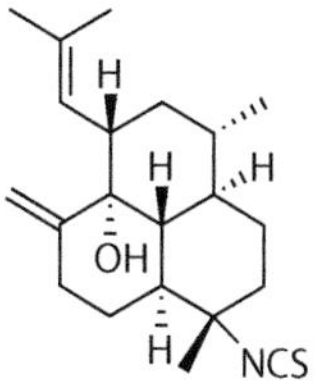

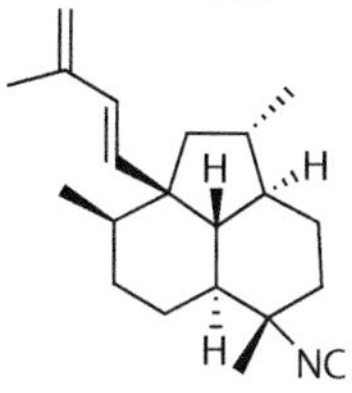

In more recent years, Wattanapiromsakul et al.[216] isolated four tricyclic diterpene isonitriles (**115–118**) from the *Ciocalapata* sp. sponge collected off the coast of Kho-Tao, Thailand, which showed antiplasmodial activity against the Laotian chloroquine resistance strain of *P. falciparm* K1, with IC_{50} values of 0.09, 1.07, 0.44, and 0.98 µM, respectively. A recent study of a Puerto Rican sponge, *Hymeniacidon* sp., has yielded monamphilectine A (**119**), a new β-lactam containing isonitrile, which shows promising antiplasmodial activity (IC_{50}, W2 = 40 nM).[217] A series of kili-hinane diterpenoids (**120–124**) were discovered from the Okinawan sponge *Acanthella* sp. These kilihinane diterpenoids are known for their anthelmintic, antibacterial, and antifouling abilities, in addition to potent antiplasmodial capabilities.[209,218] The most active in the series was kalihinol A (**120**), which exhibited a very significant antiplasmodial activity (EC_{50} = 1 nM against the FCR-03 *P. falciparum* clone). The bicyclic isothiocyanate (**125**), isolated from a *Halichondria* sp. sponge collected off the Hawaiian island Oahu, displayed moderate antiplasmodial activity, with an IC_{50} value of 5.7 µM.[214]

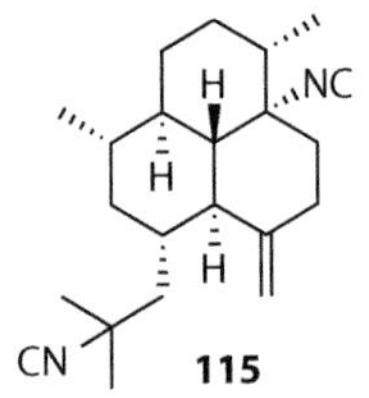

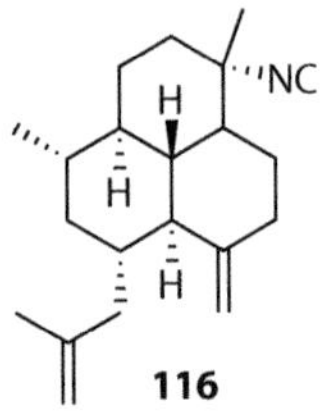

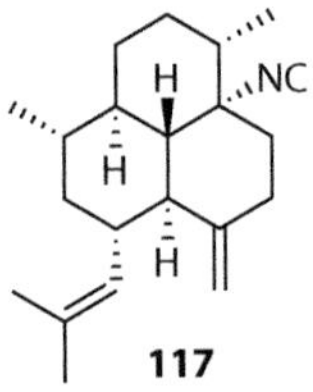

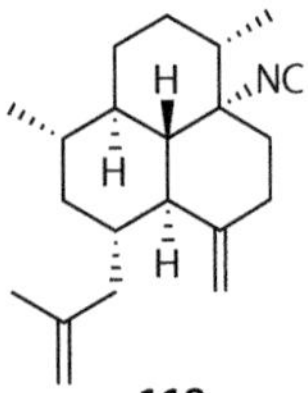

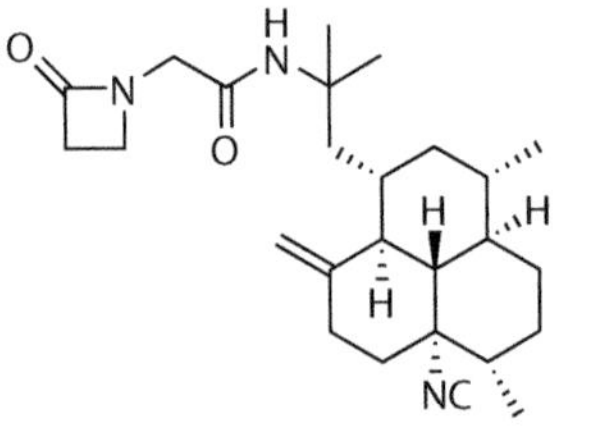

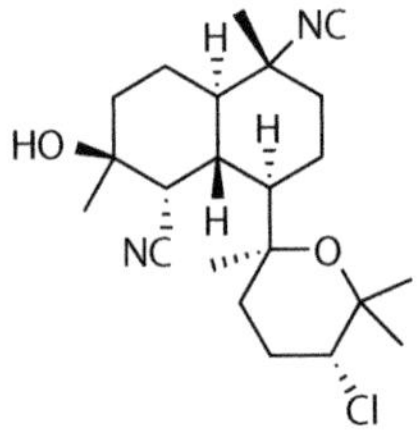

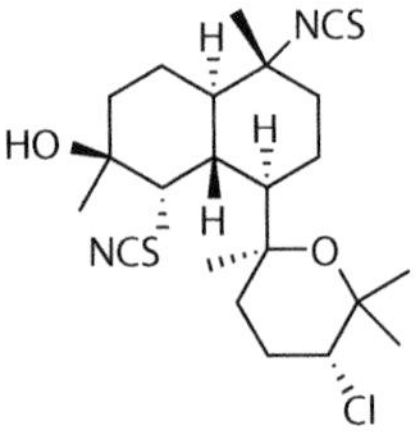

The fungal metabolite 5-carboxymellein (**126**) was isolated from the marine fungus *Hypoxylon oceanicum*[219] and later tested for antiplasmodial activity.[220] This compound showed moder-ate activity against the multi-drug-resistant K1 strain of *P. falciparum*. The indole derivative, 6-bromoaplysinopsin (**127**), was first isolated in 1985[221] and then reisolated years later from the sponge *Smenospongia aurea*[213] and was evaluated for its antiplasmodial activity (IC_{50}, D6 = 1.02 µM). In that same year, the *Xestospongia sapra* metabolite, xestoquinone (**128**), was isolated[222] and 21 years later proved to be a selective inhibitor of Pfnek-1, a never in mitosis A(NIMA) related protein kinase of *P. falciparum*.[223] The previously described **57** and its 8-hydroxy derivative (**129**) displayed antiplasmodial activity, with IC_{50} values of 8.8 and 8.6 nM, respectively, against the D6 strain of *P. falciparum*.[224] The latter also showed promising *in vivo* activity against *Plasmodium*

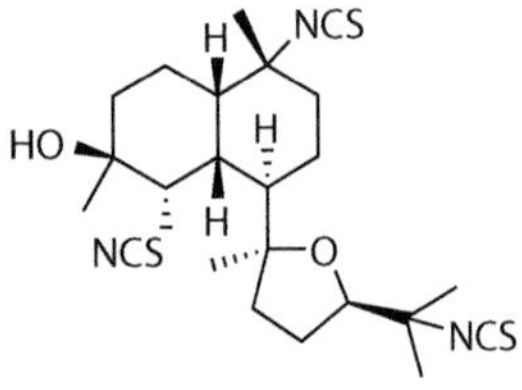

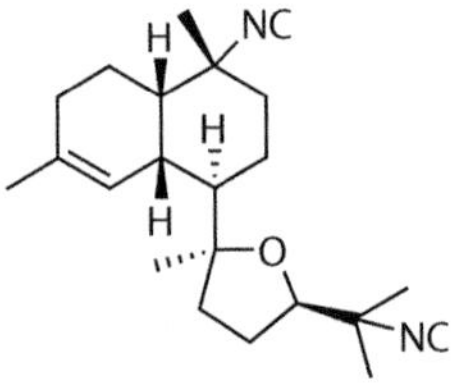

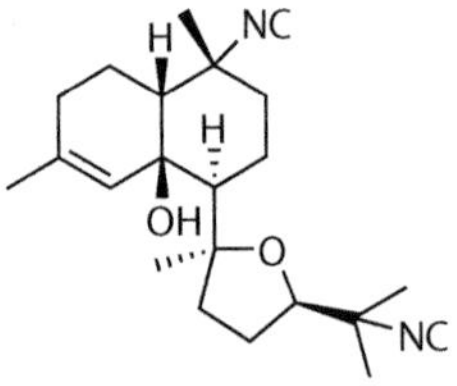

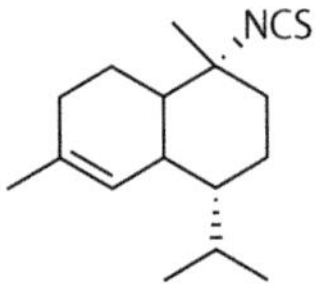

berghei.[225] The decahydroquinoline derivatives, lepadins, were isolated from two Australian tunicates, *Clavelina lepadiformis*[226] and *Didemnum* sp.[227], with lepadin E (**130**) exhibiting the greatest antiplasmodial activity (IC$_{50}$, K1 = 0.95 μM).[227] A series of phloeodictines were isolated from the New Caledonian sponges *Phloeodictyon* sp.[228] and *Oceanapia fistulosa*.[229] The most potent antiplasmodial of this series of phloeodictines was **131**, with an IC$_{50}$ value of 0.69 μM against the FGB1 strain of the malaria parasite.[229] A series of peroxyketal endoperoxides was isolated from a species of marine sponge belonging to the genus *Plakortis*. These compounds displayed good antiplasmodial activity (e.g., **132**, IC$_{50}$ = 0.15 μM against FCR3).[230,231] The norsesterpene sigmosceptrellin A (**133**), isolated from the marine sponge *Sigmosceptrella laevis*,[232] showed good antiplasmodial activity, with an IC$_{50}$ value of 1.4 μM.[233] The γ-lactam alkaloid salinosporamide A (**134**), from the bacterium of the new genus *Salinispora*,[234] exhibited significant antiplasmodial activity with IC$_{50}$ values of 11.4 and 19.6 nM, respectively, against the 3D7 and FCB strains of *P. falciparum*.[235]

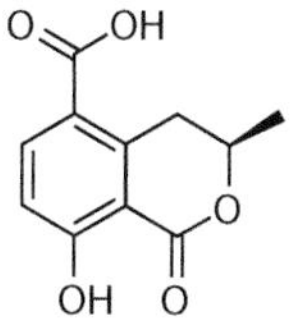

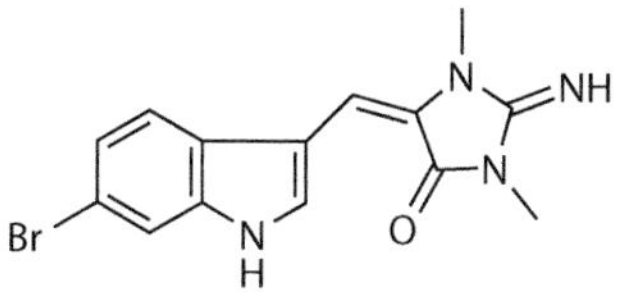

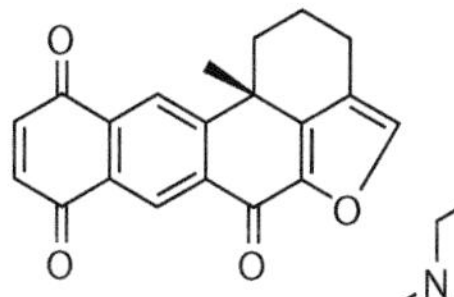

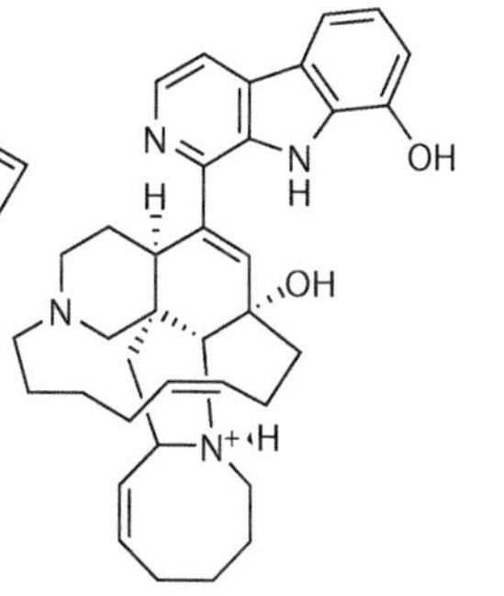

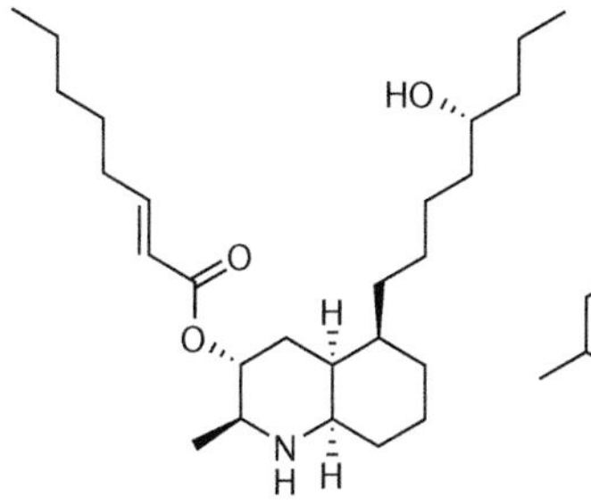

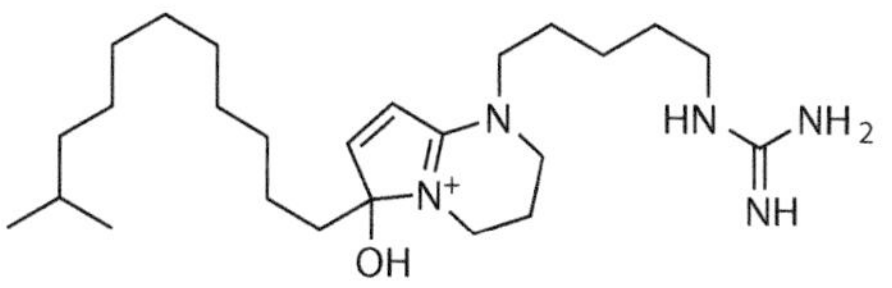

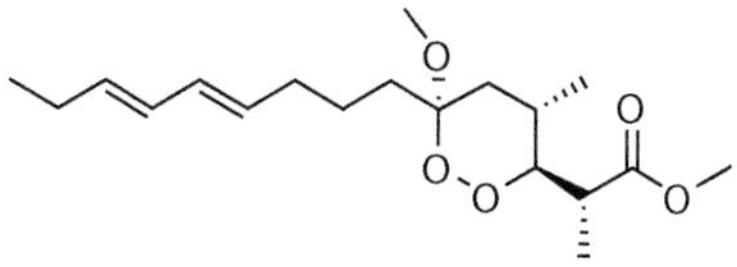

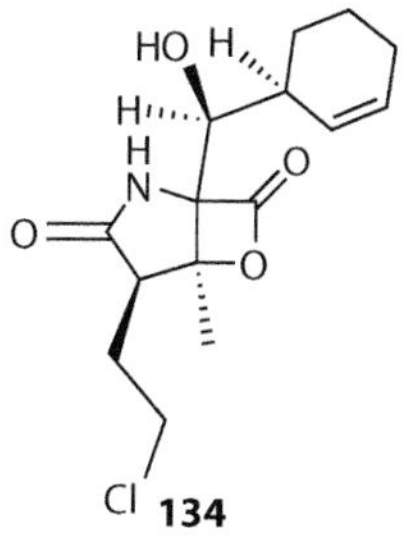

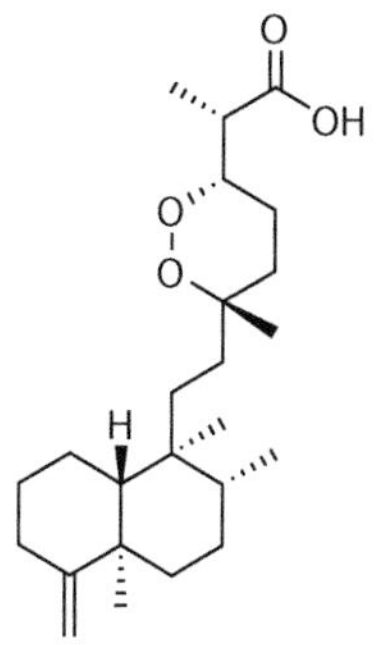

The mangrove fungus, *Aigialus parus*, yielded a series of new resorcylic macrolides as well as the known compound hypothemycin[236] (**135**). The most active compounds of these metabolites were aigialomycin D (**136**) and **135**, with antiplasmodial IC$_{50}$ values against the K1 strain of *P. falciparum* of 19.2 and 5.8 μM, respectively.[237] Several rare sesquiterpenes (e.g., **137**), as well as the known isochochlioquinones A (**138**) and cochlioquinone B[238] (**139**), were isolated from the red algae endophatic fungus *Drechslera dematioidea*.[239] The fungal metabolites **137–139** were the most promising antiplasmodials, with IC$_{50}$ values of 11.5/16.8, 2.7/6.2, and 5.5/7.2 μM, respectively, against the K1/NF54 strains of *P. falciparum*. Four new sesquiterpenes were obtained from the Turkish red algae *Laurencia obtusa*. However, only one sesquiterpene, **140**, showed moderate antiplasmodial activity, with IC$_{50}$ values of 7.1 and 10.5 μM against the D6 and W2 clones of *P. falciparum*.[240] The gorgonian coral *Briarium polyanthes* yielded 10 new briarellins, though only one, briarellin L (**141**), showed moderate activity against *P. falciparum*, with an IC$_{50}$ value of 17.7 μM.[241] A suite of trioxacarcins were isolated from a species of *Streptomyces*,[242] which proved to be potent antiplasmodials. Triotoxins A (**142**) and D (**143**) were the most active, with IC$_{50}$ values of 1.8/1.7 and 2.7/2.0 nM, respectively, for the K1 and NF54 strains.[243]

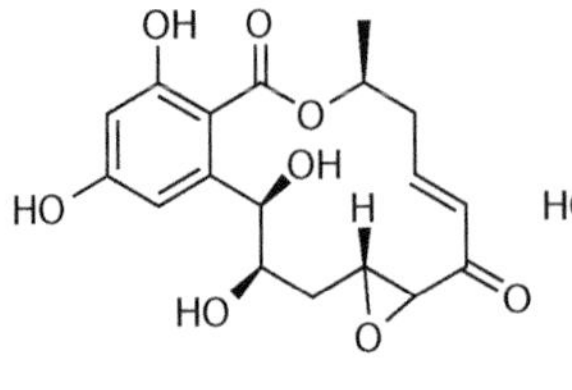

135 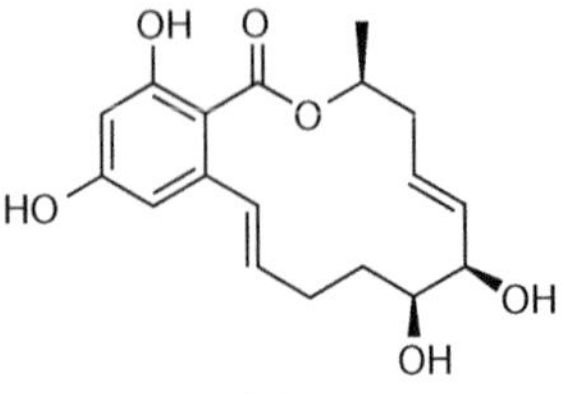　　　　136 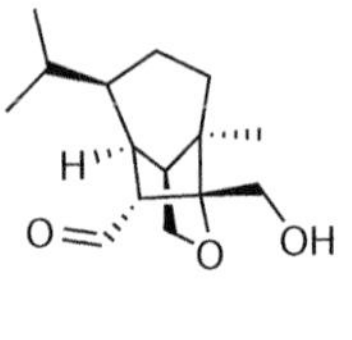　　　　137

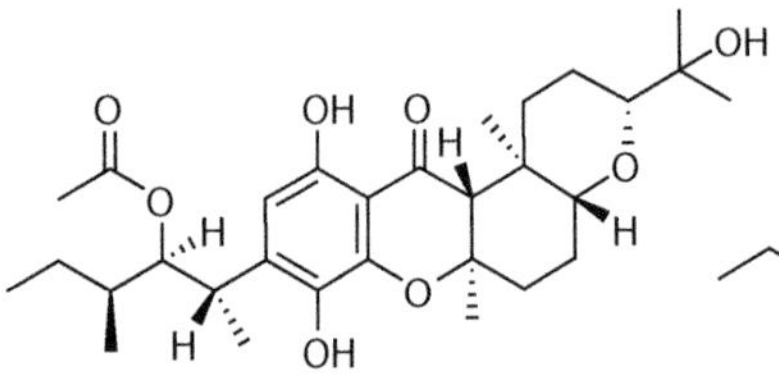

138 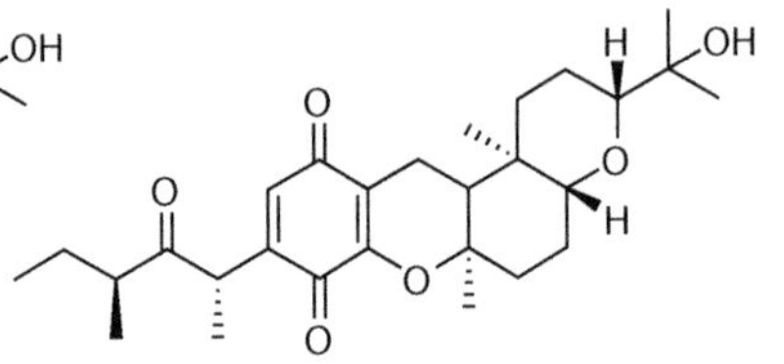　　　　139 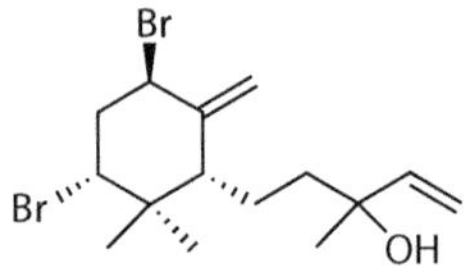　　　　140

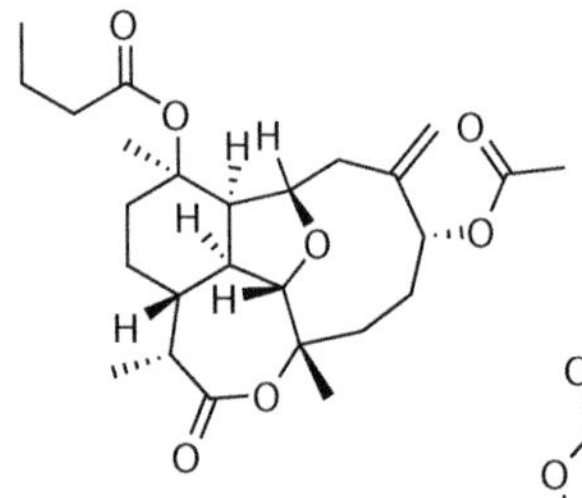

141 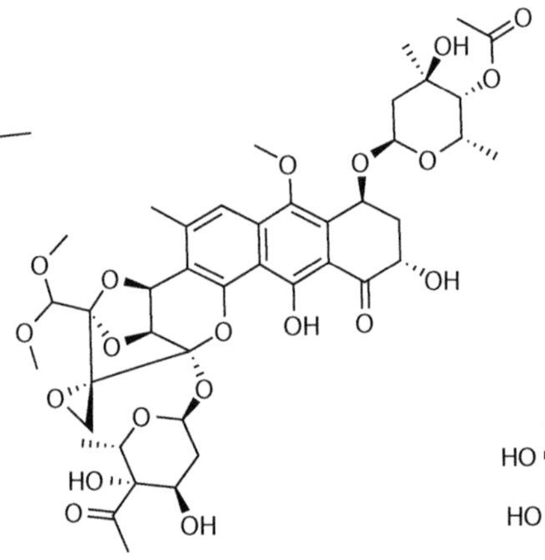　　　　142 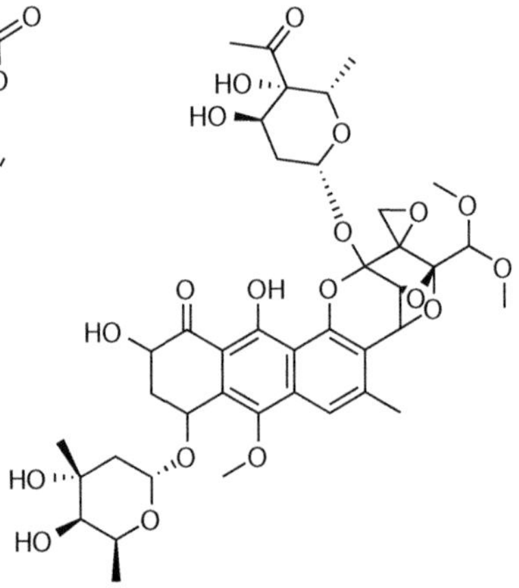　　　　143

A species of marine fungus of the genus *Phoma* afforded pycnidione (**144**). Pycnidione was tested for antiplasomodial activity against three strains of *P. falciparum*, FCR3F86, W2, and D6, with IC$_{50}$ values of 0.28, 0.37, and 0.56 μM, respectively.[244] The Panamanian cyanobacterium *Oscillatoria* sp. yielded the cyclic hexapeptides venturamides A (**145**) and B (**146**). These compounds showed good selective antiplasmodial activity, with IC$_{50}$ values of 8.2 and 5.2 μM.[245] The brown algal metabolites **147** and **148** showed moderate activity against the multi-drug-resistant strain K1 of *P. falciparum*, with IC$_{50}$ values of 10 and 4.4 μM.[246] Bioguided fractionation of the sponge *Agelas oroides* yielded (*E*)-oroidin (**149**). This compound proved to be a potent inhibitor of the *P. falciparum* enoyl-ACP reductase (*Pf*FabI), with an IC$_{50}$ value of 0.77 μM.[247] The cyanobacterium *Lyngbya majuscula* yielded three linear lipopeptides. The peptide dragomabin (**150**) showed the best differential toxicity between the parasite (W2) and human cells, with IC$_{50}$ values of 6.0 and 182.3 μM, respectively.[248] The very simple 2-aminoimidazole derivative, girolline (**151**), isolated from *Cymbastela cantharella*, showed potent antiplasmodial activity, with an IC$_{50}$ range of 77–215 nM. Due to its potency, the compound was also tested in an *in vivo* animal model, which was found to be active at a dose of 1 mg/kg/day (orally and intraperitoneally).[249] The actinomyce (*Streptomyces* sp.) polyether **152** showed effective activity toward both chloroquine-sensitive and -resistant strains of *P. falciparum* without displaying any cytotoxicity toward normal cells.[250] The algal endophyte *Chaetomium* sp. yielded **70**, which displayed an IC$_{50}$ value of 1.4 μM against the K1 strain of *P. falciparum*.[188]

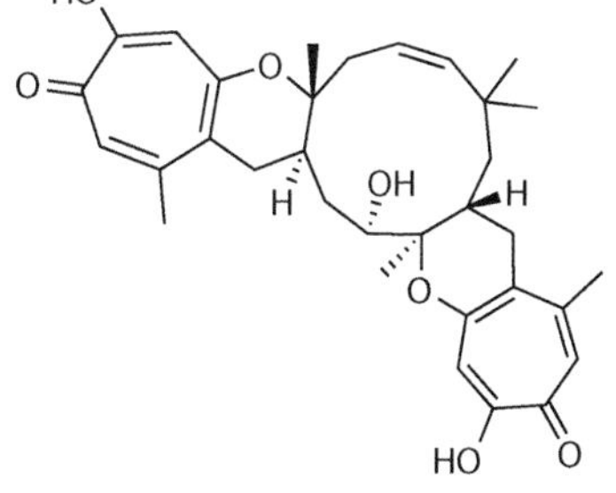

144 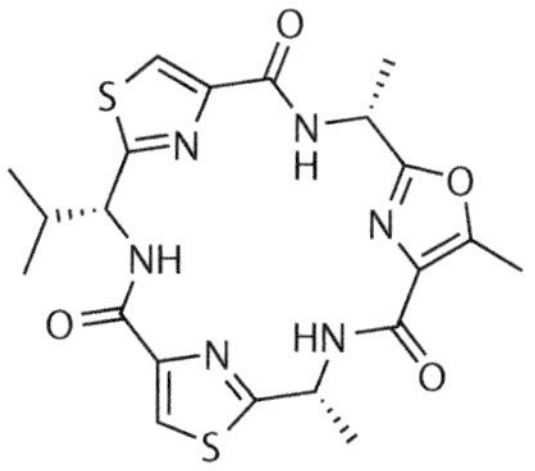   145 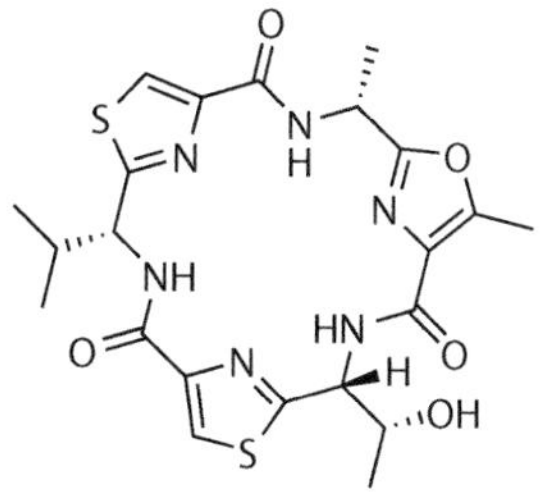   146

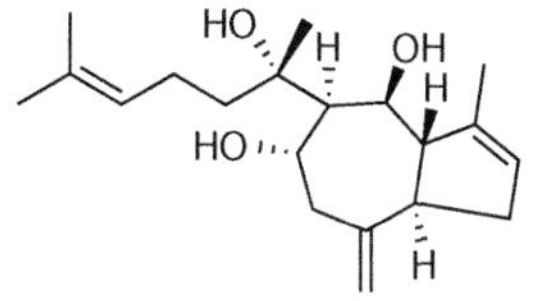

147 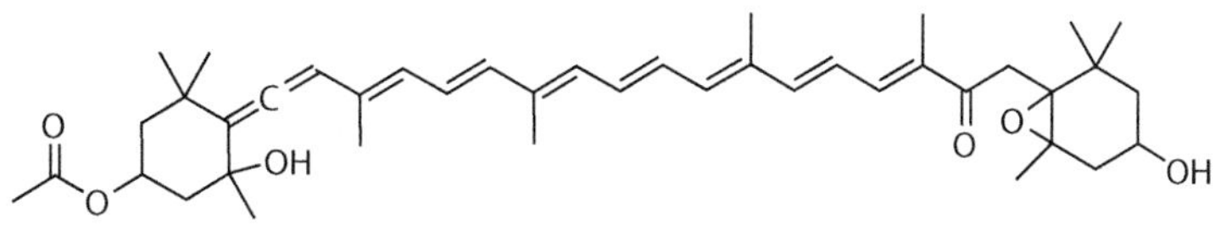   148 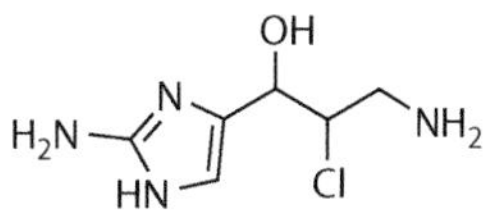   152

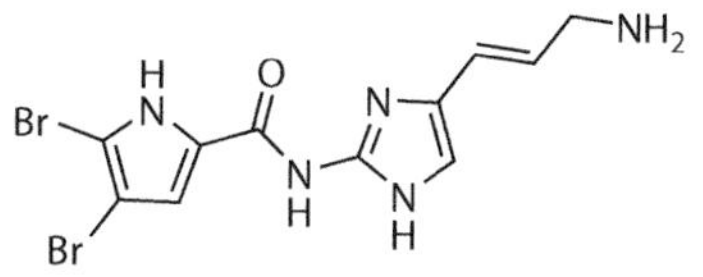

149 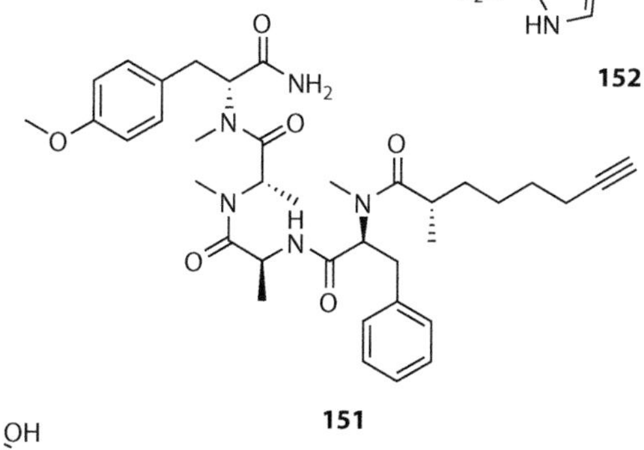   151

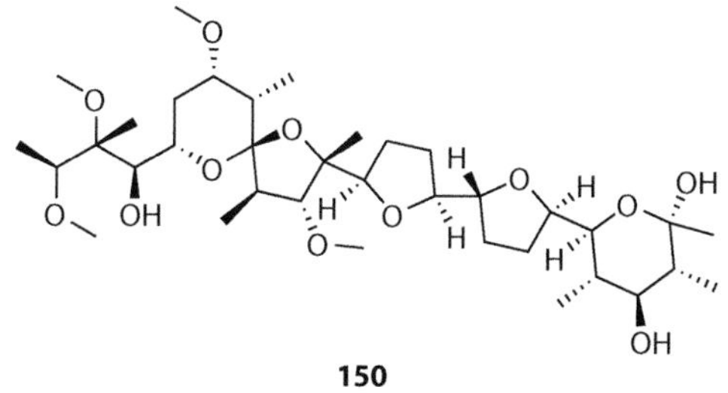

150

The marine fungal metabolites from *Nodulisporium* sp. **153** and **154** showed moderate antiplasmodial activity, with IC$_{50}$ values within the range 1–10 μM.[251] Three new diterpenes, agelasine J (**155**), K (**156**), and L (**157**), were isolated from the sponge *Agelas* cf. *mauritiana* collected in the Solomon Islands. These compounds were tested *in vitro* against the Columbia strain of *P. falciparum* (FcB1) and showed mild activity, with IC$_{50}$ values of 6.6, 8.3, and 18 μM, respectively.[252] Four new meroterpenes, alisiaquinones A–C (**158–160**) and alisiaquinol (**161**), were extracted from an unidentified deep-sea sponge collected from New Caledonia. These compounds were tested against several strains of *P. falciparum*, as well as in an assay to test the inhibition of protein farnesyl transferase (PFTase). The most potent of these was **160**, which showed good activity against *Pf*FcMC29, *Pf*FcB1, and *Pf*F32, with IC$_{50}$ values of 0.08, 0.21, and 0.15 µM, respectively, as well as inhibition of PFTase, with an IC$_{50}$ value of 1.9 µM. Compounds **158** and **160** were also tested in a murine model infected with *P. vinckei petteri*. The parasitemia was reduced after 5 days at a concentration of 20 mg/kg/day (intraperitoneal).[253] The previously mentioned cyanobacterium metabolite **73** also showed moderate activity (IC$_{50}$ = 8.4 µM) against the W2 strain of the malaria parasite.[189]

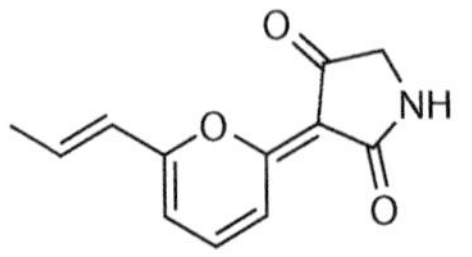

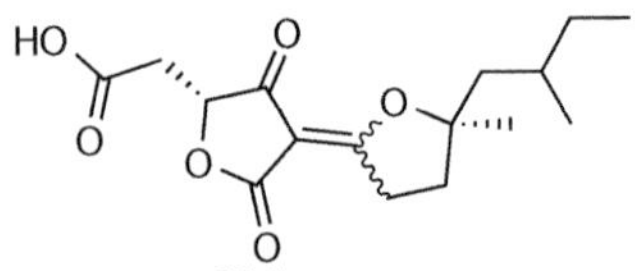

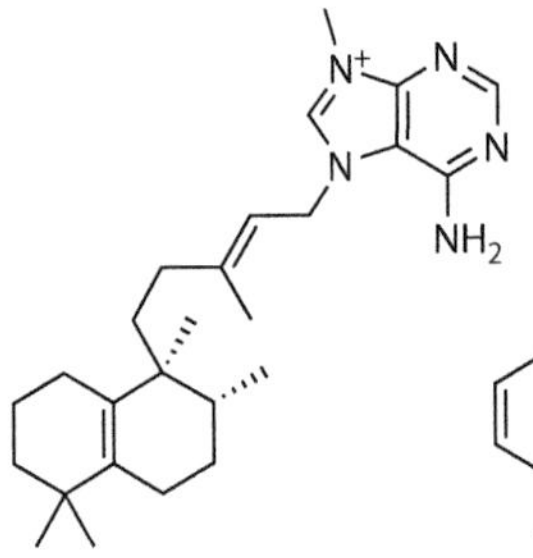

The South African red alga *Plocamium cornutum* afforded a series of polyhalogenated monoterpenes, with compounds **162** and **163** showing moderate activity against the *P. falciparum* chloroquine-sensitive strain D10, with IC$_{50}$ values of 16 and 17 µM, respectively.[254] The cyanobacterial metabolite symplostatin 4 (**164**), isolated from *Symploca* sp.,[255] has recently been tested for antiplasmodial activity. This study identified **164** as a potent inhibitor of the *P. falciparum* falcipains, suggesting that **164**'s mode of action may lie in the disruption of the hemoglobin degradation pathway.[256]

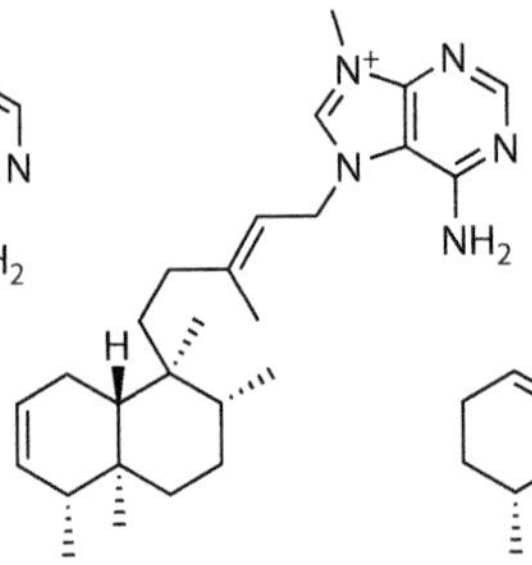

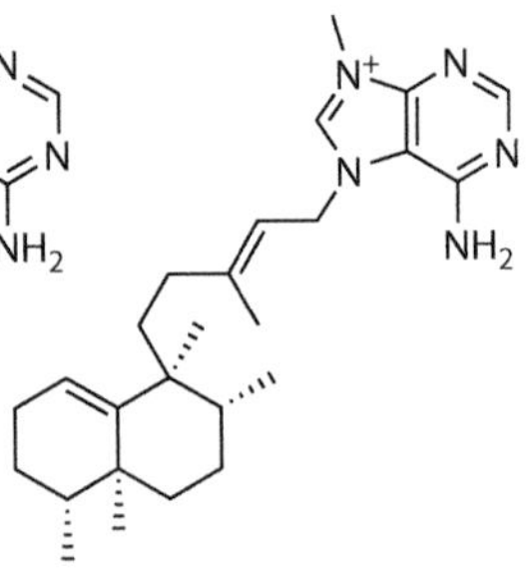

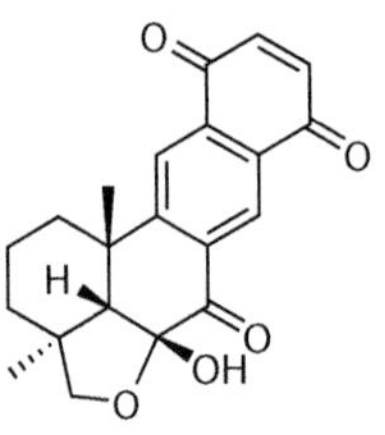

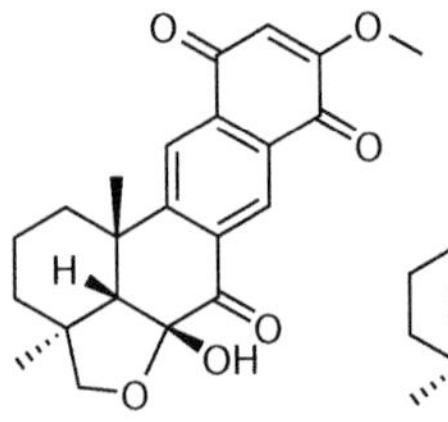

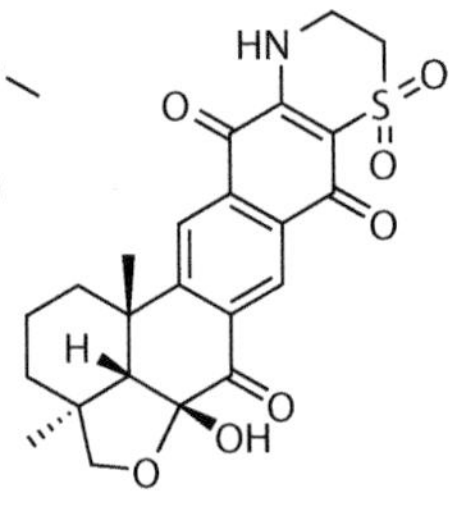

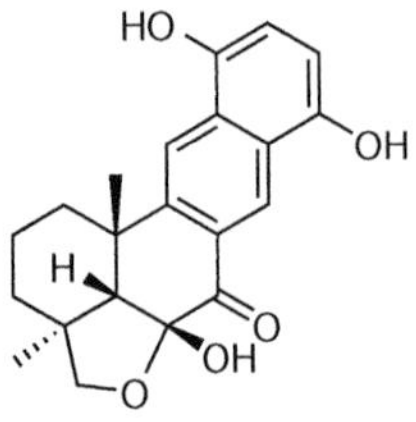

The tricyclic guanidines merobatzelladines A (**165**) and B (**166**) were isolated from the marine sponge *Monanchora* sp. They showed activity against *P. falciparum*, with IC$_{50}$ values of 1.3 and 3.2 µM, respectively.[257,258] The hard coral of the genus *Trbastraea* yielded a new bis(indole) alkaloid **167**. This compound presented micromolar activity against two strains of *P. falciparum* (IC$_{50}$, F32 = 2.7 µM; IC$_{50}$, FcB1 = 1.8 µM).[259] The compound-rich cyanobacterium *Lyngbya majuscula* afforded the cyclic peptides lagunamides A (**168**) and B (**169**), which showed potent antiplasmodial activity against *P. falciparum*, with IC$_{50}$ values of 190 and 910 nM, respectively.[260] Several compounds of the Fijian red alga *Callophycus serratus*, the bromophycolides, showed submicromolar activity of *P. falciparum*. The most effective of these were the bromophycolides A (**170**), D (**171**), E (**172**), H (**173**), M (**174**), and S (**175**), with IC_{50} values of 0.9, 0.3, 0.8, 0.9, 0.5, and 0.9 μM, respectively.[261–263] In a recent publication, Kubanek and coworkers evaluated the pharmacokinetics, metabolism, and efficacy of **170** in a murine model. The compound was administered both intravenously and intraperitoneally and showed good potential for further development as a lead compound.[264] The bis-piperidine alkaloid, neopetrosiamine A (**176**), isolated from the Caribbean marine sponge *Neopetrosia proxima*, had mild antiplasmodial activity against the *P. falciparum* (IC_{50} = 2.3 μM).[265] The β-carboline (**177**) from the Australian sponge *Acorina* sp. displayed moderate activity against the chloroquine-resistant (Dd2) and chloroquine-sensitive (3D7) strains of *P. falciparum*, with IC_{50} values of 5.4 and 3.5 μM, respectively.[266]

162

163

164

165

166

167

168

169

170

171

172

173

174

175

176

177

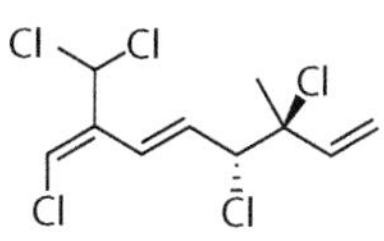

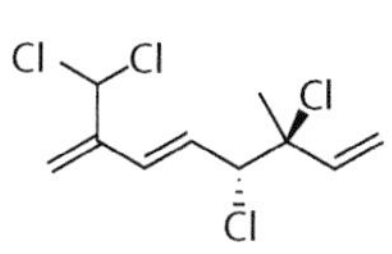

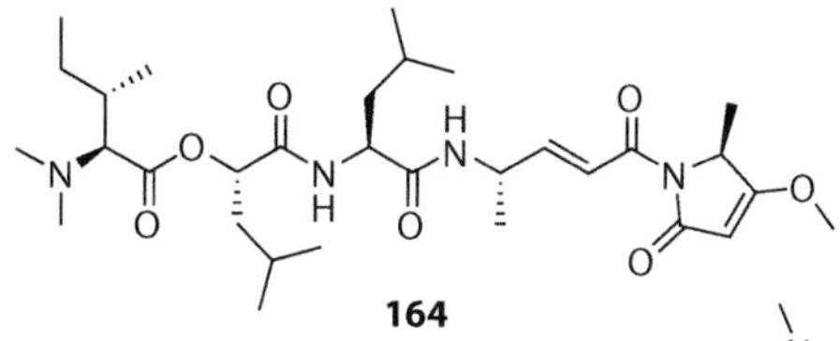

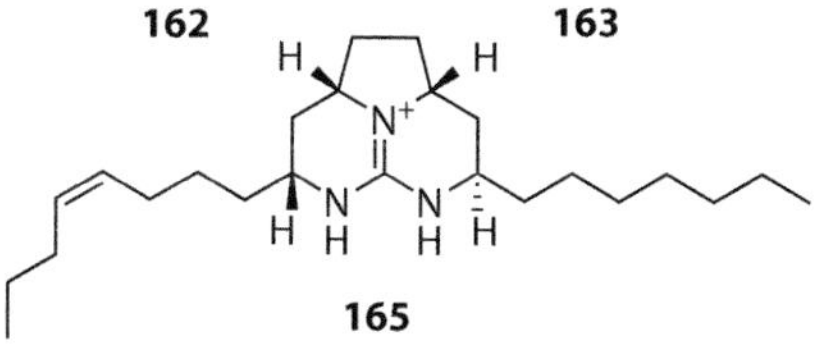

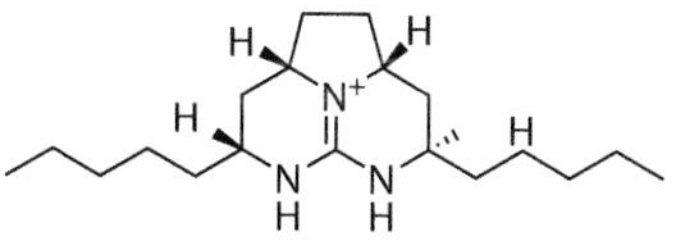

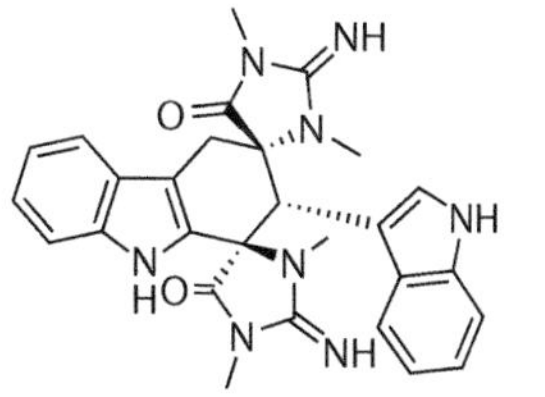

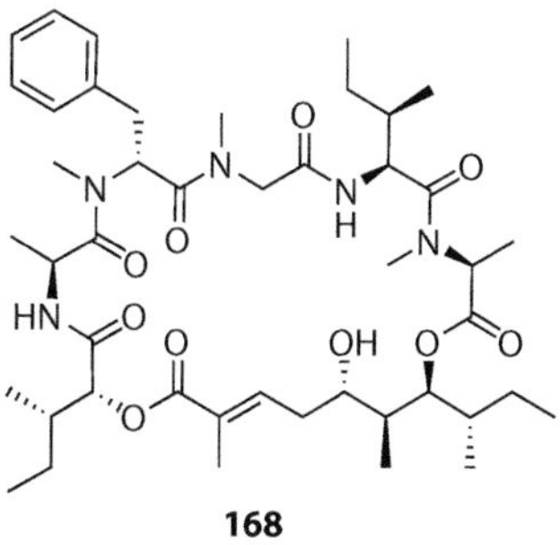

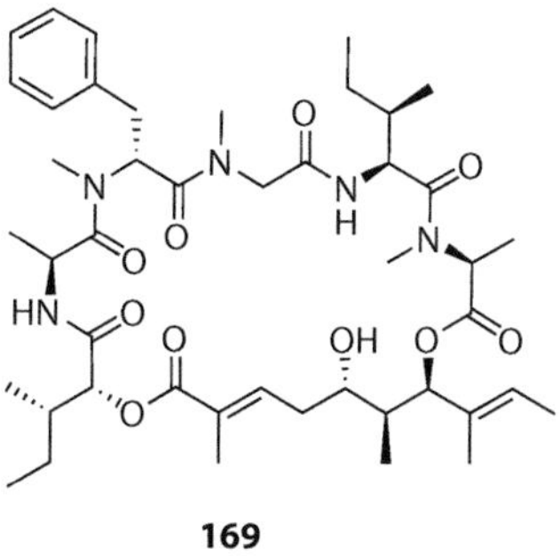

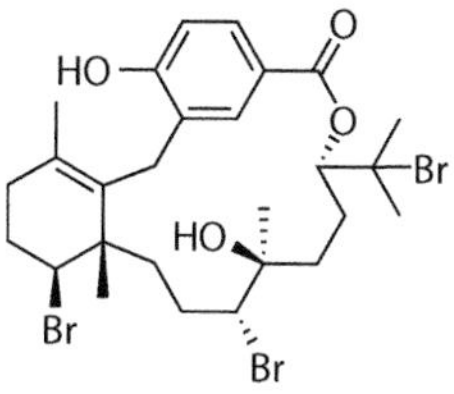

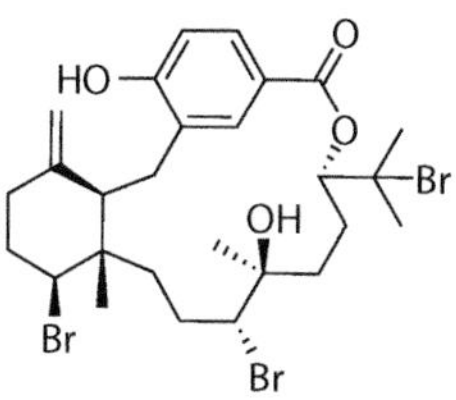

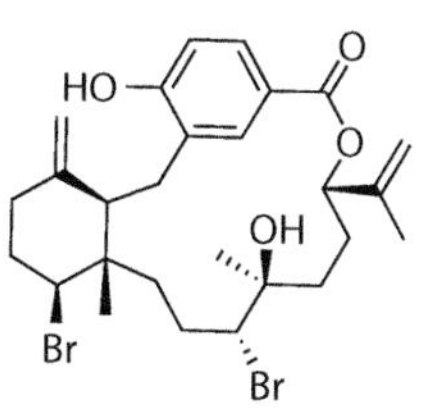

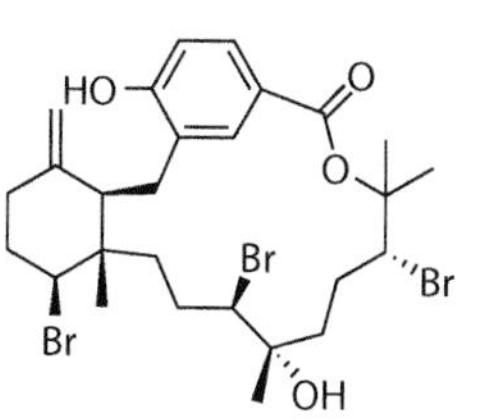

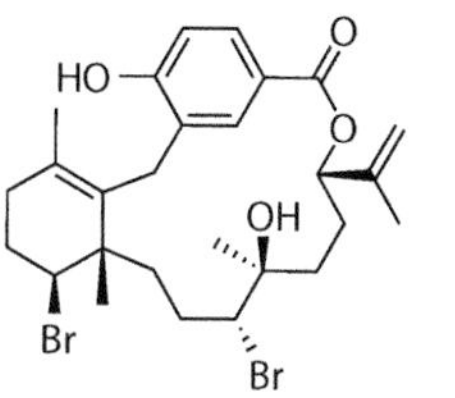

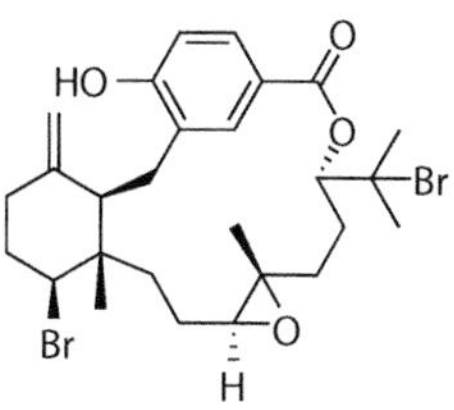

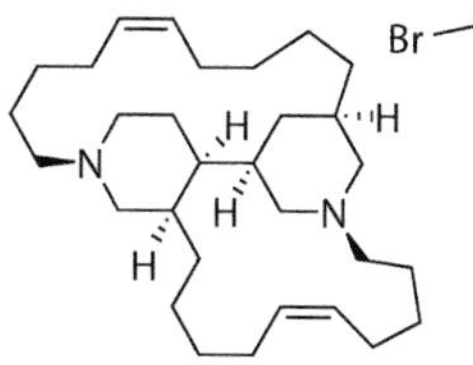

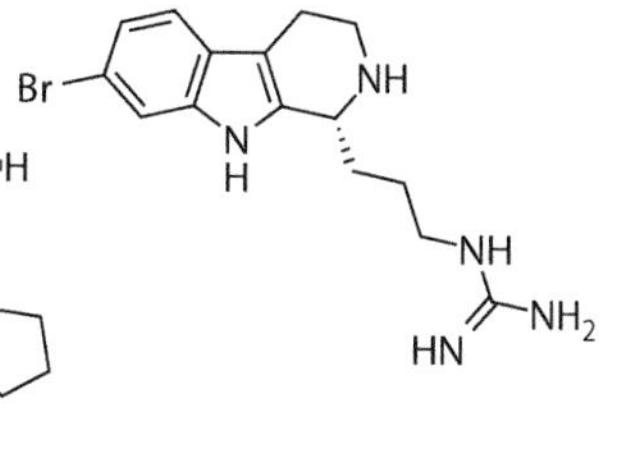

The suite of dolabellane-type diterpenes were isolated from the gorgonian octocoral of the genus *Eunicea*, which exhibited some antiplasmodial activity. The most active of these was the compound dolabellane (**178**), which inhibited the growth of 50% of the malaria parasite at a concentration of 9.4 μM.[267] A series of β-carboline and indolactam alkaloids, isolated from the marine actinomyces *Marinactinospora thermotolerans*, were tested for their ability to inhibit the growth of *P. falciparum*. The most active of these compounds were two β-carboline alkaloids, marinacarbolines C (**179**) and D (**180**), which displayed IC_{50} values of 3.09/3.38 and 5.39/3.59 μM, respectively, against the 3D7/Dd2 strains.[268] The marine fungal metabolite marilone A (**181**) exhibited antiplasmodial activity against the liver stage of *P. berghei* (IC_{50} = 12.1 μM).[269] A species of Australian sponge from the genus *Zyzzya* afforded a suite of potent antiplasmodial pyrroloiminoquinones. Tsitsikammamine C (**182**) and makaluvamines J (**183**), G (**184**), L (**185**), K (**186**), and damirone A (**187**) showed nanomolar activity against the *P. falciparum* 3D7/Dd2, with IC_{50} values of 13/18, 25/22, 36/39, 40/21, 396/300, and 1880/360 nM. Compounds **183** and **184** were evaluated in an *in vivo* mouse model, with the latter showing no mouse toxicity and suppression of parasite growth in *P. berghei*.[270]

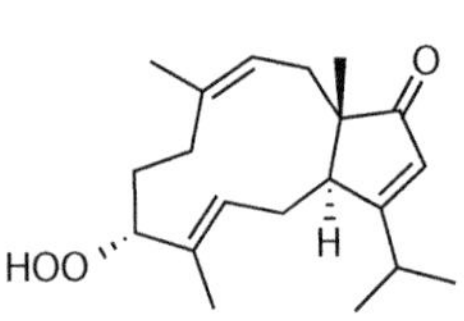

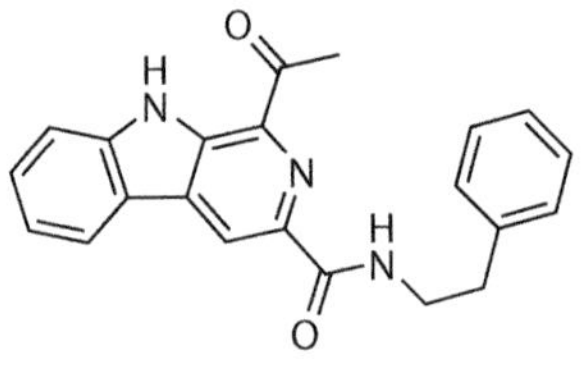

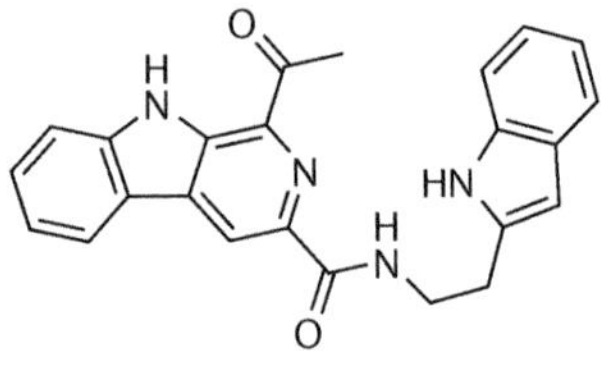

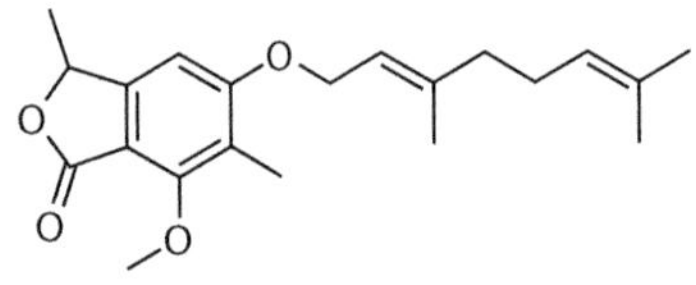

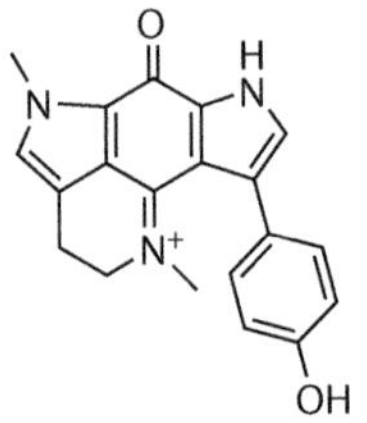

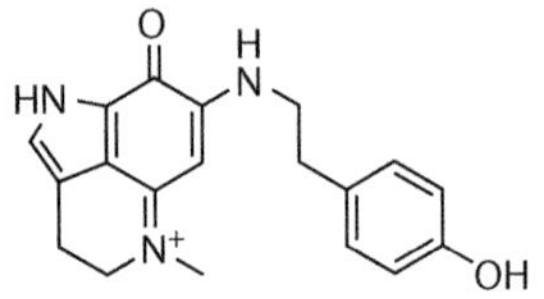

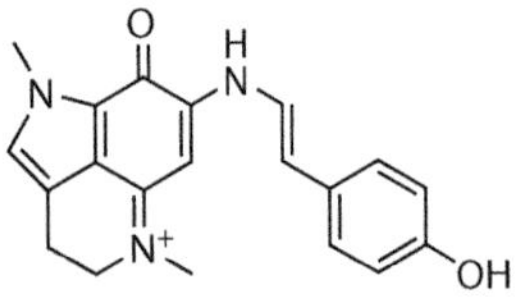

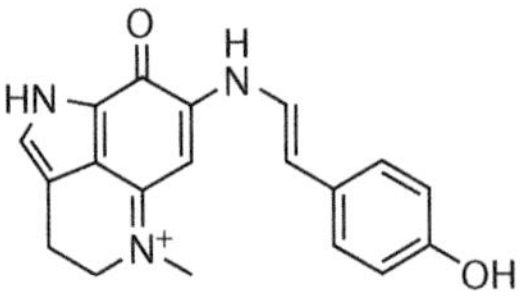

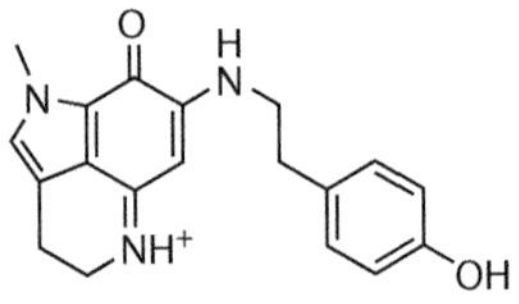

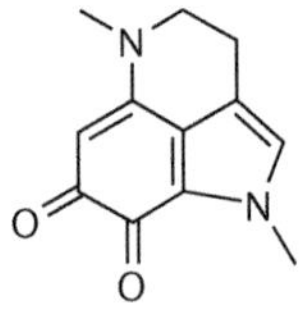

The Australian sponges from the genera *Hyattella* and *Pseudoceratina* produced the bromotyrosine alkaloids psammaplysin F (**188**), G (**189**), and H (**190**), which showed potent *in vitro* antiplasmodial activity against *P. flaciparum* 3D7, with IC_{50} values of 1.92, 5.22, and 0.41 μM.[271,272] The Indonesian marine sponge *Aplysinella strongylata* yielded a series of psammaplysins, as well as 21 new psammaplysin derivatives. Unfortunately, only one, **191**, showed moderate antiplasmodial activity against *P. falciparum* (IC_{50} = 6.4 μM).[273] The New Zealand *Didemnum* sp. afforded the indole spermidine alkaloid, didemnidine B (**192**), which displayed moderate activity toward *P. falciparum* (IC_{50}, K1 = 15 μM).[274] Another species of ascidian from this genus, *D. albopunctatum*, yielded albopunctatone (**193**), which showed moderate activity against the Dd2 and 3D7 strains of *P. falciparum*, with IC$_{50}$ values of 5.3 and 4.4 µM, respectively.[275] The anti-inflammatory compound orthidine F (**194**), isolated from the New Zealand ascidian *Aplidium orthium*,[276] was recently tested for its antiplasmodial activity, resulting in an IC$_{50}$ value of 0.89 µM against *P. falciparum* K1.[277] Synthetic alteration of the aromatic ring at the terminal ends of **194** resulted in a 100-fold increase in activity, with **195** having an antiplasmodial activity of 8.6 nM. This suite of synthetic anologues was screened against *Trypanosoma brucei* (*T. b.*) *rhodesiense*, *T. b. gambiense*, and *Leishmania donovani*, showing moderate activity against the former, with little or no activity against the latter two parasites.[277]

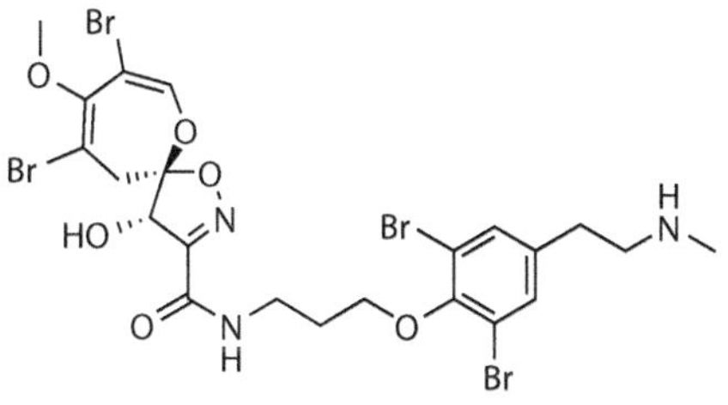

188

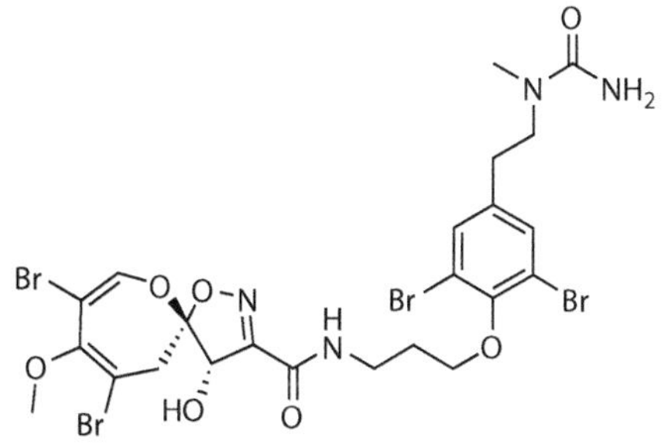

189

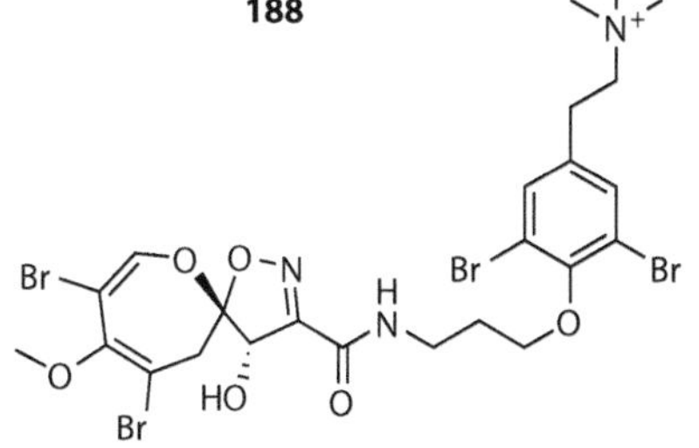

190

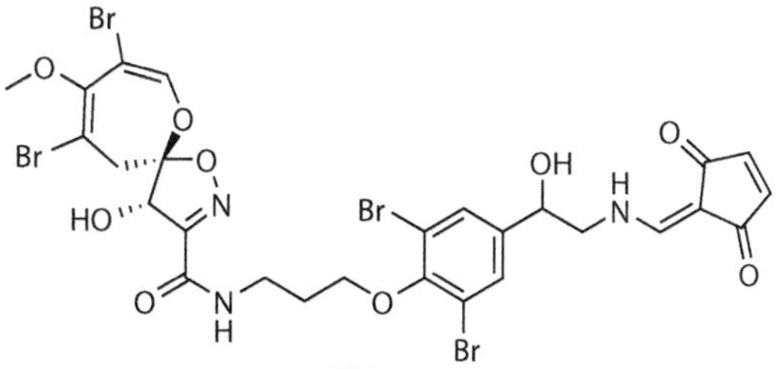

191

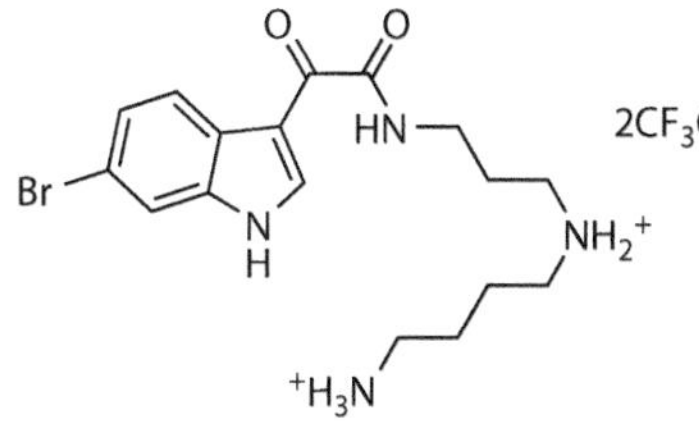

192

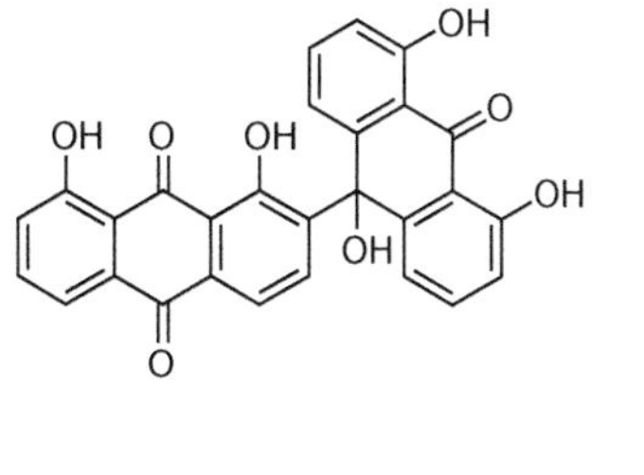

193

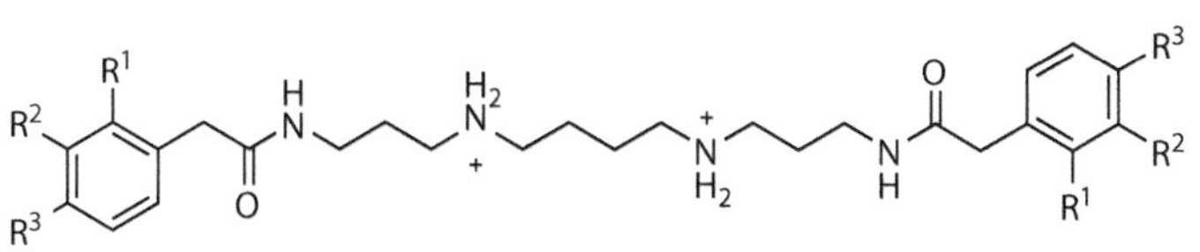

	R^1	R^2	R^3
194	H	OMe	OH
195	OH	H	H

9.4.4 *Toxoplasma*

The Jamaican sponge *Plakinastrella onkodes* afforded the cyclic peroxylactone plakortide (**196**), which showed potent *in vitro* activity toward *Toxoplasma gondii*, with an IC$_{50}$ value of 64 nM.[278] The new norsesterterpene acid muqubilone (**197**) and sigmosceptrellin B (**198**) were isolated from the Red Sea sponge *Diacarnus erythraeanus*. **197** and **198** had potent antitoxoplasmosis activity, with an IC$_{50}$ value of 0.1 μM for both compounds, with no significant cytotoxicity.[279] An undescribed genus of Indo-Pacific sponge (family Petrosiidae, order Haplosclerida) yielded three complex cyclic alkaloids: 8-hydroxymanzamine A (**129**), manzamine F (**199**), and **57**. Compound **199** showed the best activity against *T. gondii*; however, it exhibited higher toxicity to host cells, and therefore **57** was taken forward into the murine model.[217]

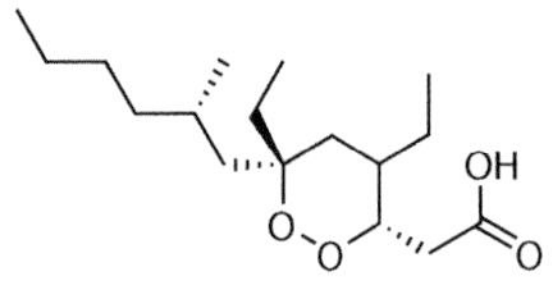

196

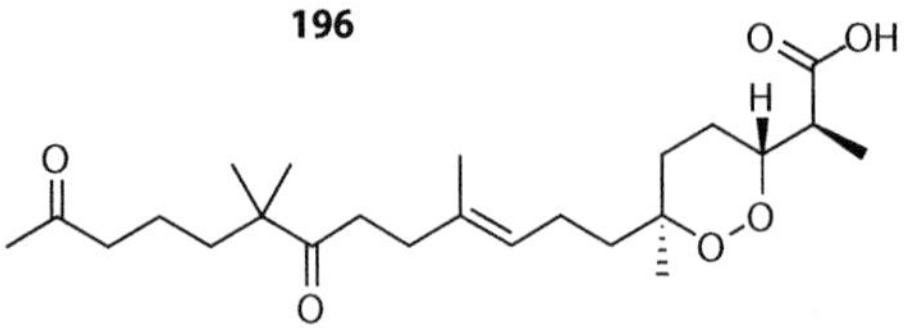

197

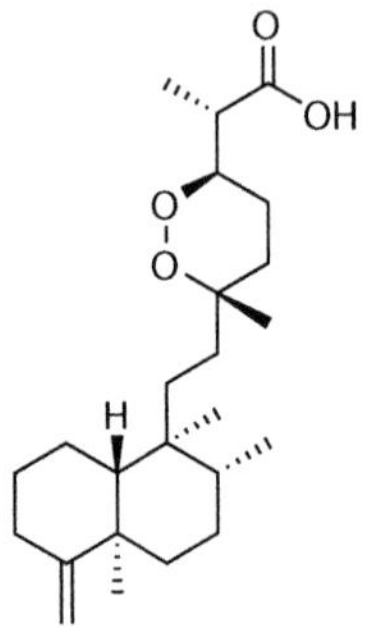

198

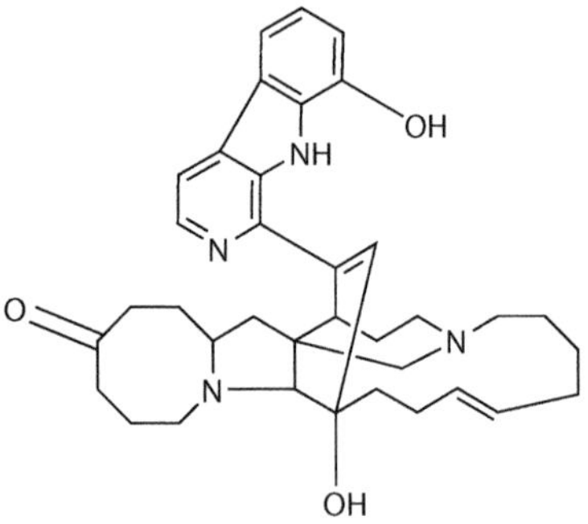

199

9.4.5 *Trypanosoma*

The ascidian of the *Didemnum* sp. yielded the metabolite ascididemin (**200**).[280] Many years later, this compound was proved to be a potent inhibitor of *Trypanosoma brucei* (*T. b.*) *rhodesiense* (IC$_{50}$ = 7.4 nM), as well as displaying moderate activity against *T. cruzi* (IC$_{50}$ = 2.4 μM). This led to a structure–activity relationship study resulting in a series of alkaloids, the most potent of which were **201** and **202**, with IC$_{50}$ values of 102.9/64.3 and 240.1/64.3 nM, respectively, against the species of *T. b. cruzi*/*T. b. rhodesiense*.[281] The cyanobacterium metabolites **145** and **146** also displayed mild antitrypanasomal activity against *T. cruzi*, with IC$_{50}$ values of 14.6 and 15.8 μM.[237] This same group isolated another peptide, viridamide A (**203**), from the cyanobacterium *Oscillatoria nigroviridis*, which showed antitrypanosomal activity against *T. cruzi*, with an IC$_{50}$ value of 1.3 μM.[184] The marine fungus *Ascochyta salicorniae*, isolated from the green alga *Ulva* sp., yielded the tetramic acid ascosalipyrrolidinone A (**204**). This compound showed activity against *T. cruzi* at a minimal inhibitory concentration (MIC) of 2.6 μM.[282] The Brazilian marine sponge *Plakortis angulospiculatus* afforded the polyketide plakortide P (**54**), which displayed mild activity against the Chagas parasite *T. cruzi*, with an IC$_{50}$ value of 6.3 μM.[176] Many years later, the bioguided fractionation of the marine sponge *Plakortis* sp. led to the discovery of 11,12-didehydro-13-oxo-plakortide Q (**205**), a potent inhibitor of *T. b. brucei* (IC$_{50}$ = 49 nM).[283] The marine fungal metabolite (*Chaetomium* sp.) chaetoxanthone C (**206**) displayed moderate activity against *T. cruzi*, with an IC$_{50}$ value of 3.1 μM.[188] The marine sponge metabolite agelasine D (**207**), isolated from *Agelas nakamuri*[284] and *A. linnaei*,[285] led to a comprehensive study on compounds active against Chagas disease. The compounds that displayed the highest selective indexes were analogues **208** and **209**, with potent antitrypanosomal activity at IC$_{50}$ values of 1.2 and <0.31 μM.[191] The previously described alkaloids **165** and **166** also showed moderate activity against *T. b. brucei*, with IC$_{50}$ values of 1.3 and 3.2 μM.[249]

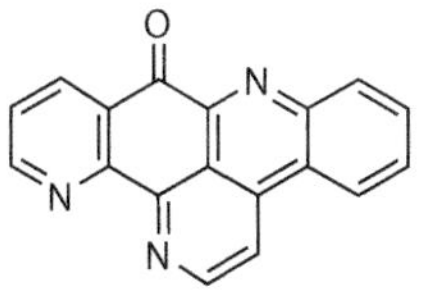

The marine sponge *Diacarnus bismarckensis* yielded several new and known peroxiterpenes, which were assessed for their ability to inhibit *T. brucei* growth. Sigmosceptrellins A (**133**) and B (**198**) and epi-muqubilin (**210**) showed the greatest activity, with IC_{50} values of 2.55, 0.51, and 2.29 µM, respectively.[286] Aignopsanoic acid A (**211**) was isolated from the Papua New Guinea sponge *Cacospongia mycofijiensis* and showed moderate activity against *T. brucei*, with an IC_{50} value of 24 µM.[287] The leishmanicidal compound **89** and dibromopalau'amine (**90**) have also shown activity against *T. b. rhodesiense*, with IC_{50} values of 4.35 and 0.80 µM, respectively.[198] The Australian ascidian *Polysyncraton echinatum* yielded three new pyridoacridine alkaloids (**212–214**), which had potent activity against the parasite of human African trypanosomiasis. Compounds **212–214** had IC_{50} values of 0.077, 0.032, and 1.33 µM, respectively.[288] These compounds are remarkably similar to ascididemin (**200**), which also exhibited potent antitrypanosomal activity. A series of steroidal saponins were isolated from the marine sponge *Pandaros acanthifolium*, the most effective of which are pandaroside G (**215**) and its methyl ester **216**, with IC_{50} values of 780 and 38 nM against *T. b. rhodesiense*.[289] In a study, 13 sponge-derived terpenoids were tested for their ability to inhibit *T. b. rhodesiense* and *T. cruzi*. The linear meroterpene 4-hydroxy-3-tetraprenylphenyl-acetic acid (**217**) showed the greatest activity toward *T. b. rhodesiense* (IC_{50} = 1.37 µM), while heptaprenyl-*p*-quinol (**218**) and 11β-acetoxyspongi-12-en-16-one (**219**) displayed the greatest activity toward *T. cruzi* (IC_{50} = 6.79 and 12.51 µM, respectively).[196] The xenican diterpenoid **96** also displayed potent antitrypanosomal activity against *T. congolense*, with an IC_{50} value of 0.25 µM.[201] The marine bacterium *Bacillus pumilus* yielded the simple compound 3-hydroxyacetylindole (**220**), which exhibited mild activity against *T. cruzi*, with an IC_{50} value of 19.4 µM.[290] The bioassay-guided isolation of Australian species of sponges from the *Iotrochoata* genus yielded two new N-cinnamoyl amino acids, iotrochamides A (**221**) and B (**222**). These compounds proved to inhibit the parasite

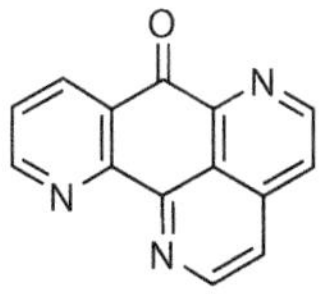

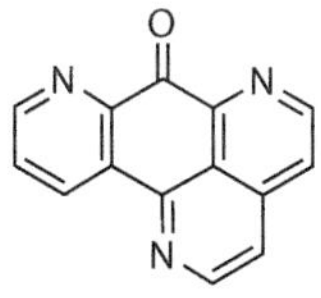

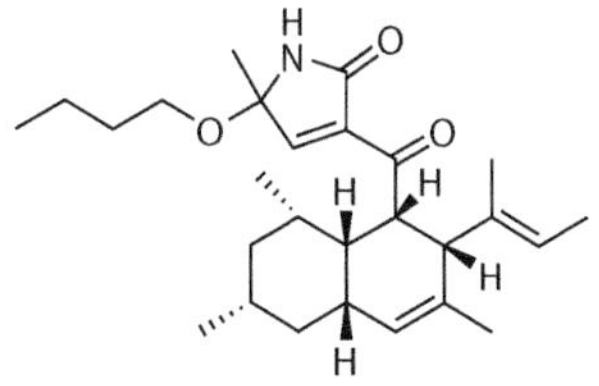

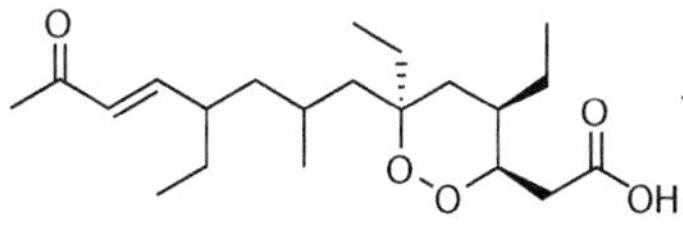

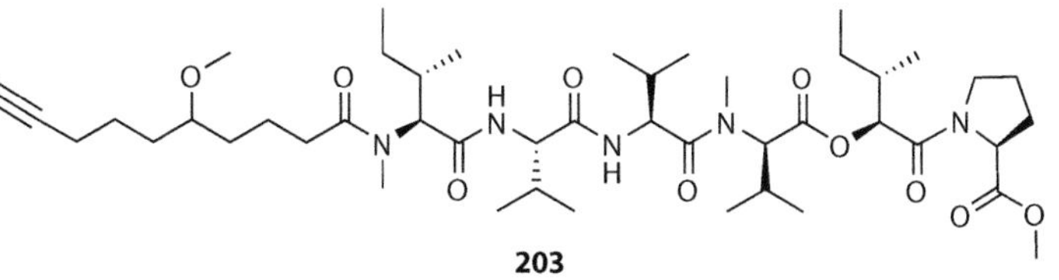

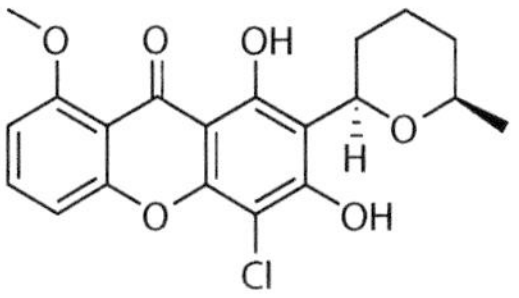

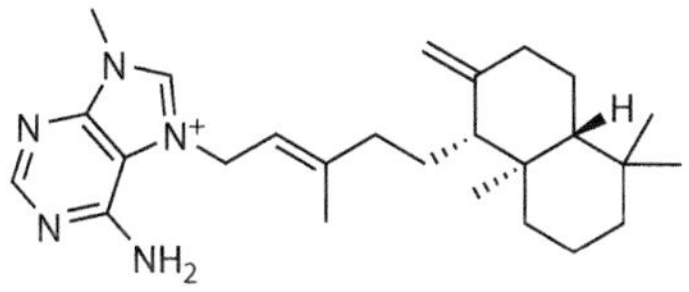

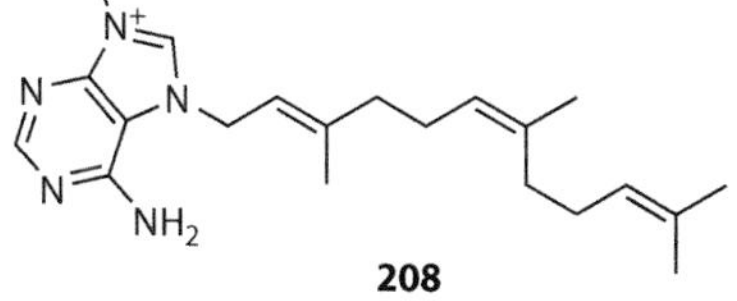

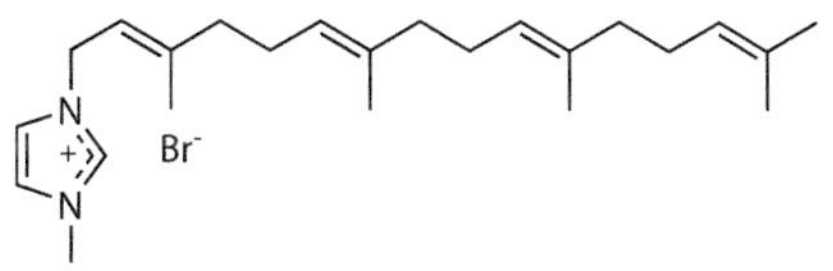

T. b. brucei, with IC$_{50}$ values of 3.4 and 4.7 μM, respectively.[291] The high-throughput screening of a natural product library (433 extracts and 428 pure compounds) out of Vietnam yielded three active compounds (**223–225**) of marine origin against the parasite *T. b. brucei*. The cembranoid laevigatol B (**223**)[292] extracted from the Vietmanese soft corals *Lobophytum cassum* and *L. laevigatum* displayed moderate antitrypanasomal activity, with an EC$_{50}$ value of 5.3 μM. Two steroids, astropectenol A (**224**)[293] and **225**, isolated from the echinoderms *Astropecten polyacanthus* and *Diadema*, resulted in EC$_{50}$ values of 1.6 and 14.6 μM, respectively.[294]

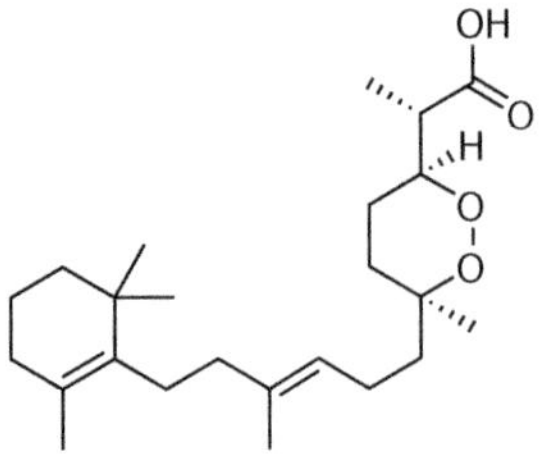

210

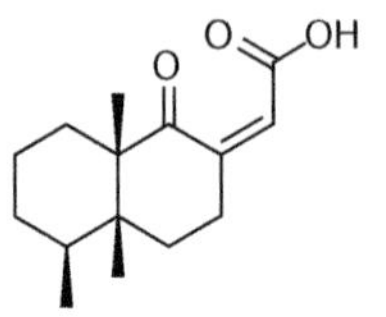

211

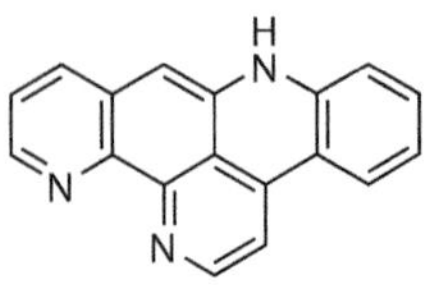

212

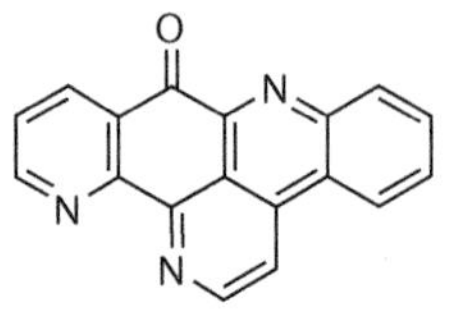

213

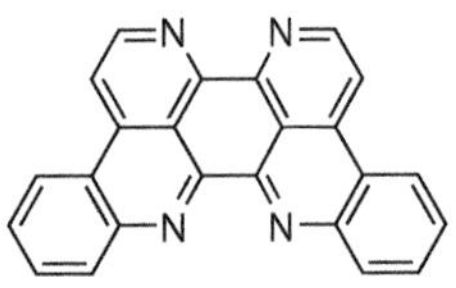

214

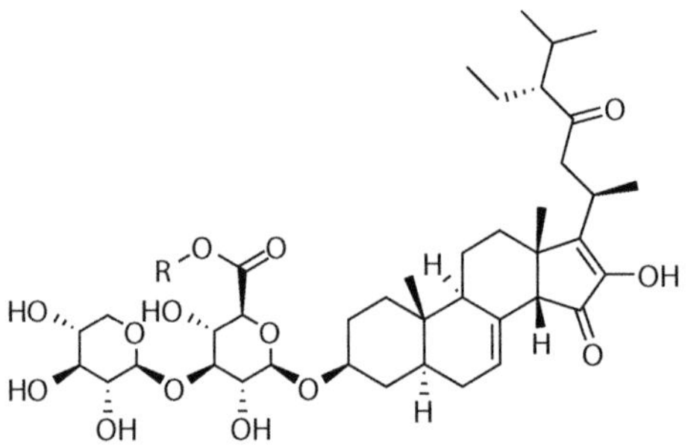

215 R = H
216 R = CH$_3$

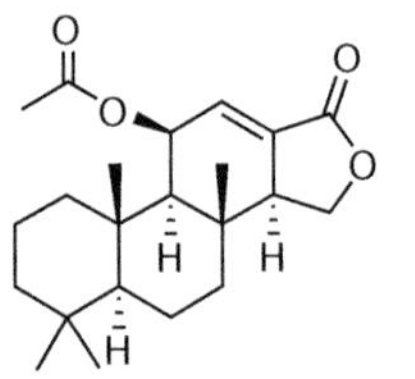

219

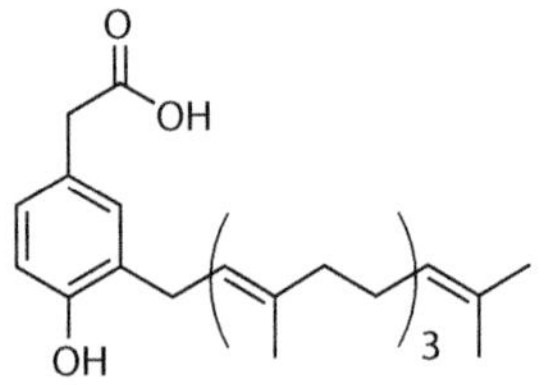

217

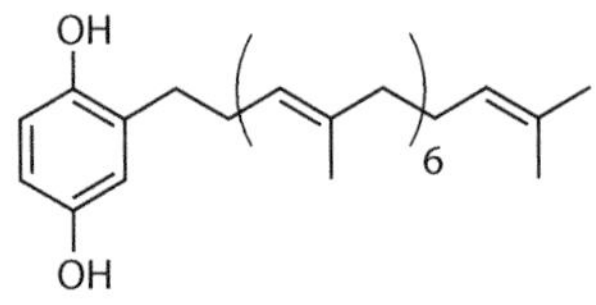

218

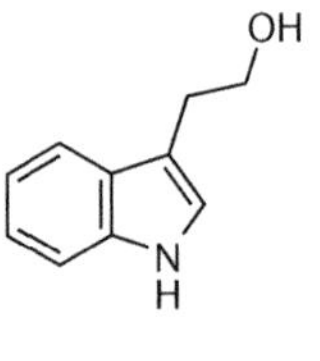

220

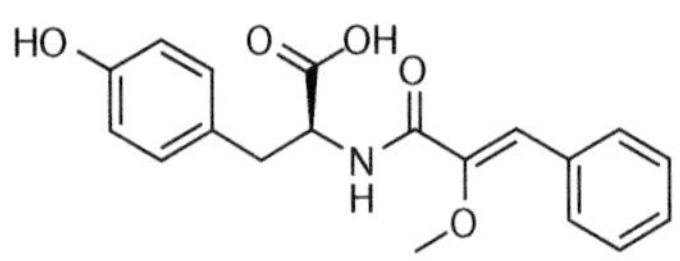

221

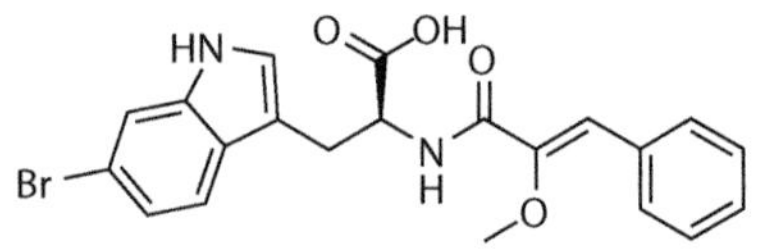

222

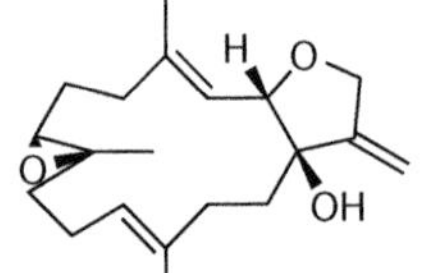

223

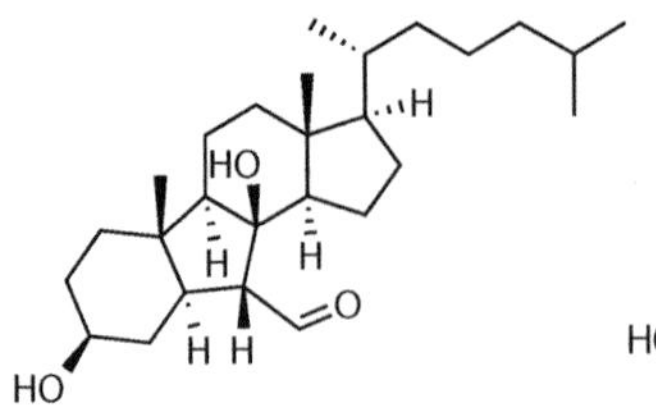

224

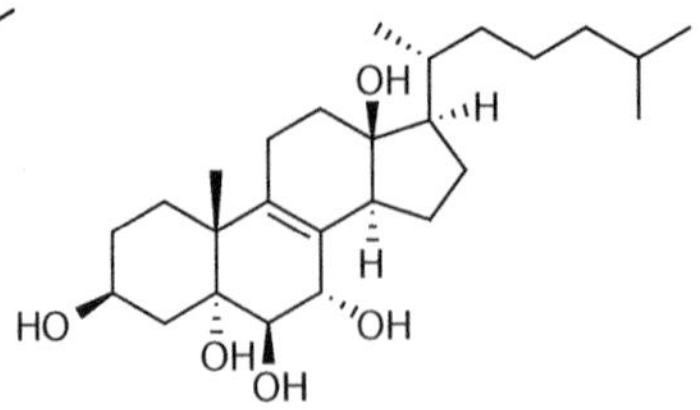

225

9.4.6 Amoeba

Currently, there are very few natural products reported in the literature screened for biological activity against amoebas. To the best of our knowledge, all of the marine isolated compounds active against amoebas are summarized here. The screening of the marine sponge metabolite **73** led to the development of a series of analogues, some of which proved active against the acanthamoebas *A. castellanii* and *A. polyphaga*.[295] The crude methanol extracts of the Malaysian sponge *Aoptos aaptos* collected from various islands were screened against the pathogenic *A. castellanii*. These crude extracts proved to be a promising lead, as their IC_{50} values ranged from 0.62 to 0.88 µg/ml.[296] Unfortunately, the active compounds were never isolated and screened and just assumed to be aaptamine-related compounds (**226–228**) previously isolated from *A. aaptos*.[297]

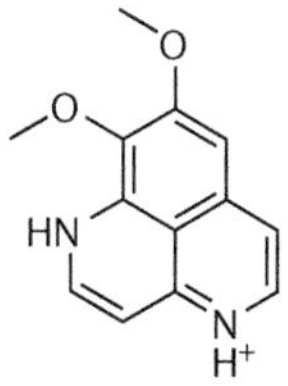

226 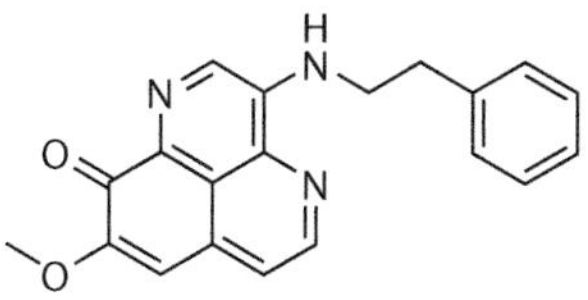   227 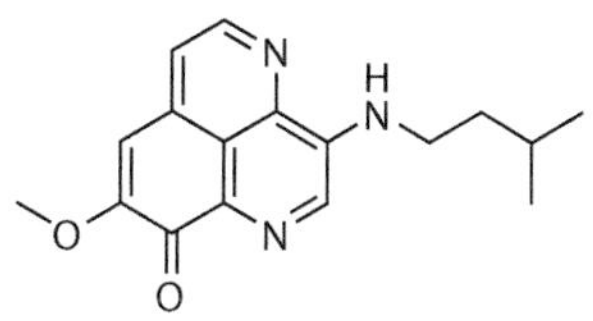   228

9.4.7 *Cryptosporidium*

Once again, very few screenings of marine natural products against *Cryptosporidium parvum* have been reported in the literature. However, one study, by Kiderlen et al.,[298] tested a series of aurones. These compounds were synthesized based on the natural products **229–231** isolated from the brown alga *Spatoglossum variabile*.[299,300] The most active analogue, **231**, exhibited a 76% growth inhibition at a concentration of 25 µM.

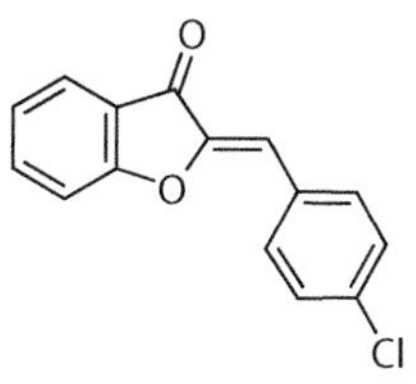

229 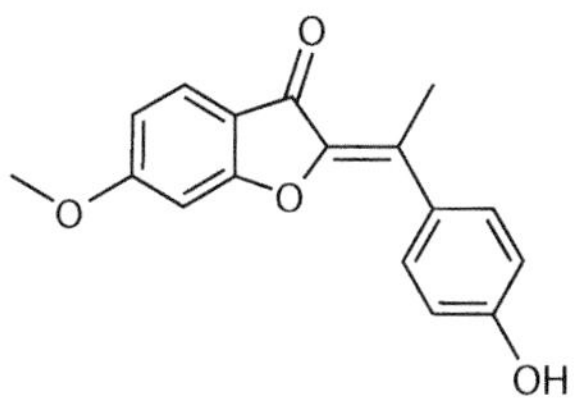   230 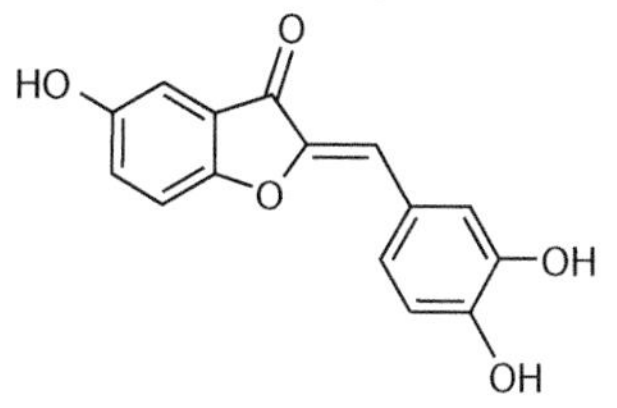   231

9.5 CONCLUSION

With more than 1 billion people at risk of infection and an annual death rate of more than 1 million people due to parasitic infections, it remains a moral obligation to develop new drugs to treat these diseases. Historically, the discovery and development of antiparasitic compounds from the natural realm has been successful, and therefore natural products still remain a reservoir of undiscovered drug candidates. The identification of new druggable targets requires that many previously isolated marine compounds should be revisited to gauge their ability to inhibit these newly discovered targets. With the range of unique scaffolds discussed in this chapter, marine natural products still represent a relevant source for drug discovery for a variety of parasites, with the hope that they may lead to the cure for many of these parasitic diseases.

REFERENCES

1. World Health Organization. *Monitoring and Evaluation of Preventative Chemotherapy.* World Health Organization, Geneva, 2013, pp. 17–28.
2. Foley, M., Tilley, L. Quinoline antimalarials: Mechanisms of action and resistance. *Int. J. Parasitol.* 1997, 27 (2), 231–240.
3. Ridley, R. G. Medical need, scientific opportunity and the drive for antimalarial drugs. *Nature* 2002, 415 (6872), 686–693.
4. Pasvol, G., Phil, D. Management of severe malaria: Interventions and controversies. *Infect. Dis. Clin. North Am.* 2005, 19 (1), 211–240.
5. Cowman, A. F., Foote, S. J. Chemotherapy and drug-resistance in malaria. *Int. J. Parasitol.* 1990, 20 (4), 503–513.
6. Schlitzer, M. Antimalarial drugs: What is in use and what is in the pipeline? *Arch. Pharm.* 2008, 341 (3), 149–163.
7. Loeb, R. F. Activity of a new antimalarial agent, pentaquine (Sn 13,276): Statement approved by the Board for Coordination of Malarial Studies. *JAMA* 1946, 132 (6), 321–323.
8. Weinke, T., Trautmann, M., Held, T., Weber, G., Eichenlaub, D., Fleischer, K., Kern, W., Pohle, H. D. Neuropsychiatric side-effects after the use of mefloquine. *Am. J. Trop. Med. Hyg.* 1991, 45 (1), 86–91.
9. Taylor, W. R. J., White, N. J. Antimalarial drug toxicity: A review. *Drug Saf.* 2004, 27 (1), 25–61.
10. Ezzet, F., van Vugt, M., Nosten, F., Looareesuwan, S., White, N. J. Pharmacokinetics and pharmaco-dynamics of lumefantrine (benflumetol) in acute falciparum malaria. *Antimicrob. Agents Chemother.* 2000, 44 (3), 697–704.
11. Winstanley, P. A. Chemotherapy for falciparum malaria: The armoury, the problems and the prospects. *Parasitol. Today* 2000, 16 (4), 146–153.
12. Kokwaro, G., Mwai, L., Nzila, A. Artemether/lumefantrine in the treatment of uncomplicated falci-parum malaria. *Exp. Opin. Pharmacother.* 2007, 8 (1), 75–94.
13. Foote, S. J., Cowman, A. F. The mode of action and the mechanism of resistance to antimalarial drugs. *Acta Trop.* 1994, 56 (2–3), 157–171.
14. Ramya, T. N. C., Mishra, S., Karmodiya, K., Surolia, N., Surolia, A. Inhibitors of nonhousekeeping func-tions of the apicoplast defy delayed death in *Plasmodium falciparum. Antimicrob. Agents Chemother.* 2007, 51 (1), 307–316.
15. Dahl, E. L., Shock, J. L., Shenai, B. R., Gut, J., DeRisi, J. L., Rosenthal, P. J. Tetracyclines specifically target the picoplast of the malaria parasite *Plasmodium falciparum. Antimicrob. Agents Chemother.* 2006, 50 (9), 3124–3131.
16. Ashley, E. A., White, N. J. Artemisinin-based combinations. *Curr. Opin. Infect. Dis.* 2005, 18 (6), 531–536.
17. Finlay, A. C., Hobby, G. L., Pan, S. Y., Regna, P. P., Routien, J. B., Seeley, D. B., Shull, G. M., Sobin, B. A., Solomons, I. A., Vinson, J. W., Kane, J. H. Terramycin, a new antibiotic. *Science* 1950, 111 (2874), 85.
18. Taylor, M. J., Makunde, W. H., McGarry, H. F., Turner, J. D., Mand, S., Hoerauf, A. Macrofilaricidal activity after doxycycline treatment of *Wuchereria bancrofti*: A double-blind, randomised placebo-controlled trial. *Lancet* 2005, 365 (9477), 2116–2121.
19. van Hellemond, J. J., Molhoek, N., Koelewijn, R., Wismans, P. J., van Genderen, P. J. J. Is paromomy-cin the drug of choice for eradication of *Dientamoeba fragilis* in adults? *Int. J. Parasitol. Drugs Drug Resist.* 2012, 2, 162–165.
20. Haskell, T. H., French, J. C., Bartz, Q. R. Paromomycin. 4. Structural Studies. *J. Am. Chem. Soc.* 1959, 81 (13), 3482–3483.
21. Haskell, T. H., French, J. C., Bartz, Q. R. Paromomycin. 3. The structure of Paromobiosamine. *J. Am. Chem. Soc.* 1959, 81 (13), 3481–3482.
22. Haskell, T. H., French, J. C., Bartz, Q. R. Paromomycin. 2. Paromobiosamine, a diaminohexosyl-D-ribose. *J. Am. Chem. Soc.* 1959, 81 (13), 3481–3481.
23. Haskell, T. H., French, J. C., Bartz, Q. R. Paromomycin. 1. Paromamine, a glycoside of D-glucosamine. *J. Am. Chem. Soc.* 1959, 81 (13), 3480–3481.

24. Oura, M., Sternberg, T. H., Wright, E. T. A new antifungal antibiotic, amphotericin B. *Antibiot. Annu.* 1955, 3, 566–573.
25. Burg, R. W., Miller, B. M., Baker, E. E., Birnbaum, J., Currie, S. A., Hartman, R., Kong, Y. L., Monaghan, R. L., Olson, G., Putter, I., Tunac, J. B., Wallick, H., Stapley, E. O., Oiwa, R., Omura, S. Avermectins, new family of potent anthelmintic agents: Producing organism and fermentation. *Antimicrob. Agents Chemother.* 1979, 15 (3), 361–367.
26. Herr, R. R., Slomp, G. Lincomycin. 2. Characterization and gross structure. *J. Am. Chem. Soc.* 1967, 89 (10), 2444–2447.
27. Hoeksema, H., Magerlein, B. J., Kagan, F., Herr, R. R., Slomp, G., Birkenmeyer, R. D., Bannister, B., Schroeder, W., Mackellar, F. A. Chemical studies on lincomycin. I. Structure of lincomycin. *J. Am. Chem. Soc.* 1964, 86 (19), 4223–4224.
28. Magerlei, B., Birkenme, R., Herr, R. R., Kagan, F. Lincomycin. V. Amino acid fragment. *J. Am. Chem. Soc.* 1967, 89 (10), 2459–2564.
29. Waksman, S. A., Lechevalier, H. A. Neomycin, a new antibiotic active against streptomycin-resistant bacteria, including tuberculosis organisms. *Science* 1949, 109 (2830), 305–307.
30. Campbell, W. C., Fisher, M. H., Stapley, E. O., Albersschonberg, G., Jacob, T. A. Ivermectin: A potent new anti-parasitic agent. *Science* 1983, 221 (4613), 823–828.
31. Yates, D. M., Wolstenholme, A. J. *Dirofilaria immitis*: Identification of a novel ligand-gated ion channel-related polypeptide. *Exp. Parasitol.* 2004, 108 (3–4), 182–185.
32. Magerlein, B. J. Modification of lincomycin. *Adv. Appl. Microbiol.* 1971, 14, 185–229.
33. Kremsner, P. G., Winkler, S., Brandts, C., Neifer, S., Bienzle, U., Graninger, W. Clindamycin in combination with chloroquine or quinine is an effective therapy for uncomplicated *Plasmodium-falciparum* malaria in children from Gabon. *J. Infect. Dis.* 1994, 169 (2), 467–470.
34. McGready, R., Cho, T., Villegas, S. L., Samuel, Villegas, L., Brockman, A., van Vugt, M., Looareesuwan, S., White, N. J., Nosten, F. Randomized comparison of quinine-clindamycin versus artesunate in the treatment of falciparum malaria in pregnancy. *Trans. R. Soc. Trop. Med. Hyg.* 2001, 95 (6), 651–656.
35. Berger, S. T., Mondino, B. J., Hoft, R. H., Donzis, P. B., Holland, G. N., Farley, M. K., Levenson, J. E. Successful medical-management of acanthamoeba keratitis. *Am. J. Ophthalmol.* 1990, 110 (4), 395–403.
36. Sensi, P., Timbal, M. T., Maffii, G. Rifomycin-Ix 2 new antibiotics of rifomycin family: Rifomycin-S and rifomycin-Sv. *Experientia* 1960, 16 (9), 412–412.
37. Lancini, G. C., Thiemann, J. E., Sartori, G., Sensi, P. Biogenesis of rifamycins. Conversion of rifamycin B into rifamycin Y. *Experientia* 1967, 23 (11), 899–900.
38. Vargas-Zepeda, J., Gomez-Alcala, A. V., Vazquez-Morales, J. A., Licea-Amaya, L., De Jonckheere, J. F., Lares-Villa, F. Successful treatment of *Naegleria fowleri* meningoencephalitis by using intravenous amphotericin B, fluconazole and rifampicin. *Arch. Med. Res.* 2005, 36 (1), 83–86.
39. Goswick, S. M., Brenner, G. M. Activities of azithromycin and amphotericin B against *Naegleria fowleri in vitro* and in a mouse model of primary amoebic meningoencephalitis. *Antimicrob. Agents Chemother.* 2003, 47 (2), 524–528.
40. Dionisio, D., Orsi, A., Sterrantino, G., Meli, M., Di Lollo, S., Manneschi, L. I., Trotta, M., Pozzi, M., Sani, L., Leoncini, F. Chronic cryptosporidiosis in patients with AIDS: Stable remission and possible eradication after long term, low dose azithromycin. *J. Clin. Pathol.* 1998, 51 (2), 138–142.
41. Vargas, S. L., Shenep, J. L., Flynn, P. M., Pui, C. H., Santana, V. M., Hughes, W. T. Azithromycin for treatment of severe *Cryptosporidium* diarrhea in 2 children with cancer. *Journal of Pediatrics* 1993, 123 (1), 154–156.
42. Robiquet, P. J. Expériences sur les cantharides. *Annales de Chimie* 1810, 76, 302–322.
43. Ghaffarifar, F. *Leishmania major*: *In vitro* and *in vivo* anti-leishmanial effect of cantharidin. *Exp. Parasitol.* 2010, 126 (2), 126–129.
44. Brusca, R. C., Brusca, G. J. *Invertebrates*. Sinauer Associates, Sunderland, MA, 1990.
45. DPDx Parasite Image Library, Centers for Disease Control, Atlanta, GA, 2014.
46. Centers for Disease Control. www.cdc.gov (accessed January 14, 2014).
47. World Health Organization. *Lymphatic Filariasis: Managing Morbidity and Preventing Disability*. World Health Organization, Geneva, 2013, pp. 1–53.

48. Gardon, J., GardonWendel, N., DemangaNgangue, D., Kamgno, J., Chippaux, J. P., Boussinesq, M. Serious reactions after mass treatment of onchocerciasis with ivermectin in an area endemic for *Loa loa* infection. *Lancet* 1997, 350 (9070), 18–22.

49. World Health Organization. *World Health Statistics 2014*. World Health Organization, Geneva, 2014, pp. 93–103.

50. World Health Organization. *World Malaria Report 2013*. World Health Organization, Geneva, 2013.

51. Sachs, J., Malaney, P. The economic and social burden of malaria. *Nature* 2002, 415 (6872), 680–685.

52. Breman, J. G. The ears of the hippopotamus: Manifestations, determinants, and estimates of the malaria burden. *Am. J. Trop. Med. Hyg.* 2001, 64 (1–2), 1–11.

53. Winstanley, P. Modern chemotherapeutic options for malaria. *Lancet Infect. Dis.* 2001, 1 (4), 242–50.

54. Kantele, A., Jokiranta, T. S. Review of cases with the emerging fifth human malaria parasite, *Plasmodium knowlesi*. *Clin. Infect. Dis.* 2011, 52 (11), 1356–1362.

55. King, C. A. Cell motility of sporozoan protozoa. *Parasitol. Today* 1988, 4 (11), 315–319.

56. Sidjanski, S., Vanderberg, J. P. Delayed migration of *Plasmodium* sporozoites from the mosquito bite site to the blood. *Am. J. Trop. Med. Hyg.* 1997, 57 (4), 426–429.

57. Meis, J. F., Verhave, J. P., Jap, P. H., Meuwissen, J. H. Transformation of sporozoites of *Plasmodium berghei* into exoerythrocytic forms in the liver of its mammalian host. *Cell Tissue Res.* 1985, 241 (2), 353–360.

58. Muhindo, M. K., Kakuru, A., Jagannathan, P., Talisuna, A., Osilo, E., Orukan, F., Arinaitwe, E., Tappero, J. W., Kaharuza, F., Kamya, M. R., Dorsey, G. Early parasite clearance following artemisinin-based combination therapy among Ugandan children with uncomplicated *Plasmodium falciparum* malaria. *Malar. J.* 2014, 13, 32.

59. Dondorp, A. M., Nosten, F., Yi, P., Das, D., Phyo, A. P., Tarning, J., Lwin, K. M., Ariey, F., Hanpithakpong, W., Lee, S. J., Ringwald, P., Silamut, K., Imwong, M., Chotivanich, K., Lim, P., Herdman, T., An, S. S., Yeung, S., Singhasivanon, P., Day, N. P. J., Lindegardh, N., Socheat, D., White, N. J. Artemisinin resistance in *Plasmodium falciparum* malaria. *New Engl. J. Med.* 2009, 361 (5), 455–467.

60. Yoder, J. S., Eddy, B. A., Visvesvara, G. S., Capewell, L., Beach, M. J. The epidemiology of primary amoebic meningoencephalitis in the USA, 1962–2008. *Epidemiol. Infect.* 2010, 138 (7), 968–975.

61. Schuster, F. L., Guglielmo, B. J., Visvesvara, G. S. In-vitro activity of miltefosine and voriconazole on clinical isolates of free-living amoebas: *Balamuthia mandrillaris*, *Acanthamoeba* spp., and *Naegleria fowleri*. *J. Eukaryot. Microbiol.* 2006, 53 (2), 121–126.

62. Kim, J. H., Jung, S. Y., Lee, Y. J., Song, K. J., Kwon, D., Kim, K., Park, S., Im, K. I., Shin, H. J. Effect of therapeutic chemical agents *in vitro* and on experimental meningoencephalitis due to *Naegleria fowleri*. *Antimicrob. Agents Chemother.* 2008, 52 (11), 4010–4016.

63. Upcroft, P., Upcroft, J. A. Drug targets and mechanisms of resistance in the anaerobic protozoa. *Clin. Microbiol. Rev.* 2001, 14 (1), 150–164.

64. Sangster, N., Batterham, P., Chapman, H. D., Duraisingh, M., Le Jambre, L., Shirley, M., Upcroft, J., Upcroft, P. Resistance to antiparasitic drugs: The role of molecular diagnosis. *Int. J. Parasitol.* 2002, 32 (5), 637–653.

65. Dunne, R. L., Dunn, L. A., Upcroft, P., O'Donoghue, P. J., Upcroft, J. A. Drug resistance in the sexually transmitted protozoan *Trichomonas vaginalis*. *Cell Res.* 2003, 13 (4), 239–249.

66. Schneider, G. Virtual screening: An endless staircase? *Nat. Rev. Drug Discov.* 2010, 9 (4), 273–276.

67. Stahl, M., Rarey, M. Detailed analysis of scoring functions for virtual screening. *J. Med. Chem.* 2001, 44 (7), 1035–1042.

68. Clardy, J., Walsh, C. Lessons from natural molecules. *Nature* 2004, 432 (7019), 829–837.

69. Lipinski, C. A., Lombardo, F., Dominy, B. W., Feeney, P. J. Experimental and computational approaches to estimate solubility and permeability in drug discovery and development settings. *Adv. Drug Del. Rev.* 1997, 23 (1–3), 3–25.

70. Venkatesan, S. K., Saudagar, P., Shukla, A. K., Dubey, V. K. Screening natural products database for identification of potential antileishmanial chemotherapeutic agents. *Interdiscip. Sci. Comput. Life Sci.* 2011, 3 (3), 217–231.

71. Ogungbe, I. V., Setzer, W. N. Comparative molecular docking of antitrypanosomal natural products into multiple *Trypanosoma brucei* drug targets. *Molecules* 2009, 14 (4), 1513–1536.

72. Srivastava, M., Singh, H., Naik, P. K. Molecular modeling evaluation of the antimalarial activity of artemisinin analogues: Molecular docking and rescoring using prime/MM-GBSA approach. *Curr. Res. J. Biol. Sci.* 2010, 2 (2), 83–102.

73. Rollinger, J. M., Stuppner, H., Langer, T. Virtual screening for the discovery of bioactive natural products. In *Progress in Drug Research*, vol. 65: *Natural Compounds as Drugs*, ed. F. Petersen, R. Amstutz. 1st ed. Birkhäuser Verlag, Basel, 2008, pp. 211–249.

74. Geary, T. G., Mackenzie, C. D. Progress and challenges in the discovery of macrofilaricidal drugs. *Expert Rev. Anti-Infect. Ther.* 2011, 9 (8), 681–695.

75. Burns, A. R., Wallace, I. M., Wildenhain, J., Tyers, M., Giaever, G., Bader, G. D., Nislow, C., Cutler, S. R., Roy, P. J. A predictive model for drug bioaccumulation and bioactivity in *Caenorhabditis elegans*. *Nat. Chem. Biol.* 2010, 6 (7), 549–557.

76. Ruiz-Lancheros, E., Viau, C., Walter, T. N., Francis, A., Geary, T. G. Activity of novel nicotinic anthelmintics in cut preparations of *Caenorhabditis elegans*. *Int. J. Parasitol.* 2011, 41 (3–4), 455–461.

77. Klein, R. D., Geary, T. G. Recombinant microorganisms as tools for high throughput screening for non-antibiotic compounds. *J. Biomol. Screen.* 1997, 2 (1), 41–49.

78. Graf, B., Rojas-Silva, P., Ayeni, A., Raskin, I. The Global Institute for BioExploration (GIBEX): Building research and development capacity worldwide. *Planta Med.* 2012, 78 (11), 1124.

79. Junio, H. A., Sy-Cordero, A. A., Ettefagh, K. A., Burns, J. T., Micko, K. T., Graf, T. N., Richter, S. J., Cannon, R. E., Oberlies, N. H., Cech, N. B. Synergy-directed fractionation of botanical medicines: A case study with goldenseal (*Hydrastis canadensis*). *J. Nat. Prod.* 2011, 74 (7), 1621–1629.

80. Shivkar, Y. M., Kumar, V. Anthelmintic activity of latex of *Calotropis procera*. *Pharm. Biol.* 2003, 41 (4), 263–265.

81. Jain, M. L., Jain, S. R. Therapeutic utility of *Ocimum-basilicum* var *album*. *Planta Med.* 1972, 22 (1), 66–70.

82. Sollmann, T. Anthelmintics: Their efficiency as tested on earthworms. *J. Pharmacol. Exp. Ther.* 1918, 12 (3), 129–170.

83. Geary, T. G., Thompson, D. P. *Caenorhabditis elegans*: How good a model for veterinary parasites? *Vet. Parasitol.* 2001, 101 (3–4), 371–386.

84. Li, C. The ever-expanding neuropeptide gene families in the nematode *Caenorhabditis elegans*. *Parasitology* 2005, 131, S109–S127.

85. Angstadt, J. D., Donmoyer, J. E., Stretton, A. O. W. Retrovesicular ganglion of the nematode ascaris. *J. Comp. Neurol.* 1989, 284 (3), 374–388.

86. Geary, T. G., Woods, D. J., Williams, T., Nwaka, S. Target identification and mechanism-based screening for anthelmintics: Application of veterinary antiparasitic research programs to search for new antiparasitic drugs for human indications. In *Drug Discovery in Infectious Diseases*, ed. P. M. Selzer. Wiley-VCH, Weinheim, Germany, 2009, pp. 1–16.

87. Kron, M., Marquard, K., Hartlein, M., Price, S., Leberman, R. An immunodominant antigen of *Brugia-malayi* is an asparaginyl-transfer-RNA synthetase. *FEBS Lett.* 1995, 374 (1), 122–124.

88. Nilsen, T. W., Maroney, P. A., Goodwin, R. G., Perrine, K. G., Denker, J. A., Nanduri, J., Kazura, J. W. Cloning and characterization of a potentially protective antigen in lymphatic filariasis. *Proc. Natl. Acad. Sci. U.S.A.* 1988, 85 (10), 3604–3607.

89. Danel, F., Caspers, P., Nuoffer, C., Hartlein, M., Kron, M. A., Page, M. G. Asparaginyl-tRNA synthetase pre-transfer editing assay. *Curr. Drug Discov. Technol.* 2010, 8, 66–75.

90. Cromley, J. C. W., Rees, M. J., Turner, C. H., Jenkins, D. C. Colorimetric quantitation of filarial viability. *Int. J. Parasitol.* 1989, 19, 77–83.

91. Tada, H., Shiho, O., Kuroshima, K., Koyama, M., Tsukamoto, K. An improved colorimetric assay for interleukin-2. J. *Immunol. Methods* 1986, 93 (2), 157–165.

92. Sereno, D., Lemesre, J. L. Axenically cultured amastigote forms as an *in vitro* model for investigation of antileishmanial agents. *Antimicrob. Agents Chemother.* 1997, 41 (5), 972–976.

93. Dutta, A., Bandyopadhyay, S., Mandal, C., Chatterjee, M. Development of a modified MTT assay for screening antimonial resistant field isolates of Indian visceral leishmaniasis. *Parasitol. Int.* 2005, 54 (2), 119–122.

94. Mikus, J., Steverding, D. A. A simple colorimetric method to screen drug cytotoxicity against *Leishmania* using the dye Alamar Blue. *Parasitol. Int.* 2000, 48, 265–269.

95. Davis, A. J., Murray, H. W., Handman, E. Drugs against leishmaniasis: A synergy of technology and partnerships. *Trends Parasitol.* 2004, 20 (2), 73–76.

96. Davis, A. J., Perugini, M. A., Smith, B. J., Stewart, J. D., Ilg, T., Hodder, A. N., Handman, E. Properties of GDP-mannose pyrophosphorylase, a critical enzyme and drug target in *Leishmania mexicana*. *J. Biol. Chem.* 2004, 279 (13), 12462–12468.

97. Moreno, M. A., Abramov, A., Abendroth, J., Alonso, A., Zhang, S., Alcolea, P. J., Edwards, T., Lorimer, D., Myler, P. J., Larraga, V. Structure of tyrosine aminotransferase from *Leishmania infantum*. *Acta Crystallogr. F Struct. Biol. Commun.* 2014, 70, 583–587.

98. Efstathiou, A., Gaboriaud-Kolar, N., Smirlis, D., Myrianthopoulos, V., Vougogiannopoulou, K., Alexandratos, A., Kritsanida, M., Mikros, E., Soteriadou, K., Skaltsounis, A. L. An inhibitor-driven study for enhancing the selectivity of indirubin derivatives towards leishmanial glycogen synthase kinase-3 over leishmanial cdc2-related protein kinase 3. *Parasit. Vectors* 2014, 7, 234.

99. Schmidt, A., Krauth-Siegel, R. Enzymes of the trypanothione metabolism as targets for antitrypanosomal drug development. *Curr. Top. Med. Chem.* 2002, 2 (11), 1239–1259.

100. Nicoll-Griffith, D. A. Use of cysteine-reactive small molecules in drug discovery for trypanosomal disease. *Expert Opin. Drug Discov.* 2012, 7 (4), 353–366.

101. Sajid, M., McKerrow, J. H. Cysteine proteases of parasitic organisms. *Mol. Biochem. Parasitol.* 2002, 120 (1), 1–21.

102. Lepesheva, G. I., Waterman, M. R. Sterol 14alpha-demethylase (CYP51) as a therapeutic target for human trypanosomiasis and leishmaniasis. *Curr. Top. Med. Chem.* 2011, 11 (16), 2060–2071.

103. Sharma, M., Chauhan, P. M. S. Dihydrofolate reductase as a therapeutic target for infectious diseases: Opportunities and challenges. *Future Med. Chem.* 2012, 4 (10), 1335–1365.

104. Birkholtz, L. M., Williams, M., Niemand, J., Louw, A. I., Persson, L., Heby, O. Polyamine homoeostasis as a drug target in pathogenic protozoa: Peculiarities and possibilities. *Biochem. J.* 2011, 438, 229–244.

105. Colotti, G., Ilari, A. Polyamine metabolism in *Leishmania*: From arginine to trypanothione. *Amino Acids* 2011, 40 (2), 269–285.

106. Chatterji, B. P., Jindal, B., Srivastava, S., Panda, D. Microtubules as antifungal and antiparasitic drug targets. *Expert Opin. Ther. Pat.* 2011, 21 (2), 167–186.

107. Zeeman, A. M., der Wel, A. V., Kocken, C. H. *Ex vivo* culture of *Plasmodium vivax* and *Plasmodium cynomolgi* and *in vitro* culture of *Plasmodium knowlesi* blood stages. *Methods Mol. Biol.* 2013, 923, 35–49.

108. Desjardins, R. E., Canfield, C. J., Haynes, J. D., Chulay, J. D. Quantitative assessment of antimalarial activity *in vitro* by a semiautomated microdilution technique. *Antimicrob. Agents Chemother.* 1979, 16 (6), 710–718.

109. Manandhar, M. S. P., Vandyke, K. Detailed purine salvage metabolism in and outside free malarial parasite. *Exp. Parasitol.* 1975, 37 (2), 138–146.

110. Lebar, M. D., Hahn, K. N., Mutka, T., Maignan, P., McClintock, J. B., Amsler, C. D., van Olphen, A., Kyle, D. E., Baker, B. J. CNS and antimalarial activity of synthetic meridianin and psammopemmin analogs. *Bioorg. Med. Chem.* 2011, 19 (19), 5756–5762.

111. Vaughan, A. M., Mikolajczak, S. A., Camargo, N., Lakshmanan, V., Kennedy, M., Lindner, S. E., Miller, J. L., Hume, J. C., Kappe, S. H. A transgenic *Plasmodium falciparum* NF54 strain that expresses GFP-luciferase throughout the parasite life cycle. *Mol. Biochem. Parasitol.* 2012, 186 (2), 143–147.

112. Fidock, D. A., Rosenthal, P. J., Croft, S. L., Brun, R., Nwaka, S. Antimalarial drug discovery: Efficacy models for compound screening. *Nat. Rev. Drug Discov.* 2004, 3 (6), 509–520.

113. Slater, A. F. G., Cerami, A. Inhibition by chloroquine of a novel heme polymerase enzyme-activity in malaria trophozoites. *Nature* 1992, 355 (6356), 167–169.

114. Kurosawa, Y., Dorn, A., Kitsuji-Shirane, M., Shimada, H., Satoh, T., Matile, H., Hofheinz, W., Masciadri, R., Kansy, M., Ridley, R. G. Hematin polymerization assay as a high-throughput screen for identification of new antimalarial pharmacophores. *Antimicrob. Agents Chemother.* 2000, 44 (10), 2638–2644.

115. Ambele, M. A., Egan, T. J. Neutral lipids associated with haemozoin mediate efficient and rapid beta-haematin formation at physiological pH, temperature and ionic composition. *Malar. J.* 2012, 11.

116. Egan, T. J., Ross, D. C., Adams, P. A. Quinoline antimalarial-drugs inhibit spontaneous formation of beta-hematin (malaria pigment). *FEBS Lett.* 1994, 352 (1), 54–57.

117. Ncokazi, K. K., Egan, T. J. A colorimetric high-throughput beta-hematin inhibition screening assay for use in the search for antimalarial compounds. *Anal. Biochem.* 2005, 338 (2), 306–319.

118. Parapini, S., Basilico, N., Pasini, E., Egan, T. J., Olliaro, P., Taramelli, D., Monti, D. Standardization of the physicochemical parameters to assess *in vitro* the beta-hematin inhibitory activity of antimalarial drugs. *Exp. Parasitol.* 2000, 96 (4), 249–256.

119. Gorka, A. P., Alumasa, J. N., Sherlach, K. S., Jacobs, L. M., Nickley, K. B., Brower, J. P., de Dios, A. C., Roepe, P. D. Cytostatic versus cytocidal activities of chloroquine analogues and inhibition of hemozoin crystal growth. *Antimicrob. Agents Chemother.* 2013, 57 (1), 356–364.

120. Sullivan, A. D., Ittarat, I., Meshnick, S. R. Patterns of haemozoin accumulation in tissue. *Parasitology* 1996, 112, 285–294.

121. Berger, B. J., Bendrat, K., Cerami, A. High-performance liquid-chromatographic analysis of biological and chemical heme polymerization. *Anal. Biochem.* 1995, 231 (1), 151–156.

122. Njunge, J. M., Ludewig, M. H., Boshoff, A., Pesce, E. R., Blatch, G. L. Hsp70s and J proteins of *Plasmodium* parasites infecting rodents and primates: Structure, function, clinical relevance, and drug targets. *Curr. Pharm. Des.* 2013, 19 (3), 387–403.

123. Pesce, E. R., Cockburn, I. L., Goble, J. L., Stephens, L. L., Blatch, G. L. Malaria heat shock proteins: Drug targets that chaperone other drug targets. *Infect. Disord. Drug Targets* 2010, 10 (3), 147–157.

124. Cockburn, I. L., Boshoff, A., Pesce, E. R., Blatch, G. L. Selective modulation of plasmodial Hsp70s by small molecules with antimalarial activity. *Biol. Chem.* 2014, 395 (11), 1353–1362.

125. Cockburn, I. L., Pesce, E. R., Pryzborski, J. M., Davies-Coleman, M. T., Clark, P. G. K., Keyzers, R. A., Stephens, L. L., Blatch, G. L. Screening for small molecule modulators of Hsp70 chaperone activity using protein aggregation suppression assays: Inhibition of the plasmodial chaperone PfHsp70-1. *Biol. Chem.* 2011, 392 (5), 431–438.

126. Pesce, E. R., Blatch, G. L. Plasmodial Hsp40 and Hsp70 chaperones: Current and future perspectives. *Parasitology* 2014, 141 (9), 1167–1176.

127. Proteau, P. J. 1-Deoxy-D-xylulose 5-phosphate reductoisomerase: An overview. *Bioorg. Chem.* 2004, 32 (6), 483–493.

128. Takahashi, S., Kuzuyama, T., Watanabe, H., Seto, H. A 1-deoxy-D-xylulose 5-phosphate reductoisomerase catalyzing the formation of 2-C-methyl-D-erythritol 4-phosphate in an alternative nonmevalonate pathway for terpenoid biosynthesis. *Proc. Natl. Acad. Sci. U.S.A.* 1998, 95 (17), 9879–9884.

129. Okuhara, M., Kuroda, Y., Goto, T., Okamoto, M., Terano, H., Kohsaka, M., Aoki, H., Imanaka, H. Studies on new phosphonic acid antibiotics. III. Isolation and characterization of FR-31564, FR-32863 and FR-33289. *J. Antibiot.* (Tokyo) 1980, 33 (1), 24–28.

130. Argyrou, A., Blanchard, J. S. Kinetic and chemical mechanism of *Mycobacterium tuberculosis* 1-deoxy-D-xylulose-5-phosphate isomeroreductase. *Biochemistry* 2004, 43 (14), 4375–4384.

131. Olliaro, P. Mode of action and mechanisms of resistance for antimalarial drugs. *Pharmacol. Ther.* 2001, 89 (2), 207–219.

132. Hudson, A. T. Atovaquone: A novel broad-spectrum anti-infective drug. *Parasitol. Today* 1993, 9 (2), 66–68.

133. Krungkrai, J. Purification, characterization and localization of mitochondrial dihydroorotate dehydrogenase in *Plasmodium falciparum*, human malaria parasite. *Biochim. Biophys. Acta* 1995, 1243 (3), 351–360.

134. Vaidya, A. B., Lashgari, M. S., Pologe, L. G., Morrisey, J. Structural features of *Plasmodium* cytochrome b that may underlie susceptibility to 8-aminoquinolines and hydroxynaphthoquinones. *Mol. Biochem. Parasitol.* 1993, 58 (1), 33–42.

135. Ziegler, H. L., Staerk, D., Christensen, J., Hviid, L., Hagerstrand, H., Jaroszewski, J. W. *In vitro Plasmodium falciparum* drug sensitivity assay: Inhibition of parasite growth by incorporation of stomatocytogenic amphiphiles into the erythrocyte membrane. *Antimicrob. Agents Chemother.* 2002, 46 (5), 1441–1446.

136. McFadden, D. C., Seeber, F., Boothroyd, J. C. Use of *Toxoplasma gondii* expressing beta-galactosidase for colorimetric assessment of drug activity *in vitro*. *Antimicrob. Agents Chemother.* 1997, 41 (9), 1849–1853.

137. Gubbels, M. J., Li, C., Striepen, B. High-throughput growth assay for *Toxoplasma gondii* using yellow fluorescent protein. *Antimicrob. Agents Chemother.* 2003, 47 (1), 309–316.

138. Hirumi, H., Hirumi, K. Continuous cultivation of *Trypanosoma brucei* blood stream forms in a medium containing a low concentration of serum protein without feeder cell layers. *J. Parasitol.* 1989, 75 (6), 985–989.

139. Raz, B., Iten, M., GretherBuhler, Y., Kaminsky, R., Brun, R. The Alamar Blue® assay to determine drug sensitivity of African trypanosomes (*T-b-rhodesiense* and *T-b-gambiense*) in vitro. *Acta Trop.* 1997, 68 (2), 139–147.

140. Buckner, F. S., Verlinde, C. L. M. J., LaFlamme, A. C., vanVoorhis, W. C. Efficient technique for screening drugs for activity against *Trypanosoma cruzi* using parasites expressing beta-galactosidase. *Antimicrob. Agents Chemother.* 1996, 40 (11), 2592–2597.

141. Engel, J. C., Ang, K. K. H., Chen, S., Arkin, M. R., McKerrow, J. H., Doyle, P. S. Image-based high-throughput drug screening targeting the intracellular stage of *Trypanosoma cruzi*, the agent of Chagas' disease. *Antimicrob. Agents Chemother.* 2010, 54 (8), 3326–3334.

142. Jacobs, R. T., Ding, C. Recent advances in drug discovery for neglected tropical diseases caused by infective kinetoplastid parasites. *Annu. Rep. Med. Chem.* 2010, 45, 277–294.

143. Connor, C. G., Hopkins, S. L., Salisbury, R. D. Effectivity of contact-lens disinfection systems against *Acanthamoeba-culbertsoni*. *Optom. Vis. Sci.* 1991, 68 (2), 138–141.

144. Khunkitti, W., Avery, S. V., Lloyd, D., Furr, J. R., Russell, A. D. Effects of biocides on *Acanthamoeba castellanii* as measured by flow cytometry and plaque assay. *J. Antimicrob. Chemother.* 1997, 40 (2), 227–233.

145. Beattie, T. K., Seal, D. V., Tomlinson, A., McFadyen, A. K., Grimason, A. M. Determination of amoebicidal activities of multipurpose contact lens solutions by using a most probable number enumeration technique. *J. Clin. Microbiol.* 2003, 41 (7), 2992–3000.

146. Henriquez, F. L., Ingram, P. R., Muench, S. P., Rice, D. W., Roberts, C. W. Molecular basis for resistance of *Acanthamoeba tubulins* to all major classes of antitubulin compounds. *Antimicrob. Agents Chemother.* 2008, 52 (3), 1133–1135.

147. McBride, J., Mullen, A. B., Carter, K. C., Roberts, C. W. Differential cytotoxicity of phospholipid analogues to pathogenic *Acanthamoeba* species and mammalian cells. *J. Antimicrob. Chemother.* 2007, 60 (3), 521–525.

148. Limouze, J., Straight, A. F., Mitchison, T., Sellers, J. E. Specificity of blebbistatin, an inhibitor of myosin II. *J. Muscle Res. Cell Motil.* 2004, 25 (4–5), 337–341.

149. Kim, B. G., Sobota, A., Bitonti, A. J., Mccann, P. P., Byers, T. J. Polyamine metabolism in acanthamoeba: Polyamine content and synthesis of ornithine, putrescine, and diaminopropane. *J. Protozool.* 1987, 34 (3), 278–284.

150. Henriquez, F. L., McBride, J., Campbell, S. J., Ramos, T., Ingram, P. R., Roberts, F., Tinney, S., Roberts, C. W. Acanthamoeba alternative oxidase genes: Identification, characterisation and potential as antimicrobial targets. *Int. J. Parasitol.* 2009, 39 (13), 1417–1424.

151. Crews, P., Manes, L. V., Boehler, M. Jasplakinolide, A cyclodepsipeptide from the marine sponge, *Jaspis* sp. *Tetrahedron Lett.* 1986, 27 (25), 2797–2800.

152. Capon, R. J., Barrow, R. A., Rochfort, S., Jobling, M., Skene, C., Lacey, E., Gill, J. H., Friedel, T., Wadsworth, D. Marine nematocides: Tetrahydrofurans from a southern Australian brown alga, *Notheia anomala*. *Tetrahedron* 1998, 54 (10), 2227–2242.

153. Ovenden, S. P. B., Capon, R. J., Lacey, E., Gill, J. H., Friedel, T., Wadsworth, D. Amphilactams A–D: Novel nematocides from southern Australian marine sponges of the genus *Amphimedon*. *J. Org. Chem.* 1999, 64 (4), 1140–1144.

154. Capon, R. J., Skene, C., Lacey, E., Gill, J. H., Wadsworth, D., Friedel, T. Geodin A magnesium salt: A novel nematocide from a southern Australian marine sponge, *Geodia*. *J. Nat. Prod.* 1999, 62 (9), 1256–1259.

155. Omura, S., Miyadera, H., Ui, H., Shiomi, K., Yamaguchi, Y., Masuma, R., Nagamitsu, T., Takano, D., Sunazuka, T., Harder, A., Kolbl, H., Namikoshi, M., Miyoshi, H., Sakamoto, K., Kita, K. An anthelmintic compound, nafuredin, shows selective inhibition of complex I in helminth mitochondria. *Proc. Natl. Acad. Sci. U.S.A.* 2001, 98 (1), 60–62.

156. Takano, D., Nagamitsu, T., Ui, H., Shiomi, K., Yamaguchi, Y., Masuma, R., Kuwajima, I., Omura, S. Total synthesis of nafuredin, a selective NADH-fumarate reductase inhibitor. *Org. Lett.* 2001, 3 (15), 2289–2291.

157. Vuong, D., Capon, R. J., Lacey, E., Gill, J. H., Heiland, K., Friedel, T. Onnamide F: A new nematocide from a southern Australian marine sponge, *Trachycladus laevispirulifer. J. Nat. Prod.* 2001, 64 (5), 640–642.

158. Capon, R. J., Skene, C., Liu, E. H. T., Lacey, E., Gill, J. H., Heiland, K., Friedel, T. Nematocidal thiocyanatins from a southern Australian marine sponge *Oceanapia* sp. *J. Nat. Prod.* 2004, 67 (8), 1277–1282.

159. Cros, F., Pelotier, B., Piva, O. Microwave-assisted cross-metathesis of unsaturated thiocyanates: Application to the synthesis of thiocyanatins A and B and analogues. *Synthesis* (Stuttgart) 2010, (2), 233–238.

160. Singh, A., Sharma, M. L., Singh, J. New synthesis of nematocidal natural products dithiocynates thiocyanatin A and 1,8,16-trihydroxyhexadecane. *Nat. Prod. Res.* 2009, 23 (11), 1029–1034.

161. Davyt, D., Fernandez, R., Suescun, L., Mombru, A. W., Saldana, J., Dominguez, L., Coll, J., Fujii, M. T., Manta, E. New sesquiterpene derivatives from the red alga *Laurencia scoparia*. Isolation, structure determination, and anthelmintic activity. *J. Nat. Prod.* 2001, 64 (12), 1552–1555.

162. Davyt, D., Fernandez, R., Suescun, L., Mombru, A. W., Saldana, J., Dominguez, L., Fujii, M. T., Manta, E. Bisabolanes from the red alga *Laurencia scoparia. J. Nat. Prod.* 2006, 69 (7), 1113–1116.

163. Juagdan, E. G., Kalidindi, R., Scheuer, P. Two new chamigranes from an Hawaiian red alga, *Laurencia cartilaginea. Tetrahedron* 1997, 53 (2), 521–528.

164. Rungprom, W., Chavasiri, W., Kokpol, U., Kotze, A., Garson, M. J. Bioactive chromodorolide diterpenes from an Aplysillid sponge. *Mar. Drugs* 2004, 2 (3), 101–107.

165. Franks, A., Haywood, P., Holmstrom, C., Egan, S., Kjelleberg, S., Kumar, N. Isolation and structure elucidation of a novel yellow pigment from the marine bacterium *Pseudoalteromonas tunicata. Molecules* 2005, 10 (10), 1286–1291.

166. Ballestriero, F., Thomas, T., Burke, C., Egan, S., Kjelleberg, S. Identification of compounds with bioactivity against the nematode *Caenorhabditis elegans* by a screen based on the functional genomics of the marine bacterium *Pseudoalteromonas tunicata* D2. *Appl. Environ. Microbiol.* 2010, 76 (17), 5710–5717.

167. Rungprorn, W., Siwu, E. R. O., Lambert, L. K., Dechsakulwatana, C., Barden, M. C., Kokpol, U., Blanchfield, J. T., Kita, M., Garson, M. J. Cyclic tetrapeptides from marine bacteria associated with the seaweed *Diginea* sp and the sponge *Halisarca ectofibrosa. Tetrahedron* 2008, 64 (14), 3147–3152.

168. Rajiv, D., Hemendra, G. Synthesis and pharmacological studies on a cyclooligopeptide from marine bacteria. *Chin. J. Chem.* 2011, 29 (9), 1911–1916.

169. Rao, C. B., Anjaneyula, A. S. R., Sarma, N. S., Venkatateswarlu, Y., Rosser, R. M., Faulkner, D. J., Chen, M. H. M., Clardy, J. Zoanthamine: A novel alkaloid from a marine zoanthid. *J. Am. Chem. Soc.* 1984, 106 (25), 7983–7984.

170. Giner, J. L., Djerassi, C. Biosynthetic-studies of marine lipids. 33. Biosynthesis of dinosterol, peridinosterol, and gorgosterol: Unusual patterns of bioalkylation in dinoflagellate sterols. *J. Org. Chem.* 1991, 56 (7), 2357–2363.

171. Lakshmi, V., Saxena, A., Pandey, K., Bajpai, P., Misra-Bhattacharya, S. Antifilarial activity of *Zoanthus* species (phylum Coelenterata, class Anthzoa) against human lymphatic filaria, *Brugia malayi. Parasitol. Res.* 2004, 93 (4), 268–273.

172. Lakshmi, V., Kumar, R., Gupta, P., Varshney, V., Srivastava, M. N., Dikshit, M., Murthy, P. K., Misra-Bhattacharya, S. The antifilarial activity of a marine red alga, *Botryocladia leptopoda*, against experimental infections with animal and human filariae. *Parasitol. Res.* 2004, 93 (6), 468–474.

173. Misra, S., Verma, M., Mishra, S. K., Srivastava, S., Lakshmi, V., Misra-Bhattacharya, S. Gedunin and photogedunin of *Xylocarpus granatum* possess antifilarial activity against human lymphatic filarial parasite *Brugia malayi* in experimental rodent host. *Parasitol. Res.* 2011, 109 (5), 1351–1360.

174. Gupta, J., Misra, S., Mishra, S. K., Srivastava, S., Srivastava, M. N., Lakshmi, V., Misra-Bhattacharya, S. Antifilarial activity of marine sponge *Haliclona oculata* against experimental *Brugia malayi* infection. *Exp. Parasitol.* 2012, 130 (4), 449–455.

175. Lakshmi, V., Srivastava, S., Mishra, S. K., Misra, S., Verma, M., Misra-Bhattacharya, S. *In vitro* and *in vivo* antifilarial potential of marine sponge, *Haliclona exigua* (Kirkpatrick), against human lymphatic filarial parasite *Brugia malayi. Parasitol. Res.* 2009, 105 (5), 1295–1301.

176. Encarnacion, R., Carrasco, G., Espinoza, M., Anthoni, U., Nielsen, P. H., Christophersen, C. Neothyoside-A, proposed structure of a triterpenoid tetraglycoside from the Pacific sea-cucumber, *Neothyone-Gibbosa. J. Nat. Prod.* 1989, 52 (2), 248–251.

177. Kitagawa, I., Kobayashi, M., Son, B. W., Suzuki, S., Kyogoku, Y. Marine natural products. 19. Pervicosides A, B, and C, lanostane-type triterpene-oligoglycoside sulfates from the sea cucumber *Holothuria Pervicax. Chem. Pharm. Bull.* (Tokyo) 1989, 37 (5), 1230–1234.

178. Encarnacion, R., Murillo, J. I., Nielsen, J., Christophersen, C. Neothyoside B, a triterpenoid diglycoside from the Pacific sea cucumber *Neothyone gibbosa. Acta Chem. Scand.* 1996, 50 (9), 848–849.

179. Encarnacion-Dimayuga, R., Murillo-Alvarez, J. I., Christophersen, C., Chan-Bacab, M., Garcia Reiriz, M. L., Zacchino, S. Leishmanicidal, antifungal, and cytotoxic activity of triterpenoid glycosides isolated from the sea cucumber *Neothyone gibbosa. Nat. Prod. Commun.* 2006, 1 (7), 541–547.

180. Rangel, H. R., Dagger, F., Compagnone, R. S. Antiproliferative effect of illimaquinone on *Leishmania mexicana. Cell Biol. Int.* 1997, 21 (6), 337–339.

181. Capon, R. J., Macleod, J. K. A revision of the absolute stereochemistry of ilimaquinone. *J. Org. Chem.* 1987, 52 (22), 5059–5060.

182. Kossuga, M. H., Nascimento, A. M., Reimao, J. Q., Tempone, A. G., Taniwaki, N. N., Veloso, K., Ferreira, A. G., Cavalcanti, B. C., Pessoa, C., Moraes, M. O., Mayer, A. M. S., Hajdu, E., Berlinck, R. G. S. Antiparasitic, antineuroinflammatory, and cytotoxic polyketides from the marine sponge *Plakortis angulospiculatus* collected in Brazil. *J. Nat. Prod.* 2008, 71 (3), 334–339.

183. Compagnone, R. S., Piña, I. C., Rangel, H. R., Dagger, F., Suarez, A. I., Reddy, M. V. R., Faulkner, D. J. Antileishmanial cyclic peroxides from the Palauan sponge *Plakortis* aff. *angulospiculatus. Tetrahedron* 1998, 54 (13), 3057–3068.

184. Rao, K. V., Donia, M. S., Peng, J. N., Garcia-Palomero, E., Alonso, D., Martinez, A., Medina, M., Franzblau, S. G., Tekwani, B. L., Khan, S. I., Wahyuono, S., Willett, K. L., Hamann, M. T. Manzamine B and E and ircinal A related alkaloids from an Indonesian *Acanthostrongylophora* sponge and their activity against infectious, tropical parasitic, and Alzheimer's diseases. *J. Nat. Prod.* 2006, 69 (7), 1034–1040.

185. Rao, K. V., Santarsiero, B. D., Mesecar, A. D., Schinazi, R. F., Tekwani, B. L., Hamann, M. T. New manzamine alkaloids with activity against infectious and tropical parasitic diseases from an indonesian sponge. *J. Nat. Prod.* 2003, 66 (6), 823–828.

186. Savoia, D., Avanzini, C., Allice, T., Callone, E., Guella, G., Dini, F. Antimicrobial activity of euplotin C, the sesquiterpene taxonomic marker from the marine ciliate *Euplotes crassus. Antimicrob. Agents Chemother.* 2004, 48 (10), 3828–3833.

187. Nakao, Y., Shiroiwa, T., Murayama, S., Matsunga, S., Goto, Y., Matsumoto, Y., Fusetani, N. Identification of renieramycin A as an antileishmanial substance in a marine sponge *Neopetrosia* sp. *Mar. Drugs* 2004, 2, 55–62.

188. Gul, W., Hammond, N. L., Yousaf, M., Bowling, J. J., Schinazi, R. F., Wirtz, S. S., Andrews, G. D. C., Cuevas, C., Hamann, M. T. Modification at the C9 position of the marine natural product isoaaptamine and the impact on HIV-1, mycobacterial, and tumor cell activity. *Biorg. Med. Chem.* 2006, 14 (24), 8495–8505.

189. Gray, C. A., de Lira, S. P., Silva, M., Pimenta, E. F., Thiemann, O. H., Oliva, G., Hajdu, E., Andersen, R. J., Berlinck, R. G. S. Sulfated meroterpenoids from the Brazilian sponge *Callyspongia* sp. are inhibitors of the antileishmaniasis target adenosine phosphoribosyl transferase. *J. Org. Chem.* 2006, 71 (23), 8685–8690.

190. Kossuga, M. H., de Lira, S. P., Nascimento, A. M., Gambardella, M. T. P., Berlinck, R. G. S., Torres, Y. R., Nascimento, G. G. E., Pimenta, E. E., Silva, M., Thiemann, O. H., Oliva, G., Tempone, A. G., Melhem, M. S. C., de Souza, A. O., Galetti, F. C. S., Silva, C. L., Cavalcanti, B., Pessoa, C. O., Moraes, M. O., Hajdu, E., Peixinho, S., Rocha, R. M. Isolation and biological activities of secondary metabolites from the sponges *Monanchora* aff. *arbuscula, Aplysina* sp., *Petromica ciocalyptoides* and *Topsentia ophiraphidites,* from the ascidian *Didemnum ligulum* and from the octocoral *Carijoa rusei. Quim. Nova* 2007, 30 (5), 1194–1202.

191. Simmons, T. L., Engene, N., Urena, L. D., Romero, L. I., Ortega-Barria, E., Gerwick, L., Gerwick, W. H. Viridamides A and B, lipodepsipeptides with antiprotozoal activity from the marine cyanobacterium *Oscillatoria nigro-viridis. J. Nat. Prod.* 2008, 71 (9), 1544–1550.

192. Gul, W., Hammond, N. L., Yousaf, M., Peng, J., Holley, A., Hamann, M. T. Chemical transformation and biological studies of marine sesquiterpene (S)-(+)-curcuphenol and its analogs. *Biochim. Biophys. Acta Gen. Subj.* 2007, 1770 (11), 1513–1519.

193. Nakao, Y., Kawatsu, S., Okamoto, C., Okamoto, M., Matsumoto, Y., Matsunaga, S., van Soest, R. W. M., Fusetani, N. Ciliatamides A–C, bioactive lipopeptides from the deep-sea sponge *Aaptos ciliatao. J. Nat. Prod.* 2008, 71 (3), 469–472.

194. Giddens, A. C., Nielsen, L., Boshoff, H. I., Tasdemir, D., Perozzo, R., Kaiser, M., Wang, F., Sacchettini, J. C., Copp, B. R. Natural product inhibitors of fatty acid biosynthesis: Synthesis of the marine microbial metabolites pseudopyronines A and B and evaluation of their anti-infective activities. *Tetrahedron* 2008, 64 (7), 1242–1249.

195. Pontius, A., Krick, A., Kehraus, S., Brun, R., Konig, G. M. Antiprotozoal activities of heterocyclic-substituted xanthones from the marine-derived fungus *Chaetomium* sp. *J. Nat. Prod.* 2008, 71 (9), 1579–1584.

196. Cruz, L. J., Luque-Ortega, J. R., Rivas, L., Albericio, F. Kahalalide F, an antitumor depsipeptide in clinical trials, and its analogues as effective antileishmanial agents. *Mol. Pharm.* 2009, 6 (3), 813–824.

197. Linington, R. G., Clark, B. R., Trimble, E. E., Almanza, A., Urena, L. D., Kyle, D. E., Gerwick, W. H. Antimalarial peptides from marine cyanobacteria: Isolation and structural elucidation of gallinamide A. *J. Nat. Prod.* 2009, 72 (1), 14–17.

198. Vik, A., Proszenyak, A., Vermeersch, M., Cos, P., Maes, L., Gundersen, L. L. Screening of agelasine D and analogs for inhibitory activity against pathogenic protozoa; identification of hits for visceral leishmaniasis and Chagas disease. *Molecules* 2009, 14 (1), 279–288.

199. Ma, W. S., Mutka, T., Vesley, B., Amsler, M. O., McClintock, J. B., Amsler, C. D., Perman, J. A., Singh, M. P., Maiese, W. M., Zaworotko, M. J., Kyle, D. E., Baker, B. J. Norselic acids A–E, highly oxidized anti-infective steroids that deter mesograzer predation, from the Antarctic sponge *Crella* sp. *J. Nat. Prod.* 2009, 72 (10), 1842–1846.

200. Sanchez, L. M., Lopez, D., Vesely, B. A., Della Togna, G., Gerwick, W. H., Kyle, D. E., Linington, R. G. Almiramides A–C: Discovery and development of a new class of leishmaniasis lead compounds. *J. Med. Chem.* 2010, 53 (10), 4187–4197.

201. Sanchez, L. M., Knudsen, G. M., Helbig, C., De Muylder, G., Mascuch, S. M., Mackey, Z. B., Gerwick, L., Clayton, C., McKerrow, J. H., Linington, R. G. Examination of the mode of action of the Almiramide family of natural products against the kinetoplastid parasite *Trypanosoma brucei. J. Nat. Prod.* 2013, 76 (4), 630–641.

202. Balunas, M. J., Linington, R. G., Tidgewell, K., Fenner, A. M., Urena, L. D., Della Togna, G., Kyle, D. E., Gerwick, W. H. Dragonamide E, a modified linear lipopeptide from *Lyngbya majuscula* with antileishmanial activity. *J. Nat. Prod.* 2010, 73 (1), 60–66.

203. Pimentel-Elardo, S. M., Kozytska, S., Bugni, T. S., Ireland, C. M., Moll, H., Hentschel, U. Anti-parasitic compounds from *Streptomyces* sp strains isolated from Mediterranean sponges. *Mar. Drugs* 2010, 8 (2), 373–380.

204. Orhan, I., Sener, B., Kaiser, M., Brun, R., Tasdemir, D. Inhibitory activity of marine sponge-derived natural products against parasitic protozoa. *Mar. Drugs* 2010, 8 (1), 47–58.

205. Singh, N., Kumar, R., Gupta, S., Dube, A., Lakshmi, V. Antileishmanial activity *in vitro* and *in vivo* of constituents of sea cucumber *Actinopyga lecanora. Parasitol. Res.* 2008, 103 (2), 351–354.

206. Scala, F., Fattorusso, E., Menna, M., Taglialatela-Scafati, O., Tierney, M., Kaiser, M., Tasdemir, D. Bromopyrrole alkaloids as lead compounds against protozoan parasites. *Mar. Drugs* 2010, 8 (7), 2162–2174.

207. dos Santos, A. O., Britta, E. A., Bianco, E. M., Ueda-Nakamura, T., Dias, B. P., Pereira, R. C., Nakamura, C. V. 4-Acetoxydolastane diterpene from the Brazilian brown alga *Canistrocarpus cervicornis* as antileishmanial agent. *Mar. Drugs* 2011, 9 (11), 2369–2383.

208. Balunas, M. J., Grosso, M. F., Villa, F. A., Engene, N., McPhail, K. L., Tidgewell, K., Pineda, L. M., Gerwick, L., Spadafora, C., Kyle, D. E., Gerwick, W. H. Coibacins A–D, antileishmanial marine cyanobacterial polyketides with intriguing biosynthetic origins. *Org. Lett.* 2012, 14 (15), 3878–3881.

209. Ishigami, S. T., Goto, Y., Inoue, N., Kawazu, S. I., Matsumoto, Y., Imahara, Y., Tarumi, M., Nakai, H., Fusetani, N., Nakao, Y. Cristaxenicin A, an antiprotozoal xenicane diterpenoid from the deep sea gorgonian *Acanthoprimnoa cristata. J. Org. Chem.* 2012, 77 (23), 10962–10966.

210. von Salm, J. L., Wilson, N. G., Vesely, B. A., Kyle, D. E., Cuce, J., Baker, B. J. Shagenes A & B, tricyclic seqquiterpenes produced by an undescribed Antarctic octocoral. *Planta Med.* 2014, 80 (10), 767–767.

211. Cafieri, F., Fattorus, E., Magno, S., Santacro, C., Sica, D. Isolation and structure of axisonitrile-1 and axisothiocyanate-1: 2 unusual sesquiterpenoids from marine sponge *Axinella-cannabina*. *Tetrahedron* 1973, 29 (24), 4259–4262.

212. Fattorusso, E., Magno, S., Mayol, L., Santacro, C., Sica, D. Isolation and structure of axisonitrile-2: New sesquiterpenoid isonitrile from sponge *Axinella-cannabina*. *Tetrahedron* 1974, 30 (21), 3911–3913.

213. Angerhofer, C. K., Pezzuto, J. M., Konig, G. M., Wright, A. D., Sticher, O. Antimalarial activity of sesquiterpenes from the marine sponge *Acanthella-klethra*. *J. Nat. Prod.* 1992, 55 (12), 1787–1789.

214. Burreson, B. J., Christophersen, C., Scheuer, P. J. Co-occurrence of two terpenoid isocyanide-formamide pairs in a marine sponge (*Halichondria* sp). *Tetrahedron* 1975, 31 (17), 2015–2018.

215. Wright, A. D., Wang, H. Q., Gurrath, M., Konig, G. M., Kocak, G., Neumann, G., Loria, P., Foley, M., Tilley, L. Inhibition of heme detoxification processes underlies the antimalarial activity of terpene isonitrile compounds from marine sponges. *J. Med. Chem.* 2001, 44 (6), 873–885.

216. Wattanapiromsakul, C., Chanthathamrongsiri, N., Bussarawit, S., Yuenyongsawad, S., Plubrukarn, A., Suwanborirux, K. 8-Isocyanoamphilecta-11(20),15-diene, a new antimalarial isonitrile diterpene from the sponge *Ciocalapata* sp. *Can. J. Chem.* 2009, 87 (5), 612–618.

217. Aviles, E., Rodriguez, A. D. Monamphilectine A, a potent antimalarial beta-lactam from marine sponge *Hymeniacidon* sp: Isolation, structure, semisynthesis, and bioactivity. *Org. Lett.* 2010, 12 (22), 5290–5293.

218. Miyaoka, H., Shimomura, M., Kimura, H., Yamada, Y., Kim, H. S., Wataya, Y. Antimalarial activity of kalihinol A and new relative diterpenoids from the Okinawan sponge, *Acanthella* sp. *Tetrahedron* 1998, 54 (44), 13467–13474.

219. Anderson, J. R., Edwards, R. L., Whalley, A. J. S. Metabolites of the higher fungi. 21. 3-Methyl-3,4-dihydroisocoumarins and related-compounds from the ascomycete family Xylariaceae. *J. Chem. Soc.-Perkin Trans. 1* 1983, (9), 2185–2192.

220. Chinworrungsee, M., Kittakoop, P., Isaka, M., Rungrod, A., Tanticharoen, M., Thebtaranonth, Y. Antimalarial halorosellinic acid from the marine fungus *Halorosellinia oceanica*. *Bioorg. Med. Chem. Lett.* 2001, 11 (15), 1965–1969.

221. Tymiak, A. A., Rinehart, K. L., Bakus, G. J. Constituents of morphologically similar sponges: *Aplysina* and *Smenospongia* species. *Tetrahedron* 1985, 41 (6), 1039–1047.

222. Nakamura, H., Kobayashi, J., Kobayashi, M., Ohizumi, Y., Hirata, Y. Physiologically active marine natural product from Porifera. 7. Xestoquinone: A novel cardiotonic marine natural product isolated from the Okinawan sea sponge *Xestospongia-sapra*. *Chem. Lett.* 1985, (6), 713–716.

223. Laurent, D., Jullian, V., Parenty, A., Knibiehler, M., Dorin, D., Schmitt, S., Lozach, O., Lebouvier, N., Frostin, M., Alby, F., Maurel, S., Doerig, C., Meijer, L., Sauvain, M. Antimalarial potential of xestoquinone, a protein kinase inhibitor isolated from a Vanuatu marine sponge *Xestospongia* sp. *Biorg. Med. Chem.* 2006, 14 (13), 4477–4482.

224. Ang, K. K. H., Holmes, M. J., Higa, T., Hamann, M. T., Kara, U. A. K. *In vivo* antimalarial activity of the beta-carboline alkaloid manzamine A. *Antimicrob. Agents Chemother.* 2000, 44 (6), 1645–1649.

225. El Sayed, K. A., Kelly, M., Kara, U. A. K., Ang, K. K. H., Katsuyama, I., Dunbar, D. C., Khan, A. A., Hamann, M. T. New manzamine alkaloids with potent activity against infectious diseases. *J. Am. Chem. Soc.* 2001, 123 (9), 1804–1808.

226. Steffan, B. Lepadin-A, a decahydroquinoline alkaloid from the tunicate *Clavelina lepadiformis*. *Tetrahedron* 1991, 47 (41), 8729–8732.

227. Wright, A. D., Goclik, E., Konig, G. M., Kaminsky, R. Lepadins D–F: Antiplasmodial and antitrypanosomal decahydroquinoline derivatives from the tropical marine tunicate *Didemnum* sp. *J. Med. Chem.* 2002, 45 (14), 3067–3072.

228. Kouranylefoll, E., Pais, M., Sevenet, T., Guittet, E., Montagnac, A., Fontaine, C., Guenard, D., Adeline, M. T., Debitus, C. Phloeodictine A and phloeodictine B: New antibacterial and cytotoxic bicyclic amidinium salts from the New Caledonian sponge, *Phloeodictyon* sp. *J. Org. Chem.* 1992, 57 (14), 3832–3835.

229. Mancini, I., Guella, G., Sauvain, M., Debitus, C., Duigou, A. G., Ausseil, F., Menou, J. L., Pietra, F. New 1,2,3,4-tetrahydropyrrolo[1,2-a]pyrimidinium alkaloids (phloeodictynes) from the New Caledonian shallow-water haplosclerid sponge *Oceanapia fistulosa*: Structural elucidation from mainly LC-tandem-MS-soft-ionization techniques and discovery of antiplasmodial activity. *Org. Biomol. Chem.* 2004, 2 (5), 783–787.

230. Kobayashi, M., Kondo, K., Kitagawa, I. Antifungal peroxyketal acids from an Okinawan marine sponge of *Plakortis* sp. *Chem. Pharm. Bull.* (Tokyo) 1993, 41 (7), 1324–1326.
231. Murakami, N., Kawanishi, M., Itagaki, S., Horii, T., Kobayashi, M. New readily accessible peroxides with high anti-malarial potency. *Bioorg. Med. Chem. Lett.* 2002, 12 (1), 69–72.
232. Capon, R. J., Macleod, J. K., Coote, S. J., Davies, S. G., Gravatt, G. L., Dordorhedgecock, I. M., Whittaker, M. Stereochemical studies on marine cyclic peroxides: An unequivocal assignment of absolute stereochemistry by asymmetric-synthesis. *Tetrahedron* 1988, 44 (6), 1637–1650.
233. El Sayed, K. A., Dunbar, D. C., Goins, D. K., Cordova, C. R., Perry, T. L., Wesson, K. J., Sanders, S. C., Janus, S. A., Hamann, M. T. The marine environment: A resource for prototype antimalarial agents. *J. Nat. Toxins* 1996, 5 (2), 261–285.
234. Feling, R. H., Buchanan, G. O., Mincer, T. J., Kauffman, C. A., Jensen, P. R., Fenical, W. Salinosporamide A: A highly cytotoxic proteasome inhibitor from a novel microbial source, a marine bacterium of the new genus *Salinospora. Angew. Chem. Int. Ed.* 2003, 42 (3), 355–357.
235. Prudhomme, J., McDaniel, E., Ponts, N., Bertani, S., Fenical, W., Jensen, P., Le Roch, K. Marine actinomycetes: A new source of compounds against the human malaria parasite. *PLoS One* 2008, 3 (6).
236. Nair, M. S. R., Carey, S. T. Metabolites of pyrenomycetes. 13. Structure of (+) hypothemycin, an antibiotic macrolide from hypomyces-trichothecoides. *Tetrahedron Lett.* 1980, 21 (21), 2011–2012.
237. Isaka, M., Suyarnsestakorn, C., Tanticharoen, M., Kongsaeree, P., Thebtaranonth, Y. Aigialomycins A–E, new resorcylic macrolides from the marine mangrove fungus *Aigialus parvus. J. Org. Chem.* 2002, 67 (5), 1561–1566.
238. Lim, C. H., Miyagawa, H., Akamatsu, M., Nakagawa, Y., Ueno, T. Structures and biological activities of phytotoxins produced by the plant pathogenic fungus *Bipolaris cynodontis* cynA. *J. Pestic. Sci.* 1998, 23 (3), 281–288.
239. Osterhage, C., Konig, G. M., Holler, U., Wright, A. D. Rare sesquiterpenes from the algicolous fungus *Drechslera dematioidea. J. Nat. Prod.* 2002, 65 (3), 306–313.
240. Topcu, G., Aydogmus, Z., Imre, S., Goren, A. C., Pezzuto, J. M., Clement, J. A., Kingston, D. G. I. Brominated sesquiterpenes from the red alga *Laurencia obtusa. J. Nat. Prod.* 2003, 66 (11), 1505–1508.
241. Ospina, C. A., Rodriguez, A. D., Ortega-Barria, E., Capson, T. L. Briarellins J–P and polyanthellin A: New eunicellin-based diterpenes from the gorgonian coral *Briareum polyanthes* and their antimalarial activity. *J. Nat. Prod.* 2003, 66 (3), 357–363.
242. Maskey, R. P., Helmke, E., Fiebig, H. H., Laatsch, H. Parimycin: Isolation and structure elucidation of a novel cytotoxic 2,3-dihydroquinizarin analogue of gamma-indomycinone from a marine streptomycete isolate. *J. Antibiot.* (Tokyo) 2002, 55 (12), 1031–1035.
243. Maskey, R. P., Helmke, E., Kayser, O., Fiebig, H. H., Maier, A., Busche, A., Laatsch, H. Anti-cancer and antibacterial trioxacarcins with high anti-malaria activity from a marine streptomycete and their absolute stereochemistry. *J. Antibiot.* 2004, 57 (12), 771–779.
244. Wright, A. D., Lang-Unnasch, N. Potential antimalarial lead structures from fungi of marine origin. *Planta Med.* 2005, 71 (10), 964–966.
245. Linington, R. G., Gonzalez, J., Urena, L. D., Romero, L. I., Ortega-Barria, E., Gerwick, W. H. Venturamides A and B: Antimalarial constituents of the Panamanian marine cyanobacterium *Oscillatoria* sp. *J. Nat. Prod.* 2007, 70 (3), 397–401.
246. Jongaramruong, J., Kongkam, N. Novel diterpenes with cytotoxic, anti-malarial and anti-tuberculosis activities from a brown alga *Dictyota* sp. *J. Asian Nat. Prod. Res.* 2007, 9 (8), 743–751.
247. Tasdemir, D., Topaloglu, B., Perozzo, R., Brun, R., O'Neill, R., Carballeira, N. M., Zhang, X. J., Tonge, P. J., Linden, A., Ruedi, P. Marine natural products from the Turkish sponge *Agelas oroides* that inhibit the enoyl reductases from *Plasmodium falciparum*, *Mycobacterium tuberculosis* and *Escherichia coli. Biorg. Med. Chem.* 2007, 15 (21), 6834–6845.
248. McPhail, K. L., Correa, J., Linington, R. G., Gonzalez, J., Ortega-Barria, E., Capson, T. L., Gerwick, W. H. Antimalarial linear lipopeptides from a Panamanian strain of the marine cyanobacterium *Lyngbya majuscula. J. Nat. Prod.* 2007, 70 (6), 984–988.
249. Benoit-Vical, F., Salery, M., Soh, P. N., Ahond, A., Poupat, C. Girolline: A potential lead structure for antiplasmodial drug research. *Planta Med.* 2008, 74 (4), 438–444.
250. Na, M., Meujo, D. A. F., Kevin, D., Hamann, M. T., Anderson, M., Hill, R. T. A new antimalarial polyether from a marine *Streptomyces* sp. H668. *Tetrahedron Lett.* 2008, 49 (44), 6282–6285.

251. Kasettrathat, C., Ngamrojanavanich, N., Wiyakrutta, S., Mahidol, C., Ruchirawat, S., Kittakoop, P. Cytotoxic and antiplasmodial substances from marine-derived fungi, *Nodulisporium* sp. and CRI247-01. *Phytochemistry* 2008, 69 (14), 2621–2626.

252. Appenzeller, J., Mihci, G., Martin, M. T., Gallard, J. F., Menou, J. L., Boury-Esnalllt, N., Hooper, J., Petek, S., Chevalley, S., Valentin, A., Zaparucha, A., Al Mourabit, A., Debitus, C. Agelasines J, K, and L from the Solomon Islands marine sponge *Agelas* cf. *mauritiana*. *J. Nat. Prod.* 2008, 71 (8), 1451–1454.

253. Desoubzdanne, D., Marcourt, L., Raux, R., Chevalley, S., Dorin, D., Doerig, C., Valentin, A., Ausseil, F., Debitus, C. Alisiaquinones and alisiaquinol, dual inhibitors of *Plasmodium falciparum* enzyme targets from a New Caledonian deep water sponge. *J. Nat. Prod.* 2008, 71 (7), 1189–1192.

254. Afolayan, A. F., Mann, M. G. A., Lategan, C. A., Smith, P. J., Bolton, J. J., Beukes, D. R. Antiplasmodial halogenated monoterpenes from the marine red alga *Plocamium cornutum*. *Phytochemistry* 2009, 70 (5), 597–600.

255. Taori, K., Liu, Y. X., Paul, V. J., Luesch, H. Combinatorial strategies by marine cyanobacteria: Symplostatin 4, an antimitotic natural dolastatin 10/15 hybrid that synergizes with the coproduced HDAC inhibitor largazole. *ChemBioChem* 2009, 10 (10), 1634–1639.

256. Stolze, S. C., Deu, E., Kaschani, F., Li, N., Florea, B. I., Richau, K. H., Colby, T., van der Hoom, R. A. L., Overkleeft, H. S., Bogyo, M., Kaiser, M. The antimalarial natural product symplostatin 4 is a nanomolar inhibitor of the food vacuole falcipains. *Chem. Biol.* 2012, 19 (12), 1546–1555.

257. Takishima, S., Ishiyama, A., Iwatsuki, M., Otoguro, K., Yamada, H., Omura, S., Kobayashi, H., van Soest, R. W. M., Matsunaga, S. Merobatzelladines A and B, anti-infective tricyclic guanidines from a marine sponge *Monanchora* sp. *Org. Lett.* 2009, 11 (12), 2655–2658.

258. Takishima, S., Ishiyama, A., Watsuki, M., Otoguro, K., Yamada, H., Omura, S., Kobayashi, H., van Soest, R. W. M., Matsunaga, S. Merobatzelladines A and B, Anti-infective tricyclic guanidines from a marine sponge *Monanchora* sp. (vol 11, pg 2655, 2009). *Org. Lett.* 2010, 12 (4), 896–896.

259. Meyer, M., Delberghe, F., Liron, F., Guillaume, M., Valentin, A., Guyot, M. An antiplasmodial new (bis) indole alkaloid from the hard coral *Tubastraea* sp. *Nat. Prod. Res.* 2009, 23 (2), 178–182.

260. Tripathi, A., Puddick, J., Prinsep, M. R., Rottmann, M., Tan, L. T. Lagunamides A and B: Cytotoxic and antimalarial cyclodepsipeptides from the marine cyanobacterium *Lyngbya majuscula*. *J. Nat. Prod.* 2010, 73 (11), 1810–1814.

261. Kubanek, J., Prusak, A. C., Snell, T. W., Giese, R. A., Fairchild, C. R., Aalbersberg, W., Hay, M. E. Bromophycolides C–I from the Fijian red alga *Callophycus serratus*. *J. Nat. Prod.* 2006, 69 (5), 731–735.

262. Lane, A. L., Stout, E. P., Lin, A. S., Prudhomme, J., Le Roch, K., Fairchild, C. R., Franzblau, S. G., Hay, M. E., Aalbersberg, W., Kubanek, J. Antimalarial bromophycolides J–Q from the Fijian red alga *Callophycus serratus*. *J. Org. Chem.* 2009, 74 (7), 2736–2742.

263. Lin, A. S., Stout, E. P., Prudhomme, J., Le Roch, K., Fairchild, C. R., Franzblau, S. G., Aalbersberg, W., Hay, M. E., Kubanek, J. Bioactive bromophycolides R-U from the Fijian red alga *Callophycus serratus*. *J. Nat. Prod.* 2010, 73 (2), 275–278.

264. Teasdale, M. E., Prudhomme, J., Torres, M., Braley, M., Cervantes, S., Bhatia, S. C., La Clair, J. J., Le Roch, K., Kubanek, J. Pharmacokinetics, metabolism, and *in vivo* efficacy of the antimalarial natural product bromophycolide A. *ACS Med. Chem. Lett.* 2013, 4 (10), 989–993.

265. Wei, X. M., Nieves, K., Rodriguez, A. D. Neopetrosiamine A, biologically active bis-piperidine alkaloid from the Caribbean sea sponge *Neopetrosia proxima*. *Bioorg. Med. Chem. Lett.* 2010, 20 (19), 5905–5908.

266. Davis, R. A., Duffy, S., Avery, V. M., Camp, D., Hooper, J. N. A., Quinn, R. J. (+)-7-Bromotrypargine: An antimalarial beta-carboline from the Australian marine sponge *Ancorina* sp. *Tetrahedron Lett.* 2010, 51 (4), 583–585.

267. Wei, X. M., Rodriguez, A. D., Baran, P., Raptis, R. G. Dolabellane-type diterpenoids with antiprotozoan activity from a southwestern Caribbean gorgonian octocoral of the genus *Eunicea*. *J. Nat. Prod.* 2010, 73 (5), 925–934.

268. Huang, H. B., Yao, Y. L., He, Z. X., Yang, T. T., Ma, J. Y., Tian, X. P., Li, Y. Y., Huang, C. G., Chen, X. P., Li, W. J., Zhang, S., Zhang, C. S., Ju, J. H. Antimalarial beta-carboline and indolactam alkaloids from *Marinactinospora thermotolerans*, a deep sea isolate. *J. Nat. Prod.* 2011, 74 (10), 2122–2127.

269. Almeida, C., Kehraus, S., Prudencio, M., Konig, G. M. Marilones A–C, phthalides from the sponge-derived fungus *Stachylidium* sp. *Beilstein J. Org. Chem.* 2011, 7, 1636–1642.

270. Davis, R. A., Buchanan, M. S., Duffy, S., Avery, V. M., Charman, S. A., Charman, W. N., White, K. L., Shackleford, D. M., Edstein, M. D., Andrews, K. T., Camp, D., Quinn, R. J. Antimalarial activity of pyrroloiminoquinones from the Australian marine sponge *Zyzzya* sp. *J. Med. Chem.* 2012, 55 (12), 5851–5858.

271. Xu, M., Andrews, K. T., Birrell, G. W., Tran, T. L., Camp, D., Davis, R. A., Quinn, R. J. Psammaplysin H, a new antimalarial bromotyrosine alkaloid from a marine sponge of the genus *Pseudoceratina*. *Bioorg. Med. Chem. Lett.* 2011, 21 (2), 846–848.

272. Yang, X. Z., Davis, R. A., Buchanan, M. S., Duffy, S., Avery, V. M., Camp, D., Quinn, R. J. Antimalarial bromotyrosine derivatives from the Australian marine sponge *Hyattella* sp. *J. Nat. Prod.* 2010, 73 (5), 985–987.

273. Mudianta, I. W., Skinner-Adams, T., Andrews, K. T., Davis, R. A., Hadi, T. A., Hayes, P. Y., Garson, M. J. Psammaplysin derivatives from the Balinese marine sponge *Aplysinella strongylata*. *J. Nat. Prod.* 2012, 75 (12), 2132–2143.

274. Finlayson, R., Pearce, A. N., Page, M. J., Kaiser, M., Bourguet-Kondracki, M. L., Harper, J. L., Webb, V. L., Copp, B. R. Didemnidines A and B, indole spermidine alkaloids from the New Zealand ascidian *Didemnum* sp. *J. Nat. Prod.* 2011, 74 (4), 888–892.

275. Carroll, A. R., Nash, B. D., Duffy, S., Avery, V. M. Albopunctatone, an antiplasmodial anthrone-anthraquinone from the Australian ascidian *Didemnum albopunctatum*. *J. Nat. Prod.* 2012, 75 (6), 1206–1209.

276. Pearce, A. N., Chia, E. W., Berridge, M. V., Maas, E. W., Page, M. J., Harper, J. L., Webb, V. L., Copp, B. R. Orthidines A–E, tubastrine, 3,4-dimethoxyphenethyl-beta-guanidine, and 1,14-sperminedihomovanillamide: Potential anti-inflammatory alkaloids isolated from the New Zealand ascidian *Aplidium orthium* that act as inhibitors of neutrophil respiratory burst. *Tetrahedron* 2008, 64 (24), 5748–5755.

277. Liew, L. P. P., Kaiser, M., Copp, B. R. Discovery and preliminary structure-activity relationship analysis of 1,14-sperminediphenylacetamides as potent and selective antimalarial lead compounds. *Bioorg. Med. Chem. Lett.* 2013, 23 (2), 452–454.

278. Perry, T. L., Dickerson, A., Khan, A. A., Kondru, R. K., Beratan, D. N., Wipf, P., Kelly, M., Hamann, M. T. New peroxylactones from the Jamaican sponge *Plakinastrella onkodes*, with inhibitory activity against the AIDS opportunistic parasitic infection *Toxoplasma gondii*. *Tetrahedron* 2001, 57 (8), 1483–1487.

279. El Sayed, K. A., Hamann, M. T., Hashish, N. E., Shier, W. T., Kelly, M., Khan, A. A. Antimalarial, antiviral, and antitoxoplasmosis norsesterterpene peroxide acids from the Red Sea sponge *Diacarnus erythraeanus*. *J. Nat. Prod.* 2001, 64 (4), 522–524.

280. Kobayashi, J., Cheng, J. F., Nakamura, H., Ohizumi, Y., Hirata, Y., Sasaki, T., Ohta, T., Nozoe, S. Ascididemin, a novel pentacyclic aromatic alkaloid with potent antileukemic activity from the Okinawan tunicate *Didemnum* sp. *Tetrahedron Lett.* 1988, 29 (10), 1177–1180.

281. Copp, B. R., Kayser, O., Brun, R., Kiderlen, A. F. Antiparasitic activity of marine pyridoacridone alkaloids related to the ascididemins. *Planta Med.* 2003, 69 (6), 527–531.

282. Osterhage, C., Kaminsky, R., König, G. M., Wright, A. D. Ascosalipyrrolidinone A, an antimicrobial alkaloid, from the obligate marine fungus *Ascochyta salicorniae*. *J. Org. Chem.* 2000, 65 (20), 6412–6417.

283. Feng, Y. J., Davis, R. A., Sykes, M., Avery, V. M., Camp, D., Quinn, R. J. Antitrypanosomal cyclic polyketide peroxides from the Australian marine sponge *Plakortis* sp. *J. Nat. Prod.* 2010, 73 (4), 716–719.

284. Nakamura, H., Wu, H. M., Ohizumi, Y., Hirata, Y. Agelasine-A, agelasine-B, agelasine-C and agelasine-D, novel bicyclic diterpenoids with a 9-methyladeninium unit possessing inhibitory effects on Na,K-atpase from the Okinawan sea sponge *Agelas* sp 1. *Tetrahedron Lett.* 1984, 25 (28), 2989–2992.

285. Hertiani, T., Edrada-Ebel, R., Ortlepp, S., van Soest, R. W. M., de Voogd, N. J., Wray, V., Hentschel, U., Kozytska, S., Muller, W. E. G., Proksch, P. From anti-fouling to biofilm inhibition: New cytotoxic secondary metabolites from two Indonesian *Agelas* sponges. *Biorg. Med. Chem.* 2010, 18 (3), 1297–1311.

286. Rubio, B. K., Tenney, K., Ang, K. H., Abdulla, M., Arkin, M., McKerrow, J. H., Crews, P. The marine sponge *Diacarnus bismarckensis* as a source of peroxiterpene inhibitors of *Trypanosoma brucei*, the causative agent of sleeping sickness. *J. Nat. Prod.* 2009, 72 (2), 218–222.

287. Johnson, T. A., Amagata, T., Sashidhara, K. V., Oliver, A. G., Tenney, K., Matainaho, T., Ang, K. K. H., McKerrow, J. H., Crews, P. The Aignopsanes, a new class of sesquiterpenes from selected chemotypes of the sponge *Cacospongia mycofijiensis*. *Org. Lett.* 2009, 11 (9), 1975–1978.

288. Feng, Y. J., Davis, R. A., Sykes, M. L., Avery, V. M., Carroll, A. R., Camp, D., Quinn, R. J. Antitrypanosomal pyridoacridine alkaloids from the Australian ascidian *Polysyncraton echinatum*. *Tetrahedron Lett.* 2010, 51 (18), 2477–2479.

289. Regalado, E. L., Tasdemir, D., Kaiser, M., Cachet, N., Amade, P., Thomas, O. P. Antiprotozoal steroidal saponins from the marine sponge *Pandaros acanthifolium*. *J. Nat. Prod.* 2010, 73 (8), 1404–1410.

290. Martinez-Luis, S., Gomez, J. F., Spadafora, C., Guzman, H. M., Gutierrez, M. Antitrypanosomal alkaloids from the marine bacterium *Bacillus pumilus*. *Molecules* 2012, 17 (9), 11146–11155.

291. Feng, Y. J., Davis, R. A., Sykes, M. L., Avery, V. M., Quinn, R. J. Iotrochamides A and B, antitrypanosomal compounds from the Australian marine sponge *Iotrochota* sp. *Bioorg. Med. Chem. Lett.* 2012, 22 (14), 4873–4876.

292. Quang, T. H., Ha, T. T., Van Minh, C., Van Kiem, P., Huong, H. T., Nguyen, T. T. N., Nhiem, N. X., Tung, N. H., Tai, B. H., Dinh, T. T. T., Song, B., Kang, H. K., Kim, Y. H. Cytotoxic and anti-inflammatory cembranoids from the Vietnamese soft coral *Lobophytum laevigatum*. *Biorg. Med. Chem.* 2011, 19 (8), 2625–2632.

293. Thao, N. P., Cuong, N. X., Luyen, B. T. T., Nam, N. H., Cuong, P. V., Thanh, N. V., Nhiem, N. X., Hanh, T. T. H., Kim, E. J., Kang, H. K., Kiem, P. V., Minh, C. V., Kim, Y. H. Steroidal constituents from the starfish *Astropecten polyacanthus* and their anticancer effects. *Chem. Pharm. Bull.* (Tokyo) 2013, 61 (10), 1044–1051.

294. Thao, N. P., No, J. H., Luyen, B. T. T., Yang, G., Byun, S. Y., Goo, J., Kim, K. T., Cuong, N. X., Nam, N. H., Minh, C. V., Schmidt, T. J., Kang, J. S., Kim, Y. H. Secondary metabolites from Vietnamese marine invertebrates with activity against *Trypanosoma brucei* and *T. cruzi*. *Molecules* 2014, 19 (6), 7869–7880.

295. Proszenyak, A., Charnock, C., Hedner, E., Larsson, R., Bohlin, L., Gundersen, L. L. Synthesis, antimicrobial and antineoplastic activities for agelasine and agelasimine analogs with a beta-cyclocitral derived substituent. *Arch. Pharm.* 2007, 340 (12), 625–634.

296. Nakisah, M. A., Muryany, M. Y. I., Fatimah, H., Fadilah, R. N., Zalilawati, M. R., Khamsah, S., Habsah, M. Anti-amoebic properties of a Malaysian marine sponge *Aaptos* sp on *Acanthamoeba castellanii*. *World J. Microbiol. Biotechnol.* 2012, 28 (3), 1237–1244.

297. Shaari, K., Ling, K. C., Rashid, Z. M., Jean, T. P., Abas, F., Raof, S. M., Zainal, Z., Lajis, N. H., Mohamad, H., Ali, A. M. Cytotoxic aaptamines from Malaysian *Aaptos aaptos*. *Mar. Drugs* 2009, 7 (1), 1–8.

298. Kayser, O., Waters, W. R., Woods, K. M., Upton, S. J., Keithly, J. S., Kiderlen, A. F. Evaluation of *in vitro* activity of aurones and related compounds against *Cryptosporidium parvum*. *Planta Med.* 2001, 67 (8), 722–725.

299. Atta, U. R., Choudhary, M. I., Hayat, S., Khan, A. M., Ahmed, A. Two new aurones from marine brown alga *Spatoglossum variabile*. *Chem. Pharm. Bull.* (Tokyo) 2001, 49 (1), 105–107.

300. Pelter, A., Ward, R. S., Heller, H. G. C-13 nuclear magnetic-resonance spectra of (Z)-aurones and (E)-aurones. *J. Chem. Soc. Perkin Trans. 1* 1979, (2), 328–329.

CHAPTER 10

Central Nervous System Modulators from the Oceans

Kh. Tanvir Ahmed,[1] **Neil Lax,**[2] **and Kevin Tidgewell**[1]

[1]Division of Medicinal Chemistry, Graduate School of Pharmaceutical Sciences,
Duquesne University, Pittsburgh, Pennsylvania
[2]Department of Biological Sciences, Bayer School of Natural and Environmental Sciences,
Duquesne University, Pittsburgh, Pennsylvania

CONTENTS

10.1 INTRODUCTION

Today much of what we know about the neurochemistry and function of the central nervous system (CNS) has been aided by the use of natural products.[1] Many of the receptors within the CNS were discovered by studying their interactions with natural products; for example, research on opium alkaloids from *Papaver somniferum* L. (Papaveraceae) guided researchers to the discovery of the opioid receptors and has helped to elucidate the mechanism of antinociception mediated by these receptors.[1–3] In this case, ethnopharmacological knowledge has aided the identification and study of a terrestrial natural product. In contrast, relatively little ethnopharmacological data are available for marine natural products.[4,5] Due to the high biological diversity of marine ecosystems, marine natural products are an obvious source of novel compounds that can modulate CNS activity. Marine organisms have evolved the ability to biosynthesize CNS-active molecules as a means of chemical defense.[6] Notable targets for these molecules are CNS ligand-gated ion channels, voltage-gated ion channels, and G protein-coupled receptors (GPCRs). They regulate a number of physiological functions that are linked to human disorders, such as sleep, anxiety, seizures, pain, mood, appetite, cognition, and memory.[7-9] Molecules with the ability to modulate ion channels or GPCRs can be used to understand the pharmacology of and treatment for these CNS disorders. There are a wide variety of compounds that have been discovered from marine sources with activity against these key targets within the CNS.[10] This chapter gives a history of marine natural products discovered with CNS modulatory activity, with specific emphasis on compounds with affinity and activity against ion channels and GPCRs.

Ion channels and GPCRs are two of the most important drug targets of the human body.[11] An analysis of all drugs approved by the Food and Drug Administration (FDA) over the past three decades has shown that GPCRs, ligand-gated ion channels, and voltage-gated ion channels are the top three drug targets, respectively.[12] All these targets are known to control important functions of the CNS, and alteration of their function can lead to various disease states like Parkinson's, Alzheimer's, schizophrenia, depression, anxiety, memory loss, addiction, neuropathy, epilepsy, brain or spinal cord ischemia, migraine, and sleep disorders.[7,13-16]

More than 22,000 natural products have been discovered from the marine environment to date, with the majority coming from microbial sources.[5] The marine environment has immense potential for drug discovery. There are already approved drugs on the market, and others are in clinical trials for neuropathic pain, cancer, Alzheimer's disease, and schizophrenia.[5,10,17,18] This chapter discusses the historical context and current drug discovery efforts in marine natural products for CNS ion channel and GPCR modulators.

10.2 ION CHANNELS

Ion channels are transmembrane proteins that facilitate ion flux across the membranes of cells. Most ion channels are either voltage or ligand gated, meaning they open and close due to changes in electrical potential of the membrane or ligand binding, respectively. Ion channels are the second most common target for currently approved drugs, behind only GPCRs.[11] Disruption in ion channels leads to a number of important cardiovascular and neural "channelopathies." Drugs that target ion channels in the CNS are indicated for sleep disorders, anxiety, epilepsy, pain, Parkinson's, Alzheimer's, and schizophrenia.[7] One of the most difficult tasks in drug discovery is identification of a bioactive lead molecule with acceptable pharmacokinetic and pharmacodynamic properties. Fortunately, most marine natural products by their shear existence are bioactive. Marine natural products have already provided scientists with considerable tools to study ion channel pharmacology (e.g., tetrodotoxin and kainic acid).[19,20] Future analysis of marine natural products will likely be a valuable source for ion channel drug discovery. Due to the large number of ion channels, more than 400,[21] this chapter focuses on known marine natural products that target ion channels relevant to the CNS.

10.2.1 Voltage-Gated Channels

10.2.1.1 Sodium Channel Modulators

Voltage-gated sodium channels (Na_Vs) are transmembrane proteins responsible for sodium influx in response to membrane depolarization in excitable cells.[22] Na_Vs belong to the family of six-transmembrane ion channels, which are composed of one α subunit and one or more connected β subunits.[17] The α

subunit is composed of four identical domains (I–IV), with each domain containing six α-helical transmembrane segments (S1–S6) and a nonhelical P segment between S5 and S6. The ion pore is composed of S5, P, and S6 segments and confers ion selectivity to the channel, while the S1–S4 segments act as the voltage sensors that allow the channel to open.[23] Auxiliary β subunits are connected with α subunits via covalent or noncovalent bonds and act to stabilize α subunits, modulate gating properties, and contribute to aggregation, migration, and cell surface expression of the Na_Vs.[23] Na_Vs are classified according to the specific α subunit incorporated. α subunits are divided into nine classes (Na_V 1.1–Na_V1.9) according to the nine functional α subunits known to date. There is an additional $Na_X\alpha$ subunit identified that lacks the amino acid for voltage gating and is instead gated by extracellular sodium concentration.[23] There are nine known binding sites for neurotoxins within Na_V channels, and a number of marketed drugs act by targeting the Na_V channels. These include carbamazepine, valproic acid, and phenytoin for epilepsy; lidocaine and bupivacaine in local anesthesia; lacosamide in seizure and pain management; and lamotrigine in epilepsy and bipolar disorder.[7,21,24] Recent research in Na_V channel drug discovery has focused on treating pain related to neuropathy, cancer, and inflammation, and a number of potential candidates are currently in different phases of clinical trials.[17] Marine natural products that target Na_V channels have been found in a variety of organisms, from puffer fish to cyanobacteria.

10.2.1.1.1 Puffer Fish Toxins

1   2

Tetrodotoxin (TTX, **1**) is currently in phase III clinical trials to treat neuropathic and cancer-related pain.[17] TTX is a water-soluble heterocyclic guanidine molecule, found in various species of puffer fish as well as mollusks, crabs, octopi, and some species of terrestrial amphibians.[25,26] Though TTX is known as a puffer fish toxin, it is likely a symbiotic bacterial metabolite. *Vibrio harveyi* and *Vibrio alginolyticus* have been found to produce TTX and several analogs, and these bacteria have been isolated from different species of puffer fish.[27] The discovery of TTX's ability to block Na_V channels in the early 1960s brought the attention of physiologists and pharmacologists worldwide to the study of TTX. It has been shown that Na_V channels can be classified as TTX-sensitive (TTX-S; Na_V1.1, 1.2, 1.3, 1.6, and 1.7) or TTX-resistant (TTX-R; Na_V1.8) sodium channels.[19] TTX blocks ion transport by binding on the α subunit of Na_V channels and is classified as a pore blocker.[28] TTX's toxicity is a result of nerve and muscle paralysis caused by inhibition of action potential generation.[29] Another guanidine neurotoxin, saxitoxin (STx, **2**), was first isolated from the mollusk *Saxidomus giganteus* and is also a pore blocker.[28] STx was later found in cyanobacteria, dinoflagellates, and their associated bacteria.[30] STx and its congeners are also called paralytic shellfish toxins, since they accumulate in filter-feeding shellfish and cause poisoning in higher organisms that consume those shellfish.[31]

Both TTX and STx have been used as pharmacological tools because of their ability to block neuronal Na_V channels. Some TTX-S Na_V channels are involved in pain generation and are upregulated after nerve injury, leading to neuropathic pain. Selective inhibition of these channels represents a potential approach to treat different pain conditions.[29] A recent human clinical study of intramuscularly injected TTX was effective in alleviating cancer pain.[32] STx has been studied for its potential as a local anesthetic, but systemic toxicity ruled out clinical use of current formulations.[33] Researchers are trying to bypass the systemic toxicity by designing a drug delivery system using liposomes.[34]

10.2.1.1.2 Conopeptides

Conopeptides are the peptide toxins found in cone snails that are used for attacking prey and defense. These mollusks belong to the Conidae family, which has roughly 500 species of cone snails.[35,36] There have been more than 1000 conopeptides discovered from cone snails. However, only a small number have been evaluated for pharmacological activity. Most conopeptides act on pain pathways and have been used to understand ion channel pharmacology involved in nociception. Conopeptides are classified into 12 pharmacological classes that come from 16 genetically distinct superfamilies of cone snails.[37] Researchers have identified four classes of conotoxins, μ, μO, δ, and ι, with Na_V channel modulatory activity. Despite their affinity for the Na_V channels, structural diversity allows them to act as either Na_V channel inhibitors (μ and μO) or activators (δ and ι).[37]

μ-Conotoxins are composed of 16 to 26 amino acid residues, and their conformation is stabilized by three disulfide linkages. The first neuronally active conotoxin, PIIA, was isolated from *Conus purpurascens*.[38] μ-Conotoxins bind to the same site as TTX and show selectivity toward TTX-S Na_V channels. Most μ-conotoxins preferentially target skeletal Na_V1.4 and neuronal Na_V1.2. However, differential activity at other Na_V subtypes makes them interesting leads for the development of analogs at therapeutically relevant Na_V channels.[37]

The μO-conotoxins display the ability to produce analgesia in animal models.[37] The most interesting μO-conotoxins are MrVIA and MrVIB, isolated from *Conus marmoreus*,[39] because they have been shown to inhibit Na_V1.8 TTX-R channels by gating modification in rat dorsal root ganglion (DRG) neurons.[37,40,41] Na_V1.8 channels are found in nociceptors and are responsible for the generation of transient TTX-R Na^+ current in DRG neurons. They are upregulated during inflammatory pain and may be a useful target for therapeutic analgesia.[41] Both δ- and ι-conotoxins are activators of Na_V channels, but through different mechanisms. δ-Conotoxins activate Na_V channels by inhibition of inactivation, resulting in continuous neuronal firing,[42] while the ι-conotoxins activate the Na_V channels without any substantial influence on channel inhibition.[37,42]

Among the four Na_V modulatory conotoxins, μO-conotoxins are the most promising in terms of Na_V channel subtype selectivity and pain therapeutics. But their detailed structural–activity relationship (SAR) analysis is not done yet because of the difficulties in synthesis, appropriate folding, and purification of analogs. Current approaches are focused on effective synthesis of the analogs, and thus identification of amino acid residues responsible for subtype selectivity.[37]

10.2.1.1.3 Sea Anemone Toxins

Sea anemones are marine organisms that have characteristic stinging organelles used to discharge venom into their prey.[43] Sea anemone toxins are an important group of marine toxins that target both Na_V and voltage-gated potassium (K_V) channels; most studies have focused on the Na_V channel effects.[28,44] Na_V channel toxins from sea anemones bind at the neurotoxin receptor site 3 of Na_V channels during depolarization and prolong channel inactivation[45] with affinities ranging from 5 nM to >10 μM.[46] These toxins are polypeptides and are divided into four types based on the number of amino acid residues. Types I and II have 46 to 51 amino acids; type III is shorter, containing 27–32 amino acids; and type IV includes all other anemone toxins.[45] The anemone toxin APETx2 is a 42–amino acid peptide isolated from *Anthopleura elegantissima*[47] and is unique due to its ability to inhibit the H^+-gated Na_V channel "acid-sensing ion channel-3" (IC_{50} = 63 nM).[47] Activation of these acid-sensing ion channels (ASICs) in sensory neurons is involved in nociception. Genetic knockdown of ASIC-3 or antagonism by APETx2 was found to be effective in pain relief in animals.[48] Recent discovery of Na_V1.8 channel blocking activity for APETx2, has increased interest in this compound for analgesic drug development.[49]

10.2.1.1.4 Cyanobacterial Toxins

Kalkitoxin (**3**), the jamaicamides (**4–6**), the antillatoxins (**7** and **8**), and the hoiamides (**9** and **10**) are Na_V modulators isolated from *Lyngbya majuscula*, an important marine cyanobacterium. Kalkitoxin and the jamaicamides are Na_V channel blockers, whereas the antillatoxins and hoiamides are activators.[50–53] Kalkitoxin is a thiazoline-containing lipopeptide that blocks the activity of TTX-S Na_V channels.[50] The mixed polyketide–peptide jamaicamides A (**4**), B (**5**), and C (**6**) are structurally unique due to their uncommon alkynyl bromide, vinyl chloride, β-methoxyeneone system, and pyrrolinone ring structural features. They were found to block Na_V channels at 5 μM.[54] Antillatoxin A (**7**) is a cyclic lipopeptide initially characterized due to its ichthyotoxic activity,[55] which was later determined to be caused by gating modification of Na_V channels and stimulation of neuronal sodium influx[56] causing activation.[52] The cyclic depsipeptide hoiamide A (**9**) acts as a partial agonist at neurotoxin site 2 on Na_V channels and was isolated from the inseparable mixture of two cyanobacteria, *Lyngbya majuscula* and *Phormidium gracile*. Later, a mixture of *Symploca* sp. and *Oscillatoria* cf. sp. yielded hoiamide B (**10**), which was found to activate the Na_v channel at the same site as hoiamide A.[57] Cyanobacteria produce a number of Na_V channel modulators that are useful as pharmacological tools and could be used in the future to develop therapeutics.

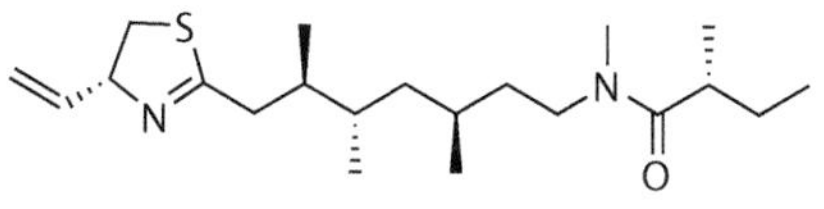

3

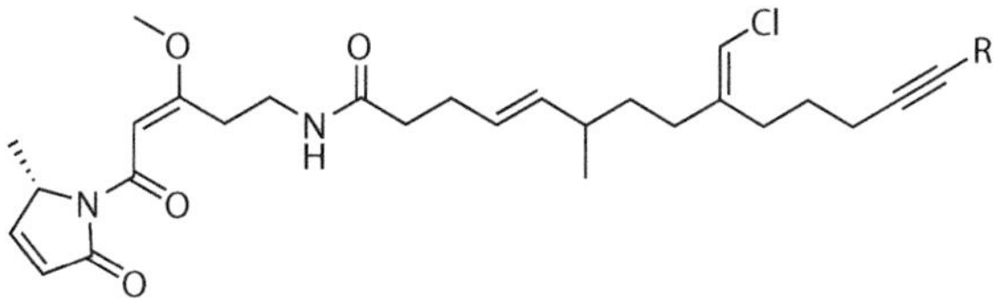

4 R = Br
5 R = H

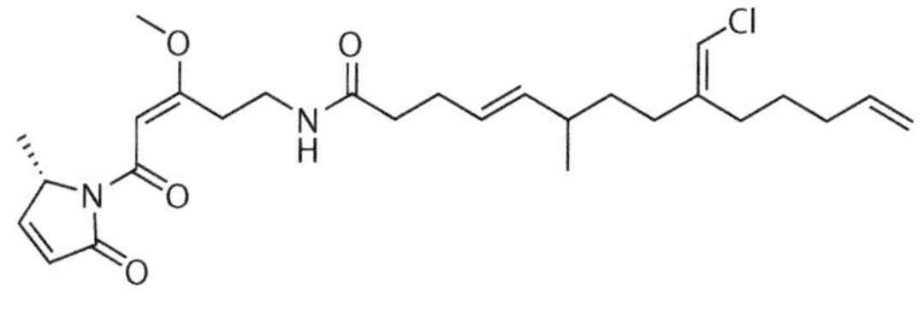

6

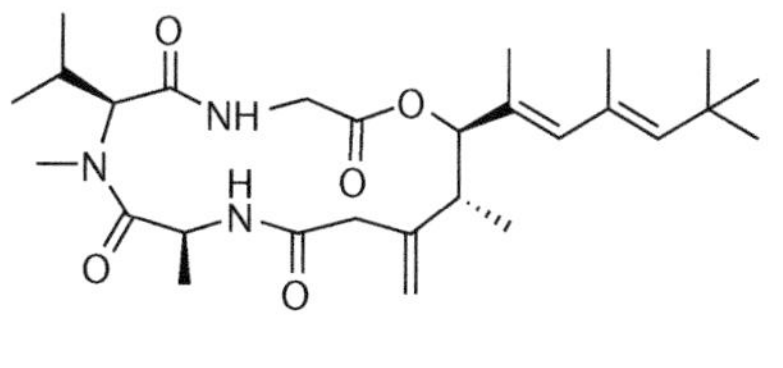

7

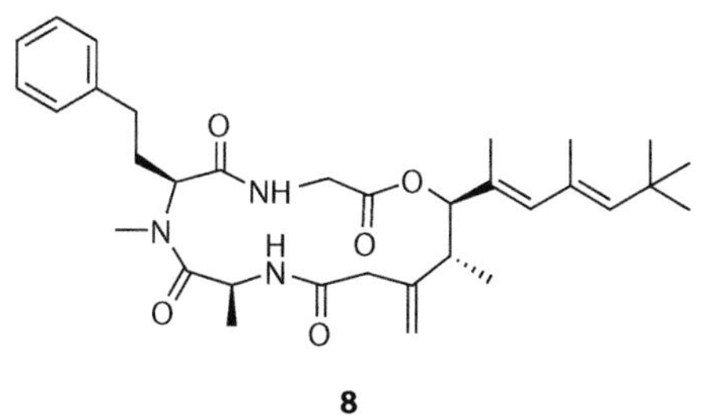

8

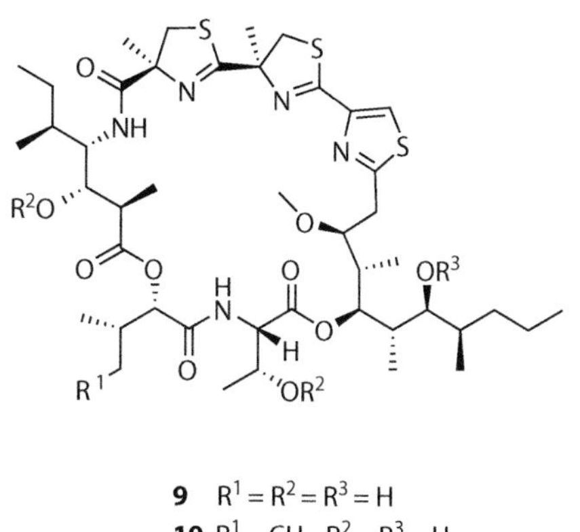

9 $R^1 = R^2 = R^3 = H$
10 $R^1 = CH_3, R^2 = R^3 = H$

10.2.1.1.5 Dinoflagellate Toxins

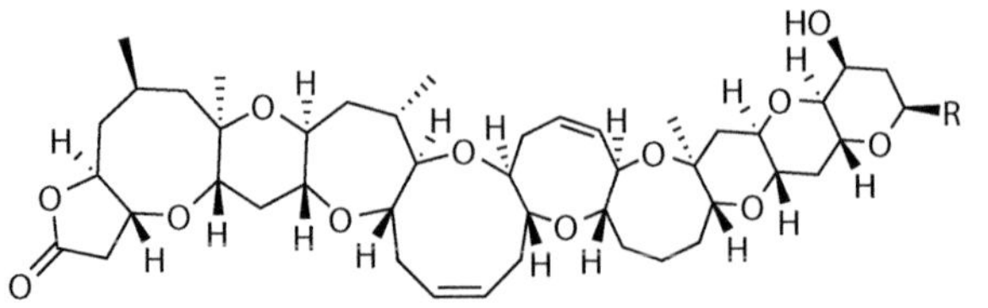

11

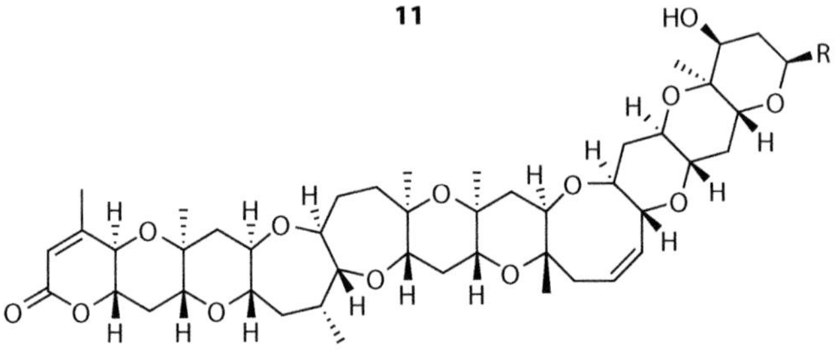

12

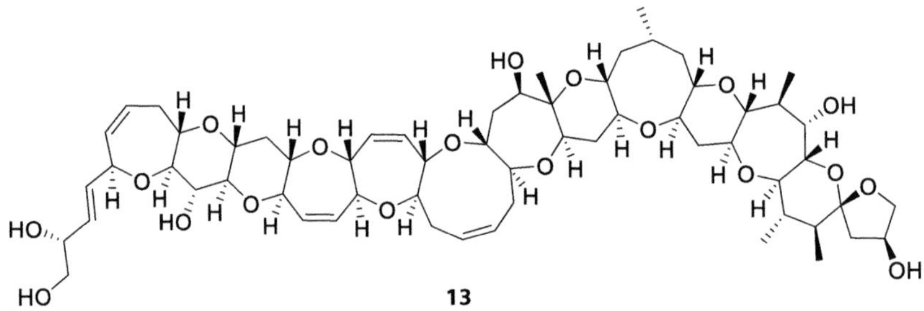

13

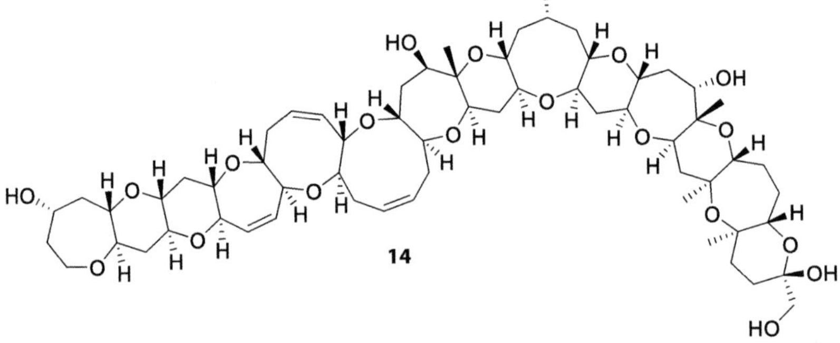

14

Two groups of ladder-shaped polyethers (LSPs), the brevetoxins (PbTx, **11** and **12**) and ciguatoxins (CTXs, **13** and **14**), have drawn considerable interest from researchers because of their involvement in neurotoxic shellfish poisoning and ciguatera fish poisoning.[58] Both of these toxins come from marine dinoflagellates and are composed of 10–13 connected cyclic ethers.[59] PbTxs are a group of neurotoxins found in the marine dinoflagellate species *Karenia brevis*.[60] All naturally occurring PbTxs identified to date belong to two types, type A PbTx (**11**) and type B PbTx (**12**), based on their backbone structures.[28] PbTxs bind with the neurotoxin receptor site 5 on the α subunit of Na_V channels and modify channel gating by maintaining an open channel at resting membrane potential and inhibiting channel inactivation.[28] CTXs were originally isolated from the dinoflagellate *Gambierdiscus toxicus* and are classified into two groups according to their activity. P-CTX (**13**) produces neurological symptoms, whereas C-CTX (**14**) results in gastrointestinal symptoms[61]; they are typically isolated from the Pacific Ocean and Caribbean Sea, respectively. CTXs bind with the Na_V channel at the same binding site as PbTx and are extremely potent, with EC_{50} (half-maximal effective concentration) values in the nanomolar to picomolar ranges.[28] LSPs are extremely potent in inhibiting Na_V channels and thus cannot be safely used therapeutically. However, they can be used as leads to develop novel compounds.

Na_V channels are targeted to offer wide therapeutic advantages, including local anesthesia, analgesia, anticonvulsant, and chronic pain management. But the effectiveness of targeting them by the currently available small molecules is restricted by severe side effects resulting from nonselectivity of the molecules.[37] Marine natural products can solve this issue because of their Na_V channel subtype selectivity. These molecules are highly potent, and their chemical diversity ranges from simple alkaloids to complex LSPs and peptides. Therefore, medicinal chemists have opportunities to consider them as leads and develop subtype-selective Na_V modulators with increased efficacy and safety profiles.

10.2.1.2 Calcium Channel Modulators

Voltage-gated calcium channels (Ca_V) are on the membranes of neurons and allow Ca^{2+} transport into the cell. Neuronal Ca^{2+} influx results in initiation of synaptic transmission.[62] After their identification from crustacean muscle tissue, subclasses of Ca_V channels have been identified in plants, invertebrates, and vertebrates.[63] Structurally, Ca_V channels are composed of α_1, α_2, β, δ, and γ subunits. The α_1 subunit is the largest subunit and forms the pore, voltage sensor, and gate. Its organization is similar to that of the α subunit of Na_V channels; it is divided into four domains (I–IV), each containing six transmembrane segments (S1–S6). Most interestingly, changing only three amino acids in the pore loop between S5 and S6 in domains I, III, and IV of the Na_V channel alters the selectivity to Ca^{2+} ions.[64] Variations of the α_1 subunit are the key determinant of subtypes for Ca_V channels, making this subunit the target of most Ca_V channel ligands.[64] γ subunits are found in skeletal and cardiac muscle Ca_V channels. The role of γ subunits in neuronal Ca_V channels is unknown.[62] There are 10 subtypes of Ca_V channels that are classified according to their voltage activation. High-voltage-activated channels include the L- (Ca_V 1.1–1.4), N- (Ca_V2.2), P/Q (Ca_V 2.1), and R-type (Ca_V 2.3) channels. T-type channels are the only low-voltage-activated channels (Ca_V 3.1–3.3).[64,65] These subtypes are critically involved in different pathological states related to CNS disorders and represent an opportunity for CNS drug development. Ca_V2.1 and 2.2 channels have been implicated in hyperexcitability disorders like pain and epilepsy.[66] Ca_V1.2 and 1.3 are involved in major neuropsychiatric conditions like mood disorders, drug dependency, and Parkinson's disease, making them potential therapeutic targets.[67] Marine organisms are a promising source of Ca_V channel modulatory compounds.

Ziconotide, a synthetic form of ω-conotoxin MVIIA, was the first marine-derived U.S. FDA approved drug. Ziconotide is one of three approved analgesics that act on N-type calcium channels (Ca_V2.2).[68,69] MVIIA, isolated from *Conus magus* in 1979, is a 25-amino acid linear peptide containing three disulfide cross-links that inhibit Ca_V2.2.[69,70] During chronic pain states, Ca_V2.2 channels are upregulated and contribute to nociception. Selective inhibition of Ca_V2.2 with MVIIA is analgesic,[71] as demonstrated with the intrathecal delivery of ziconotide for patients with severe chronic pain.[69] Other ω-conotoxins selectively inhibit neuronal Ca_V2.1 or 2.2 channels. MVIIC and

MVIID are $Ca_V2.1$ selective, while conotoxins CVID, CVIE, CVIF, and GVIA are $Ca_V2.2$ selective. The interactions between the ω-conotoxins with their binding sites are well understood, and important amino acid residues have been identified in the S5–S6 region of domain III that are required for binding.[72] Researchers are currently working to develop orally active ω-conotoxin mimics to improve the accessibility of $Ca_V2.2$ as an analgesic target.[37]

10.2.1.3 Potassium Channel Modulators

Potassium (K^+) channels are the largest, most diverse and widely distributed ion channel family, with 78 members, each containing a unique pore-forming segment. Based on their activation pattern and transmembrane segments, they are classified into four structural families: (1) inwardly rectifying two-transmembrane K^+ channels (K_{ir}: $K_{ir}1$–7); (2) background/leak or tandem two-pore four-transmembrane K^+ channels (K_{2P}: $K_{2P}1$–7, 9, 10, 12, 13, 15–18); (3) Ca^{2+}-activated six-transmembrane or seven-transmembrane K^+ channels (K_{Ca}: $K_{Ca}1$–5); and (4) voltage-gated six-transmembrane K^+ channels (K_V: K_V1–12).[13,73] Excitability of neurons, neuronal signaling, muscle contraction, and release of neurotransmitters are the fundamental actions mediated by these K^+ channels.[7] These channels represent important targets in the CNS for the development of drugs to treat multiple sclerosis, Alzheimer's, Parkinson's, anxiety, epilepsy, schizophrenia, bipolar disorder, CNS ischemia, and pain, including migraine.[13,73]

Conotoxins also target K^+ channels.[74] Conotoxin ViTx is a κ-conotoxin and was the first conotoxin found to modulate vertebrate K^+ channels. ViTx is a 35-amino acid linear peptide with four disulfide cross-links and was first isolated from *Conus virgo*. It has been shown to selectively block $K_V1.1$ and 1.3 channels.[74] A more recently discovered conotoxin from the same class, conotoxin sr11a, selectively inhibits $K_V1.2$ and 1.6.[75] Martel et al. identified the involvement of the $K_V1.2$, 1.3, and 1.6 channels in regulating dopamine (DA) release in the CNS.[76] So, these conotoxin compounds may turn out to be useful in the treatment of disorders of dopamine regulation, such as schizophrenia and Parkinson's disease.

Peptide toxins from sea anemones are also known to block current through potassium ion channels of excitable cells. The first potassium channel blockers isolated from marine sources were two peptide toxins. These toxins, BgK and ShK, were found in the sea anemones *Bunodosoma granulifera* and *Stichodactyla helianthus*, respectively. Sea anemone toxins with K^+ channel activity are classified into four types: type I peptides (35–37 amino acid residues, three disulfide bonds); type II peptides (58–59 amino acid residues, three disulfide bonds); type III peptides (41–42 amino acid residues, three disulfide bonds); and type IV peptides (8 amino acid residues, two disulfide bonds).[77] Besides K_V channels, sea anemone toxins also target the small-conductance and intermediate-conductance Ca^{2+}-activated K^+ channels (K_{Ca}).[78] Both BgK and ShK are type I peptides and are selective for $K_V1.1$, 1.2, 1.3, and 1.6 channels. BgK also blocks $K_V3.2$ channels; ShK blocks $K_V1.4$ channels. There are a number of other sea anemone toxins that also target the K_V1 (shaker) channels.[45] Type II toxins block the K_V1 channels with lower potency than type I toxins, but they also possess protease inhibition activity that protects them from degradation from proteolytic cleavage inside of prey.[79] Such protease protection can improve the pharmacokinetic profile of the compound and may be a useful attribute in the development of this sort of toxins for therapeutic use. Two type III toxins, BDS-I and BDS-II, were isolated from *Anemonia sulcata* and are the first specific blockers of the $K_V3.4$ family. These toxins act by modifying channel gating, unlike other anemone toxins.[77,79] The type IV structural class was introduced after the discovery of SHTX II from *S. haddoni*, which blocks K^+ channels in synaptosomal membranes. Channel subtype selectivity of SHTX II is unknown.[77]

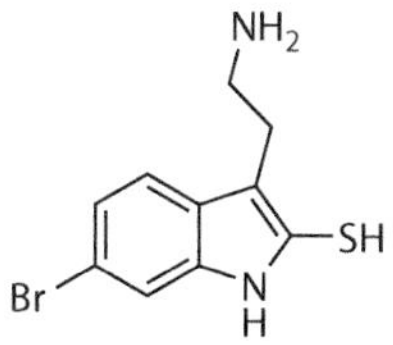

A nonpeptidic K⁺ channel inhibitor, 6-bromo-2-mercaptotryptamine (BrMT, **15**) was isolated from the marine snail *Calliostoma canaliculatum* off the Pacific coast of North America.[80] It was suggested that a dimer of BrMT binds potassium channels in the closed state and inhibits gating by stabilizing the closed conformation.[81] Gao et al. reported the first synthesis of the disulfane BrMT dimer (**16**), along with sulfane (**17**) and trisulfane (**18**) analogs. Electrophysiological assay of mouse $K_V1.1$ channels revealed that these compounds reduce K+ current amplitude and slow down the activation kinetics of the $K_v1.1$ channel. The study established that a small change in the distance between the two indole moieties does not change the activity of the compound significantly.[82]

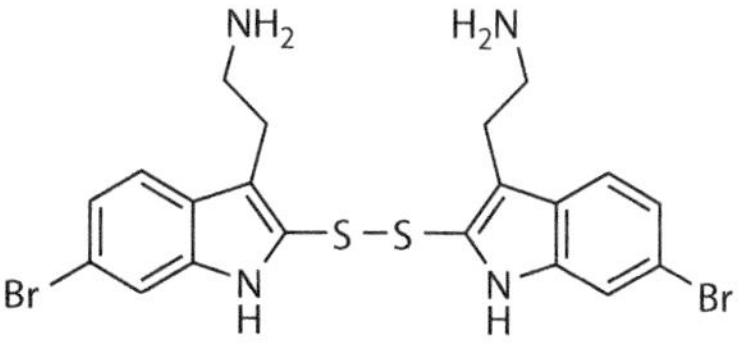

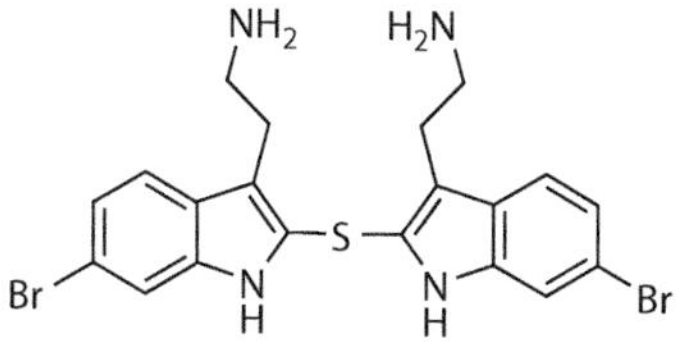

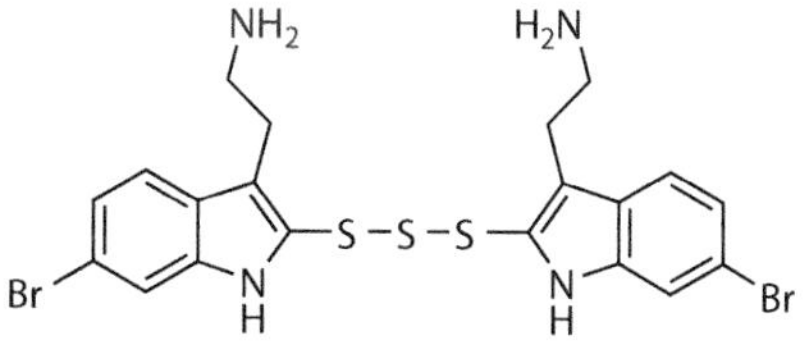

Involvement in a variety of cellular mechanisms has made K_V channels an important therapeutic target in a number of diseases. Activation of K_V channels is known to be beneficial in conditions like depression, neuronal conduction, and cognition disorders, while inhibition is useful in CNS hyper-excitability, seizures, pain, anxiety, bipolar disease, and schizophrenia.[73] In fact, K_V channels are so versatile that they can be targeted in cancer, inflammation, and cardiovascular and autoimmune diseases as well.[73] Marine organisms have provided both large and small molecules that can selectively modulate the subtypes of K_V channels. Therefore, they offer excellent leads for targeting therapeutically important K_V channels.

10.2.2 Ligand-Gated Ion Channels

10.2.2.1 nAChR Modulators

Nicotinic acetylcholine receptors (nAChRs) are Cys-loop, ligand-gated, cation-selective ion channels widely distributed in the central and peripheral nervous system. As a canonical ionotropic receptor, the neurotransmitter acetylcholine or exogenous ligands (e.g., nicotine) bind and permit transport of mono- or divalent cations (e.g., Na⁺, K⁺, Ca²⁺).[83] The channel is composed of five sub-units (α, β, γ, δ, and ε) in a homo- or heteromeric arrangement. Each subunit is divided into four transmembrane segments, M1–M4.[84] To date, 12 neuronal nAChRs have been described based on 12 variable subunits, nine α ($α_2$–$α_{10}$) and three β ($β_2$–$β_4$). The receptor function depends on the subunits forming the receptor.[85] In the brain, nAChRs are located both pre- and postsynaptically. Presynaptic nAChRs are responsible for neurotransmitter release, whereas postsynaptic nAChRs are involved in synaptic transmission and can excite or inhibit the postsynaptic cell, depending on neuronal cell type.[83] Widespread distribution, coupled with versatile activity and involvement in different disease states like Parkinson's, Alzheimer's, schizophrenia, neuropathic pain, and memory loss, makes them an important therapeutic target for drug discovery.[14] Marine natural products targeting the nAChR include a variety of chemical classes from diverse sources like cone snails, mollusks, corals, and sponges.[15,58]

10.2.2.1.1 Conotoxins and nAChRs

Conotoxins with nAChR-modulating activity are divided into seven different families of conotoxins: α-conotoxins, αC-conotoxins, αD-conotoxins, Ψ-conotoxins, αS-conotoxins, αL-conotoxins, and αJ-conotoxins.[37] The first isolated α-conotoxins were GI, GIA, and GII from *Conus geographus*. Although they lacked CNS activity,[86] they were used to explain nAChR pharmacology at the neuromuscular junction.[15] α-Conotoxin IMI, isolated from *Conus imperialis*, was the first conotoxin with mammalian CNS nAChR activity, and it is an antagonist of the α3β2, α7, and α9 subtypes.[87–89] α-Conotoxin Vc1.1, from *Conus victoriae*, is a potent and selective inhibitor of α9α10 nAChRs with an IC$_{50}$ value of 19 nM and was shown to decrease mechanical hyperalgesia in the chronic constriction injury and the partial nerve ligation models of neuropathic pain in rats.[8,15] It has now entered phase II clinical trials as a therapeutic entity for neuropathic pain.[90]

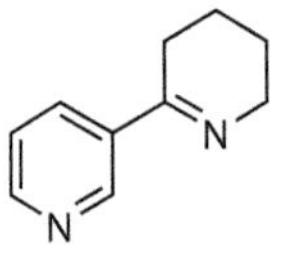

19 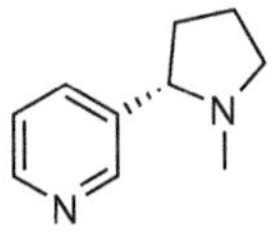   **20** 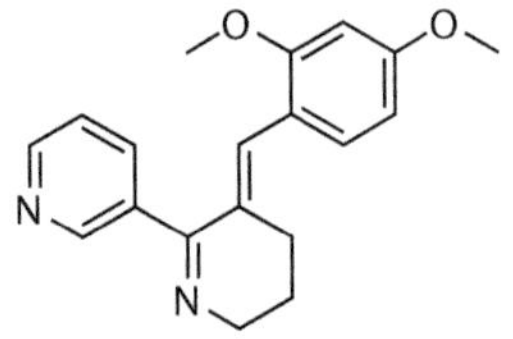   **21**

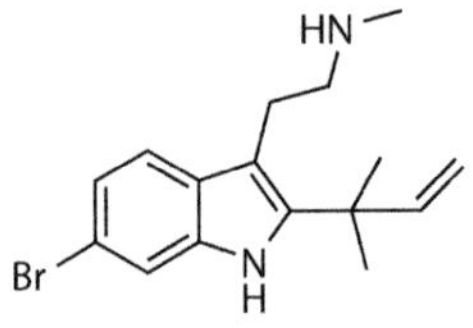

22 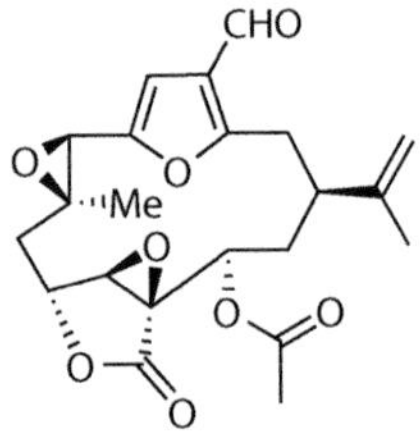   **23**

10.2.2.1.2 Nemertine Alkaloids

Anabasein (**19**) is a small-molecule nAChR ligand isolated from the Pacific nemertine (ribbon worms) *Paranemertes peregrine*. Structurally, it is related to nicotine (**20**), but unlike nicotine, it selectively stimulates the α7 subtype of nAChR.[91] The α7 nAChR has been identified as a potential therapeutic target to reverse the cognitive deficits in schizophrenia.[92] GTS-21 or 3-[2,4-dimethoxybenzylidene]anabaseine or DMXBA (**21**), derived from anabasein, has shown partial agonism at α7 nAChRs. DMXBA has reached the phase II clinical trial for the treatment of schizophrenia.[92,93]

10.2.2.1.3 Bryozoan Compounds

Another nAChR subtype-selective alkaloid, deformylflustrabromine (dFBr, **22**), was isolated from the North Sea bryozoan *Flustra foliacea*. dFBR was found to stimulate α4β2 and α2β2 receptors while suppressing the $\alpha 7$ receptor.[94–96] dFBR selectively stimulates $\alpha 4\beta 2$ by allosteric modification of channel gating. It represents an interesting lead for nAChR modulator development.[58,97] Researches have shown that agonism of the $\alpha 4\beta 2$ nACHRs improves reward, cognition, attention, and mood in animal models of addiction, depression, cognition, and attention.[98]

10.2.2.1.4 Gorgonian Compounds

Cembranoids are diterpenes with 14-membered carbocyclic skeletons capable of modulating nAChRs. The structural diversity of marine cembranoids allows them to act with receptors through different mechanisms, including irreversible inhibition at agonist sites, noncompetitive inhibition, or positive modulation.[99] This diversity makes cembranoids an interesting class of natural products in the search for treatments of neurological diseases. The marine cembranoid, lophotoxin (**23**), is found in few species of genera *Lophogorgia* and *Pseudopterogorgia*[100] and was the first reported member of this class of compounds. Studies on lophotoxin using neuronal nicotinic receptors in autonomic ganglia of chicks and rats revealed that it is a selective, high-affinity irreversible antagonist.[100–102]

The binding of agonists, antagonists, and allosteric modulators can shift the equilibrium of the channels' interconvertible states. nAChR agonists are studied to treat and understand diseases like Alzheimer's disease, Parkinson's disease, attention-deficit hyperactivity disorder, and various neuropathic pain syndromes (e.g., diabetic neuropathy, shingles, and cancer pain) where currently available therapeutics are not effective or possess undesirable side effects.[15] Antagonists are being developed to treat nicotine dependence, neuroendocrine, neuropsychiatric, and neurological diseases; memory and learning disabilities; eating disorders; pain; cardiovascular disease; and gastrointestinal disorders.[52] Compounds that act as positive allosteric modulators (PAMs) of nAChRs can enhance agonist effects and may be beneficial in conditions with decreased nicotinic tone. For example, in Alzheimer's disease, reduced nicotinic tone results from the decrease in expression of several nAChRs subtypes.[103] Cholinesterase inhibitors are one of the few approved treatments for Alzheimer's symptoms. Additional acetylcholine modulation with nAChR PAMs may further improve symptoms.[103] Of course, subtype selectivity is important since nonselectivity of PAMs in pathological conditions with variable nAChR densities may have differential effects on different populations of nAChRs.[103]

10.2.2.2 Glycine Receptor Modulators

The glycine receptor (GlyR) also belongs to the Cys-loop family of ligand-gated ion channels; it is a transmembrane protein that binds glycine on the extracellular surface and allows chloride ions to pass through the channel. Classically, GlyRs mediate inhibitory neurotransmission in the brain stem, spinal cord, and retina, where chloride concentration is normally higher extracellularly, and thus chloride influx causes the equilibrium potential to become more negative and inhibits signal transmission.[104] Structurally, GlyRs are composed of five subunits, and there are five subtypes that have been identified based on their subunit variation (α_1–α_4 with β), with the most common form found being the assembly of α_1 and β subunits.[105] GlyRs are being explored for therapeutic potential against pain and movement disorders, but detailed pharmacology is not well understood. There is an opportunity to develop subtype selective molecules that can differentiate between the different heteromeric GlyRs to better understand their specific roles in disease.[104] There is evidence for the role of the α_3 subtype in inflammatory pain. α_3 knockout mice show reduced pain sensation compared to wild-type mice. The α_3 subunit is the target for prostaglandin E_2–mediated pain sensation.[106] Missense point mutations in the α_1 subunit are responsible for hyperekplexia startle disease, and it has been shown that presynaptic GlyRs are a promising therapeutic target to treat hyperekplexia.[107] Library screening of more than 2500 extracts of southern Australian and

Antarctic marine invertebrates and algae for GlyR receptor modulatory activity has resulted in the discovery of the sesterterpene lactams (ircinialactams and ircinianin lactams) and indole-alkaloids as GlyR modulators.[108–110] Three sponges of the Irciniidae family produce three known metabolites, (12*E*,20*Z*,18*S*)-8-hydroxyvariabilin (**24**), along with the new class of glycinyl lactam sesterterpenes, 8-hydroxyircinialactam A (**25**) and 8-hydroxyircinialactam B (**26**). Sponges are sessile organisms that have evolved chemical defenses that protect them from potential grazers and allow them to reduce competition with other sessile organisms in their environment.[111] Oftentimes, sponges exist in very close symbiotic relationships with bacteria and other microorganisms; this creates difficulty in identifying the true producer of secondary metabolites.[112] Screening by whole-cell patch-clamp electrophysiology against recombinant α1 and α3 GlyRs has revealed that compounds **24–26** are subunit-specific modulators of GlyRs in the micromolar range. The glycinyl lactam sesterterpenes are α1-selective PAMs with no effects on the α3 GlyR. **24** has been found to be a PAM of the α1 and an antagonist of the α3 GlyR; this was the first report of a selective α3 GlyR antagonist.[108]

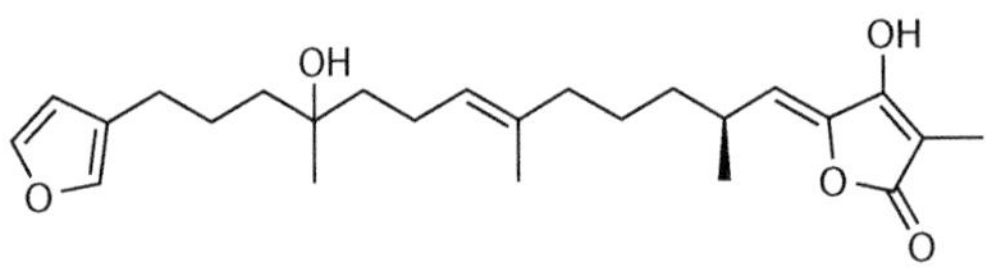

24

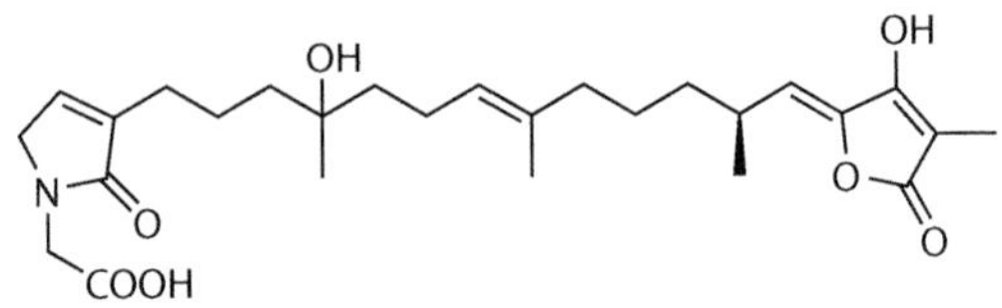

25

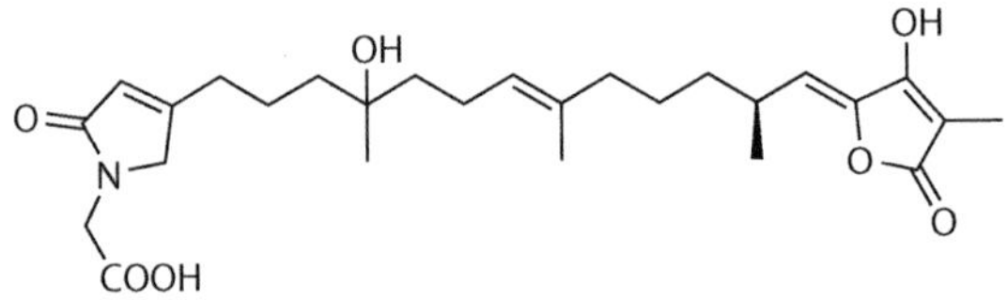

26

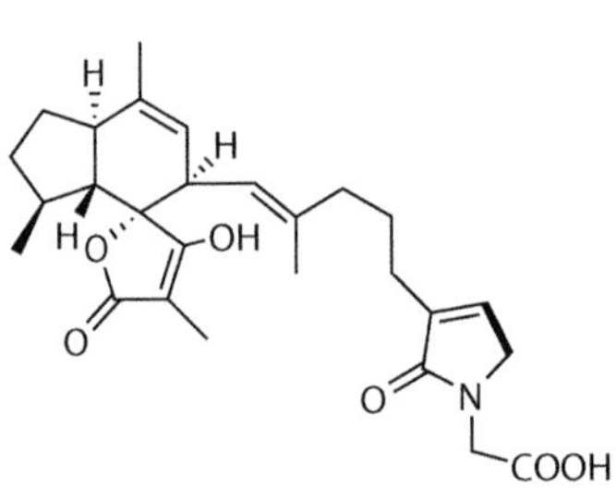

27

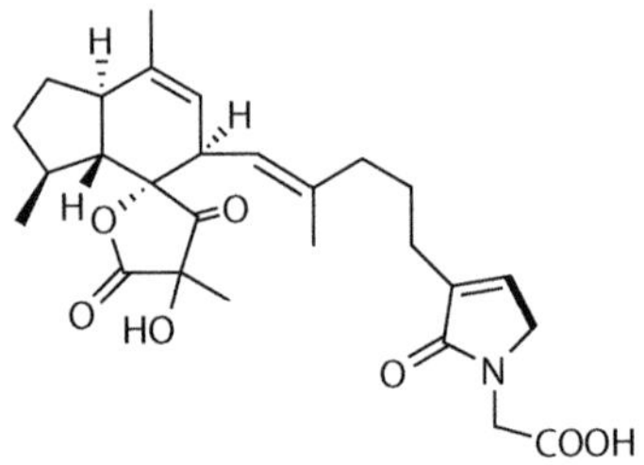

28

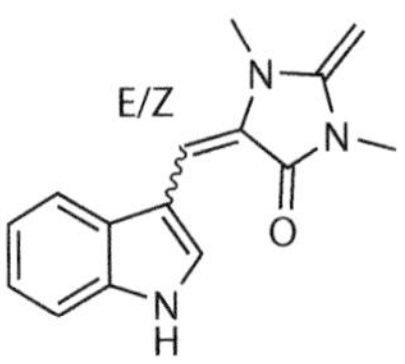

29(E)/30(Z)

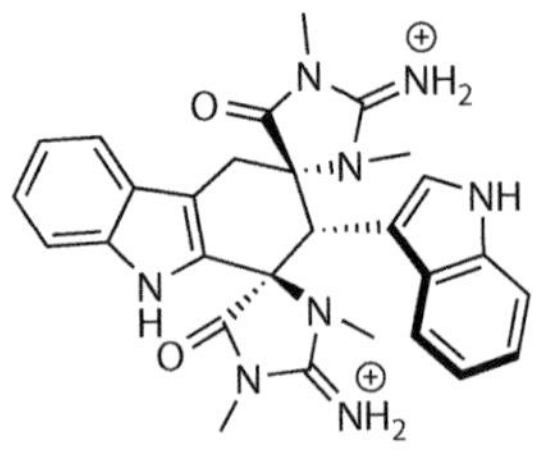

31 (TFA salt)

Later studies on sponges of the genus *Psammocinia* from southern Australia have resulted in the isolation of the ircinianinglycinyl lactams, (−)-ircinianin lactam A (**27**) and (−)-oxoircinianin lactam A (**28**), and their identification as GlyR modulators. Patch-clamp electrophysiology analysis of the ircinianinglycinyl lactams has shown that **27** is a selective α3 GlyR PAM, while **28** is an α1-selective PAM. SAR analysis of the ircinianinglycinyl lactams and other related metabolites provides evidence that the glycinyl lactam moiety is required for PAM activity in this class of compounds.[109]

8*E*-30-Deimino-30-oxoaplysinopsin (**29**), 8*Z*-30-deimino-30-oxoaplysinopsin (**30**), tubastrindole B (**31**), and indole alkaloids have been isolated from the sponge *Ianthella* cf. *flabelliformi* and have been found to show GlyR modulatory activity. A mixture of **29** and **30** antagonizes the α3 GlyR. Compound **31** shows selective antagonism at low concentrations and acts as a PAM at higher concentration for the α1 GlyR.[110]

Analysis of the marine sesterterpene lactams and indole alkaloids on GlyR pharmacology has resulted in the discovery of new classes of GlyR modulators. Currently available GlyR activators or antagonists either are weakly active or bind with multiple targets, resulting in undesirable off-target effects.[104,113] Selective activation or inhibition of GlyRs could be used in the development of therapeutic agents for muscle relaxation, epilepsy, inflammatory pain, and opioid-induced apnea and other breathing disturbances with increased potency and decreased side effects.[114–117]

10.2.2.3 Ionotropic GABA Receptor Modulators

Gamma-aminobutyric acid (GABA) is the principal inhibitory neurotransmitter of the brain; there are two types of GABA receptors: ionotropic receptors that are ligand-gated ion channels (GABA$_A$ and GABA$_C$) and metabotropic receptors that are GPCRs (GABA$_B$).[7] Ionotropic GABA receptors belong to the Cys-loop family of ligand-gated ion channels, which allows transport of chloride ions through a central pore to hyperpolarize the cell.[118] GABA$_A$ and GABA$_C$ are classified based on their subunit composition and functions.[119] They are pentameric proteins made up of a combination of subunits α_{1-6}, β_{1-3}, γ_{1-3}, δ, ε, π, θ, and ρ_{1-3}. The GABA$_A$ receptor is composed of two α, two β, and one γ subunit, while the GABA$_C$ receptor is formed by ρ_1, ρ_2, and ρ_3 in a homomeric or heteromeric fashion.[106] Each subunit has three essential parts: an extracellular N-terminal domain where ligands bind, four α-helical transmembrane domains (M1–M4), and an intracellular loop between M3 and M4. Ligand binding to the extracellular domain causes a conformational change in the transmembrane segments and opens the ion channel.[120]

Balance between neuronal excitation and inhibition is essential for normal brain function. Excessive inhibition or decreased excitation can cause coma, depression, low blood pressure, sedation, and sleep. On the other hand, excessive excitation or decreased inhibition can cause convulsions, anxiety, high blood pressure, restlessness, and insomnia. The principal strategy for drugs that target the GABAergic pathway is to regain homeostasis of the system.[121]

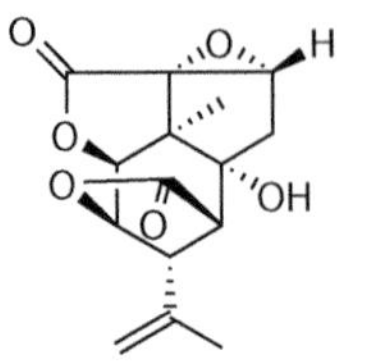

32a

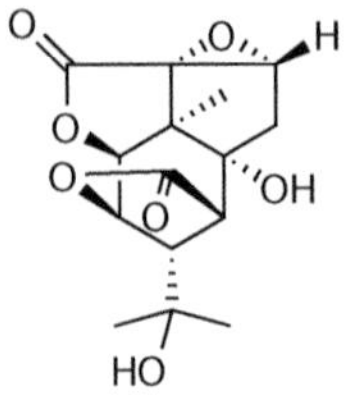

32b

32 (1:1)

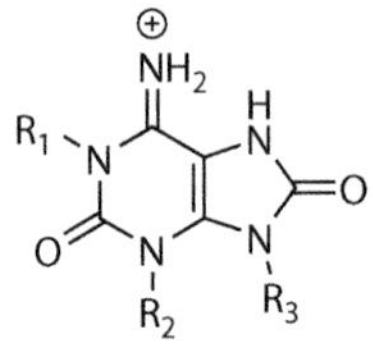

33 $R_1 = CH_3, R_2 = H, R_3 = CH_3$
34 $R_1 = CH_3, R_2 = CH_3, R_3 = H$
35 $R_1 = H, R_2 = CH_3, R_3 = CH_3$
36 $R_1 = H, R_2 = H, R_3 = CH_3$

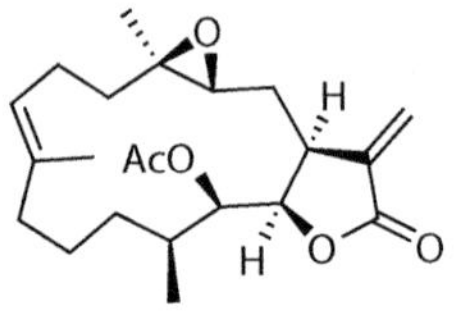

37

Picrotoxin (PTX, **32**) is an equimolar mixture of picrotoxinin (**32a**) and picrotin (**32b**), originally isolated from the berries of *Anamirta cocculus*,[122] and it has been used to study $GABA_A$ receptor pharmacology. PTX is a prototypical noncompetitive $GABA_A$ antagonist that binds at the intracellular end of the M2 helix and blocks the channel.[123] This plant-derived toxin was later isolated from a marine sponge of the genus *Spirastrella*.[124] Another group of GABAergic molecules from marine sponges are the 8-oxogunanines. A collection of several Palauan sponges, including *Cribrochalina olemda*, *Haliclona* sp., and *Amphimedon* sp., yielded four neurologically active oxoguanines (**33–36**). *In vivo* characterization of **33** showed convulsant effects, which were later correlated with inhibition of $GABA_A$ receptor-mediated inhibitory currents.[125]

Eupalmerin acetate (EPA, **37**) has activity as an agonist at the $GABA_A$ receptor and a noncompetitive inhibitor of peripheral nAChR.[126,127] EPA was isolated from the gorgonian octocorals *Eunicea succinea* and *Eunicea mammosa*. Studies have shown that EPA binds at the same site as that of the $GABA_A$ potentiator neurosteroids (allopregnanolone and tetrahydro-deoxycorticosterone) and enhances GABA or pentobarbital-generated current of the $GABA_A$ receptor.[127,128]

Both $GABA_A$ and $GABA_C$ are promising therapeutic targets. $GABA_A$ receptors are the target of several classes of therapeutic agents like benzodiazepines, anesthetics (propofol, etomidate), and barbiturates.[123] $GABA_C$ receptors are also linked to several disease states, including Alzheimer's disease and visual, sleep, and cognitive disorders. $GABA_A$ ligands with positive, negative, and neutralizing allosteric modulation each have potential therapeutic advantages, and multiple binding sites within the receptor allow for a variety of targets to explore.[129]

10.2.2.4 Ionotropic Glutamate Receptor Modulators

Ionotropic glutamate receptors (iGluRs) are ligand-gated ion channels that mediate fast excitatory synaptic transmission in the mammalian CNS.[130] They are cation-selective, homo- or heterotetrameric transmembrane proteins, with each subunit being composed of an extracellular amino terminal domain, an extracellular ligand binding domain, three transmembrane domains with a pore-forming P-loop, and an extracellular carboxy-terminal domain.[20,130] iGluRs are divided into three subclasses on the basis of their response to AMPA (α-amino-3-hydroxyl-5-methyl-4-isoxazolepropionic acid), kainate, and NMDA (N-methyl-D-aspartate). Sixteen iGluRs are classified into these subclasses as AMPA (GluA1–GluA4), kainate (GluK1–GluK5), and NMDA receptors (GluN1, GluN2A–GluN2D, and GluN3A and GluN3B).[130]

Marine organisms produce a number of compounds that selectively modulate the different subtypes of iGluRs. The kainate iGluRs are named because they were first characterized by their activation with kainic acid (**38**). Kainic acid is a small molecule first isolated from a Japanese marine algae, *Digenea simplex*.[20] Domoic acid (**39**), a kainate receptor agonist, contains the same 3-(carboxymethyl)pyrrolidine-2-carboxylic acid backbone as kainic acid, and was isolated from the marine red algae of the genus *Chondria* and diatoms of the genus *Pseudo-nitzschia*.[131] Kainic and domoic acid and their related congeners are called kainoids. They are pyrrolidine dicarboxylates and vary in their structures and activities based on substitution at the C4 positon.[132] Kainic and domoic acid evoke depolarizing currents as agonists of the kainate and AMPA receptors. Of note, they have higher binding affinity for the kainate receptor and produce excitatory and excitotoxic effects *in vivo*. SAR studies of the pyrrolidine dicarboxylate scaffold have revealed that receptor selectivity depends greatly on the stereochemistry of C2 and C4 and the structure of the C4 side chain.[20]

The kainate and AMPA receptor agonists, dysiherbaine (**40**) and neodysiherbaine (**41**), were isolated from a marine sponge *Lendenfeldia chondrodes*. They contain a *cis*-fused hexahydrofuro[3,2,-b] pyran ring substituted with a 3-[2-aminopropanoic acid] side chain. Later studies by Sakai et al. suggested that the compounds are actually the metabolite of an endosymbiotic cyanobacteria from *Synechosystis* sp.[133] Intraperitoneal injection of dysiherbaine in mice causes neurotoxic symptoms resembling the neuroexcitatory effects of the kainoids,[134] suggesting a common mechanistic pathway.

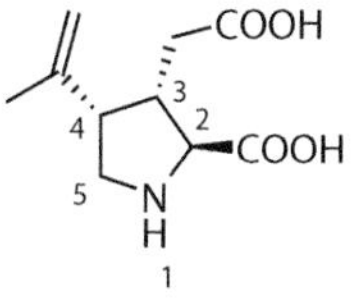

38 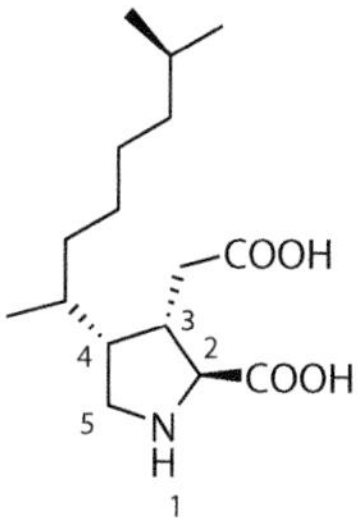 **39** 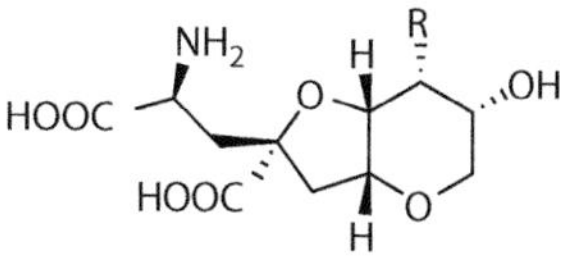 **40** R = CH₃NH **41** R = OH

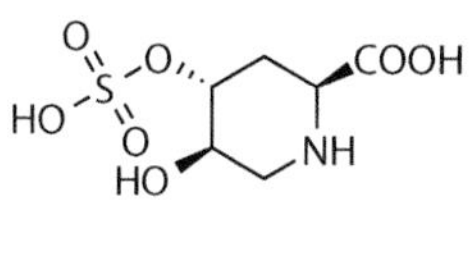

42 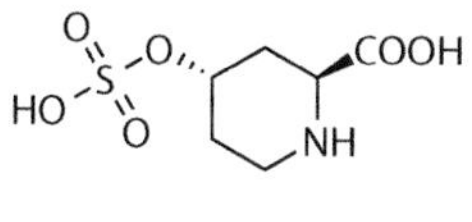 **43**

NMDA is present in a number of marine organisms.[135,136] In addition, other secondary metabolites with NMDA modulatory activity have been isolated from marine sponges. Two 4,5-substituted analogs of pipecolic acid, cribronic acid [(2*S*,4*R*,5*R*)-5-hydroxy-4-sulfooxy-piperidine2-carboxylic acid] (**42**) and (2*S*,4*S*)-4-sulfooxy-piperidine-2-carboxylic acid (*trans*-4-hydroxypipecolic acid sulfate [*t*-HPIS]) (**43**), have been isolated from the Palauan sponge *Cribrochalin aolemda* and from the Micronesian sponges *Axynella carteri* and *Stylotella aurantium*, respectively.[137] In addition to the small molecules, peptides from the venom of cone snails have been isolated that act on NMDA receptors. The conantokins are 17- to 27-amino acid peptides with a high prevalence of γ-carboxyglutamic acid residues that selectively and potently inhibit NMDA receptors.[37] NMDA receptors have been utilized as a potential target in neuropathic pain management.[138] Intrathecal administration of conantokins, which act as antagonists of the NR1 and NR2B subunits of NMDA receptors, decreases neuropathic pain in animal models.[37,139] Though one of the most effective conantokins, Con-G, was suspended in the preclinical stage due to motor-related side effects, other conantokins with NR2B subtype selectivity show potential for development as drugs for neuropathic pain.[37]

10.3 G PROTEIN-COUPLED RECEPTORS

GPCRs are a class of proteins found on the membranes of cells throughout the body, including the CNS. On an evolutionary timescale, GPCRs originated very early and can be found in organisms as diverse as plants, fungi, and protozoans.[140] Due to their long history, CNS GPCRs have diversified into a variety of different subtypes and are involved in physiological processes ranging from vision and smell to behavior and mood. While different classifications exist, human GPCRs can be classified into five main structural families: glutamate, rhodopsin, adhesion, frizzled/taste2, and secretin.[141] Of these groups, by far the largest is the rhodopsin family, which includes many receptors that are common drug targets in the central nervous system, such as the opioid, serotonin, and dopamine receptors.[142] Though GPCRs are extremely diverse, they share many common structural and functional components. GPCRs consist of seven transmembrane helices and have an extracellular N-terminal domain and an intracellular C-terminal domain. The extracellular domain is where ligand binding occurs, while the intracellular domain is where the receptor interacts with the G protein. The G protein is a heterotrimeric protein consisting of G_α, G_β, and G_γ subunits. In its inactive state, the three subunits of the G protein remain bound to the intracellular domain of the GPCR and guanosine diphosphate (GDP). Upon ligand binding, GDP is exchanged for guanosine triphosphate (GTP), and this leads to a conformational change in which the G protein splits into two parts: the G_α/GTP subunit and the G_β/G_γ subunit. These two subunits are released from the intracellular domain of the GPCR and act as secondary messengers inside of the cell. The distinct G_α, G_β, and G_γ subunits with which a GPCR interacts are what determine that receptor's specific function.[143]

Due to their ubiquitous nature, malfunctioning GPCRs are the cause of many different disease conditions, and it is estimated that 36% of drugs on the market today target GPCRs.[12] Some of these drugs and knowledge of the receptors that they target were originally developed using ethnopharmacological information. As described in the introduction, the discovery of morphine from poppy plants and its targeting of the μ-opioid receptor is a classic example of this process. It is presumed that compounds that act as exogenous ligands to GPCRs evolved for their antifeedant activity against potential herbivores.[144] Recently, researchers have moved beyond the diversity of terrestrial natural products and have started to uncover the biological diversity that exists in marine ecosystems for GPCR drug discovery. Several important modulators of GPCRs in the CNS are known to have been isolated from organisms in marine environments. Here, we review several GPCRs in the CNS for which marine-derived ligands have been found. These include the serotonin, opioid, cannabinoid, and adenosine receptors.

10.3.1 Serotonin Receptor Modulators

Serotonin (5-HT) is one of the principal monoamine neurotransmitters of the central nervous system. This neurotransmitter, along with the family of seven serotonin receptors (5-HT$_{1-7}$), with which it binds, has been found to play a role in many physiological processes and disease states, including mood, sleep, learning and memory, appetite, depression, anxiety, and disorders like schizophrenia.[16] Naturally occurring ligands with high affinity for the serotonin receptors have been derived from marine organisms, with several notable compounds originating from sponges. Compounds derived from sponges and their symbionts have been found to have a variety of bioactivities, and many of them have been found to act on serotonin receptors.

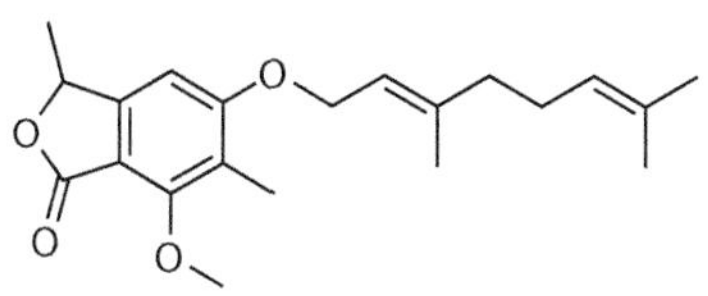

44

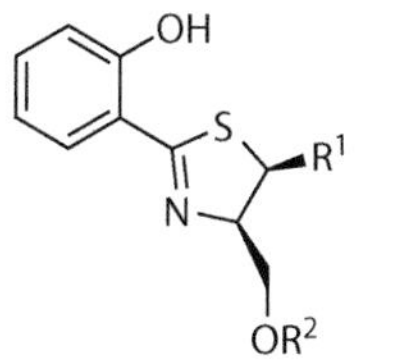

45 R^1 = CH$_3$, R^2 = H
54 R^1 = H ; R^2 = H

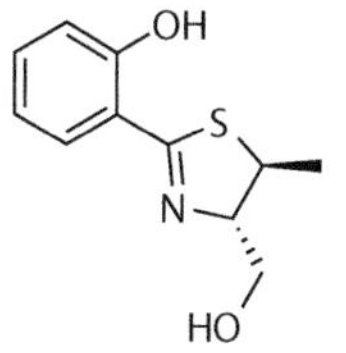

46

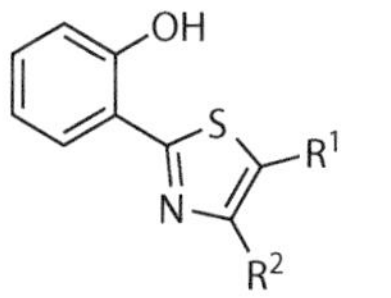

47 R^1 = CH$_3$, R^2 = CH$_2$OH
48 R^1 = CH$_3$, R^2 = CHO
49 R^1 = CH$_3$, R^2 = CONH$_2$
50 R^1 = H, R^2 = CONH$_2$

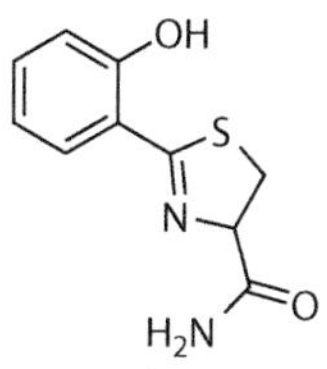

51

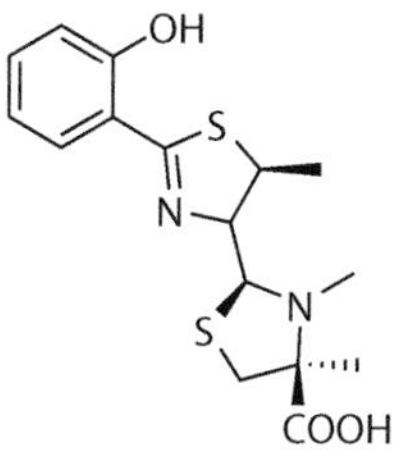

52

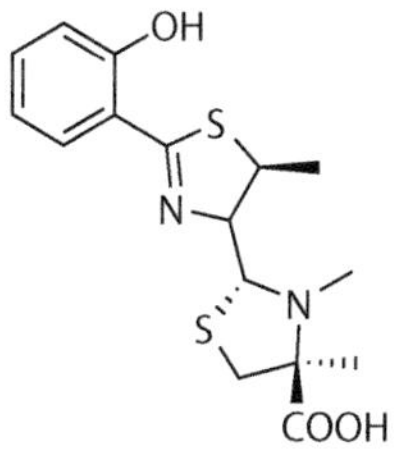

53

Marilone B (**44**) is a selective ligand for the serotonin 5-HT$_{2B}$ receptor and was isolated from a marine sponge and its symbiont. This ligand was obtained from a fungus (*Stachylidium* sp.) that lives on the sponge *Callyspongia* sp. cf. *C. flammea*. Marilone B is a phthalide [2-benzofuran-1(3H)-one] derivative.[145] Phthalide-containing compounds having been shown to have a variety of activities, including modulation in the CNS. Marilone B has the strongest affinity for 5-HT$_{2B}$ (K$_i$ = 7.7 μM) and acts as an antagonist.[145] In the CNS, the 5-HT$_{2B}$ receptor plays a role in anxiety, appetite, and sleep.[16] Antagonists of this receptor have anxiolytic properties[146] and are involved in the development of mechanical hyperalgesia in animal models.[147] It has also been shown that antagonizing 5-HT$_{2B}$ receptors in the periphery and spinal cord prevent formalin-induced flinching behavior in rodents, meaning that antagonists of this receptor could be used to develop analgesic drugs.[148] Another ligand selective for the 5-HT$_{2B}$ receptor and found in the marine environment comes from an actinomycete, *Streptomyces* sp. CP32, found living on a species of cone snail, *Conus pulicarius*.[149] Five novel compounds known as the pulicatins A–E (**45–49**) have been isolated from these bacteria. These symbiotic bacteria have also yielded several known compounds, including pulicatins F (**50**) and G (**51**) (two compounds now classified as pulicatins based on their structural similarity to the novel compounds described), watasemycins A (52) and B (53), and aerugine (54). All of these compounds are a part of the larger family of benzothiazole- and thiazoline-containing natural products. Of these, 45, 48, and 50–54 bind very strongly to the 5-HT$_{2B}$ receptor, with K$_i$ values ranging from 505 to 4965 nM. The thiazoline compounds, such as pulicatin A (45), have the highest affinity for the receptor, while the bisthiazolines, such as 52 and 53, have weaker affinities in comparison. Searching for new ligands of the 5-HT$_{2B}$ receptor is a challenge due to the known selective cardiac effects produced by manipulating this receptor.[150] Therefore, finding new compounds in the marine environment that are unlike any known ligands may allow for the discovery of compounds that lack these undesired effects via biased agonism.

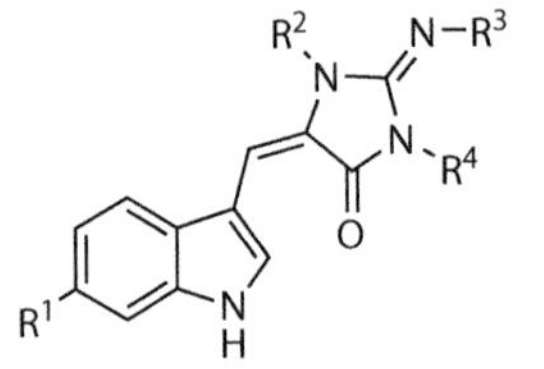

55 R^1= Br, R^2=R^3= H, R^4= CH$_3$
56 R^1 = Br, R^2 = R^4 = CH$_3$, R^3 = H
57 R^1 = H, R^2 = R^3 = R^4 = CH$_3$

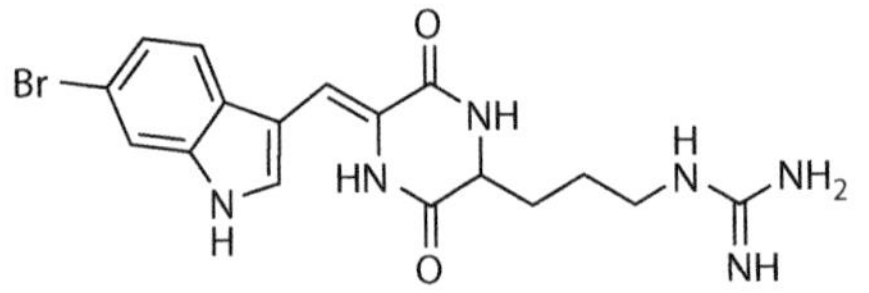

58

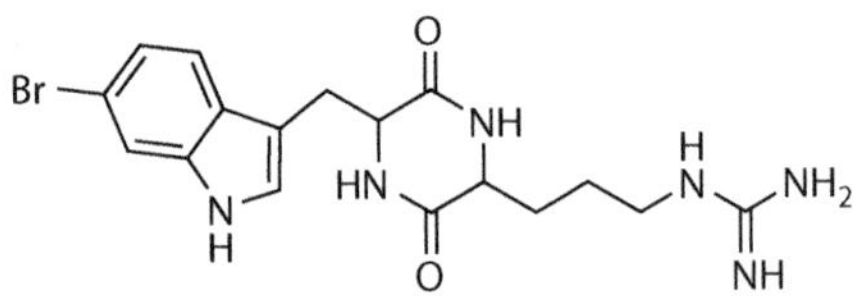

59

In addition to compounds that bind to the 5-HT$_{2B}$ receptor, marine sponges have also yielded compounds that bind to other serotonin receptor subtypes. For example, the Jamaican sponge *Smenospongia aurea* has been found to produce compounds that interact with the 5-HT$_{2A}$ as well as the 5-HT$_{2C}$ receptors.[151] Both of these receptor subtypes are involved with anxiety, appetite, sleep, mood, thermoregulation, and sexual behavior.[16] Three compounds isolated from this Jamaican sponge (6-bromo-2′-de-N-methylplysinopsin [55], 6-bromoaplysinopsin [56], and N-3′-ethylaplysinopsin [57]) all contain an indolamine pharmacophore identical to that found in the endogenous ligand serotonin. They also show affinity for the 5-HT$_{2A}$ and 5-HT$_{2C}$ receptor subtypes, with 6-bromoaplysinopsin in particular showing the highest affinity. Because these three indolamine compounds are structurally similar to serotonin, they may have been evolutionarily selected for because of these similarities. By contrast, compounds (44–54) isolated from *Stachylidium* sp. and *Streptomyces* sp. have no structural similarities to serotonin. They represent novel chemotypes to be further explored. These nonindolamine compounds could provide a basis for future drug development to develop compounds that interact with serotonin receptors differently from traditional ligands.

After screening and identifying structures of marine natural products that resemble serotonin, medicinal chemists have modified some of them to synthesize ligands with higher selectivity and affinity. An example of this is barretin (58), originally isolated from the marine sponge *Geodia barretti*, and the analogs synthesized from barretin.[152] Originally, barretin was found to be an antifouling agent that inhibited the deposition of settling stage larvae of the barnacle *Balanus improvises*. Barretin was later found to bind with specificity to serotonin receptors. This is most likely due to the homology between mammalian serotonin receptors and the 5-HT$_1$ receptors present in many species of barnacle. Testing the ability of several barretin derivatives to inhibit larval growth has shown that minor modification of its chemical structure can significantly change the ability of the compound to bind serotonin receptors and ultimately inhibit barnacle growth. In particular, with barretin and its analogs, much has been learned about the importance of the location of the bromine atom in giving this family of compounds activity. When bromine is replaced with Cl, NO_2, CH_3, or MeO at the fifth or sixth position of the indole ring, no significant bioactivity is displayed. Barretin and its derivative 8,9-dihydrobarretin (**59**) were shown to have selective serotonin binding activity. Barretin shows affinity for three serotonin receptor subtypes, $5\text{-}HT_{2A}$, $5\text{-}HT_{2C}$, and $5\text{-}HT_4$ (K_i = 1.93, 0.34, and 1.91 μM, respectively), while the dihydro analog shows exclusive affinity for the $5\text{-}HT_{2C}$ receptor (K_i = 4.63 μM). These compounds differ in that 8,9-dihydrobarretin contains a double bond between the tryptophan and arginine residues, whereas barretin contains a single bond in the same location. The double bond therefore plays an important role in determining selectivity among serotonin receptor subtypes. This research involving barretin derived from sponges provides evidence that marine products can be used to develop new pharmacological tools.[153]

10.3.2 Opioid Receptor Modulators

Opioid receptors are involved in a broad range of processes that include pain, addiction, and mood-related behavior.[3] These receptors fall into three families: the mu, delta, and kappa opioid receptors (MOR, DOR, and KOR, respectively). Unlike the serotonin receptors that have only one primary ligand found in mammals, the opioid receptors each interact with different endogenous opioid peptides, including endorphins, enkephalins, and dynorphins, and have different affinities for each. Due to the diversity of ligands able to bind with these receptors, many natural products from marine environments have been found to act as highly selective agonists and antagonists to all three of these receptor subtypes. Extracts obtained from vertebrates, invertebrates, and algae have been found to produce these types of compounds.

One notable example of a marine natural product derived from a vertebrate comes from an extract of the skin of the planehead filefish, *Stephanolepis hispidus*, originating in the marine waters of Brazil.[154] Fishermen in this part of the world are known to make water infusions of dried and powdered skins of this fish for treatment of mild inflammatory conditions of the respiratory system. Upon closer examination, it was discovered that these skin extracts have antinociceptive activity that is associated with activation of opioid receptors. The extracts, when administered intraperitoneally to mice, act as an analgesic in several behavioral models of pain, including the acetic acid writhing model, Hargreaves' test, and tail-flick assay. [154] Furthermore, when the skin extracts are coadministered with the known opioid receptor antagonist naloxone, some of these analgesic effects are blocked. This suggests that one or more compounds found in this skin extract induce analgesia through the activation of opioid receptors.[154] However, further studies are needed to determine which compounds are bioactive, their structures, and their selectivity.

Opioid ligands from invertebrates, such as the phidianidines (**60** and **61**) from the sea slug *Phidiana militaris*,[155] have been much better studied as to structure and pharmacological profile. Phidianidines A (**60**) and B (**61**) contain an unusual 1,2,4-oxadiazole ring linked to an indolamine structure. It was later found that the phidianidines could be synthesized from known indole acetic acids in relatively high yields,[156] making the problem of supply of these interesting compounds a nonissue. Further tests with the phidianidines have shown that they are selective MOR agonists (K_i = 310 and 680 nM, respectively), as well as inhibitors of the dopamine receptor (K_i = 230 and 340 nM, respectively). Selective ligands for opioid receptors have also been found in other marine invertebrates, including the Australian marine sponge, *Ianthella flabelliformis*.[157] The most potent ligand derived from this organism, bastadin-26 (**62**), has a high selectivity for the DOR (K_i = 100 nM). This compound is unusual in that it is a macrocyclic molecule in which two tyra-mino tyrosine groups are oxidatively coupled to yield a macrocyclic ether. Insufficient yields of extracts and the inability to synthesize this compound mean that it is still unknown whether basta-din-26 behaves as an agonist or antagonist at the DOR, and further studies need to be conducted.[157]

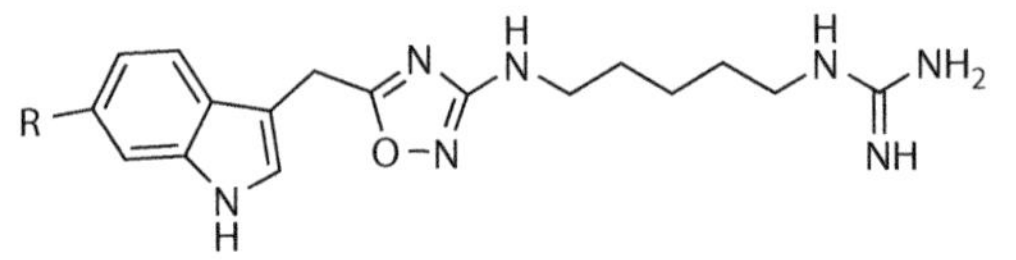

60 R = Br
61 R= H

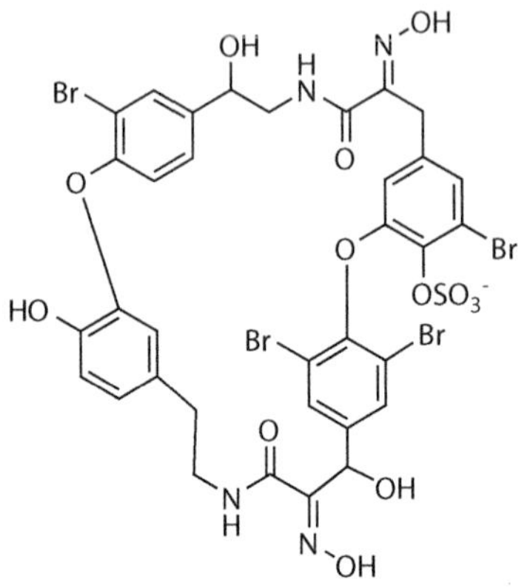

62

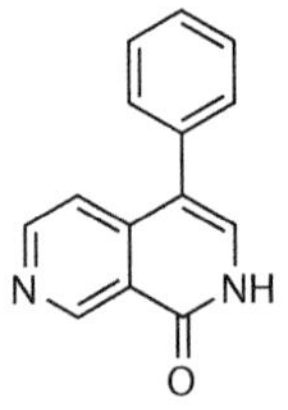

63

Ligands for the opioid receptors have also been found in species of red algae, including *Lophocladia* sp., a genus of algae found throughout the tropics and subtropics.[158] In one particular sample of this genera collected off of the coast of the Fiji islands, a compound known as lophocla-dine A (**63**) has been found to have DOR antagonist activity. Lophocladine A is an alkaloid with a 2,7-naphthrydine ring system. Until its discovery, the only naturally occurring example of a com-pound with this type of structure, dipyridylmethyl ketone, was found in the roots of the valerian plant, *Valeriana officinalis*.[158] Despite the vastly different sources, plant root versus red marine algae, these sources produce two similar chemical structures, demonstrating the relatedness of bio-synthetic pathways across species.

10.3.3 Cannabinoid Receptor Modulators

Cannabinoid receptors are activated by the endocannabinoids (endogenous cannabinoids) and are involved in processes including pain, anxiety, metabolic regulation, feeding behavior, and immune function. There are two main groups of these receptors: CB1 and CB2. CB1 recep-tors are of more interest to this discussion because they are primarily found in the CNS; CB2 receptors, on the other hand, are found primarily in immune cells.[159] Ligands for the CB1 can-nabinoid receptor have been obtained in marine environments, with the most notable originating from cyanobacteria.

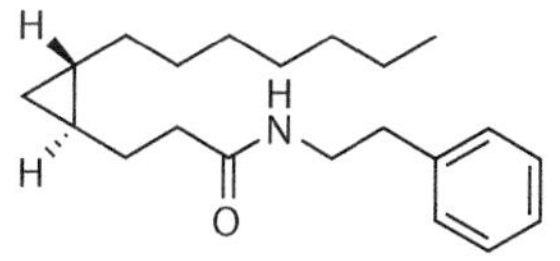

64

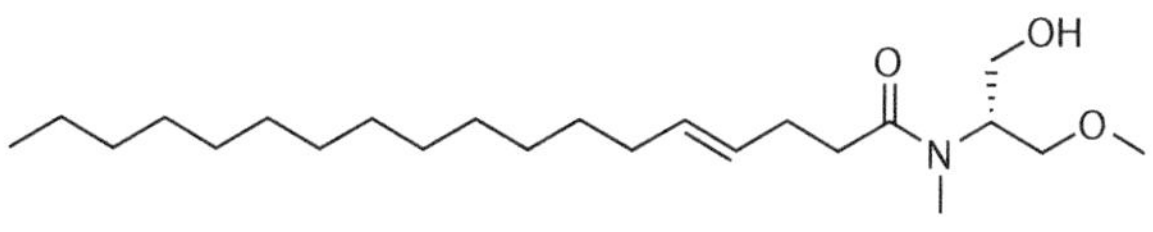

65

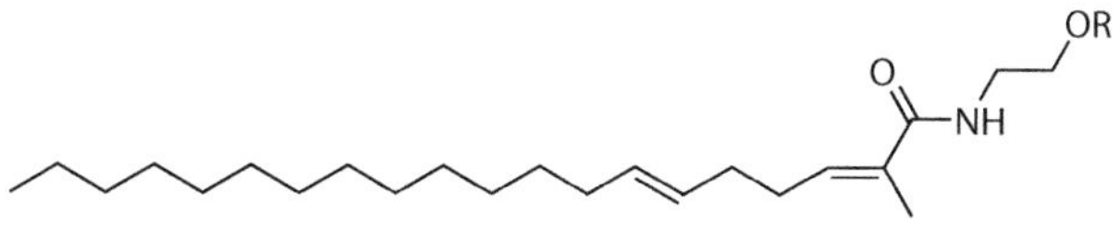

66 R=H
67 R= Ac

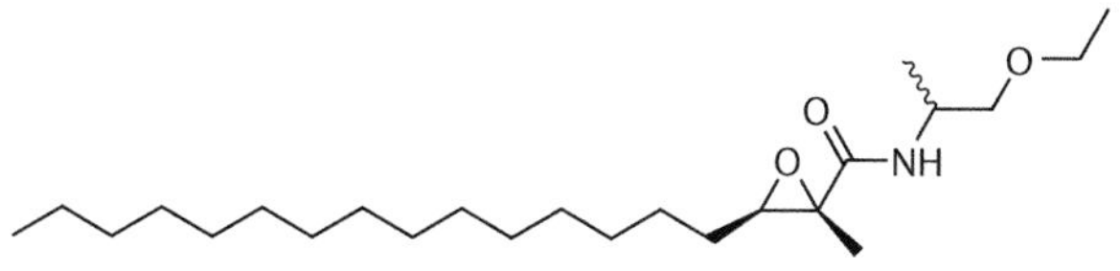

68

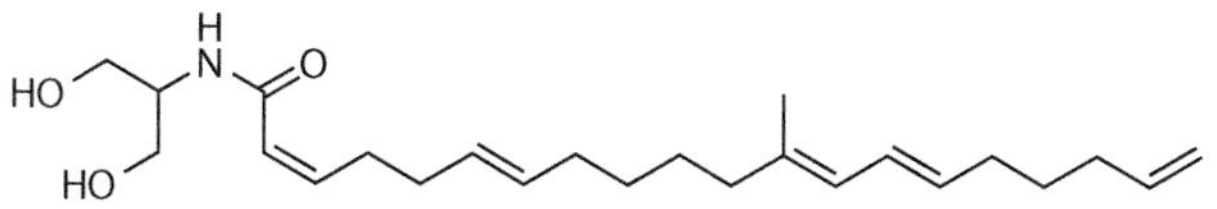

69

Many of the natural products derived from cyanobacteria have originated from one particular genera, *Lyngbya*, but from recent papers, this is likely an artifact of misidentification of samples by researchers.[160] Two compounds that both interact with cannabinoid receptors, grenadamide (**64**)[161] and serinolamide A (**65**),[162] were reported from collections of cyanobacteria obtained in different parts of the world. Grenadamide, obtained off of the island of Grenada in the Caribbean, is a cyclopropyl-containing fatty acid derivative with a β-phenylethylamine substructure. Since this substructure is common in many sympathomimetic drugs, it may be able to work in a similar fashion. Grenadamide shows some affinity for the CB1 receptor (K_i = 4.7 μM), providing additional evidence that this may be the case. Serinolamide A, also isolated from *Lyngbya majuscula*, shows agonist activity at the CB1 receptor (K_i = 1.3 μM). Serinolamide A is also a fatty acid derivative, but lacks the cyclopropyl-containing component of grenadamide. Serinolamide instead has a serinol unit, most likely derived from the incorporation of the amino acid serine. The sample of *Lyngbya majuscula* that was found to produce serinolamide was isolated from a sample collected from the coast of Papua New Guinea, displaying how similar, yet not identical compounds can be produced by related species depending on the environment in which they live. Varying ecological signals or factors could activate or express different biochemical pathways in the same species, leading to the production of different compounds. Understanding the circumstances in which these biosynthetic pathways are activated or finding alternative ways to synthesize compounds produced in the natural environment is an important aspect of marine natural product research.

Other species of cyanobacteria have been found to produce CB ligands as well. *Lyngbya semiplena* collected in Papua New Guinea produces three CB1 receptor binding agonists, semi-plenamides A (**66**), B (**67**), and G (**68**).[163] These compounds are fatty acid amides that resemble the structure of the endogenous cannabinoid anandamide (N-arachidonoylethanolamine). Their K_i values are 19.5, 18.7, and 17.9 μM, respectively, while anandamide has a K_i value of 0.4 μM at the CB1 receptor. The interactions of the three semiplenamides (**66–68**) are hypothesized to have a weaker interaction with CB1 because of the structure of their aliphatic carbon chains. All three compounds have linear aliphatic carbon chains; the chains of **67** and **68** are saturated, whereas **66** has one *trans*-double bond in the chain. SARs for fatty acid ligands at the CB1 receptor have shown that one or more *cis*-double bonds in the fatty acid chain are required to cause a U shape and allow tight binding with the receptor. Moreover, three conjugated double bonds are required for optimal interaction with the CB1 receptor, something that all three semiplenamides lack.

One final example of a cyanobacterium that produces cannabinoid receptor ligands comes from a species of a different genera, *Moorea bouillonii*.[164] This species, also collected from Papua New Guinea, produces a compound called mooreamide A (**69**), which has a structure very similar to that of anandamide and another endocannabinoid, 2-arachidonoylglycerol. Mooreamide A is a marine natural product that has shown the strongest CB1 receptor affinity to date, with a K_i value of 0.47 μM. This high affinity is likely due to the fact that mooreamide A has five double bonds in its fatty acid chain (one *trans*, four *cis*). It has also been suggested that the higher affinity may be due to the fact that both of the oxygen functional groups in the amine head group are free alcohols, unlike those in serinolamide. Regardless, mooreamide A offers a contrast to serinolamide and shows how molecular structure directly relates to the affinity of the molecules for the cannabinoid receptors. A comparison of these differences can be used in the synthesis of new compounds or generation of natural product derivatives with better pharmacological properties.

10.3.4 Adenosine Receptor Modulators

Adenosine receptors are found throughout the body, including the CNS, and are involved in cardiac rhythm, circulation, immunity, sleep regulation, angiogenesis, ischemia-reperfusion, and neurodegenerative disorders.[165] There are four types, A_1, A_{2A}, A_{2B}, and A_3, with the main endog-enous ligand for each of these receptors being the nucleic acid and adenosine triphosphate (ATP) intermediate, adenosine. Several ligands of the A_1 receptor, in particular, have been found in species of marine invertebrates.

The A_1 receptor exists throughout the body, but the highest amounts are located in the CNS, especially in the brain at excitatory nerve endings.[165] An example of a marine compound found to contain a ligand for the A_1 receptor is psammaplin A 11′-sulfate (**70**), isolated from the marine sponge *Aplysinella rhax*.[166] This compound is a nearly symmetrical bromotyrosine disulfide that is able to block the binding of the highly selective A_1 receptor antagonist DPCPX (8-cyclopentyl-1,3-dipropylxanthine) with an IC_{50} value of 90 μM. Similar A_1 receptor binding is seen with the compounds anthoptilides B (**71**) and C (**72**), isolated from the Australian sea pen *Anthoptilum* Cf. *kukanthali*.[167] These two compounds are diterpenes that block DPCPX binding to A_1 with high potency; their IC_{50} values are 45 and 3.1 μM, respectively. A final example of compounds able to block the binding of DPCPX is cembranes 1 (**73**) and 2 (**74**), isolated from a *Sarcophyton* coral.[168] These compounds are also diterpenes, which were previously identified synthetically, but were not known to occur naturally. Cembranes 1 and 2 inhibit the binding of DPCPX with an IC_{50} value of 300 μM for both. While all of these marine invertebrate-derived compounds have been shown to block the binding of the selective A_1 antagonist DPCPX, further testing is needed to determine their exact affinities and whether they act as agonists or antagonists to this receptor.

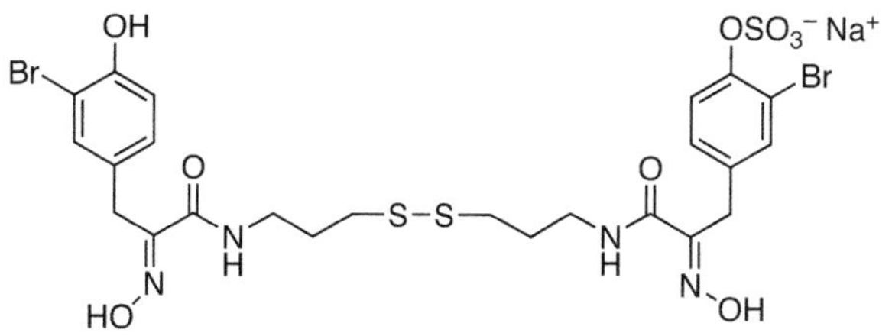

70

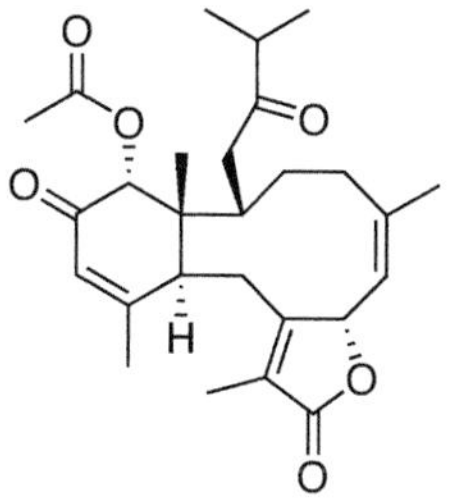

71

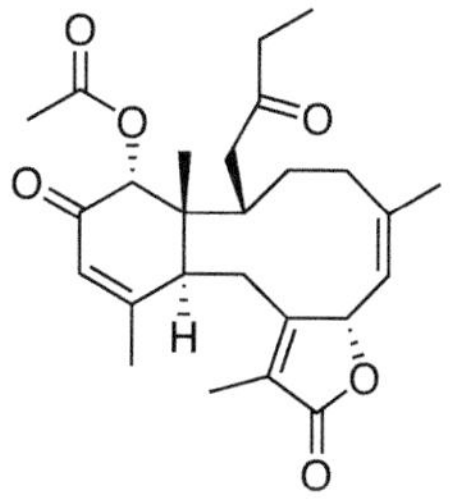

72

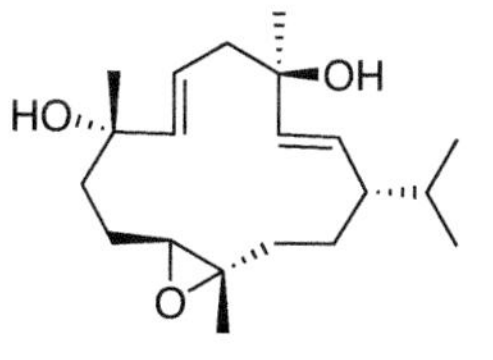

73

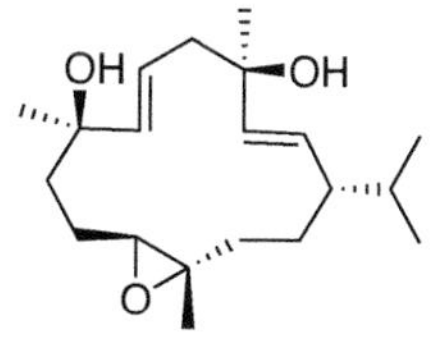

74

Other examples of marine-derived adenosine receptor binding compounds have been found to interact with multiple receptor subtypes, and more information is known about these compounds. One example, 5′-deoxy-5′-methylthioadenosine-2′,3′-diester (**75**), was isolated from the sea squirt *Atrolium robustum*.[169] This compound is a nucleoside that contains a rare methylthioadenosine moiety (a methylthio group at the 5′ position on the ribose in place of a hydroxyl group). The ribose moiety is also esterified with 3-(4-hydroxyphenyl)-2-meth-oxyacrylic acid at the 2′ position and urocanic acid at the 3′ position. The compound shows micromolar affinity for the A_1, A_{2A}, and A_3 receptors (K_i = 3.26, 17.1, and 6.94 µM, respectively). Additional assays have shown that this compound acts as a partial agonist of both the A_1 and A_3 receptors. Partial agonists of the A_1 receptor can be analgesic or sedative or act as cardiac depressants, implying that this compound could have a variety of potential therapeutic uses. Partial agonists of the A_1 receptor, because of their limited side effects, often have better therapeutic profiles than traditional full agonists. The addition of the two esterified groups makes this molecule more lipophilic, allowing it to be absorbed and cross the blood–brain barrier more easily. This implies that it could enter the CNS to exert its effects. Testing of this compound would first need to be done in animal models to see if any of these predicted effects exist, but to date this has not been done.

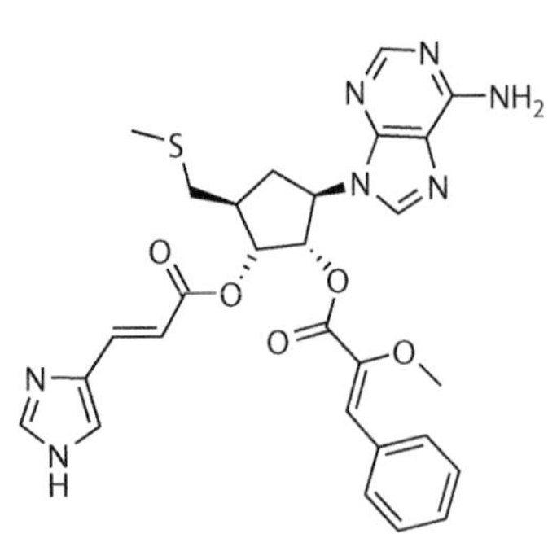

The β-carboline alkaloid, eudistomin D (**76**), originally isolated from the marine tunicate *Eudistoma olivaceum*,[170] provides an especially interesting story. Originally, this compound was found to be a potent inducer of calcium release from the sarcoplasmic reticulum (SR). While this is seemingly unrelated to effects on any receptors in the CNS, it is known that caffeine (**77**), a competitive antagonist to both the A_1 and A_{2A} receptors, is also an inducer of SR calcium release. Therefore, a series of analogs of caffeine and eudistomin D were made to examine the mechanism of action on these receptors. Indeed, a hybrid compound (**78**) made of both eudistomin and caffeine shows high affinity for the A_1 and A_{2A} receptors, with K_i values of 0.38 and 0.89 μM, respectively. These values are much better than those of caffeine for these same receptors ($K_i = 49$ and 18.1 μM, respectively), showing once again how modification of marine-derived products can lead to the synthesis of potent ligands for GPCRs.[170]

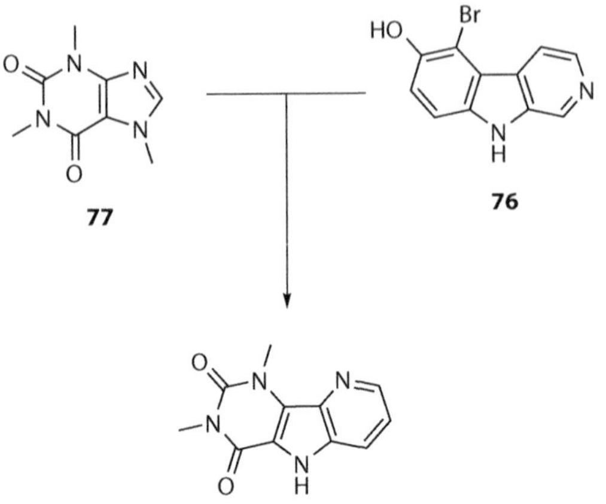

10.4 CONCLUSIONS

A vast number of compounds derived from marine organisms have been found to have bioactivity on mammalian ion channels and receptors. Researchers have only scratched the surface on what these compounds have to offer. As technology, collection, and screening methods improve, more and more compounds will be discovered and modified to make more potent and selective ligands. Some of these are already being used as therapeutic agents. In fact, several known pharmaceuticals on the market today were originally derived as marine natural products. These include trabectedin (originally found in the sea squirt *Ecteinascidia turbinata*) for the treatment of cancer,[171] vidarabine (originally synthesized from a compound from the sponge *Tethya crypta*) for the treatment of viral infection,[172] and ziconotide (originally found in the sea snail *Conus magus*) for the treatment of moderate to severe pain.[173] None of these compounds target GPCRs despite the fact that nearly 36% of all prescription drugs on the market today target a GPCR of some type.[12] Ion channels are found to be more frequently targeted by these compounds. But this trend does not negate the potential of GPCRs since discovery of ion channel modulatory activities results from more emphasis of ion channel research with marine natural products. Therefore, in addition to simply screening compounds for receptor binding activity, the next major step of marine product research will be to test GPCR modulatory compounds in biological assays that look for potential therapeutic effects. With the vast array of physiological processes that GPCRs regulate, some of these ligands could be developed into analgesics, anxiolytics, antidepressants, and more.

REFERENCES

1. Prisinzano, T. E. Natural products as tools for neuroscience: Discovery and development of novel agents to treat drug abuse. *J Nat Prod* 2009, 72 (3), 581–87.
2. Calixto, J. B., Scheidt, C., Otuki, M., Santos, A. R. Biological activity of plant extracts: Novel analgesic drugs. *Expert Opin Emerg Drugs* 2001, 6 (2), 261–79.
3. Waldhoer, M., Bartlett, S. E., Whistler, J. L. Opioid receptors. *Annu Rev Biochem* 2004, 73, 953–90.
4. Proksch, P., Edrada, R. A., Ebel, R. Drugs from the seas: Current status and microbiological implications. *Appl Microbiol Biotechnol* 2002, 59 (2–3), 125–34.
5. Gerwick, W. H., Fenner, A. M. Drug discovery from marine microbes. *Microbial Ecol* 2013, 65 (4), 800–6.
6. Haefner, B. Drugs from the deep: Marine natural products as drug candidates. *Drug Discov Today* 2003, 8 (12), 536–44.
7. Waszkielewicz, A. M., Gunia, A., Szkaradek, N., Sloczynska, K., Krupinska, S., Marona, H. Ion channels as drug targets in central nervous system disorders. *Curr Med Chem* 2013, 20 (10), 1241–85.
8. Lim, W. K, GPCR drug discovery: Novel ligands for CNS receptors. *Recent Pat CNS Drug Discov* 2007, 2 (2), 107–12.
9. Kaczorowski, G. J., McManus, O. B., Priest, B. T., Garcia, M. L. Ion channels as drug targets: The next GPCRs. *J Gen Physiol* 2008, 131 (5), 399–405.
10. Martins, A., Vieira, H., Gaspar, H., Santos, S. Marketed marine natural products in the pharmaceutical and cosmeceutical industries: Tips for success. *Mar Drugs* 2014, 12 (2), 1066–101.
11. Alexander, S. P. H., Mathie, A., Peters, J. A. Guide to receptors and channels (GRAC), 5th edition. *Br J Pharmacol* 2011, 164, S1–S2.
12. Rask-Andersen, M., Almén, M. S., Schiöth, H. B. Trends in the exploitation of novel drug targets. *Nat Rev Drug Discov* 2011, 10 (8), 579–90.
13. Judge, S. I. V., Smith, P. J., Stewart, P. E., Bever Jr., C. T. Potassium channel blockers and openers as CNS neurologic therapeutic agents. *Recent Pat CNS Drug Discov* 2007, 2 (3), 200–28.
14. Yu, R., Kompella, S. N., Adams, D. J., Craik, D. J., Kaas, Q. Determination of the alpha-conotoxin Vc1.1 binding site on the alpha9alpha10 nicotinic acetylcholine receptor. *J Med Chem* 2013, 56 (9), 3557–67.
15. Livett, B. G., Sandall, D. W., Keays, D., Down, J., Gayler, K. R., Satkunanathan, N., Khalil, Z. Therapeutic applications of conotoxins that target the neuronal nicotinic acetylcholine receptor. *Toxicon* 2006, 48 (7), 810–29.
16. Pytliak, M., Vargova, V., Mechirova, V., Felsoci, M. Serotonin receptors—from molecular biology to clinical applications. *Physiol Res* 2011, 60 (1), 15–25.
17. Bagal, S. K., Chapman, M. L., Marron, B. E., Prime, R., Storer, R. I., Swain, N. A. Recent progress in sodium channel modulators for pain. *Bioorg Med Chem Lett* 2014, 24 (16), 3690–99.
18. Olincy, A., Stevens, K. E. Treating schizophrenia symptoms with an α7 nicotinic agonist, from mice to men. *Biochem Pharmacol* 2007, 74 (8), 1192–1201.
19. Narahashi, T. Tetrodotoxin: A brief history. *Proc Jpn Acad Ser B Phys Biol Sci* 2008, 84 (5), 147–54.
20. Swanson, G. T., Sakai, R. Ligands for ionotropic glutamate receptors. *Prog Mol Subcell Biol* 2009, 46, 123–57.
21. Bagal, S. K., Brown, A. D., Cox, P. J., Omoto, K., Owen, R. M., Pryde, D. C., Sidders, B., Skerratt, S. E., Stevens, E. B., Storer, R. I., Swain, N. A. Ion channels as therapeutic targets: A drug discovery perspective. *J Med Chem* 2013, 56 (3), 593–624.
22. Cummins, T. R., Sheets, P. L., Waxman, S. G. The roles of sodium channels in nociception: Implications for mechanisms of pain. *Pain* 2007, 131 (3), 243–257.
23. M. Waszkielewicz, A., Gunia, A., Szkaradek, N., Sloczynska, K., Krupinska, S., Marona, H. Ion channels as drug targets in central nervous system disorders. *Current Med Chem* 2013, 20 (10), 1241–85.
24. Stafstrom, C. E. Mechanisms of action of antiepileptic drugs: The search for synergy. *Curr Opin Neurol* 2010, 23 (2), 157–63. DOI: 10.1097/WCO.0b013e32833735b5.
25. Cestele, S., Catterall, W. A. Molecular mechanisms of neurotoxin action on voltage-gated sodium channels. *Biochimie* 2000, 82 (9–10), 883–92.
26. Hanifin, C. T. The chemical and evolutionary ecology of tetrodotoxin (TTX) toxicity in terrestrial vertebrates. *Mar Drugs* 2010, 8 (3), 577–93.

27. Mansson, M., Gram, L., Larsen, T. O. Production of bioactive secondary metabolites by marine vibrio-naceae. *Mar Drugs* 2011, 9 (9), 1440–68.

28. Al-Sabi, A., McArthur, J., Ostroumov, V., French, R. Marine toxins that target voltage-gated sodium channels. *Mar Drugs* 2006, 4 (3), 157–92.

29. Nieto, F. R., Cobos, E. J., Tejada, M. A., Sanchez-Fernandez, C., Gonzalez-Cano, R., Cendan, C. M. Tetrodotoxin (TTX) as a therapeutic agent for pain. *Mar Drugs* 2012, 10 (2), 281–305.

30. Vasquez, M., Gruttner, C., Moeller, B., Moore, E. R. Limited selection of sodium channel blocking toxin-producing bacteria from paralytic shellfish toxin-contaminated mussels (*Aulacomya ater*). *Res Microbiol* 2002, 153 (6), 333–38.

31. Gallacher, S., Flynn, K. J., Franco, J. M., Brueggemann, E. E., Hines, H. B. Evidence for production of paralytic shellfish toxins by bacteria associated with *Alexandrium* spp. (Dinophyta) in culture. *Appl Environ Microbiol* 1997, 63 (1), 239–45.

32. Hagen, N. A., du Souich, P., Lapointe, B., Ong-Lam, M., Dubuc, B., Walde, D., Love, R., Ngoc, A. H. Tetrodotoxin for moderate to severe cancer pain: A randomized, double blind, parallel design multi-center study. *J Pain Symptom Manage* 2008, 35 (4), 420–29.

33. Kohane, D. S., Lu, N. T., Gokgol-Kline, A. C., Shubina, M., Kuang, Y., Hall, S., Strichartz, G. R., Berde, C. B. The local anesthetic properties and toxicity of saxitonin homologues for rat sciatic nerve block *in vivo*. *Reg Anesth Pain Med* 2000, 25 (1), 52–59.

34. Chorny, M., Levy, R. J. Site-specific analgesia with sustained release liposomes. *Proc Natl Acad Sci USA* 2009, 106 (17), 6891–92.

35. McIntosh, J. M., Jones, R. M. Cone venom: From accidental stings to deliberate injection. *Toxicon* 2001, 39 (10), 1447–51.

36. Shen, G. S., Layer, R. T., McCabe, R. T. Conopeptides: From deadly venoms to novel therapeutics. *Drug Discov Today* 2000, 5 (3), 98–106.

37. Lewis, R. J., Dutertre, S., Vetter, I., Christie, M. J. *Conus* venom peptide pharmacology. *Pharmacol Rev* 2012, 64 (2), 259–98.

38. Knapp, O., McArthur, J. R., Adams, D. J. Conotoxins targeting neuronal voltage-gated sodium channel subtypes: Potential analgesics? *Toxins* (Basel) 2012, 4 (11), 1236–60.

39. McIntosh, J. M., Hasson, A., Spira, M. E., Gray, W. R., Li, W., Marsh, M., Hillyard, D. R., Olivera, B. M. A new family of conotoxins that blocks voltage-gated sodium channels. *J Biol Chem* 1995, 270 (28), 16796–802.

40. Wilson, M. J., Zhang, M. M., Azam, L., Olivera, B. M., Bulaj, G., Yoshikami, D. Navbeta subunits mod-ulate the inhibition of Nav1.8 by the analgesic gating modifier muO-conotoxin MrVIB. *J Pharmacol Exp Ther* 2011, 338 (2), 687–93.

41. Daly, N. L., Ekberg, J. A., Thomas, L., Adams, D. J., Lewis, R. J., Craik, D. J. Structures of muO-conotoxins from *Conus marmoreus*: Inhibitors of tetrodotoxin (TTX)-sensitive and TTX-resistant sodium channels in mammalian sensory neurons. *J Biol Chem* 2004, 279 (24), 25774–82.

42. Terlau, H., Olivera, B. M. *Conus* venoms: A rich source of novel ion channel-targeted peptides. *Physiol Rev* 2004, 84 (1), 41–68.

43. Bosmans, F., Tytgat, J. Sea anemone venom as a source of insecticidal peptides acting on voltage-gated Na+ channels. *Toxicon* 2007, 49 (4), 550–60.

44. Moran, Y., Gordon, D., Gurevitz, M. Sea anemone toxins affecting voltage-gated sodium channels: Molecular and evolutionary features. *Toxicon* 2009, 54 (8), 1089–101.

45. Frazão, B., Vasconcelos, V., Antunes, A. Sea anemone (Cnidaria, Anthozoa, Actiniaria) toxins: An over-view. *Mar Drugs* 2012, 10 (8), 1812–51.

46. Smith, J. J., Blumenthal, K. M. Site-3 sea anemone toxins: Molecular probes of gating mechanisms in voltage-dependent sodium channels. *Toxicon* 2007, 49 (2), 159–70.

47. Diochot, S., Baron, A., Rash, L. D., Deval, E., Escoubas, P., Scarzello, S., Salinas, M., Lazdunski, M. A new sea anemone peptide, APETx2, inhibits ASIC3, a major acid-sensitive channel in sensory neurons. *EMBO J* 2004, 23 (7), 1516–25.

48. Wemmie, J. A., Taugher, R. J., Kreple, C. J. Acid-sensing ion channels in pain and disease. *Nat Rev Neurosci* 2013, 14 (7), 461–71.

49. Blanchard, M. G., Rash, L. D., Kellenberger, S. Inhibition of voltage-gated Na(+) currents in sensory neurones by the sea anemone toxin APETx2. *Br J Pharmacol* 2012, 165 (7), 2167–77.

50. LePage, K. T., Goeger, D., Yokokawa, F., Asano, T., Shioiri, T., Gerwick, W. H., Murray, T. F. The neurotoxic lipopeptide kalkitoxin interacts with voltage-sensitive sodium channels in cerebellar granule neurons. *Toxicol Lett* 2005, 158 (2), 133–39.

51. Edwards, D. J., Marquez, B. L., Nogle, L. M., McPhail, K., Goeger, D. E., Roberts, M. A., Gerwick, W. H. Structure and biosynthesis of the jamaicamides, new mixed polyketide-peptide neurotoxins from the marine cyanobacterium *Lyngbya majuscula*. *Chem Biol* 2004, 11 (6), 817–33.

52. Dwoskin, L. P., Crooks, P. A. Competitive neuronal nicotinic receptor antagonists: A new direction for drug discovery. *J Pharmacol Exp Ther* 2001, 298 (2), 395–402.

53. Vinothkumar, S., Parameswaran, P. S. Recent advances in marine drug research. *Biotechnol Adv* 2013, 31 (8), 1826–45.

54. Araoz, R., Molgo, J., Tandeau de Marsac, N. Neurotoxic cyanobacterial toxins. *Toxicon* 2010, 56 (5), 813–28.

55. Orjala, J., Nagle, D. G., Hsu, V., Gerwick, W. H. Antillatoxin: An exceptionally ichthyotoxic cyclic lipo-peptide from the tropical cyanobacterium *Lyngbya majuscula*. *J Am Chem Soc* 1995, 117 (31), 8281–82.

56. Cao, Z., Gerwick, W. H., Murray, T. F. Antillatoxin is a sodium channel activator that displays unique efficacy in heterologously expressed rNav1.2, rNav1.4 and rNav1.5 alpha subunits. *BMC Neurosci* 2010, 11, 154.

57. Choi, H., Pereira, A. R., Cao, Z., Shuman, C. F., Engene, N., Byrum, T., Matainaho, T., Murray, T. F., Mangoni, A., Gerwick, W. H. The hoiamides, structurally intriguing neurotoxic lipopeptides from Papua New Guinea marine cyanobacteria. *J Nat Prod* 2010, 73 (8), 1411–21.

58. Sakai, R., Swanson, G. T. Recent progress in neuroactive marine natural products. *Nat Prod Rep* 2014, 31 (2), 273–309.

59. Dechraoui, M.-Y., Naar, J., Pauillac, S., Legrand, A.-M. Ciguatoxins and brevetoxins, neurotoxic poly-ether compounds active on sodium channels. *Toxicon* 1999, 37 (1), 125–43.

60. Watkins, S., Reich, A., Fleming, L., Hammond, R. Neurotoxic shellfish poisoning. *Mar Drugs* 2008, 6 (3), 431–55.

61. Vernoux, J. P., Lewis, R. J. Isolation and characterisation of *Caribbean ciguatoxins* from the horse-eye jack (*Caranx latus*). *Toxicon* 1997, 35 (6), 889–900.

62. Catterall, W. A. Voltage-gated calcium channels. *Cold Spring Harb Perspect Biol* 2011, 3 (8), a003947.

63. Arias, H. Marine toxins targeting ion channels. *Mar Drugs* 2006, 4 (3), 37–69.

64. Catterall, W. A., Striessnig, J., Snutch, T. P., Perez-Reyes, E. International union of pharmacology. XL. Compendium of voltage-gated ion channels: Calcium channels. *Pharmacol Rev* 2003, 55 (4), 579–81.

65. Rahman, W., Dickenson, A. H. Voltage gated sodium and calcium channel blockers for the treatment of chronic inflammatory pain. *Neurosci Lett* 2013, 557 (Pt A), 19–26.

66. Lee, M. S. Recent progress in the discovery and development of N-type calcium channel modulators for the treatment of pain. In *Progress in Medicinal Chemistry*, ed. G. Lawton, D. R. Witty. Vol. 53. Elsevier, Amsterdam, the Netherlands, 2014, pp. 147–86.

67. Belardetti, F., Zamponi, G. W. Calcium channels as therapeutic targets. *Wiley Interdiscip Rev Membr Transp Signal* 2012, 1 (4), 433–51.

68. Tranberg, C. E., Yang, A., Vetter, I., McArthur, J. R., Baell, J. B., Lewis, R. J., Tuck, K. L., Duggan, P. J. ω-Conotoxin GVIA mimetics that bind and inhibit neuronal Cav2.2 ion channels. *Mar Drugs* 2012, 10 (10), 2349–68.

69. Malmberg, A. B., Yaksh, T. L. Effect of continuous intrathecal infusion of omega-conopeptides, N-type calcium-channel blockers, on behavior and antinociception in the formalin and hot-plate tests in rats. *Pain* 1995, 60 (1), 83–90.

70. Molinski, T. F., Dalisay, D. S., Lievens, S. L., Saludes, J. P. Drug development from marine natural products. *Nat Rev Drug Discov* 2009, 8 (1), 69–85.

71. Lewis, R. J., Garcia, M. L. Therapeutic potential of venom peptides. *Nat Rev Drug Discov* 2003, 2 (10), 790–802.

72. Ellinor, P. T., Zhang, J. F., Horne, W. A., Tsien, R. W. Structural determinants of the blockade of N-type calcium channels by a peptide neurotoxin. *Nature* 1994, 372 (6503), 272–75.

73. Wulff, H., Castle, N. A., Pardo, L. A. Voltage-gated potassium channels as therapeutic targets. *Nat Rev Drug Discov* 2009, 8 (12), 982–1001.

74. Kauferstein, S., Huys, I., Lamthanh, H., Stocklin, R., Sotto, F., Menez, A., Tytgat, J., Mebs, D. A novel conotoxin inhibiting vertebrate voltage-sensitive potassium channels. *Toxicon* 2003, 42 (1), 43–52.

75. Aguilar, M. B., Pérez-Reyes, L. I., López, Z., de la Cotera, E. P. H., Falcón, A., Ayala, C., Galván, M., Salvador, C., Escobar, L. I. Peptide srl1a from *Conus spurius* is a novel peptide blocker for Kv1 potassium channels. *Peptides* 2010, 31 (7), 1287–91.

76. Martel, P., Leo, D., Fulton, S., Bérard, M., Trudeau, L.-E. Role of Kv1 potassium channels in regulating dopamine release and presynaptic D2 receptor function. *PLoS ONE* 2011, 6 (5), e20402.

77. Castañeda, O., Harvey, A. L., Castañeda, O., Harvey, A. L. Discovery and characterization of cnidarian peptide toxins that affect neuronal potassium ion channels. *Toxicon* 2009, 54 (8), 1119–24.

78. Norton, R. S. Structures of sea anemone toxins. *Toxicon* 2009, 54 (8), 1075–88.

79. Honma, T., Shiomi, K, Peptide toxins in sea anemones: Structural and functional aspects. *Mar Biotechnol* (NY) 2006, 8 (1), 1–10.

80. Kelley, W. P., Wolters, A. M., Sack, J. T., Jockusch, R. A., Jurchen, J. C., Williams, E. R., Sweedler, J. V., Gilly, W. F. Characterization of a novel gastropod toxin (6-bromo-2-mercaptotryptamine) that inhibits Shaker K channel activity. *J Biol Chem* 2003, 278 (37), 34934–42.

81. Sack, J. T., Aldrich, R. W., Gilly, W. F. A gastropod toxin selectively slows early transitions in the Shaker K channel's activation pathway. *J Gen Physiol* 2004, 123 (6), 685–96.

82. Gao, D., Sand, R., Fu, H., Sharmin, N., Gallin, W. J., Hall, D. G. Synthesis of the non-peptidic snail toxin 6-bromo-2-mercaptotryptamine dimer (BrMT)2, its lower and higher thio homologs and their ability to modulate potassium ion channels. *Bioorg Med Chem Lett* 2013, 23 (20), 5503–6.

83. Gotti, C., Clementi, F., Fornari, A., Gaimarri, A., Guiducci, S., Manfredi, I., Moretti, M., Pedrazzi, P., Pucci, L., Zoli, M. Structural and functional diversity of native brain neuronal nicotinic receptors. *Biochem Pharmacol* 2009, 78 (7), 703–11.

84. Miyazawa, A., Fujiyoshi, Y., Unwin, N. Structure and gating mechanism of the acetylcholine receptor pore. *Nature* 2003, 423 (6943), 949–55.

85. Gotti, C., Moretti, M., Gaimarri, A., Zanardi, A., Clementi, F., Zoli, M. Heterogeneity and complexity of native brain nicotinic receptors. *Biochem Pharmacol* 2007, 74 (8), 1102–11.

86. Gray, W. R., Luque, A., Olivera, B. M., Barrett, J., Cruz, L. J. Peptide toxins from *Conus geographus* venom. *J Biol Chem* 1981, 256 (10), 4734–40.

87. McIntosh, J. M., Yoshikami, D., Mahe, E., Nielsen, D. B., Rivier, J. E., Gray, W. R., Olivera, B. M. A nicotinic acetylcholine receptor ligand of unique specificity, alpha-conotoxin ImI. *J Biol Chem* 1994, 269 (24), 16733–39.

88. Loughnan, M., Bond, T., Atkins, A., Cuevas, J., Adams, D. J., Broxton, N. M., Livett, B. G., Down, J. G., Jones, A., Alewood, P. F., Lewis, R. J. α-Conotoxin EpI, a novel sulfated peptide from *Conus episcopatus* that selectively targets neuronal nicotinic acetylcholine receptors. *J Biol Chem* 1998, 273 (25), 15667–74.

89. Nicke, A., Samochocki, M., Loughnan, M. L., Bansal, P. S., Maelicke, A., Lewis, R. J. Alpha-conotoxins EpI and AuIB switch subtype selectivity and activity in native versus recombinant nicotinic acetylcholine receptors. *FEBS Lett* 2003, 554 (1–2), 219–23.

90. Vincler, M., Wittenauer, S., Parker, R., Ellison, M., Olivera, B. M., McIntosh, J. M. Molecular mechanism for analgesia involving specific antagonism of α9α10 nicotinic acetylcholine receptors. *Proc Natl Acad Sci USA* 2006, 103 (47), 17880–84.

91. Kem, W., Soti, F., Wildeboer, K., LeFrancois, S., MacDougall, K., Wei, D.-Q., Chou, K.-C., Arias, H. The nemertine toxin anabaseine and its derivative DMXBA (GTS-21): Chemical and pharmacological properties. *Mar Drugs* 2006, 4 (3), 255–73.

92. Olincy, A., Harris, J. G., Johnson, L. L., Pender, V., Kongs, S., Allensworth, D., Ellis, J., Zerbe, G. O., Leonard, S., Stevens, K. E., Stevens, J. O., Martin, L., Adler, L. E., Soti, F., Kem, W. R., Freedman, R. Proof-of-concept trial of an α7 nicotinic agonist in schizophrenia. *Arch Gen Psychiatry* 2006, 63 (6), 630–38.

93. Wallace, T. L., Porter, R. H. P. Targeting the nicotinic alpha7 acetylcholine receptor to enhance cognition in disease. *Biochem Pharmacol* 2011, 82 (8), 891–903.

94. Peters, L., König, G. M., Terlau, H., Wright, A. D. Four new bromotryptamine derivatives from the marine bryozoan *Flustra foliacea*. *J Nat Prod* 2002, 65 (11), 1633–37.

95. Pandya, A., Yakel, J. L. Allosteric modulator desformylflustrabromine relieves the inhibition of alpha-2beta2 and alpha4beta2 nicotinic acetylcholine receptors by beta-amyloid(1–42) peptide. *J Mol Neurosci* 2011, 45 (1), 42–47.

96. Peters, L., Wright, A. D., Kehraus, S., Gündisch, D., Tilotta, M. C., König, G. M. Prenylated indole alkaloids from *Flustra foliacea* with subtype specific binding on NAChRs. *Planta Med* 2004, 70 (10), 883–86.

97. Sala, F., Mulet, J., Reddy, K. P., Bernal, J. A., Wikman, P., Valor, L. M., Peters, L., Konig, G. M., Criado, M., Sala, S. Potentiation of human alpha4beta2 neuronal nicotinic receptors by a *Flustra foliacea* metabolite. *Neurosci Lett* 2005, 373 (2), 144–49.

98. Rollema, H., Hajós, M., Seymour, P. A., Kozak, R., Majchrzak, M. J., Guanowsky, V., Horner, W. E., Chapin, D. S., Hoffmann, W. E., Johnson, D. E., McLean, S., Freeman, J., Williams, K. E. Preclinical pharmacology of the α4β2 nAChR partial agonist varenicline related to effects on reward, mood and cognition. *Biochem Pharmacol* 2009, 78 (7), 813–24.

99. Ferchmin, P. A., Pagan, O. R., Ulrich, H., Szeto, A. C., Hann, R. M., Eterovic, V. A. Actions of octocoral and tobacco cembranoids on nicotinic receptors. *Toxicon* 2009, 54 (8), 1174–82.

100. Groebe, D. R., Abramson, S. N. Lophotoxin is a slow binding irreversible inhibitor of nicotinic acetylcholine receptors. *J Biol Chem* 1995, 270 (1), 281–86.

101. Sorenson, E. M., Culver, P., Chiappinelli, V. A. Lophotoxin: Selective blockade of nicotinic transmission in autonomic ganglia by a coral neurotoxin. *Neuroscience* 1987, 20 (3), 875–84.

102. Abramson, S. N., Trischman, J. A., Tapiolas, D. M., Harold, E. E., Fenical, W., Taylor, P. Structure/activity and molecular modeling studies of the lophotoxin family of irreversible nicotinic receptor antagonists. *J Med Chem* 1991, 34 (6), 1798–804.

103. Weltzin, M. M., Schulte, M. K. Pharmacological characterization of the allosteric modulator desformylflustrabromine and its interaction with alpha4beta2 neuronal nicotinic acetylcholine receptor orthosteric ligands. *J Pharmacol Exp Ther* 2010, 334 (3), 917–26.

104. Lynch, J. W. Native glycine receptor subtypes and their physiological roles. *Neuropharmacology* 2009, 56 (1), 303–9.

105. Cascio, M. Structure and function of the glycine receptor and related nicotinicoid receptors. *J Biol Chem* 2004, 279 (19), 19383–86.

106. Verkman, A. S., Galietta, L. J. V. Chloride channels as drug targets. *Nat Rev Drug Discov* 2009, 8 (2), 153–71.

107. Xiong, W., Chen, S.-R., He, L., Cheng, K., Zhao, Y.-L., Chen, H., Li, D.-P., Homanics, G. E., Peever, J., Rice, K. C., Wu, L.-G., Pan, H.-L., Zhang, L. Presynaptic glycine receptors as a potential therapeutic target for hyperekplexia disease. *Nat Neurosci* 2014, 17 (2), 232–39.

108. Balansa, W., Islam, R., Fontaine, F., Piggott, A. M., Zhang, H., Webb, T. I., Gilbert, D. F., Lynch, J. W., Capon, R. J. Ircinialactams: Subunit-selective glycine receptor modulators from Australian sponges of the family Irciniidae. *Bioorg Med Chem* 2010, 18 (8), 2912–19.

109. Balansa, W., Islam, R., Fontaine, F., Piggott, A. M., Zhang, H., Xiao, X., Webb, T. I., Gilbert, D. F., Lynch, J. W., Capon, R. J. Sesterterpene glycinyl-lactams: A new class of glycine receptor modulator from Australian marine sponges of the genus *Psammocinia*. *Organic Biomol Chem* 2013, 11 (28), 4695–701.

110. Balansa, W., Islam, R., Gilbert, D. F., Fontaine, F., Xiao, X., Zhang, H., Piggott, A. M., Lynch, J. W., Capon, R. J. Australian marine sponge alkaloids as a new class of glycine-gated chloride channel receptor modulator. *Bioorg Med Chem* 2013, 21 (14), 4420–25.

111. Balskus, E. P. Natural products: Sponge symbionts play defense. *Nat Chem Biol* 2014, 10 (8), 611.

112. Piel, J. Metabolites from symbiotic bacteria. *Nat Prod Rep* 2009, 26 (3), 338–62.

113. Lynch, J. W., Chen, X. Subunit-specific potentiation of recombinant glycine receptors by NV-31, a bilobalide-derived compound. *Neurosci Lett* 2008, 435 (2), 147–51.

114. Laube, B., Maksay, G., Schemm, R., Betz, H. Modulation of glycine receptor function: A novel approach for therapeutic intervention at inhibitory synapses? *Trends Pharmacol Sci* 2002, 23 (11), 519–27.

115. Eichler, S. A., Kirischuk, S., Jüttner, R., Schafermeier, P. K., Legendre, P., Lehmann, T.-N., Gloveli, T., Grantyn, R., Meier, J. C. Glycinergic tonic inhibition of hippocampal neurons with depolarizing GABAergic transmission elicits histopathological signs of temporal lobe epilepsy. *J Cell Mol Med* 2008, 12 (6b), 2848–66.

116. Li, D., Romain, G., Flamar, A.-L., Duluc, D., Dullaers, M., Li, X.-H., Zurawski, S., Bosquet, N., Palucka, A. K., Le Grand, R., O'Garra, A., Zurawski, G., Banchereau, J., Oh, S. Targeting self- and foreign antigens to dendritic cells via DC-ASGPR generates IL-10–producing suppressive CD4+ T cells. *J Exp Med* 2012, 209 (1), 109–21.

117. Manzke, T., Niebert, M., Koch, U. R., Caley, A., Vogelgesang, S., Hülsmann, S., Ponimaskin, E., Müller, U., Smart, T. G., Harvey, R. J., Richter, D. W. Serotonin receptor 1A–modulated phosphorylation of glycine receptor α3 controls breathing in mice. *J Clin Investig* 2010, 120 (11), 4118–28.

118. Le Novere, N., Changeux, J. P. The ligand gated ion channel database: An example of a sequence database in neuroscience. *Philos Trans R Soc Lond B Biol Sci* 2001, 356 (1412), 1121–30.

119. Chebib, M., Johnston, G. A. GABA-activated ligand gated ion channels: Medicinal chemistry and molecular biology. *J Med Chem* 2000, 43 (8), 1427–47.

120. Miller, P. S., Aricescu, A. R. Crystal structure of a human GABAA receptor. *Nature* 2014, 512 (7514), 270–75.

121. Johnston, G. A. GABA(A) receptor channel pharmacology. *Curr Pharm Des* 2005, 11 (15), 1867–85.

122. Verpoorte, R., Siwon, J., Tieken, M. E. M., Svendsen, A. B. Studies on Indonesian medicinal plants. V. The alkaloids of *Anamirta cocculus*. *J Nat Prod* 1981, 44 (2), 221–24.

123. Carpenter, T. S., Lau, E. Y., Lightstone, F. C. Identification of a possible secondary picrotoxin-binding site on the GABAA receptor. *Chem Res Toxicol* 2013, 26 (10), 1444–54.

124. Shin, B. A., Kim, Y. R., Lee, I.-S., Sung, C. K., Hong, J., Sim, C. J., Im, K. S., Jung, J. H. Lyso-PAF analogues and lysophosphatidylcholines from the marine sponge *Spirastrella abata* as inhibitors of cholesterol biosynthesis. *J Nat Prod* 1999, 62 (11), 1554–1557.

125. Sakurada, T., Gill, M. B., Frausto, S., Copits, B., Noguchi, K., Shimamoto, K., Swanson, G. T., Sakai, R. Novel N-methylated 8-oxoisoguanines from Pacific sponges with diverse neuroactivities. *J Med Chem* 2010, 53 (16), 6089–99.

126. Eterović, V., Hann, R., Ferchmin, P. A., Rodriguez, A., Li, L., Lee, Y.-H., McNamee, M. Diterpenoids from Caribbean gorgonians act as noncompetitive inhibitors of the nicotinic acetylcholine receptor. *Cell Mol Neurobiol* 1993, 13 (2), 99–110.

127. Li, P., Reichert, D. E., Rodríguez, A. D., Manion, B. D., Evers, A. S., Eterović, V. A., Steinbach, J. H., Akk, G. Mechanisms of potentiation of the mammalian GABAA receptor by the marine cembranoid eupalmerin acetate. *Br J Pharmacol* 2008, 153 (3), 598–608.

128. Hosie, A. M., Wilkins, M. E., da Silva, H. M., Smart, T. G. Endogenous neurosteroids regulate GABAA receptors through two discrete transmembrane sites. *Nature* 2006, 444 (7118), 486–89.

129. Johnston, G. A., Hanrahan, J. R., Chebib, M., Duke, R. K., Mewett, K. N. Modulation of ionotropic GABA receptors by natural products of plant origin. *Adv Pharmacol* 2006, 54, 285–316.

130. Stawski, P., Janovjak, H., Trauner, D. Pharmacology of ionotropic glutamate receptors: A structural perspective. *Bioorg Med Chem* 2010, 18 (22), 7759–72.

131. Sawant, P. M., Weare, B. A., Holland, P. T., Selwood, A. I., King, K. L., Mikulski, C. M., Doucette, G. J., Mountfort, D. O., Kerr, D. S. Isodomoic acids A and C exhibit low KA receptor affinity and reduced *in vitro* potency relative to domoic acid in region CA1 of rat hippocampus. *Toxicon* 2007, 50 (5), 627–38.

132. Carcache, L. M., Rodriguez, J., Rein, K. S. The structural basis for kainoid selectivity at AMPA receptors revealed by low-mode docking calculations. *Bioorg Med Chem* 2003, 11 (4), 551–59.

133. Sakai, R., Yoshida, K., Kimura, A., Koike, K., Jimbo, M., Koike, K., Kobiyama, A., Kamiya, H. Cellular origin of dysiherbaine, an excitatory amino acid derived from a marine sponge. *Chembiochem* 2008, 9 (4), 543–51.

134. Sakai, R., Kamiya, H., Murata, M., Shimamoto, K. Dysiherbaine: A new neurotoxic amino acid from the Micronesian marine sponge *Dysidea herbacea*. *J Am Chem Soc* 1997, 119 (18), 4112–16.

135. Sato, M., Inoue, F., Kanno, N., Sato, Y. The occurrence of N-methyl-D-aspartic acid in muscle extracts of the blood shell, *Scapharca broughtonii*. *Biochem J* 1987, 241 (1), 309–11.

136. D'Aniello, A., Spinelli, P., De Simone, A., D'Aniello, S., Branno, M., Aniello, F., Fisher, G. H., Di Fiore, M. M., Rastogi, R. K. Occurrence and neuroendocrine role of D-aspartic acid and N-methyl-D-aspartic acid in *Ciona intestinalis*. *FEBS Lett* 2003, 552 (2–3), 193–98.

137. Sakai, R., Matsubara, H., Shimamoto, K., Jimbo, M., Kamiya, H., Namikoshi, M. Isolations of N-methyl-D-aspartic acid-type glutamate receptor ligands from Micronesian sponges. *J Nat Prod* 2003, 66 (6), 784–87.

138. Parsons, C. G. NMDA receptors as targets for drug action in neuropathic pain. *Eur J Pharmacol* 2001, 429 (1–3), 71–78.

139. Layer, R. T., Wagstaff, J. D., White, H. S. Conantokins: Peptide antagonists of NMDA receptors. *Curr Med Chem* 2004, 11 (23), 3073–84.

140. Schoneberg, T., Schulz, A., Biebermann, H., Hermsdorf, T., Rompler, H., Sangkuhl, K. Mutant G-protein-coupled receptors as a cause of human diseases. *Pharmacol Ther* 2004, 104 (3), 173–206.

141. Schiöth, H. B., Fredriksson, R. The GRAFS classification system of G-protein coupled receptors in comparative perspective. *Gen Comp Endocrinol* 2005, 142 (1–2), 94–101.

142. Lagerstrom, M. C., Schioth, H. B. Structural diversity of G protein-coupled receptors and significance for drug discovery. *Nat Rev Drug Discov* 2008, 7 (4), 339–57.

143. Wettschureck, N., Offermanns, S. Mammalian G proteins and their cell type specific functions. *Physiol Rev* 2005, 85 (4), 1159–204.

144. de Nys, R., Dworjanyn, S. A., Steinberg, P. D. A new method for determining surface concentrations of marine natural products on seaweeds. *Mar Ecol Prog Ser* 1998, 162, 79–87.

145. Almeida, C., Kehraus, S., Prudencio, M., Konig, G. M. Marilones A–C, phthalides from the sponge-derived fungus *Stachylidium* sp. *Beilstein J Org Chem* 2011, 7, 1636–42.

146. Kennett, G. A., Wood, M. D., Glen, A., Grewal, S., Forbes, I., Gadre, A., Blackburn, T. P. *In vivo* properties of SB 200646A, a 5-HT2C/2B receptor antagonist. *Br J Pharmacol* 1994, 111 (3), 797–802.

147. Lin, S. Y., Chang, W. J., Lin, C. S., Huang, C. Y., Wang, H. F., Sun, W. H. Serotonin receptor 5-HT2B mediates serotonin-induced mechanical hyperalgesia. *J Neurosci* 2011, 31 (4), 1410–18.

148. Cervantes-Duran, C., Vidal-Cantu, G. C., Barragan-Iglesias, P., Pineda-Farias, J. B., Bravo-Hernandez, M., Murbartian, J., Granados-Soto, V. Role of peripheral and spinal 5-HT2B receptors in formalin-induced nociception. *Pharmacol Biochem Behav* 2012, 102 (1), 30–35.

149. Lin, Z., Antemano, R. R., Hughen, R. W., Tianero, M. D. B., Peraud, O., Haygood, M. G., Concepcion, G. P., Olivera, B. M., Light, A., Schmidt, E. W. Pulicatins A–E, neuroactive thiazoline metabolites from cone snail-associated bacteria. *J Nat Prod* 2010, 73 (11), 1922–26.

150. Brea, J., Castro-Palomino, J., Yeste, S., Cubero, E., Parraga, A., Dominguez, E., Loza, M. I. Emerging opportunities and concerns for drug discovery at serotonin 5-HT2B receptors. *Curr Top Med Chem* 2010, 10 (5), 493–503.

151. Hu, J. F., Schetz, J. A., Kelly, M., Peng, J. N., Ang, K. K., Flotow, H., Leong, C. Y., Ng, S. B., Buss, A. D., Wilkins, S. P., Hamann, M. T. New antiinfective and human 5-HT2 receptor binding natural and semisynthetic compounds from the Jamaican sponge *Smenospongia aurea*. *J Nat Prod* 2002, 65 (4), 476–80.

152. Sjogren, M., Johnson, A. L., Hedner, E., Dahlstrom, M., Goransson, U., Shirani, H., Bergman, J., Jonsson, P. R., Bohlin, L. Antifouling activity of synthesized peptide analogs of the sponge metabolite barettin. *Peptides* 2006, 27 (9), 2058–64.

153. Hedner, E., Sjogren, M., Frandberg, P. A., Johansson, T., Goransson, U., Dahlstrom, M., Jonsson, P., Nyberg, F., Bohlin, L. Brominated cyclodipeptides from the marine sponge *Geodia barretti* as selective 5-HT ligands. *J Nat Prod* 2006, 69 (10), 1421–24.

154. Carvalho, V., Fernandes, L., Conde, T., Zamith, H., Silva, R., Surrage, A., Frutuoso, V., Castro-Faria-Neto, H., Amendoeira, F. Antinociceptive activity of *Stephanolepis hispidus* skin aqueous extract depends partly on opioid system activation. *Mar Drugs* 2013, 11 (4), 1221–34.

155. Carbone, M., Li, Y., Irace, C., Mollo, E., Castelluccio, F., Di Pascale, A., Cimino, G., Santamaria, R., Guo, Y. W., Gavagnin, M. Structure and cytotoxicity of phidianidines A and B: First finding of 1,2,4-oxadiazole system in a marine natural product. *Org Lett* 2011, 13 (10), 2516–19.

156. Brogan, J. T., Stoops, S. L., Lindsley, C. W. Total synthesis and biological evaluation of phidianidines A and B uncovers unique pharmacological profiles at CNS targets. *ACS Chem Neurosci* 2012, 3 (9), 658–64.

157. Carroll, A. R., Kaiser, S. M., Davis, R. A., Moni, R. W., Hooper, J. N., Quinn, R. J. A bastadin with potent and selective delta-opioid receptor binding affinity from the Australian sponge *Ianthella flabelliformis*. *J Nat Prod* 2010, 73 (6), 1173–76.

158. Gross, H., Goeger, D. E., Hills, P., Mooberry, S. L., Ballantine, D. L., Murray, T. F., Valeriote, F. A., Gerwick, W. H. Lophocladines, bioactive alkaloids from the red alga *Lophocladia* sp. *J Nat Prod* 2006, 69 (4), 640–44.

159. Mackie, K. Cannabinoid receptors as therapeutic targets. *Annu Rev Pharmacol Toxicol* 2006, 46 (1), 101–22.

160. Engene, N., Gunasekera, S. P., Gerwick, W. H., Paul, V. J. Phylogenetic inferences reveal a large extent of novel biodiversity in chemically rich tropical marine cyanobacteria. *Appl Environ Microbiol* 2013, 79 (6), 1882–88.

161. Sitachitta, N., Gerwick, W. H. Grenadadiene and grenadamide, cyclopropyl-containing fatty acid metabolites from the marine cyanobacterium *Lyngbya majuscula*. *J Nat Prod* 1998, 61 (5), 681–84.

162. Gutierrez, M., Pereira, A. R., Debonsi, H. M., Ligresti, A., Di Marzo, V., Gerwick, W. H. Cannabinomimetic lipid from a marine cyanobacterium. *J Nat Prod* 2011, 74 (10), 2313–17.

163. Han, B., McPhail, K. L., Ligresti, A., Di Marzo, V., Gerwick, W. H. Semiplenamides A–G, fatty acid amides from a Papua New Guinea collection of the marine cyanobacterium *Lyngbya semiplena*. *J Nat Prod* 2003, 66 (10), 1364–68.

164. Mevers, E., Matainaho, T., Allara, M., Di Marzo, V., Gerwick, W. H. Mooreamide A: A cannabinomimetic lipid from the marine cyanobacterium *Moorea bouillonii*. *Lipids* 2014, 49, 1127–32.

165. Chen, J. F., Eltzschig, H. K., Fredholm, B. B. Adenosine receptors as drug targets: What are the challenges? *Nat Rev Drug Discov* 2013, 12 (4), 265–86.

166. Pham, N. B., Butler, M. S., Quinn, R. J. Isolation of psammaplin A 11′-sulfate and bisaprasin 11′-sulfate from the marine sponge *Aplysinella rhax*. *J Nat Prod* 2000, 63 (3), 393–95.

167. Pham, N. B., Butler, M. S., Healy, P. C., Quinn, R. J. Anthoptilides A–E, new briarane diterpenes from the Australian sea pen *Anthoptilum* cf. *kukenthali*. *J Nat Prod* 2000, 63 (3), 318–21.

168. Pham, N. B., Butler, M. S., Quinn, R. J. Naturally occurring cembranes from an Australian sarcophyton species. *J Nat Prod* 2002, 65 (8), 1147–50.

169. Kehraus, S., Gorzalka, S., Hallmen, C., Iqbal, J., Muller, C. E., Wright, A. D., Wiese, M., Konig, G. M. Novel amino acid derived natural products from the ascidian *Atriolum robustum*: Identification and pharmacological characterization of a unique adenosine derivative. *J Med Chem* 2004, 47 (9), 2243–55.

170. Ohshita, K., Ishiyama, H., Oyanagi, K., Nakata, H., Kobayashi, J. Synthesis of hybrid molecules of caffeine and eudistomin D and its effects on adenosine receptors. *Bioorg Med Chem* 2007, 15 (9), 3235–40.

171. Newman, D. J., Cragg, G. M. Marine-sourced anti-cancer and cancer pain control agents in clinical and late preclinical development. *Mar Drugs* 2014, 12 (1), 255–78.

172. Sagar, S., Kaur, M., Minneman, K. P. Antiviral lead compounds from marine sponges. *Mar Drugs* 2010, 8 (10), 2619–38.

173. Williams, J. A., Day, M., Heavner, J. E. Ziconotide: An update and review. *Expert Opin Pharmacother* 2008, 9 (9), 1575–83.

CHAPTER 11

Marine Organisms in Cancer Chemoprevention

Eun-Jung Park,[1] **Anam Shaikh,**[2] **Brian T. Murphy,**[2] **and John M. Pezzuto**[1]

[1]Daniel K. Inouye College of Pharmacy, University of Hawaii at Hilo, Hilo, Hawaii
[2]Department of Medicinal Chemistry and Pharmacognosy, University of Illinois at Chicago, Chicago, Illinois

CONTENTS

As reported by Ferlay et al.,[1] on a worldwide basis, in 2008, the cases of cancer incidence and mortality were 12.7 million and 7.6 million, respectively. In 2010, it was estimated that there were about 1.63 million cancer cases in the United States, in addition to 3.5 million cases of nonmelanoma skin cancer (NMSC).[2] Fortunately, the mortality rate from NMSC is relatively low (0.69 deaths per 100,000 population a year from 1969 to 2000),[3] but definitive treatment of malignant metastatic cancer remains elusive.

Largely as a result of lifestyle changes (e.g., smoking cessation) and advances in early detection and cancer treatment, during the period of 2005–2009, the cancer mortality rate in the United States decreased 1.8% and 1.5% per year in males and females, respectively. Nonetheless, cancer remains the second overall leading cause of death. In fact, if non-disease-based causes of death are excluded (e.g., accidents/unintentional injuries, assault/homicide, and intentional self-harm/suicide), cancer is the leading cause of death in individuals under the age of 80.[4] In addition, the estimated cost associated with cancer care in the United States was $124.57 billion in 2010, and this is expected to reach at least $157.77 billion (normalized to the value of the dollar in 2010) by 2020 (http://costpro-jections.cancer.gov/, accessed August 22, 2013). Based on statistics such as these, it seems clear a prophylactic approach could appreciably improve public health, as well as lessen financial burden.

Inspection of trends in incidence rates for major cancers in the United States during the period of 1975–2009 indicates the rates of colorectal cancer are declining, mainly due to early detection and surgical removal of neoplastic polyps (polypectomy), whereas those of skin, liver, and thyroid are increasing in both genders. In view of cancer incidence as a whole, the overall rate in males decreased by 0.6% a year during the period of 2005–2009, while that in females was relatively unaffected.[4] Given the level of effort focused on reducing cancer occurrence, such an outcome is rather disappointing, yet not surprising since carcinogenesis is a complicated and variable process that is difficult to entirely control. For example, during the developmental stages of cancer, altera-tions may occur at genetic, epigenetic, proteomic, and metabolic levels. Another paradoxical factor is the gradual increase in the life span of humans. For instance, in 2050, the life expectancy of males in the United States is projected to fall in the range of 80–81, 83–86, 78–113, or 82.2–86.4 years, depending on the methodology used for the extrapolation. Most importantly, however, all of these projected life spans are significantly greater than 75.6 years, the 2008 life expectancy (http://www.cdc.gov/nchs/data/nvsr/nvsr61/nvsr61_03.pdf, accessed August 22, 2013).[5] Accordingly, as the risk for developing cancer increases considerably with age, it is expected the rate of cancer incidence will escalate in the future. In fact, it is proposed that the incidence of all cancer cases will increase from 12.7 million in 2008 to 22.2 million by 2030, based on projected demographic and trend-based changes for selected cancer sites.[6] These types of considerations further bolster the logic of increas-ing efforts to lower the occurrence of cancer, with prevention remaining at the forefront.

There are several strategies of cancer prevention, including vaccinations (e.g., Cervarix or Gardasil for preventing cervical cancer), surgical resection (e.g., mastectomy for preventing breast cancer and polypectomy for preventing colon cancer), and lifestyle changes (e.g., smoking cessation). To assess the ability of potential chemopreventive agents to reduce the incidence of cancer, large-scale clinical trials have been performed. Negative results for lung cancer prevention were obtained with the α-Tocopherol, β-Carotene (ATBC) Prevention Study (1985–1993), involving 29,133 total participants who were Finnish male smokers from 50 to 69 years of age,[7] and the Carotene and Retinol Efficacy Trial (CARET, 1985–1997), with 18,314 total participants who were smokers and workers exposed to asbestos.[8] More recently, in the Selenium and Vitamin E Cancer Prevention Trial (SELECT) conducted from 2001 to 2008 with 35,533 relatively healthy men (African American men age 50 or older and all other men age 55 or older), it was found that selenium or vitamin E, alone or in combination, did not prevent prostate cancer.[9] Moreover, extended postintervention follow-up of SELECT participants demonstrated that vitamin E consumption could even lead to a 17% increase in prostate cancer incidence.[10]

In contrast to these studies, some desirable outcomes have been reported for other cancer chemopreventive drug candidates. For instance, in the Breast Cancer Prevention Trial (BCPT, 1992–1997), with 13,388 participants, it was found that tamoxifen could reduce the occurrence of estrogen receptor-positive tumors by 69% in women at high risk of developing breast cancer with a history of lobular carcinoma *in situ*.[11] In the Study of Tamoxifen and Raloxifene (STAR) (1999–2006), conducted with 19,747 participants, it was found that tamoxifen and raloxifene, which are selective estrogen receptor modulators (SERMs), reduced the risk of invasive breast cancer in postmenopausal women with an increased 5-year predicted breast cancer risk. Notably, raloxifene exerted less adverse effects than tamoxifen; for example, raloxifene induced less endometrial cancer than tamoxifen.[12,13] The Prostate Cancer Prevention Trial (PCPT) on healthy men more than 55 years of age with prostate-specific antigen levels of less than 3 ng/ml showed that finasteride, an inhibitor of 5α-reductase, reduced the risk of developing prostate cancer by 25% in comparison with placebo. An increased risk of high-grade prostate cancer was observed as an adverse effect,[14] but after 18 years of follow-up, there was no significant between-group difference in the rates of overall survival or survival after the diagnosis of prostate cancer.[15]

Hitherto, few pharmaceutical agents have obtained approval from the Food and Drug Administration (FDA) for the purpose of reducing cancer risk. A notable exception is tamoxifen, which is used by women with ductal carcinoma *in situ* (DCIS) or high-risk women (e.g., women at least 35 years of age with a 5-year predicted risk of breast cancer of ≥1.67%, as calculated by the Gail Model). However, associated adverse effects, including an increased incidence of thromboembolic events, cataracts, and endometrial cancer, remain problematic.[16] On the other hand, recent results with aspirin suggested promising chemopreventive therapy with fewer or less severe side effects. For example, epidemiologic studies with individuals who were involved in a cardiovascular disease prevention study showed that daily intake of aspirin for around 5 years reduced colorectal cancer mortality by 34%. Another study demonstrated that aspirin decreased the recurrence of advanced adenomas by 28% in patients with a history of colorectal adenoma or cancer. The most frequently reported unwanted effect with aspirin is gastrointestinal bleeding, which is not ideal, but may be considered preferable to the adverse effects mediated by SERMs and 5α-reductase inhibitors.[17]

In principle, combination chemoprevention therapy warrants greater attention due to therapeutic synergism and ameliorated undesirable effects as the result of using lower doses of individual agents. Notably, the value of this approach has been demonstrated with a combination of difluoromethylornithine (DFMO) and sulindac for patients with a history of resected adenomas. Treatment with the combination resulted in reduced colorectal adenoma recurrence with few adverse effects.[18]

In addition to synthetic agents, as mentioned above, largely based on chemoprevention strategies spearheaded by Lee Wattenberg in the 1960s,[19,20] a number of studies have been performed to examine the potential of dietary regimens and supplements from natural sources (mainly from terrestrial botanicals) to prevent cancer. For instance, sulfur-containing compounds (e.g., allicin, diallyl disulfide, and isothiocyanates) occurring in the genus *Allium* plants (e.g., garlic and onion) and the genus *Brassica* plants (e.g., cabbage and broccoli that produce isothiocyanates); polyphenols (e.g., stilbenoids and flavonoids) abundantly found in berries, citrus fruits, and green tea; carotenoids (e.g., lycopene) responsible for red pigment in vegetables and fruits (e.g., tomatoes); and curcuminoids or gingerols isolated from the Zingiberaceae family (e.g., turmeric and ginger) have shown cancer chemopreventive potential.[21,22] It is thereby reasonable to assume active principles are associated with some dietary materials or nonedible natural products, and some studies have been performed to isolate and characterize these agents. For example, polyphenolic compounds have been well studied in terms of cancer chemoprevention as bioactive components. Various flavonoids, a group of polyphenols, with radical scavenging, anti-inflammatory, cytoprotective, and antiproliferative activities, are reported as potential cancer chemopreventive agents.

Examples we have studied include zapotin, isoliquiritigenin, rotenones, deguelin, and abyssinone. Not only *in vitro* and *in vivo* studies, but also clinical and epidemiological studies, indicate that a dietary intake of flavonoid-rich fruits and vegetables is inversely correlated with the incidence of cancers.[23] A prominent example is resveratrol. Ever since the first report on the cancer chemopreventive potential of resveratrol in 1997,[24] studies to elucidate the preventive effects of resveratrol have dramatically increased, with a wide range of biological activities being reported, such as inhibition of pro-inflammatory mediators, aromatase, and induction of NAD(P)H–quinone reductase 1 (QR1).[25,26]

In general, however, the majority of cancer chemopreventive candidates selected from numerous *in vitro* and *in vivo* studies do not fulfill the rigorous standards of mediating an efficacious response with an acceptable range of adverse effects. Some natural products have been tested in clinical trials, and the results have been negative. For example, none of the following have demonstrated beneficial effects and reproducible results in phase 3 trials for lung cancer prevention: β-carotene, retinol, 13-*cis*-retinoic acid, α-tocopherol, *N*-acetylcysteine, or selenium.[27] On the other hand, it was reported that supplementation with calcium was effective for the prevention of adenoma recurrence in patients with a history of adenomas.[28] In addition, eicosapentaenoic acid showed promise as a colorectal cancer chemoprevention agent with a favorable safety profile in familial adenomatous polyposis.[29] Finally, berry consumption might be a feasible approach for preventing esophageal squamous cell carcinoma in places like China and other high-risk regions for this disease.[30]

In sum, whereas more than 140 agents have been approved by the FDA for the treatment and palliative care of cancer, only approximately a dozen agents have obtained approval for the treatment of precancerous lesions or for the reduction of cancer risk.[31] Given the obvious advantage of disease prevention, and the systemic advantage of cancer chemoprevention, it is clear that additional work in this area is warranted. In this chapter we focus on marine organisms in cancer chemoprevention.

11.1 TARGETS FOR CANCER CHEMOPREVENTION

11.1.1 Detoxification

An excessive amount of free radicals or reactive species generated by deregulated enzymes, including cytochrome P450 (CYP), NADPH oxidase, nitric oxide synthase (NOS), and xanthine oxidase, might result in various pathological conditions, including cancer via DNA damage, protein modification, and lipid peroxidation.[32,33] In order to neutralize free radicals or reactive species, phase 2 detoxification enzymes exist that can be induced by transcriptional activation of antioxidant-response element (ARE) through nuclear factor E2–related factor 2 (Nrf2). Theoretically, monofunctional agents that induce phase 2 enzymes selectively appear to be more desirable candidates for cancer chemoprevention.[34]

11.1.2 Inflammation

Cumulating evidence supports the idea of chronic inflammation being involved in the development of various cancer types.[35] The overexpression of cyclooxygenase 2 (COX-2), a pro-inflammatory protein, plays a fundamental role in gastrointestinal cancers, including esophageal, gastric, and colorectal cancers.[36] Also, anti-inflammatory phytochemicals can ameliorate hepatocellular carcinoma.[37]

11.1.3 Estrogen Metabolism

SERMs have been clinically used for breast cancer prevention. In addition, with women bearing a high risk of breast cancer, aromatase inhibitors have shown promise in reducing risk with more acceptable side effects than SERMs. For instance, exemestane showed a 65% relative reduction in total breast cancers in the intervention compared to the placebo group.[38]

11.1.4 Upregulated Biomarkers/Signaling Pathways during Carcinogenesis

Deregulated expression and activity of tyrosine kinase receptors, including epidermal growth factor receptor (EGFR) (ErbB-1) and human epidermal growth factor receptor 2 (HER2) (Neu/ErbB-2), have been reported in human neoplasia.[39] Overexpression of EGFR has been observed in human tumors, including glioblastoma, breast, esophageal, gastric, lung, and prostate tumors.[40] Also, amplification of HER2 has been found in various tumors, including bladder, breast, endometrial, non-small-cell lung cancer, ovarian, pancreatic, and salivary gland tumors.[41]

During carcinogenesis, various protein kinases can be deregulated, including Akt, Fyn, Janus kinase 1, mitogen-activated protein kinase kinase 1 (MEK1), and phosphoinositide 3-kinase (PI3K).[42] The Akt-related pathway might be a promising target for preventing cancer since Akt is frequently upregulated during carcinogenesis and in various preneoplastic lesions.[43] Another potential target, the PI3K/Akt/mTOR pathway, has been suggested to contribute to development of hepatocellular carcinoma (HCC)[44] and colon cancer.[45]

11.1.5 Diabetes/Obesity

It has been reported that both cancer incidence and mortality are increased among diabetic patients, possibly due to increased insulin levels, which can promote cell proliferation and excessive amounts of blood glucose, which is an energy source for malignant cells.[46] Epidemiological studies demonstrate a positive correlation between obesity and multiple types of cancer. Although the molecular mechanisms remain to be unraveled, chronic inflammation by obesity can facilitate cancer development via reactive oxygen species (ROS)-induced cell damage and mutagenesis, and pro-inflammatory cytokine/transcription factor-induced tumor growth and its malignancy (e.g., invasiveness).[47] It was reported that calorie restriction attenuated tumor promotion in an epithelial carcinogenesis model accompanied by insulin-like growth factor I receptor (IGF-1R)/EGFR cross talk and a downstream signaling pathway such as protein kinase B (Akt)-mammalian target of rapamycin (mTOR).[48] From a mouse skin carcinogenesis model, circulating levels of IGF-1, insulin, and leptin were reduced by weight loss that resulted in subsequent downregulation of signaling pathways, including Ras mitogen-activated protein kinase (MAPK), PI3K-Akt-mTOR, and AMP-activated protein kinase (AMPK) pathways.[49] Indeed, it was reported that deguelin, known as a potential cancer chemopreventive agent, activates AMPK, a cellular energy sensor, demonstrating a possible role for AMPK in cancer chemoprevention.[50] In addition, elevated levels of IGF-1 are associated with an increased risk of breast cancer, and the COX-2/prostaglandin E_2 (PGE_2)/EP3 signaling pathway is involved in IGF-1-stimulated mammary tumorigenesis. In line with this, COX-2-selective inhibitors are suggested to be useful in the prevention or treatment of human breast cancer associated with elevated IGF-1 levels.[51]

11.2 MARINE ORGANISMS THAT PRODUCE THERAPEUTIC AGENTS

The search for novel drug candidates from natural sources has gradually increased over the decades. Notably, intensive and massive work has been done by the Developmental Therapeutics Program of the National Cancer Institute to establish a Natural Products Repository that includes approximately 170,000 extracts from more than 70,000 plants and 10,000 marine organisms (including invertebrates and algae), in addition to more than 30,000 extracts from bacteria and fungi. Through testing cytotoxicity against human cancer cell lines in culture, around 4,000 extracts have shown activity toward the panel (http://dtp.nci.nih.gov/branches/npb/repository.html, accessed on August 22, 2013). Although the majority of natural product research has been performed with terrestrial resources, since the late 1960s, an increasing number of studies have been performed with marine products, leading to the discovery of approximately 2,500 novel marine metabolites and the isolation of more than 10,000 compounds from 1977 to 1987.[52]

Evaluation of terrestrial plants may be advantageous since some of these materials have been used as traditional or folk medicines. However, since ancient times, there are records and medical literature on the medicinal application of marine materials in America, China, Europe, India, and the Near East. Evidence has been found suggesting inhabitants in southern Chile may have used seaweed for dietary and medicinal purposes approximately 12,310 to 12,290 years ago.[53] In the fifth to fourth centuries BCE, Hippocrates recommended treating wounds using a sea sponge.[54] In the Qin and Han dynasties (221 BC–220 AD), people included marine materials in prescriptions.[52] In early Islamic and Crusader periods, seashell (*Angulus* sp.) was used as a mild purgative and for ailments affecting women.[54] Also, coastal residents, especially East Asians, have been using seaweed in their cuisines. In recent times, marine-derived anticancer agents, including cytarabine (Cytosar-U1), eribulin (E7389 or Halaven®), and trabectedin (ET-743 or Yondelis), have been developed.[32]

Although people have used marine materials from the days of yore for various medicinal purposes, including cancer treatment, application of marine natural products as cancer chemopreventive agents has not received much attention relative to terrestrial materials. According to a comprehensive review by Appeltans et al., the number of currently known marine species is estimated to be from 150,000 to 274,000, but depending on expert opinions, inventory, and literature, the number of marine species in existence may range from 300,000 to more than 10 million. About 222,000–230,000 eukaryotic marine species with pharmaceutical potential have been recognized, including ~7,600 species of Plantae, ~19,500 of Chromista, ~550 of Protozoa, ~1,050 of Fungi, and nearly 200,000 of Animalia.[55]

In this review, a select group of known marine materials with antioxidant, anti-inflammatory, and anticarcinogenic activities, along with inhibitors of kinases associated with carcinogenesis, will be described. We exclude the description of extracts or macromolecules (e.g., polysaccharides) and materials mediating moderate or weak activities (e.g., IC_{50} [concentration that inhibits binding or activity by 50%] values in the mM range).

11.3 *IN VITRO* CANCER CHEMOPREVENTIVE ACTIVITIES

11.3.1 Marine-Derived Bacteria

Since the discovery of penicillin by Alexander Fleming in 1929, terrestrial microorganisms have been a major focal point of the global drug discovery effort. This effort was massively successful, producing more than 120 drugs that are used to treat a broad array of diseases (infectious diseases, cancers, immunomodulation, etc.).[56] However, in previous decades we have observed a marked decrease in the discovery of new chemical scaffolds from terrestrial strains, since extensive screening efforts and similar screening platforms have led to the constant rediscovery of known bioactive structural classes. To overcome this barrier, researchers have begun to explore the ocean as a source for new microbial-derived drug leads, as some selection pressures are fundamentally different than those existing in the terrestrial environment.[56,57]

Marine-derived bacteria is broadly defined as bacteria that have adapted to conditions in the ocean, but does not imply that these conditions are required for growth. Of these bacteria, some of the most prolific bioactive small-molecule-producing phyla to date are Actinobacteria and Cyanobacteria,[58,59] though this in part represents the extent to which these taxa have been studied and does not preclude the potential of other phyla (Firmicutes, Proteobacteria, etc.) to produce equally useful biologically active secondary metabolites. In the following sections, the chemopreventive activities of secondary metabolites produced by these two phyla will be discussed.

11.3.1.1 Marine-Derived Actinomycetes

Actinobacteria is a phylum of Gram-positive, high-G+C-content bacteria that are morphologically and physiologically diverse. Although Actinobacteria comprise over 195 genera, *Streptomyces* and *Micromonospora* are the major producers of secondary metabolites reported to date.[60] However, with increasing investment in cultivation technologies, particularly toward samples collected in the marine environment, researchers are beginning to discover taxa with unexplored secondary metabolite-producing potential. Using chemopreventive assays to guide these discovery efforts is a relatively young strategy and has led to the identification of several unique marine-derived small molecules. Interestingly, the majority of studies have occurred in only the past 10 years. Herein we attempt to summarize this body of work to date, focusing on those molecules that *do not* exhibit significant cytotoxicity effects.

11.3.1.1.1 Anti-Inflammatory Metabolites of Marine-Derived Actinomycetes

A number of phenazine derivatives were isolated from the fermentation broth of a marine-derived *Streptomyces* sp. and inhibited tumor necrosis factor (TNF)-α-induced NF-κB activity. In particular, lavanducyanin (**1**) and a brominated terpenoid phenazine (**2**) exhibited IC_{50} values of 16.3 and 4.1 μM, respectively. The compounds also inhibited lipopolysaccharide (LPS)-induced nitric oxide (NO) production with IC_{50} values of 8.0 and > 48.6 μM, respectively.[61] In follow-up studies, phenazine analogs were biosynthesized in order to further explore the effects of structure manipulation on their chemopreventive potential. Several derivatives were synthesized and exhibited a range of improved activities, including nanomolar induction of QR1 and inhibition of QR2, and micromolar NF-κB and inducible nitric oxide synthase (iNOS) inhibition. Particularly, a synthetic phenazine analog (**3**) exhibited a QR1 CD (concentration required to double QR1 activity) value of 4.7 nM.[62]

Two cyclohexadepsipeptides, arenamides A and B (**4**), were isolated from the fermentation broth of the first marine obligate genus to be described, *Salinispora arenicola*. They blocked TNF-induced NF-κB activation in a dose- and time-dependent manner with IC_{50} values of 3.7 and 1.7 μM, respectively.[63] The compounds also displayed inhibitory activity against iNOS and PGE_2.

Streptochlorin (**5**) was isolated from the fermentation broth of a *Streptomyces* sp. and displayed potent antiangiogenic activity. In the presence of vascular endothelial growth factor (VEGF) cells, streptochlorin inhibited cell invasion and tube formation in a dose-dependent manner; it was postulated that streptochlorin exhibited this activity by acting on the NF-κB signaling pathway. To support this claim, **5** exhibited inhibitory activity against NF-κB in a dose-dependent manner, suggesting that it may have an antiproliferative effect on cell growth, and thus may be effective in controlling the progression of angiogenesis.[64]

Fijiolides A (**6**) and B were isolated from a marine sediment-derived *Nocardiopsis* sp. from Beqa Lagoon, Fiji. They exhibited a significant reduction in TNF-α-induced NF-κB activation by 70.3% (IC_{50} value of 0.57 μM) and 46.5%, respectively.[65] Similarly, lawsonone (**7**) was isolated from the fermentation broth of a *Streptomyces* sp. (collected in Karnataka, India). It inhibited NO and pro-inflammatory cytokine production, namely, interleukin (IL)-1β, IL-6, and TNF-α, in LPS-induced macrophages in a dose-dependent manner.[66]

The measurement of NO in LPS-stimulated RAW 264.7 cells is a mechanism by which the anti-inflammatory potential of a molecule is assessed. Thienodolin (**8**), isolated from a marine-derived *Streptomyces* sp. (Chilean marine sediment), inhibited NO production in LPS-stimulated RAW 264.7 cells with an IC_{50} value of 17.2 μM.[67] Similar to other dietary phytochemicals, such as genistein and kaempferol, thienodolin showed multiple inhibitory mechanisms in LPS-induced iNOS expression. In particular, this was the first report to postulate that halogenated tryptophan analogs acted on iNOS expression and NO production by downregulating NF-κB and signal transducer and activator of transcription (STAT1) signaling pathways.

11.3.1.1.2 Detoxification Effects of Metabolites from Marine-Derived Actinomycetes

Ultrafiltration liquid chromatography–mass spectrometry was employed in an effort to develop a screen where inhibitors of QR2 could be readily identified from complex samples. In this study, both botanical extracts and bacterial extracts (the latter derived from marine sediment) were analyzed. This unique screening process led to the identification of tetrangulol methyl ether (**9**), an anthraquinone-type polyketide isolated from an *Actinomyces* sp. It exhibited an IC_{50} value of 0.16 µM and served as an example of utilizing rapid screening approaches to identify potential QR2 inhibitors from complex metabolic mixtures.[68] Given the sensitivity of MS detectors and their amenability to high-throughput analyses, coupled with the advent of extensive secondary metabolite databases, MS-based screens show much promise as attractive options to identify potential chemopreventive compounds without the need for traditional large-scale chemical isolation studies.

In a continuation of the study by Nam et al., compound **6** induced QR1 enzyme activity with an IC_{50} value of 28.4 µM, whereas fijiolide B exhibited no observable QR1 activity, which suggests that the nitrogen substitution pattern is essential for biological activity.[65] Two macrolide antibiotics, 5-*O*-α-l-rhamnosyltylactone and juvenimicin C (**10**), were isolated from the fermentation broth of a sediment-derived *Micromonospora* sp. collected from Palau. Compound **10** enhanced QR1 enzyme activity and glutathione levels by twofold, with CD values of 10.1 and 27.7 µM, respectively. In addition, glutathione reductase and glutathione peroxidase activities were elevated.[69] This is the first member of the macrolide class of antibiotics found to mediate these responses.

The ammosamides are a group of chlorinated quinolines produced by a *Streptomyces* sp. derived from sediment samples in the Bahamas. Ammosamides A and B exhibited potent cytotoxicity toward HCT-116 colon tumor cells with IC_{50} values of 320 nM each.[70] Later studies examined the fermentation broth of marine-derived *Streptomyces variabilis*, and ammosamides E (**11**) and F were discovered. Interestingly, **11** exhibited an IC_{50} value of 40 nM in a QR2 assay using human recombinant protein and, unlike the A and B analogs, exhibited no significant cytotoxicity toward cancer cell lines HCC44, HCC4017, and Calu-3, and the noncancerous human bronchial epithelial cell line HBEC30KT.[71] Further studies indicated that the addition of amines to the culture medium of *S. variabilis* resulted in the production of a wide range of ammosamide analogs, all with varying levels of biological activity.

11.3.1.1.3 Other Activities of Metabolites from Marine-Derived Actinomycetes

Among six compounds isolated from the fermentation broth of a marine-derived *Streptomyces* sp., *N*-carbamoyl-2,3-dihydroxybenzamide (**12**) and 2-acetamido-3-(2,3-dihydroxybenzoylthio) propanoic acid, as well as 2,3-dihydroxybenzoic acid and 2,3-dihydroxybenzamide, showed ED_{50} (median effective dose) values of 14.4, 13.0, 10.3, and 14.6 µM, respectively, in the DPPH free-radical (DPPH·) scavenging assay.[72]

Four α-methoxy-γ-pyrones were isolated from the extract of a marine-derived *Streptomyces albus* (Iberian peninsula), and one (**13**) exhibited potent EGFR inhibition. EGFR is a ligand-activated receptor tyrosine kinase; binding to this receptor results in a cascade of steps that ultimately ends in MAPK transducing mitogenic signals to the nucleus. Compound **13** inhibited EGF-induced activator protein 1 (AP-1) transcriptional activity with an IC_{50} value of less than 7 nM in HeLa cells.[11,73]

11.3.1.2 Marine-Derived Cyanobacteria

Cyanobacteria are commonly referred to under the misnomer blue-green algae. However, they are photosynthetic bacteria that exhibit a wide range of morphologies, from single cells to branched and unbranched filaments. They have the capacity to live in any environment that has sufficient access to sunlight, and play important roles in the ecosystem (toxin production, ability to fix nitrogen, etc.).[74] The vast majority of secondary metabolites that have been isolated from cyanobacteria belong to the genus *Lyngbya*. However, taxonomy within this phylum is in need of extensive modification.[74] In the past, researchers have utilized morphological and growth characteristics to distinguish between genera, whereas phylogenetic analyses that employ genomic methods (16S rRNA gene sequences or multilocus sequence typing) suggest that taxonomic diversity is much greater than originally thought. As is the case with marine-derived actinomycete natural products, relatively few studies explore the capacity of cyanobacterial secondary metabolites to exhibit chemopreventive activities; summarized here are studies that focus on noncytotoxic molecules.

11.3.1.2.1 Anti-Inflammatory Metabolites of Marine-Derived Cyanobacteria

In a screening program designed to discover anti-inflammatory molecules from a library of marine natural products, malyngamide F acetate (**14**) was identified. Researchers postulated that prokaryotes (such as cyanobacteria) evolved to produce small molecules that would inhibit the innate immune system and inflammatory response in vertebrates. Compound **14**, from the strain *Lyngbya majuscula*, inhibited NO production in a concentration-dependent manner in LPS-stimulated murine macrophage cells (RAW 264.7) with an IC_{50} value of 3.4 μg/ml.[75] It was determined that **14** suppressed IL-1β, IL-6, IL-10, and iNOS transcription, while it also increased TNF-α transcription.

A study was designed to screen marine natural products against both solution-based DPPH and cell-based 2′,7′-dichlorodihydrofluorescein diacetate (DCFH-DA) assays. Among the compounds screened in the assays, scytonemin (**15**), isolated from a *Scytonema* sp., showed strong activity in the DPPH assay but was virtually inactive in the DCFH-DA assay.[76] However, in a screen designed to identify compounds that inhibit kinases related to chronic hyperproliferative conditions, **15** inhibited human pololike kinase in a concentration-dependent manner with an IC_{50} value of 2 μM.[77] It inhibited the mitogen-induced proliferation of three cell types related to inflammatory hyperproliferation. The active compounds are illustrated in Figure 11.1.

11.3.2 Marine-Derived Fungi

Various secondary metabolites from marine-derived fungi have been reported, including terpenes, steroids, polyketides, peptides, alkaloids, and polysaccharides. In this section, the cancer chemopreventive potential of small molecules (not polysaccharides, proteins, etc.) from this source will be presented.

Fungi living in the marine environment can be largely classified into two categories defined by ecologic origin: obligate marine-derived fungi that originate from the ocean and exclusively grow and sporulate in a marine habitat, and facultative marine-derived fungi that originate from freshwater and terrestrial ecosystems but have adapted to the marine environment.[78,79] In fact, numerous marine-derived fungi were identified as genera, including *Aspergillus*, *Penicillium*, *Fusarium*, *Cladosporium*, and *Phoma*, which exist in terra.[80]

The use of fungi for commercial and medical purposes can be a double-edged sword. For instance, *Aspergillus oryzae* has been widely used for food fermentation, while *Aspergillus flavus* produces mycotoxins (e.g., aflatoxin) that are fatal and highly carcinogenic. Therefore, the core task prior to the evaluation of biological activities of fungal products should be a test for safety.[81]

For instance, some antioxidant compounds described below were reported to possess toxic effects, including gliotoxin,[82] tryptophol,[83] and citrinin H2.[84] Thus, these molecules should not be taken into consideration as drug candidates.

During a search of the literature, the majority of biological effects were found to be cytotoxic and antimicrobial, which seems reasonble since fungi can produce allelochemicals as a defense mechanism.[85] In fact, the first reported bioactive compound obtained from marine-derived fungus (*Acremonium chrysogenum*) was cephalosporin C, which is a prototype of broad-spectrum antibiotics, the cephalosporins.[86] The total numbers of cataloged and predicted species of marine-derived

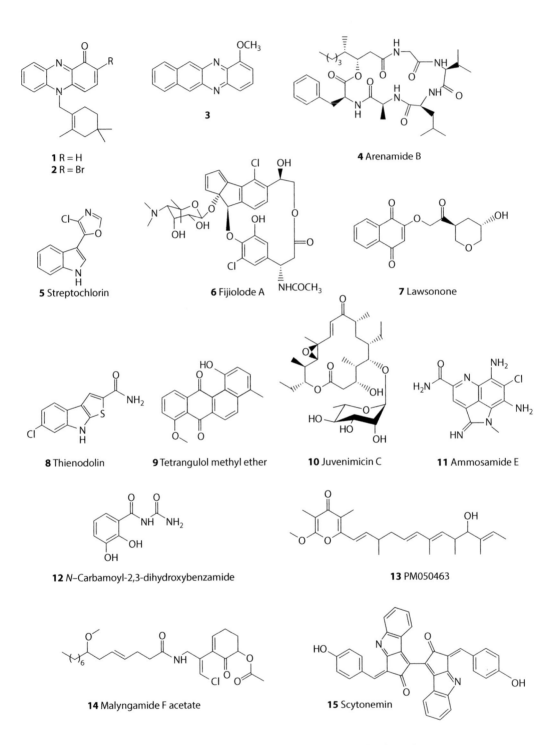

1 R = H
2 R = Br

3

4 Arenamide B

5 Streptochlorin

6 Fijiolode A

7 Lawsonone

8 Thienodolin

9 Tetrangulol methyl ether

10 Juvenimicin C

11 Ammosamide E

12 *N*–Carbamoyl-2,3-dihydroxybenzamide

13 PM050463

14 Malyngamide F acetate

15 Scytonemin

Figure 11.1 Structures of cancer chemopreventive compounds from marine bacteria.

fungi are 1,097 and 5,320, respectively, which are much fewer than those of terrestrial fungi, standing at 43,271 for the cataloged species and 611,000 for the predicted. However, the unique chemical diversity produced by marine-derived fungi leads more scientists to explore the ocean.[87] Other reports estimate that more than 1,500 species of marine-derived fungi occur in various taxonomic groups. Most marine-derived fungi are ascomycetes and basidiomycetes.[88]

Marine-derived fungi were found in a variety of marine habitats and living organisms, including plants (e.g., algae, driftwoods, and mangroves), invertebrates (e.g., tunicates, mollusks, sponges, ascidians, and corals), vertebrates (e.g., fish), and even inorganic substances (e.g., sediments and synthetic substances).[78] According to Hu et al.,[89] approximately 300 diverse novel compounds were isolated from marine-derived fungi between 1985 and 2008.

Considering there are still numerous unexplored marine-derived fungi, and these species have the ability to produce chemically diverse compounds based on surrounding environments, it is expected that the number of compounds will eventually increase. It is estimated that polyketides comprise the largest portion of molecules isolated from marine-derived fungi, followed by prenylated polyketides (meroterpenes), peptides, and alkaloids.[90] In this review, metabolites in both terrestrial and marine organisms are excluded from the illustrations.

11.3.2.1 Antioxidant Marine-Derived Fungi

The most frequently reported biological properties of isolated marine-derived fungi are antimicrobial activities and cytotoxicity toward cancer cell lines. In terms of antioxidant activities, the DPPH$^\bullet$ scavenging assay is the most commonly used method to evaluate the potential of isolated marine metabolites. In this review, small molecules showing DPPH$^\bullet$ scavenging activities with IC_{50} values under millimolar concentrations are included. Compounds that are described as weak or moderately active are excluded from this review. A considerable number of antioxidant metabolites were isolated from marine-derived fungi from various sources, especially marine sponges, algae, and sediments. So far, a rich profile of biologically active metabolites is described from filamentous fungi of terrestrial origin, especially from three genera: *Aspergillus*, *Penicillium*, and *Fusarium*.[52] The description of antioxidant marine-derived fungi metabolites is modified and expanded from a previous review.[32]

Several gentisyl alcohol derivatives showed potent DPPH$^\bullet$ scavenging activities. Gentisyl alcohol also occurs as a metabolic product of soil fungi *Penicillium patulum*. Gentisyl alcohol from *Aspergillus parasiticus* on the surface of red alga *Carpopeltis cornea*, from *Aspergillus* sp. on the surface of brown alga *Sargassum ringgoldium*, and from *Dothideomycete* sp. isolated from the red algae *Chondria crassicualis* showed DPPH$^\bullet$ scavenging activity, with slightly different IC_{50} values of 1.4[91], 1.0[92], and 7.0 μM,[93] respectively.

Gentisyl alcohol polymer terrestrol A (**16**) and monomeric derivatives, including 2-chloro-6-(methoxymethyl)-1,4-benzenediol and 2-(methoxymethyl)-1,4-benzenediol from *Penicillium terrestre* C. N. Jenson in marine sediments, showed DPPH$^\bullet$ scavenging activity with IC_{50} values of 2.6, 8.5, and 9.8 μM, respectively.[94]

Two toluhydroquinone derivatives, 5-bromotoluhydroquinone (**17**) and 4-*O*-methyltoluhydroquinone, and toluhydroquinone from the fermentation medium of a *Dothideomycete* sp. from the red alga *C. crassicualis* showed DPPH$^\bullet$ scavenging with IC_{50} values of 11.0, 17.0, and 12.0 μM, respectively.[93]

Brominated dihydroxyphenylacetic acid derivatives, including methyl 2-(6-bromo-3,4-dihydroxyphenyl)acetate and methyl 2-(2,5-dibromo-3,4-dihydroxyphenyl)acetate (**18**), and 2-(3,4-dihydroxyphenyl)acetic acid from the broth of the fungus *Aspergillus* sp. incubated with NaBr and CaBr$_2$ induced production of scavenged DPPH$^\bullet$ with IC_{50} values of 14.2, 12.1, and 11.0 μM, respectively.[95]

A *seco*-anthraquinone, 2-*O*-methyleurotinone, and a bisdihydroanthracenone derivative, eurorubrin (**19**), from *Eurotium rubrum*, an endophytic fungal strain from the inner tissue of the stem of the marine mangrove plant *Hibiscus tiliaceus*, showed DPPH[•] scavenging activities with IC_{50} values of 74.0 and 44.0 μM, respectively.[96]

A gabosine derivative, parasitenone (**20**), and 3-chloro-4,5-dihydroxybenzyl alcohol from *A. parasiticus* on the surface of red alga *Carpopeltis cornea* showed DPPH[•] scavenging activity with IC_{50} values of 57.0 and 0.6, respectively.[91] Later, **20** was found to be identical with an epoxy-cyclohexenone (polyketide), (+)-epoxydon.[97] The comparison of DPPH[•] scavenging activity showed that 3-chlorogentisyl alcohol was more potent than (+)-epoxydon monoacetate, gentisyl alcohol, and (+)-epoxydon from *Aspergillus* sp., exhibiting IC_{50} values of 1.0, 15.0, 7.0, and 6.0 μM, respectively.[98]

3-Methyl-8-hydroxy-3,4-dihydroisocoumarin [(–)-mellein], first isolated from terrestrial *Aspergillus ochraceushas* in 1933,[99] was also produced by both *Cladosporium* sp. and *A. ochraceushas*, found on the surface of the red alga *Chondria crassicualis*. (–)-Mellein showed DPPH[•] scavenging activity, with similar IC_{50} values of 34[100] and 25 μM,[101] respectively.

A halogenated isocoumarin, (*R*)-(–)-5-bromomellein (**21**), along with clavatol and circumdatin A found in the fermented medium containing the *A. ochraceus* from the red alga *C. crassicualis* and NaBr, showed DPPH[•] scavenging with IC_{50} values of 24, 30, and 32 μM, respectively.[101] Also, a benzodiazepine, 2-hydroxycircumdatin C (**22**), produced by *A. ochraceus* from the brown alga *Sargassum kjellmanianum*, showed DPPH[•] scavenging activity with an IC_{50} value of 9.9 μM.[102]

2,5-Diketopiperazines are cyclodipeptides obtained by the condensation of two α-amino acids and occur in numerous natural sources, including fungi. These compounds showed chitinase inhibitory, 5-phosphodiesterase inhibitory, oxytocin antagonistic, anticancer, anti-inflammatory, antimicrobial, antifungal, neuroprotective, anxiolytic, and herbicidal activities.[103] A diketopiperazine alkaloid, golmaenone (**23**), and an indole-containing diketopiperazine alkaloid, neoechinulin A, from an *Aspergillus* sp. on the surface of the red alga *Lomentaria catenata*, exhibited significant DPPH[•] scavenging activity with IC_{50} values of 20 and 24 μM, respectively.[104] Neoechinulin A was first isolated from the mycelium of *Aspergillus amstelodami*, grown on molasses beet cultures[105] and found in various terrestrial sources, including other fungi (e.g., *Aspergillus* sp., *Microsporum* sp., *Penicillium griseofulvum*, ascomysete *Xylaria eulossa*, endophytic fungus *Chaetomium globosum*, and *Eurotium* sp.) and plants (e.g., *Bridelia ferruginea*, *Plumbago zeylanica*, and *Pinellia cordate*).[106,107] Based on a PubMed search (data accessed on December 30, 2013), neoechinulin A has been reported to show neuroprotective activities.

New indole-containing diketopiperazine alkaloids, named variecolorins M–O (**24**), from *Penicillium griseofulvum* in a deep-ocean sediment, showed IC_{50} values of 135, 120, and 91 μM, respectively.[108]

A diketopiperazine, brevianamide W (**25**), and diketopiperazine alkaloids, including diketopiperazine V, brevianamide Q, brevianamide R, brevianamide K, and brevianamide E, isolated from the the fermentation broth of a deep-sea-inhabiting fungus *Aspergillus versicolor* CXCTD-06-6a, exhibited DPPH[•] scavenging activity, with inhibition of 55.0, 55.0, 53.7, 46.2, 61.4, and 19.3%, respectively, at a concentration of 13.9 μM.[109]

A dioxopiperazine, gliotoxin, from *Pseudallescheria* sp. found in the brown alga *Agarum cribrosum*, exhibited significant DPPH[•] scavenging activity with an IC_{50} value of 5.2 μM.[110] However, gliotoxin is a mycotoxin produced by various fungal species, including *Aspergillus fumigatus*, *Eurotium chevalieri*, *Gliocladium fimbriatum*, *Trichoderma* sp., and *Penicillium* sp., and may be a severe hazard for humans.[82]

Known tryptophol (tryptophan degrading product in *Saccharomyces cerevisiae*) and 2-(1*H*-indol-3-yl)ethyl 5-hydroxypentanoate from the yeast *Pichia membranifaciens* found in the marine sponge *Halichondria okadai* exhibited 52% and 47% DPPH[•] scavenging activity at 300 μM.[111] However, tryptophol showed genotoxicity by damaging DNA in HepG2, A549, and THP-1 cells in an alkaline comet assay.[83]

Naphtho-γ-pyrones, aurasperone B[112] and nigerasperone C, from *Aspergillus niger* EN-13 from the brown alga *Colpomenia sinuosa*, showed 41.6% and 48.1% DPPH[•] scavenging activities at

50 μg/ml.[113] Aurasperone B was also found in terrestrial fungi *Aspergilli brasiliensis* and *Aspergilli ibericus* from grapes.[114]

Sorbicillinoid derivatives, JBIR-124 and JBIR-59 (**26**), from *Penicillium citrinum* SpI080624G1f01 in a marine sponge, demonstrated DPPH$^{\bullet}$ scavenging activity with IC$_{50}$ values of 30 and 25 μM, respectively.[115]

An orcinol tetramer, tetraorcinol A (**27**), from *A. versicolor* LCJ-5-4 in the coral *Cladiella* sp., showed DPPH$^{\bullet}$ scavenging activity with an IC$_{50}$ value of 67 μM.[116] A dihydroxanthenone, 1,7-dihydroxy-3-hydroxymethyl-7,8-dihydroxanthenone-8-carboxylic acid methyl ester (AGI-B4), which was first isolated from a terrestrial source, *Aspergillus* sp. Y80118, and also found in *Aspergillus sydowii* PSU-F154 from a gorgonian sea fan of the genus *Annella*, showed DPPH$^{\bullet}$ scavenging activity with an IC$_{50}$ value of 17 μM.[117]

4-Hydroxyphenethyl methyl succinate, 4-hydroxyphenethyl 2-(4-hydroxyphenyl)acetate (**28**), and 4-methylpyrocatechol, from *Penicillium griseofulvum* Y19-07 found in mangrove *Lumnitzera racemosa*, showed DPPH$^{\bullet}$ scavenging activities with IC$_{50}$ values of 58.6, 56.2, and 7.1 μM, respectively.[118]

Polyketides, including redoxcitrinin (biogenetic precursor of citrinin) and two mycotoxins (phenol A and citrinin H2) from *Penicillium* sp. found in the green alga *Ulva pertusa*, exhibited DPPH$^{\bullet}$ scavenging activity with IC$_{50}$ values of 27.7, 23.4, and 27.2 μM, respectively.[119] Citrinin H2 is a main component after heating from citrinin, a mycotoxin found in *Penicillium* or *Aspergillus*.[84]

Two polybrominated diphenyl ethers, 4,4′,6,6′-tetrabromo-3,3′-dihydroxy-5,5′-dimethyldiphenyl ether and 2′,4,4′,6,6′-pentabromo-3,3′-dihydroxy-5,5′-dimethyldiphenyl ether (**29**), and diphenyl ethers, including 3,3′-dihydroxy-5,5′-dimethyldiphenyl ether and an inseparable mixture of violacerol-I and violacerol-II, from *Penicillium chrysogenum* in the red alga *Hypnea complex*, exhibited DPPH$^{\bullet}$ scavenging activities with IC$_{50}$ values of 18, 15, 42, and 6 μM, respectively.[120]

Farnesylhydroquinone, reported as the reduction product from terrestrial *Seseli* species and later isolated from the mycelium of genus *Penicillium* on the surface of the polymeric cord, demonstrated DPPH$^{\bullet}$ scavenging activity with an IC$_{50}$ value of 12.5 μM.[121]

Hydroquinone derivatives, 7-isopropenylbicyclo[4.2.0]octa-1,3,5-triene-2,5-diol (**30**), 2-(1-hydroxy-1-methylehtyl)-2,3-dihydrobenzofuran-5-ol, 2-dimethyl-2,3-dihydrobenzopyran-3,6-diol, and 2-(2,3-dihydroxy-3-methylbutyl)benzene-1,4-diol, from *Acremonium* sp. found in the brown alga *Cladostephus spongius* (Hudson) C. Agardh, exerted DPPH$^{\bullet}$ scavenging activities with IC$_{50}$ values of 41.2, 31.9, 55.1, and under 117.8 μM.[122] 4,5,6-Trihydroxy-7-methylphthalide (epicoccone), isolated from *Epicoccum* sp. found in the brown alga *Fucus vesiculosus*, exhibited a 95% DPPH$^{\bullet}$ scavenging effect at 25 μg/ml and 62% inhibition in the peroxidation of linolenic acid in a thiobarbituric acid reactive substances (TBARS) assay at 37 μg/ml.[123]

Benzaldehydes, chaetopyranin, 2-(2′,3-epoxy-1′,3′-heptadienyl)-6-hydroxy-5-(3-methyl-2-butenyl) benzaldehyde, and 2-[(*E*)-hept-5-enyl]-3,6-dihydroxy-5-(3-methylbut-2-enyl)benzaldehyde (isotetrahydroauroglaucin), and an anthraquinone, 1,4,5-trihydroxy-7-methoxy-2-methyl-9,10-anthraquinone (erythroglaucin), from the endophytic fungus *Chaetomium globosum* found in inner tissue of red alga *Polysihonia urceolata*, showed DPPH$^{\bullet}$ scavenging activity with IC$_{50}$ values of 110.6, 294.9, 86.0, and 206.5 μM, respectively.[124]

An anthracene glycoside, asperflavin ribofuranoside (**31**), and polyketides flavoglaucin and iso-dihydroauroglaucin, from *Microsporum* sp., exhibited DPPH$^{\bullet}$ scavenging activity with IC$_{50}$ values of 14.2, 11.3, and 11.5 μM, respectively.[125]

2,3,6,8-Tetrahydroxy-1-methylxanthone (**32**) (molecular weight [MW] = 274.23 Da), 3,6,8-trihydroxy-1-methylxanthone, and 5-(hydroxymethyl)-2-furanocarboxylic acid from *Wardomyces anomalus* showed DPPH$^{\bullet}$ scavenging activities with IC$_{50}$ values of < 91.2, 895.0, and 289.3 μM, respectively.[126]

Depsidones, aspergillusidones A and B (**33**), from *Aspergillus unguis* CRI282-03, showed radical scavenging activity in the xanthine–xanthine oxidase assay with IC$_{50}$ values of 16.0 and less than 15.6 μM, respectively.[127]

Xyloketal B was isolated from a mangrove fungus *Xylaria* sp. (no. 2508) and protects against oxidized low-density lipoprotein (oxLDL)-induced endothelial oxidative injury, probably through inhibiting NADPH oxidase-derived ROS generation, promoting NO production, and restoring Bcl-2 expression.[128] Thereafter, xyloketal B exhibited a neuroprotective effect against 1-methyl-4-phenylpyridinium-induced neurotoxicity in the C57BL/6 mouse Parkinson's disease model.[129]

The antioxidant compounds found from DPPH assay are listed in Table 11.1. Compounds newly found in marine-derived fungi when each paper was published, along with IC_{50} values under 100 μM in the DPPH assay, are illustrated in Figure 11.2.

11.3.2.2 Anti-Inflammatory Marine-Derived Fungi

The above-mentioned compound **20** inhibited LPS-induced NO production and iNOS protein expression in RAW 264.7 cells through modulating NF-κB, likely by binding covalently to a cysteine residue.[130] A highly oxygenated polyketide penilactone A (**34**) that contains a novel carbon skeleton formed from two 3,5-dimethyl-2,4-diol-acetophenone units and a γ-butyrolactone moiety was isolated from an Antarctic deep-sea-derived fungus *Penicillium crustosum* PRB-2. **34** showed 40% inhibition at the concentration of 10 μM on NF-κB transcriptional activation in RAW 264.7 cells transiently transfected with a pNF-κB-Luc expression plasmid.[131]

A hexacyclic dipeptide, azonazine (**35**), from a Hawaiian marine sediment-derived strain *Aspergillus insulicola*, inhibited NF-κB luciferase activity (IC_{50} = 8.4 μM) and nitrite production (IC_{50} = 13.7 μM) in RAW 264.7 murine macrophage cells.[132] Isolated from the marine-derived fungus *Aspergillus* sp. SF-5044, asperlin inhibited LPS-induced expression of iNOS and COX-2; production of NO, PGE$_2$, TNF-α, and IL-1β; and degradation of inhibitor of κB (IκB)-α. In addition, asperlin induced hemeoxygenase-1 (HO-1) in murine macrophages.[133] A styrylpyrone-type metabolite, penstyrylpyrone (**36**), produced by *Penicillium* sp. JF-55 from an unidentified sponge off the shores of Jeju Island, Korea, inhibited protein tyrosine phosphatase 1B activity, and inhibited expression of iNOS and COX-2; production of NO, PGE$_2$, TNF-α, and IL-1β; degradation of IκB-α; and nuclear translocation and DNA binding of NF-κB in LPS-stimulated macrophages.[134] Two hexylitaconic acid derivatives (α,β-dicarboxylic acid derivatives), 2-[(5R)-5-hydroxyhexyl]-3-methylene-1,4-dimethyl ester(2S)-butanedioic acid (**37**) and 2-(6-hydroxyhexyl)-3-methylene-1,4-dimethyl ester(2S)-butanedioic acid (**38**), from a sponge-derived fungus *Penicillium* sp., showed more than 50% inhibition of IL-1β production at the concentration of 200 μM in LPS-stimulated RAW 264.7 cells.[135] A anthraquinone, questinol (**39**), was isolated from the marine-derived fungus *Eurotium amstelodami* and exhibited anti-inflammatory capacities, as illustrated by inhibition of the production of NO, PGE$_2$, TNF-α, IL-1β, IL-6, and iNOS in LPS-stimulated RAW 264.7 cells.[136] Pyrenocine A (**40**) from the marine-derived fungus *Penicillium paxilli* Ma(G)K was found to inhibit the production of NO, PGE$_2$, and TNF-α in LPS-stimulated RAW 264.7 cells probably by interfering with the MyD88-dependent pathway.[137] *bis*-*N*-Norgliovictin (**41**) from marine-derived endophytic fungus *S3-1-c* suppressed TNF-α, IL-6, interferon-β, and monocyte chemoattractant protein 1 production in LPS-treated RAW 264.7 cells.[138] The chemical structures of novel anti-inflammatory compounds isolated from marine-derived fungi are illustrated in Figure 11.3.

11.3.2.3 Anticarcinogenic Marine-Derived Fungi

11.3.2.3.1 Induction of Detoxification and Cytoprotection by Marine-Derived Fungi

A chromone derivative, 2-(hydroxymethyl)-8-methoxy-3-methyl-4*H*-chromen-4-one (**42**, chromanone A), from *Penicillium* sp., inhibited CYP1A activity up to 60% and induced glutathione *S*-transferase (GST) and epoxide hydrolase (mEH) activity in Hepa1c1c7 cells.[139] New monomeric xanthones monodictysins B (**43**) and C (**44**), from *Monodictys putredinis* found in the inner tissue

Table 11.1 DPPH Scavengers from Marine-Derived Fungi

Compounds	Fungi	Source	Type	Structure Class	DPPH Scavenging Activity (IC$_{50}$, μM)	Positive Control: IC$_{50}$ (μM)	References
Terrestrols A (**16**)–H	Penicillium terrestre C. N. Jenson	Sediment	Sediment	Gentisyl alcohol polymer	2.6–6.3	L-Ascorbic acid: 17.4	94
5-Bromotoluhydroquinone (**17**), 4-O-methyltoluhydroquinone	Dothideomycete sp.	Chondria crassicualis	Red alga	Toluhydroquinone	11.0, 17.0	L-Ascorbic acid: 20.0	93
Eurorubrin (**19**)	Eurotium rubrum	Hibiscus tiliaceus	Mangrove	Bisdihydroanthracenone	44.0	BHT: 82.6	96
Parasitenone (**20**)	Aspergillus parasiticus	Carpopeltis cornea	Red alga	Gabosine derivative	57.0		91
(R)-(−)-5-Bromomellein (**21**)	Aspergillus ochraceus	Chondria crassicualis	Red alga	Brominated mellein	24	L-Ascorbic acid: 20.0	
2-Hydroxycircumdatin C (**22**)	Aspergillus ochraceus	Sargassum kjellmanianum	Brown alga	Benzodiazepine analog	9.9	BHT: 88.2	102
Golmaenone (**23**)	Aspergillus sp.	Lomentaria catenata	Red alga	Diketopiperazine	20		104
Variecolorins M–O (**24**)	Penicillium griseofulvum	Ocean sediment	Sediment	Diketopiperazine	135, 120, 91	L-Ascorbic acid: 26	108
Nigerasperone C	Aspergillus niger EN-13	Colpomenia sinuosa	Brown alga	Naphtho-γ-pyrone	41.6% scavenging at 50 μg/ml	BHT, 80.4% at 50 μg/ml	113

(Continued)

Table 11.1 (Continued) DPPH Scavengers from Marine-Derived Fungi

Compounds	Fungi	Source	Type	Structure Class	DPPH Scavenging Activity (IC$_{50}$, μM)	Positive Control: IC$_{50}$ (μM)	References
JBIR-124, JBIR-59 (**26**)	*Penicillium citrinum* SpI080624G1f01	Offshore of Ishigaki Island	Sponge	Sorbicillinoid derivative	30, 25	α-Tocopherol: 9.0	115
Tetraorcinol A (**27**)	*Aspergillus versicolor* LCJ-5-4	*Cladiella* sp.	Soft coral	Orcinol tetramer	67		116
4-Hydroxyphenethyl methyl succinate, 4-hydroxyphenethyl 2-(4-hydroxyphenyl)acetate (**28**)	*Penicillium griseofulvum* Y19-07	*Lumnitzera racemosa*	Mangrove	Phenolic	58.6, 56.2	Vitamin E: 26.5	118
4,4′,6,6′-Tetrabromo-3,3′-dihydroxy-5,5′-dimethyldiphenyl ether, 2′-bromo-3,3′-dihydroxy-5,5′-dimethyldiphenyl ether (**29**)	*Penicillium chrysogenum*	*Hypnea complex*	Red alga	Diphenyl ether	18, 15	L-Ascorbic acid: 20	120
7-Isopropenylbicyclo[4.2.0] octa-1,3,5-triene-2,5-diol (**30**)	*Acremonium* sp.	*Cladostephus spongius* (Hudson) C. Agardh	Brown alga	Hydroquinone	41.2	BHT: 282.2	122
4,5,6-Trihydroxy-7-methylphthalide (epicoccone)	*Epicoccum* sp.	*Fucus vesiculosus*	Brown alga	Phthalide	95% scavenging at 25 μg/ml		123
Chaetopyranin	*Chaetomium globosum*	*Polysihonia urceolata*	Red alga	Benzaldehyde	110.6	BHT: 81.69	124
Asperflavin ribofuranoside	*Microsporum* sp.	*Lomentaria catenata*	Red alga	Anthracene glycoside	14.2	L-Ascorbic acid: 20	125
2,3,6,8-Tetrahydroxy-1-methylxanthone (**32**)	*Wardomyces anomalus*	*Enteromorpha* sp.	Green alga	Xanthone	<91.2	BHT: <25 μg/ml	126

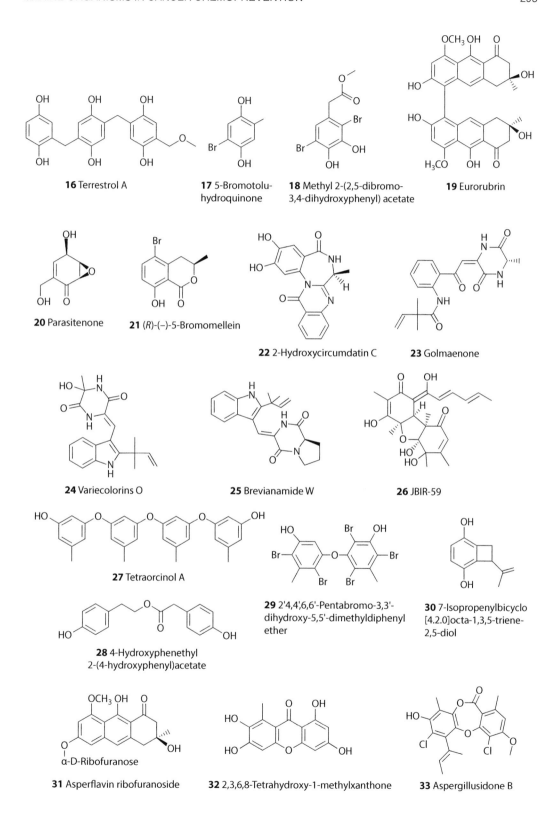

Figure 11.2 Antioxidant compounds from marine-derived fungi.

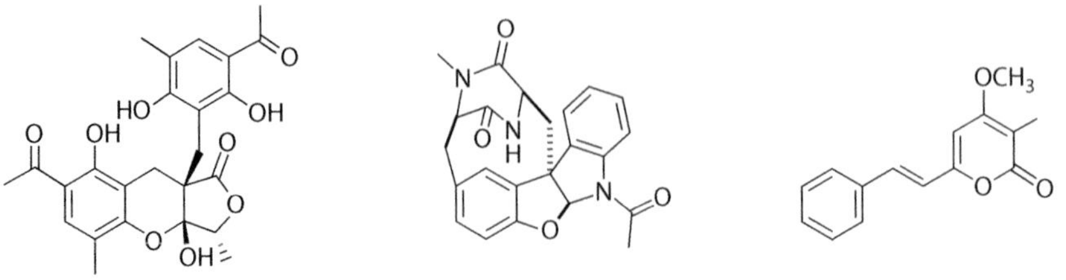

34 Penilactone A   **35** Azonazine   **36** Penstyrylpyrone

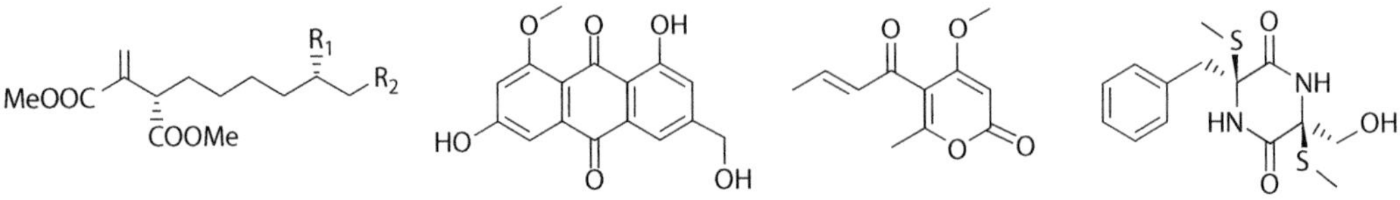

37 R₁ = OH   R₂ = H   **39** Questinol   **40** Pyrenocine A   **41** *bis-N*-Norgliovictin
38 R₁ = H   R₂ = OH

Figure 11.3   Novel anti-inflammatory compounds from marine-derived fungi.

of a marine green alga collected in Tenerife, Spain, induced QR activity in Hepa 1c1c7 cells, with CD values of 12.0 and 12.8 µM, respectively. In addition, monodictysin B inhibited CYP1A activity with an IC$_{50}$ value of 3.0 µM.[140] Two dimeric chromanones, monodictyochromones A (**45**) and B (**46**) from *M. putredinis*, inhibited CYP1A activity with IC$_{50}$ values of 5.3 and 7.5 µM, respectively. In addition, both compounds displayed moderate activity as inducers of QR in mouse Hepa1c1c7 cells, with CD values of 22.1 and 24.8 µM, respectively.[141]

A polyketide, noduliprevenone (**47**), from an endophytic *Nodulisporium* sp., competitively inhibited CYP1A activity with an IC$_{50}$ value 6.5 µM, and induced QR activity in Hepa 1c1c7 mouse culture cells, with a CD value of 5.3 µM.[142]

Emodin and chrysophanol were isolated from an *Aspergillus* sp. on the surface of a marine brown alga. They decreased γ-glutamyl transpeptidase activity and increased glutathione and cell viability with ethanol-induced HepG2/CYP2E1 cells.[143] In a dextran sulfate sodium-induced mouse model for colitis, chrysophanol significantly attenuated body weight loss, colon shortening, diarrhea, and rectal bleeding. The responses were accompanied by inhibition of COX-2 and NF-κB protein expression and caspase-1 activation. Additionally, chrysophanol inhibited the production of TNF-α and IL-6 and the expression of COX-2 levels, induced by LPS through suppression of the activation of NF-κB and caspase-1.[144] A meroterpenoid, penicillipyrone B (**48**), isolated from the marine-derived fungus *Penicillium* sp., induced QR activity by 1.9 times at a concentration of 40 µM.[145]

These newly found cancer chemopreventive marine compounds in each published paper are illustrated in Figure 11.4.

11.3.2.3.2 Other Chemopreventive Marine-Derived Fungi

A dopamine metabolite, 3,4-dihydroxyphenyl acetic acid from an *Aspergillus* sp. found in marine brown alga *Ishige okamurae*, and **20** from *Phoma herbarum* found in marine red alga *Hypnea saidana*, inhibited the activity of EGFR in a dose-dependent manner; the IC$_{50}$ values for 3,4-dihydroxyphenyl acetic acid and (+)-epoxydon were 2.8 and 0.6 µg/ml, respectively.[146] A indole alkaloid, shearinine E (**49**), from *Penicillium janthinellum* Biourge, inhibited EGF-induced malignant transformation of JB6 P⁺ Cl41 cells in a soft agar.[147] A depsidone, aspergillusidone C (**50**), showed potent aromatase inhibitory activity with an IC$_{50}$ value of 0.74 µM.[127] A monomeric

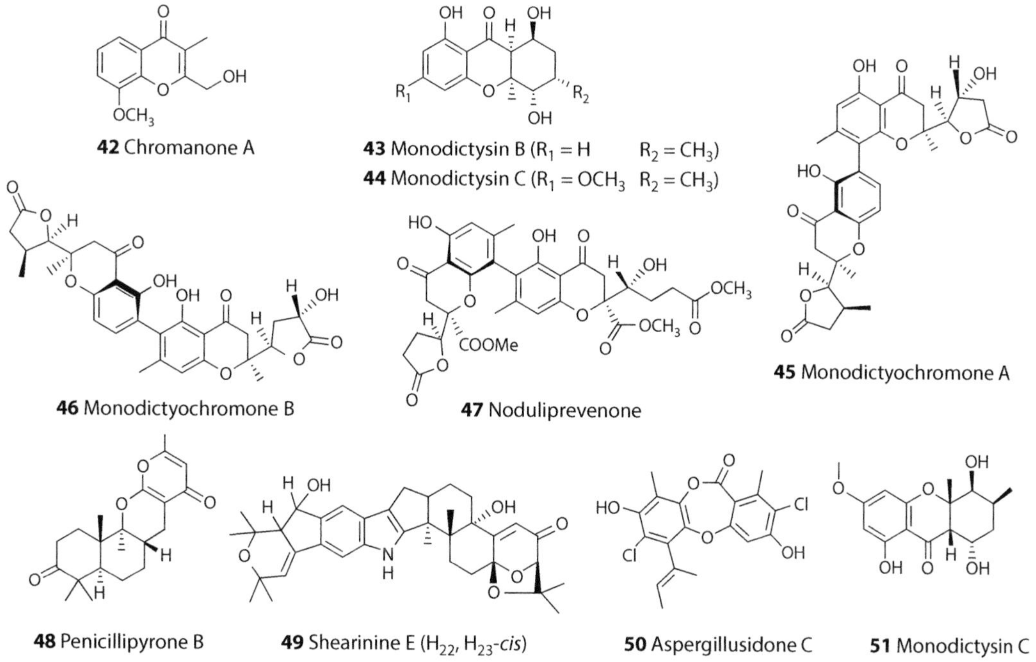

Figure 11.4 Novel cancer chemopreventive compounds from marine-derived fungi.

xanthone, monodictysin C (**51**), and two dimeric chromanones, monodictyochromones A and B, showed aromatase inhibitory activity with IC$_{50}$ values of 28.3,[140] 24.4, and 16.5 μM,[141] respectively.

11.3.3 Macroalgae

Marine macroalgae, commonly known as seaweed, are chlorophyll-containing photosynthetic algae. There are three types of seaweed, depending on pigmentation: brown algae (Phaeophyceae), including edible kombu (*Laminaria* or *Saccharina*), wakame (*Undaria pinnatifida*), and hijiki (*Hijikia fusiformis*); red algae (Rhodophyta), including edible nori (*Porphyra* sp.), Irish moss (*Chondrus crispus*), and dulse (*Palmaria palmata*); and green algae (Chlorophyceae), including edible limu palahalaha (*Ulva fasciata*).[32]

11.3.3.1 Antioxidant Marine Algae

A number of seaweed contain considerable amounts of polyphenols, including catechin, epicatechin, epigallocatechin gallate, and gallic acid. Extracts from seaweed have been reported to exert potent antioxidant activities. However, the antioxidant components of extracts from seaweed can vary, depending on environmental factors and seasons. Therefore, we describe exclusively pure compounds that possess antioxidant properties based on our previous review.[32]

11.3.3.1.1 Phlorotannins

Phlorotannins are hydrophilic oligomers of phloroglucinol (1,3,5-tryhydroxybenzene) with abundant hydroxyl groups, which are biosynthesized via the acetate–malonate pathway in brown algae.[148,149] There are subclasses of phlorotannins, including fuhalols and phlorethols with ether linkages, fucols with phenyl linkages, fucophloroethols with ether and phenyl linkages, and eckols with dibenzodioxin linkages.[149]

Phloroglucinol (**52**), from *Ecklonia stolonifera*, inhibited total ROS generation[150] and demonstrated ·OH, O_2^-·, and peroxyl radical scavenging activities with IC_{50} values of 392.5, 115.2, and 128.9 µM, respectively.[151] Commercially available phloroglucinol reduced intracellular ROS levels measured by the oxidation of DCFH-DA and inhibited membrane protein oxidation in H_2O_2-induced oxidative stress with HT1080 cells.[152]

Fucodiphloroethol G (**53**) from *Ecklonia cava* exhibited antioxidant activity by inhibiting DCFH-DA oxidation. Also, **53** showed DPPH·, ·OH, O_2^-·, and peroxyl radical scavenging activities with IC_{50} values of 14.7, 33.5, 18.6, and 18.1 µM, respectively.[151]

Trifucodiphlorethol A (**54**), trifucotriphlorethol A (**55**), and fucotriphlorethol A (**56**), from *Fucus vesiculosus* L., were identified as strong DPPH· scavengers with IC_{50} values of 14.4, 13.8, and 10.0 µg/ml, respectively. Also, these compounds displayed potent peroxyl radical scavenging activity in the oxygen radical absorbance capacity (ORAC) assay. In addition, the compounds were shown to inhibit CYP1A activity with IC_{50} values in the range of 17.9–33.7 µg/ml and aromatase activity with IC_{50} values in the range of 1.2–5.6 µg/ml.[153]

Eckol (**57**), found in *Eisenia bicyclis* (Kjellman) Setchell and *E. stolonifera*, showed a cytoprotective effect against *tert*-butyl hyperoxide (*t*-BHP)-injured HepG2 cells[154] and inhibited total ROS generation.[150] Eckstolonol (**58**), isolated from *E. stolonifera*, also inhibited total ROS generation.[150]

7-Phloroeckol (**59**) from *E. cava* presented antioxidant activity by inhibiting DCFH-DA oxidation. Also, **59** showed DPPH·, ·OH, O_2^-·, and peroxyl radical scavenging activities with IC_{50} values of 18.6, 39.6, 21.9, and 22.7 µM, respectively.[151]

Phlorofucofuroeckol A (**60**) was found in *E. cava*,[151] *E. bicyclis* (Kjellman) Setchell,[154] and *E. stolonifera*.[155] This compound showed inhibition of total ROS generation,[150] hepatoprotective effects in *t*-BHP-injured HepG2 cells,[154] DPPH·, ·OH, O_2^-·, and peroxyl radical scavenging activities with IC_{50} values of 4.7[155] (or 17.7[151]), 39.2, 21.6, and 21.4 µM, respectively.[151]

Dieckol (**61**), found in *E. bicyclis* (Kjellman) Setchell and *E. stolonifera*, showed hepatoprotective effects in *t*-BHP-injured HepG2 cells[154] and inhibited total ROS generation.[150] Compound **61** showed potent total antioxidant activity as well as DPPH·, ·OH, O_2^-·, and peroxyl radical scavenging activities with IC_{50} values of 6.2[155] (or 8.3[151]), 28.6, 16.2, and 14.5 µM, respectively.[151] Also, **61** inhibited DCFH-DA oxidation and myeloperoxidase activity in cells.[151,155]

6,6'-Bieckol (**62**) was found in *Ishige okamurae*,[156] *E. cava*,[151] and *E. bicyclis* (Kjellman) Setchell.[154] Compound **56** was reported to possess radical scavenging activities against DPPH·, hydroxyl radical (·OH), O_2^-·, and alkyl radical with IC_{50} values of 9.1, 28.7, 15.4, and 17.3 µM, respectively. Also, compound **56** markedly reduced the level of intracellular ROS determined by the DCFH-DA assay in RAW 264.7 cells, dose-dependently inhibited myeloperoxidase activity in HL-60 cells, and inhibited radical-mediated oxidation of cell membrane proteins in RAW 264.7 cells.[151,156] In line with these results, **62** exhibited similar DPPH· scavenging activities, as well as ·OH, O_2^-·, and peroxyl radical scavenging activities with IC_{50} values of 8.7, 29.7, 15.9, and 17.1 µM, respectively.[151] Moreover, **62** showed hepatoprotective effects with *t*-BHP-injured HepG2 cells; the defense was greater than that of 10 µM quercetin.[154]

8,8'-Bieckol (**63**) was isolated from *E. bicyclis* (Kjellman) Setchell and showed hepatoprotective effects with *t*-BHP-injured HepG2 cells.[154] Dioxinodehydroeckol, from *E. stolonifera*, showed DPPH· scavenging activity with an IC_{50} value of 8.8 µM.[155]

Diphlorethohydroxycarmalol (**64**) was isolated from *I. okamurae* and exhibited antioxidant activity showing radical scavenging activities against DPPH·, ·OH, alkyl radical, and superoxide radical (O_2^-·) with IC_{50} values of 10.5, 27.1, 18.8, and 16.7 µM, respectively. Also, **64** reduced intracellular ROS levels assessed by the DCFH-DA assay and radical-mediated oxidation of membrane proteins in RAW 264.7 cells, inhibited myeloperoxidase activity in HL-60 cells,[156] prevented cellular

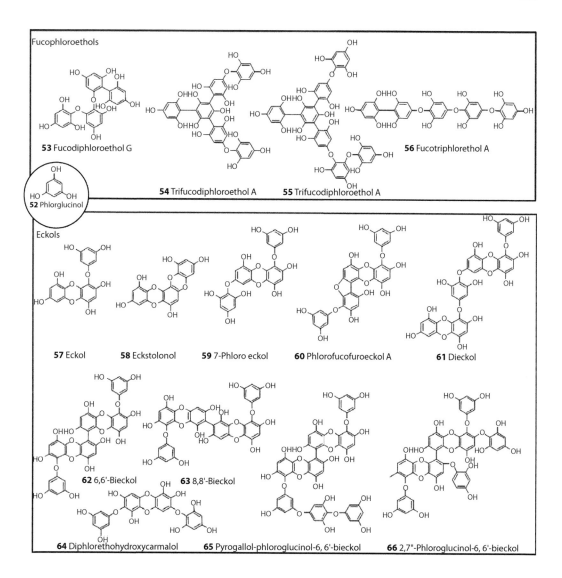

Figure 11.5 Antioxidant phlorotannins from marine algae.

damage in high-glucose (30 mM)-induced oxidative stress with human umbilical vein endothelial cells,[157] and protected human keratinocytes (HaCaT) against UVB-induced damage.[158]

Pyrogallol-phloroglucinol-6,6′-bieckol (**65**) was isolated from *E. cava* and showed scavenging activities against DPPH•, alkyl, •OH, and $O_2^{-•}$ radicals with IC_{50} values of 0.9, 2.5, 62.9, and 109.1 μM, respectively. Compound **65** inhibited DNA damage induced by H_2O_2.[159] Also, 2,7″-phloroglucinol-6,6′-bieckol (**66**) from *E. cava* showed scavenging activities with DPPH•, alkyl, •OH, and $O_2^{-•}$ radicals with IC_{50} values of 0.5, 2.1, 75.6, and 57.2 μM, respectively.[160] The structures of antioxidant phlorotannins are illustrated in Figure 11.5.

11.3.3.1.2 Bromophenols

Bromophenols are naturally occurring secondary metabolites found in marine algae. Considering antioxidant and radical scavenging activities, bromophenols isolated from red alga, including *Symphyocladia latiuscula*,[161–164] *Rhodomela confervoides*,[165–169] and *Polysiphonia urceolata*,[169] have been intensively studied. Herein, the potent antioxidant and radical scavenging bromophenols with IC$_{50}$ values under 10 μM will be described.

Symphyoketone (**67**),[161] 2,3,6-tribromo-4,5-dihydroxybenzyl alcohol (**68**),[170] *bis*(2,3,6-tribromo-4,5-dihydroxyphenyl)methane (**69**),[171] and *bis*(2,3,6-tribromo-4,5-dihydroxybenzyl) ether (**70**)[172] from *Symphyocladia latiuscula* showed DPPH˙ scavenging activities with IC$_{50}$ values of 8.5, 7.5, 8.1, and 8.5 μM, respectively.[161,163]. 2,3,6-Tribromo-4,5-dihydroxy benzyl methyl ether (**71**)[173] from *S. latiuscula* showed peroxynitrite, O$_2$$^-$˙, and ˙NO scavenging activities with IC$_{50}$ values of 0.013, 16.4, and 13.8 μM, respectively. Also, **71** reduced cell damage induced by 3-morpholinosydnonimine in bovine smooth muscle cells.[162]

3,4-Dibromo-5-[(methylsulfonyl)methyl]benzene-1,2-diol (**72**) and four other derivatives (**73–76**) isolated from either red algae *P. urceolata* or *R. confervoides*[165,169] showed DPPH˙ scavenging activities with IC$_{50}$ values of 9.5, 7.4, 8.7, 7.6, and 9.4 μM, respectively.[166] Also, nitrogen-containing bromophenols, including 3-(2,3-dibromo-4,5-dihydroxybenzyl)pyrrolidine-2,5-dione (**77**), and three additional derivatives (**78–80**)[168] found in *R. confervoides* showed DPPH˙ scavenging activities with IC$_{50}$ values of 5.2, 5.7, 5.4, and 8.9 μM, respectively.[168]

6-Bromo-1-(3-bromo-4,5-dihydroxybenzyl)phenanthro[4,5-bcd]furan-2,3,5-triol (urceolatin, **81**),[174] 7-bromo-9,10-dihydrophenanthrene-2,3,5,6-tetraol (**82**), 4,7-dibromo-9,10-dihydrophenanthrene-2,3,5,6-tetraol (**83**), and 1,8-dibromo-5,7-dihydrodibenzo[c,e]oxepine-2,3,9,10-tetraol (**84**)[175] from *P. urceolata* showed DPPH˙ scavenging activities with IC$_{50}$ values of 7.9,[174] 6.8, 6.1, and 8.1 μM.[175]

Both cymopol (**85**) from *Cymopolia barbata* and avrainvilleol (**86**) from several *Avrainvillea* spp. showed strong DPPH˙ scavenging activities and antioxidant capacities by inhibiting the production of oxidative products with DCFH-DA.[76] The chemical structrures of bromophenols from marine-derived fungi are illustrated in Figure 11.6.

11.3.3.1.3 Fucoxanthin

A xanthophyll carotenoid, fucoxanthin (**87**), found in edible brown algae (e.g., *Hijikia fusiformis*, *Undaria pinnatifida*, *Sargassum fulvellum*, *Sargassum siliquastrum*, and *Turbinaria ornate*) showed antioxidant activities *in vitro*[32] and *in vivo*.[176,177] For example, fucoxanthin exerted a protective effect on UVB oxidative stress-induced cell injury in human fibroblasts,[178] reduced cadmium-induced oxidative renal dysfunction in rats,[177] and alleviated oxidative stress via the reduction of blood glucose levels and hepatic lipid content in an obese and diabetes mouse model.[176]

11.3.3.1.4 Chromenes

A novel chromene, mojabanchromanol (**88**), from *Sargassum siliquastrum*, exhibited nearly equivalent antioxidant activity to known antioxidants, including butylated hydroxytoluene (BHT), tocopherol, and ascorbic acid in TBARS analysis and the DPPH˙ scavenging assay.[179] Other chromenes also showed antioxidant activities: Sargachromenol (**89**) from *Sargassum micracanthum* showed DPPH˙ scavenging activity with an IC$_{50}$ value of 100.2 μM.[180] A mixture of diastereomers of (3*R*,6*E*,10*E*)-13-(2,8-dimethyl-6-hydroxy-2*H*-chromen-2-yl)-2,6,10-trimethyltrideca-6,10-diene-2,3-diol from *S. micracanthum* showed inhibition of lipid peroxidation and DPPH˙ scavenging activity with IC$_{50}$ values of 0.65 and 25.0 μM, respectively (IC$_{50}$ value of L-ascorbic acid: lipid peroxidation, 159.0 μM; DPPH˙, 14.2 μM).[181]

11.3.3.1.5 Alkaloids

Fragilamide (**90**), from *Martensia fragilis*, showed moderate antioxidant activity in the DPPH˙ solution-based chemical assay and the DCFH-DA assay (IC$_{50}$ = 11 μM).[76] A 3-oxazolylindole

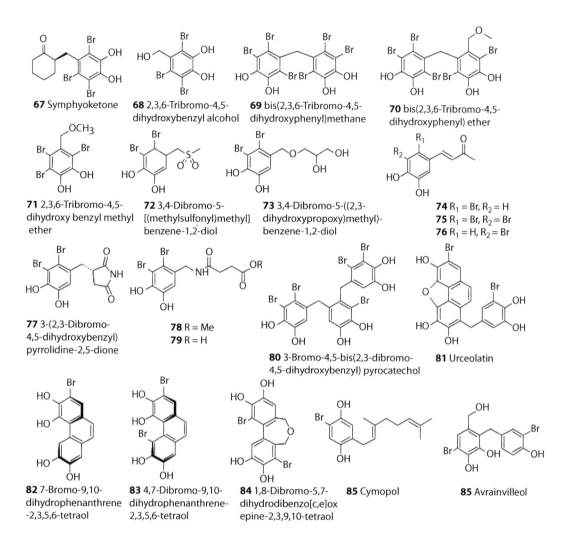

Figure 11.6 Antioxidant marine bromophenols.

alkaloid, martefragin A (**91**) from *M. fragilis*, inhibited NADPH-dependent lipid peroxidation in rat liver microsomes with an IC_{50} value of 2.8 μM (IC_{50}, α-tocopherol = 87 μM; IC_{50}, L-ascorbic acid = 200 μM).[182] Dendrodoine (**92**), an alkaloid possessing a 1,2,4-thiadiazole unit from the marine alga *Dendrodoa grossularia*, showed antioxidant activity-inhibited 2,2′-azino-bis(3-ethylbenzothiazoline-6-sulphonic acid (ABTS) radical formation to the extent of 0.18 times that of Trolox, lipid peroxidation activity, and scavenged NO radical and peroxyl radicals.[183]

11.3.3.1.6 Other Classes of Molecules

A monoterpene lactone, loliolide (**93**), protected against cell damage and inhibited apoptosis in H_2O_2-exposed Vero cells.[184] New meroterpenoids, sargachromenals A–P from *Sargassum siliquastrum*, exhibited DPPH· scavenging activities (87%–91%, the most potent of which was sargachromenal C [**94**], with 90.5% scavenging) at 100 μg/ml.[185] Besides meroterpenoids, other antioxidant compounds have been isolated from *S. siliquastrum*, including nahocols (nahocols A, A1, D1, and D2) and isonahocols (isonahocols D1 and D2 [**95**]) and sargahydroquinoic acid (**96**) (also found in

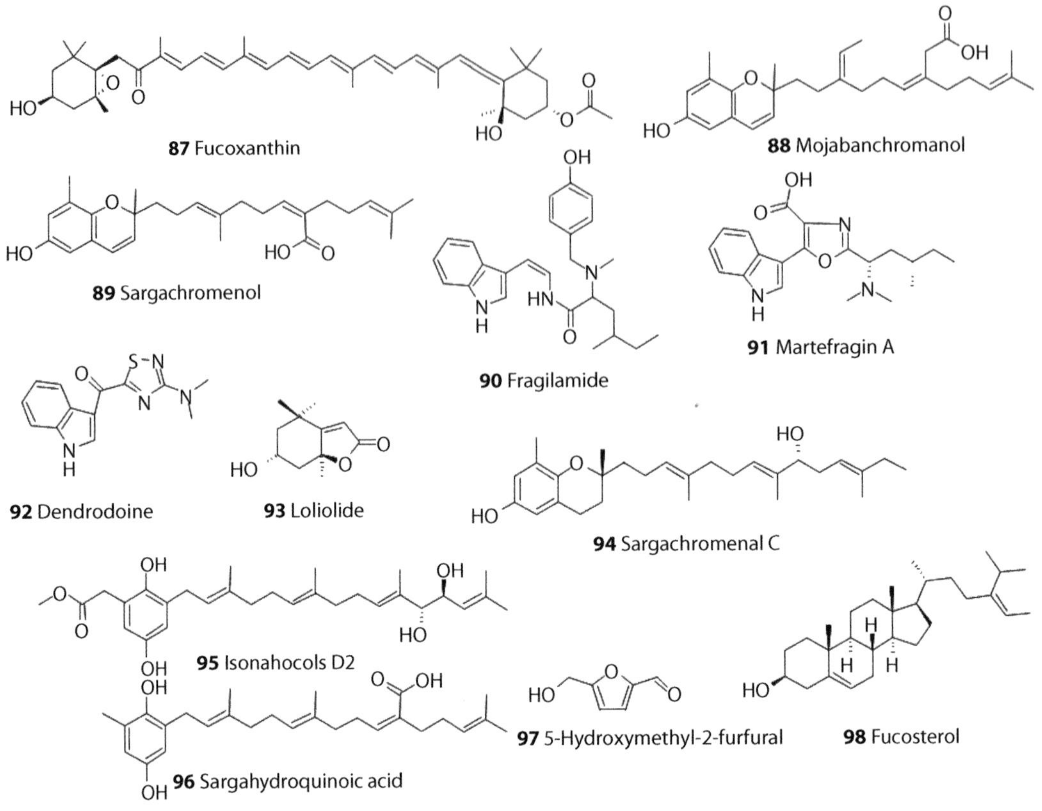

Figure 11.7   Other antioxidant compounds from marine algae.

other *Sargassum* sp. *yezoense*[186]), which showed DPPH· scavenging activities with IC_{50} values of 11.4, 12.7, 12.4, 18.5, 0.13, 0.12, and 0.17 µg/ml, respectively.[187]

2-[(2*E*,6*E*,10*E*,14*R*)-14,15-Dihydroxy-3,7,11,15-tetramethyl-2,6,10-hexadecatrien-1-yl]-6-methyl-1,4-benzenediol from *Sargassum micracanthum* showed inhibitory activity toward lipid peroxidation and DPPH· scavenging activity with IC_{50} values of 0.26 and 25.5 µM, respectively.[181]

Sargaquinoic acid was found in both *Sargassum yezoense*[186] and *S. micracanthum* and showed DPPH· scavenging activity with an IC_{50} value of 49.3 µM.[180]

5-Hydroxymethyl-2-furfural (**97**), isolated from *Laurencia undulate*, showed various antioxidant activities: free-radical scavenging with IC_{50} values of 27.1, 22.8, 45.0, and 33.5 µM on DPPH, ·OH, alkyl, and $O_2^{-·}$ radicals, respectively; intracellular ROS scavenging; inhibition of membrane protein oxidation and oxidative enzyme myeloperoxidase; and induction of superoxide dismutase (SOD) gene expression.[188]

A marine sterol, fucosterol (**98**) from *Pelvetia siliquosa*, showed hepatoprotective activities in CCl_4-intoxicated rats through the reduction of serum transaminase, sGOT, and sGPT activities, and induction of hepatic cytosolic SOD, catalase, and glutathione peroxidase activities.[189]

Other antioxidant compounds from marine algae are illustrated in Figure 11.7.

11.3.3.2 Anti-Inflammatory Marine Algae

A thiazole-containing cyclic heptapeptide, *trans,trans*-ceratospongamide (**99**), from the symbiotic marine red alga (Rhodophyta) *Ceratodictyon spongiosum*, collected from Biaro Island,

Indonesia, exhibited potent inhibition of secreted phospholipase A_2 ($sPLA_2$) expression in a cell-based model for anti-inflammation (ED_{50} = 32 nM) and also inhibited the expression of a human-$sPLA_2$ promoter-based reporter by 90%.[190]

New bromophenolic metabolites, vidalols A (**100**) and B (**101**) from the Caribbean red alga *Vidalia obtusaloba*, inhibited PLA_2.[191]

A novel marine substance, 6-*n*-tridecylsalicylic acid (**102**) from a brown alga, has similar anti-inflammatory activity but less ulcerogenic activity on a molar basis than salicylic acid.[192]

A norisoprenoid, apo-9'-fucoxanthinone (**103**), from brown algae *Sargassum muticum* (Yendo) Fensholt in Jeju Island, Korea, suppressed LPS-induced NO and PGE_2 production and NF-κB activation via suppression of IκB-α in macrophages.[193]

A phlorotannin, 6,6'-bieckol (**62**), from *Ecklonia cava*, inhibited the generation of iNOS, COX-2, TNF-α, and IL-6 through the negative regulation of the NF-κB pathway in LPS-stimulated macrophages.[194] Fucosterol (**98**), from *Eisenia bicyclis*, inhibited the expression of iNOS and COX-2 in RAW 264.7 cells.[195] Similarly, eckol (**57**), dieckol (**61**), and 7-phloroeckol (**59**) from *E. bicyclis* inhibited NO production with IC_{50} values of 52.9, 51.4, and 26.9 µg/ml, respectively.[195]

Phlorofucofuroeckol A (**60**), from *Ecklonia stolonifera*, inhibited both NO and PGE_2 production, as well as the protein expression of iNOS and COX-2[155] through NF-κB-dependent transcriptional control associated with inhibition of multiple signaling proteins, including Akt and p38 MAPK in LPS-induced RAW 264.7 cells.[196]

Phloroglucinol (**52**) inhibited the production of inflammatory mediators (e.g., TNF-α, IL-1β, IL-6, and PGE_2) in LPS-stimulated RAW 264.7 cells.[152]

Sargachromanol G (**104**) was isolated from *Sargassum siliquastrum* and inhibited the production of inflammatory markers (NO, iNOS, PGE_2, and cyclooxygenase-2) and pro-inflammatory cytokines (TNF-α, IL-1β, and IL-6) induced by LPS treatment via inhibition of the phosphorylation IκB-α and NF-κB (p65 and p50) and MAPKs (extracellular-signal-regulated kinase [ERK1/2], c-Jun N-terminal kinase [JNK], and p38 MAPK).[197]

Sargachromenol (**89**) was isolated from *Sargassum micracanthum* and suppressed the LPS-induced production of NO and PGE_2 with IC_{50} values of 81.7 and 30.2 µM, respectively, along with inhibition of the protein expression of iNOS and cyclooxygenase-2 (COX-2), by inhibiting activation of the NF-κB signaling pathway in RAW 264.7 cells.[198]

From the strain *Sargassum siliquastrum*, 9'-*cis*-(6'*R*)-fucoxanthin significantly inhibited NO production and showed slightly reduced PGE_2 production with a dose-dependent reduction in iNOS and COX-2 proteins, as well as mRNA expression associated with the attenuation of TNF-α and IL-6 formation.[199]

3-*O*-β-D-Glucopyranosyl clerosterol, from the green alga *Ulva lactuca*, *Codium decorticatum*, and *Codium iyengarii*, inhibited topical inflammation (62.3% inhibition at 1000 µg/ear using the mouse ear edema assay, equivalent to a reference, indomethacin [66% inhibition at 1000 µg/ear]).[200] Representative anti-inflammatory compounds from marine algae are illustrated in Figure 11.8.

11.3.3.3 QR1 Induction by Marine Algae

A monounsaturated keto-type C_{18} fatty acid, 7(*E*)-9-keto-octadec-7-enoic acid from *Ulva lactuca*, activated ARE via stabilization of Nrf2 from Keap1-mediated ubiquitination, and consequently induced the expression of ARE-regulated cytoprotective genes, including QR1, HO-1, thioredoxin reductase 1, catalytic and modifier subunits of glutamate-cysteine ligase, and the cystine–glutamate exchange transporter, in IMR-32 human neuroblastoma cells. In follow-up studies, 7(*E*)-9-keto-octadec-7-enoic acid upregulated *Nqo1* transcript levels in mouse heart, brain, lung, and stomach tissues.[201]

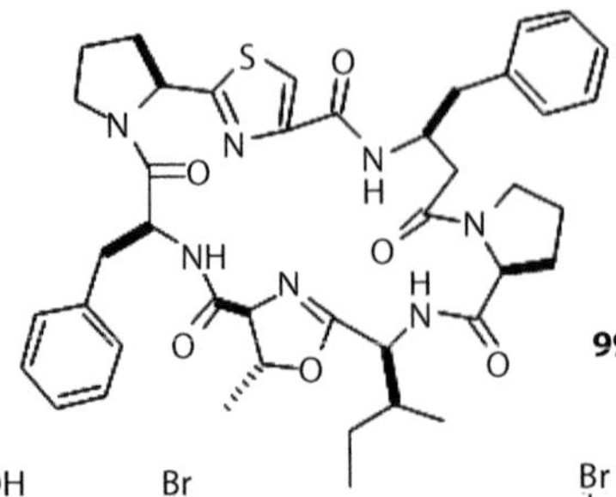

99 *Trans, trans*-Ceratospongamide

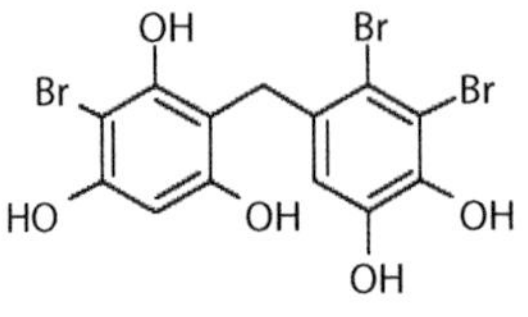

100 Vidalol A

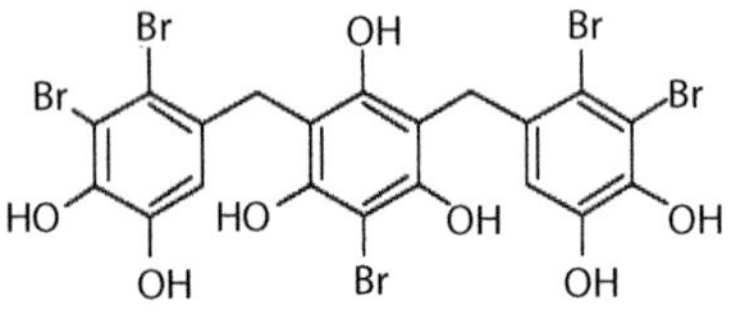

101 Vidalol B

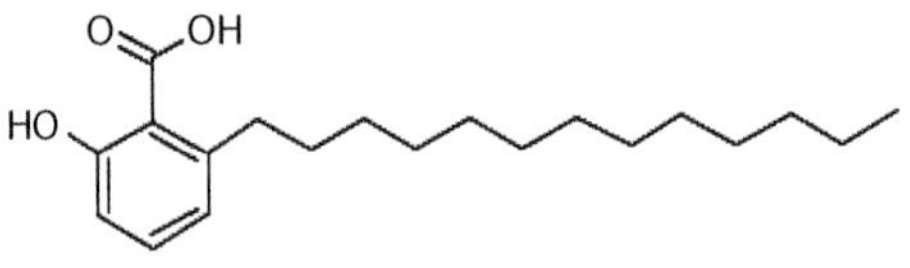

102 6-*n*-Tridecylsalicylic acid

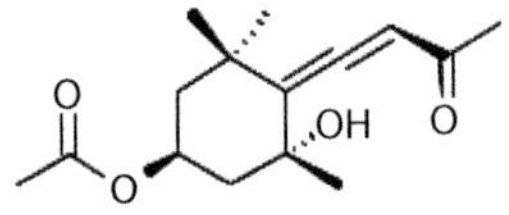

103 *apo*-9'-Fucoxanthinone

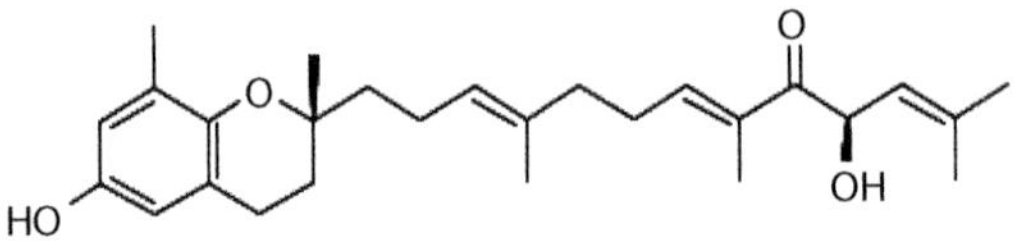

104 Sargachromanol G

Figure 11.8   Anti-inflammatory compounds from marine algae.

Fucoxanthin (**87**), isolated as a major carotenoid in brown sea algae, increased the nuclear accumulation and ARE binding of Nrf2, likely by the phosphorylation of upstream ERK and p38 MAPK, thereby inducing mRNA and protein expression of HO-1 and NQO1.[202]

Prenylated bromohydroquinones, 7-hydroxycymopochromanone and 7-hydroxycymopolone, from the marine algae *Cymopolia barbata*, had potent inhibitory activity against CYP1A1 with IC$_{50}$ values of 0.39 and 0.93 µM, respectively. Similarly, 7-hydroxycymopolone potently inhibited the activity of CYP1B1 (IC$_{50}$ = 0.14 µM).[203]

Eckol, found in *Ecklonia cava*, induced the expression of HO-1 by activating Nrf2 through upstream signals, including Erk and PI3K/Akt in Chinese hamster lung fibroblast (V79-4) cells.[204]

11.3.4 Marine Invertebrates

Marine invertebrates, which consist of about 60% of marine animal diversity, are also a major source of bioactive compounds. Approximately 66% of novel marine natural products reported between 1990 and 2005 were from marine invertebrates.[205] Upon analysis of 9812 marine invertebrate compounds reported from 1990 to 2009, the phylum Porifera was the major source (48.8%), followed by Cnidaria (28.6%), Echinodermata (8.2%), Chordata (6.9%), and Mollusca (5.8%).[205]

Marine sponges produce various metabolites, largely due to being inhabited by diverse microbial communities from more than nine bacterial phyla (e.g., Proteobacteria, Actinobacteria, Chloroflexi, Bacteroidetes, Cyanobacteria, Firmicutes, Acidobacteria, Planctomycetes, and Nitrospira).[206]

To date, marine invertebrate-derived compounds in clinical use or with therapeutic potential for cancer treatment include ecteinascidin 743 and dehydrodidemnin B from ascidians, bryostatin-1 from a bryozoan, and discodermolide, dictyostatin-1, fijianolide B, spongistatin 1, and psymberin from sponges.[207] However, the cancer chemopreventive potential of marine invertebrates has not been well studied.

Herein, we describe pure natural products (not extracts or fractions) from marine invertebrates with antioxidant, anti-inflammatory, and anticarcinogenic potential. Numerous secondary metabolites were found to show promising cytotoxic effects toward a variety of cancer cell lines. However, since cytotoxicity is more closely related to anticancer agents, cytotoxic molecules are excluded in this review.

11.3.4.1 Antioxidant Metabolites

The majority of active metabolites that possess DPPH[•] scavenging activities were found in marine sponges, including 2-octaprenyl-1,4-hydroquinone,[208] 2-(24-hydroxy)-octaprenyl-1,4-hydroquinone,[208] puupehenone,[76] (1S)-(+)-curcuphenol 76, (1S)-curcudiol,[76] aaptamine,[76] isoaaptamine,[76] 9-demethylaaptamine (**105**),[209] siphonodictyals B1 and B2,[210] avarol,[211] purpurone,[212] subereaphenols B, C,[213] and D,[214] popolohuanone A,[215] popolohuanone F,[215] arenarol,[215] halioxepine (**106**),[216] verongiaquinol,[217] siphonellinol A,[217] akadisulfate B,[218] and siphonodictyal B3 (**107**).[218]

In addition, ophiuroidine from a brittle star,[219] comaparvin and 6-methoxycomaparvin-5-methyl ether from marine echinoderm,[220] and echinochrome A (**108**) from a sea urchin[221] possess DPPH[•] scavenging activities. The chemical class, origin, and potency of each compound are described in Table 11.2.

In addition to DPPH[•] scavenging activities, marine invertebrate metabolites displayed antioxidant activities in different experimental settings. Renierol inhibited xanthine oxidase with IC_{50} values of 1.85 µg/ml with uric acid measurement and 1.4 µg/ml with superoxide radical measurement, along with hypouricemic activity in mice.[222] Dithiocarbamate-related tridentatol A from the marine hydroid *Tridentata marginata* inhibited lipid peroxidation of human LDL.[223] Farnesylhydroquinone glycosides, euplexides A–C and E, showed antioxidizing activities of 3.4, 3.6, 3.5, and 3.1 times higher than SOD at a concentration of 33.3 µg/ml.[224] The puupehenone from the *Hyrtios* sp. mentioned above showed inhibition of DCFH-DA oxidation with an IC_{50} value of 27 µM.[76] Diterpene glycosides, pseudopterosins PsQ, PsS, and PsU, showed NO scavenging activities at 42%, 31%, and 38% at 10 µM, respectively, which were higher than that of the reference, curcumin (30% scavenging activity at 14 µM).[225]

Through ORAC assay, cladiellin A from *Cladiella* sp., 1S*,4R*,5S*,6R*,7S*,10S*-1(5),6(7)-diepoxy-4-guaiol (**109**) and 1S*,4S*,5S*,10R*-4,10-guaianediol from *Sinularia* sp.[226]; 5,6-dibromo-L-hypaphorine and aureol (**110**) from *Hyrtios* sp.[227]; lobocompactols A (**111**) and B (**112**) from *Lobophytum compactum*; and phorbatopsins A (**113**), B, and C and astaxanthin from *Phorbas topsenti*[228] were found to have antioxidant activities.

The extract of sea cucumber *Holothuria atra*, which contains phenolic components, including chlorogenic acid, pyrogallol, rutin, coumaric acid, catechin, and ascorbic acid, showed hepatoprotective activity against thioacetamide-induced liver injury in rats.[229]

Antioxidant compounds from marine invertebrates that showed the most potent activity in each study are illustrated in Figure 11.9.

Table 11.2 Antioxidant Marine Invertebrate Metabolites (DPPH Assay)

Compound	Chemical Class	Origin	Type	Activity	References
2-Octaprenyl-1,4-hydroquinone	Polyprenylated hydroquinone	*Ircinia spinosula*	Sponge	IC$_{50}$: ~50 µM (lipid peroxidation, IC$_{50}$: 220 µM)	208
2-(24-Hydroxy)-octaprenyl-1,4-hydroquinone	Polyprenylated hydroquinone	*Ircinia spinosula*	Sponge	IC$_{50}$: <100 µM (lipid peroxidation, IC$_{50}$: 350 µM)	208
(+)-Puupehenone	Unsaturated tetracyclic merosesquiterpene	*Hyrtios sp.*	Sponge	Strong	76
Siphonodictyal B1	Sulfated sesquiterpene-hydroquinone	*Aka coralliphagum*	Sponge	80.8% scavenging at 200 µM (L-Ascorbic acid: 95.3% scavenging at 200 µM)	210
Siphonodictyal B2	Sulfated sesquiterpene-hydroquinone	*Aka coralliphagum*	Sponge	88.8% scavenging at 200 µM (L-Ascorbic acid: 95.3% scavenging at 200 µM)	210
Purpurone	Aromatic alkaloid	*Iotrochota sp.*	Sponge	IC$_{50}$: 19 µM	212
Subereaphenol B	Brominated phenolic compound	*Suberea mollis*	Sponge	Very strong bright yellow color (1 mg/ml)	213
Subereaphenol C	Brominated phenolic compound	*Suberea mollis*	Sponge	Very strong bright yellow color (1 mg/ml)	213
Avarol	Sesquiterpenoid hydroquinone	*Dysidea avara*	Sponge	IC$_{50}$: 18.0 µM	211
Aaptamine	Alkaloid	*Aaptos aaptos* Schmidt	Sponge	IC$_{50}$: 5.63 µM (OH• scavenging activity, IC$_{50}$: 5.1 µM)	209

(Continued)

Table 11.2 (Continued)   Antioxidant Marine Invertebrate Metabolites (DPPH Assay)

Compound	Chemical Class	Origin	Type	Activity	References
Isoaaptamine	Alkaloid	*Aaptos aaptos* Schmidt	Sponge	IC_{50}: 2.50 µM (OH• scavenging activity, IC_{50}: 1.5 µM)	209
9-Demethylaaptamine (**105**)	Alkaloid	*Aaptos aaptos* Schmidt	Sponge	IC_{50}: 1.25 µM (OH• scavenging activity, IC_{50}: 1.3 µM)	209
Popolohuanone A	Sesquiterpenoid quinone	*Dysidea* sp.	Sponge	IC_{50}: 35 µM (Trolox IC_{50}: 16 µM)	215
Popolohuanone F	Sesquiterpenoid quinone	*Dysidea* sp.	Sponge	IC_{50}: 35 µM (Trolox IC_{50}: 16 µM)	215
Arenarol	Sesquiterpenoid quinone	*Dysidea* sp.	Sponge	IC_{50}: 19 µM (Trolox IC_{50}: 16 µM)	215
Halioxepine (**106**)	Meroditerpene	*Haliclona* sp.	Sponge	IC_{50}: 3.2 µg/ml (= 7.7 µM)	216
Siphonellinol A	Sipholane	*Callyspongia siphonella*	Sponge	81.7% scavenging at 100 µg/ml	217
Verongiaquinol	Bromoquinol	*Callyspongia siphonella*	Sponge	61.25% scavenging at 100 µg/ml	217
Subereaphenol D	Brominated phenolic compound	*Suberea mollis*	Sponge	Very strong bright yellow color (1 mg/ml)	214
Akadisulfate B	Sulfated meroterpenoid	*Aka coralliphaga*	Sponge	45.5% scavenging at 200 µM (comparable with BHT)	218
Siphonodictyal B3 (**107**)	Sulfated meroterpenoid	*Aka coralliphaga*	Sponge	60.2% scavenging at 200 µM (comparable with BHT)	218
Ophiuroidine	Indoloquinazoline alkaloid	*Ophiocoma riisei*	Brittle star	IC_{50}: 240 µM (BHT IC_{50}: 360 µM)	219
Comaparvin	Naphthopyrone	*Comanthus* sp.	Echinoderm	IC_{50}: 79 µM	220
6-Methoxycomaparvin-5-methyl ether	Naphthopyrone	*Comanthus* sp.	Echinoderm	IC_{50}: 130.7 µM	220
Echinochrome A (**108**)	Polyhydoxylated naphthoquinone	*Stomopneustes variolaris* Lamarck, *Arbacia pustulosa*	Sea urchin	IC_{50}: 7.0 µM (Trolox IC_{50}: 16.0 µM)	221

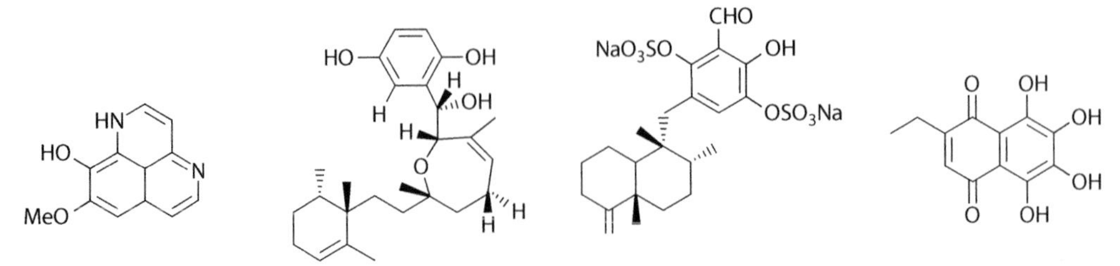

105 9-Demethylaaptamine   **106** Halioxepine   **107** Siphonodictyal B3   **108** Echinochrome A

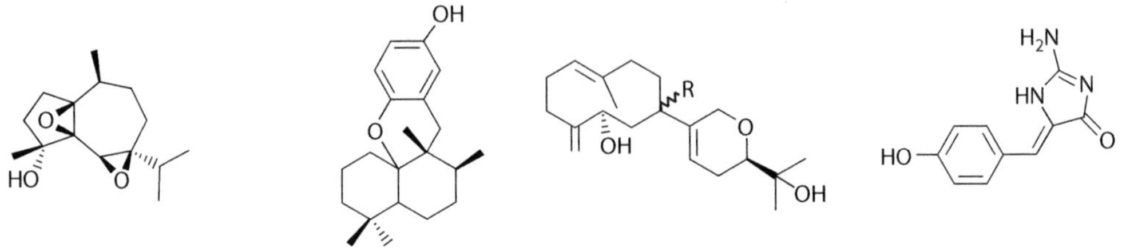

109   **110** Aureol   **111** Lobocompactol A (R=βH)
112 Lobocompactol B (R=αH)   **113** Phorbatopsin A

Figure 11.9 Antioxidant compounds from marine invertebrates.

11.3.4.2 Anti-Inflammatory Metabolites

11.3.4.2.1 Inhibitory Effects of Metabolites on Pro-Inflammatory Protein Expression (iNOS/COX-2)

11.3.4.2.1.1 Terpenoids — From the analysis of 5286 new marine invertebrate compounds discovered from 2000 to 2009, terpenoids are the most commonly reported (40.5%), followed by alkaloids (22.1%), aliphatic compounds (13.0%), steroids (7.5%), carbohydrates (6.3%), and amino acids/peptides (5.4%).[230] Terpenes are primary and secondary metabolites biosynthesized from five carbon isoprene building units; starter unit combination, structural modification, and combination with other biosynthetic pathways result in an enormous diversity of terpenoid and terpenoid-like small molecules.[231]

Based on a literature search, the majority of experiments have been performed using the RAW 264.7 cell line treated with test compounds at a concentration of 10 μM and stimulated with LPS (0.01 μg/ml) for 16 h. Potential to inhibit the inflammatory protein expression levels, including iNOS and COX-2, was determined. Caffeic acid phenethyl ester (CAPE) has been commonly used as a positive control.[232–241] Compounds found to be active in this process are listed in Table 11.3.

In regard to the anti-inflammatory potential of marine invertebrate compounds, various cembrane-type structures (diterpenoids containing a 14-membered carbon cembrane ring) isolated from soft corals (phylum Cnidaria), including the genera *Lobophytum*, *Sarcophyton*, and *Sinularia*, have been reported to inhibit protein expression of inflammatory mediators, including COX-2 and iNOS. In the soft coral *Lobophytum durum*, various cembrane-based metabolites have been reported to exhibit anti-inflammatory activities. Notably, sinularolide C, durumolide C (cembranoids with a *trans*-fused α-methylene-γ-lactone), durumolide F, and durumhemiketalolide C showed more potent inhibition than the positive control CAPE at a concentration of 10 μM in both iNOS and COX-2 protein expression.[233,242,243] The existence of α-methylene-γ-lactone moiety of cembranoids is required for the inhibitory activities toward iNOS and COX-2 expressions in LPS-stimulated

Table 11.3 Anti-Inflammatory Compounds from Marine Invertebrates (iNOS and COX-2 Protein Expression)

Compounds	Class	LPS, h[a]	Concentration[b]	iNOS (%)[c]	COX-2 (%)	Species	References
Sinularolide B	Cembrane-based diterpenoid	0.01, 16	10	0.9	63.7	*Lobophytum durum*	233
Sinularolide C	Cembrane-based diterpenoid	0.01, 16	10	1.4	42.9	*Lobophytum durum*	233
Durumolide C	Cembrane-based diterpenoid	0.01, 16	10	0.0	42.5	*Lobophytum durum*	233
20-Acetylsinularolide B	Cembrane-based diterpenoid	0.01, 16	10	0.2	N/S[d]	*Lobophytum durum*	233
Durumolide A	Cembranoid	0.01, 16	10	34.7	62.5	*Lobophytum durum*	233
Durumolide B	Cembranoid	0.01, 16	10	0.0	N/S	*Lobophytum durum*	233
Durumolide D	Cembranoid	0.01, 16	10	0.5	N/S	*Lobophytum durum*	233
Durumolide E	Cembranoid	0.01, 16	10	0.1	N/S	*Lobophytum durum*	233
Durumolide F	Cembranolide	0.01, 16	10	0.8	47.8	*Lobophytum durum*	242
Durumolide K	Cembranolide	0.01, 16	10	5.7	71.6	*Lobophytum durum*	242
Durumhemiketalolide A	Hemiketal cembranolide	0.01, 16	10	11.0	66.7	*Lobophytum durum*	243
Durumhemiketalolide C	Hemiketal cembranolide	0.01, 16	10	0.0	34.7	*Lobophytum durum*	243
Crassumolide A	Cembranoid	0.01, 16	10	26.8	50.2	*Lobophytum crassum*	232
Lobohedleolide	Cembranoid	0.01, 16	10	21.5	39.2	*Lobophytum crassum*	232
(1S,2R,7S,8R,11S)-7-Acetyloxy-8,11-epidioxy-cembra-3E,12E,15(17)-trien-16,2-olide	Cembranoid	0.01, 16	10	10.2	34.3	*Lobophytum crassum*	232
Sarcocrassocolide A	Cembranoid	0.01, 16	10	13.7	N/S	*Sarcophyton crassocaule*	240
Sarcocrassocolide B	Cembranoid	0.01, 16	10	13.3	N/S	*Sarcophyton crassocaule*	240
Sarcocrassocolide C	Cembranoid	0.01, 16	10	4.6	N/S	*Sarcophyton crassocaule*	240
Sarcocrassocolide D	Cembranoid	0.01, 16	10	7.0	N/S	*Sarcophyton crassocaule*	240
Sarcocrassolide	Cembranoid	0.01, 16	10	1.1	3.9	*Sarcophyton crassocaule*	240
13-Acetoxysarcocrassolide	Cembranoid	0.01, 16	10	6.2	N/S	*Sarcophyton crassocaule*	240
Sarcocrassocolide F	Cembranoid	0.01, 16	10	<10	N/S	*Sarcophyton crassocaule*	244
Sarcocrassocolide G	Cembranoid	0.01, 16	10	<10	N/S	*Sarcophyton crassocaule*	244
Sarcocrassocolide H	Cembranoid	0.01, 16	10	<10	N/S	*Sarcophyton crassocaule*	244
Sarcocrassocolide I	Cembranoid	0.01, 16	10	<10	>50	*Sarcophyton crassocaule*	244
Sarcocrassocolide J	Cembranoid	0.01, 16	10	<10	N/S	*Sarcophyton crassocaule*	244

(Continued)

Table 11.3 (Continued)   Anti-Inflammatory Compounds from Marine Invertebrates (iNOS and COX-2 Protein Expression)

Compounds	Class	LPS, h[a]	Concentration[b]	iNOS (%)[c]	COX-2 (%)	Species	References
Sarcocrassocolide K	Cembranoid	0.01, 16	10	<10	N/S	*Sarcophyton crassocaule*	244
Sarcocrassocolide L	Cembranoid	0.01, 16	10	<10	N/S	*Sarcophyton crassocaule*	244
Sarcocrassocolide M	Cembranoid	0.01, 16	10	4.2	62.8	*Sarcophyton crassocaule*	245
Sarcocrassocolide N	Cembranoid	0.01, 16	10	52.9	N/S	*Sarcophyton crassocaule*	245
Sarcocrassocolide O	Cembranoid	0.01, 16	10	22.7	N/S	*Sarcophyton crassocaule*	245
Flexibilisolide A	Cembranoid	0.01, 16	10	19.4	N/S	*Sinularia querciformis*	320
Flexilarin	Cembranoid	0.01, 16	10	13.8	N/S	*Sinularia querciformis*	320
Grandilobatin D	Cembranoid	0.01, 16	50	28.7	N/S	*Sinularia grandilobata.*	321
Crassarine F	Cembranoid	0.01, 16	10	N/S	65.6	*Sinularia crassa*	322
Crassarine H	Cembranoid	0.01, 16	10	35.8	N/S	*Sinularia crassa*	322
11-Dehydrosinulariolide	Cembranoid	0.01, 16	10	~30	~70	*Sinularia discrepans*	323
Sinulariolide	Cembranoid	0.01, 16	10	~50	~80	*Sinularia discrepans*	323
11-*epi*-Sinulariolide acetate	Cembranoid	0.01, 16	10	~10	~50	*Sinularia discrepans*	323
Sinularin	Cembrane-based diterpenoid	0.01, 16	10	1.2	N/S	*Sinularia triangular*	246
Dihydrosinularin (114)	Cembrane-based diterpenoid	0.01, 16	10	5.1	24.9	*Sinularia triangular*	246
(−)-14-Deoxycrassin (115)	Cembrane-based diterpenoid	0.01, 16	10	0.9	5.9	*Sinularia triangular*	246
Querciformolide C	Cembranoid	0.01, 16	10	23.7	N/S	*Sinularia querciformis*	324
Sinulariolone	Cembranoid	0.01, 16	10	N/S	17.4	*Sinularia querciformis*	324
3-Acetyloxy sinulariolone (116)	Cembranoid	0.01, 16	10	20.3	14.0	*Sinularia querciformis*	324
(+)-11,12-Epoxysarcophytol A	Oxygenated cembranoid	0.01, 16	10	76.2	54.0	*Sinularia gibberosa*	324
Gyrosanolide A	Norcembranolide	0.01, 16	10	55.2	N/S	*Sinularia gyrosa*	325
Gyrosanolide B	Norcembranolide	0.01, 16	10	18.6	N/S	*Sinularia gyrosa*	325
Gyrosanolide C	Norcembranolide	0.01, 16	10	10.6	N/S	*Sinularia gyrosa*	325
Gyrosanin A	Norcembranolide	0.01, 16	10	66.9	N/S	*Sinularia gyrosa*	325
Sinuleptolide (128)	Norditerpenoid	0.01, 16	10	10.2	N/S	*Sinularia gyrosa*	325
5-*epi*-Sinuleptolide	Norditerpenoid	0.01, 16	10	17.4	N/S	*Sinularia gyrosa*	325
Scabrolide D	Norditerpenoid	0.01, 16	10	47.2	N/S	*Sinularia gyrosa*	325
Scabrolide G	Norditerpenoid	0.01, 16	10	56.3	N/S	*Sinularia gyrosa*	325

(Continued)

Table 11.3 (Continued)   Anti-Inflammatory Compounds from Marine Invertebrates (iNOS and COX-2 Protein Expression)

Compounds	Class	LPS, h[a]	Concentration[b]	iNOS (%)[c]	COX-2 (%)	Species	References
Gyrosanol A	Diterpenoid	0.01, 16	10	N/S	16.9	*Sinularia gyrosa*	325
Gyrosanol B	Diterpenoid	0.01, 16	10	N/S	29.1	*Sinularia gyrosa*	325
Lochmolin A	Aromadendrane-type sesquiterpenoid	1, 20	10	N/S	8.7	*Sinularia lochmodes*	326
Lochmolin C	Aromadendrane-type sesquiterpenoid	1, 20	10	N/S	61.0	*Sinularia lochmodes*	326
Lochmolin D	Aromadendrane-type sesquiterpenoid	1, 20	10	N/S	83.4	*Sinularia lochmodes*	326
Scabralin A	Cadinane-type sesquiterpenoid	0.01, 16	10	39.1	N/S	*Sinularia scabra*	327
Klysimplexin sulfoxide A	Eunicellin-based diterpenoid	1, 20	10	8.8	N/S	*Klyxum simplex*	238
Klysimplexin sulfoxide B	Eunicellin-based diterpenoid	1, 20	10	17.8	N/S	*Klyxum simplex*	238
Klysimplexin sulfoxide C (117)	Eunicellin-based diterpenoid	1, 20	10	11.3	7.2	*Klyxum simplex*	238
Klysimplexin J	Eunicellin-based diterpenoid	1, 20	10	50–60	N/S	*Klyxum simplex*	241
Klysimplexin K	Eunicellin-based diterpenoid	1, 20	10	30–40	N/S	*Klyxum simplex*	241
Klysimplexin L	Eunicellin-based diterpenoid	1, 20	10	~30	N/S	*Klyxum simplex*	241
Klysimplexin M	Eunicellin-based diterpenoid	1, 20	10	~20	N/S	*Klyxum simplex*	241
Klysimplexin N	Eunicellin-based diterpenoid	1, 20	10	30–40	N/S	*Klyxum simplex*	241
Klysimplexin R	Eunicellin-based diterpenoid	1, 20	10	~20	~50	*Klyxum simplex*	241
Klysimplexin S	Eunicellin-based diterpenoid	1, 20	10	10–20	~40	*Klyxum simplex*	241
Simplexin A	Eunicellin-based diterpenoid	1, 20	10	37.7		*Klyxum simplex*	237
Simplexin D	Eunicellin-based diterpenoid	1, 20	10	15.9		*Klyxum simplex*	237
Simplexin E	Eunicellin-based diterpenoid	1, 20	10	4.8	37.7	*Klyxum simplex*	237
Krempfielin B	Eunicellin-based diterpenoid	0.01, 16	10	~40	N/S	*Cladiella krempfi*	328
Krempfielin C	Eunicellin-based diterpenoid	0.01, 16	10	50–60	N/S	*Cladiella krempfi*	328
Krempfielin D	Eunicellin-based diterpenoid	0.01, 16	10	~60	N/S	*Cladiella krempfi*	328
Litophynol B	Eunicellin-based diterpenoid	0.01, 16	10	30–40	N/S	*Cladiella krempfi*	328

(Continued)

Table 11.3 (Continued) Anti-Inflammatory Compounds from Marine Invertebrates (iNOS and COX-2 Protein Expression)

Compounds	Class	LPS, h[a]	Concentration[b]	iNOS (%)[c]	COX-2 (%)	Species	References
(1R*,2R*,3R*,6S*,7S*, 9R*,10R*,14R*)- 3-Butanoyloxycladiell- 11(17)-en-6,7-diol	Eunicellin-based diterpenoid	0.01, 16	10	~10	N/S	*Cladiella krempfi*	328
Krempfielin E	Eunicellin-based diterpenoid	0.01, 16	10	50–60	N/S	*Cladiella krempfi*	247
Krempfielin G	Eunicellin-based diterpenoid	0.01, 16	10	~30	N/S	*Cladiella krempfi*	247
Krempfielin I	Eunicellin-based diterpenoid	0.01, 16	10	30–40	N/S	*Cladiella krempfi*	247
Litophynol B	Eunicellin-based diterpenoid	0.01, 16	10	6.4	52.5	*Cladiella krempfi*	247
Sclerophytin A	Eunicellin-based diterpenoid	0.01, 16	10	~70	N/S	*Cladiella krempfi*	247
Sclerophytin B	Eunicellin-based diterpenoid	0.01, 16	10	~50	N/S	*Cladiella krempfi*	247
Litophynin I monoacetate	Eunicellin-based diterpenoid	0.01, 16	10	~80	N/S	*Cladiella krempfi*	247
6-Acetoxy litophynin E	Eunicellin-based diterpenoid	0.01, 16	10	12.8	N/S	*Cladiella krempfi*	247
Litophynin F	Eunicellin-based diterpenoid	0.01, 16	10	30–40	48.1	*Cladiella krempfi*	247
Klymollin C	Eunicellin-based diterpenoid	N/A[e]	10	42.1	~70	*Klyxum molle*	248
Klymollin D	Eunicellin-based diterpenoid	N/A	10	25.1	~70	*Klyxum molle*	248
Klymollin E	Eunicellin-based diterpenoid	N/A	10	5.7	N/S	*Klyxum molle*	248
Klymollin F (**118**)	Eunicellin-based diterpenoid	N/A	10	6.0	8.5	*Klyxum molle*	248
Klymollin G (**119**)	Eunicellin-based diterpenoid	N/A	10	5.2	4.4	*Klyxum molle*	248
Klymollin H	Eunicellin-based diterpenoid	N/A	10	32.6	70–80	*Klyxum molle*	248
Hirsutalin B	Eunicellin-based diterpenoid	0.01, 16	10	6.8	49.0	*Cladiella hirsuta*	249
Hirsutalin C	Eunicellin-based diterpenoid	0.01, 16	10	43.6	N/S	*Cladiella hirsuta*	249
Hirsutalin D	Eunicellin-based diterpenoid	0.01, 16	10	3.3	N/S	*Cladiella hirsuta*	249
Hirsutalin H	Eunicellin-based diterpenoid	0.01, 16	10	32.3	N/S	*Cladiella hirsuta*	249
Hirsutalin K	Eunicellin-based diterpenoid	1, 24	2.5–20 µg/ml	Significant	N/A	*Cladiella hirsuta*	329
Lemnalol (**120**)	Ylangene-type sesquiterpenoid	0.01, 16	10	2.0	25.0	*Lemnalia flava*	249
Dihydroxycapnellene	Capnellane-based sesquiterpenoid	0.01, 16	10	13.1	67.6	*Paralemnalia thyrsoides*	239

(Continued)

Table 11.3 (Continued)   Anti-Inflammatory Compounds from Marine Invertebrates (iNOS and COX-2 Protein Expression)

Compounds	Class	LPS, h[a]	Concentration[b]	iNOS (%)[c]	COX-2 (%)	Species	References
Erectathiol	Calamenene-type sesquiterpene	0.01, 16	10	58.0	N/S	*Nephthea erecta*	242
Δ⁹⁽¹²⁾-Capnellene-8β, 10α-diol (**121**)	Capnellane-based sesquiterpenoid	0.01, 16	10	1.2	24.8	*Capnella imbricate*	251
8α-Acetoxy-Δ⁹⁽¹²⁾-capnellene-10α-ol	Capnellane-based sesquiterpenoid	0.01, 16	10	54.4	62.9	*Capnella imbricate*	251
Δ⁹⁽¹²⁾-Capnellene-10α-ol-8-one	Capnellane-based sesquiterpenoid	0.01, 16	10	34.8	N/S	*Capnella imbricate*	251
Paraminabeolide B	Withanolide	N/A	10	7.3	N/S	*Paraminabea acronocephala*	252
Paraminabeolide C	Withanolide	N/A	10	37.9	N/S	*Paraminabea acronocephala*	252
Paraminabeolide D	Withanolide	N/A	10	43.4	N/S	*Paraminabea acronocephala*	252
Minabeolide-1 (**122**)	Withanolide	N/A	10	9.6	18.3	*Paraminabea acronocephala*	252
Minabeolide-2	Withanolide	N/A	10	45.7	51.2	*Paraminabea acronocephala*	252
Minabeolide-4 (**123**)	Withanolide	N/A	10	23.2	22.4	*Paraminabea acronocephala*	252
Minabeolide-5	Withanolide	N/A	10	6.3	31.3	*Paraminabea acronocephala*	252
Chabrosterol	Norergosterol	0.01, 16	10	12.4	45.2	*Nephthea chabroli*	242
Nebrosteroid A	4-Methylated steroid	0.01, 16	10	10.6	N/S	*Nephthea chabroli Audouin*	236
Nebrosteroid B	5-Methylated steroid	0.01, 16	10	9.6	N/S	*Nephthea chabroli Audouin*	236
Nebrosteroid C	6-Methylated steroid	0.01, 16	10	0	N/S	*Nephthea chabroli Audouin*	236
Nebrosteroid D	7-Methylated steroid	0.01, 16	10	0	63.9	*Nephthea chabroli Audouin*	236
Nebrosteroid E	8-Methylated steroid	0.01, 16	10	43.1	57.4	*Nephthea chabroli Audouin*	236
Nebrosteroid F	9-Methylated steroid	0.01, 16	10	N/S	N/S	*Nephthea chabroli Audouin*	236
Nebrosteroid G	10-Methylated steroid	0.01, 16	10	76.9	72.0	*Nephthea chabroli Audouin*	236
Nebrosteroid H	11-Methylated steroid	0.01, 16	10	32.8	N/S	*Nephthea chabroli Audouin*	236
Nebrosteroid I	Steroid	0.01, 16	10	20.2	75.3	*Nephthea chabroli Audouin*	234
Nebrosteroid J	Steroid	0.01, 16	10	65.2	86.0	*Nephthea chabroli Audouin*	234
Nebrosteroid K	Steroid	0.01, 16	10	79.2	63.8	*Nephthea chabroli Audouin*	234
Nebrosteroid L	Steroid	0.01, 16	10	61.0	60.1	*Nephthea chabroli Audouin*	234
N-Acetyldihydrosphingosine	Ceramide	0.01, 16	10	46.9	77.2	*Sarcophyton ehrenbergi Marenzeller*	235

(Continued)

Table 11.3 (Continued)   Anti-Inflammatory Compounds from Marine Invertebrates (iNOS and COX-2 Protein Expression)

Compounds	Class	LPS, h[a]	Concentration[b]	iNOS (%)[c]	COX-2 (%)	Species	References
Sarcoehrenoside A	Cerebroside	0.01, 16	10	47.3	N/S	*Sarcophyton ehrenbergi Marenzeller*	235
Sarcoehrenoside B	Cerebroside	0.01, 16	10	46.5	N/S	*Sarcophyton ehrenbergi Marenzeller*	235
Ircicerebroside	Cerebroside	0.01, 16	10	25.8	N/S	*Sarcophyton ehrenbergi Marenzeller*	235
1-O-β-d-Glucopyranosyl-2-N-(20-d-hydroxyhexadecanoyl)-9-methylsphinga-4(E),8(E)-dienine	Cerebroside	0.01, 16	10	20.3	64.3	*Sarcophyton ehrenbergi Marenzeller*	235
Griffinisterone F	Steroid	0.01, 16	10	13.4	61.7	*Dendronephthya griffini*	253
Griffinisterone G	Steroid	0.01, 16	10	6.5	31.5	*Dendronephthya griffini*	253
Griffinipregnone	Steroid	0.01, 16	10	59.6	52.3	*Dendronephthya griffini*	253
Griffinisterone H	Steroid	0.01, 16	10	15.4	N/S	*Dendronephthya griffini*	253
Gibberoketosterol	Steroid	0.01, 16	10	44.5	68.3	*Sinularia gibberosa*	254
Paraminabic acid B	Steroidal carboxylic acid	0.01, 16	10	63.9	N/S	*Paraminabea acronocephala*	255
Paraminabic acid C	Steroidal carboxylic acid	0.01, 16	10	53.5	N/S	*Paraminabea acronocephala*	255
Capilloquinone	Tetraprenylbenzoquinone	0.01, 16	10	39.6	N/S	*Sinularia capillosa*	256
Capillobenzopyranol	Furanobenzosesquiterpenoid	0.01, 16	10	36.7	N/S	*Sinularia capillosa*	256
2-[(2E,6E)-3,7-dimethyl-8-(4-methylfuran-2-yl)octa-2,6-dienyl]-5-methylcyclohexa-2,5-diene-1,4-dione	Quinone	0.01, 16	10	0.3	48.7	*Sinularia capillosa*	256
2-[(2E,6E)-3,7-dimethyl-8-(4-methylfuran-2-yl)octa-2,6-dienyl]-5-methylbenzene-1,4-diol	Hydroquinone (Quinol)	0.01, 16	10	39.5	N/S	*Sinularia capillosa*	256

[a] LPS, hour: The concentration of LPS (μg/ml), incubation time with LPS.
[b] Concentration (μM) of treated compound.
[c] iNOS (%): Percentage of iNOS protein expression in comparison with LPS-treated control.
[d] N/S: Not significantly active.
[e] N/A: Not available.

RAW 264.7 cells.[233] Cembranoids from the soft coral *Lobophytum crassum*, including crassumolide A, lobohedleolide, and (1*S*,2*R*,7*S*,8*R*,11*S*)-7-acetyloxy-8,11-*epi*-dioxy-cembra-3*E*,12*E*,15(17)-trien-16,2-olide, inhibited the expression of iNOS and COX-2 protein. The inhibitory effects against COX-2 protein expression were stronger than those of CAPE.[232]

Sarcocrassocolides A–D, F–M, and O, cembranoids isolated from the soft coral *Sarcophyton crassocaule*, reduced iNOS protein levels to less than 25% in comparison to LPS-stimulated control (100%). However, these metabolites either did not affect or moderately inhibited COX-2 expression.[240,244,245]

Of the soft coral genus *Sinularia*, inhibition of iNOS or COX-2 protein expression has been observed with various terpenoids, including cembrane-based diterpenoids, norditerpenoids, aromadendrane-type sesquiterpenoids, and cardinane-type sesquiternoids. Notably, dihydrosinularin (**114**) and (–)14-deoxycrassin (**115**) from *Sinularia triangular* and 3-acetyloxy sinulariolone (**116**) from *Sinularia querciformis* reduced both iNOS and COX-2 protein levels to less than 25% compared to the LPS-treated control, when the expression level of the LPS-treated control was set as 100%.[246]

In addition to cembranoids, various eunicellin-based diterpenoids isolated from soft corals, including klysimplexin sulfoxide C (**117**),[238] klysimplexin S, and simplexin E from *Klyxum simplex*[241]; litophynin F from *Cladiella krempfi*[247]; klymollins F (**118**) and G (**119**) from *Klyxum molle*[248]; and hirsutalin B from *Cladiella hirsute*,[249] inhibited both iNOS and COX-2 protein expression to less than 50% compared to the LPS-treated control.

Various sesquiterpenoids, including ylangene, nardosinane, capnellane, and calamenene types from soft corals including *Lemnalia flava*,[239] *Paralemnalia thyrsoides*,[239] *Nephthea erecta*,[250] and *Capnella imbricate*,[251] exerted inhibitory effects on iNOS or COX-2 protein expression. Lemnalol (**120**) from *L. flava*[239] and Δ^9(12)^-capnellene-8β,10α-diol (**121**) from *C. imbricate*[251] reduced iNOS and COX-2 protein levels below 25% in comparison to the LPS-treated control (100%).

11.3.4.2.1.2 Lipids — Steroids, including withanolides from *Paraminabea acronocephala*,[252] chabrosterol and nebrosteroids from *Nephthea chabroli*,[234,236,250] other steroids from *Dendronephthya griffini*[253] and *Sinularia gibberosa*,[254] steroidal carboxylic acids from *P. acronocephala*, and *N*-acetyldihydrosphingosine and cerebrosides from *Sarcophyton ehrenbergi* Marenzeller, showed inhibition of iNOS and COX-2 protein expression. Also, carotenoids from *Halocynthia roretzi*, including all *trans*- and 9-*cis*-alloxanthins and all *trans*- and 9-*cis*-diatoxanthins, demonstrated inhibitory capacity on iNOS and COX-2 mRNA expression. Notably, minabeolides-1 (**122**) and -4 (**123**) from *P. acronocephala* suppressed LPS-induced iNOS and COX-2 protein levels to less than 25% relative to LPS-treated control.[255]

11.3.4.2.1.3 Miscellaneous Compounds — Quinone structures, including capilloquinone; 2-[(2*E*,6*E*)-3,7-dimethyl-8-(4-methylfuran-2-yl)octa-2,6-dienyl]-5-methylcyclohexa-2,5-diene-1,4-dione; 2-[(2*E*,6*E*)-3,7-dimethyl-8-(4-methylfuran-2-yl)octa-2,6-dienyl]-5-methylbenzene-1,4-diol, and a pyran structure capillobenzopyranol, from *Sinularia capillosa*, suppressed iNOS or COX-2 expression to less than 50%.[256] Several marine invertebrate secondary metabolites that exhibit activity against pro-inflammatory mediators are illustrated in Figure 11.10.

11.3.4.2.2 Inhibition of Pro-Inflammatory Mediators and Cytokines In Vitro

During inflammation, various cytokines and mediators are released, including NO, prostaglandins, interleukins, and TNF-α. Here we describe compounds from marine invertebrates showing IC$_{50}$ values of 25 µM or less, since known positive controls show IC$_{50}$ values in this range (e.g., reference aminoguanidine IC$_{50}$ = 25.0 µM).

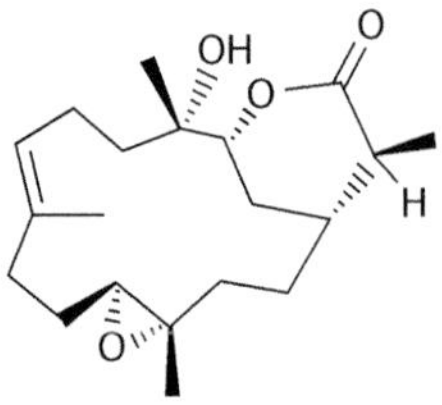

114 Dihydrosinularin 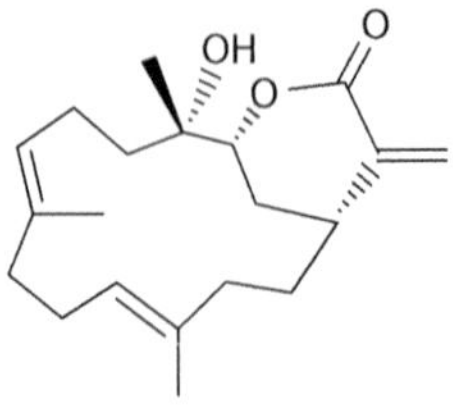   **115** (–)14-Deoxycrassin 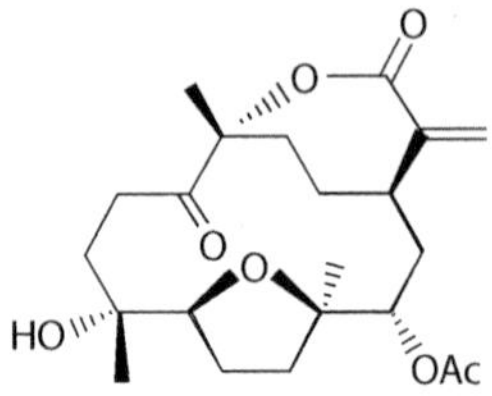   **116** 3-Acetyloxy sinulariolone

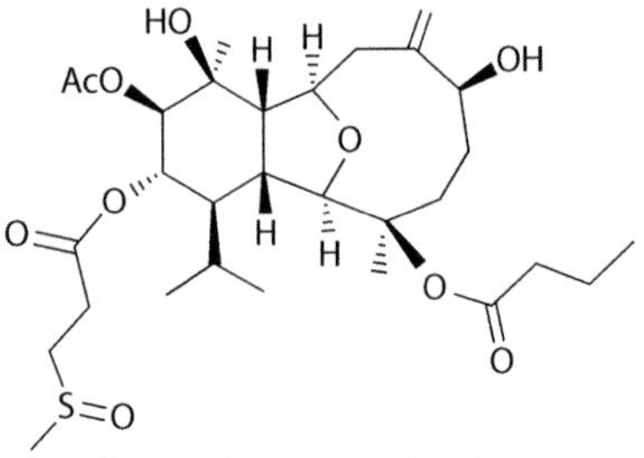

117 Klysimplexin sulfoxide C 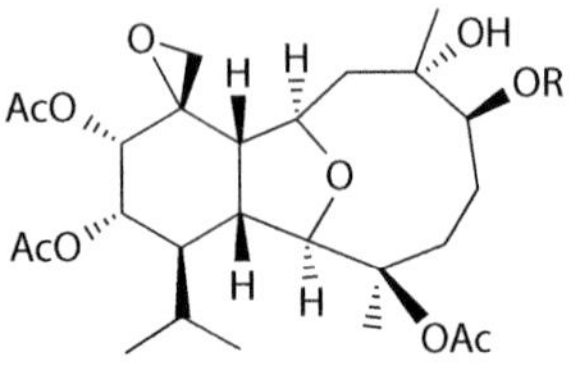   **118** Klymollin F (R=CH$_3$(CH$_2$)$_{12}$CO   **120** Lemnalol
                                          119 Klymollin G (R=CH$_3$(CH$_2$)$_{14}$CO

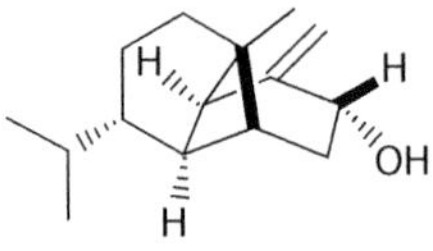

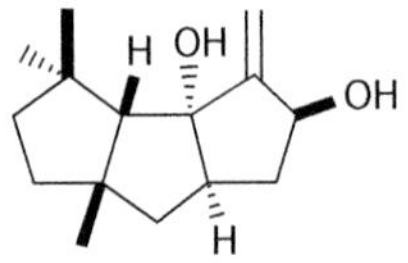

121 Δ$^{9(12)}$-Capnellene-8β, 10α-diol 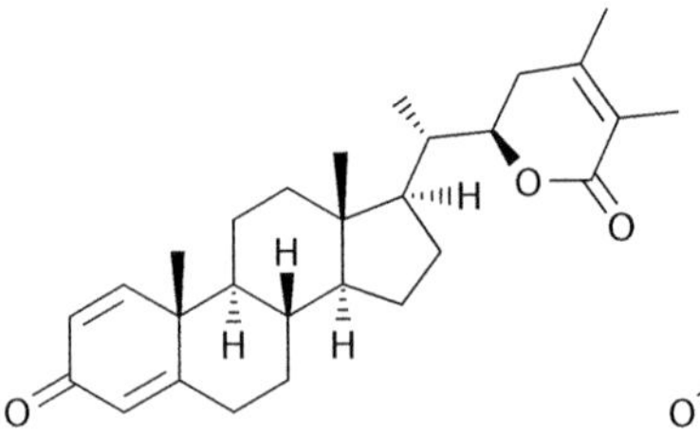   **122** Minabeolide-1 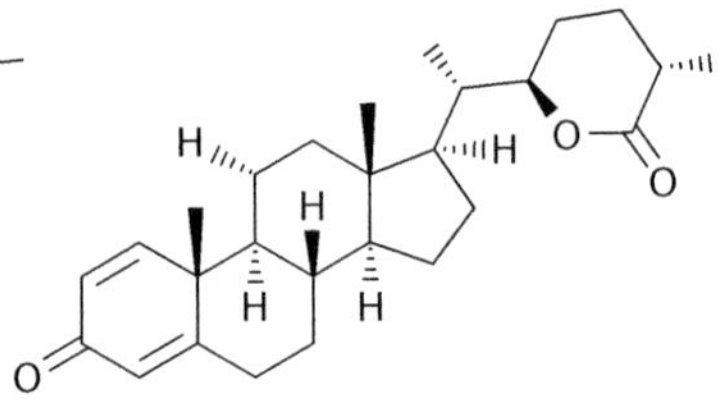   **123** Minabeolide-4

Figure 11.10 Anti-inflammatory metabolites from marine invertebrates (iNOS and COX-2).

11.3.4.2.2.1 Terpenoids — Some diterpenoids isolated from soft corals showed inhibition of inflammation mediators and cytokines. Recently, Yan et al. reported a series of biscembranoids, lobophytones A–Z, from the soft coral *Lobophytum pauciflorum*. Biscembranoids are unique tetraterpenoids with a 14-6-14-membered tricyclic backbone, possibly derived from biogenetic Diels–Alder cycloaddition of cembranoid-diene and cembranoid-dienophile. The oxygenation and cyclization of the C ring provides structural diversity of biscembranoids.[257] Lobophytones A–Z have been evaluated for anti-inflammatory activities in LPS-stimulated mouse peritoneal macrophages by measuring NO production. Lobophytones D (**124**), Q (**125**), and Z (**126**) showed potent activities with IC$_{50}$ values of 4.7, 2.8, and 2.6 μM, respectively.[257–259] With regard to seven lobophytones, H–N, tests were performed at a single low concentration of 1 μM; thus, IC$_{50}$ values are not available.[260]

As mentioned above,[233] cembranoids from the soft coral *Lobophytum crassum* contain an α-methylene-γ-lactone moiety, which is critical for inhibition of NO production[261]: crassocolide P, lobophytol acetate, cembranolide B [1a*S*-(1a*R**,4*E*,8*E*,9a*S**,12a*S**,14a*R**)], (9CI)1a,3,6,7,9a,12,12a, 13,14,14a-decahydro-4,8,14a-trimethyl-12-methylene-oxireno[5,6]cyclotetradeca[1,2-b]furan-11(2*H*)-one, cembranolide A (**127**), lobophytolide, denticulatolide, and (+)-sinularial A exhibited a significant effect on NO production in LPS (1 μg/ml)-stimulated RAW 264.7 cells with IC$_{50}$ values of 3.8, 3.8, 5.7, 7.9, 2.4, 4.9, 6.4, and 16.6 μM, respectively.

Norditerpenoids from the soft coral *Clavularia viridis*, including norcembrenolide and sinuleptolide (**128**), inhibited NO production in LPS-stimulated RAW 264.7 cells with IC_{50} values of 2.8 and 0.86 µM, respectively.[262] Additional norditerpenoids, including scabrolide A (**129**) and 13-*epi*-scabrolide C from the soft coral *Sinularia maxima*, significantly inhibited IL-12/IL-6 production in LPS-stimulated bone marrow-derived dendritic cells (BMDCs) with IC_{50} values of 5.3/13.1 and 23.5/69.9 µM, respectively.[263]

Diterpenoids, including sinumaximols B and C (**130**), and isomandapamate from the soft coral *S. maxima* significantly inhibited IL-12/IL-6 production in LPS-stimulated BMDCs with IC_{50} values of 15.2/59.8, 4.4/17.7, and 18.0/54.3 µM, respectively. In addition, sinumaximol C (**130**) inhibited TNF-α production with an IC_{50} value of 62.3 µM.[264]

Eunicellin-based diterpenes *seco*-briarellinone (**131**) and briarellin S from the octocoral *Briareum asbestinum* inhibited LPS-induced NO production in mouse peritoneal macrophages with IC_{50} values of 4.7 and 20.4 µM, respectively.[265]

In addition to corals, marine sponges were found to contain inhibitors of pro-inflammatory cytokines and mediators. Cycloamphilectenes (amphilectene-based diterpenes) from the Vanuatu sponge *Axinella* sp. inhibited the production of NO, PGE_2, and TNF-α in murine peritoneal macrophages. Notably, *N*-[(1*S*,3a*R*,4*S*,6*R*,6a*S*,9a*S*,9b*S*)-dodecahydro-1,4-dimethyl-7-methylene-6-(2-methyl-1-propen-1-yl)-1*H*-phenalen-1-yl]-formamide (**132**), containing a unique exocyclic methylene group, inhibited the production of NO most potently with an IC_{50} value of 0.1 µM without affecting TNF-α release.[266]

Marine sponges are also a source of sesterterpenoids, including ansellone B and phorbasone A acetate (**133**) from *Nephthea chabroli*,[267] and norsesterterpene peroxides, including epimuqubilin A (**134**), muqubilone B, and sigmosceptrellin A from *Latrunculia* sp.[268] These compounds inhibited NO production in LPS (1 µg/ml)-stimulated RAW 264.7 cells with IC_{50} values of 4.5, 2.8, 7.4, 23.8, and 9.9 µM, respectively. In the case of epimuqubilin A, a follow-up study demonstrated that the response was due to inhibition of the phosphorylation of IκB kinase.[269] Cacospongionolide B (**135**) was isolated from the Mediterranean sponge *Fasciospongia cavernosa* and decreased the production of NO and TNF-α with IC_{50} values of 0.33 and 0.26 µM, respectively.[270]

Sarcophytol A, a cembrane-type diterpene first isolated from the soft coral *Sarcophyton glaucum*,[271] inhibited mRNA expression and release of TNF-α in BALB/3T3 cells induced by a tumor promoter, okadaic acid.[272]

Diterpene glycosides, pseudopterosins PsQ (**136**), PsS, PsT, and PsU from *Pseudopterogorgia elisabethae* collected at the Providenciah, showed 59%, 49%, 52%, and 52% inhibition in the release of myeloperoxidase with human polymorphonuclear neutrophils, respectively, at a concentration of 10 µM, comparable to dexamethasone and indomethacin. Pseudopterosins PsP (**137**) and PsT from *P. elisabethae* showed 58% and 52%, respectively, at 10 µM in LPS (100 µg/ml)-stimulated J774 cells.[225] Pseudopterosins have a substantial commercial market as an additive in personal care products or as topical anti-inflammatory agents. The terpenoids from marine inverteberates that showed the most potent activities in each study are illustrated in Figure 11.11.

11.3.4.2.2.2 Lipids — (3*S*,14*S*)-Petrocortyne A, a lipid compound (a C_{46} polyacetylenic alcohol) from a marine sponge *Petrosia* sp., blocked TNF-α production strongly in a concentration-dependent manner with LPS-activated RAW 264.7 cells.[273]

Terpioside B, a unique glycolipid containing two fucose residues in the furanose form in its pentasaccharide chain, was isolated from the marine sponge *Terpios* sp. and is a potent inhibitor of NO release in J774 cells stimulated with LPS (1 µg/ml) for 24 h.[274]

A sterol glycoside, 3β-*O*-(3′,4′-di-*O*-acetyl-β-D-arabinopyranosyl)-25ξ-cholestane-3β,5α,6β,26-tetrol-26-acetate) (carijoside A), isolated from an octocoral *Carijoa* sp., displayed significant inhibitory effects on superoxide anion generation and elastase release by human neutrophils with IC_{50} values of 1.8 and 6.8 µg/ml, respectively.[275]

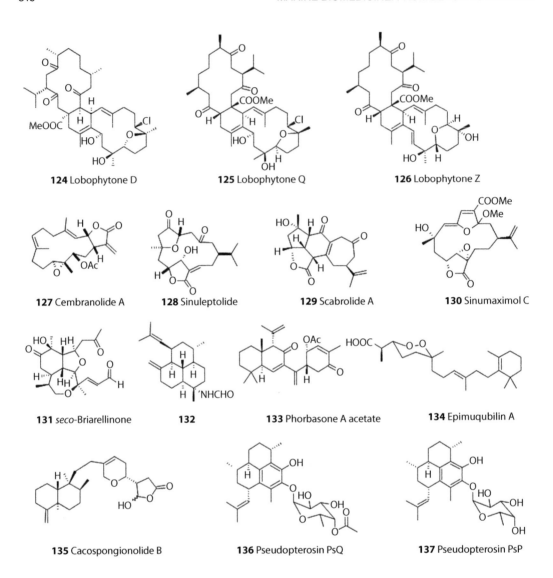

Figure 11.11 Inhibitory marine invertebrate terpenoids against pro-inflammatory mediators and cytokines.

Sinularioside, a triacetylated glycolipid containing two α-D-arabinopyranosyl residues and a myristyl alcohol unit from the Indonesian soft coral *Sinularia* sp., moderately inhibited LPS (1 μg/ml)-induced NO release in J774 macrophages.[276]

11.3.4.2.2.3 Other Classes of Molecules — *N,N*-Didesmethylgrossularine-1, a compound with a rare α-carboline structure (alkaloid), was isolated from the Indonesian ascidian *Polycarpa aurata*. The molecule was responsible for the observed inhibitory activity against IL-6 production (24 h, 10%–20% production at 10 μM) in LPS-stimulated RAW 264.7 cells.[277]

A D-form of arginine (amino acid derivatives), herdmanine D from the ascidian *Herdmania momus*, inhibited NO production in LPS (1 μg/ml)-stimulated RAW 264.7 with an IC_{50} value of 9 μM.[278]

Bolinaquinone and dysidine, isolated from a sponge *Dysidea* sp., modulated human leukocyte functions with significant inhibitory effects on elastase release by human neutrophils with IC_{50} values of 5.2 and 1.3 μM, respectively.[279]

Two naphthopyrones, comaparvin and 6-methoxycomaparvin, from the Philippine echinoderm *Comanthus* sp., exhibited inhibitory activity against TNF-α-induced NF-κB activation in rat hepatoma cells and human breast cancer cells associated with blockade of the nuclear translocation of NF-κB and IκB phosphorylation.[220]

11.3.4.2.3 Other Anti-Inflammatory Properties

A diterpene-pentoseglycoside, pseudopterosin A from the Caribbean Sea whip *Pseudoptergorgia elisabethae*, inhibited 12-*O*-tetradecanoylphorbol-13-acetate (TPA)-induced topical inflammation in mouse skin.[280,281]

Diterpene glycosides, fuscosides B and E, from the soft coral *Eunicea fusca*, showed strong anti-inflammatory activities with 80.5% and 81.5% inhibition, respectively, which is slightly higher than that of indomethacin when topically applied at a dose of 0.5 mg/ear in a TPA-induced mouse ear edema model.[282]

Lobohedleolide, from the soft corals of *Sinularia crassa* and *Lobophytum* sp., respectively, inhibited chronic inflammation in the cotton pellet-induced granuloma rat model.[283]

Dendalone 3-hydroxybutyrate was isolated from a Dictyoceratid sponge and inhibited prostaglandin synthetase activity *in vitro*.[192]

11.3.4.3 Anticarcinogenic Metabolites

11.3.4.3.1 Induction of Detoxification

A *para*-hydroquinone, strongylophorine-8 (5 μM), from the sponge *Petrosia* (*Strongylophora*) *corticata* collected in Papua New Guinea, induced mRNA expression of phase 2 enzymes (*nqo1*, *gsta4*, *ho-1*, *gclc*, and *gsta1*), along with ARE activation, and increased glutathione in HT22 cells.[284]

A steroidal glycoside, acanthifolioside G, from the Caribbean marine sponge *Pandaros acanthifolium*, increased ARE activity 2.3-fold at 25 μg/ml. In addition, it induced SOD and glutathione peroxidase associated with a decrease in lipid peroxidation and cellular oxidative damage upon H_2O_2 exposure in U373 cells.[285]

Two diterpenes, 12(*S*)-hydroperoxylsarcoph-10-ene and 8-*epi*-sarcophinone from the Red Sea soft coral *Sarcophyton glaucum*, showed chemopreventive activities with inhibition of CYP1A activity and induction of GST and QR activities. Also, a cembranolide diterpene, *ent*-sarcophine, from the coral *S. glaucum*, inhibited CYP1A activity and induced mEH, establishing chemopreventive and tumor anti-initiating activity for these metabolites.[286]

11.3.4.3.2 Tyrosine Kinase Inhibitors

A meroterpenoid, liphagal (**138**), from the marine sponge *Aka coralliphaga*, inhibited PI3Kα enzyme activity with an IC_{50} value of 100 nM.[287] Analogs of the sponge meroterpenoid liphagal (**138**) have been biosynthesized and evaluated for inhibition of PI3Kα and PI3Kγ as part of a program aimed at developing new isoform-selective PI3K inhibitors. One of the analogs (**139**) demonstrated IC_{50} values of 66 nM against PI3Kα and 1840 nM against PI3Kγ (representing a 27-fold preference for PI3Kα), exhibited enhanced chemical stability, and modestly enhanced potency and selectivity compared to the natural product liphagal.[288]

Nakijiquinone C (**140**), from an Okinawan marine sponge of the family Spongiidae, inhibited EGFR, c-erbB-2 kinase, and protein kinase C enzyme activity with IC_{50} values of 170, 26, and 23 μM, respectively.[289] Nakijiquinones G–I (**141–143**), from Okinawan marine sponges of the family Spongiidae, were also reported to inhibit HER2 kinase activity.[290]

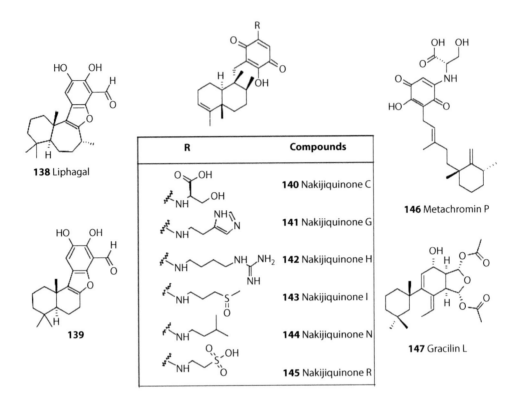

Figure 11.12   Tyrosine kinase inhibitors from marine invertebrates.

Among sesquiterpenoid quinones, nakijiquinones J–R were isolated from three collections of Okinawan marine sponges of the family Spongiidae; of these, nakijiquinones N (**144**), O, and R showed inhibitory activities against HER2 (66%, 59%, and 52% inhibition, respectively), while naki-jiquinones P and R (**145**) exhibited 76% and >99% inhibitory activity against EGFR, respectively.[291]

Sesquiterpenoid quinones, metachromins L–Q from Okinawan marine sponges, showed greater inhibition of HER2 enzyme activity (IC_{50} values of 18–190 μg/ml). The most potent of these was metachromin P (**146**).[292]

Gracilins A and L, tetrahydroaplysulfurin-1, and 3′-norspongiolactone, from the marine sponge *Spongionella* sp., inhibited EGFR tyrosine kinase greater than 50% at 100 μM. Gracilin L (**147**) was the most potent, with 75% inhibition, being nearly as active as the positive control, genistein, with 80% inhibition.[293]

12-*epi*-Scalaradial, found in the marine sponges *Cacospongia* sp., inhibited EGF-stimulated Akt phosphorylation, membrane translocation of 3-phosphoinositide-dependent protein kinase 1, and PI3K activity.[294] Tyrosine kinase inhibitors from marine invertebrates are illustrated in Figure 11.12.

11.3.4.3.3 Inhibitors of Malignant Transformation

Known cembrane-type diterpenes, sarcophytol A and sarcophine, from *Sarcophyton glaucum* (Alcyonaria, Alcyoniidae), inhibited TPA-induced JB6 cell transformation by 86.4% and 58.0%, respectively, at a concentration of 50 μM.[295] Also, sarcophytol A and sarcophine inhibited Epstein–Barr virus early antigen activation.[296]

3-Demethylubiquinone Q2 from the ascidian *Aplidium glabrum* reduced transformation of JB6 P$^+$ Cl41 cells induced by EGF (10 ng/ml); the value for inhibiting the number of the colonies by 50% (INCC$_{50}$) was 11.4 µM in the anchorage-independent transformation assay with soft agar. 3-Demethylubiquinone Q2 significantly induced AP-1- and NF-κB-dependent transcriptional activation by 190% and 217%, respectively, whereas it inhibited p53-dependent transcriptional activity by 72%.[297]

A triterpene oligoglycoside, frondoside A, from the sea cucumber *Cucumaria okhotensis*, decreased the AP-1-dependent transcriptional activity induced by UVB, TPA, or EGF in JB6-LucAP-1 cells by 41% (at 0.7 µg/ml), 20%, and 51% (at 1.1 µg/ml), respectively. Also, frondoside A inhibited EGF-induced NF-κB-dependent transcriptional activity in JB6-LucNF-κB cells by 37% (at 0.7 µg/ml) and increased p53-dependent transcriptional activity by 39% (at 1.8 µg/ml) in JB6-Lucp53 cells. The compound also inhibited colony formation of JB6 P$^+$ Cl41 cells activated with EGF (INCC$_{50}$ = 0.8 µg/ml).[298]

Leviusculoside G, isolated from the starfish *Henricia leviuscula*, decreased cell transformation in an anchorage-independent soft agar assay.[299]

A two-headed sphingolipid-like compound, rhizochalin, from the sponge *Rhizochalina incrustata*, inhibited EGF-induced malignant transformation of the mouse skin epidermal JB6 P$^+$ Cl41 cell line with an INCC$_{50}$ value of 8.6 µM. The compound induced p53-dependent transcriptional activity by approximately 1.6-fold at 10 µM with a 6 h incubation time.[300]

A asterosaponin, archasterosides B, isolated from the Vietnamese starfish *Archaster typicus*, induced basal AP-1 (1.7-fold at 12.5 µM) and p53 (1.3-fold at 12.5 µM), but not NF-κB transcriptional activation in JB6 Cl41 cells.[301]

Actinoporins are water-soluble pore-forming toxins produced by sea anemones. Actinoporin (RTX-A) from the tropical sea anemone *Heteractis crispa* (*Radianthus macrodactylus*) showed cancer-preventive activity with mice bearing Ehrlich ascites carcinoma. RTX-A suppressed the malignant transformation of mouse JB6 P$^+$ Cl41 cells stimulated by EGF in soft agar with an INCC$_{50}$ of 0.034 nM. RTX-A also induced apoptosis and inhibited basal AP-1-, NF-κB-, and p53-dependent transcriptional activity in JB6 Cl41 cells with IC$_{50}$ values of 0.1–0.4 nM.[302]

A sulfated steroidal glycoside, hippasterioside D, from the starfish *Hippasteria kurilensis* collected in the Sea of Okhotsk near Kuril Islands, exerted an inhibitory effect on HT-29 colony formation in a soft-agar clonogenic assay.[303]

Mycalamide A, isolated from the ascidian *Polysincraton* sp., inhibited EGF-induced neoplastic transformation of JB6 P$^+$ Cl41 cells with an INCC$_{50}$ of 0.14 nM, and induced apoptosis at subnanomolar or nanomolar concentrations. Mycalamide A inhibited transcriptional activity of AP-1, NF-κB, and p53 with IC$_{50}$ values of 15.6–31.2 nM, ~15.6 nM, and ~7.8 nM, respectively, after 6 h of incubation. Induction of phosphorylation of the MAPKs (ERK1/2, JNK, and p38 MAPK) was also observed at higher concentrations of mycalamide A.[304]

11.3.4.3.4 In Vivo Cancer Prevention

A marine cembranoid, sarcophytol A, significantly reduced the incidence of large bowel tumors in *N*-methyl-*N*-nitrosourea-induced colon carcinogenesis with female CD Fischer rats.[305] Sarcophytols A and B inhibited tumor promotion by teleocidin in a two-stage carcinogenesis experiment on mouse skin.[306] Also, sarcophytol A lowered the incidence of atypical ductal hyperplasia and carcinoma *in situ* in *N*-nitrobis-(2-hydroxypropyl) amine-induced pancreatic carcinogenesis in Syrian golden hamsters.[307]

In addition, semisynthetic sarcophine derivatives, including sarcodiol, sarcotriol α, sarcotriol β, sarcodiolone, and sarcodiolal, significantly inhibited tumor incidence and multiplicity in the mouse two-stage skin model using 7,12-dimethylbenz[*a*]anthracene (DMBA) as an initiator and TPA as promoter.[308]

Sarcophine-diol significantly decreased skin papilloma development in both DMBA-initiated and TPA-promoted, as well as UVB-induced, *in vivo* skin carcinogenesis models in mice.[309–311]

Frondanol A5, which consists of frondoside A, fucosylated chondroitin sulfate, and 12-methyl-tetradecanoic acid from the sea cucumber *Cucumaria frondosa*, significantly suppressed azoxymethane-induced total colonic aberrant crypt foci formation, and multicrypt aberrant foci, in a rat colon carcinogenesis model.[312]

DL15 decreased multiple plaque lesions in the 1,2-dimethylhydrazine-induced colon carcinogenesis model; activity was associated with inhibition of Bcl-2 and iNOS.[313,314]

11.4 CONCLUSIONS (CURRENT ISSUES AND FUTURE PERSPECTIVES)

Natural products have made an indelible contribution to all types of FDA-approved drugs.[315] Interestingly, for the discovery of novel cancer chemopreventive agents, the influence of natural products has been equally significant. In part, this relates to epidemiological work that has suggested inverse relationships between cancer and the consumption of phytochemical-rich fruits and vegetables. Accordingly, we and others have undertaken systematic efforts to unearth natural product cancer chemopreventive agents, and an interesting array of lead molecules have emerged.

Beyond terrestrial sources, it is rational to assume that marine-based organisms would serve as a viable source for novel cancer chemopreventive agents. However, unlike terrestrial plants, with few exceptions (e.g., dietary consumption of seaweed), epidemiological surveys and traditional uses generally do not provide assistance for the selection of marine starting materials. Nonetheless, given our current understanding of molecular targets and mechanisms of carcinogenesis, as well as high-throughput screening capacity, progress can be achieved in a rapid and efficient manner.

Once leads are discovered, in order to enable more advanced testing, the issue of compound supply must be resolved. The approach for procuring adequate quantities of a lead molecule depends on the complexity of the chemical structure, as well as the nature of the source material. However, like terrestrial materials, direct isolation may be fraught with problems. Yields may be extremely low, and the amount and composition of metabolites from identical organisms can be different, depending on the surrounding environment, including temperature and adjacent organisms (species interactions). With sponges, metabolites can vary as a result of adaptation to survive from environmental conditions, including predation or competition for limited space.[316] For example, *Oscarella balibaloi* showed a significant seasonal variation in bioactivity, with the lowest values at the end of spring and in early summer, followed by the highest values in late summer and autumn. The bioactivity was negatively correlated to the reproductive effort; minimal bioactivities coincided with the period of embryogenesis and larval development.[317]

Nevertheless, in some cases, harvest from the wild may be possible, depending on the type of natural source and the concentration of bioactive compounds found within. As a more environmentally sensitive alternative, genetic modification or cultivation can be attempted. With corals, various approaches have been employed to obtain large-scale quantities of products, including mariculture, *ex situ* culture, cell culture, and *in toto* aquaculture.[318] When the source material is a microorganism, the potential of large-scale cultivation for the production of the active principle is a distinct advantage, and precludes the need for source material recollection.

In other situations, synthetic chemistry is a plausible approach for obtaining sufficient quantities of active leads. Speculation regarding the biosynthetic assembly of a molecule can be used as a guide for the development of biomimetic cascade reactions that have been applied in very short syntheses of complex natural products. For example, the total synthesis of *ent*-penilactone A and penilactone B has been conducted through biomimetic Michael reactions between tetronic acids and *o*-quinone methides.[319]

When active chemopreventive agents are isolated from edible materials, there is at least a tacit implication of safety. However, when derived from nonedible sources, biologically active metabolites possessing antioxidant, anti-inflammatory, or detoxificating effects can also be lethal. An example of this is mycotoxins from fungal sources. Therefore, it is necessary to examine the safety of drug candidates and determine lethal and sublethal ranges. Also, metabolites could initiate and promote carcinogenesis. Potential genotoxicity and mutagenicity of active biomaterials need to be taken into account as well. In general, these factors may be considered a drawback in attempting to develop marine-based chemopreventive agents, but efficacy demonstrated in preclinical models can engender good confidence for proceeding with more advanced stages of testing.

In sum, as illustrated by substances such as cembrane and eunicellin-type terpenoids, marine materials have shown great pharmaceutical potential with chemical diversity and uniqueness. However, in the context of cancer chemoprevention, this has been largely underdeveloped. There is no doubt that efficacious cancer chemopreventive agents are present in the marine environment, and that these agents could be discovered with experimental tools that are currently available. The development of improved bioassays, coupled with a greater understanding of the biological mechanisms of chemoprevention, will facilitate the discovery of leads from the marine environment and help streamline the lengthy current discovery process.

REFERENCES

1. Ferlay, J., Shin, H., Bray, F., Forman, D., Mathers, C., Parkin, D. GLOBOCAN 2008 v2.0: Cancer incidence and mortality worldwide: IARC CancerBase no. 10 [Internet]. Lyon, France: International Agency for Research on Cancer, 2010. Available from http://globocan.iarc.fr (accessed on August 22, 2013).
2. Perera, E., Sinclair, R. An estimation of the prevalence of nonmelanoma skin cancer in the U.S. *F1000Res* 2013, 2.
3. Lewis, K. G., Weinstock, M. A. Trends in nonmelanoma skin cancer mortality rates in the United States, 1969 through 2000. *J Invest Dermatol* 2007, 127 (10), 2323–27.
4. Siegel, R., Naishadham, D., Jemal, A. Cancer statistics, 2013. *CA Cancer J Clin* 2013, 63 (1), 11–30.
5. Raftery, A. E., Chunn, J. L., Gerland, P., Sevčíková, H. Bayesian probabilistic projections of life expectancy for all countries. *Demography* 2013, 50 (3), 777–801.
6. Bray, F., Jemal, A., Grey, N., Ferlay, J., Forman, D. Global cancer transitions according to the Human Development Index (2008–2030): A population-based study. *Lancet Oncol* 2012, 13 (8), 790–801.
7. Heinonen, O. P., Huttunen, J. K., Haapakoski, J., Palmgren, J., Pietinen, P., Pikkarainen, J., Rautalahti, M., Virtamo, J., Edwards, B. K., Greenwald, P., Hartman, A. M., Taylor, P. R. The effect of vitamin E and beta carotene on the incidence of lung cancer and other cancers in male smokers. The Alpha-Tocopherol, Beta Carotene Cancer Prevention Study Group. *N Engl J Med* 1994, 330 (15), 1029–35.
8. Omenn, G. S., Goodman, G. E., Thornquist, M. D., Balmes, J., Cullen, M. R., Glass, A., Keogh, J. P., Meyskens, F. L., Valanis, B., Williams, J. H., Barnhart, S., Cherniack, M. G., Brodkin, C. A., Hammar, S. Risk factors for lung cancer and for intervention effects in CARET, the Beta-Carotene and Retinol Efficacy Trial. *J Natl Cancer Inst* 1996, 88 (21), 1550–59.
9. Lippman, S. M., Klein, E. A., Goodman, P. J., Lucia, M. S., Thompson, I. M., Ford, L. G., Parnes, H. L., Minasian, L. M., Gaziano, J. M., Hartline, J. A., Parsons, J. K., Bearden, J. D., Crawford, E. D., Goodman, G. E., Claudio, J., Winquist, E., Cook, E. D., Karp, D. D., Walther, P., Lieber, M. M., Kristal, A. R., Darke, A. K., Arnold, K. B., Ganz, P. A., Santella, R. M., Albanes, D., Taylor, P. R., Probstfield, J. L., Jagpal, T. J., Crowley, J. J., Meyskens, F. L., Baker, L. H., Coltman, C. A. Effect of selenium and vitamin E on risk of prostate cancer and other cancers: The Selenium and Vitamin E Cancer Prevention Trial (SELECT). *JAMA* 2009, 301 (1), 39–51.
10. Klein, E. A., Thompson, I. M., Tangen, C. M., Crowley, J. J., Lucia, M. S., Goodman, P. J., Minasian, L. M., Ford, L. G., Parnes, H. L., Gaziano, J. M., Karp, D. D., Lieber, M. M., Walther, P. J., Klotz, L., Parsons, J. K., Chin, J. L., Darke, A. K., Lippman, S. M., Goodman, G. E., Meyskens, F. L., Baker, L. H. Vitamin E and the risk of prostate cancer: The Selenium and Vitamin E Cancer Prevention Trial (SELECT). *JAMA* 2011, 306 (14), 1549–56.

11. Fisher, B., Costantino, J. P., Wickerham, D. L., Redmond, C. K., Kavanah, M., Cronin, W. M., Vogel, V., Robidoux, A., Dimitrov, N., Atkins, J., Daly, M., Wieand, S., Tan-Chiu, E., Ford, L., Wolmark, N. Tamoxifen for prevention of breast cancer: Report of the National Surgical Adjuvant Breast and Bowel Project P-1 Study. *J Natl Cancer Inst* 1998, 90 (18), 1371–88.

12. Vogel, V. G., Costantino, J. P., Wickerham, D. L., Cronin, W. M., Cecchini, R. S., Atkins, J. N., Bevers, T. B., Fehrenbacher, L., Pajon, E. R., Wade, J. L., Robidoux, A., Margolese, R. G., James, J., Lippman, S. M., Runowicz, C. D., Ganz, P. A., Reis, S. E., McCaskill-Stevens, W., Ford, L. G., Jordan, V. C., Wolmark, N., National Surgical Adjuvant Breast and Bowel Project (NSABP). Effects of tamoxifen vs raloxifene on the risk of developing invasive breast cancer and other disease outcomes: The NSABP Study of Tamoxifen and Raloxifene (STAR) P-2 trial. *JAMA* 2006, 295 (23), 2727–41.

13. Guilford, J. M., Pezzuto, J. M. Natural products as inhibitors of carcinogenesis. *Expert Opin Investig Drugs* 2008, 17 (9), 1341–52.

14. Thompson, I. M., Goodman, P. J., Tangen, C. M., Lucia, M. S., Miller, G. J., Ford, L. G., Lieber, M. M., Cespedes, R. D., Atkins, J. N., Lippman, S. M., Carlin, S. M., Ryan, A., Szczepanek, C. M., Crowley, J. J., Coltman, C. A. The influence of finasteride on the development of prostate cancer. *N Engl J Med* 2003, 349 (3), 215–24.

15. Thompson, I. M., Goodman, P. J., Tangen, C. M., Parnes, H. L., Minasian, L. M., Godley, P. A., Lucia, M. S., Ford, L. G. Long-term survival of participants in the Prostate Cancer Prevention Trial. *N Engl J Med* 2013, 369 (7), 603–10.

16. Nelson, H. D., Smith, M. E., Griffin, J. C., Fu, R. Use of medications to reduce risk for primary breast cancer: A systematic review for the U.S. Preventive Services Task Force. *Ann Intern Med* 2013, 158 (8), 604–14.

17. Chan, A. T., Arber, N., Burn, J., Chia, W. K., Elwood, P., Hull, M. A., Logan, R. F., Rothwell, P. M., Schrör, K., Baron, J. A. Aspirin in the chemoprevention of colorectal neoplasia: An overview. *Cancer Prev Res* (Phila) 2012, 5 (2), 164–78.

18. Meyskens, F. L., McLaren, C. E., Pelot, D., Fujikawa-Brooks, S., Carpenter, P. M., Hawk, E., Kelloff, G., Lawson, M. J., Kidao, J., McCracken, J., Albers, C. G., Ahnen, D. J., Turgeon, D. K., Goldschmid, S., Lance, P., Hagedorn, C. H., Gillen, D. L., Gerner, E. W. Difluoromethylornithine plus sulindac for the prevention of sporadic colorectal adenomas: A randomized placebo-controlled, double-blind trial. *Cancer Prev Res* (Phila) 2008, 1 (1), 32–38.

19. Wattenberg, L. W. Chemoprophylaxis of carcinogenesis: A review. *Cancer Res* 1966, 26 (7), 1520–26.

20. Guilford, J. M., Pezzuto, J. M. Cancer chemoprevention. In *Chemical Sciences, Engineering and Technology Resources.* Encyclopedia of Life Support Systems (EOLSS). Oxford: EOLSS, 2011.

21. Park, E. J., Pezzuto, J. M. Botanicals in cancer chemoprevention. *Cancer Metastasis Rev* 2002, 21 (3–4), 231–55.

22. Gullett, N. P., Ruhul Amin, A. R., Bayraktar, S., Pezzuto, J. M., Shin, D. M., Khuri, F. R., Aggarwal, B. B., Surh, Y. J., Kucuk, O. Cancer prevention with natural compounds. *Semin Oncol* 2010, 37 (3), 258–81.

23. Park, E. J., Pezzuto, J. M. Flavonoids in cancer prevention. *Anticancer Agents Med Chem* 2012, 12 (8), 836–51.

24. Jang, M., Cai, L., Udeani, G. O., Slowing, K. V., Thomas, C. F., Beecher, C. W., Fong, H. H., Farnsworth, N. R., Kinghorn, A. D., Mehta, R. G., Moon, R. C., Pezzuto, J. M. Cancer chemopreventive activity of resveratrol, a natural product derived from grapes. *Science* 1997, 275 (5297), 218–20.

25. Kondratyuk, T. P., Park, E. J., Marler, L. E., Ahn, S., Yuan, Y., Choi, Y., Yu, R., van Breemen, R. B., Sun, B., Hoshino, J., Cushman, M., Jermihov, K. C., Mesecar, A. D., Grubbs, C. J., Pezzuto, J. M. Resveratrol derivatives as promising chemopreventive agents with improved potency and selectivity. *Mol Nutr Food Res* 2011, 55 (8), 1249–65.

26. Pezzuto, J. M. The phenomenon of resveratrol: Redefining the virtues of promiscuity. *Ann NY Acad Sci* 2011, 1215, 123–30.

27. Szabo, E., Mao, J. T., Lam, S., Reid, M. E., Keith, R. L. Chemoprevention of lung cancer: Diagnosis and management of lung cancer, 3rd ed: American College of Chest Physicians evidence-based clinical practice guidelines. *Chest* 2013, 143 (5 Suppl), e40S–60S.

28. Carroll, C., Cooper, K., Papaioannou, D., Hind, D., Pilgrim, H., Tappenden, P. Supplemental calcium in the chemoprevention of colorectal cancer: A systematic review and meta-analysis. *Clin Ther* 2010, 32 (5), 789–803.

29. West, N. J., Clark, S. K., Phillips, R. K., Hutchinson, J. M., Leicester, R. J., Belluzzi, A., Hull, M. A. Eicosapentaenoic acid reduces rectal polyp number and size in familial adenomatous polyposis. *Gut* 2010, 59 (7), 918–25.

30. Stoner, G. D., Wang, L. S. Chemoprevention of esophageal squamous cell carcinoma with berries. *Top Curr Chem* 2013, 329, 1–20.

31. Patterson, S. L., Colbert Maresso, K., Hawk, E. Cancer chemoprevention: Successes and failures. *Clin Chem* 2013, 59 (1), 94–101.

32. Park, E. J., Pezzuto, J. M. Antioxidant marine products in cancer chemoprevention. *Antioxid Redox Signal* 2013, 19 (2), 115–38.

33. Pezzuto, J. M., Park, E. J., Park, E. J. Autoxidation and antioxidants. In *Encyclopedia of Pharmaceutical Technology*. 3rd ed. Boca Raton, FL: Taylor & Francis, 2013, pp. 139–54.

34. Cuendet, M., Oteham, C. P., Moon, R. C., Pezzuto, J. M. Quinone reductase induction as a biomarker for cancer chemoprevention. *J Nat Prod* 2006, 69 (3), 460–63.

35. Schottenfeld, D., Beebe-Dimmer, J. Chronic inflammation: A common and important factor in the pathogenesis of neoplasia. *CA Cancer J Clin* 2006, 56 (2), 69–83.

36. Wang, R., Guo, L., Wang, P., Yang, W., Lu, Y., Huang, Z., Tang, C. Chemoprevention of cancers in gastrointestinal tract with cyclooxygenase 2 inhibitors. *Curr Pharm Des* 2013, 19 (1), 115–25.

37. Bishayee, A., Thoppil, R. J., Waghray, A., Kruse, J. A., Novotny, N. A., Darvesh, A. S. Dietary phytochemicals in the chemoprevention and treatment of hepatocellular carcinoma: *In vivo* evidence, molecular targets, and clinical relevance. *Curr Cancer Drug Targets* 2012, 12 (9), 1191–232.

38. Dunn, B. K., Cazzaniga, M., DeCensi, A. Exemestane: One part of the chemopreventive spectrum for ER-positive breast cancer. *Breast* 2013, 22 (3), 225–37.

39. Howe, L. R., Brown, P. H. Targeting the HER/EGFR/ErbB family to prevent breast cancer. *Cancer Prev Res* (Phila) 2011, 4 (8), 1149–57.

40. Rae, J. M., Scheys, J. O., Clark, K. M., Chadwick, R. B., Kiefer, M. C., Lippman, M. E. EGFR and EGFRvIII expression in primary breast cancer and cell lines. *Breast Cancer Res Treat* 2004, 87 (1), 87–95.

41. Scholl, S., Beuzeboc, P., Pouillart, P. Targeting HER2 in other tumor types. *Ann Oncol* 2001, 12 (Suppl 1), S81–87.

42. Hou, D. X., Kumamoto, T. Flavonoids as protein kinase inhibitors for cancer chemoprevention: Direct binding and molecular modeling. *Antioxid Redox Signal* 2010, 13 (5), 691–719.

43. Crowell, J. A., Steele, V. E., Fay, J. R. Targeting the AKT protein kinase for cancer chemoprevention. *Mol Cancer Ther* 2007, 6 (8), 2139–48.

44. Buitrago-Molina, L. E., Vogel, A. mTor as a potential target for the prevention and treatment of hepatocellular carcinoma. *Curr Cancer Drug Targets* 2012, 12 (9), 1045–61.

45. Temraz, S., Mukherji, D., Shamseddine, A. Potential targets for colorectal cancer prevention. *Int J Mol Sci* 2013, 14 (9), 17279–303.

46. Sciacca, L., Vigneri, R., Tumminia, A., Frasca, F., Squatrito, S., Frittitta, L., Vigneri, P. Clinical and molecular mechanisms favoring cancer initiation and progression in diabetic patients. *Nutr Metab Cardiovasc Dis* 2013, 13 (9), 808–15.

47. Khandekar, M. J., Cohen, P., Spiegelman, B. M. Molecular mechanisms of cancer development in obesity. *Nat Rev Cancer* 2011, 11 (12), 886–95.

48. Moore, T., Beltran, L., Carbajal, S., Hursting, S. D., DiGiovanni, J. Energy balance modulates mouse skin tumor promotion through altered IGF-1R and EGFR crosstalk. *Cancer Prev Res* (Phila) 2012, 5 (10), 1236–46.

49. Xie, L., Wang, W. Weight control and cancer preventive mechanisms: Role of insulin growth factor-1-mediated signaling pathways. *Exp Biol Med* (Maywood) 2013, 238 (2), 127–32.

50. Gills, J. J., Kosmeder, J., Moon, R. C., Lantvit, D. D., Pezzuto, J. M. Effect of deguelin on UVB-induced skin carcinogenesis. *J Chemother* 2005, 17 (3), 297–301.

51. Tian, J., Lambertz, I., Berton, T. R., Rundhaug, J. E., Kiguchi, K., Shirley, S. H., Digiovanni, J., Conti, C. J., Fischer, S. M., Fuchs-Young, R. Transgenic insulin-like growth factor-1 stimulates activation of COX-2 signaling in mammary glands. *Mol Carcinog* 2012, 51 (12), 973–83.

52. Sithranga Boopathy, N., Kathiresan, K. Anticancer drugs from marine flora: An overview. *J Oncol* 2010, Article, 214186.

53. Dillehay, T. D., Ramírez, C., Pino, M., Collins, M. B., Rossen, J., Pino-Navarro, J. D. Monte Verde: Seaweed, food, medicine, and the peopling of South America. *Science* 2008, 320 (5877), 784–86.
54. Lev, E. Traditional healing with animals (zootherapy): Medieval to present-day Levantine practice. *J Ethnopharmacol* 2003, 85 (1), 107–18.
55. Appeltans, W., Ahyong, S. T., Anderson, G., Angel, M. V., Artois, T., Bailly, N., Bamber, R., Barber, A., Bartsch, I., Berta, A., Błażewicz-Paszkowycz, M., Bock, P., Boxshall, G., Boyko, C. B., Brandão, S. N., Bray, R. A., Bruce, N. L., Cairns, S. D., Chan, T. Y., Cheng, L., Collins, A. G., Cribb, T., Curini-Galletti, M., Dahdouh-Guebas, F., Davie, P. J., Dawson, M. N., De Clerck, O., Decock, W., De Grave, S., de Voogd, N. J., Domning, D. P., Emig, C. C., Erséus, C., Eschmeyer, W., Fauchald, K., Fautin, D. G., Feist, S. W., Fransen, C. H., Furuya, H., Garcia-Alvarez, O., Gerken, S., Gibson, D., Gittenberger, A., Gofas, S., Gómez-Daglio, L., Gordon, D. P., Guiry, M. D., Hernandez, F., Hoeksema, B. W., Hopcroft, R. R., Jaume, D., Kirk, P., Koedam, N., Koenemann, S., Kolb, J. B., Kristensen, R. M., Kroh, A., Lambert, G., Lazarus, D. B., Lemaitre, R., Longshaw, M., Lowry, J., Macpherson, E., Madin, L. P., Mah, C., Mapstone, G., McLaughlin, P. A., Mees, J., Meland, K., Messing, C. G., Mills, C. E., Molodtsova, T. N., Mooi, R., Neuhaus, B., Ng, P. K., Nielsen, C., Norenburg, J., Opresko, D. M., Osawa, M., Paulay, G., Perrin, W., Pilger, J. F., Poore, G. C., Pugh, P., Read, G. B., Reimer, J. D., Rius, M., Rocha, R. M., Saiz-Salinas, J. I., Scarabino, V., Schierwater, B., Schmidt-Rhaesa, A., Schnabel, K. E., Schotte, M., Schuchert, P., Schwabe, E., Segers, H., Self-Sullivan, C., Shenkar, N., Siegel, V., Sterrer, W., Stöhr, S., Swalla, B., Tasker, M. L., Thuesen, E. V., Timm, T., Todaro, M. A., Turon, X., Tyler, S., Uetz, P., van der Land, J., Vanhoorne, B., van Ofwegen, L. P., van Soest, R. W., Vanaverbeke, J., Walker-Smith, G., Walter, T. C., Warren, A., Williams, G. C., Wilson, S. P., Costello, M. J. The magnitude of global marine species diversity. *Curr Biol* 2012, 22 (23), 2189–202.
56. Murphy, B. T., Jensen, P. R., Fenical, W. *Handbook of Marine Natural Products*. New York: Springer, 2012, pp. 153–90.
57. Fenical, W., Jensen, P. R. Developing a new resource for drug discovery: Marine actinomycete bacteria. *Nat Chem Biol* 2006, 2 (12), 666–73.
58. Williams, P. G. Panning for chemical gold: Marine bacteria as a source of new therapeutics. *Trends Biotechnol* 2009, 27 (1), 45–52.
59. Gerwick, W. H., Moore, B. S. Lessons from the past and charting the future of marine natural products drug discovery and chemical biology. *Chem Biol* 2012, 19 (1), 85–98.
60. Bérdy, J. Bioactive microbial metabolites. *J Antibiot* (Tokyo) 2005, 58 (1), 1–26.
61. Kondratyuk, T. P., Park, E. J., Yu, R., van Breemen, R. B., Asolkar, R. N., Murphy, B. T., Fenical, W., Pezzuto, J. M. Novel marine phenazines as potential cancer chemopreventive and anti-inflammatory agents. *Mar Drugs* 2012, 10 (2), 451–64.
62. Conda-Sheridan, M., Marler, L., Park, E. J., Kondratyuk, T. P., Jermihov, K., Mesecar, A. D., Pezzuto, J. M., Asolkar, R. N., Fenical, W., Cushman, M. Potential chemopreventive agents based on the structure of the lead compound 2-bromo-1-hydroxyphenazine, isolated from *Streptomyces* species, strain CNS284. *J Med Chem* 2010, 53 (24), 8688–99.
63. Asolkar, R. N., Freel, K. C., Jensen, P. R., Fenical, W., Kondratyuk, T. P., Park, E. J., Pezzuto, J. M. Arenamides A–C, cytotoxic NFkappaB inhibitors from the marine actinomycete *Salinispora arenicola*. *J Nat Prod* 2009, 72 (3), 396–402.
64. Choi, I. K., Shin, H. J., Lee, H. S., Kwon, H. J. Streptochlorin, a marine natural product, inhibits NF-kappaB activation and suppresses angiogenesis *in vitro*. *J Microbiol Biotechnol* 2007, 17 (8), 1338–43.
65. Nam, S. J., Gaudêncio, S. P., Kauffman, C. A., Jensen, P. R., Kondratyuk, T. P., Marler, L. E., Pezzuto, J. M., Fenical, W. Fijiolides A and B, inhibitors of TNF-alpha-induced NFkappaB activation, from a marine-derived sediment bacterium of the genus *Nocardiopsis*. *J Nat Prod* 2010, 73 (6), 1080–86.
66. Ali, A., Khajuria, A., Sidiq, T., Kumar, A., Thakur, N. L., Naik, D., Vishwakarma, R. A. Modulation of LPS induced inflammatory response by Lawsonyl monocyclic terpene from the marine derived *Streptomyces* sp. *Immunol Lett* 2013, 150 (1–2), 79–86.
67. Park, E. J., Pezzuto, J. M., Jang, K. H., Nam, S. J., Bucarey, S. A., Fenical, W. Suppression of nitric oxide synthase by thienodolin in lipopolysaccharide-stimulated RAW 264.7 murine macrophage cells. *Nat Prod Commun* 2012, 7 (6), 789–94.
68. Choi, Y., Jermihov, K., Nam, S. J., Sturdy, M., Maloney, K., Qiu, X., Chadwick, L. R., Main, M., Chen, S. N., Mesecar, A. D., Farnsworth, N. R., Pauli, G. F., Fenical, W., Pezzuto, J. M., van Breemen, R. B., van Breemen, R. R. Screening natural products for inhibitors of quinone reductase-2 using ultrafiltration LC-MS. *Anal Chem* 2011, 83 (3), 1048–52.

69. Carlson, S., Marler, L., Nam, S. J., Santarsiero, B. D., Pezzuto, J. M., Murphy, B. T. Potential chemo-preventive activity of a new macrolide antibiotic from a marine-derived *Micromonospora* sp. *Mar Drugs* 2013, 11 (4), 1152–61.

70. Hughes, C. C., MacMillan, J. B., Gaudêncio, S. P., Fenical, W., La Clair, J. J. Ammosamides A and B target myosin. *Angew Chem Int Ed Engl* 2009, 48 (4), 728–32.

71. Pan, E., Oswald, N. W., Legako, A. G., Life, J. M., Posner, B. A., Macmillan, J. B. Precursor-directed generation of amidine containing ammosamide analogs: Ammosamides E–P. *Chem Sci* 2013, 4 (1), 482–88.

72. Sugiyama, Y., Hirota, A. New potent DPPH radical scavengers from a marine-derived actinomycete strain USF-TC31. *Biosci Biotechnol Biochem* 2009, 73 (12), 2731–34.

73. Schleissner, C., Pérez, M., Losada, A., Rodríguez, P., Crespo, C., Zúñiga, P., Fernández, R., Reyes, F., de la Calle, F. Antitumor actinopyranones produced by *Streptomyces albus* POR-04-15-053 isolated from a marine sediment. *J Nat Prod* 2011, 74 (7), 1590–96.

74. Choi, H., Pereira, A., Gerwick, W. *Handbook of Marine Natural Products.* New York: Springer, 2012, pp. 55–152.

75. Villa, F. A., Lieske, K., Gerwick, L. Selective MyD88-dependent pathway inhibition by the cyanobacte-rial natural product malyngamide F acetate. *Eur J Pharmacol* 2010, 629 (1–3), 140–46.

76. Takamatsu, S., Hodges, T. W., Rajbhandari, I., Gerwick, W. H., Hamann, M. T., Nagle, D. G. Marine natural products as novel antioxidant prototypes. *J Nat Prod* 2003, 66 (5), 605–8.

77. Stevenson, C. S., Capper, E. A., Roshak, A. K., Marquez, B., Eichman, C., Jackson, J. R., Mattern, M., Gerwick, W. H., Jacobs, R. S., Marshall, L. A. The identification and characterization of the marine natural product scytonemin as a novel antiproliferative pharmacophore. *J Pharmacol Exp Ther* 2002, 303 (2), 858–66.

78. Bugni, T. S., Ireland, C. M. Marine-derived fungi: A chemically and biologically diverse group of microorganisms. *Nat Prod Rep* 2004, 21 (1), 143–63.

79. Kohlmeyer, J., Volkmann-Kohlmeyer, B. Fungi from coral reefs: A commentary. *Mycol Res* 2003, 107 (Pt 4), 386–87.

80. Höller, U., Wright, A. D., Matthee, G. F., Konig, G. M., Draeger, S., Aust, H. J., Schulz, B. Fungi from marine sponges: Diversity, biological activity and secondary metabolites. *Mycol Res* 2000, 104 (11), 1354–65.

81. Zain, M. E. Impact of mycotoxins on humans and animals. *J Saudi Chem Soc* 2011, 15 (2), 129–44.

82. Scharf, D. H., Heinekamp, T., Remme, N., Hortschansky, P., Brakhage, A. A., Hertweck, C. Biosynthesis and function of gliotoxin in *Aspergillus fumigatus*. *Appl Microbiol Biotechnol* 2012, 93 (2), 467–72.

83. Kosalec, I., Ramić, S., Jelić, D., Antolović, R., Pepeljnjak, S., Kopjar, N. Assessment of tryptophol geno-toxicity in four cell lines *in vitro*: A pilot study with alkaline comet assay. *Arh Hig Rada Toksikol* 2011, 62 (1), 41–49.

84. Hirota, M., Mehta, A., Yoneyama, K., Kitabatake, N., Menta Alka, B. A major decomposition product, citrinin H2, from citrinin on heating with moisture. *Biosci Biotechnol Biochem* 2002, 66 (1), 206–10.

85. Lee, Y. M., Kim, M. J., Li, H., Zhang, P., Bao, B., Lee, K. J., Jung, J. H. Marine-derived *Aspergillus* species as a source of bioactive secondary metabolites. *Mar Biotechnol* (NY) 2013, 15 (5), 499–519.

86. Proksch, P., Ebel, R., Edrada, R., Riebe, F., Hongbing, L., Diesel, A., Bayer, M., Li, X., Lin, W. H., Grebenyuk, V., Müller, W. E. G., Draeger, S., Zuccaro, A., Schulz, B. Sponge-associated fungi and their bioactive compounds: The *Suberites* case. *Bot Mar* 2008, 51 (3), 209–18.

87. Mora, C., Tittensor, D. P., Adl, S., Simpson, A. G., Worm, B. How many species are there on earth and in the ocean? *PLoS Biol* 2011, 9 (8), e1001127.

88. Blackwell, M. The fungi: 1, 2, 3 … 5.1 million species? *Am J Bot* 2011, 98 (3), 426–38.

89. Hu, G. P., Yuan, J., Sun, L., She, Z. G., Wu, J. H., Lan, X. J., Zhu, X., Lin, Y. C., Chen, S. P. Statistical research on marine natural products based on data obtained between 1985 and 2008. *Mar Drugs* 2011, 9 (4), 514–25.

90. Ebel, R. Terpenes from marine-derived fungi. *Mar Drugs* 2010, 8 (8), 2340–68.

91. Son, B. W., Choi, J. S., Kim, J. C., Nam, K. W., Kim, D. S., Chung, H. Y., Kang, J. S., Choi, H. D. Parasitenone, a new epoxycyclohexenone related to gabosine from the marine-derived fungus *Aspergillus parasiticus*. *J Nat Prod* 2002, 65 (5), 794–95.

92. Yun, K., Kondempudi, C. M., Choi, H. D., Kang, J. S., Son, B. W. Microbial mannosidation of bioactive chlorogentisyl alcohol by the marine-derived fungus *Chrysosporium synchronum*. *Chem Pharm Bull* (Tokyo) 2011, 59 (4), 499–501.

93. Leutou, A. S., Yun, K., Choi, H. D., Kang, J. S., Son, B. W. New production of 5-bromotoluhydroquinone and 4-O-methyltoluhydroquinone from the marine-derived fungus *Dothideomycete* sp. *J Microbiol Biotechnol* 2012, 22 (1), 80–83.

94. Chen, L., Fang, Y., Zhu, T., Gu, Q., Zhu, W. Gentisyl alcohol derivatives from the marine-derived fungus *Penicillium terrestre*. *J Nat Prod* 2008, 71 (1), 66–70.

95. Leutou, A. S., Yun, K., Kang, J. S., Son, B. W. Induced production of methyl bromodihydroxyphenyl acetates by the marine-derived fungus *Aspergillus* sp. *Chem Pharm Bull* (Tokyo) 2013, 61 (4), 483–85.

96. Li, D. L., Li, X. M., Wang, B. G. Natural anthraquinone derivatives from a marine mangrove plant-derived endophytic fungus *Eurotium rubrum*: Structural elucidation and DPPH radical scavenging activity. *J Microbiol Biotechnol* 2009, 19 (7), 675–80.

97. Mehta, G., Pujar, S. R., Ramesh, S. S., Islam, K. Enantioselective total synthesis of polyoxygenated cyclohexanoids: (+)-Streptol, *ent*-RKTS-33 and putative '(+)-parasitenone'. Identity of parasitenone with (+)-epoxydon. *Tetrahedron Lett* 2005, 46 (19), 3373–76.

98. Li, Y., Li, X., Son, B. W. Antibacterial and radical scavenging epoxycyclohexenones and aromatic polyols from a marine isolate of the fungus *Aspergillus*. *Nat Prod Sci* 2005, 11 (3), 136–38.

99. Cole, R. J., Moore, J. H., Davis, N. D., Kirksey, J. W., Diener, U. L. 4-Hydroxymellein: New metabolite of *Aspergillus ochraceus*. *J Agr Food Chem* 1971, 19 (5), 909–11.

100. Feng, Z., Nenkep, V., Yun, K., Zhang, D., Choi, H. D., Kang, J. S., Son, B. W. Biotransformation of bioactive (–)-mellein by a marine isolate of bacterium *Stappia* sp. *J Microbiol Biotechnol* 2010, 20 (6), 985–87.

101. Yun, K., Feng, Z., Choi, H. D., Kang, J. S., Son, B. W. New production of (*R*)-(–)-5-bromomellein, a dihydroisocoumarin derivative from the marine-derived fungus *Aspergillus ochraceus*. *Chem Nat Compd* 2013, 49 (1).

102. Cui, C. M., Li, X. M., Li, C. S., Sun, H. F., Gao, S. S., Wang, B. G. Benzodiazepine alkaloids from marine-derived endophytic fungus *Aspergillus ochraceus*. *Helv Chim Acta* 2009, 92 (7), 1366–70.

103. Borthwick, A. D. 2,5-Diketopiperazines: Synthesis, reactions, medicinal chemistry, and bioactive natural products. *Chem Rev* 2012, 112 (7), 3641–716.

104. Li, Y., Li, X., Kim, S. K., Kang, J. S., Choi, H. D., Rho, J. R., Son, B. W. Golmaenone, a new diketopiperazine alkaloid from the marine-derived fungus *Aspergillus* sp. *Chem Pharm Bull* (Tokyo) 2004, 52 (3), 375–76.

105. Marchelli, R., Dossena, A., Pochini, A., Dradi, E. The structures of five new didehydropeptides related to neoechinulin, isolated from *Aspergillus amstelodami*. *J Chem Soc Perkin 1* 1977, (7), 713–17.

106. Kuramochi, K. Synthetic and structure-activity relationship studies on bioactive natural products. *Biosci Biotechnol Biochem* 2013, 77 (3), 446–54.

107. Dewapriya, P., Li, Y. X., Himaya, S. W., Pangestuti, R., Kim, S. K. Neoechinulin A suppresses amyloid-β oligomer-induced microglia activation and thereby protects PC-12 cells from inflammation-mediated toxicity. *Neurotoxicology* 2013, 35, 30–40.

108. Zhou, L. N., Zhu, T. J., Cai, S. X., Gu, Q. Q., Li, D. H. Three new indole-containing diketopiperazine alkaloids from a deep-ocean sediment derived fungus *Penicillium griseofulvum*. *Helv Chim Acta* 2010, 93, 1758–63.

109. Kong, X., Cai, S., Zhu, T., Gu, Q., Li, D., Luan, Y. Secondary metabolites of a deep sea derived fungus *Aspergillus versicolor* CXCTD-06-6a and their bioactivity. *J Ocean Univ China* 2014, 13 (4), 691–95.

110. Li, X., Kim, S. K., Nam, K. W., Kang, J. S., Choi, H. D., Son, B. W. A new antibacterial dioxopiperazine alkaloid related to gliotoxin from a marine isolate of the fungus *Pseudallescheria*. *J Antibiot* (Tokyo) 2006, 59 (4), 248–50.

111. Sugiyama, Y., Ito, Y., Suzuki, M., Hirota, A. Indole derivatives from a marine sponge-derived yeast as DPPH radical scavengers. *J Nat Prod* 2009, 72 (11), 2069–71.

112. Priestap, H. A. New naphthopyrones from *Aspergillus fonsecaeus*. *Tetrahedron* 1984, 40, 3617–24.

113. Zhang, Y., Li, X. M., Wang, B. G. Nigerasperones A–C, new monomeric and dimeric naphtho-gamma-pyrones from a marine alga-derived endophytic fungus *Aspergillus niger* EN-13. *J Antibiot* (Tokyo) 2007, 60 (3), 204–10.

114. Somma, S., Perrone, G., Logrieco, A. F. Diversity of black *Aspergilli* and mycotoxin risks in grape, wine and dried vine fruits. *Phytopathol Mediterr* 2012, 51 (1), 131–47.

115. Kawahara, T., Takagi, M., Shin-ya, K. JBIR-124: A novel antioxidative agent from a marine sponge-derived fungus *Penicillium citrinum* SpI080624G1f01. *J Antibiot* (Tokyo) 2012, 65 (1), 45–47.

116. Zhuang, Y., Teng, X., Wang, Y., Liu, P., Wang, H., Li, J., Li, G., Zhu, W. Cyclopeptides and polyketides from coral-associated fungus, *Aspergillus versicolor* LCJ-5-4. *Tetrahedron* 2011, 67 (37), 7085–89.

117. Trisuwan, K., Rukachaisirikul, V., Kaewpet, M., Phongpaichit, S., Hutadilok-Towatana, N., Preedanon, S., Sakayaroj, J. Sesquiterpene and xanthone derivatives from the sea fan-derived fungus *Aspergillus sydowii* PSU-F154. *J Nat Prod* 2011, 74 (7), 1663–67.

118. Wang, Y. N., Tian, L., Hua, H. M., Lu, X., Sun, S., Wu, H. H., Pei, Y. H. Two new compounds from the broth of the marine fungus *Penicillium griseofulvum* Y19-07. *J Asian Nat Prod Res* 2009, 11 (11), 912–17.

119. Zhang, D., Li, X., Kang, J. S., Choi, H. D., Jung, J. H., Son, B. W. Redoxcitrinin, a biogenetic precursor of citrinin from marine isolate of fungus *Penicillium* sp. *J Microbiol Biotechnol* 2007, 17 (5), 865–67.

120. Yang, G., Yun, K., Nenkep, V. N., Choi, H. D., Kang, J. S., Son, B. W. Induced production of halogenated diphenyl ethers from the marine-derived fungus *Penicillium chrysogenum*. *Chem Biodivers* 2010, 7 (11), 2766–70.

121. Son, B. W., Kim, J. C., Choi, H. D., Kang, J. S. A radical scavenging farnesylhydroquinone from a marine-derived fungus *Penicillium* sp. *Arch Pharm Res* 2002, 25 (1), 77–79.

122. Abdel-Lateff, A., König, G. M., Fisch, K. M., Höller, U., Jones, P. G., Wright, A. D. New antioxidant hydroquinone derivatives from the algicolous marine fungus *Acremonium* sp. *J Nat Prod* 2002, 65 (11), 1605–11.

123. Abdel-Lateff, A., Fisch, K. M., Wright, A. D., König, G. M. A new antioxidant isobenzofuranone derivative from the algicolous marine fungus *Epicoccum* sp. *Planta Med* 2003, 69 (9), 831–34.

124. Wang, S., Li, X. M., Teuscher, F., Li, D. L., Diesel, A., Ebel, R., Proksch, P., Wang, B. G. Chaetopyranin, a benzaldehyde derivative, and other related metabolites from *Chaetomium globosum*, an endophytic fungus derived from the marine red alga *Polysiphonia urceolata*. *J Nat Prod* 2006, 69 (11), 1622–25.

125. Li, Y., Li, X., Lee, U., Kang, J. S., Choi, H. D., Sona, B. W. A new radical scavenging anthracene glycoside, asperflavin ribofuranoside, and polyketides from a marine isolate of the fungus microsporum. *Chem Pharm Bull* (Tokyo) 2006, 54 (6), 882–83.

126. Abdel-Lateff, A., Klemke, C., König, G. M., Wright, A. D. Two new xanthone derivatives from the algicolous marine fungus *Wardomyces anomalus*. *J Nat Prod* 2003, 66 (5), 706–8.

127. Sureram, S., Wiyakrutta, S., Ngamrojanavanich, N., Mahidol, C., Ruchirawat, S., Kittakoop, P. Depsidones, aromatase inhibitors and radical scavenging agents from the marine-derived fungus *Aspergillus unguis* CRI282-03. *Planta Med* 2012, 78 (6), 582–88.

128. Chen, W. L., Qian, Y., Meng, W. F., Pang, J. Y., Lin, Y. C., Guan, Y. Y., Chen, S. P., Liu, J., Pei, Z., Wang, G. L. A novel marine compound xyloketal B protects against oxidized LDL-induced cell injury *in vitro*. *Biochem Pharmacol* 2009, 78 (8), 941–50.

129. Li, S., Shen, C., Guo, W., Zhang, X., Liu, S., Liang, F., Xu, Z., Pei, Z., Song, H., Qiu, L., Lin, Y., Pang, J. Synthesis and neuroprotective action of xyloketal derivatives in Parkinson's disease models. *Mar Drugs* 2013, 11 (12), 5159–89.

130. Saitoh, T., Suzuki, E., Takasugi, A., Obata, R., Ishikawa, Y., Umezawa, K., Nishiyama, S. Efficient synthesis of (+/−)-parasitenone, a novel inhibitor of NF-kappaB. *Bioorg Med Chem Lett* 2009, 19 (18), 5383–86.

131. Wu, G., Ma, H., Zhu, T., Li, J., Gu, Q., Li, D. Penilactones A and B, two novel polyketides from Antarctic deep-sea derived fungus *Penicillium crustosum* PRB-2. *Tetrahedron* 2012, 68 (47), 9745–49.

132. Wu, Q. X., Crews, M. S., Draskovic, M., Sohn, J., Johnson, T. A., Tenney, K., Valeriote, F. A., Yao, X. J., Bjeldanes, L. F., Crews, P. Azonazine, a novel dipeptide from a Hawaiian marine sediment-derived fungus, *Aspergillus insulicola*. *Org Lett* 2010, 12 (20), 4458–61.

133. Lee, D. S., Jeong, G. S., Li, B., Lee, S. U., Oh, H., Kim, Y. C. Asperlin from the marine-derived fungus *Aspergillus* sp. SF-5044 exerts anti-inflammatory effects through heme oxygenase-1 expression in murine macrophages. *J Pharmacol Sci* 2011, 116 (3), 283–95.

134. Lee, D. S., Jang, J. H., Ko, W., Kim, K. S., Sohn, J. H., Kang, M. S., Ahn, J. S., Kim, Y. C., Oh, H. PTP1B inhibitory and anti-inflammatory effects of secondary metabolites isolated from the marine-derived fungus *Penicillium* sp. JF-55. *Mar Drugs* 2013, 11 (4), 1409–26.

135. Li, J. L., Zhang, P., Lee, Y. M., Hong, J., Yoo, E. S., Bae, K. S., Jung, J. H. Oxygenated hexylitaconates from a marine sponge-derived fungus *Penicillium* sp. *Chem Pharm Bull* (Tokyo) 2011, 59 (1), 120–23.

136. Yang, X., Kang, M. C., Li, Y., Kim, E. A., Kang, S. M., Jeon, Y. J. Anti-inflammatory activity of questinol isolated from marine-derived fungus *Eurotium amstelodami* in lipopolysaccharide-stimulated RAW 264.7 macrophages. *J Microbiol Biotechnol* 2014, 24 (10), 1346–53.

137. Toledo, T. R., Dejani, N. N., Monnazzi, L. G., Kossuga, M. H., Berlinck, R. G., Sette, L. D., Medeiros, A. I. Potent anti-inflammatory activity of pyrenocine A isolated from the marine-derived fungus *Penicillium paxilli* Ma(G)K. *Mediators Inflamm* 2014, 2014, 767061.

138. Song, Y., Dou, H., Gong, W., Liu, X., Yu, Z., Li, E., Tan, R., Hou, Y. Bis-N-norgliovictin, a small-molecule compound from marine fungus, inhibits LPS-induced inflammation in macrophages and improves survival in sepsis. *Eur J Pharmacol* 2013, 705 (1–3), 49–60.

139. Gamal-Eldeen, A. M., Abdel-Lateff, A., Okino, T. Modulation of carcinogen metabolizing enzymes by chromanone A: A new chromone derivative from algicolous marine fungus *Penicillium* sp. *Environ Toxicol Pharmacol* 2009, 28 (3), 317–22.

140. Krick, A., Kehraus, S., Gerhäuser, C., Klimo, K., Nieger, M., Maier, A., Fiebig, H. H., Atodiresei, I., Raabe, G., Fleischhauer, J., König, G. M. Potential cancer chemopreventive *in vitro* activities of monomeric xanthone derivatives from the marine algicolous fungus *Monodictys putredinis*. *J Nat Prod* 2007, 70 (3), 353–60.

141. Pontius, A., Krick, A., Mesry, R., Kehraus, S., Foegen, S. E., Müller, M., Klimo, K., Gerhäuser, C., König, G. M. Monodictyochromes A and B, dimeric xanthone derivatives from the marine algicolous fungus *Monodictys putredinis*. *J Nat Prod* 2008, 71 (11), 1793–99.

142. Pontius, A., Krick, A., Kehraus, S., Foegen, S. E., Müller, M., Klimo, K., Gerhäuser, C., König, G. M. Noduliprevenone: A novel heterodimeric chromanone with cancer chemopreventive potential. *Chemistry* 2008, 14 (32), 9860–63.

143. Qian, Z. J., Zhang, C., Li, Y. X., Je, J. Y., Kim, S. K., Jung, W. K. Protective effects of emodin and chrysophanol isolated from marine fungus *Aspergillus* sp. on ethanol-induced toxicity in HepG2/ CYP2E1 cells. *Evid Based Complement Alternat Med* 2011, 2011, 452621.

144. Kim, S. J., Kim, M. C., Lee, B. J., Park, D. H., Hong, S. H., Um, J. Y. Anti-inflammatory activity of chrysophanol through the suppression of NF-kappaB/caspase-1 activation *in vitro* and *in vivo*. *Molecules* 2010, 15 (9), 6436–51.

145. Liao, L., Lee, J. H., You, M., Choi, T. J., Park, W., Lee, S. K., Oh, D. C., Oh, K. B., Shin, J. Penicillipyrones A and B, meroterpenoids from a marine-derived *Penicillium* sp. fungus. *J Nat Prod* 2014, 77 (2), 406–10.

146. Jo, M. J., Bae, S. J., Son, B. W., Kim, C. Y., Kim, G. D. 3,4-Dihydroxyphenyl acetic acid and (+)-epoxydon isolated from marine algae-derived microorganisms induce down regulation of epidermal growth factor activated mitogenic signaling cascade in Hela cells. *Cancer Cell Int* 2013, 13 (1), 49.

147. Smetanina, O. F., Kalinovsky, A. I., Khudyakova, Y. V., Pivkin, M. V., Dmitrenok, P. S., Fedorov, S. N., Ji, H., Kwak, J. Y., Kuznetsova, T. A. Indole alkaloids produced by a marine fungus isolate of *Penicillium janthinellum* Biourge. *J Nat Prod* 2007, 70 (6), 906–9.

148. Ragan, M. A., Glombitza, K. W. Phlorotannins, brown algal polyphenols. *Prog Phycol Res* 1986, 4, 129–241.

149. Vo, T. S., Kim, S. K. Potential anti-HIV agents from marine resources: An overview. *Mar Drugs* 2010, 8 (12), 2871–92.

150. Kang, H. S., Chung, H. Y., Kim, J. Y., Son, B. W., Jung, H. A., Choi, J. S. Inhibitory phlorotannins from the edible brown alga *Ecklonia stolonifera* on total reactive oxygen species (ROS) generation. *Arch Pharm Res* 2004, 27 (2), 194–98.

151. Li, Y., Qian, Z. J., Ryu, B., Lee, S. H., Kim, M. M., Kim, S. K. Chemical components and its antioxidant properties *in vitro*: An edible marine brown alga, *Ecklonia cava*. *Bioorg Med Chem* 2009, 17 (5), 1963–73.

152. Kim, M. M., Kim, S. K. Effect of phloroglucinol on oxidative stress and inflammation. *Food Chem Toxicol* 2010, 48 (10), 2925–33.

153. Parys, S., Kehraus, S., Krick, A., Glombitza, K. W., Carmeli, S., Klimo, K., Gerhäuser, C., König, G. M. *In vitro* chemopreventive potential of fucophlorethols from the brown alga *Fucus vesiculosus* L. by antioxidant activity and inhibition of selected cytochrome P450 enzymes. *Phytochemistry* 2010, 71 (2–3), 221–29.

154. Kim, S. M., Kang, K., Jeon, J. S., Jho, E. H., Kim, C. Y., Nho, C. W., Um, B. H. Isolation of phlorotannins from *Eisenia bicyclis* and their hepatoprotective effect against oxidative stress induced by tert-butyl hyperoxide. *Appl Biochem Biotechnol* 2011, 165 (5–6), 1296–307.

155. Kim, A. R., Shin, T. S., Lee, M. S., Park, J. Y., Park, K. E., Yoon, N. Y., Kim, J. S., Choi, J. S., Jang, B. C., Byun, D. S., Park, N. K., Kim, H. R. Isolation and identification of phlorotannins from *Ecklonia stolonifera* with antioxidant and anti-inflammatory properties. *J Agric Food Chem* 2009, 57 (9), 3483–89.

156. Zou, Y., Qian, Z. J., Li, Y., Kim, M. M., Lee, S. H., Kim, S. K. Antioxidant effects of phlorotannins isolated from *Ishige okamurae* in free radical mediated oxidative systems. *J Agric Food Chem* 2008, 56 (16), 7001–9.

157. Heo, S. J., Hwang, J. Y., Choi, J. I., Lee, S. H., Park, P. J., Kang, D. H., Oh, C., Kim, D. W., Han, J. S., Jeon, Y. J., Kim, H. J., Choi, I. W. Protective effect of diphlorethohydroxycarmalol isolated from *Ishige okamurae* against high glucose-induced-oxidative stress in human umbilical vein endothelial cells. *Food Chem Toxicol* 2010, 48 (6), 1448–54.

158. Piao, M. J., Kang, K. A., Kim, K. C., Chae, S., Kim, G. O., Shin, T., Kim, H. S., Hyun, J. W. Diphlorethohydroxycarmalol attenuated cell damage against UVB radiation via enhancing antioxidant effects and absorbing UVB ray in human HaCaT keratinocytes. *Environ Toxicol Pharmacol* 2013, 36 (2), 680–88.

159. Kang, S. M., Lee, S. H., Heo, S. J., Kim, K. N., Jeon, Y. J. Evaluation of antioxidant properties of a new compound, pyrogallol-phloroglucinol-6,6′-bieckol isolated from brown algae, *Ecklonia cava*. *Nutr Res Pract* 2011, 5 (6), 495–502.

160. Kang, S. M., Heo, S. J., Kim, K. N., Lee, S. H., Jeon, Y. J. Isolation and identification of new compound, 2,7″-phloroglucinol-6,6′-bieckol from brown algae, *Ecklonia cava* and its antioxidant effect. *J Funct Foods* 2012, 4 (1), 158–66.

161. Choi, J. S., Park, H. J., Jung, H. A., Chung, H. Y., Jung, J. H., Choi, W. C. A cyclohexanonyl bromophenol from the red alga *Symphyocladia latiuscula*. *J Nat Prod* 2000, 63 (12), 1705–6.

162. Chung, H. Y., Choi, H. R., Park, H. J., Choi, J. S., Choi, W. C. Peroxynitrite scavenging and cytoprotective activity of 2,3,6-tribromo-4,5-dihydroxybenzyl methyl ether from the marine alga *Symphyocladia latiuscula*. *J Agric Food Chem* 2001, 49 (8), 3614–21.

163. Duan, X. J., Li, X. M., Wang, B. G. Highly brominated mono- and bis-phenols from the marine red alga *Symphyocladia latiuscula* with radical-scavenging activity. *J Nat Prod* 2007, 70 (7), 1210–13.

164. Xu, X., Yin, L., Gao, L., Gao, J., Chen, J., Li, J., Song, F. Two new bromophenols with radical scavenging activity from marine red alga *Symphyocladia latiuscula*. *Mar Drugs* 2013, 11 (3), 842–47.

165. Zhao, J., Fan, X., Wang, S., Li, S., Shang, S., Yang, Y., Xu, N., Lü, Y., Shi, J. Bromophenol derivatives from the red alga *Rhodomela confervoides*. *J Nat Prod* 2004, 67 (6), 1032–35.

166. Li, K., Li, X. M., Gloer, J. B., Wang, B. G. Isolation, characterization, and antioxidant activity of bromophenols of the marine red alga *Rhodomela confervoides*. *J Agric Food Chem* 2011, 59 (18), 9916–21.

167. Javan, A. J., Javan, M. J., Tehrani, Z. A. Theoretical investigation on antioxidant activity of bromophenols from the marine red alga *Rhodomela confervoides*: H-atom vs electron transfer mechanism. *J Agric Food Chem* 2013, 61 (7), 1534–41.

168. Li, K., Li, X. M., Gloer, J. B., Wang, B. G. New nitrogen-containing bromophenols from the marine red alga *Rhodomela confervoides* and their radical scavenging activity. *Food Chem* 2012, 135 (3), 868–72.

169. Li, K., Li, X. M., Ji, N. Y., Wang, B. G. Natural bromophenols from the marine red alga *Polysiphonia urceolata* (Rhodomelaceae): Structural elucidation and DPPH radical-scavenging activity. *Bioorg Med Chem* 2007, 15 (21), 6627–31.

170. Globitza, K. W., Stoffelen, H., Murawski, U., Bielaczek, J., Egge, H. [Antibiotics from algae. 9. Bromphenols from Rhodemelaceae (author's transl.)]. *Planta Med* 1974, 25 (2), 105–14.

171. Wang, W., Okada, Y., Shi, H., Wang, Y., Okuyama, T. Structures and aldose reductase inhibitory effects of bromophenols from the red alga *Symphyocladia latiuscula*. *J Nat Prod* 2005, 68 (4), 620–22.

172. Kurata, K., Amiya, T. Bis(2,3,6-tribromo-4,5-dihydroxybenzyl) ether from the red alga, *Symphyocladia latiuscula*. *Phytochemistry* 1980, 19 (1), 141–42.

173. Park, H. J., Chung, H. Y., Kim, J., Choi, J. S. Antioxidant activity of 2,3,6-tribromo-4,5-dihydroxybenzyl methyl ether from *Symphyocladia latiuscula*. *J Fish Sci Technol* 1999, 2 (1), 1–7.

174. Li, K., Li, X. M., Ji, N. Y., Gloer, J. B., Wang, B. G. Urceolatin, a structurally unique bromophenol from *Polysiphonia urceolata*. *Org Lett* 2008, 10 (7), 1429–32.

175. Li, K., Li, X. M., Ji, N. Y., Wang, B. G. Bromophenols from the marine red alga *Polysiphonia urceolata* with DPPH radical scavenging activity. *J Nat Prod* 2008, 71 (1), 28–30.

176. Iwasaki, S., Widjaja-Adhi, M. A. K., Koide, A., Kaga, T., Nakano, S., Beppu, F., Hosokawa, M., Miyashita, K. *In vivo* antioxidant activity of fucoxanthin on obese/diabetes KK-Ay mice. *Food Nutr Sci* 2012, 3, 1491–99.

177. Bharathiraja, K., Hari Babu, L., Vijayaprakash, S., Tamilselvan, P., Balasubramanian, M. P. Fucoxanthin, a marine carotenoid protects cadmium-induced oxidative renal dysfunction in rats. *Biomed Prev Nutr* 2013, 3 (3), 201–7.

178. Heo, S. J., Jeon, Y. J. Protective effect of fucoxanthin isolated from *Sargassum siliquastrum* on UV-B induced cell damage. *J Photochem Photobiol B* 2009, 95 (2), 101–7.

179. Cho, S. H., Cho, J. Y., Kang, S. E., Hong, Y. K., Ahn, D. H. Antioxidant activity of mojabanchromanol, a novel chromene, isolated from brown alga *Sargassum siliquastrum*. *J Environ Biol* 2008, 29 (4), 479–84.

180. Ham, Y. M., Kim, K. N., Lee, W. J., Lee, N. H., Hyun, C. G. Chemical constituents from *Sargassum micracanthum* and antioxidant activity. *Int J Pharmacol* 2010, 6 (2), 147–51.

181. Iwashima, M., Mori, J., Ting, X., Matsunaga, T., Hayashi, K., Shinoda, D., Saito, H., Sankawa, U., Hayashi, T. Antioxidant and antiviral activities of plastoquinones from the brown alga *Sargassum micracanthum*, and a new chromene derivative converted from the plastoquinones. *Biol Pharm Bull* 2005, 28 (2), 374–77.

182. Takahashi, S., Matsunaga, T., Hasegawa, C., Saito, H., Fujita, D., Kiuchi, F., Tsuda, Y. Martefragin A, a novel indole alkaloid isolated from red alga, inhibits lipid peroxidation. *Chem Pharm Bull* (Tokyo) 1998, 46 (10), 1527–29.

183. De, S., Devasagayam, T. P. A., Adhikari, S., Menon, V. P. Antioxidant properties of a novel marine analogue of dendrodoine. *BARC Newsl* 2006, 273, 123–33.

184. Yang, X., Kang, M.-C., Lee, K.-W., Kang, S.-M., Lee, W.-W., Jeon, Y.-J. Antioxidant activity and cell protective effect of loliolide isolated from *Sargassum ringgoldianum* subsp. *coreanum*. *Algae* 2011, 26 (2), 201–8.

185. Jang, K. H., Lee, B. H., Choi, B. W., Lee, H. S., Shin, J. Chromenes from the brown alga *Sargassum siliquastrum*. *J Nat Prod* 2005, 68 (5), 716–23.

186. Lee, E. H., Ham, J., Ahn, H. R., Kim, M. C., Kim, C. Y., Pan, C. H., Um, B. H., Jung, S. H. Inhibitory effects of the compounds isolated from *Sargassum yezoense* on α-glucosidase and oxidative stress. *Saengyak Hakhoechi* 2009, 40 (2), 150–54.

187. Jung, M., Jang, K. H., Kim, B., Lee, B. H., Choi, B. W., Oh, K. B., Shin, J. Meroditerpenoids from the brown alga *Sargassum siliquastrum*. *J Nat Prod* 2008, 71 (10), 1714–19.

188. Li, Y. X., Li, Y., Qian, Z. J., Kim, M. M., Kim, S. K. *In vitro* antioxidant activity of 5-HMF isolated from marine red alga *Laurencia undulata* in free-radical-mediated oxidative systems. *J Microbiol Biotechnol* 2009, 19 (11), 1319–27.

189. Lee, S., Lee, Y. S., Jung, S. H., Kang, S. S., Shin, K. H. Anti-oxidant activities of fucosterol from the marine algae *Pelvetia siliquosa*. *Arch Pharm Res* 2003, 26 (9), 719–22.

190. Tan, L. T., Williamson, R. T., Gerwick, W. H., Watts, K. S., McGough, K., Jacobs, R. cis,cis- and Trans,Trans-Ceratospongamide, new bioactive cyclic heptapeptides from the Indonesian red alga *Ceratodictyon spongiosum* and symbiotic sponge *Sigmadocia symbiotica*. *J Org Chem* 2000, 65 (2), 419–25.

191. Wiemer, D. F., Idler, D. D., Fenical, W. Vidalols A and B, new anti-inflammatory bromophenols from the Caribbean marine red alga *Vidalia obtusaloba*. *Experientia* 1991, 47 (8), 851–53.

192. Buckle, P. J., Baldo, B. A., Taylor, K. M. The anti-inflammatory activity of marine natural products: 6-n-Tridecylsalicylic acid, flexibilide and dendalone 3-hydroxybutyrate. *Agents Actions* 1980, 10 (4), 361–67.

193. Yang, E. J., Ham, Y. M., Lee, W. J., Lee, N. H., Hyun, C. G. Anti-inflammatory effects of apo-9′-fucoxanthinone from the brown alga, *Sargassum muticum*. *Daru* 2013, 21 (1), 62.

194. Yang, Y. I., Shin, H. C., Kim, S. H., Park, W. Y., Lee, K. T., Choi, J. H. 6,6′-Bieckol, isolated from marine alga *Ecklonia cava*, suppressed LPS-induced nitric oxide and PGE_2 production and inflammatory cytokine expression in macrophages: The inhibition of NFκB. *Int Immunopharmacol* 2012, 12 (3), 510–17.

195. Jung, H. A., Jin, S. E., Ahn, B. R., Lee, C. M., Choi, J. S. Anti-inflammatory activity of edible brown alga *Eisenia bicyclis* and its constituents fucosterol and phlorotannins in LPS-stimulated RAW264.7 macrophages. *Food Chem Toxicol* 2013, 59, 199–206.

196. Kim, A. R., Lee, M. S., Shin, T. S., Hua, H., Jang, B. C., Choi, J. S., Byun, D. S., Utsuki, T., Ingram, D., Kim, H. R. Phlorofucofuroeckol A inhibits the LPS-stimulated iNOS and COX-2 expressions in macrophages via inhibition of NF-κB, Akt, and p38 MAPK. *Toxicol In Vitro* 2011, 25 (8), 1789–95.

197. Yoon, W. J., Heo, S. J., Han, S. C., Lee, H. J., Kang, G. J., Kang, H. K., Hyun, J. W., Koh, Y. S., Yoo, E. S. Anti-inflammatory effect of sargachromanol G isolated from *Sargassum siliquastrum* in RAW 264.7 cells. *Arch Pharm Res* 2012, 35 (8), 1421–30.

198. Yang, E. J., Ham, Y. M., Yang, K. W., Lee, N. H., Hyun, C. G. Sargachromenol from *Sargassum micracanthum* inhibits the lipopolysaccharide-induced production of inflammatory mediators in RAW 264.7 macrophages. *Sci World J* 2013, Article, 712303.

199. Heo, S. J., Yoon, W. J., Kim, K. N., Oh, C., Choi, Y. U., Yoon, K. T., Kang, D. H., Qian, Z. J., Choi, I. W., Jung, W. K. Anti-inflammatory effect of fucoxanthin derivatives isolated from *Sargassum siliquastrum* in lipopolysaccharide-stimulated RAW 264.7 macrophage. *Food Chem Toxicol* 2012, 50 (9), 3336–42.

200. Awad, N. E. Biologically active steroid from the green alga *Ulva lactuca*. *Phytother Res* 2000, 14 (8), 641–43.

201. Wang, R., Paul, V. J., Luesch, H. Seaweed extracts and unsaturated fatty acid constituents from the green alga *Ulva lactuca* as activators of the cytoprotective Nrf2-ARE pathway. *Free Radic Biol Med* 2013, 57, 141–53.

202. Liu, C. L., Chiu, Y. T., Hu, M. L. Fucoxanthin enhances HO-1 and NQO1 expression in murine hepatic BNL CL.2 cells through activation of the Nrf2/ARE system partially by its pro-oxidant activity. *J Agric Food Chem* 2011, 59 (20), 11344–51.

203. Badal, S., Gallimore, W., Huang, G., Tzeng, T. R., Delgoda, R. Cytotoxic and potent CYP1 inhibitors from the marine algae *Cymopolia barbata*. *Org Med Chem Lett* 2012, 2 (1), 21.

204. Kim, K. C., Kang, K. A., Zhang, R., Piao, M. J., Kim, G. Y., Kang, M. Y., Lee, S. J., Lee, N. H., Surh, Y. J., Hyun, J. W. Up-regulation of Nrf2-mediated heme oxygenase-1 expression by eckol, a phlorotannin compound, through activation of Erk and PI3K/Akt. *Int J Biochem Cell B* 2010, 42 (2), 297–305.

205. Leal, M. C., Puga, J., Serôdio, J., Gomes, N. C., Calado, R. Trends in the discovery of new marine natural products from invertebrates over the last two decades: Where and what are we bioprospecting? *PLoS One* 2012, 7 (1), e30580.

206. Webster, N. S., Taylor, M. W. Marine sponges and their microbial symbionts: Love and other relationships. *Environ Microbiol* 2012, 14 (2), 335–46.

207. Radjasa, O. K., Vaske, Y. M., Navarro, G., Vervoort, H. C., Tenney, K., Linington, R. G., Crews, P. Highlights of marine invertebrate-derived biosynthetic products: Their biomedical potential and possible production by microbial associants. *Bioorg Med Chem* 2011, 19 (22), 6658–74.

208. Tziveleka, L. A., Kourounakis, A. P., Kourounakis, P. N., Roussis, V., Vagias, C. Antioxidant potential of natural and synthesised polyprenylated hydroquinones. *Bioorg Med Chem* 2002, 10 (4), 935–39.

209. Larghi, E. L., Bohn, M. L., Kaufman, T. S. Aaptamine and related products: Their isolation, chemical syntheses, and biological activity. *Tetrahedron* 2009, 65 (22), 4257–82.

210. Grube, A., Assmann, M., Lichte, E., Sasse, F., Pawlik, J. R., Kock, M. Bioactive metabolites from the Caribbean sponge *Aka coralliphagum*. *J Nat Prod* 2007, 70 (4), 504–9.

211. Pejin, B., Iodice, C., Tommonaro, G., De Rosa, S. Synthesis and biological activities of thio-avarol derivatives. *J Nat Prod* 2008, 71 (11), 1850–53.

212. Liu, Y., Ji, H., Dong, J., Zhang, S., Lee, K. J., Matthew, S. Antioxidant alkaloid from the South China Sea marine sponge *Iotrochota* sp. *Z Naturforsch C* 2008, 63 (9–10), 636–38.

213. Abou-Shoer, M. I., Shaala, L. A., Youssef, D. T., Badr, J. M., Habib, A. A. Bioactive brominated metabolites from the red sea sponge *Suberea mollis*. *J Nat Prod* 2008, 71 (8), 1464–67.

214. Shaala, L. A., Bamane, F. H., Badr, J. M., Youssef, D. T. Brominated arginine-derived alkaloids from the red sea sponge *Suberea mollis*. *J Nat Prod* 2011, 74 (6), 1517–20.

215. Utkina, N. K., Denisenko, V. A., Krasokhin, V. B. Sesquiterpenoid aminoquinones from the marine sponge *Dysidea* sp. *J Nat Prod* 2010, 73 (4), 788–91.

216. Trianto, A., Hermawan, I., de Voogd, N. J., Tanaka, J. Halioxepine, a new meroditerpene from an Indonesian sponge *Haliclona* sp. *Chem Pharm Bull* (Tokyo) 2011, 59 (10), 1311–13.

217. Aqil, F., Zahin, M., El Sayed, K. A., Ahmad, I., Orabi, K. Y., Arif, J. M. Antimicrobial, antioxidant, and antimutagenic activities of selected marine natural products and tobacco cembranoids. *Drug Chem Toxicol* 2011, 34 (2), 167–79.

218. Shubina, L. K., Kalinovsky, A. I., Makarieva, T. N., Fedorov, S. N., Dyshlovoy, S. A., Dmitrenok, P. S., Kapustina, I. I., Mollo, E., Utkina, N. K., Krasokhin, V. B., Denisenko, V. A., Stonik, V. A. New meroterpenoids from the marine sponge *Aka coralliphaga*. *Nat Prod Commun* 2012, 7 (4), 487–90.

219. Utkina, N. K., Denisenko, V. A. Ophiuroidine, the first indolo[2,1-b]quinazoline alkaloid from the Caribbean brittle star *Ophiocoma riisei*. *Tetrahedron Lett* 2007, 48 (25), 4445–47.

220. Chovolou, Y., Ebada, S. S., Wätjen, W., Proksch, P. Identification of angular naphthopyrones from the Philippine echinoderm *Comanthus* species as inhibitors of the NF-κB signaling pathway. *Eur J Pharmacol* 2011, 657 (1–3), 26–34.

221. Utkina, N. K., Pokhilo, N. D. Free radical scavenging activities of naturally occurring and synthetic analogues of sea urchin naphthazarin pigments. *Nat Prod Commun* 2012, 7 (7), 901–4.

222. Wang, G. F., Shang, Y. J., Jiao, B. H., Huang, C. G. Renierol from marine sponge *Haliclona* sp.: A natural inhibitor of xanthine oxidase with hypouricemic effects. *J Enzyme Inhib Med Chem* 2008, 23 (3), 406–10.

223. Johnson, M. K., Alexander, K. E., Lindquist, N., Loo, G. Potent antioxidant activity of a dithiocarbamate-related compound from a marine hydroid. *Biochem Pharmacol* 1999, 58 (8), 1313–19.

224. Shin, J., Seo, Y., Cho, K. W., Moon, S. S., Cho, Y. J. Euplexides A–E: Novel farnesylhydroquinone glycosides from the gorgonian *Euplexaura anastomosans*. *J Org Chem* 1999, 64 (6), 1853–58.

225. Correa, H., Valenzuela, A. L., Ospina, L. F., Duque, C. Anti-inflammatory effects of the gorgonian *Pseudopterogorgia elisabethae* collected at the Islands of Providencia and San Andrés (SW Caribbean). *J Inflamm* (Lond) 2009, 6, 5.

226. Zhang, G. W., Ma, X. Q., Su, J. Y., Zhang, K., Kurihara, H., Yao, X. S., Zeng, L. M. Two new bioactive sesquiterpenes from the soft coral *Sinularia* sp. *Nat Prod Res* 2006, 20 (7), 659–64.

227. Longeon, A., Copp, B. R., Quévrain, E., Roué, M., Kientz, B., Cresteil, T., Petek, S., Debitus, C., Bourguet-Kondracki, M. L. Bioactive indole derivatives from the South Pacific marine sponges *Rhopaloeides odorabile* and *Hyrtios* sp. *Mar Drugs* 2011, 9 (5), 879–88.

228. Nguyen, T. D., Nguyen, X. C., Longeon, A., Keryhuel, A., Le, M. H., Kim, Y. H., Chau, V. M., Bourguet-Kondracki, M. L. Antioxidant benzylidene 2-aminoimidazolones from the Mediterranean sponge *Phorbas topsenti*. *Tetrahedron* 2012, 68, 9256–59.

229. Esmat, A. Y., Said, M. M., Soliman, A. A., El-Masry, K. S., Badiea, E. A. Bioactive compounds, antioxidant potential, and hepatoprotective activity of sea cucumber (*Holothuria atra*) against thioacetamide intoxication in rats. *Nutrition* 2013, 29 (1), 258–67.

230. Leal, M. C., Madeira, C., Brandão, C. A., Puga, J., Calado, R. Bioprospecting of marine invertebrates for new natural products: A chemical and zoogeographical perspective. *Molecules* 2012, 17 (8), 9842–54.

231. Ebada, S. S., Lin, W., Proksch, P. Bioactive sesterterpenes and triterpenes from marine sponges: Occurrence and pharmacological significance. *Mar Drugs* 2010, 8 (2), 313–46.

232. Chao, C. H., Wen, Z. H., Wu, Y. C., Yeh, H. C., Sheu, J. H. Cytotoxic and anti-inflammatory cembranoids from the soft coral *Lobophytum crassum*. *J Nat Prod* 2008, 71 (11), 1819–24.

233. Cheng, S. Y., Wen, Z. H., Chiou, S. F., Hsu, C. H., Wang, S. K., Dai, C. F., Chiang, M. Y., Duh, C. Y. Durumolides A–E, anti-inflammatory and antibacterial cembranolides from the soft coral *Lobophytum durum*. *Tetrahedron* 2008, 64 (41), 9698–704.

234. Cheng, S. Y., Huang, Y. C., Wen, Z. H., Hsu, C. H., Wang, S. K., Dai, C. F., Duh, C. Y. New 19-oxygenated and 4-methylated steroids from the Formosan soft coral *Nephthea chabroli*. *Steroids* 2009, 74 (6), 543–47.

235. Cheng, S. Y., Wen, Z. H., Chiou, S. F., Tsai, C. W., Wang, S. K., Hsu, C. H., Dai, C. F., Chiang, M. Y., Wang, W. H., Duh, C. Y. Ceramide and cerebrosides from the octocoral *Sarcophyton ehrenbergi*. *J Nat Prod* 2009, 72 (3), 465–68.

236. Huang, Y. C., Wen, Z. H., Wang, S. K., Hsu, C. H., Duh, C. Y. New anti-inflammatory 4-methylated steroids from the Formosan soft coral *Nephthea chabroli*. *Steroids* 2008, 73 (11), 1181–86.

237. Wu, S. L., Su, J. H., Wen, Z. H., Hsu, C. H., Chen, B. W., Dai, C. F., Kuo, Y. H., Sheu, J. H. Simplexins A–I, eunicellin-based diterpenoids from the soft coral *Klyxum simplex*. *J Nat Prod* 2009, 72 (6), 994–1000.

238. Chen, B. W., Chao, C. H., Su, J. H., Wen, Z. H., Sung, P. J., Sheu, J. H. Anti-inflammatory eunicellin-based diterpenoids from the cultured soft coral *Klyxum simplex*. *Org Biomol Chem* 2010, 8 (10), 2363–66.

239. Cheng, S. Y., Lin, E. H., Huang, J. S., Wen, Z. H., Duh, C. Y. Ylangene-type and nardosinane-type sesquiterpenoids from the soft corals *Lemnalia flava* and *Paralemnalia thyrsoides*. *Chem Pharm Bull* (Tokyo) 2010, 58 (3), 381–85.

240. Lin, W. Y., Su, J. H., Lu, Y., Wen, Z. H., Dai, C. F., Kuo, Y. H., Sheu, J. H. Cytotoxic and anti-inflammatory cembranoids from the Dongsha atoll soft coral *Sarcophyton crassocaule*. *Bioorg Med Chem* 2010, 18 (5), 1936–41.

241. Chen, B. W., Chao, C. H., Su, J. H., Tsai, C. W., Wang, W. H., Wen, Z. H., Huang, C. Y., Sung, P. J., Wu, Y. C., Sheu, J. H. Klysimplexins I–T, eunicellin-based diterpenoids from the cultured soft coral *Klyxum simplex*. *Org Biomol Chem* 2011, 9 (3), 834–44.

242. Cheng, S. Y., Wen, Z. H., Wang, S. K., Chiou, S. F., Hsu, C. H., Dai, C. F., Duh, C. Y. Anti-inflammatory cembranolides from the soft coral *Lobophytum durum*. *Bioorg Med Chem* 2009, 17 (11), 3763–69.

243. Cheng, S. Y., Wen, Z. H., Wang, S. K., Chiou, S. F., Hsu, C. H., Dai, C. F., Chiang, M. Y., Duh, C. Y. Unprecedented hemiketal cembranolides with anti-inflammatory activity from the soft coral *Lobophytum durum*. *J Nat Prod* 2009, 72 (1), 152–55.

244. Lin, W. Y., Lu, Y., Su, J. H., Wen, Z. H., Dai, C. F., Kuo, Y. H., Sheu, J. H. Bioactive cembranoids from the Dongsha atoll soft coral *Sarcophyton crassocaule*. *Mar Drugs* 2011, 9 (6), 994–1006.

245. Lin, W. Y., Lu, Y., Chen, B. W., Huang, C. Y., Su, J. H., Wen, Z. H., Dai, C. F., Kuo, Y. H., Sheu, J. H. Sarcocrassocolides M–O, bioactive cembranoids from the Dongsha atoll soft coral *Sarcophyton crassocaule*. *Mar Drugs* 2012, 10 (3), 617–26.

246. Su, J. H., Wen, Z. H. Bioactive cembrane-based diterpenoids from the soft coral *Sinularia triangular*. *Mar Drugs* 2011, 9 (6), 944–51.

247. Tai, C. J., Su, J. H., Huang, C. Y., Huang, M. S., Wen, Z. H., Dai, C. F., Sheu, J. H. Cytotoxic and anti-inflammatory eunicellin-based diterpenoids from the soft coral *Cladiella krempfi*. *Mar Drugs* 2013, 11 (3), 788–99.

248. Hsu, F. J., Chen, B. W., Wen, Z. H., Huang, C. Y., Dai, C. F., Su, J. H., Wu, Y. C., Sheu, J. H. Klymollins A–H, bioactive eunicellin-based diterpenoids from the formosan soft coral *Klyxum molle*. *J Nat Prod* 2011, 74 (11), 2467–71.

249. Chen, B. W., Chang, S. M., Huang, C. Y., Chao, C. H., Su, J. H., Wen, Z. H., Hsu, C. H., Dai, C. F., Wu, Y. C., Sheu, J. H. Hirsutalins A–H, eunicellin-based diterpenoids from the soft coral *Cladiella hirsuta*. *J Nat Prod* 2010, 73 (11), 1785–91.

250. Cheng, S. Y., Huang, Y. C., Wen, Z. H., Chiou, S. F., Wang, S. K., Hsu, C. H., Dai, C. F., Duh, C. Y. Novel sesquiterpenes and norergosterol from the soft corals *Nephthea erecta* and *Nephthea chabroli*. *Tetrahedron Lett* 2009, 50 (7), 802–6.

251. Chang, C. H., Wen, Z. H., Wang, S. K., Duh, C. Y. Capnellenes from the Formosan soft coral *Capnella imbricata*. *J Nat Prod* 2008, 71 (4), 619–21.

252. Chao, C. H., Chou, K. J., Wen, Z. H., Wang, G. H., Wu, Y. C., Dai, C. F., Sheu, J. H. Paraminabeolides A–F, cytotoxic and anti-inflammatory marine withanolides from the soft coral *Paraminabea acronocephala*. *J Nat Prod* 2011, 74 (5), 1132–41.

253. Chao, C. H., Wen, Z. H., Su, J. H., Chen, I. M., Huang, H. C., Dai, C. F., Sheu, J. H. Further study on anti-inflammatory oxygenated steroids from the octocoral *Dendronephthya griffini*. *Steroids* 2008, 73 (14), 1353–58.

254. Ahmed, A. F., Hsieh, Y. T., Wen, Z. H., Wu, Y. C., Sheu, J. H. Polyoxygenated sterols from the Formosan soft coral *Sinularia gibberosa*. *J Nat Prod* 2006, 69 (9), 1275–79.

255. Chao, C. H., Wu, Y. C., Wen, Z. H., Sheu, J. H. Steroidal carboxylic acids from soft coral *Paraminabea acronocephala*. *Mar Drugs* 2013, 11 (1), 136–45.

256. Cheng, S. Y., Huang, K. J., Wang, S. K., Wen, Z. H., Chen, P. W., Duh, C. Y. Antiviral and anti-inflammatory metabolites from the soft coral *Sinularia capillosa*. *J Nat Prod* 2010, 73 (4), 771–75.

257. Yan, P., Deng, Z., van Ofwegen, L., Proksch, P., Lin, W. Lobophytones U–Z$_1$, biscembranoids from the Chinese soft coral *Lobophytum pauciflorum*. *Chem Biodivers* 2011, 8 (9), 1724–34.

258. Yan, P., Lv, Y., van Ofwegen, L., Proksch, P., Lin, W. Lobophytones A–G, new isobiscembranoids from the soft coral *Lobophytum pauciflorum*. *Org Lett* 2010, 12 (11), 2484–87.

259. Yan, P., Deng, Z., van Ofwegen, L., Proksch, P., Lin, W. Lobophytones O–T, new biscembranoids and cembranoid from soft coral *Lobophytum pauciflorum*. *Mar Drugs* 2010, 8 (11), 2837–48.

260. Yan, P., Deng, Z., van Ofwegen, L., Proksch, P., Lin, W. Lobophytones H–N, biscembranoids from the Chinese soft coral *Lobophytum pauciflorum*. *Chem Pharm Bull* (Tokyo) 2010, 58 (12), 1591–95.

261. Wanzola, M., Furuta, T., Kohno, Y., Fukumitsu, S., Yasukochi, S., Watari, K., Tanaka, C., Higuchi, R., Miyamoto, T. Four new cembrane diterpenes isolated from an Okinawan soft coral *Lobophytum crassum* with inhibitory effects on nitric oxide production. *Chem Pharm Bull* (Tokyo) 2010, 58 (9), 1203–9.

262. Takaki, H., Koganemaru, R., Iwakawa, Y., Higuchi, R., Miyamoto, T. Inhibitory effect of norditerpenes on LPS-induced TNF-alpha production from the Okinawan soft coral, *Sinularia* sp. *Biol Pharm Bull* 2003, 26 (3), 380–82.

263. Thao, N. P., Nam, N. H., Cuong, N. X., Quang, T. H., Tung, P. T., Dat, l. D., Chae, D., Kim, S., Koh, Y. S., Kiem, P. V., Minh, C. V., Kim, Y. H. Anti-inflammatory norditerpenoids from the soft coral *Sinularia maxima*. *Bioorg Med Chem Lett* 2013, 23 (1), 228–31.

264. Thao, N. P., Nam, N. H., Cuong, N. X., Quang, T. H., Tung, P. T., Tai, B. H., Luyen, B. T., Chae, D., Kim, S., Koh, Y. S., Kiem, P. V., Minh, C. V., Kim, Y. H. Diterpenoids from the soft coral *Sinularia maxima* and their inhibitory effects on lipopolysaccharide-stimulated production of pro-inflammatory cytokines in bone marrow-derived dendritic cells. *Chem Pharm Bull* (Tokyo) 2012, 60 (12), 1581–89.

265. Gómez-Reyes, J. F., Salazar, A., Guzmán, H. M., González, Y., Fernández, P. L., Ariza-Castolo, A., Gutiérrez, M. *seco*-Briarellinone and briarellin S, two new eunicellin-based diterpenoids from the Panamanian octocoral *Briareum asbestinum*. *Mar Drugs* 2012, 10 (11), 2608–17.

266. Lucas, R., Casapullo, A., Ciasullo, L., Gomez-Paloma, L., Payá, M. Cycloamphilectenes, a new type of potent marine diterpenes: Inhibition of nitric oxide production in murine macrophages. *Life Sci* 2003, 72 (22), 2543–52.

267. Wang, W., Lee, Y., Lee, T. G., Mun, B., Giri, A. G., Lee, J., Kim, H., Hahn, D., Yang, I., Chin, J., Choi, H., Nam, S. J., Kang, H. Phorone A and isophorbasone A, sesterterpenoids isolated from the marine sponge *Phorbas* sp. *Org Lett* 2012, 14 (17), 4486–89.

268. Cheenpracha, S., Park, E. J., Rostama, B., Pezzuto, J. M., Chang, L. C. Inhibition of nitric oxide (NO) production in lipopolysaccharide (LPS)-activated murine macrophage RAW 264.7 cells by the nors-esterterpene peroxide, epimuqubilin A. *Mar Drugs* 2010, 8 (3), 429–37.

269. Park, E. J., Cheenpracha, S., Chang, L. C., Pezzuto, J. M. Suppression of cyclooxygenase-2 and inducible nitric oxide synthase expression by epimuqubilin A via IKK/IκB/NF-κB pathways in lipopolysaccharide-stimulated RAW 264.7 cells. *Phytochem Lett* 2011, 4 (4), 426–431.

270. Posadas, I., De Rosa, S., Terencio, M. C., Paya, M., Alcaraz, M. J. Cacospongionolide B suppresses the expression of inflammatory enzymes and tumour necrosis factor-alpha by inhibiting nuclear factor-kappa B activation. *Br J Pharmacol* 2003, 138 (8), 1571–79.

271. Kobayashi, M., Nakagawa, T., Mitsuhashi, H. Marine terpenes and terpenoids. I. Structures of four cembrane-type diterpenes: Sarcophytol-A, sarcophytol-A acetate, sarcophytol-B, and sarcophytonin-A, from the soft coral, *Sarcophyton glaucum*. *Chem Pharm Bull* 1979, 27 (10), 2382–87.

272. Suganuma, M., Okabe, S., Sueoka, E., Iida, N., Komori, A., Kim, S. J., Fujiki, H. A new process of cancer prevention mediated through inhibition of tumor necrosis factor alpha expression. *Cancer Res* 1996, 56 (16), 3711–15.

273. Hong, S., Kim, S. H., Rhee, M. H., Kim, A. R., Jung, J. H., Chun, T., Yoo, E. S., Cho, J. Y. *In vitro* anti-inflammatory and pro-aggregative effects of a lipid compound, petrocortyne A, from marine sponges. *Naunyn Schmiedebergs Arch Pharmacol* 2003, 368 (6), 448–56.

274. Costantino, V., Fattorusso, E., Mangoni, A., Teta, R., Panza, E., Ianaro, A. Terpioside B, a difucosyl GSL from the marine sponge *Terpios* sp. is a potent inhibitor of NO release. *Bioorg Med Chem* 2010, 18 (14), 5310–15.

275. Liu, C. Y., Hwang, T. L., Lin, M. R., Chen, Y. H., Chang, Y. C., Fang, L. S., Wang, W. H., Wu, Y. C., Sung, P. J. Carijoside A, a bioactive sterol glycoside from an octocoral *Carijoa* sp. (Clavulariidae). *Mar Drugs* 2010, 8 (7), 2014–20.

276. Putra, M. Y., Ianaro, A., Panza, E., Bavestrello, G., Cerrano, C., Fattorusso, E., Taglialatela-Scafati, O. Sinularioside, a triacetylated glycolipid from the Indonesian soft coral *Sinularia* sp., is an inhibitor of NO release. *Bioorg Med Chem Lett* 2012, 22 (8), 2723–25.

277. Oda, T., Lee, J. S., Sato, Y., Kabe, Y., Sakamoto, S., Handa, H., Mangindaan, R. E., Namikoshi, M. Inhibitory effect of N,N-didesmethylgrossularine-1 on inflammatory cytokine production in lipopolysaccharide-stimulated RAW 264.7 cells. *Mar Drugs* 2009, 7 (4), 589–99.

278. Li, J. L., Han, S. C., Yoo, E. S., Shin, S., Hong, J., Cui, Z., Li, H., Jung, J. H. Anti-inflammatory amino acid derivatives from the ascidian *Herdmania momus*. *J Nat Prod* 2011, 74 (8), 1792–97.

279. Giannini, C., Debitus, C., Lucas, R., Ubeda, A., Payá, M., Hooper, J. N., D'Auria, M. V. New sesquiterpene derivatives from the sponge *Dysidea* species with a selective inhibitor profile against human phospholipase A2 and other leukocyte functions. *J Nat Prod* 2001, 64 (5), 612–15.

280. Look, S. A., Fenical, W., Jacobs, R. S., Clardy, J. The pseudopterosins: Anti-inflammatory and analgesic natural products from the sea whip *Pseudopterogorgia elisabethae*. *Proc Natl Acad Sci USA* 1986, 83 (17), 6238–40.

281. Look, S. A., Fenical, W., Matsumoto, G. K., Clardy, J. The pseudopterosins: A new class of antiinflammatory and analgesic diterpene pentosides from the marine sea whip *Pseudopterogorgia elisabethae* (Octocorallia). *J Org Chem* 1986, 51 (26), 5140–45.

282. Reina, E., Puentes, C., Rojas, J., García, J., Ramos, F. A., Castellanos, L., Aragón, M., Ospina, L. F. Fuscoside E: A strong anti-inflammatory diterpene from Caribbean octocoral *Eunicea fusca*. *Bioorg Med Chem Lett* 2011, 21 (19), 5888–91.

283. Radhika, P., Rao, P. R., Archana, J., Rao, N. K. Anti-inflammatory activity of a new sphingosine derivative and cembrenoid diterpene (lobohedleolide) isolated from marine soft corals of *Sinularia crassa* Tixier-Durivault and *Lobophytum* species of the Andaman and Nicobar Islands. *Biol Pharm Bull* 2005, 28 (7), 1311–13.

284. Sasaki, S., Tozawa, T., Van Wagoner, R. M., Ireland, C. M., Harper, M. K., Satoh, T. Strongylophorine-8, a pro-electrophilic compound from the marine sponge *Petrosia* (*Strongylophora*) *corticata*, provides neuroprotection through Nrf2/ARE pathway. *Biochem Biophys Res Commun* 2011, 415 (1), 6–10.

285. Berrué, F., McCulloch, M. W., Boland, P., Hart, S., Harper, M. K., Johnston, J., Kerr, R. Isolation of steroidal glycosides from the Caribbean sponge *Pandaros acanthifolium*. *J Nat Prod* 2012, 75 (12), 2094–100.

286. Hegazy, M. E., Gamal Eldeen, A. M., Shahat, A. A., Abdel-Latif, F. F., Mohamed, T. A., Whittlesey, B. R., Paré, P. W. Bioactive hydroperoxyl cembranoids from the Red Sea soft coral *Sarcophyton glaucum*. *Mar Drugs* 2012, 10 (1), 209–22.

287. Marion, F., Williams, D. E., Patrick, B. O., Hollander, I., Mallon, R., Kim, S. C., Roll, D. M., Feldberg, L., Van Soest, R., Andersen, R. J. Liphagal, a selective inhibitor of PI3 kinase alpha isolated from the sponge akacoralliphaga: Structure elucidation and biomimetic synthesis. *Org Lett* 2006, 8 (2), 321–24.

288. Pereira, A. R., Strangman, W. K., Marion, F., Feldberg, L., Roll, D., Mallon, R., Hollander, I., Andersen, R. J. Synthesis of phosphatidylinositol 3-kinase (PI3K) inhibitory analogues of the sponge meroterpenoid liphagal. *J Med Chem* 2010, 53 (24), 8523–33.

289. Kobayashi, J., Madono, T., Shigemori, H. Nakijiquinones C and D, new sesquiterpenoid quinones with a hydroxy amino acid residue from a marine sponge inhibiting c-erbB-2 kinase. *Tetrahedron* 1995, 51 (40), 10867–74.

290. Takahashi, Y., Kubota, T., Ito, J., Mikami, Y., Fromont, J., Kobayashi, J. Nakijiquinones G–I, new sesquiterpenoid quinones from marine sponge. *Bioorg Med Chem* 2008, 16 (16), 7561–64.

291. Takahashi, Y., Ushio, M., Kubota, T., Yamamoto, S., Fromont, J., Kobayashi, J. Nakijiquinones J–R, sesquiterpenoid quinones with an amine residue from okinawan marine sponges. *J Nat Prod* 2010, 73 (3), 467–71.

292. Takahashi, Y., Kubota, T., Yamamoto, S., Kobayashi, J. Inhibitory effects of metachromins L–Q and its related analogs against receptor tyrosine kinases EGFR and HER2. *Bioorg Med Chem Lett* 2013, 23 (1), 117–18.

293. Rateb, M. E., Houssen, W. E., Schumacher, M., Harrison, W. T., Diederich, M., Ebel, R., Jaspars, M. Bioactive diterpene derivatives from the marine sponge *Spongionella* sp. *J Nat Prod* 2009, 72 (8), 1471–76.

294. Xie, Y., Liu, L., Huang, X., Guo, Y., Lou, L. Scalaradial inhibition of epidermal growth factor receptor-mediated Akt phosphorylation is independent of secretory phospholipase A2. *J Pharmacol Exp Ther* 2005, 314 (3), 1210–17.

295. El Sayed, K. A., Hamann, M. T., Waddling, C. A., Jensen, C., Lee, S. K., Dunstan, C. A., Pezzuto, J. M. Structurally novel bioconversion products of the marine natural product sarcophine effectively inhibit JB6 cell transformation. *J Org Chem* 1998, 63 (21), 7449–55.

296. Katsuyama, I., Fahmy, H., Zjawiony, J. K., Khalifa, S. I., Kilada, R. W., Konoshima, T., Takasaki, M., Tokuda, H. Semisynthesis of new sarcophine derivatives with chemopreventive activity. *J Nat Prod* 2002, 65 (12), 1809–14.

297. Fedorov, S. N., Radchenko, O. S., Shubina, L. K., Balaneva, N. N., Bode, A. M., Stonik, V. A., Dong, Z. Evaluation of cancer-preventive activity and structure-activity relationships of 3-demethylubiquinone Q2, isolated from the ascidian *Aplidium glabrum*, and its synthetic analogs. *Pharm Res* 2006, 23 (1), 70–81.

298. Silchenko, A. S., Avilov, S. A., Kalinin, V. I., Kalinovsky, A. I., Dmitrenok, P. S., Fedorov, S. N., Stepanov, V. G., Dong, Z., Stonik, V. A. Constituents of the sea cucumber *Cucumaria okhotensis*. Structures of okhotosides B1–B3 and cytotoxic activities of some glycosides from this species. *J Nat Prod* 2008, 71 (3), 351–56.

299. Fedorov, S. N., Shubina, L. K., Kicha, A. A., Ivanchina, N. V., Kwak, J. Y., Jin, J. O., Bode, A. M., Dong, Z., Stonik, V. A. Proapoptotic and anticarcinogenic activities of leviusculoside G from the starfish *Henricia leviuscula* and probable molecular mechanism. *Nat Prod Commun* 2008, 3 (10), 1575–80.

300. Fedorov, S. N., Makarieva, T. N., Guzii, A. G., Shubina, L. K., Kwak, J. Y., Stonik, V. A. Marine two-headed sphingolipid-like compound rhizochalin inhibits EGF-induced transformation of JB6 P+ Cl41 cells. *Lipids* 2009, 44 (9), 777–85.

301. Kicha, A. A., Ivanchina, N. V., Huong, T. T., Kalinovsky, A. I., Dmitrenok, P. S., Fedorov, S. N., Dyshlovoy, S. A., Long, P. Q., Stonik, V. A. Two new asterosaponins, archasterosides A and B, from the Vietnamese starfish *Archaster typicus* and their anticancer properties. *Bioorg Med Chem Lett* 2010, 20 (12), 3826–30.

302. Fedorov, S., Dyshlovoy, S., Monastyrnaya, M., Shubina, L., Leychenko, E., Kozlovskaya, E., Jin, J. O., Kwak, J. Y., Bode, A. M., Dong, Z., Stonik, V. The anticancer effects of actinoporin RTX-A from the sea anemone *Heteractis crispa* (= *Radianthus macrodactylus*). *Toxicon* 2010, 55 (4), 811–17.

303. Kicha, A. A., Kalinovsky, A. I., Ivanchina, N. V., Malyarenko, T. V., Dmitrenok, P. S., Ermakova, S. P., Stonik, V. A. Four new asterosaponins, hippasteriosides A–D, from the Far Eastern starfish *Hippasteria kurilensis*. *Chem Biodivers* 2011, 8 (1), 166–75.

304. Dyshlovoy, S. A., Fedorov, S. N., Kalinovsky, A. I., Shubina, L. K., Bokemeyer, C., Stonik, V. A., Honecker, F. Mycalamide A shows cytotoxic properties and prevents EGF-induced neoplastic transformation through inhibition of nuclear factors. *Mar Drugs* 2012, 10 (6), 1212–24.

305. Narisawa, T., Takahashi, M., Niwa, M., Fukaura, Y., Fujiki, H. Inhibition of methylnitrosourea-induced large bowel cancer development in rats by sarcophytol A, a product from a marine soft coral *Sarcophyton glaucum*. *Cancer Res* 1989, 49 (12), 3287–89.

306. Fujiki, H., Suganuma, M., Suguri, H., Takagi, K., Yoshizawa, S., Ootsuyama, A., Tanooka, H., Okuda, T., Kobayashi, M., Sugimura, T. New antitumor promoters: (–)-Epigallocatechin gallate and sarcophytols A and B. *Basic Life Sci* 1990, 52, 205–12.

307. Yokomatsu, H., Satake, K., Hiura, A., Tsutsumi, M., Suganuma, M. Sarcophytol A: A new chemotherapeutic and chemopreventive agent for pancreatic cancer. *Pancreas* 1994, 9 (4), 526–30.

308. Fahmy, H., Zjawiony, J. K., Konoshima, T., Tokuda, H., Khan, S., Khalifa, S. Potent skin cancer chemopreventing activity of some novel semi-synthetic cembranoids from marine sources. *Mar Drugs* 2006, 4 (2), 28–36.

309. Zhang, X., Kundoor, V., Khalifa, S., Zeman, D., Fahmy, H., Dwivedi, C. Chemopreventive effects of sarcophine-diol on skin tumor development in CD-1 mice. *Cancer Lett* 2007, 253 (1), 53–59.

310. Zhang, X., Bommareddy, A., Chen, W., Hildreth, M. B., Kaushik, R. S., Zeman, D., Khalifa, S., Fahmy, H., Dwivedi, C. Chemopreventive effects of sarcophine-diol on ultraviolet B-induced skin tumor development in SKH-1 hairless mice. *Mar Drugs* 2009, 7 (2), 153–65.

311. Guillermo, R. F., Zhang, X., Kaushik, R. S., Zeman, D., Ahmed, S. A., Khalifa, S., Fahmy, H., Dwivedi, C. Dose-response on the chemopreventive effects of sarcophine-diol on UVB-induced skin tumor development in SKH-1 hairless mice. *Mar Drugs* 2012, 10 (9), 2111–25.

312. Janakiram, N. B., Mohammed, A., Zhang, Y., Choi, C. I., Woodward, C., Collin, P., Steele, V. E., Rao, C. V. Chemopreventive effects of Frondanol A5, a *Cucumaria frondosa* extract, against rat colon carcinogenesis and inhibition of human colon cancer cell growth. *Cancer Prev Res* (Phila) 2010, 3 (1), 82–91.

313. Piplani, H., Vaish, V., Rana, C., Sanyal, S. N. Up-regulation of p53 and mitochondrial signaling pathway in apoptosis by a combination of COX-2 inhibitor, celecoxib and dolastatin 15, a marine mollusk linear peptide in experimental colon carcinogenesis. *Mol Carcinog* 2012, 52 (11), 845–58.

314. Piplani, H., Vaish, V., Sanyal, S. N. Dolastatin 15, a mollusk linear peptide, and celecoxib, a selective cyclooxygenase-2 inhibitor, prevent preneoplastic colonic lesions and induce apoptosis through inhibition of the regulatory transcription factor NF-κB and an inflammatory protein, iNOS. *Eur J Cancer Prev* 2012, 21 (6), 511–22.

315. Newman, D. J., Cragg, G. M. Natural products as sources of new drugs over the 30 years from 1981 to 2010. *J Nat Prod* 2012, 75 (3), 311–35.

316. Grasela, J. J., Pomponi, S. A., Rinkevich, B., Grima, J. Efforts to develop a cultured sponge cell line: Revisiting an intractable problem. *In Vitro Cell Dev Biol Anim* 2012, 48 (1), 12–20.

317. Ivanisevic, J., Thomas, O. P., Pedel, L., Penez, N., Ereskovsky, A. V., Culioli, G., Perez, T. Biochemical trade-offs: Evidence for ecologically linked secondary metabolism of the sponge *Oscarella balibaloi*. *PLoS One* 2011, 6 (11), e28059.

318. Leal, M. C., Calado, R., Sheridan, C., Alimonti, A., Osinga, R. Coral aquaculture to support drug discovery. *Trends Biotechnol* 2013, 31 (10), 555–61.

319. Spence, J. T., George, J. H. Biomimetic total synthesis of ent-penilactone A and penilactone B. *Org Lett* 2013, 15 (15), 3891–93.

320. Lu, Y., Su, J. H., Huang, C. Y., Liu, Y. C., Kuo, Y. H., Wen, Z. H., Hsu, C. H., Sheu, J. H. Cembranoids from the soft corals *Sinularia granosa* and *Sinularia querciformis*. *Chem Pharm Bull* (Tokyo) 2010, 58 (4), 464–66.

321. Ahmed, A. F., Tai, S. H., Wen, Z. H., Su, J. H., Wu, Y. C., Hu, W. P., Sheu, J. H. A C-3 methylated isocembranoid and 10-oxocembranoids from a formosan soft coral, *Sinularia grandilobata*. *J Nat Prod* 2008, 71 (6), 946–51.

322. Chao, C. H., Chou, K. J., Huang, C. Y., Wen, Z. H., Hsu, C. H., Wu, Y. C., Dai, C. F., Sheu, J. H. Bioactive cembranoids from the soft coral *Sinularia crassa*. *Mar Drugs* 2011, 9 (10), 1955–68.

323. Lu, Y., Su, H. J., Chen, Y. H., Wen, Z. H., Sheu, J. H., Su, J. H. Anti-inflammatory cembranoids from the Formosan soft coral *Sinularia discrepans*. *Arch Pharm Res* 2011, 34 (8), 1263–67.

324. Lu, Y., Huang, C. Y., Lin, Y. F., Wen, Z. H., Su, J. H., Kuo, Y. H., Chiang, M. Y., Sheu, J. H. Anti-inflammatory cembranoids from the soft corals *Sinularia querciformis* and *Sinularia granosa*. *J Nat Prod* 2008, 71 (10), 1754–59.

325. Cheng, S. Y., Chuang, C. T., Wen, Z. H., Wang, S. K., Chiou, S. F., Hsu, C. H., Dai, C. F., Duh, C. Y. Bioactive norditerpenoids from the soft coral *Sinularia gyrosa*. *Bioorg Med Chem* 2010, 18 (10), 3379–86.

326. Tseng, Y. J., Shen, K. P., Lin, H. L., Huang, C. Y., Dai, C. F., Sheu, J. H. Lochmolins A–G, new sesquiterpenoids from the soft coral *Sinularia lochmodes*. *Mar Drugs* 2012, 10 (7), 1572–81.

327. Su, J. H., Huang, C. Y., Li, P. J., Lu, Y., Wen, Z. H., Kao, Y. H., Sheu, J. H. Bioactive cadinane-type compounds from the soft coral *Sinularia scabra*. *Arch Pharm Res* 2012, 35 (5), 779–84.

328. Tai, C. J., Su, J. H., Huang, M. S., Wen, Z. H., Dai, C. F., Sheu, J. H. Bioactive eunicellin-based diterpenoids from the soft coral *Cladiella krempfi*. *Mar Drugs* 2011, 9 (10), 2036–45.

329. Chen, B. W., Wang, S. Y., Huang, C. Y., Chen, S. L., Wu, Y. C., Sheu, J. H. Hirsutalins I–M, eunicellin-based diterpenoids from the soft coral *Cladiella hirsuta*. *Tetrahedron* 2013, 69 (10), 2296–2301.

CHAPTER 12

Off the Beaten Path
Natural Products from Extreme Environments

Samantha M. Gromek, Ashley M. West, and Marcy J. Balunas
Division of Medicinal Chemistry, Department of Pharmaceutical Sciences,
School of Pharmacy, University of Connecticut, Storrs, Connecticut

CONTENTS

12.1 INTRODUCTION

Extreme environments such as glaciers, hot springs, caves, deserts, and deep seas are habitats with special classes of organisms, generally termed extremophiles. To classify an organism as an extremophile, the preferred growth conditions must be outside the range of the optimal growth parameters of *Escherichia coli* (37 °C, pH 7.4, salinity 0.9%–3%, and 1 atm).[1] Thus, extremophiles comprise several major classes dependent upon the environments in which they tolerate or achieve optimal growth, including psychrophiles (temperatures below 15 °C), thermophiles (temperatures above 50 °C), barophiles (high pressure), halophiles (high salt concentration), acidophiles (low pH conditions), and xerophiles (low water).[1] Additional classes of extremophiles include organisms that live in low light conditions (e.g., caves), also known as photophobic organisms, and those that live in toxic environments (e.g., Superfund sites).[2,3] There are also organisms that tolerate more than one extreme condition in an environment, known as polyextremophiles.[4] Thermoacidophiles are one type of polyextremophile that grow optimally under conditions with low pH and high temperature.[5] For the purposes of this review, the following extremophile definitions will be utilized:

Extremophile	Optimal Growth Conditions
Psychrophile	≤15 °C
Psychrotolerant	20–25 °C
Thermophile	≥50 °C
Thermotolerant	40–50 °C
Barophile	>35 MPa
Halophile	>3% NaCl
Acidophile	pH < 5
Xerophile	Dry desert

Although various extremophilic organisms have been investigated for biologically active compounds, these organisms comprise a largely unstudied biological and chemical resource for further natural product drug discovery efforts. The extreme regions of the earth's surface that these organisms inhabit provide potential targets for novel drug discovery that have yet to be mined.[5] For example, sources for psychrophilic and psychrotolerant organisms are widely distributed in marine (e.g., deep sea, snow, and glaciers) and terrestrial (e.g., Arctic soils and permafrost) regions.[6] Thermophiles can be found in hot springs, hydrothermal vents, and volcanoes. Halophiles live in hypersaline habitats such as the Red Sea. Acidophiles can be found in habitats with high sulfur content (e.g., hydrothermal vents and acid mine drainage).

Under the harsh conditions that these extreme environments present, extremophiles are able to adapt and maintain essential functions by utilizing a variety of biological and chemical adaptations, such as modulation and production of unique classes of enzymes, ability to adapt to temperature changes, and transcription and translation under extreme conditions. For example, psychrophilic and psychrotolerant bacteria respond to decreased extracellular temperature by stimulating the production of genes that code for key functional proteins to withstand cold temperatures.[7] Other examples include the dependence of halophiles on compatible solutes and an influx of inorganic ions such as potassium and chloride to change the osmotic potential to prevent excessive water loss from the cell[8] and synthesis of compact proteins that contain a high content of small amino acid residues, small loops, disulfide bonds, and hydrogen bonds by thermophiles.[9] These biological characteristics necessary for adaptation to their extreme environment may consequently result in new and potentially biologically active molecules.

This review summarizes the most recent literature regarding biologically active natural product compounds from the aforementioned extremophile classes: psychrophile, thermophile, barophile, halophile, acidophile, and xerophile. Organisms that fit into more than one class are discussed based on the primary environmental factor limiting their growth. We have not included alkaliphiles, organisms that grow optimally in pH > 8.5, because to our knowledge there has been no report of novel bioactive compounds from organisms from highly basic environments since the Wilson and Brimble review.[5]

12.2 PSYCHROPHILES AND PSYCHROTOLERANT ORGANISMS

A majority of the earth's biosphere is held constantly below 5 °C[1,10,11] and spans across aquatic (e.g., deep sea, lakes, and glaciers) and terrestrial (e.g., Arctic soils and permafrost) terrain.[1,6] Organisms that live in these cold environments are broadly categorized as either psychrophilic or psychrotolerant. Psychrophiles have an optimal growth temperature below 15 °C with a maximal temperature of ≤20 °C,[1,4,12] whereas psychrotolerant organisms have been shown to grow optimally between 20 °C and 25 °C[12] but can tolerate lower temperatures. Because of easier accessibility, organisms living in shallow Arctic and Antarctic marine environments have been examined more frequently than those in deeper waters.[13]

Psychrophiles exhibit several interesting molecular adaptations to their environment that may influence their production of secondary metabolites. They have remarkable alterations that allow them to regulate membrane fluidity, continue transcription and translation, adapt to sudden temperature changes, inhibit formation of intracellular ice crystals, and continue chemical reactions with cold-adapted enzymes.[14] Furthermore, psychrophiles need to overcome issues with nutrient and waste transport, as well as improper protein folding. Psychrophiles maintain membrane fluidity through various structural changes, such as increasing the content of *cis* unsaturated fatty acids and branched fatty acid chains in the cell membrane. The production of polar carotenoids also aids in stabilizing the cell membrane.[6,7] Additionally, these organisms protect themselves with specialized cold-shock proteins, cold-acclimation proteins, cold active enzymes, antifreeze proteins, and compatible solutes, which are small compounds with a low molecular weight, such as sugars and amino acid derivatives.[6,7,12,14]

The chemistry and biological activity of organisms that live in cold habitats have been the subject of several reviews,[5,15–19] and so we will primarily discuss more recent reports of the chemistry and biological activity of psychrophilic and psychrotolerant organisms. More specifically, our interest is focused on novel compounds derived from soft corals, sponges, tunicates, and bacteria that live in the aquatic environments.

12.2.1 Corals

Until recently, *Alcyonium* soft corals have been largely understudied, despite long-standing knowledge of their habitat in Antarctic waters.[16] As a defense mechanism, soft corals often compensate for their lack of mobility by producing strong chemical metabolites that ultimately serve to ward off predators.[20] In addition to chemical defenses, corals have structural defenses, including muco-polysaccharide lipids, which cover the surface layer of the coral and contain sterols, wax esters, and terpenoid toxins.[20] Lebar et al.[15] had previously discussed chemical compounds originating from cold-water corals, including steroids, nonpolar sesquiterpenes, sesquiterpenoids, 10-hydroxydocosapolyenoic acids, 10-hydroxydocosapolyenoic acids, 8-hydroxyeicosapolyenic acid, eicosanoids, sterols, and diterpenoids. Herein we focus on the new reports of chemical and biological investigations of cold-water corals.

Alcyonium paessleri, a species of sub-Antarctic soft coral, was collected from the South Georgia Islands at a depth of 200 m.[21] Fifteen new compounds were isolated from this collection, including the alcyopterosins (**1–15**), within the illudalane class.[21] Illudalanes have been reported to play a role in DNA binding and demonstrate cytotoxic and antispasmodic activity.[20,22] Four of the alcyopterosins, compounds **1**, **3**, **5**, and **8**, exhibited cytotoxicity in human cancer cell lines.[21] Compounds **1**, **3**, and **8** had IC_{50} (concentration that inhibits binding or activity by 50%) values of 10 µg/ml against human colon carcinoma (HT-29).[21] Compound **5** had an IC_{50} value of 13.5 µM against human larynx carcinoma (Hep-2).[21]

Another species of Antarctic *Alcyonium* soft coral, *A. grandis*, was collected from the Weddell Sea at a depth of 597.6 m.[22] Initially, the soft coral was extracted with acetone and a diethyl ether soluble fraction was further purified to afford nine novel alcyopterosins (**16–24**). The ether fraction demonstrated strong repellent activity against the sea star *Odontaster validus*.

More recently, Nunez-Pons et al.[20] evaluated the chemical ecology of five species of Antarctic *Alcyonium* soft corals (*A. antarcticum*, *A. grandis*, *A. haddoni*, *A. paucilobulatum*, *A. roseum* [deep sample and shallow sample]) from the Weddell Sea from depths between 308.8 and 622 m and from

Deception Island at a depth of 9 m. The diethyl ether extracts of these species were investigated for novel metabolites, their potential repellent activity against the seastar *O. validus* and the amphipod *Cheirimedon femoratus*, and their inhibition of sympatric marine bacteria.[20]

A. *roseum* was collected from the Weddell Sea at two depths (416 and 308.8 m; samples not pooled), with the deeper specimen affording two novel compounds with a similar aromatic motif as the alcyopterosins (**25** and **26**), while the sample collected at the more shallow depth did not contain any illudalane sesquiterpenoids.[20] A. *grandis* produced the previously reported alcyopterosins (**16–24**). Upon further analysis, all of the *Alcyonium* extracts showed a major constituent of two wax esters: C34:1 (**27**) and C32:1 (**28**).[20] It has been suggested that these waxes may be an additional layer of protection for the soft coral by increasing their unpalatability toward predators.[20]

Alcyopterosins (**16–24**) and five of the six diethyl ether extracts mentioned above were tested and found to be effective in deterring the sea star *O. validus*, with the A. *roseum* sample that did not contain alcyopterosins being the only inactive sample. Wax esters (**27** and **28**) were tested at various concentrations and found to deter predation at concentrations higher than 1 mg/g.[20] Due to a limited quantity of alcyopterosins **25** and **26**, these compounds were not tested. Three of the extracts, A. *antarcticum*, A. *haddoni*, and A. *roseum* (deep sample), and the wax esters (**27** and **28**) were not consumed by the amphipod *Cheirimedon femoratus* at varying concentrations during a feeding preference assay.[20] The six extracts and the wax esters were also tested for antibacterial activity against an unidentified marine Antarctic bacterium, with all extracts found to be active (zones of inhibition greater than 7 mm), while all the wax esters were inactive.[20]

12.2.2 Sponges

Porifera (sponges) are found throughout the world, ranging from polar to tropical marine environments. Sponges are sessile organisms that utilize their physical structures and synthesize bioactive compounds to protect themselves against predators, foulers, and grazers.[23a] Initially, research was heavily focused on sponges inhabiting temperate and tropical climates.[24] However, the discovery of novel bioactive compounds produced by cold-water sponges demonstrates that sponges inhabiting polar environments also have the capacity to produce potentially interesting metabolites.[20,24] For this section, we focus on two different classes, Hexactinellida and Demospongiae, due to the discovery of novel compounds or report of new biological activity from existing compounds belonging to these classes.

Within the phylum Porifera is the class Hexactinellida, commonly known as glass sponges, which, until recently, remained largely understudied due to past interests in another class, Demospongiae.[25] Glass sponges are typically found in marine environments at depths ranging from 200 to 6000 m, but have also been found to live at shallower depths.[26] Antarctic glass sponges, in particular, have been found at depths ranging from as deep as 100–600 m to as shallow as 20 m.[25]

The investigation of diethyl ether extracts from 31 Antarctic glass sponges (family Rossellidae), 5 Antarctic demosponges, and 4 non-Antarctic glass sponges revealed 5-α-cholestan-3-one (**29**), which had previously been discovered from the Antarctic demosponge *Artemisina apollinis*, and a mixture of glycoceramides (**30a** and **30b**).[25] While none of the compounds were detected in the Antarctic demosponge extracts, **29** was prevalent in a majority of the Antarctic glass sponge species. The mixture of glycoceramides (**30a** and **30b**), C24 and C22 fatty acid homologues, were detected in all Antarctic glass sponge species with the exception of the internal portion of *Rossella villosa*.[25] Out of the four non-Antarctic glass sponges, only two extracts from distinct collections of *Caulophacus arcticus* contained compounds **29** and **30a** or **30b**. The chemical profiles of Antarctic glass sponges and Antarctic demosponges are intriguing, given that both classes are associated with similar diatom species, but appear to produce different compounds.[25]

Consistent with assays used to study the coral compounds described above, 19 glass sponge and 5 demosponge extracts were impregnated into shrimp cubes that were fed to the sea star *Odontaster validus.*[25] 5-α-Cholestan-3-one (**29**) was present in 12 of the extracts, but only 3 of the 12 extracts with **29** were not consumed by *O. validus.* Furthermore, there was no feeding deterrence of the glycoceramide mixture (**30a** and **30b**) when given to *O. validus* or *Cheirimedon femoratus.*[25]

Also, within the phylum Porifera is the family Latrunculiidae (class Demospongiae), which has been shown to contain compounds from the discorhabdin class (discorhabdin alkaloids) that demonstrate cytotoxic and antibacterial activity.[15,27] Discorhabdin R (**31**) was isolated from two marine sponges, *Negombata* sp. from the Southern Ocean and Antarctic *Latrunculia* sp.[15] Additionally, discorhabdins C (**32**) and G (**33**) were isolated from the Antarctic sponge *Latrunculia apicalis.*[15] The genus *Latrunculia* is more commonly found in colder waters, including Antarctica, North Pacific, and Alaska, and is known to produce cytotoxic pyrroloiminoquinine alkaloids within the discorhabdin class.[27]

A new unidentified Alaskan *Latrunculia* species was recently collected near the Aleutian Islands.[27] A total of eight discorhabdins were isolated from the unidentified *Latrunculia* species.[27] There were two novel compounds, dihydrodiscorhabdin B (**34**) and discorhabdin Y (**35**), and six known compounds, discorhabdins A (**36**), C (**32**), E (**37**), and L (**38**), dihydroscorhabdin C (**39**), and a benzene derivative (**40**).[27]

Discorhabdin A (**36**), discorhabdin C (**32**), and dihydroscohabdin C (**39**) were tested *in vitro* for antiviral, antimalarial, and antimicrobial activity.[27] All three discorhabdins (**32**, **36**, **39**) were active against the hepatitis C virus (HCV) with EC_{50} (half-maximal effective concentration) values of less than 10 μM.[27] However, the compounds were also found to be cytotoxic to Huh-7 cells at 10 μM. The three compounds (**32**, **36**, **39**) showed antimalarial activity against *Plasmodium falciparum*; **36** demonstrated an IC_{50} value of 53 nM against both chloroquine-susceptible and chloroquine-resistant strains, while **39** exhibited IC_{50} values of 170 and 130 nM against the chloroquine-susceptible and chloroquine-resistant strains, respectively.[27] Compounds **36** and **39** were further explored for *in vivo* antimalarial activity using *Plasmodium berghei* mice.[27] Because mice that were administered 10 mg/kg of **39** died prior to assessing reduction of parasitemia, only the *in vivo* antimalarial activity of **36** could be determined. After two doses of 10 mg/kg each, compound **36** caused a 50% reduction in parasitemia in comparison to the control group. The three discorhabdins (**32**, **36**, **39**) were also tested against methicillin-resistant *Staphylococcus aureus* (MRSA), *Mycobacterium intracellulare*, and *Mycobacterium tuberculosis* (H37Rv).[27] Compound **32** displayed the greatest antimicrobial activity with IC_{50} values of 3.2, 0.13, and 6.8 μM against MRSA, *M. intracellulare*, and *M. tuberculosis*, respectively.

Further investigation of the Alaskan *Latrunculia* species discussed above[27] revealed a novel pyrroloiminoquinone alkaloid known as atkamine (**41**), with a chemical structure consisting of a unique heterocyclic ring system containing tryptophan, tyrosine, and a monosaturated fatty acid.[28] Due to a limited quantity of the pure isolated compound, **41** was not assessed for biological activity.[28]

Others have previously reviewed the chemical compounds found in Arctic sponges of the *Haliclona* genus, which are a promising source of novel biologically active 3-alkyl pyridinium alkaloids.[15,17,29] In a more recent report, a study of a *n*BuOH fraction of the Arctic sponge *Haliclona viscosa* collected near Kongsfjorden, Norway (ranging from 15 to 25 m in depth), afforded four new 3-alkyl pyridinium alkaloids: viscosalines B_1 (**42**), B_2 (**43**), E_1 (**44**), and E_2 (**45**).[30] From mass spectra, there appeared to be different alkyl chain lengths associated with viscosalines B (C_{12} and C_{13}) and E (C_{13} and C_{14}). Therefore, the compounds were named according to the association of the β-alanine motif with a specific alkyl chain. Viscosalines B_1 (**42**) and E_1 (**44**) have the β-alanine motif attached to a longer alkyl chain, while viscosalines B_2 (**43**) and E_2 (**45**) have the β-alanine motif attached to a shorter alkyl chain. There was no reported biological activity of the four novel 3-alkyl pyridinium alkaloids.

As noted in previous review articles, within the class Demospongiae is the species *Geodia barretti*, a cold-water marine sponge that has been found in the North Atlantic Ocean near Koster Fjord, Sweden.[15,17,31] Although *G. barretti* is sessile, the sponge appears to protect itself from undesired settlement of other marine organisms (e.g., fouling) by secreting defensive chemical compounds.[31] Two known brominated cyclodipeptides, barretin (46) and 8,9-dihydrobarettin (47), and one new dibrominated cyclopeptide, bromobenzisoxazolone barettin (48), were isolated from *G. barretti*.[31] These three compounds (46–48) prevented the settlement of the barnacle *Balanus improvises* with EC_{50} values of 0.9 µM, 7.9 µM, and 15 nM, respectively.[31] Additionally, at concentrations below 10 µM, 46 demonstrated an affinity for $5\text{-}HT_{2A}$, $5\text{-}HT_{2C}$, and $5\text{-}HT_4$, whereas 47 only exhibited affinity for $5\text{-}HT_4$ at the same concentration.[31] No affinity for any serotonin receptors was observed for compound 48. Compounds 46 and 48 are structurally similar due to the core structure composed of arginine and tryptophan, although 48 has a larger substituent at C-8, which may cause the compound to elicit a stronger inhibition against *B. improvises* and prevent binding to the serotonin receptors.[31]

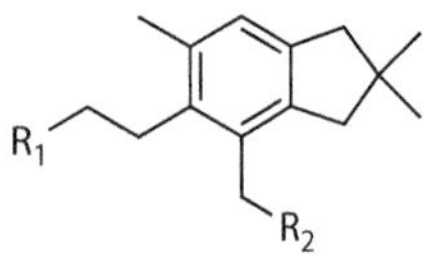

1: $R_1 = Cl; R_2 = H$
2: $R_1 = ONO_2; R_2 = H$
4: $R_1 = Cl; R_2 = OH$
7: $R_1 = ONO_2; R_2 = OH$
15: $R_1 = OH; R_2 = OH$

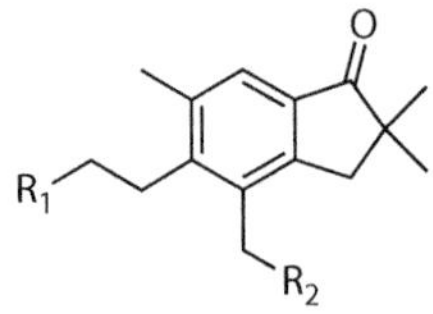

3: $R_1 = ONO_2; R_2 = H$
10: $R_1 = ONO_2; R_2 = OH$
14: $R_1 = OH; R_2 = H$

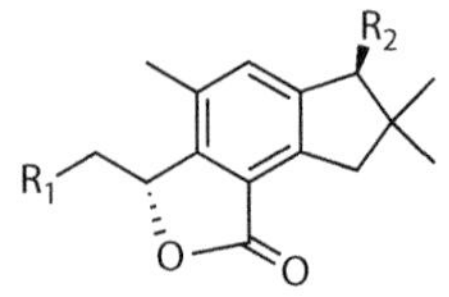

5: $R_1 = ONO_2; R_2 = H$
12: $R_1 = Cl; R_2 = OH$
13: $R_1 = ONO_2; R_2 = OH$

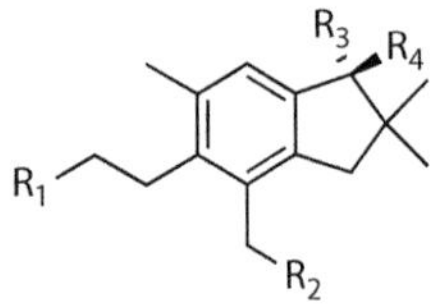

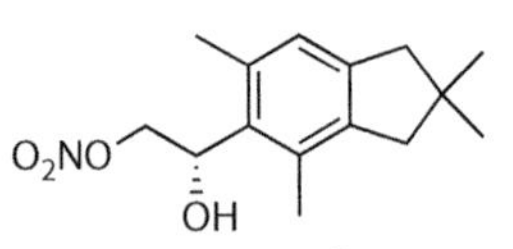

8: $R_1 = ONO_2; R_2 = H; R_3 = OH; R_4: H$
11: $R_1 = Cl; R_2 = OH; R_3 = H; R_4: OH$

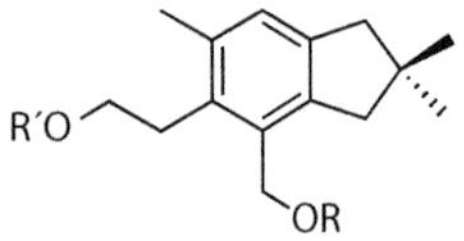

6

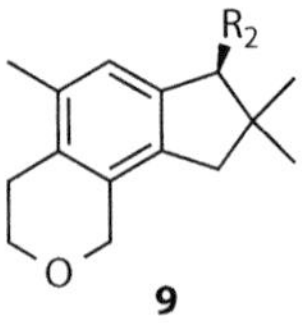

9

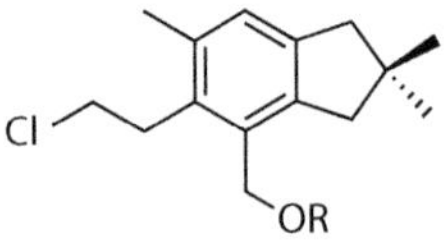

16: $R = COCH_3$
17: $R = COCH_2CH_2CH_3$

18: $R = COCH_2CH_2CH_3; R' = COCH_2CH_2CH_3$
19: $R = COCH_3; R' = COCH_3$
20: $R = COCH_3; R' = COCH_2CH_2CH_3$

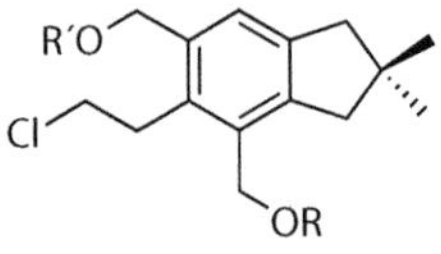

21: $R = COCH_3; R': COCH_3$
22: $R = COCH_3; R'; COCH_2CH_2CH_3$
23: $R = H; R' = H$

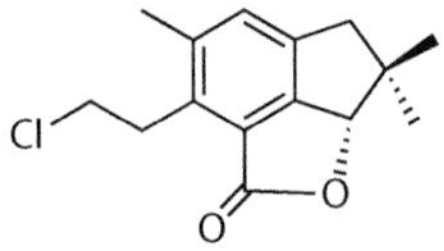

24

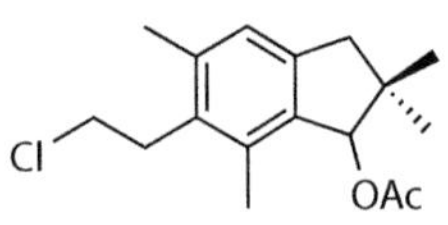

25

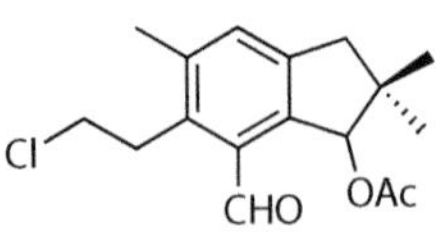

26

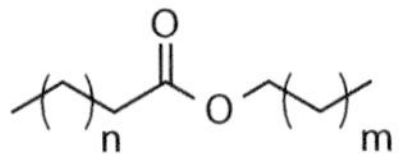

27: n = 15:1; m = 14
28: n = 15:1; m = 12

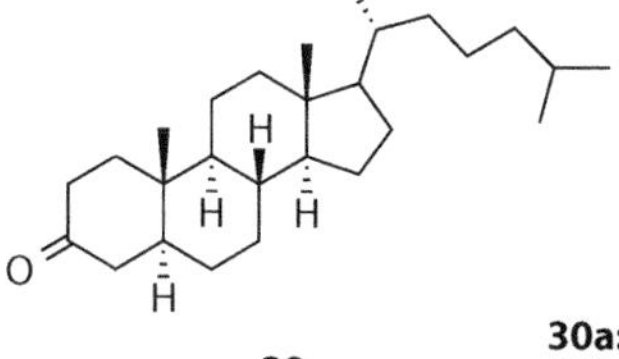

29

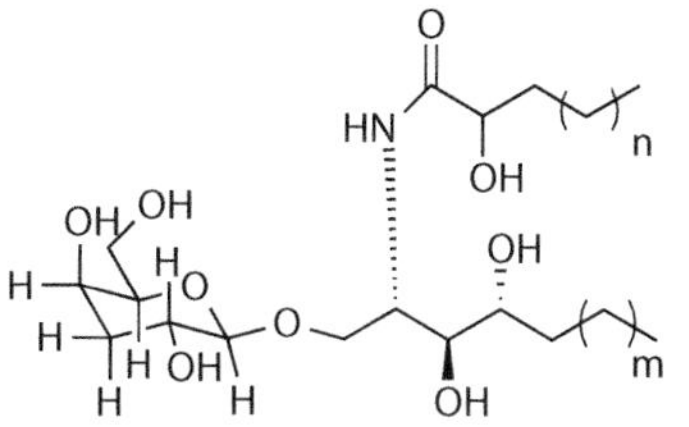

30a: fatty acid residue n = 18:1; sphingoid base residue m = 20:1
30b: fatty acid residue n = 20:1; sphingoid base residue m = 20:1

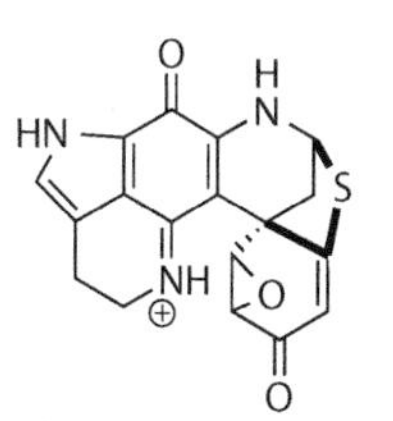

31

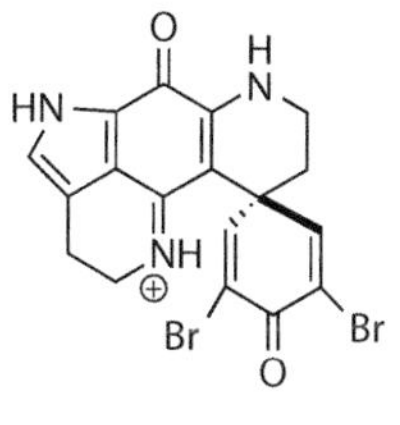

32

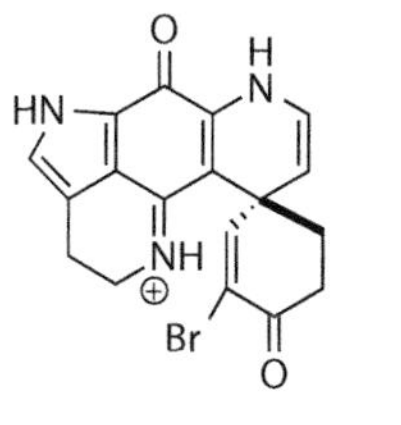

33

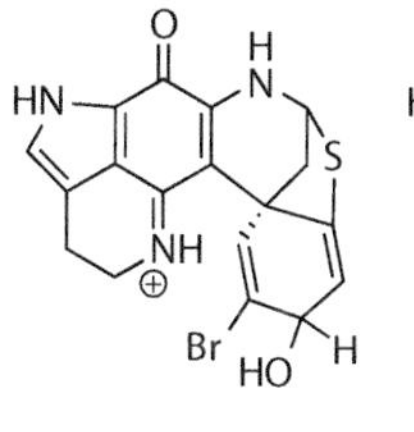

34

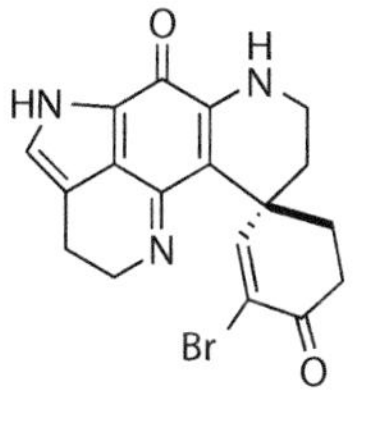

35

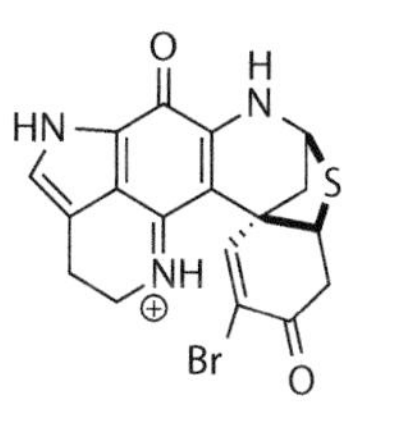

36

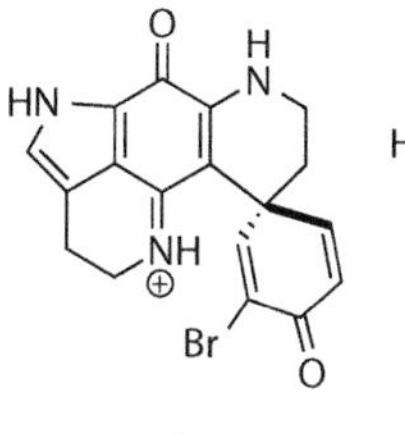

37

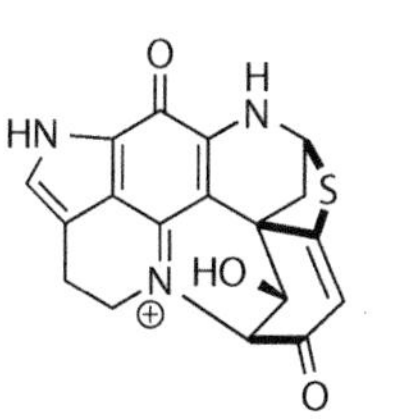

38

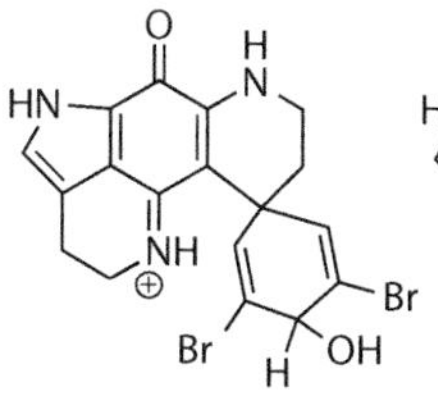

39

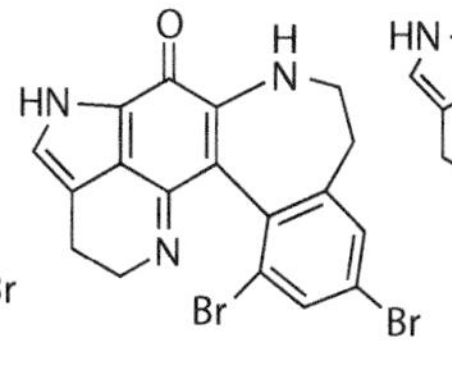

40

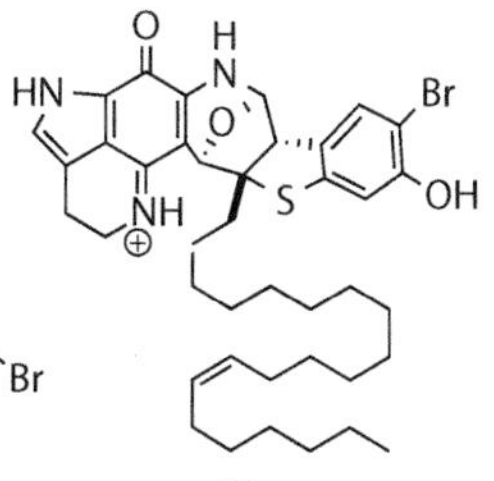

41

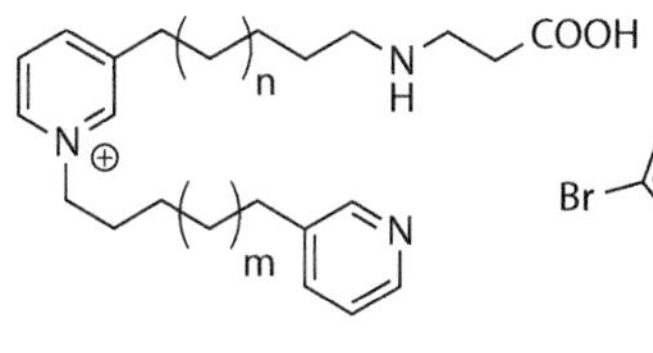

42: m = 8; n = 9
43: m = 9; n = 8
44: m = 9; n = 10
45: m = 10; n = 9

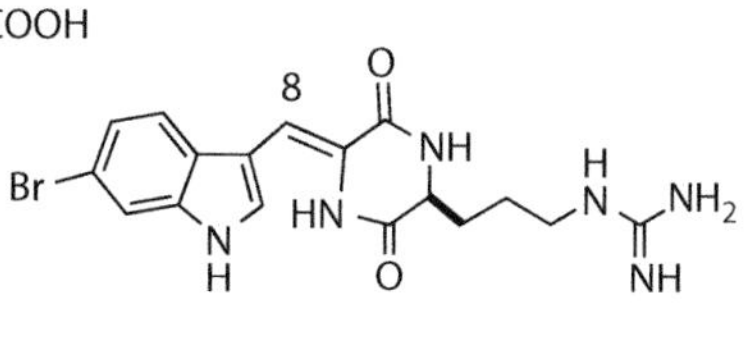

46

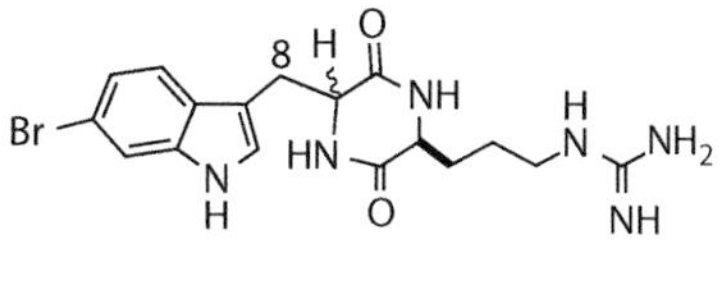

47

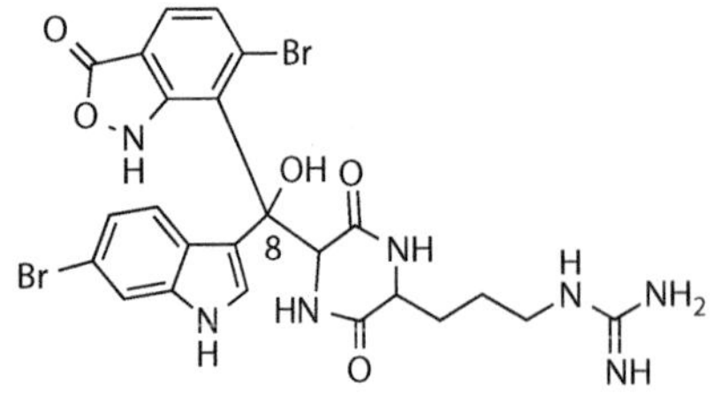

48

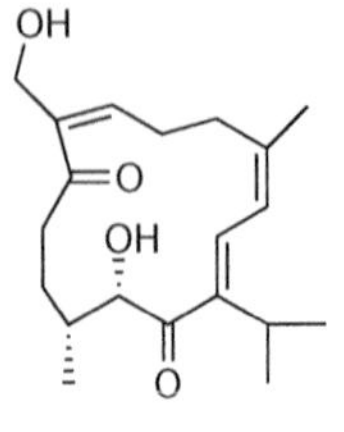

49

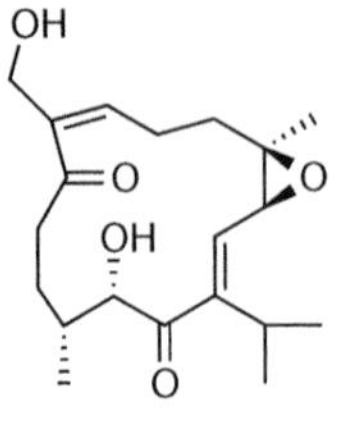

50

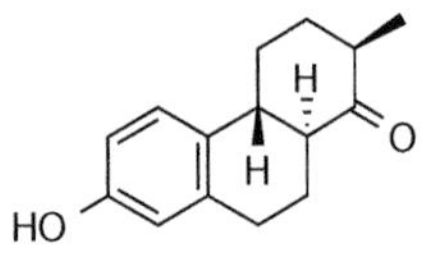

51

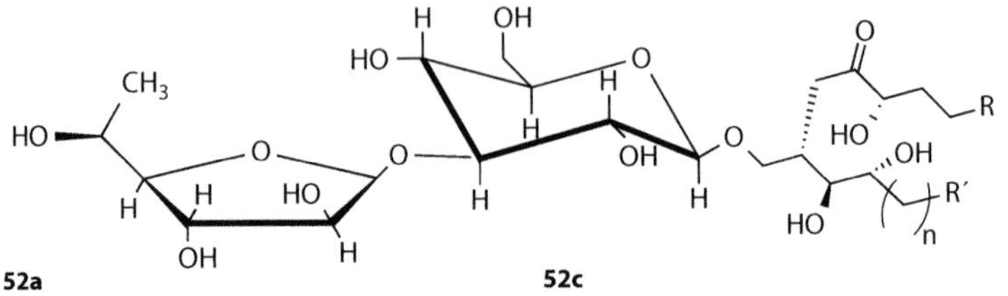

52a
fatty acid residue = 2-OH C 22:1
sphingoid base residue n = 11; 12; 12:1 13; 15

52b
fatty acid residue = 2-OH C 24:1
sphingoid base residue n = 11; 12; 12:1 13; 15

52c
fatty acid residue = 2-OH C 25:1
sphingoid base residue n = 11; 12; 13; 13:1; 14; 15

52d
fatty acid residue = 2-OH C 26:1
sphingoid base residue n = 11; 12; 13; 13:1; 14; 15

52e
fatty acid residue = 2-OH C 27:1
sphingoid base residue n = 11; 13;

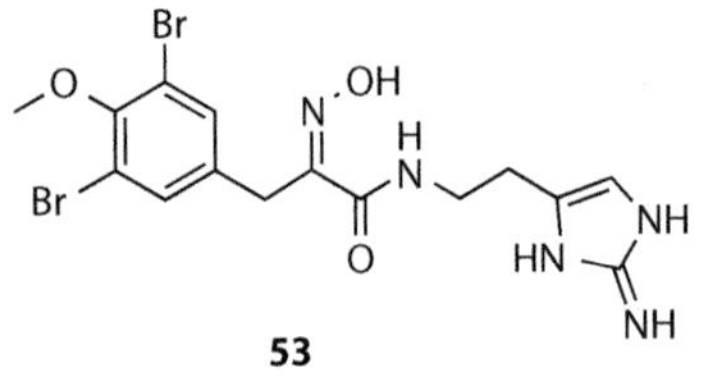

53

The two brominated cyclodipeptides (**46** and **47**) were further evaluated for their synergistic properties.[32] Four different concentrations of **46** (0.3, 0.6, 1.2, and 2.4 μM) were combined with either 1.2 or 2.4 μM of **47** and tested for their inhibitory effect against *B. improvises*. When tested alone at 2.4 μM, **46** demonstrated 100% inhibition of fouling, while the combination of **46** at 2.4 μM with either 1.2 or 2.4 μM of **47** also resulted in 100% inhibition. With the exception of **46** at 2.4 μM, a combination of compounds **46** and **47** exhibited greater inhibition than **46** alone. An isobologram was used to determine minimum concentrations to inhibit 50% settlement (0.3 μM **46** and 1.2 μM **47**). The investigators also quantified the amount of **46** and **47** secreted by *G. barretti* into the surrounding water in the laboratory and natural environment and found a 7:1 secretion ratio in the laboratory and a 14:1 ratio in the natural environment. In further testing using hermit crabs, at a ratio of 7:1 of **46** to **47**, 10 out of 12 crabs were deterred from obtaining bait.

In 1994, an Antarctic sponge *Lissodendoryx flabellata* was collected near Cape Russell at a depth of 70 m.[33] The sponge was subjected to a chemical investigation and afforded two novel cembranes, flabellatenes A (**49**) and B (**50**).[33] The major constituent, **49**, exhibited cytotoxic activity at 0.16 μM against mouse neuroblastoma cells (N18-T62) and suppressed the growth of human breast adenocarcinoma (MCF-7) and human prostate carcinoma (DU-145) cells.[33] Continued investigation of this same Antarctic sponge species, *Lissodendoryx flabellata*, from a fresh collection in Terra Nova Bay, revealed a novel polycyclic compound flabellone (**51**) and a glycospingolipid mixture (**52a–e**).[34] The biological role of **51** in the sponge has yet to be determined, but a synthetic derivative was shown to alleviate estrogen disorders.[34] Further testing revealed that at concentrations ranging from 0.15 to 0.9 µg/ml, the mixture **52a–e** suppressed cell growth in a mixed lymphocyte reaction assay with human T cells.

Interestingly, a known secondary metabolite, ianthelline (**53**), has been discovered in two sponge species living in drastically different marine environments.[23a] Ianthelline (**53**) was originally isolated by Litaudon and Guyot[23b] from a Bahamian sponge, *Ianthella ardis*.[23a] More recently, the same metabolite was isolated from an Arctic sponge, *Stryphnus fortis*—an interesting discovery since it is uncommon for species from separate orders that exist in different habitats (e.g., cold and warm water) to produce the same compound.[23a]

Due to the limited knowledge of biological activity of **53**, the metabolite was further investigated for potential cytotoxic activity, enzymatic activity, kinase inhibition and selectivity, and the effects of the compound on molecular mechanisms.[23a] Compound **53** was tested in a range of concentrations between 52 and 210 µM against 10 malignant cell lines and 1 nonmalignant cell line and caused a decrease in the percentage of cell survival as the concentration of the metabolite increased, with no selectivity toward a single cell line. When **53** was tested against the human melanoma cell line A2058, the IC_{50} value was found to be 91.57 µM.[23a] Further testing of **53** included examination of mitotic cell division in sea urchin embryos, although no division was seen at concentrations as high as 200 µM. In addition, **53** did not affect DNA synthesis. However, **53** seemed to cause an interruption in pronuclear migration and disrupted microtubule organization. Compound **53** also weakly inhibited the phosphorylation activity of cAMP-dependent protein kinase A, tyrosine kinase ABL, and protein-tyrosine phosphatase 1B.[23a] The compound (**53**) was then screened against 131 kinases and found to be moderately selective as indicated by a Gini coefficient of 0.22.[23a]

12.2.3 Tunicates

Tunicates (ascidians, sea squirts) are sedentary, sessile organisms that have modified their physical structure and chemical defense mechanisms according to their surrounding environment and thus provide myriad opportunities for novel bioactive compound discovery.[35] Antarctic tunicates have been collected at various depths along the ocean floor. Similar to marine sponges, the structural features of tunicates create a microhabitat for marine invertebrates and microorganisms.[35] The close association between tunicates and their associated invertebrates or microorganisms causes debate as to the origin of the biologically active compounds that have been reported.[30] For this section, we discuss recently reported metabolites from tunicates in the genera *Aplidium* and *Synoicum*.

Consistent with extracts from soft coral and sponges described above, 25 diethyl ether extracts from deep-water Antarctic tunicates of the *Aplidium* and *Synoicum* genera were evaluated for their chemical composition and deterrence activity against *Odontaster validus* and *Cheirimedon femoratus*.[35] Specimens were collected near the Eastern Weddell Sea at depths ranging from 280 to 340 m, and internal and external portions of the tunicates were separated to assess the allocation of chemical defense compounds.[35] The internal extract of *Aplidium fuegiense* contained rossinone B (**54**), 2,3-epoxy-rossinone B (**55**), 3-epi-rossinone B (**56**), and 5,6-epoxy-rossinone B (**57**),[35] while the external extract afforded only a limited amount of rossinone B (**54**). The known compounds, meridianins A–G (**58–64**), were detected in all extracts (internal and external sections) from *Aplidium falklandicum* and *Aplidium meridianum*.[35] One of the samples of *A. falklandicum* was further investigated and revealed the presence of 12 novel minor meridianins with proposed chemical structures (**65–76**), determined via orbitrap liquid chromatography–high-resolution mass spectrometry–mass spectrometry (LC-HRMS-MS).

O. validus and *C. femoratus* rejected food treated with rossinone B (**54**) at a concentration of 4.8 mg/g and meridianins A–G (**58–64**) at 19.11 mg/g, with these concentrations similar to what is found in the natural environment.[35] All of the extracts, meridianins A–G (**58–64**) and rossinone B (**54**), were tested against a marine bacterium, with activity found only in the mixture of meridianins A–G (**58–64**), which had an inhibition zone (>10 mm) comparable to that of the positive control.[35]

Meridianins A (**58**), B (**59**), C (**60**), and E (**62**) were isolated from an Antarctic tunicate *Synoicum* sp.[36a] and subsequently compared to psammonpemmins A–C (**84–86**; isolated from an Antarctic sponge *Psammopemma* sp.).[36b] Comparison of the nuclear magnetic resonance (NMR) data of meridianin A (**58**) isolated from *Synoicum* sp. collected near the Palmer Station, Antarctica, and synthetic psammopemmin A (**84**) strongly supports the structural revision of **84** to **58**.[36a] Also, due to the similarity in NMR data, psammopemmin C (**86**) should be corrected to meridianin B (**59**) and psammopemmin B (**85**) should be corrected to meridianin E (**62**).[36a]

In another study, investigators explored the chemical composition of a sub-Arctic ascidian *Synoicum pulmonaria* collected near Troms, Norway.[37] Further purification of a crude acetonitrile extract afforded two novel compounds: synoxazolidinones A (**77**) and B (**78**), each with a unique 4-oxazolidinone moiety similar to that found in the linezolid class of antimicrobial drugs.[37] Compounds **77** and **78** displayed antimicrobial and antifungal activity, with **77** demonstrating potent antimicrobial and antifungal activity against *Staphylococcus aureus*, MRSA, *Corynebacterium glutamicum*, and *Saccharomyces cerevisiae*.[37] The only structural difference between compounds **77** and **78** is the chlorine located on C-11, suggesting that this chlorine may be necessary to elicit a biological response.[37]

Within the same genus is the Antarctic ascidian *Synoicum adareanum*, collected near the Palmer Station.[38] Purification resulted in the isolation of four novel macrolides, palmerolides D–G (**79–82**), and the known palmerolide A (**83**).[38] Previous evidence has shown that palmerolide A (**83**) is a V-ATPase inhibitor.[39] Palmerolides A (**83**) and D–G (**79–82**) were assessed for V-ATPase inhibition and cytotoxicity against the UACC-62 human melanoma cell line, with **83**, **79**, and **82** exhibiting the strongest activity (IC$_{50}$ values of 0.002, 0.025, and 0.007 μM, respectively), possibly due to the presence of C2′/C3′ olefin that does not exist in **81**.[38] Additionally, compounds **83**, **79**, and **81** displayed the most potent cytotoxic activity in this same cell line, with IC$_{50}$ values of 0.024, 0.002, and 0.758 μM, respectively.[38] Interestingly, palmerolide G (**82**) exhibited similar potency as a V-ATPase inhibitor as palmerolide A (**83**), except palmerolide G (**82**) demonstrated weaker cytotoxic activity (IC$_{50}$ value of 1.207 μM), possibly attributed to the Z-geometry of C-21/C-22 trisubstituted alkene.[38]

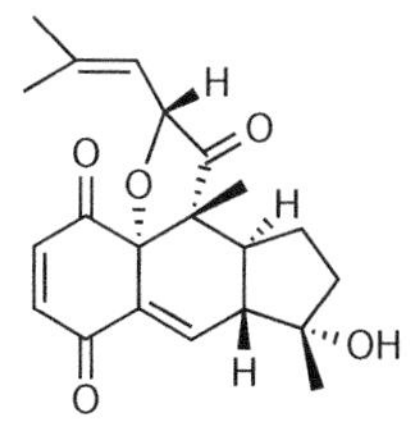

54

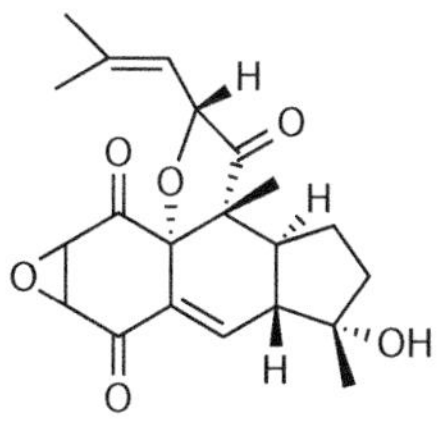

55

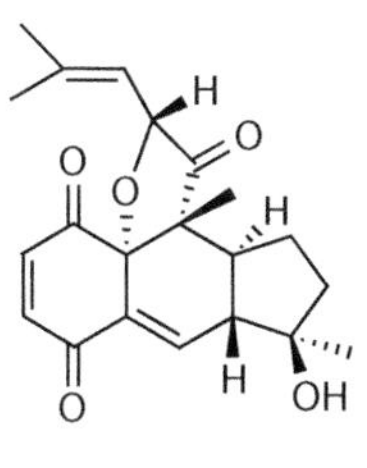

56

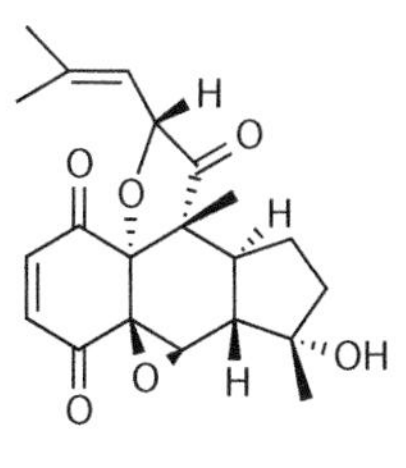

57

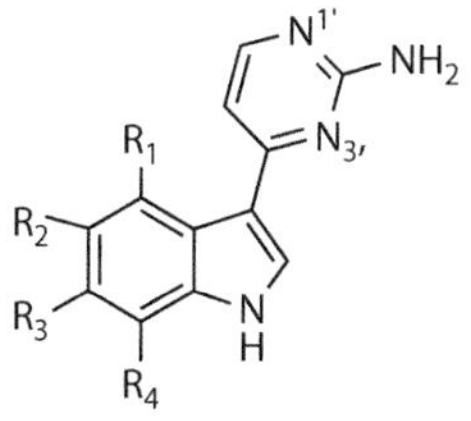

58: R_1 = OH; R_2 = R_3 = R_4 = H
59: R_1 = OH; R_2 = R_4 = H; R_3 = Br
60: R_1 = R_3 = R_4 = H; R_2 = Br
61: R_1 = R_2 = R_4 = H; R_3 = Br
62: R_1 = OH; R_2 = R_3 = H; R_4 = Br

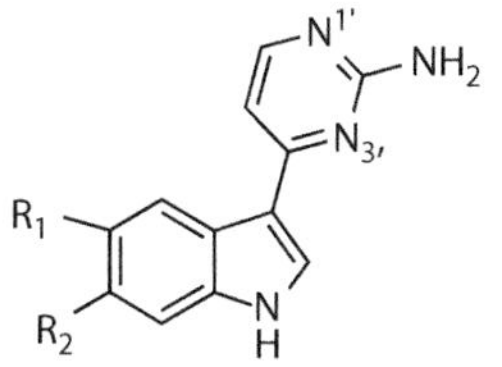

63: R_1 = R_2 = Br
64: R_1 = R_2 = H

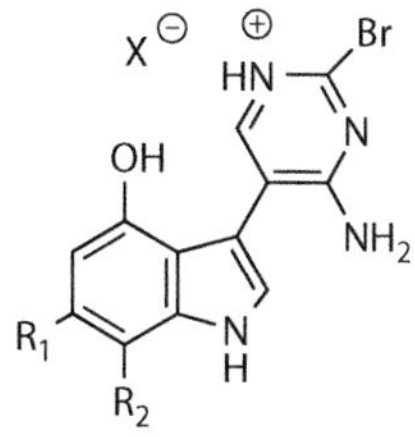

84: R^1 = H; R^2 = H
85: R^1 = H; R^2 = Br
86: R^1 = Br; R^2 = H

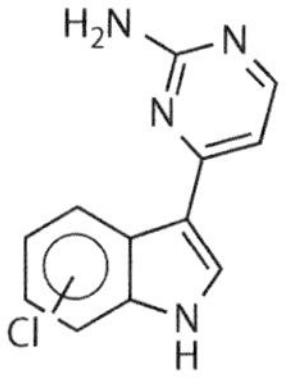

65

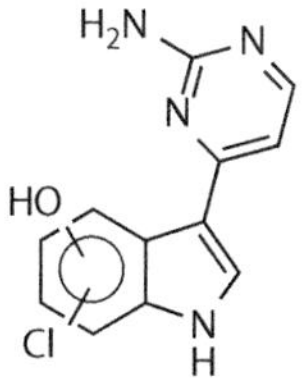

66
67

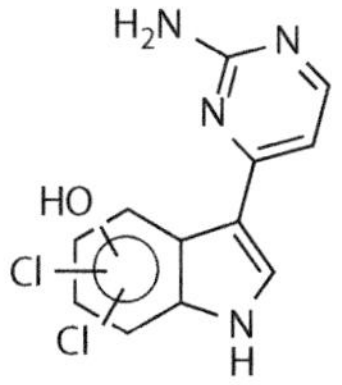

68

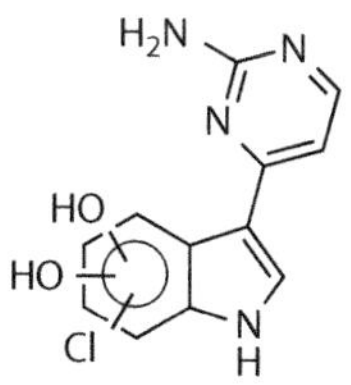

69

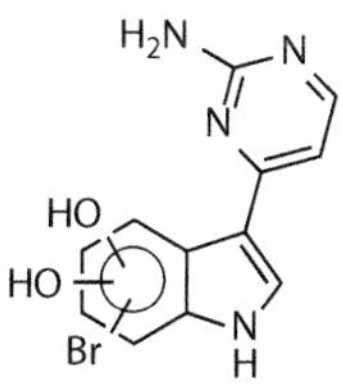

70

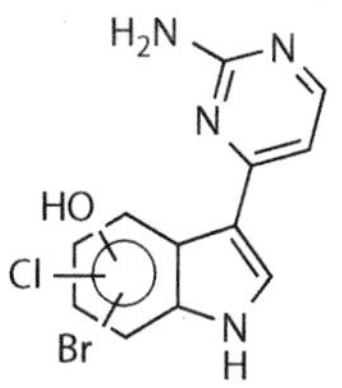

71

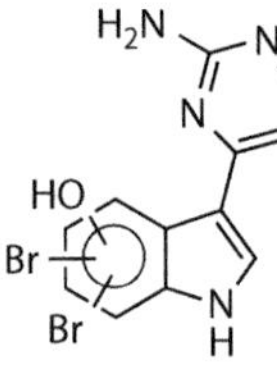

72
73   **74** 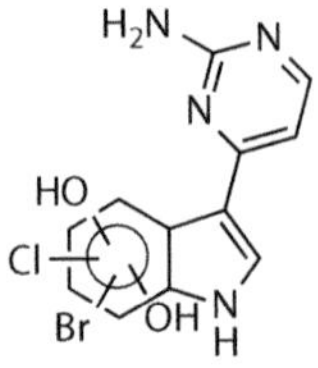   **75** 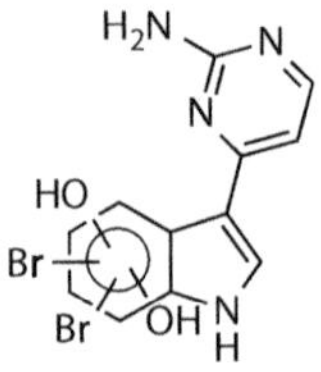   **76** 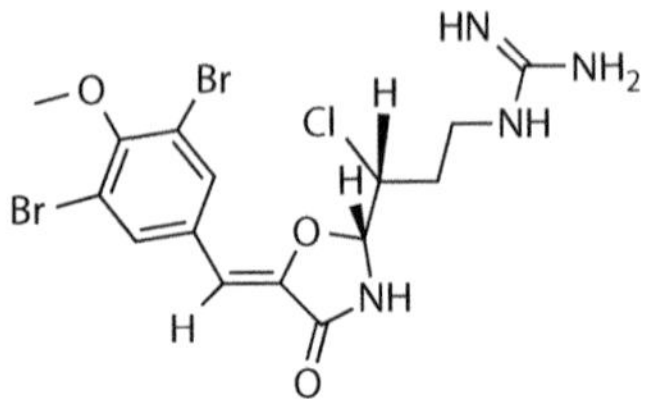   **77**

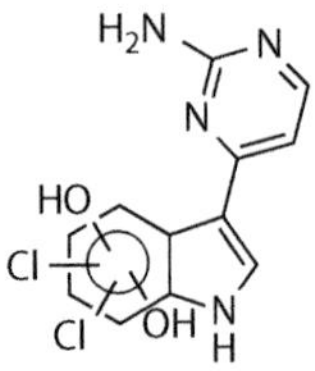

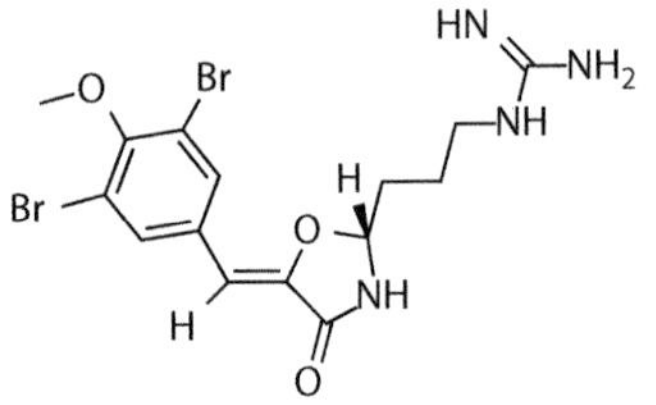

78

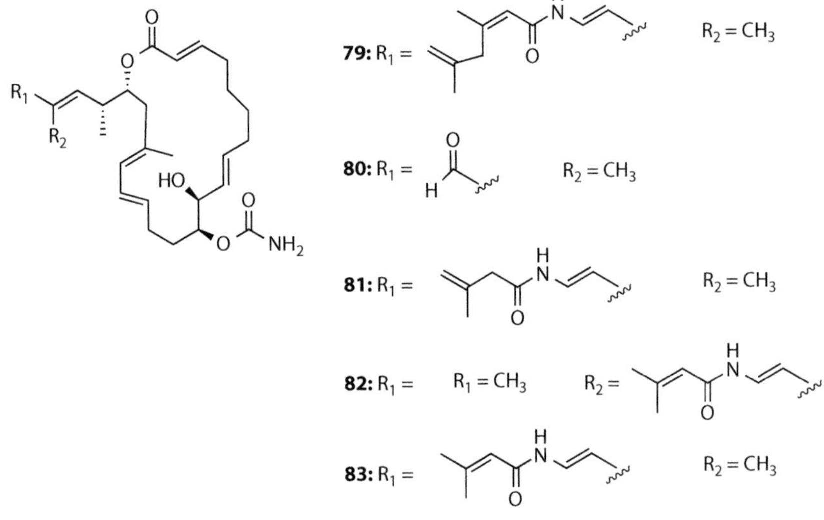

79: $R_1 =$ ... $R_2 = CH_3$

80: $R_1 =$ ... $R_2 = CH_3$

81: $R_1 =$ ... $R_2 = CH_3$

82: $R_1 =$ $R_1 = CH_3$ $R_2 =$ ...

83: $R_1 =$ ... $R_2 = CH_3$

12.2.4 Bacteria

Unlike sponges and soft corals, which often have more restricted niches, microorganisms are typically dispersed throughout cold marine environments and span a broad range of niches.[40] In addition, bacterial strains in cold marine environments are understudied, and therefore their capability to produce biological active compounds is not widely understood.[40] Recently, 16 Arctic bacterial isolates belonging to the *Actinobacteria* and γ-proteobacteria demonstrated antibacterial activity against *Vibrio anguillarum*.[40] These strains were isolated from sea ice, copepods, meltwater, surface water, and deep water collected in the Arctic Ocean. These isolates were then subjected to 16S rRNA sequencing, with seven isolates (PP12, MB182, SS14, TT4, ZZ3, LM7, WX11) from the genus *Arthrobacter* found to inhibit *Staphylococcus aureus*. In addition, two isolates from the genus *Vibrio* exhibited antibacterial activity against *S. aureus*.[40] Upon further analysis of the *Arthrobacter* sp. WX11 ethanol extract, arthrobacillins A–C (**87–89**) were provisionally confirmed using LC-MS compared with the published metabolites. The same extract revealed two unidentified compounds with molecular formulas of $C_{50}H_{90}O_{21}$ and $C_{52}H_{94}O_{21}$.[40] Because **87–89** are cyclic glycolipids composed of lipophilic and hydrophilic moieties, assessment of their antibiotic activity was difficult.[40]

An Arctic bacterium, *Salegentibacter* strain T436, was collected from the Arctic Ocean.[41] Initially, a crude ethyl acetate extract was purified to yield 19 compounds (**90–108**), each containing a nitro group, including four novel compounds.[41] All 19 compounds were tested against seven fungi (*Candida albicans, Paecilomyces variotii, Penicillium notatum, Mucor miehei, Magnaporthe grisea, Nematospora coryli, Ustilago nuda*), five bacteria (*Bacillus brevis, Bacillus subtilis, Micrococcus luteus, Escherichia coli, Proteus vulgaris*), and six cell lines (mouse lymphocytic leukemia L1210, human acute T cell leukemia Jurkat, human colorectal adenocarcinoma Colo-320, human breast adenocarcinoma MDA-MB-131, human promyelocytic leukemia HL-60, and human breast adenocarcinoma MCF-7).[41] A previously reported compound (**105**) displayed the most potent antimicrobial and cytotoxic activity with minimum inhibitory concentration (MIC) values from 0.5 to 50 µg/ml against the fungi and from 12.5 to 20 µg/ml against the bacteria.[41] Also, **105** had IC_{50} values of 20, 10, 30, >100, and 17 µg/ml for L1201, Jurkat, MDA-MB-321, MCF-7, and Colo-320, respectively.[41] The four novel compounds (**97, 100, 101, 103**) did not display significant antimicrobial or cytotoxic activity. Additional analysis of this *Salegentibacter* strain T436 revealed new nitro-containing compounds (**109–115**), with further discussion of the known nitro-containing compounds (**90–108**).[42] Furthermore, 2,6 dimethoxy-1,4-benzoquinone (**116**) and tryptophol methyl ether (**117**), known plant metabolites with cytotoxic activity, were both isolated from this strain.[42]

Another study examined the bacterial strains *Pseudomonas* sp. RG-6, *Pseudomonas* sp. RG-8, and *Yersinia* sp. RP-3, isolated from a cold freshwater river in Chilean Patagonia.[43] The results suggest that the bacterial strains have implemented an acclimation strategy because they were able to successfully grow at 4 °C and 30 °C. After analysis of the 16S RNA gene, RG-6 was found to be most similar to *Pseudomonas brenneri* and RG-8 was most similar to *Pseudomonas trivialis*. RP-3 was identified as *Yersinia aldovae*.[43] These strains were tested for biological activity, with the supernatant of RG-8 and RP-3 exhibiting the strongest activity against both Gram-positive and Gram-negative bacteria with zones of inhibition ranging from 10 to 42 mm and from 12 to 38 mm, respectively. Further analysis of the supernatant from RG-6, RG-8, and RP-3 with gas chromatography–mass spectrometry (GC-MS) revealed the presence of sesquiterpene derivatives, in addition to phthalate (**118**, possibly a contaminant), δ-selinene (**119**), and τ-endesmol. RG-6 and RG-8 also had the presence of hexadecanoic acid (**120**), while RP-3 and RG-6 afforded eicosane (**121**). β-Maailene was only present in RP-3.[43]

There has been recent interest in mining *Streptomyces* from the marine environment to discover novel compounds. A recent study examined 49 marine sediment samples collected at Bamfield, Indian Arm, Georgia Strait, and Howe Sound (British Columbia coast) at depths ranging from 20 to 200 m.[44] From the samples, 186 *Streptomyces* strains were isolated, and their ethyl acetate extracts were tested for antibacterial and antifungal properties, with 47 isolates exhibiting activity.[44] Four bacterial isolates (RJA2961, RJA2926, RJA2895, RJA3265) that demonstrated the strongest antimicrobial activity were further investigated for chemical composition.[44] Purification of RJA2961 revealed four novel novobiocins (**122–125**) and two known compounds, including the antibiotic novobiocin (**126**) and desmethyldescarbamoylnovobiocin (**127**). RJA2926 afforded two previously described compounds, elaiophylin (**128**) and nigericin (**129**), and both of these compounds are known to exhibit strong antibacterial activity against Gram-positive bacteria. RJA2895 contained the previously reported antifungal compounds, the butenolides (**130** and **131**), while known antifungal compound antimycin A2B (**132**) was isolated from RJA3265.[44] All six novobiocins (**122–127**) isolated from RJA2961 were tested against MRSA.[44] Novobiocin (**126**) displayed the most potent activity against MRSA with a MIC value of 0.25 µg/ml. Compounds **122** and **123** exhibited activity against MRSA with MIC values of 16 and 8 µg/ml, respectively. The strength of the activity against MRSA may be attributed to the presence or absence of substituents at positions 3″ and 4″ of the noviose motif and position 5 on the hydroxybenzoate ring.[44]

One of the actinomycete bacteria isolated from marine sediment samples collected near Trondheimfjord, Norway (MP53-27; depth 4.5 m), was pursued for biologically active compounds due to the strong antifungal activity against polyene-resistant *Candida glabrata* and polyene-sensitive *C. albicans*.[45] MP53-27 also exhibited cytotoxicity and selectivity toward IPC-81 rat leukemia cells. MP53-27 was identified as a *Streptosporangium* sp., and isolation from this sample afforded the known compound iodinin (**133**), consisting of a phenazine moiety that is structurally similar to DNA intercalators.[45] Compound **133** was found to have MIC values of 0.36, 0.36, 0.71, and 0.35 µg/ml against *C. glabrata*, *C. albicans*, *Enterococcus faecium* 37832, and *E. faecium* CTC49245, respectively.[45] Moreover, **133** was tested against 10 mammalian cell lines and demonstrated the greatest LC_{50} value in IPC-81 wild-type (0.33 µM) and NB4 human leukemia (0.34 µM) cell lines, with evidence of apoptosis found after treating the NB4 leukemia cells.[45]

In another investigation of Antarctic soil, 4 samples afforded 39 actinobacteria isolates.[46] These isolates were screened against *C. albicans*, *S. aureus*, MRSA, and *Pseudomonas aeruginosa*, with 15 out of 39 isolates demonstrating antibacterial or antifungal activity.[46] Bacteria from the genus *Breribacterium* exhibited the strongest antimicrobial properties, with the second highest bioactive genera being *Arthrobacter* and *Micromonospora*. One isolate from *Brevibacterium* and one from *Kocuria* were active against *S. aureus* and MRSA, while two isolates from the *Gordonia* genus were active against *C. albicans* and *S. aureus*.[46]

Many freshwater cyanobacterial communities in the polar regions flourish during the summer and construct benthic mats, and in a manner similar to that of the invertebrates discussed above, the formation of these benthic mats creates a microhabitat for marine invertebrates and microorganisms.[47] Some of these phototrophic organisms have evolved to withstand the harsh polar environment, such as cold temperatures and high levels of UV radiation, and although these adaptations require high energy, many of these cyanobacteria continue to biosynthesize the known toxins saxitoxins (**134**) and microcystins (**135**) (generic structures of **134** and **135** are presented).[47]

Five Arctic cyanobacterial communities, samples A–E, were screened for the presence of gene clusters specific to compounds **134** and **135** to gauge the potential of the organism to produce the toxins.[47] Samples were subjected to enzyme-linked immunosorbent assay (ELISA) assays and LC-MS analysis to explore the presence of compounds **134** and **135**, and a microcystin variant (Asp³, ADMADDA⁵, Dhb⁷) MC-RR (**136**) was found in sample A.[47] This was the first time the MC-RR variant (**136**) was reported in the Arctic, but the variant had been previously described from a *Nostoc* strain (DUN901) in the United Kingdom.[48] ELISA and LC-MS confirmed that samples B–E did not contain microcystins, while saxitoxins (**134**) were found only in sample E. Interestingly, the authors found there were lower quantities of toxins detected in the Arctic samples than in cyanobacteria living in warmer climates, although Arctic toxin concentrations were similar to cyanobacterial mats in the Antarctic.[47]

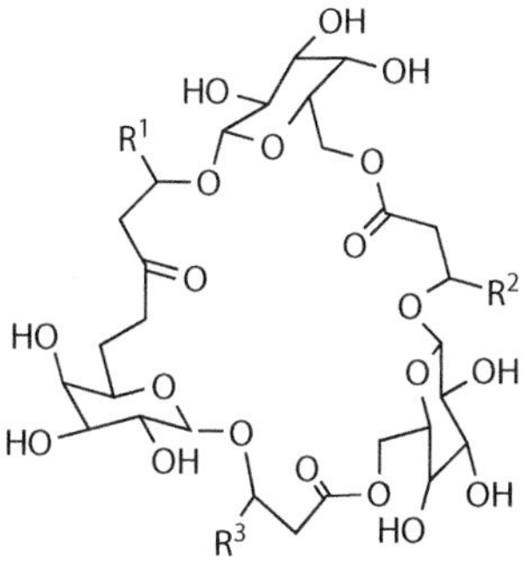

87: $R^1 = R^2 = R^3 = (CH_2)_3CH_3$
88: $R^1 = R^2 = (CH_2)_8CH_3$; $R^3 = (CH_2)_{10}CH_3$
89: $R^1 = (CH_2)_8CH_3$; $R^2 = R^3(CH_2)_{10}CH_3$

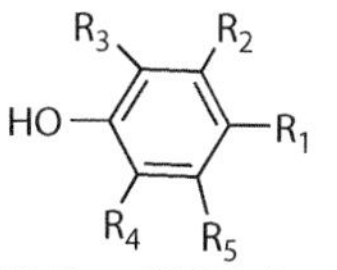

90: R_1 = COOH; R_2 = H;
R_3 = H; R_4 = NO$_2$; R_5 = H
91: R_1 = NO$_2$; R_2 = H;
R_3 = NO$_2$; R_4 = OCH$_3$;
R_5 = H
92: R_1 = NO$_2$; R_2 = NO$_2$;
R_3 = H; R_4 = H; R_5 = OCH$_3$

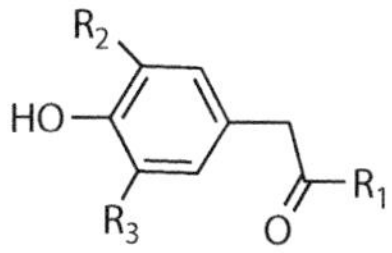

93: R_1 = OCH$_3$; R_2 = NO$_2$;
R_3 = H
94: R_1 = OCH$_3$; R_2 = NO$_2$;
R_3 = NO$_2$
95: R_1 = OH; R_2 = NO$_2$;
R_3 = H
96: R_1 = OH; R_2 = NO$_2$;
R_3 = NO$_2$

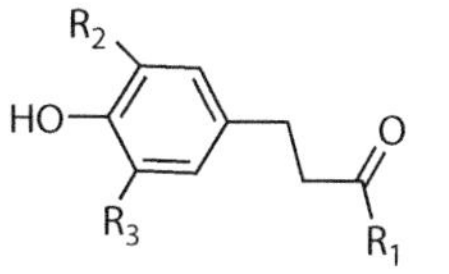

97: R_1 = OCH$_3$; R_2 = NO$_2$;
R_3 = NO$_2$
98: R_1 = OH; R_2 = NO$_2$;
R_3 = H
99: R_1 = OH; R_2 = NO$_2$;
R_3 = NO$_2$

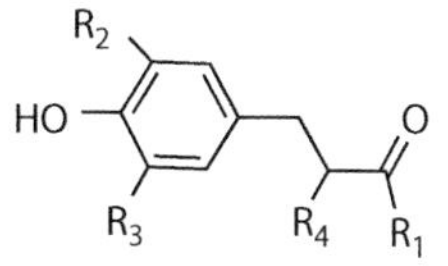

100: R_1 = OCH$_3$; R_2 = NO$_2$;
R_3 = NO$_2$; R_4 = Cl
101: R_1 = OCH$_3$; R_2 = NO$_2$;
R_3 = H; R_4 = OH

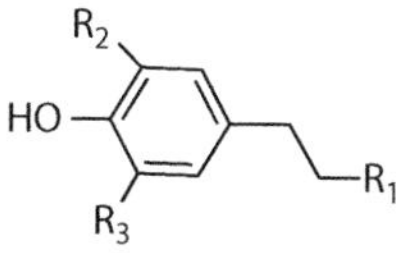

102: R_1 = OH; R_2 = NO$_2$; R_3 = H
103: R_1 = Cl; R_2 = NO$_2$; R_3 = NO$_2$
104: R_1 = OH; R_2 = NO$_2$; R_3 = NO$_2$

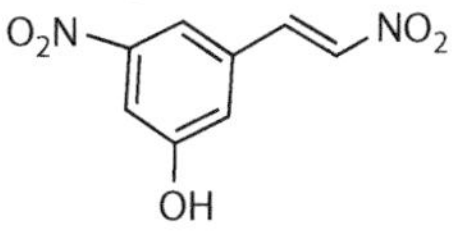

105

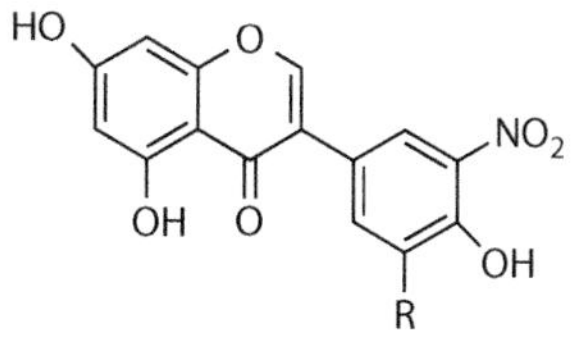

106: R = NO$_2$
107: R = H

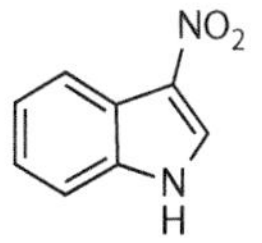

108

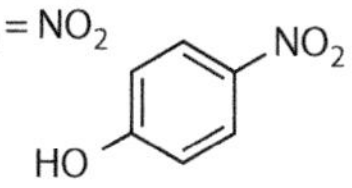

109

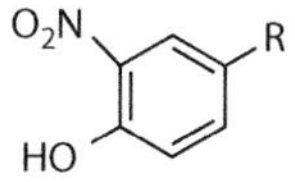

110: R = CH$_2$CHClCOOCH$_3$
111: R = CH$_2$CH$_2$NHAc

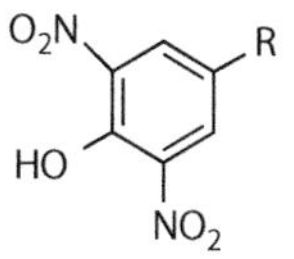

112: R = CH$_2$CH$_2$NHCOCH$_3$
113: R = CH = CHNO$_2$

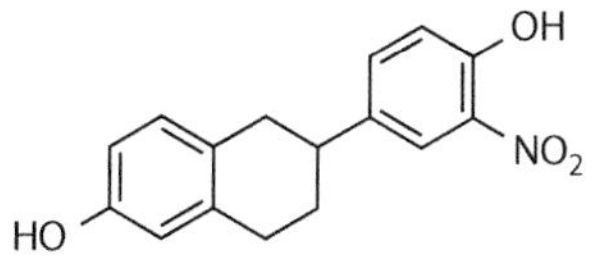

114

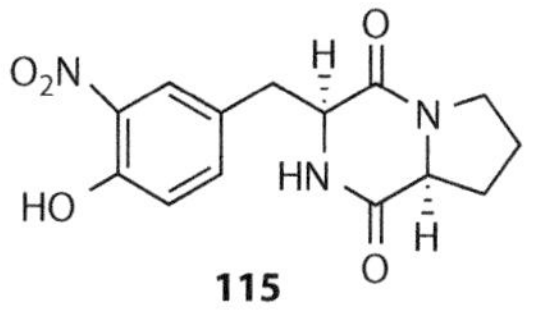

115

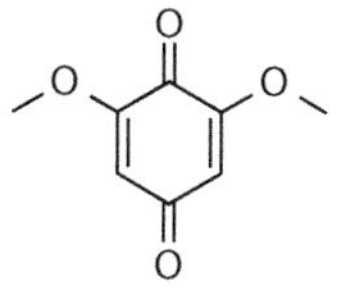

116

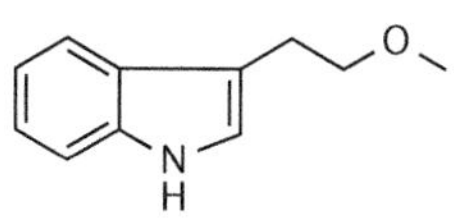

117

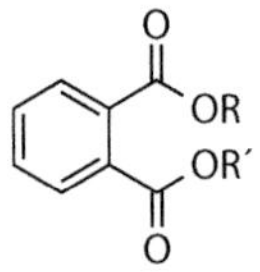

118

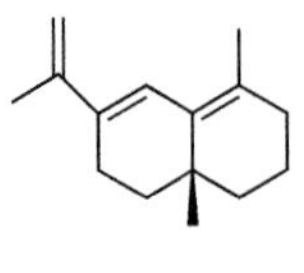

119

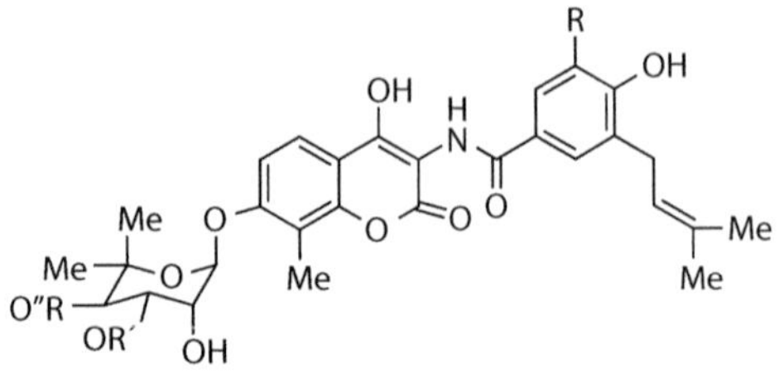

120

121

122: R = H; R′ = C(O)NH$_2$; R″ = H
123: R = OH; R′ = C(O)NH$_2$; R″ = Me
124: R = OH; R′ = H; R″ = H
125: R = OH; R′ = C(O)NH$_2$; R″ = H
126: R = H; R′ = C(O)NH$_2$; R″ = Me
127: R = H; R′ = H; R″ = H

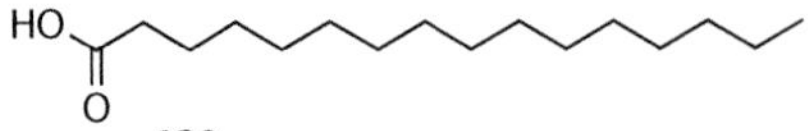

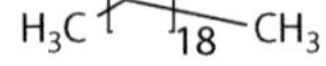

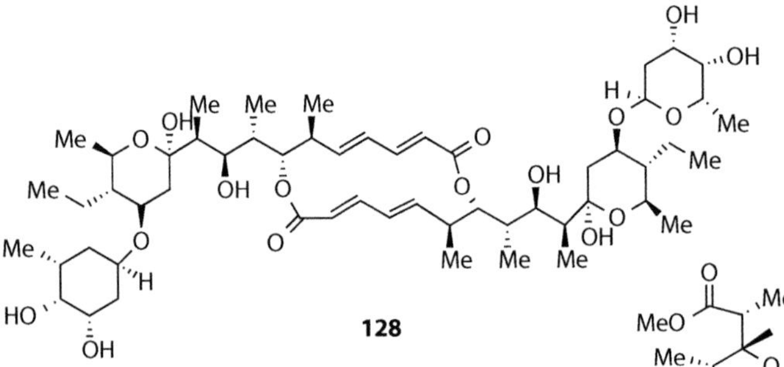

128

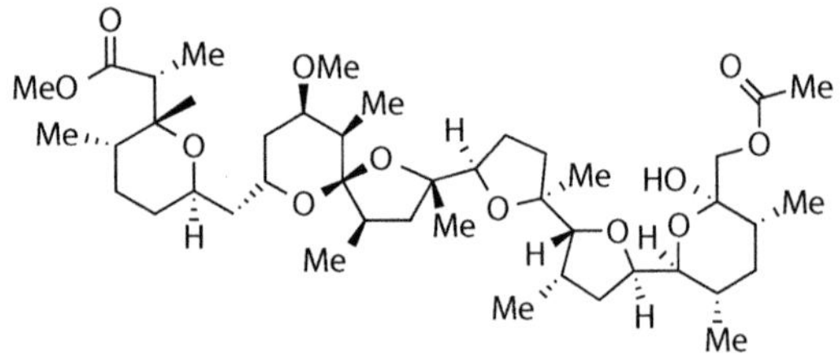

129

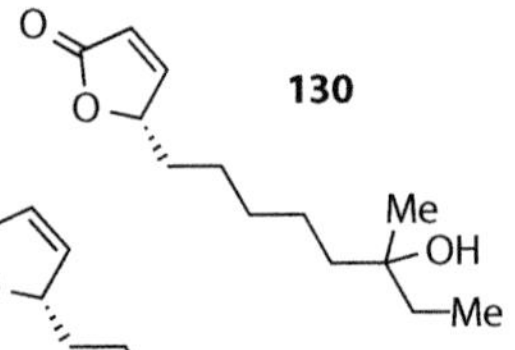

130

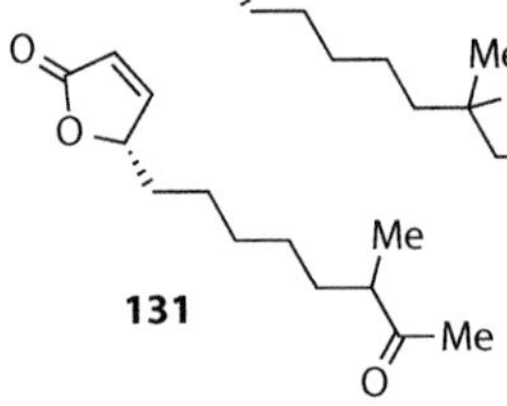

131

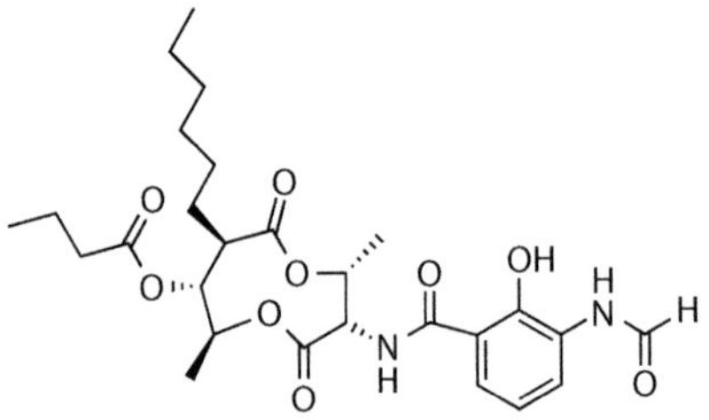

132

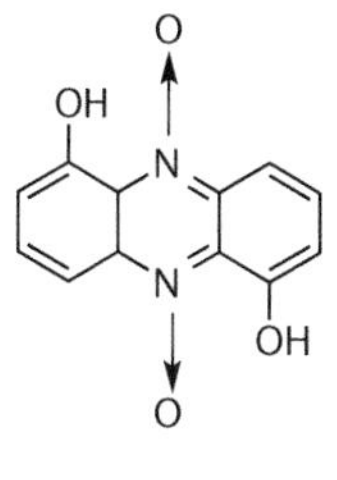

133

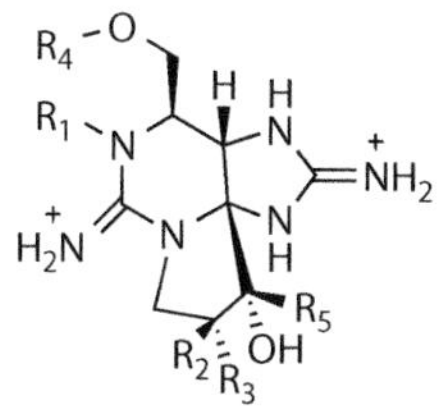

134:
R_1 = H or OH_3
R_2 = H or SO_3
R_3 = H or SO_3
R_4 = H, $COOH_3$, $CONH_3$, $COHNHSO_3$, or COC_6H_4OH
R_5 = H or OH_3

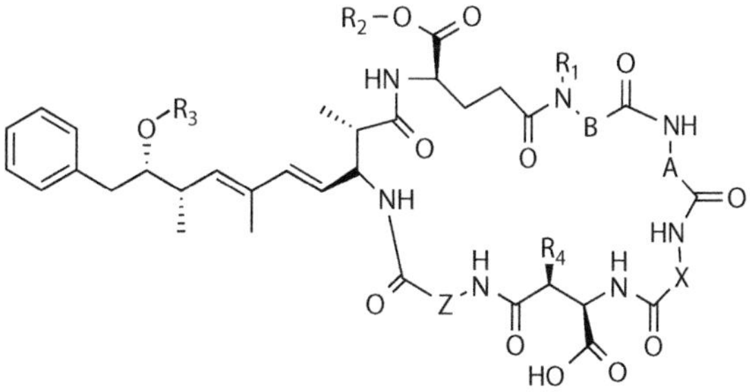

135:
R_1 = H or CH_3
R_2 = H, CH_3, or C_3H_6OH
R_3 = H, CH_3, or $COOH_3$
R_4 = H or CH_3
A = D-alanine, D-leucine, D-serine or glycine
B = Dehydroalanine, Dehydrobutyrine, Serine or Lanthionine
X = Variable L-amino acids
Z = Variable L-amino acids

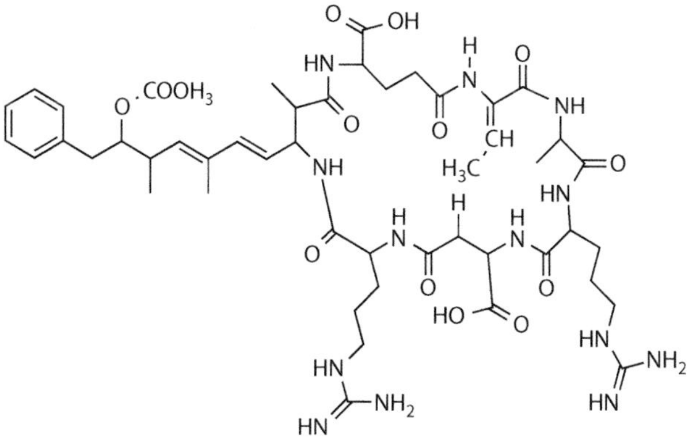

136

12.3 THERMOPHILES

Thermophiles are another potential source of novel natural product compounds. Thermophilic organisms grow and thrive in environments with temperatures above 50 °C, while thermotolerant organisms can tolerate high temperatures, but their optimum growth temperatures are below 50 °C.[1,5] To date, these organisms have largely been explored for their unique enzymes, with less work on isolation and identification of bioactive metabolites from these organisms (see Thornburg et al.,[49] Wilson and Brimble,[5] and Pettit[50] for recent reviews). Thermophilic organisms have been collected from deep-sea hydrothermal vents, hot springs, and geysers, as well as products of biodegradation (e.g., compost heaps and manure piles).

Thornburg et al.[49] reviewed the prior literature related to natural products from deep-sea hydrothermal vents, with another review of extremophile metabolites as well as deep-sea hydrothermal vent metabolites by Pettit.[50] Organisms living in or near these vents must survive large changes in chemical concentrations, pH, and temperature (4 °C–400 °C) between the fluids discharged from the vent and the surrounding seawater.[49,50] In addition, these organisms must tolerate high pressure and absence of light, and are thus considered polyextremophiles, including thermophilic, barophilic, and photophobic characteristics. Those two reviews report only a very few deep-sea hydrothermal vent secondary metabolites from the siderophore and chroman classes (**137**, **138**, **139–144**).[49,50] Two other chroman derivatives, ammonificins C and D (**145–146**), were more recently reported from *Thermovibrio ammonificans*, collected from a hydrothermal vent on the East Pacific Rise.[51] Compounds **145** and **146** were found to induce apoptosis of more than 20% of W2 (apoptosis-competent) cells at 2 and 3 μM, respectively.[51]

Wilson and Brimble[5] and Pettit[50] also reviewed other secondary metabolites from thermophilic organisms, including samples from hot springs, soil, penguin excrement, sewage sludge, yogurt and cheese cultures, and smoldering coal piles. Other thermophilic reports of secondary metabolites include the discovery of a new molecular architecture for the chelation of iron in *Thermobifida fusca*, a thermophilic model organism that is found in decaying organic material.[52] *T. fusca* was found to produce three siderophores with original structures for iron chelation, fuscachelins A–C (**147–149**), that were not present when *T. fusca* was grown in the presence of iron.[52] In a study of hot spring microorganisms, Pednekar et al. collected 73 strains of bacteria from Vajreshwari—Ganeshpuri Hot Springs in Western India.[53] They tested extracts against several strains of infectious microorganisms and found one strain that inhibited both MRSA and vancomycin-resistant enterococci (VRE).[53] In another study of a known thermophile, Hu et al.[54] studied the benzodiazepine biosynthetic pathway of *Streptomyces refuineus*, the producing organisms of anthramycin (**150**). In addition to these reports of secondary metabolites, numerous locations known to house thermophilic organisms have been largely unexplored for their bioactive metabolites; these include volcanoes as well as the underground mine fire in Centralia, Pennsylvania.

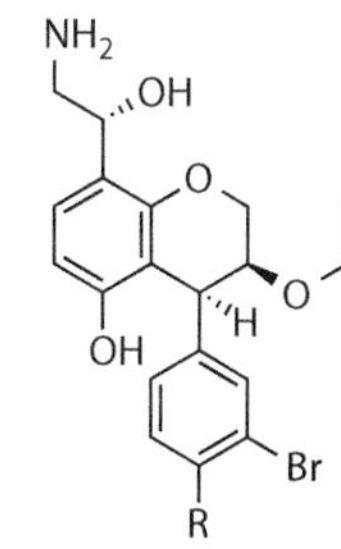

137: R = OH
138: R = Br

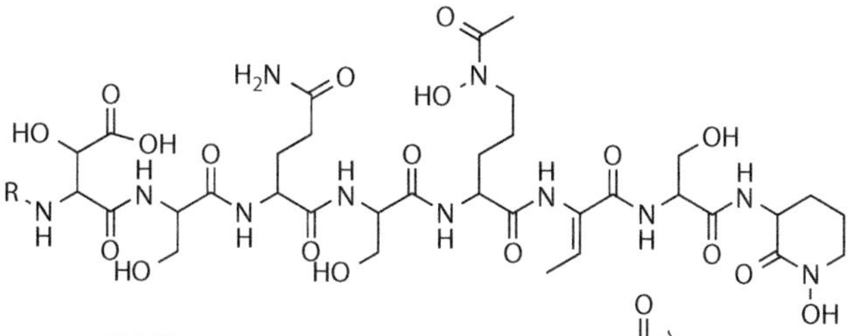

145: R = OH
146: R = Br

139: R =

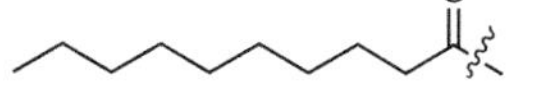

140: R =

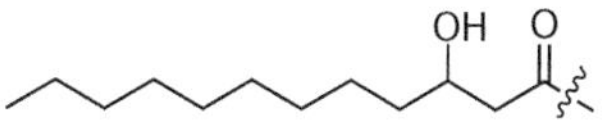

141: R =

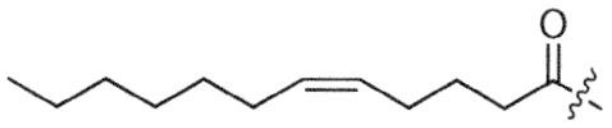

142: R =

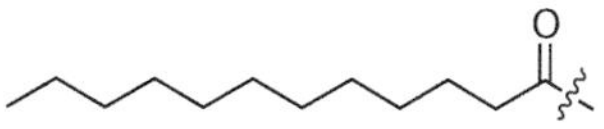

143: R =

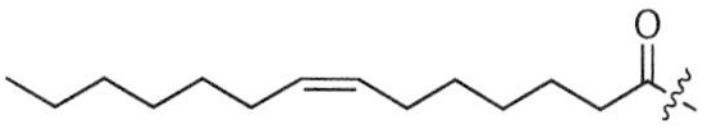

144: R =

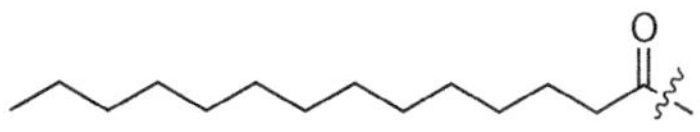

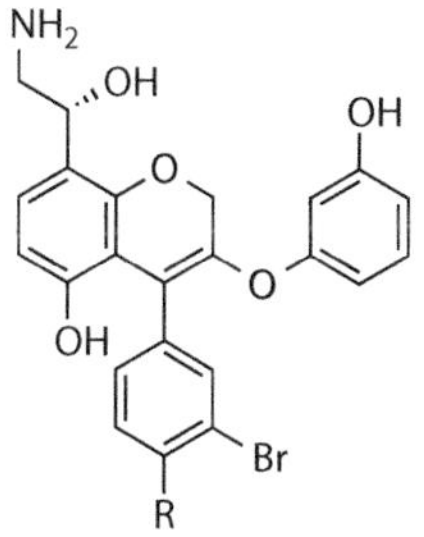

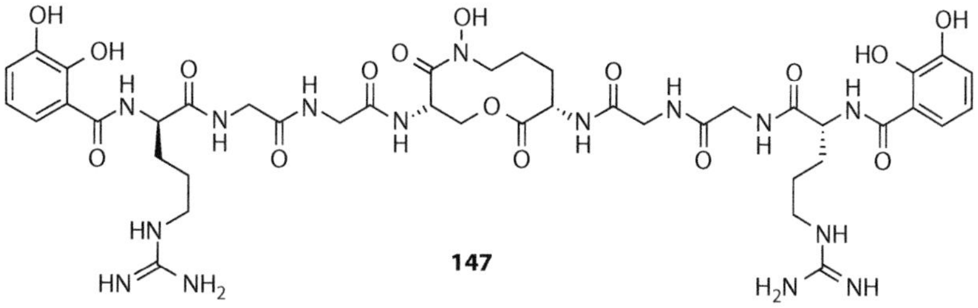

147

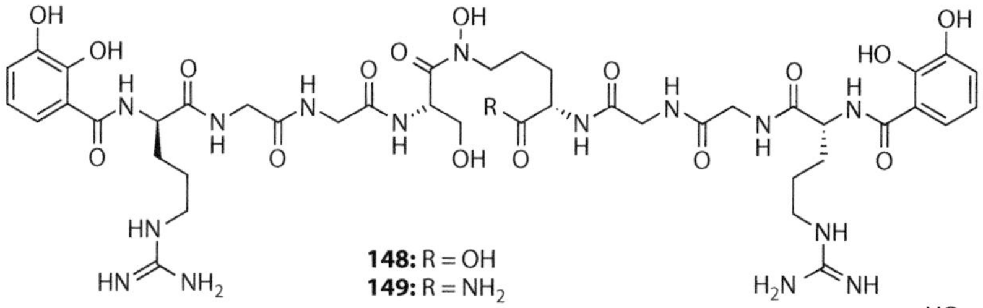

148: R = OH
149: R = NH₂

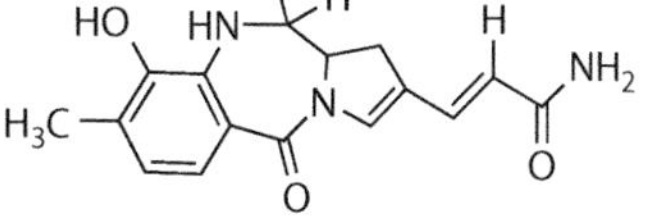

150

12.4 BAROPHILES

The deepest point on earth has been recorded to be 10,915–10,920 m deep.[55] It can be found within the Challenger Deep, the southernmost end of the Mariana Trench. Organisms living in these and other deep-sea waters must be able to endure the high pressure at these depths. Barophiles (also known as piezophiles) are organisms that grow under high pressure (>35 MPa).[5] Deep-sea sponges, echinoderms, and microorganisms are the most commonly collected barophilic samples for the isolation of secondary metabolites.[56] Many barophilic organisms are also polyextremophiles, being acclimated to extreme cold (baropsychrophiles) or extremely high temperatures (e.g., deep-sea hydrothermal vents and thus barothermophiles).[56] Skropeta[56] reviewed natural product secondary metabolites from deep-sea fauna. Various groups have investigated the secondary metabolites produced by barotolerant microorganisms from deep-sea sediments, but to the best of our knowledge, there have been no reports of metabolites isolated from true barophiles, possibly because these organisms require high pressure for collection, transportation, and culturing. Thus, for the purpose of this review, we will focus on the secondary metabolites reported from barotolerant deep-sea fauna reported since.[56]

Three reports regarding metabolites from deep-sea sponges have recently been published. A new compound citharoxazole (**151**) and the known compound batzelline C (**152**) were isolated from the deep-sea sponge *Latrunculia citharistae* collected in the Mediterranean Sea.[57] Compound **151** is closely related to the batzelline class of compounds but contains a benzoxazole moiety, a structural element not previously found in the batzelline derivatives. No biological testing of these compounds was reported.[57] Another group[58] isolated a cyclic peptide (**153**) from a deep-sea sponge, *Discodermia japonica*. Compound **153** was determined to have spectroscopic data identical to the data of the previously reported compound cyclolithistide A.[59] Tajima et al.[58] determined that the original structural assignment for cyclolithistide A was incorrect and provided the structural revision of the compound (**153**). In another recent report, Pinchuk et al.[60] used two hamacanthins, *cis*-3,4-dihydrohamacanthin B (**154**) and 5(R)-hamacanthin B (**155**), to make derivatives with platelet-derived growth factor receptor beta (PDGFRβ) binding activity. Their main lead derivative (**156**) was tested for protein kinase inhibition, PDGFRβ binding, and biological activity in several cancer cell lines, as well as used for molecular modeling studies with PDGFRβ.[60] In addition to inhibiting PDGFRβ with an IC_{50} value of 0.5 μM, compound **156** had an IC_{50} value of 26 nM against HL-60 cells, a cell line that has been shown to be dependent on PDGFRβ signaling.

Two additional reports of secondary metabolites from other deep-sea fauna have recently been published. Proisocrinins A–F (**157–162**) are brominated anthraquinones that were isolated from a stalked crinoid *Proisocrinus ruberrimus*, an echinoderm collected in the Okinawa Trough in Japan.[61] No biological data were reported, although the compounds were described as the first report of tribromo and tetrabromo anthraquinones from natural sources.[61] Another new xenicane diterpenoid, cristaxenicin A (**163**), was isolated from the deep-sea gorgonian coral *Acanthoprimnoa cristata*.[62] Compound **163** was tested for activity against *Leishmania amazonensis*, *Trypanosoma congolese*, and *Plasmodium falciparum*, in addition to being testing for cytotoxicity against P388 and HeLa cells. The most potent activity was against *L. amazonensis* (IC_{50} value of 88 nM), with activity against *T. congolese* as well (IC_{50} value of 250 nM) and low activity against *P. falciparum* (IC_{50} value of 11 μM). Very little cytotoxicity was found with IC_{50} values of 4.7 and 2.1 μM in P388 and HeLa cells, respectively, thus exhibiting a fairly high selectivity index for the trypanosomid diseases.[62]

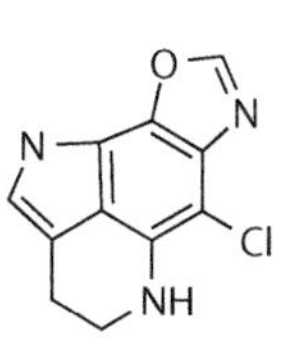

151

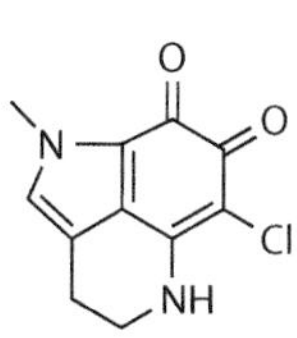

152

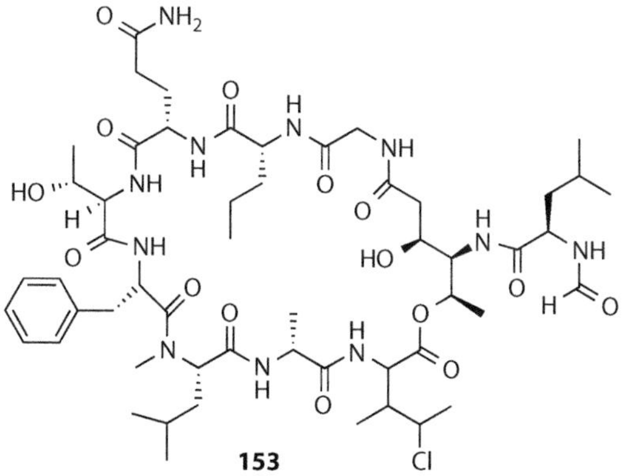

153

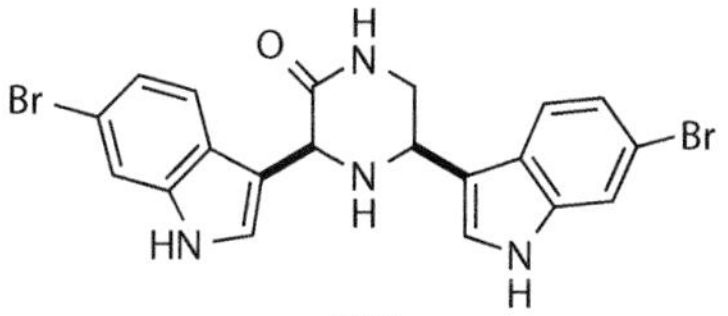

154

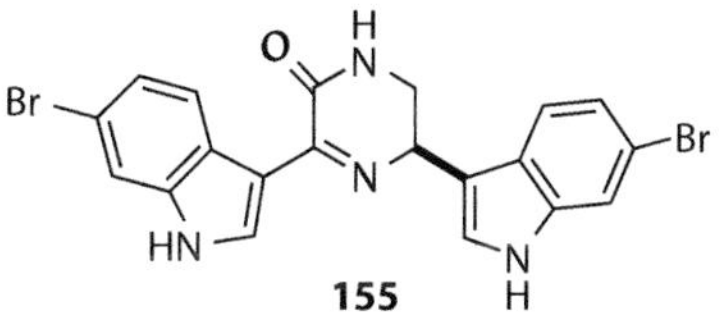

155

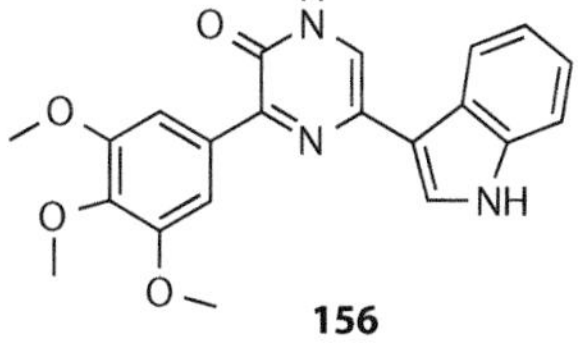

156

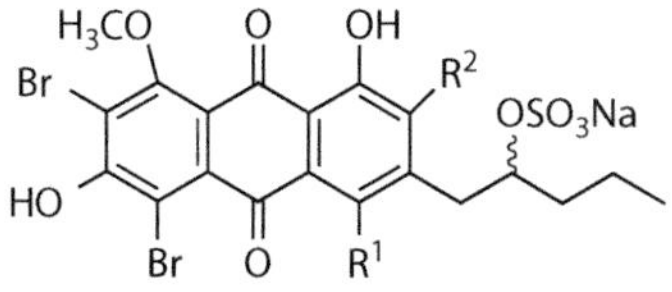

157: R^1 = R^2 = Br
158: R^1 = Br; R^2 = H
159: R^1 = H; R^2 = Br

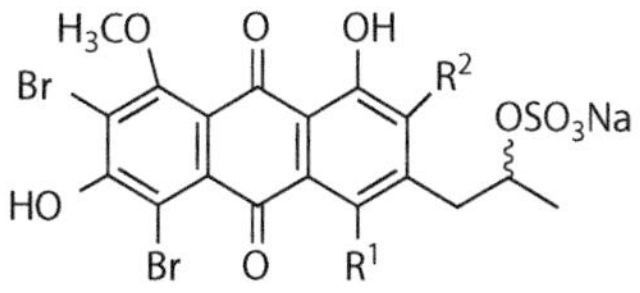

160: R^1 = R^2 = Br
161: R^1 = Br; R^2 = H
162: R^1 = H; R^2 = Br

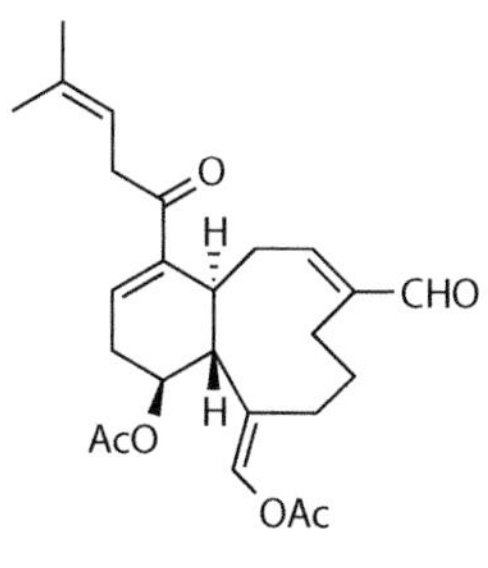

163

12.5 HALOPHILES

Halophiles are organisms that live in hypersaline habitats, including the Red Sea, Dead Sea, and solar salterns. Extreme halophiles have optimum growth at 20%–30% (wt/vol) sodium chloride concentrations, while moderate halophiles grow in 5%–20% (wt/vol) NaCl, and slight halophiles are found in 2%–5% (wt/vol) NaCl.[1] One of the key cellular strategies of organisms that live in this extreme environment is to increase the salt concentration in the cell to prevent excessive water loss from the cell. The cells rely on compatible solutes or an influx of inorganic ions, including potassium and chloride ions, to maintain homeostasis.[8]

Several recent investigations of halophilic microorganisms have been conducted. From a leaf piece of a hard coral on Akajima Island in Okinawa, Japan, Osawa et al.[63] cultured a strain of a halophilic bacterium, *Micrococcus yunnanensis* strain AOY-1, from which three unusual carotenoids were isolated, including sarcinaxanthin (**164**), sarcinaxanthin monoglucoside (**165**), and sarcinaxanthin diglucoside (**166**). These compounds were tested for their antioxidant activity using a singlet oxygen ($^1O_2$) quenching model and found to have IC_{50} values of 57, 54, and 74 μM, respectively (the positive control in this assay was astaxanthin, whose IC_{50} value was 8.9 μM).[63] The authors hypothesize that the antioxidant activity of these compounds might contribute to the organism's ability to tolerate the high levels of ultraviolet irradiation of surface seawater.

An actinomycete, *Actinopolyspora erythraea* YIM 90600, was collected in the Baicheng salt field of Xingjiang Province, China.[64] The strain was cultured in 10% sodium chloride at pH 7.8, which is considered to be in the range of optimal growth conditions for moderately halophilic bacteria. Zhao et al.[64] isolated and characterized three novel compounds, actinopolysporins A–C (**167–169**), and one previously described compound, tubercidin (**170**). All four compounds were evaluated for their ability to stabilize Pdcd4, a tumor suppressor that inhibits the translation of RNA helicase eIF4A. Actinopolysporins A–C (**167–169**) did not demonstrate the ability to stabilize Pdcd4. Tubercidin (**170**) did stabilize Pdcd4 with an IC_{50} value of 0.88 μM.[64]

An extremely halophilic species of Archaea, *Haloterrigena hispanica*, was isolated from the Fuente de Piedra saline lake in Spain.[65] The strain requires a minimum of 15% NaCl, pH 7.0, and temperature of 50 °C for optimum growth. Five diketopiperazine compounds were isolated from this strain, including cyclo-(D-prolyl-L-tyrosine) (**171**), cyclo-(L-prolyl-L-tyrosine) (**172**), cyclo-(L-prolyl-L-valine) (**173**), cyclo-(L-prolyl-L-phenylalanine) (**174**), and cyclo-(L-prolyl-L-isoleucine) (**175**). Although not an *N*-acyl homoserine lactone (AHL) signaling molecule, compound **173** was found to activate the AHL bioreporter (minimum concentration to activate was 2.5 mM).[65] The authors hypothesize that these compounds are produced to help the organism adapt to its hypersaline environment.

Ectoine (**176**) is a zwitterionic compound originally isolated from the halophilic bacterium *Ectothiorhodospira halochloris*.[66] A commercially available compound, ectoine (**176**), was used for several studies related to colitis, including inflammatory bowel diseases (IBDs) such as Crohn's disease and ulcerative colitis.[67] Previous studies suggested that compound **176** can prevent inflammatory responses related to IBD; thus, *in vivo* follow-up studies were carried out using adult male Wistar rats. Following treatment with compound **176**, several symptoms related to the induction of colitis were ameliorated and levels of inflammation-related biomarkers TNF-α, IL-1β, ICAM-1, PGE_2, and LTB_4 were reduced.[67]

The compounds 8-*O*-methyltetrangulol (**177**) and naphthomycin A (**178**) were isolated from a moderately halophilic strain of *Streptomyces* sp. nov. WH26.[68] This bacterium was collected from the Weihai Solar Saltern in China and determined to belong to the *Streptomyces* genus and is likely a new species. The two compounds (**177** and **178**) were tested for cytotoxicity against four cell lines

(A549, HeLa, BEL-740, HT-29), with both compounds exhibiting the strongest activity in BEL-7402 cells (IC$_{50}$ values of 47.1 µM and 7.9 µM, respectively).[68] A halophilic fungus, from the *Aspergillus* genus, designated as *Aspergillus* sp. nov. F1, was isolated from the same marine solar saltern in Weihai, China.[69] Three known compounds, ergosterol (**179**), rosellichalasin (**180**), and cytochalasin E (**181**), were isolated and identified. These were tested for cytotoxic activity against A549, HeLa, BEL-7402, and RKO cells, with the strongest activity found in RKO cells (IC$_{50}$ values of 3.3, 62.3, and 37.3 µM, respectively).

Another halophilic actinomycete, *Nocardiopsis gilva* YIM 90087, was isolated from a saline soil sample in China.[70] Fifteen compounds (**182–196**) were isolated from this strain, including a new *p*-terphenyl compound (**182**), a new benzothiazole *p*-terphenyl derivative (**184**), and 13 known compounds (*p*-terphenyl compound [**183**], novobiocin [**185**], 9 cyclodipeptides [**186–194**], and 2 aromatic acids [**195** and **196**]). These were tested against several strains of pathogenic *Fusarium* species known to cause *Fusarium* head blight, a fungal disease affecting several important agricultural crops. Compounds were also tested against other fungal and bacterial pathogens and for their 2,2′-azino-bis-3-ethylbenzothiazoline-6-sufonic acid radical (ABTS$^+$.) scavenging capacity. One of the new *p*-terphenyl compounds (**182**) was found to have MIC values of 8, 16, and 128 µg/ml against *Fusarium avenaceum*, *Fusarium graminearum*, and *Fusarium culmorum*, respectively, and also exhibited 68.6% ABTS$^+$. scavenging capacity.[70]

Another halophilic *Nocardiopsis* species, *N. terrae* YIM 90022, was isolated from saline soil in China by the same group.[71] Six compounds, including one new natural product, a quinolone alkaloid, 4-oxo-1,4-dihydroquinoline-3-carboxamide (**197**), and five known compounds [*p*-hydroxybenzoic acid (**198**), *N*-acetyl-anthranilic acid (**199**), indoly-3-carboxylic acid (**200**), cyclo(Trp-Gly) (**201**), and cyclo(Leu-Ala) (**202**)] were isolated. The new natural product (**197**) exhibited antimicrobial activity against *Staphylococcus aureus* (MIC 64 µg/ml), *Bacillus subtilis* (MIC 64 µg/ml), *Escherichia coli* (MIC 128 µg/ml), and *Pyricularia oryzae* (MIC 256 µg/ml). Compound **199** also exhibited some antimicrobial activity against several fungal and bacterial pathogens, although with weaker potency.[71]

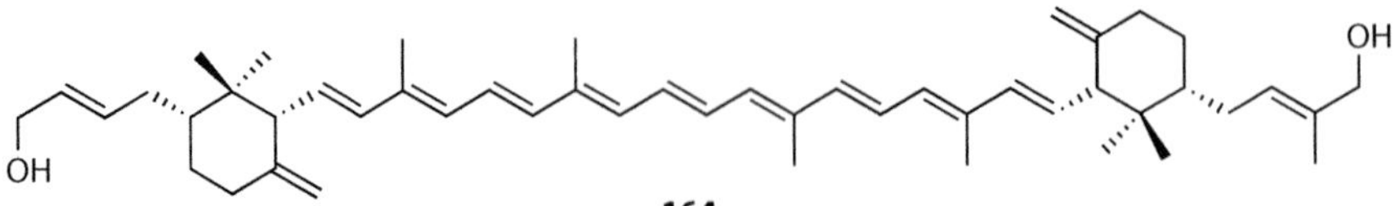

164

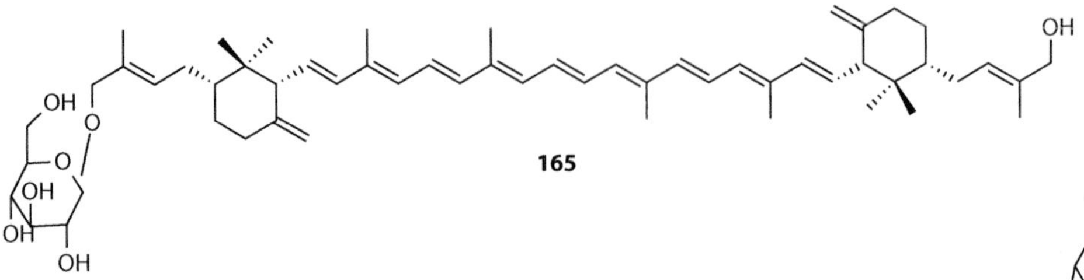

165

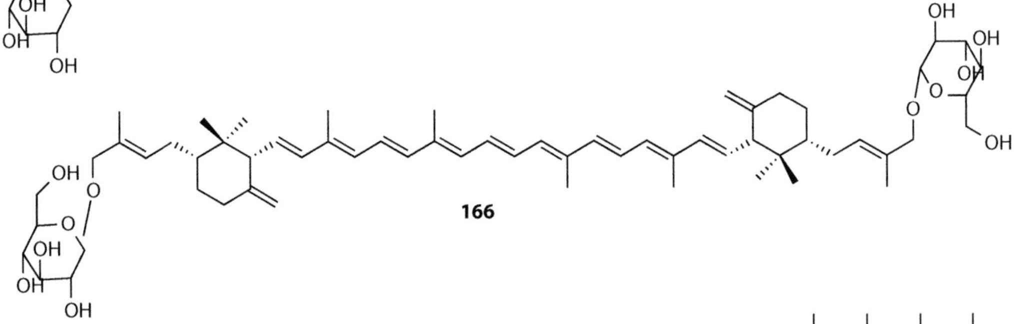

166

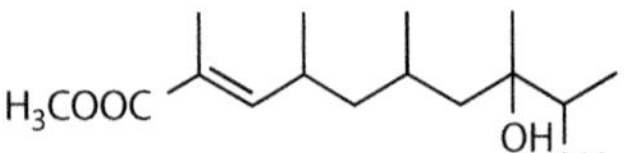

167

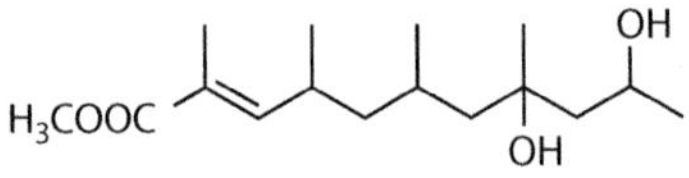

168

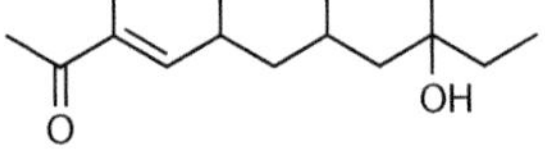

169

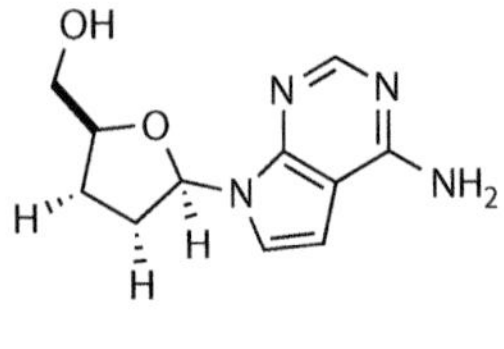

170

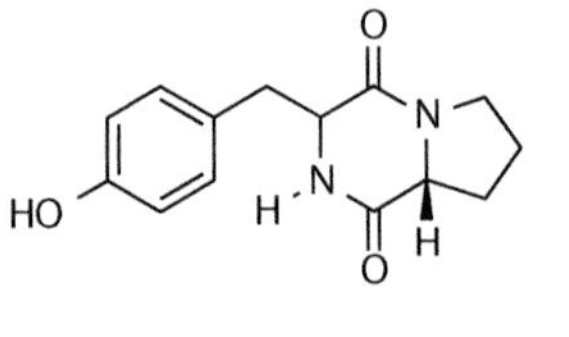

171

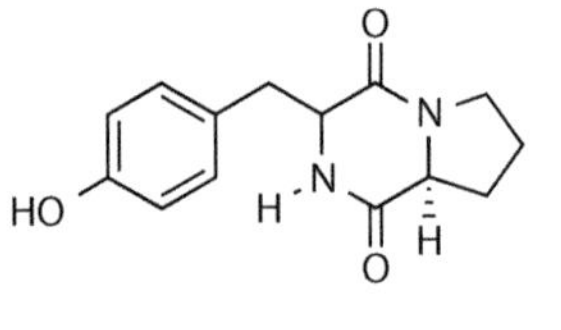

172

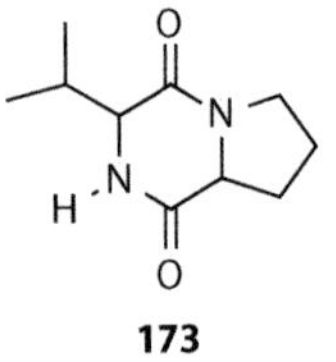

173

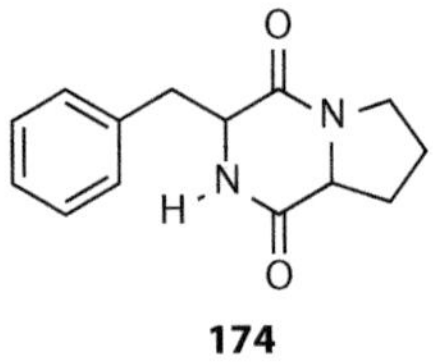

174

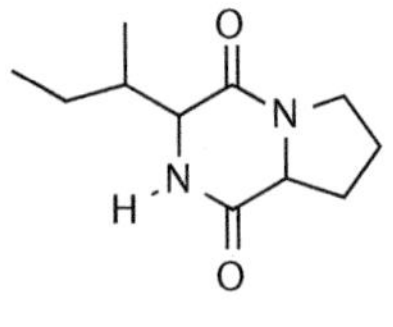

175

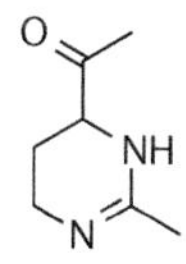

176

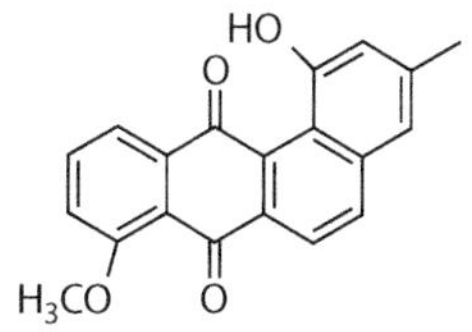

177

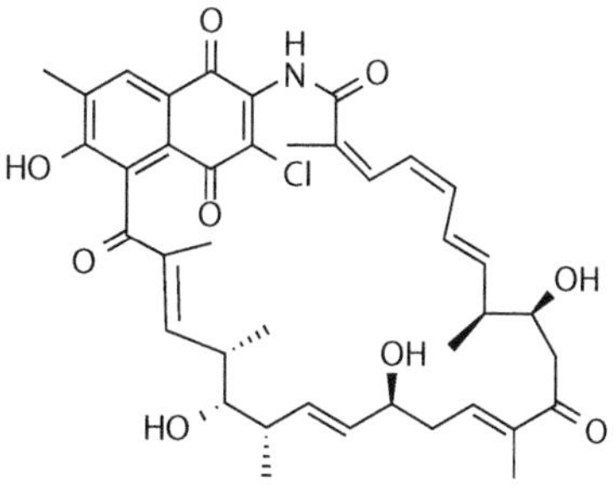

178

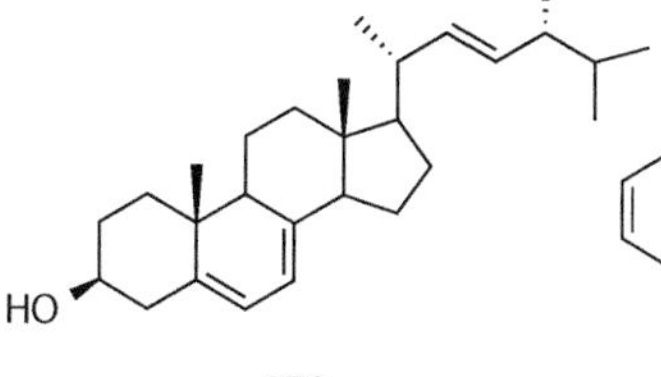

179

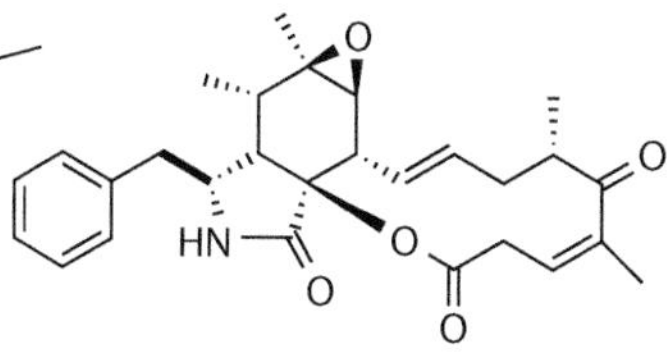

180

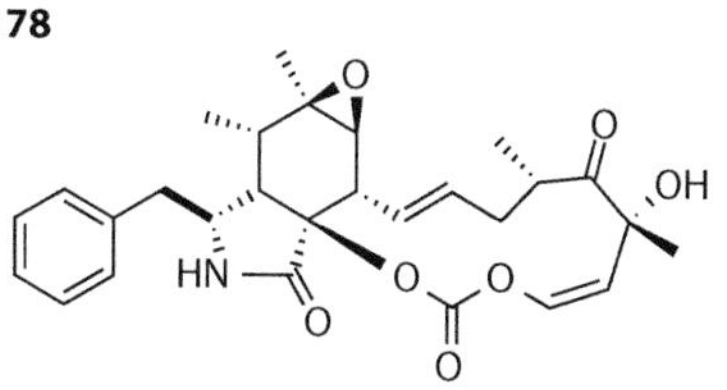

181

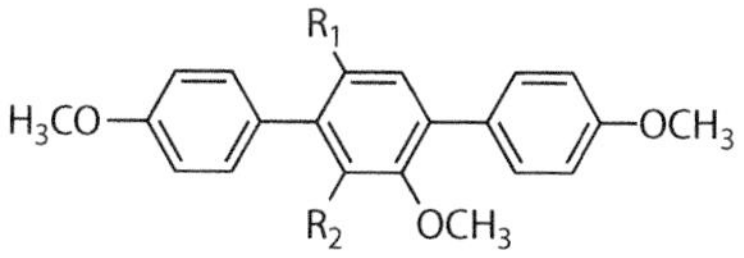

182: R$_1$ = OH; R$_2$ = OCH$_3$
183: R$_1$ = OCH$_3$; R$_2$ = OH

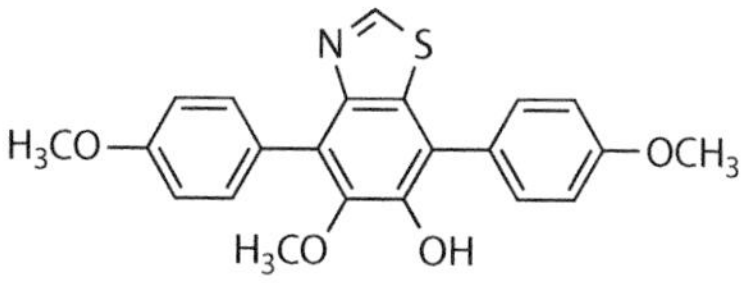

184

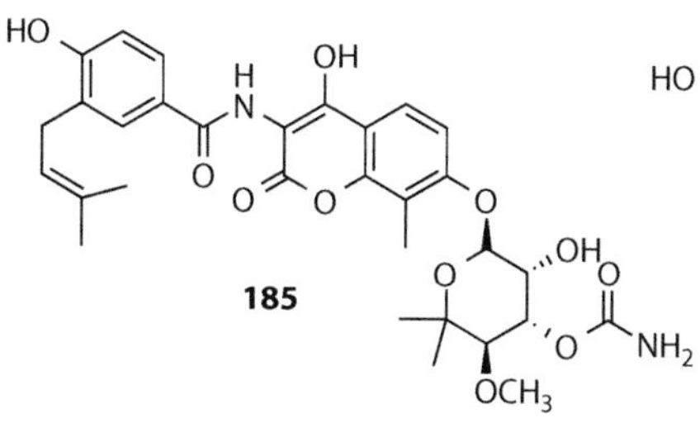

185

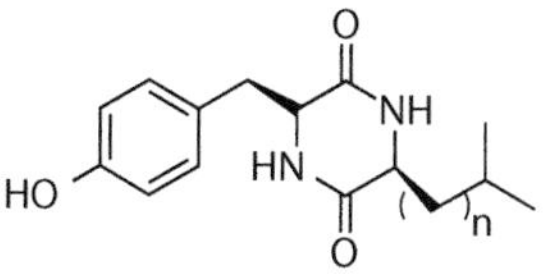

186: n = 1
187: n = 0

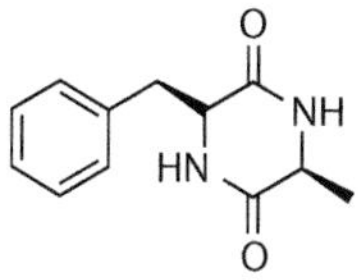

188

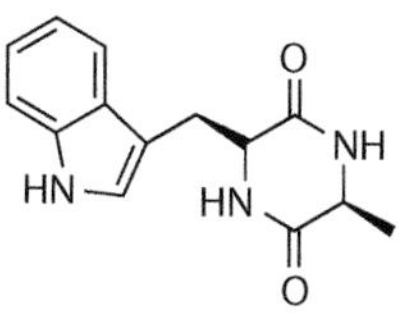

189

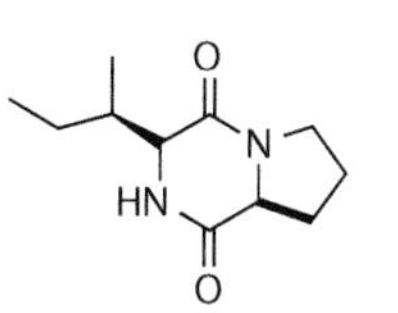

190

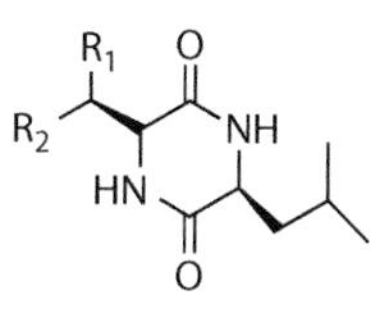

191: R$_1$ = R$_2$ = H
193: R$_1$ = R$_2$ = CH$_3$

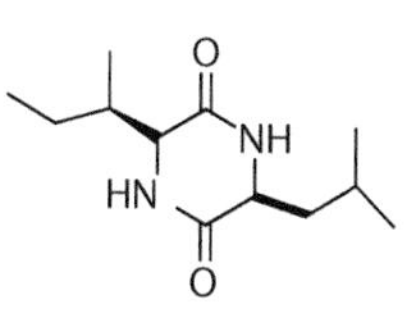

192

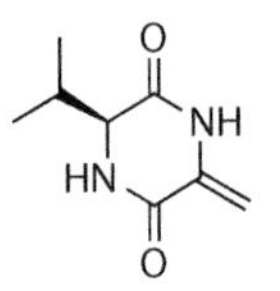

194

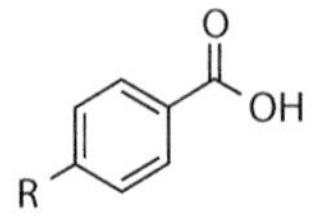

195: R = OH
196: R = H

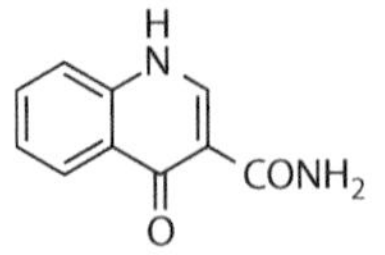

197

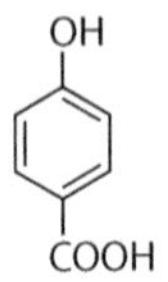

198

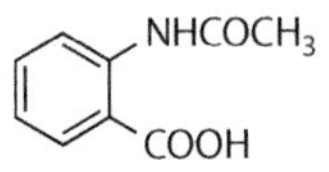

199

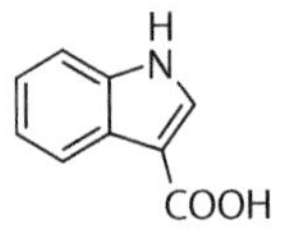

200

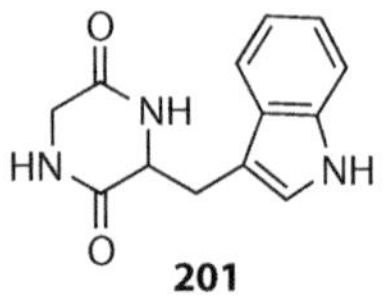

201

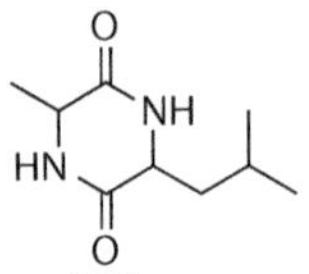

202

12.6 ACIDOPHILES

As mentioned in the introduction, acidophiles are considered to be organisms that can live in highly acidic environments (pH < 5).[1] Specifically, optimal growth is achieved at pH values between 2 and 4. These organisms have been found in habitats with high sulfur content, including abandoned mines, Berkley Pit in Montana, and hydrothermal vents.[1] Acidophiles have made several key modifications to their molecular mechanisms to survive in the harsh conditions. One of the important adaptations is the maintenance of a neutral pH in the cytoplasm of cells by constricting the pore size in the membrane channels, producing a cell membrane that is highly resistant to the entry of protons, and utilizing the ability to expel protons from the cell.[72] The section discusses biologically active compounds or extracts originating from acidophilic or acid-tolerant fungi and bacteria.

12.6.1 Fungi

There has been ongoing isolation of biologically active compounds from the Berkeley Pit Lake in Butte, Montana, since 1995.[3] The abandoned open copper mine is filled with acidified water (pH 2.5) containing metal sulfates with depths upwards of 540 m. Wilson and Brimble[5] summarized compounds discovered from the Berkley Pit until 2008. *Penicillium* species from the Berkley Pit are known to be a source of novel bioactive compounds.[3,5,73,74]

Three recent reports of *Penicillium rubrum* and *Penicillium solitum* afforded a total of 10 novel compounds (**203–205, 210–213, 220–222**).[3,73,74] In the first of these reports, three new meroterpenes, berkelyones A–C (**203–205**), were isolated from *P. rubrum* along with the known compounds preaustinoids A (**206**) and A1 (**207**).[74] All compounds were evaluated for their ability to inhibit caspase-1 and to mitigate interleukin-1β (IL-1β) production, with the strongest activity found in each assay for berkeleyone A with IC$_{50}$ values of 68 and 2.7 μM, respectively. Additionally, docking studies were performed to observe the orientation of compounds **203–207**, berkleydione (**208**), and berkeleytrione (**209**) in the catalytic pocket (caspase-1).[74]

Further work on *P. rubrum*[3] resulted in the isolation of new compounds berkazaphilones A (**210**) and B (**211**), berkedienoic acid (**212**), and berkedienolactone (**213**), as well as the known compounds berkazaphilone C (azaphilone) (**214**), penisimplicissin (**217**), aldehyde (**218**), vermistatin (**215**), dihydrovermistatin (**216**), and methylparaconic acid (**219**). Seven of the compounds (**210, 211, 214–218**) were tested for caspase-1 inhibition, with compounds **211** and **214** demonstrating IC$_{100}$ values (total enzyme inhibition) of 25 μM, while compounds **210, 217**, and **218** had IC$_{100}$ values of 250 μM.[3] Compounds **211, 214**, and **215–217** were further tested for IL-1β inhibition and exhibited inhibition at concentrations of 250 μM, with compounds **211** and **214** having the strongest inhibition

(5 and 50 μM, respectively). Compounds **211**, **214**, **216**, and **217** were tested against the NCI 60 cell line panel with selective activity against leukemia cell lines by several of the compounds, with the exception of **216**.[3]

A different species, *P. solitum*, was isolated from the surface water of the Berkley Pit.[73] The crude extract was found to have strong inhibitory activity against caspase-1, caspase-3, and MMP-3. Three new compounds were isolated, including berkedrimane A (**220**), berkedrimane B (**221**), and trimethyl 4-hydroxy-4,5-dicarboxy-14-pentadecenoate (**222**). Moreover, two previously reported compounds, secospiculisporic acid (**223**) and spiculisporic acid (**224**), were also isolated. Berkedrimanes A (**220**) and B (**221**) were tested for caspase-1 and caspase-3 inhibition and found to be moderately active (IC$_{50}$ values ranging from 50 to 200 μM), with further testing for inhibition of the production of IL-1β resulting in low micromolar inhibition.[73]

Penicillium purpurogenum JS03-21 (an acid-tolerant fungal strain) was isolated from red soil in Jianshui, Yunnan, China.[75] Because of the promising novel chemical scaffolds isolated from the Berkley Pit, the investigators grew the fungus under acidic conditions (pH 2). Six novel compounds were isolated, including purpurqinones A–C (**225–227**), purpuresters A and B (**228** and **229**), and 2,6,7-trihydroxy-3-methyl-napthalene-1,4-dione (**230**). Also, three previously described compounds, TAN-931 (**231**), (–)-mitorubrin (**232**), and orsellinic acid, were isolated. All of the compounds (**225–232**) were screened for cytopathic effect against the influenza A virus. Compounds purpurqinones B (**226**) and C (**227**), purpuresters A (**228**), and TAN-931 (**231**) demonstrated IC$_{50}$ values of 61.3, 64.0, 85.3, and 58.6 μM, respectively. The positive control, ribavirin, exhibited an IC$_{50}$ value of 100.8 μM, suggesting that compounds **226–228** and **231** have strong antiviral activity.[75]

12.6.2 Bacteria

Acid mine drainage is another natural source of organisms that may have the ability to produce new secondary metabolites. Recently, *Lysinibacillus fusiformis* (a Gram-positive bacterium) was isolated from an abandoned coal mine in Yong-Dong, South Korea.[76] This highly acidic environment (pH 3.0) is saturated with heavy metals and iron. Six compounds were isolated, including the two new compounds spirobacillenes A (**233**) and B (**234**) and the four known compounds (Z)-3-hydroxy-4-(3-indolyl)-hydroxyphenyl-2-butenone (**235**), (Z)-3-hydroxy-4-(3-indolyl)-L-phenyl-2-butenone (**236**), soraphinol A (**237**), and kurasoin B (**238**). Compounds **235** and **236** were described for the first time as natural compounds. All six compounds were tested for antimicrobial activity against 11 pathogenic strains of bacteria and fungi. Compound **235** demonstrated the strongest antimicrobial activity (MIC value of 3.13 μg/ml) against *Micrococcus luteus* and *Enterococcus hirae*, which are both Gram-positive bacteria. Further investigation of the potential biological activity was explored through inhibition studies of nitric oxide production and reactive oxygen species generation.[76]

From the same abandoned mine in Young-dong, South Korea, a *Streptomyces* sp. KMC004 was isolated at pH 3.0. The species afforded two novel angucyclic quinones: angumycinones A (**239**) and B (**240**). In addition, six previously reported angucyclinone metabolites were isolated: MM 47755 (**241**), (+)-rubiginone B$_2$ (**242**), (+)-ochromycinone (**243**), (+)-hatomarubigin A (**244**), (+)-rubiginone D$_2$ (**245**), and X-14881 E (**246**). All 8 compounds were tested against 10 pathogenic strains of bacteria and fungi. Both of the new angucyclic quinones (**239** and **240**) exhibited antimicrobial activity. Angumycinone A (**239**) demonstrated MIC values of 6.25 and 12.5 μg/ml against *M. luteus* and *E. hirae*, respectively. Angumycinone B (**240**) displayed MIC values of 0.78, 1.56, and 12.5 μg/ml against *M. luteus*, *E. hirae*, and MRSA. Compounds **241**, **243**, and **246** showed antimicrobial activity, whereas **242**, **244**, and **245** were considered not active against any of the pathogenic strains tested. When tested in the antimicrobial assays, compounds **239–246** had no activity against the fungi or Gram-negative bacteria.[77]

An actinomycete, *Streptomyces* sp. KMA-001, was isolated from a similar habitat (presence of low pH and heavy metals).[76] The species was collected in an abandoned mine in Yeonhwa, Korea, from a heat-treated soil sample. This strain afforded one new 16-membered macrolide, aldgamycin I (**247**), as well as four previously described macrolides, aldgamycins E–G (**248–250**) and chalcomycin (**251**). All five compounds (**247–251**) were tested against *M. luteus*, *Bacillus subtilis*, *Proteus vulgaris*, *SOalmonella typhimurium*, *Aspergillus fumigatus*, *Trichophyton rubrum*, and MRSA. Compounds **247–251** exhibited the strongest antimicrobial activity against *M. luteus* and *S. typhimurium*, with MIC values ranging from 0.78 to 12.5 µg/ml. However, compounds **247–251** demonstrated no activity against *A. fumigatus*, *T. rubrum*, or MRSA, with MIC values of >100 µg/ml.[76]

12.6.3 Co-Culture of Fungi and Bacteria

A bacterial strain (*Sphingomonas* KMK-001) and a fungal strain (*Aspergillus fumigatus* KMC-901) were isolated from an abandoned coal mine (Young-dong) in Gangneung, South Korea, at an elevation of 750 m.[76] The two species were isolated from acid mine drainage at pH 3.0. The bacterial and fungal strains were co-cultured to determine if new secondary metabolites would be biosynthesized compared to the two strains individually cultured. A novel compound, glionitrin A (**252**), was isolated after further purification of a crude ethyl acetate extract. To ensure compound **252** was produced as a result of the co-culturing, the bacterial and fungal strains were cultured separately. *A. fumigatus* KMC-901 afforded previously reported compounds gliotoxin (**253**) and dehydrogliotoxin (**254**), while no secondary metabolites were isolated from *Sphingomonas* KMK-001.

Glionitrin A (**252**) was screened against six cancer cell lines to evaluate potential cytotoxic properties. Compound **252** demonstrated strong cytotoxic activity with IC_{50} values of 0.82, 0.55, 0.45, and 0.24 µM against the cell lines HCT-116, A549, AGS, and DU145, respectively, with additional activity against MCF-7 and HepG2 cells resulting in IC_{50} values of 2.0 and 2.3 µM, respectively. In addition, an antimicrobial array of nine pathogenic strains was used to test glionitrin A (**252**). Compound **252** exhibited strong antibacterial activity against three different MRSA strains and a Gram-positive strain (*M. luteus*).[76]

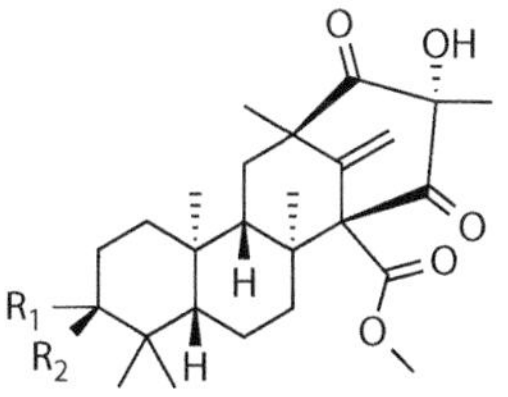

203: R$_1$ = H; R$_2$ = OH
206: R$_1$ = R$_2$ = O

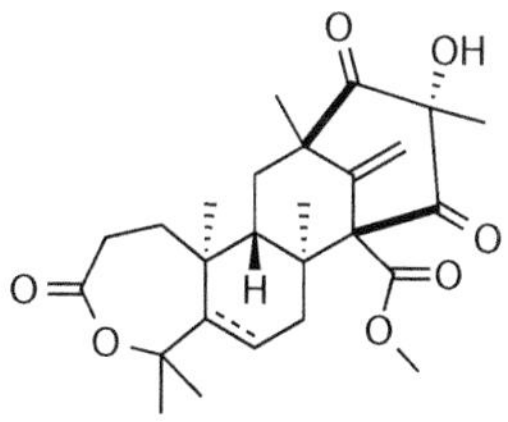

204
207

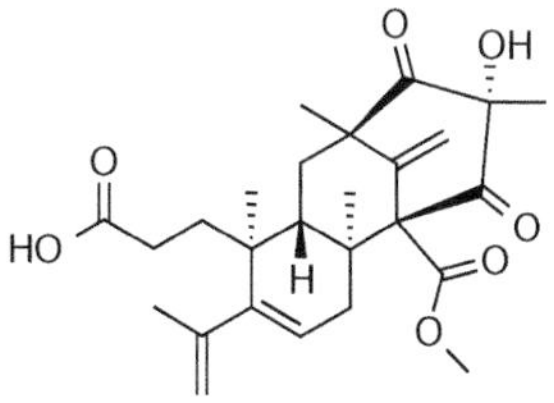

205

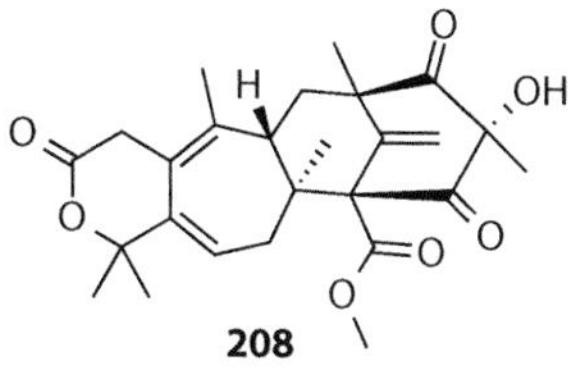

208

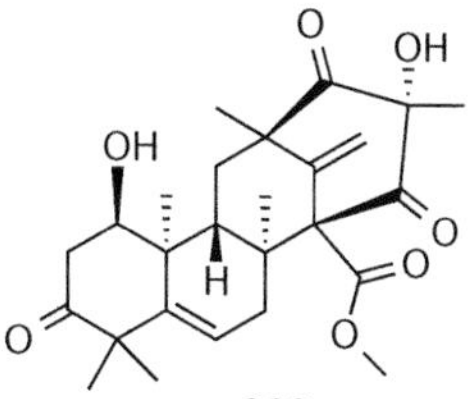

209

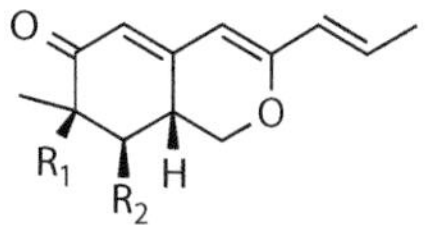

210: R$_1$ = H; R$_2$ = OH

211: R$_1$ =
R$_2$ = OH
214: R$_1$ = OH
R$_2$ =

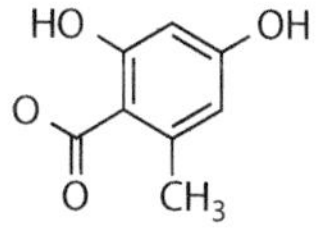

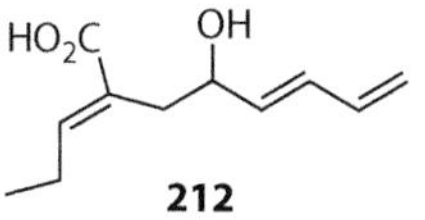

212

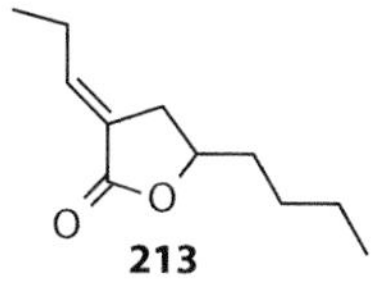

213

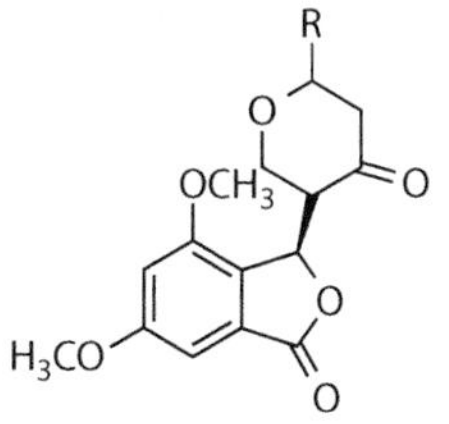

215: R = CH = CH-CH$_3$
216: R = CH$_2$-CH$_2$-CH$_3$
217: R = CH$_3$

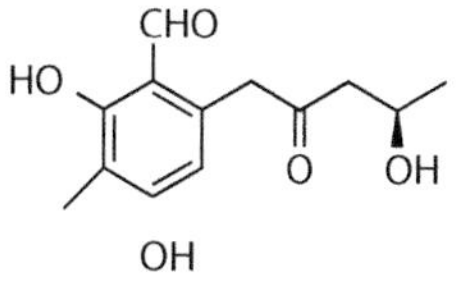

218

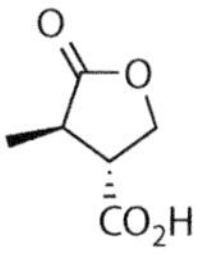

219

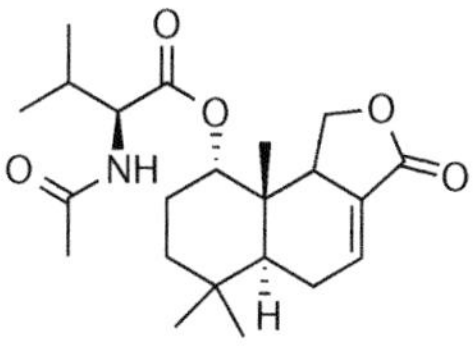

224

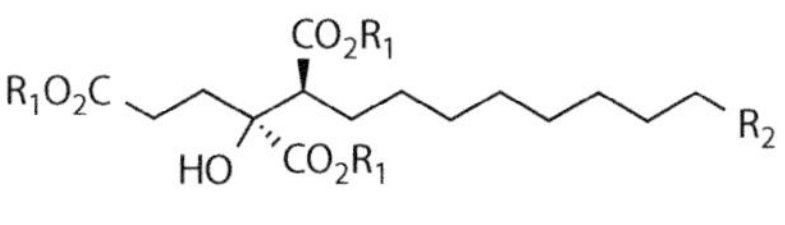

222: R$_1$ = CH$_3$; R$_2$ = CH = CH$_2$
223: R$_1$ = H; R$_2$ = CH$_2$CH$_3$

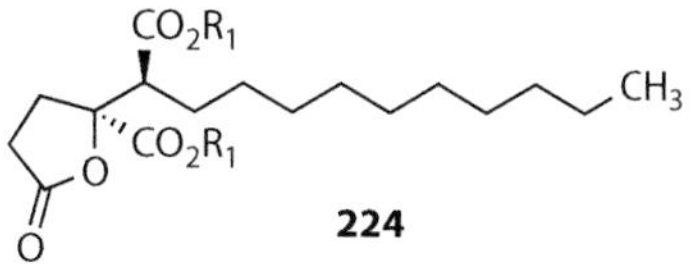

220: R = H
221: R = OH

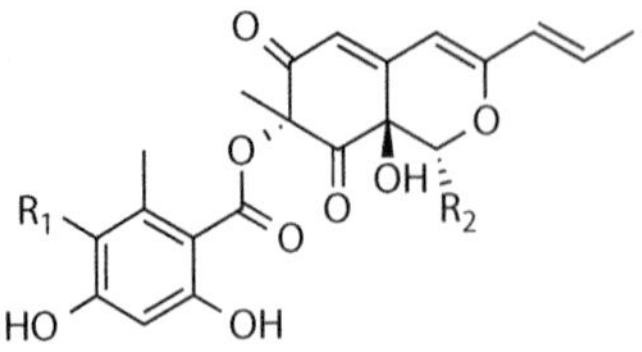

225: R$_1$ = H; R$_2$ = OH
226: R$_1$ = OH; R$_2$ = OH
227: R$_1$ = H; R$_2$ = H

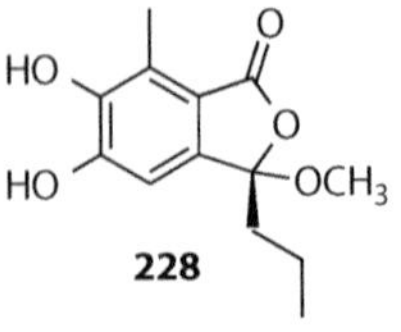

228

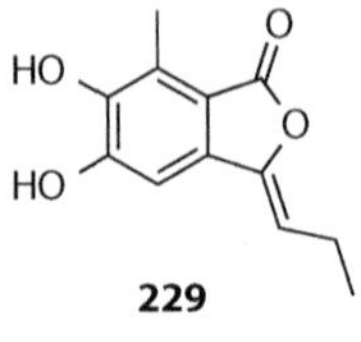

229

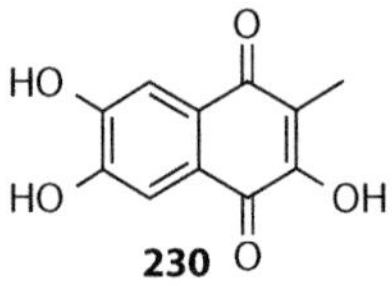

230

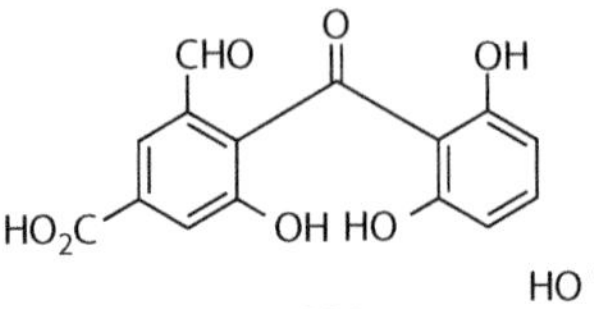

231

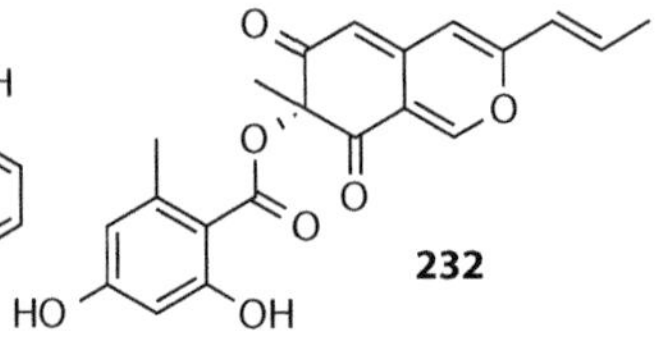

232

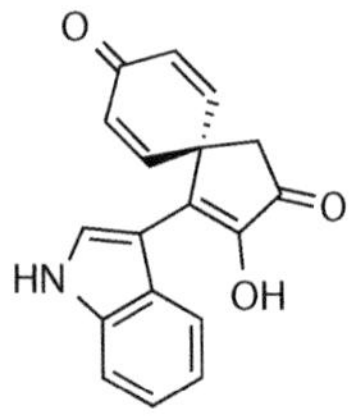

233

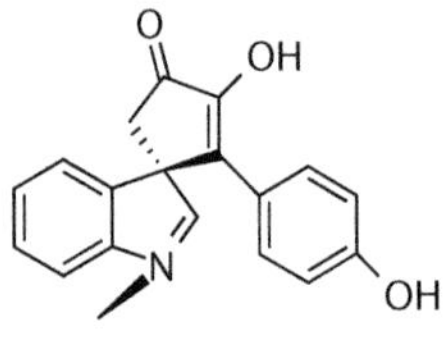

234

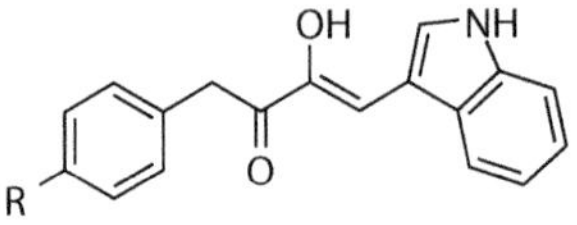

235: R = OH
236: R = H

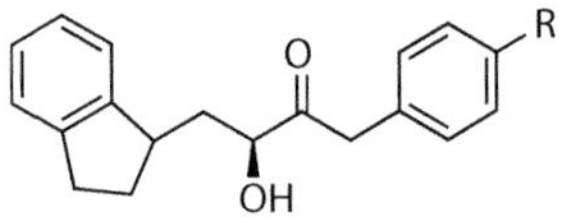

237: R = OH
238: R = H

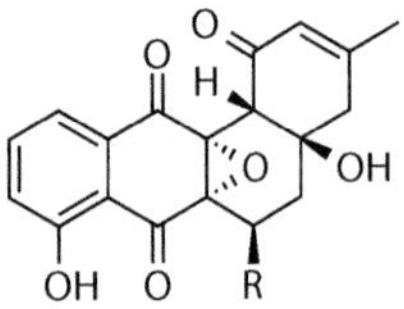

239: R = H
240: R = OH

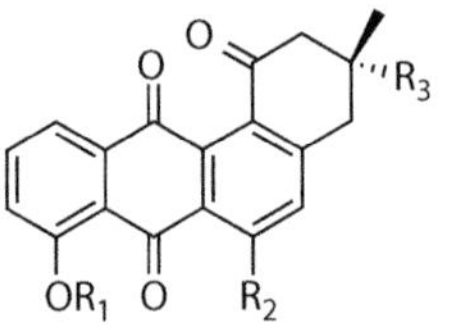

241: R$_1$ = CH$_3$; R$_2$ = H; R$_3$ = OH
242: R$_1$ = CH$_3$; R$_2$ = H; R$_3$ = H
243: R$_1$ = H; R$_2$ = H; R$_3$ = H
244: R$_1$ = CH$_3$; R$_2$ = OH; R$_3$ = H

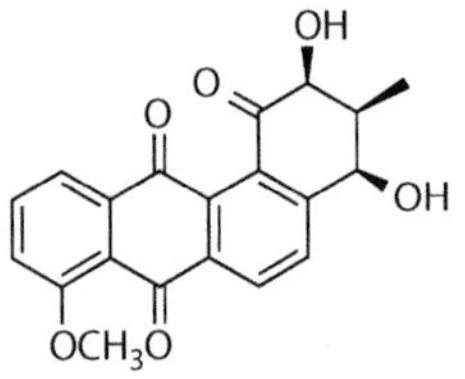

245

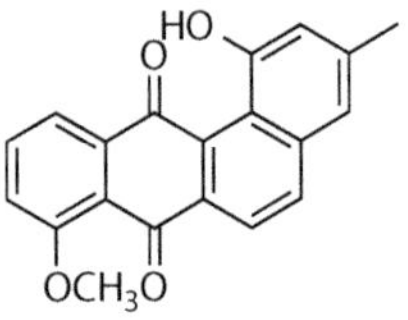

246

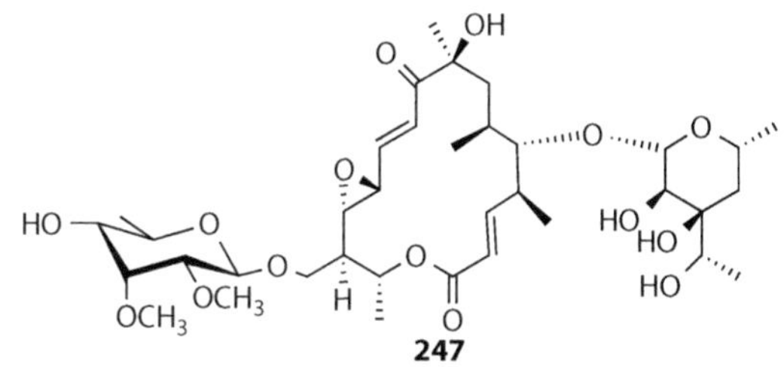

247

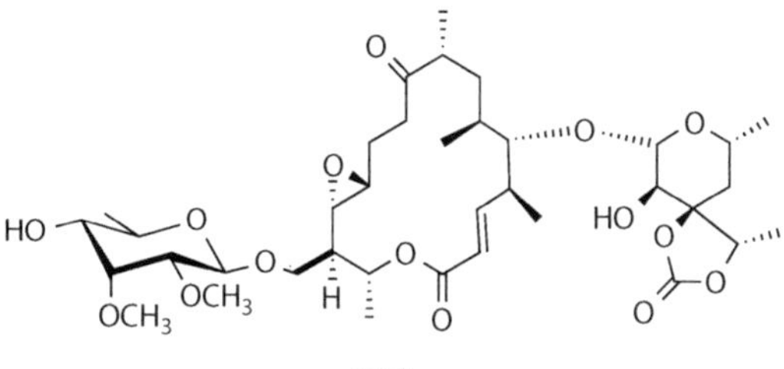

248

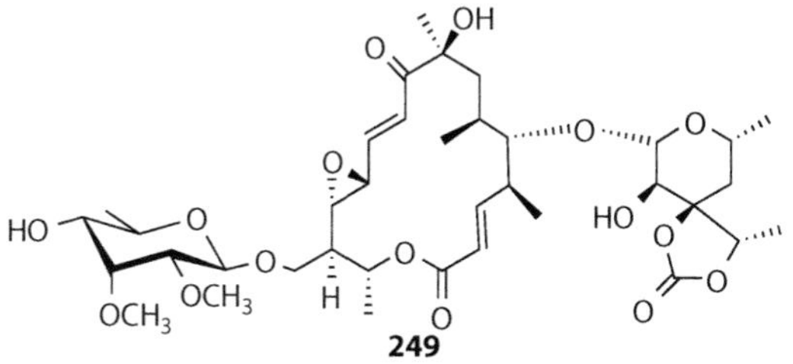

249

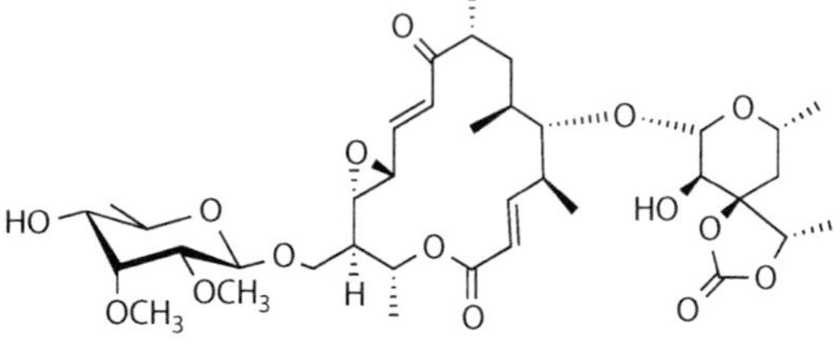

250

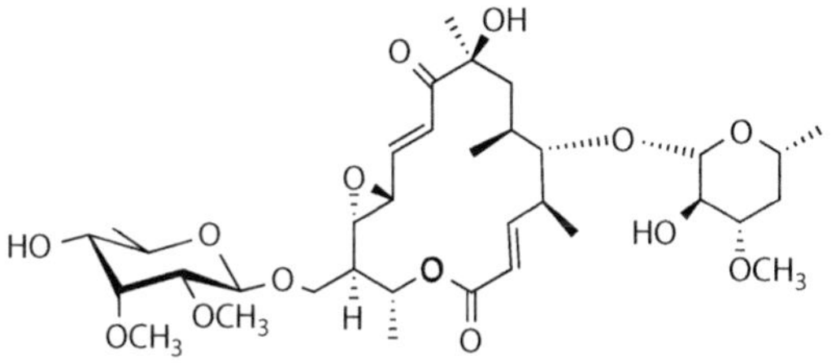

251

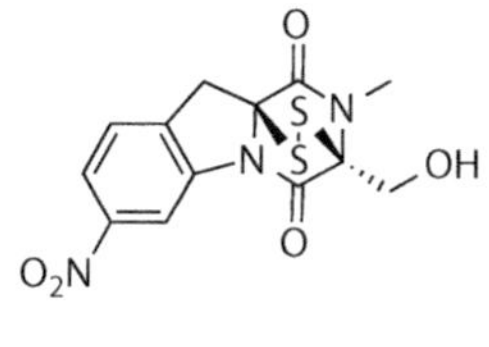

252

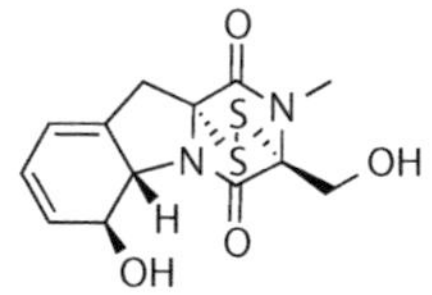

253

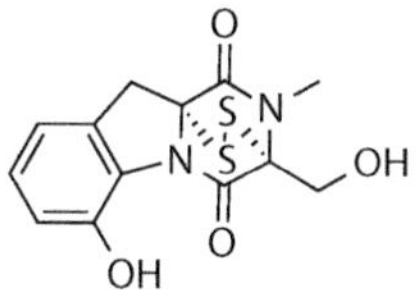

254

12.7 XEROPHILES

In response to the scarce availability of water, organisms become anhydrobiotic (in a desiccation state) and rely on several molecular adaptations to protect themselves from cellular damage.[78] Similar to the other organisms that live in extreme conditions, prokaryotes appear to depend on compatible solutes to replace water, synthesize key proteins (e.g., heat shock proteins), and acquire essential molecules (e.g., Mn^{2+} and sugar).[4,78,79] Xerophiles may be a potential source of novel compounds with intriguing biotechnological applications due to their ability to enter a desiccation state and resume normal cellular processes after rehydration.[78,80]

Rateb et al.[81] explored bacteria isolated from the Atacama Desert (Laguna de Chaxa) in northern Chile. After genome scanning 21 isolates, they further examined the chemical composition of *Streptomyces* sp. C34, which afforded four new compounds in the ansamycin class: chaxamycins A–D (**255–258**). Ansamycins are known to selectively interact with shock protein 90 (Hsp90).[82] Furthermore, compounds within this class have been shown to exhibit antibacterial and antitumor properties.[81] Thus, compounds **255–258** were tested in Hsp90 and antibacterial assays. At concentrations of 100 μM, chaxamycins A–C (**255–257**) inhibited the intrinsic ATPase activity of human Hsp90α by 46%, 45%, and 41%, respectively.[81] All four compounds were tested against *Staphylococcus aureus* and *Escherichia coli*. Compound **258** demonstrated the strongest antibacterial activity with MIC values of 1.21 and 0.05 μg/ml against *E. coli* and *S. aureus*, respectively. Compound **258** was further evaluated for activity against a methicillin-sensitive strain *S. aureus* (MIC 0.06–0.13 μg/ml) and 10 methicillin-resistant clinical isolates, with activity against 9 of the 10 clinical isolates (MIC < 1 μg/ml).[81]

255: R^1 = OH; R^2 = CH$_3$
256: R^1 = H; R^2 = CH$_3$
257: R^1 = OH; R^2 = CH$_2$OH

258

Another desert organism, a plant from the Crassulaceae family, *Rhodiola imbricate* (also known as rose root, Arctic root, and golden root), was collected at Chang La-Top (5330 m above sea level) in the Indian trans-Himalayan cold desert.[83] The plant is consumed as food and frequently used in traditional medicine.[83] Previous reports of biologically active compounds from this species have generally focused on isolation from the roots, specifically with extraction using polar solvents. Therefore, Tayade et al.[83] focused on characterizing compounds extracted using a range of nonpolar and polar solvents. GC-MS analysis of the root extracts revealed a total of 63 chemotypes: 22 from the n-hexane extract, 18 from the chloroform extract, 25 from the dichloromethane extract, 19 from the ethyl acetate extract, 18 from the methanol extract, and 12 from the 60% ethanol extract (some chemotypes found in more than one type of extract). The various chemotypes characterized via GC-MS included steroids, alkanes, esters, ethers, fatty acids, and terpenoids (structures not included in this review due to space limitations).

Although there has not been a significant amount of compounds isolated and characterized from xerophiles, extracts from various organisms have been reported to exhibit biological activity that may be promising candidates for natural product discovery.[84–86] Moreover, gene sequencing of soil samples from the Sonoran Desert of Arizona, the Anza-Borrego region of the Sonoran Desert in California, and the Great Basin Desert of Utah suggested that there may be minimal similarity in the secondary metabolites produced by these xerophilic organisms.[87] Thus, these studies support the need for further investigation of desert organisms for novel compounds.

12.8 CONCLUSIONS AND FUTURE DIRECTIONS

The extremophiles reported herein comprise several classes, including psychrophiles, thermophiles, barophiles, halophiles, acidophiles, and xerophiles, most of which have been studied since the previous reviews on metabolites from extreme environments (e.g., Skropeta,[56] Wilson and Brimble,[5] and Pettit[50]). Given the large number of new compounds reported thereafter, many with unique structural moieties, research on these intriguing organisms is gaining attention. Herein we have reviewed the reports of 258 compounds isolated from extremophiles, including 136 from psychophiles (28 from soft corals, 25 from sponges, 33 from tunicates, and 50 from bacteria), 14 from thermophiles, 13 barotolerant compounds from deep-sea environments, 39 from halophiles, 52 from acidophiles, and 4 from xerophiles. Many of these metabolites exhibit potent biological activities against various targets, suggesting that further investigation of these extremophiles is warranted.

Although there have been significant discoveries in both chemical diversity and biological activity from extremophilic organisms, there remain many underexplored and unexplored regions (e.g., alkaliphiles, photophobic organisms from caves, organisms adapted to toxic environments, and volcanic microbes), thus suggesting a strong potential for future contributions from these environments to the field of natural product drug discovery. As scientists continue to explore new regions and identify new organisms, natural product chemistry will continue to expand into these new realms.

REFERENCES

1. Canganella, F., Wiegel, J. Extremophiles: From abyssal to terrestrial ecosystems and possibly beyond. *Naturwissenschaften* 2011, 98 (4), 253–279.
2. Bhullar, K., Waglechner, N., Pawlowski, A., Koteva, K., Banks, E. D., Johnston, M. D., Barton, H. A., Wright, G. D. Antibiotic resistance is prevalent in an isolated cave microbiome. *PLoS One* 2012, 7 (4).
3. Stierle, A. A., Stierle, D. B., Girtsman, T. Caspase-1 inhibitors from an extremophilic fungus that target specific leukemia cell lines. *Journal of Natural Products* 2012, 75 (3), 344–350.
4. Rothschild, L. J., Mancinelli, R. L. Life in extreme environments. *Nature* 2001, 409 (6823), 1092–1101.
5. Wilson, Z. E., Brimble, M. A. Molecules derived from the extremes of life. *Natural Product Reports* 2009, 26 (1), 44–71.
6. Margesin, R., Miteva, V. Diversity and ecology of psychrophilic microorganisms. *Research in Microbiology* 2011, 162 (3), 346–361.
7. Shivaji, S., Prakash, J. S. S. How do bacteria sense and respond to low temperature? *Archives of Microbiology* 2010, 192 (2), 85–95.
8. Averhoff, B., Muller, V. Exploring research frontiers in microbiology: Recent advances in halophilic and thermophilic extremophiles. *Research in Microbiology* 2010, 161 (6), 506–514.
9. Meruelo, A. D., Han, S. K., Kim, S., Bowie, J. U. Structural differences between thermophilic and mesophilic membrane proteins. *Protein Science* 2012, 21 (11), 1746–1753.
10. Cavicchioli, R., Thomas, T., Curmi, P. M. G. Cold stress response in Archaea. *Extremophiles Review* 2000, 4 (6), 321–331.
11. Feller, G., Gerday, C. Psychrophilic enzymes: Hot topics in cold adaptation. *Nature Reviews Microbiology* 2003, 1 (3), 200–208.
12. Deming, J. W. Extremophiles: Cold environments. In *Encyclopedia of Microbiology*, ed. M. Schaechter. Vol. 1. Academic Press, Oxford, UK. 2009, pp. 147–158.

13. Taboada, S., Nunez-Pons, L., Avila, C. Feeding repellence of Antarctic and sub-Antarctic benthic inver-tebrates against the omnivorous sea star *Odontaster validus*. *Polar Biology* 2013, 36 (1), 13–25.

14. D'Amico, S., Collins, T., Marx, J.-C., Feller, G., Gerday, C. Psychrophillic microorganisms: Challenges for life. *EMBO Reports* 2006, 7 (4), 385–389.

15. Lebar, M. D., Heimbegner, J. L., Baker, B. J. Cold-water marine natural products. *Natural Product Reports* 2007, 24 (4), 774–797.

16. McClintock, J. B., Amsler, C. D., Baker, B. J. Overview of the chemical ecology of benthic marine invertebrates along the western Antarctic peninsula. *Integrative and Comparative Biology* 2010, 50 (6), 967–980.

17. Abbas, S., Kelly, M., Bowling, J., Sims, J., Waters, A., Hamann, M. Advancement into the Arctic region for bioactive sponge secondary metabolites. *Marine Drugs* 2011, 9 (11), 2423–2437.

18. Boustie, J., Tomasi, S., Grube, M. Bioactive lichen metabolites: Alpine habitats as an untapped source. *Phytochemistry Reviews* 2011, 10 (3), 287–307.

19. Bratchkova, A., Ivanova, V. Bioactve metabolites produced by microorganisms collected in Antarctica and the Arctic. *Biotechnology and Biotechnological Equipment* 2011, 24 (4), 1–7.

20. Nunez-Pons, L., Carbone, M., Vazquez, J., Gavagnin, M., Avila, C. Lipophilic defenses from *Alcyonium* soft corals of Antarctica. *Journal of Chemical Ecology* 2013, 39 (5), 675–685.

21. Palermo, J. A., Rodriguez Brasco, M. F., Spagnuolo, C., Seldes, A. M. Illudalane sesquiterpenoids from the soft coral *Alcyonium paessleri*: The first natural nitrate esters. *Journal of Organic Chemistry* 2000, 65 (15), 4482–4486.

22. Carbone, M., Nunez-Pons, L., Castelluccio, F., Avila, C., Gavagnin, M. Illudalane sesquiterpenoids of the alcyopterosin series from the Antarctic marine soft coral *Alcyonium grandis*. *Journal of Natural Products* 2009, 72 (2), 1357–1360.

23. (a) Hanssen, K. O., Andersen, J. H., Stiberg, T., Engh, R. A., Svenson, J., Geneviere, A.-M., Hansen, E. Antitumoral and mechanistic studies of ianthelline isolated from the Arctic sponge *Stryphnus fortis*. *Anticancer Research* 2012, 32 (10), 4287–4297. (b) Litaudon, M., Guyot, M. Ianthalline, un nouveau derive de 1fl dibromo-3,5 tyrosine, isole de l'eponge *Ianthella ardis* (Bahamas). *Tetrahedron Letters* 1986, 27 (37), 4455–4456.

24. Turk, T., Avgustin, J. A., Batista, U., Strugar, G., Kosmina, R., Civovic, S., Janussen, D., Kauferstein, S., Mebs, D., Sepcic, K. Biological activities of ethanolic extracts from deep-sea Antarctic marine sponges. *Marine Drugs* 2013, 11 (4), 1126–1139.

25. Nunez-Pons, L., Carbone, M., Paris, D., Melck, D., Rios, P., Cristobo, J., Castelluccio, F., Gavagnin, M., Avila, C. Chemo-ecological studies on hexactinellid sponges from the Southern Ocean. *Naturwissenschaften* 2012, 99 (5), 353–368.

26. Soest, R. W. M. V., Boury-Esnault, N., Vacelet, J., Dohrmann, M., Erpenbeck, D., De Voogd, N. J., Santodomingo, N., Vanhoorne, B., Kelly, M., Hooper, J. N. A. Global diversity of sponges (Porifera). *PLoS One* 2012, 7 (4), 1–23.

27. Na, M., Ding, Y., Wang, B., Tekwani, B., Schinazi, R. F., Franzblau, S., Kelly, M., Stone, R., Li, X.-C., Ferreira, D., Hamann, M. Anti-infective discorhabdins from a deep-water Alaskan sponge of the genus *Latrunculia*. *Journal of Natural Products* 2010, 73 (3), 383–387.

28. Zou, Y., Hamann, M. Atkamine: A new pyrroloiminoquinone scaffold from the cold water Aleutian Islands *Latrunculia* sponge. *Organic Letters* 2013, 15 (7), 1516–1519.

29. Timm, C., Mordhorst, T., Kock, M. Synthesis of 3-alkyl pyridinium alkaloids from the Arctic sponge *Haliclona viscosa*. *Marine Drugs* 2010, 8 (3), 483–497.

30. Schmidt, G., Timm, C., Grube, A., Volk, C. A., Kock, M. Viscosalines B(1,2) and E(1,2): Challenging new 3-alkyl pyridinium alkaloids from the marine sponge *Haliclona viscosa*. *Chemistry* 2012, 18 (26), 8180–8189.

31. Hedner, E., Sjogren, M., Andersson, R., Goransson, U., Jonsson, P. R., Bohlin, L. Antifouling activity of a dibrominated cyclopeptide from the marine sponge *Geodia barretti*. *Journal of Natural Products* 2008, 71 (3), 330–333.

32. Sjogren, M., Jonsson, P. R., Dahlstrom, M., Lundalv, T., Burman, R., Goransson, U., Bohlin, L. Two brominated cyclic dipeptides released by the coldwater marine sponge *Geodia barretti* act in synergy as chemical defense. *Journal of Natural Products* 2011, 74 (3), 449–454.

33. Fontana, A., Ciavatta, M. L., Amodeo, P., Cimino, G. Single solution phase conformation of new anti-proliferative cembranes. *Tetrahedron* 1999, 55 (4), 1143–1152.

34. (a) Cutignano, A., De Palma, R., Fontana, A. A chemical investigation of the Antarctic sponge *Lyssodendoryx flabellata*. *Natural Product Research* 2012, 26 (13), 1240–1248.

35. Nunez-Pons, L., Carbone, M., Vazquez, J., Rodriguez, J., Nieto, R. M., Varela, M. M., Gavagnin, M., Avila, C. Natural products from Antarctic colonial ascidians of the genera *Aplidium* and *Synoicum*: Variability and defensive role. *Marine Drugs* 2012, 10 (8), 1741–1764.

36. Lebar, M. D., Baker, B. J. Synthesis and structure reassessment of psammopemmin A. *Australian Journal of Chemistry* 2010, 63 (6), 862–866. (b) Butler, H. S., Cupon, R. J., Lu, C. C. Psammopemmins (A–C), novel brominated 4-hydroxyindole alkaloids from an Anarctic sponge. *Australian Journal of Chemistry* 1992, 45, 1871–1877.

37. Tadesse, M., Strom, M. B., Svenson, J., Jaspars, M., Milne, B. F., Torfoss, V., Andersen, J. H., Hansen, E., Stensvag, K., Haug, T. Synoxazolidinones A and B: Novel bioactive alkaloids from the Ascidian *Synoicum pulmonaria*. *Organic Letters* 2010, 12 (21), 4752–4755.

38. Noguez, J. H., Diyabalanage, T., Miyata, Y., Xie, X.-S., Valeriote, F. A., Amsler, C. D., McClintock, J. B., Baker, B. J. Palmerolide macrolides from the Antarctic tunicate *Synoicum adareanum*. *Bioorganic and Medicinal Chemistry* 2011, 19 (22), 6608–6614.

39. Diyabalanage, T., Amsler, C. D., McClintock, J. B., Baker, B. J. Palmerolide A, a cytotoxic macrolide from the Antarctic tunicate *Synoicum adareanum*. *Journal of the American Chemical Society* 2006, 128 (17), 5630–5631.

40. Wietz, M., Mansson, M., Bowman, J. S., Blom, N., Ng, Y., Gram, L. Wide distribution of closely related, antibiotic-producing *Arthrobacter* strains throughout the Arctic Ocean. *Applied and Environmental Microbiology* 2012, 78 (6), 2039–2042.

41. Al-Zereini, W., Schuhmann, I., Laatsch, H., Helmke, E., Anke, H. New aromatic nitro compounds from *Salegentibacter* sp. T436 an Arctic sea ice bacterium: Taxonomy, fermentation, isolation and biological activities. *Journal of Antibiotics* 2007, 60 (5), 301–308.

42. Schuhmann, I., Fotso-Fondja Yao, C. B., Al-Zereini, W., Anke, H., Helmke, E., Laatsch, H. Nitro derivatives from the Arctic ice bacterium *Salegentibacter* sp. isolate T436. *Journal of Antibiotics* 2009, 62 (8), 453–460.

43. Barros, J., Becerra, J., Gonzalez, C., Martinez, M. Antibacterial metabolites synthesized by psychrotrophic bacteria isolated from cold-freshwater environments. *Folia Microbiologica* 2012, 58 (2), 127–133.

44. Dalisay, D. S., Williams, D. E., Wang, X. L., Centko, R., Chen, J., Andersen, R. J. Marine sediment-derived *Streptomyces* bacteria from British Columbia, Canada are a promising microbiota resource for the discovery of antimicrobial natural products. *PLoS One* 2013, 8 (10).

45. Sletta, H., Degnes, K. F., Herfindal, L., Klinkenberg, G., Fjaervik, E., Zahlsen, K., Brunsvik, A., Nygaard, G., Aachmann, F. L., Ellingsen, T. E., Doskeland, S. O., Zotchev, S. B. Anti-microbial and cytotoxic 1,6-dihydroxyphenazine-5,10-dioxide (iodinin) produced by *Streptosporangium* sp. DSM 45942 isolated from the fjord sediment. *Applied Microbiology and Biotechnology* 2013, 98 (2), 603–610.

46. Lee, L. H., Cheah, Y. K., Sidik, S. M., Ab Mutalib, N. S., Tang, Y. L., Lin, H. P., Hong, K. Molecular characterization of Antarctic actinobacteria and screening for antimicrobial metabolite production. *World Journal of Microbiology and Biotechnology* 2012, 28 (5), 2125–2137.

47. Kleinteich, J., Wood, S. A., Puddick, J., Schleheck, D., Kupper, F. C., Dietrich, D. R. Potent toxins in Arctic environments: Presence of saxitoxins and an unusual microcystin variant in Arctic freshwater ecosystems. *Chemico-Biological Interactions* 2013, 206 (2), 423–431.

48. Beattie, K. A., Kaya, K., Sano, T., Codd, G. A. Three dehydrobutyrine-containing microcystins from Nostoc. *Phytochemistry* 1998, 47 (7), 1289–1292.

49. Thornburg, C., Zabriskie, M., McPhail, K. Deep-sea hydrothermal vents: Potential hot spots for natural products discovery. *Journal of Natural Products* 2010, 73 (3), 489–499.

50. Pettit, R. K. Culturability and secondary metabolite diversity of extreme microbes: Expanding contribution of deep sea and deep-sea vent microbes to natural product discovery. *Marine Biotechnology* (New York) 2011, 13 (1), 1–11.

51. Andrianasolo, E. H., Haramaty, L., Roasrio-Passapera, R., Vetriani, C., Falkowski, P., White, E., Lutz, R. Ammonificins C and D, hydroxyethylamine chromene derivatives from a cultured marine hydrothermal vent bacterium, *Thermovibrio ammonificans*. *Marine Drugs* 2012, 10 (10), 2300–2311.

52. Dimise, E. J., Widboom, P. F., Bruner, S. D. Structure elucidation and biosynthesis of fuscachelins, peptide siderophores from the moderate thermophile *Thermobifida fusca*. *Proceedings of the National Academy of Sciences* 2008, 105 (40), 15311–15316.

53. Pednekar, P., Jain, R., Mahajan, G. Anti-infective potential of hot-spring bacteria. *Journal of Global Infectious Diseases* 2011, 3 (3), 241–245.

54. Hu, Y. F., Phelan, V., Ntai, I., Famet, C. M., Zazopoulos, E., Bachmann, B. O. Benzodiazepine biosynthesis in *Streptomyces refuineus* (vol 14, pg 691, 2007). *Chemistry and Biology* 2007, 14 (7), 870.

55. Abdel-Mageed, W. M., Milne, B. F., Wagner, M., Schumacher, M., Sandor, P., Pathom-aree, W., Goodfellow, M., Bull, A. T., Horikoshi, K., Ebel, R., Diederich, M., Fiedler, H.-P., Jaspars, M. Dermacozines, a new phenazine family from deep-sea dermacocci isolated from a Mariana Trench sediment. *Organic and Biomolecular Chemistry* 2010, 8 (10), 2352–2362.

56. Skropeta, D. Deep-sea natural products. *Natural Product Reports* 2008, 25 (6), 1131–1166.

57. Genta-Jouve, G., Francezon, N., Puissant, A., Auberger, P., Vacelet, J., Perez, T., Fontana, A., Al Mourabit, A., Thomas, O. P. Structure elucidation of the new citharoxazole from the Mediterranean deep-sea sponge *Latrunculia (Biannulata) citharistae*. *Magn. Reson. Chem.* 2011, 49 (8), 533–536.

58. Tajima, H., Wakimoto, T., Takada, K., Ise, Y., Abe, I. Revised structure of cyclolithistide A, a cyclic depsipeptide from the marine sponge *Discodermia japonica*. *Journal of Natural Products* 2014, 77 (1), 154–158.

59. Clark, D. P., Carroll, J., Naylor, S., Crews, P. An antifungal cyclodepsipeptide, cyclolithistide A, from the sponge *Theonella swinhoei*. *Journal of Organic Chemistry* 1998, 63 (24), 8757–8764.

60. Pinchuk, B., Johannes, E., Gul, S., Schlosser, J., Schaechtele, C., Totzke, F., Peifer, C. Marine derived hamacanthins as lead for the development of novel PDGFRbeta protein kinase inhibitors. *Marine Drugs* 2013, 11 (9), 3209–3223.

61. Wolkenstein, K., Schoefberger, W., Muller, N., Oji, T. Proisocrinins A–F, brominated anthraquinone pigments from the stalked crinoid *Proisocrinus ruberrimus*. *Journal of Natural Products* 2009, 72 (11), 2036–2039.

62. Ishigami, S. T., Goto, Y., Inoue, N., Kawazu, S., Matsumoto, Y., Imahara, Y., Tarumi, M., Nakai, H., Fusetani, N., Nakao, Y. Cristaxenicin A, an antiprotozoal xenicane diterpenoid from the deep sea gorgonian *Acanthoprimnoa cristata*. *Journal of Organic Chemistry* 2012, 77 (23), 10962–10966.

63. Osawa, A., Ishii, Y., Sasamura, N., Morita, M., Kasai, H., Maoka, T., Shindo, K. Characterization and antioxidative activities of rare C(50) carotenoids—sarcinaxanthin, sarcinaxanthin monoglucoside, and sarcinaxanthin diglucoside—obtained from *Micrococcus yunnanensis*. *Journal of Oleo Science* 2010, 59 (12), 653–659.

64. Zhao, L.-X., Huang, S.-Z., Tang, S.-K., Jiang, C.-L., Duan, Y., Beutler, J. A., Henrich, C. J., McMahon, J. B., Schmid, T., Blees, J. S., Colburn, N. H., Rajski, S. R., Shen, B. Actinopolysporins A–C and tubercidin as a Pdcd4 stabilizer from the halphilic actinomycete *Actinopolyspora erythraea* YIM 90600. *Journal of Natural Products* 2011, 74 (9), 1990–1995.

65. Tommonaro, G., Abbamondi, G. R., Iodice, C., Tait, K., De Rosa, S. Diketopiperazines produced by the halophilic archaeon, *Haloterrigena hispanica*, activate AHL bioreporters. *Microbial Ecology* 2012, 63 (3), 490–495.

66. Galinski, E. A., Pfeiffer, H. P., Truper, H. G. 1,4,5,6-Tetrahydro-2-methyl-4-pyrimidinecarboxylic acid: A novel cyclic amino acid from halophilic phototrophic bacteria of the genus *Ectothiorhodospira*. *European Journal of Biochemistry* 1985, 149 (1), 135–139.

67. Abdel-Aziz, H., Wadie, W., Abdallah, D. M., Lentzen, G., Khayyal, M. T. Novel effects of ectoine, a bacteria-derived natural tetrahydropyrimidine, in experimental colitis. *Phytomedicine* 2013, 20 (7), 585–591.

68. Liu, H., Xiao, L., Wei, J., Schmitz, J. C., Liu, M., Wang, C., Cheng, L., Wu, N., Chen, L., Zhang, Y., Lin, X. Identification of *Streptomyces* sp. nov. WH26 producing cytotoxic compounds isolated from marine solar saltern in China. *World Journal of Microbiology and Biotechnology* 2013, 29 (7), 1271–1278.

69. Xiao, L., Liu, H., Wu, N., Liu, M., Wei, J., Zhang, Y., Lin, X. Characterization of the high cytochalasin E and rosellichalasin producing *Aspergillus* sp. nov. F1 isolated from marine solar saltern in China. *World Journal of Microbiology and Biotechnology* 2013, 29 (1), 11–17.

70. Tian, S. Z., Pu, X., Luo, G., Zhao, L. X., Xu, L. H., Li, W. J., Luo, Y. Isolation and characterization of new p-terphenyls with antifungal, antibacterial, and antioxidant activities from halophilic actinomycete *Nocardiopsis gilva* YIM 90087. *Journal of Agricultural and Food Chemistry* 2013, 61 (12), 3006–3012.

71. Tian, S., Yang, Y., Liu, K., Xiong, Z., Xu, L., Zhao, L. Antimicrobial metabolites from a novel halophilic actinomycete *Nocardiopsis terrae* YIM 90022. *Natural Product Research* 2014, 28 (5), 344–346.

72. Baker-Austin, C., Dopson, M. Life in acid: pH homeostasis in acidophiles. *Trends in Microbiology* 2007, 15 (4), 165–171.

73. Stierle, D. B., Stierle, A. A., Girtsman, T., McIntyre, K., Nichols, J. Caspase-1 and –3 inhibiting drimane sesquiterpenoids from the extremophilic fungus *Penicillium solitum*. *Journal of Natural Products* 2012, 75 (2), 262–266.

74. Stierle, D. B., Stierle, A. A., Patacini, B., McIntyre, K., Girtsman, T., Bolstad, E. Berkeleyones and related meroterpenes from a deep water acid mine waste fungus that inhibit the production of interleukin 1-beta from induced inflammasomes. *Journal of Natural Products* 2011, 74 (10), 2273–2277.

75. Wang, H., Wang, Y., Wang, W., Fu, P., Liu, P., Zhu, W. Anti-influenza virus polyketides from the acid-tolerant fungus *Penicillium purpurogenum* JS03-21. *Journal of Natural Products* 2011, 74 (9), 2014–2018.

76. Park, J. S., Yang, H. O., Kwon, H. C. Aldgamycin I, an antibacterial 16-membered macrolide from the abandoned mine bacterium, *Streptomyces* sp. KMA-001. *Journal of Antibiotics* (Tokyo) 2009, 62 (3), 171–175.

77. Park, H. B., Lee, J. K., Lee, K. R., Kwon, H. C. Angumycinones A and B, two new angucyclic quinones from *Streptomyces* sp KMC004 isolated from acidic mine drainage. *Tetrahedron Letters* 2014, 55 (1), 63–66.

78. Garcia, A. H. Anhydrobiosis in bacteria: From physiology to applications. *Journal of Bioscience* (Bangalore) 2011, 36 (5), 939–950.

79. Rampelotto, P. H. Extremophiles and extreme environments. *Life* 2013, 3 (3), 482–485.

80. Alpert, P. The limits and frontiers of desiccation-tolerant life. *Integrative and Comparative Biology* 2005, 45 (5), 685–695.

81. Rateb, M. E., Houssen, W. E., Arnold, M., Abdelrahman, M. H., Deng, H., Harrison, W. T. A., Okoro, C. K., Asenjo, J. A., Andrews, B. A., Ferguson, G., Bull, A. T., Goodfellow, M., Ebel, R., Jaspars, M. Chaxamycins A–D, bioactive ansamycins from a hyper-arid desert *Streptomyces* sp. *Journal of Natural Products* 2011, 74 (6), 1491–1499.

82. Porter, J. R., Ge, J., Lee, J., Normant, E., West, K. Ansamycin inhibitors of Hsp90: Nature's prototype for anti-chaperone therapy. *Current Topics in Medicinal Chemistry* 2009, 9 (15), 1386–1418.

83. Tayade, A. B., Dhar, P., Kumar, J., Sharma, M., Chauhan, R. S., Chaurasia, O. P., Srivastava, R. B. Chemometric profile of root extracts of *Rhodiola imbricata* Edgew. with hyphenated gas chromatography mass spectrometric technique. *PLoS One* 2013, 8 (1), 1–15.

84. Elmann, A., Mordechay, S., Erlank, H., Telerman, A., Rindner, M., Ofir, R. Anti-neuroinflammatory effects of the extract of *Achillea fragrantissima*. *BMC Complementary and Alternative Medicine* 2011, 11 (98), 1–10.

85. Wang, Q., Li, F., Zhang, X., Zhang, Y., Hou, Y., Zhang, S., Wu, Z. Purification and characterization of a CkTLP protein from *Cynanchum komarovii* seeds that confers antifungal activity. *PLoS One* 2011, 6 (2), 1–10.

86. Ahmad, M., Ghafoor, N., Aamir, M. N. Antibacterial activity of mother tinctures of cholistan desert plants in Pakistan. *Indian Journal of Pharmaceutical Sciences* 2012, 74 (5), 465–468.

87. Reddy, B. V. B., Kallifidas, D., Kim, J. H., Charlop-Powers, Z., Feng, Z., Brady, S. F. Natural product biosynthetic gene diversity in geographically distinct soil microbiomes. *Applied and Environmental Microbiology* 2012, 78 (10), 3744–3752.

CHAPTER 13

Bioprospecting Fungi and the Labyrinthulomycetes at the Ocean–Land Interface

Ka-Lai Pang,[1] **Clement K. M. Tsui,**[2] **E. B. Gareth Jones,**[3] **and Lilian L. P. Vrijmoed**[4,5]

[1]Institute of Marine Biology and Centre of Excellence for the Oceans,
National Taiwan Ocean University, Keelung, Taiwan (ROC)

[2]Department of Pathology and Laboratory Medicine, University of British Columbia,
Vancouver, British Columbia, Canada

[3]Department of Botany and Microbiology, College of Science, King Saud University,
Riyadh, Kingdom of Saudi Arabia

[4]Department of Biology and Chemistry, City University of Hong Kong, Kowloon, Hong Kong SAR

[5]Beijing Normal University–Hong Kong Baptist University, United International College,
Tangjiawan, Zhuhai, Guangdong, People's Republic of China

CONTENTS

13.1 INTRODUCTION

With the increasing threats of new infections and antibiotic resistance of microbes, finding new drugs has never been so pressing. While organic synthesis continues to improve the effectiveness of available drugs, finding new backbones of chemical structures is crucial in the long-term development of effective drugs. Traditionally, soil microorganisms from terrestrial habitats have been explored for chemical diversity, and many compounds have been isolated and used in the market.[1] However, repeated isolation of phylogenetically related organisms may have resulted in compounds with similar physical, chemical, biochemical, or pharmacological properties. For the past couple of decades, new chemical structures have been isolated from marine organisms, implicating a niche for drug discovery research.

Fungi, the second largest group of living eukaryotic organisms after insects, were estimated to have numbers ranging from 1.5 million[2] to more than 5 million.[3] They are mostly saprobes, but many form symbiotic relationships with other organisms, including mutualists, commensals, and parasites. Fungi produce secondary metabolites to initiate infection or fight off competitors in the environment.[4] Many secondary metabolites from terrestrial fungi are of human interest, such as paclitaxel by *Taxomyces* spp.[5,6] and camptothecin by an endophytic zygomycete *Nothapodytes foetida*.[7]

Apart from filamentous fungi, the diversity of labyrinthulomycetes, a group of fungus-like protists, is also high. Recently, they have gained increased attention due to their spectacular biotechnological potentials.[8] They produce omega-3 (ω-3) polyunsaturated fatty acids (PUFAs), such as docosahexaenoic acid (DHA), eicosapentaenoic acid (EPA), and several other bioactive metabolites, known to have nutritional implications in human health.

The land–ocean boundary/zone creates unique environments, which foster distinct ecological groups of fungi and microbes. In the past two decades, efforts have been made to uncover the chemical diversity of natural products of fungi isolated from this zone. The labyrinthulomycetes are also common inhabitants in this zone, growing on organic debris, such as fallen mangrove leaves and seaweeds. In this chapter, we introduce the diverse habitats and various ecological groups of fungi and the labyrinthulomycetes, as well as the chemical structures or compounds discovered so far from this land–ocean coastal zone.

13.2 WHERE THE SEA MEETS THE LAND

Marine environments differ significantly from the terrestrial milieu in terms of physical (temperature variance, oxygen) and chemical (salinity, density and viscosity of seawater) characteristics.[9] At the boundary, there is a wide variety of estuarine and coastal ecosystems that are economically important to provide three ecological system services: fisheries, nursery habitats, and filtering and detoxification of pollutants.[10,11] These diverse ecosystems are listed below:

- Seagrass beds: Flowering plants colonizing shallow marine and estuarine waters.
- Mangals (Figure 13.1a): Salt-tolerant tree species along sheltered bays, estuaries, and inlets in tropical and subtropical latitudes. Mangrove foliage provides nutrients, vitamins, amino acids, proteins, and fatty acids (FA) as food for herbivores.[91] The fallen leaves and detritus of mangroves are the gross primary food source for various ecological animal groups, such as shredders and detritivores.
- Algal beds (Figure 13.1b): Subtidal or intertidal photic habitats of the marine and estuarine coastlines where seaweeds grow.
- Coral reefs: Dead coral calcium carbonate skeletons covered by a layer of live corals nearshore or in shallow offshore environments.
- Salt marshes (Figure 13.1c): Intertidal grassland with a high productivity in wave-protected shorelines.
- Sandy beaches and dunes (Figure 13.1d): Low-lying coastal areas where sand is deposited through wave action.

Salinity varies greatly among these environments with fully saline water (seagrass and algal beds, coral reefs, sandy beaches and dunes) or brackish water (seagrass and algal beds, mangals, salt marshes). In sandy beaches and dunes, mangals, and salt marshes, organisms are exposed to an environment similar to that of their terrestrial counterparts during low tide, while during high tide,

Figure 13.1 Coastal habitats. (a) Sai-Keng mangrove in Hong Kong. (b) An algal bed at Keelung, Taiwan. (c) *Salicornia* growing in a salt marsh at Langstone Harbour, Portsmouth, UK. (d) A sandy beach at Tung Ping Chau, Hong Kong.

the habitats are marine and the organisms are subject to the physical and chemical stresses experienced in a typical marine environment.[12]

13.3 FUNGI AT OCEAN–LAND INTERFACE

Several ecological groups of fungi inhabit the ocean–land junction: saprobic fungi on intertidal wood and senescent leaves, marine zoosporic organisms, endophytic fungi associated with marine-related plants, seagrasses and seaweeds, sediment fungi, and planktonic species and lichens that form an integral part of marine ecosystems as saprobes or by forming symbiotic relationships with other organisms.[13] These fungi are mostly saprobes, causing decay of cellulosic materials, and thus involved in the recycling of nutrients in the environment.[13] Others form symbiotic relationships with different marine or marine-related fungi, such as lichens, or cause diseases of animals, plants, and seaweeds.[14]

Diversity of terrestrial fungi differs from that of marine fungi; salinity and pH are the key factors that govern their occurrence.[15] However, salinity of these habitats varies dramatically from low salinity (<10%) in some mangals to full saline water (3.5%) or higher in the Red Sea.[15] Marine-derived fungi have also evolved unique morphological and ecological features to overcome the viscosity of water and scarcity of substrata for growth, such as a passive release of spores and appendaged spores for entanglement or attachment to substrata.[16]

Jones and Pang[13] documented 560 marine, mostly saprobic, fungi, which were mainly isolated from intertidal wood, seaweeds, and dead, submerged parts of grasses in mangals, algal beds, salt marshes, sandy beaches, and rocky shores. Jones[17] states that this is an underestimate, projecting there may as many as 12,500 species in marine habitats, These fungi decompose woody and leafy substrata by the production of various oxidative enzymes and therefore play an essential role in nutrient recycling in marine and estuarine environments.[13] The majority of these fungi belong to the Ascomycota and asexual morphs, while there are few basidiomycetes.[14] Eleven phylogenetic lineages of marine fungi occur in six classes: Ustilaginomycetes, Agaricomycetes (Basidiomycota), Dothideomycetes, Eurotiomycetes, Lecanoromycetes, and Sordariomycetes (Ascomycota).[13]

Many fungi have been reported from dead and living corals. Kohlmeyer and Volkmann-Kohlmeyer[18] established the family Koralionastetaceae in the Ascomycota to include species growing on coral slabs. This family was subsequently referred to a new order, Koralionastetales.[19] Many terrestrial fungal genera such as *Aspergillus* and *Penicillium* have also been reported growing in the organic matrix of the carbonate skeletons inside corals as endolithic saprotrophs.[20] Fungi have also been reported from other organisms in the coral reef environment, with aspergillosis on *Gorgonia ventalina* caused by *Aspergillus sydowii* being the most famous example.[21] *Penicillium*, *Cladosporium*, *Gloeotinia*, *Rhodotorula*, *Stachybotrys*, and *Xylaria* species have also been found in the sea fan.[22]

Increased interests have emerged in recent years on endophytes of marine and marine-adapted plants and algae. Sakayaroj et al.[23] summarized the diversity of fungal endophytes of marine and salt-affected plants over the last decade. Fifty-two plant host species in 23 plant families were examined for fungal endophytes in mostly tropical and subtropical countries. Terrestrial fungal genera dominated, including *Aspergillus*, *Penicillium*, *Cladosporium*, *Guignardia/Phyllosticta*, *Alternaria*, and *Diaporthe/Phomopsis*.[24] Suryanarayanan et al.[25] investigated the diversity of endophytic fungi in 25 algal species (11 brown algae, 6 green algae, and 8 red algae) and found that brown algae supported the highest diversity of endophytic fungi and green algae supported the lowest. The majority were asexual fungi, with *Aspergillus* the dominant genus among the algae. Asexual fungi of terrestrial origin, belonging predominantly to the Ascomycota in the orders Hypocreales, Eurotiales, and Capnodiales, also dominated in seagrasses with many mycelial morphotypes, but some marine taxa have been found, such as *Trichocladium alopallonellum*, *Trichocladium achrasporum* and *Cumulospora marina*.[26,27] More data are available for fungal endophytes of mangrove plants. Some endophytic fungi are ubiquitous, such as *Colletotrichum*, *Phomopsis*, *Phyllosticta*, and *Sporormiella*, while some are specific to a host plant species.[28] Tissue specificity of endophytic fungi occurred in *Rhizophora apiculata*[29] and *Kandelia obovata*.[24] These terrestrial-like fungi need to be sequenced and compared to those described from terrestrial habitats to determine if they are specially adapted to the marine environment.

13.3.1 Secondary Metabolites of Fungi from Marine and Marine-Related Habitats

All major groups of chemical structures are documented for marine and marine-derived fungi: polyketides (41%), alkaloids (20%), peptides (12%), terpenoids (14%), prenylated polyketids (8%), shikimates (2%), and lipids (1%).[30] Major sources of fungi studied for their ability to produce new metabolites include mangrove fungi,[31–34] endophytic fungi,[35–37] marine animals,[38] marine-derived fungi,[39] and seaweeds.[40] Of the different host substrates for fungi, algae (21%) yielded the highest number of hits, followed by sponges (19%), sediments (16%), and mangrove species (6%).[41] Many marine fungi produce a wide range of compounds, as illustrated by the results of Pan et al.,[34] who documented 42 new or novel compounds and 35 known compounds from 25 fungal strains, an average of 3 from each species. *Dendryphiella salina* and *Halorosellinia oceanica* are prolific fungi in the production of novel compounds.[31,42] New compounds of *H. oceanica* include 15G 256 ALPHA (**1**) and 15G 256 BETA (**2**), novel depsipeptides that affect fungal cell wall formation,[43] and macrocyclic polylactones sesquiterpenoid lactone (**3**)[44] and halorosellinic acid (**4**), with antimalarial activity (IC_{50} = 13 µg/ml) and weak antibacterial activity (minimum inhibitory concentration of 200 mg/ml).[31,34] *Dendryphiella salina* yielded dendryphiellin A, an unprecedented sesquiterepene esterified with a branched C9 carboxylic acid; dendryphiellins B–G; dendryphiellic acids A and B; glyceol dendryphiellate A; and dedryphiellins A1, E1, E2, F, and G.[42,45]

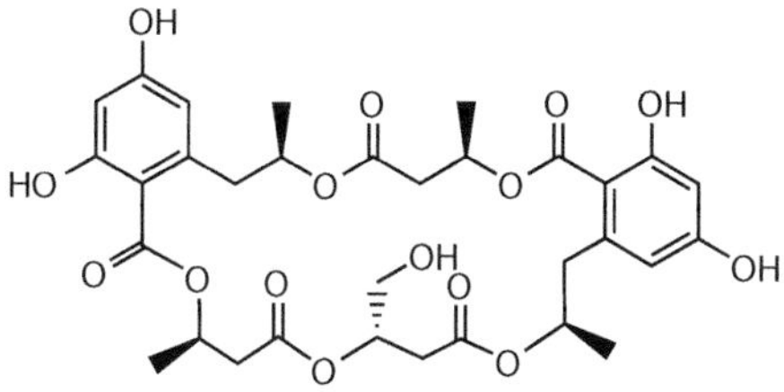

It is difficult to evaluate how many marine and marine-derived fungi have been screened for new or novel bioactive compounds; a conservative number would be 15,000 to 20,000 isolates (E. B. G. Jones, unpublished data). Of the compounds derived from marine fungi, at least three are commercially available or in advanced clinical trials: cephalosporin C (and many derivatives), isolated from a near sewage outlet on the Sardinian coast; phomactins, isolated from a *Phoma* sp. found growing on a crab shell of *Chionecetes opilio*, yielding a fungal diterpene with an antagonist of platelet-activating factor; and Plinabulin, a diketopipeerazine halimide, with activity against cancer and in advanced clinical trials.[41] Many compounds derived from marine fungi, which were shown to possess activity, are either too toxic (cytochalasins) or insufficiently active at low concentrations, or the basic backbone of the compound is already well known. Target medical conditions for new compounds from marine fungi have included cancer,[36,37] antimalarial,[33] and platelet-activating factor (PAF)[46,47] antagonists. However, the key sector requiring more intensive search and discovery is for antimicrobials. With the growing concern over drug resistance, there is an urgent need to invest in research to discover new antibiotics against common bacterial infections. The medical profession has drawn attention to the over-prescription of antibiotics and the concern that future surgery may be compromised due to a lack of suitable drugs to ward off postinfection recovery of patients.

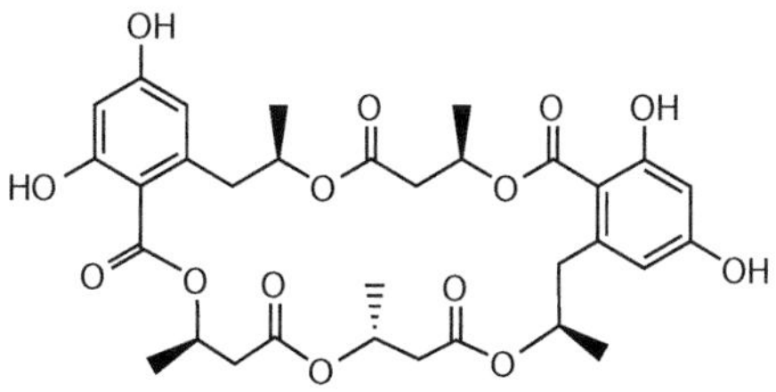

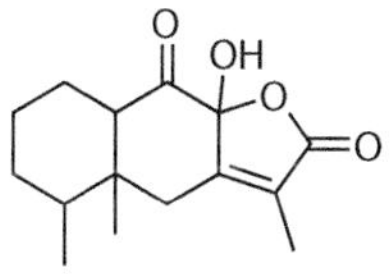

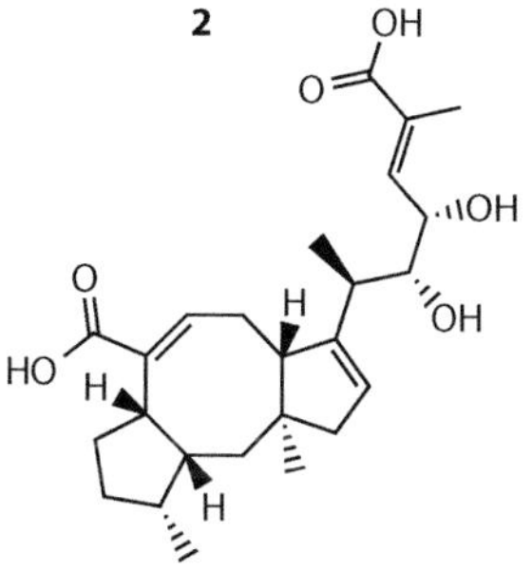

13.4 LABYRINTHULOMYCETES IN MANGROVES

The Labyrinthulomycetes are a ubiquitous but poorly understood group of marine protists classified under the Stramenopiles (also called Heterokonta).[48,49] They include organisms known as labyrinthulids, thraustochytrids, and aplanochytrids.[48] Major genera include *Thraustochytrium*, *Aplanochytrium*, *Japonochytrium*, *Ulkenia*, and *Schizochytrium*.[50] Recent taxonomic revisions have divided *Schizochytrium* into multiple genera.[51] The vegetative cells of the labyrinthulomycetes range from spindle shaped to spherical, and they are unique by producing an ectoplasmic net (EN), which is a branched network of plasma membrane extensions associated with a specialized organelle called sagenogenetosome.[50] EN attaches the vegetative cells to surfaces and secretes hydrolytic enzymes in contact with their substrates.[50] Representatives of the labyrinthulomycetes can easily be isolated from mangrove leaves, sediment, open water, and guts of marine invertebrates using pine pollen baits and yeast extract peptone (YEP) agar.[52]

Initially grouped as fungi due to morphological similarities, the labyrinthulomycetes play an ecological role similar to that of saprobes in the decomposition of organic matter of plant, algal, or animal origin in the water column, sediments, and mangroves.[53] However, some representatives are parasitic or pathogenic, causing serious diseases such as eelgrass wasting disease and the hard clam disease Quahog Parasite Unknown (QPX) in North America and Europe.[54,55]

Due to their ability to synthesize *de novo* and to accumulate a high level of essential omega-3 (ω-3) and omega-6 (ω-6) polyunsaturated fatty acids (PUFAs), such as docosahexaenoic acid (DHA) (C22:6, **5**), docosapentaenoic acid (DPA) (C22:5), and eicosapentaenoic acid (EPA) (C20:5, **6**),[56] labyrinthulomycetes play an important role in the nutrition of marine metazoans by improving detritus food quality (trophic upgrading).[53] The high PUFA content of labyrinthulomycetes has resulted in increasing biotechnological application and interest in these organisms over the past two decades because the application of microbial DHA in human health has been expanding. Also, biodiesel from lipids produced by microbes may overcome many of the sustainability challenges previously derived from petroleum-based fuels. Other metabolites, such as enzymes, carotenoids, squalene, thraustochytrosides A–C, and extracellular polysaccharides, can also be obtained from these marine protists.[8,57,58]

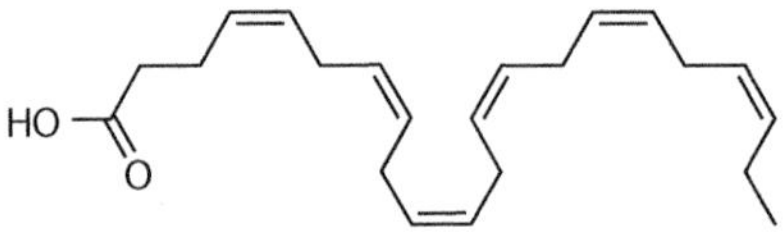

5

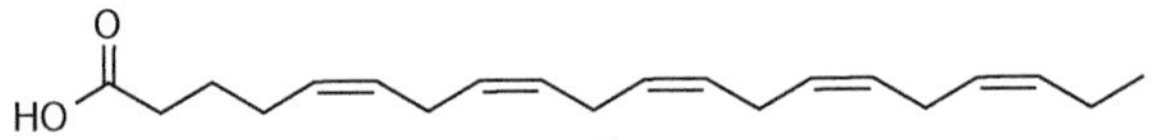

6

13.4.1 Omega-3 Polyunsaturated Fatty Acids

Polyunsaturated fatty acids are essential components of higher eukaryotes, and they provide flexibility and selective permeability properties to cell membranes.[59] However, higher eukaryotes cannot synthesize most PUFAs, but have to obtain them through daily diets. Docosahexaenoic acid (DHA) is a PUFA essential for human beings because it is one of the major fatty acids in the brain's gray matter and in the eyes' retina.[60] DHA has been demonstrated to be important in brain and eye development in infants. Also, DHA generates positive effects on cardiovascular diseases such as coronary heart disease, stroke, and hypertension[61] and neural disorders such as dementia, Alzheimer's, and depression.[62] The current commercial source of DHA is fish and fish oil, but DHA has been increasingly manufactured from thraustochytrids and marine algae, and numerous publications and patents have described the latest pathway and technologies in the synthesis of these compounds.[56]

Many representatives of labyrinthulomycetes have been reported to produce PUFA. Particularly 22:6 ω-3 is found at very high levels in *Aurantiochytrium limacinum* (formerly *Schizochytrium limacinum*) isolated from Japan mangroves.[63] *Aurantiochytrium mangrovei* is another common thraustochytrid in Asia with a high level of DHA production.[52] Various species in *Thraustochytrium* can also produce large amounts of DHA and EPA in defined media conditions (see review in Barclay et al.[56]). The production of DHA in the thraustochytrids involves two different biochemical pathways: the traditional desaturation and elongation pathway and the independent polyketide synthase pathway.[64,65] The improvement of genetic modified strains and the development of an efficient large-scale cultivation system for the commercial production of these nutrients from thraustochytrids would address a major global need in the FA market.

13.4.2 Squalene

Squalene (2,6,10,15,19,23-hexamethyltetracosa-6,6,10,14,18,20-tetracosahexaene; **7**) is a polyunsaturated triterpenic hydrocarbon ($C_{30}H_{50}$). Squalene is a key precursor to the biosynthesis of cholesterol and steroids in plants and animals, and it has several beneficial properties.[66] Squalene has attracted attention in the medical and pharmaceutical industries because it has been reported to be an effective inhibitor of chemically induced skin, lung, and colon tumorigenesis.[66,67] Squalene has also been demonstrated to be effective in reducing total cholesterol, low-density lipoprotein cholesterol, and triglyceride levels in people with hypercholesterolemia.[67]

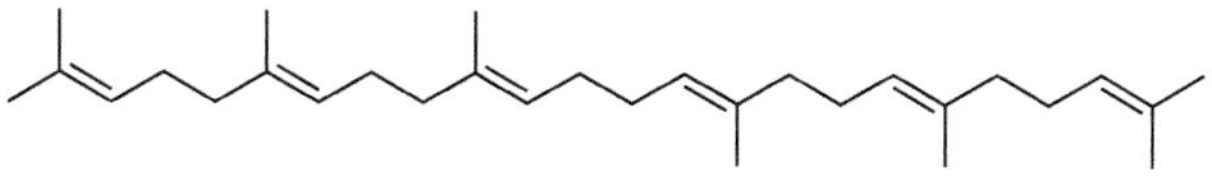

7

The major source of squalene is from the shark liver oil, but the availability of squalene may not be sustainable due to international concern for marine wildlife conservation. Squalene is also found in olive oil, palm oil, wheat germ oil, and rice bran oil.[67] Plant sources have been under investigation, but the scale of production is limited. Thraustochytrids can accumulate a high level of squalene, for instance, thraustochytrid ACEM 6063 (0.1 mg/g biomass),[68] *Aurantiochytrium mangrovei* FB 1 (0.162 mg/g biomass),[69] and *Aurantiochytrium* sp. 18W-13a (171 mg/g dry weight).[70]

13.4.3 Carotenoids

Carotenoids are one of the most widely distributed groups of naturally occurring pigments. The colors of these pigments range from yellow to red. Carotenoids are also known as sources of the pinkish red pigmentation in salmonids and crustaceans.[71] The most well-known examples are astaxanthin (3,3'-dihydroxy-β,β'-carotene-4,4'-dione, **8**) and canthaxanthin (β,β'-carotene-4,4'-dione, **9**), which are widely reported in plants, algae, bacteria, fungi, and animals. These pigments are important because they are widely used as dyes in the food and cosmetic industries.

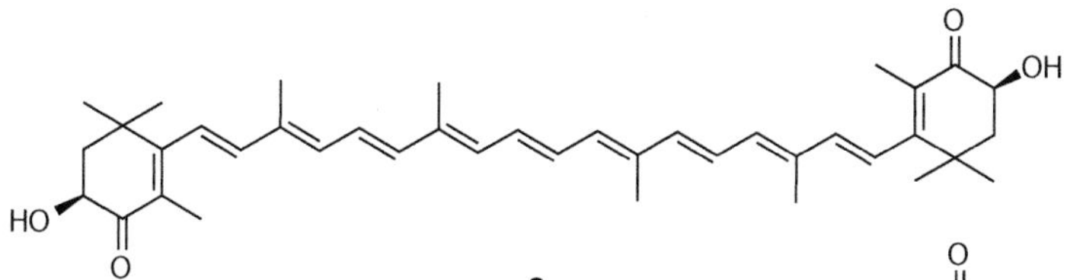

8

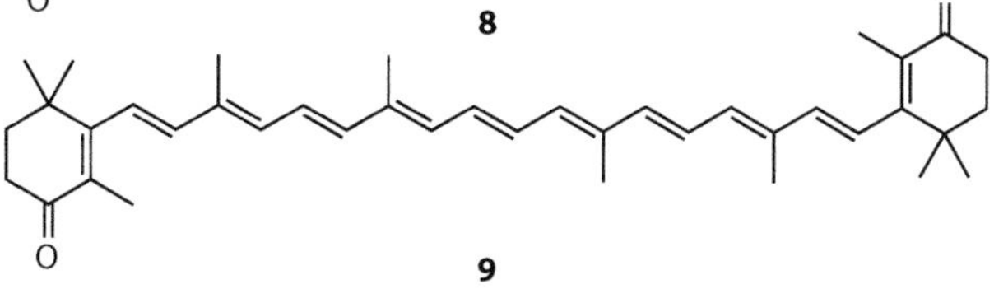

9

Owing to their unique chemical structure, these xanthophylls play important physiological roles in humans, including activation of the immune system[72] and protection against lipid peroxidation and oxidative damage of low-density lipoprotein cholesterol, as well as induced carcinogenesis,[72] most likely by their antioxidant and chemoprotective properties.[71] In addition, xanthophylls are important in the maintenance of the visual system. Since β-carotene acts as a precursor for vitamin A, sufficient uptake of β-carotene can help to prevent diseases caused by vitamin A deficiency, including blindness, immune dysfunction, and skin disorders.[73] Since mammals can metabolize but not synthesize carotenoids, they need to obtain these nutrients through their diets and as supplements.

Traditionally, carotenoids were extracted from plants such as paprika and saffron. However, microbial carotenoids have attracted significant attention lately because of the ease of increasing production by environmental and genetic manipulation. Thraustochytrids are also a promising source of carotenoids. The production of carotenoid pigments such as astaxanthin, zeaxanthin, canthaxanthin, echinenone, phoenicoxanthin, and β-carotene by *Thraustochytrium* sp. has been reported.[74,75] Thraustochytrid strain KH-105 was reported as β-carotene and xanthophyll accumulator (astaxanthin and canthaxanthin).[76]

13.5 AREAS FOR FURTHER RESEARCH

Marine yeasts, deep-sea fungi, planktonic fungi, endophytes of marine-related plants, and algicolous fungi are the understudied groups, and these fungi can bring an estimated number of marine fungi to 12,000.[13] With this discrepancy in the current and estimated numbers, there is still a vast unknown diversity of fungi in the sea, possibly forming unreported phylogenetic lineages. Fungi in the deep-sea are largely unknown, although environmental sequences suggest the existence of a number of ascomycetes and basidiomycetes with unknown ecological roles in such hostile environments (low temperature and high pressure).[77] Bacteria and Archaea are abundant in the same environment, so competition is intense with the low nutrient level. Whether deep-sea fungi produce novel secondary metabolites to fight off competition requires further research.

Endophytic fungi form symbiotic relationships with marine-related plants and seaweeds, and they might produce chemical structures to accelerate or delay senescence of the host plants,[78–80] play a protective role against insect herbivory,[81–84] and protect hosts against pathogens.[85] These fungal chemicals could be interesting to screen for anticancer, anti-insect, and antimicrobial activity. The additional stress of salinity might induce structurally diverse compounds. However, chemical structures produced by fungi *in vitro* are possibly different, implying direct extraction from tissues of plants or seaweeds.

Saprobic fungi in terrestrial environments have been the main source of bioactive compounds, while marine fungi have only been explored for secondary metabolites since the discovery of siccayne in the marine basidiomycete *Halocyhphina villosa*.[86] Since that time, some 1100 new compounds have been published in more than 500 publications.[87] Perhaps microbial antagonism on wood is severe since only a proportion of fungi can produce wood-modifying enzymes for nutrition.[92,93] Marine sediment is rich in fungal diversity; while fungi on seaweeds are basically understudied, their isolation and fermentation would definitely be meaningful. Unfortunately, few laboratories are currently studying marine fungi.[94]

13.6 CONCLUSIONS

Marine fungi and the labyrinthulomycetes are diverse and thus untapped sources of novel chemical structures and compounds for drug discovery and nutraceutical research. The recent increase and availability of whole genome sequences allow mining of genes that are involved in the metabolism of microorganisms. A large number of orphan biosynthesis pathways have been identified by bioinformatics, showing that microbial genomes contain gene clusters responsible for biosynthesis of secondary metabolites. Fungi are known to produce several classes of secondary metabolites, including polyketides (PKs) and nonribosomal peptides (NRPs). Genes involved in PK and NRP biosynthesis are identified by genomic analysis in *Aspergillus nidulans*, and in particular, 27 polyketide synthases (PKSs) and 14 nonribosomal peptide synthetases (NRPSs) were recognized.[88–90] Recently, genome sequencing and sequence mining have revealed that genes involved in the biosynthesis of secondary metabolites are abundant, and there are still a vast number of novel compounds with new chemical structures that could be cloned, engineered, and produced from microorganisms. Therefore, overexpression of biosynthetic genes in the production of secondary metabolites using a heterologous bacterial expression system is key to creating combinatorial biosynthesis for deigned manufacture of novel compounds that are of medical applications.

ACKNOWLEDGMENTS

Ka-Lai Pang thanks the Ministry of Science and Technology, Taiwan (grant no. 101–2621-B-019–001-MY3) for financial support. Gareth Jones is supported by the Distinguished Scientist Fellowship Program (DSFP), King Saud University, Riyadh, Saudi Arabia. Alice Chan from City University of Hong Kong is thanked for her technical assistance.

REFERENCES

1. Peláez, F. The historical delivery of antibiotics from microbial natural products—Can history repeat? *Biochem. Pharmacol.* 2006, 71 (7), 981–990.
2. Hyde, K. D. Where are the missing fungi? Does Hong Kong have any answers? *Mycol. Res.* 2001, 105 (12), 1514–1518.
3. Blackwell, M. The fungi: 1, 2, 3 … 5.1 million species? *Am. J. Bot.* 2011, 98 (3), 426–438.
4. Fox, E. M., Howlett, B. J. Secondary metabolism: Regulation and role in fungal biology. *Curr. Opin. Microbiol.* 2008, 11 (6), 481–487.
5. Stierle, A., Strobel, G., Stierle, D. Taxol and taxane production by *Taxomyces andreanae*, an endophytic fungus of Pacific yew. *Science* 1993, 260 (5105), 214–216.
6. Stierle, A., Strobel, G., Stierle, D., Grothaus, P., Bignami, G. The search for a taxol-producing microorganism among the endophytic fungi of the pacific yew, *Taxus brevifolia. J. Nat. Prod.* 1995, 58 (9), 1315–1324.
7. Puri, S. C., Verma, V., Amna, T., Qazi, G. N., Spiteller, M. An endophytic fungus from *Nothapodytes foetida* that produces camptothecin. *J. Nat. Prod.* 2005, 68, 1717–1719.
8. Lewis, T.E., Nichols, P. D., McMeekin, T. A. The biotechnological potential of thraustochytrids. *Mar. Biotechnol.* 1999, 1 (6), 580–587.
9. Raffaelli, D., Solan, M., Webb, T. J. Do marine and terrestrial ecologists do it differently? *Mar. Ecol. Prog. Ser.* 2005, 304, 283–289.
10. Gray, J. S. Marine biodiversity: Patterns, threats and conservation needs. *Biodivers. Conserv.* 1997, 6 (1), 153–175.
11. Barbier, E. B., Hacker, S. D., Kennedy, C., Koch, E. W., Stier, A. C., Silliman, B. R. The value of estuarine and coastal ecosystem services. *Ecol. Monogr.* 2011, 81 (2), 169–193.
12. Underwood, A. J. Intertidal ecologists work in the 'gap' between marine and terrestrial ecology: Bridging the gap between aquatic and terrestrial ecology. *Mar. Ecol. Prog. Ser.* 2005, 304, 297–302.
13. Jones, E. B. G., Pang, K. L. Tropical aquatic fungi. *Biodivers. Conserv.* 2012, 21 (9), 2403–2423.
14. Jones, E. B. G., Sakayaroj, J., Suetrong, S., Somrithipol, S., Pang, K. L. Classification of marine Ascomycota, anamorphic taxa and Basidiomycota. *Fungal Divers.* 2009, 35, 1–203.
15. Jones, E. B. G. Marine fungi: Some factors influencing biodiversity. *Fungal Divers.* 2000, 4, 53–73.
16. Jones, E. B. G. Fungal adhesion. Presidential address 1992. *Mycol. Res.* 1994, 98, 961–981.
17. Jones, E. B. G. Are there more marine fungi to be described? *Bot. Mar.* 2011, 54 (4), 343–354.
18. Kohlmeyer, J., Volkmann-Kohlmeyer, B. Marine fungi from Belize with a description of two new genera of ascomycetes. *Bot. Mar.* 1987, 30 (3), 195–204.
19. Campbell, J., Inderbitzin, P., Kohlmeyer, J., Volkmann-Kohlmeyer, B. Koralionastetales, a new order of marine *Ascomycota* in the *Sordariomycetes. Mycol. Res.* 2009, 113 (3), 373–380.

20. Raghukumar, C., Ravindran, J. Fungi and their role in corals and coral reef ecosystems. In *Biology of Marine Fungi.* Springer-Verlag, Berlin-Heidelberg, 2012, pp. 89–113.

21. Smith, G. W., Ives, L. D., Nagelkerken, I. A., Ritchie, K. B. Caribbean sea-fan mortalities. *Nature* 1996, 383, 487.

22. Toledo-Hernández, C., Sabat, A. M., Zuluaga-Montero, A. Density, size structure and aspergillosis prevalence in *Gorgonia ventalina* at six localities in Puerto Rico. *Mar. Biol.* 2007, 152 (3), 527–535.

23. Sakayaroj, J., Preedanon, S., Phongpaichit, S., Buatong, J., Chaowalit, P., Rukachaisirikul, V. Diversity of endophytic and marine-derived fungi associated with marine plants and animals. In *Marine Fungi and Fungal-Like Organisms.* De Gruyter, Berlin, 2012, pp. 291-328.

24. Pang, K. L., Vrijmoed, L. L. P., Goh, T. K., Plaingam, N., Jones, E. B. G. *Fungal Endophytes Associated with Kandelia candel (Rhizophoraceae).* Vol. 51. Botanica Marina, Hong Kong, 2008, pp. 171–178.

25. Suryanarayanan, T. S., Venkatachalam, A., Thirunavukkarasu, N., Ravishankar, J. P., Doble, M., Geetha, V. Internal mycobiota of marine macroalgae from the Tamilnadu coast: Distribution, diversity and biotechnological potential. *Bot. Mar.* 2010, 53 (5), 457–468.

26. Sakayaroj, J., Preedanon, S., Supaphon, O., Jones, E. B. G., Phongpaichit, S. Phylogenetic diversity of endophyte assemblages associated with the tropical seagrass *Enhalus acoroides* in Thailand. *Fungal Divers.* 2010, 42 (1), 27–45.

27. Mata, J. L., Cebrián, J. Fungal endophytes of the seagrasses *Halodule wrightii* and *Thalassia testudinum* in the north-central Gulf of Mexico. *Bot. Mar.* 2013, 56 (5–6), 541–545.

28. Kumaresan, V., Suryanarayanan, T. S. Occurrence and distribution of endophytic fungi in a mangrove community. *Mycol. Res.* 2001, 105 (11), 1388–1391.

29. Kumaresan, V., Suryanarayanan, T. S., Johnson, J. A. Ecology of mangrove endophytes. In *Fungi in Marine Environments.* Fungal Diversity Press, Hong Kong, 2002, pp. 145–166.

30. Rateb, M. E., Ebel, R. Secondary metabolites of fungi from marine habitats. *Nat. Prod. Rep.* 2011, 28 (2), 290–344.

31. Chinworrungsee, M., Kittakoop, P., Isaka, M., Rungrod, A., Tanticharoen, M., Thebtaranonth, Y. Antimalarial halorosellinic acid from the marine fungus *Halorosellinia oceanica. Bioorg. Med. Chem. Lett.* 2001, 11 (15), 1965–1969.

32. Lin, Y., Wu, X., Feng, S., Jiang, G., Luo, J., Zhou, S., Vrijmoed, L. L. P., Jones, E. B. G., Krohn, K., Steingröver, K. Five unique compounds: Xyloketals from mangrove fungus *Xylaria* sp. from the South China Sea coast. *J. Org. Chem.* 2001, 66 (19), 6252–6256.

33. Isaka, M., Suyarnsestakorn, C., Tanticharoen, M., Kongsaeree, P., Thebtaranonth, Y. Aigialomycins A–E, new resorcylic macrolides from the marine mangrove fungus *Aigialus parvus. J. Org. Chem.* 2002, 67 (5), 1561–1566.

34. Pan, J. Y., Jones, E. B. G., She, Z. Y., Ling, Y. C. Review of bioactive compounds from fungi in the South China Sea. *Bot. Mar.* 2008, 51 (3), 179–190.

35. Schulz, B., Draeger, S., Rheinheimer, J., Siems, K., Loesgen, S., Bitzer, J., Schloerke, O., Zeeck, A., Kock, I., Hussain, H. Screening strategies for obtaining novel, biologically active, fungal secondary metabolites from marine habitats. *Bot. Mar.* 2008, 51 (3), 219–234.

36. Chaeprasert, S., Piapukiew, J., Whalley, A. J. S., Sihanonth, P. Endophytic fungi from mangrove plant species of Thailand: Their antimicrobial and anticancer potentials. *Bot. Mar.* 2010, 53 (6), 555–564.

37. Debbab, A., Aly, A. H., Proksch, P. Endophytes and associated marine derived fungi: Ecological and chemical perspectives. *Fungal Divers.* 2012, 57 (1), 45–83.

38. Trisuwan, K., Khamthong, N., Rukachaisirikul, V., Phongpaichit, S., Preedanon, S., Sakayaroj, J. Anthraquinone, cyclopentanone, and naphthoquinone derivatives from the sea fan-derived fungi *Fusarium* spp. PSU-F14 and PSU-F135. *J. Nat. Prod.* 2010, 73 (9), 1507–1511.

39. Trisuwan, K., Rukachaisirikul, V., Sukpondma, Y., Preedanon, S., Phongpaichit, S., Rungjindamai, N., Sakayaroj, J. Epoxydons and a pyrone from the marine-derived fungus *Nigrospora* sp. PSU-F5. *J. Nat. Prod.* 2008, 71 (8), 1323–1326.

40. Jones, E. B. G. Bioactive compounds in marine organisms. *Bot. Mar.* 2008, 51 (3), 161–162.

41. Ebel, R. Natural products from marine-derived fungi. In *Marine Fungi and Fungal-Like Organisms.* De Gruyter, Berlin, 2012, pp. 411–440.

42. Guerriero, A., D'Ambrosio, M., Pietra, F., Cuomo, V., Vanzanella, F. Dendryphiellin A, the first fungal trinor—eremophilane. Isolation from the marine deuteromycete *Dendryphiella salina* (SUTHERLAND) PUGHet NICOT. *Helv. Chim. Acta* 1988, 71 (1), 57–61.

43. Schlingmann, G., Milne, L., Williams, D. R., Carter, G. T. Cell wall active antifungal compounds produced by the marine fungus *Hypoxylon oceanicum* LL-15G256. II. Isolation and structure determination. *J. Antibiot.* 1998, 51 (3), 303–316.

44. Li, H. J., Lin, Y. C., Wang, L., Zhou, S. N., Vrijmoed, L. L. P. Metabolites of marine fungus *Hypoxylon oceanicum* from the South China Sea. *Acta Scientiarum Naturalium Universitatis Sunyatseni* 2000, 40 (4), 70–72, 80.

45. Guerriero, A., D'Ambrosio, M., Cuomo, V., Vanzanella, F., Pietra, F. Novel trinor-eremophilanes (dedryphiellin B, C, and D), eremophilanes (dendryphiellin E, F and G) and branched C9-carboxylic acids (dendryphiellic acid A and B) from the marine deuteromycete *Dendryphiella salina* (Sutherland) Pugh et Nicot. *Helv. Chim. Acta* 1989, 72, 438–446.

46. Sugano, M., Sato, A., Iijima, Y., Furuya, K., Kuwano, H., Hata, T. Phomactin E, F, and G: New phomactin-group PAF antagonists from a marine fungus *Phoma* sp. *J. Antibiot.* 1995, 48 (10), 1188.

47. Ishino, M., Kiyomichi, N., Takatori, K., Sugita, T., Shiro, M., Kinoshita, K., Takahashi, K., Koyama, K. Phomactin I, 13-*epi*-phomactin I, and phomactin J, three novel diterpenes from a marine-derived fungus. *Tetrahedron* 2010, 66 (14), 2594–2597.

48. Honda, D., Yokochi, T., Nakahara, T., Raghukumar, S., Nakagiri, A., Schaumann, K., Higashihara, T. Molecular phylogeny of labyrinthulids and thraustochytrids based on the sequencing of 18S ribosomal RNA gene. *J. Eukaryot. Microbiol.* 1999, 46 (6), 637–647.

49. Tsui, C. K. M., Marshall, W., Yokoyama, R., Honda, D., Lippmeier, J. C., Craven, K. D., Peterson, P. D., Berbee, M. L. Labyrinthulomycetes phylogeny and its implications for the evolutionary loss of chloroplasts and gain of ectoplasmic gliding. *Mol. Phylogen. Evol.* 2009, 50 (1), 129–140.

50. Porter, D. Phylum Labyrinthulomycota. In *Handbook of Protoctista*. Jones and Barlette Publishers, Boston, 1990, pp. 388–398.

51. Yokoyama, R., Honda, D. Taxonomic rearrangement of the genus *Schizochytrium* sensu lato based on morphology, chemotaxonomic characteristics, and 18S rRNA gene phylogeny (Thraustochytriaceae, Labyrinthulomycetes): Emendation for *Schizochytrium* and erection of *Aurantiochytrium* and *Oblongichytrium* gen. nov. *Mycoscience* 2007, 48 (4), 199–211.

52. Wong, M. K. M., Tsui, C. K. M., Au, D. W. T., Vrijmoed, L. L. P. Docosahexaenoic acid production and ultrastructure of the thraustochytrid *Aurantiochytrium mangrovei* MP2 under high glucose concentrations. *Mycoscience* 2008, 49 (4), 266–270.

53. Raghukumar, S. Ecology of the marine protists, the labyrinthulomycetes (thraustochytrids and labyrinthulids). *Eur. J. Protistol.* 2002, 38 (2), 127–145.

54. Muehlstein, L. K., Porter, D., Short, F. T. *Labyrinthula* sp., a marine slime mold producing the symptoms of wasting disease in eelgrass, *Zostera marina*. *Mar. Biol.* 1988, 99 (4), 465–472.

55. Stokes, N. A., Calvo, L. M. R., Reece, K. S., Burreson, E. M. Molecular diagnostics, field validation, and phylogenetic analysis of Quahog Parasite Unknown (QPX), a pathogen of the hard clam *Mercenaria*. *Dis. Aquat. Org.* 2002, 52, 233–247.

56. Barclay, W., Weaver, C., Metz, J., Hansen, J., Cohen, Z., Ratledge, C. Development of a docosahexaenoic acid production technology using *Schizochytrium*: Historical perspective and update. In *Single cell Oils: Microbial and Algal Oils*, ed. Z. Cohen, C. Ratledge. 2nd ed. AOCS Publishing, Urbana, IL, 2010, pp. 75–96.

57. Jenkins, K. M., Jensen, P. R., Fenical, W. Thraustochytrosides AC: New glycosphingolipids from a unique marine protist, *Thraustochytrium globosum*. *Tetrahedron Lett.* 1999, 40 (43), 7637–7640.

58. Sijtsma, L., de Swaaf, M. E. Biotechnological production and applications of the ω-3 polyunsaturated fatty acid docosahexaenoic acid. *Appl. Microbiol. Biotechnol.* 2004, 64, 146–153.

59. Ward, O. P., Singh, A. Microbial production of polyunsaturated fatty acids. In *Biotechnological Applications of Microbes*. I. K. International, New Delhi, 2005, pp. 199–220.

60. Muskiet, F. A., Fokkema, M. R., Schaafsma, A., Boersma, E. R., Crawford, M. A. Is docosahexaenoic acid (DHA) essential? Lessons from DHA status regulation, our ancient diet, epidemiology and randomized controlled trials. *J. Nutr.* 2004, 134 (1), 183–186.

61. Das, U. N. Folic acid and polyunsaturated fatty acids improve cognitive function and prevent depression, dementia, and Alzheimer's disease—But how and why? *Prostaglandins Leukot. Essent. Fatty Acids* 2008, 78 (1), 11–19.

62. Kris-Etherton, P. M., Harris, W. S., Appel, L. J. Fish consumption, fish oil, omega-3 fatty acids, and cardiovascular disease. *Circulation* 2002, 106 (21), 2747–2757.

63. Pyle, D. J., Garcia, R. A., Wen, Z. Producing docosahexaenoic acid (DHA)-rich algae from biodiesel-derived crude glycerol: Effects of impurities on DHA production and algal biomass composition. *J. Agric. Food. Chem.* 2008, 56 (11), 3933–3939.

64. Qiu, X., Hong, H. P., Mackenzie, T. A. Identification of a Δ4 fatty acid desaturase from *Thraustochytrium* sp. involved in the biosynthesis of docosahexanoic acid by heterologous expression in *Saccharomyces cerevisiae* and *Brassica juncea. J. Biol. Chem.* 2001, 276, 31561–31566.

65. Metz, J. G., Roessler, P., Facciotti, D., Levering, C., Dittrich, M., Lassner, R., Valentine, R., Lardizabal, K., Domergue, F., Yamada, A., Yazawa, K., Knauf, V., Browse, J. Production of polyunsaturated fatty acids by polyketide synthases in both prokaryotes and eukaryotes. *Science* 2001, 293 (5528), 290–293.

66. Smith, T. J. Squalene: Potential chemopreventive agent. *Expert Opin. Investig. Drugs* 2000, 9 (8), 1841–1848.

67. Spanova, M., Daum, G. Squalene biochemistry, molecular biology, process biotechnology, and applications. *Eur. J. Lipid Sci. Technol.* 2011, 113 (11), 1299–1320.

68. Lewis, T.E., Nichols, P. D., McMeekin, T. A. Sterol and squalene content of a docosahexaenoic-acid-producing thraustochytrid: Influence of culture age, temperature, and dissolved oxygen. *Mar. Biotechnol.* 2001, 3 (5), 439–447.

69. Jiang, Y., Fan, K.-W., Tsz-Yeung Wong, R., Chen, F. Fatty acid composition and squalene content of the marine microalga *Schizochytrium mangrovei. J. Agric. Food. Chem.* 2004, 52 (5), 1196–1200.

70. Nakazawa, A., Matsuura, H., Kose, R., Kato, S., Honda, D., Inouye, I., Kaya, K., Watanabe, M. M. Optimization of culture conditions of the thraustochytrid *Aurantiochytrium* sp. strain 18W-13a for squalene production. *Bioresour. Technol.* 2012, 109, 287–291.

71. Goodwin, T. W. Metabolism, nutrition, and function of carotenoids. *Annu. Rev. Nutr.* 1986, 6 (1), 273–297.

72. Tanaka, T., Makita, H., Ohnishi, M., Mori, H., Satoh, K., Hara, A. Chemoprevention of rat oral carcinogenesis by naturally occurring xanthophylls, astaxanthin and canthaxanthin. *Cancer Res.* 1995, 55 (18), 4059–4064.

73. Fierce, Y., de Morais Vieira, M., Piantedosi, R., Wyss, A., Blaner, W. S., Paik, J. *In vitro* and *in vivo* characterization of retinoid synthesis from β-carotene. *Arch. Biochem. Biophys.* 2008, 472 (2), 126–138.

74. Carmona, M. L., Naganuma, T., Yamaoka, Y. Identification by HPLC-MS of carotenoids of the *Thraustochytrium* CHN-1 strain isolated from the Seto Inland Sea. *Biosci. Biotechnol. Biochem.* 2003, 67 (4), 884–888.

75. Armenta, R. E., Burja, A., Radianingtyas, H., Barrow, C. J. Critical assessment of various techniques for the extraction of carotenoids and co-enzyme Q10 from the thraustochytrid strain ONC-T18. *J. Agric. Food Chem.* 2006, 54 (26), 9752–9758.

76. Aki, T., Hachida, K., Yoshinaga, M., Katai, Y., Yamasaki, T., Kawamoto, S., Kakizono, T., Maoka, T., Shigeta, S., Suzuki, O., Ono, K. Thraustochytrid as a potential source of carotenoids. *J. Am. Oil Chem. Soc.* 2003, 80 (8), 789–794.

77. Xu, W., Pang, K.-L., Luo, Z.-H. High fungal diversity and abundance recovered in the deep-sea sediments of the Pacific Ocean. *Microb. Ecol.* 2014, 1–11.

78. Petrini, O., Sieber, T. N., Toti, L., Viret, O. Ecology, metabolite production, and substrate utilization in endophytic fungi. *Nat. Toxins* 1993, 1 (3), 185–196.

79. Saikkonen, K., Faeth, S. H., Helander, M., Sullivan, T. J. Fungal endophytes: A continuum of interactions with host plants. *Annu. Rev. Ecol. Syst.* 1998, 319–343.

80. Jones, E. B. G., Stanley, S. J., Pinruan, U. Marine endophyte sources of new chemical natural products: A review. *Bot. Mar.* 2008, 51 (3), 163–170.

81. Clay, K. Endophytes as antagonists of plant pests. In *Microbial Ecology of Leaves.* Springer-Verlag, New York, 1991, pp. 331–357.

82. Carroll, G. C. Fungal associates of woody plants as insect antagonists in leaves and stems. In *Microbial Mediation of Plant-Herbivore Interactions.* Wiley, New York, 1991, pp. 253–271.

83. Wilson, D., Carroll, G. C. Avoidance of high-endophyte space by gall-forming insects. *Ecology* 1997, 78 (7), 2153–2163.

84. Azevedo, J. L., Maccheroni Jr, W., Pereira, J. O., de Araújo, W. L. Endophytic microorganisms: A review on insect control and recent advances on tropical plants. *Electron. J. Biotechnol.* 2000, 3 (1), 15–16.

85. Rodriguez, R., Redman, R. More than 400 million years of evolution and some plants still can't make it on their own: Plant stress tolerance via fungal symbiosis. *J. Exp. Bot.* 2008, 59 (5), 1109–1114.

86. Kupta, J., Anke, T., Steglich, W., Zechlin, L. Antibiotics from basidiomycetes. XI. The biological activity of siccayne, isolated from the marine fungus *Halocyphina villosa* J. & E. Kohlmeyer. *J. Antibiot.* 1981, 34 (3), 298–304.

87. Overy, D.P., Bayman, P., Kerr, R. G., Bills, G. F. An assessment of natural product discovery from marine (sensu stricto) and marine-derived fungi. *Mycology* 2014, 5, 145–167.

88. Koglin, A., Walsh, C. T. Structural insights into nonribosomal peptide enzymatic assembly lines. *Nat. Prod. Rep.* 2009, 26 (8), 987–1000.

89. Gao, X., Wang, P., Tang, Y. Engineered polyketide biosynthesis and biocatalysis in *Escherichia coli*. *Appl. Microbiol. Biotechnol.* 2010, 88 (6), 1233–1242.

90. Galm, U., Wendt-Pienkowski, E., Wang, L., Huang, S.-X., Unsin, C., Tao, M., Coughlin, J. M., Shen, B. Comparative analysis of the biosynthetic gene clusters and pathways for three structurally related antitumor antibiotics: Bleomycin, tallysomycin, and zorbamycin. *J. Nat. Prod.* 2011, 74 (3), 526–536.

91. Tanaka S., Shinjo, T., Hoshino, M. Nutritive value and in vitro digestibility of some mangrove species. In *Proceedings of VII Pacific Science Inter-Congress Mangrove Session, 97-100*. ISME/VII Pacific Science Inter-Congress, Okinawa, 1994, pp. 97–100.

92. Mouzouras, R., Jones, E. B. G., Venkatasamy, R., Moss, S. T. Decay of wood by microorganisms in marine environments. In *Record of the Annual Convention B.W.P.A.* British Wood Preserving Association, Huntingdon, 1986, pp. 27–44.

93. Velmurrugan, N., Lee, Y. S. Enzymes from marine fungi: current research and future prospects. In *Marine Fungi and Fungal-Like Organisms*. De Gruyter, Berlin, 2012, pp. 441–474.

94. Pang, K. L., Jones, E. B. G. Epilogue: importance and impact of marine mycology and fungal-like organisms: challenges for the future. In *Marine Fungi and Fungal-Like Organisms*. De Gruyter, Berlin, 2012, pp. 509–517.

CHAPTER 14

Medicinal Chemistry and Lead Optimization of Marine Natural Products

James W. Leahy
Department of Chemistry and Center for Drug Discovery and Innovation,
University of South Florida, Tampa, Florida

CONTENTS

14.1 INTRODUCTION

Natural products have had a profound impact on the discovery and development of medicinal agents throughout history.[1] In the two centuries since Friedrich Sertürner first isolated morphine (**1**) from opium,[2] there have been a significant number of novel natural products revealed and elucidated that have played a crucial role in preventing disease and treating symptoms. From anti-infective agents such as penicillin (**2**)[3] and erythromycin (**3**)[4] to anticancer compounds such as vincristine (**4**)[5] and Taxol (**5**),[6] countless human lives have been saved through the auspices of natural product drugs. Even today, some diseases, such as malaria, are predominantly treated with natural products like artemisinin (**6**).[7] In addition, some compounds play a prominent role in the elucidation of critical biological pathways that can be exploited in the treatment of a number of conditions. For example, staurosporine (**7**) is used to study protein kinases[8] and has facilitated the discovery of inhibitors like Gleevec (**8**)[9] and Iressa (**9**).[10]

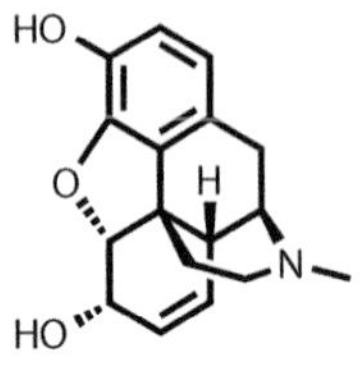

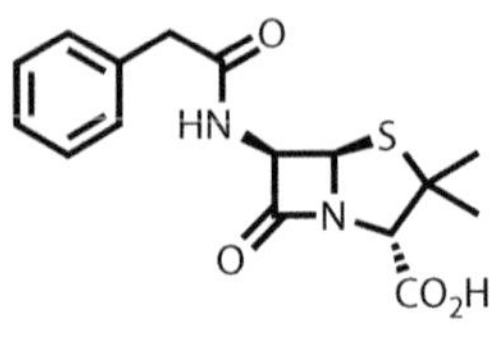

Given that more than 70% of the earth's surface is covered by the ocean, which accounts for 99% of the living space on the planet, it is surprising that all of the aforementioned natural products emanate from terrestrial sources.[11] In fact, the vast majority of medicinally relevant natural products have historically come from terrestrial sources. Part of this can be attributed to the fact that less than 10% of the ocean has ever been explored by humans, which suggests that there are a large number of possibilities for the discovery of new agents.[11]

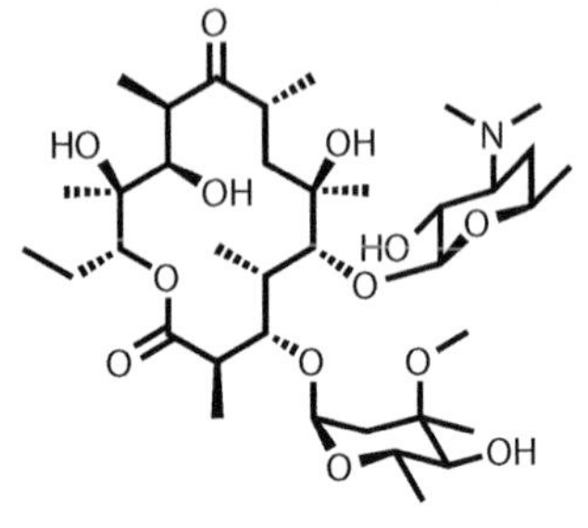

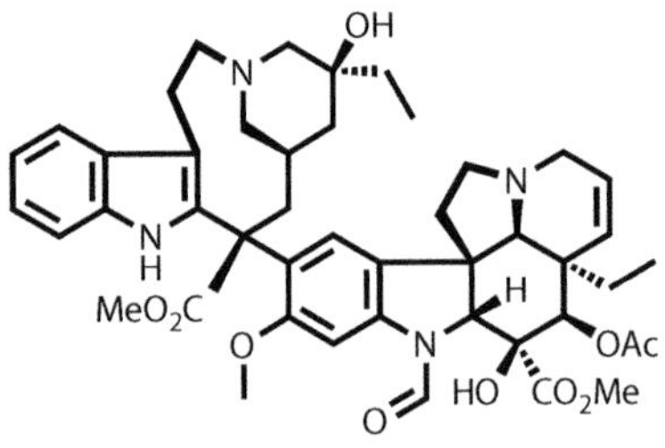

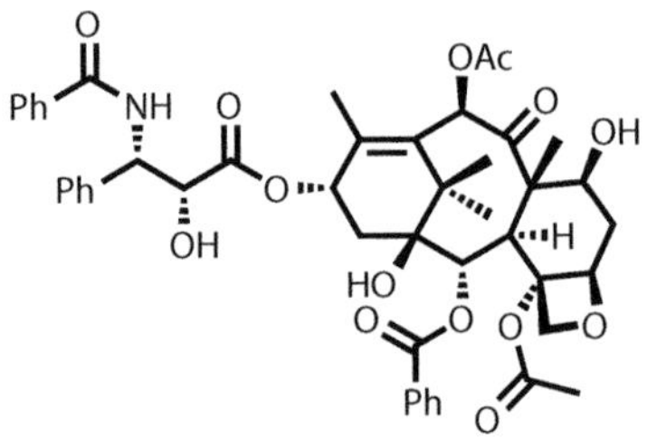

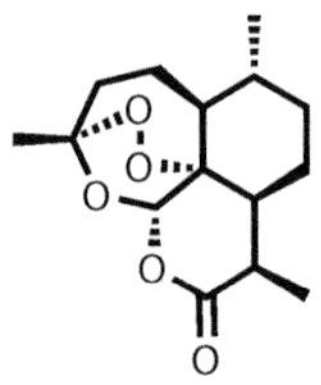

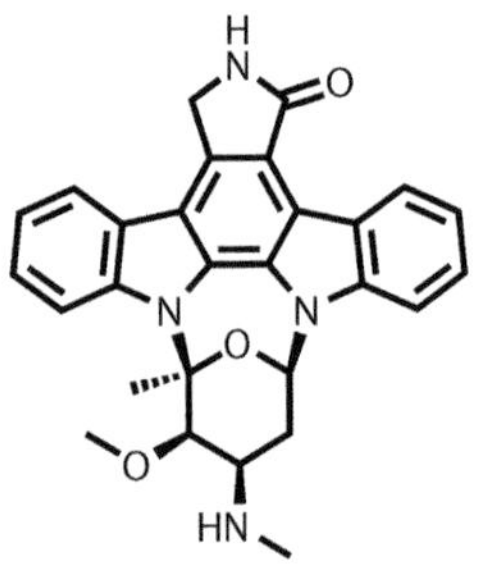

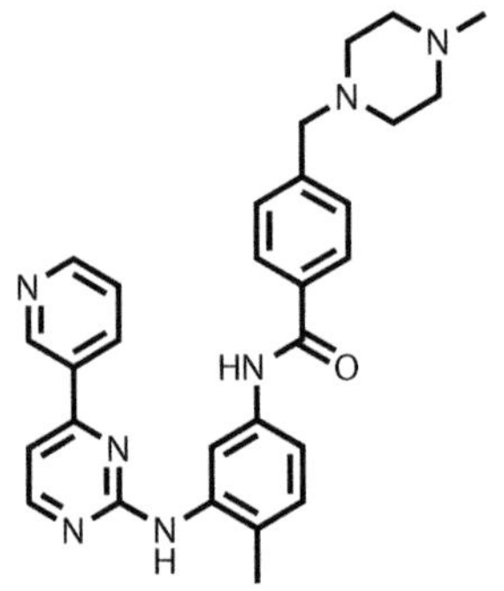

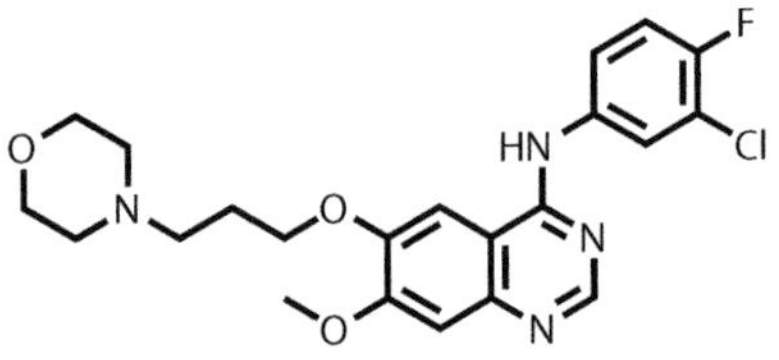

There have been a number of transformative innovations that have had a dramatic impact on modern drug discovery. One has been the concurrent development of automation and high-throughput screening capabilities, which has allowed for the rapid analysis of large collections of compounds.[12] Coupled with the advent of parallel lead optimization of activity and pharmacokinetic properties,[13] there has been a marked decrease in the amount of time (and patience) provided for a drug discovery project to progress. Toward that end, phrases such as "fail early and fail cheaply" have become the predominant philosophy behind many discovery campaigns,[14] and they are not always congruous with the historical timelines required for the development of natural products that have become drugs.

14.2 LEAD OPTIMIZATION

At its core, drug discovery is the search for a molecule that safely elicits a desired biological response without causing any harmful side effects. This has often been accomplished through the use of medicinal chemistry, where a systematic evaluation of a biologically active compound of interest through the generation of new molecules allows for the determination of what modifications improve or diminish the characteristics of interest.[15] Improvements in synthetic methodology, purification, and structural analysis have contributed greatly to modern medicinal chemistry, in terms of both the generation of sizable screening collections and accelerating access to analogs.[16] This approach is the same regardless of whether the initial compound is a natural product. Two of the most well-documented examples of this are aspirin (11) and heroin (12), discoveries that were first made in the nineteenth century. Acetylation of salicylic acid, a natural product initially isolated from willow bark, transformed a poorly tolerated compound into arguably the most successful drug in history, which remains a widely used pain reliever around the world.[17] Similarly, acetylation of morphine extracted from poppy plants provides heroin, which was marketed as a superior and nonaddictive alternative to morphine.[18] This approach to using natural products as the foundation for medicinal chemistry ventures has continued to provide a successful avenue for the discovery of novel therapeutic agents. The statin drugs used to lower blood cholesterol levels owe their origins to compactin (13)[19] and lovastatin (14),[20] natural products isolated from fungi and subsequently found to inhibit 3-hydroxy-3-methylglutaryl-Coenzyme A (HMG-CoA) reductase. Compounds such as Pravachol (15)[21] and Zocor (16)[22] are the result of direct chemical modification of the original natural products, while Lipitor (17)[23] is a totally synthetic molecule that contains critical structural features identified from the initial lead compounds. Given that Lipitor, which only went off patent in 2011, is generally considered to be the best-selling drug in the history of pharmaceuticals,[24] natural products clearly remain an important resource for the discovery of new drugs.

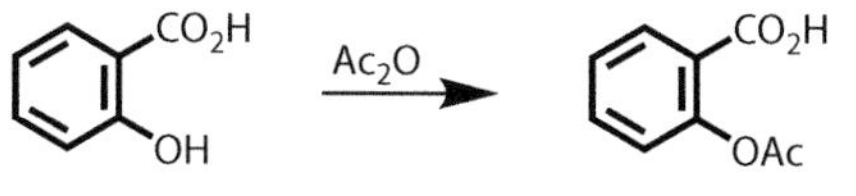

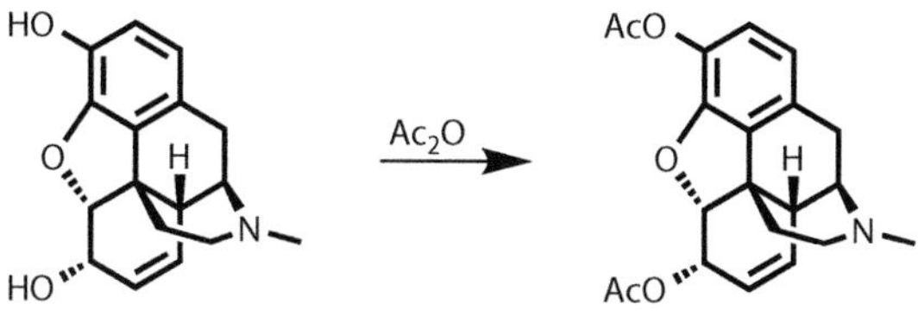

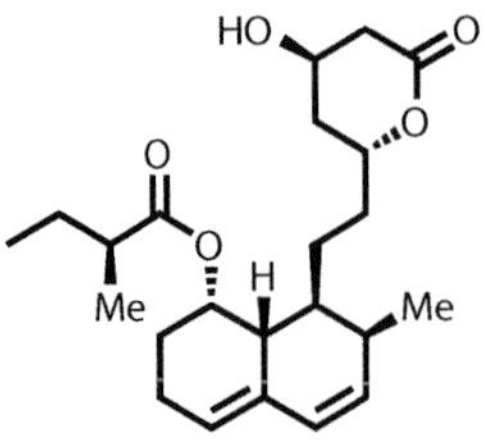

13

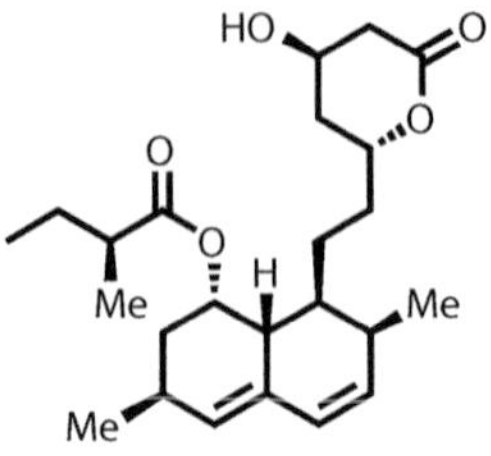

14

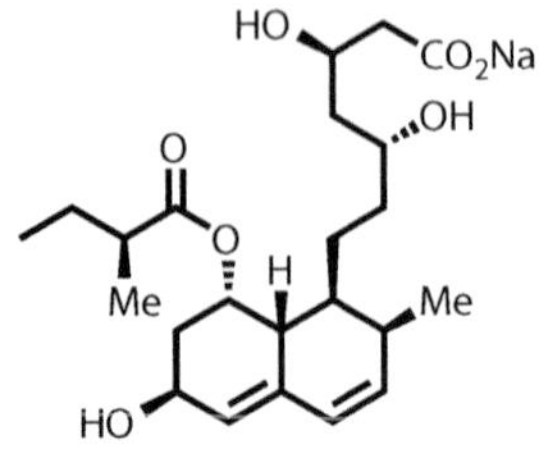

15

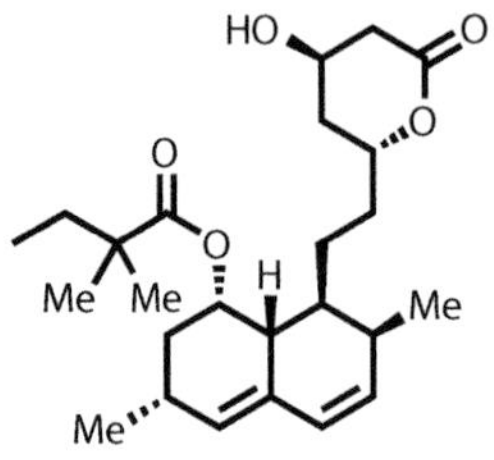

16

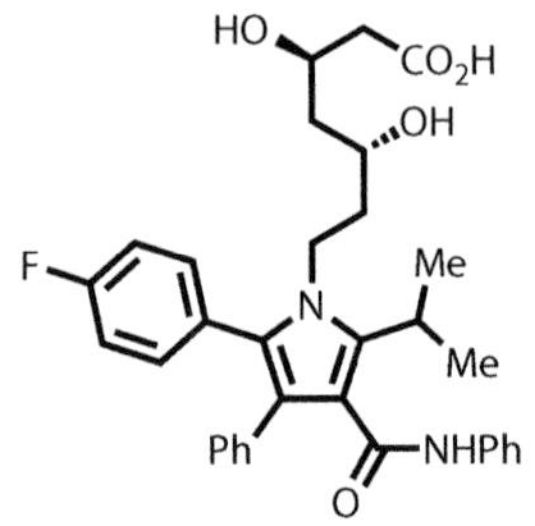

17

Each of these examples used natural products that led to the identification of compounds that became marketed drugs. Interestingly, the modifications in each case were not aimed at improving the activity of the initial natural product. Instead, they resulted in an improvement in other parameters that made them more suitable as drug candidates. A common misconception is that the bulk of the lead optimization process is occupied by searching for potency instead of other considerations, such as selectivity, physicochemical properties, pharmacokinetics, toxicity, or even reliable production. Improvements in screening capabilities have enabled the practice of parallel lead optimization,[13] where all of these issues can be considered at the same time, instead of optimizing the activity of a particular class of compounds only to discover that the entire series exhibits poor pharmacokinetic characteristics. Toward that end, a number of *in vitro* assays have been developed in an effort to model the various processes that contribute to the overall pharmacokinetic profile of compounds and are now routinely performed as part of the overall parallel lead optimization process.

14.2.1 Pharmacokinetics in Lead Optimization

The iterative process of medicinal chemistry is predicated on the concept of synthesizing a molecule, evaluating its properties, and using that information to guide the selection of new analogs to be synthesized (Figure 14.1). One of the more difficult tasks in this regard is the evaluation of the pharmacokinetics of a particular compound, primarily because there are so many interconnected activities that contribute to the overall fate of a molecule after it is administered.[25] Additionally, performing a full pharmacokinetic analysis in an animal model such as a mouse is a fairly time-consuming operation and requires a relatively significant amount of material (typically more than 10 mg). In the case of natural products, that can exceed the amount of compound obtained from the isolation process.

Since the overall pharmacokinetic fate of a xenobiotic compound can be largely accounted for through the processes of absorption, distribution, metabolism, and excretion (ADME) (Figure 14.2),[26] it is beneficial to try to consider each of these individually. Identification of which of these processes represents the biggest obstacle to a favorable pharmacokinetic allows for the design of novel analogs that might circumvent the impediment. This has prompted the development of assays designed to mimic each of these individual processes, as well as computational models to help guide the

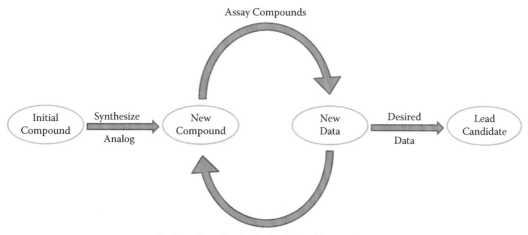

Figure 14.1 A representative decision loop for the lead optimization process. While a more detailed version of this can be developed that incorporates structural information to design analogs to address affinity or selectivity or automation to speed up synthesis, ultimately every lead optimization campaign utilizes a loop like this to identify a development candidate.

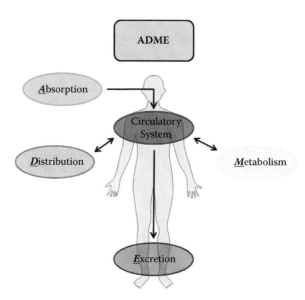

Figure 14.2 Graphical depiction of the individual ADME components (absorption, metabolism, distribution, and excretion) that contribute to the overall pharmacokinetic profile of a compound.

selection of analogs with the greatest likelihood of having improved pharmacokinetic properties. In many instances, these assays have been optimized into a format suitable for high-throughput analyses that require relatively little material. Since marine organisms are unlikely to have evolved chemical defenses in response to human predation, it should not be expected that they would create compounds that would have optimized human pharmacokinetics. It is therefore helpful to consider the ADME parameters of natural products, as well as any novel analogs, in order to ascertain any problems in advance of actual pharmacokinetic analyses.

14.2.1.1 Absorption

The first process that is generally considered for a xenobiotic that has been administered into the body is absorption, which refers to the progression of the compound from its initial administration into the circulatory system.[25] This process therefore differs depending on the route of administration, since intravenous dosing would bypass any need to consider the absorption of the compound. In fact, absorption is the primary process that contributes to the bioavailability of a compound, which is simply the ratio of the concentration of the compound that reaches the bloodstream via a particular administration route to the concentration after intravenous administration.[27]

All nonintravenous administration routes of a xenobiotic require the compound to pass through one or more membranes in order to reach the bloodstream and enter into circulation. A number of factors not entirely related to the structure of the compound can contribute to this, including formulation, particle size, crystallinity, and salt counterion selection, and they are beyond the scope of discussion here.[28] However, there are several aspects of the molecule that can be rapidly assessed in order to maximize the likelihood of favorable absorption characteristics.

The first of these is the solubility of the compound,[29] which can be considered under a variety of conditions, including variables such as the composition of the solvent and whether the solubility is being measured under kinetic or thermodynamic circumstances. Poor solubility is an obvious impediment to favorable pharmacokinetic behavior and can also hide downstream issues that the compound could face by artificially limiting the amount of material that enters circulation.[29] In order to assess compounds as quickly as possible, a determination of the kinetic solubility in buffered media from a standard dimethyl sulfoxide (DMSO) solution can be performed using microtiter plates in a turbidimetric assay.[30] This format allows for solubility analysis to be performed on hundreds of samples at a time, with the data quickly available and thus able to guide the design and synthesis of novel analogs. There has been some discussion about whether there is more value in obtaining kinetic or thermodynamic solubility data, but the determination of thermodynamic solubility is sufficiently time-consuming that it is not performed on a routine basis.[31] Furthermore, the solubility can be determined with various media, including at different pH levels designed to mimic the conditions in the stomach or small intestine.

Another measurement of solubility that is used to guide drug discovery efforts is the partition coefficient P,[32] which considers the relative solubility of a compound in octanol and water such that P = [compound in octanol]/[compound in water]. In this case, the lipophilic octanol is intended to mimic the hydrophobicity of the phospholipid bilayer of a cell membrane and thus approximate the membrane permeability of a compound. In his seminal work that led to the formulation of the rule of five, Lipinski examined a number of drug candidates and noted that compounds were more likely to have favorable pharmacokinetic characteristics if the logP of that compound was less than five, suggesting that overly lipophilic compounds are unlikely to pass through membranes and thus become available to the organism.[33] While it is possible to measure the partition coefficient of compounds using straightforward protocols, considerable effort has been devoted to developing methods to calculate an approximate value based on computational models.[34] These have become so prevalent that programs such as ChemDraw® provide a calculated logP (often referred to as clogP) for any drawn structure as part of a default display of chemical properties. There has been a substantial effort to find additional methods for predicting favorable pharmacokinetic properties based on computational models, such as polar surface area, which extend beyond the scope of this discussion.[35] It should be noted that Lipinski has pointed out that many natural products are exceptions to the rule of five.[36]

While solubility is an important feature of a compound, a more direct measure of its probability of having favorable absorption in humans relates to its membrane permeability. There are a number of methods for rapidly evaluating this. Immobilized artificial membrane (IAM) chromatography,

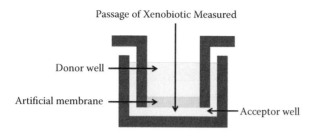

Passage of Xenobiotic Measured

Donor well

Artificial membrane

Acceptor well

Figure 14.3   Schematic diagram of a parallel artificial membrane permeability assay (PAMPA). A solution of the compound to be evaluated with a known concentration is added to the donor well that is affixed with an artificial semipermeable membrane. The amount of the compound that passes through the membrane can then be detected in the acceptor well.

for example, allows for the high-throughput analysis of samples via a simple high-performance liquid chromatography (HPLC) protocol, although this is in reality a measure of the lipophilicity of a compound more than actual permeability.[37] Caco-2 human epithelial cells can be grown as a monolayer in a multiwell filter plate, and the passage of compounds through that monolayer can be measured as a direct analysis of their permeability.[38] Since the use of cell lines is somewhat more labor-intensive, as well as prone to greater experimental variability, there has been a shift toward the use of a parallel artificial membrane permeability assay (PAMPA).[39] This assay is also conducted in a multiwell format (Figure 14.3), but since the artificial membranes allow for advanced and automated preparation of the filter plates, there is a decreased opportunity for variability between plates.[40] With such a wealth of assays available, the vast majority of lead optimization campaigns include a thorough investigation of *in vitro* and *in silico* absorption profiling.

14.2.1.2 Distribution

The next step in the pharmacokinetic process is the distribution of xenobiotic compounds from the bloodstream to the intended site of interest. Much as the absorption process requires nonintravenous compounds to pass through membranes in order to enter the circulatory system, they have to pass through membranes again in order to leave the bloodstream. Therefore, membrane permeability assays serve as a predictor of whether compounds can be expected to have favorable distribution characteristics as well. Unlike with absorption, however, the permeability assays can be tailored to consider the target area of interest. For example, the blood–brain barrier protects the brain and the central nervous system from invasion by a number of large molecules and microorganisms.[41] It also prevents many smaller molecules from passing through, which can be an impediment to the discovery of drugs targeting the brain. PAMPA assays have been used to mimic the blood–brain barrier *in vitro* to help select compounds with the greatest chance of demonstrating activity *in vivo*.[40]

In addition to the exploration of simple membrane permeability, distribution can also be studied through the use of permeability in modified cell lines. The aforementioned Caco-2 cell line can be made to overexpress p-glycoprotein (Pgp), which actively transports many different compounds across membranes that would not otherwise pass through.[42] By performing assays under standard conditions as well as in the presence of Pgp inhibitors, a reasonable picture of the overall distribution of any compound, including via simple membrane permeability or drug transport, can be determined.

Interestingly, membrane permeability is not necessarily required for xenobiotic compounds to leave the circulatory system. Fenestrated capillaries have small pores that allow small molecules to pass through them and thus enter the interstitial fluid surrounding cells (Figure 14.4).[43] Essential compounds such as nutrients pass through these pores and directly into the interstitial fluids to facilitate access to cells. The pores are sufficiently small that larger proteins do not pass through and thus

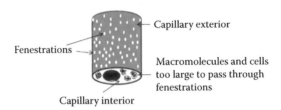

Figure 14.4 An illustration of a fenestrated capillary. The small perforations allow for the passage of fluid as well as small molecules between the capillary and the interstitial fluid, but larger molecules, such as albumin or entire cells, cannot leave the plasma.

are retained in the bloodstream.[27] For this reason, the proteins contained in the blood plasma play a critical role in the distribution of potential drugs. Compounds that aggregate with plasma proteins are unlikely to pass through the capillary pores, impeding their ability to reach the desired target. While there are a number of proteins that exist in the blood, one of the most abundant is albumin, which makes up more than half of the total protein in human plasma.[44] Albumin plays a number of physiological roles, but one of its primary functions is as a transporter of lipophilic compounds such as fatty acids and hormones.[45] Since it is a fairly nonselective transporter, it is not surprising that it also binds many potential drug candidates, and thus can retard their distribution out of the bloodstream. In fact, it is not uncommon to run some cell-based assays in the presence of albumin in order to avoid overly positive data that can result when lipophilic compounds aggregate with cell membranes.[46] Therefore, a common and straightforward assay that is routinely performed during the drug discovery phase is a protein binding assay.[47,48] Although high protein binding (>99%) does not necessarily preclude a compound from serving as a viable drug (and in some cases can be beneficial), it can dramatically reduce the distribution and pharmacodynamic effect of the compound.[44]

Finally, not all organic molecules are stable in plasma for a significant period of time. Blood contains a number of proteins that can have an adverse effect on molecules, and it is therefore useful to evaluate the stability of compounds over time in plasma.[49]

14.2.1.3 Metabolism

While absorption and distribution deal with the movement of a xenobiotic compound from its site of administration to its site of interest, the remaining aspects of the pharmacokinetic process relate to removal of the compound from the body. One of the most prominent mechanisms for this is the metabolism of xenobiotics into compounds that can be cleared more rapidly.[50] Most of this metabolism happens in the liver, which is the first organ to encounter molecules that enter the bloodstream via oral administration.[27] Blood in the portal vein carries all of these compounds to the liver before going to the heart via the hepatic vein and thus entering the circulatory system. While in the liver, a series of biotransformations can occur to xenobiotics. Oxidation and bioconjugation are the two most significant of these biotransformations from a drug discovery perspective. Oxidations can proceed through a number of different enzymatic processes, such as monoamine oxidase and alcohol dehydrogenase, but one of the most prolific enzyme classes that conduct biological oxidations is the cytochrome P450 (CYP) enzymes. There are a number of different CYP isoforms that exist, each of which has its own type of substrate specificity, and the levels of these isoforms can vary between individuals.[51] This can be best illustrated by the observation that tamoxifen (**18**) must undergo oxidation by CYP 2D6 to hydroxytamoxifen (**19**), which is an active antagonist of the estrogen receptor and can be effective in the treatment of breast cancer.[52] Some patients do not respond to treatment with tamoxifen, and it has been found that most of these patients had a deficiency in their CYP 2D6. This wide variability helps illustrate the challenge of drug discovery and points to the virtual impossibility of finding a drug that is universally tolerable across an entire patient population.

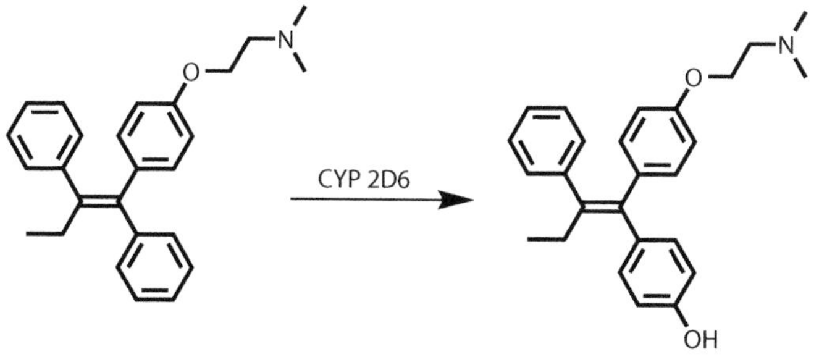

18   CYP 2D6   **19**

From a basic pharmacokinetics perspective, the greatest concern about metabolism is the extent to which the compound in question undergoes chemical modification. This can be readily assessed by examining the stability of the compound in the presence of liver microsomes, where the bulk of xenobiotic metabolism takes place *in vivo*.[53] The ready availability of microsomes from different animals that are commonly used in pharmacokinetic and pharmacodynamic evaluation also removes some of the doubt that can arise from trying to infer data between species. Compounds are routinely exposed to liver microsomes of the species of interest and evaluated by HPLC at regularly scheduled intervals to determine stability of the compound under these conditions.[53] Compounds that are rapidly metabolized in this manner would be expected to demonstrate high clearance and a short half-life in an actual pharmacokinetic study.

Since metabolism is such a critical function for overall survival, a more thorough consideration of how a compound interacts with this process is crucial. Toward this end, it is common to examine the activity of the most common CYP isoforms in the presence of new compounds during the lead optimization process.[44] By exposing a known substrate to the appropriate CYP in the presence and absence of the compound being evaluated, it is possible to determine if it inhibits the isoform from functioning properly. If a compound inhibits its own metabolism by CYP, it might artificially appear to have a favorable pharmacokinetic profile by maintaining a desirable concentration through an undesirable mechanism. Furthermore, drug–drug interactions represent one of the biggest concerns in the development of a new therapeutic agent,[54] and one of the most common mechanisms for this phenomenon is the inhibition of a CYP isoform by one drug that is required for the metabolism of another. As shown above, a compound that inhibits the activity of CYP 2D6 would not be compatible with the use of tamoxifen, which requires 2D6 for the generation of an active species. Recently, the cytotoxic natural product austocystin D (**20**) was found to require CYP activation in order to demonstrate *in vivo* antitumor activity.[55] Alternatively, Coumadin (**21**) is a widely used anticoagulant with a very small therapeutic index that is cleared primarily via CYP 2C9 oxidation.[56] A compound that inhibits 2C9 could lead to a dangerous increase in Coumadin concentration *in vivo*. It is therefore quite common to investigate the inhibition of compounds against some of the most common CYP isoforms (phenacetin [**22**] for 1A2, tolbutamide [**24**] for 2C9, dextromethorphan [**27**] for 2D6, and both midazolam [**29**] and testosterone [**31**] for 3A4).[57] It should be noted that the metabolic process is far more complex than can be appreciated from a simple consideration of whether a particular compound inhibits a particular isoform. For example, a compound that is not a CYP inhibitor might be a substrate for one isoform, and the corresponding product that is formed might be an inhibitor for a different isoform. For this reason, it is now common to study compounds in a cocktail of multiple CYP isoforms in the presence of their substrates in order to evaluate any inhibition.[58] This allows for the detection of this type of cross-reactivity, at least among the isoforms that are included in the cocktail.

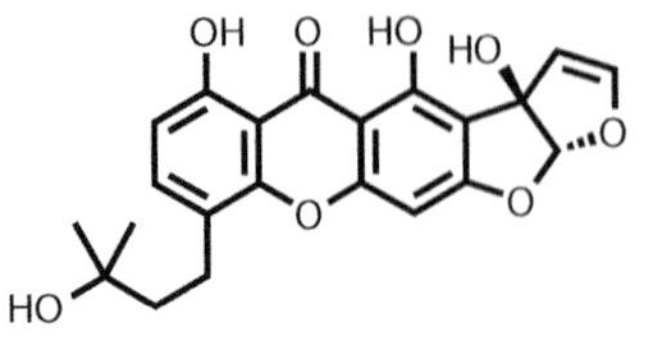

20 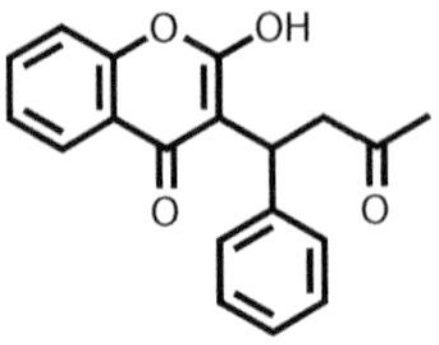   **21**

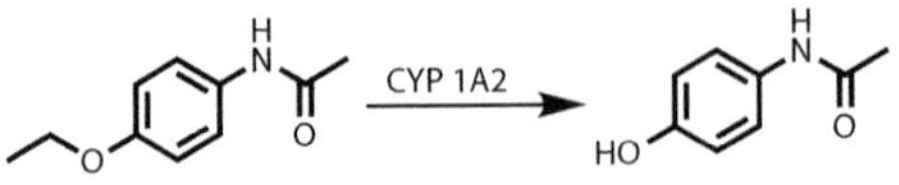

22   CYP 1A2   **23**

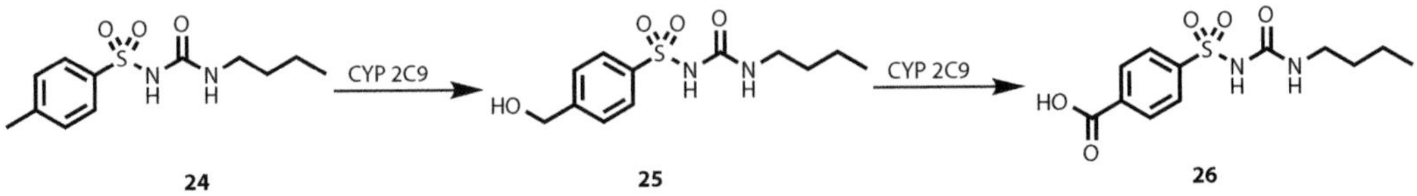

24   CYP 2C9   **25**   CYP 2C9   **26**

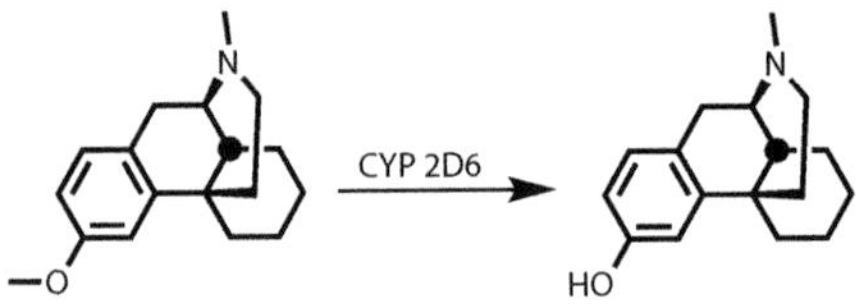

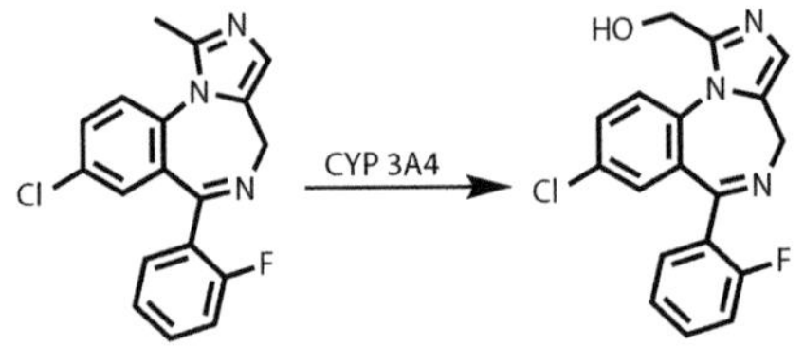

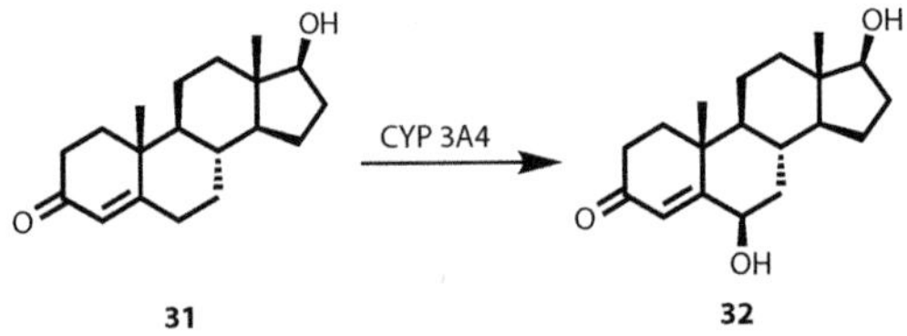

27   CYP 2D6   **28**

29   CYP 3A4   **30**

31   CYP 3A4   **32**

A full discussion of metabolism and xenobiotics should also include a consideration of the mechanism-based inhibition of cytochrome P450 isoforms.[59] This type of inhibition involves the irreversible inhibition of the enzyme that ultimately requires a *de novo* synthesis of the isoform in question. While there is not currently a widely used high-throughput method for investigating mechanism-based inhibition of CYPs, it is often explored prior to a full pharmacokinetic analysis to expose compounds that might exhibit artificially high exposure.

14.2.1.4 Excretion

Of the four processes that contribute to the overall pharmacokinetic profile of a compound, excretion tends to draw the least attention during the lead optimization phase of drug discovery. Generally, by the time a xenobiotic has undergone metabolism to the form that is eliminated from the body, it has outlived its biological intention. However, there are *in vitro* assays that help to predict aspects of the excretion process. One such example is to evaluate the multidrug resistance-associated protein 2 (mrp2) transport pathway to predict biliary excretion.[60]

14.2.2 Toxicity

Although not inherently part of traditional pharmacokinetic analysis, it is common to consider the toxicity of a compound as part of the lead optimization process. In fact, it is not uncommon to see toxicity lumped in with the remainder of the aforementioned *in vitro* pharmacokinetic optimization, such that ADMET and ADME/Tox are commonly used terms. A number of cytotoxicity assays exist that can be used to rapidly assess compounds, and a selectivity profile could be generated with little difficulty.[61]

While there are a number of assays to examine toxicity to whole cells, there are not many toxicity mechanisms that have been fully developed. One that has been is the hERG (human Ether-ā-go-go Related Gene) ion channel, which regulates potassium ion gradients across muscle cells.[62] This is critical in heart muscle, where the hERG channel contributes to the restoration of the appropriate intracellular potassium concentration during ventricular repolarization. Compounds that inhibit this channel can lead to a pronounced elongation of the QT interval in electrocardiograms and potentially fatal cardiac arrhythmia (Figure 14.5).[63] The hERG channel has been shown to be somewhat promiscuous in that it can be inhibited by a number of different potential drugs. Examples of approved drugs that have been pulled from the market due to hERG interactions are terfenadine (**33**)[64] and astemizole (**34**).[65] An electrophysiology patch-clamp assay has been developed to evaluate whether compounds inhibit the hERG channel and is now routinely performed as part of the lead optimization process.[66]

33   **34**

Another common toxicity assay that is run on advanced compounds during the lead optimization process is for mutagenicity. The Ames test allows for a rapid assessment of mutagenicity for compounds using several bacterial strains that can determine their capacity for inducing mutations and, presumably, potential for being carcinogenic.[67] While not routinely run on every compound during the lead optimization process, virtually every compound that is considered for an investigational new drug application goes through a genotoxicity screening that typically starts with an Ames test.

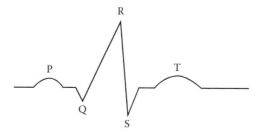

Figure 14.5 Typical electrocardiogram of a heartbeat. When the hERG ion channel is inhibited, it leads to a significant increase in the time between the initiation of the QRS sequence (which represents the ventricular depolarization that triggers contraction) and the completion of the T wave (ventricular repolarization). Elongated QT intervals have been linked to ventricular fibrillation and an increased incidence of sudden death.

14.2.3 Structure–Activity Relationship

While the pharmacokinetics and ADME/Tox of a compound are critical and often underappreciated components of any lead optimization efforts, the parallel drive to maximize the desired biological activity while minimizing any undesired interactions is still of paramount importance. In this regard, the search for a discernible structure–activity relationship (SAR) pattern can become the primary focus of the medicinal chemistry efforts. Full discussions of various strategies for the general optimization of target interactions with small molecules, including the use of quantitative methods, are available in the literature.[68]

One method that permeates historical SAR work is the use of bioisosteres.[69] Since hydrogen bonds and dipolar interactions are a crucial part of the interaction between ligands and receptors, it is typically important that they be retained or mimicked in analogs in order to display the desired biological activity. In fact, nature is not above using the same tactic in the creation of bioactive natural products. Balanol (**35**) is a potent inhibitor of protein kinase C that can be obtained from fungal sources, and crystallographic studies of this compound with cAMP-dependent protein kinase show that the benzophenone portion of the molecule acts as a bioisostere for the triphosphate portion of ATP (adenosine triphosphate) (**36**).[70]

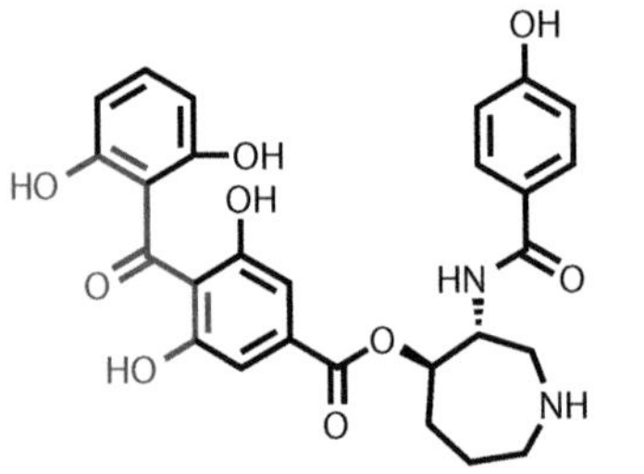

35 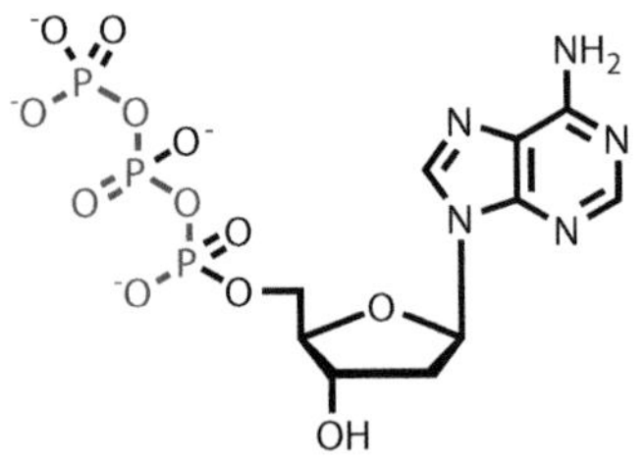   36

There are numerous examples of the use of bioisosteres in medicinal chemistry that vary from simple use of fluorine to block metabolism[71] to heterocycles that can replace the potentially chemically sensitive functionality of molecules.[69] In this regard, lead optimization campaigns based on marine natural products are no different. The only limitation in this regard is access to the materials required to investigate their viability.

14.3 MATERIAL ACCESS

There is no question that natural products have been invaluable in the pursuit of novel therapeutic agents. The molecular architecture provided from nature that has been optimized through various evolutionary pressures has produced many chemical scaffolds that may have never been considered without their discovery. However, that molecular complexity can also restrict their use as viable chemical leads for practical drug discovery campaigns. Isolation of the material from natural sources can be both costly and impractical. For example, a large-scale operation to isolate pure bryostatin 1 (**37**) to be used in clinical trials required 10 months, 10,000 gallons of wet *Bugula neritina* that was estimated to represent 12,700 kg of wet animal, and considerable amounts of solvent and silica gel in order to provide 18 g of the marine natural product for further studies.[72] Cultivation of such a large amount of the bryozoan represents a tremendous challenge from both a cost and a logistical standpoint. However, the total synthesis of bryostatin 1 has also proven to be a formidable challenge. There has been considerable interest in the development of synthetic approaches to this natural product, and Keck recently reported the first total synthesis of bryostatin 1 nearly 30 years after the isolation was initially reported.[73] This optimized and highly convergent synthesis still proceeds in 0.4% overall yield from methyl isobutyrate, providing 2 mg of the product in the final step, and would present a tremendous challenge to scale up to give grams of bryostatin 1. More importantly, both of these approaches afford a single compound, which presupposes that the compound has already been optimized. Performing lead optimization on material obtained in this manner restricts the number of options available, partially because of the limited amount of the starting material, but also due to the lack of synthetic options available once the compound has undergone the full complement of biotransformations that lead up to its natural synthesis. Conversely, introducing a modification early in the synthesis requires the redundant performance of all of the remaining steps in the total synthesis before the change can truly be evaluated from a biological perspective. There have therefore been considerable efforts expended to explore alternative methods to provide access to natural products as well as analogs that might otherwise be difficult to obtain.

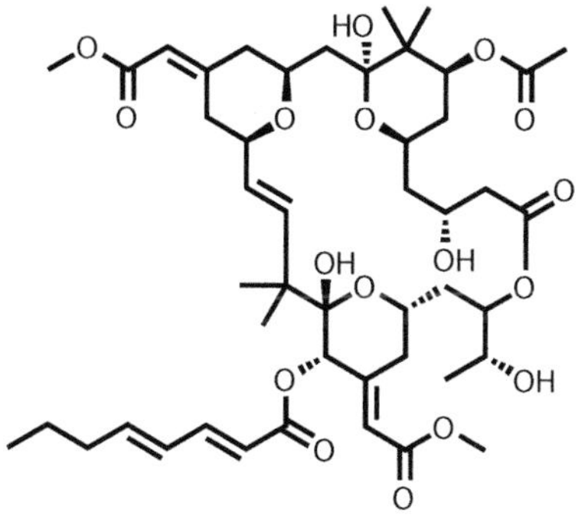

37

One example of a bioactive natural product that attracted considerable attention due to material access is Taxol (38).[74] The compound was initially isolated from the bark of the Pacific yew tree,[75] and an effort to cultivate a useful supply resulted in the collection of 1200 kg of the bark that yielded just 10 g of pure Taxol.[76] This prompted a number of groups to pursue studies aimed at the total synthesis of Taxol.[77] However, the significant breakthrough resulted from the discovery that 10-deacetylbaccatin could be isolated in reasonable quantities from the needles of *Taxus baccata*, a different species of yew tree.[78] With a viable source of a starting material that contained the Taxol core, Holton demonstrated a practical semisynthesis of the natural product that ultimately paved the way for the generation of sufficient material to complete the clinical trials necessary for commercial approval.[79] Meanwhile, efforts to develop an alternative source of the compound without the need to conduct semisynthesis resulted in the discovery of a fermentation process that could be conducted with *Taxus* cells that directly provided Taxol.[80]

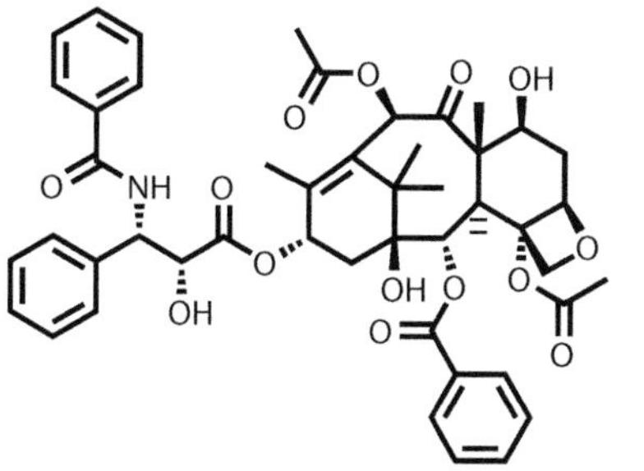

38

More recently, engineered cell lines for the production of specific chemicals have been used effectively to overcome many of the shortcomings of traditional natural product isolation.[81] For example, microbial cells have been engineered to produce not only Taxol itself, but also compounds along the biosynthetic pathway that could be used for the exploration of new analogs.[82] Exploration of the polyketide synthase pathway, as well as some of the downstream processes, has allowed for the development of a method that manipulates the enzymes in a manner that unprecedented analogs can be isolated.[83] In this manner, the field of combinatorial biosynthesis has become viable.[84] With these types of efforts still in the stages of early exploration, most lead optimization of marine natural products requires considerable effort for isolation and synthesis to move forward.

14.4 CASE STUDIES

While there are a relatively large number of natural products that have served as the launching pad for drug discovery and lead optimization campaigns that have resulted in approved drugs, most of them were based on compounds that were isolated from sources that fall outside of what would be considered marine. There are, however, several examples of marine natural products that have gone through successful clinical trials and become approved drugs. There are also a number of examples of compounds that have undergone lead optimization that can help illustrate the role of medicinal chemistry in their development to date, and several thorough discussions have previously been published.[1,85–91]

14.4.1 Approved Drugs from Marine Natural Products

The search for drug candidates from marine sources can largely be traced to the post-World War II pursuit of natural products in general. Werner Bergmann extracted several novel nucleoside analogs from a sponge that he had collected off the Florida Keys in 1945.[92] Subsequent structural elucidation of these nucleosides revealed that the carbohydrate component was different from the ribofuranosyl or 2-deoxyribofuranosyl portions typically found on compounds of this type. Specifically, the compounds isolated from the sponge were spongothymidine (**39**) and spongouridine (**40**), which contain an arabino-furanosyl carbohydrate portion.[93] Prior to that discovery, any efforts that had been explored in the area of modified nucleosides had contained novel base portions, and no modifications to the carbohydrates had ever been reported. In the face of this discovery, research endeavors using this carbohydrate modification resulted in the synthesis of the arabinofuranosyl nucleosides, including cytidine arabinoside (**41**)[94] and adenine arabinoside (**42**).[95] Ultimately, cytidine arabinoside became the first of these compounds to be approved by the U.S. Food and Drug Administration (FDA) as cytarabine (also known as ara-C) in 1969, and it is still used today as an injectable treatment for leukemia.[96] Similarly, adenine arabinoside, known as Vira-A or vidarabine, was approved in 1976 as the first nucleoside herpes drug.[97] These two drugs were developed before many of the current approaches to maximizing pharmacokinetics were put into practice, although it is noteworthy that both fall within the Lipinski parameters, with clogP values below 0. Neither appears to pose a CYP inhibition concern, although cytarabine is a substrate for CYP 3A4.[98] Given the number of carbohydrate-modified nucleosides that have subsequently been approved and widely used, such as AZT (azidothymidine) (**43**) and acyclovir (**44**), a case could be made that these were the ultimate extension of lead optimization efforts from marine natural products.

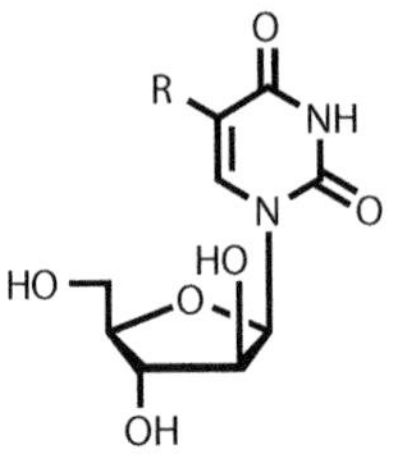

39 (R = Me)
40 (R = H)   **41** 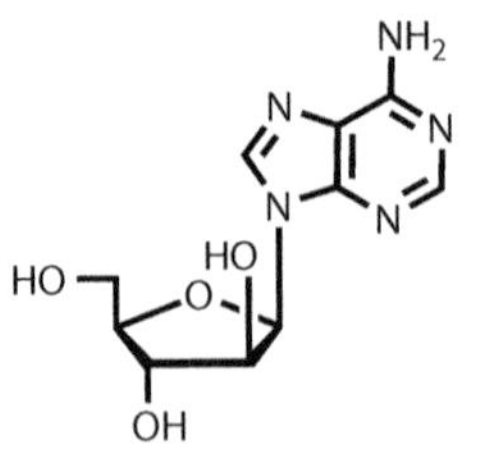   **42**

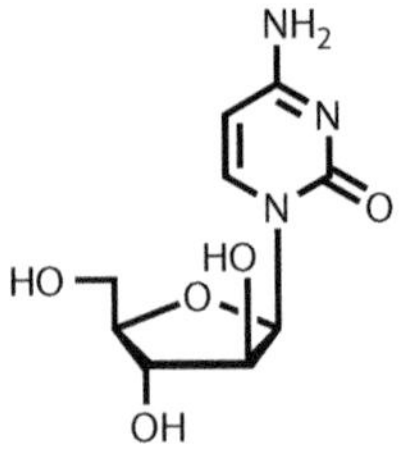

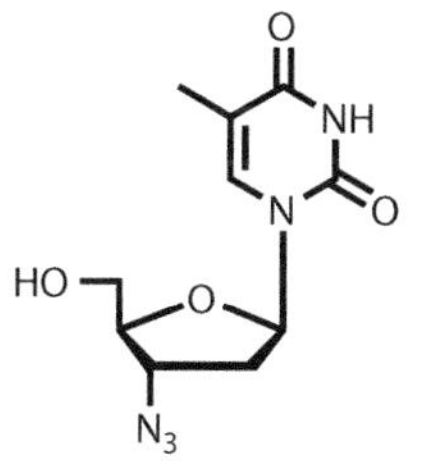

43 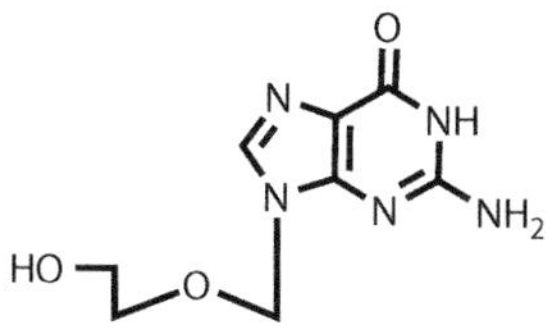   **44**

14.4.1.1 Ziconotide and Eribulin

In the intervening years since the approval of vidarabine, there have only been three drugs that have been approved as a direct result of marine natural product lead optimization studies. Ziconotide (**45**) is a peptide that was isolated from cone snails that is approved as an intrathecal analgesic agent.[99,100] As a water-soluble peptide with a molecular weight above 2500 g/mol, it falls far outside of the Lipinski parameters and has a clogP of –15. Eribulin (**46**) is an antimitotic spiro-ketal that has been developed by Eisai and approved for the treatment of breast cancer.[101] Eribulin is the end result of a lead optimization effort based on halichondrin B (**47**), which was isolated from the marine sponge *Halichondria okadai*.[102] It has a clogP of 2.31 and a molecular weight of 729.9 g/mol. Detailed accounts of the development of these drugs are discussed in Chapters 16 (ziconotide) and 17 (eribulin) of this book.

A case can also be made for the inclusion of Lovaza as a marine drug since it has been approved by the FDA for hypertriglyceridemia and is a prescription drug.[103] However, since Lovaza is comprised of the ethyl esters of a mixture of omega-3 fatty acids rather than a single agent, it falls beyond the scope of this discussion.

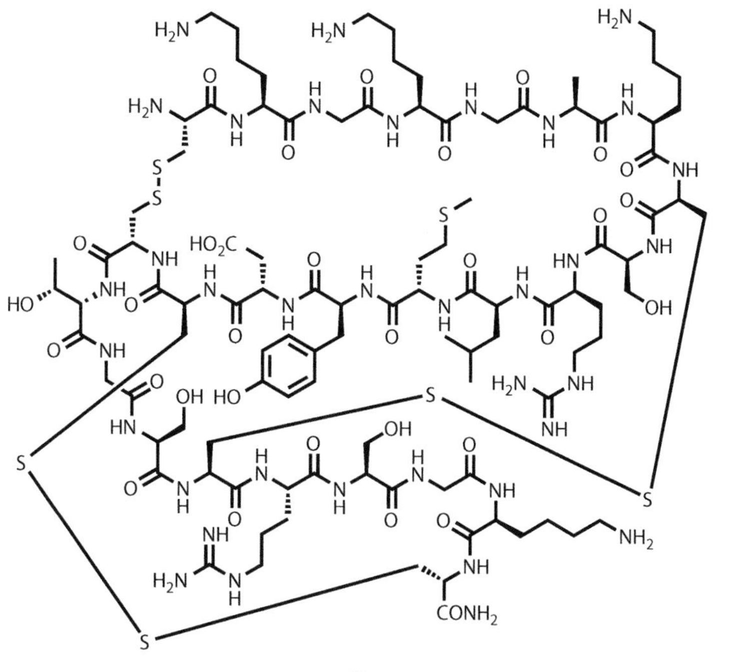

45

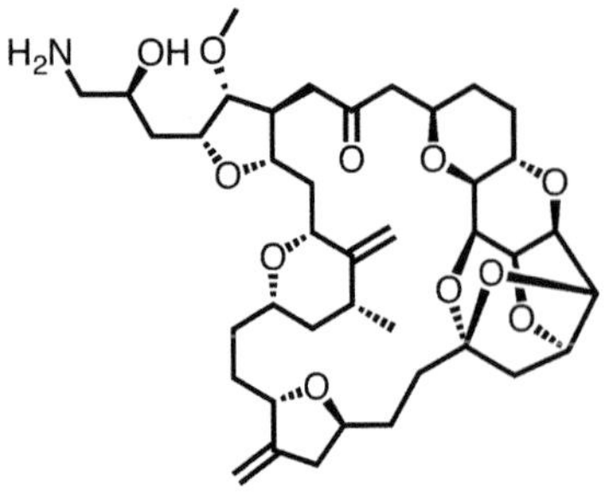

46 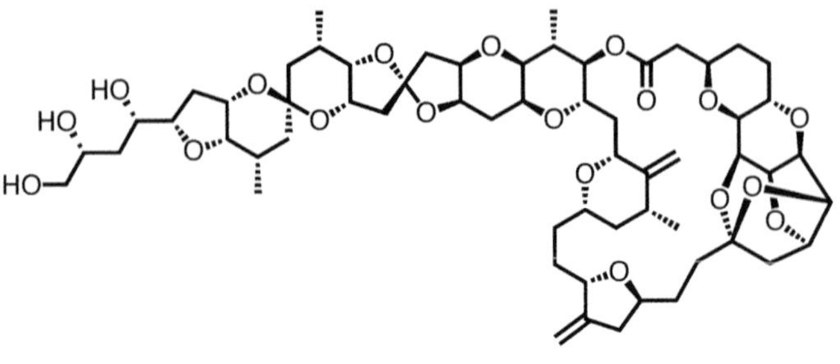   47

14.4.1.2 Yondelis

With the discovery of bioactive agents from sponges that resulted in the discovery and development of approved drugs, attention turned to the examination of other marine organisms. One type of organism that has shown promise in this regard is tunicates, which have yielded a number of bioactive natural products.[104] The presence of antitumor activity in extracts of *Ecteinascidia turbinata* was first reported in 1969 in a screen using a P388 murine leukemia cell line,[105] and subsequent efforts to identify the active components led to the first publication of the relative structure of ecteinascidin 743 (**48**) in 1992.[106,107] The absolute stereochemistry was confirmed in 1996.[108]

The ecteinascidins possess an unprecedented scaffold that contains multiple tetrahydroisoquinoline subunits. Extensive analysis of nuclear magnetic resonance (NMR) and mass spectrometry data provided most of the information required to establish the structure of the natural products, but chemical degradation of ecteinascidin 597 (**49**) gave methyl cysteinate (**50**), and chiral GC analysis was able to be used to demonstrate the absolute stereochemistry about this center.[108] With this stereocenter in hand, the absolute stereochemistry of the entire ecteinascidin family of natural products was able to be inferred. While the structural complexity of the ecteinascidins contributed to the difficulty in establishing the structure, the lack of material available was also a significant factor. Ecteinascidin 743 was the most abundant member of the family that was found, but it was only isolated in 0.0001% yield.[109] The level of bioactivity (IC_{50} [concentration that inhibits binding or activity by 50%] of 0.5 ng/ml in the L1210 assay), however, warranted additional study. PharmaMar therefore developed a comprehensive sea farm system to cultivate sufficient quantities of ecteinascidin 743 necessary to complete these studies and facilitate clinical trials.[110]

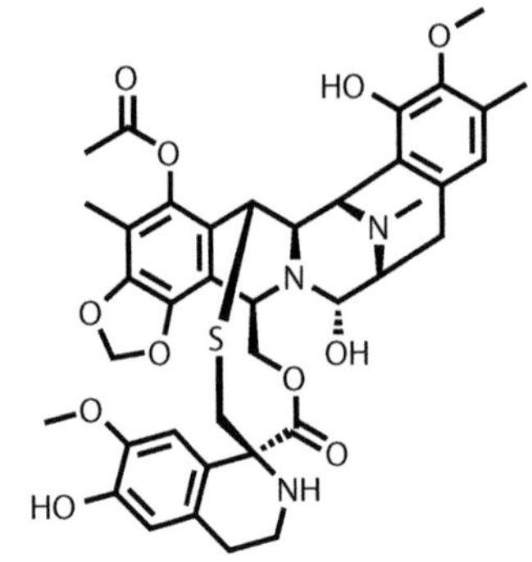

1) HgCl$_2$
2) NaBH$_4$
3) HCl, Δ
4) MeOH, H$^+$

48 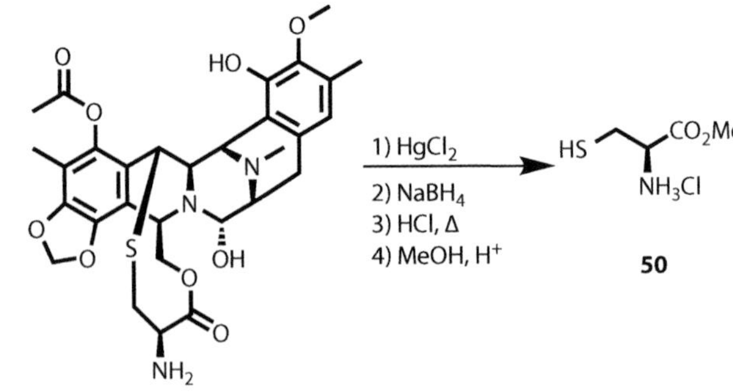   49   50

There are a number of compounds that have been approved for various cancer treatments whose activity can be traced to their ability to bind to DNA.[111] Some compounds, such as the platinum-based agents, cause DNA damage that leads to the apoptosis of rapidly proliferating cells,[112] while others, such as etoposide (**51**), interfere with the DNA–topoisomerase interaction, resulting in improper DNA synthesis.[113] Ecteinascidin 743 binds to the minor groove of DNA in a manner that inhibits the activity of the transcription-coupled nucleotide excision repair system, which results in double-strand DNA cleavage and cell death. In this manner, ecteinascidin 743 represents a new class of anticancer agents that is active against a variety of solid tumors, including prostate, renal, breast, ovarian, and non-small-cell lung cancer cell lines.[114]

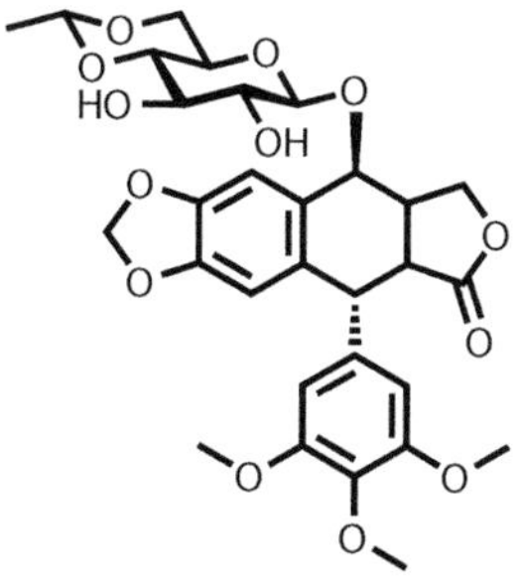

51

With the material access to the ecteinascidins limited to extraction from the tunicate and the supplies obtained used in the preclinical evaluation of the compound, no serious efforts were initially entertained to explore ecteinascidin analogs to improve on the pharmacokinetic properties. The molecular weight of ecteinascidin 743 is sufficiently high and solubility sufficiently low that there was no expectation of oral bioavailability. With a relatively high clogP of 4.20, intravenous administration was explored from the outset.[114] It is more than 94% protein bound with an affinity for albumin site I,[115] leading to a relatively high volume of distribution (V_{ss} = 959 L/m^2), and indicates significant distribution.[116] Following an infusion of ecteinascidin 743, a biphasic decrease in the concentration of the compound in plasma suggests multicompartmental distribution. Following the initial rapid decrease in plasma concentration, ecteinascidin 743 has a half-life of 43.8 h. Treatment of the compound with human microsomes led to a significant degradation, with 68% remaining after incubation for 4 h.[117] The primary isoform responsible for ecteinascidin 743 metabolism is CYP 3A4, although it is also metabolized by several other isoforms, including 2C9 and 2D6. It does not appear to significantly inhibit cytochrome P450. As a DNA binding agent, ecteinascidin displays genotoxicity as expected.[118]

The acceptable pharmacokinetics displayed by ecteinascidin 743 warranted further clinical evaluation of the compound, but the cultivation efforts were not viewed as a viable ongoing source of material for expanding these studies. Meanwhile, the unique molecular architecture of the ecteinascidins had attracted synthetic interest, resulting in the first total synthesis by Corey based on a biomimetic approach.[119] Among the notable transformations in this synthesis were the installation of the medium-sized 10-membered macrolide ring via a thiol addition to the quinone methide (**53** to **54**) and the subsequent generation of the spirotetrahydroisoquinoline C subunit through a selective Pictet–Spengler reaction (**54** to **48**). This revelation resulted in the development of a semisynthetic route by Manzanares and coworkers at PharmaMar.[120] Their approach started with cyanosafracin B (**55**), an analog of the natural antibiotic safracin B (**56**). Optimization of a fermentation process from *Pseudomonas fluorescens* allowed for the production of cyanosafracin B on a kilogram scale.[120,121] Following reduction of the quinone into the hydroquinone with suitable protecting group manipulation (**57**), the pendant alanine was removed through an Edman degradation to give an alcohol (**58**) that essentially intersected Corey's initial synthesis. In total, this semisynthetic approach proceeds in 0.996% overall yield from cyanosafracin B over 19 isolated and purified steps, which was enough of an improvement that it has supplanted the cultivation process for the supply of ecteinascidin 743.

With an adequate supply of the drug candidate secured, ecteinascidin 743 has been examined in a number of clinical trials.[122] Most significantly, PharmaMar was granted orphan drug status for the compound on May 30, 2001, by the European Medicines Association (EMEA) for the indication of soft-tissue sarcoma under the trade name Yondelis and the generic name trabectedin.[123] The FDA granted the same designation on April 7, 2005.[124] The EMEA has since approved the use of Yondelis as a single agent for advanced soft-tissue sarcoma, as well as a second-line indication for ovarian cancer in conjunction with doxorubicin.[123] On November 20, 2008, Ortho Biotech submitted to the FDA a new drug application for this combination treatment of ovarian cancer, but it voluntarily withdrew the application on April 29, 2011. While Yondelis is currently approved for use in 73 countries, it has not yet been approved for use in the United States.

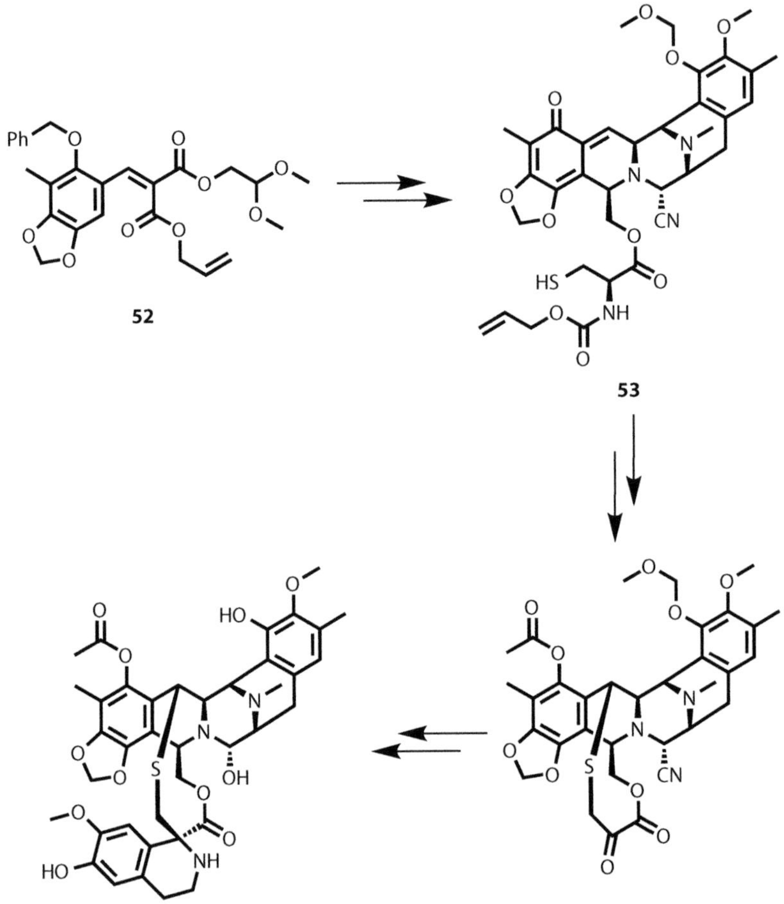

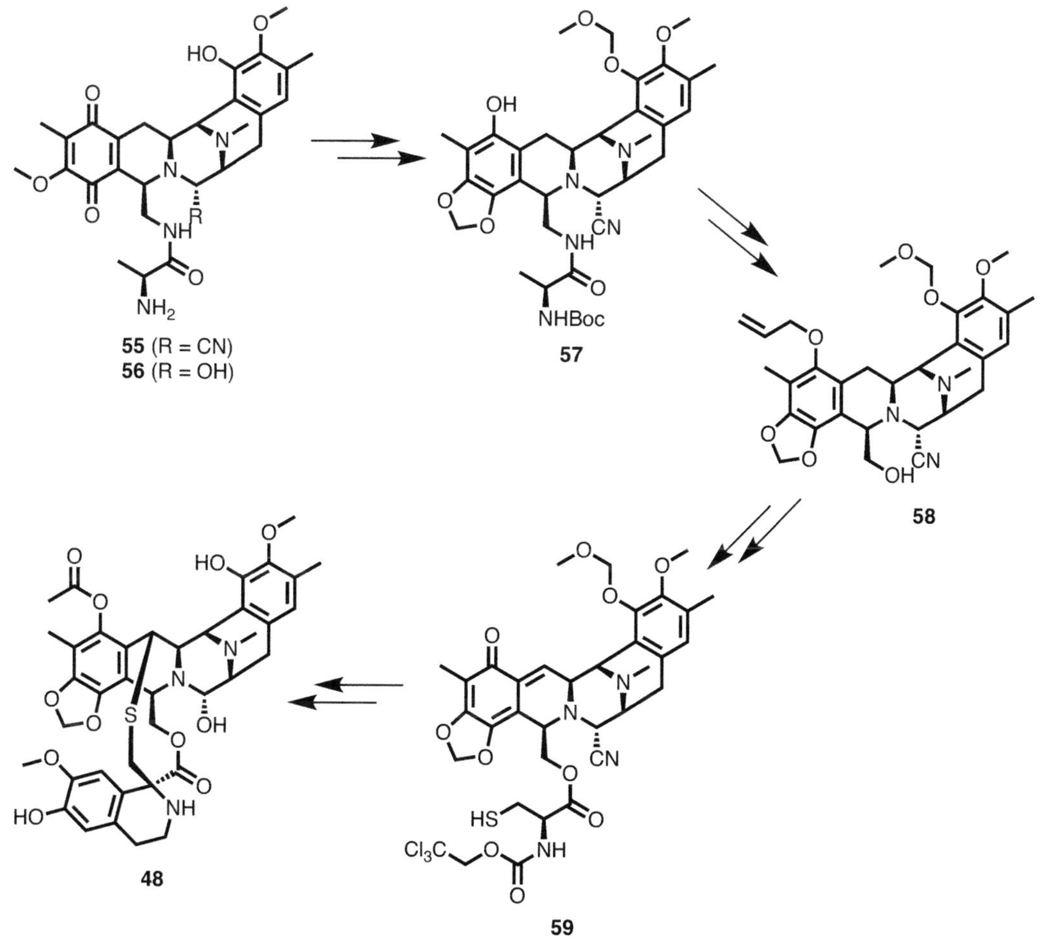

55 (R = CN)
56 (R = OH)

57

58

48

59

Since Yondelis has already been approved for human use and an optimized route that supplies the material has been realized, there has not been a pronounced push to pursue lead optimization studies. This is in part due to the fact that even the semisynthetic process requires 19 synthetic steps that are largely protecting group manipulations, which does not lend itself to a streamlined library approach to generating analogs. After completing his initial total synthesis, Corey was able to prepare a number of truncated analogs that did not contain the macrolide ring or the spirotetrahydroisoquinoline portions that were effectively as active as ecteinascidin 743 in a variety of cell lines.[125] Using molecular modeling to predict that a phthalide would serve as a reasonable surrogate for the C subunit as it protrudes from the DNA-bound complex, Corey correctly predicted that this analog would be active, and in fact, it was the most active analog prepared. Liu was able to demonstrate potent activity with greatly simplified analogs such as **63**,[126] and in conjunction with synthetic studies aimed at the total synthesis of saframycin A, Myers developed a scalable synthesis of an intermediate that could be used to explore a series of analogs, including a parallel synthetic approach.[127] Ultimately, he was able to identify several compounds, such as **65** and **66**, that had potent *in vitro* and *in vivo* activity.[128]

PharmaMar has reported an ecteinascidin analog (PM00104, **67**) that differs from Corey's phthalascidin through the use of a trifluoromethylcinnamyl amide.[129] This new analog is also a DNA binding agent, and the compound demonstrates antimyeloma activity *in vitro* and *in vivo*. Under the trade name Zalypsis, it has been investigated clinically as a potential treatment for a number of different cancer indications and is currently in a European clinical trial for multiple myeloma.[130,131]

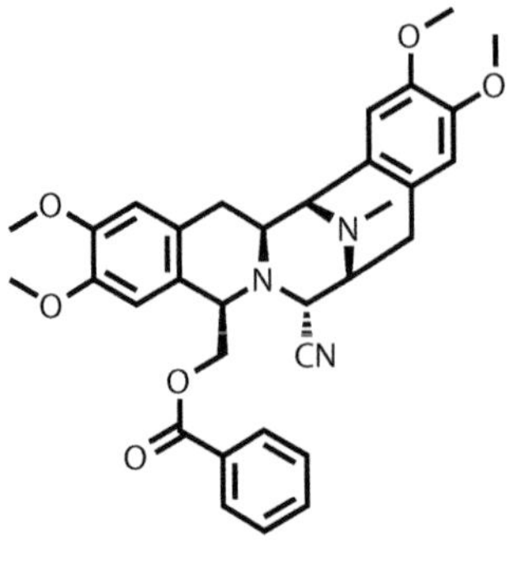

63

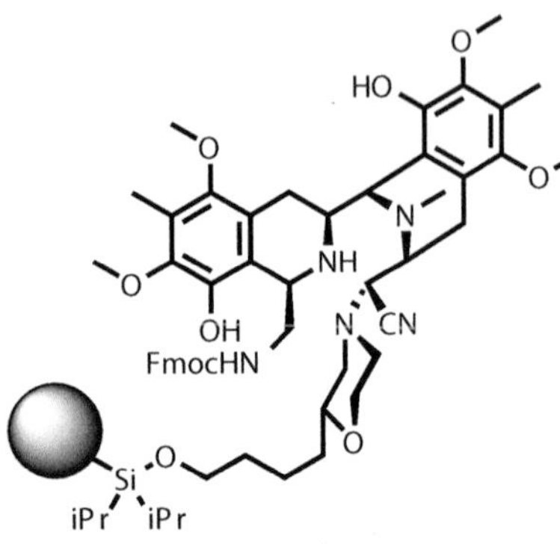

64

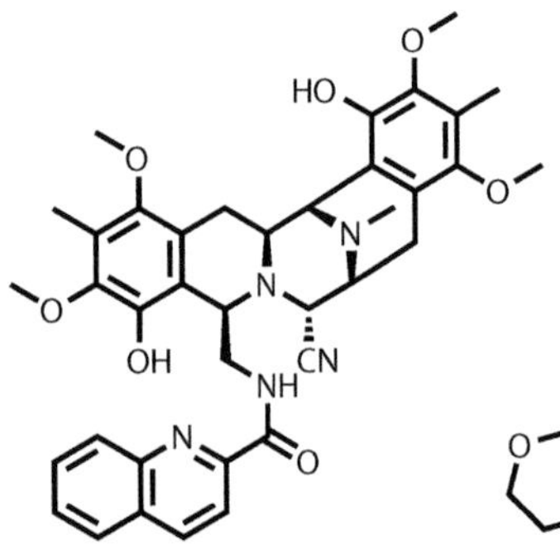

65

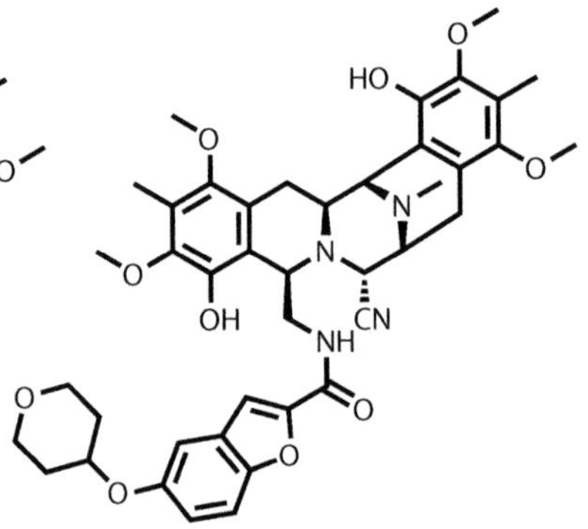

66

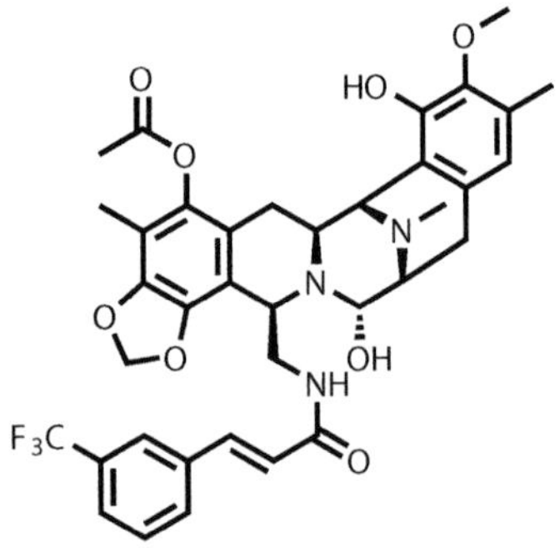

67

14.4.2 Other Marine Natural Product Clinical Candidates

The aforementioned compounds that have already been approved are validation that marine natural products can serve as the foundation for a drug discovery program. Considerable efforts have been undertaken to identify additional compounds that could serve as the basis for novel therapeutic agents that have resulted in projects at virtually every stage of the drug discovery process. Although these efforts are still in progress, a discussion of a few notable examples is included here.

14.4.2.1 Didemnin

The ecteinascidins are far from the only drug candidates derived from tunicates.[132] One of the extensively studied examples is the didemnins.[107,133] This family of cyclic depsipeptides was also initially identified by the Rinehart group at Illinois. Isolated from the Caribbean tunicate *Trididemnum solidum*, the didemnins appear to serve as a chemical defense against predation. Studies have shown that these natural products can be found in the larvae of the tunicates and confer resistance to feeding fish.[134]

The two first members of this family of natural products that were identified were didemnins A (**68**) and B (**69**), which differed only by the presence of a lactylproline appendage.[133] Both displayed activity in the initial antiherpetic assay that was used as a guide to identify compounds of interest, with the less abundant didemnin B demonstrating significantly pronounced activity over the truncated didemnin A. Over time, a number of additional related natural products have been isolated, primarily involving modifications to the peripheral *N*-methylleucine residue, including pyruvate analog dehydrodidemnin B (**70**).[135] Each of the members of this family of depsipeptides is comprised of a macrolide containing several atypical residues, including an α-(α-hydroxyisovaleryl)propionyl (Hip) and either an isostatine or norstatine group (**71**). More recently, the tamandarin family of related depsipeptides has been isolated from a Brazilian tunicate; they differ from the didemnins by the replacement of the Hip residue with an α-hydroxyisovaleric acid (Hiv) group (**72, 73**).[136]

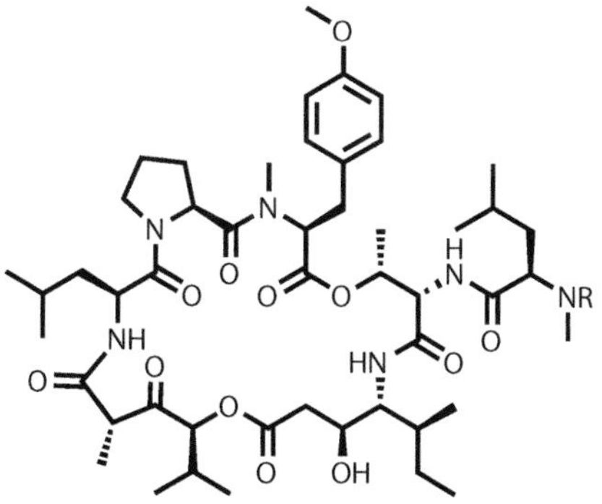

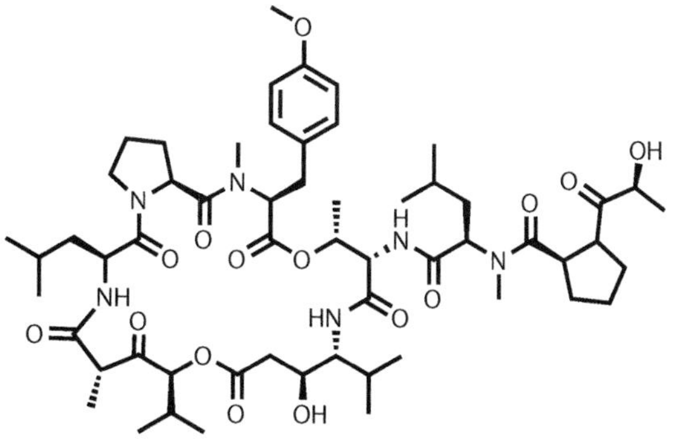

68 (R = H)

69 (R = 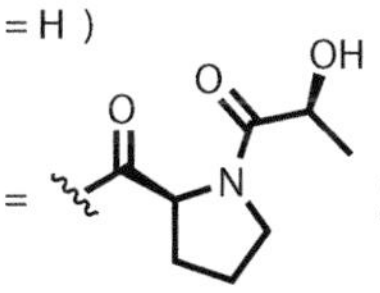)

70 (R =)

71

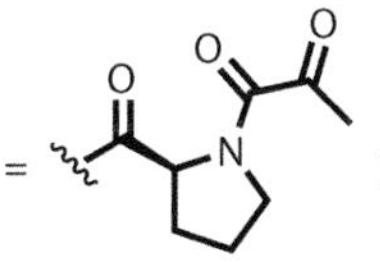

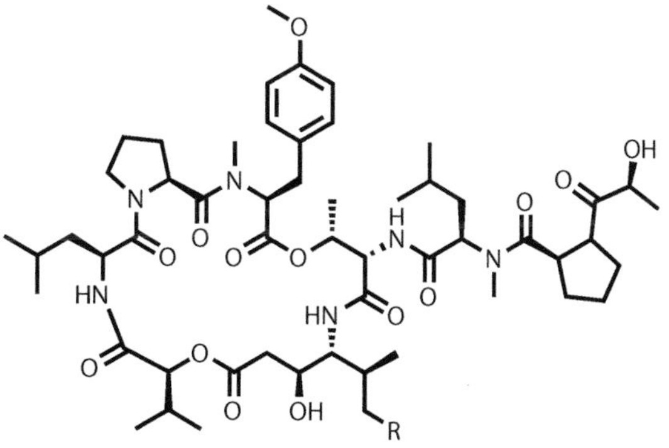

72 (R = Me)
73 (R = H)

While the initial bioassays that led the Rinehart group to focus on the didemnins were a herpes simplex virus assay and a human cytotoxicity assay, subsequent evaluation demonstrated remarkable antiviral, antitumor, and immunosuppressive activity.[91] Despite considerable efforts to elucidate the mechanism of action for the didemnins, a full understanding of how they elicit a biological response and induce apoptosis has not yet been achieved. They have been linked to the inhibition of protein biosynthesis,[137] and they bind to elongation factor 1α (EF-1α),[138] but it is not yet clear that this binding is sufficient to account for the pronounced apoptotic activity of the natural products.

Based on the activity of didemnin B against a number of different human tumor cell lines, it was considered to be a viable clinical candidate. From an ADME standpoint, didemnin B is far from ideal. It has a molecular weight of 1112 g/mol, more than 15 rotatable bonds, and is virtually insoluble in water below a pH of 10, so even with a clogP below 5, it is expected to present pharmacokinetic challenges.[139] Ultimately, didemnin B required formulation that included Cremophor,[140] which is known to introduce complications and side effects in some patients[141] but is used clinically in the formulation of some drug treatments, such as Taxol.[142] It was examined in at least three different phase II clinical trials for renal, breast, and ovarian cancer but was not advanced to phase III consideration due in part to adverse reactions and neuromuscular toxicity that were observed.[143] It was noted that didemnin B quickly disappeared from the plasma after administration, with less than 50% remaining after 1 h.[144] Furthermore, less than 10% of the intact compound was found to undergo biliary excretion, suggesting significant metabolism.[145]

The toxicities that led to the discontinuation of didemnin B as a clinical candidate suggested that any further studies would require either a compound that was less toxic or a compound that was more active in order to open up an adequate therapeutic window.[146] Rinehart had already reported that dehydrodidemnin B was approximately 10 times more active than didemnin B, with IC_{50} values below 1 nM against several different tumor cell lines,[147] suggesting that it could represent a viable option. As with didemnin B, material access to the drug candidate was realized through the development of a synthetic route rather than through the isolation of the compound from cultivation of the tunicate, taking advantage of highly efficient peptide coupling reactions to produce the material in good yields.[148] Pharmacokinetically, dehydrodidemnin B appears to be superior to didemnin B, with a mean half-life of more than 20 h and a suitably high volume of distribution.[149] PharmaMar is currently pursuing multiple clinical trials of dehydrodidemnin B under the trade name of Aplidin, including a phase III trial in refractory myeloma.[150]

From a SAR standpoint, the depsipeptide nature of the molecule allows for synthesis of a number of analogs in a straightforward manner.[146] These efforts have revealed a few interesting structural features. For example, ester linkages such as those in cyclic depsipeptides render the molecule sensitive to hydrolysis, either through hydrolysis under acidic conditions such as experienced in the stomach or through lipase-mediated hydrolysis.[151] Replacement of either ester linkage with a similarly substituted amide in the macrolide (**74, 75**) leads to an approximately 10-fold decrease in activity. However, simultaneous replacement of both esters as amides (**76**) actually leads to a significant increase in activity compared to the didemnin B analog. Curiously, small changes to the isostatine (**77–79**) and Hip (**80**) portions of the macrolide generally led to a significant drop in biological activity,[146] which makes the significant activity of the tamandarins somewhat surprising.[136] With the pronounced improvement in activity of Aplidin, side chain modifications have attracted a greater level of attention, and analogs with improved bioactivity have been identified (e.g., **81–83**).[143] Meanwhile, new bioactive natural analogs of this family continue to be found (**84**),[152,153] suggesting that the didemnin story will continue to evolve.

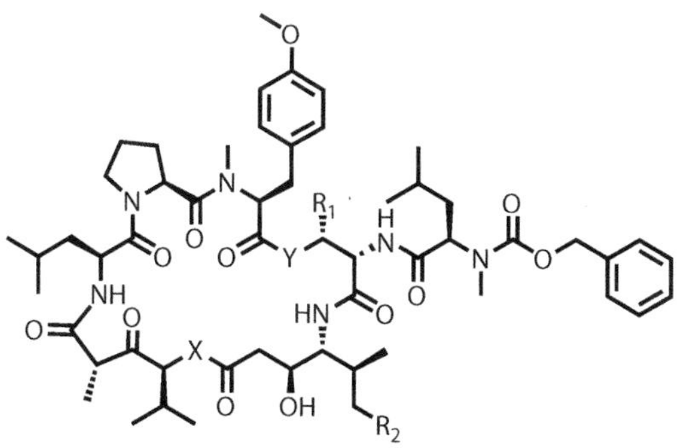

	R$_1$	R$_2$	X	Y	Lung	Colon
73	Me	Me	O	O	ND	1
74	H	H	NH	O	10	10
75	Me	H	O	NH	10	10
76	H	H	NH	NH	0.01	0.01

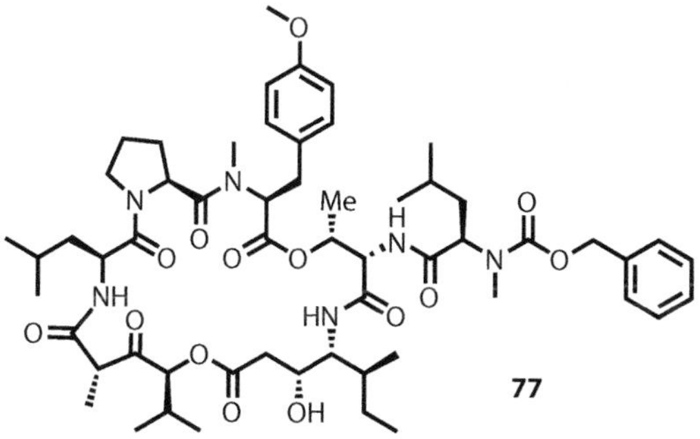

Cancer Cell Line IC$_{50}$ (nM)

77

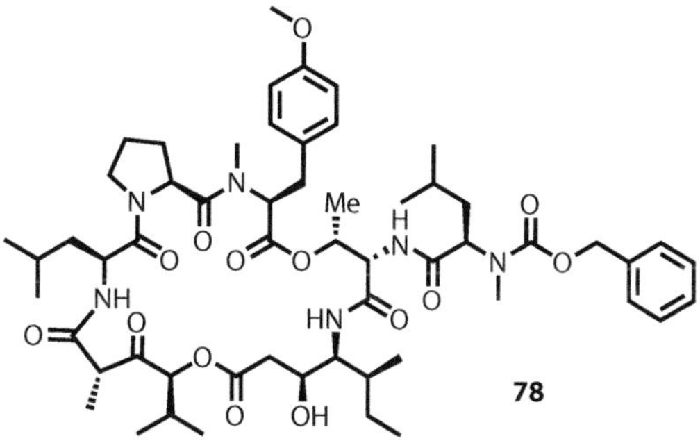

78

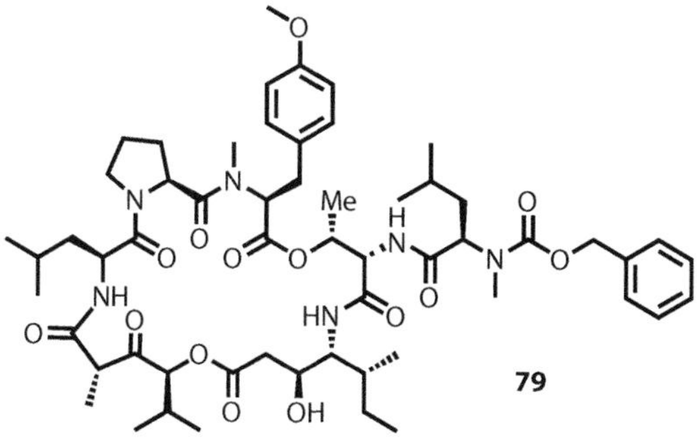

79

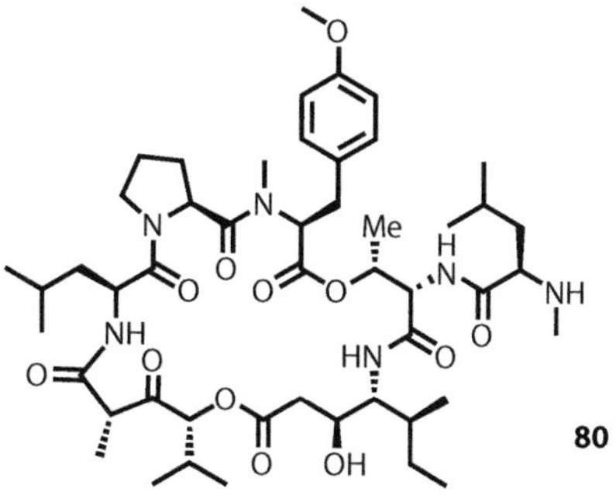

80

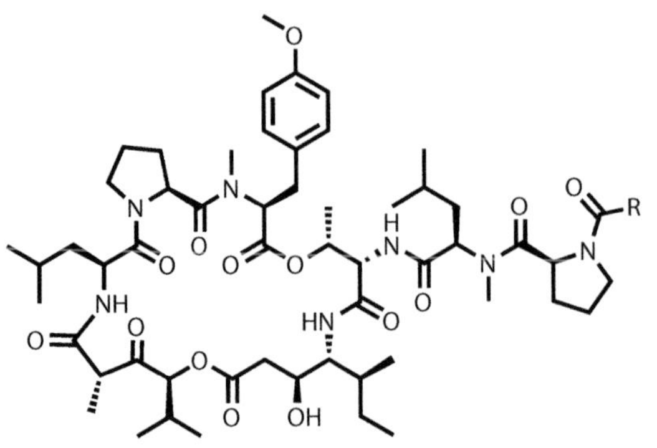

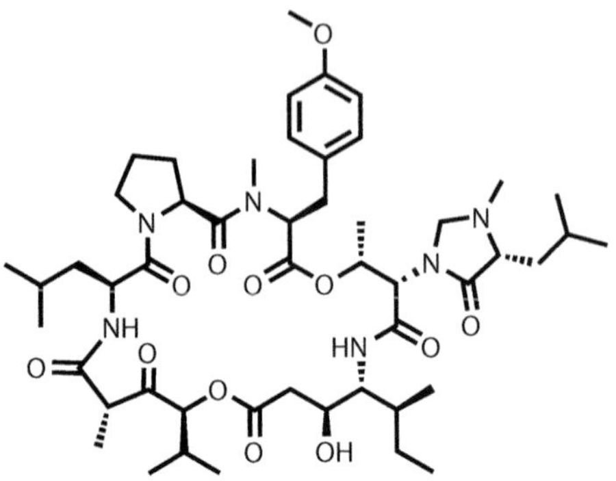

	R	Lung	Colon
		Cancer Cell Line IC$_{50}$ (nM)	
69		10	10
70		0.2	0.5
81	Me	0.2	0.5
82	Et	0.2	0.5
83	iBu	0.2	0.5

14.4.2.2 Bryostatin

Tunicates are not the only marine filter feeders that produce medicinally interesting natural products. In fact, Pettit recognized that *Bugula neritina* (Linnaeus) bryozoans contain anticancer properties concurrent with Rinehart's earliest tunicate efforts.[154] This activity led to the discovery and structural elucidation of the bryostatins, unprecedented polyketide macrolides that afforded significant life extension in a murine P388 lymphocytic leukemia model.[155,156] The first member of this family of natural products to be identified was bryostatin 1 (**37**), and to date the structures of 20 different natural bryostatins have been reported.[157] Bryostatin 1 is a potent modulator of protein kinase C (PKC),[158] activating several of the isoforms of the enzyme.[159] PKC plays a critical role in a number of cell signaling pathways, so anything that modifies its activity would be expected to have a significant biological effect.[160] Additionally, bryostatin 1 modulates P-glycoprotein in several cancer cell lines,[161] which is known to control multidrug resistance.

The activity of bryostatin 1 prompted an interest in its potential use as a clinical candidate, but material access was recognized as a legitimate challenge. A large-scale isolation of bryostatin 1 under good manufacturing practices was undertaken that resulted in the isolation of 18 g of pure bryostatin 1 from 14 tons of damp animal.[72] This was sufficient to allow for early clinical trials to commence, which initially explored the use of bryostatin 1 as a single agent for the induction of interleukin 6 and tumor necrosis factor alpha.[162] It does not have favorable Lipinski rule of five parameters, with a logP just above 5, a molecular weight of more than 900 g/mol, and 13 rotatable bonds, leading to little expectation of oral bioavailability.[163] Further hampering pharmacokinetic analysis is the fact that bryostatin 1 is given in such small doses that observing the material in humans is difficult using conventional methods of detection.[163] To date, the available pharmacokinetic data have been generated in animal models using radiolabeled material, which has hampered clinical trials.[164] Bryostatin 1 is delivered via an intravenous infusion and is widely distributed with an acceptable half-life of 23 h in mice.[165] It does not appear to be significantly broken down via metabolism.

The successful phase I safety data, where the most significant adverse event was myalgia, suggested that bryostatin 1 should be examined for clinical efficacy.[166] Eleven different phase II trials using bryostatin 1 as a single agent were pursued, including melanoma,[167] renal cell carcinoma,[168] colorectal carcinoma,[169] sarcoma,[170] and relapsed myeloma.[171] Unfortunately, none of these trials were sufficiently successful to warrant further clinical evaluation as a single agent. Subsequent clinical efforts have focused on the use of bryostatin 1 in conjunction with other agents in the hopes of taking advantage of its multidrug resistance activity. Toward that end, it has been evaluated with vincristine,[172] Taxol,[173] temsirolimus,[174] cisplatin,[175] fludrabine,[176] gemcitabine,[177] and cladribine[178] in a number of cancer indications, but to date none have resulted in the initiation of a phase III trial.

As the number of clinical evaluations of bryostatin 1 expanded, the need for additional material became a critical driver. It was clear from the outset that simple harvesting of the bryozoans would not be feasible, prompting efforts to develop a cultivation protocol for production of an adequate supply. Ultimately, an aquaculture was optimized that allowed for the generation of more than 100 g of bryostatin 1 per year.[179] Meanwhile, the unique scaffold attracted considerable synthetic interest, with a number of successful syntheses of different members of the bryostatin family of natural products.[157] Among the members to succumb to total synthesis were bryostatins 2 (**85**),[180] 3 (**86**),[181] 7 (**87**),[182] 9 (**88**),[183] and 16 (**89**)[184]; several excellent reviews of these synthetic efforts have been published.[157,185] As mentioned above, Keck has now completed a total synthesis of bryostatin 1.[73] Concurrent with these synthetic efforts, a number of bryostatin analogs have been prepared that allow for some insights into the required structural features for activity. These have been contemplated essentially from the start of the earliest synthetic efforts utilizing modeling strategies that suggested similarities with phorbol (**90**),[186] another known PKC modulator. While the bryostatins suppress tumor growth, phorbol esters stimulate growth.[187]

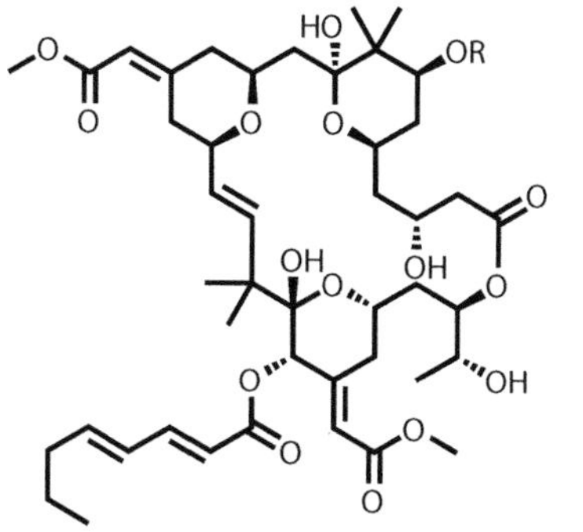

37 (R = Ac)
85 (R = H)

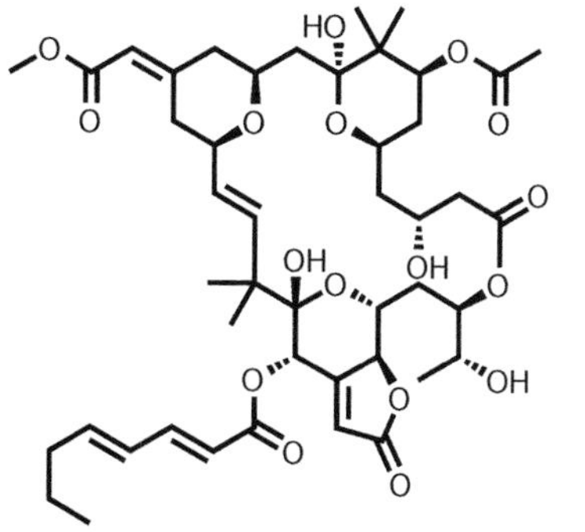

86

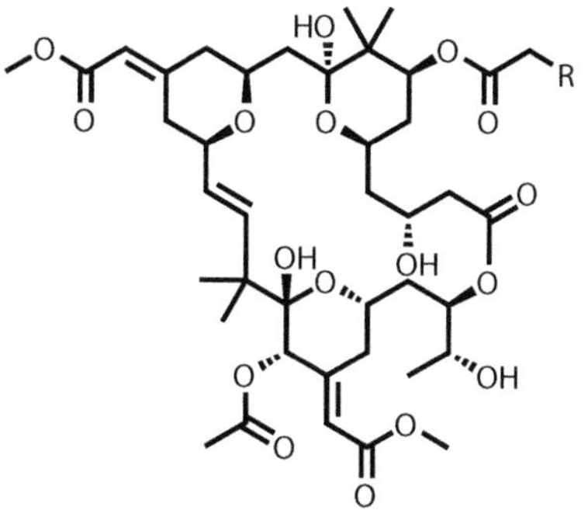

87 (R = H)
88 (R = Et)

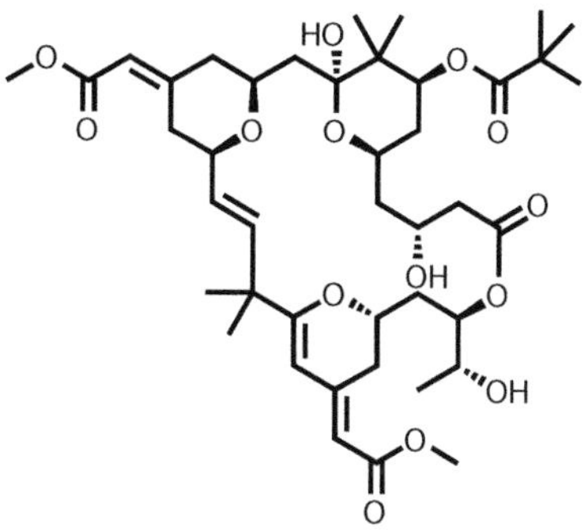

89

Some of the structural changes created in this analog work represent significant modifications to the parent scaffold. Trost, for example, has generated ring-expanded bryostatin analogs such as **91** that would only be accessible through synthetic efforts.[188] The amount of effort required to synthesize these analogs points to the difficulty associated with attempting to perform extensive structure–activity relationships in this manner. Two groups that have undertaken extensive studies of bryostatin analogs are those of Wender and Keck. The Keck analogs, dubbed Merle compounds after Merle Haggard,[189] have helped to illustrate some critical features to the A and B rings of bryostatin 1. All of the Merle analogs bind to PKC, with K_i's in the same 1 nM range as bryostatin 1.[190] However, some analogs, such as Merle 23 (**92**) and Merle 32 (**93**), behave like phorbol 12-myristate-13-acetate, while others, such as Merle 28 (**94**) and Merle 30 (**95**), behave like bryostatin 1.[191] To date, there is no single structural feature that appears to switch the type of behavior displayed by these analogs, suggesting a much more subtle interaction that regulates tumor suppression or tumor activation.

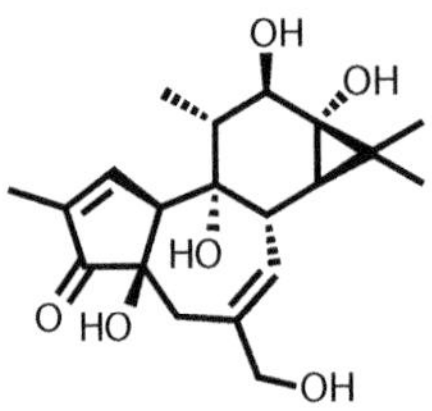

90

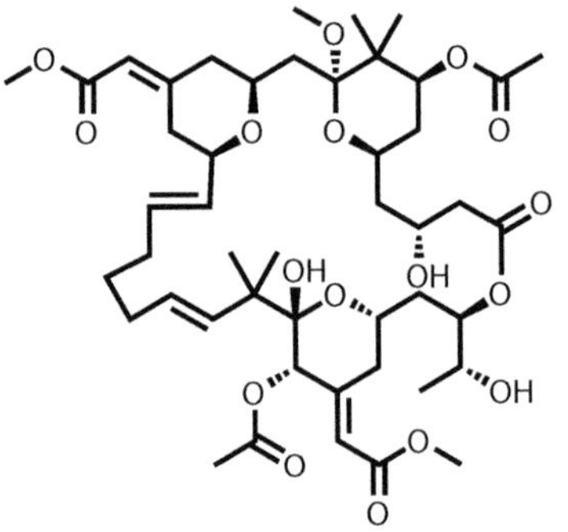

91

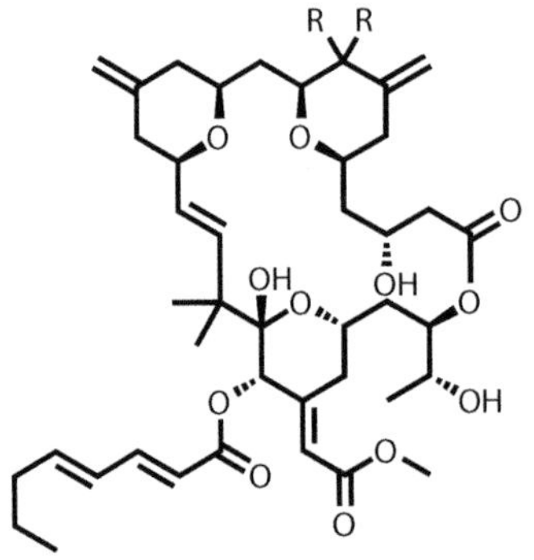

92 (R = H)
93 (R = Me)

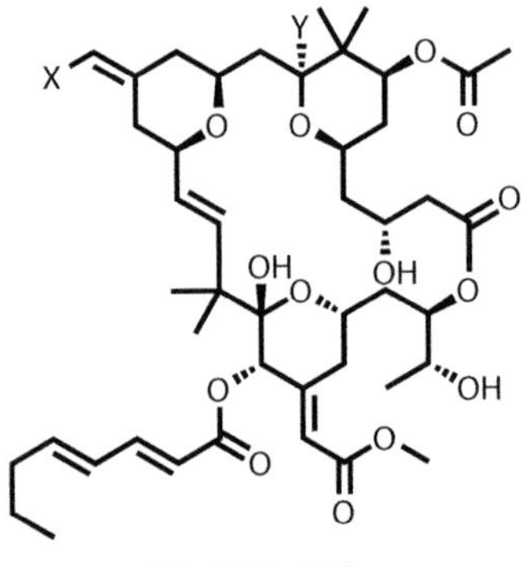

94 (X = H, Y = OH)
95 (X = CO$_2$Me, Y = H)

Wender's efforts aimed at the synthesis of bryostatin analogs, or "bryologs," has been the most extensive, with more than 100 different analogs reported to date.[189] These efforts have revealed several intriguing features about the bottom portion of the molecule, which has been suggested as the binding region of the bryostatins to PKC.[192] A critical observation was that simplified analogs at the A and B ring positions were found to be nearly equipotent with respect to PKC inhibition. For example, bryolog **96** has a PKC K_i of 3.4 nM, which is within a factor of three of bryostatin 1 (K_i = 1.35 nM).[193] This set the stage for the observation that truncation of the exocyclic side chain from a chiral hydroxyethyl substituent to the achiral hydroxymethyl group led to a marked increase in potency against PKC (K_i = 0.2 nM), and this compound, which has been given the name picolog (**97**), is more potent in a number of *in vitro* and *in vivo* assays.[194,195] The importance of this alcohol was further demonstrated via oxidation of the secondary alcohol to the corresponding ketone (**98**), which rendered the analog effectively inactive.[196] More recently, Wender has developed analogs that take advantage of the observation that bryostatin 1 is capable of inducing HIV expression from latent viral reservoirs.[197] Toward this end, he has identified analogs such as **99** that are nearly 10 times more active than bryostatin 1 and more than 1000 times more active than prostratin, which is currently under preclinical evaluation for this indication.[198]

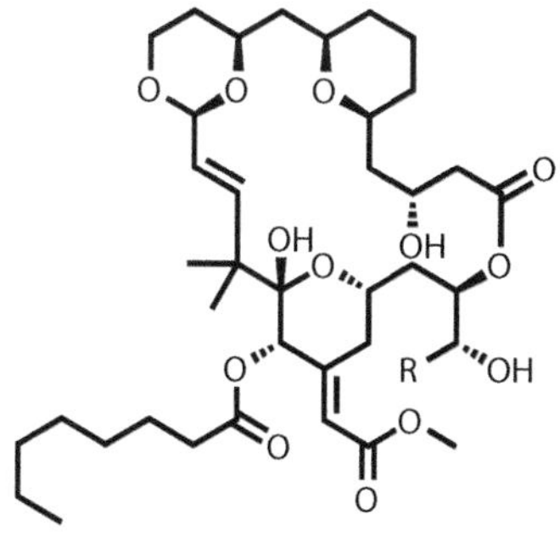

96 (R = Me)
97 (R = H)

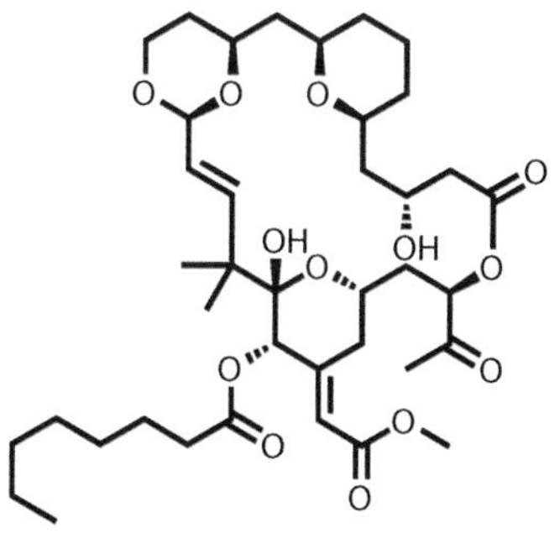

98

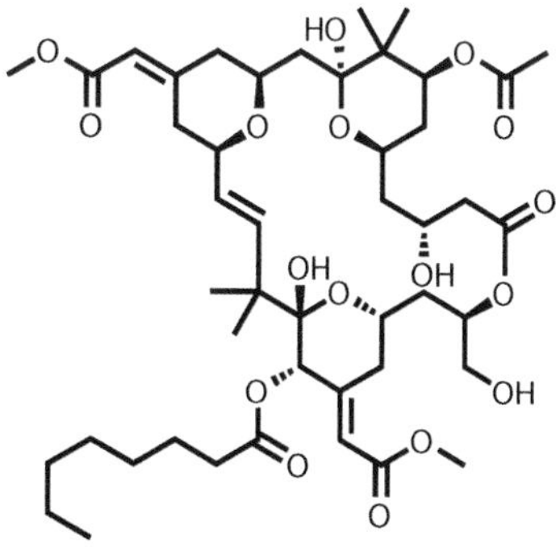

99

An alternative to direct synthesis may be available since the molecule is prepared through a polyketide biosynthetic route.[199] In fact, the biosynthesis of the bryostatins has been attributed to the bacterial symbiont *Candidatus* Endobugula sertula, an as yet uncultivated gamma-proteobacterium. Work is underway to elucidate the requisite polyketide modules that would open the door to performing combinatorial biosynthesis. Alternatively, cultivation of *Candidatus* Endobugula sertula would allow for the development of a fermentation process. This would facilitate expansion of clinical trials in burgeoning areas, such as for Alzheimer's disease.[200]

14.4.2.3 Dolastatins 10 and 15

In addition to initiating the incredible amount of research discussed above with his discovery of the bryostatins, Pettit was also responsible for the discovery of another prolific research area based on his work with sea hares.[201] Extracts from the gastropod *Dolabella auricularia* were found to possess potent inhibitory properties against P388 leukemia.[202] Among the active compounds identified in these efforts were a series of peptides called the dolastatins. Including the cumulative efforts of Pettit and Yamada, dozens of dolastatins have been characterized to date,[203] but the one that has attracted the greatest amount of attention is dolastatin 10 (**100**).[204] This linear pentapeptide is comprised of uncommon amino acids and appears to exert its considerable antitumor activity through an antimitotic mechanism by inhibiting the polymerization of tubulin.[205] There is an extensive collection of compounds that have been found to interfere with the dynamic instability of microtubules and thus are potentially useful in the search for chemotherapeutic agents,[206] but most of these are architecturally complex molecules that do not lend themselves to lead optimization. By contrast, the peptide nature of the dolastatins makes them ideally suited to the rapid generation of synthetic analogs. Moore has since found that dolastatin 10 can be isolated from marine *Symploca* cyanobacteria, suggesting that its presence in sea hares is likely of dietary origin.[207]

The ability to form amide bonds with good reliability and in high yields has provided access to modified peptides and helped launch the area of parallel organic synthesis.[208] Much of the methodological development that led to these improvements was undertaken at the same time that the structures of the individual dolastatins were being elucidated, which delayed some of the initial lead optimization efforts. In the case of dolastatin 10, the amino acid composition was determined to include valine, *N*,*N*-dimethylvaline, and three unique amino acids: dolaisoleucine (Dil), dolaproine (Dap), and dolaphenine (Doe).[205] Each of these novel components contains stereocenters, and it was determined early on that the stereocenter on the Doe unit was not critical for activity.[209] Furthermore, the thiazole itself was not essential, as the phenethylamine derivative proved to be nearly equipotent with dolastatin 10 and also received some clinical attention (*vide infra*).

Not long after the initial disclosure of dolastatin 10, another very active member of this family of natural products was reported that was given the name dolastatin 15 (**101**).[210] Like dolastatin 10, this compound contained a linear arrangement of residues, but dolastatin 15 is a depsipeptide that does not contain the Dil, Dap, or Doe components. Instead, it has two proline subunits along with 2-hydroxyisovaleric acid (Hiva) and dolapyrrolidone (Dpy) groups on the C-terminus. Dolastatin 15 has a significantly decreased affinity for tubulin compared to dolastatin 10, even though they both are potent antiproliferative agents.[211] Furthermore, dolastatin 15 is active against prokaryotic cells that do not contain tubulin, suggesting that there may be an alternative and as yet undiscovered target for these compounds.[203]

The discovery of the activity of two structurally unique dolastatins opened the door to extensive analog synthesis that examined hybridized versions of each individual member, as well as completely novel analogs. A full analysis of the SAR of the dolastatins would be beyond the scope of this discussion, and there are several reviews that have been published on the subject.[203,212] Among the features that have been uncovered are that the unique Dil residue of dolastatin 10 appears to play a critical role in binding to tubulin (and may explain the decreased tubulin affinity of dolastatin 15, which does not have this group).[213] Also, as with dolastatin 10, the C-terminus appears less critical, as simplified analogs that don't contain either the Hiva or Dpy groups could be used with very little impact on the overall activity.[214] In total, these efforts have resulted in the clinical exploration of several different candidates.

Dolastatin 10 itself was the first compound to enter the clinic.[215] As a peptide, it was administered as an intravenous bolus, as expected. It is somewhat lipophilic with a clogP above 5, but it was well distributed with a plasma half-life of more than 5 h and did not appear to present any metabolic problems, although high doses resulted in neuropathy for some patients. Similarly, the phenethylamine dolastatin 10 analog that is missing the thiazole of the Doe subunit (**102**) was also studied in the clinic by Daiichi-Sankyo under the names of soblidotin and TXT-1027.[216] Soblidotin has improved physicochemical parameters, including improved kinetic solubility and decreased clogP. It was also administered intravenously and demonstrated similar pharmacokinetic characteristics, and although the neutropenia was less frequent in the case of soblidotin, it was still present.

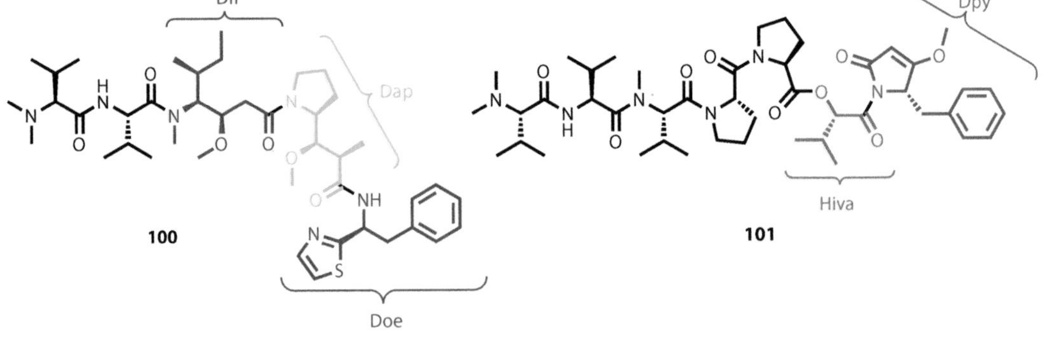

100   **101**

While dolastatin 15 itself has not been pursued in a clinical setting, analogs have been used in human testing. The first of these was cemadotin (LU103793, **103**), which is comprised of three valines and two prolines.[217] This peptide, which has quite reasonable physicochemical characteristics as well, was dosed as an intravenous infusion that varied in length between 1 and 5 days. Although neutropenia was observed in some patients, hypertension was the dose-limiting toxicity for cemadotin. Synthadotin (tasidotin hydrochloride, ILX651, **104**) differs from cemadotin through the use of a *t*-butylamide instead of a benzylamide, and it has also been explored in the clinic.[218] It has the most favorable physicochemical properties of any of the dolastatin analogs that have gone into the clinic, and like the others, it was administered intravenously. For both cemadotin and synthadotin, the same primary metabolite was found through hydrolysis of the C-terminal amide (**105**).[219] Despite a large number of clinical evaluations, none of these dolastatin clinical candidates were sufficiently efficacious to warrant further evaluation as a single agent, and Daiichi has withdrawn a proposed study of soblidotin as part of a combination therapy.

Although the use of dolastatin analogs as a small-molecule drug has not proven to be effective to date, research into the discovery of new analogs that might possess improved pharmacokinetic properties is still in progress.[220] However, Seattle Genetics has explored the use of a dolastatin 10 analog as part of an antibody–drug conjugate with brentuximab (**106**), an antibody that specifically targets CD30.[221] CD30 is a membrane protein associated with tumor necrosis factor and is considered a marker for Hodgkin's lymphoma.[222] Using an analog of soblidotin where the C-terminal phenethylamine has been exchanged for norephedrine (an analog previously known as auristatin E, **107**),[223] the brentuximab antibody was covalently attached in place of one of the N-terminal methyl groups.[224] In this manner, the antimitotic dolastatin analog given the name of monomethyl auristatin E or vedotin is delivered to the cell in question, decreasing the toxicity that can be associated with inhibiting the mitosis of noncancerous cells.[225] Brentuximab vedotin, under the trade name Adcetris, was approved by the FDA for use in the United States as a biologic for the treatment of Hodgkin's lymphoma and anaplastic large-cell lymphoma.[226]

14.4.2.4 Other Marine Natural Products

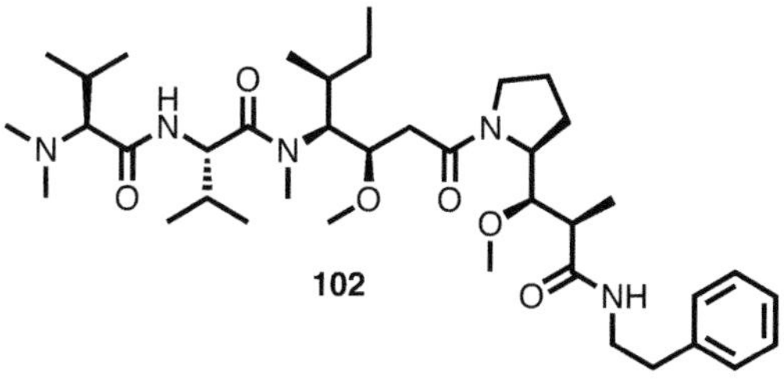

102

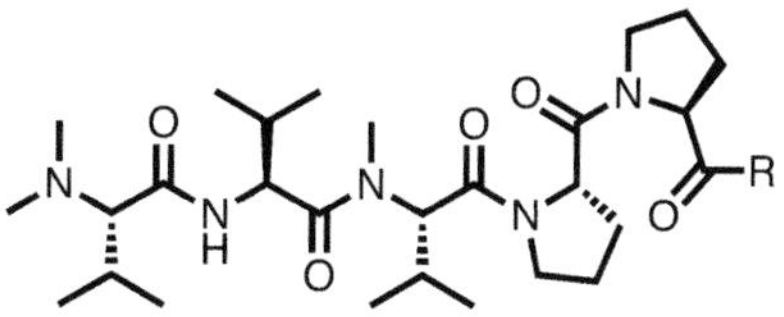

103 (R = NHCH$_2$Ph)
104 (R = NHt-Bu)
105 (R = OH)

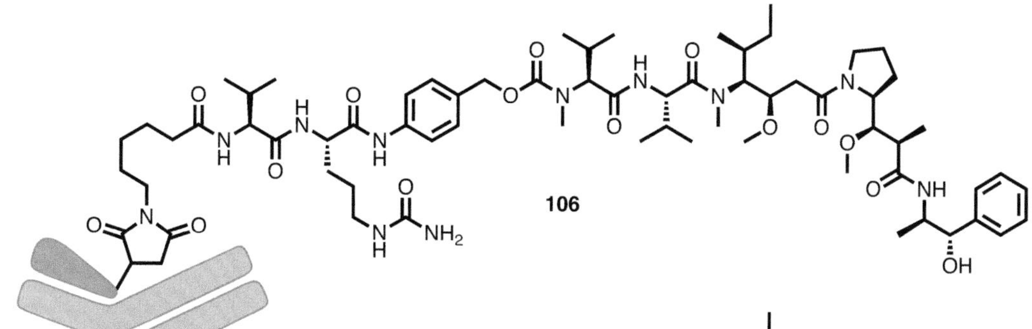

106

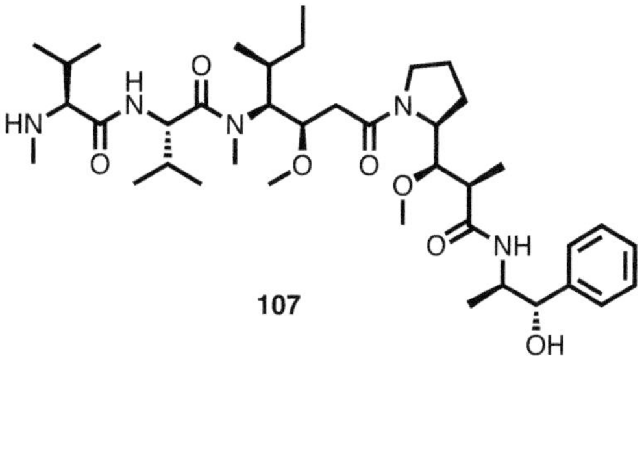

Brentuximab

107

There are a number of other examples of exciting possibilities for lead optimization based on natural products. Discodermolide (**108**), a microtubule-stabilizing polyketide from the Caribbean sponge *Discodermia dissoluta* that is not a substrate for the p-glycoprotein drug efflux pump,[227–229] was able to be pursued in the clinic after the development of a synthetic route that could be used to provide the natural product in gram quantities.[230] This route, along with research into the biosynthesis through polyketide synthase, could lead to new and viable analogs with improved pharmacokinetic properties. Cryptophycin 52 (**109**) is a synthetic analog of the arenastatins and cryptophycins, cyclic depsipeptides that have been isolated from sources ranging from the Okinawan marine sponge *Dysidea arenaria* to *Nostoc* cyanobacteria.[231–234] This class of compounds has attracted substantial interest in the clinic, as the peptide nature of the molecules allowed for considerable lead optimization that resulted in **109** becoming the first clinical candidate in this series.[235] It was selected for further investigation in large part because the dimethyl-β-alanine modification provided improved hydrolytic stability over the monomethylated cryptophycin 1 natural product.[236] Cryptophycin 52 (LY355703) was examined in a phase II metastatic colorectal cancer trial,[237] and investigation into new cryptophycin analogs continues today.[238] The spongistatins (**110** and **111**), spiroketal-containing macrolides found in several different genera of sponges,[239–241] are extremely potent antimitotic agents that could prove clinically useful if sufficient material could be made available, and they represent a challenging opportunity as a starting point for lead optimization. Salinosporamide A (NPI-0052, **112**), a cytotoxic proteasome inhibitor that was initially isolated from a marine *Salinospora* bacterium,[242] is currently in active clinical trials for a number of indications, including melanoma, pancreatic cancer, lung cancer, and multiple myeloma.[243,244] It induces apoptosis through a mechanism unique from bortezomib, suggesting the prospect of utility as a combination treatment.[245] PharmaMar has recently published two novel tubulin binding natural products from the sponge *Lithoplocamia lithistoides*.[246] One of these compounds, PM060184 (**113**), is already in a phase I clinical trial for patients with advanced solid tumors. Since the requisite material for this trial was generated by total synthesis, lead optimization studies focusing on ADME and pharmacokinetic properties of this series are viable. Undoubtedly, the development of novel drugs based on lead structures from marine sources will continue to serve as a worthwhile venture.

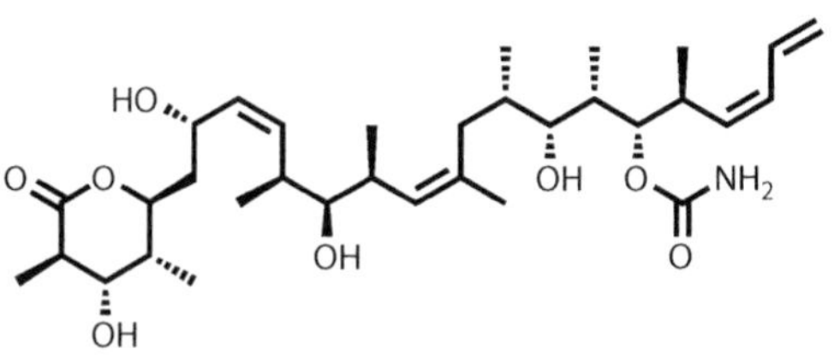

108

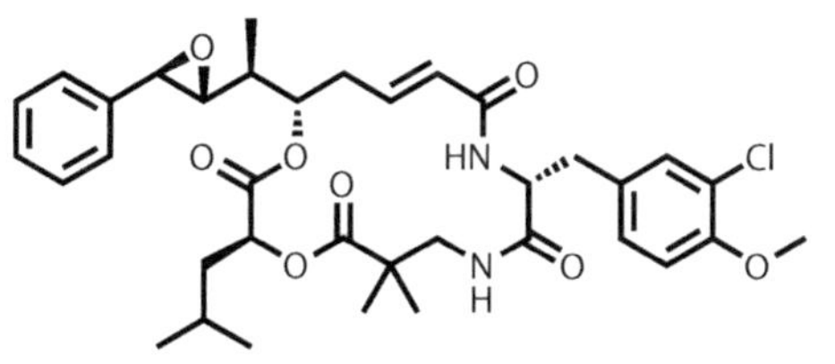

109

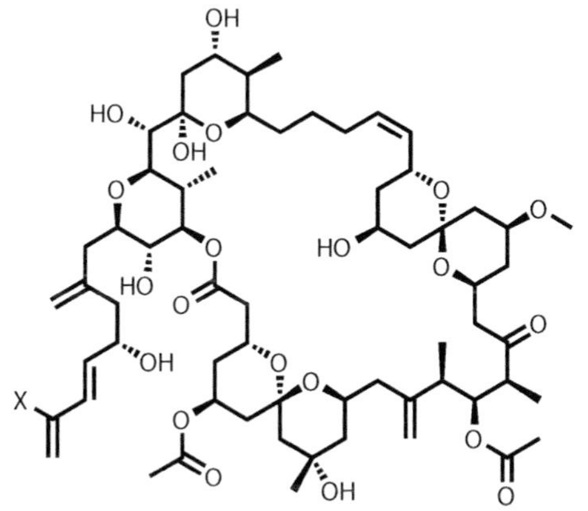

112

110 (X = Cl)
111 (X = H)

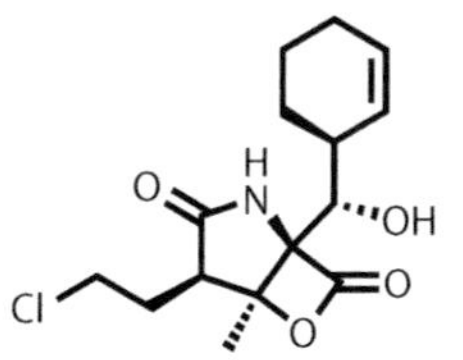

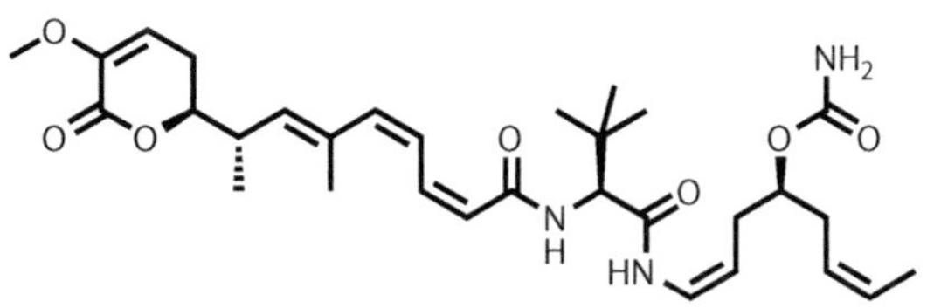

113

REFERENCES

1. Newman, D. J., Cragg, G. M. Natural products as sources of new drugs over the 30 years from 1981 to 2010. *J. Nat. Prod.* 2012, 75 (3), 311–335.
2. Lockemann, G. Friedrich Wilhelm Serturner, the discoverer of morphine. *J. Chem. Educ.* 1951, 28, 277–279.
3. Kardos, N., Demain, A. L. Penicillin: The medicine with the greatest impact on therapeutic outcomes. *Appl. Microbiol. Biotechnol.* 2011, 92 (4), 677–687.
4. McCowen, M. C., Callender, M. E., Lawlis, J. F., Jr., Brandt, M. C. The effects of erythromycin (ilotycin, lilly) against certain parasitic organisms. *Am. J. Trop. Med. Hyg.* 1953, 2 (2), 212–218.
5. Whitelaw, D. M., Cowan, D. H., Cassidy, F. R., Patterson, T. A. Clinical experience with vincristine. *Cancer Chemother. Rep.* 1963, 30, 13–20.
6. Schiff, P. B., Fant, J., Horwitz, S. B. Promotion of microtubule assembly *in vitro* by Taxol. *Nature* (London) 1979, 277 (5698), 665–667.
7. Antimalaria studies on Qinghaosu. *Chin. Med. J.* (Peking, Engl. Ed.) 1979, 92 (12), 811–816.
8. Tamaoki, T., Nomoto, H., Takahashi, I., Kato, Y., Morimoto, M., Tomita, F. Staurosporine, a potent inhibitor of phospholipid/Ca++dependent protein kinase. *Biochem. Biophys. Res. Commun.* 1986, 135 (2), 397–402.
9. Deininger, M. W. N., Goldman, J. M., Lydon, N., Melo, J. V. The tyrosine kinase inhibitor CGP57148B selectively inhibits the growth of BCR-ABL-positive cells. *Blood* 1997, 90 (9), 3691–3698.
10. Baselga, J., Averbuch, S. D. ZD1839 ('Iressa') as an anticancer agent. *Drugs* 2000, 60 (Suppl. 1), 33–40; discussion, 41–42.

11. Appeltans, W., Ahyong, S. T., Anderson, G., Angel, M. V., Artois, T., Bailly, N., Bamber, R., Barber, A., Bartsch, I., Berta, A., Blazewicz-Paszkowycz, M., Bock, P., Boxshall, G., Boyko, C. B., Brandao, S. N., Bray, R. A., Bruce, N. L., Cairns, S. D., Chan, T.-Y., Cheng, L., Collins, A. G., Cribb, T., Curini-Galletti, M., Dahdouh-Guebas, F., Davie, P. J. F., Dawson, M. N., De'Clerck, O., Decock, W., De'Grave, S., de'Voogd, N. J., Domning, D. P., Emig, C. C., Erseus, C., Eschmeyer, W., Fauchald, K., Fautin, D. G., Feist, S. W., Fransen, C. H. J. M., Furuya, H., Garcia-Alvarez, O., Gerken, S., Gibson, D., Gittenberger, A., Gofas, S., Gomez-Daglio, L., Gordon, D. P., Guiry, M. D., Hernandez, F., Hoeksema, B. W., Hopcroft, R. R., Jaume, D., Kirk, P., Koedam, N., Koenemann, S., Kolb, J. B., Kristensen, R. M., Kroh, A., Lambert, G., Lazarus, D. B., Lemaitre, R., Longshaw, M., Lowry, J., MacPherson, E., Madin, L. P., Mah, C., Mapstone, G., McLaughlin, P. A., Mees, J., Meland, K., Messing, C. G., Mills, C. E., Molodtsova, T. N., Mooi, R., Neuhaus, B., Ng, P. K. L., Nielsen, C., Norenburg, J., Opresko, D. M., Osawa, M., Paulay, G., Perrin, W., Pilger, J. F., Poore, G. C. B., Pugh, P., Read, G. B., Reimer, J. D., Rius, M., Rocha, R. M., Saiz-Salinas, J. I., Scarabino, V., Schierwater, B., Schmidt-Rhaesa, A., Schnabel, K. E., Schotte, M., Schuchert, P., Schwabe, E., Segers, H., Self-Sullivan, C., Shenkar, N., Siegel, V., Sterrer, W., Stohr, S., Swalla, B., Tasker, M. L., Thuesen, E. V., Timm, T., Todaro, M. A., Turon, X., Tyler, S., Uetz, P., van der Land, J., Vanhoorne, B., van Ofwegen, L. P., van Soest, R. W. M., Vanaverbeke, J., Walker-Smith, G., Walter, T. C., Warren, A., Williams, G. C., Wilson, S. P., Costello, M. J. The magnitude of global marine species diversity. *Curr. Biol.* 2012, 22 (23), 2189–2202.
12. Macarron, R., Banks, M. N., Bojanic, D., Burns, D. J., Cirovic, D. A., Garyantes, T., Green, D. V. S., Hertzberg, R. P., Janzen, W. P., Paslay, J. W., Schopfer, U., Sittampalam, G. S. Impact of high-throughput screening in biomedical research. *Nat. Rev. Drug Discov.* 2011, 10 (3), 188–195.
13. Wang, J. Comprehensive assessment of ADMET risks in drug discovery. *Curr. Pharm. Des.* 2009, 15 (19), 2195–2219.
14. Ekins, S., Ring, B. J., Grace, J., McRobie-Belle, D. J., Wrighton, S. A. Present and future *in vitro* approaches for drug metabolism. *J. Pharmacol. Toxicol. Methods* 2001, 44 (1), 313–324.
15. Foye, W. O., Lemke, T. L., Williams, D. A. *Principles of Medicinal Chemistry.* 4th ed. Williams & Wilkins, Baltimore, MD, 1995.
16. Dolle, R. E., Worm, K. *Role of Chemistry in Lead Discovery.* John Wiley & Sons, Hoboken, NJ, 2010, pp. 259–290.
17. Mahdi, J. G., Mahdi, A. J., Mahdi, A. J., Bowen, I. D. The historical analysis of aspirin discovery, its relation to the willow tree and antiproliferative and anticancer potential. *Cell Prolif.* 2006, 39 (2), 147–155.
18. Sneader, W. The discovery of heroin. *Lancet* 1998, 352 (9141), 1697–1699.
19. Endo, A., Kuroda, M., Tsujita, Y. ML-236A, ML-236B, and ML-236C, new inhibitors of cholesterogenesis produced by *Penicillium citrinum. J. Antibiot.* 1976, 29 (12), 1346–1348.
20. Endo, A., Kuroda, M., Tanzawa, K. Competitive inhibition of 3-hydroxy-3-methylglutaryl coenzyme A reductase by ML-236A and ML-236B fungal metabolites, having hypocholesterolemic activity. *FEBS Lett.* 1976, 72 (2), 323–326.
21. Nakaya, N., Homma, Y., Tamachi, H., Goto, Y. The effect of CS-514, an inhibitor of HMG-CoA reductase, on serum lipids in healthy volunteers. *Atherosclerosis* (Shannon, Irel.) 1986, 61 (2), 125–128.
22. Mol, M. J., Erkelens, D. W., Leuven, J. A., Schouten, J. A., Stalenhoef, A. F. Effects of synvinolin (MK-733) on plasma lipids in familial hypercholesterolaemia. *Lancet* 1986, 2 (8513), 936–939.
23. Baumann, K. L., Butler, D. E., Deering, C. F., Mennen, K. E., Millar, A., Nanninga, T. N., Palmer, C. W., Roth, B. D. The convergent synthesis of CI-981, an optically active, highly potent, tissue-selective inhibitor of HMG-CoA reductase. *Tetrahedron Lett.* 1992, 33 (17), 2283–2284.
24. Johnson, L. A. "Against Odds, Lipitor Became World's Top Seller," USA Today, 28 Dec., 2011. Web. 30 June 2015. <http://usatoday30.usatoday.com/news/health/medical/health/medical/treatments/story/2011-12-28/Against-odds-Lipitor-became-worlds-top-seller/52250720/1>
25. Fan, J., de Lannoy, I. A. M. Pharmacokinetics. *Biochem. Pharmacol.* (Amsterdam, Neth.) 2014, 87 (1), 93–120.
26. Hawking, F. Pharmacology of sulfonamides: Recent work. *Br. Med. J.* 1945, 1945, I, 505–509.
27. Brunton, L., Chabner, B., Knollman, B. *Goodman & Gilman's The Pharmacological Basis of Therapeutics.* 12th ed. McGraw-Hill, New York, 2010, p. 1808.
28. Stahl, P. H., Wermuth, C. G., eds. *Handbook of Pharmaceutical Salts: Properties, Selection, and Use.* 2nd, rev. ed. Verlag Helvetica Chimica Acta, Weinheim, Germany, 2011, p. 446.

29. Alsenz, J. The impact of solubility and dissolution assessment on formulation strategy and implications for oral drug disposition. In *Encyclopedia of Drug Metabolism and Interactions*. John Wiley & Sons, New York, 2012, p. 493–562.

30. Lipinski, C. A., Lombardo, F., Dominy, B. W., Feeney, P. J. Experimental and computational approaches to estimate solubility and permeability in drug discovery and development settings. *Adv. Drug Delivery Rev.* 2001, 46 (1–3), 3–26.

31. Alsenz, J., Meister, E., Haenel, E. Development of a partially automated solubility screening (PASS) assay for early drug development. *J. Pharm. Sci.* 2007, 96 (7), 1748–1762.

32. Kubinyi, H. Drug partitioning: Relationships between forward and reverse rate constants and partition coefficient. *J. Pharm. Sci.* 1978, 67 (2), 262–263.

33. Lipinski, C. A., Lombardo, F., Dominy, B. W., Feeney, P. J. Experimental and computational approaches to estimate solubility and permeability in drug discovery and development settings. *Adv. Drug Deliv. Rev.* 1997, 23 (1–3), 3–25.

34. Gaillard, P., Carrupt, P. A., Testa, B., Boudon, A. Molecular lipophilicity potential, a tool in 3D QSAR: Method and applications. *J. Comput. Aided Mol. Des.* 1994, 8 (2), 83–96.

35. Lu, J. J., Crimin, K., Goodwin, J. T., Crivori, P., Orrenius, C., Xing, L., Tandler, P. J., Vidmar, T. J., Amore, B. M., Wilson, A. G. E., Stouten, P. F. W., Burton, P. S. Influence of molecular flexibility and polar surface area metrics on oral bioavailability in the rat. *J. Med. Chem.* 2004, 47 (24), 6104–6107.

36. Lipinski, C. A. Chris Lipinski discusses life and chemistry after the rule of five. *Drug Discov. Today* 2003, 8 (1), 12–16.

37. Alvarez, F. M., Bottom, C. B., Chikhale, P., Pidgeon, C. In *Immobilized Artificial Membrane Chromatography: Prediction of Drug Transport across Biological Barriers*. Plenum, New York, 1993, pp. 151–67.

38. van Breemen, R. B., Li, Y. Caco-2 cell permeability assays to measure drug absorption. *Expert Opin. Drug Metab. Toxicol.* 2005, 1 (2), 175–185.

39. Kansy, M., Senner, F., Gubernator, K. Physicochemical high throughput screening: Parallel artificial membrane permeation assay in the description of passive absorption processes. *J. Med. Chem.* 1998, 41 (7), 1007–1010.

40. Faller, B. Artificial membrane assays to assess permeability. *Curr. Drug Metab.* 2008, 9 (9), 886–892.

41. Friedemann, U., Elkeles, A. The blood-brain barrier in infectious diseases: Its permeability to toxins in relation to their electrical charges. *Lancet* 1934, I, 719-24, 775-7.

42. Luo, F. R., Paranjpe, P. V., Guo, A., Rubin, E., Sinko, P. Intestinal transport of irinotecan in Caco-2 cells and MDCK II cells overexpressing efflux transporters Pgp, cMOAT, and MRP1. *Drug Metab. Dispos.* 2002, 30 (7), 763–770.

43. Stan, R. V., Tse, D., Deharvengt, S. J., Smits, N. C., Xu, Y., Luciano, M. R., McGarry, C. L., Buitendijk, M., Nemani, K. V., Elgueta, R., Kobayashi, T., Shipman, S. L., Moodie, K. L., Daghlian, C. P., Ernst, P. A., Lee, H.-K., Suriawinata, A. A., Schned, A. R., Longnecker, D. S., Fiering, S. N., Noelle, R. J., Gimi, B., Shworak, N. W., Carriere, C. The diaphragms of fenestrated endothelia: Gatekeepers of vascular permeability and blood composition. *Dev. Cell* 2012, 23 (6), 1203–1218.

44. Kerns, E. H., Di, L. Pharmaceutical profiling in drug discovery. *Drug Discov. Today* 2003, 8 (7), 316–323.

45. van der Vusse, G. J. Albumin as fatty acid transporter. *Drug Metab. Pharmacokinet.* 2009, 24 (4), 300–307.

46. Conant, C. G., Schwartz, M. A., Ionescu-Zanetti, C. Well plate-coupled microfluidic devices designed for facile image-based cell adhesion and transmigration assays. *J. Biomol. Screening* 2010, 15 (1), 102–106.

47. Vuignier, K., Schappler, J., Veuthey, J.-L., Carrupt, P.-A., Martel, S. Drug-protein binding: A critical review of analytical tools. *Anal. Bioanal. Chem.* 2010, 398 (1), 53–66.

48. Vuignier, K., Veuthey, J.-L., Carrupt, P.-A., Schappler, J. Global analytical strategy to measure drug-plasma protein interactions: From high-throughput to in-depth analysis. *Drug Discov. Today* 2013, 18 (21–22), 1030–1034.

49. Di, L., Kerns, E. H., Hong, Y., Chen, H. Development and application of high throughput plasma stability assay for drug discovery. *Int. J. Pharm.* 2005, 297 (1–2), 110–119.

50. Kirchmair, J., Williamson, M. J., Tyzack, J. D., Tan, L., Bond, P. J., Bender, A., Glen, R. C. Computational prediction of metabolism: Sites, products, SAR, P450 enzyme dynamics, and mechanisms. *J. Chem. Inf. Model* 2012, 52 (3), 617–648.

51. Bertz, R. J., Granneman, G. R. Use of *in vitro* and *in vivo* data to estimate the likelihood of metabolic pharmacokinetic interactions. *Clin. Pharmacokinet.* 1997, 32 (3), 210–258.

52. Jin, Y., Desta, Z., Stearns, V., Ward, B., Ho, H., Lee, K.-H., Skaar, T., Storniolo, A. M., Li, L., Araba, A., Blanchard, R., Nguyen, A., Ullmer, L., Hayden, J., Lemler, S., Weinshilboum, R. M., Rae, J. M., Hayes, D. F., Flockhart, D. A. CYP2D6 genotype, antidepressant use, and tamoxifen metabolism during adjuvant breast cancer treatment. *J. Natl. Cancer Inst.* 2005, 97 (1), 30–39.

53. Di, L., Kerns, E. H., Hong, Y., Kleintop, T. A., McConnell, O. J., Huryn, D. M. Optimization of a higher throughput microsomal stability screening assay for profiling drug discovery candidates. *J. Biomol. Screening* 2003, 8 (4), 453–462.

54. Bjornsson, T. D., Callaghan, J. T., Einolf, H. J., Fischer, V., Gan, L., Grimm, S., Kao, J., King, S. P., Miwa, G., Ni, L., Kumar, G., McLeod, J., Obach, R. S., Roberts, S., Roe, A., Shah, A., Snikeris, F., Sullivan, J. T., Tweedie, D., Vega, J. M., Walsh, J., Wrighton, S. A. The conduct of *in vitro* and *in vivo* drug-drug interaction studies: A Pharmaceutical Research and Manufacturers of America (PhRMA) perspective. *Drug Metab. Dispos.* 2003, 31 (7), 815–832.

55. Marks, K. M., Park, E. S., Arefolov, A., Russo, K., Ishihara, K., Ring, J. E., Clardy, J., Clarke, A. S., Pelish, H. E. The selectivity of austocystin D arises from cell-line-specific drug activation by cytochrome P450 enzymes. *J. Nat. Prod.* 2011, 74 (4), 567–573.

56. Gage, B. F., Eby, C., Milligan, P. E., Banet, G. A., Duncan, J. R., McLeod, H. L. Use of pharmacogenetics and clinical factors to predict the maintenance dose of warfarin. *Thromb. Haemostasis* 2004, 91 (1), 87–94.

57. Walsky, R. L., Obach, R. S. Validated assays for human cytochrome P450 activities. *Drug Metab. Dispos.* 2004, 32 (6), 647–660.

58. Di, L., Kerns, E. H., Li, S. Q., Carter, G. T. Comparison of cytochrome P450 inhibition assays for drug discovery using human liver microsomes with LC-MS, rhCYP450 isozymes with fluorescence, and double cocktail with LC-MS. *Int. J. Pharm.* 2007, 335 (1–2), 1–11.

59. Kalgutkar, A. S., Obach, R. S., Maurer, T. S. Mechanism-based inactivation of cytochrome P450 enzymes: Chemical mechanisms, structure-activity relationships and relationship to clinical drug-drug interactions and idiosyncratic adverse drug reactions. *Curr. Drug Metab.* 2007, 8 (5), 407–447.

60. Colombo, F., Armstrong, C., Duan, J., Rioux, N. A high throughput *in vitro* mrp2 assay to predict *in vivo* biliary excretion. *Xenobiotica* 2012, 42 (2), 157–163.

61. Szymanski, P., Markowicz, M., Mikiciuk-Olasik, E. Adaptation of high-throughput screening in drug discovery: Toxicological screening tests. *Int. J. Mol. Sci.* 2012, 13, 427–452.

62. Vandenberg, J. I., Walker, B. D., Campbell, T. J. HERG K+ channels: Friend and foe. *Trends Pharmacol. Sci.* 2001, 22 (5), 240–246.

63. Cavalli, A., Poluzzi, E., De Ponti, F., Recanatini, M. Toward a pharmacophore for drugs inducing the long QT syndrome: Insights from a CoMFA study of HERG K+ channel blockers. *J. Med. Chem.* 2002, 45 (18), 3844–3853.

64. Roy, M.-L., Dumaine, R., Brown, A. M. HERG, a primary human ventricular target of the nonsedating antihistamine terfenadine. *Circulation* 1996, 94 (4), 817–823.

65. Zhou, Z., Vorperian, V. R., Gong, Q., Zhang, S., January, C. T. Block of HERG potassium channels by the antihistamine astemizole and its metabolites desmethylastemizole and norastemizole. *J. Cardiovasc. Electrophysiol.* 1999, 10 (6), 836–843.

66. Dubin, A. E., Nasser, N., Rohrbacher, J., Hermans, A. N., Marrannes, R., Grantham, C., Van Rossem, K., Cik, M., Chaplan, S. R., Gallacher, D., Xu, J., Guia, A., Byrne, N. G., Mathes, C. Identifying modulators of hERG channel activity using the PatchXpress planar patch clamp. *J. Biomol. Screening* 2005, 10 (2), 168–181.

67. Mortelmans, K., Zeiger, E. The Ames *Salmonella*/microsome mutagenicity assay. *Mutat. Res.* 2000, 455 (1–2), 29–60.

68. Tropsha, A. Best practices for QSAR model development, validation, and exploitation. *Mol. Inf.* 2010, 29 (6–7), 476–488.

69. Meanwell, N. A. Synopsis of some recent tactical application of bioisosteres in drug design. *J. Med. Chem.* 2011, 54 (8), 2529–2591.

70. Narayana, N., Diller, T. C., Koide, K., Bunnage, M. E., Nicolaou, K. C., Brunton, L. L., Xuong, N.-H., Ten Eyck, L. F., Taylor, S. S. Crystal structure of the potent natural product inhibitor balanol in complex with the catalytic subunit of cAMP-dependent protein kinase. *Biochemistry* 1999, 38 (8), 2367–2376.

71. Hagmann, W. K. The many roles for fluorine in medicinal chemistry. *J. Med. Chem.* 2008, 51 (15), 4359–4369.

72. Schaufelberger, D. E., Koleck, M. P., Beutler, J. A., Vatakis, A. M., Alvarado, A. B., Andrews, P., Marzo, L. V., Muschik, G. M., Roach, J., Ross, J. T., Lebherz, W. B., Reeves, M. P., Eberwein, R. M., Rodgers, L. L., Testerman, R. P., Snader, K. M., Forenza, S. The large-scale isolation of bryostatin 1 from *Bugula neritina* following current good manufacturing practices. *J. Nat. Prod.* 1991, 54 (5), 1265–1270.

73. Keck, G. E., Poudel, Y. B., Cummins, T. J., Rudra, A., Covel, J. A. Total synthesis of bryostatin 1. *J. Am. Chem. Soc.* 2011, 133 (4), 744–747.

74. Guenard, D., Gueritte-Voegelein, F., Potier, P. Taxol and taxotere: Discovery, chemistry, and structure-activity relationships. *Acc. Chem. Res.* 1993, 26 (4), 160–167.

75. Wani, M. C., Taylor, H. L., Wall, M. E., Coggon, P., McPhail, A. T. Plant antitumor agents. VI. Isolation and structure of Taxol, a novel antileukemic and antitumor agent from *Taxus brevifolia*. *J. Am. Chem. Soc.* 1971, 93 (9), 2325–2327.

76. Goodman, J., Walsh, V. *The Story of Taxol: Nature and Politics in the Pursuit of an Anti-Cancer Drug.* Cambridge University Press, Cambridge, 2001, p. 304.

77. Kingston, D. G. I. The shape of things to come: Structural and synthetic studies of Taxol and related compounds. *Phytochemistry* 2007, 68 (14), 1844–1854.

78. Gaullier, J. C., Mandard, B., Margraff, R. Method of obtaining 10-desacetylbaccatin III. WO9407882A1, 1994.

79. Holton, R. A. Method for preparation of Taxol. EP400971A2, 1990.

80. Fett-Neto, A. G., Melanson, S. J., Sakata, K., DiCosmo, F. Improved growth and Taxol yield in developing calli of *Taxus cuspidata* by medium composition modification. *Biotechnology* (N.Y.) 1993, 11 (6), 731–734.

81. Ajikumar, P. K., Xiao, W.-H., Tyo, K. E. J., Wang, Y., Simeon, F., Leonard, E., Mucha, O., Phon, T. H., Pfeifer, B., Stephanopoulos, G. Isoprenoid pathway optimization for Taxol precursor overproduction in *Escherichia coli*. *Science* (Washington, D.C.) 2010, 330 (6000), 70–74.

82. Onrubia, M., Cusido, R. M., Ramirez, K., Hernandez-Vazquez, L., Moyano, E., Bonfill, M., Palazon, J. Bioprocessing of plant *in vitro* systems for the mass production of pharmaceutically important metabolites: Paclitaxel and its derivatives. *Curr. Med. Chem.* 2013, 20 (7), 880–891.

83. Cane, D. E., Walsh, C. T., Khosla, C. Harnessing the biosynthetic code: Combinations, permutations, and mutations. *Science* (Washington, D.C.) 1998, 282 (5386), 63–68.

84. Fischbach, M. A., Walsh, C. T. Assembly-line enzymology for polyketide and nonribosomal peptide antibiotics: Logic, machinery, and mechanisms. *Chem. Rev.* (Washington, D.C.) 2006, 106 (8), 3468–3496.

85. Gerwick, W. H., Moore, B. S. Lessons from the past and charting the future of marine natural products drug discovery and chemical biology. *Chem. Biol.* (Oxford, U.K.) 2012, 19 (1), 85–98.

86. Montaser, R., Luesch, H. Marine natural products: A new wave of drugs? *Future Med. Chem.* 2011, 3 (12), 1475–1489.

87. Mayer, A. M. S., Glaser, K. B., Cuevas, C., Jacobs, R. S., Kem, W., Little, R. D., McIntosh, J. M., Newman, D. J., Potts, B. C., Shuster, D. E. The odyssey of marine pharmaceuticals: A current pipeline perspective. *Trends Pharmacol. Sci.* 2010, 31 (6), 255–265.

88. Molinski, T. F., Dalisay, D. S., Lievens, S. L., Saludes, J. P. Drug development from marine natural products. *Nat. Rev. Drug Discov.* 2009, 8 (1), 69–85.

89. Proksch, P., Edrada-Ebel, R., Ebel, R. Drugs from the sea: Opportunities and obstacles. *Mar. Drugs* 2003, 1 (1), 5–17.

90. Newman, D. J., Cragg, G. M. Marine natural products and related compounds in clinical and advanced preclinical trials. *J. Nat. Prod.* 2004, 67 (8), 1216–1238.

91. Nuijen, B., Bouma, M., Manada, C., Jimeno, J. M., Schellens, J. H. M., Bult, A., Beijnen, J. H. Pharmaceutical development of anticancer agents derived from marine sources. *Anti-Cancer Drugs* 2000, 11 (10), 793–811.

92. Bergmann, W., Feeney, R. J. Marine products. XXXII. The nucleosides of sponges. I. *J. Org. Chem.* 1951, 16, 981–987.

93. Bergmann, W., Burke, D. C. Marine products. XXXIX. The nucleosides of sponges. III. Spongothymidine and spongouridine. *J. Org. Chem.* 1955, 20, 1501–1507.

94. Walwick, E. R., Roberts, W. K., Dekker, C. A. Cyclization during the phosphorylation of uridine and cytidine by polyphosphoric acid: A new route to the O2,2′-cyclonucleosides. *Proc. Chem. Soc.* (London) 1959, 84.

95. Lee, W. W., Benitez, A., Goodman, L., Baker, B. R. Potential anticancer agents. XL. Synthesis of the β-anomer of 9-(D-arabinofuranosyl)adenine. *J. Am. Chem. Soc.* 1960, 82, 2648–2649.

96. Loewenberg, B., Pabst, T., Vellenga, E., van Putten, W., Schouten, H. C., Graux, C., Ferrant, A., Sonneveld, P., Biemond, B. J., Gratwohl, A., de Greef, G. E., Verdonck, L. F., Schaafsma, M. R., Gregor, M., Theobald, M., Schanz, U., Maertens, J., Ossenkoppele, G. J. Cytarabine dose for acute myeloid leukemia. *N. Engl. J. Med.* 2011, 364 (11), 1027–1036.

97. Whitley, R. J., Soong, S. J., Dolin, R., Galasso, G. J., Ch'ien, L. T., Alford, C. A. Adenine arabinoside therapy of biopsy-proved herpes simplex encephalitis. National Institute of Allergy and Infectious Diseases collaborative antiviral study. *N. Engl. J. Med.* 1977, 297 (6), 289–294.

98. Colburn, D. E., Giles, F. J., Oladovich, D., Smith, J. A. *In vitro* evaluation of cytochrome P450-mediated drug interactions between cytarabine, idarubicin, itraconazole and caspofungin. *Hematology* (Abingdon, U.K.) 2004, 9 (3), 217–221.

99. Jain, K. K. An evaluation of intrathecal ziconotide for the treatment of chronic pain. *Expert Opin. Invest. Drugs* 2000, 9 (10), 2403–2410.

100. Yu, M. J., Kishi, Y., Littlefield, B. A. In *Discovery of E7389, a Fully Synthetic Macrocyclic Ketone Analog of Halichondrin B*. CRC Press, Boca Raton, FL, 2012, pp. 317–345.

101. Gradishar, W. J. The place for eribulin in the treatment of metastatic breast cancer. *Curr. Oncol. Rep.* 2011, 13 (1), 11–16.

102. Hirata, Y., Uemura, D. Halichondrins: Antitumor polyether macrolides from a marine sponge. *Pure Appl. Chem.* 1986, 58 (5), 701–710.

103. Koski, R. R. Omega-3-acid ethyl esters (Lovaza) for severe hypertriglyceridemia. *Pharm. Ther.* 2008, 33 (5).

104. Sings, H. L., Rinehart, K. L. Compounds produced from potential tunicate-blue-green algal symbiosis: A review. *J. Ind. Microbiol. Biotechnol.* 1996, 17 (5–6), 385–396.

105. Sigel, M. M. W., L. L., Lichter, W., Dudeck, L. E., Gargus, J. L., Lucas, A. H. In *Food-Drugs from the Sea*, ed. H. W. Youngken Jr. Marine Technology Society, Washington, D.C., 1969, p. 281.

106. Sakai, R., Rinehart, K. L., Guan, Y., Wang, A. H. J. Additional antitumor ecteinascidins from a Caribbean tunicate: Crystal structures and activities *in vivo*. *Proc. Natl. Acad. Sci. U.S.A.* 1992, 89 (23), 11456–11460.

107. Cuevas, C., Francesch, A., Galmarini, C. M., Aviles, P., Munt, S. Ecteinascidin-743 (Yondelis), Aplidin, and Irvalec. In *Anticancer Agents from Natural Products*, ed. D. G. I. Kingston, G. M. Cragg, D. J. Newman. 2nd ed. CRC Press, Boca Raton, FL, 2012, pp. 291–316, 2 plates.

108. Sakai, R., Jares-Erijman, E. A., Manzanares, I., Elipe, M. V. S., Rinehart, K. L. Ecteinascidins: Putative biosynthetic precursors and absolute stereochemistry. *J. Am. Chem. Soc.* 1996, 118 (38), 9017–9023.

109. Rinehart, K. L., Holt, T. G., Fregeau, N. L., Stroh, J. G., Keifer, P. A., Sun, F., Li, L. H., Martin, D. G. Ecteinascidins 729, 743, 745, 759A, 759B, and 770: Potent antitumor agents from the Caribbean tunicate *Ecteinascidia turbinata*. *J. Org. Chem.* 1990, 55 (15), 4512–4515.

110. Cuevas, C., Francesch, A. Development of Yondelis (trabectedin, ET-743): A semisynthetic process solves the supply problem. *Nat. Prod. Rep.* 2009, 26 (3), 322–337.

111. Martinez, R., Chacon-Garcia, L. The search of DNA-intercalators as antitumoral drugs: What worked and what did not work. *Curr. Med. Chem.* 2005, 12 (2), 127–151.

112. Zhang, J., Zhang, F., Li, H., Liu, C., Xia, J., Ma, L., Chu, W., Zhang, Z., Chen, C., Li, S., Wang, S. Recent progress and future potential for metal complexes as anticancer drugs targeting G-quadruplex DNA. *Curr. Med. Chem.* 2012, 19 (18), 2957–2975.

113. Capranico, G., Zunino, F. Antitumor inhibitors of DNA topoisomerases. *Curr. Pharm. Des.* 1995, 1 (1), 1–14.

114. Takebayashi, Y., Pourquier, P., Zimonjic, D. B., Nakayama, K., Emmert, S., Ueda, T., Urasaki, Y., Kanzaki, A., Akiyama, S.-I., Popescu, N., Kraemer, K. H., Pommier, Y. Antiproliferative activity of ecteinascidin 743 is dependent upon transcription-coupled nucleotide-excision repair. *Nat. Med.* (N.Y.) 2001, 7 (8), 961–966.

115. Beumer, J. H., Lopez-Lazaro, L., Schellens, J. H. M., Beijnen, J. H., van Tellingen, O. Evaluation of human plasma protein binding of trabectedin (Yondelis, ET-743). *Curr. Clin. Pharmacol.* 2009, 4 (1), 38–42.

116. Lau, L., Supko, J. G., Blaney, S., Hershon, L., Seibel, N., Krailo, M., Qu, W., Malkin, D., Jimeno, J., Bernstein, M., Baruchel, S. A phase I and pharmacokinetic study of ecteinascidin-743 (Yondelis) in children with refractory solid tumors. A Children's Oncology Group study. *Clin. Cancer Res.* 2005, 11 (2, Pt. 1), 672–677.

117. Brandon, E. F. A., Sparidans, R. W., Guijt, K.-J., Loewenthal, S., Meijerman, I., Beijnen, J. H., Schellens, J. H. M. *In vitro* characterization of the human biotransformation and CYP reaction phenotype of ET-743 (Yondelis, trabectidin), a novel marine anti-cancer drug. *Invest. New Drugs* 2006, 24 (1), 3–14.

118. Cavallo, D., Ursini, C. L., Omodeo-Sale, E., Iavicoli, S. Micronucleus induction and FISH analysis in buccal cells and lymphocytes of nurses administering antineoplastic drugs. *Mutat. Res. Genet. Toxicol. Environ. Mutagen.* 2007, 628 (1), 11–18.

119. Corey, E. J., Gin, D. Y., Kania, R. S. Enantioselective total synthesis of ecteinascidin 743. *J. Am. Chem. Soc.* 1996, 118 (38), 9202–9203.

120. Cuevas, C., Perez, M., Martin, M. J., Chicharro, J. L., Fernandez-Rivas, C., Flores, M., Francesch, A., Gallego, P., Zarzuelo, M., de la Calle, F., Garcia, J., Polanco, C., Rodriguez, I., Manzanares, I. Synthesis of ecteinascidin ET-743 and phthalascidin Pt-650 from cyanosafracin B. *Org. Lett.* 2000, 2 (16), 2545–2548.

121. Ikeda, Y., Idemoto, H., Hirayama, F., Yamamoto, K., Iwao, K., Asao, T., Munakata, T. Safracins, new antitumor antibiotics. I. Producing organism, fermentation and isolation. *J. Antibiot.* 1983, 36 (10), 1279–1283.

122. Fayette, J., Boyle, H., Chabaud, S., Favier, B., Engel, C., Cassier, P., Thiesse, P., Meeus, P., Sunyach, M.-P., Vaz, G., Ray-Coquard, I., Ranchere, D., Decouvelaere, A.-V., Alberti, L., Perol, D., Blay, J.-Y. Efficacy of trabectedin for advanced sarcomas in clinical trials versus compassionate use programs: Analysis of 92 patients treated in a single institution. *Anti-Cancer Drugs* 2010, 21 (1), 113–119.

123. Yondelis. http://www.ema.europa.eu/docs/en_GB/document_library/EPAR_-_Summary_for_the_public/human/000773/WC500045833.pdf (accessed January 21, 2014).

124. YONDELIS® granted orphan drug designation by the U.S. FDA for the treatment of ovarian cancer. http://www.zeltia.com/media/docs/dgjiaour.pdf?ie=UTF-8&oe=UTF-8&q=prettyphoto&iframe=true&width=100%25&height=100%25 (accessed January 21, 2014).

125. Martinez, E. J., Owa, T., Schreiber, S. L., Corey, E. J. Phthalascidin, a synthetic antitumor agent with potency and mode of action comparable to ecteinascidin 743. *Proc. Natl. Acad. Sci. U.S.A.* 1999, 96 (7), 3496–3501.

126. Liu, Z.-Z., Wang, Y., Tang, Y.-F., Chen, S.-Z., Chen, X.-G., Li, H.-Y. Synthesis and antitumor activity of simplified ecteinascidin-saframycin analogs. *Bioorg. Med. Chem. Lett.* 2006, 16 (5), 1282–1285.

127. Myers, A. G., Lanman, B. A. A solid-supported, enantioselective synthesis suitable for the rapid preparation of large numbers of diverse structural analogues of (–)-saframycin A. *J. Am. Chem. Soc.* 2002, 124 (44), 12969–12971.

128. Spencer, J. R., Sendzik, M., Oeh, J., Sabbatini, P., Dalrymple, S. A., Magill, C., Kim, H. M., Zhang, P., Squires, N., Moss, K. G., Sukbuntherng, J., Graupe, D., Eksterowicz, J., Young, P. R., Myers, A. G., Green, M. J. Evaluation of antitumor properties of novel saframycin analogs *in vitro* and *in vivo*. *Bioorg. Med. Chem. Lett.* 2006, 16 (18), 4884–4888.

129. De Bono, J., Paz-Ares, L., Tabernero, J., Smyth, J., De las Heras, B. PM00104 compound for use in cancer therapy. WO2008135792A1, 2008.

130. Massard, C., Margetts, J., Amellal, N., Drew, Y., Bahleda, R., Stevens, P., Armand, J. P., Calvert, H., Soria, J. C., Coronado, C., Kahatt, C., Alfaro, V., Siguero, M., Fernandez-Teruel, C., Plummer, R. Phase I study of PM00104 (Zalypsis) administered as a 1-hour weekly infusion resting every fourth week in patients with advanced solid tumors. *Invest. New Drugs* 2013, 31 (3), 623–630.

131. Martin, L. P., Krasner, C., Rutledge, T., Ibanes, M. L., Fernandez-Garcia, E. M., Kahatt, C., Gomez, M. S., McMeekin, S. Phase II study of weekly PM00104 (ZALYPSIS) in patients with pretreated advanced/metastatic endometrial or cervical cancer. *Med. Oncol.* (N.Y.) 2013, 30 (3), 1–4.

132. Rinehart, K. L. Antitumor compounds from tunicates. *Med. Res. Rev.* 2000, 20 (1), 1–27.

133. Rinehart, K. L., Jr., Gloer, J. B., Cook, J. C., Jr., Mizsak, S. A., Scahill, T. A. Structures of the didemnins, antiviral and cytotoxic depsipeptides from a Caribbean tunicate. *J. Am. Chem. Soc.* 1981, 103 (7), 1857–1859.

134. Joullie, M. M., Leonard, M. S., Portonovo, P., Liang, B., Ding, X., La Clair, J. J. Chemical defense in ascidians of the Didemnidae family. *Bioconjugate Chem.* 2003, 14 (1), 30–37.

135. Vera, M. D., Joullie, M. M. Natural products as probes of cell biology: 20 years of didemnin research. *Med. Res. Rev.* 2002, 22 (2), 102–145.

136. Vervoort, H., Fenical, W., de Epifanio, R. Tamandarins A and B: New cytotoxic depsipeptides from a Brazilian ascidian of the family Didemnidae. *J. Org. Chem.* 2000, 65 (3), 782–792.

137. SirDeshpande, B. V., Toogood, P. L. Mechanism of protein synthesis inhibition by didemnin B *in vitro*. *Biochemistry* 1995, 34 (28), 9177–9184.

138. Marco, E., Martin-Santamaria, S., Cuevas, C., Gago, F. Structural basis for the binding of didemnins to human elongation factor eEF1A and rationale for the potent antitumor activity of these marine natural products. *J. Med. Chem.* 2004, 47 (18), 4439–4452.

139. http://www.chemicalize.org/structure/#!mol=77327–05–0&source=fp (accessed January 21, 2014).

140. Hochster, H., Oratz, R., Ettinger, D. S., Borden, E. A phase II study of didemnin B (NSC 325319) in advanced malignant melanoma: An Eastern Cooperative Oncology Group study (PB687). *Invest. New Drugs* 1999, 16 (3), 259–263.

141. Gelderblom, H., Verweij, J., Nooter, K., Sparreboom, A. Cremophor EL: The drawbacks and advantages of vehicle selection for drug formulation. *Eur. J. Cancer* 2001, 37 (13), 1590–1598.

142. Sparreboom, A., Scripture, C. D., Trieu, V., Williams, P. J., De, T., Yang, A., Beals, B., Figg, W. D., Hawkins, M., Desai, N. Comparative preclinical and clinical pharmacokinetics of a Cremophor-free, nanoparticle albumin-bound paclitaxel (ABI-007) and paclitaxel formulated in Cremophor (Taxol). *Clin. Cancer Res.* 2005, 11 (11), 4136–4143.

143. Lee, J., Currano, J. N., Carroll, P. J., Joullie, M. M. Didemnins, tamandarins and related natural products. *Nat. Prod. Rep.* 2012, 29 (3), 404–424.

144. Benvenuto, J. A., Newman, R. A., Bignami, G. S., Raybould, T. J., Raber, M. N., Esparza, L., Walters, R. S. Phase II clinical and pharmacological study of didemnin B in patients with metastatic breast cancer. *Invest. New Drugs* 1992, 10 (2), 113–117.

145. Beasley, V. R., Bruno, S. J., Burner, J. S., Choi, B. W., Rinehart, K. L., Koritz, G. D., Levengood, J. M. Fate of tritiated didemnin B in mice: Excretion and tissue concentrations after an intraperitoneal dose. *Biopharm. Drug Dispos.* 2005, 26 (8), 341–351.

146. Sakai, R., Kishore, V., Kundu, B., Faircloth, G., Gloer, J. B., Carney, J. R., Namikoshi, M., Sun, F., Hughes, R. G., Jr., et al. Structure-activity relationships of the didemnins. *J. Med. Chem.* 1996, 39 (14), 2819–2834.

147. Rinehart, K. L., Lithgow-Bertelloni, A. M. Dehydrodidemnin B. WO9104985A1, 1991.

148. Lloyd-Williams, P., Jou, G., Gonzalez, I., Caba, J. M., Albericio, F., Giralt, E. Total synthesis of dehydrodidemnin B: Use of phosphonium and uronium salts in peptide synthesis in solution. *Peptides 1996, Proceedings of the European Peptide Symposium, Mayflower Scientific, Kingswinford, UK*, 1998, pp. 589–590.

149. Faivre, S., Chieze, S., Delbaldo, C., Ady-vago, N., Guzman, C., Lopez-Lazaro, L., Lozahic, S., Jimeno, J., Pico, F., Armand, J. P., Martin, J. A. L., Raymond, E. Phase I and pharmacokinetic study of aplidine, a new marine cyclodepsipeptide in patients with advanced malignancies. *J. Clin. Oncol.* 2005, 23 (31), 7871–7880.

150. Mateos, M. V., Cibeira, M. T., Richardson, P. G., Prosper, F., Oriol, A., de la Rubia, J., Lahuerta, J. J., Garcia-Sanz, R., Extremera, S., Szyldergemajn, S., Corrado, C., Singer, H., Mitsiades, C. S., Anderson, K. C., Blade, J., San Miguel, J. Phase II clinical and pharmacokinetic study of plitidepsin 3-hour infusion every two weeks alone or with dexamethasone in relapsed and refractory multiple myeloma. *Clin. Cancer Res.* 2010, 16 (12), 3260–3269.

151. Schmidt, U., Griesser, H., Haas, G., Kroner, M., Riedl, B., Schumacher, A., Sutoris, F., Haupt, A., Emling, F. Amino acids and peptides, part 105. Cyclopeptides, part 34. Synthesis and cytostatic activities of didemnin derivatives. *J. Pept. Res.* 1999, 54 (2), 146–161.

152. Gloer, J. B. *Structures of the Didemnins* (doctoral thesis), University of Illinois. Diss. Abst. Int. B 1984, 45, 188.

153. Molinski, T. F., Ko, J., Reynolds, K. A., Lievens, S. C., Skarda, K. R. N,N′-Methyleno-didemnin A from the ascidian *Trididemnum solidum*: Complete NMR assignments and confirmation of the imidazolidinone ring by strategic analysis of 1JCH. *J. Nat. Prod.* 2011, 74 (4), 882–887.

154. Pettit, G. R., Day, J. F., Hartwell, J. L., Wood, H. B. Antineoplastic components of marine animals. *Nature* 1970, 227 (5261), 962–963.

155. Pettit, G. R., Herald, C. L., Doubek, D. L., Herald, D. L., Arnold, E., Clardy, J. Isolation and structure of bryostatin 1. *J. Am. Chem. Soc.* 1982, 104 (24), 6846–6848.

156. Newman, D. J. In *The bryostatins*, CRC Press, Boca Raton, FL, 2012; pp 199–218.

157. Hale, K. J., Manaviazar, S. New approaches to the total synthesis of the bryostatin antitumor macrolides. *Chem. Asian J.* 2010, 5 (4), 704–754.

158. Ramsdell, J. S., Pettit, G. R., Tashjian, A. H., Jr. Three activators of protein kinase C, bryostatins, dioleins, and phorbol esters, show differing specificities of action on GH4 pituitary cells. *J. Biol. Chem.* 1986, 261 (36), 17073–17080.

159. Grant, S., Turner, A. J., Freemerman, A. J., Wang, Z., Kramer, L., Jarvis, W. D. Modulation of protein kinase C activity and calcium-sensitive isoform expression in human myeloid leukemia cells by bryostatin 1: Relationship to differentiation and Ara-C-induced apoptosis. *Exp. Cell Res.* 1996, 228 (1), 65–75.

160. Nishizuka, Y. The role of protein kinase C in cell surface signal transduction and tumor promotion. *Nature* (London) 1984, 308 (5961), 693–698.

161. Scala, S., Dickstein, B., Regis, J., Szallasi, Z., Blumberg, P. M., Bates, S. E. Bryostatin 1 affects P-glycoprotein phosphorylation but not function in multidrug-resistant human breast cancer cells. *Clin. Cancer Res.* 1995, 1 (12), 1581–1587.

162. Philip, P. A., Rea, D., Thavasu, P., Carmichael, J., Stuart, N. S., Rockett, H., Talbot, D. C., Ganesan, T., Pettit, G. R., Balkwill, F. Phase I study of bryostatin 1: Assessment of interleukin 6 and tumor necrosis factor alpha induction *in vivo*. The Cancer Research Campaign Phase I Committee. *J. Natl. Cancer Inst.* 1993, 85 (22), 1812–1818.

163. Bryostatin 1. http://www.chemicalize.org/structure/#!mol=83314–01–6&source=calculate (accessed January 21, 2014).

164. Jayson, G. C., Crowther, D., Prendiville, J., McGown, A. T., Scheid, C., Stern, P., Young, R., Brenchley, P., Chang, J. A phase I trial of bryostatin 1 in patients with advanced malignancy using a 24 hour intravenous infusion. *Br. J. Cancer* 1995, 72 (2), 461–468.

165. Zhang, X., Zhang, R., Zhao, H., Cai, H., Gush, K. A., Kerr, R. G., Pettit, G. R., Kraft, A. S. Preclinical pharmacology of the natural product anticancer agent bryostatin 1, an activator of protein kinase C. *Cancer Res.* 1996, 56 (4), 802–808.

166. Cragg, L. H., Andreeff, M., Feldman, E., Roberts, J., Murgo, A., Winning, M., Tombes, M. B., Roboz, G., Kramer, L., Grant, S. Phase I trial and correlative laboratory studies of bryostatin 1 (NSC 339555) and high-dose 1-B-D-arabinofuranosylcytosine in patients with refractory acute leukemia. *Clin. Cancer Res.* 2002, 8 (7), 2123–2133.

167. Bedikian, A. Y., Plager, C., Stewart, J. R., O'Brian, C. A., Herdman, S. K., Ross, M., Papadopoulos, N., Eton, O., Ellerhorst, J., Smith, T. Phase II evaluation of bryostatin-1 in metastatic melanoma. *Melanoma Res.* 2001, 11 (2), 183–188.

168. Pagliaro, L., Daliani, D., Amato, R., Tu, S.-M., Jones, D., Smith, T., Logothetis, C., Millikan, R. A phase II trial of bryostatin-1 for patients with metastatic renal cell carcinoma. *Cancer* (N.Y.) 2000, 89 (3), 615–618.

169. Zonder, J. A., Shields, A. F., Zalupski, M., Chaplen, R., Heilbrun, L. K., Arlauskas, P., Philip, P. A. A phase II trial of bryostatin 1 in the treatment of metastatic colorectal cancer. *Clin. Cancer Res.* 2001, 7 (1), 38–42.

170. Brockstein, B., Samuels, B., Humerickhouse, R., Arietta, R., Fishkin, P., Wade, J., Sosman, J., Vokes, E. E. Phase II studies of bryostatin-1 in patients with advanced sarcoma and advanced head and neck cancer. *Invest. New Drugs* 2001, 19 (3), 249–254.

171. Varterasian, M. L., Pemberton, P. A., Hulburd, K., Rodriguez, D. H., Murgo, A., Al-Katib, A. M. Phase II study of bryostatin 1 in patients with relapsed multiple myeloma. *Invest. New Drugs* 2001, 19 (3), 245–247.

172. Barr, P. M., Lazarus, H. M., Cooper, B. W., Schluchter, M. D., Panneerselvam, A., Jacobberger, J. W., Hsu, J. W., Janakiraman, N., Simic, A., Dowlati, A., Remick, S. C. Phase II study of bryostatin 1 and vincristine for aggressive non-Hodgkin lymphoma relapsing after an autologous stem cell transplant. *Am. J. Hematol.* 2009, 84 (8), 484–487.

173. Lam, A. P., Sparano, J. A., Vinciguerra, V., Ocean, A. J., Christos, P., Hochster, H., Camacho, F., Goel, S., Mani, S., Kaubisch, A. Phase II study of paclitaxel plus the protein kinase C inhibitor bryostatin-1 in advanced pancreatic carcinoma. *Am. J. Clin. Oncol.* 2010, 33 (2), 121–124.

174. Wong, Y., Hudes., G. R., von Mehren, M., Malizzia, L., Roethke, S. K., Litwin, S., Haas, N. B. Phase I study of temsirolimus (TEM) and bryostatin (BRYO) in patients with metastatic renal cell carcinoma (RCC). In *Genitourinary Cancers Symposium*, Orlando, FL, 2009, Abstr. 313.

175. Morgan, R. J., Jr., Leong, L., Chow, W., Gandara, D., Frankel, P., Garcia, A., Lenz, H.-J., Doroshow, J. H. Phase II trial of bryostatin-1 in combination with cisplatin in patients with recurrent or persistent epithelial ovarian cancer: A California cancer consortium study. *Invest. New Drugs* 2012, 30 (2), 723–728.

176. Roberts, J. D., Smith, M. R., Feldman, E. J., Cragg, L., Millenson, M. M., Roboz, G. J., Honeycutt, C., Thune, R., Padavic-Shaller, K., Carter, W. H., Ramakrishnan, V., Murgo, A. J., Grant, S. Phase I study of bryostatin 1 and fludarabine in patients with chronic lymphocytic leukemia and indolent (non-Hodgkin's) lymphoma. *Clin. Cancer Res.* 2006, 12 (19), 5809–5816.

177. El-Rayes, B. F., Gadgeel, S., Shields, A. F., Manza, S., Lorusso, P., Philip, P. A. Phase I study of bryostatin 1 and gemcitabine. *Clin. Cancer Res.* 2006, 12 (23), 7059–7062.

178. Ahmad, I., Al-Katib, A. M., Beck, F. W., Mohammad, R. M. Sequential treatment of a resistant chronic lymphocytic leukemia patient with bryostatin 1 followed by 2-chlorodeoxyadenosine: Case report. *Clin. Cancer Res.* 2000, 6 (4), 1328–1332.

179. Mendola, D. Aquaculture of three phyla of marine invertebrates to yield bioactive metabolites: Process developments and economics. *Biomol. Eng.* 2003, 20 (4–6), 441–458.

180. Evans, D. A., Carter, P. H., Carreira, E. M., Charette, A. B., Prunet, J. A., Lautens, M. Total synthesis of bryostatin 2. *J. Am. Chem. Soc.* 1999, 121 (33), 7540–7552.

181. Ohmori, K., Ogawa, Y., Obitsu, T., Ishikawa, Y., Nishiyama, S., Yamamura, S. Total synthesis of bryostatin 3. *Angew. Chem. Int. Ed.* 2000, 39 (13), 2290–2294.

182. Lu, Y., Woo, S. K., Krische, M. J. Total synthesis of bryostatin 7 via C–C bond-forming hydrogenation. *J. Am. Chem. Soc.* 2011, 133 (35), 13876–13879.

183. Wender, P. A., Schrier, A. J. Total synthesis of bryostatin 9. *J. Am. Chem. Soc.* 2011, 133 (24), 9228–9231.

184. Trost, B. M., Dong, G. Total synthesis of bryostatin 16 using a Pd-catalyzed diyne coupling as macrocyclization method and synthesis of C20-epi-bryostatin 7 as a potent anticancer agent. *J. Am. Chem. Soc.* 2010, 132 (46), 16403–16416.

185. Hale, K. J., Hummersone, M. G., Manaviazar, S., Frigerio, M. The chemistry and biology of the bryostatin antitumour macrolides. *Nat. Prod. Rep.* 2002, 19 (4), 413–453.

186. Wender, P. A., Cribbs, C. M., Koehler, K. F., Sharkey, N. A., Herald, C. L., Kamano, Y., Pettit, G. R., Blumberg, P. M. Modeling of the bryostatins to the phorbol ester pharmacophore on protein kinase C. *Proc. Natl. Acad. Sci. U.S.A.* 1988, 85 (19), 7197–7201.

187. Castagna, M., Takai, Y., Kaibuchi, K., Sano, K., Kikkawa, U., Nishizuka, Y. Direct activation of calcium-activated, phospholipid-dependent protein kinase by tumor-promoting phorbol esters. *J. Biol. Chem.* 1982, 257 (13), 7847–7851.

188. Trost, B. M., Yang, H., Thiel, O. R., Frontier, A. J., Brindle, C. S. Synthesis of a ring-expanded bryostatin analogue. *J. Am. Chem. Soc.* 2007, 129 (8), 2206–2207.

189. Halford, B. The bryostatins' tale. *Chem. Eng. News* 2011, 89 (43), 10.

190. Keck, G. E., Poudel, Y. B., Rudra, A., Stephens, J. C., Kedei, N., Lewin, N. E., Blumberg, P. M. Role of the C8gem-dimethyl group of bryostatin 1 on its unique pattern of biological activity. *Bioorg. Med. Chem. Lett.* 2012, 22 (12), 4084–4088.

191. Kedei, N., Telek, A., Michalowski, A. M., Kraft, M. B., Li, W., Poudel, Y. B., Rudra, A., Petersen, M. E., Keck, G. E., Blumberg, P. M. Comparison of transcriptional response to phorbol ester, bryostatin 1, and bryostatin analogs in LNCaP and U937 cancer cell lines provides insight into their differential mechanism of action. *Biochem. Pharmacol.* (Amsterdam, Neth.) 2013, 85 (3), 313–324.

192. Wender, P. A., Clarke, M. O., Horan, J. C. Role of the A-ring of bryostatin analogues in PKC binding: Synthesis and initial biological evaluation of new A-ring-modified bryologs. *Org. Lett.* 2005, 7 (10), 1995–1998.

193. Wender, P. A., Debrabander, J., Harran, P. G., Jimenez, J.-M., Koehler, M. F. T., Lippa, B., Park, C.-M., Siedenbiedel, C., Pettit, G. R. The design, computer modeling, solution structure, and biological evaluation of synthetic analogs of bryostatin 1. *Proc. Natl. Acad. Sci. U.S.A.* 1998, 95 (12), 6624–6629.

194. Wender, P. A., Baryza, J. L., Bennett, C. E., Bi, F. C., Brenner, S. E., Clarke, M. O., Horan, J. C., Kan, C., Lacote, E., Lippa, B., Nell, P. G., Turner, T. M. The practical synthesis of a novel and highly potent analogue of bryostatin. *J. Am. Chem. Soc.* 2002, 124 (46), 13648–13649.

195. Wender, P. A., DeChristopher, B. A., Schrier, A. J. Efficient synthetic access to a new family of highly potent bryostatin analogues via a prins-driven macrocyclization strategy. *J. Am. Chem. Soc.* 2008, 130 (21), 6658–6659.

196. Wender, P. A., Hilinski, M. K., Mayweg, A. V. W. Late-stage intermolecular CH activation for lead diversification: A highly chemoselective oxyfunctionalization of the C-9 position of potent bryostatin analogues. *Org. Lett.* 2005, 7 (1), 79–82.

197. DeChristopher, B. A., Loy, B. A., Marsden, M. D., Schrier, A. J., Zack, J. A., Wender, P. A. Designed, synthetically accessible bryostatin analogues potently induce activation of latent HIV reservoirs *in vitro*. *Nat. Chem.* 2012, 4 (9), 705–710.

198. Archin, N. M., Margolis, D. M. Emerging strategies to deplete the HIV reservoir. *Curr. Opin. Infect. Dis.* 2014, 27 (1), 29–35.

199. Trindade-Silva, A. E., Lim-Fong, G. E., Sharp, K. H., Haygood, M. G. Bryostatins: Biological context and biotechnological prospects. *Curr. Opin. Biotechnol.* 2010, 21 (6), 834–842.

200. Sun, M.-K., Alkon, D. L. Bryostatin-1: Pharmacology and therapeutic potential as a CNS drug. *CNS Drug Rev.* 2006, 12 (1), 1–8.

201. Pettit, G. R., Herald, C. L., Ode, R. H., Brown, P., Gust, D., Michel, C. The isolation of loliolide from an Indian Ocean opisthobranch mollusc. *J. Nat. Prod.* 1980, 43 (6), 752–755.

202. Pettit, G. R., Kamano, Y., Fujii, Y., Herald, C. L., Inoue, M., Brown, P., Gust, D., Kitahara, K., Schmidt, J. M., Doubek, D. L., Michael, C. Antineoplastic agents. Part 72. Marine animal biosynthetic constituents for cancer chemotherapy. *J. Nat. Prod.* 1981, 44 (4), 482–785.

203. Flahive, E., Srirangam, J. In *The dolastatins: Novel antitumor agents from Dolabella auricularia.* CRC Press, Boca Raton, FL, 2012, pp. 263–289, 3 plates.

204. Pettit, G. R., Kamano, Y., Herald, C. L., Tuinman, A. A., Boettner, F. E., Kizu, H., Schmidt, J. M., Baczynskyj, L., Tomer, K. B., Bontems, R. J. The isolation and structure of a remarkable marine animal antineoplastic constituent: Dolastatin 10. *J. Am. Chem. Soc.* 1987, 109 (22), 6883–6885.

205. Bai, R., Pettit, G. R., Hamel, E. Dolastatin 10, a powerful cytostatic peptide derived from a marine animal: Inhibition of tubulin polymerization mediated through the vinca alkaloid binding domain. *Biochem. Pharmacol.* 1990, 39 (12), 1941–1949.

206. Wilson, L., Panda, D., Jordan, M. A. Modulation of microtubule dynamics by drugs: A paradigm for the actions of cellular regulators. *Cell Struct. Funct.* 1999, 24 (5), 329–335.

207. Luesch, H., Moore, R. E., Paul, V. J., Mooberry, S. L., Corbett, T. H. Isolation of dolastatin 10 from the marine cyanobacterium *Symploca* species VP642 and total stereochemistry and biological evaluation of its analogue symplostatin 1. *J. Nat. Prod.* 2001, 64 (7), 907–910.

208. Gallop, M. A., Barrett, R. W., Dower, W. J., Fodor, S. P. A., Gordon, E. M. Applications of combinatorial technologies to drug discovery. 1. Background and peptide combinatorial libraries. *J. Med. Chem.* 1994, 37 (9), 1233–1251.

209. Bai, R., Roach, M. C., Jayaram, S. K., Barkoczy, J., Pettit, G. R., Luduena, R. F., Hamel, E. Differential effects of active isomers, segments, and analogs of dolastatin 10 on ligand interactions with tubulin: Correlation with cytotoxicity. *Biochem. Pharmacol.* 1993, 45 (7), 1503–1515.

210. Pettit, G. R., Kamano, Y., Dufresne, C., Cerny, R. L., Herald, C. L., Schmidt, J. M. Isolation and structure of the cytostatic linear depsipeptide dolastatin 15. *J. Org. Chem.* 1989, 54 (26), 6005–6006.

211. Hamel, E. Natural products which interact with tubulin in the vinca domain: Maytansine, rhizoxin, phomopsin A, dolastatins 10 and 15 and halichondrin B. *Pharmacol. Ther.* 1992, 55 (1), 31–51.

212. Flahive, E., Srirangam, J. The dolastatins: Novel antitumor agents from *Dolabella aricularia*. In *Anticancer Agents from Natural Products*, ed. G. M. Cragg, D. G. I. Kingston, D. J. Newman. CRC Press, Boca Raton, FL, 2005, pp. 191–213, 1 plate.

213. Bai, R., Pettit, G. R., Hamel, E. Structure-activity studies with chiral isomers and with segments of the antimitotic marine peptide dolastatin 10. *Biochem. Pharmacol.* 1990, 40 (8), 1859–1864.

214. Roux, F., Galeotti, N., Poncet, J., Jouin, P., Cros, S., Zenke, G. Synthesis and *in vitro* cytotoxicity of diastereoisomerically modified dolastatin 15 analogs. *Bioorg. Med. Chem. Lett.* 1994, 16 (4), 1947–1950.

215. Pitot, H. C., McElroy, E. A., Jr., Reid, J. M., Windebank, A. J., Sloan, J. A., Erlichman, C., Bagniewski, P. G., Walker, D. L., Rubin, J., Goldberg, R. M., Adjei, A. A., Ames, M. M. Phase I trial of dolastatin-10 (NSC 376128) in patients with advanced solid tumors. *Clin. Cancer Res.* 1999, 5 (3), 525–531.

216. Horti, J., Juhasz, E., Monostori, Z., Maeda, K., Eckhardt, S., Bodrogi, I. Phase I study of TZT-1027, a novel synthetic dolastatin 10 derivative, for the treatment of patients with non-small cell lung cancer. *Cancer Chemother. Pharmacol.* 2008, 62 (1), 173–180.

217. Mross, K., Berdel, W. E., Fiebig, H. H., Velagapudi, R., von, B. I. M., Unger, C. Clinical and pharmacologic phase I study of cemadotin-HCl (LU103793), a novel antimitotic peptide, given as 24-hour infusion in patients with advanced cancer. A study of the Arbeitsgemeinschaft Internistische Onkologie (AIO) Phase I Group and Arbeitsgruppe Pharmakologie in der Onkologie und Haematologie (APOH) Group of the German Cancer Society. *Ann. Oncol.* 1998, 9 (12), 1323–1330.

218. Ebbinghaus, S., Rubin, E., Hersh, E., Cranmer, L. D., Bonate, P. L., Fram, R. J., Jekunen, A., Weitman, S., Hammond, L. A. A phase I study of the dolastatin-15 analogue tasidotin (ILX651) administered intravenously daily for 5 consecutive days every 3 weeks in patients with advanced solid tumors. *Clin. Cancer Res.* 2005, 11 (21), 7807–7816.

219. Bai, R., Edler, M. C., Bonate, P. L., Copeland, T. D., Pettit, G. R., Luduena, R. F., Hamel, E. Intracellular activation and deactivation of tasidotin, an analog of dolastatin 15: Correlation with cytotoxicity. *Mol. Pharmacol.* 2009, 75 (1), 218–226.

220. Gajula, P. K., Asthana, J., Panda, D., Chakraborty, T. K. A synthetic dolastatin 10 analogue suppresses microtubule dynamics, inhibits cell proliferation, and induces apoptotic cell death. *J. Med. Chem.* 2013, 56 (6), 2235–2245.

221. Pro, B., Advani, R., Brice, P., Bartlett, N. L., Rosenblatt, J. D., Illidge, T., Matous, J., Ramchandren, R., Fanale, M., Connors, J. M., Yang, Y., Sievers, E. L., Kennedy, D. A., Shustov, A. Brentuximab vedotin (SGN-35) in patients with relapsed or refractory systemic anaplastic large-cell lymphoma: Results of a phase II study. *J. Clin. Oncol.* 2012, 30 (18), 2190–2196.

222. Smith, C. A., Gruss, H. J., Davis, T., Anderson, D., Farrah, T., Baker, E., Sutherland, G. R., Brannan, C. I., Copeland, N. G., Jenkins, N. A., Grabstein, K. H., Gliniak, B., McAlister, I. B., Fanslow, W., Alderson, M., Falk, B., Gimpel, S., Gillis, S., Din, W. S., Goodwin, R. G., Armitage, R. J. CD30 antigen, a marker for Hodgkin's lymphoma, is a receptor whose ligand defines an emerging family of cytokines with homology to TNF. *Cell* (Cambridge, Mass.) 1993, 73 (7), 1349–1360.

223. Pettit, G. R., Srirangam, J. K., Barkoczy, J., Williams, M. D., Boyd, M. R., Hamel, E., Pettit, R. K., Hogan, F., Bai, R., Chapuis, J.-C., McAllister, S. C., Schmidt, J. M. Antineoplastic agents 365. Dolastatin 10 SAR probes. *Anti-Cancer Drug Des.* 1998, 13 (4), 243–277.

224. Doronina, S. O., Toki, B. E., Torgov, M. Y., Mendelsohn, B. A., Cerveny, C. G., Chace, D. F., DeBlanc, R. L., Gearing, R. P., Bovee, T. D., Siegall, C. B., Francisco, J. A., Wahl, A. F., Meyer, D. L., Senter, P. D. Development of potent monoclonal antibody auristatin conjugates for cancer therapy. *Nat. Biotechnol.* 2003, 21 (7), 778–784.

225. Junutula, J. R., Raab, H., Clark, S., Bhakta, S., Leipold, D. D., Weir, S., Chen, Y., Simpson, M., Tsai, S. P., Dennis, M. S., Lu, Y., Meng, Y. G., Ng, C., Yang, J., Lee, C. C., Duenas, E., Gorrell, J., Katta, V., Kim, A., McDorman, K., Flagella, K., Venook, R., Ross, S., Spencer, S. D., Wong, W. L., Lowman, H. B., Vandlen, R., Sliwkowski, M. X., Scheller, R. H., Polakis, P., Mallet, W. Site-specific conjugation of a cytotoxic drug to an antibody improves the therapeutic index. *Nat. Biotechnol.* 2008, 26 (8), 925–932.

226. FDA. FDA approves Adcetris to treat two types of lymphoma. http://www.fda.gov/NewsEvents/Newsroom/PressAnnouncements/ucm268781.htm (accessed January 22, 2014).

227. Gunasekera, S. P., Gunasekera, M., Longley, R. E., Schulte, G. K. Discodermolide: A new bioactive polyhydroxylated lactone from the marine sponge *Discodermia dissoluta. J. Org. Chem.* 1990, 55 (16), 4912–4915.

228. Gunasekera, S. P., Gunasekera, M., Longley, R. E., Schulte, G. K. Discodermolide: A new bioactive polyhydroxylated lactone from the marine sponge *Discodermia dissoluta* [Erratum to document cited in CA113(9):75187b]. *J. Org. Chem.* 1991, 56 (3), 1346.

229. Gunasekera, S. P., Wright, A. E. Chemistry and biology of the discodermolides, potent mitotic spindle poisons. In *Anticancer Agents from Natural Products*, ed. D. G. I. Kingston, G. M. Cragg, D. J. Newman. 2nd ed. CRC Press, Boca Raton, FL, 2012, pp. 241–262, 1 plate.

230. Smith, A. B., III, Kaufman, M. D., Beauchamp, T. J., LaMarche, M. J., Arimoto, H. Gram-scale synthesis of (+)-discodermolide. *Org. Lett.* 1999, 1 (11), 1823–1826.

231. Kobayashi, M., Aoki, S., Ohyabu, N., Kurosu, M., Wang, W., Kitagawa, I. Arenastatin A, a potent cytotoxic depsipeptide from the Okinawan marine sponge *Dysidea arenaria. Tetrahedron Lett.* 1994, 35 (43), 7969–7972.

232. Schwartz, R. E., Hirsch, C. F., Sesin, D. F., Flor, J. E., Chartrain, M., Fromtling, R. E., Harris, G. H., Salvatore, M. J., Liesch, J. M., Yudin, K. Pharmaceuticals from cultured algae. *J. Ind. Microbiol.* 1990, 5 (2–3), 113–123.

233. Smith, C. D., Zhang, X., Mooberry, S. L., Patterson, G. M., Moore, R. E. Cryptophycin: A new antimicrotubule agent active against drug-resistant cells. *Cancer Res.* 1994, 54 (14), 3779–3784.

234. Al-awar, R. S., Shih, C. *The Isolation, Characterization, and Development of a Novel Class of Potent Antimitotic Macrocyclic Depsipeptides: The Cryptophycins.* CRC Press, Boca Raton, FL, 2012, pp. 219–240, 2 plates.

235. Moore, R. E., Tius, M. A., Barrow, R. A., Liang, J., Corbett, T. H., Valeriote, F. A., Hemscheidt, T. K. Isolation, characterization, and synthesis of new cryptophycin compounds as anticancer agents. WO9640184A1, 1996.

236. Eggen, M., Georg, G. I. The cryptophycins: Their synthesis and anticancer activity. *Med. Res. Rev.* 2002, 22 (2), 85–101.

237. Edelman, M. J., Gandara, D. R., Hausner, P., Israel, V., Thornton, D., DeSanto, J., Doyle, L. A. Phase 2 study of cryptophycin 52 (LY355703) in patients previously treated with platinum based chemotherapy for advanced non-small cell lung cancer. *Lung Cancer* 2003, 39 (2), 197–199.

238. Weiss, C., Sammet, B., Sewald, N. Recent approaches for the synthesis of modified cryptophycins. *Nat. Prod. Rep.* 2013, 30 (7), 924–940.

239. Pettit, G. R., Chicacz, Z. A., Gao, F., Herald, C. L., Boyd, M. R., Schmidt, J. M., Hooper, J. N. A. Antineoplastic agents. 257. Isolation and structure of spongistatin 1. *J. Org. Chem.* 1993, 58 (6), 1302–1304.

240. Kobayashi, M. A., S., Sakai, H., Kawazoe, K., Kihara, N., Sasaki, T., Kitagawa, I. Altohyrtin A, a potent anti-tumor macrolide from the Okinawan marine sponge *Hyrtios altum. Tetrahedron Lett.* 1993, 34 (17), 2795–2798.

241. Fusetani, N., Shinoda, K., Matsunaga, S. Bioactive marine metabolites. 48. Cinachyrolide A: A potent cytotoxic macrolide possessing two spiro ketals from marine sponge *Cinachyra* sp. *J. Am. Chem. Soc.* 1993, 115 (10), 3977–3981.

242. Feling, R. H., Buchanan, G. O., Mincer, T. J., Kauffman, C. A., Jensen, P. R., Fenical, W. Salinosporamide A: A highly cytotoxic proteasome inhibitor from a novel microbial source, a marine bacterium of the new genus *Salinospora. Angew. Chem. Int. Ed.* 2003, 42 (3), 355–357.

243. Millward, M., Price, T., Townsend, A., Sweeney, C., Spencer, A., Sukumaran, S., Longenecker, A., Lee, L., Lay, A., Sharma, G., Gemmill, R. M., Drabkin, H. A., Lloyd, G. K., Neuteboom, S. T. C., McConkey, D. J., Palladino, M. A., Spear, M. A. Phase 1 clinical trial of the novel proteasome inhibitor marizomib with the histone deacetylase inhibitor vorinostat in patients with melanoma, pancreatic and lung cancer based on *in vitro* assessments of the combination. *Invest. New Drugs* 2012, 30 (6), 2303–2317.

244. Miller, C. P., Ban, K., Dujka, M. E., McConkey, D. J., Munsell, M., Palladino, M., Chandra, J. NPI-0052, a novel proteasome inhibitor, induces caspase-8 and ROS-dependent apoptosis alone and in combination with HDAC inhibitors in leukemia cells. *Blood* 2007, 110 (1), 267–277.
245. Chauhan, D., Catley, L., Li, G., Podar, K., Hideshima, T., Velankar, M., Mitsiades, C., Mitsiades, N., Yasui, H., Letai, A., Ovaa, H., Berkers, C., Nicholson, B., Chao, T.-H., Neuteboom, S. T. C., Richardson, P., Palladino, M. A., Anderson, K. C. A novel orally active proteasome inhibitor induces apoptosis in multiple myeloma cells with mechanisms distinct from bortezomib. *Cancer Cell* 2005, 8 (5), 407–419.
246. Martín, M. J., Coello, L., Fernández, R., Reyes, F., Rodríguez, A., Murcia, C., Garranzo, M., Mateo, C., Sánchez-Sancho, F., Bueno, S., de Eguilior, C., Francesch, A., Munt, S., Cuevas, C. Isolation and first total synthesis of PM050489 and PM060184, two new marine anticancer compounds. *J. Am. Chem. Soc.* 2013, 135 (27), 10164–10171.

PART **3**

Marine Products in Biomedicine

CHAPTER 15

The Travails of Clinical Trials with Marine-Sourced Compounds

David J. Newman and Gordon M. Cragg
Natural Products Branch, Developmental Therapeutics Program, Division of Cancer Treatment and Diagnosis, Frederick National Laboratory for Cancer Research, Frederick, Maryland

CONTENTS

15.1 INTRODUCTION

Though there are now a number of agents either in clinical use or in clinical trials for a variety of diseases, from antitumor agents to pain management, the route to marine (or aqueous)-sourced agents is in fact a long one, with very significant detours due to problems that were not at all obvious in the early days of studying materials isolated from marine sources. The success stories of antibiotics, which in the early days almost always came directly from microbial broths, and the centuries of use of plant-derived materials, where we should note that even today, 80+% of the world's population have only plants as a source of medication, led to the idea in the late 1960s to early 1970s that the sea might be as productive in terms of pharmacologically active materials as drugs or leads thereto. It is interesting to note at this point that the original isolation of the microbe that produced cephalosporins actually came from a "quasi-marine source" off Cagliari in Sardinia.[1]

Although a fair number of marine-derived agents have entered preclinical trials and, in a number of cases, clinical trials for a number of diseases of humans, at the original time of this writing, the second quarter of 2013 and updated in late September 2014 for clinical trials, only a very few agents derived from marine organisms have actually become drugs. Though a larger number have entered clinical trials, mainly at this moment, in cancer or cancer-related diseases (and the major reason for the emphasis on cancer is very simple), almost the only funder of marine-derived compound discovery for the past 30 years has been the U.S. National Cancer Institute (NCI), using a variety of funding methodologies and also providing materials from their own collections for use by academics. In addition to the large number in relative terms that entered the "cancer route," a small number of other agents were tested in areas such as pain control and other neurological areas.

The thrust of this chapter is not what might be expected from the authors' normal approach to marine compounds and their usage in drug discovery. What we will be doing, at the editor's request, is discussing compounds from the aspect of the "almost heroic measures" that had to be undertaken in the drug development process, coupled with, in most cases, why and where they failed, rather than simply stating that few marine-sourced materials have made it across what, in the drug developer's parlance, is known as the "valley of death." This is a somewhat flexible term, but we are using it to show the development space that covers the time from the initial discovery that a compound might have interesting *in vitro* and early *in vivo* activity to the time that the compound, or its chemical progeny, might enter advanced preclinical trials, such as investigational new drug application (INDA)-directed toxicology and the areas beyond that that lead to the filing of the INDA.

In Figure 15.1, we show an idealized representation of the route from discovery through development to clinical trials and approval as a drug for a particular disease entity. We will cover some aspects of discovery as the process leads to preclinical candidates, but we will use what might be considered a modified case study approach to show where compounds have come to a stop, and then comment on the reasons why, in cases where we can find published data, or draw reasonable conclusions from publications in the public domain.

15.2 FUNDAMENTAL PROBLEM IN MARINE NATURAL PRODUCT AND DRUG DEVELOPMENT

For many years, the major problem could be described in a very few words by modifying the phrase that was immortalized in later years when discussing real estate, that the important aspects were "location, location, location." Here, we would simply change this to "supply, supply, supply." In the early days of marine natural product chemistry (MNPC), when the major participants were predominantly chemists who liked to dive, usually in warm waters in the Caribbean Sea and Pacific tropical areas, most samples were obtained from relatively close-in waters to a shoreline and at 10 to 30 m depths. The source invertebrates were organisms that were easily collectable and ranged from large colonies to those organisms that were not in great abundance.

Very interesting chemical compounds with pharmacologic activities were discovered from these earlier collections, including the arabinosides that ultimately led to both antiviral and antitumor nucleosides that entered and, in some cases, still are in clinical use.[2–4] A lot of the initial work was funded by the NCI in the mid-1960s, both as direct collections and in the form of R01 investigator-initiated grants. To these must be added the very significant early work performed by the Scheuer group and its successors in Hawaii, which was funded almost exclusively by the National Science Foundation (NSF). The multiplicity of structures found by these efforts can be seen by inspection,

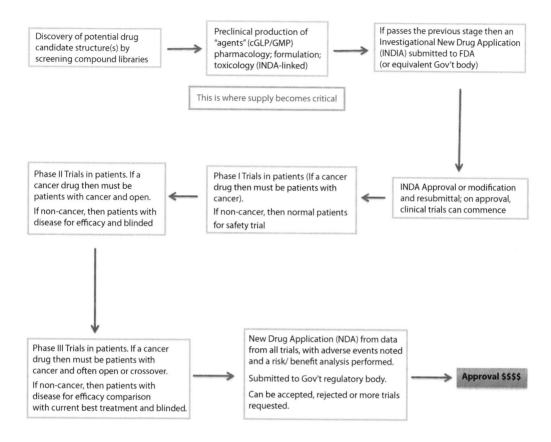

Figure 15.1 Idealized scheme for drug discovery and development.

though written after the fact, of the review articles in *Natural Product Reports* by Faulkner on marine natural products that commenced in 1986,[5] continued for 12 years until his untimely death, and then recommenced with the New Zealand group in 2003,[6] with a new review every year covering the compounds reported in the literature the preceding year. This series of reviews comprises the most highly cited articles in that journal, showing the interest in the field.

The different emphases of these two funding agencies led to different but complementary approaches, in that the NCI-funded work was designed to look for bioactivity-driven discoveries (with NCI providing most of the assays) and the NSF-funded work was designed for basic marine chemistry discovery, with bioactivity being a secondary process, at least in the 1960–1980 time frame.

The first three major potential antitumor agent compound classes discovered from the early NCI programs are excellent examples of the supply problem as seen in the 1970–1980 time frame. The compounds in question were isolated from the encrusting ascidian *Trididemnum solidum* (the didemnins), the nudibranch *Dolabella auricularia* (the dolastatins), and the encrusting bryozoan *Bugula neritina* (the bryostatins). In all cases, there were a multiplicity of compounds isolated and subsequently identified, but in the case of the first two, enough of the organism could not be collected, and in the third case, herculean efforts were required (details later under the compound).

In the succeeding sections we will deal first with materials that have been successful in going through the system to become approved drugs or used in comparable areas, for example, in agriculture, showing the problems that had to be overcome (where we know them). Then we will follow with materials that are currently in clinical trials or that failed at some stage in their clinical trials, and we will give an idea of why if the reason is in the public domain. Due to our familiarity with the field and the lack of comparable data in most other disease areas, the majority of examples will be drawn from compounds directed against cancer.

15.3 COMPOUNDS THAT ARE NOW APPROVED AS DRUGS OR COMPARABLE AGENTS IN OTHER FIELDS

15.3.1 Nereistoxin (NTX)

The marine annelid worm *Lumbriconereis heteropoda* Marenz and another similar animal, *Lumbrineris brevicerria*, both produce the disulfide compound known as nereistoxin (NTX) (**1**) (Figure 15.2).[7] This material, centuries before, was used by African tribes in the Lake Victoria area (in the form of the worm) as a fish poison or bait. Subsequently, it became the base molecule for the insecticide Cartap™ (**2**) (Figure 15.2), which is effectively a prodrug of NTX with the metabolism reported in 1971.[8] Although what could be considered traditional usage led to this series of commercialized agents, no benefits flowed back to the Lake Victoria tribesmen, as the original discovery occurred decades before the Convention on Biodiversity (CBD).

15.3.2 Ziconotide

This compound (**3**) (Figure 15.2), a peptidic cone snail toxin, has the distinction of being the first "direct from the sea" agent to be approved for any disease. In late December 2004, the Food and Drug Administration (FDA) approved this peptidic compound for treatment of intractable neuropathic pain, with phantom limb pain being a particular area for this drug; but since it has to be delivered via an intrathecal syringe from a reservoir in the peritoneum of the patient, the number of patients willing to tolerate the intricate delivery system is low, but even today, there are studies in the literature demonstrating its value in very specific situations.[9]

The compound is the exact equivalent of the 25-residue peptide isolated from the venom of the cone snail *Conus magus* under the name of ω-conopeptide MVIIA.[10] A small biotech company was set up by Olivera that was subsequently sold to Neurex in California, and between them, they made more than 200 variations on the structure, but finally came to realize that the native peptide was the most effective. The Irish company Elan purchased the compound and rights, and it was approved as stated earlier. Two companies currently are involved: Eisai, which purchased the European rights to Elan's ziconotide, and Azur Pharma, which purchased the rest of the world rights in 2010. Jazz Pharmaceuticals merged with Azur Pharma in late 2011 and is based in Ireland.

15.3.3 Trabectedin, Yondelis®, or ET-743

The tetrahydroisoquinoline alkaloid trabectedin (ET-743) (**4**) (Figure 15.2) was the first compound directly from the sea to be approved for the treatment of cancer. The initial work on this compound class, though not yet identified, was in 1969, when ethanolic extracts of the Caribbean tunicate *Ecteinascidia turbinata* were first reported by Sigel and colleagues to have antiproliferative properties.[11] It took approximately 17 years for the first report of the isolation and structural characterization of ecteinascidin derivatives to be reported in a 1986 PhD thesis by Holt[12a] in Rinehart's group at the Univesity of Illinois, Champaign-Urbana.[12b] This work was followed by two papers published back-to-back on what became known as trabectedin, by groups at the University of Illinois[12] and Harbor Branch Oceanographic Institute.[13]

The compound was later licensed to the Spanish pharmaceutical company PharmaMar by Harbor Branch, where its isolation was optimized using a variety of techniques—initially by large-scale

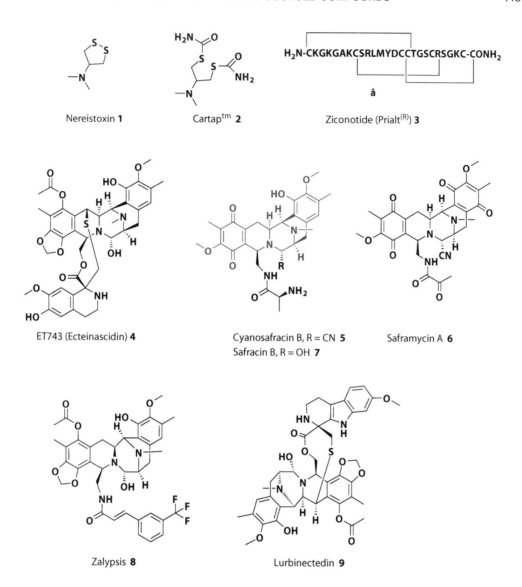

Figure 15.2 Compounds **1** to **9**.

collections from marine environments and then from large-scale in-sea and on-land aquaculture of the source tunicate. Enough material for the initial and early clinical trials came from several years of aquaculture and isolation.[14] Realizing that this technology could not produce enough material if development was successful, synthetic efforts were also made to obtain more trabectedin. The first total synthesis was reported by Corey's research group, inspired by the probable biosynthetic route to trabectedin.[15] Several other syntheses of trabectedin were published but ultimately did not meet the requirements for the commercial production of trabectedin. However, by working in conjunction with a companion fermentation group under the same corporate structure (Zeltia SA), PharmaMar developed a short, semisynthetic synthesis capable of producing multigram quantities of trabectedin[16] by starting from the antibiotic cyanosafracin B (**5**) (Figure 15.2),[17] a fermentation product of a marine-derived *Pseudomonas fluorescens*.[18]

Trabectedin was approved by the European Medicines Agency (EMEA) in September 2007 for the treatment of sarcoma and in 2009 for the treatment of ovarian cancer. More detailed information about the discovery and mechanism of action of trabectedin can be found in the recent reviews by Patel[19] and D'Incalci and Galmarini.[20] Following the launches in the EU, in 2009, PharmaMar launched this drug in the Philippines for the treatment of ovarian cancer, and in 2011, it was approved in Japan for the treatment of malignant soft tissue tumors accompanied with chromosomal transloca-tion. In the United States, an application was filed for the approval of the use of trabectedin in com-bination with docetaxel (a semisynthetic analog of paclitaxel) for the treatment of recurrent ovarian cancer. However, the FDA recommended that an additional phase III trial be conducted, which led to the voluntary withdrawal of the new drug application by Johnson & Johnson. We should point out, however, that as is usual with any approved antitumor drug, trabectedin is still in many phase I–III clinical trials in many areas of the world for a variety of cancers. Thus, using the International Clinical Trials Registry Platform (ICTRP) (http://apps.who.int/trialsearch/default.aspx), as of the end of September 2014, there were 13 clinical trials covering 25 different sites, as some trials are in multiple countries, with most at phase II levels of this drug with different combinations of other agents against a variety of tumor types, starting from May 2010 as the approval date.

From the perspective of actual source, trabectedin is structurally related to the saframycin (**6**) (Figure 15.2) and safracin (**7**) (Figure 15.2) classes of known antibiotics, which share similar struc-tural tetrahydroisoquinoline frameworks. Thus, trabectedin and other compounds were thought to probably have microbial components involved in their production. In 2011, Rath and coworkers demonstrated via metagenomic sequencing that a microbial consortium derived from *Ecteinascidia turbinata*[21] was the probable producer of trabectedin. By using the known gene clusters of the safra-mycin[22] and safracin[23] metabolites as markers, the "contig" encoding the nonribosomal peptide synthetase (NRPS) biosynthetic enzymes involved in trabectedin production was identified, as well as the probable producing organism, a γ-proteobacterium named *Candidatus Endoecteinascidia frumentensis* (AY054370).

These results confirmed previous reports speculating that bacterial candidates from Mediterranean and Caribbean *E. turbinata* isolates[24,25] were involved in trabectedin production, as *C. E. frumentensis* was found in *E. turbinata* from both geographic locations. Thus, identification of these clusters may lead to expression in heterologous hosts and the possibility of producing novel compounds by metabolic engineering or a combination of fermentation and chemical modification. It is also tempting to speculate about as yet unrecognized cryptic clusters in the genome of the potential producing organism.

15.3.4 Derivatives of the Base ET-743 Structure

Although these compounds could be discussed under a later section, for continuity, we consider that they are best commented on immediately following their parent.

PharmaMar has developed the closely related compound zalypsis or PM00104 (**8**) (Figure 15.2), which bears structural similarity to both ET-743 and the jorumycins, compounds originally isolated from a nudibranch[26] and subsequently modified with a trifluoromethyl-phenyl substituent on the southern edge. Almost all publications on this drug candidate are either abstracts at meetings or reports of phase I clinical trials; but in 2009, NCI scientists demonstrated that its mechanism of action differed in its interaction with DNA when compared to ET-743, and established that gamma-H2AX could serve as a pharmacodynamics marker for trials of this compound.[27] The agent recently completed a phase II trial against Ewing's sarcoma (NCT01222767), but no details have been pub-lished yet. This compound was also in other European trials not currently listed in the NCT data-base, but listed in the ICTRP portal. The first (EudraCT 2010-020994-18) was a phase II trial for the treatment of cervical carcinoma (completed but no published details yet), and the second is a phase II trial against refractory melanoma, which is still ongoing (EudraCT 2009-016054-40). Since these data became available, a phase II trial in Italy for Ewing's sarcoma was completed (EUCTR2010-022221-15-IT), but no results have yet been published.

Another closely related drug candidate is lurbinectedin, PM01183 (**9**) (Figure 15.2), which has a tetrahydro-β-carboline moiety instead of the tetrahydroisoquinoline present on the northern edge. The structure of this compound was first reported by Leal et al. in 2010.[28] A later report in 2012 demonstrated that it too was a DNA-minor groove binder.[29] No details of the synthetic routes have been published other than in the patent literature. Currently, as of September 2014, there are six trials listed in the ICTRP, with the first commencing in February 2010 as a phase II study of pancreatic cancer in the UK (completed) and Spain, though no details have yet been published (EUCTR2010-0294292-30-GB/ES), with five others listed on the ICTRP and also on the NCT site. In September 2013, four were registered covering solid tumors (NCT01951157 [phase II], 01980667 [phase I], 01970553 [phase I], and 01970540 [phase I]), and one, the latest, in August 2014 (NCT02210364), was a phase I study against metastatic breast cancer, pancreatic cancer, and metastatic colorectal cancer. Two others are also listed only on the NCT site, NCT01314599, a phase I for acute leukemia from 2011, and NCT01525589, a phase II trial against BRCA 1 or 2 or unselected metastatic breast cancer. To date, late September 2014, no data have yet been reported. However, what needs to be emphasized from the listings given is that no one registry list is comprehensive; all have to be cross-checked at the same time.

15.3.5 Halichondrin B and Eribulin

In 1986, Hirata and Uemura reported the isolation of the structurally complex natural product halichondrin B (**10**) (Figure 15.3) along with several other halichondrin derivatives from a collection of 600 kg of the marine sponge *Halichondria okadai*. This sponge was reported to be associated with a number of symbionts, and thus the halichondrins may well be produced by a single-celled organism or a combination of different single-celled organisms,[30] and exhibited potent cytotoxic activity against the murine B-16 melanoma cell line (IC$_{50}$ [concentration that inhibits binding or activity by 50%] = 0.09 nM). Concomitantly, other groups had isolated the same series of compounds from different sponges in different areas, ranging from the Central Pacific to the Indian Ocean and waters off New Zealand, but the report from Hirata and Uemura was the first to fully

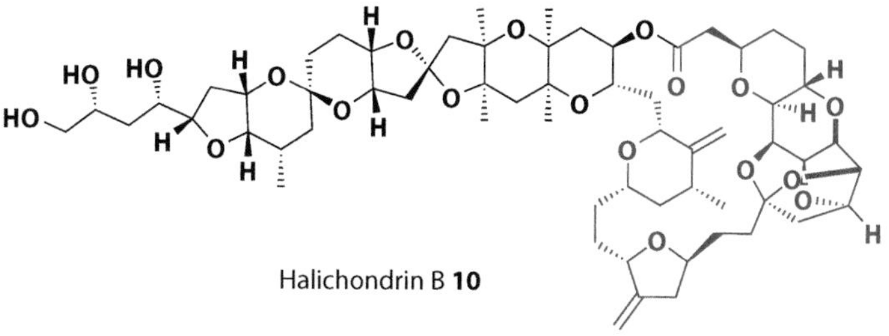

Halichondrin B **10**

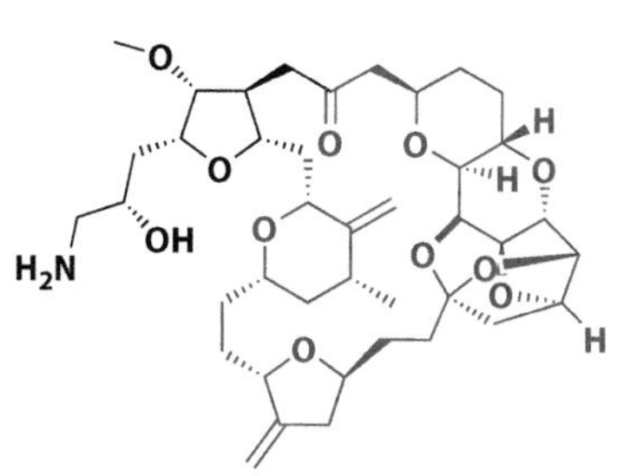

Ebribulin **11**

Figure 15.3 Halichondrin B and its relationship to eribulin.

identify the series. Its mechanism of action as a tubulin destabilizing agent using materials provided by the Pettit group was reported by an NCI group in 1991,[31] and the compound was approved for preclinical development by the NCI in early 1992, thus generating a major problem: How should the material be sourced?

Using the skills of the New Zealand group under Blunt and Munro at the University of Canterbury and other New Zealand-based processing chemists, plus full cooperation from the National Institute of Water and Atmospheric Research (NIWA) in New Zealand, one metric tonne of deep-water *Lissodendoryx* sp. was harvested, and after herculean efforts, 300 mg of >98% pure halichondrin B was produced for use by the NCI in its preclinical evaluations. In addition, successful in-sea aquaculture was performed that demonstrated that the compound could be produced from sponges grown at 10 m, rather than collected at 200+ m by dredging.[32]

The aquaculture of sponges for such production is still alive and well, particularly in academia,[33] and recent work has demonstrated that the microbial consortia in the sponge is maintained during such manipulations.[34]

Before the original Japanese report of the halichondrins, Yoshito Kishi at Harvard in Cambridge, Massachusetts, became aware of the exquisite chemical structures of the parent molecules and wished to demonstrate that the basic backbone could be synthesized by the then available (late 1980s) chemical techniques. As a result, he applied for an R01 investigator-initiated grant to investigate the synthesis of the halichondrins. The work was successful, and in 1992, Kishi and coworkers reported[35] that they had synthesized halichondrin B and norhalichondrin B. During the syntheses of the various components of the complex molecule, Kishi worked very closely with the then Eisai Research Institute that was close to Cambridge, Massachusetts. Their biologists tested the modular components of the synthesis as it progressed and were the first to recognize that the biological activity predominantly resided in the ring portion of the molecule and not, as was thought from the naturally occurring compounds, in the tail portion. As a result of their investigations, close to 200 derivatives of the truncated natural product were made and evaluated by Eisai in close collaboration with the Kishi group.

Following subsequent discussions between NCI and Eisai scientists at two American Association for Cancer Research (AACR) meetings in the middle 1990s, where parts of each other's stories were presented, Eisai and NCI collaborated in testing the best synthetic analogs against the pure New Zealand halichondrin B in a series of *in vitro* time course assays and *in vivo* studies in mice with human xenografts. The upshot of these head-to-head comparisons was that the truncated halichondrin B analog, now known as eribulin (**11**) (Figure 15.3) (with structural similarities to **10** shown in red in Figure 15.3), showed significantly more potent activity in the *in vivo* studies.

This compound was chosen for advanced preclinical and then clinical studies in humans, with the first clinical trial approved under NCI auspices in 2000, using materials made under current good manufacturing practice (cGMP) conditions by total synthesis. The compound, under the generic name eribulin (and proprietary name of Halaven®), was approved for refractory breast cancer by the FDA in 2010, following a very convoluted path through clinical trials. We should reemphasize that eribulin is by far the most complex drug ever produced by total synthesis and is a testament to all of the investigators in three countries and multiple organizations that cooperated to make it a success.

The chemists and biologists involved in the derivation at both Eisai (in the United States) and Harvard published a case study on Halaven® in 2011 in *the Annual Reports in Medicinal Chemistry*, which should be consulted for further information.[36] In addition, very recently, three papers covering the details of the process development of the compound were published by the Eisai chemical teams involved in the journal *Synlett*, which should be consulted to see the methodologies necessary to produce this very complex compound under cGMP conditions.[37–39] As of the end of September 2014, there were more than 80 clinical trials listed as "recruiting" since the approval of eribulin as a drug in 2010, from phase II through phase III in the ICTRP registry.

What is of note in this case is that this is probably the second example in the marine area where a total synthesis of the compound started in an academic laboratory and the clinical-grade material used the academic synthesis as its starting point. The other one is covered in Section 15.4.6, under discodermolide, where Paterson's base synthesis was used by Novartis as its route to the clinical candidate. Since we do not have accurate dates of process development in either case, the orders might be reversed if these are ever published.

15.4 SELECTED COMPOUNDS THAT ENTERED HUMAN CLINICAL TRIALS

As mentioned in the introduction, a number of compounds with a marine background have entered clinical trials in the past 20+ years. When a compound fails in a clinical trial, in probably the majority of cases in the cancer area, it ends up being too toxic or the risk–benefit analyses from trial data rule out further work, or in at least one case, a business decision by the company involved moved emphasis away from the agent.

Due to the costs involved in clinical trials, which are measured in the end in the hundreds of million dollars (included are the preclinical toxicology efforts, provision of cGMP drug candidates, etc.), decisions are made regularly whether to continue trials, and the decisions are frequently based on the apparent return on investment (ROI), rather than on scientific grounds. So when a compound is not continued with, the reasons given may or may not be the complete story; this needs to be borne in mind when reading the literature.

In this section, we will cover the following agents. Obviously, there are many others that we could choose from, but in these cases, the pharmacological reasons for not continuing were known, but not always formally reported in the cases of bryostatin, dolastatin 10, didemnin B, and discodermolide. In the case of hemiasterlin, Wyeth elected to discontinue for business reasons at the entry to phase II clinical trials.

Each of these will be discussed from the basis of the problems involved in obtaining enough of the compound and, in the case of dolastatin 10, how a successful derivative did finally become an approved antitumor drug, but as a warhead on an antibody. Since it did not become an agent as the isolated or modified compound, we are including it here rather than under the earlier heading of approved marine drugs.

15.4.1 Didemnin B

This cyclic peptide (**12**) (Figure 15.4) was, as mentioned earlier, the first marine-derived compound to go into human clinical trials as an antitumor agent. It was isolated from the encrusting ascidian *Trididemnum solidum* during a collection trip in the Caribbean by Rinehart from the

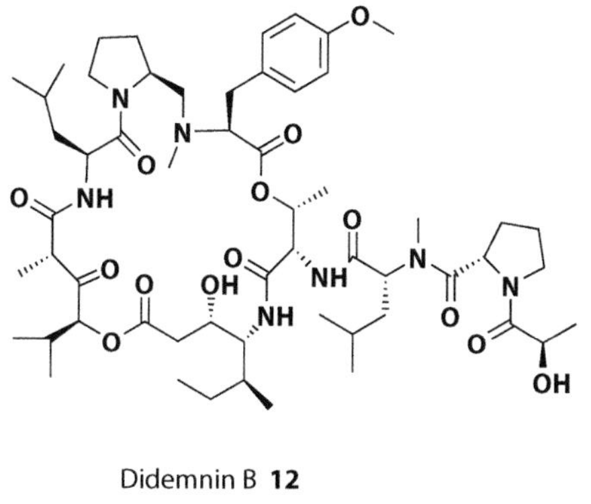

Didemnin B **12** 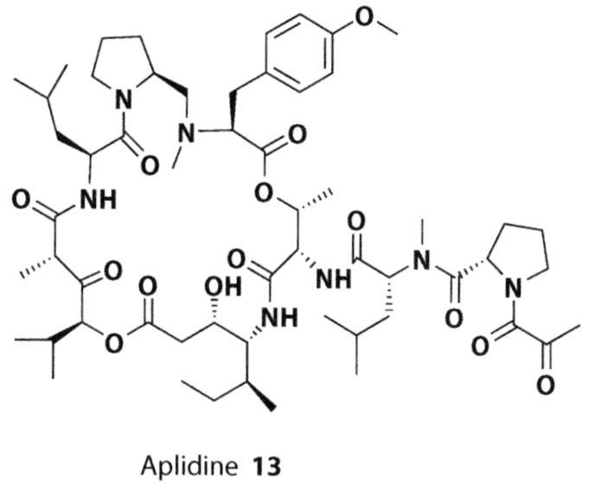   Aplidine **13**

Figure 15.4 Didemnin and aplidine.

University of Illinois.[40–42] Due to the complexity of the molecule and the very small amounts available of the producing organism, Rinehart had to synthesize the molecule and other analogs in order to obtain enough material for preclinical and then clinical trials.[43] The compound entered clinical trials under the auspices of the NCI[44] and subsequently failed at the phase II level due to toxicity. In addition, it also exhibited significant immunomodulating, predominantly suppressive properties. The toxicity observed was possibly due to the standard methodology used for dosing in the early 1980s, that is, a single bolus of the compound at the maximum tolerated dose. This possibility was alluded to by Joullie in her review in 2012.[45]

As will be seen, quite a common phenomenon with these early compounds was that although they failed as the initial agent, years later, a modification, sometimes quite simple and sometimes much more complex in nature (total synthesis of smaller or simpler molecules based on the initial structure), or a slight variation on the natural products that were isolated from other organisms sometimes leads to successful clinical results.

15.4.2 Aplidine (Dehydrodidemnin B)

In the case of didemnin B, in the original reports by the Rinehart group, a very close relative was isolated and identified, with details initially given in a patent application in 1991.[46] Subsequently, investigators at the Spanish company PharmaMar were able to clinically investigate this very closely related substance, Aplidine or didehydrodidemnin B (DDB), which only differed from the original by the conversion of a lactyl side chain to a pyruvyl side chain (13) (Figure 15.4). PharmaMar rediscovered this compound from the Mediterranean invertebrate *Aplidium albicans*, and as with its very close relative, it had to be synthesized, as the natural source was too sparse to be able to obtain the material in any quantity.

The details of the PharmaMar studies and their methodologies and other background information, including the synthesis of the molecule story around aplidine, which is currently in phase III clinical trials for multiple myeloma and in phase II trials for T-cell lymphomas, were given in the recent review by Cuevas et al.,[18] which should be consulted for more information on specific points. Currently (September 2014), there are 17 trials from phase I to phase III listed on the ICTRP website with this compound.

15.4.3 True Source of Didemnins (Perhaps also Aplidine)

As mentioned above, PharmaMar used chemical synthesis[47] to produce the aplidine used for clinical studies. For other synthetic routes to these agents or the closely related tamandarins A and B (14) (Figure 15.5), a truncated version of the base molecule, the reader should consult the excellent recent review by Joullie's group at the University of Pennsylvania.[45] It is possible, however, that bacterial fermentation may eventually be used to produce this compound, as other tunicate-derived didemnins have recently been determined to be produced by free-living and potentially symbiotic bacteria.[48]

In 2012, Xu and coworkers[49] reported the sequence of the producing cluster for didemnin B from the genome of the marine α-proteobacteria *Tistrella mobilis*, an organism similar to the Japanese organism reported by Tsukimoto et al.,[48] but isolated from the Red Sea.[49] By subsequent use of imaging mass spectrometry of the producing culture, real-time conversion of the (projected) didemnin X and Y precursors to didemnin B was observed.

With the identification of the didemnin B gene cluster, it may be feasible to use genetic engineering to create renewable supplies of aplidine via microbial fermentation. Furthermore, metagenomic analyses of *Aplidium albicans* using *Tistrella mobilis* gene clusters as markers may lead to the identification of the bacterial genes involved in aplidine biosynthesis in the tunicate, as it has not yet been proven that the same free-living microbe produces both aplidine and its reduced congener,

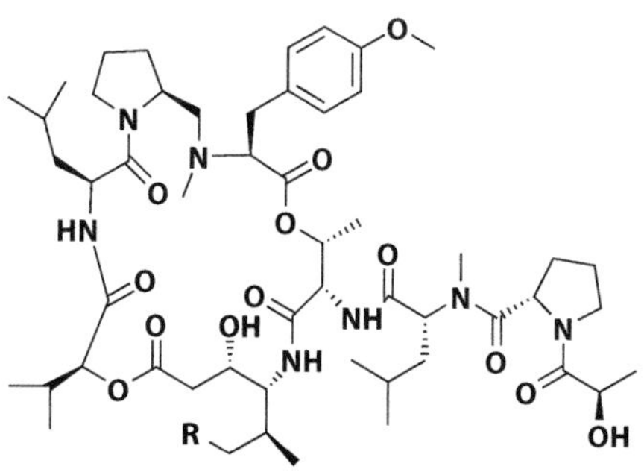

Tamandarins **14**
A, R = CH$_3$
B, R = H

Figure 15.5 Tamandarins A and B.

didemnin B. In addition, recently a group of investigators who have been instrumental in the discovery and biosynthetic processes related to tunicate-derived peptides published an excellent review demonstrating that multiple biosynthetic and genetic processes could be involved in the production of these metabolites, not the invertebrate itself.[50]

15.4.4 Bryostatins

Bryostatin 1 and its related compounds (**15–34**) (Figure 15.6) and dolastatin 10 (**35**) (Figure 15.7), which will be discussed in Section 15.4.5, are, with didemnin B (see Section 15.4.1), excellent exemplars of the heroic efforts that had to be used in order to isolate enough marine-derived material for initial clinical trials.

The initial discovery of bryostatin 3 (**17**) (Figure 15.6) was indirectly reported in 1970.[51] Subsequent developments leading to the report of the isolation and x-ray structure of bryostatin 1 (**15**) (Figure 15.6) in 1982,[52] and the multiyear program that culminated in the isolation and purification of 18 bryostatins, have been well documented by a variety of authors over the years.[53–60]

Structurally, all of the molecules possess a 20-membered macrolactone ring. Modifying the description by Hale et al.,[60,61] all of the known bryostatins possess a 20-membered macrolactone ring in which there are three remotely substituted pyran rings that are linked by a methylene bridge and an (*E*)-disubstituted alkene; all have geminal dimethyls at C$_8$ and C$_{18}$ and a four-carbon side chain (carbons 4–1) from the A ring to the lactone oxygen, with another four-carbon chain (C$_{24-27}$) on the other side of the lactone oxygen to the C ring. Most have an exocyclic methyl enoate in their B and C rings, though bryostatin 3, in particular, has a butenolide rather than the C-ring methyl enoate, and bryostatins 16 (**30**) (Figure 15.6) and 17 (**31**) (Figure 15.6) have glycals in place of the regular C$_{19}$ and C$_{20}$ hydroxyl moieties.

In the early reviews by Hale[60] and Mutter,[58] 18 structures (bryostatins 1–18, **15–32**) (Figure 15.6) were listed, with the remaining 2 structures presumed to be desoxy-bryostatin 4 and desoxy-bryostatin 5 isolated from *Lissodendoryx isodictyalis*.[62a] However, work subsequently reported from the People's Republic of China in 1998 in Chinese[62a] and in 2004 in English[62b] gave the structure for bryostatin 19 purified from a South China Sea collection of *B. neritina* (**33**) (Figure 15.6). Then in the same year, this report was followed by the publication by Lopanik et al. reporting the isolation of bryostatin 20 (**34**) (Figure 15.6) from an Atlantic-sourced *B. neritina*.[64] Comparison with the structures of the other 18 bryostatins shows that these are closely related to bryostatin 3 in terms of their basic ring components.

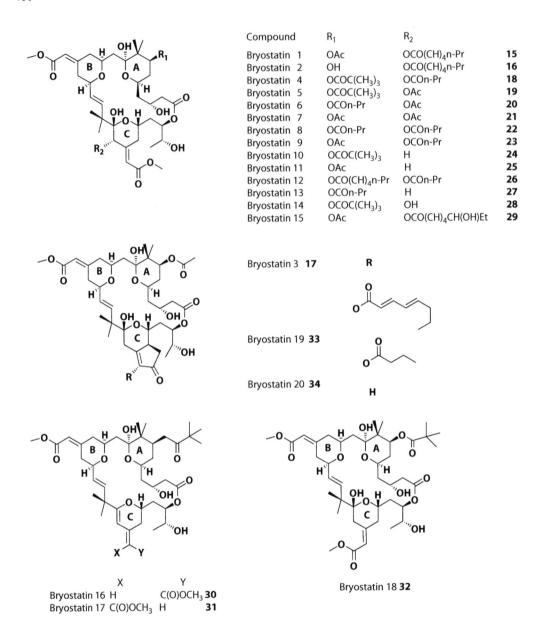

Compound	R_1	R_2	
Bryostatin 1	OAc	OCO(CH)$_4$n-Pr	15
Bryostatin 2	OH	OCO(CH)$_4$n-Pr	16
Bryostatin 4	OCOC(CH$_3$)$_3$	OCOn-Pr	18
Bryostatin 5	OCOC(CH$_3$)$_3$	OAc	19
Bryostatin 6	OCOn-Pr	OAc	20
Bryostatin 7	OAc	OAc	21
Bryostatin 8	OCOn-Pr	OCOn-Pr	22
Bryostatin 9	OAc	OCOn-Pr	23
Bryostatin 10	OCOC(CH$_3$)$_3$	H	24
Bryostatin 11	OAc	H	25
Bryostatin 12	OCO(CH)$_4$n-Pr	OCOn-Pr	26
Bryostatin 13	OCOn-Pr	H	27
Bryostatin 14	OCOC(CH$_3$)$_3$	OH	28
Bryostatin 15	OAc	OCO(CH)$_4$CH(OH)Et	29

Bryostatin 3 **17**

Bryostatin 19 **33**

Bryostatin 20 **34**

R

H

Bryostatin 18 **32**

	X	Y	
Bryostatin 16	H	C(O)OCH$_3$	**30**
Bryostatin 17	C(O)OCH$_3$	H	**31**

Figure 15.6 Bryostatins **1** to **20**.

In order to obtain enough bryostatin 1 (the compound of choice from the original three reported by Pettit), herculean efforts had to be made initially by the Pettit group, working in conjunction with collectors in the Gulf of Mexico in U.S. waters, in order to obtain enough material for the initial phase I clinical trials in the UK in the early 1990s.

All of the clinical work reported to date has been with material from wild collections, and except for the initial work by Pettit, the source waters have been either the Gulf of California (small collections) or the Pacific Ocean off Palos Verdes, California. The initial material used by NCI for further clinical trials came from a collection of close to 13 metric tonnes collected predominantly in the area around Palos Verdes in California and yielded 18 g of cGMP material.[57]

NCI used the Small Business Innovation Research (SBIR) mechanism to investigate, with the Massachusetts company Aphios, the potential for supercritical extraction of bryostatins from *B. neritina*, utilizing both wild collections and some of the material from aquaculture. This was successful, and the potential for such methods was shown in the review by Newman,[57] where a comparison was made of the methods used for the earlier cGMP purification versus the Aphios technique. The earlier method used for cGMP production at NCI–Frederick involved a four-stage extraction–concentration process, and then a six-step process repeated 15 times in order to produce the 18 g of bryostatin 1 from 13 metric tonnes of *B. neritina*, referred to above. That process was multiweek in duration and used massive amounts of solvents. In contrast, the supercritical technique used basically carbon dioxide as the extraction medium on the wet animal mass, followed by supercritical chromatography, and the process was reduced to six simple operations performed within days, yielding material of similar purity.

However, it became obvious, even as the initial trials were beginning in the early 1990s, that wild collections would not suffice and production via chemical synthesis would probably not be viable at that time. Thus, in the early 1990s, NCI, utilizing the SBIR program, established a phase I and a subsequent phase II aquaculture contract with a small company, CalBioMarine, with the aim of investigating on-land and in-sea aquaculture for production of *B. neritina* under conditions that were not affected by the vagaries of nature. This series of projects was successful, culminating in the proof by CalBioMarine that the organism could be grown under both conditions, and also that bryostatin could be isolated from the aquacultured animal in quantities and at costs less than those incurred in wild collections.[65]

Bryostatin has been through more than 80 clinical trials at the phase I and II levels, with or without the addition of a cytotoxic agent in the protocol. One trial using bryostatin and temsirolimus at the phase I level was shown as "actively recruiting" as of the beginning of May 2013 (NCT00112476) on the NCT website, but as of September 2014, the trial had closed, but no information had yet been published. There are three listed as "unknown status," including a potentially interesting use in Alzheimer's disease at the phase II level, but no information has been posted since 2008 (NCT00606164). However, very recently, a phase I study with a new company called Neurotrope was listed as recruiting among Alzheimer's patients for a phase I study under NCT02221947 from July 2014 under double-blinded conditions covering pharmacokinetics, safety, and efficacy. The major reason for not continuing with bryostatin aside from the trials mentioned against Alzheimer's disease appears to be very significant myalgia in patients. Whether this side effect can be ameliorated by alteration of dosing regimens is not clear.

Condensed versions of all published results from clinical trials of bryostatin 1 (the only compound of the 20 known used as a drug candidate) from the early 1990s through August 2010 are given in two review chapters by Newman, one published in 2005[66] and the other in 2012.[67] These should be consulted for information of the manifold variations used in scheduling regimens over the years.

15.4.4.1 Chemical Syntheses of Bryostatin and Truncated Structures

However, the lack of success of bryostatin as an antitumor drug has not stopped some excellent chemistry groups from synthesizing some of the bryostatins, or devising simplified molecules based on the basic skeleton that have much higher *in vitro* activities and can be produced by total synthesis. From the initial reports on the structures of the bryostatins, synthetic chemists recognized the base molecule as a challenge for their skills, with a significant number of published syntheses of parts of the bryostatin 1 molecule being presented or published, with an early example being a discussion of routes to bryostatin 1 by Masamune in 1988.[68] It should be pointed out, however, that with the exception of the Trost synthesis referred to later in Section 15.4.4.1, none were substitutes for the isolation and purification of bryostatin 1 from natural sources.

The first total synthesis was the enantiomeric synthesis of bryostatin 7 (**21**) (Figure 15.6) in 1990 by Masamune's group,[69] followed in 1998–1999 by details of an enantiomeric synthesis of bryostatin 2 (**16**) (Figure 15.6) from the Evans group.[70,71] The synthesis of bryostatin 3 (**19**) (Figure 15.6) was reported by Nishiyama and coworkers in 2000,[72] with an explanation of the strategies developed by Ohmori in 2004.[73] These earlier syntheses, together with the reported partial syntheses of other bryostatins, were reviewed in detail through 2002 by Hale et al.[60]

The problems involved in the syntheses of these complex molecules were exemplified by the discussion in the 2006 report by Hale's group on the formal total synthesis of bryostatin 7 (**21**) (Figure 15.6). This synthesis was based on the original Masamune route, but with variations that could permit syntheses of other novel analogs that would not have been accessible via the original Masamune synthetic process. They discovered that some of the original nuclear magnetic resonance (NMR) data in the Masamune synthesis appeared to be incorrect, and it took 3 years for this to be proven. However, in 2006 they published the full results, leading to a synthesis of bryostatin 7 that could be modified en route.[74]

Two years later, in 2008, Trost and Dong published their elegant synthesis of bryostatin 16 (**30**) (Figure 15.6) [75] involving some novel metal-linked catalysis steps[76] that included a ruthenium tandem alkyne–enone coupling, and then a palladium catalyzed alkyne—ynoate macrocyclization to give the cyclized precursor of bryostatin 16.

A truly excellent compendium and thorough discussion of the chemistry efforts around the synthesis of the bryostatins was published by Hale and Manaviazar covering the published results up through 2010. It should be read by any chemists interested in the manifold methods and specific methodological differences that can be, and have been, used in both successful and unsuccessful syntheses of these agents.[61] However, in spite of all of these methods, up to 2011, no *de novo* synthesis of bryostatin 1 had been published, though in the early days of studying the bryostatins, Pettit et al. demonstrated that bryostatin 2 could be converted to bryostatins 1 and 12 (**28**) (Figure 15.6).[54] Thus, one can argue that a formal synthesis of bryostatin 1 was achieved by using the bryostatin 2 asymmetric synthesis of Evans et al.[70,71] and then applying the Pettit conversion method. However, in 2011, the Keck group reported the first complete total synthesis of this agent.[77] This report was rapidly followed by a paper from Manaviazar and Hale with details of a shorter route[78] to the same compound. Also in the same year, Trost et al. published on a ring-expanded version obtained by total synthesis,[79,80] so the synthetic story has not yet finished.

A number of bryostatin analogs (often called bryologs) have been synthesized using methods such as function-oriented synthesis. This was employed by Wender and other workers to develop simplified analogs with comparable or improved activities.[81–84] Further information is given in the review by Newman[67] and in a recent paper by the Keck group.[85]

What is also of significant import is that the Wender group reported *in vitro* anti-HIV activity for some of their newer analogs in 2012. It will be interesting to see if this can be further developed with *in vivo* activity.[86]

15.4.4.2 True Source of the Bryostatins?

These metabolites, just as in the case of the ET-743-related agents, are now thought to be produced by the as yet uncultured microbe *Candidatus Endobugula sertula*, but definitive proof is lacking. The most promising piece of direct evidence was the reduction of the amount of bryostatin 1 in *B. neritina* colonies obtained from antibiotic-treated larvae.[87,88] Later work by Haygood and coworkers led to the isolation and cloning of the gene cluster involved in the biosynthesis of the base ring structure of bryostatins, but the large, acyl-transferase (AT) polyketide synthases (PKSs) have deterred their heterologous expression.[89] A relatively up-to-date account on the bryostatin source story and its potential was published in 2010 by the Haygood group and should be consulted for further details.[90]

15.4.5 Dolastatins and Analogs

The dolastatins are a series of cytotoxic peptides that were originally isolated in very low yield from the Indian Ocean mollusk *Dolabella auricularia* by Pettit's group as part of its work on marine invertebrates with the NCI. The early aspects of the work and the very significant amount of synthetic methodology that had to be performed to obtain any significant quantities of these linear and cyclic peptides, due to the inability to use the mollusk as the source, were covered in detail by Flahive and Srirangam in a 2005 book chapter,[91] with details of the cGMP production being published by Pettit in 1997.[92] Due to the potency of the linear depsipeptide dolastatin 10 (**35**) (Figure 15.7) and its mechanism of action as a tubulin interactive agent binding close to the vinca domain at a site where other peptidic agents bound,[93–95] the compound entered phase I clinical trials in the 1990s under the auspices of the NCI.

Dolastatin 10 progressed through to phase II trials as a single agent, and although tolerated at the doses used, which were high enough to give the expected levels *in vivo* to inhibit cell growth, it did not demonstrate significant antitumor activity in a phase II trial against prostate cancer in man.[96] Similarly, no significant activity was seen in a phase II trial against metastatic melanoma, even though again, bloodstream levels high enough to affect cells were demonstrated.[97]

Subsequently, dolastatins and related compounds were isolated from a free-living cyanophyte (*Symploca* sp.).[98] Thus, in contrast to the bryostatins, the actual producing organism could be isolated and identified. Although dolastatins of varying structures were synthesized, with some going into clinical trials, for example, auristatin PE (**36**) (Figure 15.7), which completed phase II clinical trials for the treatment of non-small-cell lung cancer (NCT00061854) and metastatic soft tissue sarcoma (NCT00064220), no recent work has been reported on this agent as a single entity in spite of being tested by a number of companies as ownership changed.

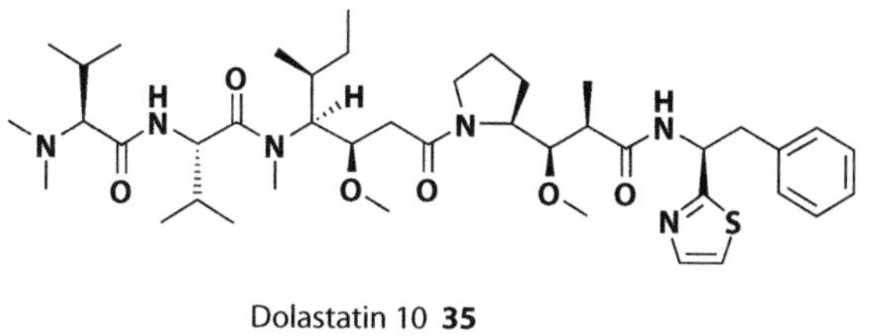

Dolastatin 10 **35**

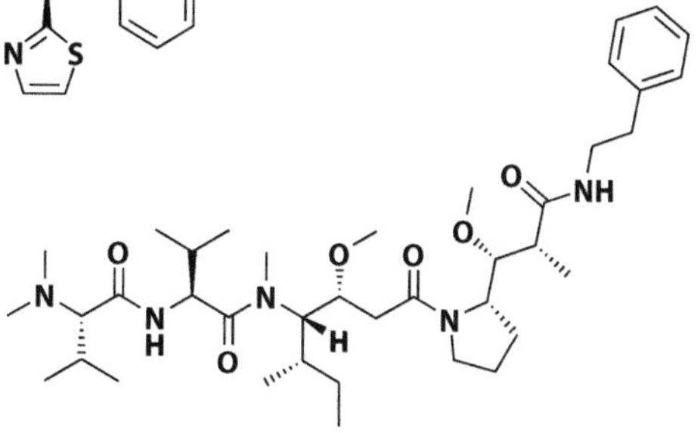

Auristatin PE **36**

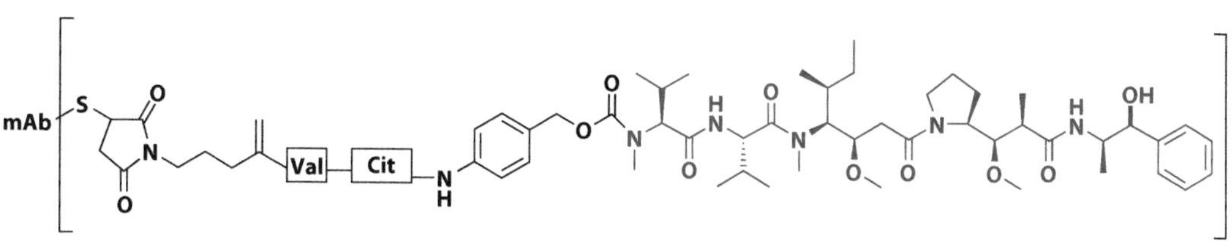

Brentuximab vedotin; Red color is momomethylauristatin E **37**

Figure 15.7 Dolastatin 10 and its derivatives.

However, just to demonstrate that old compounds can have a new lease on life, in 2011, the FDA approved an anti-CD30 antibody-conjugated monomethyl auristatin E (brentuximab vedotin or SGN-35, **37**) (Figure 15.7) for the treatment of various lymphomas. Some details of the development of brentuximab vedotin can be found in a recent perspective by Younes and colleagues,[99] which should be read in conjunction with a recent review by the inventors[100] and a 2013 clinical review.[101] This antibody-directed combination is the "tip of the iceberg" as far as derivatives of auristatin are concerned, as there are more than 10 clinical trials listed on variations of antibody–linker auristatin (or derivative) as the warhead listed.

Thus, the concept of linking auristatin and derivatives to monoclonal antibodies targeted at specific epitopes will lead to an increase in the number of agents and derivatives in cancer therapy, even though the original lead compound and number of variations failed to proceed beyond phase II trials for either toxicity or lack of formal efficacy in human patients. As of September 2014, there were 133 records covering 69 trials (multiple records under a trial usually in different EU countries) just on a search using vedotin (monomethylauristatin E) that covers the initial approved molecule. There are many others in phase I–II trials with this warhead or its close chemical cousin, mono-methylauristatin F, where the C-terminal amino acid has changed, but with different monoclonal targeting agents shown in the Thomson Reuters Integrity™ database.

15.4.6 Discodermolide

This polyhydroxylated lactone (**38**) (Figure 15.8) was first reported by the Harbor Branch group in 1990 following isolation from the Caribbean sponge *Discodermia dissoluta*, originally collected at a depth of 33 m off the Bahamas. In the early reports, the compound was considered to be a new immunosuppressive and an incidental cytotoxin, but in 1996, it was reported that discodermolide bound to microtubules more potently than Taxol™, a discovery that confirmed *in silico* studies at the University of Pittsburgh.[102] Concomitantly with these reports, a variety of chemical synthetic groups had seen discodermolide as a good candidate for total synthesis, with Paterson and Florence publishing an excellent review in 2003 of the various synthetic schemas in use.[103] The compound had been licensed by Harbor Branch to Ciba-Geigy prior to its incorporation into Novartis, and then

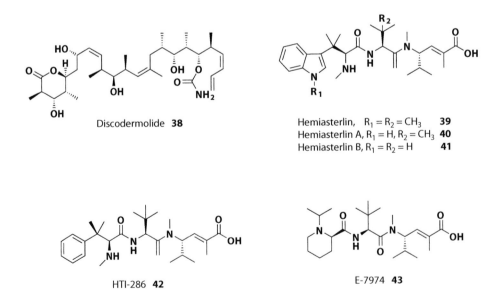

Discodermolide **38**

Hemiasterlin, $R_1 = R_2 = CH_3$ **39**
Hemiasterlin A, $R_1 = H, R_2 = CH_3$ **40**
Hemiasterlin B, $R_1 = R_2 = H$ **41**

HTI-286 **42**

E-7974 **43**

Figure 15.8 Discodermolide, hemiasterlin, and derivatives.

the Novartis development group, using a variation on the Paterson techniques, published its cGMP process synthetic scheme.[104]

Novartis placed discodemolide into phase I clinical trials as a potential treatment against solid tumors but withdrew the compound due to toxicity in 2004. However, in the next seven years, there were publications from a number of groups in academia still interested in this molecule and other variations. New synthetic methods for discodermolide were reported by de Lemos et al.,[105,106] as well as those for analogs.[107,108] In addition, the formations of hybrid combinations of discodermolide and dictyostatin were reported in the same time frame.[109] Recently, an updated review of discodermolide was published by Gunasekera and Wright, two of the original discoverers; this should be consulted for more detail as to similar molecules isolated under similar conditions and the variations in structure made by synthesis.[110]

15.4.7 HTI-286 (Hemiasterlin Derivative, Taltobulin)

The base molecule for this compound class, hemiasterlin (**39**) (Figure 15.8), was originally isolated by Kashman's group[111] from the South African sponge *Hemiasterella minor*, an organism that also contained jaspamide and geodiamolide TA. The following year, Andersen's group at the University of British Columbia (UBC) published the structures of a number of peptides, including geodiamolides A–F, the hemiasterlin as described by Kashman, plus two novel hemiasterlins, A (**40**) (Figure 15.8) and B (**41**) (Figure 15.8), and other geodiamolides and criamides isolated from a Papua New Guinea sponge now reclassified as a *Cymbastela* sp.[112]

In 1997, following testing of all the hemiasterlins, it was discovered that these agents interacted with tubulin to produce microtubule depolymerization in a manner similar to that reported for nocodazole and vinblastine.[113] Further investigations by Hamel's group at NCI, using a sample of hemiasterlin isolated at NCI, confirmed the effects on tubulin assembly and demonstrated that the base molecule probably bound at what is being called the peptide binding site.[114]

Prior to the formal publication of the mechanism of action, in order to determine structural–activity relationship (SAR) requirements, Andersen at UBC began a synthetic program to produce the original hemiasterlin, using a scheme that would permit variations on the overall structure.[115] The hemiasterlins, including the analogs made by Andersen's group, which included HTI-286 (**42**)[116] (Figure 15.8), were licensed by UBC to Wyeth (now Pfizer), and following significant amounts of synthetic work by Wyeth, they reported that the original agent was still superior.[117] HTI-286 entered phase I clinical trials, and a full paper giving details of the *in vitro* and *in vivo* animal data was published by Loganzo et al. in 2003.[118] Although the compound was scheduled for phase II trials and may even have begun recruiting patients, Wyeth made a conscious business decision not to proceed further with the agent in the 2004–2005 time frame. Fuller details of the work involved are given in a recent review from Andersen's group, which should be consulted for information on this particular compound as of late 2011.[119]

However, other groups have proceeded with variations on the hemiasterlins, with Eisai's E7974 (**43**) (Figure 15.8) entering phase I trials with a recent report demonstrating efficacy in this setting.[120] Three trials are listed at the phase I level in the NCT database, but no others have been added since 2005. Quite recently, a multinational group has reported novel agents based on the hemiasterlin tripeptide structure that show *in vitro* and *in vivo* activities.[121,122] Thus, close to 20 years after the original reports of this "simple" tripeptide, work is still ongoing around the basic structure.

15.5 CONCLUSIONS

As stated in the introduction, we have attempted to show how compounds from the marine environment have led to drugs directly, but mainly after long gestation periods and the development of complex sourcing strategies that involve the talents of a multitude of scientists in areas ranging from marine biology through aquaculture to semi- and total synthesis.

We have also attempted to show how failure of a structure with initial promise has led to the development of analogs that, using different techniques, led to a successful drug where the original structure led to a warhead, a methodology not even on the horizon when the original discovery was made. Obviously, we could have included a large number of compounds that were in the running to become clinical candidates, but we did not intend that this chapter would become a litany of structures with cytotoxic activity.

What is important, and has been emphasized to a varying extent in each section, is that, as now widely recognized, the marine invertebrate from which the initial agent was found and from which development was begun was almost in all cases not the actual producer. However, at the time that these compounds entered preclinical and clinical trials, the technology did not exist to be able to perform genomic analyses either rapidly or cheaply. It should be realized that it has been just over a decade since the first complete genomic sequence of any streptomycete was published, when Ōmura's group in 2001 reported the sequence of *Streptomyces avermitilis* and identified the potential of what are now known as cryptic clusters.[123] Thus, the basic information and subsequent technologies that enabled the actual organism that produced didemnin B to be identified[49] were not invented until well after the initial clinical trials of the compound had finished.

REFERENCES

1. Brotzu, G. Ricerche su di un nuovo antibiotico. *Lav. Ist. Ig. Cagliari* 1948, 1–11.
2. Bergmann, W., Burke, D. C. Marine products. XXXIX. The nucleosides of sponges. III. Spongothymidine and spongouridine. *J. Org. Chem.* 1955, 20, 1501–1507.
3. Bergmann, W., Feeney, R. J. Isolation of a new thymine pentoside from sponges. *J. Am. Chem. Soc.* 1950, 72, 2809–2810.
4. Bergmann, W., Feeney, R. J. Marine products. XXXII. The nucelosides of sponges. I. *J. Org. Chem.* 1951, 16, 981–987.
5. Faulkner, D. J. Marine natural products. *Nat. Prod. Rep.* 1986, 3 (1), 1–33.
6. Blunt, J. W., Copp, B. R., Munro, M. H. G., Northcote, P. T., Prinsep, M. R. Marine natural products. *Nat. Prod. Rep.* 2003, 20 (1), 1–48.
7. Yu, X., Lane, W. V., Loring, R. H. Nereistoxin: A naturally occurring toxin with redox effects on neuronal nicotinic acetylcholine receptors in chick retina. *J. Pharm. Exp. Ther.* 1993, 264 (2), 689–694.
8. Sakai, M., Sato, Y. Metabolic conversion of the nereistoxin-related cornpounds into nereistoxin as a factor of their insecticidal action. In *2nd International Pesticides and Chemical Conference*, Tel A'viv, 1971, pp. 455–467.
9. Alicino, I., Giglio, M., Manca, F., Bruno, F., Puntillo, F. Intrathecal combination of ziconotide and morphine for refractory cancer pain: A rapidly acting and effective choice. *Pain* 2012, 153 (1), 245–249.
10. Olivera, B. M., Cruz, L. J., de Santos, V., LeCheminant, G. W., Griffin, D., Zeikus, R., McIntosh, J. M., Galyean, R., Varga, J., Gray, W. R., Rivier, J. Neuronal calcium channel subtypes using omega-conotoxin from *Conus magus* venom. *Biochemistry* 1987, 26, 2086–2090.
11. Sigel, M. M., Wellham, L. L., Lichter, W., Dudeck, L. E., Gargus, J. L., Lucas, L. H. *Food-Drugs from the Sea: Proceedings 1969*. Marine Technology Society, Washington, DC, 1970.
12. (a) Rinehart, K., Ryuichi, S., Holt, T. G. Compositions comprising ecteinascidins and a method of treating herpes simplex virus infections therewith. US Patent 5256663A, 1986. (b) Rinehart, K., Holt, T. G., Fregeau, N. L., Stroh, J. G., Kiefer, P. A., Sun, F., Li, L. H., Martin, D. G. Ecteinascidins 729, 743, 745, 759A, 759B and 770: Potent antitumor agents from the Caribbean tunicate *Ecteinascidia turbinata*. *J. Org. Chem.* 1990, 55, 4512–4515.
13. Wright, A. E., Forleo, D. A., Gunawardana, G. P., Gunasekera, S. P., Koehn, F. E., McConnell, O. J. Antitumor tetrahydroisoquinoline alkaloids from the colonial ascidian *Ecteinascidia turbinata*. *J. Org. Chem.* 1990, 55, 4508–4512.

14. Cuevas, C., Francesch, A. Development of Yondelis® (trabectedin, ET-743). A semisynthetic process solves the supply problem. *Nat. Prod. Rep.* 2009, 26 (3), 322–337.

15. Corey, E. J., Gin, D. Y., Kania, R. S. Enantioselective total synthesis of ecteinascidin 743. *J. Am. Chem. Soc.* 1996, 118 (38), 9202–9203.

16. Cuevas, C., Pérez, M., Martín, M. J., Chicharro, J. L., Fernández-Rivas, C., Flores, M., Francesch, A., Gallego, P., Zarzuelo, M., de la Calle, F., García, J., Polanco, C., Rodríguez, I., Manzanares, I. Synthesis of ecteinascidin ET-743 and phthalascidin Pt-650 from cyanosafracin B. *Org. Lett.* 2000, 2 (16), 2545–2548.

17. Ikeda, Y., Idemoto, H., Hirayama, F., Yamamoto, K., Iwao, K., Asao, T., Munakata, T. Safracins, new antitumor antibitics. I. Producing organism, fermentation and isolation. *J. Antibiot.* 1983, 36 (10), 1279–1283.

18. Cuevas, C., Francesch, A., Galmarini, C. M., Aviles, P., Munt, S. Ecteinascidin-743 (Yondelis®), Aplidin®, and Irvalec®. In *Anticancer Agents from Natural Products*, ed. G. M. Cragg, D. G. I. Kingston, D. J. Newman. 2nd ed. Taylor & Francis, Boca Raton, FL, 2012, pp. 291–316.

19. Patel, R. M. Trabectedin: A novel molecular therapeitic in cancer. *Int. J. Curr. Pharm. Res.* 2011, 3 (3), 65–70.

20. D'Incalci, M., Galmarini, C. M. A review of trabectedin (ET-743): A unique mechanism of action. *Mol. Cancer Ther.* 2010, 9 (8), 2157–2163.

21. Rath, C. M., Janto, B., Earl, J., Ahmed, A., Hu, F. Z., Hiller, L., Dahlgren, M., Kreft, R., Yu, F., Wolff, J. J., Kweon, H. K., Christiansen, M. A., Håkansson, K., Williams, R. M., Ehrlich, G. D., Sherman, D. H. Meta-omic characterization of the marine invertebrate microbial consortium that produces the chemotherapeutic natural product ET-743. *ACS Chem. Biol.* 2011, 6 (11), 1244–1256.

22. Li, L., Deng, W., Song, J., Ding, W., Zhao, Q.-F., Peng, C., Song, W.-W., Tang, G.-L., Liu, W. Characterization of the saframycin A gene cluster from *Streptomyces lavendulae* NRRL 11002 revealing a nonribosomal peptide synthetase system for assembling the unusual tetrapeptidyl skeleton in an iterative manner. *J. Bacteriol.* 2008, 190 (1), 251–263.

23. Velasco, A., Acebo, P., Gomez, A., Schleissner, C., Rodriguez, P., Aparicio, T., Conde, S., Munoz, R., de la Calle, F., Garcia, J. L., Sanchez-Puelles, J. M. Molecular characterization of the safracin biosynthetic pathway from *Pseudomonas fluorescens* A2-2: Designing new cytoxic compounds. *Mol. Microbiol.* 2005, 56 (1), 144–154.

24. Moss, C., Green, D. H., Perez, B., Velasco, A., Henriquez, R., McKenzie, J. D. Intracellular bacteria associated with the ascidian *Ecteinascidia turbinata*: Phylogenic and *in situ* hybridization analysis. *Mar. Biol.* 2003, 143, 99–110.

25. Perez-Matos, A. E., Rosado, W., Govind, N. S. Bacterial diversity associated with the Carbbean tunicate *Ecteinascidia turbinata*. *Anton. van Leeuwen.* 2007, 92, 155–164.

26. Fontana, A., Cavaliere, P., Wahidulla, S., Naik, C. G., Cimino, G. A new antitumor isoquinoline alkaloid from the marine nudibranch *Jorunna funebris*. *Tetrahedron* 2000, 56 (37), 7305–7308.

27. Guirouilh-Barbat, J., Antony, S., Pommier, Y. Zalypsis (PM00104) is a potent inducer of gamma-H2AX foci and reveals the importance of the C ring of trabectedin for transcription-coupled repair inhibition. *Mol. Cancer Ther.* 2009, 8 (7), 2007–2014.

28. Leal, J. F., Martínez-Díez, M., García-Hernández, V., Moneo, V., Domingo, A., Bueren-Calabuig, J. A., Negr, I. A., Gago, F., Guillén-Navarro, M. J., Avilés, P., Cuevas, C., García-Fernández, L. F., Galmarini, C. M. PM01183, a new DNA minor groove covalent binder with potent *in vitro* and *in vivo* anti-tumour activity. *Br. J. Pharmacol.* 2010, 161 (5), 1099–1110.

29. Vidal, A., Muñoz, C., Guillén, M. J., Moretó, J., Puertas, S., Martínez-Iniesta, M., Figueras, A., Padullés, L., García-Rodriguez, F. J., Berdiel-Acer, M., Pujana, M. A., Salazar, R., M., G.-M., Mart, í. L., Ponce, J., Molleví, D. G., Capella, G., Condom, E., Viñals, F., Huertas, D., Cuevas, C., Esteller, M., Avilés, P., Villanueva, A. Lurbinectedin (PM01183), a new DNA minor groove binder, inhibits growth of orthotopic primary graft of cisplatin-resistant epithelial ovarian cancer. *Clin. Cancer Res.* 2012, 18 (19), 5399–5411.

30. Hirata, Y., Uemura, D. Halichondrins: Antitumor polyether macrolides from a marine sponge. *Pure Appl. Chem.* 1986, 58 (5), 701–710.

31. Bai, R. L., Paull, K. D., Herald, C. L., Malspeis, L., Pettit, G. R., Hamel, E. Halichondrin B and homo-halichondrin B, marine natural products binding in the vinca domain of tubulin: Discovery of tubulin-based mechanism of action by analysis of differential cytotoxicity data. *J. Biol. Chem.* 1991, 266 (24), 15882–15889.

32. Munro, M. H. G., Blunt, J. W., Dumdei, E. J., Hickford, S. J. H., Lill, R. E., Li, S., Battershill, C. N., Duckworth, A. R. The discovery and development of marine compounds with pharmaceutical potential. *J. Biotechnol.* 1999, 70 (1–3), 15–25.

33. Schippers, K. J., Sipkema, D., Osinga, R., Smidt, H., Pomponi, S. A., Martens, D. E., Wijffels, R. H. Cultivation of sponges, sponge cells and symbionts: Achievements and future prospects. *Adv. Marine Biol.* 2012, 62, 273–337.

34. Bergman, O., Haber, M., Mayzel, B., Anderson, M. A., Shpigel, M., Hill, R. T., Ilan, M. Marine-based cultivation of *Diacarnus* sponges and the bacterial community composition of wild and maricultured sponges and their larvae. *Marine Biotechnol.* 2011, 13 (6), 1169–1182.

35. Aicher, T. D., Buszek, K. R., Fang, F. G., Forsyth, C. J., Jung, S. H., Kishi, Y., Matelich, M. C., Scola, P. M., Spero, D. M., Yoon, S. K. Total synthesis of halichondrin B and norhalichondrin B. *J. Am. Chem. Soc.* 1992, 114, 3162–3164.

36. Yu, M. J., Zheng, W., Seletsky, B. M., Littlefield, B. A., Kishi, Y. Case history: Discovery of eribulin (HALAVEN™), a halichondrin B analogue that prolongs overall survival in patients with metastatic breast cancer. In *Annual Reports in Medicinal Chemistry*, ed. J. E. Macor. Vol. 46. Elsevier, Amsterdam, 2011, pp. 227–241.

37. Austad, B. C., Benayoud, F., Calkins, T. L., Campagna, S., Chase, C. E., Choi, H.-W., Christ, W., Costanzo, R., Cutter, J., Endo, A., Fang, F. G., Hu, Y., Lewis, B. M., Lewis, M. D., McKenna, S., Noland, T. A., Orr, J. D., Pesant, M., Schnaderbeck, M. J., Wilkie, G. D., Abe, T., Asai, N., Asai, Y., Kayano, A., Kimoto, Y., Komatsu, Y., Kubota, M., Kuroda, H., Mizuno, M., Nakamura, T., Omae, T., Ozeki, N., Suzuki, T., Takigawa, T., Watanabe, T., Yoshizawa, K. Process development of Halaven®: Synthesis of the C14-C35 fragment via iterative Nozaki-Hiyama-Kishi reaction-Williamson ether cyclization. *Synlett* 2013, 24 (3), 327–332.

38. Austad, B. C., Calkins, T. L., Chase, C. E., Fang, F. G., Horstmann, T. E., Hu, Y., Lewis, B. M., Niu, X., Noland, T. A., Orr, J. D., Schnaderbeck, M. J., Zhang, H., Asakawa, N., Asai, N., Chiba, H., Hasebe, T., Hoshino, Y., Ishizuka, H., Kajima, T., Kayano, A., Komatsu, Y., Kubota, M., Kuroda, H., Miyazawa, M., Tagami, K., Watanabe, T. Commercial manufacture of Halaven®: Chemoselective transformations en route to structurally complex macrocyclic ketones. *Synlett* 2013, 24 (3), 333–337.

39. Chase, C. E., Fang, F. G., Lewis, B. M., Wilkie, G. D., Schnaderbeck, M. J., Zhu, X. Process development of Halaven®: Synthesis of the C1-C13 fragment from d-(–)-gulono-1,4-lactone. *Synlett* 2013, 24, 323–326.

40. Rinehart Jr, K. L., Gloer, J. B., Cook Jr, J. C., Mizsak, S. A., Scahill, T. A. Structures of the didemnins, antiviral and cytotoxic depsipeptides from a Carribean tunicate. *J. Am. Chem. Soc.* 1981, 103, 1857–1859.

41. Rinehart Jr, K. L., Gloer, J. B., Hughes Jr, R. G., Renis, H. E., McGovren, J. P., Swynenberg, E. B., Stringfellow, D. A., Kuentzel, S. L., Li, L. H. Didemnins: Antiviral and antitumor depsipeptides from a Caribbean tunicate. *Science* 1981, 212, 933–935.

42. Rinehart Jr, K. L., Shaw, P. D., Shield, L. S., Gloer, J. B., Harbour, G. C., Koker, M. E. S., Samain, D., Schwartz, R. E., Tymiak, A. A., Weller, D. L., Carter, G. T., Munro, M. H. G., Hughes Jr, R. G., Renis, H. E., Swynenberg, E. B., Stringfellow, D. A., Vavra, J. J., Coats, J. H., Zurenko, G. E., Kuentzel, S. L., Li, L. H., Bakus, G. J., Brusca, R. C., Craft, L. L., Young, D. N., Connor, J. L. Marine natural products as sources of antiviral, antimicrobial, and antineoplastic agents. *Pure. Appl. Chem.* 1981, 53, 795–817.

43. Rinehart Jr, K. L., Kishore, V., Nagarajan, S., Lake, R. J., Gloer, J. B., Bozich, F. A., Li, K.-M., Maleczka Jr, R. E., Todsen, W. L., Munro, M. H. G., Sullins, D. W., Sakai, R. Total synthesis of didemnins A, B, and C. *J. Am. Chem. Soc.* 1987, 109, 6846–6848.

44. Chun, H. G., Davies, B., Hoth, D., Suffness, M., Plowman, J., Flora, K., Grieshaber, C., Leyland-Jones, B. Didemnin B. *Investig. New Drugs* 1986, 4 (3), 279–284.

45. Lee, J., Currano, J. N., Carroll, P. J., Joullie, M. M. Didemnins, tamandarins and related natural products. *Nat. Prod. Rep.* 2012, 29 (3), 404–424.

46. Rinehart Jr, K. L., Lithgow-Bertelloni, A. M. Novel Antiviral and Cytotoxic Agent. WO 9104985 A1, 19 April, 1991.

47. Jou, G., González, I., Albericio, F., Lloyd-Williams, P., Giralt, E. Total synthesis of dehydrodidemnin B: Use of uronium and phosphonium salt coupling reagents in peptide synthesis in solution. *J. Org. Chem.* 1997, 62 (2), 354–366.

48. Tsukimoto, M., Nagaoka, M., Shishido, Y., Fujimoto, J., Nishisaka, F., Matsumoto, S., Harunari, E., Imada, C., Matsuzaki, T. bacterial production of the tunicate-derived antitumor cyclic depsipeptide didemnin B. *J. Nat. Prod.* 2011, 74 (11), 2329–2331.

49. Xu, Y., Kersten, R. D., Nam, S.-J., Lu, L., Al-Suwailem, A. M., Zheng, H., Fenical, W., Dorrestein, P. C., Moore, B. S., Qian, P.-Y. Bacterial biosynthesis and maturation of the didemnin anti-cancer agents. *J. Am. Chem. Soc.* 2012, 134 (20), 8625–8632.

50. Schmidt, E. W., Donia, M. S., McIntosh, J. A., Fricke, W. F., Ravel, J. Origin and variation of tunicate secondary metabolites. *J. Nat. Prod.* 2012, 75 (2), 295–304.

51. Pettit, G. R., Day, J. F., Hartwell, J. L., Wood, H. B. Antineoplastic components of marine animals. *Nature* 1970, 227, 962–963.

52. Pettit, G. R., Herald, C. L., Doubeck, D. L., Herald, D. L., Arnold, E., Clardy, J. Isolation and structure of bryostatin 1. *J. Am. Chem. Soc.* 1982, 104, 6846–6848.

53. Suffness, M., Newman, D. J., Snader, K. M. Discovery and development of antineoplastic agents from natural sources. In *Bioorganic Marine Chemistry*, ed. P. Scheuer. Vol. 3. Springer-Verlag, Berlin, 1989, pp. 131–167.

54. Pettit, G. R. The bryostatins. In *Progress in the Chemistry of Organic Natural Products*, ed. W. Hertz, G. W. Kirby, W. Steglich, C. Tamm. Vol. 57. Springer-Verlag, New York, 1991, pp. 153–195.

55. Pettit, G. R., Sengupta, D., Herald, C. L., Sharkey, N. A., Blumberg, P. M. Synthetic conversion of bryostatin 2 to bryostatin 1 and related bryopyrans. *Can. J. Chem.* 1991, 69 (5), 856–860.

56. Pettit, G. R. Progress in the discovery of biosynthetic anticancer drugs. *J. Nat. Prod.* 1996, 59, 812–821.

57. Newman, D. J. Bryostatin: From bryozan to cancer drug. In *Bryozoans in Space and Time*, ed. D. P. Gordon, A. M. Smith, J. A. Grant-Mackie. NIWA, Wellington, NZ, 1996, pp. 9–17.

58. Mutter, R., Wills, M. Chemistry and clinical biology of the bryostatins. *Bioorg. Med. Chem.* 2000, 8, 1841–1860.

59. Pettit, G. R., Herald, C. L., Hogan, F. Biosynthetic products for anticancer drug design and treatment— the bryostatins. In *Anticancer Drug Development*, ed. B. C. Baguley, D. J. Kerr. Academic Press, San Diego, 2002, pp. 203–235.

60. Hale, K. J., Hummersone, M. C., Manaviazar, S., Frigerio, M. The chemistry and biology of the bryostatin antitumour macrolides. *Nat. Prod. Rep.* 2002, 19, 413–453.

61. Hale, K. J., Manaviazar, S. New approaches to the total synthesis of the bryostatin antitumor macrolides. *Chem. Asian J.* 2010, 5, 704–754.

62a. Petit, G. R., Kamano, Y., Herald, C. L., Schmidt, J. M., Zubrod, C. G. Relationship of *Bugula neritina* (Bryozoa) antieoplastic constitutents to the yellow sponge *Lyssodendoryx isodictyalis*. *Pure Appl Chem.* 1986, 58, 415–421.

62b. Lin, H., Yao, X., Yi, Y., Li, X., Wu, H. Bryostatin 19: A new antineoplastic compound from *Bugula neritina* in South China sea. *Zhongguo Haiyang Yaowu* 1998, 17, 1–3.

63. Lin, H., Liu, G., Yi, Y., Yao, X., Wu, H. Studies on antineoplastic constituents from marine bryozoan *Bugula neritina* inhabiting South China sea: Isolation and structural elucidaiton of a novel macrolide. *Dier Junyi Daxue Xuebao* 2004, 25 (5), 473–478.

64. Lopanik, N., Gustafson, K. R., Lindquist, N. Structure of bryostatin 20: A symbiont-produced chemical defense for larvae of the host bryozoan, *Bugula neritina*. *J. Nat. Prod.* 2004, 67, 1412–1414.

65. Mendola, D. Aquaculture of three phyla of marine invertebrates to yield bioactive metabolites: Process development and economics. *Biomol. Engin.* 2003, 20, 441–458.

66. Newman, D. J. The Bryostatins. In *Anticancer Agents from Natural Products*, ed. G. M. Cragg, D. G. I. Kingston, D. J. Newman. 1st ed. Taylor & Francis, Boca Raton, FL, 2005, pp. 137–150.

67. Newman, D. J. The Bryostatins. In *Anticancer Agents from Natural Products*, ed. G. M. Cragg, D. G. I. Kingston, D. J. Newman. 2nd ed. Taylor & Francis, Boca Raton, FL, 2012, pp. 199–218.

68. Masamune, S. Asymmetric synthesis and its applications: Towards the synthesis of bryostatin 1. *Pure Appl. Chem.* 1988, 60, 1587–1596.

69. Kageyama, M., Tamura, T., Nantz, M. H., Roberts, J. C., Somfai, P., Whritenour, D. C., Masamune, S. Synthesis of bryostatin 7. *J. Am. Chem. Soc.* 1990, 112, 7407–7408.

70. Evans, D. A., Carter, P. H., Carreira, E. M., Charette, A. B., Prunet, J. A., Lautens, M. Total synthesis of bryostatin 2. *J. Am. Chem. Soc.* 1999, 121, 7540–7552.

71. Evans, D. A., Carter, P. H., Carreira, E. M., Prunet, J. A., Charette, A. B., Lautens, M. Asymmetric synthesis of bryostatin 2. *Angew. Chem. Int. Ed.* 1998, 37 (17), 2354–2359.

72. Ohmori, K., Ogawa, Y., Obitsu, T., Ishikawa, Y., Nishiyama, S., Yamamura, S. Total synthesis of bryostatin 3. *Angew. Chem. Int. Ed.* 2000, 39 (13), 2290–2294.

73. Ohmori, K. Evolution of synthetic strategies for highly functionalized natural products: A successful route to bryostatin 3. *Bull. Chem. Soc. Japan* 2004, 77 (5), 875–885.

74. Manaviazar, S., Frigerio, M., Bhatia, G. S., Hummersone, M. G., Aliev, A. E., Hale, K. J. Enantioselective formal total synthesis of the antitumor macrolide bryostatin 7. *Org. Lett.* 2006, 8 (20), 4477–4480.

75. Trost, B. M., Dong, G. Total synthesis of bryostatin 16 using atom-economical and chemoselective approaches. *Nature* 2008, 456, 485–488.

76. Miller, A. K. Catalysis in the total synthesis of bryostatin 16. *Angew. Chem. Int. Ed.* 2009, 48 (18), 3221–3223.

77. Keck, G. E., Poudel, Y. B., Cummins, T. J., Rudra, A., Covel, J. A. Total synthesis of bryostatin 1. *J. Am. Chem. Soc.* 2011, 133 (4), 744–747.

78. Manaviazar, S., Hale, K. J. Total synthesis of bryostatin 1: A short route. *Angew. Chem. Int. Ed.* 2011, 50 (38), 8786–8789.

79. Trost, B. M., Yang, H., Brindle, C. S., Dong, G. Atom-economic and stereoselective syntheses of the ring A and B subunits of the bryostatins. *Chem. Eur. J.* 2011, 17 (35), 9777–9788.

80. Trost, B. M., Yang, H., Dong, G. Total syntheses of bryostatins: Synthesis of two ring-expanded bryostatin analogues and the development of a new-generation strategy to access the C7–C27 fragment. *Chem. Eur. J.* 2011, 17 (35), 9789–9805.

81. Trost, B. M., Dong, G. Total synthesis of bryostatin 16 using a Pd-catalyzed diyne coupling as macrocyclization method and synthesis of C20-epi-bryostatin 7 as a potent anticancer agent. *J. Am. Chem. Soc.* 2010, 132 (46), 16403–16416.

82. Wender, P. A., Baryza, J. L., Brenner, S. E., DeChristopher, B. A., Loy, B. A., Schrier, A. J., Verma, V. A. Design, synthesis, and evaluation of potent bryostatin analogs that modulate PKC translocation selectivity. *Proc. Natl. Acad. Sci. USA* 2011, 108 (17), 6721–6726.

83. Wender, P. A., Reuber, J. Function oriented synthesis: Preparation and initial biological evaluation of new A-ring-modified bryologs. *Tetrahedron* 2011, 67 (51), 9998–10005.

84. Keck, G. E., Kraft, M. B., Truong, A. P., Li, W., Sanchez, C. C., Kedei, N., Lewin, N. E., Blumberg, P. M. Convergent assembly of highly potent analogues of bryostatin 1 via pyran annulation: Bryostatin look-alikes that mimic phorbol ester function. *J. Am. Chem. Soc.* 2008, 130 (21), 6660–6661.

85. Kedei, N., Lewin, N. E., Géczy, T., Selezneva, J., Braun, D. C., Chen, J., Herrmann, M. A., Heldman, M. R., Lim, L., Mannan, P., Garfield, S. H., Poudel, Y. B., Cummins, T. J., Rudra, A., Blumberg, P. M., Keck, G. E. Biological Profile of the less lipophilic and synthetically more accessible bryostatin 7 closely resembles that of bryostatin 1. *ACS Chem. Biol.* 2013, 8 (3), 767–777.

86. DeChristopher, B. A., LoyBrian, A., Marsden, M., D., Schrier, A. J., Zack, J. A., Wender, P. A. Designed, synthetically accessible bryostatin analogues potently induce activation of latent HIV reservoirs *in vitro*. *Nat. Chem.* 2012, 4 (9), 705–710.

87. Davidson, S. K., Allen, S. W., Lim, G. E., Anderson, C. M., Haygood, M. G. Evidence for the biosynthesis of bryostatins by the bacterial symbiont "*Candidatus Endobugula sertula*" of the Bryozoan *Bugula neritina*. *Appl. Environ. Microbiol.* 2001, 67 (10), 4531–4537.

88. Lopanik, N., Lindquist, N., Targett, N. Potent cytotoxins produced by a microbial symbiont protect host larvae from predation. *Oecologia* 2004, 139 (1), 131–139.

89. Lopanik, N. B., Shields, J. A., Buchholz, T. J., Rath, C. M., Hothersall, J., Haygood, M. G., Hakansson, K., Thomas, C. M., Sherman, D. H. *In vivo* and *in vitro* trans-acylation by BryP, the putative bryostatin pathway acyltransferase derived from an uncultured marine symbiont. *Chem. Biol.* 2008, 15 (11), 1175–1186.

90. Trindade-Silva, A. E., Lim-Fong, G. E., Sharp, K. H., Haygood, M. G. Bryostatins: Biological context and biotechnological prospects. *Curr. Opin. Biotechnol.* 2010, 21 (6), 834–842.

91. Flahive, E., Srirangam, J. The dolastatins: Novel antitumor agents from *Dolabella auricularia*. In *Anticancer Agents from Natural Products*, ed. G. M. Cragg, D. G. I. Kingston, D. J. Newman. Taylor & Francis, Boca Raton, FL, 2005, pp. 191–213.

92. Pettit, G. R. The dolastatins. *Fortschr. Chem. Org. Naturst.* 1997, 70, 1–79.

93. Bai, R., Pettit, G. R., Hamel, E. Dolastatin 10, a powerful cytostatic peptide derived from a marine animal: Inhibition of tubulin polymerization mediated throgh the vinca alkaloid binding domain. *Biochem. Pharmacol.* 1990, 39, 1941–1949.

94. Bai, R., Pettit, G. R., Hamel, E. Binding of dolastatin 10 to tubulin at a distinct site for peptide antimitotic agents near the exchangeable nucleotide and vinca alkaloid sites. *J. Biol. Chem.* 1990, 265 (28), 17141–17149.

95. Bai, R., Friedman, S. J., Pettit, G. R., Hamel, E. Dolastatin 15, a potent antimitotic depsipeptide derived from *Dolabella auricularia*: Interactions with tubulin and effects on cellular microtubules. *Biochem. Pharmacol.* 1992, 43, 2637–2645.

96. Vaishampayan, H., Glode, M., Du, W., Kraft, A., Hudes, G., Wright, J., Hussain, M. Phase II study of dolastatin-10 in patients with hormone-refractory metastatic prostate adenocarcinoma. *Clin. Canc. Res.* 2000, 6, 4205–4208.

97. Margolin, K., Longmate, J., Synold, T. W., Gandara, D. R., Weber, J., Gonzalez, R., Johansen, M. J., Newman, R., Doroshow, J. H. Dolastatin-10 in metastatic melanoma: A phase II and pharmacokinetic trial of the California Cancer Consortium. *Invest. New Drugs* 2001, 19 (4), 335–340.

98. Luesch, H., Moore, R. E., Paul, V. J., Mooberry, S. L., Corbett, T. H. Isolation of dolastatin 10 from the marine cyanobacterium *Symploca* species VP642 and total stereochemistry and biological evaluation of its analogue symplostatin 1. *J. Nat. Prod.* 2001, 64 (7), 907–910.

99. Younes, A., Yasothan, U., Kirkpatrick, P. Brentuximab vedotin. *Nat. Rev. Drug. Discov.* 2012, 11 (1), 19–20.

100. Sievers, E. L., Senter, P. D. Antibody-drug conjugates in cancer therapy. *Annu. Rev. Med.* 2013, 64, 15–29.

101. Newland, A. M., Li, J. X., Wasco, L. E., Aziz, M. T., Lowe, D. K. Brentuximab vedotin: A CD30-directed antibody-cytotoxic drug conjugate. *Pharmacother.* 2013, 33 (1), 93–104.

102. ter Haar, E., Kowalski, R. J., Hamel, E., Lin, C. M., Longley, R. E., Gunasekera, S. P., Rosenkranz, H. S., Day, B. W. Discodermolide, a cytotoxic marine agent that stabilizes microtubules more potently than Taxol. *Biochemistry* 1996, 35, 243–250.

103. Paterson, I., Florence, G. J. The development of a practical total synthesis of discodermolide, a promising microtubule-stabilizing anticancer agent. *Eur. J. Org. Chem.* 2003, 2193–2208.

104. Mickel, S. J., Neiderer, D., Daeffler, R., Osmani, A., Kuesters, E., Schmid, E., Schaer, K., Gamboni, R., Chen, W., Loeser, E., Kinder Jr, F. R., Konigsberger, K., Prasad, K., Ramsey, T. M., Repic, O., Wang, R.-M., Florence, G. J., Paterson, I. Large-scale synthesis of the anti-cancer marine natural product (+)-discodermolide. Part 5: Linkage of fragments C_{1-6} and C_{7-24} and finale. *Org. Proc. Res. Dev.* 2004, 8 (1), 122–130.

105. de Lemos, E., Poree, F.-H., Commercon, A., Betzer, J.-F., Pancrazi, A., Ardisson, J. a-Oxygenated crotyltitanium and dyotropic rearrangement in the total synthesis of discodermolide. *Angew. Chem. Int. Ed.* 2007, 46, 1917–1921.

106. Roche, C., Le Roux, R., Haddad, M., Phansavath, P., Genêt, J.-P. A ruthenium-mediated asymmetric hydrogenation approach to the synthesis of discodermolide subunits. *Synlett* 2009, (4), 573–576.

107. Du Jourdin, X. M., Noshi, M., Fuchs, P. L. Designer discodermolide segments via ozonolysis of vinyl phosphonates. *Org. Lett.* 2009, 11 (3), 543–546.

108. de Lemos, E., Agouridas, E., Sorin, G., Guerreiro, A., Commerçon, A., Pancrazi, A., Betzer, J.-F., Lannou, M.-I., Ardisson, J. Conception, synthesis, and biological evaluation of original discodermolide analogues. *Chem. Eur. J.* 2011, 17 (36), 10123–10134.

109. Paterson, I., Naylor, G. J., Gardner, N. M., Guzmán, E., Wright, A. E. Total synthesis and biological evaluation of a series of macrocyclic hybrids and analogues of the antimitotic natural products dictyostatin, discodermolide, and Taxol. *Chem. Asian J.* 2011, (2), 459–473.

110. Gunasekera, S. P., Wright, A. E. Chemistry and biology of the discodermolides, potent mitotic spindle poisons. In *Anticancer Agents from Natural Products*, ed. G. M. Cragg, D. G. I. Kingston, D. J. Newman. 2nd ed. Taylor & Francis, Boca Raton, FL, 2012, pp. 241–262.

111. Talpir, R., Benayahu, Y., Kashman, Y., Pannell, L., Schleyer, M. Hemiasterlin and geodiamolide TA: Two new cytotoxic peptides from the marine sponge *Hemiasterella minor* (Kirkpatrick). *Tetrahedron Lett.* 1994, 35, 4453–4456.

112. Coleman, J. E., de Silva, E. D., Kong, F., Andersen, R. J., Allen, T. M. Cytotoxic peptides from the marine sponge *Cymbastela* sp. *Tetrahedron* 1995, 51 (39), 10653–10662.

113. Anderson, H. J., Coleman, J. E., Andersen, R. J., Roberge, M. Cytotoxic peptides hemiasterlin, hemiasterlin A and hemiasterlin B induce mitotic arrest and abnormal spindle formation. *Cancer Chemother. Pharmacol.* 1997, 39, 223–226.

114. Bai, R., Durso, N. A., Sackett, D. L., Hamel, E. Interactions of the sponge-derived antimitotic tripeptide hemiasterlin with tubulin: Comparison with dolastatin 10 and cryptophycin 1. *Biochemistry* 1999, 38, 14302–14310.

115. Andersen, R. J., Coleman, J. E., Piers, E., Wallace, D. J. Total synthesis of (–)-hemiasterlin, a structurally novel tripeptide that exhibits potent cytotoxic activity. *Tetrahedron Lett.* 1997, 38 (3), 317–320.

116. Nieman, J. A., Coleman, J. E., Wallace, D. J., Piers, E., Lim, L. Y., Roberge, M., Andersen, R. J. Synthesis and antimitotic/cytotoxic activity of hemiasterlin analogues. *J. Nat. Prod.* 2003, 66, 183–199.

117. Loganzo, F., Discafani, C., Annable, T., Beyer, C., Musto, S., Hardy, C., Hernandez, R., Baxter, M., Ayral-Kaloustian, S., Zask, A., Singanallore, T., Khafizova, G., Asselin, A., Poruchynsky, M. S., Fojo, T., Nieman, J., Andersen, R. J., Greenberger, L. M. HTI-286, a synthetic analog of the anti-microtubule tripeptide hemiasterlin, potently inhibits growth of cultured tumor cells, overcomes resistance to paclitaxel mediated by various mechanisms, and demonstrates intravenous and oral *in vivo* efficacy. *Proc. Am. Assoc. Cancer Res.* 2002, 43, abstract 1316.

118. Loganzo, F., Discafani, C., Annable, T., Beyer, C., Musto, S., Hari, M., Tan, X., Hardy, C., Hernandez, R., Baxter, M., Singanallore, T., Khafizova, G., Poruchynsky, M. S., Fojo, T., Nieman, J. A., Ayral-Kaloustian, S., Zask, A., Andersen, R. J., Greenberger, L. M. HTI-286, a synthetic analogue of the tripeptide hemiasterlin, is a potent antimicrotubule agent that circumvents P-glycoprotein-mediated resistance *in vitro* and *in vivo*. *Cancer Res.* 2003, 63, 1838–1845.

119. Andersen, R. J., Williams, D. E., Strangman, W. K., Roberge, M. HTI-286 (taltobulin), a synthetic analog of the antimitotic natural product hemiasterlin. In *Anticancer Agents from Natural Products*, ed. G. M. Cragg, D. G. I. Kingston, D. J. Newman. 2nd ed. Taylor & Francis, Boca Raton, FL, 2012, pp. 347–362.

120. Rocha-Lima, C. M., Bayraktar, S., MacIntyre, J., Raez, L., Flores, A. M., Ferrell, A., Rubin, E. H., Poplin, E. A., R., T. A., Lucarelli, A., Zojwalla, N. A phase 1 trial of E7974 administered on day 1 of a 21-day cycle in patients with advanced solid tumors. *Cancer* 2012, 118 (17), 4262–4270.

121. Simoni, D., Lee, R. M., Durrant, D. E., Chi, N.-W., Baruchello, R., Rondanin, R., Rullo, C., Marchetti, P. Versatile synthesis of new cytotoxic agents structurally related to hemiasterlins. *Bioorg. Med. Chem. Lett.* 2010, 20 (11), 3431–3435.

122. Hsu, L.-C., Durrant, D. E., Huang, C.-C., Chi, N.-W., Baruchello, R., Rondanin, R., Rullo, C., Marchetti, P., Grisolia, G., Simoni, D., Lee, R. M. Development of hemiasterlin derivatives as potential anticancer agents that inhibit tubulin polymerization and synergize with a stilbene tubulin inhibitor. *Invest. New Drugs* 2012, 30 (4), 1379–1388.

123. Ōmura, S., Ikeda, H., Ishikawa, J., Hanamoto, A., Takahashi, C., Shinose, M., Takahashi, Y., Horikawa, H., Nakazawa, H., Osonoe, T., Kikuchi, H., Shiba, T., Sakaki, Y., Hattori, M. Genome sequence of an industrial microorganism *Streptomyces avermitilis*: Deducing the ability of producing secondary metabolites. *Proc. Nat. Acad. Sci. USA* 2001, 98, 12215–12220.

CHAPTER 16

Conopeptides, Marine Natural Products from Venoms
Biomedical Applications and Future Research Applications

Baldomero M. Olivera, Helena Safavi-Hemami, Martin P. Horvath, and **Russell W. Teichert**
Department of Biology, University of Utah, Salt Lake City, Utah

CONTENTS

In this chapter, we present an overview of cone snails, their venoms, and the biomedical potential of individual venom components that fully spans the theme of this book, from beach to bedside. We start off by providing an overview of the interactions of humans with cone snails in a temporal sequence. From the earliest times, people picked up cone shells from the beach, an activity that continues to this day. Gradually, it became clear that the animals that made the pretty shells that washed up on the beach were venomous and could even be deadly. In recent years, cone snails are increasingly valued for their biomedical potential. A direct manifestation of this is that one compound from a cone snail venom is now an approved commercial drug for intractable pain.

Like other marine natural products, venom components from cone snails have been developed individually for biomedical applications. The standard strategy of biochemically characterizing each compound, followed by exploring biomedical applications, has been routinely followed. However, we have probably come to an important transition point for *Conus* venom peptides. Although the standard characterization of the individual venom components will undoubtedly continue, combinations of conopeptides will increasingly be used for a broad spectrum of research and biomedical purposes. This transition to combination-based approaches is a particularly important theme that we develop in the chapter. Combinations of venom peptides that we refer to as toolkit cabals should have utility for basic research purposes, and in the longer term could lead to effective combination drug therapies, particularly for the more intractable pathological conditions of the nervous system.

This chapter has five sections. The first is the background, which presents the history of human interactions with cone snails, as well as an overview of their evolution, phylogeny, and general biology. The second section is an account of the early work characterizing cone snail venoms, followed by a review of some of the direct biomedical applications that emerged from these initial studies. The third section is an overview of the basic science that underlies the biomedical applications of *Conus* peptides. In the fourth section, we describe a new combination pharmacology approach, constellation pharmacology, which applies insights from cone snail biology as a conceptual framework and utilizes the highly selective venom components from cone snails as the key building blocks for enabling this pharmacological platform. In the fifth and final section, we discuss perspectives and potential future applications of cone snail peptides.

16.1 BACKGROUND

16.1.1 From the Beach: A History of Humans and Cone Snails

There is a long historical record of humans collecting cone shells from beaches, with abundant evidence that their unique and beautiful shell patterns have always been admired. This has occurred since the beginning of human civilization, in a variety of different cultural contexts. More than 5000 years ago in Mesopotamia, inhabitants of the first urban settlement, Uruk, fabricated necklaces by stringing cone shells together with precious stones such as carnelian;[1] such a necklace is shown in Figure 16.1. The intriguing aspect of the archeological findings is that one of the cone shells in the Uruk necklaces was *Conus ebraeus*, the "Hebrew cone," which is not found in the closest coastal waters (the Persian Gulf). The shells presumably had to be imported from somewhere in the Arabian Peninsula that faces the Indian Ocean, or from the Indian subcontinent itself. Today, we see evidence for the continuing fascination with cone shells (Figure 16.1). The Himba tribe in northern Namibia imports a cone shell, *Conus litteratus*, from East Africa for married women to wear as a symbol of status (D. Nebs, personal communication). The same shell is used by tribes along the Sepak River in New Guinea; the spire of the shell is cut out and worn as a pendant.

In Europe, a craze for collecting cone shells, particularly in Holland, was noted by many observers in the seventeenth through nineteenth centuries.[2] Cone shells were included in the paintings of the period. Rembrandt so admired the marble cone, *Conus marmoreus* (the type species of the genus

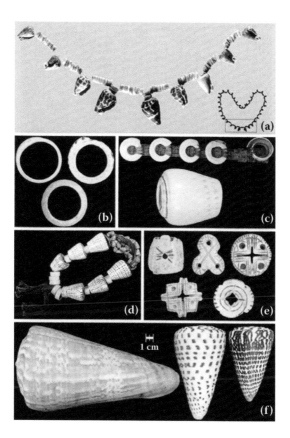

Figure 16.1 Cone shells in diverse cultural contexts. (a) Necklace recovered at the site of Urak in Mesopotamia, the first human urban settlement. Two *Conus* species are used: the larger shells are *Conus ebraeus*, the "Hebrew cone," and the smaller shells are *Conus parvatus*. The disks are probably made from the spires of the latter species. This necklace is at the Pergamon Museum in Berlin. (b) Three bracelets that were found together from a fourteenth-century Philippine gravesite. Two are made by sectioning the widest part of the "Alphabet cone" shells, *Conus litteratus* and *Conus leopardus*; the third one is made from blue glass—the glass bracelet was probably imported from China. (c) A traditional belt made of cone shells worn by the Codillera mountain tribes in the Philippines. The larger shells are *C. litteratus* and *C. leopardus*. The smaller shells are *C. ebraeus* and *Conus coronatus*. (d) Belts of Berber tribesmen in Morocco, strung with cone shell discs, and part of a shell that was used as money several centuries ago by the tribesmen. (e) A carved cone shell used as money in Mauritiana in the Middle Ages. These are all made from specimens of *Conus pulcher* found off the Atlantic coast of North Africa. (f) Shells of *Conus pulcher* (left), *C. leopardus* (middle), and *C. litteratus* (right). The specimen of *C. pulcher* is 208 mm in length—this is the largest living cone snail species.

Conus), that he made a wood block of the shell that he continually revised over a long period. Two shells in particular were legendary: *Conus cedonulli* and *Conus gloriamaris*. Shells of *C. cedonulli*, the "matchless cone," were apparently collected from the island of St. Vincent in the Caribbean when the first Europeans arrived. It is likely that the shells were obtained from a colony that had been buried and killed by the eruption of the La Soufrière volcano many centuries earlier (W. Old, personal communication). Three of these shells reached European collections. They were so highly valued that one of these shells outsold a painting by Vermeer by fourfold in an auction in Amsterdam in 1796. The most famous cone shell of all was the "glory of the sea cone," *C. gloriamaris*. At one point in time, it was regarded the most valuable natural history object on earth (these cone shells are shown in Figure 16.2).

Fascination with the beauty of these shells increased with the growing awareness that the living animals that made the shells were dangerous and capable of killing people. Since the time of

Figure 16.2 Highly prized cone shells. Two specimens of *Conus cedonulli*, the "matchless cone," are shown on the left-hand side of the figure. These shells were recovered from dredge hauls from the harbor in the island of St. Vincent in the Caribbean. One specimen outsold the painting by Vermeer shown in the inset. It is thought that these shells were buried during an eruption of the La Soufrière volcano years earlier. A specimen of *Conus gloriamaris*, the "glory of the sea cone," from Panglao Island, Bohol, in the Central Philippines, is shown on the right-hand side of the figure. This was one of the most famous seashell treasures from the sixteenth through eighteenth centuries. By the middle of the twentieth century, it was even thought to be extinct; no specimen had been found in the twentieth century until 1957.

Rumphius at the end of the seventeenth century in Indonesia, it was known by zoologists that people could be killed by the sting of a cone snail.[3] A missionary who went to Vanuatu wrote that the natives knew that a cone snail sting could be fatal. There were a few reports, including one in the 1930s from Queensland,[4] that detailed human fatality from cone snail stings. This only added to their mystique—beautiful but deadly. However, essentially nothing was known about the biology of cone snails, the animals inside the shells.

The present framework for the biology of cone snails was single-handedly erected by the field-work of a young graduate student at Yale who did his PhD thesis research in Hawaii in the 1950s, Alan Kohn.[5] The careful analysis of the ecology of living cone snails carried out by Kohn remains a monumental contribution to our knowledge of the diverse prey that different species of cone snails envenomate (Figure 16.3). Kohn's work established for the first time that some lineages of cone snails specialize on fish as prey. He recorded the envenomation of fish by the large fish-hunting species, *Conus striatus*, the "striated cone." Kohn also showed that some species devour other marine snails, as well as species that eat a diversity of different types of worms, primarily polychaetes.[6]

16.1.2 Insights into Cone Snail Biology Based on Phylogenetic Data

Insights into the evolution of *Conus* and the relationship between different species of cone snails have been obtained through genetic data from >300 species.[7] A phylogenetic tree of cone snail species from standard genetic markers is shown in Figure 16.4. From this phylogenetic analysis, it

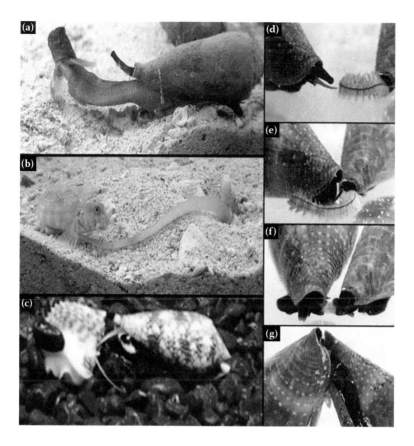

Figure 16.3 Live cone snails and their prey. Left-hand panels: (a) *Conus magus*, the "magician's cone," which has just envenomated and swallowed a fish. (b) *Conus striatus*, the "striated cone," a fish-hunting species, is shown with its extended yellowish proboscis, ready to strike its fish prey. (c) *Conus textile*, a molluscivorous species, is shown about to inject venom into its snail prey. Right-hand panels: Two specimens of *Conus lividus* compete for prey. (d) The snail on the right has envenomated a worm (*Noreis* sp., a polychaete), but the snail on the left is about to inject venom through its red proboscis. (e–g) A continuation of the time-lapsed progression is shown. (g) In their struggle to suck up the worm, both snails are levitated.

is clear that there are many distinctive clades within the genus *Conus* and, surprisingly, that some cone snails are only distantly related to the majority of *Conus* species. In a recent proposal revising the taxonomy of cone snails, the family Conidae (including all cone snails) was divided into four different genera. The great majority of cone snail species (ca. 500) remain within the genus *Conus*. However, three smaller groups were separated from *Conus* on the basis of phylogentic analysis: *Conasprella*, *Profundiconus*, and *Californiconus*, with the last genus being monospecific, only including the species *Conus californicus* (now to be called *Californiconus californicus*) found off the coast of Southern California and the Baja California peninsula. Almost all of the work on cone snail toxins has been published for species in the genus *Conus*, although some limited data have been obtained from *Californiconus* and *Conasprella*.

Within the genus *Conus*, there is a diverse set of distinct lineages with all of the species in each lineage specializing in a distinctive spectrum of prey. Most clades from the phylogenetic tree shown in Figure 16.4 specialize in hunting various types of marine worms. At around the Miocene period, two major shifts in prey occurred, generating a number of fish-hunting cone snails, as well as mollusc-hunting lineages. Molecular changes that led to the evolution of fish hunting from a worm-hunting ancestor have recently been reconstructed.[8]

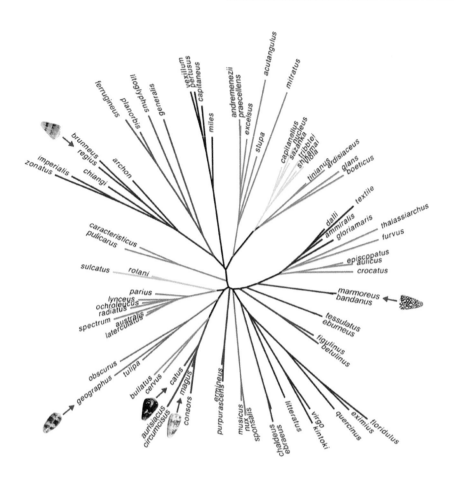

Figure 16.4 Molecular phylogeny of *Conus*. Different clades of cone snails are shown. The tree was established using standard mitochondrial markers and one nuclear marker adapted from Aman et al.[8]
Arrows indicate *Conus* species from which venom peptides that have undergone development
as drug leads were derived. *Conus marmoreus* is a snail-hunting species. *Conus regius* preys
on fireworms. *Conus geographus*, *Conus catus*, and *Conus magus* are each fish hunters. The
venom peptide from *C. magus*, ω-conotoxin MVIIA, has become an approved drug for intractable
pain (see Table 16.2).

Detailed knowledge of cone snail phylogeny has facilitated the discovery of biomedically relevant peptides from cone snail venoms. Access to the phylogenetic information (Figure 16.4) can lead
to the quick elucidation of peptide toxins that share the same or related molecular-targeting specificity. For example, the first peptide from a clade of *Conus* species discovered to target a particular ion
channel or receptor family may be used as a starting point for discovering functionally similar peptides from related cone snails. Using the phylogenetic tree as a guide, molecular cloning techniques
can be used to obtain similar peptide sequences from closely related *Conus* species. Such peptides
can be synthesized (or expressed), and the molecular-targeting selectivity of each individual venom
peptide can then be assessed. Typically, venom peptide homologues from closely related species target the same family of receptors or ion channels, but often with contrasting selectivity for particular
subtypes of receptors or ion channels.[9] In this way, knowing the detailed phylogeny of cone snails
can greatly facilitate the development of a diverse pharmacological toolkit containing compounds
that target different molecular isoforms within a receptor or ion channel family.

In the immediate future, efforts to discover novel venom peptides will undoubtedly continue
and likely expand. Although the venoms of cone snails will be a continuing source of bioactive

peptides, bioactive components from other venomous molluscs, such as the turrids and terebrids, will also be characterized[10] (for a phylogenetic overview of all venomous marine snails, see Olivera et al.[11] and Puillandre et al.[12]).

16.2 CONE SNAIL VENOMS: BIOMEDICAL APPLICATIONS

16.2.1 History of *Conus* Venom Research

The first experimental characterization of *Conus* venoms was aimed at discovering the pharmacological basis of the physiological activity elicited by the venoms. Early insights into the bioactive components of these venoms were provided by several studies, particularly by Endean and coworkers.[13] The work of Spence and Quinn provided an early harbinger of the novelty of *Conus* venom components.[14] These workers demonstrated that the venom of *Conus geographus*, the "geography cone," had an activity that specifically inhibited muscle action potentials but not axonal action potentials.[14] At the time, different isoforms of voltage-gated sodium channels were unknown, including those expressed in neurons and muscle. Consequently, the pharmacological basis of the phenomenological effect evoked by *C. geographus* venom temporarily remained a mystery. The work of Spence and Quinn was also pioneering because, for the first time, a purified venom component was characterized that specifically inhibited action potentials in muscle.[14,15]

The biochemical characterization of *Conus* venoms was initiated by Kohn and coworkers.[16] Early studies identifying novel bioactive venom components have been reviewed.[17,18] In the initial period, the major strategy was to use traditional biochemical methods for separating venom components in conjunction with an assay for biological activity. In this way, individual bioactive components could be purified by following biological activity through successive fractionation steps. Each purified venom component was then characterized biochemically. This early work demonstrated that the bioactivity of these venoms could be accounted for by small, highly structured venom peptides, and that cone snail venoms were extremely complex with ~100–200 different peptides termed conopeptides or conotoxins. Eventually it also became clear that the sequences of closely related conotoxins typically diverged between even the most closely related *Conus* species. The astounding implication was that each cone snail venom contains its own distinctive complement of a large number of bioactive conopeptides.[18,19]

In this phase of initial biochemical characterization, fairly large amounts of venom were required because bioactivity-guided purification was the standard approach. Most bioassays were phenotypic, for example, by injection into mice, either intraperitoneal injection (initially) or intracranial injection directly into the central nervous system. After the individual peptides were purified and their sequences determined, many of the conopeptides could be chemically synthesized; see, for example, Cruz et al.[20] The availability of bioactive synthetic toxin enabled the detailed mechanistic characterization of each peptide by electrophysiology, typically using neuromuscular junction preparations in the early days.[21] Accordingly, peptides that blocked presynaptic calcium channels,[22] postsynaptic nicotinic receptors,[23] and muscle action potentials[21] were among the first groups of venom peptides extensively characterized.

The second stage of conopeptide discovery incorporated molecular cloning of conopeptide genes. A peptide known as ω-conotoxin MVIIC was chemically synthesized exclusively based on the amino acid sequence predicted by a cDNA sequence obtained from *Conus magus* venom duct tissue.[24] The molecular-targeting selectivity and mechanism of this peptide were elucidated using electrophysiology.[24] Through the analysis of venom duct messenger RNAs (e.g., reverse transcriptase polymer chain reaction [RT-PCR] or shotgun sequencing of cDNA libraries), a very large and continuously growing data set of venom peptide sequences was compiled.[1,25] From this large data

set of known peptide sequences, only a relatively minor fraction of the peptides have been synthesized and characterized.

The initial neuromuscular junction preparations were eventually replaced by the heterologous
expression of specific ion channel or receptor subtypes in *Xenopus* oocytes or in cultured cells.
These heterologous expression systems have been particularly valuable for elucidating conopeptide
targets. Notably, this approach has elucidated the targeting selectivity of many conopeptides that
inhibit various nicotinic acetylcholine receptor subtypes, as well as conopeptides that target other
ligand-gated ion channels and voltage-gated channels. Collectively, the conopeptides may comprise
the largest set of related natural products that target specific ion channel and receptor subtypes. A
summary of conopeptides that are widely used as research tools by the neuroscience community is
provided in Table 16.1. These early studies of individual conopeptides led to a number of biomedical
applications, as reviewed in the following sections.

Table 16.1   Subtype-Selective Conopeptides (for Mammalian Targets)

Conopeptide	Molecular Target	Published Reference
Nicotinic Receptors		
α-RgIA	α9α10 nAChR	Ellison et al.[125]
α-BuIA	nAChR-containing β4 subunit	Azam et al.[100]
α-ArIB (V11L;V16D)	α7 nAChR	Whiteaker et al.[126]
α-MII (H9A;L15A)	α6β2 nAChR	McIntosh et al.[127]
α-BuIA (T5A;P6O)	α6β4 nAChR	Azam et al.[34]
α-AuIB	α3β4 nAChR	Luo et al.[128]
α-MI	α1δ interface of nAChR	Lou and McIntosh[129]
αA-OIVA (K15N)	α1γ interface of nAChR	Teichert et al.[130,131]
Neurotensin Receptors		
Contulakin-G	NTS1	Craig et al.[132]
Glutamate Receptors		
Conantokin-RI-B	NR2B	Gowd et al.[133]
Conantokin-R	NR2A, NR2B	White et al.[134]
Conantokin-RI-A	NR2A, NR2B, NR2D	Gowd et al.[135]
Calcium Channels		
ω-GVIA	N-type calcium channel ($Ca_V2.2$)	Kerr and Yoshikami[136]
ω-MVIIC	P/Q-type calcium channel ($Ca_V2.1$)	Hillyard et al.[24]
Sodium Channels		
μ-TIIIA	$Na_V1.1$[a]	Wilson et al.[137]
μ-PIIIA	$Na_V1.1, 1.6$[a]	Wilson et al.[137]
μ-KIIIA or SmIIIA	$Na_V1.1, 1.6, 1.7$[a]	Wilson et al.[137]
μO-MrVIB	$Na_V1.8$	Bulaj et al.[45]
μO§-GVIIJ	Na_V channels with β1 or β3 subunits	Gajewiak et al.[56]
Potassium Channels		
κM-RIIIJ	$K_V1.2$	Chen et al.[138]
pl14a	$K_V1.6$	Imperial et al.[139]
Conkunitzin-S1	$K_V1.7$	Bayrhuber et al.[140]

[a] Also potently blocks $Na_V1.2$ and $Na_V1.4$.

16.2.2 Diagnostic Applications

Among the major classes of marine natural products, the peptides from cone snail venom have the most clearly demonstrable biomedical applications in both diagnostics and therapeutics. In this section and the following one, we present two specific case studies; the first is an example of diagnostic application of conopeptides, and Section 16.2.3 presents an example of the therapeutic use of a conopeptide.

An example of diagnostic applications of components from cone snail venom is the use of these venom peptides to differentiate between two autoimmune diseases, myasthenia gravis and Lambert–Eaton syndrome. Both of these conditions present as muscle weakness, but in fact, except for the autoimmune component of the diseases, they are very different, both in origin and in likely outcome.

In myasthenia gravis, autoimmunity is targeted to the postsynaptic acetylcholine receptor of the neuromuscular junction.[26] Autoantibodies that attack the nicotinic acetylcholine receptor result in a disruption of synaptic transmission and ultimately the well-known symptoms of progressive muscle weakness. Lambert–Eaton syndrome, although presenting with the same general symptomatology, has a different molecular cause, with autoantibodies elicited against the presynaptic voltage-gated calcium channels,[26] which are necessary for the exocytosis of the neurotransmitter acetylcholine. In the absence of neurotransmitter release, the same muscle weakness is observed.

In general, the consequences of Lambert–Eaton syndrome are very different from those of myasthenia gravis. The autoimmunity associated with Lambert–Eaton syndrome commonly arises because the patient also has small-cell lung carcinoma,[26] a very aggressive form of lung cancer. Apparently, the tumor cells inappropriately express the voltage-gated calcium channels that are normally only found in presynaptic termini. This out-of-context expression pattern is probably eliciting an immune response against those calcium channels found at cancer cells and also at the presynaptic terminus. The physiological result is progressive muscle weakness. However, because these patients have small-cell lung carcinoma, which is medically the more rapidly progressing pathology, the treatment for patients that have Lambert–Eaton syndrome is quite different than that for patients that have myasthenia gravis.

In order to determine whether patients have Lambert–Eaton syndrome, ω-conotoxin GVIA, which binds N-type calcium channels, and ω-conotoxin MVIIC, which binds P/Q-type calcium channels, are radiolabeled, and a preparation of solubilized calcium channels is provided as the substrate for binding (separately) these radiolabeled peptides.[27,28] If the patient has a detectable titer of autoimmune antibodies targeted to these presynaptic calcium channels, then this can be detected by immunoprecipitation of the radiolabeled *Conus* peptides. This has become a widely used diagnostic procedure for differentiating between Lambert–Eaton syndrome and myasthenia gravis.[28]

16.2.3 Therapeutic Applications of *Conus* Peptides

One conopeptide inhibitor of voltage-gated calcium channels was developed and approved as a therapeutic drug for intractable pain. This peptide, ω-conotoxin MVIIA, from *Conus magus*, the "magician's cone," was first developed by Neurex, Inc., with a team led by George Miljanich and J. Ramachandran, and later it was further developed by Elan, Inc. as an approved drug for pain. It is now known by the generic name ziconotide and the trade name Prialt®.

Notably, the commercial drug is chemically identical to the natural peptide, without a single alteration in any functional group. In order to alleviate severe intractable pain, the peptide is infused intrathecally with a pump, and for that reason, it is typically only given to patients with advanced cancer or other forms of intractable pain that are not significantly alleviated by morphine. Although not practical for more general applications, the requirement for intrathecal access is not a major obstacle because in many cases, the patients who are prescribed Prialt have already progressed in tolerance to morphine to the point that morphine infusion directly into the spinal cord has become necessary to alleviate pain. For a significant number of such patients, a pump for intrathecal drug delivery has already been installed.

The analgesic effects of Prialt are mediated by inhibition of the presynaptic N-type voltage-gated calcium channel ($Ca_V2.2$), which prevents neurotransmitter release from somatosensory neurons that synapse with dendrites of spinal neurons in the dorsal horn. Notably, the release of neurotransmitter requires an influx of calcium at the presynaptic terminus through $Ca_V2.2$. In the presence of Prialt, these $Ca_V2.2$ channels are blocked, and as a result, no neurotransmitter is released and the pain signal is not transmitted to the spinal neuron or to the rest of the central nervous system. The inhibition of pain signaling from the peripheral to the central nervous system thus alleviates severe intractable pain. In addition to Prialt, five other conopeptides have reached human clinical trials, as summarized in Table 16.2. Additionally, there are a number of *Conus* peptides that are currently being explored for therapeutic applications at the preclinical stage (also shown in Table 16.2).

Table 16.2   Conopeptides Developed for Therapeutic Applications

Conopeptide (Indication) [Molecular Target]	Stage of Human Clinical Trials	Peptide from the Venom of (*Conus* Species)	*Conus* Subgenus (Prey)	Developed by
ω-MVIIA (Prialt)[a] (neuropathic pain) [$Ca_V2.2$]	Approved in the United States and EU	*Conus magus* (magician's cone)	*Pionoconus* (fish)	Elan Corp.
ω-CVID (AM336)[b] (cancer pain) [$Ca_V2.2$]	Phase II	*Conus catus* (cat's cone)	*Pionoconus* (fish)	Amrad, Inc.
x-MrIA (XEN2174)[b] (neuropathic pain) [norepinephrine transporter]	Phase II	*Conus marmoreus* (marble cone)	*Conus* (marine snails)	Xenome, Inc.
α-Vc1.1 (ACV1)[c,h] (neuropathic pain) [α9α10 Nicotinic receptor]	Phase II	*Conus victoriae* (Victoria's cone)	*Cylindrus* (marine snails)	Metabolic, Inc.
Contulakin-G (CGX1160)[d] (neuropathic pain) [neurotensin receptor]	Phase I	*Conus geographus* (geography cone)	*Gastridium* (fish)	Cognetix, Inc.
Conantokin-G (CGX1007)[e] (epilepsy) N-methyl-D-asparate receptor (NMDA)	Phase I	*Conus geographus* (geography cone)	*Gastridium* (fish)	Cognetix, Inc.
κ-PVIIA (CGX1051)[f] (cardioprotection) [K_V1 subfamily]	Preclinical	*Conus purpurascens* (purple cone)	*Chelyconus* (fish)	Cognetix, Inc.
α-RgIA[g,h] (nerve injury pain) [α9α10 nicotinic receptor]	Preclinical	*Conus regius* (royal cone)	*Stephanoconus* (fireworms)	Kineta, Inc.

Note: Details on the molecular targets and results from preclinical and clinical studies are reviewed in the following:
a Milianich[141]
b Armishaw and Alewood[142]
c McIntosh et al.[143]
d Han et al.[25]
e Twede et al.[144]
f Koch et al.[145]
g Vincler et al.[146]
h An alternative target for these peptides has been suggested: the gamma-aminobutyric acid (GABA) B receptor. For a discussion of these alternatives, see Vetter and Lewis.[147]

16.3 MOLECULAR MECHANISMS OF CONOPEPTIDE DIVERSITY AND TARGETING SELECTIVITY

16.3.1 Overview of Conopeptide Diversity

The venoms of cone snails are remarkably diverse. Each of the 500–700 species of cone snail synthesizes its own characteristic repertoire of ~200 venom peptides, providing a library of as many as 100,000 different bioactive compounds from this genus alone. It is this rapid diversification of venom components over the course of cone snail evolution that provides the basis for the generation of compounds with improved specificity and efficacy for novel neuronal targets. The following three sections will provide an overview of the major molecular mechanisms contributing to the rapid diversification of cone snail toxins. Section 16.3.1.1 describes the accelerated evolution of conopeptide-encoding genes. Section 16.3.1.2 summarizes the extensive processing of the nascent peptide chain, including folding, modification, and proteolytic cleavage. Finally, Section 16.3.1.3 describes adaptations of the venom duct that enable the high-throughput manufacturing and secretion of conopeptides.

16.3.1.1 Molecular Basis of Conopeptide Diversity

At the present time, the genome of a cone snail has yet to be completely sequenced, and studies examining the evolution of conopeptide-encoding genes have therefore focused on sequence information obtained for a limited number of conopeptide gene superfamilies. Nevertheless, what has become clear is that conopeptide genes are among the fastest-evolving peptide-encoding genes in metazoans, and that this rapid evolution is enabled by an unusually high degree of gene duplication events.[29–31] Gene duplications allow for an important gene to maintain its function, while mutations in redundant gene copies can lead to advantageous new functions, including novel or improved targeting selectivity. This is nicely illustrated for a group of conotoxins belonging to the A-superfamily. Depending on the number of amino acids positioned between the second and third and between the third and fourth cysteine residues, α-conotoxins are classified into distinct classes (e.g., in the $\alpha_{3/5}$-conotoxins, three and five amino acids are located between these cysteine residues), each exhibiting differences in targeting specific subtypes of the muscle or neuronal nicotinic acetylcholine receptor.[32,33] Gene sequence analysis revealed that the $\alpha_{4/6}$- and $\alpha_{4/3}$-conotoxins evolved by duplication of an ancestral $\alpha_{4/7}$-conotoxin gene, followed by subsequent divergence.[29] Such events provide opportunities for a selective advantage since even a single nonsynonymous mutation within the toxin-encoding region of a gene can be sufficient to yield a shift in receptor subtype selectivity.[34,35]

In the arms race between predator and prey, neofunctionalizations provide a significant advantage for survival and can lead to speciation and the occupancy of new biological niches. In addition to gene duplications, high mutation rates (including point mutations, insertions, and deletions) are apparent in the genes encoding conopeptides. Interestingly, the high mutation rate is restricted to the region of the gene that encodes the mature toxin and is not apparent for neighboring regions, as described further here.

Conotoxin precursors can be readily divided into three distinct regions: an amino-terminal signal sequence, a propeptide region, and the region that will become the mature toxin at the carboxy-terminus (Figure 16.5). The signal sequence targets the conotoxin precursor to the endoplasmic reticulum (ER). The propeptide region, located between signal sequence and the mature toxin region, has been suggested to play a role in secretion, posttranslational modification, and folding.[36–38] Conopeptides that belong to the same gene superfamily share a conserved signal sequence, but the mature toxin region is hypervariable, with the exception of a conserved framework of cysteine

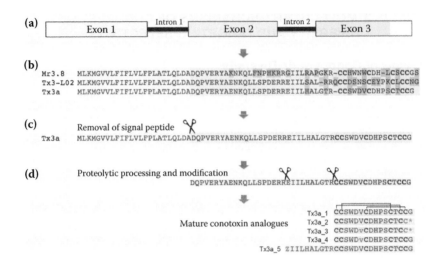

Figure 16.5 Conotoxins from gene to peptide products. (a) Genes encoding conotoxin precursors are located on separate exons. The depicted example represents an M-superfamily gene with three exons. Three exons are generally found for genes belonging to the I1-, I2-, M-, O-, P-, S-, and T-superfamilies, while two exons are characteristic for members of the A-superfamily. (b) Conotoxin transcripts and their encoding peptide products are characterized by a conserved N-terminal signal sequence (yellow), an intermediate propeptide (purple), and a single copy of the toxin at the C-terminus (blue). Alignment of M-superfamily conotoxins from *Conus marmoreus* (Mr) and *Conus textile* (Tx3-L02 and Tx3a) illustrates that signal sequences are highly conserved, whereas propeptide and toxin regions are increasingly variable (amino acid changes are shaded gray). (c) Following translocation into the ER, the nascent pre-propeptide experiences a series of modifications beginning with removal of the signal sequence by proteolysis (scissors). (d) Processing continues with removal of the propeptide sequence (scissors) and posttranslational modification of amino acids (highlighted in orange) to yield the final mature peptides. These processing events can generate several analogues originating from the same gene copy, as shown here for five analogues of Tx3a found in *C. textile* venom. The analogue Tx3a_1 is shown with its correctly formed disulfide bonds. Modified amino acid residues are as follows: C*, amidated cysteine; v, hydroxyvaline; Z, pyroglutamate.

residues. Propeptide sequences typically exhibit intermediate mutation rates. This striking feature of contrasting mutation rates for adjacent gene regions was noted in the first study that reported the cloning of conotoxin genes[39] and has been consistently reported in subsequent studies.[29,40,41]

Even though the molecular mechanisms responsible for this unusual juxtaposition of conserved and variable regions are not fully understood, studies on the gene organization of several toxin superfamilies have provided new insights. An analysis of the rate of nonsynonymous to synonymous mutations (a measure of selective pressure acting on a gene) revealed that the mature toxin region experiences strong positive selection.[29,40,41] Positive selection can explain elevated nonsynonymous mutations (amino acid changes), but curiously, the rate of synonymous mutations is also much higher than normally observed.[41] Signal sequences and mature toxin regions are generally encoded by different exons; the rate of point mutations and insertions and deletions increases two- to eightfold between signal peptide-encoding and toxin-encoding exons[42] (Figure 16.5). One mechanism that has been suggested repeatedly to account for contrasting mutation rates is that an error-prone DNA polymerase or a polymerase lacking proofreading activity is specifically involved in the replication of the toxin-encoding exons.[42,43] Indeed, mutations commonly introduced by these polymerases are frequently observed within the mature toxin regions of conopeptide genes.[42] Evidence is emerging for additional molecular mechanisms, such as gene recombination[44] and alternative splicing.[42] Ongoing genome and transcriptome sequencing projects are likely to elucidate the importance of these and other mechanisms.

16.3.1.2 Posttranslational Venom Diversification

Each of the molecular mechanisms discussed in the previous section contributes to conopeptide diversification at the gene and transcript level. Following translation on the ribosome, additional molecular mechanisms allow for further expansion of the venom peptide repertoire. Conopeptides are translocated and further processed in the ER of the venom glandular cells, where successful secretion of properly assembled peptide toxin is preceded by three biochemical processes: oxidative folding, modification, and proteolytic cleavage.

Oxidative folding describes the process by which disulfide bonds are formed between cysteine residues. Even though some conopeptides lack disulfide bonds (e.g., conantokins and contulakins), the vast majority of conopeptides are disulfide-rich. The formation of the proper disulfide bonds is a prerequisite for adopting the correct three-dimensional structure necessary for functional activity. This is not a trivial task. For instance, conotoxin MVIIA (the drug Prialt) contains six cysteine residues that are connected by three disulfide bonds. If scrambled, there are 15 permutations for the fully oxidized version of a peptide like MVIIA, only one of which is active. The disulfide bonding becomes more complex as the number of disulfide bonds increases; a peptide with 4 disulfide bonds must precisely select one of the 105 possible structural scaffolds.

These examples illustrate the high potential for misfolding during *in vitro* synthesis of conotoxins. For instance, µO-conotoxin MrVIB, a peptide with promising analgesic properties,[45] is highly abundant in the venom of *Conus marmoreus*.[46] However, *in vitro* folding of MrVIB results in very low yields of functionally active peptide,[47] rendering this conotoxin unsuitable for large-scale pharmacological studies.

Cone snails are masters of peptide folding. Over the course of 50 million years of evolution, cone snails have evolved strategies for the efficient folding of conopeptides. The quest to understand these processes has provided several important insights. Formation of disulfide bonds in the ER of the venom glandular cells may in some cases be assisted by the presence of the propeptide,[37] posttranslational modifications,[48,49] and "helper proteins."[50–52]

One of these helper proteins is protein disulfide isomerase (PDI), an enzyme that catalyzes the oxidation and isomerization of disulfide bonds in peptides and proteins. PDI is one of the most abundant proteins in the venom duct of cone snails.[48,53] The recombinantly expressed PDI enzyme from *Conus marmoreus* efficiently catalyzes the oxidation and isomerization of various conopeptide substrates *in vitro*.[54] Comprehensive transcriptome and proteome surveys have further revealed that cone snails express an unprecedented number of PDI isoforms,[53] suggesting that the molecular mechanisms underlying the diversification of conotoxin gene superfamilies (e.g., gene duplications) may also apply to the proteins important for their proper assembly.

The timing of disulfide bond formation is unknown but likely occurs co-translationally, while the nascent peptide chain is translocated into the ER. Following translocation and folding, a large number of conotoxins are further modified. The diversity and density of posttranslational modifications observed in conopeptides is one of the most remarkable characteristics of these venom peptides. Modifications are known to affect conopeptide structure, potency, and target selectivity and clearly represent another key mechanism of conopeptide diversification. Conopeptide modifications identified to date include amidation of the carboxy-terminus; hydroxylation of proline and lysine; γ-carboxylation of glutamate; bromination of tryptophan; L-to-D epimerization of leucine, phenylalanine, tryptophan, and valine; O-glycosylation of serine and threonine; sulfation of tyrosine; cyclization of N-terminal glutamate to pyroglutamate (reviewed in Buczek et al.[55]); and S-cysteinylation of cysteine.[56] It should be noted that some of these modifications require further transport from the ER to the Golgi apparatus (e.g., O-glycosylation).

Even though common in *Conus*, many of these modifications are rarely found in other biological systems. Particularly, the discovery of γ-carboxylation of glutamate residues in conopeptides by a vitamin K-dependent carboxylase was surprising at the time,[57,58] as this modification was believed to be a unique feature of the mammalian blood-clotting cascade. Another modification of interest is the bromination of tryptophan, which was first discovered in cone snails[59] and subsequently shown to be present in the mammalian nervous system.[60] In addition to this diversity, the sheer density of alterations reported for some conotoxins is astounding. For instance, in bromocontryphan-R, a peptide isolated from *Conus radiatus* that elicits a "stiff-tail" syndrome in mice, five of eight amino acids are modified: L-to-D epimerization of tryptophan 4, hydroxylation of proline 6, bromination of tryptophan 7, disulfide bond formation between cysteines 2 and 8, and C-terminal amidation of cysteine 8.[61] Similarly, TxVA, a 13-amino acid-long peptide from *Conus textile* that causes hyperactivity and spasticity in mice, contains nine modified residues: two γ-carboxyglutamates, a bromo-tryptophan, an O-glycosylated threonine, a 4-hydroxyproline residue, and four cysteines forming two disulfide bonds.[62] The high density of modifications is further exemplified by the finding that some amino acid alterations can occur sequentially at the same residue. For instance, in conophans isolated from *Conus gladiator* and *Conus mus*, the already unusual L-to-D epimerizations of valines are accompanied by hydroxylations.[63]

Advances in mass spectrometric sequencing have revealed that a single peptide sequence can be present in a venom with variable modifications, which significantly extends the number of unique peptides that originate from a single gene[64] (see Figure 16.5). For example, the α-conotoxin Vc1.1 was first identified by cDNA sequencing from the venom duct of *Conus victoriae*,[65] a method that does not allow for the identification of posttranslational modifications. Synthetic Vc1.1 accelerated functional recovery of injured neurons and showed significant analgesic effects in several animal models of neuropathic pain.[66] Vc1.1 (known as ACV1) progressed to phase IIA human clinical trials (Table 16.2) but was not developed further. After its initial discovery, several analogues of Vc1.1 that display different modifications (e.g., hydroxylation of prolines and carboxylation of glutamate) were identified using mass spectrometry.[67,68] Synthetic peptide variants containing these modifications were less potent at inhibiting neuronal nicotinic receptors when tested in bovine adrenal chromaffin cell preparations,[69] demonstrating functional differences between the various analogues. Similar observations were made for Tx3a, a conotoxin isolated from the venom of *C. textile*[64] (Figure 16.5d), suggesting that these are not isolated events but reflect a generalized adaptation of cone snails for venom diversification.

Similarly, alternative proteolytic cleavage of the propeptide at the amino terminus of the toxin contributes to the expansion of chemical and pharmacological diversity for conopeptides,[64,70] as best exemplified by Tx3a. Proteolytic processing of Tx3a at its predicted cleavage site (generally following one or two basic residues adjacent to the mature toxin sequence) produces a 16-amino acid-long peptide. However, an analogue containing 10 additional amino acids at its amino terminus is also found in the same venom. Interestingly, the N-terminal glutamate residue of this 26-residue analogue is modified to a pyroglutamate[64] (Figure 16.5d, analogue Tx3a_5). Another example is found in the venom of the fish hunter *Conus consors*. Cleavage at a common cleavage site yields the 12-residue peptide Cn1B.[71] This peptide is less abundant than its longer analogue Cn1A that contains two additional residues at the amino terminus and originates from the same peptide precursor. While both analogues are abundant in the venom,[71] it is not known whether each produces different functional effects in prey. Functional data exist for two similar analogues found in the venom of *Conus magus*, α-MI and α-MIC. The N-terminally truncated variant, α-conotoxin MI, is significantly less potent at the muscle nicotinic acetylcholine receptor than its longer counterpart, α-MIC.[70]

The *C. consors* and *C. magus* conopeptides described above were identified by Edman sequencing of venom fractions,[71] a method that requires the conopeptides to be present at relatively high levels. Advances in peptide sequencing now allow for the detection of previously overlooked, low-abundance conopeptides. Using techniques based on mass spectrometry, a large number of differentially processed conopeptides derived from a limited number of gene product precursors were reported in the venoms of *C. consors*.[72] The question of whether these are of biological importance or merely represent by-products of conopeptide biosynthesis or sample processing has not been addressed systematically and certainly warrants further investigation.

Little is known about the proteases involved in the proteolytic cleavage of conopeptides. Recombinantly expressed Tex31, a protease from the venom duct of *Conus textile*, successfully processes a number of conotoxin-like peptides *in vitro* with strong cleavage-site selectivity for basic amino acid residues at both positions P1 and P2 amino terminal to the scissile bond and leucine at position P4.[73] For a large number of conotoxins, this specificity correlates well with the *in vivo* frequency of amino acids observed at these positions, but not all conopeptides satisfy these requirements, indicating that proteases other than Tex31 must be involved in the proteolytic cleavage of conopeptides *in vivo*. The great diversity of proteases expressed in the venom duct of *Conus geographus*,[74] together with reports on the presence of alternative cleavage sites, certainly suggests that the proteolytic processing of conopeptides is complex.

16.3.1.3 Venom Duct Compartmentalization Enables High-Throughput Peptide Production

At any given time, hundreds of different conopeptides are expressed, folded, modified, and processed in the epithelial cells of the venom duct and secreted into its lumen. The epithelial cells of the duct and its lumen are densely packed with granules of different size, shape, and ultrastructure,[53,76,77] further highlighting the specialized role of this organ in conopeptide biosynthesis.

The venom duct is long and convoluted and can measure up to four times the snail's body length. The internal, proximal end of the duct merges into the muscular venom bulb. Muscular movement of the bulb pushes venom toward the distal end, where it is loaded into a harpoon-like radular tooth for injection.[50] The evolutionary origin of the venom duct was long subject to speculation. Morphological evidence obtained from different life stages of *Conus lividus* now suggests that the venom duct evolved by rapid pinching off from the epithelium of the midesophageal wall.[77] Originally comprised of a small epithelial sheet, ongoing epithelial cell remodeling gave rise to elongation of the venom duct and epithelial cell specialization. Anatomical, molecular, and biochemical interrogations into epithelial cell morphology and conopeptide gene expression and abundances concordantly show that the venom duct is compartmentalized into distinct regions of specialized function.[64,76–79]

For example, in *Conus geographus*, the majority of all sequenced conopeptide transcripts experience differential expression along the length of the venom duct (e.g., 98% of all conantokin-G reads are found in the regions close to the injection apparatus).[79] Larger polypeptides, such as pore-forming proteins, phospholipases, and accessory proteins important for conopeptide biosynthesis and folding, also exhibit region-specific changes in gene expression and protein abundance.[74] Spatial variation in conopeptide expression, abundance, and the degree of posttranslational processing is also observed in *Conus textile*.[64,78,80] Elongation and subsequent functionalization can be regarded as yet another key evolutionary innovation enabling the high-throughput manufacturing of the high diversity and high density of venom components found in cone snails.

16.3.2 Structural Biology: How Selective Targeting Is Achieved at the Molecular Level

As described in the sections above, conopeptides interact with a wide range of targets in the neuromuscular system, and individual conopeptides are each typically selective for a precise subtype of signaling molecule (i.e., ion channel, neurotransmitter transporter, or receptor). This section provides an overview of structural biology for conopeptides, with particular emphasis on the question of how selective targeting is achieved at the molecular level.

The first experimentally determined, high-resolution structures of a conopeptide were that of α-conotoxin G1 described by Pardi's group and independently by the Kobayashi lab.[81,82] This was followed by structures for a μ-conotoxin and its derivatives[83–85] and a ω-conotoxin structure.[86] Since these landmark discoveries, the number of structures related to conopeptides has steadily increased (Figure 16.6a). The majority of structures currently available are determined for isolated conopeptides in solution by nuclear magnetic resonance (NMR) spectroscopy plus a few structures determined by x-ray crystallography. Although highly revealing, these structures of individual conopeptides show only half of the molecular surface responsible for target specificity. The other half corresponds with signaling molecule targets, and structures of these are extremely challenging to obtain because the signaling molecules are membrane-embedded, multisubunit assemblies, predominantly with heteromeric subunit composition. Structures for voltage-gated ion channels, ligand-gated ion channels, and neurotransmitter transporters are nevertheless steadily emerging. The potentially most illuminating of structures, the complex of a conopeptide and its signaling molecule target, has now recently been determined for con-ikot-ikot S bound to the α-amino-3-hydroxy-5-methyl-4-isoxazole propionic acid (AMPA) receptor.[87]

A route to something closely approaching a conopeptide target structure first opened up in 2001, when the structure of the acetylcholine-binding protein (AChBP) from the freshwater snail *Lymnaea stagnalis* was described by x-ray crystallography.[88] This structure and the several derivative structures that followed, including structures of the protein in complex with agonists and conopeptide antagonists, have significantly improved our understanding of the molecular interactions that lead to subtype selectivity for certain α-conotoxins. Members of the α-conotoxin family inhibit nicotinic acetylcholine receptors, often with very exacting selectivity for certain combinations of nicotinic acetylcholine receptor subunits. Together with structure–function activity studies of nicotinic acetylcholine receptors, structures of the AChBP in various co-complexed forms suggest that α-conotoxins achieve subtype selectivity by two strategies: by direct contact with residues that are different for different nicotinic acetylcholine receptor subunits and by an indirect strategy that targets structural transitions of the nicotinic acetylcholine receptor. The goal of this section is to describe the structural biology supporting these ideas. We begin with an overview of the AChBP structure.

The nicotinic acetylcholine receptor or domains derived from this receptor have not proven amenable to analysis by structural methods, with one notable exception being the reconstruction from electron microscopy images of receptors found in the electric organ of the Torpedo ray.[89,90] Sequence alignments showed that AChBP, expressed in the freshwater snail *Lymnaea* and other mollusks, is homologous with the acetylcholine-binding domain of nicotinic acetylcholine receptors and most closely resembles the neuronal subtypes comprising exclusively α subunits.[91] Thus, the AChBP is a naturally occurring, water-soluble analogue of nicotinic acetylcholine receptors found in the vertebrate brain.

Crystals of AChBP diffract to high resolution with upper diffraction limits typically ranging from 1.8 to 2.4 Å, meaning sufficient structural detail can be obtained to define the overall fold of the protein, the positions of residues, and the nature of intermolecular interactions (hydrogen-bonds, van der Waals contacts, salt bridges, etc.) at subunit interfaces and at ligand-binding pockets. Crystal structures of AChBP show a five-subunit assembly with rotational symmetry. Each subunit contributes a principal (+) and a complementary (–) surface to the subunit interface. Agonist- and antagonist-binding sites are found at this +/– interface between subunits.[88]

Figure 16.6b highlights the quaternary structure of AChBP and also illustrates a difference in disposition for one portion of the structure in each subunit that is elicited by agonist binding. The binding of agonist (small magenta shape in Figure 16.6b) is associated with the closure of the C loop over the binding pocket, and it is believed that this conformational change in the ligand-binding domain is the key step in triggering channel opening in the transmembrane region of nicotinic acetylcholine receptors. The C loop harbors disulfide-bridged vicinal cysteine residues (hence the

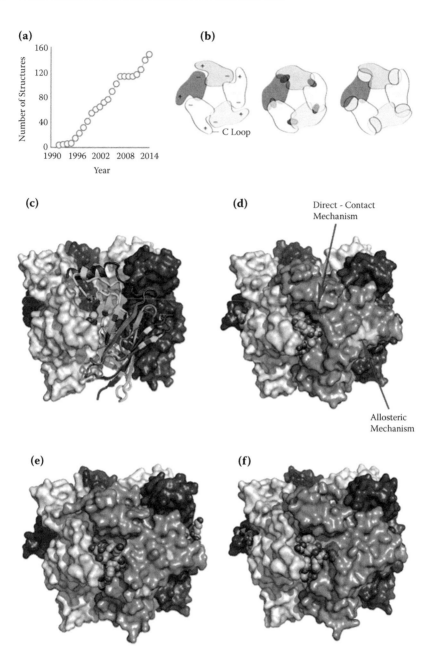

Figure 16.6 Structural biology of conopeptides. (a) The number of conopeptide structures increases each year. (b) Views from the "top" show AChBP in its apo (left), agonist-bound (middle) and α-conopeptide-bound forms (right). Two surfaces, principal (+) and complementary (−), form the agonist and antagonist-binding sites. Note how the C-loop adopts different positions. (c) Side view of AChBP shows agonist with all atoms, one subunit as a rainbow-colored ribbon, and the other four subunits as grey surfaces. Specificity determinants are color-coded magenta for those that make van der Waals (VDW) contact and lime for residues that make no direct contact with α-conotoxin. (d) AChBP in complex with α-BuIA (all atom) shows that specificity determinants include both direct-contact (magenta) and not-directly-contacting (lime) residues. (e) AChBP in complex with α-PnIA shows that all specificity determinants belong to the not-directly-contacting class (lime). (f) AChBP in complex with α-conotoxin ImI shows specificity determinants belong to the direct-contact class (magenta). Molecular models were generated on the basis of atomic coordinates found in the following PDB entries: 2byn, 2byq, 4ez1, 2br8, and 2c9t.

Table 16.3   Specificity Determinants That Impact Sensitivity to α-Conotoxins

	α-Conotoxin			
	α-BuIA	**α-BuIA**	**α-PnIA**	**α-ImI**
Receptor subunits	α6 vs. α4	β4 vs. β2	α3 vs. α2	α7 vs. α1
Specificity	60,000	60 (k-off)	1,000	50
Location[a]	C loop	Complementary surface (−)	C loop	Complementary surface (−)
Inferred strategy	Allostery	Direct contact	Allostery	Direct contact
Specificity positions (subunit)	185, 187, 188, 198, 205 (α6)	59, 111, 119 (β4)	182, 188, 198 (α3)	55, 59, 77 (α7)
Positions in AChBP[b]	181, 183, 184, 194, 201	54, 106, 114	178, 184, 194	53, 57, 75

[a] In Figure 16.6, specificity determinants are color-coded magenta if they are located on the complementary surface of a subunit and make direct contact with α-conotoxin and color-coded lime green if they are located on beta strands connected by the C loop and make no direct contact with α-conotoxin.

[b] Residue positions in AChBP are indexed as found in the sequence from *Aplysia californica*. Atomic coordinates deposited with the Protein Data Bank (PDB) for AChBP structures often include two N-terminal residues that are remnants following proteolysis of an engineered tag, meaning that residue positions for these structures (*4ez1, 2br8*, etc., but not *2c9t*) will be shifted +2 relative to the positions reported here.

name of this loop—C for cysteines) that are a distinguishing character for α subunits of nicotinic acetylcholine receptors. The vicinal cysteine moiety makes van der Waals (VDW) contact with the agonist. Direct VDW contact is also observed between C loop residues and α-conotoxin antagonists (large gray shape in Figure 16.6b), which bind to the same site found at +/− interfaces between AChBP subunits. However, in the case of antagonists, the C loop is forced into a different position, farther from the adjacent subunit, as a consequence of the bulkier nature of α-conotoxins. In the ligand-free apo-AChBP structure, the C loop adopts a position intermediate to that seen for the agonist and antagonist. The idea that agonist activity correlates directly with C loop closure is reinforced by structural analysis of partial agonists that show partial closure of the C loop in crystal structures.[92] Changes in the position of the C loop can be detected by changes in assembly volume and by the distance measured for vicinal cysteine residues relative to the center of mass of the whole assembly.

Tabulated in Table 16.3 are the positions of residues that have been shown to be critical for subtype selectivity exhibited by certain α-conotoxins. These subtype selectivity determinants were identified by measuring α-conotoxin sensitivity for acetylcholine-responsive ion channels constructed with subunit chimeras.[93–96] Because of close homology between nicotinic acetylcholine receptor subunits and AChBP subunits, it has been possible to map these subtype selectivity determinants to positions on the AChBP crystal structure. Figure 16.6c shows the structure of AChBP in complex with the agonist epibatidine,[92] with residues corresponding to specificity determinants of nicotinic acetylcholine receptor subunits highlighted. These fall into two classes on the basis of whether residues at these positions make direct VDW contact or no contact with α-conotoxin as observed for AChBP-α-conotoxin crystal structures.[92,97,98] The direct-contact specificity determinants localize to one of four antiparallel beta strands found on the complementary (−) surface (magenta stripes in Figure 16.6c). The no-contact specificity determinants are contained within the two beta strands that are connected by the C loop (lime stripes and surfaces in Figure 16.6c).

The structural disposition of the direct-contact and no-contact specificity determinants suggests that there are two molecular strategies for establishing subtype specificity. The direct-contact determinants probably work by a direct-binding recognition mechanism, while the no-contact determinants likely act through an allosteric mechanism. Figure 16.6c maps currently known specificity determinants onto the structure of AChBP in a composite form. These positions are also shown in Figure 16.6d–f, as seen for the individual AChBP-α-conotoxin structures, with only the subset of specificity determinants particular to that α-conotoxin highlighted in each case.

The complex with α-BuIA from *Conus bullatus* (J.M. McIntosh, unpublished results) exhibits both classes of specificity determinants (Figure 16.6d). BuIA appears to use the allosteric mechanism for differentiating between α6 and α4 subunits and the direct-binding mechanism for differentiating between β4 and β2 subunits. Residues in β4 that confer slow dissociation were identified by measuring the rate of ion-current recovery for nicotinic acetylcholine receptors with chimeric β subunits.[95] The direct-binding mechanism is fairly simple to envision and explains the observed 60-fold slower dissociation rate observed for nicotinic acetylcholine receptors constructed with β4 subunits than for receptors constructed with β2 subunits,[95] resulting from differences in residues making contact with α-BuIA. These specificity determinants in β4 correspond with positions in the AChBP that make direct VDW contact with α-BuIA in the co-crystal structure (color-coded magenta in Figure 16.6d). The most straightforward interpretation is that certain amino acid residues create opportunities for favorable intermolecular interactions with α-BuIA, and other amino acid resides at these positions are either less favorable or repulsive with respect to interaction with the α-conotoxin.

A more complicated model is needed to explain how α-BuIA differentiates between homologous α subunits of the nicotinic acetylcholine receptor. Receptors constructed with α6 subunits show a 60,000-fold greater sensitivity to α-BuIA than nicotinic acetylcholine receptors constructed with α4 subunits.[96,99] Five residues found in α6 were shown to be most critical for α-BuIA sensitivity since replacing residues in α4 at the corresponding positions substantially increased α-BuIA sensitivity for α4/α6* chimeric receptors (* indicates that other subunits were present in these receptors in addition to the α4/α6* chimeric subunits). Residues in AChBP corresponding to these specificity determinants make no direct contact with α-BuIA, as seen in the structure of the AChBP-α-BuIA complex (color-coded lime in Figure 16.6d). Clearly, some indirect specificity-engendering mechanism is at work. Prevalent models point to the dynamic property of the C loop and invoke an explanation rooted in the fundamentals of allostery.

Allosteric systems couple binding events with structural changes. In the current system, the C loop repositions in response to binding with the agonist and α-conotoxin. In fact, binding and release of these molecules must be accompanied by movement in the C loop since there is otherwise no route in and out of the binding pocket. What is less obvious is the idea that particular residues found on particular subunits of the nicotinic acetylcholine receptors may alter the dynamic properties of the C loop in such a way as to encourage binding with α-conotoxin in some cases and discourage binding in others. Specificity determinants on the α6 subunits that dramatically increase sensitivity to α-BuIA when transplanted to chimeric α6/α4 subunits are located exactly where one might expect to influence flexibility and motion in the C loop. It is therefore reasonable to consider that the size, shape, polarity, and other properties associated with residues at these positions could facilitate movements that make binding with α-BuIA more likely, and residues with contrasting properties may make binding more challenging and thus decrease conotoxin potency.

The structure of AChBP has been determined in complex with two other α-conotoxins with known subtype selectivity determinants: α-PnIA[97] and α-ImI.[92,98] The α-conotoxins from *Conus pennaceus* have been shown to differentiate between the neuronal nicotinic acetylcholine receptors α3β2 and α7, with α-PnIA more potently inhibiting the α3β2 subtype and α-PnIB more potently inhibiting α7.[100] Leucine at position 10 was found to be critical for α7 recognition; the Ala → Leu substitution at position 10 of α-PnIA caused a switch in subtype preference so that α-PnIA[A10L] behaved similarly with α-PnIB in inhibiting α7 nicotinic acetylcholine receptors most potently.[100] Tracking down the specificity determinants that allowed PnIA to preferentially target receptors of the α3β2 subtype over α2β2 receptors by a factor of 1000 was accomplished by comparing the sensitivity of receptors containing α2/α3* chimeric subunits.[94] The positions of these specificity determinants are reported in Table 16.3, and the corresponding surfaces on the AChBP structure are highlighted in Figure 16.6e, which shows the complex with a variant of α-PnIA, α-PnIA[A10L, D14K].

Similar to the specificity determinants that allow α-BuIA to distinguish among α subunits, the specificity determinants identified for α-PnIA map to regions nearby the C loop but do not make direct contact with α-conotoxin. The idea that specificity could be dictated by residues that do not actually contact the binding agent was first proposed on the basis of homology models of the α3β2 structure that had been constructed before the x-ray crystal structures of α-conotoxin in complex with AChBP became available.[94] These structural investigations led to the highly probable conclusion that α-PnIA recognizes particular nicotinic acetylcholine receptors via an allosteric mechanism involving dynamic properties of the C loop and structural elements closely connected with the C loop.

The α-conotoxin from *Conus imperialis*, α-ImI,[101] differentiates between α7 and α1, preferring the former by a factor of 50, and the specificity determinants have been identified through the chimera receptor approach.[93] The residues on AChBP corresponding with these specificity determinants (Table 16.3) cluster on the complementary surface (–) and make close VDW contact with α-ImI (Figure 16.6f) in crystal structures, leading to the inference that α-ImI recognizes particular nicotinic acetylcholine receptor subtypes via a direct-binding mechanism, akin to how α-BuIA differentiates between β subunits of nicotinic acetylcholine receptors.

The availability of these structures for α-conotoxins complexed with a protein that is highly comparable to their molecular signaling target allows for generalizations regarding the molecular interactions leading to subtype selectivity. It is noteworthy that the especially high degree of specificity demonstrated by α-BuIA and by α-PnIA in recognition of α subunits appears to be accomplished by an allosteric mechanism that involves no direct contact between α-conotoxin and residues on α6 (for α-BuIA) or α3 (for α-PnIA). The direct-binding mechanism inferred for α-BuIA and α-ImI seems substantially less effective at establishing specificity (compare specificity values of 50 and 60 for direct-binding examples and values of 60,000 and 1000 for allostery examples in Table 16.3). It will be interesting to see whether the two strategies described for α-conotoxins are also employed by other conotoxins for establishing selectivity in choice of molecular targets. The hypothesis that the allosteric strategy is generally more effective than the direct-binding strategy will certainly need to be tested and should motivate structure determination for other systems.

While waiting for further experimental evidence, some reasoning of a speculative nature is offered here. The signaling molecules of the neuromuscular system targeted by conopeptides are allosteric systems that undergo structural transitions in response to agonist binding or membrane depolarization. Consequently, conopeptides experiencing evolutionary pressure to improve selectivity have ample opportunities for targeting both static surfaces via the direct-binding modality and structural transitions via the allosteric modality. One reason that the allosteric modality may provide enhanced targeting specificity relates to constraints on the surfaces of conopeptide-binding pockets, which may be highly similar among subtypes of a given signaling family, especially if these binding pockets perform an essential function, such as binding of neurotransmitters. Such constraints are probably significantly relaxed for regions that establish movement and flexibility in structural elements, meaning these regions targeted by the allosteric strategy will be more different among subtypes, and thus more easily differentiable through molecular interactions with the subtype-specific conotoxin.

16.4 CONSTELLATION PHARMACOLOGY

16.4.1 Insights into Cone Snail Envenomation through Toxin Cabals

Research into cone snail venoms, spanning four decades, has elucidated the broad envenomation strategies of various cone snail species. This was first achieved for fish-hunting cone snail species, such as *Conus striatus* and *Conus geographus*. From the venoms of these fish-hunting cone snail species, many different conopeptides were purified and their molecular targets and mechanisms were eventually discovered. These various discoveries were likes pieces to a puzzle that eventually produced a more complete picture of the functional integration between the many structurally diverse peptides from cone snail venoms.

A key insight from this research is that envenomation of a targeted animal produces physiological effects that are elicited by the coordinated activity of multiple peptides within the venom. These functionally coupled groups of peptide toxins are called cabals.[102] The term *cabal* has been used to refer to secret societies that attempt to overthrow existing authority. By analogy, a toxin cabal is a group of peptides that work in concert to alter the physiology of the envenomated animal in a way that facilitates a cone snail's efforts to capture prey, defend itself against other predators, or deter competitors. We refer to the molecular targets of a cabal as a constellation. Typically, these are a set of functionally linked receptors and ion channels.

Each venom may contain peptides that belong to different cabals. Specific examples from fish-hunting cone snail venoms are well understood. The motor cabal comprises a set of venom peptides that efficiently inhibit neuromuscular transmission, thereby producing a muscular paralysis that helps fish-hunting cone snails capture their fish prey.[102] Components of the motor cabal include such well-characterized conotoxins as ω-conotoxins that block presynaptic calcium channels, α-conotoxins and ψ-conotoxins that inhibit nicotinic acetylcholine receptors, and μ-conotoxins that block voltage-gated sodium channels.[102]

In some species of fish-hunting cone snails, including *C. striatus*, a second set of venom peptides causes an immediate tetanic paralysis through the massive depolarization of axons adjacent to the injection site.[103] Collectively, these peptides are known as the lightning-strike cabal and include κ-conotoxins that block voltage-gated potassium channels and δ-conotoxins that delay inactivation of voltage-gated sodium channels.[103] This cabal facilitates fish hunting in cone snails that use a hook-and-line strategy. These are fish-hunting cone snails that harpoon their prey with a barbed radular tooth and then pull their prey into their mouth. The rapid tetanic paralysis helps to prevent the fish from struggling to break away from the radular tooth, giving components of the motor cabal time to circulate to various muscles and motor neurons.

In contrast to hook-and-line hunters such as *C. striatus* described above, *C. geographus* does not elicit the massive hyperexcitability of the nervous system; rather, it elicits a syndrome of sensory deprivation and hypoactivity through venom components collectively called the nirvana cabal.[104] This appears to facilitate fish hunting in the few cone snail species that employ a net strategy, distending their mouth to engulf their fish prey.

In essence, to achieve desired biological endpoints, cone snails use a combination drug-targeting strategy. There are multiple components in each cabal, and prey capture seems to require multiple cabals. Cumulatively, the diverse venom components that belong to multiple toxin cabals comprise an extremely sophisticated pharmacological toolkit.

The implication of the cabal strategy in cone snail envenomation is that the structurally diverse conopeptides have evolved to cooperatively interact with a large number of different molecular targets. Presumably, there has been positive selection for any bioactive peptide that makes a cabal more efficient and effective. Considering the vast biodiversity of cone snail species and the presumably greater diversity of their prey, predators, and competitors, the ion channels and receptors targeted by the venom of a particular species cannot be predicted a priori. Part of the unpredictability stems from the diverse expression patterns of receptor and ion channel subtypes across species. Although receptor and ion channel subtypes are highly conserved structurally across species, their expression patterns may not be conserved.

The range of ion channel and receptor targets that have been discovered so far has been remarkable. As just one example, it was not expected that there would be peptides in cone snail venoms that target specific isoforms of glutamate receptors because the prey of most cone snail lineages utilize acetylcholine as the neurotransmitter at their neuromuscular junctions. Nevertheless, among the very first peptides that were characterized from *C. geographus* were the "sleeper peptides," the conantokins, which are subtype-selective N-methyl-D-aspartate (NMDA) receptor antagonists.[17,105,106] These are apparently a component of the nirvana cabal. In addition, fish-hunting cone snails such as *C. striatus* have peptides such as con-ikot-ikot S that inhibit the desensitization of specific subtypes of α-amino-3-hydroxy-5-methyl-4-isoxazolepropionic acid (AMPA) receptors, presumably as part of the lightning-strike cabal.[107] These are just a few of the structurally and functionally diverse conopeptides that have been characterized so far. Even though many conopeptides have become useful tools for pharmacological research, there are far more conopeptides that remain undiscovered and uncharacterized. These have tremendous potential for both basic research and drug discovery.

16.4.2 Toxin Cabals as a Conceptual Basis for Constellation Pharmacology

Prior research on *Conus* peptides might be viewed as a general effort to deconstruct the cabals that cone snails have evolved to target functionally coupled constellations of receptors and ion channels for predation or defense. The toxin cabals have become the inspiration for a conceptual and experimental advance that we call constellation pharmacology.[108–113] Concisely, constellation pharmacology reconstitutes biomedically useful cabals comprising a diverse set of pharmacological agents, including venom peptides, with the aim of defining functionally coupled signaling components. We will refer to these as toolkit cabals. The potential experimental advances to be derived from constellation pharmacology are far-reaching and include the following: (1) the elucidation of functionally coupled components (constellations) of a signaling pathway in a single cell or a circuit; (2) a high-content phenotypic screening assay for drug discovery, where diverse cell-specific constellations in a heterogeneous culture of cells provide the high-content assay; and (3) the potential to develop toolkit cabals for direct therapeutic applications.

16.4.2.1 Elucidating Constellations through Pharmacology

As described above, many different signaling components are functionally linked, either in a neural circuit or in some other signaling pathway, and these functionally linked signaling components are the targets of the natural toxin cabals. Although the cognate constellation of a cabal may span cellular networks, our initial development of constellation pharmacology has focused on elucidating constellations within single cells,[108–113] as the cell is the basic unit of biological integration. In the case of the cells that form nervous systems, plasma membrane receptors and ion channels are some of the most important determinants of cellular identity. In order to elucidate the function of individual cell types, it is essential to identify the combination of signaling molecules that are functionally expressed in each cell and how these signaling molecules are functionally coupled to confer unique physiological properties upon each cell type. In essence, cell-specific constellations create different neuronal cell types with distinct physiological properties, functions, and roles.

In the ongoing development of constellation pharmacology, we use subtype-selective pharmacological agents to elucidate the cell-specific constellations of key signaling proteins that demarcate specific cell types within heterogeneous native cell populations. At the present time, we are utilizing calcium imaging for constellation pharmacology because we have successfully developed calcium imaging assays to interrogate a broad spectrum of receptors and ion channels across an entire cell population simultaneously. The high-content nature of calcium imaging is thus a significant advantage for this approach. However, we do not use the term *constellation pharmacology* as a synonym for calcium imaging. On the contrary, as a conceptual and experimental advance, constellation pharmacology could potentially also be accomplished by combining subtype-selective pharmacology with other methods, including high-throughput electrophysiology and imaging with voltage-sensitive dyes. As described below for genetically labeled cell types, we incorporate additional technologies and techniques into this experimental approach whenever it is advantageous to do so.

The elucidation of cell-specific constellations is important work on its own, as it is increasingly recognized that a deeper understanding of nervous systems will require a definitive and comprehensive classification scheme for neuronal cell types.[114–116] Constellation pharmacology has the potential to provide such a classification scheme on the basis of differences in cell-specific constellations.

Using the constellation pharmacology approach, we first differentiate between each of the diverse neuronal and glial cell types present in a cell population by their cell-specific expression of various ion channel and receptor subtypes. To do so, we sequentially apply various subtype-selective pharmacological agents to a large population of native neurons and glia, while monitoring the responses of 100–200 cells individually and simultaneously by calcium imaging. The output of such an experiment is a large data set that is sufficient to populate a matrix of cell types and several components of each cell-type-specific constellation.[108,110–113]

The prior work to deconstruct toxin cabals has become the basis for reconstructing toolkit cabals that elucidate cell-specific constellations. To elucidate cell-specific constellations, we may utilize a series of receptor agonists and antagonists in combination with calcium imaging, which can uncover the ligand-gated ion channel subtypes and some G protein-coupled receptor (GPCR) subtypes that are differentially expressed within specific cell types.[108–113] For example, a transient increase in cytosolic calcium concentration may be elicited by transient application of acetylcholine. The elicited calcium signal may be blocked by antagonists of acetylcholine receptors. The α-conotoxins, which inhibit specific nicotinic receptor subtypes, have proven to be useful for this type of experimental protocol.[108–110] Additionally, we may block voltage-gated sodium, calcium, or potassium channels with subtype-selective inhibitors, as a means of identifying which voltage-gated ion channel subtypes are differentially expressed across neuronal cell types. Blockers of voltage-gated sodium and calcium channels decrease the calcium signal elicited by a depolarizing stimulus, while blockers of voltage-gated potassium channels amplify the calcium signal elicited by depolarization (Figure 16.7). To elucidate cell-specific constellations of voltage-gated ion channels, we have successfully used various ω-conotoxins (calcium channel blockers), κM- and κJ-conotoxins (potassium channel blockers), and μO-conotoxins (sodium channel blockers).[110–113] Historically, the pharmacology of sodium channels was restricted to simple differentiation between tetrodotoxin-sensitive or -insensitive channels, but the recent use of μ-conotoxins revealed specific combinations of voltage-gated sodium channel subtypes expressed in small-diameter and large-diameter dorsal root ganglion (DRG) neurons.[117] In summary, we utilize various types of pharmacological agents (in combination with other physiochemical stimuli) to uncover the cell-specific expression patterns of ligand-gated ion channel subtypes (neurotransmitter receptors), GPCR subtypes (neuromodulator receptors), and voltage-gated ion channel subtypes. Cumulatively, these are key differentiators of neuronal cell types.

As more subtype-selective venom peptides are discovered, it will become increasingly easier to elucidate the combinations of receptor and ion channel subtypes expressed in specific cell types, circuits, and cellular networks. The recent report that there are not only conopeptides that can distinguish between the different types of α subunits present in voltage-gated sodium channels, but also a conopeptide that can be used to discriminate between different β subunits, illustrates the frontiers of pharmacology that can be explored using newly discovered conopeptides.[56]

In addition to pharmacological challenges, we have started to use genetically labeled cell types (in collaboration with the David Ginty lab at Harvard) in combination with constellation pharmacology, as demonstrated in Figure 16.7. These genetically labeled cells express fluorescent proteins from cell-type-specific promoters. Their fluorescent labeling allows these cells to be sorted by fluorescence-activated cell sorting (FACS) for quantitative analysis of transcript expression levels.

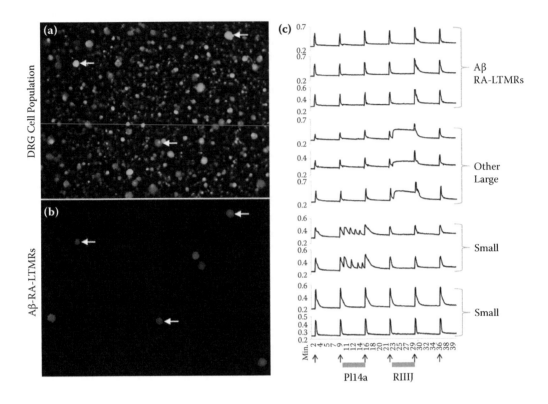

Figure 16.7 We have combined constellation pharmacology with the genetic labeling of cell types. (a) One field of view shows dissociated DRG cells loaded with Fura-2 dye. (b) The same field of view reveals a subset of large-diameter DRG neurons that express the red fluorescent protein tdTomato. The labeled cells are rapidly adapting low-threshold mechanoreceptors that have myelinated Aβ fibers (Aβ-RA-LTMRs). (c) Calcium imaging traces show the response of selected individual neurons to depolarizing stimulus (high concentration of extracellular potassium; see arrows at the bottom of the panel). These particular traces were selected from those obtained for the several hundred neurons that were simultaneously monitored. The topmost experimental group shows the responses obtained for Aβ-RA-LTMR neurons. These neurons were unaffected by a blocker of $K_V1.6$ channels (κJ-conotoxin pl14a), but a blocker of $K_V1.2$ channels (κM-conotoxin RIIIJ) modestly amplified the response to a depolarizing stimulus in these Aβ-RA-LTMRs. Other types of large- or small-diameter neurons exhibited different phenotypes in response to these stimuli. (Modified from Teichert, R. W., et al., *Annu Rev Pharmacol Toxicol* 2014, 55, 249–267. With permission from *Annual Review of Pharmacology and Toxicology*.)

We predict that the combination of genetically labeled cells and constellation pharmacology will become a powerful approach to identify neuronal cell types and elucidate their physiological roles by identifying their functionally coupled cell-specific constellations.

16.4.2.2 Constellation Pharmacology as a High-Content Screening Platform

By elucidating the spectrum of cell-specific constellations across a population of heterogeneous cell types, not only are all of the cell types differentiated from each other, but also this effort provides a basis for cell-based high-content screening. For example, we have used constellation pharmacology to discover that a novel venom peptide affects a subset of small-diameter capsaicin-sensitive somatosensory neurons, with activity that is consistent with blocking voltage-gated potassium channels.[10] This activity profile and the spectrum of responsive cells will help us to identify its molecular target. We have also used this experimental approach to purify a novel peptide from venom (by following its activity in calcium imaging assays) and identify its molecular target and mechanism of action, which is to delay inactivation of voltage-gated sodium channels.[8]

16.4.2.3 Toolkit Cabals for Therapeutic Applications

As an increasing number of subtype-selective venom peptides are discovered and characterized, these peptides can be used to create toolkit cabals not only for basic research, but also for direct diagnostic and therapeutic applications. A cabal may target a constellation of signaling components to produce a desired therapeutic endpoint. In a few cases, this principle has already been applied to clinical medicine. For example, there are effective cocktails of HIV drugs, with each cocktail component targeting one of several functionally linked proteins necessary for viral infection and replication, including integrases, proteases, and reverse transcriptases.[118,119] In principle, a cabal could be constructed to maximize a therapeutic effect because of synergies achieved by targeting multiple components of a signaling pathway, while simultaneously minimizing side effects because each component of the cabal could be administered at relatively low dosages to achieve such synergies.

16.5 PERSPECTIVES

16.5.1 Discovery and Validation of Novel Drug Targets

There are multiple examples of conopeptides that have been used to discover and validate novel drug targets and mechanisms for alleviating pain. Perhaps the best example is the development of ω-conotoxin MVIIA (ziconotide, Prialt), which showed that $Ca_V2.2$ (the N-type calcium channel) is a potential drug target for the treatment of severe intractable pain. Prior to the development of Prialt, the $Ca_V2.2$ channel had not been considered a drug target for the treatment of intractable pain. However, after Prialt was used to validate $Ca_V2.2$ as a drug target, there was an effort in both Australia and Canada focused on developing small-molecule analgesic drugs that target $Ca_V2.2$. In addition to Prialt, other conopeptides have reached clinical trials for the treatment of pain, as summarized in Table 16.2. These peptides include Vc1.1, which blocks a nicotinic acetylcholine receptor subtype ($α9α10$), and contulakin-G, which blocks the neurotensin-1 receptor. These drug targets and mechanisms for alleviating pain were discovered and validated by subtype-selective conopeptides.

The prior contributions of subtype-selective conopeptides pale in comparison with future prospects. Of the >100,000 unique conopeptides that are expressed in cone snail venoms, only a tiny fraction have been evaluated for targeting selectivity. Previously, large quantities of each peptide were required for target identification and selectivity testing because the bioassays used to identify a peptide's activity did not significantly narrow the list of possible molecular targets. Furthermore, testing each peptide on one potential molecular target at a time was slow and cumbersome. We predict that the combination of constellation pharmacology with genetically labeled cell types from reporter mice and their corresponding transcriptomes will greatly facilitate target identification when constellation pharmacology is used as a high-content screening platform for cone snail venoms, venom fractions, or pure peptides. Consequently, the targeting selectivity of novel venom peptides should be evaluated at an accelerated rate. This should in turn facilitate the discovery and validation of novel drug targets and mechanisms for a variety of nervous system diseases.

16.5.2 Discovery and Characterization of Native Heteromeric Ion Channels

Among the most underexplored molecular targets for drug discovery are heteromeric ligand- and voltage-gated ion channels. In order to form a functional pore, many ion channels require that several subunits assemble together (usually four or five), with the possibility of distinct subunits forming a heteromeric functional assembly. Presumably, all families of ion channels that are formed by multimeric assembly include heteromeric combinations of subunits (as well as homomeric combinations).[120] For example, these include all known families of ligand-gated ion channels, such as glutamate receptors, nicotinic acetylcholine receptors, gamma-aminobutyric (GABA) receptors, transient receptor potential (TRP) channels, P2X purino receptors, and cyclic nucleotide-gated channels. The potassium channels are also formed by multimeric assembly into heteromers. In contrast, sodium channels and calcium channels are each encoded by a single gene in which four domains are covalently linked in a single large α subunit. Since there are ~70 different genes that encode potassium channel subunits (in mammals), the number of possible heteromeric potassium channels is vast. The rules for which subunits may or may not assemble to form a functional channel have not been definitively established. Therefore, the potassium channel family probably represents the greatest challenge for pharmacology in the future, but may also provide one of the best opportunities to identify differentiating molecular targets for drug discovery. Conopeptides that target voltage-gated potassium channels include those that have demonstrated cardioprotection during reperfusion after an ischemic episode (such as a myocardial infarction) in preclinical animal models (Table 16.2).

Not only are heteromeric channels potential drug targets, but they are also important targets that can be functionally perturbed by pharmacology for basic research. Among ion channel families, it seems likely that potassium channels may comprise the greatest diversity of heteromeric subtypes and, as a consequence, may well be among the best ion channels for differentiating between cell types present at any particular anatomical locus. Every excitable cell and every cell type capable of regulated secretion expresses potassium channels, and in many cases, a diverse set of potassium channels is present. In most cell types, the heteromeric or homomeric potassium channel subtypes that are expressed are unknown. An accumulating body of evidence suggests that heteromeric potassium channels are expressed in a number of cell types.[112,121–123] The identification, classification, and functional characterization of specific cell types are now viewed as an emerging frontier in biology that will lead to a deeper understanding of integrated molecular and cellular functions. Subtype-selective agents that target potassium channels may be used to differentiate between cell types and selectively perturb the function of specific cell types. Thus, there is an urgent need to discover novel pharmacological agents that target various heteromeric potassium channel subtypes with a high degree of selectivity. The venom peptides from cone snails and other venomous mollusks are a rich resource for this type of future pharmacological discovery and application.

16.5.3 Future Prospects

Because of their targeting selectivity, venom peptides from cone snails have had an impressive biomedical impact, as detailed in the sections above. One has become an approved drug, and a handful of others have reached human clinical trials; others are at various preclinical stages of drug development. There are diagnostic applications as well. Studies on the individual peptides revealed novel molecular targets with potential for drug development programs.

Furthermore, many peptides from cone snail venoms have been used as important research tools in neuroscience, particularly for synaptic physiology and defining the role of different isoforms of nicotinic receptors and voltage-gated ion channels. Some of these peptides were critical in the characterization of their molecular targets in biochemical, physiological, and pathological investigations. Thus, for the targets of ω-conotoxins such as ω-GVIA or ω-MVIIC, conotoxins were critical for the biochemical purification and further characterization of the targeted calcium channel subtype and in addressing the role of the molecular target in the pathophysiology of a disease state (e.g., neuropathic pain).

A major theme of this chapter is that although the biomedical applications of cone snail venom peptides have been impressive so far, the prospects for the future are far brighter. There are several reasons for this optimism. First, using conopeptides in combination has seeded the development of a new platform for discovery described above, constellation pharmacology. The accumulation of highly subtype-selective conopeptides was a foundation for this platform, which should greatly accelerate the discovery and definition of novel conopeptides in the future. Constellation pharmacology provides a concrete road map for identifying the molecular target of each peptide to be characterized. It also makes it possible to quickly scan venoms, venom fractions, or individual peptides for their ability to have discriminating functional effects for specific cell types. Both aspects make the discovery process far more efficient.

The accelerated discovery of novel *Conus* peptides and their respective molecular targets should facilitate the assembly of more effective pharmacological combinations for neuroscience research. The applications for research should expand rapidly with more and more novel, highly selective conopeptides characterized. As noted above, a particularly exciting possibility is the discovery of conopeptides that have novel specificity for heteromeric subtypes of ligand-gated and voltage-gated ion channels.

Constellation pharmacology will initially impact the research community, but there is little question that the results will ultimately have important biomedical applications. In the long term, the direct therapeutic and diagnostic applications of conopeptides may mimic how they are used by the cone snails—as cabals targeted to constellations. However, instead of undermining normal physiology, conopeptides would then be routinely used as components of toolkit cabals for beneficial biomedical purposes.

ACKNOWLEDGMENTS

The research work of the authors has been supported by the National Institutes of General Medical Sciences (program project grant GM48677). We are grateful to Terry Merritt for typing the manuscript and My Huynh for preparing figures. Dr. Patrice Corneli prepared the phylogenetic tree shown in Figure 16.4, based on data supplied by Maren Watkins.

REFERENCES

1. Terlau, H., Olivera, B. M. *Conus* venoms: A rich source of novel ion channel-targeted peptides. *Physiol Rev* 2004, 84 (1), 41–68.
2. Dance, S. P. *Shell Collecting: An Illustrated History*. University of California Press, Berkeley CA, 1966, p. 343.
3. Rumphius, G. E. *D'Amboinsche Rariteikamer*. Fr. Halma, Amsterdam, 1705.
4. Flecker, H. Cone shell mollusc poisoning, with report of a fatal case. *Med J Aust* 1936, 1, 464–66.
5. Kohn, A. J. The ecology of *Conus* in Hawaii. *Ecol Monogr* 1959, 29 (1), 47–90.
6. Kohn, A. J. Food specialization in *Conus* in Hawaii and California. *Ecology* 1966, 47 (6), 1041–43.
7. Puillandre, N., Duda, T. F., Meyer, C., Olivera, B. M., Bouchet, P. One, four or 100 genera? A new classification of the cone snails. *J Molluscan Stud* 2014. doi: 10.1093/mollus/eyu055.
8. Aman, J. W., Imperial, J. S., Ueberheide, B., Zhang, M. M., Aguilar, M., Taylor, D., Watkins, M., Yoshikami, D., Showers-Corneli, P., Safavi-Hemami, H., Biggs, J., Teichert, R. W., Olivera, B. M. Insights into the origins of fish hunting in venomous cone snails from studies of *Conus* tessulatus. *Proc Natl Acad Sci USA* 2015, 112 (16), 5087–92.
9. Olivera, B. M., Teichert, R. W. Diversity of the neurotoxic *Conus* peptides: A model for concerted pharmacological discovery. *Mol Interv* 2007, 7 (5), 251–260.

10. Imperial, J. S., Cabang, A. B., Song, J., Raghuraman, S., Gajewiak, J., Watkins, M., Showers-Corneli, P., Fedosov, A., Concepcion, G. P., Terlau, H., Teichert, R. W., Olivera, B. M. A family of excitatory peptide toxins from venomous crassispirine snails: Using constellation pharmacology to assess bioactivity. *Toxicon* 2014, 89, 45–54.

11. Olivera, B. M., Showers Corneli, P., Watkins, M., Fedosov, A. Biodiversity of cone snails and other venomous marine gastropods: Evolutionary success through neuropharmacology. *Annu Rev Animal Biosci* 2014, 2, 487–513.

12. Puillandre, N., Kantor, Y. I., Sysoev, A., Couloux, A., Meyer, C., Rawlings, T., Todd, J. A., Bouchet, P. The dragon tamed? A molecular phylogeny of the Conoidea (Gastropoda). *J Molluscan Stud* 2011, 77 (3), 259–72.

13. Endean, R., Parish, G., Gyr, P. Pharmacology of the venom of *Conus geographus*. *Toxicon* 1974, 12 (2), 131–38.

14. Spence, I., Gillessen, D., Gregson, R. P., Quinn, R. J. Characterization of the neurotoxic constituents of *Conus geographus* (L) venom. *Life Sci* 1977, 21 (12), 1759–69.

15. McManus, O. B., Musick, J. R., Gonzalez, C. Peptides isolated from the venom of *Conus geographus* block neuromuscular transmission. *Neurosci Lett* 1981, 25 (1), 57–62.

16. Kohn, A. J., Saunders, P. R., Wiener, S. Preliminary studies on the venom of the marine snail *Conus*. *Ann NY Acad Sci* 1960, 90, 706–25.

17. Olivera, B. M., Gray, W. R., Zeikus, R., McIntosh, J. M., Varga, J., Rivier, J., de Santos, V., Cruz, L. J. Peptide neurotoxins from fish-hunting cone snails. *Science* 1985, 230 (4732), 1338–43.

18. Olivera, B. M., Rivier, J., Clark, C., Ramilo, C. A., Corpuz, G. P., Abogadie, F. C., Mena, E. E., Woodward, S. R., Hillyard, D. R., Cruz, L. J. Diversity of *Conus* neuropeptides. *Science* 1990, 249 (4966), 257–63.

19. Hillyard, D. R., Olivera, B. M., Woodward, S., Corpuz, G. P., Gray, W. R., Ramilo, C. A., Cruz, L. J. A molluscivorous *Conus* toxin: Conserved frameworks in conotoxins. *Biochemistry* 1989, 28 (1), 358–61.

20. Cruz, L. J., Kupryszewski, G., LeCheminant, G. W., Gray, W. R., Olivera, B. M., Rivier, J. mu-Conotoxin GIIIA, a peptide ligand for muscle sodium channels: Chemical synthesis, radiolabeling, and receptor characterization. *Biochemistry* 1989, 28 (8), 3437–42.

21. Cruz, L. J., Gray, W. R., Olivera, B. M., Zeikus, R. D., Kerr, L., Yoshikami, D., Moczydlowski, E. *Conus geographus* toxins that discriminate between neuronal and muscle sodium channels. *J Biol Chem* 1985, 260 (16), 9280–88.

22. Olivera, B. M., Miljanich, G. P., Ramachandran, J., Adams, M. E. Calcium channel diversity and neurotransmitter release: The omega-conotoxins and omega-agatoxins. *Annu Rev Biochem* 1994, 63, 823–67.

23. Gray, W. R., Luque, A., Olivera, B. M., Barrett, J., Cruz, L. J. Peptide toxins from *Conus geographus* venom. *J Biol Chem* 1981, 256 (10), 4734–40.

24. Hillyard, D. R., Monje, V. D., Mintz, I. M., Bean, B. P., Nadasdi, L., Ramachandran, J., Miljanich, G., Azimi-Zoonooz, A., McIntosh, J. M., Cruz, L. J., Imperial, J. S., Olivera, B. M. A new *Conus* peptide ligand for mammalian presynaptic Ca2+ channels. *Neuron* 1992, 9 (1), 69–77.

25. Han, T. S., Teichert, R. W., Olivera, B. M., Bulaj, G. *Conus* venoms: A rich source of peptide-based therapeutics. *Curr Pharm Des* 2008, 14 (24), 2462–79.

26. Mahadeva, B., Phillips, L. H., 2nd, Juel, V. C. Autoimmune disorders of neuromuscular transmission. *Semin Neurol* 2008, 28 (2), 212–27.

27. Lennon, V. A., Lambert, E. H. Autoantibodies bind solubilized calcium channel-omega-conotoxin complexes from small cell lung carcinoma: A diagnostic aid for Lambert-Eaton myasthenic syndrome. *Mayo Clin Proc* 1989, 64 (12), 1498–504.

28. Titulaer, M. J., Lang, B., Verschuuren, J. J. Lambert-Eaton myasthenic syndrome: From clinical characteristics to therapeutic strategies. *Lancet Neurol* 2011, 10 (12), 1098–107.

29. Chang, D., Duda, T. F., Jr. Extensive and continuous duplication facilitates rapid evolution and diversification of gene families. *Mol Biol Evol* 2012, 29 (8), 2019–29.

30. Duda, T. F., Jr., Remigio, E. A. Variation and evolution of toxin gene expression patterns of six closely related venomous marine snails. *Mol Ecol* 2008, 17 (12), 3018–32.

31. Puillandre, N., Watkins, M., Olivera, B. M. Evolution of *Conus* peptide genes: Duplication and positive selection in the A-superfamily. *J Mol Evol* 2010, 70 (2), 190–202.

32. McIntosh, J. M., Santos, A. D., Olivera, B. M. *Conus* peptides targeted to specific nicotinic acetylcholine receptor subtypes. *Annu Rev Biochem* 1999, 68, 59–88.

33. Tsetlin, V., Utkin, Y., Kasheverov, I. Polypeptide and peptide toxins, magnifying lenses for binding sites in nicotinic acetylcholine receptors. *Biochem Pharmacol* 2009, 78 (7), 720–31.

34. Azam, L., Maskos, U., Changeux, J. P., Dowell, C. D., Christensen, S., De Biasi, M., McIntosh, J. M. alpha-Conotoxin BuIA[T5A;P6O]: A novel ligand that discriminates between alpha6ss4 and alpha6ss2 nicotinic acetylcholine receptors and blocks nicotine-stimulated norepinephrine release. *FASEB J* 2010, 24 (12), 5113–23.

35. Van Der Haegen, A., Peigneur, S., Tytgat, J. Importance of position 8 in mu-conotoxin KIIIA for voltage-gated sodium channel selectivity. *FEBS J* 2011, 278 (18), 3408–18.

36. Bandyopadhyay, P. K., Colledge, C. J., Walker, C. S., Zhou, L. M., Hillyard, D. R., Olivera, B. M. Conantokin-G precursor and its role in gamma-carboxylation by a vitamin K-dependent carboxylase from a *Conus* snail. *J Biol Chem* 1998, 273 (10), 5447–50.

37. Buczek, O., Olivera, B. M., Bulaj, G. Propeptide does not act as an intramolecular chaperone but facilitates protein disulfide isomerase-assisted folding of a conotoxin precursor. *Biochemistry* 2004, 43 (4), 1093–101.

38. Conticello, S. G., Kowalsman, N. D., Jacobsen, C., Yudkovsky, G., Sato, K., Elazar, Z., Petersen, C. M., Aronheim, A., Fainzilber, M. The prodomain of a secreted hydrophobic mini-protein facilitates its export from the endoplasmic reticulum by hitchhiking on sorting receptors. *J Biol Chem* 2003, 278 (29), 26311–14.

39. Woodward, S. R., Cruz, L. J., Olivera, B. M., Hillyard, D. R. Constant and hypervariable regions in conotoxin propeptides. *EMBO J* 1990, 9 (4), 1015–20.

40. Duda, T. F., Jr., Palumbi, S. R. Evolutionary diversification of multigene families: Allelic selection of toxins in predatory cone snails. *Mol Biol Evol* 2000, 17 (9), 1286–93.

41. Olivera, B. M., Walker, C., Cartier, G. E., Hooper, D., Santos, A. D., Schoenfeld, R., Shetty, R., Watkins, M., Bandyopadhyay, P., Hillyard, D. R. Speciation of cone snails and interspecific hyperdivergence of their venom peptides: Potential evolutionary significance of introns. *Ann NY Acad Sci* 1999, 870, 223–37.

42. Wu, Y., Wang, L., Zhou, M., You, Y., Zhu, X., Qiang, Y., Qin, M., Luo, S., Ren, Z., Xu, A. Molecular evolution and diversity of *Conus* peptide toxins, as revealed by gene structure and intron sequence analyses. *PLoS ONE* 2013, 8 (12), e82495.

43. Conticello, S. G., Gilad, Y., Avidan, N., Ben-Asher, E., Levy, Z., Fainzilber, M. Mechanisms for evolving hypervariability: The case of conopeptides. *Mol Biol Evol* 2001, 18 (2), 120–31.

44. Espiritu, D. J., Watkins, M., Dia-Monje, V., Cartier, G. E., Cruz, L. J., Olivera, B. M. Venomous cone snails: Molecular phylogeny and the generation of toxin diversity. *Toxicon* 2001, 39 (12), 1899–916.

45. Bulaj, G., Zhang, M. M., Green, B. R., Fiedler, B., Layer, R. T., Wei, S., Nielsen, J. S., Low, S. J., Klein, B. D., Wagstaff, J. D., Chicoine, L., Harty, T. P., Terlau, H., Yoshikami, D., Olivera, B. M. Synthetic muO-conotoxin MrVIB blocks TTX-resistant sodium channel NaV1.8 and has a long-lasting analgesic activity. *Biochemistry* 2006, 45 (23), 7404–14.

46. Fainzilber, M., van der Schors, R., Lodder, J. C., Li, K. W., Geraerts, W. P., Kits, K. S. New sodium channel-blocking conotoxins also affect calcium currents in *Lymnaea* neurons. *Biochemistry* 1995, 34 (16), 5364–71.

47. de Araujo, A. D., Callaghan, B., Nevin, S. T., Daly, N. L., Craik, D. J., Moretta, M., Hopping, G., Christie, M. J., Adams, D. J., Alewood, P. F. Total synthesis of the analgesic conotoxin MrVIB through selenocysteine-assisted folding. *Angew Chem Int Ed Engl* 2011, 50 (29), 6527–29.

48. Bulaj, G., Buczek, O., Goodsell, I., Jimenez, E. C., Kranski, J., Nielsen, J. S., Garrett, J. E., Olivera, B. M. Efficient oxidative folding of conotoxins and the radiation of venomous cone snails. *Proc Natl Acad Sci USA* 2003, 100 (Suppl 2), 14562–68.

49. Lopez-Vera, E., Walewska, A., Skalicky, J. J., Olivera, B. M., Bulaj, G. Role of hydroxyprolines in the *in vitro* oxidative folding and biological activity of conotoxins. *Biochemistry* 2008, 47 (6), 1741–51.

50. Safavi-Hemami, H., Bulaj, G., Olivera, B. M., Williamson, N. A., Purcell, A. W. Identification of *Conus* peptidylprolyl cis-trans isomerases (PPIases) and assessment of their role in the oxidative folding of conotoxins. *J Biol Chem* 2010, 285 (17), 12735–46.

51. Safavi-Hemami, H., Gorasia, D. G., Steiner, A. M., Williamson, N. A., Karas, J. A., Gajewiak, J., Olivera, B. M., Bulaj, G., Purcell, A. W. Modulation of conotoxin structure and function is achieved through a multienzyme complex in the venom glands of cone snails. *J Biol Chem* 2012, 287 (41), 34288–303.

52. Wang, Z. Q., Han, Y. H., Shao, X. X., Chi, C. W., Guo, Z. Y. Molecular cloning, expression and characterization of protein disulfide isomerase from *Conus marmoreus*. *FEBS J* 2007, 274 (18), 4778–87.

53. Safavi-Hemami, H., Siero, W. A., Gorasia, D. G., Young, N. D., Macmillan, D., Williamson, N. A., Purcell, A. W. Specialisation of the venom gland proteome in predatory cone snails reveals functional diversification of the conotoxin biosynthetic pathway. *J Proteome Res* 2011, 10 (9), 3904–19.

54. Wang, C. C. Protein disulfide isomerase as an enzyme and a chaperone in protein folding. *Methods Enzymol* 2002, 348, 66–75.

55. Buczek, O., Bulaj, G., Olivera, B. M. Conotoxins and the posttranslational modification of secreted gene products. *Cell Mol Life Sci* 2005, 62 (24), 3067–79.

56. Gajewiak, J., Azam, L., Imperial, J., Walewska, A., Green, B. R., Bandyopadhyay, P. K., Raghuraman, S., Ueberheide, B., Bern, M., Zhou, H. M., Minassian, N. A., Hagan, R. H., Flinspach, M., Liu, Y., Bulaj, G., Wickenden, A. D., Olivera, B. M., Yoshikami, D., Zhang, M. M. A disulfide tether stabilizes the block of sodium channels by the conotoxin muO section sign-GVIIJ. *Proc Natl Acad Sci USA* 2014, 111 (7), 2758–63.

57. McIntosh, J. M., Olivera, B. M., Cruz, L. J., Gray, W. R. Gamma-carboxyglutamate in a neuroactive toxin. *J Biol Chem* 1984, 259 (23), 14343–46.

58. Stanley, T. B., Stafford, D. W., Olivera, B. M., Bandyopadhyay, P. K. Identification of a vitamin K-dependent carboxylase in the venom duct of a *Conus snail*. *FEBS Lett* 1997, 407 (1), 85–88.

59. Craig, A. G., Jimenez, E. C., Dykert, J., Nielsen, D. B., Gulyas, J., Abogadie, F. C., Porter, J., Rivier, J. E., Cruz, L. J., Olivera, B. M., McIntosh, J. M. A novel post-translational modification involving bromination of tryptophan: Identification of the residue, L-6-bromotryptophan, in peptides from *Conus imperialis* and *Conus radiatus* venom. *J Biol Chem* 1997, 272 (8), 4689–98.

60. Fujii, R., Yoshida, H., Fukusumi, S., Habata, Y., Hosoya, M., Kawamata, Y., Yano, T., Hinuma, S., Kitada, C., Asami, T., Mori, M., Fujisawa, Y., Fujino, M. Identification of a neuropeptide modified with bromine as an endogenous ligand for GPR7. *J Biol Chem* 2002, 277 (37), 34010–16.

61. Jimenez, E. C., Craig, A. G., Watkins, M., Hillyard, D. R., Gray, W. R., Gulyas, J., Rivier, J. E., Cruz, L. J., Olivera, B. M. Bromocontryphan: Post-translational bromination of tryptophan. *Biochemistry* 1997, 36 (5), 989–94.

62. Rigby, A. C., Lucas-Meunier, E., Kalume, D. E., Czerwiec, E., Hambe, B., Dahlqvist, I., Fossier, P., Baux, G., Roepstorff, P., Baleja, J. D., Furie, B. C., Furie, B., Stenflo, J. A conotoxin from *Conus textile* with unusual posttranslational modifications reduces presynaptic Ca2+ influx. *Proc Natl Acad Sci USA* 1999, 96 (10), 5758–63.

63. Pisarewicz, K., Mora, D., Pflueger, F. C., Fields, G. B., Mari, F. Polypeptide chains containing D-gamma-hydroxyvaline. *J Am Chem Soc* 2005, 127 (17), 6207–15.

64. Tayo, L. L., Lu, B., Cruz, L. J., Yates, J. R., 3rd. Proteomic analysis provides insights on venom processing in *Conus textile*. *J Proteome Res* 2010, 9 (5), 2292–301.

65. Sandall, D. W., Satkunanathan, N., Keays, D. A., Polidano, M. A., Liping, X., Pham, V., Down, J. G., Khalil, Z., Livett, B. G., Gayler, K. R. A novel alpha-conotoxin identified by gene sequencing is active in suppressing the vascular response to selective stimulation of sensory nerves *in vivo*. *Biochemistry* 2003, 42 (22), 6904–11.

66. Satkunanathan, N., Livett, B., Gayler, K., Sandall, D., Down, J., Khalil, Z. Alpha-conotoxin Vc1.1 alleviates neuropathic pain and accelerates functional recovery of injured neurones. *Brain Res* 2005, 1059 (2), 149–58.

67. Jakubowski, J. A., Keays, D. A., Kelley, W. P., Sandall, D. W., Bingham, J. P., Livett, B. G., Gayler, K. R., Sweedler, J. V. Determining sequences and post-translational modifications of novel conotoxins in *Conus victoriae* using cDNA sequencing and mass spectrometry. *J Mass Spectrom* 2004, 39 (5), 548–57.

68. Safavi-Hemami, H., Siero, W. A., Kuang, Z., Williamson, N. A., Karas, J. A., Page, L. R., MacMillan, D., Callaghan, B., Kompella, S. N., Adams, D. J., Norton, R. S., Purcell, A. W. Embryonic toxin expression in the cone snail *Conus victoriae*: Primed to kill or divergent function? *J Biol Chem* 2011, 286 (25), 22546–57.

69. Townsend, A., Livett, B. G., Bingham, J.-P., Truong, H.-T., Karas, J. A., O'Donnell, P., Williamson, N. A., Purcell, A. W., Scanlon, D. Mass spectral identification of Vc1.1 and differential distribution of conopeptides in the venom duct of *Conus victoriae*: Effect of post-translational modifications and disulfide isomerisation on bioactivity. *Int J Pept Res Ther* 2009, 15 (3), 195–203.

70. Kapono, C. A., Thapa, P., Cabalteja, C. C., Guendisch, D., Collier, A. C., Bingham, J. P. Conotoxin truncation as a post-translational modification to increase the pharmacological diversity within the milked venom of *Conus magus*. *Toxicon* 2013, 70, 170–78.

71. Favreau, P., Krimm, I., Le Gall, F., Bobenrieth, M. J., Lamthanh, H., Bouet, F., Servent, D., Molgo, J., Menez, A., Letourneux, Y., Lancelin, J. M. Biochemical characterization and nuclear magnetic resonance structure of novel alpha-conotoxins isolated from the venom of *Conus consors*. *Biochemistry* 1999, 38 (19), 6317–26.

72. Violette, A., Biass, D., Dutertre, S., Koua, D., Piquemal, D., Pierrat, F., Stocklin, R., Favreau, P. Large-scale discovery of conopeptides and conoproteins in the injectable venom of a fish-hunting cone snail using a combined proteomic and transcriptomic approach. *J Proteomics* 2012, 75 (17), 5215–25.

73. Milne, T. J., Abbenante, G., Tyndall, J. D., Halliday, J., Lewis, R. J. Isolation and characterization of a cone snail protease with homology to CRISP proteins of the pathogenesis-related protein superfamily. *J Biol Chem* 2003, 278 (33), 31105–10.

74. Safavi-Hemami, H., Hu, H., Gorasia, D. G., Bandyopadhyay, P. K., Veith, P. D., Young, N. D., Reynolds, E. C., Yandell, M., Olivera, B. M., Purcell, A. W. Combined proteomic and transcriptomic interrogation of the venom gland of *Conus geographus* uncovers novel components and functional compartmentalization. *Mol Cell Proteomics* 2014, 13 (4), 938–53.

75. Endean, R., Duchemin, C. The venom apparatus of *Conus magus*. *Toxicon* 1967, 4 (4), 275–84.

76. Marshall, J., Kelley, W. P., Rubakhin, S. S., Bingham, J. P., Sweedler, J. V., Gilly, W. F. Anatomical correlates of venom production in *Conus californicus*. *Biol Bull* 2002, 203 (1), 27–41.

77. Page, L. R. Developmental modularity and phenotypic novelty within a biphasic life cycle: Morphogenesis of a cone snail venom gland. *Proc Biol Sci* 2012, 279 (1726), 77–83.

78. Garrett, J. E., Buczek, O., Watkins, M., Olivera, B. M., Bulaj, G. Biochemical and gene expression analyses of conotoxins in *Conus textile* venom ducts. *Biochem Biophys Res Commun* 2005, 328 (1), 362–67.

79. Hu, H., Bandyopadhyay, P. K., Olivera, B. M., Yandell, M. Elucidation of the molecular envenomation strategy of the cone snail *Conus geographus* through transcriptome sequencing of its venom duct. *BMC Genomics* 2012, 13, 284.

80. Dobson, R., Collodoro, M., Gilles, N., Turtoi, A., De Pauw, E., Quinton, L. Secretion and maturation of conotoxins in the venom ducts of *Conus textile*. *Toxicon* 2012, 60 (8), 1370–79.

81. Pardi, A., Galdes, A., Florance, J., Maniconte, D. Solution structures of alpha-conotoxin G1 determined by two-dimensional NMR spectroscopy. *Biochemistry* 1989, 28 (13), 5494–501.

82. Kobayashi, Y., Ohkubo, T., Kyogoku, Y., Nishiuchi, Y., Sakakibara, S., Braun, W., Go, N. Solution conformation of conotoxin GI determined by 1H nuclear magnetic resonance spectroscopy and distance geometry calculations. *Biochemistry* 1989, 28 (11), 4853–60.

83. Lancelin, J. M., Kohda, D., Tate, S., Yanagawa, Y., Abe, T., Satake, M., Inagaki, F. Tertiary structure of conotoxin GIIIA in aqueous solution. *Biochemistry* 1991, 30 (28), 6908–16.

84. Ott, K. H., Becker, S., Gordon, R. D., Ruterjans, H. Solution structure of mu-conotoxin GIIIA analysed by 2D-NMR and distance geometry calculations. *FEBS Lett* 1991, 278 (2), 160–66.

85. Wakamatsu, K., Kohda, D., Hatanaka, H., Lancelin, J. M., Ishida, Y., Oya, M., Nakamura, H., Inagaki, F., Sato, K. Structure-activity relationships of mu-conotoxin GIIIA: Structure determination of active and inactive sodium channel blocker peptides by NMR and simulated annealing calculations. *Biochemistry* 1992, 31 (50), 12577–84.

86. Davis, J. H., Bradley, E. K., Miljanich, G. P., Nadasdi, L., Ramachandran, J., Basus, V. J. Solution structure of omega-conotoxin GVIA using 2-D NMR spectroscopy and relaxation matrix analysis. *Biochemistry* 1993, 32 (29), 7396–405.

87. Chen, L., Durr, K. L., Gouaux, E. X-ray structures of AMPA receptor-cone snail toxin complexes illuminate activation mechanism. *Science* 2014, 345 (6200), 1021–26.
88. Brejc, K., van Dijk, W. J., Klaassen, R. V., Schuurmans, M., van Der Oost, J., Smit, A. B., Sixma, T. K. Crystal structure of an ACh-binding protein reveals the ligand-binding domain of nicotinic receptors. *Nature* 2001, 411 (6835), 269–76.
89. Unwin, N. Nicotinic acetylcholine receptor and the structural basis of neuromuscular transmission: Insights from Torpedo postsynaptic membranes. *Q Rev Biophys* 2013, 46 (4), 283–322.
90. Unwin, N. Refined structure of the nicotinic acetylcholine receptor at 4A resolution. *J Mol Biol* 2005, 346 (4), 967–89.
91. Smit, A. B., Syed, N. I., Schaap, D., van Minnen, J., Klumperman, J., Kits, K. S., Lodder, H., van der Schors, R. C., van Elk, R., Sorgedrager, B., Brejc, K., Sixma, T. K., Geraerts, W. P. A glia-derived acetylcholine-binding protein that modulates synaptic transmission. *Nature* 2001, 411 (6835), 261–68.
92. Hansen, S. B., Sulzenbacher, G., Huxford, T., Marchot, P., Taylor, P., Bourne, Y. Structures of *Aplysia* AChBP complexes with nicotinic agonists and antagonists reveal distinctive binding interfaces and conformations. *EMBO J* 2005, 24 (20), 3635–46.
93. Quiram, P. A., Sine, S. M. Identification of residues in the neuronal alpha7 acetylcholine receptor that confer selectivity for conotoxin ImI. *J Biol Chem* 1998, 273 (18), 11001–6.
94. Everhart, D., Reiller, E., Mirzoian, A., McIntosh, J. M., Malhotra, A., Luetje, C. W. Identification of residues that confer alpha-conotoxin-PnIA sensitivity on the alpha 3 subunit of neuronal nicotinic acetylcholine receptors. *J Pharmacol Exp Ther* 2003, 306 (2), 664–70.
95. Shiembob, D. L., Roberts, R. L., Luetje, C. W., McIntosh, J. M. Determinants of alpha-conotoxin BuIA selectivity on the nicotinic acetylcholine receptor beta subunit. *Biochemistry* 2006, 45 (37), 11200–7.
96. Kim, H. W., McIntosh, J. M. alpha6 nAChR subunit residues that confer alpha-conotoxin BuIA selectivity. *FASEB J* 2012, 26 (10), 4102–10.
97. Celie, P. H., Kasheverov, I. E., Mordvintsev, D. Y., Hogg, R. C., van Nierop, P., van Elk, R., van Rossum-Fikkert, S. E., Zhmak, M. N., Bertrand, D., Tsetlin, V., Sixma, T. K., Smit, A. B. Crystal structure of nicotinic acetylcholine receptor homolog AChBP in complex with an alpha-conotoxin PnIA variant. *Nat Struct Mol Biol* 2005, 12 (7), 582–88.
98. Ulens, C., Hogg, R. C., Celie, P. H., Bertrand, D., Tsetlin, V., Smit, A. B., Sixma, T. K. Structural determinants of selective alpha-conotoxin binding to a nicotinic acetylcholine receptor homolog AChBP. *Proc Natl Acad Sci USA* 2006, 103 (10), 3615–20.
99. Azam, L., Dowell, C., Watkins, M., Stitzel, J. A., Olivera, B. M., McIntosh, J. M. Alpha-conotoxin BuIA, a novel peptide from *Conus bullatus*, distinguishes among neuronal nicotinic acetylcholine receptors. *J Biol Chem* 2005, 280 (1), 80–87.
100. Luo, S., Nguyen, T. A., Cartier, G. E., Olivera, B. M., Yoshikami, D., McIntosh, J. M. Single-residue alteration in alpha-conotoxin PnIA switches its nAChR subtype selectivity. *Biochemistry* 1999, 38 (44), 14542–48.
101. McIntosh, J. M., Yoshikami, D., Mahe, E., Nielsen, D. B., Rivier, J. E., Gray, W. R., Olivera, B. M. A nicotinic acetylcholine receptor ligand of unique specificity, alpha-conotoxin ImI. *J Biol Chem* 1994, 269 (24), 16733–39.
102. Olivera, B. M. E.E. Just Lecture. 1996. *Conus* venom peptides, receptor and ion channel targets, and drug design: 50 million years of neuropharmacology. *Mol Biol Cell* 1997, 8 (11), 2101–9.
103. Terlau, H., Shon, K. J., Grilley, M., Stocker, M., Stuhmer, W., Olivera, B. M. Strategy for rapid immobilization of prey by a fish-hunting marine snail. *Nature* 1996, 381 (6578), 148–51.
104. Olivera, B. M. *Conus* venom peptides: Reflections from the biology of clades and species. *Annu Rev Ecol Syst* 2002, 33, 25–47.
105. Platt, R. J., Curtice, K. J., Twede, V. D., Watkins, M., Gruszczynski, P., Bulaj, G., Horvath, M. P., Olivera, B. M. From molecular phylogeny towards differentiating pharmacology for NMDA receptor subtypes. *Toxicon* 2014, 81, 67–79.
106. Olivera, B. M., McIntosh, J. M., Clark, C., Middlemas, D., Gray, W. R., Cruz, L. J. A sleep-inducing peptide from *Conus geographus* venom. *Toxicon* 1985, 23 (2), 277–82.
107. Walker, C. S., Jensen, S., Ellison, M., Matta, J. A., Lee, W. Y., Imperial, J. S., Duclos, N., Brockie, P. J., Madsen, D. M., Isaac, J. T., Olivera, B., Maricq, A. V. A novel *Conus* snail polypeptide causes excitotoxicity by blocking desensitization of AMPA receptors. *Curr Biol* 2009, 19 (11), 900–8.

108. Raghuraman, S., Garcia, A. J., Anderson, T. M., Twede, V. D., Curtice, K. J., Chase, K., Ramirez, J. M., Olivera, B. M., Teichert, R. W. Defining modulatory inputs into CNS neuronal subclasses by functional pharmacological profiling. *Proc Natl Acad Sci USA* 2014, 111 (17), 6449–54.

109. Smith, N. J., Hone, A. J., Memon, T., Bossi, S., Smith, T. E., McIntosh, J. M., Olivera, B. M., Teichert, R. W. Comparative functional expression of nAChR subtypes in rodent DRG neurons. *Front Cell Neurosci* 2013, 7, 225.

110. Teichert, R. W., Memon, T., Aman, J. W., Olivera, B. M. Using constellation pharmacology to define comprehensively a somatosensory neuronal subclass. *Proc Natl Acad Sci USA* 2014, 111 (6), 2319–24.

111. Teichert, R. W., Raghuraman, S., Memon, T., Cox, J. L., Foulkes, T., Rivier, J. E., Olivera, B. M. Characterization of two neuronal subclasses through constellation pharmacology. *Proc Natl Acad Sci USA* 2012, 109 (31), 12758–63.

112. Teichert, R. W., Schmidt, E. W., Olivera, B. M. Constellation pharmacology: A new paradigm for drug discovery. *Annu Rev Pharmacol Toxicol* 2015, 55, 573–89.

113. Teichert, R. W., Smith, N. J., Raghuraman, S., Yoshikami, D., Light, A. R., Olivera, B. M. Functional profiling of neurons through cellular neuropharmacology. *Proc Natl Acad Sci USA* 2012, 109 (5), 1388–95.

114. Bernard, A., Sorensen, S. A., Lein, E. S. Shifting the paradigm: New approaches for characterizing and classifying neurons. *Curr Opin Neurobiol* 2009, 19 (5), 530–36.

115. Fishell, G., Heintz, N. The neuron identity problem: Form meets function. *Neuron* 2013, 80 (3), 602–12.

116. Wichterle, H., Gifford, D., Mazzoni, E. Neuroscience: Mapping neuronal diversity one cell at a time. *Science* 2013, 341 (6147), 726–27.

117. Zhang, M. M., Wilson, M. J., Gajewiak, J., Rivier, J. E., Bulaj, G., Olivera, B. M., Yoshikami, D. Pharmacological fractionation of tetrodotoxin-sensitive sodium currents in rat dorsal root ganglion neurons by mu-conotoxins. *Br J Pharmacol* 2013, 169 (1), 102–14.

118. Arts, E. J., Hazuda, D. J. HIV-1 antiretroviral drug therapy. *Cold Spring Harb Perspect Med* 2012, 2 (4), a007161.

119. Margolis, D. M., Hazuda, D. J. Combined approaches for HIV cure. *Curr Opin HIV AIDS* 2013, 8 (3), 230–35.

120. Sack, J. T., Shamotienko, O., Dolly, J. O. How to validate a heteromeric ion channel drug target: Assessing proper expression of concatenated subunits. *J Gen Physiol* 2008, 131 (5), 415–20.

121. Ottschytsch, N., Raes, A., Van Hoorick, D., Snyders, D. J. Obligatory heterotetramerization of three previously uncharacterized Kv channel alpha-subunits identified in the human genome. *Proc Natl Acad Sci USA* 2002, 99 (12), 7986–91.

122. Al-Sabi, A., Shamotienko, O., Dhochartaigh, S. N., Muniyappa, N., Le Berre, M., Shaban, H., Wang, J., Sack, J. T., Dolly, J. O. Arrangement of Kv1 alpha subunits dictates sensitivity to tetraethylammonium. *J Gen Physiol* 2010, 136 (3), 273–82.

123. Madrid, R., de la Pena, E., Donovan-Rodriguez, T., Belmonte, C., Viana, F. Variable threshold of trigeminal cold-thermosensitive neurons is determined by a balance between TRPM8 and Kv1 potassium channels. *J Neurosci* 2009, 29 (10), 3120–31.

124. Ellison, M., Feng, Z. P., Park, A. J., Zhang, X., Olivera, B. M., McIntosh, J. M., Norton, R. S. Alpha-RgIA, a novel conotoxin that blocks the alpha9alpha10 nAChR: Structure and identification of key receptor-binding residues. *J Mol Biol* 2008, 377 (4), 1216–27.

125. Whiteaker, P., Christensen, S., Yoshikami, D., Dowell, C., Watkins, M., Gulyas, J., Rivier, J., Olivera, B. M., McIntosh, J. M. Discovery, synthesis, and structure activity of a highly selective alpha7 nicotinic acetylcholine receptor antagonist. *Biochemistry* 2007, 46 (22), 6628–38.

126. McIntosh, J. M., Azam, L., Staheli, S., Dowell, C., Lindstrom, J. M., Kuryatov, A., Garrett, J. E., Marks, M. J., Whiteaker, P. Analogs of alpha-conotoxin MII are selective for alpha6-containing nicotinic acetylcholine receptors. *Mol Pharmacol* 2004, 65 (4), 944–52.

127. Luo, S., Kulak, J. M., Cartier, G. E., Jacobsen, R. B., Yoshikami, D., Olivera, B. M., McIntosh, J. M. alpha-Conotoxin AuIB selectively blocks alpha3 beta4 nicotinic acetylcholine receptors and nicotine-evoked norepinephrine release. *J Neurosci* 1998, 18 (21), 8571–79.

128. Luo, S., McIntosh, J. M. Iodo-alpha-conotoxin MI selectively binds the alpha/delta subunit interface of muscle nicotinic acetylcholine receptors. *Biochemistry* 2004, 43 (21), 6656–62.

129. Teichert, R. W., Lopez-Vera, E., Gulyas, J., Watkins, M., Rivier, J., Olivera, B. M. Definition and characterization of the short alphaA-conotoxins: A single residue determines dissociation kinetics from the fetal muscle nicotinic acetylcholine receptor. *Biochemistry* 2006, 45 (4), 1304–12.

130. Teichert, R. W., Rivier, J., Torres, J., Dykert, J., Miller, C., Olivera, B. M. A uniquely selective inhibitor of the mammalian fetal neuromuscular nicotinic acetylcholine receptor. *J Neurosci* 2005, 25 (3), 732–36.

131. Craig, A. G., Norberg, T., Griffin, D., Hoeger, C., Akhtar, M., Schmidt, K., Low, W., Dykert, J., Richelson, E., Navarro, V., Mazella, J., Watkins, M., Hillyard, D., Imperial, J., Cruz, L. J., Olivera, B. M. Contulakin-G, an O-glycosylated invertebrate neurotensin. *J Biol Chem* 1999, 274 (20), 13752–59.

132. Gowd, K. H., Han, T. S., Twede, V., Gajewiak, J., Smith, M. D., Watkins, M., Platt, R. J., Toledo, G., White, H. S., Olivera, B. M., Bulaj, G. Conantokins derived from the *Asprella* clade impart conRl-B, an N-methyl d-aspartate receptor antagonist with a unique selectivity profile for NR2B subunits. *Biochemistry* 2012, 51 (23), 4685–92.

133. White, H. S., McCabe, R. T., Abogadie, F., Torres, J., Rivier, J. E., Paarmann, I., Hollmann, M., Olivera, B. M., Cruz, L. J. Conantokin-R, a subtype-selective NMDA receptor antagonist and potent anticonvulsant peptide. *Soc Neurosci Abstr* 1997, 23, 2164.

134. Gowd, K. H., Watkins, M., Twede, V. D., Bulaj, G. W., Olivera, B. M. Characterization of conantokin Rl-A: Molecular phylogeny as structure/function study. *J Pept Sci* 2010, 16 (8), 375–82.

135. Kerr, L. M., Yoshikami, D. A venom peptide with a novel presynaptic blocking action. *Nature* 1984, 308 (5956), 282–84.

136. Wilson, M. J., Yoshikami, D., Azam, L., Gajewiak, J., Olivera, B. M., Bulaj, G., Zhang, M. M. mu-Conotoxins that differentially block sodium channels NaV1.1 through 1.8 identify those responsible for action potentials in sciatic nerve. *Proc Natl Acad Sci USA* 2011, 108 (25), 10302–7.

137. Chen, P., Dendorfer, A., Finol-Urdaneta, R. K., Terlau, H., Olivera, B. M. Biochemical characterization of kappaM-RIIIJ, a Kv1.2 channel blocker: Evaluation of cardioprotective effects of kappaM-conotoxins. *J Biol Chem* 2010, 285 (20), 14882–89.

138. Imperial, J. S., Bansal, P. S., Alewood, P. F., Daly, N. L., Craik, D. J., Sporning, A., Terlau, H., Lopez-Vera, E., Bandyopadhyay, P. K., Olivera, B. M. A novel conotoxin inhibitor of Kv1.6 channel and nAChR subtypes defines a new superfamily of conotoxins. *Biochemistry* 2006, 45 (27), 8331–40.

139. Bayrhuber, M., Vijayan, V., Ferber, M., Graf, R., Korukottu, J., Imperial, J., Garrett, J. E., Olivera, B. M., Terlau, H., Zweckstetter, M., Becker, S. Conkunitzin-S1 is the first member of a new Kunitz-type neurotoxin family: Structural and functional characterization. *J Biol Chem* 2005, 280 (25), 23766–770.

140. Miljanich, G. P. Ziconotide: Neuronal calcium channel blocker for treating severe chronic pain. *Curr Med Chem* 2004, 11 (23), 3029–40.

141. Armishaw, C. J., Alewood, P. F. Conotoxins as research tools and drug leads. *Curr Protein Pept Sci* 2005, 6 (3), 221–40.

142. McIntosh, J. M., Absalom, N., Chebib, M., Elgoyhen, A. B., Vincler, M. Alpha9 nicotinic acetylcholine receptors and the treatment of pain. *Biochem Pharmacol* 2009, 78 (7), 693–702.

143. Twede, V. D., Miljanich, G., Olivera, B. M., Bulaj, G. Neuroprotective and cardioprotective conopeptides: An emerging class of drug leads. *Curr Opin Drug Discov Dev* 2009, 12 (2), 231–39.

144. Koch, E. D., Olivera, B. M., Terlau, H., Conti, F. The binding of kappa-conotoxin PVIIA and fast C-type inactivation of Shaker K+ channels are mutually exclusive. *Biophys J* 2004, 86 (1 Pt 1), 191–209.

145. Vincler, M., Wittenauer, S., Parker, R., Ellison, M., Olivera, B. M., McIntosh, J. M. Molecular mechanism for analgesia involving specific antagonism of alpha9alpha10 nicotinic acetylcholine receptors. *Proc Natl Acad Sci USA* 2006, 103 (47), 17880–84.

146. Vetter, I., Lewis, R. J. Therapeutic potential of cone snail venom peptides (conopeptides). *Curr Top Med Chem* 2012, 12 (14), 1546–52.

CHAPTER 17

Development and Commercialization of a Fully Synthetic Marine Natural Product Analogue, Halaven® (Eribulin Mesylate)

Charles E. Chase,[1] **Hyeong-wook Choi,**[1] **Atsushi Endo,**[1] **Francis G. Fang,**[1] **and John D. Orr**[2]

[1]Integrated Chemistry, Eisai, Inc., Andover, Massachusetts
[2]Analytical Chemistry, Eisai, Inc., Andover, Massachusetts

CONTENTS

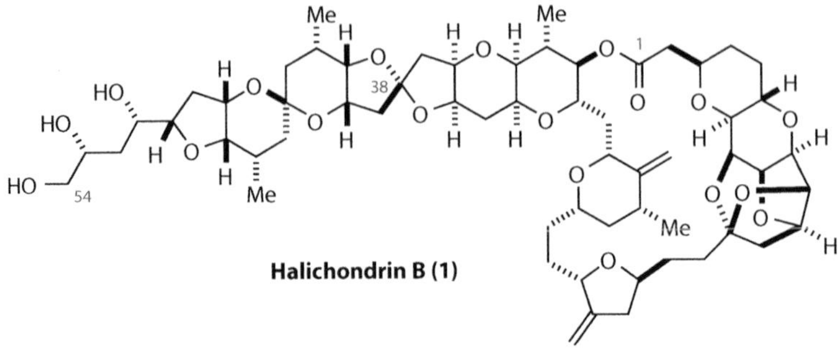

Figure 17.1 Halichondrin B: A marine natural product with potent *in vitro* and *in vivo* antitumor activity.

17.1 INTRODUCTION

Natural products represent a continuing source of inspiration for drug discovery.[1] Nevertheless, many natural products comprise a region of chemical space outside the realm of traditional small-molecule synthetic drugs typically developed by the pharmaceutical industry. Perhaps due to their structural complexity, many natural products, particularly those not readily available from nature, remain underexploited as potential medicinal agents. The innovations and advances in organic chemistry have contributed to the perception that the total synthesis of virtually any natural product is achievable in the laboratory. However, the perception also persists that natural product total synthesis on a manufacturing scale is still a significant challenge.[2]

Translating a microgram- or milligram-scale synthesis to one capable of producing kilogram quantities with acceptable cost, production time, and throughput, and which is also good manufacturing practice (GMP) compliant,[3] presents a significant challenge. Moreover, GMP compliance requires the demonstration of control of quality, including stereochemical quality, and communicating that control to gain regulatory approval of a new medicine, and this presents a significant hurdle, particularly for structurally complex, fully synthetic, natural product-based drugs.

A structurally complex marine natural product with compelling *in vitro* and *in vivo* anticancer activity, halichondrin B (**1**) (Figure 17.1), was reported by Uemura et al. in 1985.[4,5] The first and, until recently, only total synthesis of halichondrin B was reported by Kishi et al. in 1992.[6–8] These two discoveries provided the inspiration to initiate a halichondrin-based drug discovery program at Eisai.

17.2 ERIBULIN DISCOVERY

In the case of halichondrin B (**1**), availability of the natural product from the marine sponge source severely limited the ability to further evaluate the extremely promising *in vitro* and *in vivo* anticancer activity. The total synthesis of **1** provided the chemical reactions, routes, and intermediates needed to structurally modify and discover the candidate molecule that became eribulin (Halaven® [eribulin mesylate] [**3**] Figure 17.2), but additional work was necessary to realize the development, approval, and manufacture of this novel anticancer agent. Halaven®, a fully synthetic macrocyclic ketone analogue of the marine natural product halichondrin B (**1**), is now available for treatment of certain patients with advanced breast cancer, and represents Eisai's most advanced synthetic natural product program.[9–17]

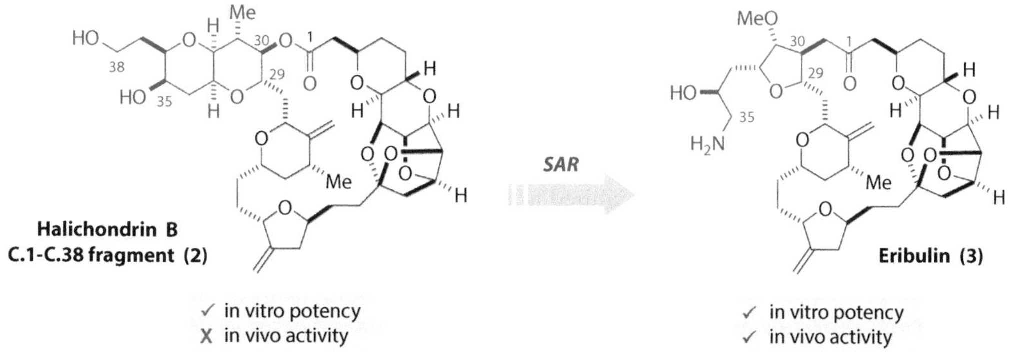

Figure 17.2 Discovery of Halaven® (eribulin) (**3**) from the halichondrin B C.1–C.38 macrolide (**2**).

The eribulin discovery program began with a synthetic intermediate in the halichondrin B total synthesis, namely, the C.1–C.38 macrolide fragment (**2**) (Figure 17.2). Importantly, **2** was found to have *in vitro* activity similar to that of halichondrin B (**1**), indicating that the macrocyclic portion of the natural product was sufficient for biological activity. Unfortunately, **2** did not exhibit *in vivo* activity, suggesting the need for structural modification and optimization prior to selecting a clinical candidate for development. A guiding principle during the lead optimization studies resulting in the discovery of eribulin (**3**) was to focus on pharmaceutical and biological considerations. Structural simplification is often stated to be an important goal of medicinal chemistry, but in fact, the eventually identified candidate compound retained all of the stereochemical complexity of the lead compound and also introduced some unique new synthetic challenges. One key structural difference between **2** and **3** is the fact that **2** is a macrolide and **3** is a macrocyclic ketone. The synthetic implication of this structural modification is that the key macrocyclization reaction would instead require making a carbon–carbon bond rather than a carbon–oxygen bond, as was the case in the total synthesis of **1**.[6] Another structural change is that the C.29–C.38 fused bicyclic ring system containing seven stereogenic centers was replaced with a C.29–C.35 monocyclic ring system containing five stereogenic centers. This structural change might appear to be a simplification, but due to the presence of a branched carbon stereogenic center at C.30 and an isolated acyclic stereogenic center at C.34, the overall change could be viewed as more synthetically challenging. Thus, eribulin (**3**) retains all of the stereochemical complexity of the halichondrin B macrocyclic core, while presenting new synthetic challenges associated with the macrocyclic ketone structural modifications.

17.3 ERIBULIN PROCESS DEVELOPMENT

17.3.1 Process Development Challenges

Due to the high potency of halichondrin B, initial biological assessment of analogues could be conducted with submilligram quantities of material. Yet, even during the early discovery phases of the program, the ability to obtain the necessary amounts of material to evaluate analogues and optimize the drug candidate structure proved quite challenging. There was the perception that even if a candidate compound was successfully identified, the synthesis would be far too complex to enable practical production. Thus, the ability to proceed into the clinic (gram scale), let alone to commercial production (kilogram scale), was viewed with some skepticism. Moreover, establishing a GMP process with documented control of stereoisomers and process impurities[*,†] to enable regulatory approval added to the process development challenges.

In order to better understand and directly address the key synthetic process issues, a detailed analysis of the initial synthesis was conducted. By far, the biggest contributors to the synthesis cost and time of production were found to be the numerous chromatographic purifications. From an operational standpoint, the tremendous quantities of solvent involved, the time needed to conduct separations, the analytical resources required to map and pool fractions, and the costs for stationary phase and multiple runs all contributed to make chromatography the single biggest obstacle in translating the eribulin synthesis to a manufacturing process.

As an illustration of the impact of chromatography on a single chemical processing step, consider the chiral resolution of a structurally simple benzazepine drug candidate (Figure 17.3).[18] The crystallization-based chiral resolution was one to two orders of magnitude more efficient than the chromatography-based process in terms of cost and production time. The cost of solvent waste disposal (green chemistry impact), while not highlighted in Figure 17.3, is similarly significant. The differences in efficiency are amplified with regard to quality control. Stereochemical (and process) quality control via chromatography comes at the expense of sample loading and throughput. Chromatography requires investment in specialized equipment and media, does not scale linearly, requires multiple injections at production scale, is prone to run-to-run variability, and necessitates the development of fraction concentration procedures to preserve isolate quality. With the need to conduct multiple chromatographic runs, collect multiple fractions, and analyze each, typically by

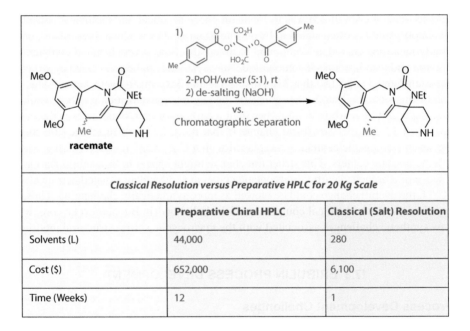

Classical Resolution versus Preparative HPLC for 20 Kg Scale		
	Preparative Chiral HPLC	**Classical (Salt) Resolution**
Solvents (L)	44,000	280
Cost ($)	652,000	6,100
Time (Weeks)	12	1

Figure 17.3 Preparative HPLC vs. classical resolution.

---

[*] The analytical control strategy necessary for GMP compliance necessitated analytical method design and validation. Method requirements included high sensitivity (0.05% impurity quantitation limits) and high specificity (for impurities, including stereoisomers). More than 100 analytical methods were required for the testing of starting materials, intermediates, and eribulin mesylate. Further, more than 200 analytical impurity reference standards required synthesis and testing.

[†] International Conference on Harmonization of Technical Requirements for Registration of Pharmaceuticals for Human Use: Validation of Analytical Procedures: Text and Methodology, Q2(R1), November 2005.

high-performance liquid chromatography (HPLC), as opposed to conducting crystallization, isolating, washing, and drying the product, the extra time needed to perform a chromatographic purification can be appreciated.

While the example in Figure 17.3 is at the higher end of chromatographic separation challenges (requiring 2200 volumes of solvent per kg substrate), nevertheless, for a molecule with multiple stereogenic centers and multiple separations, the impact of chromatography can be appreciated. A typical chromatographic purification requires 400–2500 volumes[*,†] of solvent, and thus is a dominant contributor to the cost of each step in a synthesis. It may seem intuitive that chromatography makes up the dominant cost in a complex molecule synthesis. Yet, most publications do not highlight the presence or absence of chromatographies when making claims about the merits of one synthetic route over another. In fact, the experiments include quantitative details about nearly everything except the chromatographic purification. Thus, one of the most important factors in assessing the practicality of a synthetic route is often overlooked. Based on this analysis, there was a realization that if most or all of the chromatographic purifications could be removed from the eribulin synthesis, a cost- and time-competitive practical manufacturing process might be established.

The synthetic strategy employed to construct a natural product can have a significant impact on the ability to discover crystalline intermediates. Perhaps the major practical difference between complex natural product synthesis and more traditional small-molecule drug production is the expectation that chromatography will be required for total synthesis of a natural product, while crystallization is the expectation for industrial synthesis. Given the importance of crystallization, the synthetic strategies employed to construct halichondrin (and Halaven) fragments evolved with a focus on adopting approaches that would naturally lead toward more crystallization-based routes. A priori knowledge about the crystallinity of intermediates was not available. However, there were synthetic design criteria that could logically increase the chances of obtaining crystalline intermediates. Incorporating structural features that reduce conformational degrees of freedom and enhance intermolecular associations might be expected to lead to more crystalline intermediates. For example, temporary covalent bonds, H-bond systems, and p-stacking elements could all naturally enforce conformational control and thus might be expected to improve the chances of identifying crystalline intermediates. An added benefit to having crystalline intermediates would be the ability to consistently control stereochemical and process impurities to the rigorous standards required for regulatory approval. Thus, a guiding principle in design, evaluation, and prioritization of synthetic routes to eribulin would be crystallinity.

While the structure of eribulin is substantially simplified relative to that of halichondrin B (19 vs. 32 stereogenic centers and a 36- vs. 54-carbon backbone), the development and commercialization employing total synthesis still represented a significant challenge.[19] Among the synthetic issues that needed to be addressed in order to establish a consistent and reliable commercial supply of drug were (1) large-scale production of the two fragments, representing C.1–C.13 and C.14–C.35; (2) chemoselective large-scale C–C bond formations to assemble a 27-membered macrocycle; (3) regio- and stereocontrolled transannular polycyclic ketalization and aminoalcohol formation; and (4) industrialization of the Nozaki–Hiyama–Kishi (NHK) reaction. One singularly important chemistry technology, the NHK reaction,[20–24] was indispensable to realize the total synthesis of halichondrin B. In order to commercialize the eribulin manufacturing process, challenges in the industrialization of the NHK reaction needed to be overcome. Needless to say, all of the above synthetic operations would require a high level of understanding and control to enable GMP production and documentation sufficient to obtain regulatory approval.

---

[*] For example, 400–2500 volumes is equivalent to 400–2500 L of solvent per kg of substrate.
[†] Still, W. C., Kahn, M., and Mitra, A. 1978. Rapid chromatographic technique for preparative separations with moderate resolution. *J. Org. Chem.* 43:2923–2925.

17.3.2 Eribulin Retrosynthetic Analysis

The retrosynthetic analyses of eribulin (**3**) and halichondrin B C.1–C.38 (**2**) share common themes but diverge in the final key steps (Figure 17.4). While the halichondrin B macrolide was assembled by forming an ester bond as the key ring-forming reaction, eribulin requires formation of a C–C bond in the macrocyclization step. The most obvious choice for this key C–C bond and ring-forming process would be an NHK reaction at C.13–C.14 due to the precedent established intermolecularly in the halichondrin B synthesis. In order to take advantage of an NHK macro-cyclization at C.13–C.14, polycyclic ketal formation would need to be postponed until after the macrocyclic ketone formation. As a result, enone **4** became the eribulin macrocyclic target, con-structed utilizing an intramolecular NHK reaction as the key ring-forming technology.

Retrosynthesis of eribulin (**3**) is shown in Figure 17.5. Disconnection to a C.1–C.13 aldehyde analogue of the known C.1–C.13 ester[25] revealed a bifunctional compound, vinyl iodide aldehyde

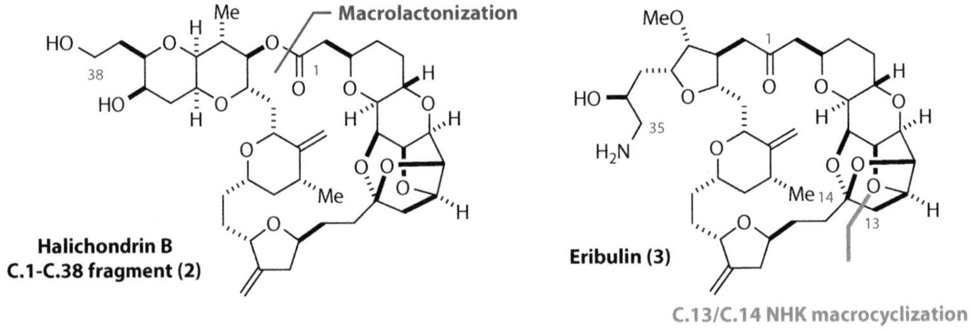

Figure 17.4 Macrocyclization of halichondrin B and eribulin.

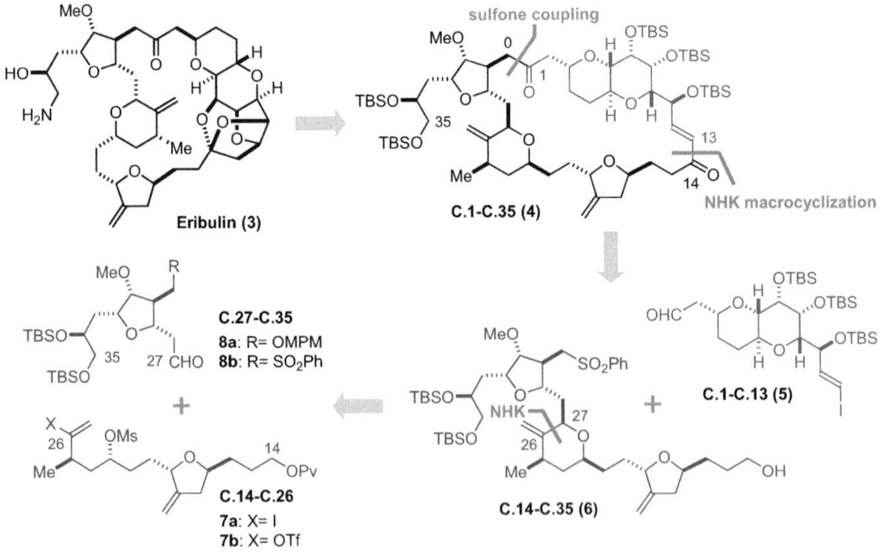

Figure 17.5 Retrosynthetic analysis.

(**5**), which was well suited for the establishment of both the C.0–C.1 and C.13–C.14 C–C bonds. The coupling partner, sulfone-alcohol C.14–C.35 (**6**), was a new fragment specifically designed for eribulin. Disconnection of C.14–C.35 by analogy with the assembly of the halichondrin C.14–C.38 fragment revealed a vinyl triflate (**7b**) analogous to the known C.14–C.26 vinyl iodide (**7a**) used in the halichondrin B synthesis.[26] The polyfunctional vinyl triflate-mesylate-pivalate C.14–C.26 fragment (**7b**) was well-suited for efficient construction of the tricyclic pyran-containing C.14–C.35 fragment (**6**) and also served to further illustrate the functional group compatibility of the NHK reaction for connecting complex fragments. The coupling partner of **7**, C.27–C.35 aldehyde (**8**), was a new fragment specifically required for eribulin and was a replacement for the natural C.27–C.38 halichondrin B moiety.

17.3.3 C.1–C.13 Fragment: L-Mannonolactone (9) Route

Three synthetic routes for the C.1–C.13 fragment from three different carbohydrate starting materials (D-galactose, L-mannonolactone (**9**), and D-ribonolactone) have been reported by Kishi et al.[25–27] The route from L-mannonolactone (**9**) was initially chosen for development due to the high degree of stereochemical overlap with the target fragment (C.8, C.9, C.10, and C.11) (Figure 17.6). Modifications to the Kishi process that were implemented at Eisai included (1) reduction of known lactone **10** with lithium aluminum hydride (LAH) at 0 °C rather than use of cryogenic diisobutyl aluminum hydride (DIBAL), (2) crystallization of the lactol **11**, (3) crystallization of diol **13**, (4) replacing the acylation solvent or base, pyridine, with triethylamine, 4-dimethylaminopyridine (DMAP), and tetrahydrofuran (THF), (5) crystallization of the diacetate **14**, and (6) replacing Triton-B(OMe) with sodium methoxide for the preparation of **16**.

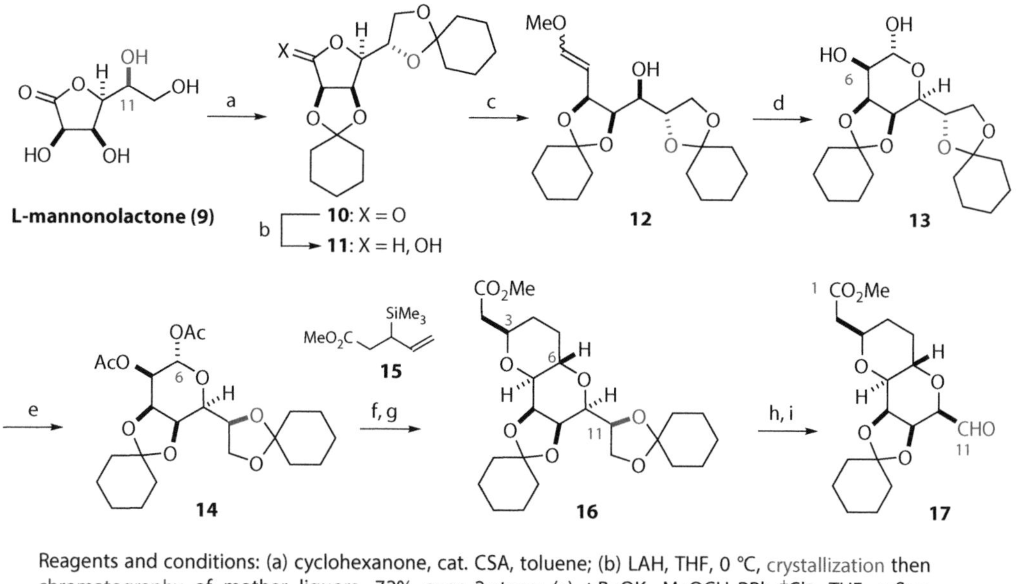

Reagents and conditions: (a) cyclohexanone, cat. CSA, toluene; (b) LAH, THF, 0 °C, crystallization then chromatography of mother liquors, 72% over 2-steps; (c) *t*-BuOK, MeOCH$_2$PPh$_3$$^+Cl^-$, THF, reflux, chromatography, 99%; (d) OsO$_4$, NMO, aq. acetone, –5 °C (slow addition), crystallization, 46%; (e) Ac$_2$O, THF, DMAP, Et$_3$N, crystallization, 92%; (f) **15**, TMSOTf, CH$_3$CN, 60 °C, chromatography, 71% (single isomer at C.6); (g) NaOMe, MeOH, THF, 0 °C to rt, chromatography, 86% (single isomer at C.3); (h) AcOH-H$_2$O, 80 °C, chromatography, 69%; (i) NaIO$_4$, THF, pH 7 buffer, quantitative.

Figure 17.6 Synthesis of C.1–C.11 aldehyde **17** from L-mannonolactone (**9**).

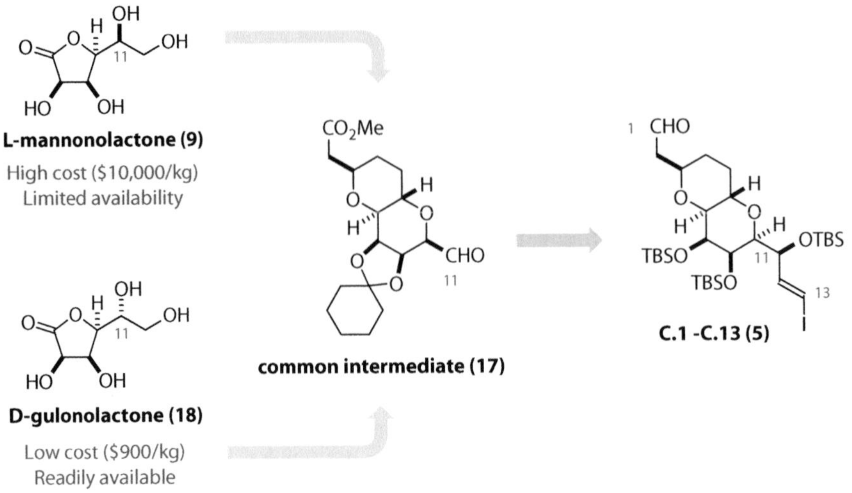

Figure 17.7 Comparison of L-mannonolactone (**9**) and D-gulonolactone (**18**).

17.3.4 C.1–C.13 Fragment: D-Gulonolactone (20) Route

Although the route to aldehyde **17** had a number of strong points, the cost and availability of L-mannonolactone (**9**) became limiting issues during development, requiring a new approach and a new source of chirality. D-Gulonolactone (**18**) contains the same three stereogenic centers that match C.8, C.9, and C.10 as **9** (Figure 17.7). With D-gulonolactone (**18**), the only differing stereogenic center (that corresponding to C.11) would be destroyed and reformed in the C.1–C.13 synthesis. Since D-gulonolactone (**18**) is readily available in bulk and also an order of magnitude cheaper than **9**, the development of a new route from **18** was investigated.

The synthesis of aldehyde **17** from D-gulonolactone (**18**)[28] is shown in Figure 17.8. Several differences in chemical reactivity and physical properties were noted during this investigation that had positive consequences for the overall efficiency of the process. First, the ancillary cyclohexylidene ring was found to be considerably more reactive than in the mannonolactone series and thus incompatible with the C-glycosidation step. This reactivity could be used to advantage by simply incorporating the removal of this cyclohexylidene moiety into the acetylation step. Second, because the C.11–C.12 diol was revealed during the annulation step to form the C.3–C.7 pyran ring, an extra deprotection step could be avoided. Third, the C.11–C.12 diol **24** (epimeric to the mannonolactone series) was found to be crystalline. Completion of the C.1–C.13 fragment (**5**) from **17** was accomplished in a fashion similar to that described by Kishi et al.[25,27] A stereoselective NHK reaction with *trans*-2-bromovinyltrimethylsilane installed the C.11 stereochemistry of **25** in a 9:1 ratio. Hydrolysis of the remaining cyclohexylidine and protection of the resulting triol **26** as the tri-TBS (tert-butylmethylsilyl) ether provided **27**. The synthesis was completed by replacement of the trimethylsilyl group with iodide[29] and DIBAL reduction of the ester to provide the aldehyde **5**. Finally, a number of crystalline late intermediates were discovered during the course of developing the late stages of the C.1–C.13 route, including the triol-vinylsilane **26** and tri-TBS-vinylsilane **27**.

17.3.5 C.14–C.26 Fragment: L-Glutamic Acid (28) and L-Arabinose (33) Route

The C.14–C.26 synthesis that was used initially for the generation of halichondrin B (**1**) is shown in Figure 17.9.[6] Because of the presence of multiple noncontiguous acyclic stereogenic centers, the initial synthesis of C.14–C.26 (**7**) depended upon the coupling of two chiral fragments

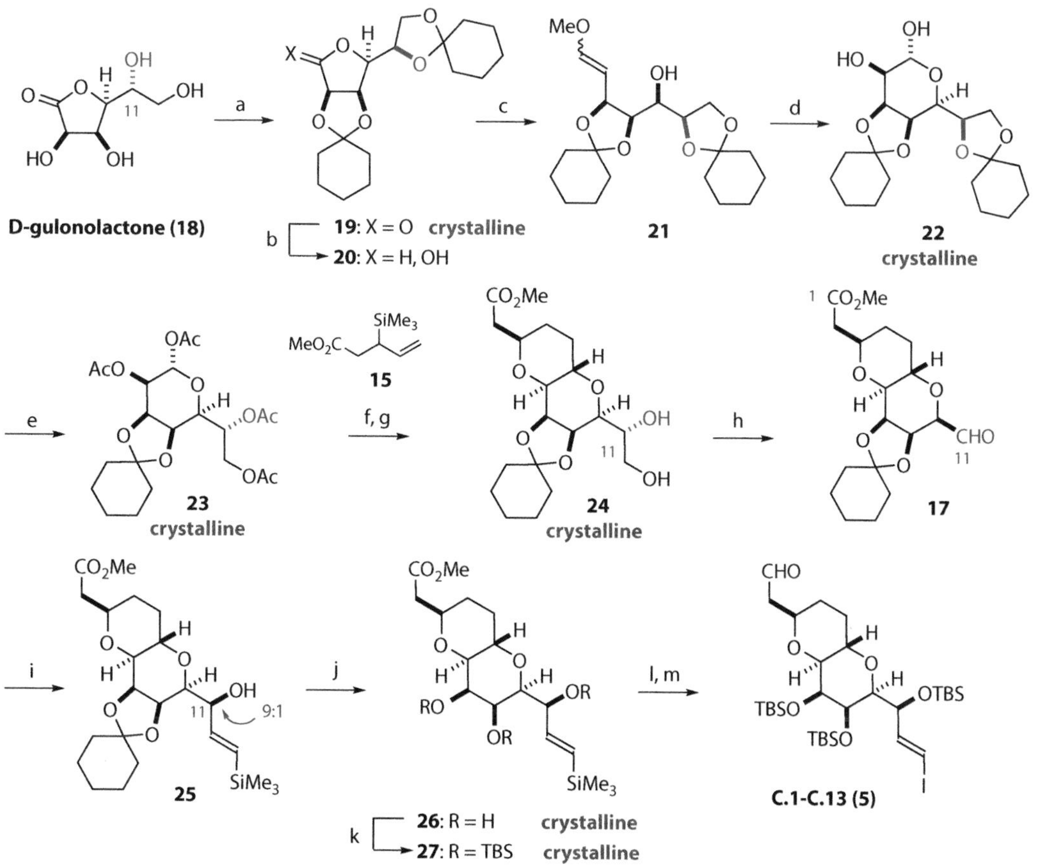

Reagents and conditions: (a) cyclohexanone, p-TsOH, toluene, 110 °C, crystallization, 60%; (b) DIBAL, toluene-THF, −5 °C, quant.; (c) t-BuOK, Ph₃P⁺CH₂OMe·Cl⁻, −5 to 30 °C, 81%; (d) K₂OsO₄·2H₂O, NMO, acetone-water, 30 °C, crystallization, 55%; (e) ZnCl₂, AcOH, Ac₂O, 30 °C, crystallization, 69%; (f) **15**, BF₃·OEt₂, CH₃CN, 15 °C, 95%; (g) NaOMe, MeOH, MTBE, 15 °C, crystallization, 62%; (h) NaIO₄, EtOAc, 15 °C, quant.; (i) CrCl₂, NiCl₂, 1-bromo-2-trimethylsilylethene, DMSO, CH₃CN, 30 °C, 45%; (j) AcOH, H₂O, 95 °C, crystallization, 70%; (k) TBSOTf, 2,6-lutidine, MTBE, 30 °C, crystallization, 75%; (l) NIS, CH₃CN, toluene, TBSCl, 35 °C, chromatography, 89%; (m) DIBAL, toluene, < −60 °C, chromatography, 80-100%.

Figure 17.8 Synthesis of C.1–C.13 fragment **5** from D-gulonolactone (**18**) via multiple crystalline intermediates.

(C.14–C.21 and C.22–C.26). The C.22–C.26 fragment is derived from L-glutamic acid (**28**), which was converted in a series of known steps into a butyrolactone, which was then methylated providing **29**, and subsequently converted into a C.22–C.26 ketophosphonate vinyl iodide **32** bearing the single C.25 methyl stereogenic center. One advantage of the C.22–C.26 route from **28** was the crystallinity of the methylated butyrolactone **29**, allowing good control of the C.25 stereochemical quality. The C.14–C.21 fragment (**37**) was derived from **33** in 18 steps. The C.20 stereogenic center originated from arabinose, while the C.17 stereogenic center was installed by a highly stereoselective C-glycosidation. Joining the C.14–C.21 (**37**) and C.22–C.26 (**32**) fragments with a Horner–Wadsworth–Emmons reaction, Stryker reagent-mediated conjugate reduction of the resultant enone and borohydride reduction gave the final stereogenic center, C.23, as a 2:1 mixture of isomers (desired–undesired). The C.23 alcohol epimers were readily separable by chromatography due to an unexpectedly large polarity difference. Presumably, hydrogen bonding between the C.23 alcohol and the C.20 THF ether oxygen contributed to different adsorption affinities. Mesylation of

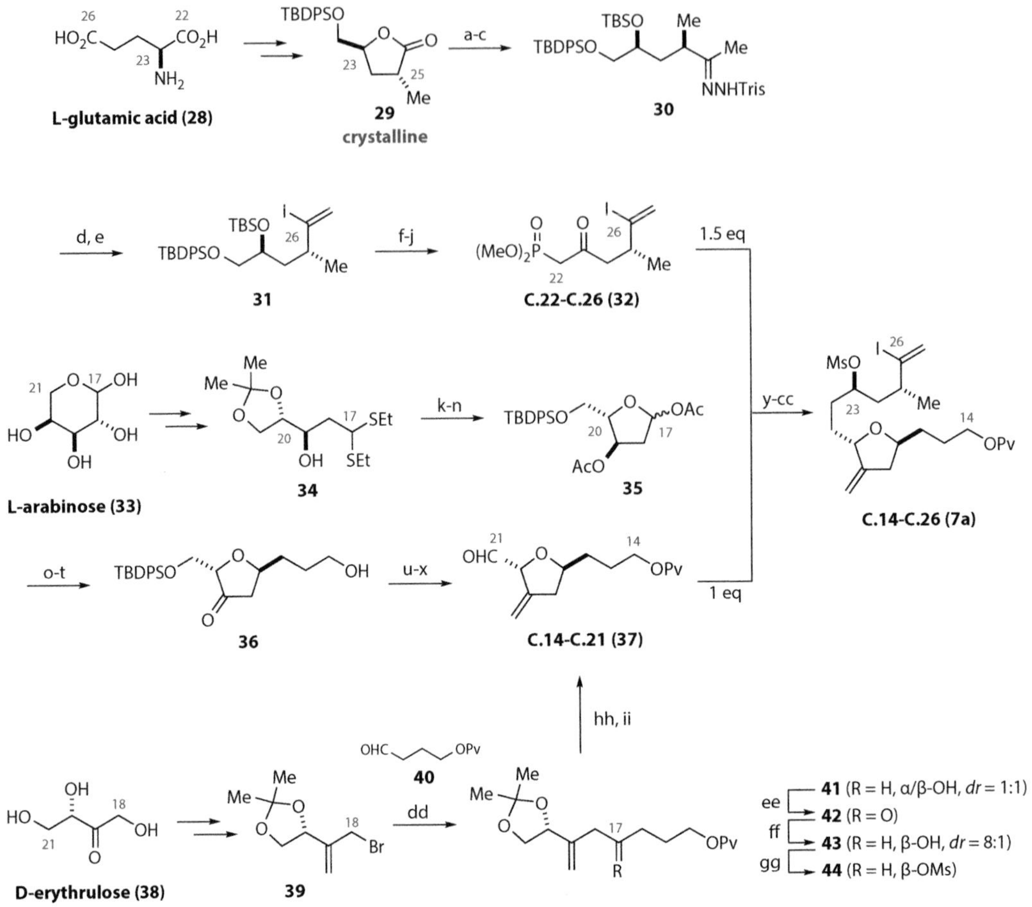

Reagents and conditions: (a) MeLi, THF; (b) TBSCl, imidazole, chromatography, 82% (2 steps); (c) TrisNHNH$_2$, conc. HCl, THF; (d) (1) n-BuLi (4 eq), TMEDA-pentane (1:4), (2) n-Bu$_3$SnCl; (e) I$_2$, CH$_2$Cl$_2$, chromatography, 78% overall; (f) HF, CH$_3$CN-THF, chromatography, 70%; (g) NaIO$_4$, THF-water; (h) Jones reagent, acetone; (i) BnOH, CH$_2$Cl$_2$, DCC, DMAP, chromatography, 94%; (j) (MeO)$_2$P(O)CH$_2$Li, THF, 78%; (k) AcOH-H$_2$O (4:1), chromatography, 94%; (l) TBDPSCl, imidazole, CH$_2$Cl$_2$, chromatography, 94%; (m) I$_2$, NaHCO$_3$, H$_2$O, acetone; (n) Ac$_2$O, pyridine, DMAP, chromatography, 94% (2 steps); (o) allyltrimethylsilane, BF$_3$·OEt$_2$, CH$_3$CN, 90%; (p) (1) 9-BBN, THF, (2) 30% H$_2$O$_2$, NaOH; (q) MMTrCl, Et$_3$N, CH$_2$Cl$_2$; (r) K$_2$CO$_3$, MeOH, chromatography, 71% (3 steps); (s) (COCl)$_2$, DMSO, Et$_3$N, CH$_2$Cl$_2$, chromatography, 97%; (t) p-TsOH, CH$_2$Cl$_2$, MeOH, chromatography, 97%; (u) Tebbe reagent, toluene-THF-pyridine (3:1:1), chromatography, 87%; (v) PvCl, pyridine, DMAP, CH$_2$Cl$_2$; (w) TBAF, THF, chromatography, 91% (2 steps); (x) Dess-Martin periodinane, CH$_2$Cl$_2$; (y) NaH (1.3 equiv), THF, chromatography, 88%; (z) [CuH(Ph$_3$P)]$_6$ (1.8 eq), benzene, water (16 eq); (aa) NaBH$_4$, MeOH, chromatography, 70% (desired, high R$_f$), 30% (C.23-epimer, low R$_f$); (bb) (1) epi-C.23 from aa, Ph$_3$P, DEAD, 4-NO$_2$-PhCO$_2$H, Et$_2$O, chromatography, 89%, (2) K$_3$CO$_3$, MeOH, chromatography, 99%; (cc) Ms$_2$O, Et$_3$N, DMAP, CH$_2$Cl$_2$, 96%; (dd) **40**, Zn, THF, chromatography, 59%; (ee) (COCl)$_2$, DMSO, Et$_3$N, CH$_2$Cl$_2$, −78 °C, chromatography, 93%; (ff) L-Selectride, chromatography; (gg) MsCl, Et$_3$N, CH$_2$Cl$_2$; (hh) HCl, chromatography, 79%; (ii) (COCl)$_2$, DMSO, Et$_3$N, CH$_2$Cl$_2$, −78 °C.

Figure 17.9 Initial synthesis of C.14–C.26.

the C.23 alcohol provided the C.14–C.26 fragment **7a**. While this synthesis had some notable challenges, it did set three of the four stereogenic centers with essentially complete chirality control. The one stereogenic center that was not selectively established, C.23, was actually the one isomer for which chromatographic separation was not so impractical due to the large R$_f$ difference. However, the overall C.14–C.26 synthesis was deemed to be rather challenging to develop because of the propensity for the intermediate β,γ-unsaturated aldehyde **37** to isomerize prior to the key C.21–C.22 bond-forming step.

17.3.6 C.14–C.26 Fragment: L-Glutamic Acid (28) and D-Erythrulose (38) Route

Another route to C.14–C.21 using D-erythrulose (**38**) as the source of chirality was devised at Eisai (Figure 17.9).[30] In a few steps, **38** was transformed into the bromide **39**. A zinc-mediated allylation with aldehyde **40** provided a 1:1 ratio of C.17 alcohol diastereomers. Oxidation and stereoselective ketone reduction with L-selectride improved the C.17 alcohol diastereoselectivity to 8:1. Activation of the C.17 alcohol as a mesylate (**44**), followed by acid-catalyzed acetonide deprotection, resulted in concomitant Williamson ether cyclization to establish the desired trans-tetrahydrofuran C.14–C.21 fragment. Swern oxidation intercepted the previous route providing aldehyde **37**. The success of the Williamson ether cyclization to establish a THF ring under acidic conditions provided an important proof of principle for establishing another approach to the C.14–C.26 fragment (**7a**).

17.3.7 C.14–C.26 Fragment via C.19–C.20 Bond Formation: Industrialization of the NHK Reaction

A key technology employed for the assembly of halichondrin B (**1**) was the NHK reaction (Figure 17.5). The NHK reaction had only been used in a laboratory setting, and thus an important practical obstacle to commercial manufacture of eribulin was to demonstrate performance and reproducibility in a kilolab and pilot plant setting. Moreover, due to the high cost and long lead times required to prepare the advanced substrates, evaluating the NHK reaction at a late stage incurred substantial risk to the overall program. A synthetic approach to the C.14–C.26 fragment (**7**) was designed that mirrored that used to construct C.14–C.35 fragment **6** (Figure 17.10). A strategic advantage of utilizing the NHK at an earlier stage of the synthesis would be the opportunity to establish the key late-stage C–C bond-forming technology in a pilot plant on a much cheaper and more stable substrate. Namely, a C.19–C.20 NHK reaction–Williamson etherification sequence coupling C.14–C.19 with C.20–C.26 was considered.

The vinyl halide precursor, **50**, necessary to demonstrate the NHK-based strategy for the C.14–C.26 fragment **7b**, is shown in Figure 17.11. Acid-catalyzed hydration of dihydrofuran (**45**), followed by tin-mediated allylation with 2,3-dibromopropene, resulted in the racemic C.14–C.19 intermediate **46**. The primary hydroxyl group was selectively protected as the TBDPS ether providing **47**, which was resolved using chiral-simulated moving-bed chromatography to provide enantiomers **48** and **49**. The undesired enantiomer, **48**, was converted to **49** by Mitsunobu inversion, and tosylation of C.17 provided the crystalline C.14–C.19 fragment, **50**.

The synthesis for the C.20–C.26 aldehyde NHK coupling partner for vinyl bromide **50** is shown in Figure 17.12. Jacobsen kinetic resolution of hexadiene monoepoxide (**51**) provided epoxide **52**

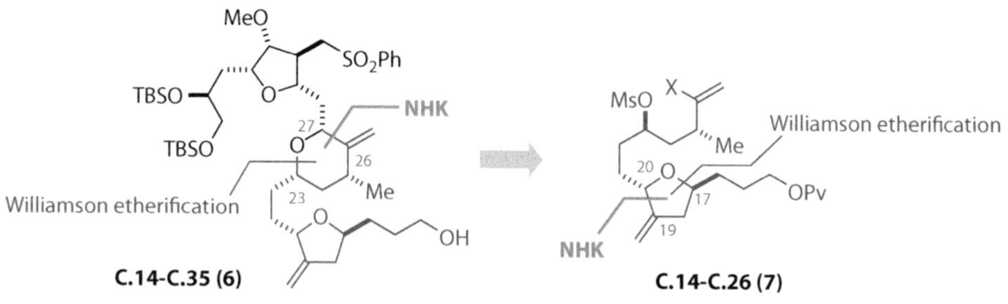

Figure 17.10 NHK reaction for key bond formation.

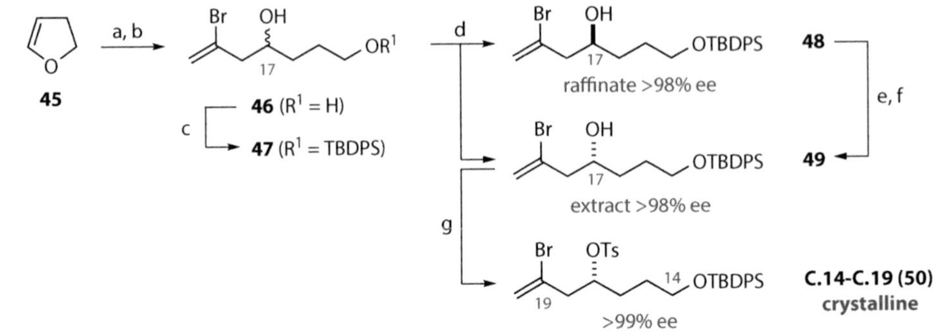

Reagents and conditions: (a) Amberlyst 15, H_2O, 5 °C; (b) 2,3-dibromo-1-propene, Sn, HBr, H_2O, 35 °C, 50% from **45**; (c) TBDPSCl, imidazole, DMF, 0–15 °C, 88%; (d) SMB chromatography, Chiralpak-OD 20 micron, 43% **48** + 44% **49**; (e) 4-NO_2-$PhCO_2H$, DEAD, PPh_3, toluene, 0 °C; (f) LiOH, THF-H_2O, 20 °C, chromatography, 83% from **48**; (g) TsCl, TEA, DMAP, CH_2Cl_2, 20 °C, crystallization, 78%.

Figure 17.11   Scale-up synthesis of C.14–C.19.

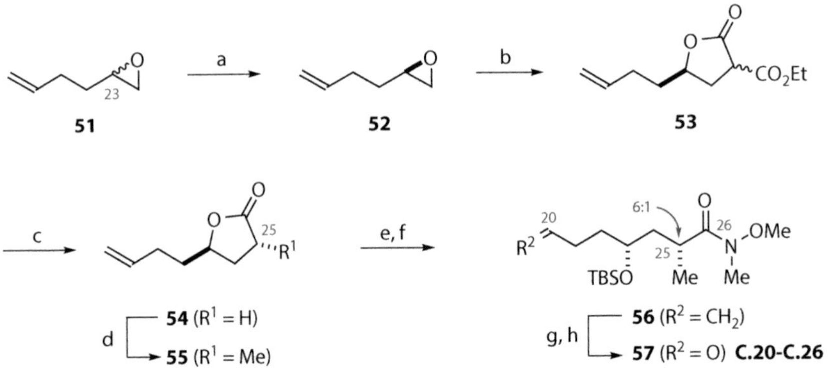

Reagents and conditions: (a) (R,R)-Jacobsen's salen(Co) catalyst, H_2O, 20 °C, distillation, 40%, >99% ee; (b) diethyl malonate, NaOEt, EtOH, 65 °C; (c) $MgCl_2(H_2O)_6$, DMF, 135 °C, distillation, 54% from **52**; (d) LHMDS, MeI, THF, –75 °C, distillation, 82%; (e) Me_3Al, (N,O)-dimethylhydroxylamine hydrochloride, CH_2Cl_2-toluene, 0 °C; (f) TBSCl, DMF, 20 °C , 53% from **55**; (g) OsO_4, NMO, CH_2Cl_2-H_2O, 20 °C; (h) $NaIO_4$, pH 7 phosphate buffer, 20 °C, chromatography, 85%.

Figure 17.12   Scale-up synthesis of C.20–C.26.

with C.23 stereoselectivity of 99% ee.[31,32] Subsequent manipulations resulted in the formation of lactone **54**, in which the C.25 center was established as a 6:1 ratio of methyl epimers by lactone enolate alkylation. The diastereomeric mixture was processed to the Weinreb amide aldehyde **57** in a four-step process.

NHK coupling of C.14–C.19 vinyl bromide **50** with the C.20–C.26 aldehyde **57** in the presence of chiral sulfonamide ligand **61**[33–35] provided allylic alcohol-homoallylic tosylate **58** as an 8:1 mixture at C.20 (Figure 17.13). By analogy with the precedent established in the erythulose-based route, Williamson ether cyclization of **58** occurred in isopropanol in the presence of silica gel to afford the C.14–C.26 tetrahydrofuran **59**. Completion of the C.14–C.26 fragment **7b** was achieved in six additional steps: (1) methyl ketone formation from the Weinreb amide, (2) vinyl triflate installation, (3) desilylation, (4) preparative HPLC to remove three significant diastereomers (epi-C.20, epi-C.25, and diepi-C.20/C.25) as well as minor diastereomers consisting of epi-C.17 and epi-C.23

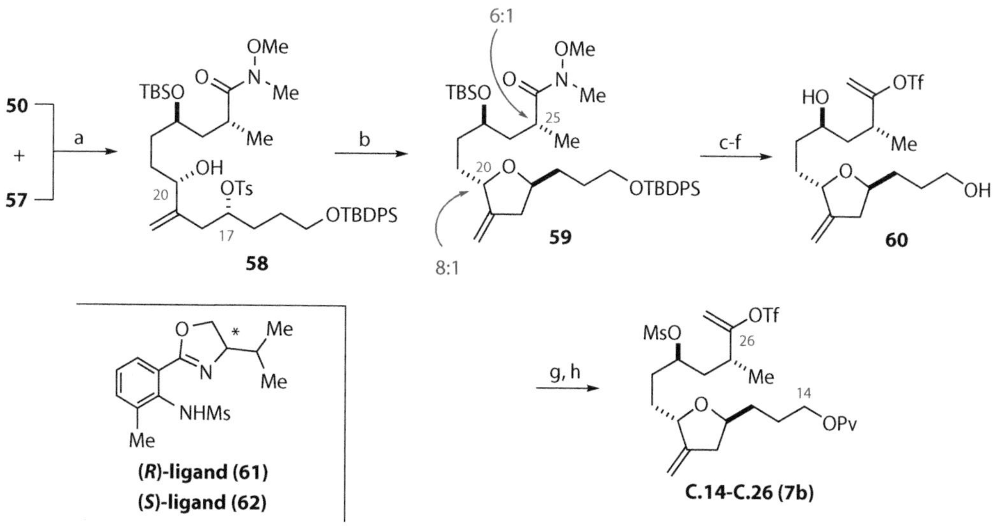

Reagents and conditions: (a) $NiCl_2$, $CrCl_2$, TEA, **61**, THF, 25 °C; (b) SiO_2, 2-PrOH, 25 °C, chromatograpy, 60% based on **50**; (c) MeMgCl, THF, 0 °C, chromatography; (d) KHMDS, $PhN(Tf)_2$, THF-toluene, –70 °C; (e) 6 M HCl in 2-PrOH, MeOH, 15 °C, chromatography, 56% from **59**; (f) preparative HPLC, 54%; (g) PvCl, 2,4,6-collidine, DMAP, CH_2Cl_2, 0 °C, chromatography, 94%; (h) MsCl, TEA, THF, 0 °C, 99%.

Figure 17.13 Scale-up synthesis of C.14–C.26: Industrialization of the Nozaki–Hiyama–Kishi reaction.

and other process-related impurities providing diol **60**, (5) regioselective C.14-OH pivaloylation, and (6) C.23-OH mesylate formation.

Having demonstrated the overall viability of the asymmetric NHK–Williamson ether approach for the C.14–C.26 fragment synthesis, the assessment and understanding of parameters required to execute the process in fixed equipment could commence on available and relatively stable substrates. Key parameters for the development of the NHK reaction in fixed equipment were evaluated, including (1) oxygen control, (2) water content, (3) reagent stoichiometry, (4) temperature profile during metal–ligand complexation, and (5) decomplexation conditions. Chromium(II) is well known to be oxygen sensitive. The asymmetric variant, being largely homogeneous, was expected to be especially sensitive to the presence of oxygen. Experimentally, contamination with oxygen at any stage of the reaction was found to be detrimental. More than 0.1 equivalent of oxygen significantly reduced the reaction conversion. Having the appropriate engineering controls to monitor and limit oxygen content throughout operations in fixed equipment turned out to be crucial to successfully conduct the NHK reaction. For example, solid additions were performed under inert conditions using appropriate solid addition techniques (e.g., alpha/beta type valves), and typically oxygen levels, measured in the reactor headspace, could be kept below 200 ppm. On the other hand, the asymmetric NHK reaction was somewhat less sensitive to water contamination, where up to 0.8 equivalent of water was not detrimental to the reaction conversion.

Temperature during chromous chloride–chiral ligand complexation was also found to be an important reaction parameter. During laboratory operations, depending on the ambient laboratory temperature, reaction conversions were found to be variable (Figure 17.14, step A). Using ReactIR™, it was determined that temperatures greater than 30 °C were optimal for achieving efficient metal–ligand complex formation in a reasonable time frame (ca. 3 h).

Proven acceptable range (PAR) studies were also performed to assess the optimal reagent stoichiometry. The final conditions, employing 3.1 equivalents of ligand and 2.3–3.7 equivalents of chromous chloride showed similar reaction conversions, although the use of less than 2.3 equivalents

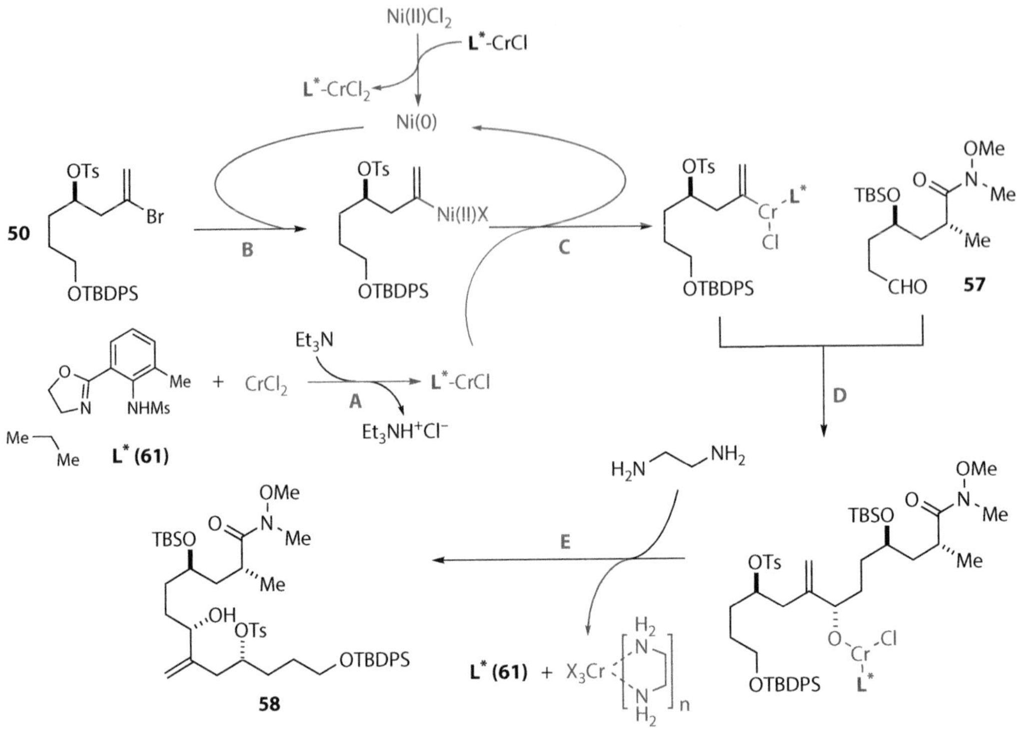

Figure 17.14 C.19–C.20 NHK reaction.

of chromous chloride resulted in a sharp decrease in the reaction conversion. In addition, the use of 2.5–3.7 equivalents of triethylamine for the complexation gave similar reaction conversions. The results seem to suggest that more than 2.3 equivalents of ligand–$CrCl_2$ complex are required to obtain acceptable reaction conversions. For the nickel insertion to the vinyl bromide (step B), the amount of nickel(II) chloride was reduced to 0.1 equivalents to suppress the homocoupling of vinyl bromides. Finally, ethylene diamine was identified as an efficient decomplexation agent to (1) dissociate the ligand and reaction product from chromium, (2) solubilize chromium in the aqueous layer (step E), and (3) facilitate extractive chromium removal.

With successful demonstration of the asymmetric NHK reaction in fixed equipment, the confidence to go forward with more critical late-stage asymmetric NHK processes (i.e., C.26–C.27 and C.13–C.14) was secured. However, the C.14–C.26 synthesis using the NHK process described had several shortcomings. Stereogenic centers C.20 and C.25 were formed as 8:1 and 6:1 mixtures, respectively, and the C.14–C.26 diol **60** required preparative HPLC for isolation of the single pure diastereomer. Additionally, because the NHK reaction was superstoichiometric and conducted at an earlier point in the synthesis, substantial amounts of the metals used in the eribulin synthesis were in the C.19–C.20 coupling.

17.3.8 C.14–C.26 Fragment from D-Quinic Acid: Crystallization-Based Route

The two previous syntheses of C.14–C.26 (**7**) employed at Eisai broke down the molecule into two chiral subfragments (Figure 17.15). Due to the presence of multiple noncontiguous acyclic stereogenic centers, a matching chirality strategy to assemble C.14–C.26 seemed to be required. In part, the need to couple two chiral fragments was what distinguished C.14–C.26 from the other two fragments of eribulin and made stereoselective synthesis relatively more complex.

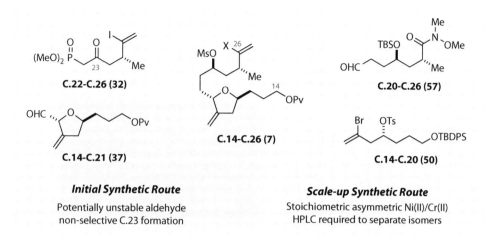

Figure 17.15   Previous C.14–C.26 syntheses use matching chirality to set stereochemistry.

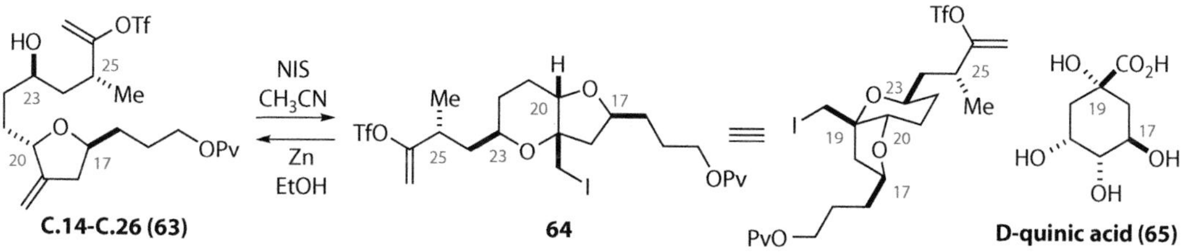

Figure 17.16   C.14–C.26 via a bicyclic template: D-Quinic acid (**65**) as a single chiral source.

Reconfiguring the monocyclic C.14–C.26 fragment **63** as a cis-fused bicyclic ether (**64**) would potentially allow for predictable stereochemical communication between C.17, C.20, and C.23 across a structurally rigid template (Figure 17.16). In order to demonstrate the feasibility of that strategy, **63** was subjected to iodoetherification reaction conditions and the product, bicyclic iodoether **64**, was subjected to reductive elimination conditions to successfully regenerate **63**. Having demonstrated the viability of a bicyclic iodoether as a precursor of C.14–C.26, further retrosynthetic analysis on a structurally rigid template could be conducted. The orientation of C.17, C.20, and C.23 substituents, all on the convex face of the cis-6,5-bicyclic ring, suggested good opportunities for effective stereochemical communication. The lone stereogenic center that remained potentially in question was C.25. Further retrosynthesis suggested that the iodoether at C.19 being *syn* to the carbon at C.17 might be reconnected to arrive at a readily available single source of chirality, D-quinic acid (**65**).[36]

Starting from D-quinic acid (**65**), the known cyclohexylidine lactone **66**[37] was converted to a glycosidation substrate **67** by temporary protection of the tertiary alcohol as a silyl ether, DIBAL reduction, and diacetate formation (Figure 17.17). The crystalline diacetate **67** was treated with the five-carbon allylsilane **15** (previously used in the C.1–C.13 synthesis) to stereoselectively establish the C.20 center. The use of this particular bulky allyl silane was crucial for obtaining high selectivity at C.20. Use of a smaller allyl silane (e.g., allyltrimethylsilane) afforded only a 2:1 ratio at C.20. Sodium methoxide treatment of the intermediate C-glycoside product deacetylated the tertiary acetate, conjugated the β,γ-unsaturated ester, and then induced an oxy-Michael reaction to stereoselectively establish the C.23 center and form the crystalline polycyclic structure **68**. Thus,

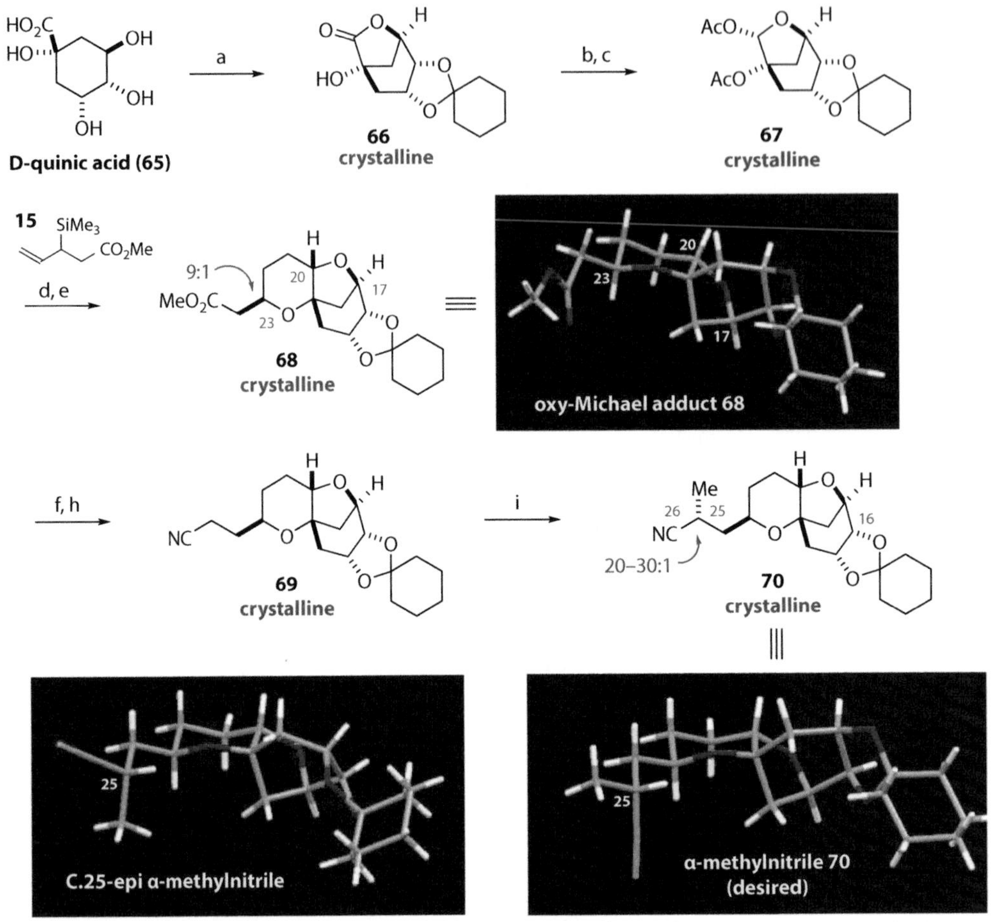

Reagents and conditions: (a) cyclohexanone, *p*-TsOH, toluene, reflux, crystallization, 95%; (b) (1) TMSCl, imidazole, toluene, 20 °C, (2) DIBAL, toluene, –78 °C; (c) (1) AcOH, H$_2$O DMAP, THF, 20 °C, (2) Ac$_2$O, TEA, 20 °C, crystallization, 72% overall from **66**; (d) **15**, BF$_3$.OEt$_2$, TFAA, CH$_3$CN, 20 °C; NaOMe, MeOH, THF, 10 °C; (f) LiAlH$_4$, THF, 15 °C; (g) MsCl, TEA, THF, 25 °C (product confirmed to be crystalline); (h) KCN, EtOH, 70 °C, crystallization, 30% overall from **67**; (i) (1) KHMDS, MeI, toluene-THF, –75 °C, (2) cat. KHMDS, toluene, –75 °C, crystallization, 72% overall from **69**.

Figure 17.17   Synthesis of a crystalline polycyclic C.16–C.26 template from D-quinic acid.

in a concise route from an available chiral building block using well-established methodology and reagents, three of the four stereocenters in the C.14–C.26 fragment, namely, C.17, C.20, and C.23, had been assembled on a crystalline polycyclic template.

The overall synthesis thus far mirrored that used for the C.1–C.13 fragment in a couple of important ways: (1) a cyclohexylidene-protected carbohydrate served as the crystalline template upon which the fragment was constructed, and (2) the same five-carbon allylsilane was used to establish two stereogenic centers on a pyran template. The crystallinity of all the intermediates in the quinic acid route thus far was an important finding, since it provided the means to control quality, throughput, cost, and so forth. Additionally, having an abundance of crystalline intermediates added flexibility to the overall manufacturing and control strategy.

Reduction of the ester **68**, activation, and cyanide displacement provided crystalline C.14–C.26 nitrile **69**. Importantly, the stereochemistry at C.23 was upgraded during the course of that synthetic process and crystallization to afford a ratio greater than 100:1 of desired-to-undesired configuration.

The crystallinity of nitrile **69** was encouraging since it suggested that incorporation of the C.25 methyl group might also result in a crystalline product. In the event, methylation of **69** resulted in a mixture of C.25 epimers (Figure 17.17). Fortunately, the desired C.25 methylated nitrile **70** was crystalline (m.p. = 123 °C), while the undesired C.25 methyl isomer possessed significantly different physical properties (m.p. = 81 °C). Thus, it became practical to separate the mixture of C.25 methyl epimers by crystallization. As a result of significant reaction optimization, the final sequence for the nitrile alkylation involved deprotonation of **69** with 1.1 equivalents of potassium bis(trimethylsilyl)amide and alkylating with methyl iodide at –78 °C. A 2:1 ratio of C.25 methyl epimers was initially formed. Equilibration by direct *in situ* retreatment and aging of the isomeric mixture by addition of another 0.15 equivalent of potassium bis(trimethylsilyl)amide improves the C.25 ratio to 4:1. Crystallization afforded **70** with a 20–30:1 diastereomeric ratio at C.25. A second crop of **70** was obtained after reequilibration of the mother liquor (0.45 equivalent of potassium bis(trimethylsilyl)amide, –78 °C).

With crystalline **70** in hand and with knowledge of the epimerizability of the C.25 center, exploration of a dynamic process was initiated (Figure 17.18). Treatment of **70** with a C.25 diastereomeric ratio of 1:1.6 (desired–undesired) with a catalytic alkoxide base in a 3:1 mixture of heptane–toluene at room temperature set in place an equilibration in which a 1:1 diastereomeric was achieved after 2.5 h. Eventually, the desired product crystallized out, with a 12:1 ratio (desired–undesired) in 64% yield.[38] Thus, crystallization-induced diastereoselective transformation was established as a useful procedure for setting the final and most challenging stereocenter in the C.14–C.26 fragment, C.25.

Having served the purpose of providing a scaffold upon which to construct a crystalline polycyclic array containing all four stereogenic centers contained in the C.14–C.26 fragment, the cyclohexane ring could now be deconstructed to unveil the iodo-ether bicyclic C.14–C.26 fragment **77** prior to Vasella fragmentation (Figure 17.19). Removal of the ketal using methanol and *p*-toluenesulfonic acid did not result in complete deketalization. Although the starting material appeared to be consumed, invariably recovered starting material was detected after work-up. It was reasoned that the intermediate hemi-ketal, enol ether, or the product cyclohexanone and diol could undergo the reverse reaction upon work-up to account for the recovery of starting material. Addition of phenylhydrazine served to drive the reaction to completion. Presumably, the phenylhydrazine helps to drive the deketalization equilibrium by converting the cyclohexanone to a phenylhydrazone that can undergo an irreversible Fischer indole synthesis.[39,40] The cis-diol **71** was then regio- and stereoselectively converted to bromoacetate **73** using Mattock's reagent (**72**). The stereochemical outcome of the bromoacetate formation can be explained by axial opening of the intermediate oxonium-bridged diol on the conformationally well-defined polycyclic system. The original plan for deconstructing the cyclohexane ring had called for converting the bromoacetate to an epoxide and relying on the bridged polycyclic system to direct allylic alcohol formation. However, with the fortuitous regioselectivity obtained in bromoacetate **73**, the bromide could be directly eliminated to provide crystalline allylic acetate **74**. Finally, in a one-pot, four-step sequence, the allylic acetate could be ozonized, reduced, deacetylated, and undergo periodate cleavage of the triol to give crystalline lactol **75**. One advantage of having numerous crystalline intermediates was the flexibility

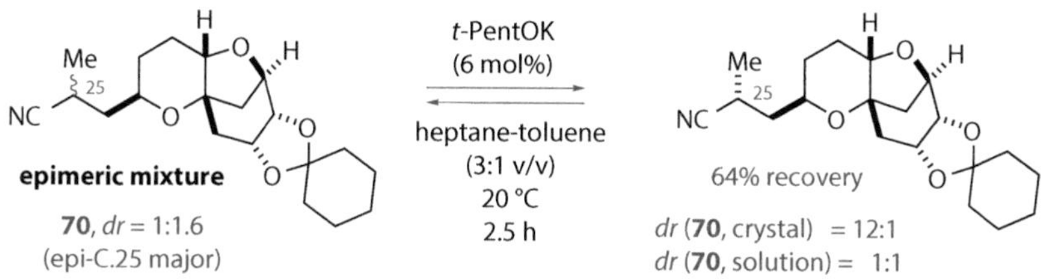

Figure 17.18 Crystallization-induced diastereoselective transformation of an α-methyl nitrile.

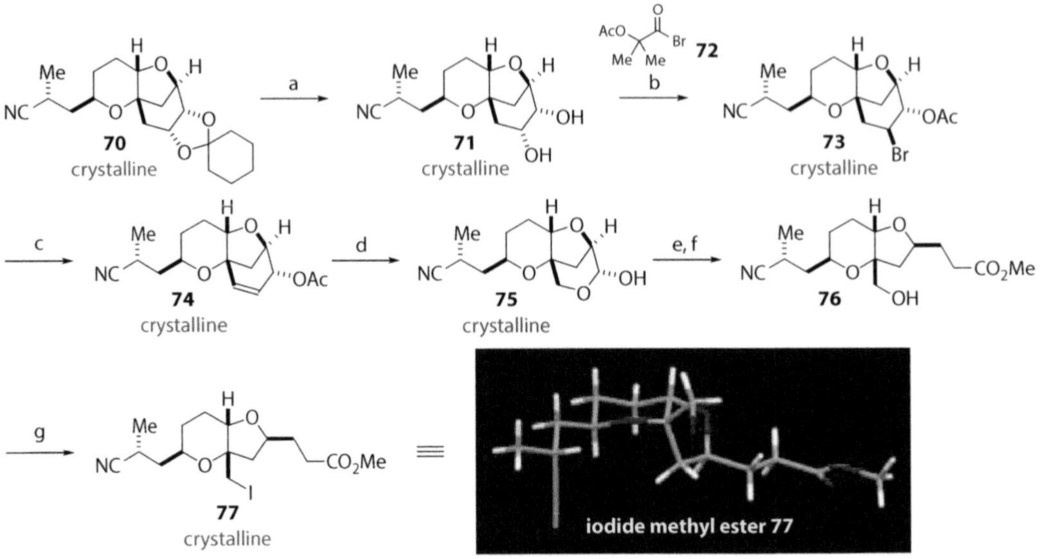

iodide methyl ester 77

Reagents and conditions: (a) phenylhydrazine hydrochloride, p-TsOH, MeOH, 55 °C, crystallization, 99%; (b) **72**, CH$_3$CN, cat. H$_2$O, 0 °C, 76%; (c) DBU, toluene, 100 °C, crystallization, 66%; (d) (1) O$_3$, i-PrOAc-MeOH, –30 °C, (2) NaBH$_4$, 0 °C, (3) K$_2$CO$_3$, 20 °C, (4) NaIO$_4$, THF-H$_2$O, 20 °C, 90% overall from **74**; (e) (MeO)$_2$POCH$_2$CO$_2$Me, (i-Pr)$_2$NEt, LiCl, CH$_3$CN, 18 °C, 98%; (f) H$_2$, Pd-C, (i-Pr)$_2$NEt, EtOAc, rt, 89%; (g) (1) Tf$_2$O, 2,6-lutidine, MTBE, 0 °C, (2) NaI, DMF, rt, crystallization, 81%.

Figure 17.19 Cyclohexane ring opening to a crystalline C.14–C.26 iodoether via multiple crystalline intermediates.

to choose which crystals would be optimal as isolation and control points. In this case, the lactol, while easily isolated, was actually directly carried forth to the final crystalline iodo-methyl ester **77** without the need for isolation.

Homologation of lactols **75** to **76** using a Horner–Wadsworth–Emmons reaction under the Masamune–Roush conditions, followed by reduction of the unsaturated ester, proved uneventful. The use of the Masamune–Roush conditions was important, as the normal, non-LiCl conditions afforded considerable amounts of oxy-Michael product resulting from the neopentyl alcohol reacting with the unsaturated ester. Conversion of the neopentyl alcohol to a neopentyl iodide was not initially straightforward. As might be expected, direct S$_N$2-type displacement processes of the neopentyl leaving group were not fruitful. However, taking advantage of the neighboring group pyran oxygen, it was reasoned that a sufficiently good leaving group (e.g., a triflate) under suitable polar aprotic solvent conditions, DMF and NaI, could generate the neopentyl iodide **77**, via anchimeric assistance by the pyran oxygen.[38] Importantly, this iodoether turned out to be crystalline. To further emphasize the interplay between conformation, stereocenters, and crystallinity, it should be noted that the corresponding C.25 normethyl analogue of the iodoether methyl ester was not crystalline. Thus, the extra methyl group at C.25, in addition to being required for the synthesis, was also important for the physicochemical properties, or crystallinity, of this fragment.

The final maneuvers to establish the C.14–C.26 fragment proceeded uneventfully, as shown in Figure 17.20.[38] Reduction of ester **77** with LiBH$_4$ and Vasella fragmentation provided the diol **78**. Lactone **79** was prepared via acid-catalyzed cyclization, forming an imidate, followed by hydrolysis. Silylation of the primary alcohol and Weinreb amide formation from the lactone provided **80**. To suppress lactone formation, silylation of the secondary alcohol was required to provide **59**. Two-step vinyl triflate installation, desilylation, pivaloylation, and mesylation completed the sequence, providing the C.14–C.26 fragment, **7b**.

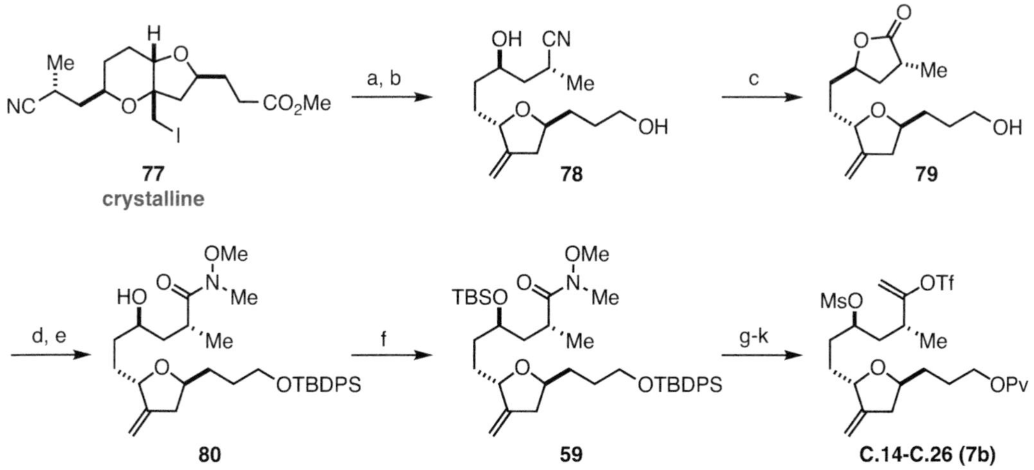

Reagents and conditions: (a) LiBH$_4$, MeOH, THF-toluene, 13 °C; (b) Zn, AcOH, THF-H$_2$O, 0 °C, 99% from **70**; (c) (1) HCl, 2-PrOH-toluene, 25 °C, (2) H$_2$O, 45 °C, 94%; (d) TBDPSCl, imidazole, DMF, 23 °C; (e) HNMe(OMe)HCl, 2.0 M Me$_3$Al in toluene, CH$_2$Cl$_2$, 0 °C; (f) TBSCl, imidazole, DMF, 23 °C; (g) MeMgCl, THF, 0 °C; (h) Tf$_2$NPh, KHMDS, THF-toluene, −70 °C; (i) HCl, 2-PrOH-MeOH, rt; (j) PvCl, 2,4,6-collidine, DMAP, DCM, 0 °C, chromatography, 45% from **79**; (k) MsCl, Et$_3$N, THF, 99%.

Figure 17.20 Intercepting the current manufacturing route.

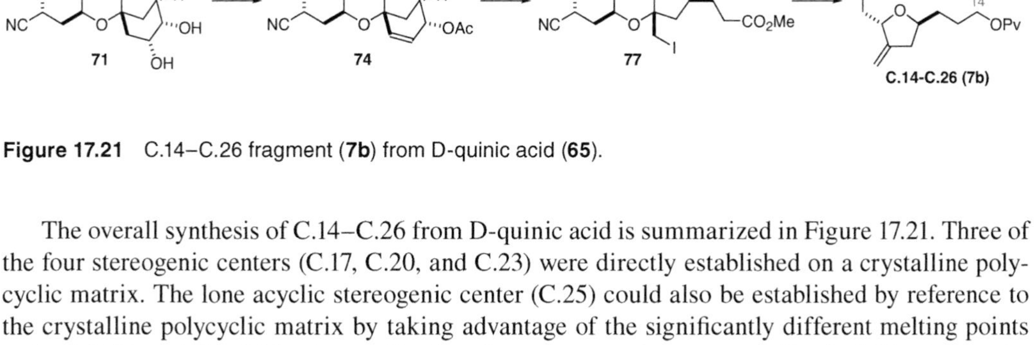

Figure 17.21 C.14–C.26 fragment (**7b**) from D-quinic acid (**65**).

The overall synthesis of C.14–C.26 from D-quinic acid is summarized in Figure 17.21. Three of the four stereogenic centers (C.17, C.20, and C.23) were directly established on a crystalline poly-cyclic matrix. The lone acyclic stereogenic center (C.25) could also be established by reference to the crystalline polycyclic matrix by taking advantage of the significantly different melting points and solubilities of the C.25 epimers. Further, from the crystal structures of both C.25 epimers, the impact of the C.25 stereogenic center on the overall conformation of the polycyclic matrix can be appreciated.

17.3.9 C.27–C.35 Fragment: L-Arabinose (81) Route

The initial synthesis of the C.27–C.35 fragment is depicted in Figure 17.22.[13] Selective silylation of the primary hydroxyl of L-arabinose (**81**), followed by triacetylation and C-glycosidation with

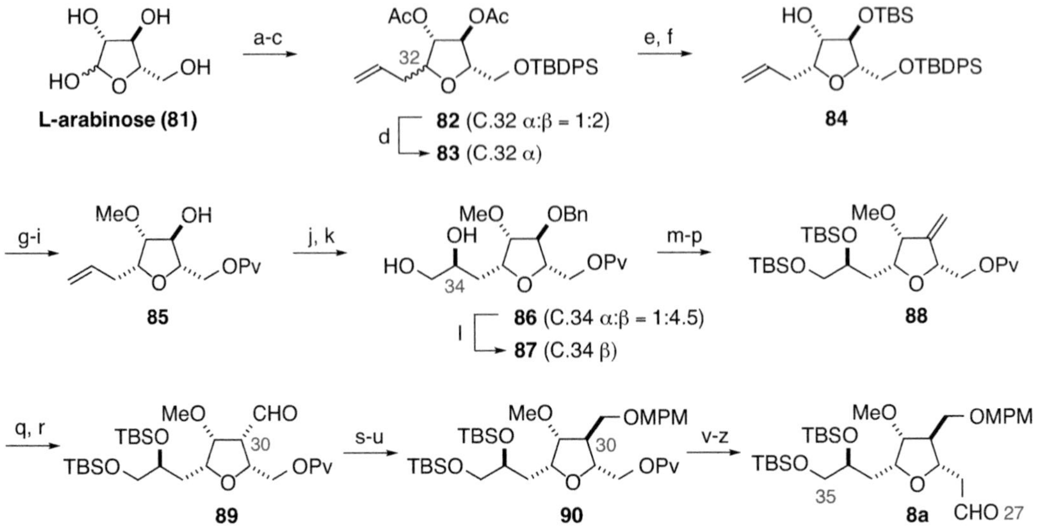

Reagents and conditions: (a) TBDPSCl, DMF, imidazole, chromatography, 61%; (b) Ac$_2$O, pyridine, chromatography, 97%; (c) allyltrimethylsilane, BF$_3$·OEt$_2$, toluene; (d) chromatography, 69%; (e) K$_2$CO$_3$, MeOH, chromatography, 37%; (f) TBSCl, imidazole, CH$_2$Cl$_2$, chromatography, 83%; (g) MeI, NaH, DMF, THF, chromatography, 96%; (h) HCl, MeOH, chromatography, 87%; (i) PvCl, pyridine, chromatography, 96%; (j) BnBr, aq. NaOH, n-Bu$_4$NHSO$_4$, CH$_2$Cl$_2$, chromatography, 98%; (k) OsO$_4$, (DHQ)$_2$PyD, K$_2$CO$_3$, K$_3$Fe(CN)$_6$, aq. t-BuOH; (l) chromatography, 75%; (m) TBSOTf, Et$_3$N, CH$_2$Cl$_2$, chromatography, 94%; (n) Pd(OH)$_2$, H$_2$, EtOAc, chromatography, 96%; (o) TPAP, NMO, CH$_2$Cl$_2$; (p) Tebbe reagent, THF, chromatography, 74% (2 steps); (q) (1) 9-BBN, THF; (2) NaBO$_3$-H$_2$O, chromatography, 91%; (r) (COCl)$_2$, DMSO, Et$_3$N, CH$_2$Cl$_2$, −78 °C; (s) Et$_3$N, CH$_2$Cl$_2$; (t) NaBH$_4$, EtOH, Et$_2$O, chromatography, 85% (3 steps); (u) MPMOC(=NH)CCl$_3$, BF$_3$·OEt$_2$, CH$_2$Cl$_2$, chromatography, 85%; (v) LiAlH$_4$, Et$_2$O, chromatography, 93%; (w) (COCl)$_2$, DMSO, Et$_3$N, CH$_2$Cl$_2$, −78 °C, chromatography; (x) Ph$_3$PCH$_3$$^+Br^-$, n-BuLi, THF, chromatography, 71% (2 steps); (y) (1) 9-BBN, THF, (2) NaBO$_3$-H$_2$O, chromatography, 92%; (z) (COCl)$_2$, DMSO, Et$_3$N, CH$_2$Cl$_2$, −78 °C, chromatography, 92%.

Figure 17.22   Initial C.27–C.35 synthesis.

allyltrimethylsilane, provided a 2:1 mixture of C.32 allyl diastereomers **82** in which the minor isomer was the desired compound. The isomers were separated by chromatography to provide the required C.32 α-diastereomer **83**. Deacetylation and chemoselective silylation of C.30 alcohol provided **84**. Methylation of the C.31 hydroxyl group followed by acidic desilylation and pivaloylation of the primary hydroxyl group afforded pivalate **85**. Benzylation of C.30 alcohol under phase transfer conditions, followed by asymmetric dihydroxylation of the C.34/C.35 olefin, provided diol **86** with a 4.5:1 diastereomeric ratio at C.34. The diastereomers were separated by chromatography, resulting in diol **87**. Protection of the diol as the bis-silylether and deprotection of the benzyl ether by transfer hydrogenolysis provided a C.30 alcohol that was oxidized to the ketone with tetrapropylammonium perruthenate (TPAP) and directly olefinated with Tebbe reagent providing **88**. Installing the MPM (4-methoxybenzyl)-protected hydroxymethyl group at C.30 began by hydroboration of the C.30 olefin of **88** to provide a single primary alcohol with exclusively the undesired configuration at C.30. Swern oxidation afforded the aldehyde **89** that was epimerized by treatment with triethylamine in dichloromethane. The aldehyde, with correct C.30 stereochemistry, was reduced with sodium borohydride and the resulting hydroxyl group protected as the methoxyphenylmethyl (MPM) ether providing **90**. A five-step sequence to extend the C.29 side chain involved LAH reduction of the pivalate ester, followed by Swern oxidation, Wittig olefination, hydroboration and oxidation, and finally Swern oxidation to provide the C.27–C.35 aldehyde **8a**.

17.3.10 C.27–C.35 Fragment from (S)-1,2,4-Butanetriol (91) and 3-Butyn-1-ol (93): Yamaguchi Coupling Route

A second-generation synthesis of a C.27–C35 fragment, featuring a C.30–C.31 C–C bond disconnection, is shown in Figure 17.23.[12] The C.31–C.35 fragment was prepared in four steps from (S)-1,2,4-butanetriol 91. Pentylidine formation to form the five-membered ketal was achieved with 3-pentanone and p-TsOH. Swern oxidation provided an aldehyde that was transformed into the C.31–C.35 acetylene **92** by the two-step Corey–Fuchs protocol.[41] The synthesis of the C.27–C.30 fragment was a five-step sequence involving (1) benzylation of 3-butyn-1-ol **93**, (2) lithiation of the acetylene and quenching with p-formaldehyde, (3) hydrogenation with Lindlar catalyst to provide the allylic alcohol, (4) Sharpless epoxidation, and (5) MPM protection of the epoxy alcohol to provide **94**. Yamaguchi coupling[42] of the acetylene **92** and epoxide **94** provided a 5:1 mixture of regioisomers that were separable by flash chromatography, providing the coupled adduct **95** in a 69% yield. Hydrogenation with Lindlar catalyst, followed by acetylation, provided the (Z)-allylic acetate **96**. Dihydroxylation provided an 8–9:1 mixture of diastereomers that was transformed to the dimesylate **97**. On a laboratory scale, **97** proved suitable for chromatographic removal of the unwanted diastereomer; however, because of concerns over mesylate stability during extended exposure to silica, the diastereomeric mixture was treated with Triton B methoxide to induce acetate deprotection and cyclization of the resulting alkoxide to the C.32 mesylate providing the tetrahydrofuran **98**. Deprotection of the C.31 mesylate was achieved by treatment with MeMgBr to provide a C.31 alcohol. Chromatographic purification to remove the unwanted C.31/C.32 diastereomer, followed by methylation of the C.31 hydroxyl group, provided the C.31 methyl ether. The final steps involved exchange of the pentylidine as the TBS ethers, selective hydrogenolysis of the benzyl ether, and Swern oxidation to provide the aldehyde **8a**.

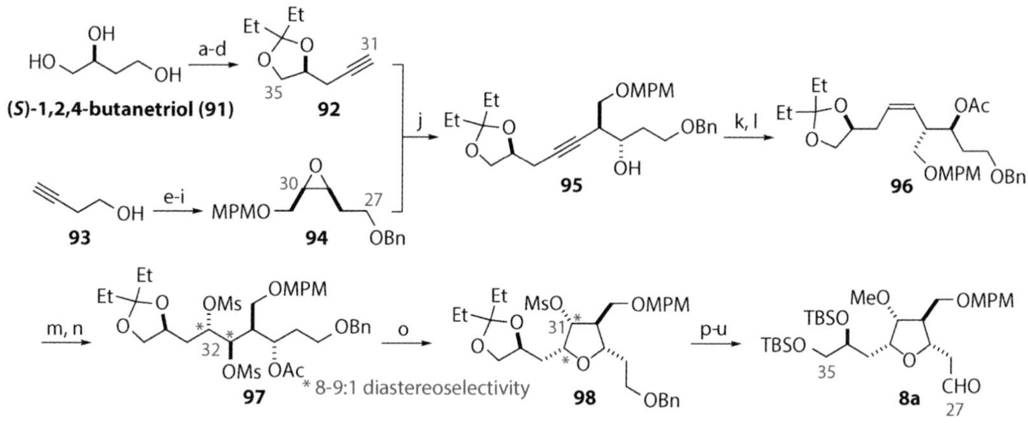

Reagents and conditions: (a) 3-pentanone, p-TsOH, 88%; (b) (COCl)$_2$, DMSO, Et$_3$N, –60 °C, wiped film distillation, 72%; (c) CBr$_4$, PPh$_3$, CH$_2$Cl$_2$, quant.; (d) n-BuLi, THF, WTF distillation, 76%; (e) t-BuOK, BnBr, THF; (f) n-BuLi, (CH$_2$O)$_n$, –35°C, chromatography, 56% (2 steps); (g) H$_2$, Lindlar's catalyst, quinoline, hexanes; wiped film distillation, quant.; (h) (–)-DIPT, Ti(OiPr)$_4$, t-BuOOH, CH$_2$Cl$_2$, 94%; (i) MPMCl, NaH, DMPU, THF, chromatography, 74%; (j) n-BuLi, BF$_3$OEt$_2$, toluene, –50 °C, chromatography, 75%; (k) H$_2$, Lindlar's catalyst, quinoline, hexanes/CH$_2$Cl$_2$, quant.; (l) Ac$_2$O, pyridine, DMAP, CH$_2$Cl$_2$, chromatography, quant.; (m) OsO$_4$, NMO, t-BuOH, H$_2$O,chromatography, 93% as 9:1 diastereomeric mixture; (n) MsCl, pyridine, 69%; (o) Triton B, MeOH, chromatography, 38% (p) MeMgBr, THF, 45°C; (q) t-BuOK, MeI, THF, 0 °C, chromatography, 81% (2 steps); (r) HCl, THF, MeOH, H$_2$O; (s) TBSCl, imidazole, DMF, chromatography, 98% (2 steps); (t) H$_2$, Raney-Ni, EtOH, chromatography, 92%; (u) (COCl)$_2$, DMSO, Et$_3$N, CH$_2$Cl$_2$, –60 °C, chromatography, 73%.

Figure 17.23 Initial scale-up route for C.27–C.35.

17.3.11 C.27–C.35 Fragment: D-Glucuronolactone (99) Route

A third route from D-glucuronolactone (**99**) (Figure 17.24) targeting a C.27–C.35 fragment **8b** that already contained the phenylsulfone moiety was reported by Kishi and coworkers.[43] This route had a number of attractive features, including an available carbohydrate source of chirality and a number of crystalline intermediates.[31,32,43] Furthermore, by incorporating the phenylsulfone into the C.27–C.35 fragment, the overall synthesis to C.14–C.35 became more convergent and there-fore more efficient in usage of C.14–C.26. D-Glucuronolactone (**99**) contains two of the stereo-genic centers (C.31 and C.32) that are required for eribulin. Acetonide formation resulted in the crystalline hydroxyl lactone **100**, and two-step deoxygenation provided the crystalline lactone **102**. DIBAL reduction and Peterson olefination provided the crystalline alcohol **105** via crystalline diol **104**. Benzylation and asymmetric dihydroxylation provided the lone acyclic stereocenter at C.34 in about a 3:1 mixture (diol **107**). The 3:1 mixture was dibenzoylated and then carried forward into a C-glycosidation with allyltrimethylsilane. This C-glycoside alcohol **109** proved to be crystalline and was able to be separated in high yield and quality from the corresponding C.34 epimer in an

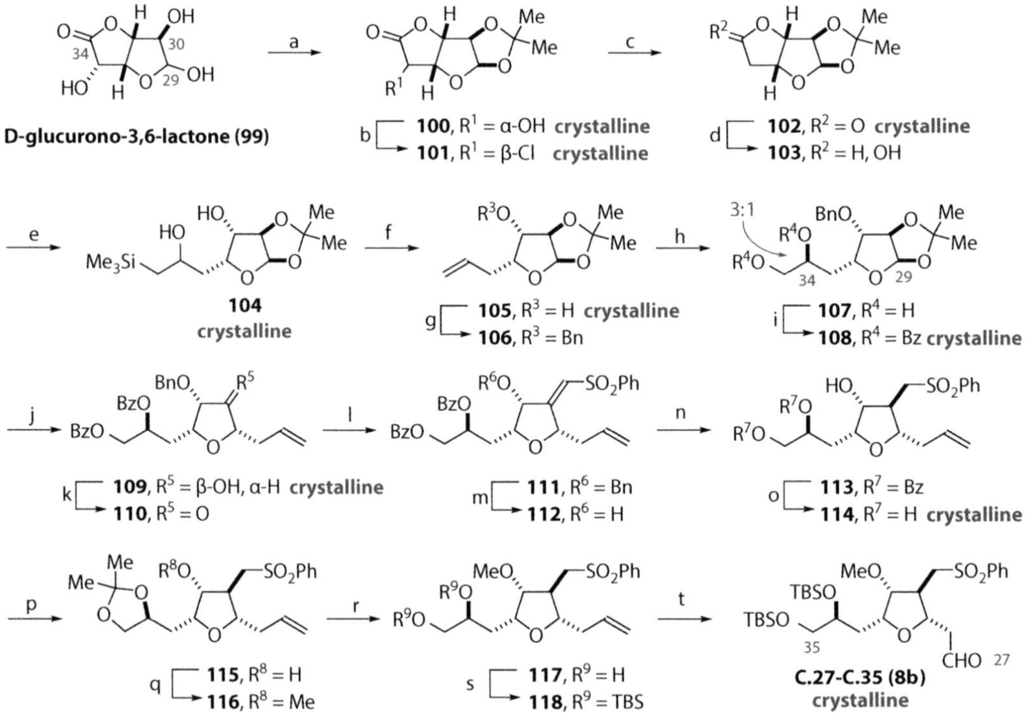

Reagents and conditions: (a) H₂SO₄, acetone-CH₃CN, 65 °C; (b) SO₂Cl₂, pyridine, CH₃CN, 0 °C, crystallization, 79% from **93**; (c) Pd-C, H₂, pyridine, THF, 25 °C, crystallization, 79%; (d) DIBAL, toluene, -20 °C, 95%; (e) Mg, TMSCH2Cl, THF, 60 °C, crystallization, quant.; (f) 15 wt% KHMDS in toluene, THF, 25 °C, crystallization, quant.; (g) BnBr, *t*-BuOK, THF, -10 °C, quant.; (h) (DHQ)₂AQN, OsO₄, K₃Fe(CN)₆, K₂CO₃, *t*-BuOH, water, 0 °C, 98%; (i) BzCl, DMAP, NMO, toluene, 75 °C, 99%; (j) allyltrimethylsilane, TiCl4, Ti(O*i*-Pr)₄, toluene, 25 °C, crystallization, 75%; (k) DMSO, trichloroacetic anhydride, TEA, toluene, -10 °C, quant.; (l) (EtO)₂P(O)CH₂SO₂Ph, LHMDS, THF-toluene, 0–20 °C, 98%; (m) TMSI, toluene-CH₃CN, 60 °C, 88%; (n) NaBH(OAc)₃, (*n*-Bu)₄NCl, DME-toluene, 80 °C, quant; (o) K₂CO₃, MeOH, 50 °C, crystallization, 79%; (p) 2-methoxypropene, H₂SO₄, DME, 40 °C, 96%; (q) *t*-BuONa, MeI, DME-NMP, 10 °C, 98%; (r) 2 M HCl, MeOH, 25 °C, 99%; (s) TBSCl, imidazole, DMF, 25 °C, 89%; (t) O₃, 2-PrOH, <50 °C; BHT in 2-PrOH, Lindlar's catalyst, H₂, 15 °C, crystallization, 89%.

Figure 17.24   C.27–C.35 from D-glucuronolactone.

approximately 70% yield. The last remaining stereocenter in the C.27–C.35 sector was the C.30 carbon center. Installation of C.30 was accomplished by a Horner–Wadsworth–Emmons reaction on C.30 ketone **110** to provide the unsaturated sulfone **111**. Deprotection of the C.31 benzyl ether and stereoselective hydroxyl-directed conjugate reduction of the unsaturated sulfone provided sulfone **113**. The sulfone **113** contains all five stereogenic centers required for the C.27–C.35 fragment; however, the last crystalline purification point was four steps earlier at the C-glycosidation product dibenzoate alcohol **109**. In order to identify a later stage process and stereochemical control point, the benzoate protecting groups were removed to reveal crystalline triol **114**. Acetonide protection of the vicinal diol and capping the C.31 alcohol as a methyl ether provided **115** and **116**, respectively. Acetonide deprotection and silylation provided **118** in a suitably protected form in advance of generating the aldehyde. Ozonolysis gave the C.27–C.35 aldehyde **8b**, which also turned out to be crystalline. The C.27–C.35 fragment synthesis was in fact the longest linear sequence of the three syntheses, yet C.27–C.35 is the cheapest, and that is in large part because there are no chromatographic purifications in this 20-step synthesis and numerous crystalline materials to control cost, environmental impact, quality, and so forth.

17.3.12 C.14–C.35 Fragment via NHK Reaction/Williamson Ether Cyclization Reaction

The NHK reaction that was employed in the course of an earlier C.14–C.26 synthesis could now be applied with some confidence to the crucial C.14–C.26 and C.27–C.35 coupling strategy (Figure 17.25).[31] The C.14–C.26 fragment, unlike the C.14–C.19 fragment used in the C.19–C.20 NHK process, was quite unstable. While the C.14–C.26 mesylate-pivalate **7b** could be manufactured

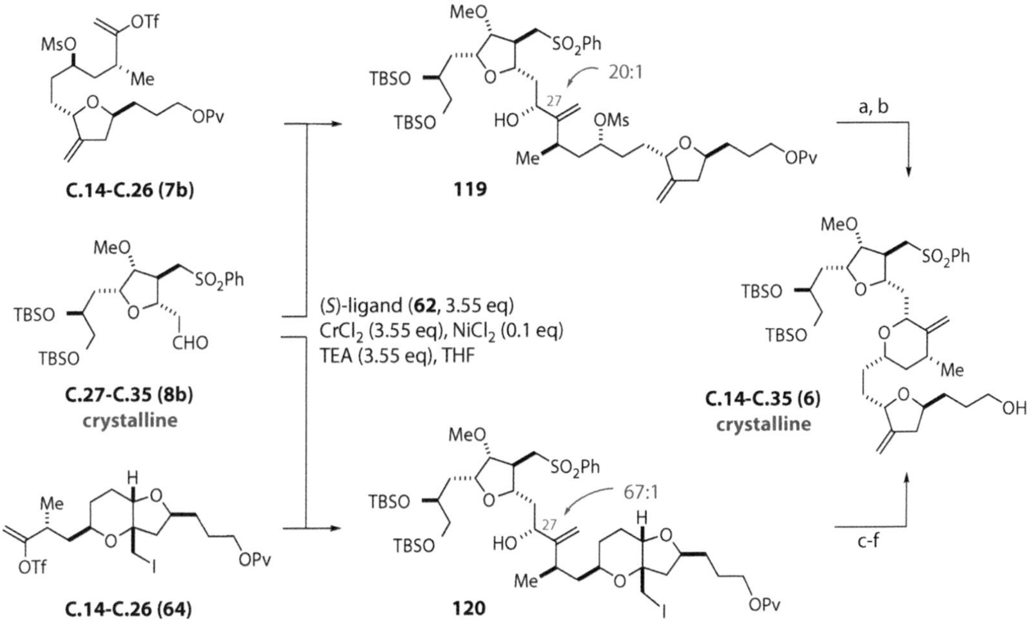

Reagents and conditions: (a) KHMDS, THF-toluene, −20 °C, chromatography, 65% from **7b**; (b) DIBAL, toluene, −70 °C, chromatography, crystallization, 79%; (c) Ac$_2$O, DMAP, pyridine, chromatography, 52% from **64**; (d) Zn, AcOH, THF- H$_2$O, chromatography, 83%; (e) Ms$_2$O, pyridine, CH$_2$Cl$_2$; (f) NaOMe, THF-MeOH, chromatography, 87% (2 steps).

Figure 17.25   Enhanced stereoselectivity in the C.26–C.27 NHK coupling on a bicyclic iodoether template.

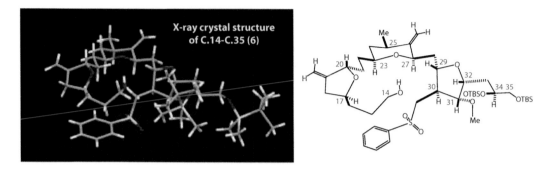

Figure 17.26 Single crystal x-ray structure of C.14–C.35 fragment (**6**).

on scale, that material could not be chromatographed or easily recovered in the event of a problem in the NHK reaction. Additionally, the NHK product **119**, which contains a mesylate, was also not entirely stable to the reaction conditions and needed to be converted to the Williamson ether product in a timely manner. The successful experience of the C.19–C.20 asymmetric NHK pilot plant implementation provided confidence that the important reaction parameters had been identified and controlled. Nevertheless, it was with some trepidation that kilogram quantities of C.14–C.26 were placed into a fixed reactor. Gratifyingly, the process performed as well as or better than observed in the laboratory and afforded 20:1 selectivity at the newly formed C.27 stereogenic center. Direct treatment of that crude product with potassium bis(trimethylsilyl)amide provided the C.14–C.35 fragment with the correct stereochemistry at C.27 and C.23. Somewhat unexpectedly, depivaloylation resulted in a crystalline C.14–C.35 fragment **6** (Figure 17.26).

Having the stable bicyclic iodoether **64**[*] (Figure 17.25) suggested an alternate and perhaps advantageous process for assembling C.14–C.35. Asymmetric NHK coupling of **64** with **8b** proceeded with ca. 67:1 diastereoselectivity at C.27.[44] This highly stereoselective process perhaps benefits from chelation of the iodoether oxygen to the incipient vinyl chromium species. The coupling product **120** can be converted in a straightforward sequence of reactions (acetylation, Vasella fragmentation, mesylation, and sodium methoxide treatment) to the crystalline C.14–C.35 fragment (**6**). This chelation-controlled C.14–C.35 synthesis has the added benefit of avoiding carrying an unstable mesylate through the NHK process, and thus minimizing risk during this complex reaction process.

The first discovery that **6** was crystalline (Figure 17.26) occurred after producing an extensively chromatographed high-purity reference standard. This example highlights the value to process development of an essential part of analytical method development and quality control strategy development: production of highly purified reference standards. Oftentimes, the first crystalline samples were obtained during the production of primary reference standards. The finding that **6** is

[*] See Figure 17.16. An alternate synthesis from **76** was also developed:

Reagents and conditions: (a) TBSCl, Imidazole; (b) LiBH$_4$, THF, (c) PvCl, Et$_3$N, 73% (3 steps); (d) benzyl 2-bromoacetate, Zn, cat MsOH, THF, 74%; (e) AcOH, THF-H$_2$O; (f) Pd-C, H$_2$; (g) toluene, D, 70% (3 steps); (h) PhNTf$_2$, KHMDS; (i) HCl-IPA, MeOH, 83% (2 steps); (j) (1) Tf$_2$O, pyridine, CH$_2$Cl$_2$, (2) NaI, DMF, one-pot, 87%.

crystalline turned out to be practically important for manufacturing eribulin and also, logistically, for handling the fragment production that was occurring at various plants around the world.

17.3.13 Final Assembly: Construction of Eribulin from C.1–C.13 (5) and C.14–C.35 (6) Fragments

With secured supplies of the two advanced building blocks, C.1–C.13 (**5**) and C.14–C.35 (**6**), assembly of the macrocycle and completion of the synthesis of eribulin could be developed for commercialization. An important aspect of the final assembly was that the overall synthetic strategy had remained consistent from the medicinal chemistry route through clinical development and commercialization. Thus, from an experience and quality control perspective, maximum value was extracted from the knowledge base developed at Eisai. An important discovery highlighted earlier was the finding that the C.14–C.35 fragment (**6**) was crystalline. The crystal structure of **6** provided valuable structural information that was helpful in understanding and optimizing the last intermolecular C–C bond formation between the sulfone-bearing carbon and the C.1 aldehyde. From the crystal structure, a hydrogen bond between the C.14 hydroxyl group and the C.17 tetrahydrofuran oxygen appeared to be present based on the interatomic bond distance (2.79 Å). Additionally, the crystal structure suggested that the sulfone-bearing carbon at C.30 was in relatively close proximity to the C.14 hydroxyl group. In as much as the C–C bond formation step would minimally involve deprotonation of both the hydroxyl group and the sulfonyl carbon, it was hoped that the proximity of the dianionic species would prove favorable by allowing an organized four-membered ring ionic interaction of the alternating charges, resulting in coulombic stabilization. Consistent with this hypothesis, use of just stoichiometric amounts of base to accomplish the double deprotonation and minimizing the solvent polarity (i.e., replacing most of the initially used ether solvents with heptanes) were found to be optimal for the reaction. The efficiency of the key intermolecular C–C bond formation in the final assembly was found to be quite high with ca. 85%–90% yield on a kilogram scale and ca. 8%–10% recovered C.14–C.35 sulfone (Figure 17.27). Oxidation of the resultant diol **121** to the keto-aldehyde **122** was accomplished in essentially quantitative yield using Dess–Martin periodinane. Selective desulfonylation of the polyfunctional molecule was accomplished with samarium diiodide in methanolic THF. The use of methanol turned out to be fortuitous and important. Use of other, bulkier alcohols (e.g., ethanol and isopropanol) led to significant amounts of aldehyde reduction. Apparently, methanol in the presence of samarium diiodide rapidly and quantitatively forms a hemiacetal with the C.14 aldehyde carbonyl, thus protecting this functionality from reduction. The stage was set for the final and most critical C–C bond formation via an intramolecular NHK reaction. At the time of initial development, the process did not employ an asymmetric ligand. Thus, the reaction times were long (7–10 days), required high dilution (ca. 350 volumes), and had not been conducted beyond the milligram scale. Toward the initial clinical batch production, approximate half-kilogram-scale macrocyclization quantities would be required. Although the process did not require stereoselectivity, the idea of utilizing the recently developed asymmetric variant of the NHK reaction seemed attractive due to the increased rates and homogeneity of the reaction. It was hoped that inverse addition of substrate to the reaction matrix, along with the efficient kinetics of the cyclization process, could simulate high dilution conditions without having to resort to large amounts of solvent. Having successfully conducted the asymmetric NHK reaction in fixed equipment on two prior C–C bond-forming reactions (C.19–C.20 and C.26–C.27), the understanding and control of the reaction parameters for the macrocyclization process were well established. In the event, the cyclization process worked remarkably well on the kilogram scale to provide an 80%–90% yield of macrocycle in less than a day with less than 60 volumes of solvent. The S-ligand **62** was found to be optimal for the reaction kinetics. Oxidation of the allylic alcohol to the corresponding enone could be conducted with a variety of oxidants, including Dess–Martin periodinane and modified Moffatt reagent. Purification of the macrocyclic enone **125** via chromatography controlled upstream process impurities. The assembly of the cage structure of eribulin required (1) removal of five *tert*-butyldimethylsilyl

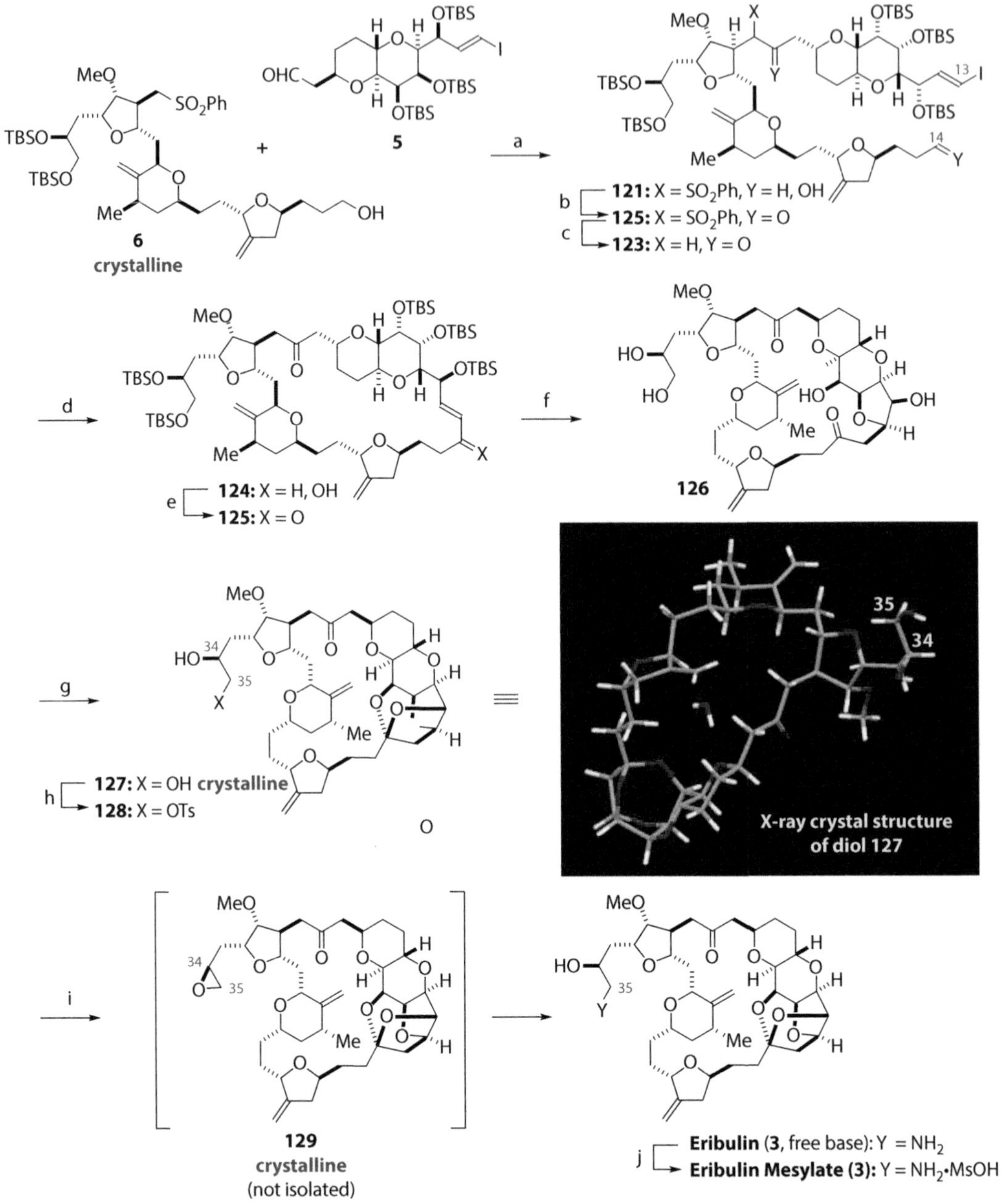

Reagents and conditions: (a) *n*-BuLi, THF/heptane, < –50 °C, 75–95%; (b) Dess-Martin periodinance, CH_2Cl_2, cat. water, chromatography, 75–95%; (c) SmI_2, THF/MeOH, < 65 °C, chromatography, 75–95%; (d) $CrCl_2$, $NiCl_2$, (*S*)-Ligand (**62**), Et_3N, CH_3CN, THF; (e) Dess-Martin periodinane, CH_2Cl_2, cat. water, chromatography, 60–80% from **123**; (f) TBAF, imidazole hydrochloride; (g) (1) PPTS, CH_2Cl_2; (2) CH_3CN water, chromatography; (3) crystallization, 55–75% from **125**; (h) Ts_2O, 2,4,6-collidine, pyridine, CH_2Cl_2; (i) NH_4OH, 2-PrOH, chromatography, 75–95%; (j) (1) MsOH, NH_4OH, CH_3CN, water, (2) [precipitation] CH_2Cl_2, *n*-pentane, 80–100%.

Figure 17.27 Final sequence.

groups, (2) regio- and stereoselective transannular oxy-Michael addition of the C.9 hydroxyl group to the C.12 position of the macrocyclic enone, and (3) polycyclic ketal formation. The initial conditions for the deprotection/oxy-Michael step utilized buffered tetrabutylammonium fluoride (TBAF) in THF and required 7–10 days for completion of the reaction. Improved kinetics (<1 day) and selectivity (20:1 at C.12) for the deprotection process and oxy-Michael reaction were obtained by switching to dimethylacetamide in place of THF. Ketalization was accomplished with standard conditions. The final intermediate diol **127** was also discovered to be crystalline and served as an important structural proof and quality control point. The x-ray structure of **127** provided the first solid-state information confirming the overall structure and stereochemistry. From this final crystalline intermediate, selective activation of the primary C.35 alcohol as a sulfonate, *in situ* epoxide formation, and regioselective aminolysis of the terminal epoxide completed the synthesis of eribulin.

17.3.14 Green Chemistry Analysis: Impact of Crystallization-Based Syntheses

A consistent theme throughout the development of the manufacturing routes for the eribulin fragments was the progression from chromatography-based to crystallization-based syntheses. Figure 17.28 summarizes the number of crystallization and chromatography steps for fragments 5,

Eribulin Fragment		Crystallization – Chromatography	
		Early Development Route	**Scale-Up Route**
C.1-C.13 (5)	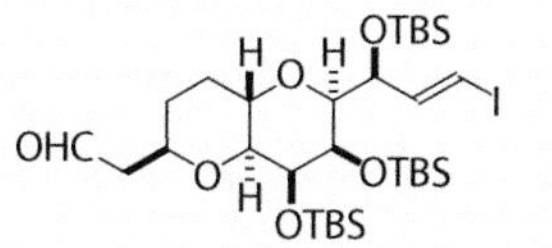	2 – 9* Figure 6	6 – 2 (# of crystals: 7) Figure 8
C.14-C.26 (7a/b)	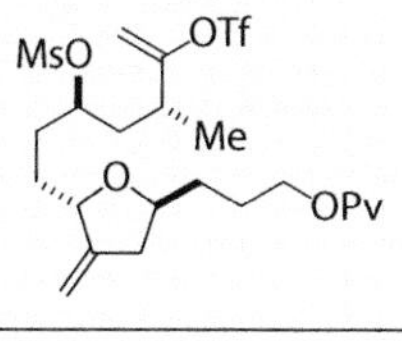	4 – 21 7a, Figure 9	1 – 8 (# of crystals: 1) 7b, Figure 11–13
C.27-C.35 (8a/b)	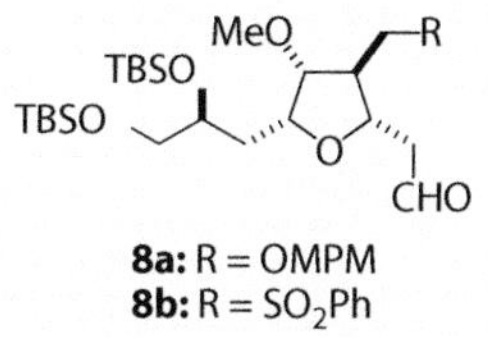 8a: R = OMPM 8b: R = SO₂Ph	0 – 9 8a, Figure 23	7 – 0 (# of crystals: 9) 8b, Figure 24

* The early development route included four column chromatographies from compounds 17 to 5 (Figure 6).

Figure 17.28 Eribulin fragment synthesis: Crystallization-based improvements.

7b, 8a, and 8b as the eribulin synthesis evolved from the early development route (early development route is defined as Figure 17.6 for C.1–C.13, Figure 17.9 for C.14–C.26, and Figure 17.23 for C.27–C.35) to the scale-up route (scale-up route is defined as Figure 17.8 for C.1–C.13, Figures 17.11 through 17.13 for C.14–C.26, and Figure 17.24 for C.27–C.35). Whereas significant progress was achieved in advancing the C.1–C.13 and C.27–C.35 fragment syntheses from chromatography-based to crystallization-based routes, the C.14–C.26 fragment synthesis still heavily depended upon chromatography as the key quality controlling technology.

To help illustrate the importance of evolving toward crystalline-based syntheses, Figure 17.29 depicts a green chemistry analysis for the scale-up route, that is, the relative waste stream per fragment per kilogram of eribulin produced. At this point in the evolution of the eribulin synthesis, the C.14–C.26 fragment synthesis generated twice as much waste as the C.27–C.35 fragment synthesis and 60% more waste than the C.1–C.13 fragment synthesis. Based strictly upon the structures and numbers of stereogenic centers, the reasons for the relative inefficiency of the C.14–C.26 fragment synthesis may not be immediately apparent. However, the seven stereogenic centers in C.1–C.13 and the five stereogenic centers in C.27–C.35 are primarily embedded in ring systems, while the C.14–C.26 fragment contains multiple, *noncontiguous*, *acyclic* stereogenic centers. As a result, the relative and absolute stereochemistry of the four stereogenic centers in the C.14–C.26 fragment were initially more challenging to establish and control. Stereochemical quality control on the C.14–C.26 fragment therefore necessitated the extensive use of chromatography in the scale-up route.

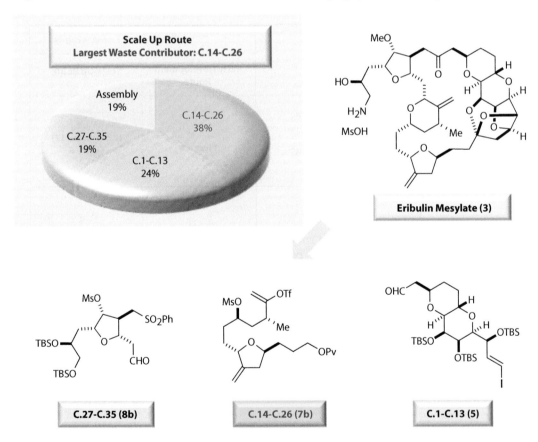

Figure 17.29 Eribulin waste stream (green chemistry) analysis.

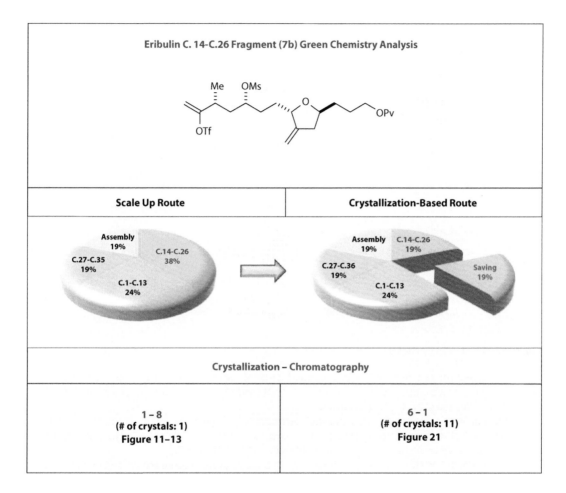

Figure 17.30   Green chemistry impact of C.14–C.26 crystallization-based route.

As the C.14–C.26 synthesis matured to a crystallization-based route, the impact of crystalliza-tion efficiency versus chromatography is reflected in the dramatic reduction of waste generated (Figure 17.30). In the scale-up route, the C.14–C.26 fragment was the major contributor to the eribulin waste stream, whereas in the C.14–C.26 crystallization-based route, C.14–C.26 is now the most efficiently produced of the three fragments, resulting in a 19% reduction of the eribulin waste stream. A similar analysis could be conducted on each of the three fragments as the synthetic routes evolved from early discovery to crystallization-based routes, and would show a similar trend. This one example focuses on how crystallization positively impacts green chemistry (i.e., minimizes waste generation); however, a similar correlation between crystallization and other key performance metrics such as throughput, time of production, and cost could be provided.

17.4 CONCLUSIONS

A summary scheme displaying the fragments used to construct eribulin is shown in Figure 17.31 to help illustrate some common themes:

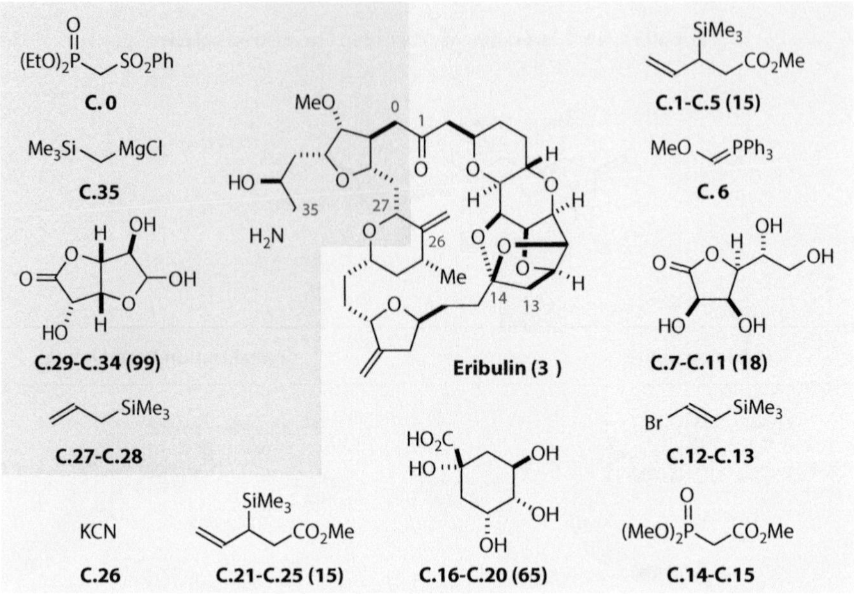

Figure 17.31 Eribulin synthesis.

1. The carbohydrate templates D-gulonolactone, D-quinic acid, and D-glucuronolactone serve as the source of chirality at C.8, C.9, C.10, C.17, C.31, and C.32 for the three fragments, C.1–C.13, C.14–C.26, and C.27–C.35, respectively.

2. A common five-carbon allylsilane, methyl-3-trimethylsilyl-4-pentenoate (**15**), is the source of 10 carbons (C.1–C.5 and C.21–C.25) and is utilized twice for stereoselective C-glycosidation/oxy-Michael pyran annulation sequences to establish C.3, C.6, C.20, and C.23.

3. Stereoselective NHK reactions established the C.11 and C.27 stereogenic centers and provided the key C–C bond construction and macrocyclization technology that made the discovery, development, and eventual commercialization of Halaven® possible.

The utility of carbohydrate templates as cheap renewable sources of chirality and crystalline templates upon which to build nonchromatographic complex fragment syntheses was demonstrated in all three sectors of the eribulin molecule. Moreover, the ability to control stereochemical quality via crystallization as stereogenic centers were introduced in each fragment allowed for a much more manageable stereochemical control strategy during manufacturing. The importance of replacing chromatography with crystallization to address issues of cost, throughput, production time, and environmental impact was demonstrated for each of the three fragments. A specific example describing the introduction of D-quinic acid as a C.14–C.26 source illustrates how replacing a chromatography-based synthetic route with a crystallization-based synthetic route can have a dramatic positive environmental impact. For example, one of the halichondrin synthetic fragments (C.14–C.26) presented unique challenges among the various building blocks due to the presence of multiple noncontiguous acyclic stereogenic centers. Reconfiguring this stereochemical network into a temporary cyclic array afforded the ability to (1) identify a crystalline carbohydrate template as a single source of chirality and (2) predictably communicate and control stereochemistry on a cyclic array. This classical cyclic stereocontrol approach required additional deconvolution steps, but these were more than compensated for by the elimination of chromatography and identification of numerous crystalline intermediates. The identification of crystalline carbohydrate sources of chirality was a consistent theme throughout the evolution of the synthetic strategies on each of the Halaven® fragments. The crystallinity of the carbohydrate templates not only benefited the control of cyclic stereogenic centers, but also invariably had benefits in establishing and controlling the quality of acyclic stereogenic centers via crystallization. By emphasizing synthetic strategies and tactics that lead to crystalline intermediates and eliminate the need for chromatography, the acquisition of practical quantities of even structurally complex, natural product-based compounds can be achieved. The versatility and functional group compatibility of the NHK reaction allowed for a high degree of convergency in the synthesis. Coupled with the crystallization-based fragment syntheses, the overall complexity of the synthesis could be reduced to several smaller, more manageable problems.

Eisai's experience in the discovery, development, and commercialization of eribulin has shown that by focusing on crystallinity, convergency, and efficient chemistry technology, structurally complex, natural product-based drug candidates can be manufactured by total synthesis, in the required quantity and quality, to achieve their full medicinal potential.

ACKNOWLEDGMENTS

We would like to acknowledge all of the dedicated scientists at Eisai, past and present, for their contributions to realize the medicinal potential of Halaven®. We would also like to acknowledge Dr. Farid Benayound for the benzazapine resolution data (Figure 17.3) and Drs. Bruce Littlefield, George Moniz, Matthew Schnaderbeck, and Ted Suh for their critical reviews and valuable feedback on this chapter.

DEDICATION

The authors wish to dedicate this work to Professor Yoshito Kishi of Harvard University for his scientific leadership in pioneering syntheses of structurally complex, biologically active marine natural products, including halichondrin B.

REFERENCES

1. Newman, D. J., and Cragg, G. M. 2012. Natural products as sources of new drugs over the 30 years from 1981 to 2010. *J. Nat. Prod.* 75:311–335.
2. Kuttruff, C. A., Eastgate, M. D., and Baran, P. S. 2014. Natural product synthesis in the age of scalability. *Nat. Prod. Rep.* 31:419–432.
3. International Conference on Harmonization of Technical Requirements for Registration of Pharmaceuticals for Human Use: Good Manufacturing Practice Guide for Active Pharmaceutical Ingredients. ICH Q7, February 2013.
4. Uemura, D., Takahashi, K., Yamamoto, T., et al. 1985. Norhalichondrin A: An antitumor polyether macrolide from a marine sponge. *J. Am. Chem. Soc.* 107:4796–4798.
5. Hirata, Y., and Uemura, D. 1986. Halichondrins: Antitumor macrolides from a marine sponge. *Pure Appl. Chem.* 58:701–710.
6. Aicher, T. D., Buszek, F. G., Fang, F. G., et al. 1992. Total synthesis of halichondrin B and norhalichondrin B. *J. Am. Chem. Soc.* 114:3162–3164.
7. Jackson, K. L., Henderson, J. A., and Phillips, A. J. 2009. The halichondrins and E7389. *Chem. Rev.* 109:3044–3079.
8. Jackson, K. L., Henderson, J. A., Motoyoshi, H., and Phillips, A. J. 2009. A total synthesis of norhalichondrin B. *Angew. Chem. Int. Ed.* 48:2346–2350.
9. Yu, M. J., Kishi, Y., and Littlefield, B. A. 2005. Discovery of E7389, a fully synthetic macrocyclic ketone analogue of halichondrin B. In *Anticancer Agents from Natural Products*, ed. G. M. Cragg, D. G. I. Kingston, and D. J. Newman. Taylor & Francis Group, CRC Press, Boca Raton, FL, pp. 241–265. (2nd ed., 2012, pp. 317–345.)

10. Towle, M. J., Salvato, K. A., Budrow, J., et al. 2001. *In vitro* and *in vivo* anticancer activities of synthetic macrocyclic ketone analogues of halichondrin B. *Cancer Res.* 61:1013–1021.

11. Yu, M. J., Zheng, W., and Seletsky, B. 2013. From micrograms to grams: Scale-up synthesis of eribulin mesylate. *Nat. Prod. Reports.* 30:1158–1164.

12. Zheng, W., Seletsky, B. M., Palme, M. H., et al. 2004. Macrocyclic ketone analogues of halichondrin B. *J. Bioorg. Med. Chem. Lett.* 14:5551–5554.

13. Littlefield, B. A., Palme, M. H., Seletsky, B. M., Towle, M. J., Yu, M. J., and Zheng, W. World Patent 9965894: Preparation of halichondrin analogs as anticancer or antimitotic agents, December 23, 1999.

14. Littlefield, B. A., Palme, M. H., Seletsky, B. M., Towle, M. J., Yu, M. J., and Zheng, W. U.S. Patent 6365759 B1: Intermediate compounds for preparing macrocylcic analogs, April 2, 2002.

15. Vahdat, L. T., Pruitt, B., Fabian, C. J., et al. 2009. Phase II study of eribulin mesylate, a halichondrin B analog, in patients with metastatic breast cancer previously treated with an anthracycline and a taxane. *J. Clin. Oncol.* 27:2954–2961.

16. Chiba, H., and Tagami, K. 2011. Research and development of Halaven™ (eribulin mesylate). *J. Synth. Org. Chem. Jpn.* 69:600–610.

17. Wang, Y., Habgood, G. J., Christ, W. J., Kishi, Y., Littlefield, B. A., and Yu, M. A. 2000. Structure activity relationships of halichondrin B analogues: Modifications at C.30–C.38. *Bioorg. Med. Chem. Lett.* 10:1029–1032.

18. Spyvee, M., Seletsky, B. M., Schiller, S., and Fang, F. World Patent WO 2009/070305 A1: Method of making imidazoazepinone compounds, June 4, 2009.

19. Fang, F. G. Synthetic studies on the halichondrin B analog, E7389. American Chemical Society 56th Southeast Regional Meeting, Research Triangle Park, NC, November 10–13, 2004.

20. Okude, Y., Hirano, S., Hiyama, T., and Nozaki, H. 1977. Grignard-type carbonyl addition of allyl halides by means of chromous salt. A chemospecific synthesis of homoallyl alcohols. *J. Am. Chem. Soc.* 99:3179–3181.

21. Hiyama, T., Kimura, K., and Nozaki, H. 1981. Chromium(II) mediated threo selective synthesis of homoallyl alcohols. *Tetrahedron Lett.* 22:1037–1040.

22. Hiyama, T., Okude, Y., Kimura, K., and Nozaki, H. 1982. Highly selective carbon-carbon bond forming reactions mediated by chromium(II) reagents. *Bull. Chem. Soc. Jpn.* 55:561–568.

23. Takai, K., Kimura, K., Kuroda, T., Hiyama, T., and Nozaki, H. 1983. Selective Grignard-type carbonyl addition of alkenyl halides mediated by chromium(II) chloride. *Tetrahedron Lett.* 24:5281–5284.

24. Smith, K. M. 2006. Perfluoraryl-substituted cyclopentadienyl complexes of transition metals. *Coord. Chem. Rev.* 250:1023–1031.

25. Stamos, D. P., and Kishi, Y. 1996. Synthetic studies on halichondrins: A practical synthesis of the C.1–C.13 segment. *Tetrahedron Lett.* 37:8643–8646.

26. Aicher, T. D., and Kishi, Y. 1987. Synthetic studies towards halichondrins. *Tetrahedron Lett.* 8:3463–3466.

27. Duan, J. J.-W., and Kishi, Y. 1993. Synthetic studies on halichondrins: A new practical synthesis of the C.1–C.12 segment. *Tetrahedron Lett.* 34:7541–7544.

28. Chase, C. E., Fang, F. G., Lewis, B. M., Schnaderbeck, M. J., Wilkie, G. D., and Zhu, X. 2013. Process development of Halaven®: Synthesis of the C1–C13 fragment from D-(–)-gulono-1,4-lactone. *Synlett* 24:323–326.

29. Stamos, D. P., Taylor, A. G., and Kishi, Y. 1996. A mild preparation of vinyliodides from vinylsilanes. *Tetrahedron Lett.* 37:8647–8650.

30. Yu, M. J., Zheng, W., and Seletsky, B. M. 2013. From micrograms to grams: Scale-up synthesis of eribulin mesylate. *Nat. Prod. Rep.* 30:1158–1164.

31. Austad, B. C., Benayoud, F., Calkins, T. L., et al. 2013. Process development of Halaven®: Synthesis of the C14–C35 fragment via iterative Nozaki–Hiyama–Kishi reaction–Williamson ether cyclization. *Synlett* 24:327–332.

32. Austad, B., Chase, C. E., and Fang, F. G. World Patent 118565: Intermediates for the preparation of halichondrin B, December 12, 2005.

33. Wan, Z.-K., Choi, H.-W., Kang, F.-A., Nakajima, K., Demeke, D., and Kishi, Y. 2002. Asymmetric Ni(II)/Cr(II)-mediated coupling reaction: Stoichiometric process. *Org. Lett.* 4:4431–4434.

34. Choi, H.-W., Nakajima, K., Demeke, D., et al. 2002. Asymmetric Ni(II)/Cr(II)-mediated coupling reaction: Catalytic process. *Org. Lett.* 4:4435–4438.

35. Kurosu, M., Lin, M.-H., and Kishi, Y. 2004. Fe/Cr- and Co/Cr-mediated catalytic asymmetric 2-halo-allylations of aldehydes. *J. Am. Chem. Soc.* 126:12248–12249.

36. Belanger, F., Chase, C. E., Endo, A., Fang, F. G., Li, J., Mathieu, S. R., Wilcoxen, A. Z., Zhang, H. 2015. Stereoselective synthesis of the halaven C14-C26 fragment from D-quinic acid: Crystallization-induced diastereoselective transformation of an α methyl nitrile. *Agnew. Chem. Int. Ed.* 54:5108–5111.

37. Hemmerle, H., Burger, H. J., Below, P., et al. 1997. Chlorogenic acid and synthetic chlorogenic acid derivatives: Novel inhibitors of hepatic glucose-6-phosphate translocase. *J. Med. Chem.* 40:137–145.

38. Chase, C., Endo, A., Fang, F. G., and Li, J. U.S. Patent 8093410 B2: Intermediates and methods for the synthesis of halichondrin B analogs, January 10, 2012.

39. Fischer, E., and Jourdan, F. 1883. Ueber die Hydrazine der Brenztraubensäure. *Ber. Dtsch. Chem. Ges.* 16:2241–2245.

40. Fischer, E., and Hess, O. 1884. Synthese von Indolderivaten. *Ber. Dtsch. Chem. Ges.* 17:559–568.

41. Corey, E. J., and Fuchs, P. L. 1972. A synthetic method for formyl→ethynyl conversion (RCHO→RC = CH or RC = CR′). *Tetrahedron Lett.* 13:3769–3772.

42. Yamaguchi, M. and Hirao, I. 1983. An efficient method for the alkynylation of oxiranes using alkynyl boranes. *Tetrahedron Lett.* 24:391–394.

43. Choi, H., Demeke, D., Kang, F.-A., et al. 2003. Synthetic studies on the marine natural product halichon-drins. *Pure Appl. Chem.* 75:1–17.

44. Endo, A., Chase, C. E., and Fang, F. G. U.S. Patent 8350067 B2: Compounds useful in the synthesis of halichondrin B analogs, January 8, 2012.

CHAPTER 18

Marine Natural Products in Pharma
How Industry Missed the Boat

Guy T. Carter and Valerie S. Bernan
Carter-Bernan Consulting, LLC, New City, New York

CONTENTS

The following account is the authors' personal perspective of events as they unfolded during their tenure (1980–2010) in Big Pharma, with the aim of reflecting on the current status of natural products in the industry.

18.1 HISTORICAL PERSPECTIVE

The modern pharmaceutical industry (Pharma) was built on a foundation of natural products derived from terrestrial actinomycetes isolated from soil. Pharma can trace its major growth phase to the 1950s and 1960s when the "golden age of antibiotics" came to fruition. The profits from marketing these "wonder drugs" propelled the growth and prosperity of Pharma for decades. The lessons learned in antibiotic discovery were translated into the various other therapeutic areas, where numerous unmet medical needs presented great opportunities for drug discovery. The majority of the alternate therapies required much simpler chemistry, and synthetic medicinal chemistry grew into the drug discovery engine that dominates the industry today. In general, Pharma moved away from natural products as a drug discovery platform to structural design and analogue synthesis (me too) as an effective means for generating new commercial products. This transition was essentially complete when the tools of molecular biology that enabled the facile production of single biochemical targets (enzymes, receptors, etc.) were paired with extensive synthetic chemical libraries generated by combinatorial chemistry. Thus began the era of programmed drug discovery that starts with "hits" generated via high-throughput screening (HTS) against single biochemical targets, followed by chemical engineering of potent and selective leads that would be optimized and developed as clinical candidates.

Marine natural product (MNP) discovery began to gain momentum in academia as a scientific discipline in the 1970s. The field was initially fueled by the uniqueness of MNP chemical structures, including a wide variety of organohalogen compounds. Basic research aimed at understanding the nature, origins, and ecological roles of the novel chemistry was soon supplanted by the quest for new drugs from the sea. Of course, it was funding by the National Cancer Institute (NCI) and other agencies that drove this shift in focus, but drugs from the sea concept, with a few notable exceptions, were not supported by Pharma and remained largely an academic pursuit.

As academic drugs from the sea programs began to uncover highly potent cytotoxic compounds directed against cancer cell lines, there was an increasing awareness of how similar some of these metabolites were to counterparts derived from terrestrial microorganisms. One of the earliest examples of this phenomenon was the isolation of renieramycin from the sponge *Reniera* sp. by Frincke and Faulkner, reported in 1982.[1] Renieramycin's structural similarity to the *Streptomyces* product saframycin was obvious, which further energized the debate as to whose biosynthetic machinery was responsible. Despite the enthusiasm engendered by the discovery of highly potent and novel chemistry, Pharma was still reluctant to devote major resources for the development of these leads. This decision was driven by the dogmatic view that structurally complex natural products could not be produced cost-effectively through *de novo* chemical synthesis. This was not an unreasonable position given the state of the art of synthetic chemistry and the prevailing economics of drug therapies circa 1980. (The evolution of synthetic chemical technologies has begun to shift this paradigm, as we have seen in the case of Halavan [see Chapter 17].)

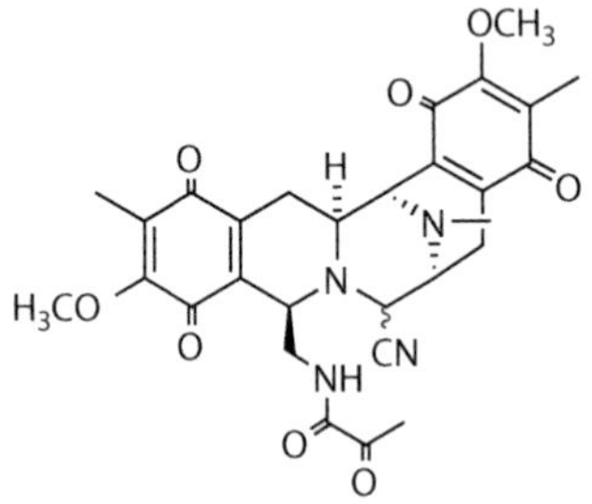

Saframycin A 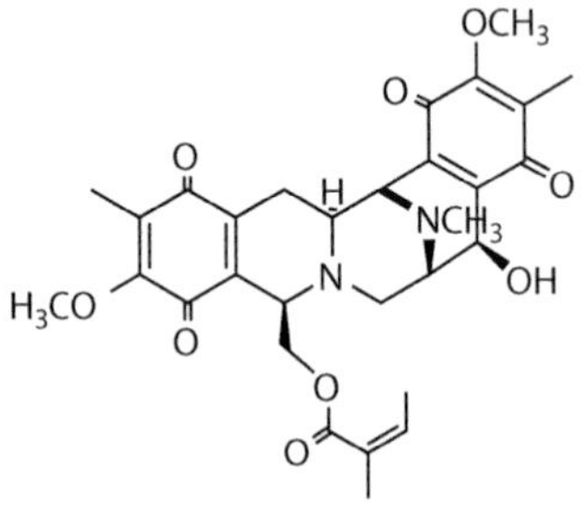   Renieramycin A

Ultimately, it was the issue of compound supply that kept Pharma from playing a major role in the development of marine natural products in the 1980s and 1990s. It was also in this period that the majority of Big Pharma companies abandoned their natural product platforms in favor of HTS— hit-to-lead approaches. Despite this trend, several enlightened Pharma companies were able to stay in the marine natural product game through the collaborative Natural Products Drug Discovery Group (NCDDG) and International Cooperative Biodiversity Group (ICBG) programs sponsored by the National Institutes of Health (NIH). Meanwhile, a revolution had begun in the marine natural product world that would reshape the field, namely, the cultivation of marine microbes as sources for new chemistry. This marine microbial revolution was driven by MNP pioneers who essentially rediscovered the vast biosynthetic potential of microorganisms—the platform that Pharma had so recently abandoned.

In the following sections we discuss important developments that have impacted Pharma's level of interest in MNP research. The perspective presented here represents our views of the Pharma industry's attitudes in general toward MNP research. There are obvious exceptions of companies who invested greater resources in MNP work; however, our emphasis is on the overall trends across the pharmaceutical industry.

18.2 NCDDG AND ICBG PROGRAMS

The National Cooperative Natural Products Drug Discovery Group (NCNPDDG) was established in 1989 as a novel public–private partnership for the development of natural product drugs.[2] Its aim was to support broad, innovative, multidisciplinary approaches to the discovery of new natural source-derived anticancer drugs. The programs consisted of multidisciplinary teams with varied research expertise from government, academic institutions, and industry. This approach was quite attractive to Pharma, as it provided access to new natural products from unexplored regions of the world, via academic partners that would not have been possible otherwise. The marriage of academia and industry complemented each other, with each supplying expertise the other did not have, and fostered high-risk research with a potential high payoff. Industry provided the screening technology and pharmaceutical development know-how and, in return, gained access to new ideas and talent for increased competiveness. Most significantly, the academic groups had a direct link to the pharmaceutical development world—a bridge that was previously difficult to build. It was a win–win situation that attracted many Pharma players, for example, BMS, Novartis, SmithKline Beecham, Lilly, Glaxo, and American Cyanamid (Lederle). Over the period of NCDDG funding, several MNP programs were supported, including collaborations with BMS, Lederle (Wyeth),[3] and Novartis.[4] The programs produced several compounds that went into clinical trials and a plethora of research results.

Wyeth's participation in the NCDDG program with Chris Ireland, Raymond Andersen, and Jon Clardy as academic partners continued from 1995 to 2005. As an industrial partner, Wyeth contributed the screening and development aspects, as noted above, but also its considerable expertise in the isolation and cultivation of microorganisms as a platform for MNP discovery.[3] One of the scientific highlights of the program was the discovery of the lomaiviticins, whose fascinating chemical scaffold continues to stimulate research by chemists and biologists.[5] The most advanced lead to come from this collaboration was the hemiasterlin derivative HTI-286 that had been prepared by the Andersen group at the University of British Columbia (UBC). Extremely limited quantities of hemiasterlins had previously hampered the development of these compounds, particularly in animal models. The total synthesis of HTI-286, which was further developed at Wyeth, provided an abundant supply of the compounds for further testing.[6] Results of extensive preclinical evaluation were quite encouraging, and HTI-286 was advanced into phase 1 clinical trials by Wyeth. The compound passed phase 1 trials and entered into phase 2; it showed positive results but was withdrawn at that phase. Nonetheless, it was a striking achievement of what the NCDDG could achieve when academia and industry united their expertise. The advancement of the hemiasterlin program also triggered milestone payments to Papua New Guinea, where the original source organism had been collected.

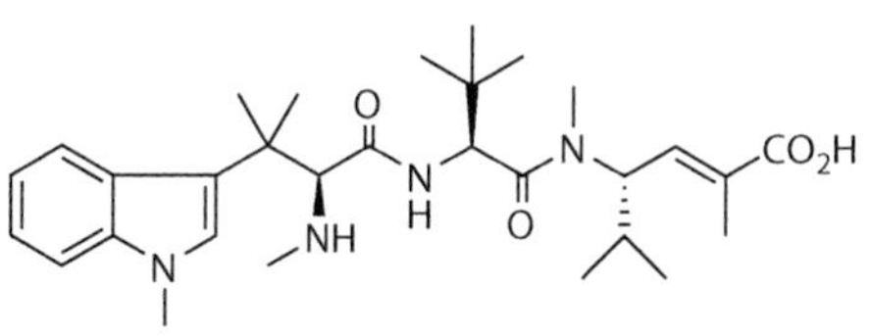

hemiasterlin

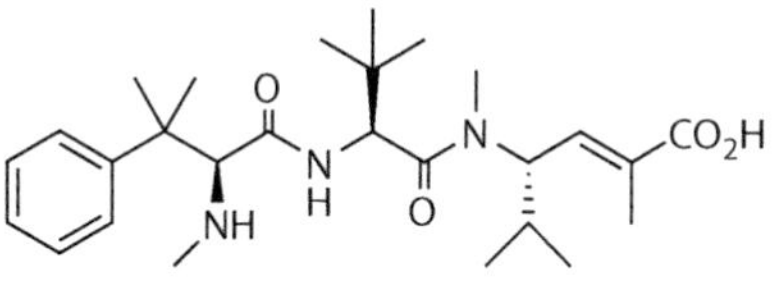

HTI-286

lomaiviticin

In a similar effort, the National Institutes of Health, the National Science Foundation, and the U.S. Agency for International Development joined together to create the International Cooperative Biodiversity Group (ICBG) program to promote the goals of biodiversity conservation, economic development, and pharmaceutical discovery.[2] This program, founded in 1993, was designed to guide natural product drug discovery so that the local communities and source countries could derive direct benefits from their own diverse biological resources. These programs are based in developing countries where most medicines are based on natural product remedies. Technology transfer and benefit sharing are integral parts of the grants that promote positive relationships and a fair exchange of technology for access to the biodiversity of host countries. The ICBG program also provides clear incentives for preservation and sustainable use of the host country's biodiversity. As was the case in the NCDDG, this program was attractive for Pharma to gain access to marine biodiversity where permits otherwise would be impossible to obtain. At issue, however, was the lack of Pharma's commitment to natural product research in the mid-1990s. Today there are a number of ICBG programs with strong MNP components, including Georgia Tech–Fiji, Scripps–Panama, Utah–Papua New Guinea, and Oregon Health Sciences University–Philippines. However, the Pharma component is less prominent.

The NCDDG and ICBG programs have had a strong positive impact on the relationship between academic groups and Pharma in the pursuit of drugs from MNPs. The general decline of Pharma's NP-based drug discovery definitely diminished its enthusiasm for these programs from the mid-1990s onward. As Pharma's attitude toward natural products evolves and with new developments promising to ease the supply issue, would such cooperative programs generate greater interest today?

18.3 CHANGING LANDSCAPE OF DRUG DISCOVERY: MOVING TARGETS

The paradigm described above for drug discovery over the past 25 years has failed to provide sufficient numbers of innovative and effective new drugs to continue to fuel the R&D engine. High-throughput screening of hundreds of thousands of small molecules against single targets, followed by medicinal chemical engineering of lead compounds as the "magic bullet," has resulted in a lower rate of clinical success than predicted by the established attrition model.[7] The reductionist approach of impacting diseases through the interaction of small molecules with single target biomolecules, such as enzymes or receptors, is being supplanted by strategies to affect multiple targets.

The involvement of several multiplexed biomolecules in a particular disease state and the process of intervention via drug therapy have been referred to as network pharmacology.[8] Effective drugs in these networks are variously known as designed multiple ligands[9] or multi-target-directed ligands,[10] and this concept is gaining momentum in Pharma. Creating compounds with these characteristics would be an ideal approach; however, current efforts are limited primarily by our lack of knowledge of the vital targets and how they interact.[11] Given these limitations in knowledge, a viable alternative that embraces the multitarget nature of disease intervention is to revisit phenotypic screening with a library of suitably complex small molecules as potential ligands.[12] Regrettably, such multifaceted compounds have been systematically purged from the screening libraries in Pharma owing to their noncompliance to Lipinski's rule of five criteria for acceptable absorption, distribution, metabolism, and excretion (ADME) properties. So here is one wide-open area for marine natural products to reclaim a vital role in drug discovery. Pharma is increasingly eager to screen libraries of multifaceted natural compounds in their newest screening systems. Lilly's Open Innovation Drug Discovery program (https://openinnovation.lilly.com/) is the premier example of such outreach efforts and highlights Pharma's new perspective on natural products.

18.4 MARINE MICROBIAL REVOLUTION

The advancement of any drug lead requires many cycles of optimization of the biological efficacy, as well as enhancement of pharmaceutical properties (druglikeness). The ultimate success of a natural product-based medicinal chemistry program will be predicated upon adequate compound supply; therefore, renewable, highly productive sources are greatly favored. Historically adequate supplies of natural product leads were obtained from a cultivable source, such as microorganisms or higher plants. For the most part, the exceptionally novel chemical entities derived from marine invertebrates were excluded from this process. However, recent research in marine microbiology and symbiosis offers a means to circumvent this roadblock.

For years there has been speculation that marine natural products isolated from sponges and other marine invertebrates actually derive from the biosynthetic machinery of symbiotic bacteria. Concrete evidence for this hypothesis has been hard won, for example, in one of the best-studied cases: the production of bryostatin in *Bugula neritina*.[13] A more recent development is the discovery of a microbial origin for the well-known MNP didemnin B. Didemnin B emerged as a promising anticancer drug candidate from the laboratory of the late Kenneth Rinehart, at the University of Illinois in 1981.[14] The compound, initially isolated from the tunicate *Trididemnum solidum*, received considerable attention in the biomedical community as fulfilling the promise of MNPs as drugs from the sea. While achieving benchmark status for MNPs by entering the clinic, it was also abundantly clear that the limited supply of the natural product had been a major obstacle to more rapid development. It was therefore quite remarkable when Tsukimoto and coworkers reported the isolation of a marine α-proteobacterium, *Tistrella mobilus*, from Japanese marine sediment that is fully capable of the production of didemnin B.[15] Upon examination of different *Tistrella* strains, *T. mobilus* and *T. bauzanensis* were also shown to produce didemnin B and nordidemnin B at varying levels. The complete genome sequence of *T. mobilis* revealed five replicons, one chromosomal and four circular plasmids. One of the megaplasmids revealed the didemnin biosynthetic gene cluster, indicating that the metabolic pathway may be acquired by horizonal gene transfer.[16] Although the yields reported in these new bacterial strains are modest, it can be expected that higher levels will be obtained. Furthermore, the information encoded in the biosynthetic gene cluster will enable biosynthetic manipulation for further optimization of this seminal class of MNPs.

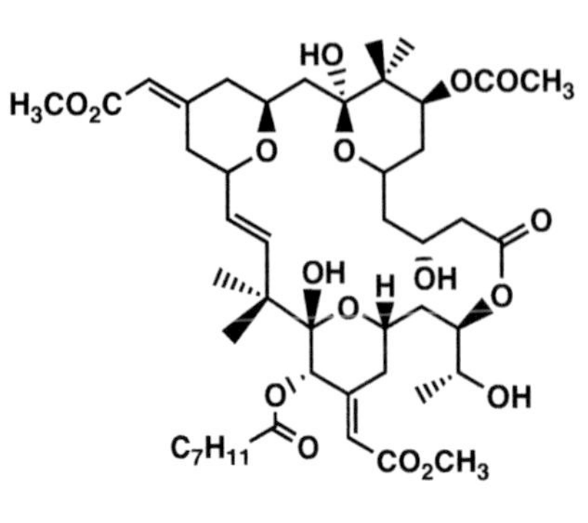

Bryostatin 1

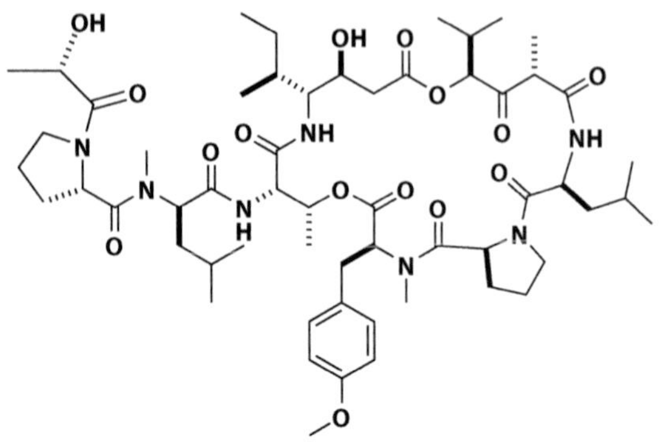

Didemnin B

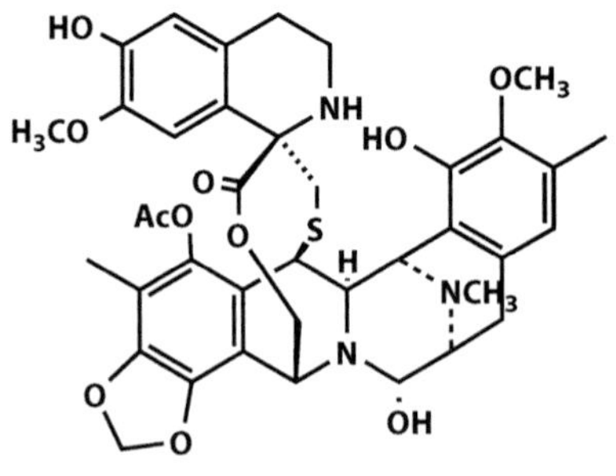

Ecteinascidin 743

The ecteinascidins (e.g., ET-743) were also isolated by the Rinehart group and were advanced by PharmaMar to the clinic for eventual approval for cancer therapy in 2007.[17] Significantly, PharmaMar's commercial production of ET-743 was made possible by co-opting a microbial metabolite (cyanosafracin B) as a key intermediate in a semisynthetic process.[18] An early result, but highly promising nonetheless, is the report from the Sherman laboratory of isolating putative genes for ecteinascidin biosynthesis from a pool of bacterial symbionts.[19] Taken together, these results suggest that such elusive metabolites could be made readily available through large-scale fermentation in the near future. Adequate supplies of the natural product and the potential for biosynthetic medicinal chemistry make microbial MNPs practical starting points for drug discovery. It is a fair assumption that had the microbial origin of such significant MNPs been known 20 years ago, Pharma's attitude toward them as leads would have been more favorable.

18.5 GENETIC ENGINEERING AND GENOMICS

As early as 1980, Pharma was seduced by the promise of genetic engineering for the creation of hybrid antibiotics as a rescue mechanism for a dry pipeline of new products. As is regrettably typical, Pharma (again with notable exceptions) charged into genetic engineering in a superficial way without allocation of adequate resources to support the development of effective tools to make such programs a reality. Consequently, as the promise of hybrid antibiotics was not realized within the fleeting time frame envisaged by Pharma management, these programs were abandoned, along with the traditional natural product platforms. This lost opportunity for Pharma was left for biotech companies and academic groups to advance. Although the original hybrid antibiotic concept has yet to be realized in a commercial sense, the tools of genetic engineering have been developed well enough to enable the practice of biosynthetic medicinal chemistry.

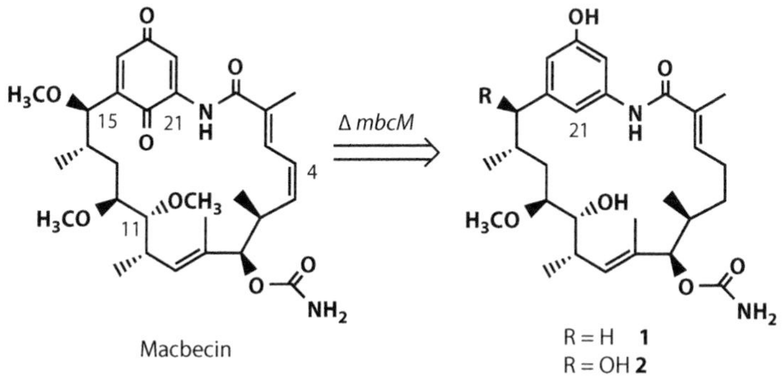

Figure 18.1 Macbecin analogues produced in mbcM knockout strain of *A. pretiosum*.

The targets of biosynthetic medicinal chemistry are not restricted by preexisting functional group "handles" on a scaffold, and thus are often complementary to the products of semisynthesis. An incisive example of biosynthetic medicinal chemistry is illustrated in Figure 18.1 for the Hsp90 inhibitor macbecin.[20] As with other ansamycins, such as geldanamycin, the quinone moiety was implicated in causing off-target effects, leading to toxicity. With a firm knowledge of the biosynthetic sequence in hand, Martin and coworkers at Biotica Technologies genetically engineered a strain of *Actinosynnema pretiosum* in which the gene responsible for oxidation at C21 (mbcM) was deleted. Eliminating this oxygenation step prevented the formation of the quinone and resulted in the exclusive production of nonquinone products **1** and **2**. Evidently, blocking quinone formation also inhibited the downstream tailoring processes that are responsible for the C4,5 desaturation, C15 hydroxylation and C11,15 O-methylation. Compound **1** was the most promising of these analogues, retaining the cellular potency of macbecin, while its toxicity was dramatically reduced. In order to enhance the overall production of **1**, the downstream tailoring gene for C15 hydroxylation was also deleted, resulting in a productive strain that yielded essentially the single component **1** at 200 mg L^{-1}. In those cases where biosynthetic pathways are available for manipulation, combining biosynthetic medicinal chemistry with semisynthetic processes can provide access to a much wider range of scaffolds and analogues.[21] Practical application of these biosynthetic technologies to challenging chemical scaffolds affords new routes for structural diversification for MNPs that enhances their chances as drug leads.

Insights provided by the sequencing and annotation of genomes, particularly of highly productive *Actinomycetes*, have led to new and greater expectations from natural product research.[22] Advances in molecular biology have dramatically changed the horizons for marine natural product research. With the dynamic infusion of genomics, proteomics, metabolomics, and transcriptomics, new areas of science are now open to inquiry. One of the most fascinating aspects of marine biology is the symbiotic production of MNPs in host organisms in cooperation with associated microorganisms. Application of the new *omic* tools promises to provide insight into the chemistry of symbiosis. While much of this promise remains to be realized, it is increasingly obvious from the genomics of marine microorganisms that there is an incredible wealth of novel MNPs remaining to be discovered.[23]

18.6 OUTLOOK

In hindsight, one has to wonder how the MNP field would have evolved if the microbial origins of MNPs had been revealed before Pharma bailed out of the natural product business. Had Pharma redeployed its expertise in microbial isolation, fermentation, and strain improvement to the production of MNPs, would there be more than a handful of drugs from the sea on the market today? It is regrettable that sponsors of drug discovery research, whether from industry or funding organizations, place such short time horizons for demonstration of success. The reality is that drug discovery remains a research-intensive process. It is still a "science project," not an exercise in engineering, in part because we do not know enough about the fundamental processes involved in disease. Natural products have had a major role in unraveling mechanisms for disease control that have led to "miracle drugs," such as antibiotics and other breakthrough therapies. Genome sequencing tells us that there is a bounty of cryptic secondary metabolites yet to be discovered whose biology may inform new directions for drug therapy. At the same time, Pharma has shown an interest in evaluating more complex, multifaceted compounds that are intrinsically functional in newer phenotypic or high-content assays. Given these trends, we should be optimistic that MNPs will have greater opportunities to inform and impact our drug discovery science going forward.

REFERENCES

1. Frincke, J. M., Faulkner, D. J. Antimicrobial metabolites of the sponge *Reniera* sp. *J Am Chem Soc* 1982, 104, 265–69.
2. Suffness, M., Cragg, G. M., Grever, M. R., Grifo, F. J., Johnson, G., Mead, J. A. R., Schepartz, S. A., Venditti, J. M., Wolpert, M. The national cooperative natural products drug discovery group and international cooperative biodiversity group programs. *Int J Pharmacognosy* 1995, 33 (Suppl), 5–16.
3. Ireland, C. M., Aalbersberg, W., Andersen, R. J., Ayral-Kaloustian, S., Berlinck, R. G. S., Bernan, V., Carter, G., Churchill, A. C. L., Clardy, J., Concepcion, G. P., De Silva, E. D., Discafani, C., Fojo, T., Frost, P., Gibson, D., Greenberger, L. M., Greenstein, M., Harper, M. K., Mallon, R., Loganzo, F., Nunes, M., Poruchynsky, M. S., Zask, A. Anticancer agents from unique natural product sources. *Pharm Biol* 2003, 41 (Suppl), 15–38.
4. Crews, P., Gerwick, W. H., Schmitz, F. J., France, D., Bair, K. W., Wright, A. E., Hallock, Y. Molecular approaches to discover marine natural products anticancer leads: An update from a drug discovery group collaboration. *Pharm Biol* 2003, 41 (Suppl), 39–52.
5. He, H., Ding, W. D., Bernan, V. S., Richardson, A. D., Ireland, C. M., Greenstein, M., Ellestad, G. A., Carter, G. T. Lomaiviticins A and B, potent antitumor antibiotics from *Micromonospora lomaivitiensis*. *J Am Chem Soc* 2001, 123 (22), 5362–63.
6. Loganzo, F., Discafani, C. M., Annable, T., Beyer, C., Musto, S., Hari, M., Tan, X., Hardy, C., Hernandez, R., Baxter, M., Singanallore, T., Khafizova, G., Poruchynsky, M. S., Fojo, T., Nieman, J. A., Ayral-Kaloustian, S., Zask, A., Andersen, R. J., Greenberger, L. M. HTI-286, a synthetic analogue of the tripeptide hemiasterlin, is a potent antimicrotubule agent that circumvents P-glycoprotein-mediated resistance *in vitro* and *in vivo*. *Cancer Res* 2003, 63 (8), 1838–45.
7. Paul, S. M., Mytelka, D. S., Dunwiddie, C. T., Persinger, C. C., Munos, B. H., Lindborg, S. R., Schacht, A. L. How to improve R&D productivity: The pharmaceutical industry's grand challenge. *Nat Rev Drug Discov* 2010, 9 (3), 203–14.
8. Hopkins, A. L. Network pharmacology: The next paradigm in drug discovery. *Nat Chem Biol* 2008, 4 (11), 682–90.
9. Morphy, R., Rankovic, Z. The physicochemical challenges of designing multiple ligands. *J Med Chem* 2006, 49 (16), 4961–70.
10. Cavalli, A., Bolognesi, M. L., Minarini, A., Rosini, M., Tumiatti, V., Recanatini, M., Melchiorre, C. Multi-target-directed ligands to combat neurodegenerative diseases. *J Med Chem* 2008, 51 (3), 347–72.
11. Morphy, R., Rankovic, Z. Designing multiple ligands: Medicinal chemistry strategies and challenges. *Curr Pharm Des* 2009, 15 (6), 587–600.
12. Lee, J. A., Uhlik, M. T., Moxham, C. M., Tomandl, D., Sall, D. J. Modern phenotypic drug discovery is a viable, neoclassic Pharma strategy. *J Med Chem* 2012, 55 (10), 4527–38.
13. Trindade-Silva, A. E., Lim-Fong, G. E., Sharp, K. H., Haygood, M. G. Bryostatins: Biological context and biotechnological prospects. *Curr Opin Biotechnol* 2010, 21 (6), 834–42.

14. Rinehart, K. L., Jr., Gloer, J. B., Hughes, R. G., Jr., Renis, H. E., McGovren, J. P., Swynenberg, E. B., Stringfellow, D. A., Kuentzel, S. L., Li, L. H. Didemnins: Antiviral and antitumor depsipeptides from a Caribbean tunicate. *Science* 1981, 212 (4497), 933–35.

15. Tsukimoto, M., Nagaoka, M., Shishido, Y., Fujimoto, J., Nishisaka, F., Matsumoto, S., Harunari, E., Imada, C., Matsuzaki, T. Bacterial production of the tunicate-derived antitumor cyclic depsipeptide didemnin B. *J Nat Prod* 2011, 74 (11), 2329–31.

16. Xu, Y., Kersten, R. D., Nam, S. J., Lu, L., Al-Suwailem, A. M., Zheng, H., Fenical, W., Dorrestein, P. C., Moore, B. S., Qian, P. Y. Bacterial biosynthesis and maturation of the didemnin anti-cancer agents. *J Am Chem Soc* 2012, 134 (20), 8625–32.

17. van Kesteren, C., Twelves, C., Bowman, A., Hoekman, K., Lopez-Lazaro, L., Jimeno, J., Guzman, C., Mathot, R. A., Simpson, A., Vermorken, J. B., Smyth, J., Schellens, J. H., Hillebrand, M. J., Rosing, H., Beijnen, J. H. Clinical pharmacology of the novel marine-derived anticancer agent ecteinascidin 743 administered as a 1- and 3-h infusion in a phase I study. *Anticancer Drugs* 2002, 13 (4), 381–93.

18. Cuevas, C., Francesch, A. Development of Yondelis (trabectedin, ET-743): A semisynthetic process solves the supply problem. *Nat Prod Rep* 2009, 26 (3), 322–37.

19. Rath, C. M., Janto, B., Earl, J., Ahmed, A., Hu, F. Z., Hiller, L., Dahlgren, M., Kreft, R., Yu, F., Wolff, J. J., Kweon, H. K., Christiansen, M. A., Hakansson, K., Williams, R. M., Ehrlich, G. D., Sherman, D. H. Meta-omic characterization of the marine invertebrate microbial consortium that produces the chemotherapeutic natural product ET-743. *ACS Chem Biol* 2011, 6 (11), 1244–56.

20. Zhang, M. Q., Gaisser, S., Nur, E. A. M., Sheehan, L. S., Vousden, W. A., Gaitatzis, N., Peck, G., Coates, N. J., Moss, S. J., Radzom, M., Foster, T. A., Sheridan, R. M., Gregory, M. A., Roe, S. M., Prodromou, C., Pearl, L., Boyd, S. M., Wilkinson, B., Martin, C. J. Optimizing natural products by biosynthetic engineering: Discovery of nonquinone Hsp90 inhibitors. *J Med Chem* 2008, 51 (18), 5494–97.

21. Kirschning, A., Hahn, F. Merging chemical synthesis and biosynthesis: A new chapter in the total synthesis of natural products and natural product libraries. *Angew Chem Int Ed Engl* 2012, 51 (17), 4012–22.

22. Walsh, C. T., Fischbach, M. A. Natural products version 2.0: Connecting genes to molecules. *J Am Chem Soc* 2010, 132 (8), 2469–93.

23. Ziemert, N., Podell, S., Penn, K., Badger, J. H., Allen, E., Jensen, P. R. The natural product domain seeker NaPDoS: A phylogeny based bioinformatic tool to classify secondary metabolite gene diversity. *PLoS One* 2012, 7 (3), e34064.

CHAPTER 19

In Vitro Model for Defining Neurotoxicity of Anticancer Agents

Balanehru Subramanian,[1] **Halina Pietraszkiewicz,**[2] **Joseph E. Media,**[2] **and Frederick A. Valeriote**[2]

[1]Central Inter-Disciplinary Research Facility, Pillayarkuppam, Puducherry, India
[2]Henry Ford Health System, Department of Internal Medicine, Division of Hematology and Oncology and Josephine Ford Cancer Center, Detroit, Michigan

CONTENTS

19.1 INTRODUCTION

Peripheral neuropathy is one of the impeding neurotoxic complications of cancer chemotherapeutic agents,[1] particularly tubulin inhibitors such as vincristine and taxanes, as well as platinum-based drugs[2] and, more recently, proteasome inhibitors.[3] As shown in the case of the chemotherapeutic agent oxaliplatin, a consequence of chemotherapy is the development of a painful peripheral neuropathy and decreases in both nerve conduction velocity and their action potential amplitudes.[4] Further, binding of cisplatin to DNA in the dorsal root ganglion neurons leads to death and atrophy of the ganglia.[5] Although management of neuropathic pain using antidepressants or anticonvulsants is available for agents such as Taxol, peripheral neurotoxicity represents a dose-limiting complication.[6] Indeed, this toxicity has been dose limiting for many new antitubulin drugs, such as cryptophycin-52.[7] Thus, there is a need for the development of new chemotherapeutic agents or analogues of present ones that have minimal or no neurotoxicity; this seems of particular importance if antitubulin, platinum-based drugs and proteasome inhibitors are to have their full potential in chemotherapy. Of the various research models available for assessing the neurotoxicity of cytotoxic drugs, PC-12 cells have been widely used,[8,9] in which assessment of nerve growth factor-dependent neurite outgrowth has been the basis for defining chemotherapy-induced neurotoxicity for a variety of compounds.[10–13]

We have modified this method to correlate the observed inhibition of neurite outgrowth with the cytotoxicity of the agents studied on the same PC-12 cells. We have defined the method for obtaining viable single-cell suspensions of PC-12 cells that can be used for both the neurite outgrowth assay and the clonogenic cell kill assay. The importance of demonstrating any difference between neurite outgrowth inhibition and cytotoxicity seems essential in understanding both the drug's mechanism of action and its potential reversal. A correlation between neurite outgrowth inhibition and the effect on clonogenic cell survival by four standard anticancer agents and two novel investigational anticancer agents is presented here.

19.2 MATERIALS AND METHODS

19.2.1 Culture Chemicals

Dulbecco's Modified Eagle Medium (DMEM), 2.5% trypsin and L-glutamine, Roswell Park Memorial Institute medium (RPMI) 1640 with L-glutamine and sodium bicarbonate, 2.5% trypsin and L-glutamine were obtained from Celox (St. Paul, Minnesota); horse serum and bovine calf serum from Hyclone (Logan, Utah); and Dulbecco's phosphate-buffered saline with calcium and magnesium, Hank's Balanced Salt Solution (HBSS) without calcium, magnesium, and penicillin-streptomycin from Gibco BRL (Rockville, Maryland). Noble agar, collagenase type XI-S, DNase I type II, ethylenetriaminetetraacetic acid (EDTA) disodium salt, dimethylsulfoxide (DMSO), 2-mercaptoethanol, and trypan blue were purchased from Sigma (St. Louis, Missouri), and nerve growth factor (NGF) (2.5S, mouse, grade II) from Roche Diagnostics Corp. (Indianapolis, Indiana). Poly-D-Lysine precoated 60 mm dishes and T-25 and T-75 culture flasks were purchased from Falcon Products (Becton Dickinson Labware, Franklin Lakes, New Jersey), and all other cultureware from Corning (Corning, New York).

19.2.2 Cytotoxic Agents

Taxol, Adriamycin, vincristine, and 5-fluorouracil were purchased from Sigma (St. Louis, Missouri). Dr, Joseph Shih at Eli Lilly Research (Indianapolis, Indiana) provided us with cryptophycin-52 in fine powder form. Stock solutions of 10 mg/ml of Adriamycin and 5-fluorouracil were prepared in sterile distilled water. Stock solutions of 10 mg/ml were prepared by dissolving Taxol, vincristine, and cryptophycin-52 in absolute ethanol. Stock solutions stored at –20 °C were stable under the described conditions. Dilutions were done in either ethanol or water.

19.2.3 PC-12 Cell Preparation

PC-12 rat pheochromocytoma cells obtained from the American Type Culture Collection (Rockville, Maryland) were cultured in a 10% CO_2 humidified incubator at 37 °C, in DMEM supplemented with 10% horse serum and 5% BCS, hereafter referred to as DMEM culture medium, at a density of 10^6 cells in 15 ml of medium per T-75 tissue culture flask. Cells were maintained in culture by passage once every week. For both weekly passage and the neurite outgrowth assay, PC-12 cells were cultured in flasks, washed once with HBSS, and incubated with 2 ml of 2.5% trypsin at 37 °C in a Shaker incubator for 15 min. Floaters from the flask, obtained by centrifuging the medium at 200g for 5 min at 5 °C, were added to the flask before the addition of trypsin. For the clonogenic assay, the incubation time was reduced to 7 min. Following trypsinization, cells were collected by centrifugation at 200g for 5 min at 5 °C, washed once with HBSS, and resuspended in HBSS. The cell suspension was passed two to three times through a 27-gauge needle to obtain single cells. Viability was assessed by trypan blue staining.

19.2.4 NGF-Induced Neurite Outgrowth Assay

PC-12 cells were plated at a density of 10^6 cells/T-75 flask in 15 ml of medium as described above. To each flask NGF was added at a final concentration of 50 ng/ml. Flasks were refed every other day with fresh medium and replenished with fresh NGF at a final concentration of 50 ng/ml. Cells were exposed to NGF for a total of 8 days prior to neurite outgrowth assay. The PC-12 cells were then rinsed with HBSS and trypsinized as described above. Cells were washed once with DMEM culture medium and viability assessed by trypan blue dye uptake. Cells (5×10^4) were plated in each 60 mm poly-D-lysine precoated dish, in a total of 3 ml DMEM culture medium containing 10 ng/ml of NGF. After 3 days in the presence of 10 ng/ml of NGF, 100 cells were randomly chosen and scored for neurite expression using a phase-contrast microscope. The percentage of neurite-forming cells, both long (>2× cell body) and short (<2× cell body), was calculated. To determine whether cytotoxic agents have neurotoxic effect, we exposed cells to appropriate concentrations of these agents and incubated at 37 °C, 10% CO_2 for 24 and 48 h. Following incubation, cells were scored for neurite expression. The percent control of neurite-expressing cells was calculated and represents an average of three experiments.

19.2.5 Neurite Recovery after Drug Treatment

After scoring the plates, the drug-containing medium was aspirated and the cells were washed, followed by a 10 min incubation at room temperature with drug-free DMEM culture medium. Fresh DMEM culture medium containing 10 ng/ml of NGF was added, and the cells were incubated at 37 °C, 10% CO_2 for another 48 h. Control plates with and without NGF were also washed and replaced with DMEM culture medium with and without NGF, respectively. The percent of neurite-expressing cells compared to the control was calculated after both 24 and 48 h and represents an average of three experiments.

19.2.6 PC-12 Clonogenic Cell Survival Assay

PC-12 cells were grown in 5 ml DMEM culture medium at 37 °C and 5% CO_2 at a starting concentration of 5×10^5 cells/T-25 flask. After 24 h, cells were exposed to different concentrations of the cytotoxic drugs. Flasks were incubated for 48 h in a 5% CO_2 incubator at 37 °C and the cells harvested with trypsin, as described above, washed once with HBSS, and resuspended in HBSS with viability determined by trypan blue staining. Cells were prepared in DMEM culture medium containing 5×10^{-5} M β-mercaptoethanol and agar at a final concentration of 0.3%. After thorough mixing, aliquots were pipetted over a hardened layer of 0.6% agar so as to yield 1000 cells per plate. The plates were incubated in a 5% CO_2 incubator maintained at 37 °C for 14 days. Colonies were scored under a stereozoom microscope (Bausch & Lamb, model BVB-125 with Nicholas Illuminator). Colonies were scored as ≥50 cells. The surviving fraction was expressed as percent control untreated and represents an average of three experiments.

19.3 RESULTS

The cell harvesting and preparation methods yielded single cells with 90%–95% viability by trypan blue dye exclusion. Following pretreatment with NGF, PC-12 cells were able to both proliferate and express neurites on poly-D-lysine-coated plates. Neurites were expressed in 70%–74% of the cells; cells that did not express neurites retained cell viability.

Four standard anticancer agents, Adriamycin, 5-fluorouracil, vincristine, and Taxol, were tested for neurite growth inhibition. All drugs showed dose-dependent inhibition of neurite outgrowth expressed as shown in Figure 19.1 (closed circles). 5-Fluorouracil showed the least inhibition, with only 90% inhibition of neurite outgrowth at the highest concentration tested, 100 μg/ml. Adriamycin and vincristine were the most effective standard agents, with complete inhibition of neurite outgrowth at a concentration of 10 μg/ml. Taxol showed complete inhibition of neurite formation at 100 μg/ml and 90% inhibition at 10 μg/ml.

The pattern of neurite inhibition changed dramatically when the drug was washed from the plates and recovery of neurite formation was assessed (Figure 19.1, open symbols). 5-Fluorouracil demonstrated a complete reversal of neurite inhibition; Adriamycin and vincristine demonstrated significant reversal. Taxol showed little, if any, recovery.

To differentiate the neurite outgrowth inhibition from general cytotoxicity, the cytotoxic effect of the agents was determined in a PC-12 clonogenic assay. Results are shown in Figure 19.1 (solid squares). No cytotoxic effect was noted for 5-fluorouracil. On the other extreme, Taxol cytotoxicity tracked nearly identical to the neurite inhibition pattern. For Adriamycin and vincristine, the effect on clonogenicity was much less than that on neurite formation; however, after washout the concentration–response for neurite inhibition was similar to that for clonogenicity.

For the investigational agent studied, cryptophycin-52 exhibited a remarkable neurite outgrowth inhibition with a 100% inhibition at 10^{-4} μg/ml, a 1 million-fold lower concentration than that found for Taxol. This value increased to 10^{-3} μg/ml following drug washout. For the clonogenic assays, PC-12 cell killing reached 10% of the control at 1 μg/ml.

19.4 DISCUSSION

Tubulin remains one of the most common, as well as important, molecular targets for both standard and new anticancer drugs, including vincristine, Taxol, and the cryptophycins discussed here, as well as combrestatins, dolastatins, halichondrins, hemiasterlins, epothilones, and many others.[14] Unfortunately, these drugs and their analogues, both in clinical use and under development, pose a serious toxic side effect often attributed to as peripheral neuropathy.[15] The neurological complications of the drugs are so severe in many instances that the drug is withdrawn from clinical trials even before the trials are completed for thorough evaluation of the agent for its chemotherapeutic efficacy. Thus, evaluation of an agent's neurotoxic property becomes imperative even before it is taken into clinical trials. While one argument is that neuropathy is a consequence of tubulin binding itself, recent studies have demonstrated that this is not a necessary consequence, as seen by the tubulin binding drug ixabepilone.[16]

Further, with combinations of such drugs, for example, platinum compounds and taxanes, the peripheral neuropathy is more severe than either agent alone.[17,18] At present, there is no simple, inexpensive *in vitro* assay available that can determine this property. Although neurotoxicity could be evaluated in animal models, it is not widely used considering the cost and the labor-intensive protocols. Thus, a simple *in vitro* assay that could deliver the same amount of information quickly would expedite the development process and potentially avoid clinical trial failures.

Finally, with new classes of agents, such as proteasome inhibitors, demonstrating peripheral neuropathy, and given that this class has demonstrated significant activity in the treatment of multiple myeloma,[3] future drug development in this area should require a comparative test for peripheral neuropathy in preclinical testing. The timeliness of such testing might be considered imperative given the number of new proteasome inhibitors being developed at present and might be expected to increase in the future (as has been the case for tubulin binding drugs). Indeed, carfilzomib,[19]

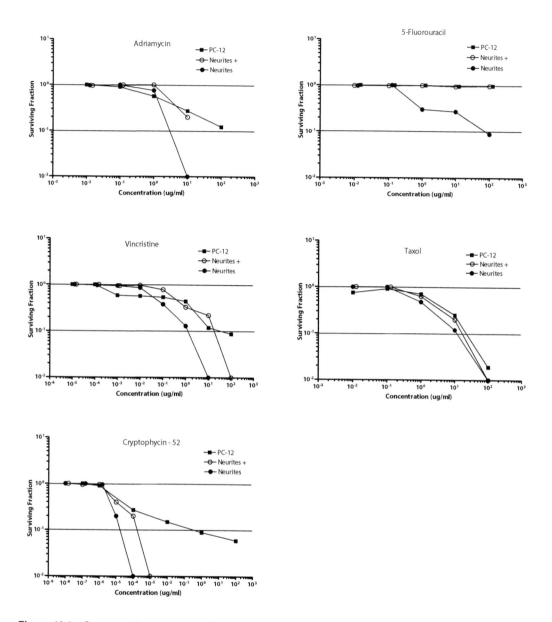

Figure 19.1 Concentration–response of Adriamycin, vincristine, Taxol, 5-fluorouracil, and cryptophycin-52 exposure for 48 h on the neurite outgrowth formation (●) and clonogenic cell survival on rat pheochromocytoma PC-12 cells (■) cultured *in vitro*. Neurite formation is also shown after drug washout and exposure to nerve growth factor for a further 48 h (O). Surviving fraction is expressed as a percentage of the control, untreated cells. Results are an average of three experiments. Arrows indicate survival levels less than 10^{-3}.

salinosporamide A,[20] and the carmaphycins[21] are examples of the next generation of these compounds. And, given that other classes of compounds, such as immunomodulators, or specific targeted compounds, such as JAK2 inhibitors, also demonstrate peripheral neuropathy, one might expect that in the future, peripheral neuropathy will replace hematological toxicity as the major dose-limiting toxicity for new anticancer drugs, thereby increasing the necessity of having a useful and effective prescreen for this toxicity.

We report here a modified, simple, inexpensive *in vitro* assay using cultured monodispersed PC-12 cells to evaluate the neurotoxic potential of anticancer agents. This assay has the advantage of comparing the inhibition of neurite outgrowth, which is a measure of neurotoxicity, to that of the agent's cytotoxic effect on the same cell population. PC-12 cells aggregate quickly and grow in clumps in culture. Obtaining single cells in culture is essential, considering the endpoint measurement, quantitation of outgrowth of short and long neurites. We have modified the existing method[10–12] using 2.5% trypsin. A very uniform, viable single-cell suspension whose growth characteristics are equivalent to those of the aggregate culture was established. This included cell viability, daily growth, plating efficiency, and cell cycle analysis by flow cytometry (data not shown). By using this single-cell culture method, highly reproducible results were obtained with both neurite outgrowth assay and clonogenic cell kill assay. Clonogenic assay was performed to confirm that the absence of neurite outgrowth was not the effect of cytotoxicity of the agent.

Inhibition of neurite formation in cultured PC-12 cells was carried out with four standard anticancer agents and an investigational agent. Results obtained with the four standard anticancer agents showed a correlation between their clonogenic cytotoxicity and neurite inhibition. Vincristine, for which there are several reports available on the neurotoxic side effects,[22,23] clearly demonstrated a differential effect on neurite inhibition that was independent of its cytotoxicity. The results suggest that the assay is capable of distinguishing the effect of the drug in cellular differentiation (neurite outgrowth) versus clonogenic toxicity and growth (cell survival and proliferation). Adriamycin is not known to exhibit peripheral neuropathy in the clinic, and its effect on PC-12 neurites was nearly completely reversible. Taxol demonstrated a dose–response pattern in which clonogenicity, neurite inhibition, and lack of reversibility by washout of drug indicted that the neurite inhibition was a direct consequence of PC-12 cell killing. 5-Fluorouracil neurite outgrowth inhibition was completely reversible in our assay system. 5-Fluorouracil-associated neurotoxicity is uncommon, and peripheral neuropathy with 5-fluorouracil therapy has only been rarely reported.[24]

Cryptophycin-52 demonstrated a remarkable neurite outgrowth inhibition totally independent of PC-12 clonogenic toxicity. The assay predicts cryptophycin-52 to be a potent neurotoxic compound and little reversibility of the effect upon washout of the drug. Evidence of such profound toxicity has been reported in patients treated with cryptophycin-52[25] and is the dose-limiting toxicity found in both phase I and phase II clinical studies.[7,26]

As an antiproliferative agent, cryptophycin-52 has been shown to be severalfold more potent than either Taxol or vinblastine.[27] Cryptophycin-52 possesses antimitotic activity by depolymerization of spindle microtubules and has also been reported to be a potent suppressor of microtubule dynamics.[28] Although it has been reported that drugs that interfere with microtubule dynamics have been found to inhibit neurite outgrowth, there were no reports on the neurotoxicity of CP-52 before it was subjected to clinical trials. Our neurite assay using a single-cell culture of PC-12 cells provides the platform to eliminate such shortcomings.

While different mechanisms have been described for the antiproliferative action of the vincristine (binding to tubulin) and Taxol (tubulin stabilization),[29,30] they both have also been correlated to the neurotoxic property. However, the involvement of several signal transduction pathways and intracellular signaling molecules in mediating the neurotoxic mechanism has recently been shown.[31–33] Further, a relationship between peripheral neuropathy and a dose-dependent decrease in nerve growth factor has been reported for patients treated with platinum- or taxane-containing regimens.[34] Indeed, nerve growth factor treatment may have a therapeutic application in the rescue of patients from cisplatin neurotoxicity.[35] Thus, simple extrapolation of microtubule interactions alone may not provide sufficient information to test for neurotoxicity.

The modified PC-12 *in vitro* model described here using single-cell suspensions is a valuable assay system for the screening of anticancer agents for their neurotoxic side effects and to differentiate these effects from antiproliferative effects. This assay has the potential to evaluate and compare the therapeutic efficacy of any new or existing standard anticancer agent and to suggest potential drugs or analogues with minimal or no neurotoxicity for subsequent clinical testing and development.

REFERENCES

1. Amato, A. A., Collins, M. P. Neuropathies associated with malignancy. *Semin Neurol* 1998, 18 (1), 125–144.
2. Bianchi, G., Vitali, G., Caraceni, A., Ravaglia, S., Capri, G., Cundari, S., Zanna, C., Gianni, L. Symptomatic and neurophysiological responses of paclitaxel- or cisplatin-induced neuropathy to oral acetyl-L-carnitine. *Eur J Cancer* 2005, 41 (12), 1746–1750.
3. Lonial, S., Mitsiades, C. S., Richardson, P. G. Treatment options for relapsed and refractory multiple myeloma. *Clin Cancer Res* 2011, 17 (6), 1264–1277.
4. Renn, C. L., Carozzi, V. A., Rhee, P., Gallop, D., Dorsey, S. G., Cavaletti, G. Multimodal assessment of painful peripheral neuropathy induced by chronic oxaliplatin-based chemotherapy in mice. *Mol Pain* 2011, 7, 29.
5. McDonald, E. S., Randon, K. R., Knight, A., Windebank, A. J. Cisplatin preferentially binds to DNA in dorsal root ganglion neurons *in vitro* and *in vivo*: A potential mechanism for neurotoxicity. *Neurobiol Dis* 2005, 18 (2), 305–313.
6. Uhm, J. H., Yung, W. K. Neurologic complications of cancer therapy. *Curr Treat Opt Neurol* 1999, 1 (5), 428–437.
7. Edelman, M. J., Gandara, D. R., Hausner, P., Israel, V., Thornton, D., DeSanto, J., Doyle, L. A. Phase 2 study of cryptophycin 52 (LY355703) in patients previously treated with platinum based chemotherapy for advanced non-small cell lung cancer. *Lung Cancer* 2003, 39 (2), 197–199.
8. Greene, L. A., Tischler, A. S. Establishment of a noradrenergic clonal line of rat adrenal pheochromocytoma cells which respond to nerve growth factor. *Proc Natl Acad Sci USA* 1976, 73 (7), 2424–2428.
9. Greene, L., Tischler, A. PC12 pheochromocytoma cells in neurobiological research. *Adv Cell Neurobiol* 1982, 3, 373–414.
10. Geldof, A. A. Nerve-growth-factor-dependent neurite outgrowth assay: A research model for chemotherapy-induced neuropathy. *J Cancer Res Clin* 1995, 121 (11), 657–660.
11. Geldof, A. A., Mastbergen, S. C., Henrar, R. E. C., Faircloth, G. T. Cytotoxicity and neurocytotoxicity of new marine anticancer agents evaluated using *in vitro* assays. *Cancer Chemoth Pharm* 1999, 44 (4), 312–318.
12. Geldof, A. A., Minneboo, A., Heimans, J. J. Vinca-alkaloid neurotoxicity measured using an *in vitro* model. *J Neuro-Oncol* 1998, 37 (2), 109–113.
13. Zielke, A., Wong, M. G., Siperstein, A. E., Clark, O. H., Rothmund, M., Duh, Q. Y. Antiproliferative effect of suramin on primary cultures of human pheochromocytomas and rat PC12 pheochromocytoma cells. *J Surg Oncol* 1997, 66 (1), 11–18.
14. Kingston, D. G. I. Tubulin-interactive natural products as anticancer agents. *J Nat Prod* 2009, 72 (3), 507–515.
15. Rowinsky, E. K., Donehower, R. C. The clinical-pharmacology and use of antimicrotubule agents in cancer chemotherapeutics. *Pharmacol Therapeut* 1991, 52 (1), 35–84.
16. Wozniak, K. M., Nomoto, K., Lapidus, R. G., Wu, Y., Carozzi, V., Cavaletti, G., Hayakawa, K., Hosokawa, S., Towle, M. J., Littlefield, B. A., Slusher, B. S. Comparison of neuropathy-inducing effects of eribulin mesylate, paclitaxel, and ixabepilone in mice. *Cancer Res* 2011, 71 (11), 3952–3962.
17. Hilkens, P. H., Pronk, L. C., Verweij, J., Vecht, C. J., van Putten, W. L., van den Bent, M. J. Peripheral neuropathy induced by combination chemotherapy of docetaxel and cisplatin. *Br J Cancer* 1997, 75 (3), 417–422.
18. Chaudhry, V., Rowinsky, E. K., Sartorius, S. E., Donehower, R. C., Cornblath, D. R. Peripheral neuropathy from Taxol and cisplatin combination chemotherapy: Clinical and electrophysiological studies. *Ann Neurol* 1994, 35 (3), 304–311.

19. Alsina, M., Trudel, S., Furman, R. R., Rosen, P. J., O'Connor, O. A., Comenzo, R. L., Wong, A., Kunkel, L. A., Molineaux, C. J., Goy, A. A phase I single-agent study of twice-weekly consecutive-day dosing of the proteasome inhibitor carfilzomib in patients with relapsed or refractory multiple myeloma or lymphoma. *Clin Cancer Res* 2012, 18 (17), 4830–4840.

20. Williamson, M. J., Blank, J. L., Bruzzese, F. J., Cao, Y., Daniels, J. S., Dick, L. R., Labutti, J., Mazzola, A. M., Patil, A. D., Reimer, C. L., Solomon, M. S., Stirling, M., Tian, Y., Tsu, C. A., Weatherhead, G. S., Zhang, J. X., Rolfe, M. Comparison of biochemical and biological effects of ML858 (salinosporamide A) and bortezomib. *Mol Cancer Ther* 2006, 5 (12), 3052–3061.

21. Pereira, A. R., Kale, A. J., Fenley, A. T., Byrum, T., Debonsi, H. M., Gilson, M. K., Valeriote, F. A., Moore, B. S., Gerwick, W. H. The carmaphycins: New proteasome inhibitors exhibiting an alpha,beta-epoxyketone warhead from a marine cyanobacterium. *Chembiochem* 2012, 13 (6), 810–817.

22. Hilkens, P. H., ven den Bent, M. J. Chemotherapy-induced peripheral neuropathy. *J Peripher Nervous Syst* 1997, 2 (4), 350–361.

23. Fiori, M. G., Schiavinato, A., Lini, E., Nunzi, M. G. Peripheral neuropathy induced by intravenous administration of vincristine sulfate in the rabbit: An ultrastructural study. *Toxicol Pathol* 1995, 23 (3), 248–255.

24. Saif, M. W., Wilson, R. H., Harold, N., Keith, B., Dougherty, D. S., Grem, J. L. Peripheral neuropathy associated with weekly oral 5-fluorouracil, leucovorin and eniluracil. *Anticancer Drugs* 2001, 12 (6), 525–531.

25. Stevenson, J. P., Sun, W. J., Gallagher, M., Johnson, R., Vaughn, D., Schuchter, L., Algazy, K., Hahn, S., Enas, N., Ellis, D., Thornton, D., O'Dwyer, P. J. Phase I trial of the cryptophycin analogue LY355703 administered as an intravenous infusion on a day 1 and 8 schedule every 21 days. *Clin Cancer Res* 2002, 8 (8), 2524–2529.

26. Sessa, C., Weigang-Kohler, K., Pagani, O., Greim, G., Mora, O., De Pas, T., Burgess, M., Weimer, I., Johnson, R. Phase I and pharmacological studies of the cryptophycin analogue LY355703 administered on a single intermittent or weekly schedule. *Eur J Cancer* 2002, 38 (18), 2388–2396.

27. Smith, C. D., Zhang, X. Mechanism of action cryptophycin. Interaction with the vinca alkaloid domain of tubulin. *J Biol Chem* 1996, 271 (11), 6192–6198.

28. Panda, D., DeLuca, K., Williams, D., Jordan, M. A., Wilson, L. Antiproliferative mechanism of action of cryptophycin-52: Kinetic stabilization of microtubule dynamics by high-affinity binding to microtubule ends. *Proc Natl Acad Sci USA* 1998, 95 (16), 9313–9318.

29. Jordan, A., Hadfield, J. A., Lawrence, N. J., McGown, A. T. Tubulin as a target for anticancer drugs: Agents which interact with the mitotic spindle. *Med Res Rev* 1998, 18 (4), 259–296.

30. Amos, L. A., Lowe, J. How Taxol® stabilises microtubule structure. *Chem Biol* 1999, 6 (3), R65–R69.

31. Daniels, R. H., Hall, P. S., Bokoch, G. M. Membrane targeting of p21-activated kinase 1 (PAK1) induces neurite outgrowth from PC12 cells. *Embo J* 1998, 17 (3), 754–764.

32. Katoh, H., Yasui, H., Yamaguchi, Y., Aoki, J., Fujita, H., Mori, K., Negishi, M. Small GTPase RhoG is a key regulator for neurite outgrowth in PC12 cells. *Mol Cell Biol* 2000, 20 (19), 7378–7387.

33. James, S. E., Burden, H., Burgess, R., Xie, Y. M., Yang, T., Massa, S. M., Longo, F. M., Lu, Q. Anticancer drug induced neurotoxicity and identification of Rho pathway signaling modulators as potential neuroprotectants. *Neurotoxicology* 2008, 29 (4), 605–612.

34. De Santis, S., Pace, A., Bove, L., Cognetti, F., Properzi, F., Fiore, M., Triaca, V., Savarese, A., Simone, M. D., Jandolo, B., Manzione, L., Aloe, L. Patients treated with antitumor drugs displaying neurological deficits are characterized by a low circulating level of nerve growth factor. *Clin Cancer Res* 2000, 6 (1), 90–95.

35. Fischer, S. J., Podratz, J. L., Windebank, A. J. Nerve growth factor rescue of cisplatin neurotoxicity is mediated through the high affinity receptor: Studies in PC12 cells and p75 null mouse dorsal root ganglia. *Neurosci Lett* 2001, 308 (1), 1–4.

PART 4

Prospects for the Future of Marine Biodiscovery

CHAPTER 20

Comments on the Past and Future of Marine Natural Product Drug Discovery

William Fenical

Center for Marine Biotechnology and Biomedicine, Scripps Institution of Oceanography, University of California, San Diego, La Jolla, California

CONTENTS

20.1 INTRODUCTION

Those who have wondered about the validity of a crystal ball to forecast the future will know that accurate prediction is unlikely, if not impossible (the predicted computer doomsday of the year 2000 is a perfect example). Who would have thought, for example, that entire genomes could be deciphered today for a few hundred dollars when only 15 years ago this was a multi-million-dollar activity. Today, we enjoy a rapidly changing science that incorporates, literally overnight, the most recently developed techniques in separation science (ultra-high-performance liquid chromatography [UHPLC]), analytical methods (imaging mass spectrometry, molecular networking), genome mining, and synthetic biology. Given the rate of development of these new technologies, their broad-scale application in marine drug discovery is already here and will be the norm in the future. Given this, the future looks exciting, especially for those who choose to invest in the numerous diverse and interactive disciplines required for effective drug discovery lead refinement and preclinical development.

What follows is my rambling and free association within this field based upon my 50+ years of involvement. These are just my thoughts, and I suspect that they may be different from many of yours. I try to provide both constructive criticism and praise where due.

20.2 THE PAST: WHAT WE HAVE DONE WELL AND NOT SO WELL

Since the beginning of the field on marine natural products in the 1960s, our science has been justified based upon the promise of developing a new source for drug discovery. However, this goal was very slow to develop, instead being replaced by a rapid race to publish structures, often with trivial bioactivities mentioned, but never even cursorily explored. Worldwide, simple structure elucidation papers, of low impact and consequence, unfortunately continue to dominate our science. This may have contributed to the view that natural product (NP) chemistry is an antiquated science held by many. But, in more recent times, many academics have recognized a greater goal and have made significant advances in chemistry and ultimately in drug development. Over the past 30+ years, unique marine natural products have provided a very strong foundation for the advancement of science and chemistry. The unprecedented large metabolites, such as maitotoxin, palytoxin, and polytheonamides, among many others, have provided the synthetic community with unparalleled challenges. Synthetic challenges were met, new nuclear magnetic resonance (NMR) tools were developed to meet these challenges, and new computer approaches in genomics have been created to locate and annotate complex biosynthetic pathways. The fact that there are some seven or more successfully approved marine drugs, virtually all of which were discovered by academic researchers, shows that academic research can be an important component in drug development. Furthermore, the trend to go past structure elucidation to truly participate in the fundamental requirements of drug development is growing, with around 25 marine natural products (MNPs), or derivatives, in various phases of development (see Chapter 15 and references therein). Overall, and in the most recent time frame, we are doing this much better, but the future will require even more effort.

It is human nature to take the easiest way first. Thus, MNP researchers have generally focused their attention on the easiest sources of marine life to examine. Sponges, without question, have been favored because they are abundant, easily collected, and chemically rich and unique. The development of the cancer drug Halaven® (see Chapter 17), a derivative designed from the sponge metabolite halichondramide, has illustrated the importance of sponge metabolites, as has the utility of several sponge metabolites as drug lead structures and molecular probes of diverse cellular processes. More MNPs are being used than ever before as probes for specific intracellular targets important in gaining a greater understanding of basic cell biology.

There are, however, many other marine life-forms that have received minimal attention. Generally, these are the more difficult to access and the small or microscopic components of the world's oceans: bacteria, microalgae, marine eukaryotes, protists, and more. Examining these organisms typically requires attention to cultivation and in general represents a much larger investment in infrastructure and biological background. So far, we have not done well in opening these areas of inquiry. The dinoflagellates (Dinophyceae), in particular, represent great potential; however, only a few have prioritized this group. Several dinos have enormous genomes in excess of 50 times the size of the human genome, and we have seen spectacular and unprecedented polyketides, such as maitotoxin, yessotoxins, and ciguatoxins, come from this source. While developing a program in this area is challenging, it remains a significant resource for the future. One could easily pinpoint a variety of eukaryotic microscopic organisms, such as the labrinthulids, thraustochytrids, foraminiferans, radiolarians, and other marine protists (12 groups in total), that have escaped study (See Chapters 12 and 13).

Despite more than two decades of investigation, marine microorganisms (mainly bacteria) remain a highly underexplored resource. Generally, this is due to the increased difficulty and costs of a cultivation-based program, but also because bacteria in the marine environment have specific requirements for nutrients and culture conditions not represented in the textbook bacterial cultivation methodologies in the literature. In deference to the decades-old proposal that environmental bacteria are largely unculturable, it is now clear that a multitude of new methods involving high dilution isolation, low-nutrient and highly marine-selective media, and long agar plate incubation times can lead to significant levels of culture success. It is time that the old proposals that bacteria are unculturable be abandoned in favor of systematic, environmentally well-designed approaches.

20.3 MARINE DRUG DISCOVERY PROGRAMS: ENLARGING THERAPEUTIC FOCI

It is abundantly clear that MNP-based drug discovery programs have been and continue to be limited by the financial resources available to academics. Cancer and infectious diseases, in particular, have been the justification and funding rationale for many programs. Today, these continue to be the areas of strong focus. There are perhaps several reasons for this, including the availability of extramural funding and societal need, but I am afraid that this has led to the perception that marine metabolites are toxic and have little applicability to drug discovery in areas such as diabetes, neurodegeneration, cardiovascular, and CNS-based diseases and treatments. This is unfortunate, and simply untrue, as the chemical and biological complexity of marine metabolites is vast; thus, these metabolites hold great potential in each of these areas. In most of my experience, less than 1%–2% of the molecules isolated were cytotoxic; the remaining were of unknown function. What remains is for each of us to develop the interest, commitment, and expertise to approach these new fields and develop the biological collaborations essential to enter unfamiliar new therapeutic areas.

It seems to me that the time is right to expand our visions of therapeutic direction in favor of a more expanded vista for drug discovery and development.

20.4 MARINE DRUG DISCOVERY, THE BIOTECH INDUSTRY, AND BIG PHARMA

I have been asked: "What is the best strategy to take an important new compound forward into development?" It seems to me that there are at least two answers to this question: (1) do sufficient preclinical research such that the compound is *clearly* more potent, safer, and more efficacious than the industry standard, and then present it to Big Pharma, or (2) on the basis of numerous quality lead molecules, work to create a new biotech company to develop your pipeline of leads. These approaches work, but they require a significant deviation from the usual activities of an academic researcher. This is a topic of massive scope that I will not delve into further here, but it is certainly one that can result in success.

Ultimately, whether Big Pharma or biotech, it is by definition that the financial resources of the large pharmaceutical industries are ultimately required to develop drugs and have them tested in humans and approved. This reality requires that we become familiar with the high standards set in Big Pharma and strive to integrate these important criteria into our programs.

20.5 SMALL MOLECULES OR BIG MOLECULES (BIOLOGICALS)?

If you have kept your eye on the pharmaceutical industry, you know that the discovery of biological drugs, vaccines, antibodies, and bacteriophages is of growing importance in therapeutic development. MNP chemists have focused almost entirely on small organic molecules, and from a chemist's viewpoint, this has been exceedingly productive. However, in the process, marine-derived macromolecules and water-soluble metabolites have received far less attention, and even been avoided. With the commercial success of the first marine peptide analgesic drug, Prialt®, from snails of the genus *Conus* (See Chapter 16), and other peptide anlaogs that are now following, it has become clear that marine organisms likely produce a vast diversity of bioactive peptides and related macromolecules, with considerable potential in drug development. Perhaps some of us should consider asking unprecedented questions of the more unfamiliar molecules found in marine life.

20.6 SCREENING VERSUS GENOMIC APPROACHES TO MARINE MICROBIAL DRUG DISCOVERY

Genomics and synthetic biology are now the keywords in the NP chemistry of marine microorganisms. With low-cost genome sequencing and the development of effective genome annotation methods, the biosynthetic pathways of most natural products can be identified. What has been a revolutionary finding is that many bacteria possess many more biosynthetic gene clusters than the number of metabolites that have been isolated in culture. It is common that only 1 or 2 metabolites can be isolated from a strain that can contain 20–30 clearly identifiable biosynthetic gene clusters. Most or many of these clusters are considered "silent" and would not be expressed under normal cultivation conditions. Many of these gene clusters are flanked by repressor genes that have been found to effectively downregulate their production. Indeed, with marine microbes, a new era has arrived that allows cloning of these new genes and expression of production in easily cultured bacterial hosts such as *Escherichia coli* or *Streptomyces coelicolor*. As this science matures, it is quite likely that the processes of cloning and expression and gene synthetic biology will lead to a massive expansion of microbial metabolites. Also, new tools such as imaging matrix-assisted laser desorption ionization (MALDI) mass spectrometry, when applied to co-cultivation of competing microbial strains, have clearly shown that competition can be a major activator of the expression of new antibiotic pathways. Perhaps the laboratory, single-strain cultivation applied for more than 50 years by Big Pharma was never the ideal way to find new antibiotic drugs.

While significant progress can be reported with marine microorganisms, the genomes of most chemically rich marine invertebrates remain unknown. In many cases, sequencing of select invertebrate genes thought to be of great stability and hence of evolutionary significance (internal transcribed spacer [ITS] regions, for example), has led to expanded understanding of the marine animal taxonomy. However, it remains to be shown that marine animals possess the genetic composition needed to produce secondary metabolites (See Chapter 3). Our fascination with the concept that microbes are solely responsible for secondary metabolite synthesis in invertebrates continues to be based on both positive and negative evidence. Only when we are fully capable of sequencing invertebrate genomes, and effectively annotating and compiling animal biosynthetic gene clusters, will we have conclusive evidence to support either a microbe- or an animal-based production.

20.7 WHAT CHANGES ARE WE LIKELY TO SEE?

Marine natural product chemistry is already expanding as young researchers with new talents infiltrate this field. Programs in new therapeutic areas are being established in successful ways. Young researchers with contemporary training are designing work that is at the cutting edge of our field. What I see as abundantly clear is that the field of marine drug discovery must be highly interdisciplinary with biological collaborations absolutely necessary for success. Marine chemists can no longer be satisfied with publishing weak papers with new structure and cursory bioactivity data. There is little science in continuing with this approach. Programs must have pharmacological depth and attend to many of the same requirements that researchers in industry face, such as analog synthesis, toxicology, concurrent absorption, distribution, metabolism, and excretion (ADME) studies, target identification, and more. I feel that it is with this new focus and dedication that marine drug discovery will flourish.

20.8 CONCLUSIONS

I have tried to outline our illustrious past and predict some components we face to create an equally fantastic future. However, it seems clear that the future will be far more difficult and demanding as our granting agencies (and review groups) establish demanding criteria for successful funding. We can no longer participate in the isolation and publication style of business that characterized our approach for the past 40 years. Rather, we will be faced with the need to develop highly collaborative, multi-PI programs that combine chemical expertise, biological sophistication, and preclinical expertise. Furthermore, for success, MNP chemists need to create links to Big Pharma or biotech to both appreciate challenges and create a pathway for success.

Index

Printed and bound by CPI Group (UK) Ltd, Croydon, CR0 4YY

24/10/2024

01778285-0009